6th International Conference on Inertial Fusion Sciences and Applications 2009

Journal of Physics: Conference Series Volume 244

San Francisco, California, USA
6-11 September 2009

ISBN: 978-1-61738-932-0
ISSN: 1742-6588

Printed from e-media with permission by:

Curran Associates, Inc.
57 Morehouse Lane
Red Hook, NY 12571

Some format issues inherent in the e-media version may also appear in this print version.

Printed by Curran Associates, Inc. (2010)

For permission requests, please contact the Institute of Physics at the address below.

Institute of Physics
Dirac House, Temple Back
Bristol BS1 6BE UK

Phone: 44 1 17 929 7481
Fax: 44 1 17 920 0979

techtracking@iop.org

Additional copies of this publication are available from:

Curran Associates, Inc.
57 Morehouse Lane
Red Hook, NY 12571 USA
Phone: 845-758-0400
Fax: 845-758-2634
Email: curran@proceedings.com
Web: www.proceedings.com

TABLE OF CONTENTS

KEYNOTE AND PLENARY PAPERS

IMPLOSION HYDRODYNAMICS AND HYDRO-INSTABILITIES

IGNITION AND HIGH GAIN PELLET DESIGN

LASER AND BEAM PLASMA INTERACTIONS

FAST IGNITION AND HIGH INTENSITY LASER MATTER INTERACTIONS

RADIATION HYDRODYNAMICS

HIGH ENERGY LASER TECHNOLOGY AND IGNITION FACILITIES

HEAVY ION BEAM DRIVERS

Z-PINCHES AND PULSED POWER

IFE POWER PLANTS

TARGET FABRICATION

ICF AND PLASMA DIAGNOSTICS

HOT DENSE PLASMA ATOMIC PROCESSES

LABORATORY ASTROPHYSICS

EOS AND CONDENSED MATTER PHYSICS

LASER ACCELERATION

X-RAY LASER

OTHER APPLICATIONS

The Sixth International Conference on Inertial Fusion Sciences and Applications IOP Publishing
Journal of Physics: Conference Series **244** (2010) 012001 doi:10.1088/1742-6596/244/1/012001

LASER-PRODUCED PLASMAS AS UNIQUE X-RAY SOUCES FOR INDUSTRY AND ASTROPHYSICS

S. Fujioka[1], H. Nishimura[1], H. Takabe[1], N. Yamamoto[1], K. Nishihara[1], D. Salzmann[1], T. Norimatsu[1], N. Miyanaga[1], K. Mima[1], H. Azechi[1], A. Sunahara[2], Y. Shimada[2], Y. Izawa[2], K. Nagai[3], F. Wang[4], J. Zhong[4], G. Zhao[4], Y. Li[5], Q. Dong[5], S. Wang[5], Y. Zhang[5,6], J. Zhang[5], Y-J Rhee[7], Y-W Lee[7], D-H. Kwon[7]

[1]Institute of Laser Engineering, Osaka University, 2-6 Yamada-oka, Suita, Osaka, Japan 565-0871
[2]Institute for Laser Technology, 2-6 Yamada-oka, Suita, Osaka, Japan 565-0871
[3]Chemical Resources Laboratory, Tokyo Institute of Technology, 4259 Nagatsuta, Midori-ku, Yokohama 226-8503, Japan
[4]National Astronomical Observatories, Chinese Academy of Sciences, Beijing 100012, China
[5]Beijing National Laboratory for Condensed Matter Physics, Institute of Physics, Chinese Academy of Sciences, Beijing 1000190, China
[6]Department of Physics, Shanghai Jiao Tong University, Shanghai 200240, China
[7]Laboratory for Quantum Optics, Korea Atomic Energy Research Institute, 1045 Daedeok Street Yuseong-gu, Daejon 305-353, Korea

E-mail: `sfujioka@ile.osaka-u.ac.jp`

Abstract. Laser produced plasma is one of the brilliant x-ray source that has unique capabilities for use in a wide range of science. Here we describe two examples of laser-produced plasma x-ray source application; one is for the semiconductor device industry and the other is for the astronomy. Extreme ultraviolet (EUV) light sources for microlithography are receiving much attention as an industrial application of laser-produced x-ray source. High power and clean EUV light source, 13.5 nm of wavelength, is developed for mass-production of next generation semiconductor devices. Highest EUV conversion efficiency of 4% has been attained by using low-density and minimum-mass tin targets produced by laser-driven explosion of micro-droplet. In addition, it was recently demonstrated that laser-produced x-ray source is very useful to simulate x-ray astronomical phenomena in the laboratory. A 0.5-keV Planckian x-ray source was created with laser driven implosion for producing non-local-thermodynamical-equilibrium (non-LTE) photoionized plasmas, which is a key to understand astronomical compact objects. Laboratory experiment of non-LTE photoionized plasma offers novel test bed for validation and verification of computational codes used in x-ray astronomy.

1. Introduction

A laser-produced plasma is an attractive radiation source since it has high brightness and is compact. Generating high-intensity radiation from laser-produced plasmas is one of the fundamental and long-term research objectives for several applications including use as a probe for experiments in high-energy-density physics, and a compact x-ray source for nondestructive inspection, radiotherapy, and nanofabrication [1]. Laser-produced x-ray source has unique capabilities for use in a wide range of interdisciplinary research tools and has great potential also in industrial production field. Application of laser-produced x-ray source have yet to be widely explored, because no intense x-ray sources have been available other than the synchrotron radiation sources.

Out of those applications, an extreme ultraviolet (EUV) light source for lithography is receiving considerable attention and is being investigated intensively as an attractive application for laser-produced plasmas in industry [2]. EUV lithography is a developing technology expected to be used in mass production of the next-generation semiconductor devices, which have a node size of less than 32 nm. Moore's law, the famous scaling law for the feature size of semiconductor devices, requires EUV lithography to be implemented in the manufacturing stage by 2011 [3]. 13.5 nm-wavelength radiation within a 2% bandwidth has been selected as suitable EUV radiation for lithography because a Mo/Si multilayer mirror has the highest reflectivity ($\sim 70\%$) for in-band light. Laser-produced tin (Sn) plasma has a high-intensity emission peak at 13.5 nm, thus much effort is being devoted to developing a Sn-based EUV light source [4] [5] [6] [7] [8] [9] [10] [11] [12] [13]. Some critical issues in the development of a Sn-based EUV source include achieving a high CE of incident laser energy into in-band light, reducing debris emanating from light source plasmas, and suppressing out-of-band light in the vacuum ultraviolet region. One of the ways to achieve these goals is to use a minimum-mass target [14] [15], which contains the minimum number of Sn atoms required for EUV radiant energy. Using a minimum-mass target substantially reduces debris [14] [15] [16] [17] and out-of-band radiation [18].

The other application to be described in this paper is the laboratory x-ray astrophysics. In a high-mass x-ray binary, a compact object sweeps up material as it orbits through the stellar wind of its companion star. As material is accreted onto the compact object, a fraction of its gravitational potential energy is converted into thermal energy, and some of this thermal energy is released as x-ray radiation, whose mean radiation temperature is on the order of 1 keV, which ionizes the surrounding stellar wind. X-ray astronomers rely on non local-thermodynamic-equilibrium (non-LTE) physics of photoionized plasma generated in such the stellar wind. Theoretical models have been developed on the basis of observed spectra [19] [20] [21] [22] and complex computer codes are written to analyze them [23] [24] [25] [26] [27] [28]. Verification and validation of photoionized plasma models and codes are difficult, because laboratory experiments of photoionized plasma are rare [29] [30], mainly owing to the lack of high flux x-ray sources. The novelty of the present experiment is to use laser-driven implosion for x-ray source production that emits a flash of brilliant, continuum x-rays which simulate those produced by an astronomical compact object. This brilliant x-ray source offers a novel test bed for validation and verification of computational codes used in x-ray astronomy.

2. Laser-driven explosion of micro-droplet for efficient and clean EUV light source generation

An EUV power of 180 W is required at the IF point to achieve EUV lithography[2]. About 33% of the EUV radiation emitted from the source plasma can be transported to the IF point [2]. Thus, a practical EUV light source requires 545 W of EUV power in 2π sr. Assuming a repetition rate of 50 kHz for EUV light radiation generation, the in-band radiant energy must be 10.9 mJ/pulse, which corresponds to 7.4×10^{14} photons/pulse. In a laser-produced plasma, Sn ions emit in-band photons while traveling from the target surface through the high-temperature, low-density corona region. The number of in-band photons emitted per Sn ion was measured to be in the range of 5 to 10 [31]. Therefore, the minimum number of Sn atoms for the required EUV radiant energy is evaluated to be in the range of 7.4×10^{13} to 1.5×10^{14}. This number of Sn atoms is contained in a pure Sn microdroplet with a diameter of 20 μm. A CO_2 laser (wavelength: 10.6 μm) is a practical driver for efficient EUV light source production [9, 32],

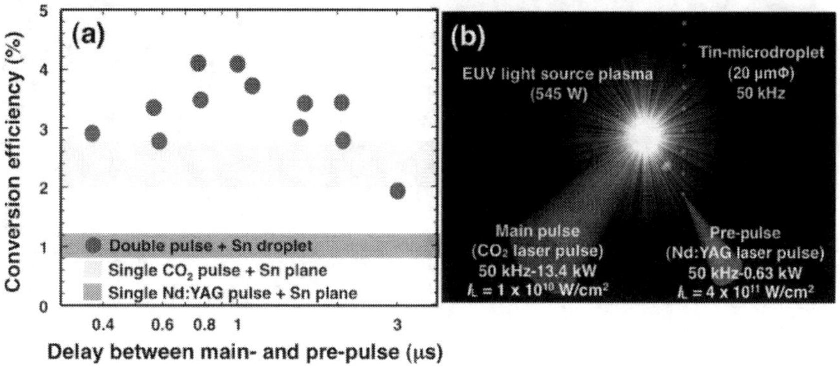

Figure 1. (a) Energy CEs of the main CO_2 laser beam to 13.5-nm-wavelength light within a 2% bandwidth. The CEs were measured by varying the temporal separation between the main CO_2 pulse and the Nd:YAG pre-pulse. An EUV-CE of 4% was attained by this scheme. (b) Schematic of a practical EUV light source based on laser-driven explosion of pure tin microdroplet.

because CO_2 laser-produced Sn plasma is optically thin for the in-band radiation so that self-absorption of the in-band radiation is negligibly small in the plasma [8]. The optimal intensity and pulse duration of a CO_2 laser were assumed to be 1×10^{10} W/cm^2 and 40 ns, respectively, and the EUV-CE was taken to be 4% (details are discussed below), so that the energy in a single laser pulse will be 273 mJ/pulse. To obtain this optimal laser intensity, the laser spot diameter must be 295 μm. This simple evaluation indicates that the diameter (20 μm) of the minimum-mass droplet is too small to be irradiated by the optimal laser intensity (1×10^{10} W/cm^2) and required laser energy (273 mJ). To resolve this considerable mismatch between the optimal laser spot size (295 μm) and droplet diameter (20 μm), the droplet must be expanded prior to laser irradiation by the main pulse. To achieve this we used instantaneous heating by a laser pre-pulse to expand a small microdroplet.

The expansion behavior of pre-pulse irradiated Sn microdroplets was observed using a laser shadowgraph technique [31]. The diameter of the droplets was 36 μm, and the droplets were supported by a 6 μm-diameter carbon fiber. The laser pre-pulse used to expand the droplets was the fundamental output of a Q-switched Nd:YAG laser. The pre-pulses were Gaussian with a pulse width of 8 ns and their intensities were in the range of 5×10^{10} to 4×10^{11} W/cm^2. The second-harmonic of a Nd:YAG laser was used as the probe laser pulse (Gaussian with a 10-ns pulse width). The expansion of the laser-heated droplet changed drastically when the laser intensity was increased and the expansion diameter approaches the optimal laser spot diameter. To expand the microdroplet, the intensity of the Nd:YAG laser pre-pulse must exceed 3×10^{11} W/cm^2. The flying velocity of the expanded target was 0.5 - 1 km/s for 2 - 4 $\times$ 10^{11} W/cm^2.

The expanded pure Sn droplet was irradiated with a CO_2 laser pulse. The incident angle of the pre-pulse was 20 degrees. from the incident axis of the CO_2 laser beam. The laser pre-pulse was the fundamental of a Q-switched Nd:YAG laser, whose output was a Gaussian with a pulse width of 8 ns, a focal spot size of 50 μm, and an intensity of 4×10^{11} W/cm^2. The main laser pulses was from a CO_2 laser, and was Gaussian with a pulse width in the range 30 - 50 ns, a focal spot diameter of 250 μm, and an intensity of 1×10^{10} W/cm^2. An in-band EUV microscope [33] coupled with a back-illuminated x-ray CCD camera was installed at 90 degrees with respect to the incident axis of the CO_2 laser beam.

The dependence of EUV-CE on the delay time between the pre-pulse and the main pulse is shown in Fig. 1. The EUV-CEs are in the range of 2.0 to 2.5 % for Sn plates irradiated with a single CO_2 pulse. This value is almost equal to that reported by Ueno *et al.* [32]. The highest

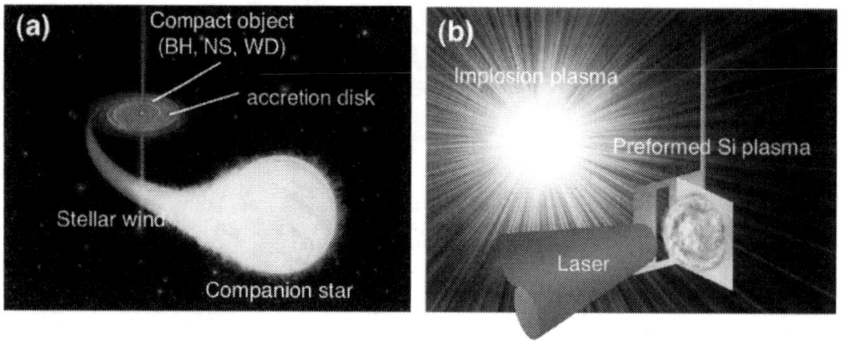

Figure 2. Illustrations of photoionizing plasma systems in the laboratory experiment and astronomy. (a) A schematic diagram of a binary system consisting of a black hole and a companion star. (b) A spherical hollow plastic shell is imploded with 12 laser beams from the GEKKO-XII facility.

CE (4%) was obtained for a delay of 1 μs. The dependence of the EUV-CE on laser intensity was also measured, and the optimal laser intensity was found to be 1×10^{10} W/cm^2 for a 1-μs delay.

An example of a practical EUV light source system design is described. The droplet diameter is assumed to be 20 μm, the feasibility of which has been demonstrated at a repetition rate of 500 kHz [34, 35]. A Nd:YAG laser is selected to generate the pre-pulses. The intensity of the laser pre-pulses is 4×10^{11} W/cm^2, the pulse width is 10 ns, the wavelength is 1.064 μm, and the spot diameter is 20 μm. Thus, the pre-pulse energy is 12.6 mJ/pulse and its power is 628 W. The intensity of the laser pre-pulse is sufficiently high to generate EUV light, and an EUV-CE of 1.5% was obtained in the above experiment. Thus, the EUV radiant energy due to pre-pulse irradiation is about 0.2 mJ/pulse. The residual radiant energy (10.7 mJ/pulse) must be generated by the main CO_2 laser irradiation. The intensity of the main pulse is 1×10^{10} W/cm^2, the pulse width is 40 ns, and the EUV-CE was 4%. Thus, the laser energy is 268 mJ/pulse and the spot size is 292 μm. In this manner, a practical EUV light source can be designed with a 50 kHz-0.63 kW Nd:YAG laser system and a 50 kHz-13.4 kW CO_2 using a 20-μm-diameter droplet. The total EUV-CE of the system is calculated to be 3.9%.

3. Laboratory x-ray astrophysics with a high-power laser facility

We report a novel laboratory simulation of photoionized plasma generated at a high-power laser facility [36]. An example of photoionized plasma is the stellar wind around compact objects. The novelty of the present experiment is a laser-driven implosion that creates a flash of brilliant, continuum x-rays which simulate those produced by an astronomical compact object. As demonstrated below, an imploded core plasma emits a nearly blackbody continuum x-ray pulse of sub-keV mean radiation temperature (T_R). It irradiates a low-density ($n_e = 1 \times 10^{20}$ cm^{-3}) and low-temperature (T_e = 30 eV) silicon plasma. Silicon is highly abundant in the universe. In fact, characteristic line emissions from silicon ions are observed in CYGNUS X-3 [37, 38] and VELA X-1 [39–42].

X-ray emission from the imploded core plasma was characterized with two types of spectrometers, an absolutely calibrated transmission-grating (TG) spectrometer and a filtered-pinhole-array (FPA) device. These were coupled to cooled x-ray CCD cameras, whose spectral response and analog-to-digital gain were calibrated with two radioisotopes (^{55}Fe and ^{109}Cd) which emit 3, 6, and 22 keV x-rays. The TG spectrometer was designed to measure the absolute

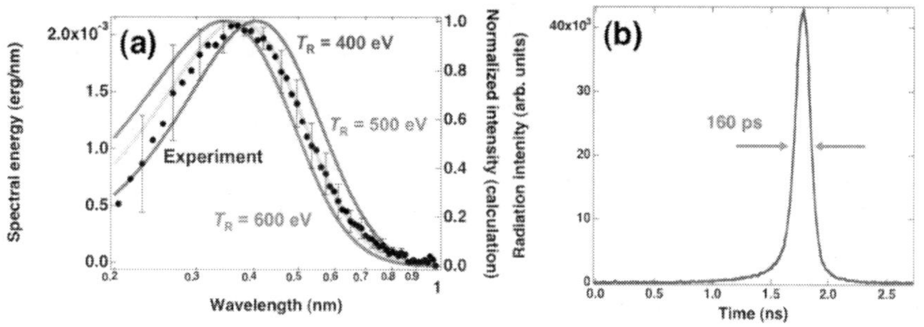

Figure 3. (a) X-ray spectrum from the core plasma measured with a transmission grating spectrometer. (b) Temporal profile of the x-ray pulse measured with an x-ray streak camera.

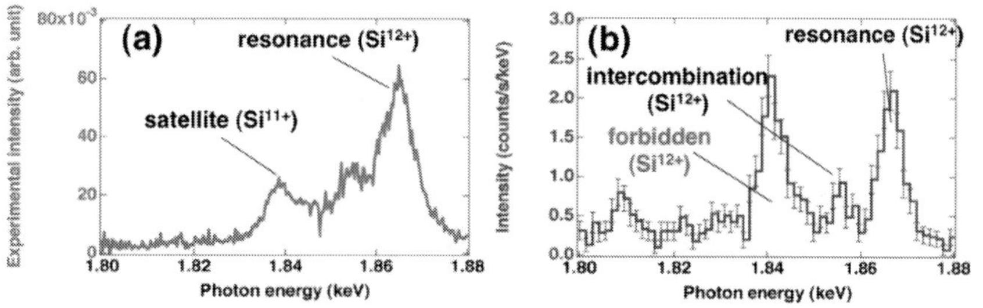

Figure 4. Comparison between the experimental x-ray spectra of photoionized Si plasma (a) and x-ray spectrum observed from VELA-X1 with the on-board Chandra spectrometer (b).

intensity and spectral shape from a core plasma in the 1 to 6 keV photon-energy range, and the FPA spectrometer measures the two-dimensional (2D) distribution of the radiation temperature (T_R) with nine x-ray filters in the 1 to 8 keV photon-energy range. Figure 3 (a) plots the x-ray spectrum from the core plasma measured with the TG spectrometer, along with Planck spectra at various radiation temperatures (T_R = 400, 500, and 600 eV). The spectral response of the grating, x-ray filters, x-ray CCD camera, and source size were all taken into account. The spectrum for T_R = 500 eV best agrees with the experiment. The maximum temperature is 810 eV, and the core has the shape of a rugby ball, with long and short axes of, respectively, 35 and 63 μm FWHM. Both the TG and FPA spectrometers give a consistent temporal- and spatial-averaged temperature of $T_R = 480 \pm 20$ eV. The duration of the blackbody x-ray pulse was measured to be 160 ± 20 ps FWHM with an x-ray streak camera as shown in Fig. 3 (b), whose temporal resolution and spectral window were 20 ps and 1 to 5 keV (10 μm aluminum filter and 300 Å-thick gold cathode), respectively.

In the astrophysical literature [19, 29], the ionization parameter is defined as $\xi = 16\pi^2 J/n_e$ (with units of erg cm/s) to measure the importance of photoionization in a plasma. Here J and n_e are the mean x-ray intensity (erg/s/cm^2) and the electron number density in the plasma (cm^{-3}), respectively. UV and x-ray satellites observed characteristic x-ray emission of highly-ionized ions from the region of photoionized plasmas with ξ = 10 - 10^4 ergs cm/s. For example, the ionization parameter for VELA X-1, a neutron star binary system, peaks around ξ = 300 [39]. In our laboratory plasma, the ionization parameter is $\xi = 5.9 \pm 3.8$, which is slightly below the lower limit of the astrophysical value. Figures 4 (a) and (b) are x-ray spectra from the laboratory plasma and from VELA X-1, respectively. Detailed analysis of the experimental results, astronomical observation, and model are described in the reference [36, 43].

4. Summary

We have demonstrated two distinguished applications of laser-produced x-ray source for industry and basic science. High efficiency of laser produced EUV light source generated from microdroplet and dual laser pulses ensures to provide ongoingly us high performance computer devices at reasonable cost. A brilliant flash of keV Planckian x-ray produced by laser-driven implosion offers a novel test bed for understanding non-LTE photoionized plasma physics relevant to x-ray astronomy. Unique capabilities of laser produced x-ray source will be continuously explored and open new horizon in the wide field of science.

5. Acknowledgement

The authors thank the dedicated technical support of the staff at the GEKKO-XII facility for the laser operation, target fabrication, and plasma diagnostics. A part of this work (EUV light source development) was performed under the auspices of a Leading Project promoted by MEXT (Japanese Ministry of Education, Culture, Sports, Science and Technology). The other part of this research (laboratory x-ray astrophysics) was supported by MEXT, Special Education and Research Expenses for "Laboratory Astrophysics with High-Power Laser" and a Grant-in-Aid for Young Scientists (A) for "Non-LTE Photoionized Plasma Generation with Laser-Produced Blackbody Radiator (Grant No. 21684034)" by the National Basic Research Program of China (973 Program, Grant No. 2007CB815102-03) and by the Korean European Union Framework Programme 7 "Collaboration on HiPER project" (Grant No. 2008-00003).

[1] Turcu I C E and Dance J B 1999 *X-rays from Laser Plasmas: Generations and Applications* (Wext Sussex, England: John Wiley & Sons)
[2] Bakshi V (ed) 2005 *EUV Sources for Lithography* (Bellingham, WA: SPIE Press) ISBN 0-8194-5845-7
[3] International technology roadmap for semiconductors URL `http://public.itrs.net/`
[4] Spitzer R C, Orzechowski T J, Phillion D W, Kauffman R L and Cerjan C 1996 *J. Appl. Phys.* **79** 2251
[5] Richardson M, Koay G S, Takenoshita K, Keyser C and Al-Rabban M 2004 *J. Vac. Sci. Technol. B* **22** 785
[6] Shimada Y, Nishimura H, Nakai M, Hashimoto K, Yamaura M, Tao Y, Shigemori K, Okuno T, Nishihara K, Kawamura T, Sunahara A, Nishikawa T, Sasaki A, Nagai K, Norimatsu T, Fujioka S and Uchida S 2005 *Appl. Phys. Lett.* **86** 051501
[7] Aota T and Tomie T 2005 *Phys. Rev. Lett.* **94** 015004
[8] Fujioka S, Nishimura H, Nishihara K, Sasaki A, Sunahara A, Okuno T, Ueda N, Ando T, Tao Y, Shimada Y, Hashimoto K, Yamaura M, Shigemori K, Nakai M, Nagai K, Norimatsu T and Nishikawa T 2005 *Phys. Rev. Lett.* **95** 235004
[9] Tanaka H, Matsumoto A, Akinaga K, Takahashi A and Okada T 2005 *Appl. Phys. Lett.* **87** 041503
[10] Nishihara K, Sasaki A, Sunahara A and Nishikawa T 2006 *EUV Sources for Lithography* vol PM149 (SPIE Press) chap 11, p 339
[11] Ando T, Fujioka S, Nishimura H, Aota T, Ueda N, Shimomura M, Sakaguchi H, Yasuda Y, Nagai K, Norimatsu T, Nishihara K, Sunahara A, Murakami M, Miyanaga N, Izawa Y and Mima K 2006 *Appl. Phys. Lett.* **89** 151501
[12] Nishihara K, Sunahara A, Sasaki A, Nunami M, Tanuma H, Fujioka S, Shimada Y, Fujima K, Furukawa H, Kato T, Koike F, More R, Murakami M, Nishikawa T, Zhakhovskii V, Gamata K, Takata A and Ueda H 2008 *Phys. Plasmas* **15** 056708
[13] White J, O'Sullivan G, Zakharov S, Choi P, Zakharov V, Nishimura H, Fujioka S and Nishihara K 2008 *Appl. Phys. Lett.* **92** 151501
[14] Fujioka S, Nishimura H, Nishihara K, Kang Y G, Shimada Y, Sunahara A, Furukawa H, Gu Q, Nagai K, Norimatsu T, Miyanaga N, Izawa Y and Mima K 2005 *Appl. Phys. Lett.* **87** 241503–1 – 241503–3
[15] Namba S, Fujioka S, Nishimura H, Yasuda Y, Nagai K, Miyanaga N, Izawa Y, Mima K and Takiyama K 2006 *Appl. Phys. Lett.* **88** 171503
[16] Tanaka H, Hashimoto Y, Tamaru K, Takahashi A and Okada T 2006 *Appl. Phys. Lett.* **89** 181109
[17] Shimomura M, Fujioka S, Ando T, Sakaguchi H, Nakai Y, Yasuda Y, Nishimura H, Nagai K, Norimatsu T, Nishihara K, Miyanaga N, Izawa Y and Mima K 2008 *Appl. Phys. Express* **1** 056001
[18] Sakaguchi H, Fujioka S, Namba S, Tanuma H, Ohashi H, Suda S, Shimomura M, Nakai Y, Kimura Y, Yasuda Y, Nishimura H, Norimatsu T, Nishihara K, Miyanaga N, Izawa Y, Mima K and Sunahara A 2008 *Appl. Phys. Lett.* **92** 111503
[19] Tarter C B, Tucker W H and Salpeter E E 1969 *Astrophys. J.* **156** 943 – 951

[20] Osterbrock D E 1989 *Astrophysics of Gaseous Nebulae and Active Galactic Nuclei* (Mill Valley, CA: University Science Books)
[21] Liedahl D A 1999 *X-ray spectroscopy in astrophysics* (Berlin: Springer-Verlag) chap The x-ray spectral properties of photoionized plasmas and transient plasmas, p 189
[22] Ballantyne D R, Ross R R and Fabian A C 2002 *Mon. Not. R. Astron. Soc.* **336** 867 – 872
[23] Kallman T R, Liedahl D, Osterheld A, Goldstein W and Kahn S 1996 *Astrophys. J.* **465** 994 – 1009
[24] Ferland G J, Korista K T, Verner D A, Ferguson J W, Kingdon J B and Verner E M 1998 *Pub. Astro. Soc. Pacific* **110** 761 – 778
[25] Rose S J 1998 *J. Phys. B* **31** 2129 – 2144
[26] Chung H K, Chen M H, Morgan W L, Ralchenko Y and Lee R W 2005 *High Energy Density Physics* **1** 3 – 12
[27] Rose S J, van Hoof P A M, Jonauskas V, Keenan F P, Kisielius R, Ramsbottom C, Foord M E, Heeter R F and Springer P T 2004 *J. Phys. B* **37** L337 – L342
[28] Porter R L and Ferland G J 2007 *Astro. Phys. J.* **664** 586 – 595
[29] Foord M E, Heeter R F, van Hoof P A M, Thoe R S, Bailey J E, Cuneo M E, Chung H K, Liedahl D A, Fournier K B, Chandler G A, Jonauskas V, Kisielius R, Mix L P, Ramsbottom C, Springer P T, Keenan F P, Rose S J and Goldstein W H 2004 *Phys. Rev. Lett.* **93** 055002
[30] Wang F, Fujioka S, Nishimura H, Kato D, Li Y, Zhao G, Zhang J and Takabe H 2008 *Phys. Plasmas* **15** 073108–1 – 073108–14
[31] Fujioka S, Shimomura M, Shimada Y, Maeda S, Sakaguchi H, Nakai Y, Aota T, Nishimura H, Ozaki N, Sunahara A, Nishihara K, Miyanaga N, Izawa Y and Mima K 2008 *Appl. Phys. Lett.* **92** 241502
[32] Ueno Y, Soumagne G, Sumitani A, Endo A and Higashiguchi T 2007 *Appl. Phys. Lett.* **91** 231501
[33] Tao Y, Nakai M, Nishimura H, Fujioka S, Okuno T, Fujiwara T, Ueda N, Miyanaga N and Izawa Y 2005 *Rev. Sci. Instrum.* **75** 5173
[34] Endo A, Hoshino H, Suganuma T, Nowak K, Yamagida T, Yabu T, Asayama T, Ueno Y, Moriya M, Nakano M, Someya H, Nishisaka T, Abe T, Soumagne G, Komori H, Mizoguchi H, Sumitani A and Toyoda K 2007 *2007 International EUVL Symposium* (Sapporo, Japan) pp SO–07
[35] Brandt D C, Fomenkov I V, Bowering N R, Ershov A I, Partlo W N, Myers D W, Vaschnko G O, Khodykin O V, Bykanov A N, Hoffman J R, Chrobak C P, Chavez J A, Simmons R D and Hsu A 2007 *2007 International EUVL Symposium* (Sapporo, Japan) pp SO–06
[36] Fujioka S, Takabe H, Yamamoto N, Salzmann D, Wang F, Nishimura H, Li Y, Dong Q, Wang S, Zhang Y, Rhee Y J, Lee Y W, Han J M, Tanabe M, Fujiwara T, Nakabayashi Y, Zhao G, Zhang J and Mima K 2009 *Nat. Phys.* **5** 821 – 825
[37] Paerels F, Cottam J, Sako M, Liedahl D, Brinkman A C, der Meer R L J V, Kaastra S and Predehl P 2000 *Astrophys. J. Lett.* **533** L135 – L138
[38] Liedahl D A and Paerels F 1996 *Astrophys. J.* **468** L33 – L36
[39] Sako M, Liedahl D A, Kahn S M and Paerels F 1999 *Astro. Phys. J.* **525** 921
[40] Schulz N S, Canizares C R, Lee J C and Sako M 2002 *Astrophys. J. Lett.* **564** L21 – L25
[41] Goldstein G, Huenemoerder D P and Blank D 2004 *Astronomical J.* **127** 2310 – 2321
[42] Watanabe S, Sako M, Ishida M, Ishisaki Y, Kahn S M, Kohmura T, Nagase F, Paerels F and Takahashi T 2006 *Astrophys. J.* **651** 421 – 437
[43] Fujioka S, Yamamoto N, Salzmann D, Wang F, Li Y, Dong Q, Wang S, Zhang Y, Rhee Y J, Lee Y W, Han J M, Kwon D H, Zhong J, Zhao G, Tanabe M, Fujiwara T, Nakabayashi Y, Zhang J, Nishimura H, Takabe H and Mima K 2009 *Plasma Phys. Controll Fusion* **51** 124032

High temperature, high density opacity measurements using short pulse lasers.

D J Hoarty[1], S F James[1], C R D Brown[1], B M Williams[1], T Guymer[1], M Hill[1], J Morton[1], D Chapman[1], R Shepherd[2], J Dunn[2], G Brown[2], M Schneider[2], P Beiersdorfer[2], H K Chung[3], J WO Harris[1], L Upcraft[1], C C Smith[1] and R W Lee[4]

1 AWE Plasma Physics Department, Building E1.1, Reading, Berkshire RG7 4PR, UK
2 Physics and Advanced Technologies, Lawrence Livermore National Laboratory, 8000 East Avenue, California 94550, USA.
3 Department of Mechanical and Aerospace Engineering, University of California, San Diego, La Jolla, California 92093, USA.
3 Department of Applied Science, University of California, Davis, Davis, California 95616, USA
Email: david.hoarty@awe.co.uk

Abstract. Heating of thin foil targets by a high power laser at intensities of 10^{17} -10^{19}W/cm^2 has been studied as a method for producing high temperature, high density samples to investigate X-ray opacity and equation of state. The targets were plastic (parylene N) foils with a buried microdot of a sample material, which was either aluminium, germanium or a mixture of germanium and titanium mixture of germanium and titanium. L-shell and K-shell spectra were taken using crystal spectrometers recording onto film and an ultrafast X-ray streak camera coupled to a conical focussing crystal with a time resolution of 1ps. The conditions in the microdot were inferred by comparing the measured spectra to synthetic spectra produced by the time-dependent collisional-radiative (CR) models FLY and FLYCHK. The data were also compared to simulated spectra from a number of opacity codes assuming local thermodynamic equilibrium (LTE). Temperature and density gradients were taken into account in the comparisons. The sample conditions, inferred from the CR modelling using FLYCHK, were 800±100eV and 1.5±0.5g/cc, in the germanium/titanium samples and 600+50/-150eV, 3-4g/cc in the pure germanium or aluminium samples. The higher densities were achieved by using a combination of long and short pulses to compress and heat the foils respectively. The experimental results and comparisons to predicted spectra are presented and discussed.

High density and high temperature plasma properties such as opacity and equation of state are important in understanding the physics of stellar interiors [1] and in research into inertial confinement fusion [2]. Recent experiments at AWE have used short pulse lasers to generate high-density (>1g/cc), high-temperature (>500eV) material in an attempt to benchmark opacity modelling codes at these extreme conditions. The experiments used the HELEN OPCPA [3] laser at AWE where previously it has been shown that, using short pulses of second harmonic light or infra-red light reflected from a plasma mirror, plastic foil targets can be heated to temperatures in excess of 500eV to a depth of up to 15μm [4]. This previous work used K-shell X-ray emission spectra from thin layers of aluminium, between 0.05 and 0.2μm thick, buried in the plastic to diagnose the conditions at the position of the buried layer. Recently data from germanium samples has been obtained from buried layer targets. The experimental setup is shown in Fig. 1. Several types of experiments were performed. In the first the sample was made from a mixture of germanium and titanium. The targets were microdot samples, 50μm in diameter 0.1μm thick and composed of 50% germanium and 50% titanium mixtures by particle number. The overcoat of parylene-N plastic between the laser and the buried dot was 1±0.5μm. The laser pulse was Gaussian, both spatially and temporally (0.5ps FWHM), at a wavelength of 0.53μm (2nd harmonic) delivering typically 20J of energy to target, producing intensities of ~10^{18}W/cm^2. In order to mitigate prepulse conversion to second harmonic was employed. That is, since the conversion process is strongly intensity dependent, the lower intensity prepulse was not converted with high efficiency and the unconverted light was rejected at four 2ω mirrors. In the second type of experiment a long pulse beam (represented in figure 1 by the dotted lines) was incident on one side of the target producing a shock compression of the buried layer. The compressed target was then irradiated with the short pulse beam thereby heating the compressed buried layer to high temperature. In this case experiments were done with the buried layer material made of either aluminium or germanium. The long pulse shape had a 100ps rise, a flat-top 700ps long and a fall of 200ps. The long pulse beam energy was typically 300J.

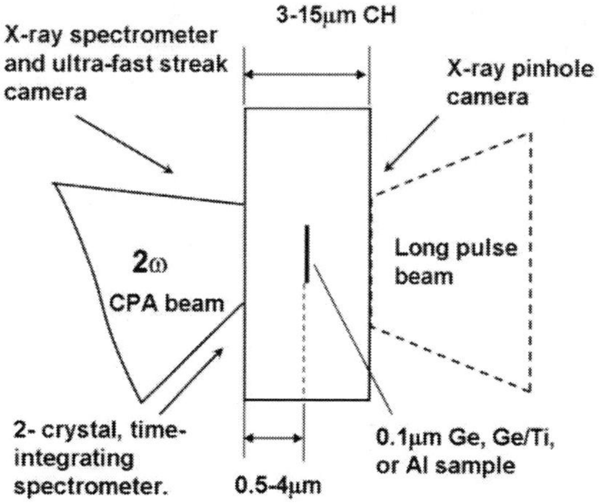

Figure 1: Schematic of the experimental setup.

The results from experiments where the target was irradiated with the short pulse beam only show germanium emission spectra in the 1.3-2.5keV spectral range covering 2p-3d and 2p-4d transitions in charge states up to 29^+. In the mixed germanium and titanium sample the plasma conditions were inferred from the titanium spectra by comparing the measurements of line ratios and widths to titanium spectra predicted by the time-dependent collisional-radiative model (CR) FLY [5] and from the germanium data by comparison to simulation predictions using the code FLYCHK [6]. In addition, to check the departure of the germanium sample from LTE the measured spectrum was compared to several LTE opacity codes with varying levels of spectral detail. In all cases experimental data were recorded using both time-integrated and time-resolved spectrometers. The time-resolved spectra show the pulse duration and emission history, whereas the time-integrated spectrometer allows absolutely calibrated data to be recorded for direct emissivity comparisons with collisional-radiative (CR) and LTE code predictions. The source size was measured with an X-ray pinhole camera with a ten micron diameter pinhole and a magnification of 30, recording onto X-ray film (Kodak Industrex C-type). The pulse duration was measured using an ultra fast X-ray streak camera –see Fig. 1 for the experimental schematic. The two time-integrated spectra were simultaneously recorded on calibrated X-ray film (Kodak Industrex CX-type) using a spectrometer that had two convex curved crystals each 200mm radius of curvature placed side by side. The crystals were PET, to cover the frequency range of the titanium K-shell (4.5-6.5keV) and CsAP to cover the germanium L-shell and aluminium K- shell (1.3-2.5keV) spectrum. Data from the time-integrating spectrometer is shown in Fig. 2a. Using curved crystals gives a greater frequency range than flat crystals and reduces source broadening. The streaked data were taken using a focussing conical CsAP crystal coupled to an ultra-fast streak camera, with a Photonis P860X tube, at a streak rate of 3ps/mm and temporal resolution of 1ps. The camera used a solid potassium iodide photocathode (0.1μm KI/0.02μm Al/ 1μm plastic). The crystal spectrometer geometry, and camera slit length limited the spectral range of the streaked germanium measurement to 1.55-1.78keV. An example of the streaked data is shown in Fig. 2b. The duration of the observed spectrum is limited by the dynamic range of the streak camera.

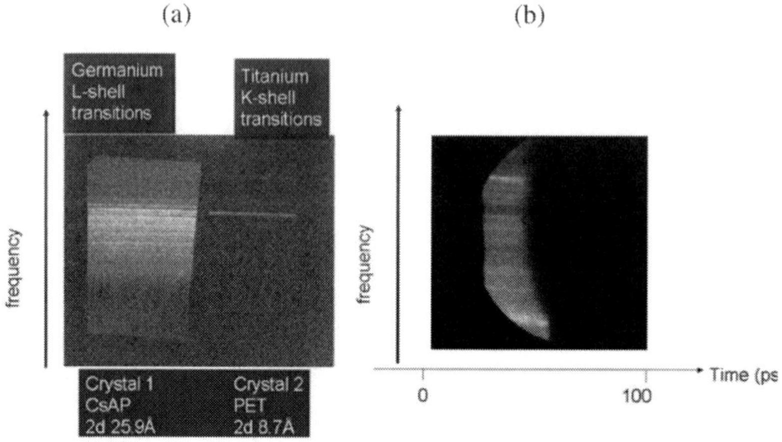

Figure 2: Spectral data for the Ge/Ti mixture: (a) Time-integrated spectrometer data and (b) time-resolved germanium data

Both the germanium and titanium spectra from mixed titanium-germanium samples were used to infer the temperature and density. The titanium spectra were compared to synthetic spectra generated by the time-dependent collisional-radiative (CR) model FLY. FLY is a low atomic number ($Z \leq 26$) K-shell spectral model and limited to H-like, He-like and Li-like ions. A recently developed generalised time-dependent CR model, FLYCHK, can model L-shell spectra and was used to analyse the germanium L-shell measurements. The germanium comparisons should ideally take the time-dependence of the plasma conditions into account but the first attempts at modelling shown here assume that the plasma is in a steady state. A measured germanium spectrum is shown in Fig. 3 plotted as emissivity, in J/kev/cm^2/sr/s, versus X-ray energy in keV. The spectral unfold takes into account the source area (measured with the X-ray pinhole camera), X-ray pulse length (measured on the streak camera), crystal reflectivity, film response (calibrated off-line with the AWE Excalibur X-ray source [7]) and spectrometer geometry. The germanium L-shell spectrum is predominantly emission from 2p-3d transitions for ions up to 29^+ (Li-like Ge). The streaked germanium spectra show a temporal variation in the emission. This late time emission which arises from the Ge/Ti mixture sample when it has decompressed was subtracted from the time-integrated data to minimize the effect on the spectrum due to under dense conditions.

A simulated spectrum calculated using FLYCHK for plasma conditions of 1g/cc, 800eV ±100eV is shown in Fig. 4. Four calculations at a density 1g/cc and electron temperatures of 700, 750, 800, 850 and 900eV were averaged to attempt to take into account the effect of spatial gradients. The FLYCHK data has been convolved with a Gaussian of FWHM 3eV to model detector resolution. The simulated spectra are in reasonable agreement with experimental data both in the spectral distribution and the absolute emissivity values. The LTE opacity code predictions using GRASP [8], CASSANDRA [9] and DAVROS [10], which incorporate different levels of detail, were also used to calculate synthetic Germanium L-shell spectra. Best fit was obtained in LTE at an electron temperature of 660 ± 60eV and density 1.5±0.5g/cc. CASSANDRA and DAVROS both use a DCA type (detailed configuration accounting) method; in the GRASP model detailed term calculations were performed to produce energy levels and oscillator strengths which were then used to construct a spectrum using Saha-Boltzmann statistics.

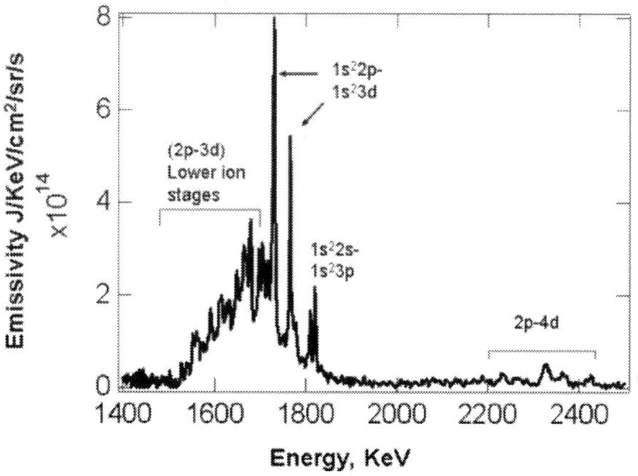

Figure 3: Measured germanium spectra from the time-integrating spectrometer converted to intensity units at the source, and corrected for the late time emission.

These LTE spectra were also convolved with the instrument resolution. The results indicate that taking gradients into account yields reasonable agreement both with the absolute emissivity and the spectral features. The gradients of about ± 10% in temperature and 30% in density are similar to those used in the fit with the CR model.

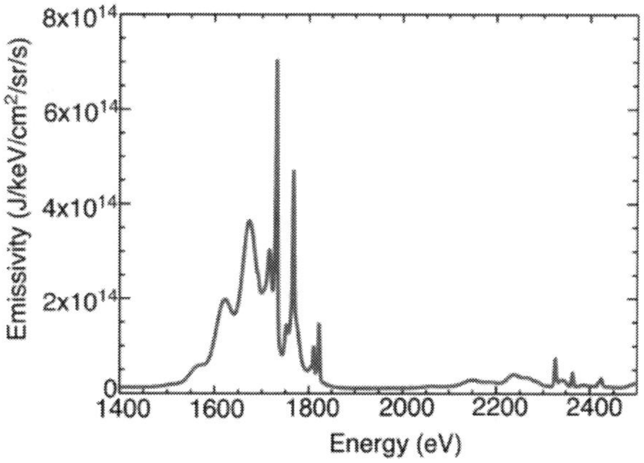

Figure 4: Simulated spectrum from the collisional-radiative equilibrium model FLYCHK. The spectrum was obtained by averaging the contributions from plasma at 1g/cc and 700, 750, 800, 850 and 900eV with equal weighting.

The titanium spectrum was also compared to model prediction to infer the sample temperature and density. Titanium was chosen due to the relative simplicity of the K-shell spectrum compared to the L-shell spectrum for germanium, the large difference in the spectral ranges of the two materials, i.e., that one spectrum would not obscure the other, and the technical ease of mixing the two materials. Comparisons of the titanium results with the synthetic spectra from FLY allowed a density of 1.5±0.5g/cc to be inferred from the width of the spectral lines and a temperature of 1±0.1 keV based on the line-intensity ratios. This is more than 100eV higher than the temperature inferred from FLYCHK comparisons to the germanium data. However, analysis of the spectra using FLYCHK showed that the Ti Ly_α/ He_β ratio used to obtain the temperature was susceptible to perturbation by background hot electrons. The germanium (and aluminium) spectra are predicted to be unaffected by the hot electron background suggesting the Ti spectra has some inaccuracies as a temperature diagnostic in this case, where the hydrogen-like population is comparatively low. The comparison of CR and LTE model calculations shows a significant departure of the germanium from LTE; with the non-LTE model giving best agreement at a temperature some 20% higher than the LTE calculations. In terms of the sample ionisation at 800eV calculations indicate a difference between non-LTE and LTE models of 0.8 of an ion stage. To obtain data at conditions closer to LTE several changes to the experiment were made. In a second set of experiments the target was compressed by a long pulse beam before being heated by the short pulse beam (see figure 1). To achieve LTE conditions the rates of collisional processes in the plasma must overwhelmingly dominate radiative processes. The sensitivity of the collisional rates and radiative rates to plasma conditions and transition energy can be shown by taking the ratio of collisional excitation to spontaneous radiative emission as in Eq. 1 [11].

$$\frac{(n_e C_{ij})}{A_{ji}} = \frac{n_e}{1.3x10^{13} \Delta E_{ij}^{\ 3} \sqrt{T_e}} \qquad (1)$$

Where $n_e C_{ij}$ is the collisional excitation rate coefficient; A_{ji} is the spontaneous radiative decay coefficient; n_e electron density (/cc); ΔE_{ij} is the transition energy (keV) and T_e is the electron temperature (keV).

In the second type of experiment the sample density was increased by shock compression as described above and the position of the buried layer was moved so that there was a plastic coating of 4µm between the buried layer and the side of the foil irradiated with the short pulse beam. The change in position lowered the temperature of the buried layer. From Eq. 1 it can be seen that increasing the density and reducing the electron temperature should produce sample conditions closer to LTE. Initially, measurements were made using aluminium as the buried layer in these experiments. The aluminium measurements confirmed that the sample was at higher density when using a combination of long and short pulses than with the short pulse alone. From comparisons of the streaked experimental spectra to predicted spectra from FLY, a peak temperature of 650 ±50eV and density 3g/cc were inferred. Radiation-hydrodynamics simulations using the NYM [12] code were used to predict the conditions in a germanium layer, buried at the same place in the foil, when subjected to shock compression and heating by the short pulse. The effect of the short pulse heating was simulated by energy deposition over the 0.5ps timescale of the laser pulse to reproduce the temperature and density history observed in the time resolved data from the streak camera. The energy deposition as a function of depth was determined in extensive measurements of peak temperature as a function of depth for a wide variety of laser parameters [13]. The resulting time history and detection threshold of the time integrating spectrometers was then used to establish the conditions in the time integrated data. The conditions predicted for the radiation-hydrodynamics simulations were a gradient in temperature of 450-650eV at 4g/cc density. Figure 5 shows the FLCHK predicted mean ionization for LTE and NLTE simulations for a 4g/cc germanium sample over a range of temperatures from 400-800eV. The sample is predicted to depart from LTE by between 0.15 and 0.35 ion stages over the 450-650eV range of temperatures. Figure 5 shows a measured emission spectrum from a shock-compressed, short-pulse heated germanium buried layer compared to predictions from FLYCHK for both LTE and NLTE taking into account gradients in the sample as predicted by the radiation-hydrodynamics simulations. Figure 6 shows the effect of the departure from LTE on the predicted spectrum of germanium. However, this difference is now similar to the uncertainty in the sample conditions.

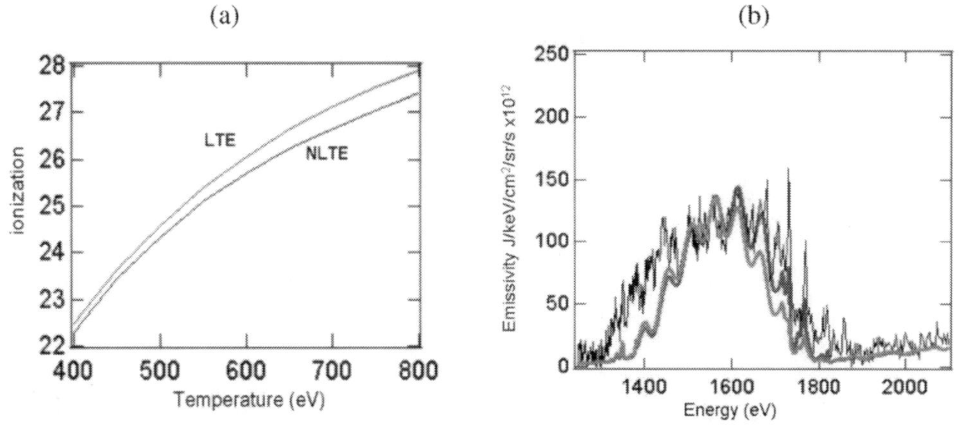

Figure 5: (a) comparison of ionization predicted by FLYCHK in LTE (top curve) and NLTE (bottom curve): (b) comparison of the experimental data from shock compressed and short pulse heated germanium to predictions of the FLYCHK code. The thin black curve is the experimental data; the grey curve the NLTE prediction and the thick black the LTE prediction.

In summary, experiments using buried layer targets have shown that sample plasmas at high temperatures and high densities can be produced to study opacity and high temperature equation of state. Germanium emission data have been obtained at high density 1.5±0.5 g/cc and high temperature 825±75 eV with the conditions inferred from comparisons of the data to the CR model FLYCHK. Reasonable agreement with the spectral shape and absolute emissivity was obtained when spatial gradients in the sample are incorporated into the analysis. Comparison of the data to the predictions of several LTE models also shows good agreement but at a temperature 20% lower (660±60) than the CR modelling, using similar gradients. To obtain data at conditions closer to LTE a combination of long and short pulse laser beams was used to shock compress and heat buried layer samples. Data from aluminium samples was obtained at 650±50eV and 3g/cc and from germanium at 550±100eV, 3.5±0.5g/cc. The higher density germanium data was within 0.4 of an ionization stage of the LTE value. Reasonable agreement was obtained with the absolute emissivity but the gradients in the sample conditions are large. Future work will extend the technique to higher densities using multiple beams to produce coalescing shocks at the buried layer and will aim to reduce gradients.

Acknowledgements
The authors would like to thank Y Ralchenko for providing access to the FLYCHK code.

References:
[1] S. J. Rose, Contemporary Physics **45**, 109 (2004)
[2] J. D. Lindl et al, Phys. Plasmas **11**, 339 (2004)
[3] M. J. Norman et al., Applied Optics **41** 3497 (2002)
[4] D. J Hoarty et al, HEDP **3**, 115, (2007)
[5] R.W. Lee, J.T. Larsen, J. Quant. Spectrosc. Radiat. Transfer, **56,** 535, (1996)
[6] H. K. Chung et al, HEDP **1**, 3, (2005)
[7] C. Bentley et al, Rev. Sci. Instrum. **72**, 1202 (2001)
[8] F.A Parpia, C Froese Fischer, IP Grant, Comput. Phys. Commun. **94**, 249, (1996)
[9] B J B Crowley and J W O Harris, J. Quant. Spectrosc. Radiat. Transfer, **71**, (2000) 257
[10] L. Upcraft private communication.
[11] J D Huba, NRL Plasma Formulary (2009)
[12] P D Roberts et al, J. Phys. D **13**, 1957 (1980)
[13] C R D Brown et al, to be published.

The LMJ program: an overview

Charles Lion

Commissariat à l'Energie Atomique, Centre DAM Ile de France, 91297 Arpajon, France

charles.lion@cea.fr

Abstract. The Mégajoule laser Program (LMJ) is one of the most important parts of the French Simulation Program. The LMJ is designed to deliver about 1.8 MJ of ultraviolet light for high energy density science. It is designed so that fusion can be achieved with it. The first experiments are to be carried out at the end of 2014. Achieving fusion with LMJ requires a coordinated program associating the facility itself, as well as optimized fusion targets, plasma diagnostics, and simulation tools. This paper presents an update about this overall program.

1. Introduction

The Laser Mégajoule (LMJ) facility will deliver up to 1.8 MJ and 550 TW of ultraviolet light. It is one of the most important pieces of the "Simulation Program" in support of the French nuclear weapons stockpile stewardship, and is devoted to laboratory experiments in a large domain of Plasma Physics. It is a key facility for the French national security program and for basic science research. The project has gone through a number of phases including completion of design and building construction, commissioned in December 2008. Presently, the laser bundles amplification section are being assembled; Laser Bay 1 shown in figure 1 is complete. Details about the buildings and the laser system can be found in [7]. The LMJ more stringent specifications are dictated by fusion experiments. Achieving ignition on LMJ requires an integrated plan including the facility itself, target systems, diagnostics, target physics as well as a preliminary experimental program on different laser facilities.

Figure 1: View of amplification sections in Laser Bay 1

2. Ignition target design

The main approach for achieving ignition on the LMJ relies on the indirect drive scheme. Ignition target design uses intensively our simulation tools; central to our approach is the use of 2D and 3D codes validated by experiments. In order to explore the large parameter space available for ICF on the LMJ, a variety of ignition targets have been designed [5]. These targets are linked to different strategies of risk mitigation. The baseline LMJ ignition design has a graded-doped CH capsule in a hohlraum made of a uranium-gold composite. High yield fusion targets requiring no more than 1.4 MJ of laser energy have been identified, allowing first attempts to achieve ignition in a 160-beam configuration, grouped in 2 cones (33.2° and 49°). Rugby-shaped hohlraums are extensively studied, as they offer better energetics and improved symmetry in a 2-cone operation.

We are still dealing with the years-long process of setting requirements for the LMJ ignition targets. For this purpose, multivariable sensitivity models are built in order to assess and manage risks due to combined technological uncertainties and laser dispersions. For instance, a neuronal network model of the 1D yield versus 22 parameters has been built, giving access to a robustness map in the shell thicknesses parameter plan, as shown in figure 2; 1D margin is defined as the highest multiple of all the nominal 1D requirements that returns at least 50 % yield.

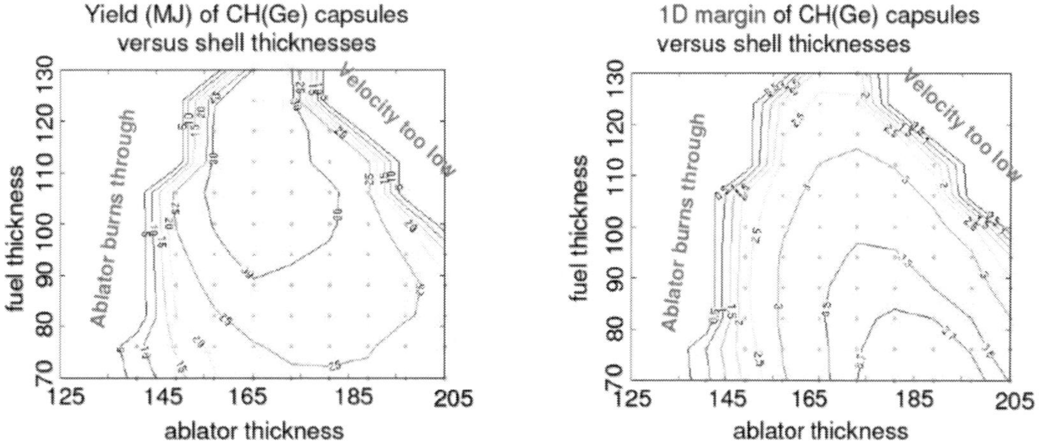

Figure 2: Robustness map of LMJ capsules in the shell thicknesses parameter plan

3. Target system

Ignition targets consist of a complex set of components requiring precise assembly and capsules with nanometer roughness surface finish. The nominal target designs are now based on cocktail hohlraums, whose fabrication has started at the CEA target lab. An important cryogenic program is under way to manufacture the Cryogenic Target Assembly, including its filling and its transportation at cryogenic temperature. The conformation of the DT ice is studied in specific cryostats to meet the required specifications. The target development program to prepare the LMJ experimental campaigns is detailed elsewhere [6].

4. Experimental campaigns

Experiments are presently done on different laser facilities to better ensure success for the fusion program. The LIL facility has been designed as a LMJ prototype, intensively used to test and improve the laser components. In addition, it is used to perform laser-plasma experiments, and is well adapted to test some issues of the experimental tuning. The first Laser Plasma Interaction experiment on LIL was devoted to test the specific longitudinal LMJ smoothing [13]. The simulated Raman spectrum given by the FCI2 code post processed with the linear gain code PIRANAH is coherent with the experimental results (figure 3). Experimental backscattered power has been successfully recovered by complex 3D simulations done with the full hydro-code HERA including a paraxial propagation package [8].

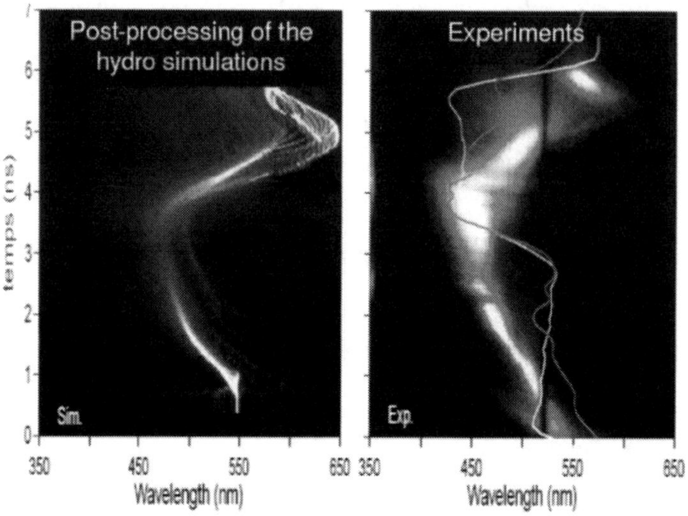

Figure 3: LIL experiment: Simulated and experimental Raman spectra

A shock timing campaign on planar targets is planned in 2010, in order to already control as much as possible this major source of uncertainty for ignition.

Controlling hohlraum energetics in the limit of high radiation temperatures is a key to target performances and has motivated a number of experiments over the past two decades (figure 4). In April 2009, radiative temperatures up to 260 eV have been reached on Omega by the CEA teams using scale ¾ cylindrical hohlraums. Valuable LPI data have then been obtained; hard X rays with energies up to 140 keV have been observed. As hydrodynamic instabilities are a known difficulty of ignition capsules, we have begun experiments on Omega to validate an innovative concept for ablator structuring: the use of a laminated ablator should dramatically reduce Rayleigh-Taylor growth at the ablation front. First experimental results are encouraging [9].

Since 2002, the rugby-shaped hohlraum concept has been tested at small scale on Omega [4], and has exhibited the expected benefit in a 2 cone-operation. It is now used for implosion experiments [11].

The CEA/LLNL/MIT implosions showed a significant drive enhancement (around 18% peak x-ray flux) with rugby hohlraums compared to cylindrical hohlraums. The highest yield for indirectly driven DD/Argon capsules was obtained during this campaign: 1.5 10^{10} neutrons. This yield was high enough to allow for a full suite of neutron and proton diagnostics, giving the first burn history measurements in indirect drive. Figure 5 shows the unfolded neutron image.

5. Plasma diagnostics

Two main types of plasma diagnostics can be identified for use on the LMJ: they are respectively dedicated to x-ray and nuclear measurements. The LMJ diagnostic designs include hardening for harsh environment. Part of the target design effort is also devoted to study diagnostic signatures in simulated experiments, in order to progress on the specifications for analysis of LMJ plasmas.

X-ray imagers and spectrometer will be first activated on LMJ for laser measurements, such as pointing stabilities, structure and power balance of beams, and for hohlraum physics experiments; they may also be used for specific radiography experiments. The imagers design take into account the environmental issues, mainly hard x-rays and debris/shrapnel during this preparation phase without significant fusion gain.

Journal of Physics: Conference Series **244** (2010) 012003 doi:10.1088/1742-6596/244/1/012003

Diagnostics deployed on LIL and OMEGA experiments will be transposed to LMJ, as the wide band spectrometer DMX or the Neutron Imaging System (NIS). A new Large NIS camera, scaled for MJ class lasers, was tested in May 2009 on Omega at 13 m from the Target Chamber Center. This large NIS allows us to consider a 7-8 μm spatial resolution for ICF experiments on the LMJ facility 40 meters line of sight.

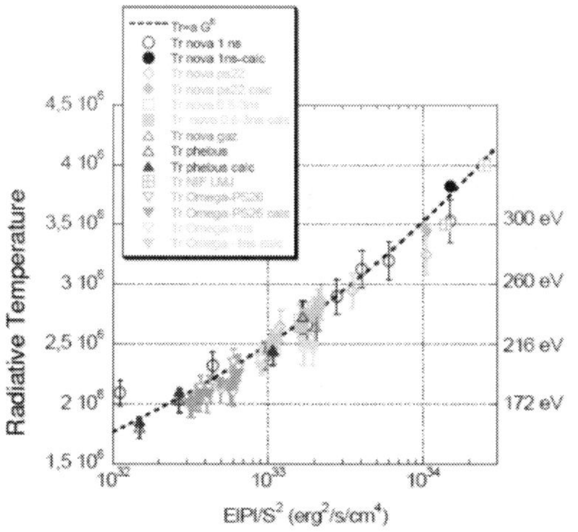

Figure 4 : Radiative temperatures reached experimentally on different laser facilities, to compare to the 300eV LMJ nominal drive

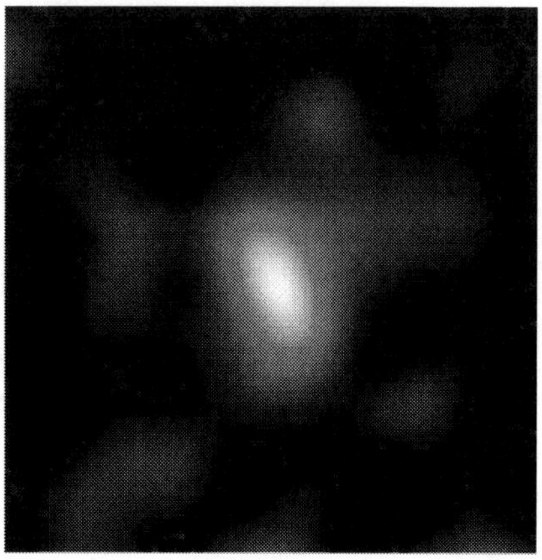

Figure 5 : Neutron image of a Omega implosion in a rugby-shaped hohlraum

6. Conclusion

LMJ will provide unique capabilities for the french Simulation program. Every component of the LMJ's ignition program is well underway. Since [1], new targets have been designed and optimized to achieve ignition with less than 1.4 MJ. The first experiments will be carried out at the end of 2014.

References

[1] Besnard D. 2008 *J. Phys. : Conf. Ser.* **112** 012004
[2] Bourgade JL et al . 2008 *Rev. Sci. Instrum.* **79** 10F301
[3] Casner A *et al., J. Phys. : Conf. Ser.*, these proceedings
[4] Casner A, Huser G, Jadaud J-P, Liberatore S, Galmiche D and Vandenboomgaerde M 2009 *Phys. Plasmas*
[5] Cherfils-Clérouin C *et al*, to be published in *Plasma Physics and Controlled Fusion*; Cherfils-Clérouin C *et al., J. Phys. : Conf. Ser.*, these proceedings
[6] Collier R *et al., J. Phys. : Conf. Ser.*, these proceedings
[7] Ebrardt J and Chaput JM, *J. Phys. : Conf. Ser.*, these proceedings
[8] Loiseau P *et al.* 2007 *37th Anomalous Absorption Conference*
[9] Masse L 2007 *Phys. Rev. Lett.* **98** 245007
[10] Masse L *et al., J. Phys. : Conf. Ser.*, these proceedings
[11] Philippe F *et al., J. Phys. : Conf. Ser.*, these proceedings
[12] Vandenboomgaerde *et al.* 2007 *Phys. Rev. Lett.* **99** 65004
[13] Videau L *et al.* 2008 *Plasma Physics and Controlled Fusion* **50** 12

Journal of Physics: Conference Series **244** (2010) 012004 doi:10.1088/1742-6596/244/1/012004

Progress in Cryogenic Target Implosions on OMEGA

R L McCrory[1,2], D D Meyerhofer[1,2], R Betti[1,2], T R Boehly[1], R S Craxton[1], J A Delettrez[1], D H Edgell[1], V Yu Glebov[1], V N Goncharov[1], D R Harding[1], S X Hu[1], J P Knauer[1], F J Marshall[1], P W McKenty[1], P B Radha[1], S P Regan[1], T C Sangster[1], W Seka[1], R W Short[1], D Shvarts[1,3], S Skupsky[1], V A Smalyuk[1], J M Soures[1], C Stoeckl[1], W Theobald[1], B Yaakobi[1], J A Frenje[4], C K Li[4], R D Petrasso[4], F H Séguin[4], and D T Casey[4]

[1] Laboratory for Laser Energetics, University of Rochester, 250 E. River Road, Rochester, NY 14623, USA

[2] Departments of Mechanical Engineering and Physics & Astronomy, University of Rochester

[3] Nuclear Research Center, Negev, Israel

[4] Plasma Science and Fusion Center, Massachusetts Institute of Technology, Cambridge, MA, USA

E-mail: rmcc@lle.rochester.edu

Abstract. Cryogenic deuterium–tritium targets are imploded on the OMEGA Laser System in a direct-drive configuration. Areal densities of approximately 200 mg/cm^2 have been measured with implosion velocities of 3×10^7 cm/s. These implosions are used to study the dynamics of cryogenic target compression and to develop areal-density diagnostics that will be used as part of the ignition campaign on the National Ignition Facility.

1. Introduction

Most inertial confinement fusion (ICF) ignition target designs rely on the high compression of a spherical shell of deuterium–tritium (DT) ice surrounded by an ablator, commonly referred to as "hot-spot" ignition [1,2]. If the appropriate compressed-target conditions are created, a central low-density region with an ion temperature of 5 to 10 keV and an areal density of 0.2 to 0.4 g/cm^2 surrounded by a cold compressed shell with an areal density of 1 to 3 g/cm^2, the α-particle heating from the DT reactions will lead to a thermonuclear instability, known as ignition, that will produce significant fusion output. The capsule can be driven with x rays generated within a hohlraum (indirect drive) [2] or directly by a laser system (direct drive) [3]. The initial ICF ignition experiments on the National Ignition Facility (NIF) will use indirect drive to compress a Cu-doped Be or Ge-doped CH ablator surrounding a DT-ice shell [4].

Cryogenic DT target implosions are carried out on the 60-beam OMEGA Laser System [5]. Direct-drive illumination is used on OMEGA because it can couple significantly more energy to the compressed capsule than indirect drive, making it possible to achieve higher areal densities with ignition-relevant implosion velocities.

Current cryogenic target implosions on OMEGA have implosion velocities of 3×10^7 cm/s and produce total areal densities of ~200 mg/cm^2 and neutron-averaged ion temperatures in excess of 2 keV. They provide the only current opportunity to study the dynamics of hot-spot formation and to develop the techniques that will be used to diagnose the areal density of ignition experiments on the NIF. This article is divided into four sections. Section 2 describes some of the diagnostic development being carried out and Sec. 3 describes progress in understanding the performance of cryogenic target experiments with ignition-relevant implosion velocities. The article is summarized in Sec. 4.

2. Measuring the areal density in cryogenic-DT implosions

A traditional technique to determine the areal density of compressed DT targets is measuring the knock-on deuteron (triton) spectrum [3,6]. The 14.1-MeV neutrons that are produced in the DT–fusion reaction collide with ions in the compressed shell, giving them MeV energies. As the energetic ions pass through the remaining fuel, they are slowed down by Coulomb scattering. The areal density of the compressed target is inferred from the resulting energy spectrum. When the areal density of the target exceeds approximately 150 mg/cm^2, the knock-on spectrum becomes insensitive to the areal density (saturation) and the knock-on particles can no longer be used to measure the areal density of the target.

New diagnostic techniques are being developed to measure areal densities of 200 mg/cm^2 and higher to understand the OMEGA cryogenic experiments and for the ignition campaigns on the NIF. There are two basic ways to diagnose high areal densities [7]: externally backlight the implosion using a high-energy petawatt laser system such as OMEGA EP [8] or the Advanced Radiographic Capability (ARC) for the NIF [9], or use "self-backlighting" by the 14.1-MeV neutrons produced in the hot spot. The use of the knock-on spectra at lower areal densities is an example of this type of "self-backlighting." Figure 1 shows a schematic of the various ways that the neutrons can be used for "self-backlighting." The neutrons can collide with the compressed-shell materials, reducing their energy so that the down-scattered neutron energy spectrum is determined by the areal density of the compressed fuel. The same collisions can produce neutrons with energies greater than 14.1 MeV through the collisional acceleration of D or T ions followed by a subsequent "in-flight" DT–fusion (tertiary) reaction [10]. A measurement of the neutron-energy spectrum, or a portion of it, makes it possible to determine the target areal density.

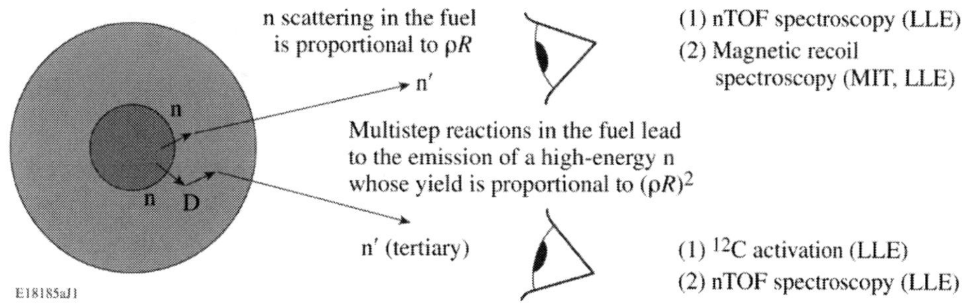

Figure 1. Various collisional processes undergone by the 14.1-MeV neutrons from the hot spot that make it possible to infer the areal density of the compressed shell.

Diagnostic techniques to measure the neutron energy spectrum and, therefore, infer the target areal density are being developed on OMEGA by LLE and MIT [6,11–14]. These include the use of gated neutron time-of-flight detectors (nTOF's), ^{12}C activation to measure the tertiary neutron yield, and the magnetic recoil spectrometer (MRS) [11–14]. On OMEGA, the tertiary yield is proportional to $(\rho R)^2$ while at higher areal densities it is proportional to ρR. The basic principle of the MRS is that the neutrons emitted from the target interact with a CH or CD foil. Nearly head-on collisions produce H or D ions that are subsequently analyzed with a magnetic spectrometer. Since the neutron collisions with H or D are understood kinematically, a measurement of the H or D spectrum, along with the instrumental response, allows the neutron spectrum to be inferred. In the following section, initial data from the MRS will be used to infer the areal density of OMEGA cryogenic-DT target implosions.

3. Performance of cryogenic-DT implosions

In results from cryogenic implosions on the OMEGA Laser System reported previously, areal densities of ~200 mg/cm^2 were measured with implosion velocities of 2.2×10^7 cm/s [15,16]. While this was an important step toward ignition-scaled implosions on OMEGA, the velocity was too low (2.2×10^7 cm/s) to allow scaling to ignition implosions on the NIF. It was found that the continuous pulse shapes (that were scaled from the NIF direct-drive baseline target design [17]) were very sensitive to the details of the pulse shape, producing two shock waves propagating through the target rather than the desired shock wave followed by a compression wave [18]. The baseline direct-drive-target design was changed to one that produces a series of four shock waves that are timed appropriately to produce the desired compressed-target conditions. A representative pulse shape is shown in Fig. 2. This "triple-picket" pulse shape produces a shock-wave structure that is similar to that used for the initial indirect-drive implosions planned for the NIF [4]. The advantage of this pulse shape is that the timing and energy of the pickets can be easily experimentally tuned to optimize target performance. The picket timing/energy has been tuned using direct measurements of the shock timing in "key-hole" targets [19]. The implosion velocity is 3×10^7 cm/s with an adiabat of 2, where the adiabat α is the ratio of the shell pressure to the Fermi-degenerate pressure at the same density. The target was driven with the pulse shape shown in Fig. 2 with a total on-target energy of 23 kJ. The peak intensity was 8×10^{14} W/cm^2. The beams were smoothed with distributed phase plates [20] and polarization smoothing [21].

A second requirement for cryogenic target implosions is that the target nonuniformity be small enough to achieve ignition-relevant performance. Figure 3 shows a shadowgraphic image of a DT target used to study hot-spot formation. The target is a 5-μm-thick CH shell containing a 65-μm-thick DT-ice layer. The inner-ice-surface roughness is 1.8-μm rms, slightly above the ignition design requirement [17].

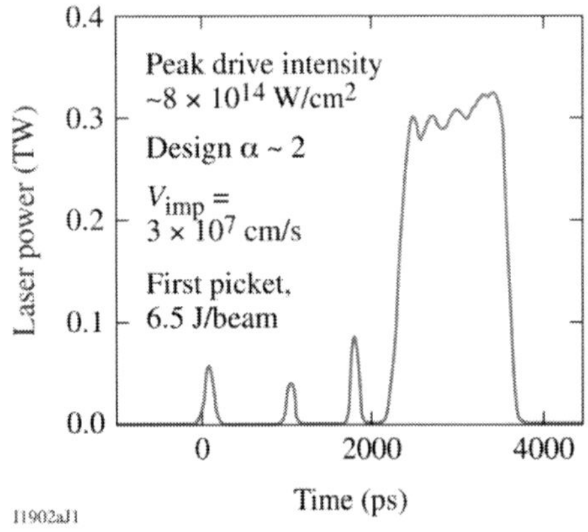

Figure 2. Laser pulse shape (per beam) used in the current OMEGA cryogenic target implosions.

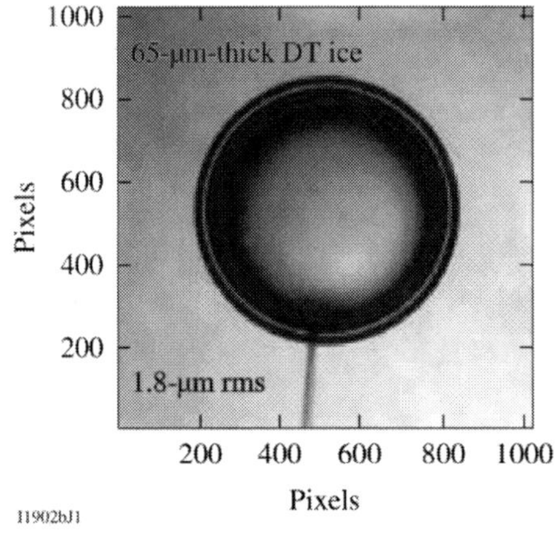

Figure 3. Shadowgraph of a stalk-mounted cryogenic-DT capsule used for these experiments with an inner-ice roughness of 1.8-μm rms.

The areal density of the compressed shell was inferred from the MRS data. Figure 4 shows the deuteron spectrum measured by the MRS, generated by collisions of the neutrons leaving the target with a diagnostic CD foil. The blue line is a fit to the data that is generated by taking test neutron spectra and convolving them with the MRS instrumental response. In this case, the inferred areal density was 179±34 mg/cm^2. The knock-on deuteron spectra measured from two other directions were saturated. This is consistent with this areal density.

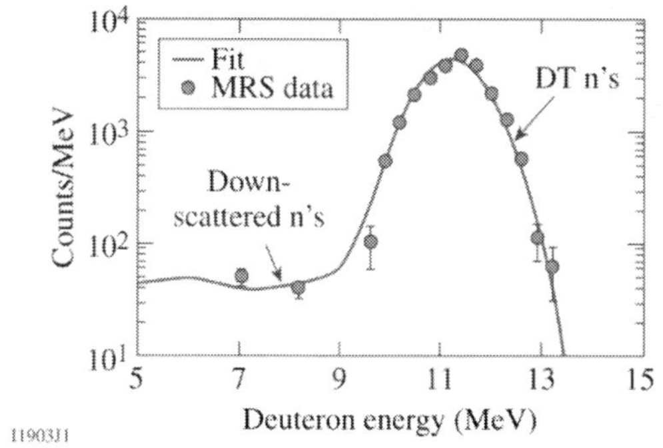

Figure 4. A deuteron-energy spectrum measured with the magnetic recoil spectrometer. The blue line is a fit to the neutron spectrum that is convolved with the instrument response to infer an areal density of 179±34 mg/cm^2.

An ion temperature of 2.2±0.5 keV was inferred using the nTOF detector [22,23]. The total neutron yield was 3.4×10^{12}, approximately 5% of that predicted by 1-D simulations.

4. Summary

Figure 5 compares the new data obtained with the triple-picket-pulse shape to that obtained previously with a continuous pulse shape having a peak intensity of 5×10^{14} W/cm^2 [15,16]. The data points are plotted versus their neutron-averaged ion-temperature and areal density. This provides a Lawson-like criterion to determine the scaling to ignition [24,25]. The blue line represents a 1-D marginal ignition threshold and the yellow shaded region is where ignition and

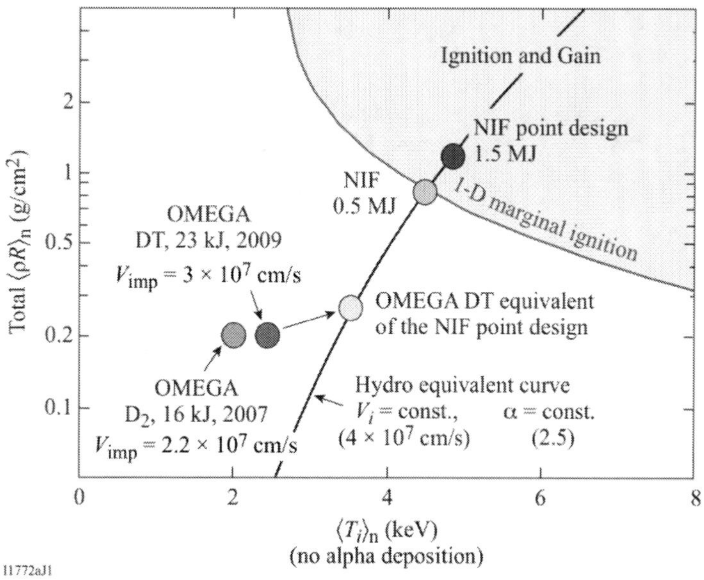

Figure 5. Comparison of the new data obtained with the triple-picket-pulse shape with that obtained previously with a continuous pulse shape. This provides a Lawson-like criterion to determine the scaling to ignition [24,25]. The blue line represents a 1-D marginal ignition threshold and the yellow shaded region is where ignition and gain would be obtained. The black line represents a set of hydrodynamically equivalent designs (varying laser energy) that have a velocity of 4×10^7 cm/s using a continuous pulse shape. The yellow point shows the NIF baseline point design [17] scaled to OMEGA energies.

gain would be obtained. The black line represents a set of hydrodynamically equivalent designs (varying laser energy) that have a velocity of 4×10^7 cm/s using a continuous pulse shape. The yellow point shows the NIF continuous pulse-shape point design [17] scaled to OMEGA energies.

The areal density measured with the triple picket pulse is similar to that of the continuous pulse shape while the triple picket design has a higher implosion velocity and a higher ion temperature, moving toward the line that is hydrodynamically scaled from the direct-drive-ignition point design for the NIF [17].

In summary, progress continues in understanding cryogenic-DT implosions on OMEGA and developing areal-density diagnostics that will be used for the ignition campaign on the NIF. A new triple-picket-pulse shape makes it possible to experimentally tune the multiple shock waves and to develop high-velocity ($>3 \times 10^7$ cm/s) implosions for OMEGA cryogenic target experiments in the near future.

Acknowledgment

This work was supported by the U.S. Department of Energy Office of Inertial Confinement Fusion under Cooperative Agreement No. DE-FC52-08NA28302, the University of Rochester, and the New York State Energy Research and Development Authority. The support of DOE does not constitute an endorsement by DOE of the views expressed in this article.

References

[1] Atzeni S and Meyer-ter-Vehn J 2004 *The Physics of Inertial Fusion: Beam Plasma Interaction, Hydrodynamics, Hot Dense Matter*, International Series of Monographs on Physics (Oxford: Clarendon Press)
[2] Lindl J D, 1998 *Inertial Confinement Fusion: The Quest for Ignition and Energy Gain Using Indirect Drive* (New York: Springer-Verlag) chapter 6 pp 61–82
[3] McCrory R L *et al.* 1988 *Nature* **335** 225–30
[4] Haan S W *et al.* 2007 *Eur. Phys. J. D* **44** 249–58
[5] Boehly T R *et al.* 1997 *Opt. Commun.* **133** 495–506
[6] Frenje J A, Li C K, Séguin F H, Casey D T, Petrasso R D, Sangster T C, Betti R, Glebov V Yu and Meyerhofer D D 2009 *Phys. Plasmas* **16** 042704
[7] Sangster T C *et al.* The nuclear diagnostics suite for the NIF, this proceeding
[8] Waxer L J *et al.* 2005 *Opt. Photonics News* **16** 30–6
[9] Barty C P J *et al.* 2004 in *Inertial Fusion Sciences and Applications 2003*, ed B A Hammel *et al.* (La Grange Park, IL: American Nuclear Society) pp 612–15
[10] Glebov V Yu, Stoeckl C, Sangster T C, Meyerhofer D D, Radha P B, Padalino C, Baumgart L, Colburn R and Fuschino J 2003 *Rev. Sci. Instrum.* **74** 1717–21
[11] Frenje J A *et al.* 2008 *Rev. Sci. Instrum.* **79** 10E502
[12] Frenje J A *et al.* 2001 *Rev. Sci. Instrum.* **72** 854–8
[13] Frenje J A *et al.* 2009 *Phys. Plasmas* **16** 022702
[14] Frenje J A *et al.* The magnetic recoil spectrometer (MRS) for absolute measurements of the neutron spectrum at OMEGA and the NIF, this proceeding
[15] Sangster T C *et al.* 2008 *Phys. Rev. Lett.* **100** 185006
[16] McCrory R L *et al.* 2008 *Phys. Plasmas* **15** 055503
[17] McKenty P W, Goncharov V N, Town R P J, Skupsky S, Betti R and McCrory R L 2001 *Phys. Plasmas* **8** 2315–22
[18] Goncharov V N, Sangster T C, Boehly T R, McCrory R L, Radha P B, Smalyuk V A, Skupsky S, Frenje J A and Petrasso R D Multiple-picket, cryogenic target designs and performance for OMEGA and the NIF, this proceeding
[19] Boehly T R *et al.* 2009 *Phys. Plasmas* **16** 056302
[20] Lin Y, Kessler T J and Lawrence G N *et al.* 1995 *Opt. Lett.* **20** 764–766
[21] Boehly T R, Smalyuk V A, Meyerhofer D D, Knauer J P, Bradley D K, Craxton R S, Guardalben M J, Skupsky S and Kessler T J 1999 *J. Appl. Phys.* **85** 3444–47

[22] Glebov V Yu, Stoeckl C, Sangster T C, Roberts S, Schmid G J, Lerhce R A and Moran M J 2004 *Rev. Sci. Instrum.* **75** 3559–62

[23] Lerche R A. Schmid G J and Glebov V Yu Ion temperature measured with radiation hardened CVD diamond detector and 14-MeV ICF fusion neutrons using short time-of-flight paths presented at ICOPS 2005, International Conference on Plasma Sciences (Monterey, CA, 20–23 June 2005) paper 10670

[24] Zhou C D and Betti R 2008 *Phys. Plasmas* **15** 102707

[25] Betti R, Anderson K S, Chang P Y, Nora R and Zhou C D A measurable Lawson criterion for inertial confinement fusion, this proceeding

Journal of Physics: Conference Series **244** (2010) 012005 doi:10.1088/1742-6596/244/1/012005

Status and Future Prospects of Laser Fusion and High Power Laser Applications

Kunioki Mima

The Graduate School for the Creation of the Photonics Industries
Institute of Nuclear Fusion, Universidad Politecnica de Madrid
Institute of laser Engineering, Osaka University
2-6, yamada-oka, Suita, Osaka, Japan

mima@ile.osaka-u.ac.jp

Abstract. In Asia, there are many institutes for the R&D of high power laser science and applications. They are 5 major institutes in Japan, 4 major institutes in China, 2 institutes in Korea, and 3 institutes in India. The recent achievements and future prospects of those institutes will be over viewed.

In the laser fusion research, the FIREX-I project in Japan has been progressing. The 10kJ short pulse LFEX laser has completed and started the experiments with a single beam. About 1kJ pulse energy will be injected into a cone target. The experimental results of the FIREX experiments will be presented. As the target design for the experiments, a new target, namely, a double cone target was proposed, in which the high energy electrons are well confined and the heating efficiency is significantly improved. Together with the fusion experiments, Osaka University has carried out laboratory astrophysics experiments on photo ionizing plasmas to observe a unique X-ray spectrum from non-LTE plasmas. In 2008, Osaka university has started a new Photon research center in relation with the new program: Consortium for Photon Science and Technology: C-PhoST, in which ultra intense laser plasmas research and related education will be carried out for 10 years.

At APRI, JAEA, the fundamental science on the relativistic laser plasmas and the applications of laser particle acceleration has been developed. The application of laser ion acceleration has been investigated on the beam cancer therapy since 2007.

In China, The high power glass laser: Shenguan-II and a peta watt beam have been operated to work on radiation hydro dynamics at SIOFM Shanghai. The laser material and optics are developed at SIOFM and LFRC. The IAPCM and the IOP continued the studies on radiation hydrodynamics and on relativistic laser plasmas interactions. At LFRC in China, the construction of Shenguan III glass laser of 200kJ in blue has progressed and will be completed in 2012.

Together with the Korean program, I will overview the above Asian programs.

1. Introduction and Asian research networks

Laser fusion and high energy density science (HEDS) together with high power laser technology are developing in the world as well as in Asia. In this lecture, the status of the Asian community of this filed, and Asian history of the research are overviewed. Namely, the status and future prospects of projects in China, Korea, India and Japan are described. In particular, for fast ignition research, the present and future of FIREX project at Osaka are discussed.

The major institutes and universities in the researches on high power laser, laser plasmas, high energy density science and laser fusion in Asia are the followings. In Japan, they are the Institute of laser Engineering and the Faculty of Engineering of Osaka University., the Kansai Photon Research Institute of JAEA, The Graduate School for the Creation of New Photonics Industries at Hamamatsu, and Institute of Chemical Research of Kyoto University, and so on. In China, there are the Shanghai Institute of Optics and Fine Mechanics, Laser Fusion Research Center at Mianyang, Institute of Physics, Institute of Applied Physics and Computational Mathematics at Beijing, and Shanghai Jao Tong University, and so on. In, Korea, Korean Atomic Energy Research Institute, Korean Advanced Institute of Science and Technlogy, and APRI of K-JIST. In India, they are Tata Institute of Fundamental Research Mumbai, and Bhabha Atomic Research Center at Mumbai, and Institute of Plasma Research at Ahmedabad.. Those institutes are connected by some research networks for laser fusion, HEDS, and High Filed Science researches which are 1) AILN (Asian Intense Laser Network) including 20 institutes and universities are included, 2) APFA (Asian Plasma and Fusion Association), 3) Asian CORE program on "Development of next generation ultra-short pulse lasers for high field science" and so on. Those net works are going to be connected to the world wide net works like EU research group under the European Physical Society (EPS) and America research group under the American Physical Society (APS). (See Fig.1)

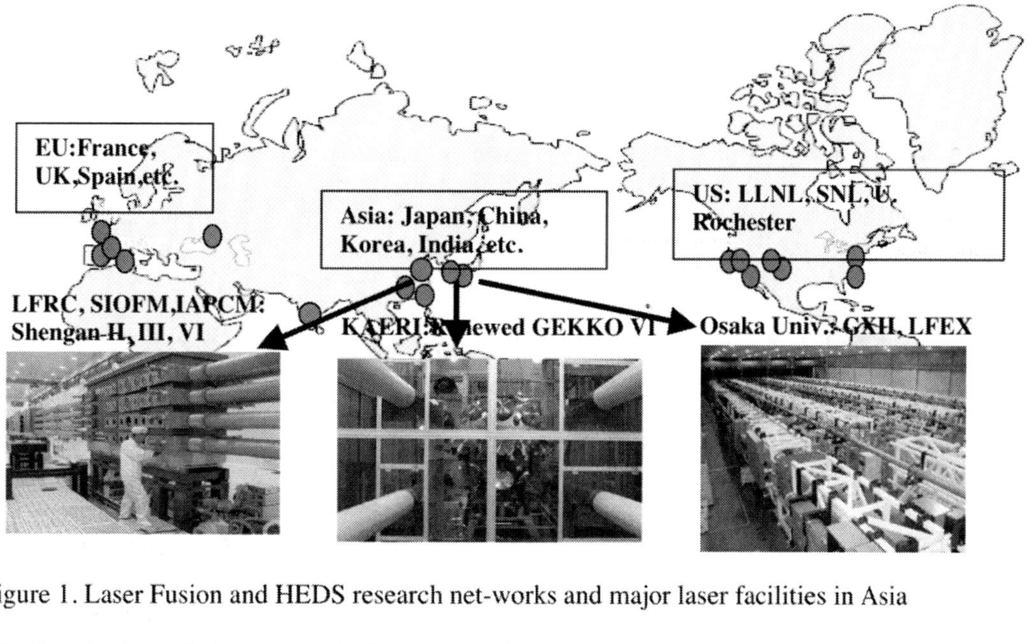

Figure 1. Laser Fusion and HEDS research net-works and major laser facilities in Asia

As for the laser fusion research, Japan and China are the leading countries. In Japan, the FIREX project has started since 2002 and the main facility, LFEX laser started the operation since 2009 March. On the other hand, the multi kJ lasers called SG laser have been developed and the 200kJ laser will be completed in 2012 at Mianyang, China. The both "SG project" and FIREX project are aiming at demonstration of ignition. The Japanese project is concentrated on the direct driven implosion and fast ignition. On the other hand, the SG project is for the indirect implosion and fast ignition. (See Fig.1)

The high field science has been widely developing in many countries. Namely, petwatt level lasers are operated at ILE and KPRI of JAEA in Japan, KPRI in Korea, and SIOFM in China. Further more, many projects related to ultra intense short pulse lasers are going on all over the Asian.

2. China and Korea Activities

2.1 China Program

Multi kJ level pulse laser facilities have been developed for the laser fusion and HEDS in SIOFM and Laser Fusion Research Centre (LFRC) at Mianyang. At SIOFM, the Shengan II laser (SG-II) are operated for laser plasma and fusion experiments. SG-II is 8 beam glass laser with 8kJ in 3ns, which is combined with the petawatt laser of 1kJ.1ps. Based upon the development of SG-II, the construction of SG-III has started and will be in operation in 2012. The SG-III is 48 beam system and will deliver 200kJ in 3ns at 0.351 μm wavelength. In China, ultra intense short pulse Ti-Saphire lasers are also operated. They are Xtreme Light III (XL-III) laser system (720 TW/ 30fs) at the Institute of Physics at Beijing and SILEX laser at LFRC (300TW). They are open for the international research community. The SG-III laboratory has three area: laser system, target chamber, laser component cleaning and assembling area. For fast ignition experiment, ultra intense short pulse laser will be also added in future. The plasma experiments with XL-III have been very successful to contribute to the relativistic laser plasma research.

Laser plasma physics and laser fusion researches have also been progressing in China. For an example, the Rayleigh-Taylor instability (RTI) of a directly driven plastic foil is measured by 13.9 nm EUV laser of a Nickel-like ion plasmas in SG-II experiments. A CH foil thickness with 50 μm wave length corrugation was irradiated with laser beams of 1-2kJ/0.53μm/2ns pulse width. As shown in the Figure 2, the bubble and spike structure as the nonlinear stage of the RTI mode was precisely measured. The experimental results have been compared with the 2-D simulation to validate the simulation code. Till the highly nonlinear stage, the experiments and simulation results agree well up to bubble and spike amplitude of 50 μm.

The ultra intense laser plasma interactions have also been investigated at IOP, Beijing and the SJTU by theory and simulation. For an example, on the high energy electron transport, the Kα X-ray images were taken for the pre-formed plasmas interacting with an obliquely injected 10TW laser pulse. The image is asymmetric for large incident angle (70°) and symmetric for 45° incident. The non-symmetrical distributions were probably induced by the electrons that escape into the vacuum and drawn back into the target again by the electrostatic fields. For more detail, see the proceeding by Y. T. Li [1]

The laser technology R&D in China has been progressing at the SIOFM and their collaboration institutes and universities. Nonlinear crystals, laser glass, large aperture deformable mirror, and so on has been developed. The deformable mirror fabricated at the Institute of Optics and Electronics, Chengdu CAS.

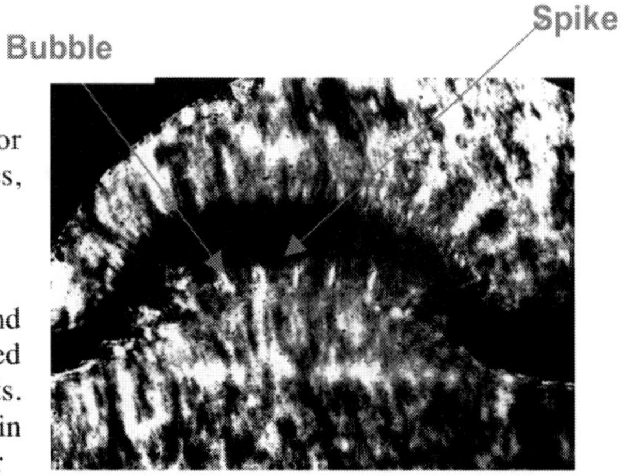

Figure 2. Bubble and spike structure measured by 13.6nm EUV laser. The wave length of the structure is 50μm. The amplitude is about 50μm.

2.2 Korean program

At KAERI, GVI laser at Osaka was transferred and renewed as KLF (KAERI Laser Facility). It started the operation recently to do fast ignition experiments. A 100 TW OPCPA short pulse laser will be added in future. At KPRI. GIST, the 100TW Ti:Sapphire laser is operated at APRI. This Ti:Sapphire laser will be up graded to a half PW and start operation for users in 2010. The pulse energy and duration of the system are 20J and 40fs.

3. Japanese program

In 2008, the Photon Frontier Network (PFN) was

organized to explore frontier of photon S&T, to foster young researchers, and to promote collaborations among institutes, universities, and industries. Under the PFN, two centers are established as it was presented by Y..Kato in the plenary of this IFSA. One of them is the " Consortium for Photon Science and Technology" (C-PhoST) lead by R.Kodama. In the C-PhoST, plasma photonics, laser acceleration, high energy density science, laser fusion, laser plasma Thz radiation and so on are carried out together with the young scientists fostering. The more detail on the net work was presented by Y.Kato [2]. At the Advanced Photon Research Center (APRC), JAEA, fundamental physics, laser ion beam cancer therapy, and laser acceleration with super powerful lasers are studied. The main facility of the institute is the J- Karen laser which can deliver a laser pulse with a few J/30fs and very high contrast ratio of 10^{-10} [3]. The project on the laser produced proton therapy started 3 years ago. The project includes the development of ultra intense short pulse laser operated with 1 kHz, the proton acceleration physics and target design, proton beam irradiation system, medical research on the proton irradiation effects and so on. Recently, the cancer cells are irradiated with a laser produced proton beam to detect the effects of the irradiation. In this experiments, proton fluence on the sample was 20 Gy with 200 laser shots. It is shown that DNA double stranded breaks in the region irradiated by a laser- driven proton beam [4].

Laser acceleration experiments also advanced at APRC. The JLITE-X Ti-Sapphire laser of 0.16 J/ 40fs irradiated the gas jet target of Argon and Helium. The monochromatic electron beams are generated. The central energy of the e-beams is 24.8 MeV for Helium and 8.5 MeV for Ar. In particular, the probability of generation of mono energy beam is very high 884%) for the case of Argon gas target. Moreover, the angular collimation of the electron beam is very high for Argon gas The divergence angle was 10mrad for Argon and 31mrad for Helium. as shown in the beam angular (after .K.Kondo Presented in ICUIL international conference) This difference of the beam divergence is interpreted by the laser beam channeling in the Argon gas target. Namely the clusters are formed in the Argon gas jet but not in the Helium gas jet. The laser beam is expected to be well guided in the cluster plasma in comparison with the usual gas plasma since the cluster plasma will have very strong positive susceptibility [5].

At ILE Osaka University, laser fusion, laser astrophysics, laser driven EUV source, laser Astronomical physics, hot dense matter physics, and so on have been explored. Because of the limitation of high reflectivity optics, the 13.5nm EUV is the only one candidate for the next generation lithography. As the 13.5nm EUV radiation source, laser produced Sn plasmas have been investigated for many years under the MEXT project at Osaka University. In 2007-2008, we discovered that very high conversion efficiency from laser energy to 13,5nm EUV radiation can be achieved. The irradiation geometry is shown in the Fig.3 where a few tens of micrometer droplets are injected with a few kHz repetition rate and irradiated with Nd-YAG pre-pulse of the intensity 10^{11}W/cm^2 and after 1 μm sec, a CO_2 laser main pulse is irradiated. In the experiment, the conversion efficiency increased up to 3.5% to 2% band width around 13.5nm. This result is very promising for using Sn laser plasmas as the lithography radiation source.

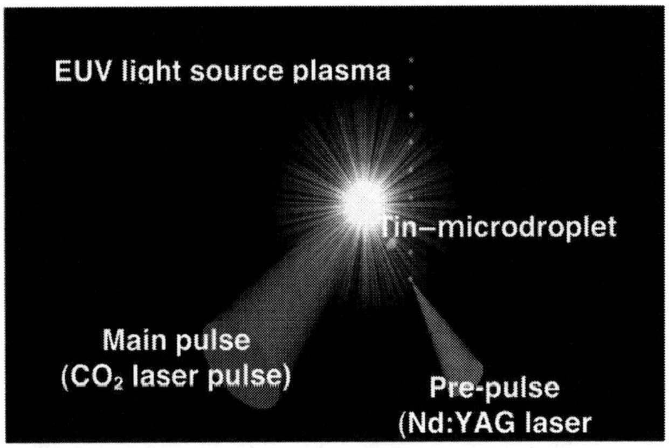

Figure 3. High efficiency EUV generation from micro Sn droplet irradiated with Nd-YAG laser and CO_2 laser double pulse

In thc lascr plasma astrophysics, a recent topic is the physics of photo ionizing plasmas. X-ray spectroscopy of non-thermal equilibrium plasmas photo-ionized by intense radiation is a key to understanding compact objects, such as black holes, based on astronomical observations[6].

The experiments carried out by the international collaboration among China Korea and Japan studied photo-ionizing plasmas in laboratory under well-efined and astrophysical relevant conditions. Photo-ionized plasma is here generated using a 0.5-keV Planckian x-ray source created by means of a laser-driven implosion.

The measured x-ray spectrum from the photo-ionized silicon plasma resembles those observed from the binary stars Cygnus X-3 6, 7 and Vela X-1 8–10 with the Chandra x-ray satellite. This experiment demonstrates that an extreme radiation field was produced in the laboratory, however, the theoretical interpretation of the laboratory spectrum significantly contradicts the generally accepted explanations in x-ray astronomy. This model experiment offers a novel test bed for validation and verification of computational codes used in x-ray astronomy. The geometry of the experiment is shown in Fig.4.

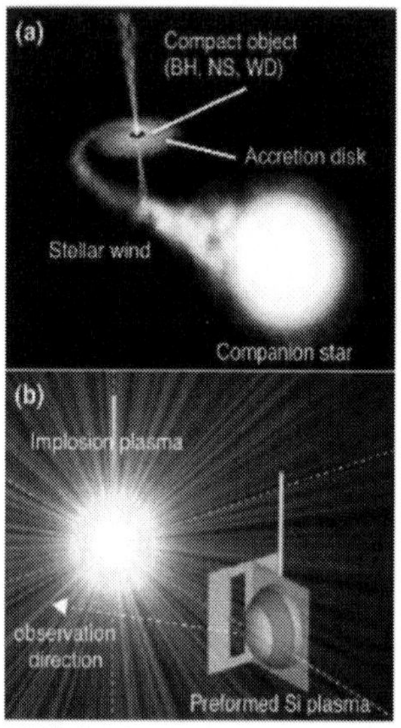

Figure 4. (a) Geometry of strong X-ray emission from compact objects like a black hole with a companion star. (b) Geometry of model experiments where a plastic shell is imploded and near the maximum compression very strong X-rays are emitted. The X-rays irradiate near by Si plasmas to create photo-ionized plasmas.(After S.Fujioka etal., Nature Physics DOI: 1038/NPHYS1402)

As for the HEDS, the properties of hydrogen at high pressure and high density are of great scientific interest. The equation of state (EOS) of hydrogen at these conditions is essential for modeling of the interior structure of gas giant planets. The large diversity in the estimation of Jupiter's core mass is resulted from the uncertainty in the EOS data especially in the region around the insulator-to-metal transition. The EOS of hydrogen isotopes has important practical applications for inertial confinement fusion, and metallic hydrogen is suggested as a prospective candidate of high-temperature superconductor [7].

4. FIREX projects

As for the fast ignition research at ILE, Osaka University, the LFEX laser for the FIREX-I project has completed and the integrated experiment on imploded plasma heating has started. The preliminary experiments indicate neutron enhancement although the input laser intensity was not high enough to realize high coupling efficiency. As shown in fig.5, when the peta watt laser shot coincides with the maximum compression timing, the neutron yield was enhanced.

For improving the coupling efficiency, a new target, so called double cone target has been proposed. Particle-in-cell simulations aimed at improving the coupling efficiency of input laser energy deposited to the compressed core by using a double cone are carried out. It is found that high energy electrons are effectively blocked by the vacuum gap inside the wing of the double cone.(see Fig. 6)

Two main mechanisms to confine high energy electrons are found, which are the sheath electric field at the rear of the inner cone wing and the quasi-static magnetic field inside the vacuum gap. However, the electrostatic retention of the high energy electrons terminates within a few hundreds of femto seconds as the vacuum gap fills with plasma. In comparison, the quasi-static magnetic fields, though arises up slowly in comparison with the electrostatic fields but continue to grow to the order of

100MG. The generation mechanism of the quasi-static magnetic fields is discussed in detail, and it is concluded that the quasi-static magnetic fields continue to confine the high energy electrons for longer than a few picoseconds. The double cones result in the confinement and focusing of 22.3% of the input energy for deposition in the compressed core [8].

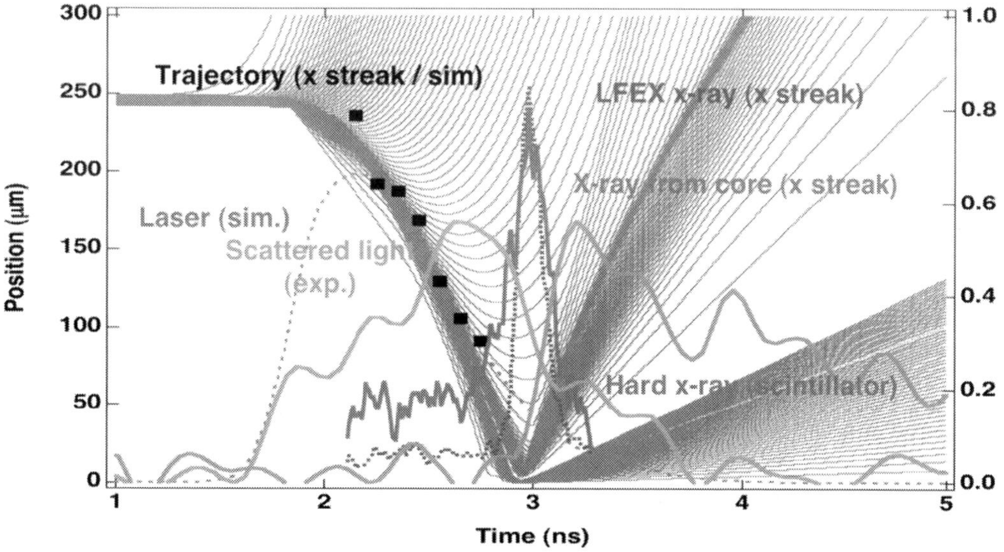

Figure 5 In this shot (shot number 3272) the shot timing is checked by the X-ray streak camera, LFEX X-ray signal and X-ray from the imploded core. The neutron yield is enhanced in this shot.

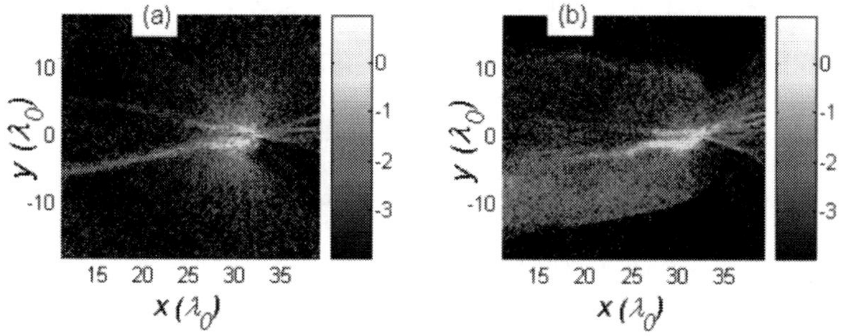

Fig.6 (Color). The electron energy density for single cone (a) and double cone (b) at timet=1500fs.Here the electron energy density is normalized by $mn_{cr}c^2$.

The multi beam LFEX laser will be completed by the April 2010 and the full beam LFEX experiment will start from Middle of 2010. The imploded high density plasma ie expected to be heated up to the ignition temperature in 2011.

References

[1] Y. T. Li etal, the proceeding for the poster 2.10.078 of IFSA2009.
[2] Y.Kato, Plenary talk of IFSA 2009
[3] H.Kiriyama, etal., Optics Communications 282, 625 (2009)
[4] A.Yogo etal., Appl.Phys. Lett 94, 181802 (2009)
[5] Alexeev PRL 90 103402 (2003)
[6] S.Fujioka etal., Nature Physics DOI: 1038/NPHYS1402
[7] T. Sano etal, Physical review letters submitted
[8] H.Cai etal Physical Review Latters 2009

Journal of Physics: Conference Series **244** (2010) 012006 doi:10.1088/1742-6596/244/1/012006

Ignition and Inertial Confinement Fusion at The National Ignition Facility

Edward I Moses

Lawrence Livermore National Laboratory, 7000 East Avenue, Livermore, CA 94551, USA

E-mail: moses1@llnl.gov

Abstract. The National Ignition Facility (NIF), the world's largest and most powerful laser system for inertial confinement fusion (ICF) and for studying high-energy-density (HED) science, is now operational at Lawrence Livermore National Laboratory (LLNL). The NIF is now conducting experiments to commission the laser drive, the hohlraum and the capsule and to develop the infrastructure needed to begin the first ignition experiments in FY 2010. Demonstration of ignition and thermonuclear burn in the laboratory is a major NIF goal. NIF will achieve this by concentrating the energy from the 192 beams into a mm3-sized target and igniting a deuterium-tritium mix, liberating more energy than is required to initiate the fusion reaction. NIF's ignition program is a national effort managed via the National Ignition Campaign (NIC). The NIC has two major goals: execution of DT ignition experiments starting in FY2010 with the goal of demonstrating ignition and a reliable, repeatable ignition platform by the conclusion of the NIC at the end of FY2012. The NIC will also develop the infrastructure and the processes required to operate NIF as a national user facility. The achievement of ignition at NIF will demonstrate the scientific feasibility of ICF and focus worldwide attention on laser fusion as a viable energy option. A laser fusion-based energy concept that builds on NIF, known as LIFE (Laser Inertial Fusion Energy), is currently under development. LIFE is inherently safe and can provide a global carbon-free energy generation solution in the 21st century. This paper describes recent progress on NIF, NIC, and the LIFE concept.

1. The National Ignition Facility

The National Ignition Facility (NIF) is the U.S. Department of Energy (DOE) and the National Nuclear Security Administration (NNSA) center to study inertial confinement fusion (ICF) and high energy density (HED) science. NIF is by far the largest scientific project ever successfully completed by the DOE. NIF, a cornerstone of the U.S. Stockpile Stewardship Program (SSP), will execute the science experiments necessary to ensure a safe, secure, and reliable nuclear weapon stockpile without underground testing. The 192-beam football-stadium sized NIF [1] is now operational at Lawrence Livermore National Laboratory (LLNL). A total 192-beam energy of 1.1 MJ 3ω was demonstrated on March 10, 2009, over 30 times more energy than ever produced in an ICF laser system. Currently, the NIF is conducting experiments to commission the laser drive, the hohlraum and the capsule and to develop the infrastructure needed to begin the first ignition experiments in FY 2010. For example, 11 physics shots were performed in the period from August 16 to September 5, 2009. These shots used 192 beams of over 500 KJ and acquired excellent data.

NIF's 192 beams are directed into a 10-meter-diameter high-vacuum target chamber containing a ~1-cm-long cylindrical hohlraum target. The NIF target chamber contains entry ports for all the laser beams and over 100 ports for diagnostic instrumentation and target insertion. Sophisticated diagnostic instruments such as x-ray and neutron spectrometers, microscopes, and streak cameras, can be mounted around the equator and at the poles of the target chamber. The laser interaction with the hohlraum will produce a radiation field with temperature of several hundreds eV. The resulting hohlraum conditions will provide the necessary environment to explore a wide range of HED science experiments, including laboratory-scale thermonuclear ignition and burn.

NIF is the most complex optical instrument ever constructed, with over 38,000 large and small optics and 60,000 points controlled by two million lines of software. The NIF 3ω energy specification of 1.8 MJ requires an order of magnitude increase in operating fluence over previous laser systems. Developing high-quality optics that can withstand the NIF environment has been a major research and development focus at LLNL. A systematic and robust approach for optics finishing improvement and optics maintenance has been developed to support the ignition demanding requirements.

2.The National Ignition Campaign

2.1 Overview

The NIF ignition program is executed via the National Ignition Campaign (NIC) [2], a national effort that includes General Atomics (GA), LLNL, Los Alamos National Laboratory (LANL), Sandia National Laboratory (SNL), and the University of Rochester Laboratory for Laser Energetics (LLE). The NIC has two major goals: execution of deuterium-tritium (DT) ignition experiments starting in FY2010 with the goal of demonstrating ignition and a reliable, repeatable ignition platform at the completion of the NIC by the end of FY2012. The NIC will also develop the infrastructure and the processes required to operate NIF as a national user facility. The scope for NIC includes the ignition physics program as well as the development of the diagnostics, targets, target cryogenic system, phase plates and other optics, and personnel and environmental protection activities required to execute ignition experiments.

NIF ignition experiments will use a centimeter-scale hohlraum containing a millimeter-scale thin-walled plastic or beryllium capsule filled with a mix of deuterium and tritium. Compression of the capsule by the ~280-eV radiation field in the ignition hohlraum drives the DT fuel to conditions under which it will ignite and burn, liberating more energy than is required to initiate the fusion reaction [3]. NIF is designed to achieve target temperatures of 100 million K, radiation temperature over 3.5 million K, density of 1,000 g/cm^3 and 100 billion times atmospheric pressure. These conditions have never been created in a laboratory and exist naturally only in the interiors of the stars and during thermonuclear burn.

Initial experiments aimed at understanding the energetics of the NIF ignition hohlraum and capsule-tuning experiments began in the summer of 2009. Cryogenic ignition targets filled with a deuterium-tritium mix capable of yields in the 10-20 MJ range will be tested in late FY2010. Experiments at other laser facilities such as OMEGA at LLE, Z at SNL, Trident at LANL and Jupiter at LLNL have been and continue to be used to develop and demonstrate tuning, shock timing, laser ablation and the diagnostics techniques needed to achieve ignition.

The NIC ignition campaigns uses the "indirect-drive" configuration, where the laser beams are directed into the ends of a cylinder coated with gold or other high-Z material mounted vertically within the target chamber. The laser irradiation of the cylinder produces a radiation field inside the cylinder that implodes a mm-scale capsule filled with a mix of deuterium and tritium. As shown in

Figure 1, the laser beams are deployed in multiple cones to control the time-dependent symmetry of the radiation drive. The resulting asymmetry is maintained to less than 1% on average. The high degree of symmetry of the NIF implosion coupled with a precisely tailored target design and corresponding laser pulse results in the high peak fuel ρR required for alpha heating of the fuel and capsule ignition.

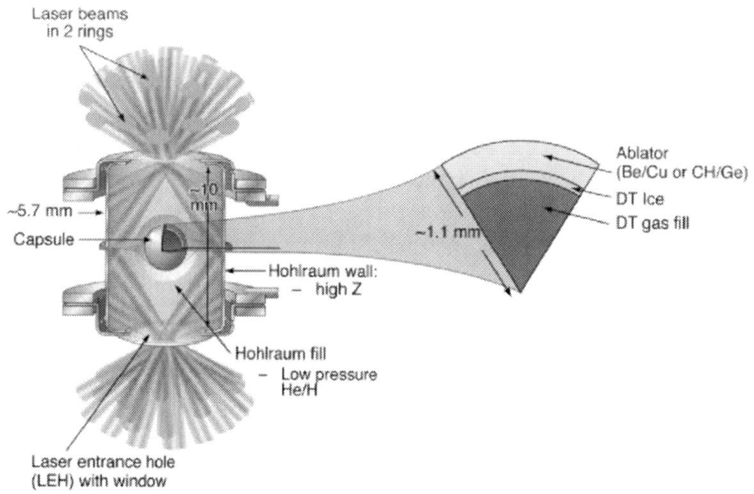

Figure 1. Schematic of NIF ignition target.

Initial ignition experiments will begin in late FY2010 at 192-beam ultraviolet laser energy of approximately 1.2 MJ. This energy level is consistent with the laser commissioning plan that ramps up to the planned 192-beam operating energy of 1.8 MJ by late 2011. The target design to be used is shown in Figure 1. Two capsule ablator materials (Be with Cu dopant and CH with Ge dopant) are currently under consideration. Both capsules are filled with DT gas via a 10-mm-diameter SiO_2 fill tube to a density of 0.3 mg/cm^3 with a DT solid layer at 18.3 degrees Kelvin. The Cu and Ge dopant density is varied through the ablator and is typically present at less than 1% concentration. Maintaining these two design options reduces risk and allows the choice of a target that is optimally configured to the precise laser and target fabrication capabilities available.

2.2 Ignition Target Developments and Fabrication

Ignition target development and fabrication has achieved considerable progress since the start of NIF construction in 1997; indeed, ignition target development and fabrication is a research and development program in its own right. NIC ignition targets must meet demanding specifications. Components must be machined to within an accuracy of 1 μm, with joints as small as 100 nm. The margin of error for target assembly is less than 8 μm. Typically, the capsule outer surface must be smooth to within 1 nm, and the thickness and corresponding opacity of the doped layers must also be carefully controlled.

A long-standing challenge for ignition target fabrication is the frozen DT fuel layer [4]. This layer must be formed at approximately 1.5 degrees below the triple point of the DT mixture. The layer temperature must fluctuate no more than 1 mK, the roughness of the inner layer surface must be maintained at 1-μm RMS roughness or better, and spherical isotherms must be maintained at the layer surface via auxiliary heating. These challenges have been addressed, and ignition targets meeting all specifications are now in production. Figure 2 shows an actual ignition target assembly featuring the thermo-mechanical package used to maintain the target at the required specification. The target is held at the center of the NIF target chamber via a cryostat attached to the NIF target positioner. The system also includes a characterization station capable of imaging the DT layer in three spatial dimensions within minutes.

2.3 NIC Diagnostics

The diagnostic suite for NIC is also a major focus of the NIC efforts in FY2009 and FY2010. By the end of FY2010, it is anticipated that approximately 35 diagnostics [5] measuring x-ray, neutron, charged particle, optical, and other emissions will be installed. Examples of diagnostics include full-aperture backscatter measurement capability for use in hohlraum energetics experiments, velocity interferometers for shock timing, absolutely calibrated soft x-ray spectrometers to measure the radiation drive, gamma-ray detectors to measure the burn history of the ignition target, and magnetic recoil spectrometers for neutron spectroscopy. Diagnostics development is a national and international effort. Further information regarding NIF diagnostics is available at the NIF website.*

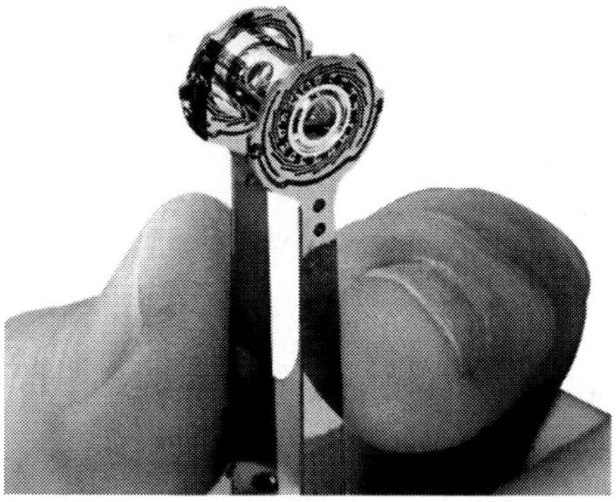

Figure 2. NIF fusion target thermo-mechanical package.

2.4 NIC Experimental Plan

The NIC experimental plan consists of four phases. The first phase will culminate with the first attempts at inertial fusion ignition in late FY2010. The subsequent three phases will refine the target and laser parameters and investigate the physics of the ignition regime, with a goal of providing a reliable and repeatable ignition platform by the conclusion of NIC at the end of FY2012.

Based on many years of experimentation and simulations, it is believed that the NIC experimental campaign requires tuning 14 laser and 3 target parameters to those required for ignition conditions. The laser and target parameters to be tuned are shown in Figure 3. The tuning of these parameters corresponds to tailoring the capsule adiabat, velocity, symmetry, and degree of hydrodynamic instability. These 17 parameters are tuned in four steps designed to tailor the laser and cryogenic DT target to ignition conditions. In the first or "drive" step, the empty hohlraum is tuned to produce the necessary radiation drive on the capsule as a function of time. In the "tuning" step, a variety of non-cryogenic and cryogenic deuterium-filled capsules are used to adjust the hohlraum symmetry and shock timing so as so produce the compressed fuel central "hot spot" conditions required for ignition. The third step consists of layered cryogenic implosions conducted with a 50%/49%/1% mix of tritium, hydrogen, and deuterium (THD) respectively. The reduced yield from these THD targets allows the full diagnostic suite to be employed and the presence of the required temperature and fuel areal density to be verified. The final step is DT ignition implosions with expected gains of 10-20. The FY2010 DT ignition experiments will be conducted with E_{laser} ~1.2 MJ. Laser energies of 1.8 MJ should be available for subsequent experiments.

* NIF web site, https://lasers.llnl.gov/programs/nic/diagnostics.php

In preparing for executing the ignition experimental series, the NIC team conducted a "simulated campaign" to exercise the experimental team and develop the ability of NIC scientists to quickly tune the ignition target to the required conditions. The "Blue Team" specified and executed simulated experiments that exercised much of the NIF laser, target, and operational infrastructure, while the "Red Team" provided synthetic data that took into account specified target fabrication errors, laser power imbalance, backscatter, and cross-beam energy transfer. Over a several month period, the Blue Team demonstrated synthetic ignition by successfully adjusting laser and target parameters to compensate for detunings specified by the Red Team and reduced from days to hours the time required to examine a data set and experimentally tune the laser and target.

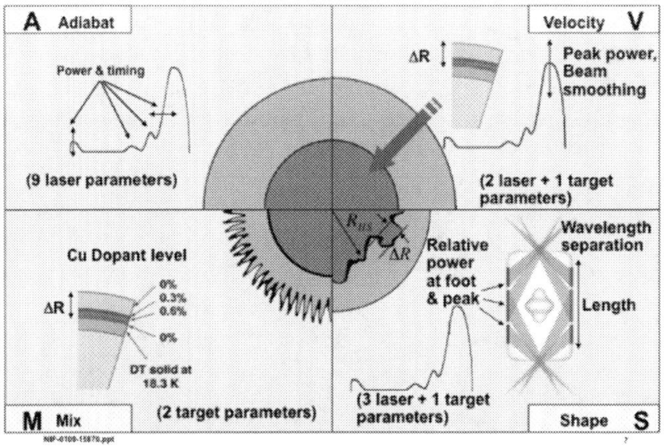

Figure 3. Laser and target parameters to be tuned in the NIF ignition campaigns.

3.0 NIF Multi-Mission Experimental Program

While the ignition experiments will be the primary focus through FY2012, NIF will execute other experiments in support of stockpile stewardship, national defense, and fundamental HED science. Weapons physics experiments conducted in this time frame will reduce uncertainties in key areas and enable improved predictive capability. Experiments conducted in collaboration with the Defense Threat Reduction Agency (DTRA) and the Missile Defense Agency (MDA) will validate NIF as an ultra-low debris soft x-ray ($E < 15$ keV) simulator. Nuclear survivability tests of specific components will also be conducted.

NIF will provide national and international researchers unparalleled opportunities to explore fundamental astrophysics, planetary physics, hydrodynamics, nonlinear optical physics, and materials science. Examples of experiments planned for NIF include investigation of the physics of planetary interiors, the formation of elements with $Z>26$ via Type II supernovae explosions, excited state nuclear reactions, and ultra-intense laser-matter interactions. NIF will ultimately be a major international center for fundamental HED science.

4.0 Ignition and Inertial Fusion Energy

The demonstration of ignition at NIF will demonstrate the scientific feasibility of ICF and will likely focus the world's attention on the possibility of an ICF energy option. NIF will have a pulse repetition frequency of only 10^{-4} Hz and estimated electrical efficiency of 0.5%, while a 2000-MW Inertial Fusion Energy (IFE) power plant would require a repetition frequency of 5-10 Hz and 20% electrical efficiency, based on current expectations of achievable target gain. Fusion energy will thus require the development of lasers, chambers, optical components, and other systems capable of operating at the ~10 Hz repetition rate required for hundreds of MW to a few GW fusion power plants. Considerable progress has been made on these technologies.

The achievement of ignition at NIF will motivate more detailed consideration of ICF as an option for clean, sustainable energy. Both pure fusion and fusion-fission hybrid schemes for energy production are under consideration. The key feature of this Laser Inertial Fusion Energy (LIFE) concept is the use of ICF-generated neutrons to induce fission reactions in the fuel, extracting virtually all its energy content and leaving only a small residue of long-lived actinide waste [6]. This approach could close the nuclear fuel cycle without the need for chemical separation and reprocessing, while generating thousands of megawatts of carbon-free electricity. Such scheme could also incinerate more than 99 percent of spent reactor fuel and extend the service life of deep geologic repositories by up to a factor of 20. The LIFE concept appears considerably attractive given the closed nature of the fuel cycle and the possibility of burning spent nuclear fuel and excess weapons plutonium and highly enriched uranium.

5.0 Conclusion

NIF is now operational, with ignition experiments expected to begin in late FY2010. After many years of R&D, most of the pieces needed for demonstrating ignition at NIF are in place, including the NIF laser, a detailed point design with adequate margin, high-quality targets meeting all specifications, and advanced diagnostics capable of precision tuning of capsule performance to ignition conditions. NIC is the centerpiece of the NIF experimental program. NIF ignition will allow access to the burning plasma regime for the first time, enabling important stockpile stewardship studies and demonstrating the scientific feasibility of ICF. More generally, NIF's ability to create extraordinarily high pressures, temperatures, and densities—as much as 1 trillion atmospheres pressure, 100 million degrees K temperature, and 1,000 g/cm^3 density—will enable major fundamental advances in support of DOE's national security, energy, and fundamental science missions. NIF and other major facilities worldwide will launch a new era in HED science and the demonstration of ignition may one day lead to an inexhaustible clean power supply for humanity.

Ignition at NIF will be a landmark scientific achievement and will open new possibilities for clean, sustainable energy production. NIF will also allow fundamental study of matter at extreme energy densities, including the possibility of examining astrophysical phenomena in the laboratory for the first time. NIF will open a truly new era in HED science.

This work performed under the auspices of the U.S. Department of Energy by Lawrence Livermore National Laboratory under Contract DE-AC52-07NA27344. LLNS-CONF-417774.

REFERENCES

[1] Haynam C A et al. 2007 National Ignition Facility Laser Performance Status *Applied Optics* **46** 3276
[2] "National Ignition Campaign Execution Plan" 2005 LLNL report UCRL-AR-213718, NIF-0111975-AA
[3] Lindl J D 1998 Inertial confinement fusion: the quest for ignition and energy gain using indirect drive *American Institute of Physics Press* New York
[4] Slater J, Kozioziemski B, Collins G W, Mapoles E R, Pipes J, Burmann J, Bernat T P 1999 Cryogenic D-T fuel layers formed in 1 mm spheres by beta-layering *Fusion Tech.* **35** 2 229–233
[5] Moses E I 2008 Ignition on the National Ignition Facility: a path toward inertial fusion energy, *Nuclear Fusion*, to be published
[6] Moses E I, Diaz de la Rubia T, Storm E, Latkowski J F, Farmer J C, Abbott R P, Kramer K J, Peterson P F, Shaw H F, and Lehman II R F 2009 A Sustainable Nuclear Fuel Cycle Based on Inertial Fusion Energy *Fusion Science and Technology* **56** 547

Grand Challenges of Inertial Fusion Energy

J H Nuckolls
Lawrence Livermore National Laboratory, P.O. Box 808, L-001, Livermore, CA 94551 USA

E-mail: nuckolls2@llnl.gov

Abstract

As soon as practical, Earth's low-cost, abundant, environmentally attractive fusion energy resources should be applied to the urgent global challenges of climate change, energy supply, economic growth, and the developing world. A National Ignition Campaign is under way at the recently completed National Ignition Facility (NIF) at Lawrence Livermore National Laboratory (LLNL) to ignite high-gain inertial fusion targets in the 2010–2012 time frame. Achieving ignition on NIF could be the catalyst for national and global leaders to support the development of inertial fusion energy (IFE) to meet the future's worldwide electric power demand. With sustained, high-priority funding could practical IFE be possible by the 2020 timeframe? The answer lies in how well can the community address and solve technical challenges in four key areas: achieving ignition, producing advanced targets and drivers, creating a practical fusion engine, and developing economical fusion power plants.

1. Introduction

At the National Ignition Facility[1], an extraordinary three-year inertial fusion ignition campaign is under way[2,3]. Ignition, advanced drivers and targets, fusion engines, and inertial fusion power plants are four grand challenges that may define a path to accelerated development of inertial fusion energy (IFE). The ultimate goal is to develop economically attractive IFE power plants that will attract large-scale commercial funding to enable global deployment.

Ignition – Ignition is a grand challenge[4]. Laser heated hohlraums energize tiny super-implosions designed to ignite propagating thermonuclear burn[5,6]. Record high velocities, densities, symmetry, and temperatures must be achieved before excessive growth of plasma and hydrodynamic instabilities occur and impede the process.

Advanced Targets and Drivers—NIF will enable the testing and development of advanced targets. High-gain spark-ignited targets can be initiated with several megajoules of 2ω laser light. Very high-gain fast-ignitor targets[7,8,9,10] can be imploded with hundreds of kilojoules of 2ω laser light or ignited with an estimated hundred-kilojoule petawatt pulse of laser light[11,12]. A small mass of high-density, fast-ignited DT may be used to ignite a 10-times larger mass of DT with density less than 10 g/cm^3. Alternatively, low-density DT may be imploded, in principle, more efficiently with a small mass of driver-ignited chemical propellant than with a 10% efficient laser, energized by a 40% power plant. Fusion yields as high as 1000 MJ may eventually be achieved.

Similarly, high repetition rate/efficient/high-average-power drivers[13,14] are being developed that can energize efficient implosions, including 2ω diode-pumped ceramic lasers capable of 10 shots per second. In addition, advanced petawatt fast ignition lasers are being developed that could reduce the cost of implosion laser systems.

Thermonuclear Engine—A burst mode, heat capacity, thermonuclear engine with fusion performance comparable to the International Thermonuclear Experimental Reactor (ITER)[15] may be feasible. Time and cost would be reduced by the use of highly modular drivers and target factories, as well as by implementing advanced lasers and targets. The repetition rate would gradually be increased to create a powerful 14-MeV neutron source. Radiation-hardened materials[16] for IFE and magnetic fusion energy (MFE) fusion reactors and Laser Inertial Fusion Energy (LIFE) fission hybrids could be developed[17]. Development of LIFE pure fusion reactors could be accelerated by using high-temperature fusion chambers with protected first walls to avoid the long-duration qualification of new 14-MeV neutron radiation-hardened materials and to achieve high thermal-electric efficiencies.

IFE Power Plants—Reduced driver and target costs will enable accelerated development of economically attractive IFE power plants. Successive generations of IFE power plants with economic advantages enabled by continuing advances in drivers and targets could well succeed at meeting the exacting criteria of future global energy requirements.

Detailed plans should be developed to assess the feasibility and estimated costs of accelerated IFE development.

2. Ignition

Ignition is the pivotal grand challenge. The National Ignition Campaign (NIC) at LLNL includes hardware and infrastructure needed to execute the initial ignition experiments and to continue research on ignition in subsequent years. Key elements of the campaign include target design code validation and equipment, such as diagnostics and the cryogenic target system.

During the ignition process, laser-heated hohlraums energize tiny super-implosions designed to ignite propagating thermonuclear burn. High velocities, densities, symmetry, and temperatures must be achieved before excessive growth of plasma and hydrodynamic instabilities occur.

In the NIC, hundreds of highly diagnosed experiments will be conducted to improve the physics approximations used in the target design codes. Unknowns will be bounded, and the target design will be optimized. "Black swan" failures modes, including phenomena that do not follow Gaussian statistics are judged to be unlikely, but cannot be precluded.

Consequently, alternative targets are being developed and may be utilized, including polar direct-drive shock-ignited targets[18,19], two-shell targets, and fast ignition targets.

Fast-ignited isochoric targets may be one of the most promising advances to the point design spark-ignited isobaric targets currently being used in the ignition campaign. In the fast ignition approach, in which compression is separated from the ignition phase, implosion velocity and density are less than those in spark ignition, so in principle, fast ignition will allow some relaxation of the requirement for maintaining precise, spherical symmetry in the implosion. The improved energy gain is estimated to be as much as a factor of 10 over the spark-ignited approach. For these reasons, enhanced performance fast-ignited targets may be a leading candidate for use in fusion engines and economically attractive power plants.

3. Advanced Targets and Drivers

After ICF target designs achieve ignition, performance may be improved in post-ignition NIF experiments, beginning with the point design indirectly driven target used in current NIF ignition experiments. Although indirectly driven target schemes were proposed for inertial confinement fusion 50 years ago, and extraordinary improvements have been made, I believe there is still high potential for performance improvement after ignition is demonstrated.

A major goal for IFE is to significantly reduce target[20] and driver costs, which would reduce costs of fusion engines and future power plants. As a first step, it has been estimated that the driver capital cost can be reduced to less than $400 M, which corresponds to 800 kJ at $500/joule, based on current driver development trends. An annual target cost of $30 M, corresponds to a 1,000-MJ-target yield with a 3-Hz repetition rate, assuming a 30¢ cost per target. Maximum target improvements may reduce these costs another factor of two (to about 5% of the capital cost and gross income of a 1-GWe power plant).

Directly driven fast ignition targets are estimated to yield ~250 MJ when imploded with an 800-kJ laser. To increase the fusion yield several fold, the isochoric imploded density distribution may be modified to create a steep density gradient (or density step). The ignition laser energy is reduced by the use of high DT densities, and the implosion laser energy is reduced (or the DT mass and yield increased) by the use of lower DT densities. With density-radius products greater than ~1 g/cm^2, thermonuclear burn may be propagated from a small mass of high-density fast-ignited DT to ignite a 10-times larger mass of low-density DT.

George Zimmerman of Lawrence Livermore National Laboratory has run a 2D LASNEX explosion calculation of a density step approximation to a density gradient. The initial configuration for the calculation is shown in Fig. 1 The calculation of a fast-ignited hemispherical DT mass, density 300 g/cm^3, $\rho R \sim 3$ g/cm^2, exploding into a 10 times larger hemispherical DT mass, density 30g/cm^3, $\rho R \sim 1.4$ g/ cm^2 is shown in Fig. 2. In the calculation, the fast-ignited mass explodes and ignites the 10-times larger mass, 10-times lower density hemisphere, and the total yield increases ~ 10 fold.

If the compression energy invested in the lower-density hemisphere is used to add high-density mass to the 300 g/cm^3 hemisphere, the explosion yield increases by about four fold. Consequently, the gain is increased ~2.5 fold by use of the density step. With 800 kJ of laser energy, a density gradient may be used in a directly driven fast ignited target to increase the yield from 250 MJ to ~600 MJ. If the implosion efficiency can be increased as the density and implosion velocity are reduced, even higher yields may be achieved.

If DT densities as low as a few g/cm^3 are ignited, dense pusher-tampers and efficient cannon-like implosion systems may be utilized (instead of ablative implosions). Inertial tampers may also increase the DT burn efficiency and fusion yield.

In another option, very small driver energies may be used to initiate ~100 grams of chemical propellant (costing ~10¢) to energize cannon-like implosions to a few g/cm^3. Use of magnetic fields should also be explored. Fusion yields of 1,000 MJ may be achieved with enhanced performance fast-ignited targets.

In light of these various alternatives, instead of developing more complex target designs to increase the target yield, it may be possible to develop simple lower-cost targets. Greatly increased efforts should be devoted to the development of advanced targets given the importance of this factor on the future of IFE.

4. The Fusion Engine

Assuming advanced target factory modules are developed in parallel with high-performance targets and high-performance driver modules, these advances would significantly reduce the cost of a fusion engine and fusion power plants and could serve as a catalyst to initiate accelerated fusion engine and IFE development programs.

The physical separation of the driver, target factory, and fusion explosion chamber in the fusion engine (and power reactor) minimizes radiation damage. This separation, the high modularity of the driver and target factory, and the small size and high improvement potential of the driver and targets may enable rapid evolution of the fusion engine.

Fusion chambers must be constructed to operate at a high temperature, more than 500 degrees Celsius, to enable high thermal-electric efficiencies for the power conversion system. The fusion explosion chamber first wall radiation damage and cyclic fatigue may be greatly reduced by using a protective shield to mitigate exposure to intense heat and radiation from the target's X rays, ions and neutrons.

An fusion engine could be constructed to operate initially as a surged heat capacity system, e.g., a shot every 100 seconds for 1,000 seconds. Performance would then evolve by increasing the repetition rate and operating time, e.g., to a greater than one-Hz, 100-MW fusion power system operating 80% of full time.

Applications of a fusion engine 100-MW, 14-MeV neutron source might include exploration of fission waste burn-up, breeding enough tritium to enable startup of many full-scale fusion reactors, development and testing of advanced radiation-hardened materials, and development of nonproliferation safeguards for fusion power plants. Demonstration of the technological feasibility of economically attractive inertial fusion power plants would be accomplished as rapidly as possible.

5. IFE Power Plants

With strong sustained government funding of an IFE power plant development program, less than 30 years (possibly 10 to 15) may be required to develop the first economically attractive IFE power plant, i.e., a power plant that would be replicated, rapidly improved, and deployed in large numbers, using large-scale commercial funding (or large-scale government funding in China and other controlled economies)

Several years would be subtracted from IFE development time by a near-term surge in funding to accelerate achievement of ignition and development of advanced targets, modules of advanced drivers and target factories, and a prototype high-temperature fusion explosion chamber.

With accelerated development, in the 2020s fusion energy may begin to contribute to solving 21st Century problems – including climate change, energy supply, economic growth, and the rise of the developing world.

IFE power plants will have powerful genetic advantages that may enable rapid improvement. With cheap, clean inexhaustible fusion fuel available to all nations, IFE may excel in a long-range competition of future energy sources.

This work performed under the auspices of the U.S. Department of Energy by Lawrence Livermore National Laboratory under Contract DE-AC52-07NA27344.

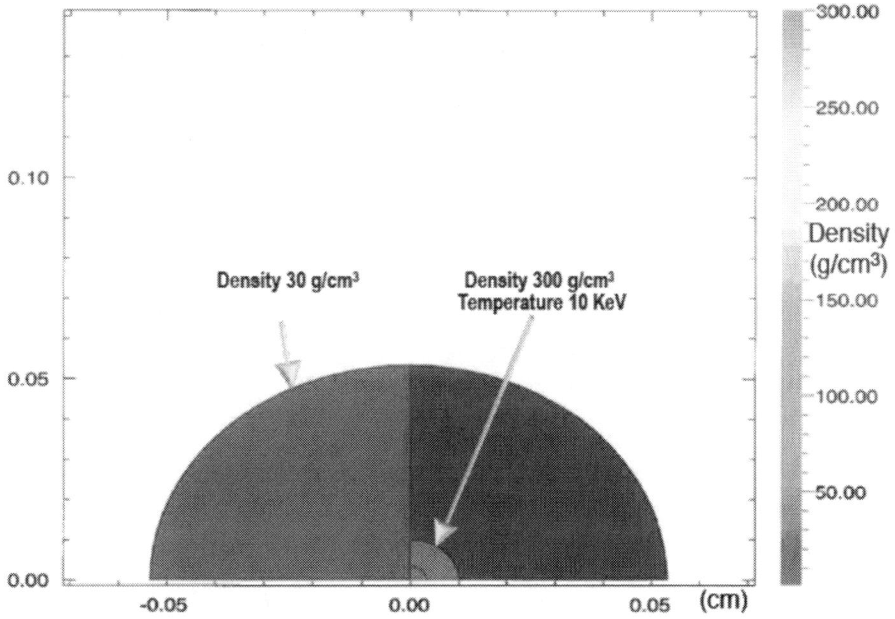

Fig. 1 Initial configuration for calculations of ignition across a 10X density step

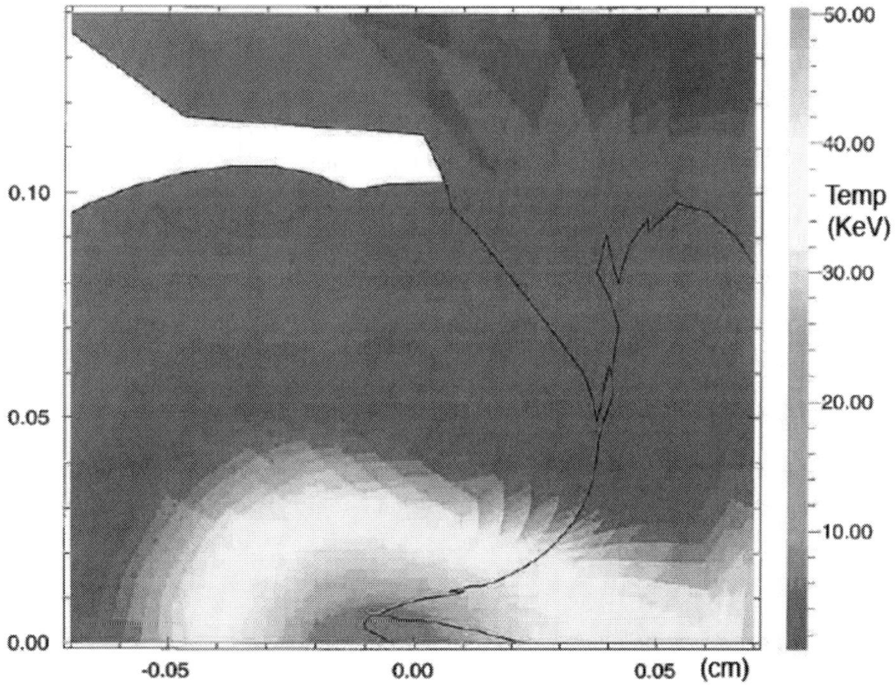

Fig. 2 Fusion driven explosion of the high-density fuel ignites the 10X lower density fuel (George Zimmerman - private communication)

REFERENCES

[1] E. I. Moses, "Ignition on the National Ignition Facility: a path towards inertial fusion energy", Nucl. Fusion 49(2009) 104022

[2] S. H. Glenzer, et al., Symmetric Inertial Confinement Fusion Implosions at Ultra-high Laser Energies, Science Express/www.science express.org/ 28 Jan 2010 page 1 science 1185634

[3] The initial results of this campaign are also documented in the IFSA 2009 proceedings

[4] J. H. Nuckolls, L. Wood, A. Thiessen, and G. B. Zimmerman, "Laser compression of matter to super-high densities: thermonuclear (CTR) applications," Nature 239, 139 (1972)

[5] J. D. Lindl, "Inertial Confinement Fusion," Springer-Verlag 1998.

[6] J. D. Lindl, "The physics basis for ignition using indirect-drive targets on the National Ignition Facility," Phys. Plasmas 11(2), 339 (2004).

[7] M. Tabak et al., "Ignition and high gain with ultrapowerful lasers," Phys. Plasmas 1 (1994) 1626-34

[8] M. Tabak et al. "Fast Ignition: Overview and Background," Fusion Science and Technology 49, 254–277 (2006).

[9] J. J. Honrubia and J. Meyer-ter-Vehn, "Fast ignition of fusion targets by laser-driven electrons, Plasma Phys. Control. Fusion 51 (2009 014008

[10] R. Kodama, et. al., "Fast Heating of ultrahigh-density plasma as a step toward fusion ignition", Nature, 412, 798 (2001)

[11] M. Dunne, "A high-power laser fusion facility for Europe", Nature Physics 2,2 (2006)

[12] K. Mima and T. Takada, "Proof of principle experiment for fast ignition and the Fast Ignition Realization Experiment (FIREX), Fus. Sci. and Tech 49,358 (2006)

[13] A. Bayramian, et al., "The Mercury project: A high average power, gas-cooled laser for inertial fusion energy development," Fusion Science and Technology 52, 383–387 (2007).

[14] J. D. Caird et al., "Nd:Glass Laser Design for Laser ICF Fission Energy (LIFE)," Fusion Science and Technology, submitted (2008).

[15] http://www.iter.org

[16] J. C. FARMER, "LIFE materials: Overview of fuels and structural materials issues," LLNL-TR-407386 Rev 1, 2008.

[17] J. F. Latkowski et al., "System Integration Issues for the Laser Inertial Confinement Fusion Fission Energy (LIFE) Engine," Fusion Science and Technology, submitted (2008).

[18] R. Betti, et. al., "Shock ignition of thermonuclear fuel with high areal density", Phys. Rev. Lett 98, 155001 (2007)

[19] L. J. Perkins, R. Betti, K. N. LaFortune, W. H. Williams, "Shock Ignition: A new approach to high gain inertial confinement fusion on the National Ignition Facility", Phys. Rev. Lett. 103, 045004 (2009)

[20] D. T. Frey et al., "Mass production methods for fabrication of inertial fusion targets," Fusion Science and Technology 51, 786–790 (2007); D. T. Goodin et al., "Developing a commercial production process for 500,000 targets per day: A key challenge for inertial fusion energy," Physics of Plasmas 13, 056305 (2006); R. MILES et al., "LIFE target fabrication research plan," LLNL report, 2008, LLNL-TR-408722.)

---- See also J. Nuckolls, L. Wood, "Future of Inertial Fusion Energy", Proceedings of 11th INCENES(2002) UCRL-36-14960

Journal of Physics: Conference Series **244** (2010) 012008 doi:10.1088/1742-6596/244/1/012008

High energy density science with tabletop terawatt lasers

G Ravindra Kumar
Tata Institute of Fundamental Research, 1 Homi Bhabha Road, Colaba, Mumbai - 400005, India

E-mail: grk@tifr.res.in

Abstract. High energy density science (HEDS) addresses some of the most fundamental questions in contemporary research. Tabletop terawatt femtosecond lasers provide a robust platform for experiments in HEDS – on the one hand, they push the limits of ultrarapid excitation and on the other they provide a perfect complement to the experiments performed with high energy, longer pulse lasers. After a general introduction we consider results in a niche regime of these lasers, namely femtosecond dynamics. Specifically, we consider pump-probe experiments with relativistic intensity 30 fs, 10 Hz, 800 nm Ti: Sapphire laser pulses. The physical processes considered are the transport of hot electrons with their associated giant magnetic fields, the spatio-temporal evolution of these magnetic fields at target front and rear and ultrafast dynamics of the plasma critical surface.

In the last twenty years, the chirped pulse amplification scheme has revolutionized the production of high intensity light in two important ways - (a) pushing the upper limits of intensity more than a thousand fold and (b) reducing the foot print of high intensity lasers by a factor as large as 100 (with concomitant cost reduction). In doing so, it also democratized the participation of researchers by offering smaller groups with limited resources an opportunity to join the ranks and make effective contributions. This last aspect of 'globalization' has led to establishing more robust methods in the area.

This paper seeks to discuss some of these features in the area of high energy density science (HEDS). It is well known that HEDS addresses some of the most fundamental questions in science [1], with its emphasis on understanding the creation and behaviour of extreme states. It is further gratifying to note that table top, femtosecond lasers can play a key role in the creation of these states as the femtosecond nature of the pulses freezes out deleterious motions and couples energy at solid densities. While many interesting HEDS experiments are already being performed at the big facilities around the world and more will definitely be undertaken at the NIF, table top experiments can provide complimentary and unique perspectives in this area. High energy laser pulses from the big facilities are necessarily limited to a few shots per experiment and our colleagues address and attempt to overcome this limitation by increasing the versatility of diagnostics in the experiment and providing enough multiplicity and redundance. The experiments are viewed in close conjunction with simulations and this all round effort has been pushing the frontiers for many years. It is obvious however that the field would benefit immensely by experiments in a more traditional mode, i.e. ease of operation, ready manipulation of beams, more repetitive collection of data in the establishment of systematic, calibration and cross checks by multiple labs with similar capabilities and possibilities for testing extended ideas with incremental efforts. Current tabletop lasers promise all these and more.

Journal of Physics: Conference Series **244** (2010) 012008 doi:10.1088/1742-6596/244/1/012008

The big question in HEDS i.e. creation and evolution of extreme states under controlled conditions, is best met by high intensity laser pulses with the shortest possible duration, provided they are truly short. The latter implies that the pulses have to have increasing contrast with background simultaneously with increasing intensity. What make tabletop femtosecond lasers very attractive is the recent improvements in producing temporally leaner and cleaner pulses. Presently, laser systems operating at the 100 TW level offers an ASE contrast of better than 10^{-9} and a contrast of 10^{-8} at the 10-20 ps level [2]. At a peak intensity of 10^{21} W cm^{-2}, prepulse levels are only at the level of $10^{12\text{-}13}$ W cm^{-2} at the nanosecond and picosecond levels. The second aspect is that of ever shortening duration. From the 100 fs level at which high intensity pulses were routinely produced a decade ago by a tabletop laser, we have advanced to the 10-20 femtosecond level and shorter few cycle pulses should be available at relativistic intensities in the near future. Some of the interesting consequences of this shrinking are the induction of nonadiabatic effects, higher levels of absorption in steeper, denser plasmas and associated physical effects. Absorption of as much as 80% of laser light has been observed with clean p-polarized 8 fs pulses in highly overdense plasma with a very steep gradient, possibly from mechanisms other than vacuum heating [3]. Similar experiments at ultrahigh intensities (close to 10^{21} W cm^{-2}) at Livermore also showed very high levels of coupling (upto 60% for normal and 85% for 45 degree incidence [4]. It is also well known now that femtosecond, high contrast pulses shining on ultrathin targets can produce the highest proton and ion energies (see for example, Maksimchuk *et al.*, IFSA09 oral 3.4.3).

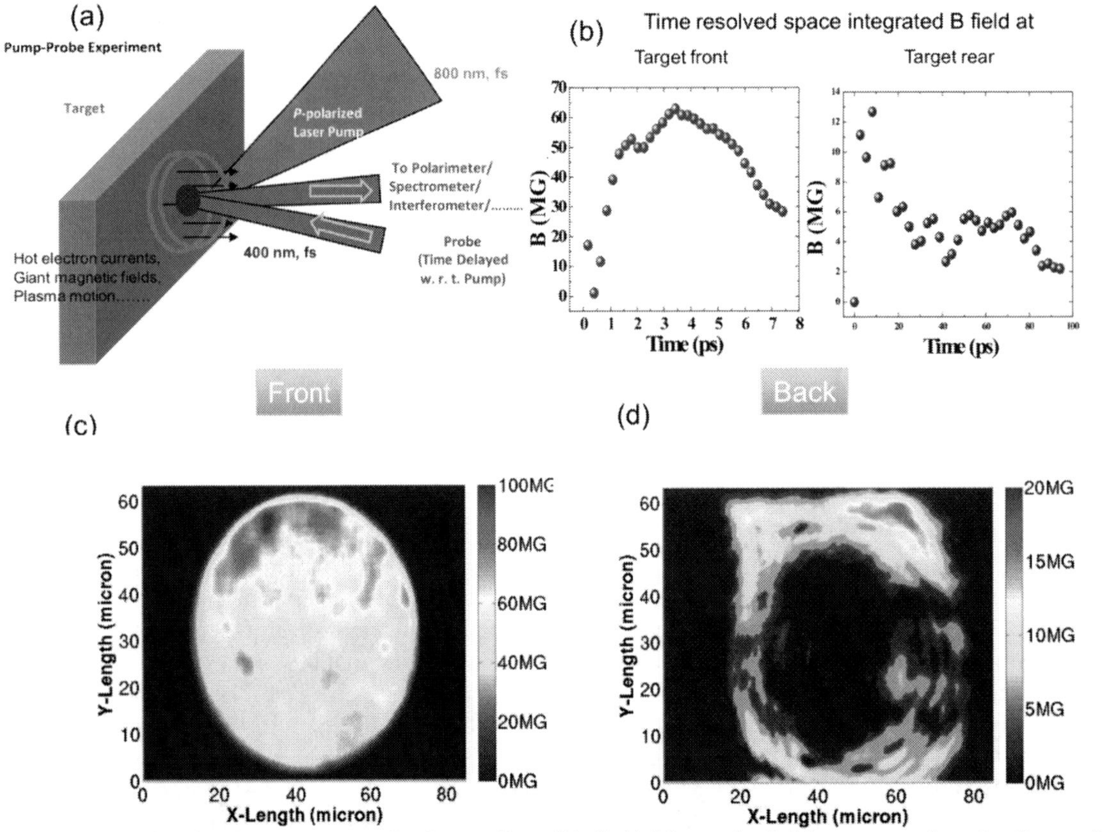

Figure 1. (a) Sketch of set up Pump-Probe studies. (b) (left) Magnetic fields measured at the front of an aluminium coated glass target and (right) the rear of a 100 micron fused silica target. (c) Spatially resolved magnetic field for 3 picosecond pump-probe delay at the critical surface at target front for an aluminium coated glass target. (d) Spatially resolved magnetic field at the critical surface of plasma at the rear of a 100 micron thick fused silica target at the same delay.

Let me now illustrate the power of the tabletop lasers in the realm of unravelling ultrafast dynamics for which they are uniquely positioned. A typical pump-probe set up that is used in our laboratory is depicted in Fig.1 (top, left). The measurement of the polarization state of the probe, reflected from its critical layer (polarimetry) gives a measure of the giant magnetic fields, depicted on the right. Magnetic fields as large as 65 MG at the front of an aluminium coated glass target and 15 MG at the back of a 100 micron fused silica target have been measured at intensities of 10^{18}- 10^{19} W cm^{-2}. These magnetic fields are caused by the hot electron currents in the target and their time evolution mirrors the electron transport process involving both forward and return currents. We have also captured spatial profiles of the magnetic fields at different instants, shown at the bottom of the figure (x-y profile of the magnetic field as captured by the CCD camera). These are the first ever measurements of the simultaneous space and time resolved giant magnetic field. The 2D profile of the magnetic field is not cylindrically symmetric but has a structure inside. This indicates that current distribution which is responsible for this magnetic field has filamentary behaviour. It is interesting to see that the field at the back has a hollow portion at the centre. Mapping of this magnetic field provides the insight of the transport of relativistic electrons in the solid dense plasma. These data are being further refined and analyzed in detail and will be published in details elsewhere along with appropriate modelling (for some details see Mondal *et al.*, IFSA09 poster 3.10.038). Preliminary analysis reveals the signatures of magnetic turbulence in these plasmas. Our recent work has also shown that crucial transport information can be derived from the measurement of giant magnetic pulses [5-7].

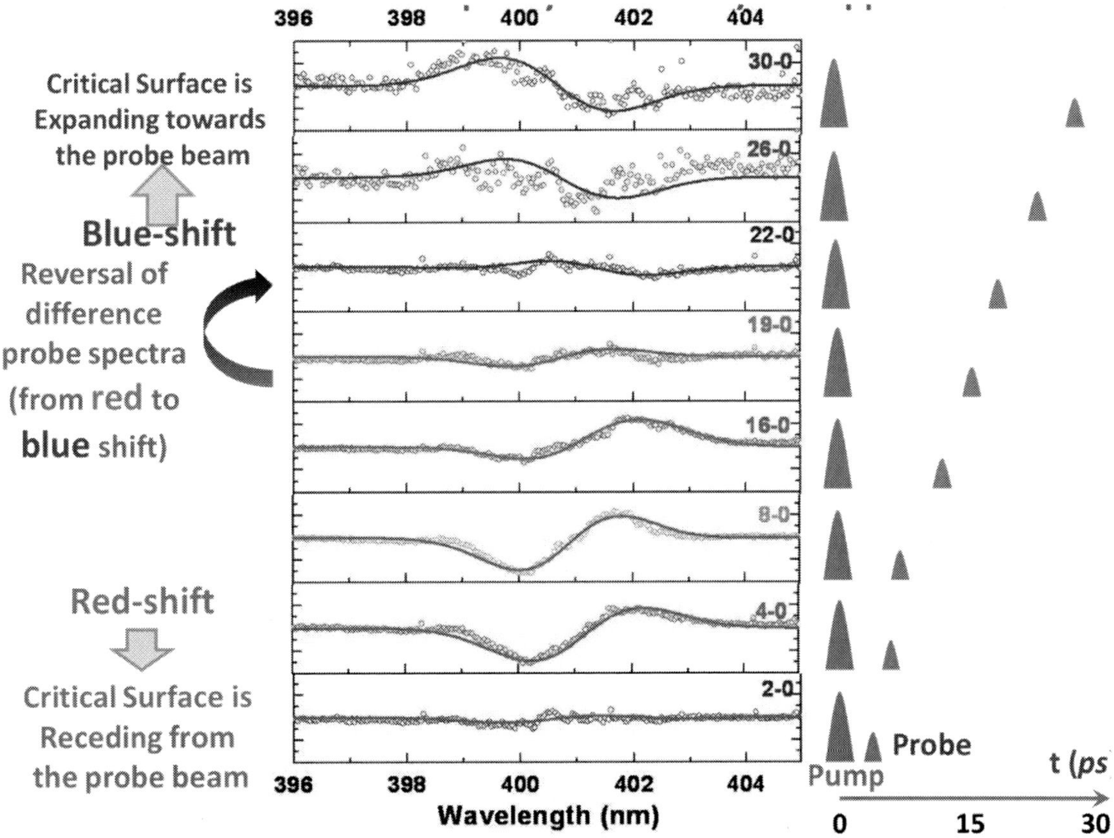

Figure 2.Time resolved Doppler shifts in the probe beam reflected From plasma crated by an 800 nm, 30 fs laser pulse at 5 × 10^{18} W cm^{-2}. The target is Al coated glass.

Another aspect that is interesting to study is plasma motion on ultrafast timescales, specifically, the motion of the critical surface, which we show in Fig. 2. The Doppler shifts in probe spectra indicate that the critical surface moves into the denser plasma portion at initial times, and the motion reverses at later time [8]. These results are important for understanding mass and energy transport in the plasma at early times. For more details, please see Oral presentation 2.1.4, Lad *et al.* in these proceedings.

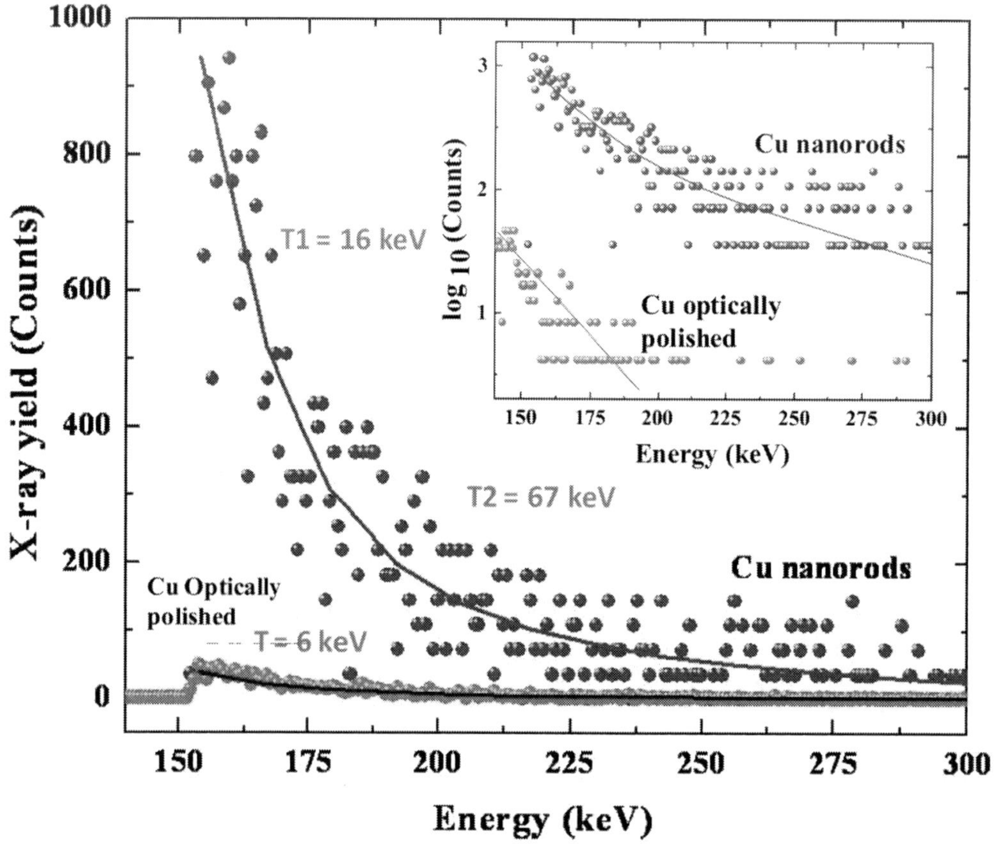

Figure 3. Comparison of bremsstrahlung emission from polished copper (red) and copper nanorod coated targets (blue) at incident laser intensity of $10^{15} - 10^{16}$ W cm^{-2}. Inset shows the logarithmic plot of bremsstrahlung emission.

We have also been interested in enhancing laser coupling to the target and enhancing hot electron production. One of the attractive ways to do this is to have nanostructuring on the target surface [9]. The structure enhances local electric fields and cause efficient absorption. We have recently studied surfaces coated with copper nanorods (150 nm diameter, 20-30 micron length, 100-150 nm spacing between rods) and obtained nearly 30 times enhancement in the hard x-ray emission caused by the hot electrons generated. The temperature of the hot electrons is increased by more than an order of magnitude (Fig. 3).

Before concluding, it is relevant to offer a few remarks about the increasing participation by the Indian community in this area. Recent Indian efforts are concentrated in HEDS, be it studying opacities, x-ray emission, shock wave propagation, hot electron generation and transport, giant magnetic fields, explosion of cluster targets, behaviour of nanostructures etc. The emphasis is on attempts to understand basic physics of the hot, dense matter that is created and applications in novel x-ray sources, particle acceleration and laser fusion, particularly fast ignition.

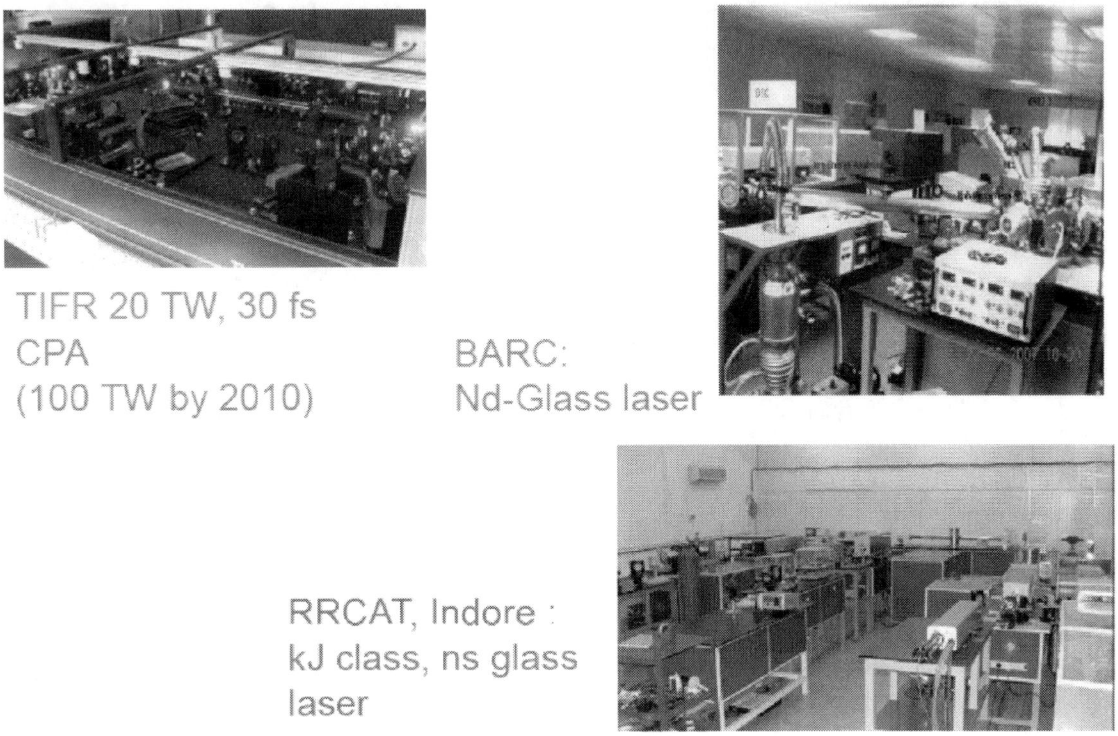

Figure 4. Indian high energy density science scenario.

The experimental areas are being pursued mainly at the Tata Institute of Fundamental Research (TIFR), Mumbai (erstwhile Bombay), Raja Ramanna Centre for Advanced Technology (RRCAT), Indore and Bhaba Atomic Research Centre (BARC), Mumbai, apart from some universities. Theoretical work, including analytical aspects and simulations is driven by efforts at the Institute for Plasma Research (IPR), Gandhinagar, Indian Institute of Technology, Delhi and the BARC, Mumbai. Figure 4 shows a few of the major facilities at different laboratories. In 2010 two high contrast, 100 TW, 'petawatt scale' laser facilities will be installed at TIFR and RRCAT. A PW laser is slated to be developed by 2013. For a summary of these aspects, please see [10].

Conclusions

Tabletop terawatt, femtosecond lasers are enabling faster progress in high energy density science by virtue of the flexibility, robustness and repeatability they offer. Interesting efforts are being made by the HEDS community to search for and understand newer ways of coupling light to dense plasma, particle emissions and their dependence on various laser and target parameters, hot electron transport, giant magnetic field generation and similar plasma processes. The feasibility to unravel ultrafast dynamics carves out a niche area for the tabletop lasers and promises new understanding of plasma dynamics and their impact on target physics.

Acknowledgement

The work reported here is done jointly with S. Mondal, Amit D. Lad. Saima Ahmed, V. Narayanan, P. Ayyub, D. Carvalho and I. Chakraborty (all from TIFR), S. Sengupta, A. Das and P. K. Kaw (all from IPR), W. M. Wang and Z. M. Sheng. I acknowledge a DAE-SRC-ORI research grant that partially facilitates this research at TIFR. A special word of gratitude to Amit D. Lad for his extensive help with the manuscript.

References

[1] Drake R P High Energy Density Physics- Fundamentals, Inertial Fusion and Experimental *Astrophysics* 2006 (Springer-Verlag, Heidelberg); Board of Physics and Astronomy, Frontiers in High Energy Density Physics; The X-games of contemporary science 2003 (National Academies Press, USA)

[2] Mourou G , Tajima T and Bulanov S V 2006 *Rev. Mod .Phys.* **78** , 309; Kiriyama H *et al.*, 2008 *Opt. Lett.* **33,** 645

[3] Cerchez M, Jung R, Osterholz J, Toncian T, Willi O, Mulser P and Ruhl H, 2008 *Phys. Rev.Lett.* **100,** 245001

[4] Ping Y, Shepherd R, Lasinski B F, Tabak M, Chen H, Chung H K, Fournier K B, Hansen S B, Kemp A, Liedahl D A, Widmann K, Wilks S C, Rozmus W and Sherlock M 2008 *Phys. Rev. Lett.* **100,** 085004

[5] Sandhu A S, Dharmadhikari A, Rajeev P P, Kumar G R, Sengupta S, Das A and Kaw P K 2002 *Phys. Rev. Lett.* **89,** 225002

[6] Sandhu A S, Kumar G R, Sengupta S, Das A and Kaw P K 2006 *Phys. Rev.* E **73,** 036409

[7] Kahaly S, Mondal S, Kumar G R, Sengupta S, Das A and Kaw P K 2009 Phys. Plasmas **16,** 043114

[8] Mondal S, Lad A D, Narayanan V, Ahmed S, Pasley J, Robinson A P L, Rajeev P P and Kumar G R 2008 *Phys. Rev. Lett.* (submitted)

[9] Kahaly S, Yadav S, Wang W M, Sengupta S, Sheng Z M, Das A, Kaw P K and Kumar G R 2008 *Phys. Rev. Lett.* **101,** 145001.

[10] Kumar G R 2009 *Pramana – J. Phys.* **73**, 113

Journal of Physics: Conference Series **244** (2010) 012009 doi:10.1088/1742-6596/244/1/012009

Transport of laser accelerated proton beams and isochoric heating of matter

M Roth[1], I Alber[1], V Bagnoud[2], C Brown[3], R Clarke[4], H Daido[5] J Fernandez[6], K Flippo[6], S Gaillard[6], C Gauthier[6], S Glenzer[7], G Gregori[3], M Günther[1], K Harres[1], R Heathcote[4], A Kritcher[7], N Kugland[7], S LePape[7], B Li[4], M Makita[8], J Mithen[3], C Niemann[9], F Nürnberg[1], D Offermann[6], A Otten[1], A Pelka[1], D Riley[8], G Schaumann[4], M Schollmeier[1], J Schütrumpf[1], M Tampo[5], A Tauschwitz[2], An Tauschwitz[10]

[1]Inst. für Kernphysik, Technische Universität Darmstadt, 64289 Darmstadt, Germany
[2]GSI Helmholtzzentrum f. Schwerionenforschung GmbH, 64291 Darmstadt, Germany
[3]Clarendon Laboratory, University of Oxford, Parks Road, Oxford OX1 3PU, UK
[4]STFC, Rutherford Appleton Laboratory, Chilton, Didcot, OX14 OQX, UK
[5]Photo Medical Research Center, JAEA, Kizugawa-City, Kyoto 619-0215, Japan
[6]Los Alamos National Laboratory, Los Alamos, New Mexico 87545, USA
[7]Lawrence Livermore National Laboratory, Livermore, California 94551, USA
[8]School of Mathematics and Physics, Queen's Univ. of Belfast, Belfast BT7 1NN, UK
[9]Physics Department, Univ. of California Los Angeles, Los Angeles, CA 90095, USA
[10]Inst. f. Theo. Phys, Johann Wolfgang Goethe-Universität, 60438 Frankfurt, Germany

E-mail: markus.roth@physik.tu-darmstadt.de

Abstract. The acceleration of intense proton and ion beams by ultra-intense lasers has matured to a point where applications in basic research and technology are being developed. Crucial for harvesting the unmatched beam parameters driven by the relativistic electron sheath is the precise control of the beam. We report on recent experiments using the PHELIX laser at GSI, the VULCAN laser at RAL and the TRIDENT laser at LANL to control and use laser accelerated proton beams for applications in high energy density research. We demonstrate efficient collimation of the proton beam using high field pulsed solenoid magnets, a prerequisite to capture and transport the beam for applications. Furthermore we report on two campaigns to use intense, short proton bunches to isochorically heat solid targets up to the warm dense matter state. The temporal profile of the proton beam allows for rapid heating of the target, much faster than the hydrodynamic response time thereby creating a strongly coupled plasma at solid density. The target parameters are then probed by X-ray Thomson scattering (XRTS) to reveal the density and temperature of the heated volume. This combination of two powerful techniques developed during the past few years allows for the generation and investigation of macroscopic samples of matter in states present in giant planets or the interior of the earth.

Introduction

The realization of intense and energetic laser pulses has resulted in enormous scientific activity over the past decade due to many potential applications including the generation of giga-electron-volt, narrow band electron pulses [1], intense x-ray pulses [2], laser-driven nuclear phenomena [3], inertial fusion energy [4,5], as well as the acceleration of protons from hydrocarbon impurities and heavy ions to mega-electron-volt energies from thin foil targets [6–9]. These ion beams (particularly protons) are generated in a very robust and reproducible way with up to 10^{13} protons by the target normal sheath acceleration mechanism [10]. The ions are accelerated, forming a quasineutral plasma with an exponential energy spectrum that exhibits a sharp cutoff at its maximum energy [11]. Unlike conventionally accelerated ion beams, they contain very high particle numbers in short, picosecond pulses and have unprecedented emittance; i.e., the beams expand in a very laminar fashion [12]. These features make them useful as a diagnostic tool (e.g., proton radiography of transient processes [13]) and they could have applications as compact particle accelerators [14] for the creation of high-energy density (HED) matter [15] or for proton fast ignition [5]. Because of the high intensity and the volumetric energy deposition of the proton beams in solid material, large samples of Warm Dense Matter (WDM) can be produced. This exotic state of matter between solid and plasma state is of high interest and relevant for astrophysical research and possible industrial applications.

Ion beam transport

The interaction between an ultra-intense laser pulse (I >10^{18} W/cm2) and a thin target foil leads to an acceleration of protons up to kinetic energies of several tens of MeV. In the short acceleration time of a few picoseconds up to 10^{13} protons and ions are accelerated. Well defined, smooth beams with transverse emittances up to hundred times smaller than in conventional accelerators are observed in experiments [12,16]. This remarkably good beam quality motivates the injection into conventional accelerators. However, such an application requires further investigations to reduce the half-opening angle of the beam which is up to 40 degree and to minimize the energy spread. An external magnetic field as a collimation and energy filtering method decouples the acceleration process and the transport providing the opportunity of independent optimization of both processes[17].

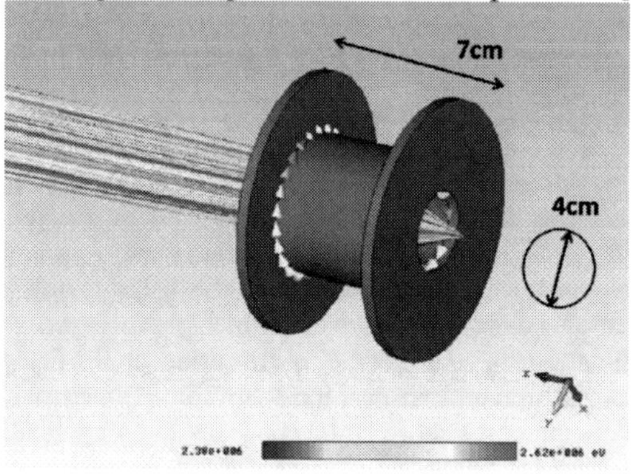

Figure 1: The figure shows a simulation with CST Studio. The solenoid magnet used in the experiments has a length of 7cm and an open inner diameter of 4 cm. A magnetic flux density of 8 T in the solenoid collimates protons with an energy of 2.5 MeV.

We report on an experiment carried out at the PHELIX laser system at GSI. PHELIX (Petawatt High Energy Laser for Ion eXperiments) is a high energy laser system operating at 1μm using large scale Nd:glass amplifiers. In its current state it delivers up to 150 J in 700 fs to reach focused intensities of 5 x 10^{19} W/cm^2. In the experiments we used a pulsed high field solenoid to collimate and focus the proton beam and further increase the proton flux density. Since the solenoid focuses the beam in all transverse directions it can be used as a compact single device. Thus, higher transmission of protons through the solenoid can be achieved in comparison to quadrupoles. The strong magnetic flux densities of up to 15 T were generated by some of the capacitors normally used to trigger the flash lamps of PHELIX. The solenoid was placed 17 mm behind the target to make sure that all protons enter the solenoid's aperture of 44 mm. The simulation with CST Particle Studio in figure 1 shows the behavior of the proton beam in the magnetic field. For this we took the actual beam divergence from experiment as input parameters for the CST simulation. A magnetic flux density of 8 T leads to a collimation of protons with a kinetic energy of 2.5 MeV. The measured particle numbers detected behind the solenoid at a magnetic flux density of 8 T was compared to the initial particle numbers. Therefore the detector was placed 240 mm behind the target and had a size of (5 x 5) cm (see Figure 2). As the simulation suggests, the 2.3 MeV protons are slightly focused, so that the beam diameter behind the solenoid is smaller than the detector size.

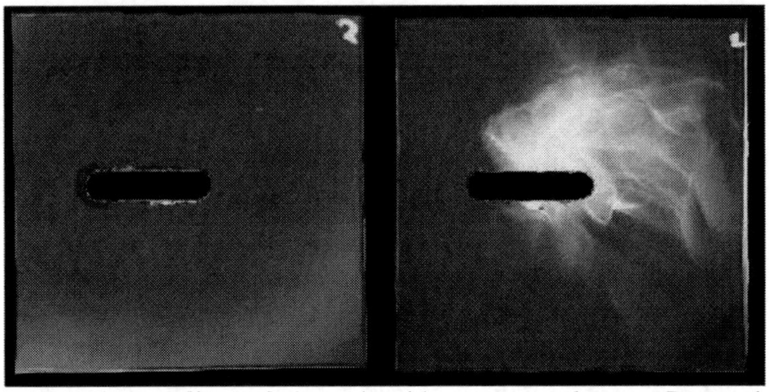

Figure 2: Example of the effect of the solenoid on proton beams. Proton beam image of two consecutive shots measured at a distance of 240 mm behind the target. Whereas no significant energy deposition is visible without energizing the coil a clear proton beam signal can be seen on the right part of the image.

About (95 +5 -19)% of the initial particles could be detected. In comparison to laser-proton-acceleration without beam parallelization by the solenoid only 23% of the initial protons can be detected in the RCF due to the strong divergence of the beam. Since the magnetic field is not strong enough to collimate protons with energies higher than 2.5 MeV the amount of detected particles in comparison to the initial particles decreases strongly with increasing proton energy.

In summary we have demonstrated collimation of laser accelerated proton beams with a pulsed high field solenoid. For the designed particle energy almost the entire beam could be detected. As can be seen in Figure 2 the beam is not really homogeneous. We verified that the proton beam accelerated off the laser target was a smooth laminar one, so the distortions in the beam were caused by the interaction of the ion with the magnet. Two main aspects are currently investigated in more detail. First, the loss of quasi neutrality, because the co-moving electrons are forced on the solenoid axis, thereby generating an electrostatic lens inside the magnet. This also is assumed to be the reason for an enhanced focusing capability, which was observed, compared to numerical simulations taking into account the influence of the B-field only. Second, there was observed an interaction of the ions (or plasma) with the inner wall of the solenoid causing local surface discharges which certainly influenced the homogeneity of the B-field. The analysis of the spatial structure of the transported ion beam is currently investigated and will be presented soon.

Driving Warm Dense Matter

The unmatched beam intensity and excellent beam quality results in important applications like ion source development, radiography, the concept of proton driven fast ignition and the generation of exotic states of matter. The latter is of particular importance due to the different interaction mechanism of ions compared to lasers. Whereas lasers only interact with the surface of a sample, ions can penetrate deep into the material of interest thereby generating large samples of homogeneously heated matter. The short pulse duration of laser produced ion beams furthermore allows for the investigation of equation of states close to the solid state density, because of the material's inertia preventing the expansion of the sample within the interaction. In addition to these unique characteristics, the interaction of ions with matter dominantly is due to collisions and does not include a high temperature plasma corona as it is present in laser matter interaction. The absence of a large radiation background is of importance to the experiment as will be explained later. Large conversion efficiencies have been observed and significant energy can be transferred from the ultra intense laser via the ion beam into the sample of interest. Because of the high beam quality, ballistic focusing has been demonstrated, allowing for an increase in local energy deposition and thus to higher temperatures.
The generation of large homogenous samples of warm dense matter is, as challenging as it is, accompanied by the even more challenging task to diagnose this state of matter, as usual diagnostic techniques fail under these conditions. The material density results in a huge opacity and the relatively low temperature does not allow traditional spectroscopic methods to be applied. Moreover the sample size, deposited energy and lifetime of the matter state are strongly interrelated and dominated by the stagnation time of the atoms in the probe. Thus high spatial and temporal resolution is required to gain quantitative data in those experiments. XRTS as the second technique developed in the recent years is ideally suited for these conditions.
The scattering of externally generated x-rays off electrons has demonstrated excellent diagnostic quality [18-22]. It is not only able to penetrate deep into the matter revealing the properties in the bulk material, but it also simultaneously results in the most wanted parameters temperature and density with highest precision. The challenge is the small cross section for the interaction which requires a powerful x-ray source as well as low background radiation level as well as high resolution spectrometers with high efficiency.
We combined these two new techniques for the first time to investigate the transformation of solid, low-Z material into the state of warm dense matter. For this we used the 60 J, 1 ps CPA beam to generate an intense proton beam to heat the sample of interest, and the 200 J, 20 ps beam 8 of the VULCAN laser to drive an x-ray source to scatter off the proton driven sample. In a second campaign we used the high contrast short pulse beam of the LANL Trident laser to drive a powerful proton beam and the remaining two long pulse beams to drive thermal Chlorine backlighter radiation. By varying the distance of the proton source to the warm dense matter sample we shaped the pulse length of the proton beam due to the dispersion caused by the energy distribution of the ions. We also shaped the ion source target in order to collimate the ion beam onto the sample for maximising the energy deposition. One advantage of the presented concept is that we can separate the ion driver from the sample and therefore can apply sufficient shielding to explore the material conditions in absence of a large radiation field.
This experiment also served as a demonstration experiment for the concept of proton driven fast ignition, where the low-Z material is to be heated to fusion conditions by a short pulse laser driven proton beam. The complex targets necessary for these experiments were produced in a collaborative effort between the RAL target laboratory and the one at Technische Universität Darmstadt.

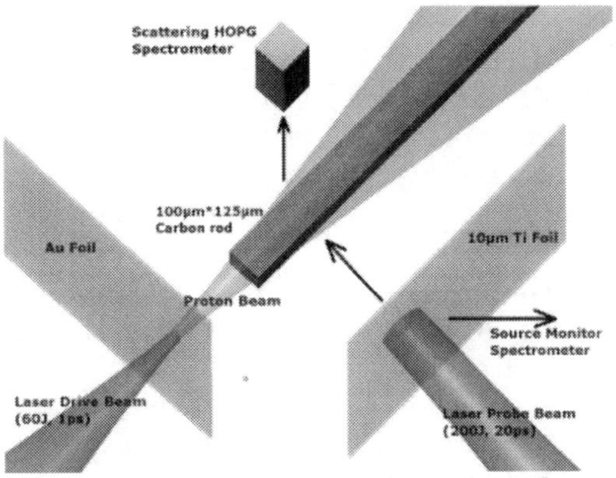

Figure 3: Target design for the RAL campaign. The 100 TW laser from the lower left part hits the gold foil to accelerate protons to heat the solid carbon rod. From the right the 20 ps 200 J backlighter beam irradiated the Titanium foil. The scattered radiation is detected at an angle of 90 degree through a slit in the shield at the bottom of the target.

The energy deposited by the proton beam was measured by the analysis of the part of the proton spectrum that missed the target. Because of the high intensity of the proton beam in addition to RCF, a new method based on nuclear activation was used to determine the proton beam energy content. Based on the deposited energy the temperature of the driven carbon sample was calculated. Figure 4 shows the temperature distribution of the RAL experiment in the carbon sample together with the measured region. The calculation was based on actual proton particle spectra from the experiment and used tabulated stopping power data and EOS to simulate the temperature distribution.

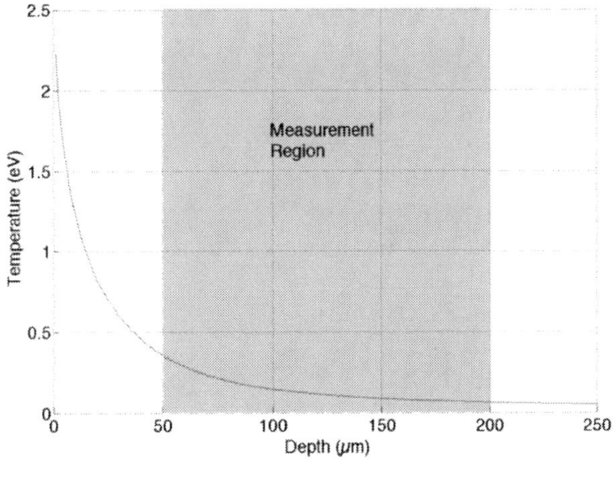

Figure 4: Temperature distribution in the solid carbon sample based on the measured proton spectrum. The shaded area represents the measured region, covering the temperature range of the solid – liquid phase transition.

The temperature in these first experiments was limited by the available proton beam energy. It covers the range between liquid carbon and the excited solid phase, a region of high interest. The change in temperature results in a strong decrease of the Thomson scattered signal elastic component. This was observed in the experiment and excellent agreement with theory could be demonstrated. Details as well as the results from the LANL experiments are still subject to analysis and will be published soon. However in the RAL experiments the predicted temperature could be demonstrated, showing the feasibility of the method for the first time.

Summary

We have successfully demonstrated the efficient capture and transport of laser accelerated proton beams using pulsed high field solenoid magnets. Almost the entire particle number could be guided through the ion beam transport system for the design particle energy. In these experiments a novel version of a nuclear temperature measurement was developed using giant resonance excitation of different isotopes. In two experimental campaigns laser accelerated proton beams were used to isochorically heat a solid carbon sample. The temperature was measured using x-ray Thomson scattering and showed good agreement with the predictions based on state of the art equation of state parameters. Finally a new diagnostic for very intense ion beams based on nuclear activation with high spatial and spectral resolution was successfully tested.

We acknowledge the expert support from the PHELIX, VULCAN and TRIDENT laser facilities. A. Otten, A. Pelka and M. Günther are supported by the BMBF.
J. Fernandez., K. Flippo., D.C. Gautier were funded under LANL Laboratory Directed Research and Development Program, LDRD-DR No. 20040064.
C. R. D. Brown was funded by AWE plc, Aldermaston.
G. Gregori was supported by EPSRC Grant No. EP/G007187/1 and by the Science and Technology Facilities Council of the United Kingdom.
The work of S.Glenzer, A.Kritcher, N.Kugland, and S.LePape was performed under the auspices of the U.S. Department of Energy by Lawrence Livermore National Laboratory under Contract No. DE-AC52-07NA27344. Work was also supported by Laboratory Directed Research and
Development Grants No. 08-ERI-002 and No. 08-LW- 004.

References

[1] W. P. Leemans, B Nagler, A. J. Gonsalves, Cs. Toth, K. Nakamura, C. G. R. Gedes, E. Esarey, C.B. Schroeder and S.M. Hooker , Nature Phys. **2**, 696 (2006).
[2] M. M. Murnane, H. C. Kapteyn, M. D. Rosen and R. Falcone, Science **251**, 531 (1991).
[3] K.W. D. Ledingham, P. McKenna, R. P. Singhal, Science **300**, 1107 (2003).
[4] M. Tabak, J. Hammer, M. E. Glinsky, W. L. Kruer, S.C. Wilks, J. Woodworth, E. M. Campbell and M. D. Perry, Phys. Plasmas **1**, 1626 (1994).
[5] M. Roth, et al. , Phys. Rev. Lett. **86**, 436 (2001).
[6] R.A. Snavely, et al., Phys. Rev. Lett. **85**, 2945 (2000).
[7] E. L. Clark, et al., Phys. Rev. Lett. **84**, 670 (2000).
[8] A. Maksimchuk, S. Gu, K. Flippo, D. Umstadter, V. Yu. Bychenkov, Phys. Rev. Lett. **84**, 4108 (2000).
[9] M. Hegelich, et al., Phys. Rev. Lett. **89**, 085002 (2002).
[10] S.C. Wilks, et al., Phys. Plasmas **8**, 542 (2001).
[11] J. Fuchs, et al., Nature Phys. **2**, 48 (2006).
[12] T. E. Cowan, et al., Phys. Rev. Lett. **92**, 204801 (2004).
[13] L. Romagnani, et al., Phys. Rev. Lett. **95**, 195001 (2005).
[14] A. Pukhov, Phys. Rev. Lett. **86**, 3562 (2001).
[15] P. K. Patel et al., Phys. Rev. Lett. **91**, 125004 (2003).
[16] F. Nürnberg, et al., Rev. Sci. Instr., Vol. 80 (3), p. 033301 (2009)
[17] M. Schollmeier, et al., Phys. Rev. Lett. **101**, 055004 (2008).
[18] G. Gregori *et al.*, Phys. Rev. E **67** 026412 (2003).
[19] S. H. Glenzer *et al.*, Phys. Rev. Lett. **90**,175002 (2003).
[20] S. H. Glenzer *et al.*, Phys. Rev. Lett. **98,** 065002 (2007).
[21] A. L. Kritcher *et al.*, Science **322**, 69 (2008).
[22] E. Garcia Saiz *et al.*, Nature Physics **4**, 940 (2008).
[23] R.J. Clarke et al., Nucl. Instr. and Meth. A -585, 117-120 (2008)
[24] P. McKenna, et al., Phys. Rev. E, 70, 036405 (2004)

Journal of Physics: Conference Series **244** (2010) 012010 doi:10.1088/1742-6596/244/1/012010

Multi kJ level Laser Concepts for HiPER Facility

J.-C. Chanteloup[1], D. Albach[1], A. Lucianetti[1], K. Ertel[2], S. Banerjee[2], P. D. Mason[2], C. Hernandez-Gomez[2], J. L. Collier[2], J.Hein[3], M. Wolf[3], J. Körner[3], B. J. Le Garrec[4]

[1]Laboratoire LULI, Ecole Polytechnique, 91120 Palaiseau, France, [2]Central Laser Facility, STFC Rutherford Appleton Laboratory, Chilton, Didcot, OX11 0QX, United-Kingdom, [3]Institute for Optics and Quantum Electronics, Max-Wien-Platz 1, 07743 Jena, Germany, [4]CEA-CESTA, 15 avenue des Sablières, BP2, 33114 Le Barp, France.

jean-christophe.chanteloup@polytechnique.fr

Abstract One of the biggest challenges the HiPER project is facing is to identify a laser architecture that meets all the demanding requirements. Among those are high wall-plug efficiency (15 to 20%) and repetition rate (5 to 10 Hz). In order to perform this task, four teams from the co-authors' institutions are working together and exploring several approaches described here.

1. Introduction

Two options rely on Yb:YAG gain medium as it can be manufactured in large sizes (ceramics), provides good thermo-mechanical properties and a reasonably long fluorescence lifetime. Active Mirror (AM) (section 2) and He-gas cooled Stack of Thin Slabs (SLS) [1] are the two amplifier architectures explored. In both cases, cryogenic temperature will improve thermal management (better thermal conductivity) but will affect Amplified Spontaneous Emission (ASE) management due to higher gain. ASE issues are addressed differently leading to an AM concept relying on multiple ~kJ beams to be ultimately phased when required whereas the SLS concept allows to reach ~10 kJ in a single aperture (a 1 kJ demonstrator is described in section 3). A third option based on fluoride laser materials with and without cryo-cooling together with pulse stacking techniques is investigated (section 4). Indeed a key factor for HiPER is the reduction of manufacturing/ownership costs: a long fluorescence life time laser material is then a solution to reduce the diode quantity. Yb doped fluorides exhibit exceptional parameters but required advanced extraction schemes due to their low emission cross sections. This third option features a very broad band amplification capability allowing a highly flexible short pulse generation (required for HiPER fast ignition scheme). Finally our fourth option (section 5) relies on the use of coherent/incoherent addition of a few million fibres as proposed in 2007 at the IFSA conference [2]. Such a design is attractive in terms of laser safety, operation and maintenance. This concept will required a large number of fibres since one single fibre can only handle a limited power [3]. Amplification relies on diode pumped fibre amplifiers, a technology known to carry limited wall-plug efficiency. Main questions to be answered are: how much power can we manage per fibre (for both ns and ps pulses), how many fibres shall we need to scale to a typical 10 kW level output at 1μm wavelength (1 kJ at 10 Hz rep-rate) and how much thermal power has to be removed at that level.

2. Yb:YAG active mirror amplifier concept

The Laboratoire pour l'Utilisation des Lasers Intenses (LULI) explores an IFE driver concept based on parallel amplification of nine ~120cm² aperture beams carrying each about one kiloJoule of laser light. The chosen host matrix is the YAG garnet whereas Yb^{3+} is selected as a doping ion. Aperture requirement will not exceed 160 mm in diagonal. Amplifiers concept is based on the so called "active mirror" architecture. This approach allows a re-circulation of the pump light improving efficiency. The thermal management approach relies on single side cooling.

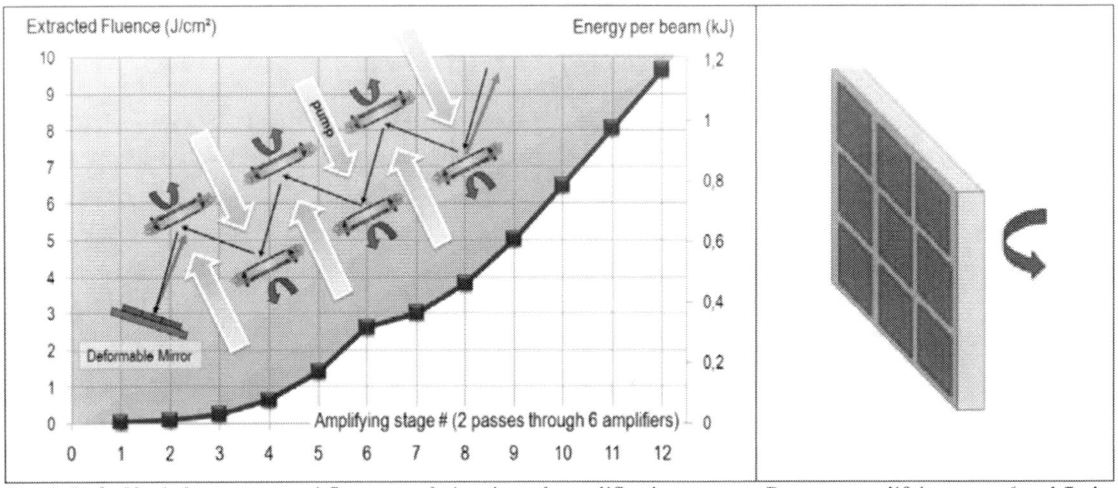

Fig. 1: Left: Single beam extracted fluence evolution through amplification process. Between amplifying stage 6 and 7, the beam is hitting a deformable mirror and is send back for a second pass through the amplification chain. Right: Mosaic active mirror amplifier distribution with single side cooling (blue arrow).

Amplifier chain design is based on parallel extraction of nine (3x3) beams propagating back and forth through a set of six active mirrors. A set of nine deformable mirrors will allow wave front correction for beam quality management. ASE, B Integral and thermal managements, together with global efficiency, will lead to the optimum set of values for doping concentration, thickness and pump brightness. A 1D numerical model was developed in order to identify the optimum set. It shows that doping level will be below 1 at%, thickness will vary between 1 and 3 cm and pump brightness will vary between 4 and 8 kW/cm².

3. Yb:YAG gas cooled amplifier concept

The Central Laser Facility develops a Yb:YAG ceramic based concept for a kJ-class DPSSL amplifier that could potentially be scaled to multi-kJ energies. Numerical modeling has shown that the required pump and extraction fluence levels for efficient operation are prohibitively high at room temperature, due to the thermal population of the lower laser level and the still rather low gain cross section. This can be overcome by cooling the medium to cryogenic temperatures. Fig. 1 shows gain and maximum achievable optical-to-optical efficiency for a Yb:YAG amplifier that is end-pumped at 940 nm from both sides with 1 ms long pulses of 5 nm FWHM spectral width. At 175 K operating temperature, the efficiency reaches > 50 % (the limit being 58%) at the moderate total pump fluence of 10 J/cm². The small signal gain is at this point 3.8 and the optimum optical depth (doping x thickness) is 3.15 % cm. To avoid excessive ASE losses, the maximum transverse gain-length product needs to stay below 3 at any point. Hence, regardless of amplifier size, the doping level needs to be chosen such that the thickness is 1.3 times the width.

With this very low aspect ratio, distributed cooling is required. The cooling approach demonstrated on the Mercury laser [1] was thus chosen for our concept. Here, the gain medium is divided into a number of thin slabs with He-gas as coolant flowing between them (see Fig. 3). Also, slabs with different doping concentrations can be employed. The concentrations can be chosen such that the overall thickness is reduced and the heat load in the individual slabs equalized. If the amplifier is sized for a 1 kJ output energy at a fluence of 5 J/cm^2, the aperture size will be 14 x14 cm^2. For a constant doping level, the required thickness would then be 18 cm, with the doping concentration being 0.18%. If ten slabs with optimized different doping levels are used, the material thickness can be reduced to 10 cm, with the doping then ranging from 0.18% in the outer slabs to 0.5 % in the two central slabs.

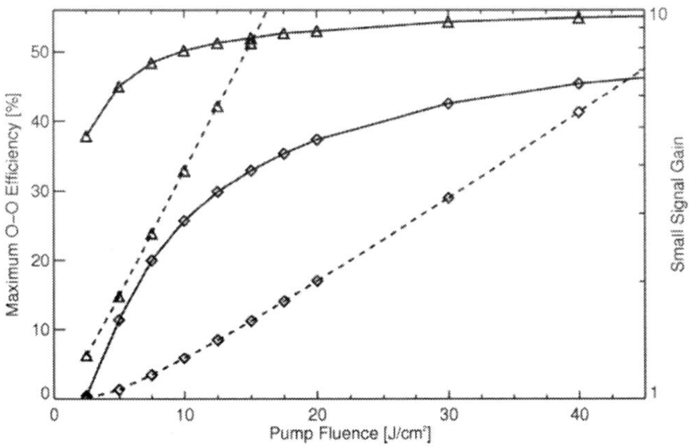

Fig. 2: Efficiency (solid lines) and gain (dashed lines) as a function of total pump fluence, calculated at room temperature (diamonds) and at 175 K (triangles).

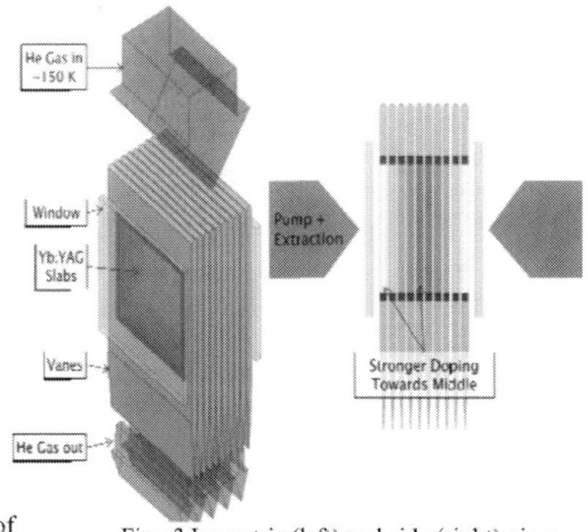

Fig.: 3 Isometric (left) and side (right) view of amplifier geometry

4. Yb:CaF2 gas cooled amplifier concept

The driving laser source for a fusion facility has to fulfil certain criteria, where of main interest are the total power consumption and the facility costs needed to provide the requested pulse parameters. The following proposal from The Institute for Optics and Quantum Electronics, addresses these constraints in the direction of minimizing the peak power requirement for the pumping diode lasers. Combined with that optimisation, the laser fluence in a beam line has to be maximized in order to reduce the required facility space consumption. Nevertheless, this approach is connected with the drawback of the higher amount of heat per volume that needs to be dissipated by the cooling system.

This proposal is based on a laser material with a long fluorescence life time, a good thermal conductivity, broad bandwidths for absorption as well as for emission, and most of all a high probability of large scale production. According to the high fluence in the laser beam line an appropriate extraction scheme is proposed together with a gas-cooled thin-disk stack as the amplifying medium.

Laser active media that fulfil the criteria above typically show a low emission cross section. The use of a low gain material has the advantage of reduced probability of transverse lasing affects, what allow an easier scale of the lasers aperture at relatively high doping concentrations. Such a scheme helps to keep the B-integral, i.e. the accumulated non-linear phase shift, on a low level if the non-linear refractive index is low for this material. CaF_2 seems to be a promising host for the laser active ions to fulfil these criteria. It is interesting to note that the first diode pumped laser incorporated a doped CaF_2 [4] and the first ceramic laser also used that type of material [5]. Pure single crystalline CaF_2 nowadays are produced for applications in lithographic systems on a tone scale. The growth of crystals with diameters of 380 mm is typical. Additionally, it was shown that highly doping with Ytterbium ions can be managed in order to ensure the required oxidation level Yb^{3+}.

Absorption and emission spectra of Yb:CaF_2 reveal a broad absorption ranging from 910 nm to 960 nm, wavelengths allowing convenient pumping with high power diode lasers [6].

Fluorides typically have low absorption in the UV. The result is a low refractive index at the emission wavelength in the near infrared. This broad transparency range results in a less likely multi-photon absorption and combined with the low refractive index the damage threshold can be assumed to be higher compared to other Yb-doped laser materials. First measurements support this assumption but further investigation is required to see how surface treatment and anti-reflective coatings influences this behaviour.

The operation of Yb^{3+}-lasers at low temperatures is favourable because of the much higher gain and extraction efficiency [7]. The reason for this behaviour is the reduced population of the lower laser level in this quasi three level system which turns into a four level system with cryogenic cooling. Both the thermal expansion coefficient, and the thermal conductivity of Yb:CaF_2 behave in a positive way at temperatures around 100K, a temperature where liquid nitrogen can be used as the coolant. Additionally, the spectral properties change in a positive direction at lower temperatures and the amplification band width will be preserved.

The advantages of using Yb:CaF_2 as the laser active material for a fusion facility can be summarized as follows: This material has a low refractive, and non-linear refractive index at the emission wavelength, therefore a low B-integral in the amplifier chain can be managed. The low refractive index allows simpler anti-reflection coatings with high performance. CaF_2 is actually produced at a large scale for lithography and doped crystal production is likely. Yb:CaF_2 shows broad absorption and emission bands offering an advantage for pumping with unstabilized high power diode lasers. The broad emission allow wavelength multiplexing (the proposed scheme for energy extraction) as well as amplification of very short pulses and therefore provide a high versatility for the fast/shock ignition laser chains of the HiPER facility. Even short pulses in the picosecond range can be wavelength multiplexed because of the femtosecond pulse length capability of this laser material. The low emission cross section prevents the amplifier from transverse lasing and the long fluorescence life time reduces the amount of required diode laser pump power, i.e. the number of diodes and the costs of the system. The broad transmission range of Yb:CaF_2 indicates a potentially high damage threshold, what allows higher fluences of the extracting laser pulse and a high extraction efficiency at cryogenic temperatures. The thermal conductivity of Yb:CaF_2 is comparable to Yb:YAG and therefore a high repetition rate operation is supported. Amplifiers will be cooled with a similar technology as the one described in section 3.

It must be noted that in order to use the advantages of Yb:CaF_2 as the laser material a more complex extraction scheme for the amplifier design has to be chosen. The low single pass gain should be compensated by high laser beam fluences. To avoid laser induced damage of optical elements including the laser material itself it is proposed to amplify a couple of pulses in one amplifier head. To keep the effort of laser beam steering and relay imaging on an acceptable level a combination of angular and wavelength multiplexing is proposed. If, for instance, four pulses with different wavelengths are sent in four beams through the amplifying head, a total of sixteen pulses can contribute to the extraction fluence, where the peak intensity is not affected if all pulses are delayed by their pulse length. After amplification pulses can be rearranged to form very powerful beams by incoherent beam combination. Because the large band width of Yb:CaF_2 this scheme is also applicable for the amplification of chirped pulses if this is necessary for fast/shock ignition laser fusion facilities.

5. Fiber approach

A new amplifying concept designed to produce high energy in either short or long pulses using coherent or incoherent addition of few millions fibres has been proposed in 2007 at the IFSA conference [2] and is currently explored by the Commissariat à l'Energie Atomique (CEA). It is clear that the fibre design looks really attractive from many different points of view: laser safety, operation and maintenance. The system will need a large number of fibres because one single fibre can handle a limited power. The system is based on diode pumped fibre lasers and this type of device has limited wall-plug efficiency: all that is not "light" will be removed as "heat". So far there are three questions to be answered: how much power can we manage per fibre, how many fibres we shall need, how much thermal power has to be removed?

5.1 Assess maximum energy per fibre.

For most high-output-power applications, several unique advantages have made ytterbium (Yb) the doping ion of choice. More specifically, ytterbium-doped fibres offer high output powers tuneable over a broad range of wavelengths, from around 975 to 1200 nanometres (typically around 1060 nanometres).

One of the most significant keys to ensuring broad marketability of fibres lasers is the development of a technique for producing ever-increasing output powers without sacrificing beam quality. Whatever the fibre design, it must not limit the total achievable output power and in pulsed laser devices the average power, peak power, and pulse energy.

One of the major advances in fibres technology in recent years has been the advent of large-mode-area (LMA) fibres, and the potential for these fibres to deliver diffraction-limited beam quality with mode-field areas greater than 10 times that for standard telecom type fibres. Companies without access to LMA fibre technology are essentially limited by nonlinear phenomenon such as stimulated Raman scattering (SRS) to the near-100-watt CW power regime. Single-fibre lasers have delivered output powers approaching 1.5 kilowatts and theoretical modelling suggests 10 kilowatts may be possible with the current fibre technology.

The key limits to power scaling [3] are: thermal rupture, melting of the core, thermal lensing, non linear effects, SBS and SRS, optical damage [8] (and how it is related to non linear effects), X-ray, γ-ray and neutron induced damages. For ns pulses, the range is 1 to 27 mJ. The typical output fluence range is 30 to 86 J/cm^2, depending on the core diameter. The damage scaling law is 475 J/cm^2 per ns (for 10 ns F_{damage}= 475 J/cm^2, etc.). Even at high repetition rates (tens of kHz), the average power does not exceed 50W, far from the CW limits indicated in figure 4. It seems obvious that multimode fibres will deliver much more energy, but the numerical aperture has to fit the overall laser design requirements, otherwise the energy will be lost.

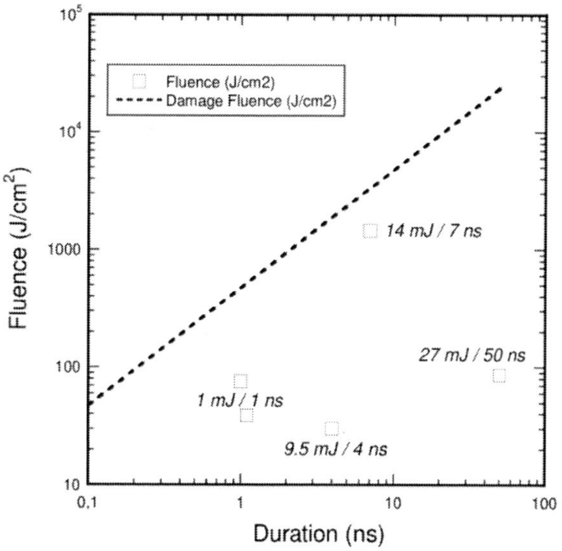

Fig. 4: Plot of the maximum fluence in fibre as a function of pulse (from ref.8). For pulses longer than 50 ps, damage occurs in single silica fibre at 475 GW/cm^2. Peak power does not exceed ≈ 2.5 MW.

5.2 Assess scalability to typical 10 kW level outputs

Both NIF and LMJ have included fibre lasers in their front-end design, with at least three purposes:
1/ Split a given function arising from the oscillator into 192 or 240 “arms” to feed the different laser beam lines; 2/Amplify the oscillator signal with high gain (10^6) and 3/ Delay the laser pulse to let fast electronic devices to implement closed-loop error signals to drive “safely” the laser system.

The system design must assess pulse shaping capability for ns and ps pulses. This principle is based on an Average Waveform Generator that is computer controlled. The input pulse shape is being computed in order to compensate for saturation in the multi stage amplifier and to compensate for non linear effects in the frequency conversion crystals.

The ability to power scale fibre devices depends on two key criteria: pump power can be efficiently and cost effectively coupled into the fibre, and the fibre core parameters are such that it can convert this pump power into bright lasing or amplification without saturation or nonlinear processes that limit the performance.

It is common for companies to power scale laser output further by incoherently combining the outputs from several fibre lasers, sacrificing beam quality as the power increases. Of course and depending on the system requirements, both coherent and incoherent additions have to be made to ensure enabling pulse synchronisation, smoothing and coherent propagation of ps ignition beams down to the target in the cone-core design.

5.3 Assess thermal management

From a 1mJ fibre laser to a 1 kJ system, at least 1 million fibre lasers would be required, certainly more. Whatever the system design, fibre lasers will be stacked and thermal management has to be assessed to decide what type of cooling is required (if any).

Because a single fibre laser is a diode pumped laser, it is easy to calculate the heat deposited in the fibre that will escape to the outside through the cladding. This calculation is made according to the distribution of the signal and pump powers along the fibre. For simplicity and without loss of accuracy, the cylindrical approximation will lead to the well-known thermal loading expressed in W/m [9]. Thermal management includes thermal, stress and thermo-optic effects in the fibre [10].

6. Acknowledgements

The authors gratefully acknowledge the support of the Funding Agencies in undertaking this work (EC FP7 project number 211737): EC, European Commission, MŠMT, Ministry of Education, Youth and Sports of the Czech Republic and STFC, Science and Technology Facility Council of the United Kingdom.

7. Conclusion

Four currently explored options for kJ level diode pumped amplifiers have been presented in the framework of the HiPER program. Down selection will be performed in HiPER program next phase.

8. References

[1] A.Bayramian et al., "The mercury project: A high average power gas-cooled laser for inertial fusion energy development", Fusion Science Technology, Vol. 52, Oct. 2007

[2] G.Mourou et al., "New amplifying concept for inertial fusion driver", J. of Physics: Conference Series 112 (2008) 032052

[3] J.Dawson et al., "Scaling limits of high average power fibre lasers", IEEE LEOS annual meeting,ThE1, nov.2008

[4] R. Keyes et al., "Injection Luminescent pumping of CaF_2:U^{3+} with GaAs diode lasers", Appl. Phys. Lett., Vol.4, pp50-51 (1964).

[5] S.Hatch et al., "Hot-pressed polycrystalline CaF_2:Dy^{2+} laser", Appl. Phys. Lett., Vol.5, pp153-154 (1964)

[6] M. Siebold et al., "Terawatt diode-pumped Yb: CaF_2 laser", Opt. Lett., Vol.33, No.23, pp2770-2772 (2008)

[7] A. Pugžlys et al., "Spectroscopic and Lasing Properties of Cryogenically Cooled Yb,Na:CaF_2", CLEO, paper MF4 (2008)

[8] A.Smith et al., IEEE J. of Selected Topics in Quantum Electronics, Vol.15, No.1, pp153-158 (2009)

[9] Y.Lee et al., IEEE J. of Selected Topics in Quantum Electronics, Vol.15, No.1, pp93-102 (2009)

[10] D. Brown et al., IEEE J. of Selected Topics in Quantum Electronics, Vol.2, pp207-217 (2001)

Modelling the Effect of ^{3}He in Direct Drive Capsule Implosions*

W J Garbett[1], C J Horsfield[1], H W Herrmann[2], J R Langenbrunner[2], J H Cooley[2], D C Wilson[2], S C Evans[2], T J Sedillo[2], M S Rubery[1], D Drew[1], J M Mack[2], C S Young[2], G A Kyrala[2], J Frenje[3], V Yu Glebov[4].

[1] AWE Aldermaston, Reading, Berkshire, RG7 4PR, UK
[2] Los Alamos National Laboratory, Los Alamos, NM 87545, USA
[3] M.I.T., Boston, MA, USA
[4] Laboratory for Laser Energetics, University of Rochester, Rochester, NY, USA

E-mail: warren.j.garbett@awe.co.uk

Abstract. D^{3}He fuels are often used in ICF implosion experiments, either as a surrogate for DT to restrict the output neutron yield, or to produce protons for use in diagnosis of core conditions. Recent experiments have suggested that capsules filled with D^{3}He do not behave as expected, but that both proton and neutron yields are anomalously degraded relative to the pure D_2 case. We have performed direct drive implosion experiments using the Omega laser to examine the effect of ^{3}He on DT-filled glass capsules. The use of DT fuel allows reaction history measurements to be obtained using the Gas Cherenkov diagnostic (GCD). It was hoped that the detailed information provided by GCD measurements would complement existing measurements to constrain modelling. We present recent modelling and analysis of the experiments using radiation-hydrocode simulations, and explore some of the hypotheses proposed to explain the results.

1. Introduction

Recent experiments have suggested that ICF capsules filled with D^{3}He do not behave as expected, with fusion yields being anomalously degraded relative to the pure D_2 case [1]. This result has consequences for experiments where D^{3}He is used either as a surrogate for DT or where it is used to generate 14.7MeV protons as an additional diagnostic of core conditions.

To further explore the anomaly, direct drive implosion experiments have been performed at Omega which examine the effect of ^{3}He on DT-filled glass capsules [2]. The use of DT fuel allows reaction history measurements to be obtained using the Gas Cerenkov diagnostic (GCD) [3]. This instrument measures reaction history without degradation by neutron time-of-flight, by observing gamma rays from a branch of the DT reaction. These experiments are amongst the first to demonstrate application of the GCD to elucidate implosion physics. GCD measurements complement existing observations to provide additional insight and help constrain capsule modelling. Modelling and analysis of the experiments is presented, with particular emphasis on the GCD and neutron yield data.

Journal of Physics: Conference Series **244** (2010) 022001 doi:10.1088/1742-6596/244/2/022001

2. Experiment

Two separate series of experiments were performed at the Omega laser using thin shell glass capsules (nominally 5μm thickness). Initial experiments used ^{3}He mixed additively to 5atm of equimolar DT. More recently the targets have used a 'hydrodynamically equivalent' gas fill [1] in which ^{3}He is substituted for D (with the tritium pressure held constant). Hydrodynamic equivalence minimises variation in the implosion dynamics as fuel composition is varied and is based on the assumption of fully ionised ideal gas. In practice simulations suggest that there is some deviation from perfect equivalence even when the pressure condition is met.

The GCD provides low noise, time resolved measurements of the fuel burn. Resolution in these experiments was ~75ps following deconvolution of the instrument response. Two yield phases are apparent in all data; identified as being due to convergence of the initial shock at the centre of the fuel (shock phase) and subsequent compression of the fuel by the capsule shell (compression phase) (Figure 1). The signal evolves appreciably as the ^{3}He fraction is increased, with the compression phase yield being clearly reduced. Excellent shot-to-shot reproducibility gives confidence that the observed features are real.

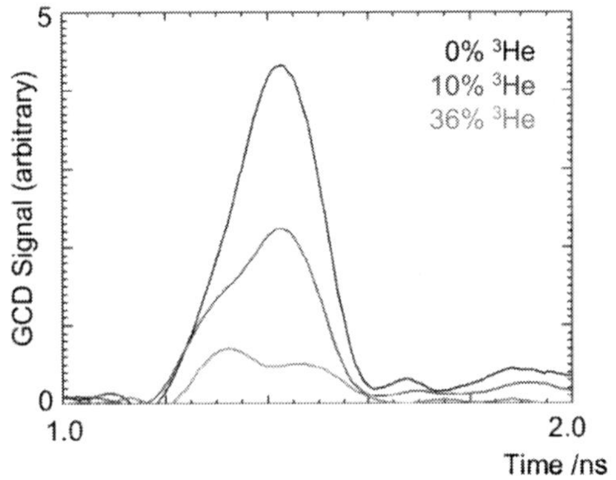

Figure 1. DT burn histories measured using the GCD. The curves show distinct shock and compression phases (colour online).

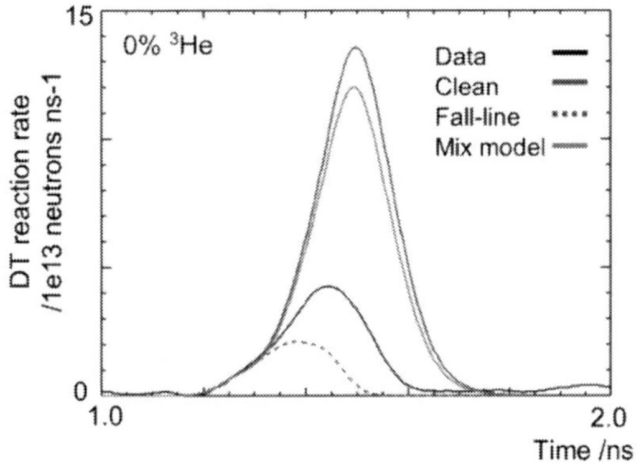

Figure 2. Comparison of GCD data with simulated reaction rates for 0% ^{3}He case. Results are shown for clean calculations, Youngs' mix model and the fall-line model (colour online).

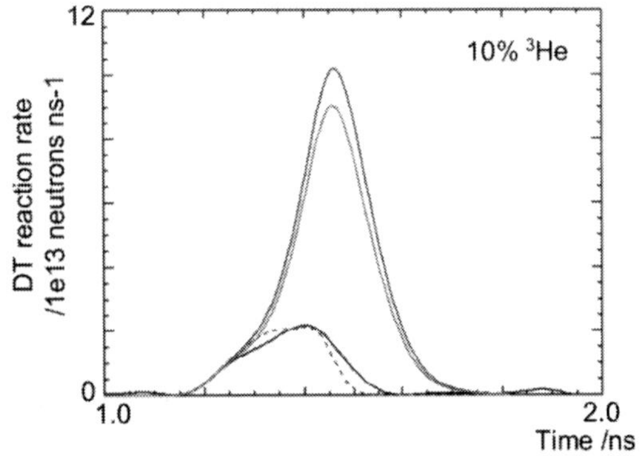

Figure 3. As figure 2 for 10% ^{3}He case. The fall-line model now gives good agreement with experimental data (colour online).

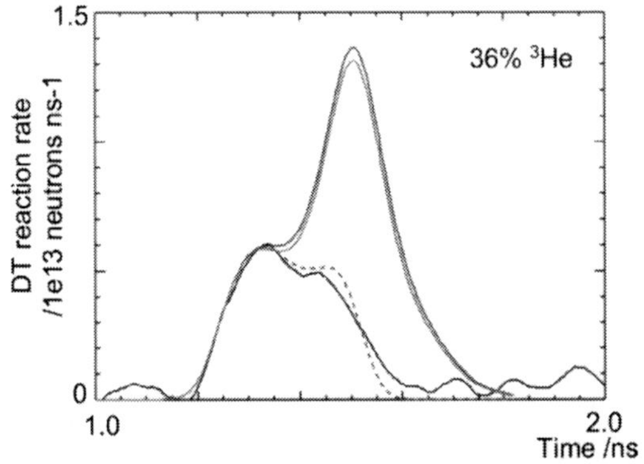

Figure 4. As figure 2 for 36% ^{3}He case. In all cases simulated reaction histories are convolved with 75ps Gaussian response (colour online).

3. Modelling

Capsules were modelled in 1D spherical geometry using the radiation-hydrocode Nym [4]. Energetics were calibrated using the rise of the GCD signal. This part of the burn history was found to be insensitive to hydrodynamic mix and so provides an unambiguous calibration provided proper account is taken of the instrument response. This improves on bang-time calibration, which is subject to assumptions about mix.

Comparison of simulated and measured burn histories shows that clean simulations match the initial shock phase, but overestimate the compression phase (Figures 2-4). The effects of mix degradation were estimated using both Youngs' model [5] and fall-line calculations [6]. Youngs' model calculates the 1D spatial extent of mixing (mix width) assuming non-linear growth. The fall-line model assumes the extent of mixing into the fuel is limited by the trajectory the gas-shell interface would take if no deceleration occurs, and provides a worst-case estimate of degradation for a 1D model. Youngs' model predicts little yield degradation. However the fall-line calculation predicts significantly greater yield degradation and gives surprisingly good agreement with the measured burn histories, particularly at 36% ^{3}He.

Yield data and calculations from the initial (additive) experiments is shown in figure 5. Despite the poor match to GCD data, clean simulations are found to match the yield trend well; i.e. yield-over-clean is approximately constant. Fall-line calculations underestimate the relative yield degradation at higher ^{3}He concentration. Thus if hydrodynamic mix is to account for the observed yields it is necessary to increase the extent of mix with increasing ^{3}He. This is consistent with LANL calculations using the Scannapieco-Cheng mix model [2]. This finding is difficult to justify physically and suggests instead that mix alone cannot account for the observed performance. Other mechanisms must act to degrade the compression phase yield.

The data from the hydro-equivalent targets (figure 6) appears consistent with that from [1]. It is clear that the results do not match either analytic scaling or simulations. Simulations also show that the departure from hydrodynamic-equivalence cannot be attributed to non-ideal gas or non-equilibrium effects. The anomaly must arise from effects not captured by the 1D code model.

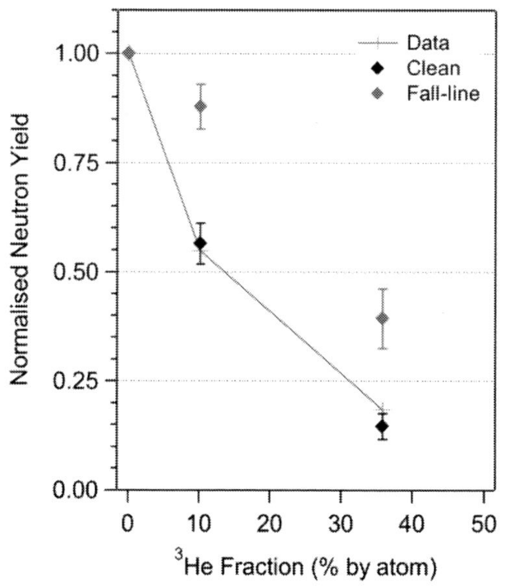

Figure 5. Observed and calculated DT neutron yields for initial experiments with ^{3}He mixed additively to DT. Yields are normalized to 0% ^{3}He case (colour online).

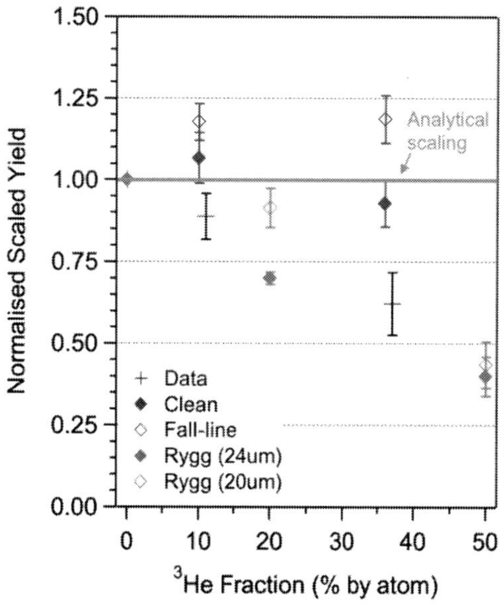

Figure 6. Scaled yields from hydro-equivalent targets. Yield scaling is detailed in [1] except that here values are normalized to the 0% ^{3}He case. (colour online).

4. Discussion

The GCD data shows that the compression phase yield is not well modelled. This is consistent with other evidence from similar experiments [7] which also suggests that capsule compression is reduced compared to simulation. A number of hypotheses have been proposed to account for the reduced compression. These include fast electron preheat from resonant absorption and LPI, non-local electron transport, shell break-up following interaction with the outgoing reflected shock, and the effect of 2D asymmetries on fuel assembly.

In order to account for the ^{3}He anomaly requires a differential effect as the gas composition is varied. One hypothesis that has been explored is uncertainty in the combination of the different equations of state (EoS) in the DT-^{3}He mixture. The experiments have been modelled using two different EoS mixing algorithms; the additive volume rule and the ideal mixing rule [8]. The capsules show little sensitivity to these algorithms, with yield variation of ~2% between the two cases. Theoretically the two methods should agree for ideal gases, for which they are exact. The lack of sensitivity thus implies that the individual EoS used are sufficiently close to ideal gases that the mixing algorithm is unimportant, and that other more sophisticated mixing rules will produce the same result.

It is also possible that the constituent EoS are in error. Although uncertainties in deuterium EoS are well documented [9], our results would require the EoS to be less compressible than the standard Sesame model, which is inconsistent with existing measurements. However, we admit the possibility of uncertainties in the derived tritium and ^{3}He EoS, which have not been examined here.

5. Summary

A series of direct drive implosion experiments has been performed at the Omega laser to explore the effect of ^{3}He on DT-filled glass capsules. Reaction histories measured using the Gas Cherenkov diagnostic (GCD) show two phases, attributed to the shock and compression. The shock phase is well matched by 1D simulation, but the compression yield is lower than calculated. The experiments are consistent with previous data, showing an anomalous effect with ^{3}He. Mix alone cannot account for the observed behaviour. However, fall-line calculations, which represent an unrealistic extreme 1D mix limit, do reproduce the GCD data surprisingly well. A number of hypotheses have been proposed to explain the reduced compression and are being actively explored. We have shown that EoS mixing uncertainties are unlikely to have a significant effect due to the near-ideal constituent EoS.

References

[1] Rygg J R *et al.* 2006 *Phys. Plasmas* **13** 052702
[2] Herrmann H W *et al.* 2009 *Phys. Plasmas* **16** 056312
[3] Caldwell S E *et al.* 2003 *Rev. Sci. Instrum.* **74** 3
[4] Roberts P D, Rose S J, Thompson P C and Wright R J 1980 *J. Phys.* D **13** 1957
[5] Youngs D L 1989 *Physica* D **37** 270
[6] Amendt P *et al.* 2002 *Phys. Plasmas* **9** 2221
Welser-Sherrill L *et al.* 2008 *Phys. Plasmas* **15** 072702
[7] Kyrala G A *et al.* 2007 *High Energy Density Physics* **3** 163
[8] Saumon D, LA-UR-08-0304
[9] Da Silva L B *et al.* 1997 *Phys. Rev. Lett.* **78** 483
Knudson M D *et al.* 2004 *Phys. Rev.* B **69** 144209

Hydrodynamic Studies of Laser Fusion using Plasma Block Ignition Driven by Nonlinear Ponderomotive Forces

H Hora[1], G H Miley[2] and Block Ignition Team*

[1]Department of Theoretical Physics, University of NSW, Sydney 2052, Australia
[2]Department of Nuclear, Plasma and Radiological Engineering, University of Illinois, Urbana IL, 61801, USA

E-mail: h.hora@unsw.edu.au

Abstract: In contrast to laser fusion with spherical plasma compression to very high densities, the alternative was studied with side-on ignition of solid state (or modestly higher) deuterium-tritium, based on hydrodynamic theory. The extremely difficult conditions may be overcome using petawatt-picosecond (PW-ps) laser pulses, however, only by applying the advantages of a discovered anomaly of plane geometry interaction without relativistic self-focusing using very high contrast ratios of the laser. The conditions of this side-on ignition could be further relaxed by inclusion of interaction effects. Ignition of solid state density or modestly compressed deuterium-tritium (DT) and of proton-boron11 (HB11) seems to be possible with laser pulses of several PW power and ps duration.

1. Introduction

Laser driven fusion aims for a clean, inexhaustible, safe and everywhere available energy source. For this aim, the largest laser on earth is now available [1] producing pulses of 1.1 MJ energy of few nanosecond duration leading for the first time on earth, to a controlled fusion reaction generating more energy than the laser pulse needed for the ignition. The fusion fuel consists of isotopes of heavy hydrogen (deuterium D) with the super-heavy hydrogen isotope tritium T, where a *spherical laser irradiation compresses* the fuel *to more than 1000 times the solid* density with heating to ignition temperatures of several ten million degrees centigrade. Following this scheme, a prototype of a power station [2] is scheduled for 2020 based on very compact and high-efficient diode pumped lasers which are now available.

Parallel to these developments are schemes for generating fusion energy by using the new type laser pulses [3] with more than PW (petawatt) power [4] and of ps (picoseconds) or shorter duration. Following the modified scheme of Nuckolls and Wood [5] it should be possible for working with power reactors using DT fuel after chemical compression to about 10 times solid state density, ignited with very intense electron beams of 5 MeV energy arriving at fusion gains of 10,000.

This paper reports on another scheme using PW-ps laser pulses following a side-on ignition of solid state density fusion fuel developed 1972 [6][7] which was not possible then because energy flux densities E^* with pulses of about ps duration above a threshold E_t^* for DT

$$E^* > E_t^* = 4\times 10^8\ J/cm^2 \quad ;\ \text{ignition temperature}\ 7.2\ \text{keV} \qquad (1)$$

were necessary. These conditions were very far away from reality and the scheme of laser fusion with spherical compression of the fuel to very high densities was followed up [1]. This situation changed after 2 PW-ps laser pulses were available [3]. Highly relativistic effects appeared as pair production and nuclear transmutation, but in contrast to these usual measurments, an anomaly was discovered with drastically different observations. If the pulses are very clean with a contrast ratio of 10^8, the x-ray emission is found to be very low instead of the usually high intensity energetic bursts, and the emission of fast ions shows much lower energies than usual but with high directionality and in numbers independent of intensity. These effects were explained with skin layer interaction predicted earlier from plane geometry analysis (see section 10 of [8])[9] and confirmed by experiments in many details [10]. Only laser beams with the high contrast ratio avoided the generation of a plasma plume in front of the target leading to relativistic self focusing [11]. The plane geometry interaction was verified for the first time in experiments by Sauerbrey [12] with the clean pulses, where the measured Doppler shift of the accelerated plasma agreed then exactly as predicted before from nonlinear (ponderomotive) force acceleration [10]. Based on these results, a come-back of the ignition of solid density or modestly compressed fusion fuel with irradiation of the laser driven ion current densities above 10^{10} Amp/cm^2 became interesting again.

2. Side-on Ignition by nonlinear force driven clean PW-ps laser pulses

The limit of Eq. (1) of the energy flux density E^* for ignition of solid state density DT by pulses of about ps duration was derived hydrodynamically [6][7]. The condition (1) can be seen from the computation of the characteristic plots of the plasma temperature in the interaction area vs time for a given parameter E^*. The results of Chu ([6] see there Fig. 2) shows where the characteristics increase (ignition) down to E_t^* where a stationary temperature is reached. For lower E^* the characteristics decay with time after a maximum with no ignition.

Other effects are also very important in this new side-on driving scheme. The threshold E_t^* decreases considerably [13] when the reduction of thermal conduction by the inhibition factor and the reduction of the stopping length by collective effects is included into the analysis of Chu [6]. A highly increased skin-layer thicknesses is created using – as an example - Rayleigh density profiles for initial conditions [8][14]. This approach also resolved the problem of internal reflection of light in inhomogeneous optical media ([8] section 7) at least during the initial stages of interaction.

For clarification it has to be underlined that this is basically no new theoretical work. It is a continuation what Chu [6] and Bobin [7] had achieved at aboiut 1972. New are only the now available PW-ps laser pulses, however with the exception that the usually measured wild relativistic effects are carefully excluded. Instead the very rare measurements of plasma block generation with extremely clean laser pulses for avoiding relativistic self-focusing is the new result [9,10] as it was predicted by computationts with the nonlinear force ([8] section 10).

3. Ignition of proton-boron(11)

A very interesting fusion fuel is the reaction of light hydrogen with the boron isotope 11 (p-^{11}B) because it produces mono-energetic 2.888 MeV alpha particles if the incident protons have an energy of less than 150 keV [15] for direct conversion into electricity. Initially no neutrons are produced and [16] less radioactivity per generated energy appears than by burning coal due to its content of 2ppm uranium. For *spherical laser ignition*, however, it was calculated from the beginning [17] that compression to 100,000 times the solid state is needed for burning p-^{11}B.

These results excluded any hope of laser fusion for p-^{11}B by high compression with *spherical* ignition. However, extending the *side-on ignition* from DT [10][13] to the case of p-^{11}B, the very surprising fact was experienced [18] that the ignition conditions are *only about 10 times more difficult* than for DT. The factor 100,000 for more difficulties for p-^{11}B at spherical compression compared with DT is given from 100 times higher density, 100 times higher laser pulse energy, and 10 times lower gain [17]. For the side-on block ignition of p-^{11}B first only the very conservative assumptions of Chu [6] are taken arriving at the characteristics (Fig. 1) similar to those for DT Chu (see his Fig. 2 [6]). In difference to the earlier reported results [18], we present here new computations for much longer times t of the temperature T of the reaction wave front at incident energy flux densities E^* as parameter. If E^* is too low, the curves decay on time t and no ingiton happens. Only if T is not decreasing as for $E^* = 2\times10^9$ J/cm^2 or a little lower, but above 1×10^9 J/cm^2, then ignition happens. From Fig. 1 the resulting threshold E^* for ignition of p-^{11}B is then within a error bar of 35% is

$$E_t^* = 1.5\times10^9\ J/cm^2 \quad ;\ \text{ignition temperature 87 keV} \qquad (2)$$

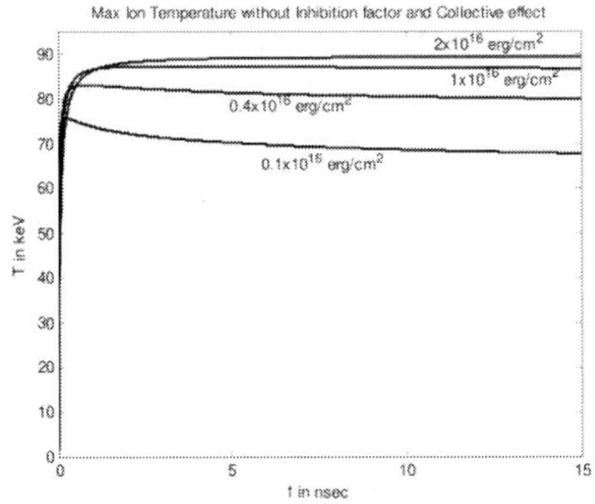

Fig. 1. Dependence of the temperature T on time t for side-on ignition of solid density p-^{11}B *working with conditions used for DT by Chu [6] (parameter E* in units used by Chu).*

Further new results were with inclusion of the inhibition factor and including Gabor's collective effect for reaction products similar to the case of DT [13] arriving at a minor decrease of the temperature in absolute values of the temperature similar to the results for DT. Bremsstrahlung losses were automatic included in the computation but again as new result confirmed by new computer outputs.

The essential result presented here – without more precise evaluations - is, that the side-on igniton of solid p-^{11}B is less than about a factor 10 more difficult than igniting DT *only*, while the spherical laser compression was about 100,000 times more difficult. Laser fusion of unompressed p-^{11}B may then work with dozen PW laser pulses of ps duration with sufficient suppression of prepulses [10][14].

4. Discuscussion and Conclusions

The results are based on the highly detailed and modified [13] hydrodynamic computations of Chu [6]. This model works basically on the generation of a hydrodynamic two-dimensional shock front and is in principle different from the three dimensional volumetric processes considered by Nuckolls and Wood [5] who needed the ρR criterion for the ignition with electron beams. It should be mentioned that the volume ignition results [19] - confirmed by the evaluation of the Wheeler modes [20] – are basically different to the two-dimensional shock-like interactions of the Chu analysis [6]. This has been clarified in details following the equivalence of Eqs. (13.7) and (13.8) of Ref. [8] about ρR.

More serious problems with our purely hydromechanical results may need to be studied with other methods, e.g. with ingeniously modified single particle models, discovered by Wilks, et al [21] as Particle In Cells (PIC) techniques. The problem of the interpenetration of plasma particles between very hot and cold plasma [22] may lead to an improvement of the conditions at least for our situation at sub-relativistic temperatures and sufficiently small Debye lengths for the block description with electric double layers [8].

Another problem is that the here reported hydrodynamic analysis is based on plane geometry presumptions. In reality there is a lateral spread of the interacting laser beam though this is of a radius in the range of thousand laser wave lengths. Several of these geometries were evaluated by Bobin [7] indicating that the better conditions for the nonlinear force driven block igniton of HB11 by several orders of magnitudes in contrast to the spherical laser compression scheme may sustain a further analysis. Experiments about the side-on ignition should guide and clarify short-cuts of the theoretical research.

A very more detailed analysis is needed but at least the basic properties for the side-on ignition are clearly visible. Most significant are the very surprising results for the fusion of uncompressed p-^{11}B fusion energy generation with laser pulses in the range of few dozens of PW power and ps duration avoiding neutron generation, nearly no radioactivity, and with a minimum of heat pollution in a power station or for space propulsion. Modest precompresion by chemical driving [5] will improve the conditions especially for HB11. The x-radiation in the reactor is around or below 100 keV and can be screened off. No secondary nuclear reactions in the power station will occure. This provides an exciting vision of a very attractive sustainable future power plant for worldwide use. Its achievement will depend on continued advances in laser optics, target physics and power conversion technology.

References:

*Plasma Block Ignition Team: Co-autors of references [9,10,14,18].

[1] Moses E, Miller G H, Kauffman R L 2006 *J. de Physique IV,* **133** 9; Moses E I 2008 *J. Physics – Conf. Ser.* **112** 012003; Moses E I 2010 Edward Teller Lecture 2009 (these proceedings)
[2] Storm E, Latkowski J F, Farmer J C, Abbott R P, Amendt P A et al. 2010 LIFE – a laser inertial fusion-based approach to high energy production *J.Physics – Conf. Ser.* (these proceedings)
[3] Mourou G and Tajima T 2002 *Inertial Fusion Science & Applications 2001* (Paris: Elsevier) 831
[4] Tabak M, Hammer J, Glinski M N, Kruer W L, Wilks S C et al.1994 *Physics of Plasmas* **1**, 1626
[5] Nuckolls J H and Wood L 2002 *Future of Inertial Fusion Energy*, LLNL Preprint UCRL-JC 149860; see also *Edward Teller Lectures* London: Imperial College Press (2005) p. 13-14
[6] Chu M S 1972 *Physics of Fluids* **15** 412
[7] Bobin J L 1974 in *Laser Interaction and Related Plasma Phenomena, H. Schwarz and H. Hora eds.* (New York: Plenum Press) Vol. 4B, 465
[8] Hora H 1991 *Plasmas at High Temperature and Densities* (Heidelberg: Springer)
[9] Hora H, Badziak J, Boody F B, Höpfl R, Jungwirth K, Kralikova B, Kraska J, Laska L, Parys P, Perina J, Pfeifer V M, Rohlena K, Skala J, Ullschmied J, Wolowski J, and Woryna E 2002 *Opt. Commun.* **207** 333; Hora H 2003 *Czechosl. J. Phys.* **53**, 199
[10] Hora H, Badziak J, Read M N, Li Yu-Tong, Liang Tian-Jiao, Cang Yu, Liu Hong, Sheng Zheng-Ming, Zhang Jie, Osman F, Miley G H, Zhang Weiyan, He Xiantu, Peng Hansheng, Glowacz S, Jablonski S, Woloswki J, Skladanowski J, Jungwirth K, Rohlena K, Ullschmied J 2007 *Physics of Plasmas* **14** 07270/1-7
[11] Hora H 1975 *J. Opt. Soc. Am.* **65** 882
[12] Sauerbrey R 1996 *Physics of Plasmas* **3** 4712
[13] Hora H, Malekynia B, Ghoranneviss M, Miley G H et al. 2008 *Appl. Phys. Letters* **93** 011101
[14] Hora H 2009 *Laser and Particle Beams* **27** 207; Yazdani E, Cang Y, Sadighi-Bonaz R, Hora H, Osman F 2009 *Laser and Part. Beams* **27** 149
[15] Stave S, Ahmed M W, France R, Henshaw S S, Müller B, Weller H R 2009 Final understanding the 11B(p,α)2α reaction at very low energies. *Phys. Rev. Lett.* (in print)
[16] Weaver S, Zimmerman G and Wood L 1973 *Exotic CTR fuel:Non-thermal effects and laser Fusion application.* Report UCRL-74938. Livermore CA: Lawrence Livermore Laboratory.
[17] Hora H 1975 *Laser Plasmas and Nuclear Energy* (New York: Plenum Press) pages vii & 95; Scheffel C, Höpfl R, Martinez-Val J M, Eliezer S, Kasotakis G, Piera M, Sarris E, and Stening R J 1997 *Laser and Particle Beams* **15**, 565
[18] Hora H, Miley G H, Azizi N, Malekynia B, He X 2009 *Laser & Part. Beams* **27**, 491; Hora H, MileyG H, Ghoranneviss M, Malekynia B, Azizi N 2009 *Optics Communications* **282**, 4124
[19] Hora H, Ray P S 1978 *Zeitschrift f. Naturforschung* 33A, 890 (1978); Hora H, Azechi H, Kitagawa Y, Mima k, Murakami M, Nakai S, Nishihara K, et al.1998 *J. Plasma Phys.* **60**, 743
[20] Kirkpatrick R C, Wheeler John A 1981 *Nuclear Fusion* 21, 389
[21] Wilks S C et al. 1992 *Phys. Rev. Lett.* **89**, 1383.
[22] Hora H 1983 *Atomkernenergie* **42**, 7.

Journal of Physics: Conference Series **244** (2010) 022003 doi:10.1088/1742-6596/244/2/022003

Rayleigh-Taylor instability growth control by an oscillating acceleration in heavy ion inertial fusion

S. Kawata[1], T. Kodera[1], Y. Hisatomi[1], A. I. Ogoyski[2], S. Koseki[1] and D. Barada[1]

[1]Utsunomiya University, Utsunomiya 321-8585, Japan
[2]Varna Technical University, Varna 9010, Bulgaria

kwt@cc.utsunomiya-u.ac.jp

Abstract. Uniformity of heavy ion beam (HIB) illumination is one of key issues in HIB inertial confinement fusion (HIF) in order to compress a fuel sufficiently and release fusion energy effectively. In this paper a new mitigation method of the Rayleigh-Taylor (R-T) instability growth is presented to make a HIF target robust against a non-uniform implosion. In this study a new mitigation method of the R-T instability growth is proposed based on an oscillating perturbed acceleration, which can be realized by a rotation or oscillation of the HIB illumination axis onto a fuel pellet. The R-T instability analyses and fluid simulations demonstrate that the oscillating acceleration reduces the R-T instability growth significantly. In this paper a baseline steady acceleration g_0 is perturbed by a perturbed oscillating acceleration δg, which is spatially non-uniform and oscillates in time ($g_0 >> \delta g$). An example result shows an 84% reduction of the R-T instability growth. In the analytical work two stratified inviscid fluids are treated under the perturbed oscillating acceleration. The linear analysis shows that the R-T instability growth rate does not change from the standard expression of γ. However, the R-T instability growth is strongly affected and mitigated by the oscillating acceleration; The transverse velocity of w is derived at the interface of the two stratified fluids. The R-T instability is induced by δg. The result presents that the perturbed oscillating acceleration reduces the R-T instability growth significantly depending on the magnitude of the HIB oscillation frequency.

1. Introduction

Heavy ion beam (HIB) serves a superior tool to study and to produce a high energy density state and HIB inertial fusion (HIF) plasma [1-12]. We have been pursuing a robust fuel target implosion mode by computer simulations in HIF [12-14]: to this end, research issues include HIB illumination uniformity, fuel target implosion uniformity, the Rayleigh-Taylor (R-T) instability mitigation, a non-uniformity smoothing by radiation, etc. To date a new HIB illumination scheme onto a fuel pellet was found, and it is robust against a target displacement in a fusion reactor [13]: the new robust HIB illumination scheme allows a fuel pellet displacement of 200-300 μm. It was also found that the radiation smoothing effect of the implosion non-uniformity is expected in a HIF direct-driven fuel target [15]. In addition, on the R-T instability control, a new method was also suggested in HIF by using an oscillating acceleration, which can be realized by the oscillating HIBs [12, 14, 16]. In general, fuel implosion in inertial fusion is attended by the R-T instability. Especially short wavelength modes of the R-T instability have a larger growth rate and should be mitigated by radiation transport, thermal conduction, ablation motion, etc.

In this paper we focus on the new method of the R-T instability growth control, and we found that an oscillating perturbed acceleration reduces the R-T instability growth drastically. The oscillating acceleration can be realized by oscillating HIBs illumination onto a target. The R-T instability under a non-uniform acceleration is studied by an analytical work and fluid simulations. In this paper a baseline steady acceleration g_0 is perturbed by an oscillating acceleration δg, which is spatially non-uniform and oscillates in time ($g_0 >> \delta g$). The HIB controllability and the well-defined HIB energy deposition may serve a well-defined perturbation of the implosion acceleration. When each HIB axis is oscillated or rotated, the wobbling HIB may serve a non-uniform and dynamically oscillating acceleration. Our analytical work and numerical computation results present that the oscillating perturbed acceleration reduces the R-T instability growth significantly. This new scenario for the R-T instability growth control can be realized by the HIB unique feature of the beam axis controllability and oscillation.

2. Rayleigh-Taylor instability under a perturbed oscillating acceleration

First we perform an analytical work to estimate an oscillating acceleration effect on the R-T instability growth:

$$\begin{aligned} g(x,y,z,t) &= g_0 + \delta g(x,y,z,t) \\ &= g_0 + g_1 f_1(x,y)\exp(-\beta|z|)\exp(i\Omega t). \end{aligned} \quad (1)$$

Here the perturbed oscillating acceleration is introduced, and we assume $f_1(x,y) = \exp(ik_x x + ik_y y)$. In this analysis two stratified inviscid fluids are treated under the perturbed oscillating acceleration. A linear analysis shows that the R-T instability growth rate does not change from the standard expression of $\gamma = \sqrt{\alpha g_0 k}$ for $g_0 >> g_1$. Here α is the Atwood number. However, the R-T instability growth is strongly affected and mitigated by the oscillating acceleration. In order to find the oscillating acceleration effect on the R-T instability growth, the transverse velocity of w is derived at the interface of the two stratified fluids as follows:

$$w = \frac{\gamma + i\Omega}{\gamma^2 + \Omega^2} g_1 \exp(ik_x x + ik_y y)[\exp(\gamma t) - \exp(i\Omega t)] \quad (2)$$

The R-T instability is induced by g_1. In realistic parameter values in the HIF target implosion, Ω can be larger than γ. for $\gamma << \Omega$, $|w| \approx (1/\Omega) g_1 \exp(\gamma t)$. For $\gamma \approx \Omega$, $|w| \approx (1/\sqrt{2}\Omega) g_1 \exp(\gamma t)$. The result presents that the perturbed oscillating acceleration reduces the R-T instability growth significantly depending on the magnitude of Ω. The reduction ratio is γ/Ω. Therefore the the R-T instability growth is reduced much for $\gamma << \Omega$.

This result suggests an interesting realistic scenario for the R-T instability control in HIF. First illuminating HIBs define the initial perturbation amplitude and phase of the non-uniformity introduced by HIBs themselves. The mode may have the largest perturbation amplitude and should be suppressed. Then the steerable HIB axes move slightly by the non-uniformity wavelength order to produce the oscillating g_1. The HIBs illumination non-uniformity scale size may be the order of the HIB diameter, and the HIBs' axes oscillation amplitude is also the order of the HIB radius. Therefore this scenario is rather

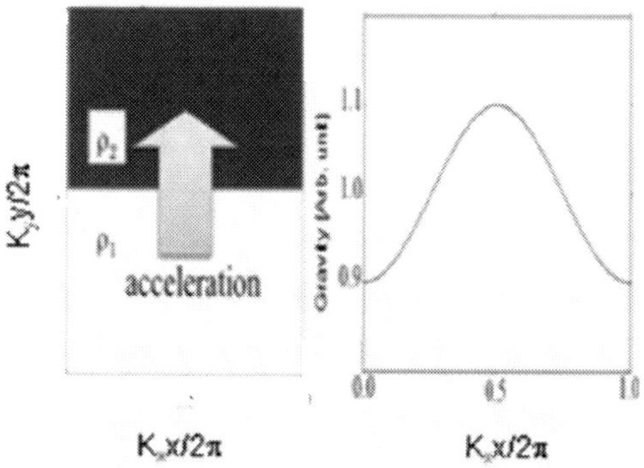

Fig. 1 Two-dimensional fluid simulation setup for the Rayleigh-Taylor Instability growth control by the acceleration oscillation. In this case 10% of the non-uniform acceleration δg oscillates.

Journal of Physics: Conference Series **244** (2010) 022003 doi:10.1088/1742-6596/244/2/022003

realistic and can be realized in a HIB accelerator system.

3. Reduction of Rayleigh-Taylor instability growth by a perturbed oscillating acceleration

In this section we perform 2-dimensional fluid simulations for the R-T instability growth control by the perturbed oscillating acceleration. Multi-HIBs illumination introduces a non-uniform acceleration. In addition, a HIB accelerator final element can give a little HIB axis oscillation or rotation. The oscillation or rotation frequency may be 100MHz ~ a few tens GHz. The HIB axis oscillation or rotation brings the HIBs deposition energy oscillation or rotation in a fuel target energy deposition layer. Therefore, the implosion acceleration consists of a baseline steady acceleration g_0 and a perturbed acceleration δg.

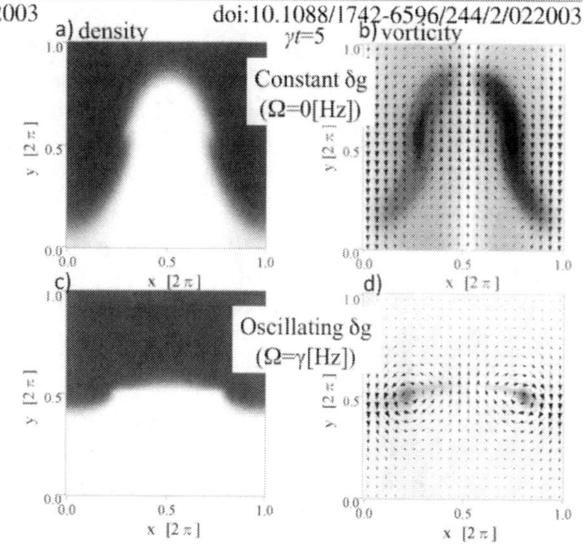

Fig. 2 The Rayleigh-Taylor instability control example results under the oscillating acceleration. In this case the oscillation frequency Ω of the oscillating part δg of the acceleration equals to γ (see c) and d)). In a) and b) Ω=0.

Figure 1 shows a simulation setup employed in this paper. Figure 2 presents the simulation results for the R-T instability growth. In Fig. 2, $\rho_2 : \rho_1 = 3:10$, $g_0=1$, $|\delta g|=0.1$ and k=1 in the normalized scale: the spatial scale is the wavelength of the R-T instability, the time scale is the inverse of $\gamma=\sqrt{\alpha g_0 k}$. In Figs. 2 a) and b) $\Omega=0$, and in Figs. 2 c) and d) $\Omega=\gamma$. For example, for $g_0=10^{13} m/s^2$ and $k=1/mm$, which are typical parameter values in inertial fusion, $\Omega \approx 100MHz$. The results in Fig. 2 demonstrate that the oscillating perturbed acceleration reduces the R-T instability growth significantly. The reduction ratio $(w(\Omega=0)-w(\Omega=\gamma))/w(\Omega=0)$ is 84.6% (see Fig. 3).

We have also checked a two-mode case: $[\exp(ikx+2ikx)]/\sqrt{2}$. Figure 4 shows another example for the R-T instability control for the two-mode case. The simulation results in Fig. 4 also show that the oscillating acceleration is very effective to mitigate the R-T instability growth as far as $\gamma \ll \Omega$. The growth reduction ratio is 76.3%.

These results demonstrate that the R-T instability growth is significantly mitigated by the oscillating perturbed acceleration.

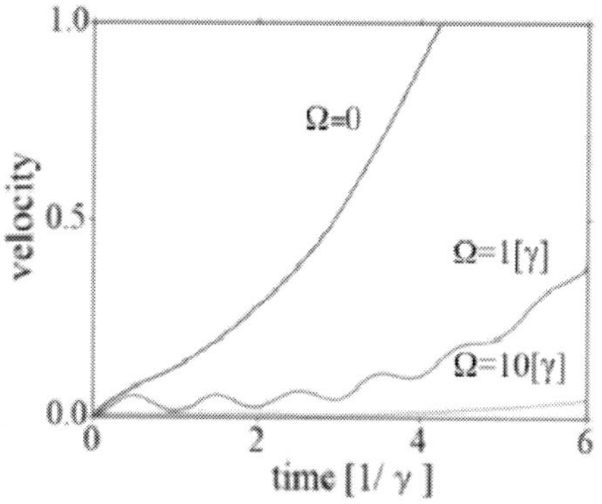

Fig. 3 Time history of velocity w induced by the R-T instability for each Ω. The oscillating perturbed acceleration contributes to the R-T instability control significantly. In this case 10% of the acceleration is perturbed.

4. Discussions and Summary

The HIB wobbling oscillation frequency may reach 1GHz ~ 100 GHz. For the R-T growth reduction presented in this paper, Eq. (2) suggests to go to the higher frequency. On the other hand, the fluid plasma should response to the HIB oscillation to produce the oscillating acceleration δg. When the acceleration oscillation frequency Ω is too high, the plasma cannot response the oscillation, so that the acceleration oscillation δg does not appear. The plasma response time τ is estimated by $\tau \sim L/Cs$. Where L should be a typical scale length of the plasma and Cs is the sound speed. Typically τ is about 1/ 1GHz~1/10GHz. Therefore, the preferable HIB wobbling frequency may be about 1~10 GHz or so.

This number may be achievable by the present accelerator technology. In this range of the oscillation the wobbling HIB deposited energy may serve a time-averaged overall energy deposition profile [16] with the oscillating part. This should be studied in the near future. In order to create the oscillating acceleration of δg, there may be other method in HIF: for example, each HIB can have a density modulation without the axis wobbling, and this density modulation phase can be also controlled so that δg appears.

In this paper we proposed the new R-T instability growth control method by the oscillating perturbed acceleration. In HIF the oscillating perturbed acceleration can be realized by the HIB axis little oscillation at a HIB accelerator final element. The oscillation frequency can be 100MHz~1GHz in HIB. The analytical and numerical works demonstrate that the proposed method mitigates the R-T instability growth effectively. The results mean that a fuel target implosion non-uniformity in HIF can be significantly smoothed by the Oscillating perturbed acceleration. The oscillating acceleration mitigation of the R-T instability growth can be applied to direct-driven, indirect-driven and hybrid targets in HIF. This new scenario is based on a unique preferable feature of the HIF concept.

In general, fuel implosion in inertial fusion is attended by the R-T instability. Especially short wavelength modes of the R-T instability have a larger growth rate and should be mitigated by radiation transport, thermal conduction, ablation motion, etc and the new method by the wobbling beams proposed in this paper. The detail studies covering all modes in the R-T instability should be performed in the near future.

Acknowledgements

This work was performed under the collaboration with Dr. B.G. Logan, J. Barnard and their colleagues in HIF-VNL, U.S.A., and was partly supported by JSPS and ILE, Osaka University.

References

[1] B. G. Logan, R. O. Bangerter, D. A. Callahan, M. Tabak, M. Roth, L. J. Perkins, G. Caporaso, Fusion Science and Tech. 49 (2006) 399.

[2] B. G. Logan, F. Bieniosek, C. Celata, et. al., Nucl. Instr. and Meth. A 544 (2005) 1.

[3] L.R. Prost, P.A. Seidl, F.M. Bieniosek, et. al., Phys. Rev. ST Accel. Beams 8 (2005) 020101.

[4] N.A. Tahir, A. Shutov, I.V. Lomonosov, et. al., High Energy Density Phys. 2 (2006) 21.

[5] N.A. Tahir, C. Deutsch, V.E. Fortov, et. al., Phys. Rev. Lett. 95 (2005) 035001.

[6] S. Medin, M. Basko, M. Churazov, et. al., Nucl. Instr. and Meth. A 544 (2005) 300.

[7] Y. Oguri, K. Kashiwagi, J. Kaneko, J. Hasegawa, M. Yoshida, M. Ogawa, Phys. Rev. ST Accel. Beams 8 (2005) 060401.

[8] M. Watanabe, M. Nakajima, K. Horioka, Nucl. Instr. and Meth. A 464 (2001) 440.

[9] K. Horioka, M. Nakajima, M. Watanabe, et.al., Laser and Particle Beams 20 (2002) 609.

[10] S. Neff, R. Knobloch, D.H.H. Hoffmann, A. Tauschwitz, S.S.Yu, Laser Part. Beams 24 (2006) 71.

[11] C. Deutsch, G. Maynard. R. Bimbot, D. Gardes, S. Della-Negra, M. Dumail, B. Kubica, A. Richard, M. F. Rivet, A. Servajean, C. Fleurier, A. Sanba, D. H. H. Hoffmann, K. Weyrich, H. Wahl, Nucl. Instr. and Meth. A 278 (1989) 38.

[12] S. Kawata, K. Horioka, M. Murakami, Y. Oguri, J. Hasegawa, K. Takayama, H. Yoneda, K. Miyazawa, T. Someya, A.I. Ogoyski, M. Seino, T. Kikuchi, T. Kawamura, M. Ogawa, Nucl. Instr. and Meth. A 577 (2007) 21.

[13] S. Miyazawa, A. I. Ogoyski, S. Kawata, T. Someya and T. Kikuchi, Phys. Plasmas 12 (2005) 122702.

[14] S. Kawata, T. Sato, T. Teramoto, E. Bandoh, Y. Masubuchi, H. Watanabe and I. Takahashi, Laser Part. Beams 11 (1993) 757.

[15] S. Kawata, K. Miyazawa, T. Kikuchi, T. Someya, Nuclear Instr. and Meth. A 577 (2007) 332.

[16] J. Runge and B. G. Logan, Phys. Plasmas 16 (2009) 033109.

Journal of Physics: Conference Series **244** (2010) 022004 doi:10.1088/1742-6596/244/2/022004

Capsule Performance Optimization in the National Ignition Campaign*

O L Landen, B J MacGowan, S W Haan, and J Edwards

Lawrence Livermore National Laboratory, Livermore, CA, USA

E-mail: landen1@llnl.gov

Abstract. A capsule performance optimization campaign will be conducted at the National Ignition Facility [1] to substantially increase the probability of ignition. The campaign will experimentally correct for residual uncertainties in the implosion and hohlraum physics used in our radiation-hydrodynamic computational models before proceeding to cryogenic-layered implosions and ignition attempts. The required tuning techniques using a variety of ignition capsule surrogates have been demonstrated at the Omega facility under scaled hohlraum and capsule conditions relevant to the ignition design and shown to meet the required sensitivity and accuracy. In addition, a roll-up of all expected random and systematic uncertainties in setting the key ignition laser and target parameters due to residual measurement, calibration, cross-coupling, surrogacy, and scale-up errors has been derived that meets the required budget.

1. Introduction

The overall goal of the capsule performance optimization campaign is to empirically correct for residual uncertainties in the implosion and hohlraum physics used in our radiation-hydrodynamic computational models [2,3] before proceeding to cryogenic-layered implosions [4] and ignition attempts. This will be accomplished using a variety of surrogate targets that will set key laser, hohlraum and capsule parameters to maximize ignition capsule implosion velocity, while minimizing fuel entropy (or adiabat), core shape asymmetry and ablator-fuel mix. This is followed by intentionally-dudded tritium-rich but deuterium-poor cryo-layered implosions to check the efficacy of the tuning through shared observables such as core symmetry and bangtime, and from implosion performance. Finally, if the chosen ignition design called for larger scale, the tuning would be checked at this scale, before proceeding to tests of alpha-heating and ignition.

Extensive computational multivariable sensitivity studies have shown that, the probability of ignition is well correlated [5] to the four key implosion parameters of 1D peak fuel implosion velocity v, 1D burn-averaged imploded fuel adiabat α, rms asymmetry $\Delta R_{hotspot}/R_{hotspot}$ at the hotspot-main fuel interface, and fraction $\Delta R_{mix}/\Delta R_{fuel}$ of fuel mixed with ablator. The product of power laws of these four parameters, for small excursions, can be used to define an Ignition Threshold Factor (ITF) given by the following equation:

$$\mathrm{ITF} = \frac{E_{fuel}}{3.2kJ}\left(\frac{v}{380km/\sec}\right)^{5.9}\left(\frac{\alpha}{1.46}\right)^{-3.9}\left(1-\frac{\Delta R_{hotspot}}{R_{hotspot}}\right)^{3.5}\left(1-\frac{0.5\Delta R_{mix}}{\Delta R_{fuel}}\right)$$

The constants 380 km/s and 1.46 in the denominators are specific to the particular 285 eV 1.2 MJ Be design [6] considered here that culminates in 3.2 kJ of stored capsule fuel energy. An ITF of 1 equates to 50% probability of ignition. Tuning is expected to increase the mean ITF from ≈0.2 to ≈1.5, with ITF widths of ≈0.2 and ≈0.5 as set by the target physics models uncertainties, and by the quadrature sum of expected residual shot-to-shot variability in laser and target parameters and residual errors in tuning, respectively.

Implosion Performance Offsets			Laser or Target Offsets			Tuning Accuracy	
Parameter	Initial	Final	Parameter	Initial	Final	Observable	Value
DT Fuel Adiabat	+10%	+3%	1st 2ns Inner Cone Energy Fraction	±25%	±10%	Reemit P2 flux asymmetry	±15%
Implosion Core Asymmetry	50% rms	15% rms	1st 2ns Inner Cone Energy Fraction	±25%	±10%	Reemit P2 flux asymmetry	±15%
DT Fuel Adiabat	+10%	+3%	1st 2ns Laser Power	±20%	±10%	1st Shock velocity	±5%
DT Fuel Adiabat	+10%	+3%	Trough Laser Power	±20%	±10%	1st Shock velocity	±5%
DT Fuel Adiabat	+10%	+3%	2nd Shock Laser Power	±10%	±4%	2nd Shock velocity	±2%
DT Fuel Adiabat	+10%	+3%	3rd Shock Laser Power	±10%	±4%	3rd Shock velocity	±2%
DT Fuel Adiabat	+10%	+3%	2nd Shock Launch Time	±200ps	±50ps	2nd Shock overtake point	±6 μm
DT Fuel Adiabat	+10%	+3%	3rd Shock Launch Time	±200ps	±50ps	3rd Shock overtake point	±6 μm
DT Fuel Adiabat	+10%	+3%	4th Shock Launch time	±200 ps	±100 ps	4th Shock breakout time	±100 ps
DT Fuel Adiabat	+10%	+3%	4th Rise Duration	±200ps	±100ps	4th rise Tr slope to peak power	±5%
Ablator Mass Remaining	±80%	±25%	Initial Ablator Thickness	±30 μm	±10 μm	StreakCap Mass Remaining	±13%
Peak Implosion Velocity	±10%	±2%	Peak Laser Power	±20%	±4%	Velocity at r = 300 μm	±2%
Implosion Core Asymmetry	50% rms	16% rms	Peak Inner Cone Energy Fraction	±20%	±5%	Symcap P2 core asymmetry	±7.5 %
Implosion Core Asymmetry	50% rms	16% rms	Hohlraum Length	±400 μm	±200 μm	Symcap P4 core asymmetry	±7,5 %
Ablator-fuel Mix	±40%	±15%	Mid-Z Ablator Dopant Fraction	±0.3%	±.0.1%	2-5 keV x-rays in hohlraum	±10%
Peak Implosion Velocity	±10%	±2%	Peak Laser Power	±20%	±4%	Symcap Bangtime	±50 ps

Table I. Expected initial and residual post-tune 1σ offset from optimum ignition implosion performance, associated initial and post-tune 1σ offsets in optimal laser and target parameters, and required accuracy for tuning associated observables.

The expected initial and final uncertainties in the four implosion parameters are given in the second and third columns in Table I. The initial uncertainties have been estimated based on a combination of level of confidence in extrapolating radiation hydrodynamics models fitting Nova, Omega and Z facility hohlraum energetics, x-ray driven planar hydrodynamics and gas-filled hohlraum implosions data and residual differences between EOS, opacity and conductivity models for the hohlraum, ablator and DT fuel plasmas. These uncertainties translate to uncertainties in capsule ablation rate affecting implosion velocities, to uncertainties in hohlraum x-ray conversion efficiency, albedo and radiation hydrodynamics affecting drive symmetry, and to uncertainties in hard x-ray preheat levels, ablator compressibility and dopant opacity affecting fuel adiabats through shock transit times, and affecting level of ablator-fuel mix through the ablator-fuel interface Atwood number.

The tuning campaign is based on the principal that these physics uncertainties can be empirically corrected for by adjusting key laser and target parameters around their nominal values, thereby increasing the ITF by increasing implosion velocity, and lowering fuel adiabat, asymmetry and mix. 16 principal adjustable parameters have been identified, schematically shown in Figure 1 and listed in the fourth column in Table I alongside the implosion parameter they affect. For the laser, they are the power levels for the 5 phases in the laser pulse, the launch time for the second, third and fourth steps, the end-point in the 4th rise of laser power (when the pulse first reaches peak power), and the power balance between inner and outer cones during the first and last phase. For the target, there are 3 parameters; the hohlraum length, capsule ablator thickness for fixed inside diameter, and capsule ablator mid-Z dopant fraction. The fifth and sixth columns show the expected initial and final 1σ uncertainties in setting these parameters that are consistent with the uncertainties quoted for the four implosion parameters

2. Tuning Techniques

Extensive sets of shots were completed at the Nova and Omega facility to demonstrate and downselect between proposed tuning techniques. The mainline tuning targets chosen are the high Z re-emission spheres [7] setting the foot cone power balance from the observed foot drive symmetry, liquid D_2-filled "keyhole" targets setting the laser power profile up to peak power from the observed shock speeds and overtake distances and times [8], x-ray imaged imploded capsules setting the peak cone power balance and hohlraum length from observed core symmetry [9] and streaked x-ray backlit imploding capsules [10] setting the initial ablator thickness and peak laser power from the radiographically-inferred ablator mass remaining [11] and implosion velocity. In addition, the soft x-ray power diagnostic "Dante" will be used to set the 4th rise launch time from the 4th rise slope and to set the ablator dopant fraction from the measured hard (> 1.8 keV) x-ray preheat levels. The last two columns of Table I list the observables and their required tuning accuracy.

3. Tuning Strategy and Accuracy

The goals of the capsule tuning campaign are to specify the optimum adjustable parameter value and its uncertainty, and to assess that shot-to-shot variability is as expected. A cluster of N shots at a nominal laser and target setting would be taken to assess the 1σ shot-to-shot variability in the observable (to $\sigma/\sqrt{[2(N-1)]}$ accuracy) and compare to expectations. For the latter, the random measurement error bars must be and are expected from scaling from current technique demonstrations to be less than the data scatter. The second step is to correct the data for known preshot shot-to-shot target variations and postshot shot-to-shot laser variations, using calculated slope sensitivities to reduce the scatter in the data to just target and laser diagnostic metrology errors and errors in measuring the observable. In general, the mean of this corrected data will be offset from the optimum value of the observable we are aiming for, precorrected for any known surrogacy offset. The third step is to gather another set of M data points, where in general M < N since data scatter has already been established, for another value of the adjustable parameter that would bracket the optimum setting. The optimum value of the adjustable parameter is then found by linear interpolation between the two datasets with a statistical accuracy = $\sigma/\sqrt{(M+N)}$/mean slope. Finally, one will have to add in quadrature systematic errors due to uncertainty in surrogacy, physics of the technique and calibrations. The various contributions to the tuning accuracy for each of the adjustable laser and target parameters is shown in Figure 1 in terms of their variance normalized to the tuning budget [12] listed in the sixth column in Table I. Many of these terms are themselves rss sums of various contributors [13]. Fig. 1 shows that we expect to meet the tuning accuracy budget for all parameters.

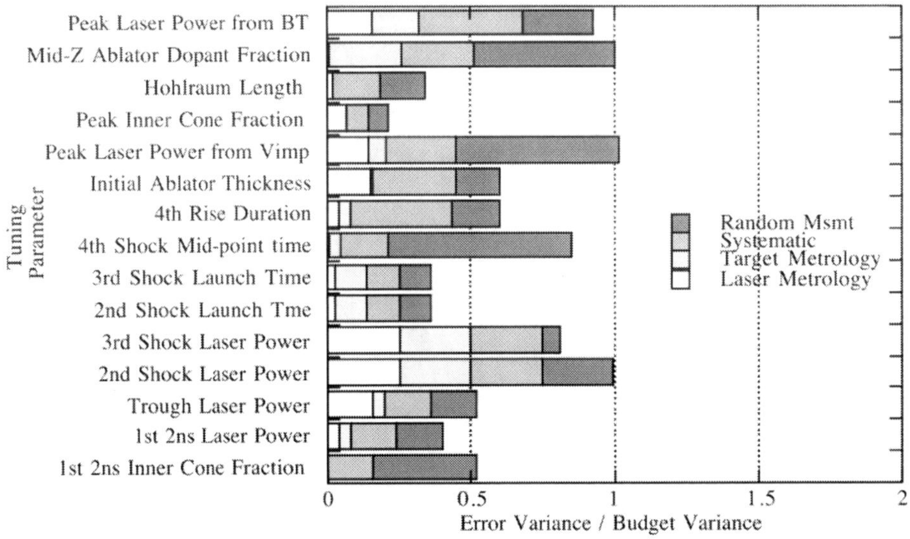

Figure 1. Residual expected variances after tuning normalized to budget for each of the laser and target adjustable parameters.

4. Summary

A capsule performance optimization campaign will experimentally correct for residual uncertainties in the implosion and hohlraum physics used in our radiation-hydrodynamic computational models before proceeding to cryogenic-layered implosions and ignition attempts. The required tuning techniques have been shown experimentally and computationally to meet the required sensitivity and accuracy. The tuning campaign plans include checks of repeatability, iterations to overcome residual cross-couplings and contingency shots. Finally, a set of additional in-flight capsule measurements to isolate capsule implosion physics issues if needed have also been conceptualized.

*Work performed under the auspices of the Department of Energy by the Lawrence Livermore National Laboratory under contract number W-7405-ENG-48.

References

[1] Miller G M, Moses E I and Wuest C R 2004 *Nucl. Fusion* **44** 228.
[2] Zimmerman G B and Kruer W L 1975 *Comments Plasma Phys. Control. Fusion* **2** 51.
[3] Marinak M M, *et. al.* 2001 *Phys. Plasmas* **8** 2275.
[4] Sangster T C, *et. al.* 2007 *Nucl. Fusion* **47** S686.
[5] Haan S W, *et al.* these proceedings.
[6] Haan S W, et. al. 2005 *Phys. Plasmas* **5** 1953.
[7] Dewald E L, *et. al.* 2008 *Rev. Sci. Instrum.* **79** 10E903.
[8] Boehly T R, *et. al. 2009 Phys. Plasmas* **16** 056302.
[9] Kyrala G, *et. al.* 2008 *BAPS* **53** 247.
[10] Hicks D, *et. al.* 2008 *BAPS* **53** 2.
[11] Spears B 2008 *J. Phys.: Conf. Ser.* **112** 022003.
[12] Haan S W, *et al.* 2007 *Eur. Phys. J. D* **44**, 249.
[13] Landen O L to be submitted to *Phys. Plasmas*.

Implosion of a large spherical void

Yian Lei, Bingqiang Sun, Jian Liu, Chao Chen
State Key Laboratory of Nuclear Physics and Technology,
School of Physics, Beijing University, Beijing, 100871, China
E-mail: yalei@pku.edu.cn

Abstract. Implosion is the most efficient way to achieve high energy density (HED) or inertial confinement fusion (ICF). Traditional ICF schemes use millimeter size fusion targets, which usually suffer from unsymmetrical driving and easy-to-rise hydrodynamic instabilities. We investigated the physics of the implosion of a thick shell spherical void, with the inside being vacuum or of very low pressure, and the radius being a few centimeters. Very high pressure and temperature will be achieved in a small region, which may lead to the thermal nuclear fusion ignition.

1. Introduction

Current HED or ICF driving methods, such as high average power laser (HAPL) [1, 2, 3], Z-pinch [4], high energy or intense particle beams [5, 6, 7, 8], or high power wire discharge in materials [9], etc., are mostly involved in some undesirable features, such as huge and expensive external devices (laser, pinch, particle beam, hypervelocity impact), non-equilibrium HED states (energy, temperature, pressure of ions, electrons, and radiation are not uniform), difficult to control, etc. On the other hand, implosion [10] or converging matter wave is very efficient in concentrating energy both temporally and spatially, which finds their applications typically in weaponry and ICF. Another live example is the sonoluminescence phenomena [11, 12]. Though the driving energy for each bubble is very small, the collapsing or implosion of the bubble can produce very high pressure and temperature [13, 14]. This even intrigued the controversial bubble fusion idea [15]. However, as the total energy of each bubble is limited (< 1 mJ), and the vapor pressure in the cavity is high (in the order of 100 Pascal), the compression ($\sim 10^3$ fold) is unlikely to produce a temperature well above 10^5 Kelvin [13].

In ICF regime, the inner DT gas has quite high initial density, which prevents a high ratio compression. Premature shockwave heating also increases the inner pressure, which contributes to the Rayleigh-Taylor instability. The amount of the gas is limited by the driving energy and hydrodynamic efficiency.

If we can dramatically raise the total implosion energy and/or greatly reduce the vapor pressure in the cavity, then we can achieve much higher compression ratio, hence much higher pressure and temperature. Particularly, if the cavity is vacuum, the inner surface of the cavity can reach a very high speed. Upon impact at the center, temperature up to ICF regime is achievable.

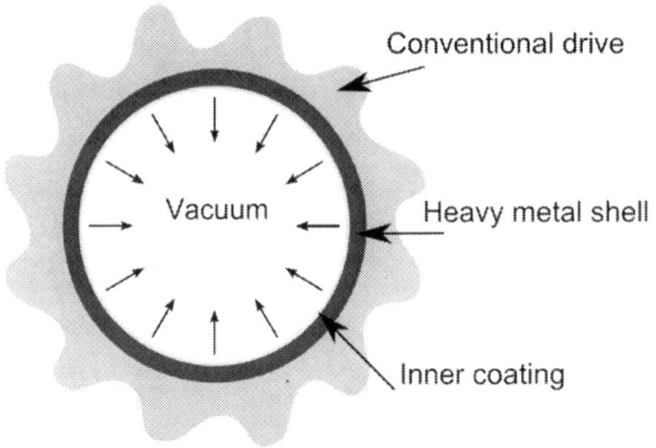

Figure 1. High energy cavity implosion by conventional drive. The cavity is vacuum or of very low pressure, and the shell is heavy metal. The inside of the shell is coated with fusion fuel.

2. High energy cavity implosion

Fig. 1 shows the sketch of the approach. A shell made of high-Z metal is used as the driving medium. Conventional means such as high explosive or high pressure is used to drive the implosion. In the case of vacuum, the fusion fuel (DT-rich solid compound such as $LiB(DT)_4$) or any material of interest is coated on the inner wall of the shell. The radius of the cavity can be a few centimeters (versus $1 \sim 2$ mm in ICF targets), to take full advantage of the converging compression effect. There is no or only very low pressure ($3 \sim 4$ orders of magnitude lower than cryogenic ICF targets) DT gas inside.

Compared with current high power short pulsed drives (HPSPD) such as lasers, pinches, or beams, this approach has some significant features:

(i) The device is much simpler and cheaper.
(ii) Conventional drive such as high explosive implosion is well understood [16], with high hydrodynamic efficiency ($20 \sim 30\%$ vs $< 10\%$ in HPSPDs), high total driving energy (up to ~ 100 MJ), and much lower cost.
(iii) Longer driving time, larger driving area, and higher driving efficiency, make the accumulation of energy more easily. The driving times in conventional means are usually 2 to 3 orders of magnitude longer than HPSPDs.
(iv) The thick and dense high-Z shell provides extra confinement in both momentum and optical opacity.
(v) Due to the vacuum inside, the pressure distribution near the inner wall creates a perfect sphere in the driving stage, which will greatly mitigate the hydrodynamic instability problem afterwards.
(vi) Easily adjustable pressure or temperature by setting different gas density in the cavity.

In ICF scenario, it is obvious that most of these features are very desirable. The downside of this approach may be the difficulty of diagnosis, due to the blocking of the thick shell.

A simple one dimensional incompressible analysis is presented in the next section to illustrate the principle features of implosion.

3. An ICF scenario example in one dimensional incompressible fluid model

Under high power driving, the heavy metal shell can be treated as fluid. To simplify the analysis and illustrate the main features of the imploding process, we assume the shell is incompressible and the viscosities are zero.

Table 1. The velocities, radii and pressures of selected moments. The bottom line is from typical ICF simulation [17].

r_1 (cm)	R_1 (cm)	u_1 (km/s)	f_h	$p_{\max}$ (Mbar)
2.00	2.05	10.0	0.49	0.004
0.50	0.90	18.7	0.23	3.186
0.125	0.85	108	0.014	238.0
0.031	0.85	815	3.1×10^{-4}	1.5×10^4
0.010	0.85	4443	1.1×10^{-5}	4.56×10^5
0.010[1]	–	350[2]	–	$\sim 10^6$

An inertial fusion oriented calculation is carried out as follows. Supposing at the end of the driving, the shell gains a speed of $u_0 = 10$ km/s, with an inner radius of $r_0 = 2$ cm, a thickness of $l = 0.5$ mm, and a density of $\rho = 10$ g/cm^3, the mass of the shell is $M = 25.7$ gram, and the total kinetic energy is $E = 1.2$ MJ.

Fig. 2 shows the fraction of the integrated kinetic energy starting from r_1. It is clear that the smaller the r_1 becomes, the more concentrated the kinetic energy will be to the center of the shell. At $r_1 = r_0/64 = 0.31$ mm, the small layer between $r_1 = 0.31$ mm and $r = 0.6$ mm will have half the total kinetic energy, but its volume (mass) fraction (f_h) is only 0.03%.

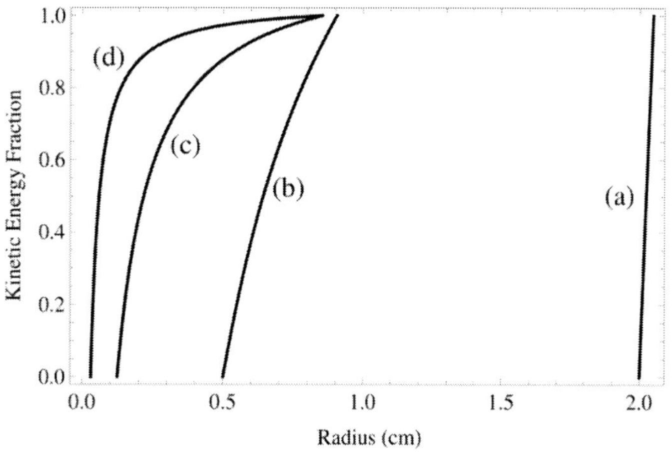

Figure 2. Energy concentration. Kinetic energy fraction integrated from the inner wall. The inner radii for (a), (b), (c) and (d) are 2.0 cm, 0.5 cm, 0.125 cm and 0.031 cm, respectively, and the half points (kinetic energy fraction 0.5) are 2.025 cm, 0.644 cm, 0.218 cm and 0.06 cm, respectively.

For comparison, the inner and outer radii r_1 and R_1, the velocities of the inner wall u_1, half kinetic energy fraction f_h, and the maximum pressures $p_{\max}$ of the imploding shell at selected moments, as well as some comparable identities in typical ICF simulation [17] are displayed in Table 1. Note the inner radius r, implosion velocity u, and pressure p have different meanings in ICF. For $r_1 = 0.01$ cm, the small half kinetic energy region has a kinetic energy density of 2×10^9 J/g, which is of the same order as in ICF ignition, but the total energy $E = 0.6$ MJ is much higher.

4. Discussions

Mathematically, the implosion center is singular, and the implosion speed tends to infinity. We did not calculate inner radii cases less 100 μm for two reasons. One is $r \sim 100$ μm is the typical ICF stagnation radius. The other is that realistically there may be a small amount of matter in the cavity, the vacuum assumption is no longer true.

The incompressibility assumption is hard to meet in high energy cases. High-Z metal such as gold or uranium is chosen to reduce the compressibility in HED. The calculations in the last section are only illustrative, not qualitatively trustworthy. However, they do reveal some of the attractive features as a HED drive: a), high hydrodynamic efficiency. Total kinetic energy concentrates to the center; b), very high pressure and velocity is achievable.

For compressible fluid, a large portion of the total hydrodynamic kinetic energy would be transferred into internal energy. At pressure as high as ~ 1 Gbar, high Z materials may have an internal energy density of ~ 1 MJ/g. However, as the ultra high pressure region is very limited as shown in Table 1, there should be still a large portion of imploding kinetic energy left.

A preliminary one-dimensional radiation hydrodynamic simulation is carried out by a modified MULTI code [18]. Simulation shows, with the shell to be gold, and the inside to be low pressure DT gas, $3 \sim 4$ MJ of imploding dynamic energy can produce a DT ion temperature of about $1 \sim 2$ keV, and DT density of about 100 g/cm. Fusion neutron yield is around 10^{12} to 10^{15}. Ignition is hard to achieve in reasonable settings. The major energy loss is the electron heat transfer of the shell and radiation.

However, there are some facts missing in current simulations: (a) as the shell is much thicker and denser than that in mainstream ICF approaches, neutron heating is not negligible, but is absent in current code. (b), as opacity is important, hybrid shell can lower the radiation loss [19]. (c) If the shell is depleted Uranium, extra heating by fast neutron induced fission will be of great help. In this case, the total energy output would be mainly fission energy.

Further investigation is worthwhile to find out if this approach is applicable in energy production.

Acknowledgement

This work is supported by National Natural Science Foundation of China (10976001, 10935001), and 973 program (2008CB717803).

References

[1] Nuckolls J, Wood L, Thiessen A, and Zimmerman G 1972 *Nature*, **239**, 139
[2] Clarke J S, Fisher H N and Mason R J 1973 *Phys. Rev. Lett.*, **30**, 89-92
[3] Pirri A N 1977 *Phys. of Fluids* **20**, 221
[4] Kalantar D H and Hammer D A 1993 *Phys. Rev. Lett.*, **71**, 3806
[5] Hoffmann D H H et al 2002 *Phys. of Plasmas* **9**, 3651
[6] Fortov V E, Hoffmann D H H, Sharkov B Yu 2008 *Phys.-Usp.* **51** 109-131
[7] Hoffmann D H H et al 2005 *Laser and Particle Beams* **23** 47-53
[8] Tahir N A et al 2005 *Nucl. Inst. Meth.* **544** 16-26
[9] Sasaki T, Yano Y, Nakajima M, Kawamura T and Horioka K 2006 *Laser and Particle Beams*, **24** 371-380
[10] Hawkins D 1961 LASL Report No. LAMS-2532 (Vol. I)
[11] Frenzel H and Schultes H 1934 *Z. Phys. Chem.* **B27**, 421
[12] Gaitan D F 1990 Ph.D. dissertation, University of Mississippi.
[13] Brenner M P, Hilgenfeldt S and Lohse D 2002 *Rev. Mod. Phys.* **74**, 425-484
[14] Moss W C, Clarke D B, White J W and Young D A 1994 *Phys. Fluids*, **6**, 2979-2985
[15] Taleyarkhan R P et al 2002 *Science*, **295**, 1868-1873
[16] Trunin R F 1998, *Shock Compression of Condensed Materials*, Cambridge University Press
[17] Atzeni S and Meyer-Ter-Vehn J 2004 *The physics of inertial fusion*, Clarendon Press-Oxford
[18] Ramis R, Schmalz R and Meyer-Ter-Vehn J 1987, *Compu. Phys. Commu.* **49**, 475-505
[19] Jones O S et al 2007, *Phys. Plasmas* **14**, 056311

Journal of Physics: Conference Series **244** (2010) 022006 doi:10.1088/1742-6596/244/2/022006

Single-shell direct-drive capsule designs to study effects of perturbations on burn

G R Magelssen, J A Cobble, I L Tregillis, M J Schmitt, S H Batha, P A Bradley, and K A Defriend Obrey
Los Alamos National Laboratory, N.M. 87545, USA

The effect of small localized perturbations, such as fill tubes and mounting tents, on the NIF ignition capsule and the effect of hemi-joints on high gain double shell capsules are an important issue in achieving ignition on NIF. Our codes have difficulty modeling small features and their effect on mix. Because of issues of symmetry, shock timing, high-z shells, mix etc. trying to understand the effect of localized perturbations ("defect") on these high gain NIF capsules will be difficult. To begin the study of defects on DT burn, a direct-drive exploding pusher is used. Experimental results are presented for capsules with and without defects. The unperturbed capsules give reproducible yields while the perturbed capsules show significant drops in yield. Both AMR and Lasnex codes over predict the unperturbed capsule yield

1. Introduction

Small capsule perturbations may impact our ability to achieve high yields on NIF. Diagnosing the hydrodynamic development and the effect of these defects on burn will be difficult. Los Alamos is developing a program to better understand the hydrodynamics of defects and how they influence burn. Much work has been done on the study of defects in planar geometry and that work will not be discussed here. Only experiments in convergent geometry will be discussed. The hydrodynamics of defects in single [1] and double-shell cylindrical configurations [2] was done on Omega earlier. Some discrepancies between simulations and experiment were found. For example, simulations suggested greater penetration of the defect jet than was seen experimentally. Spherical double-shell experiments were also done at Los Alamos [3]. These, also, concentrated on the hydrodynamics of defect-generated jets. Target fabrication and image resolution were significant issues for these experiments. The effects of fill tubes on the hydrodynamics of ignition targets were simulated by J. Edwards et al. [4] Experiments on fill tubes by G. R. Bennet et al. [5] compared simulation and experimental data and experiments by N. Izumi and calculations by S. H. Langer [6] made additional comparisons.

2. Experiments

To begin our experimental study of defects on DT burn, an exploding pusher is used. The exploding pusher is insensitive to symmetry and has a very reproducible yield. Exploding pushers with defect rings 15-17 microns wide and 2.2-2.7 microns deep were shot on the Omega laser. The capsules were CH shells 431-436 microns in radius and 8.1-8.5 microns thick. They were filled with a 50-50 ratio of 5-5.5 atmospheres of DT and were driven by direct laser absorption. The exploding pusher results are compared to Omega shots on targets with thicker shells of CH. These targets were 441-445 microns in radius and 15.1-15.2 microns thick. They were, also, driven by direct laser drive. The defects were similar to those for the exploding pusher. Because these targets were being used as a test for the NIF neutron imager the constraints on the targets were severe. The feature needed to reduce the yield enough from an unperturbed capsule to give a clear indication that the reduction was from the defect. However, the yield had to be high enough for a good neutron image. Further, the neutron image needed to have some measurable distortion over the unperturbed capsule. Calculations with both Eularian AMR and Lagrangian codes were done and compared to the experimental results. The AMR code does not have laser physics and, so, a model was developed to approximate the laser absorption.

The model assumes the laser energy is absorbed in a thin shell, typically 2-3 microns, on the surface of the capsule. The laser profile was a flat 1 ns pulse of 24 kJ. The AMR and Lasnex codes used this profile but reduced the energy by 35%. This reduction is consistent with experimental measurements made by the University of Rochester Laser Energetics team (LLE) [7]. Tables 1 and 2 give detailed experimental data for our Omega shots. In Table 1, this includes the target yield, average ion temperature for the DT, the experimental time of peak burn, the burn width, the capsule initial DT pressure and the initial capsule diameter and shell thickness. In Table 2, the capsule thickness and diameter are repeated as well as detailed information about the defect ring and the absorbed RMS energy balance that was typically about 2%. The energy for most beams was about 400 joules. For two shots there was one beam that had only 80 joules. For those shots, the RMS energy balance increased to about 10%. Figure 1 shows an image of one of the defect targets made by General Atomics and shot on Omega. The defect ring is described by three parameters, an outer defect width(width-O), an inner defect width(width-I), and a defect depth. The meaning of each of these parameters is illustrated in Figure 2. In the target chamber, the capsules were held by a stalk. The SiC stalk had a diameter of 17-20 microns and was glued to the capsule with an epoxy. The epoxy had a radius of 25 to 40 microns and a highth of 10-15 microns. The yield for our experiments was above the threshold needed ($5x10^{12}$) to obtain good neutron signals. Unfortunately, the neutron resolution (above 80 microns) was not good enough to give information on the structure of the burning core. X-ray emission images did show the presence of the stalk in the wings of the burning core. Both the exploding pusher and thick targets gave reproducible yields. Furthermore, compared to the unperturbed targets, the defect targets show a clear drop in yield.

Table 1. Experimental Results Part 1

Thickness microns	Diameter microns	Type	Yield neutrons	Ion Temperature keV	Bang time picoseconds	Burn width picoseconds	DT atms
8.2	872	perfect	2.27E+13	9.6	983	136	5.12
8.5	872	perfect	2.13E+13	9.5	1031	135	5.48
8.2	867	defect	1.28E+13	9.0	1027	169	5.43
8.2	862	defect	1.40E+13	10.0	1043	171	5.39
8.4	867	defect	1.56E+13	9.6	1035	170	5.30
8.5	868	defect	1.60E+13	8.9	980	182	5.05
15.1	885	perfect	1.27E+13	5.7	1447	170	5.41
15.2	891	perfect	1.37E+13	5.4	1504	159	5.33
15.1	884	defect	1.11E+13	5.2	1483	169	5.15
15.2	883	defect	6.35E+12	4.8	1462	206	5.48

Table 2. Experimental Results Part 2

Thickness microns	Diameter microns	Type	Width-O microns	Width-I microns	Defect Depth microns	RMS Energy Balance %
8.2	872	perfect	0	0	0	2.1
8.5	872	perfect	0	0	0	2.2
8.2	867	defect	15	9	2.8	2.1
8.2	862	defect	15	9	2.5	2.1
8.4	867	defect	16	9	2.4	2.1
8.5	868	defect	17	9	2.7	2.1
15.1	885	perfect	0	0	0	2.1
15.2	891	perfect	0	0	0	10.6
15.1	884	defect	18	8	2.2	2.1
15.2	883	defect	21	9	2.7	10.6

3. Simulations

Our calculations were done with radiation, flux-limited thermal conduction, sesame opacity and equation-of-state. They were done with three temperatures (electron, ion, and radiation) and with thermonuclear burn. Laser absorption was perpendicular to the capsule surface for Lasnex calculations that include an energy reduction based on work by LLE. Laser absorption was approximated by an energy source in the CH shell for the AMR calculations.

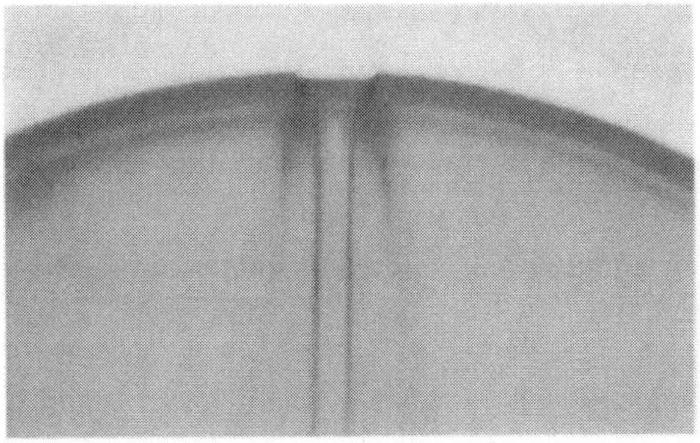

Figure 1. Typical GA defect used in our experiments.

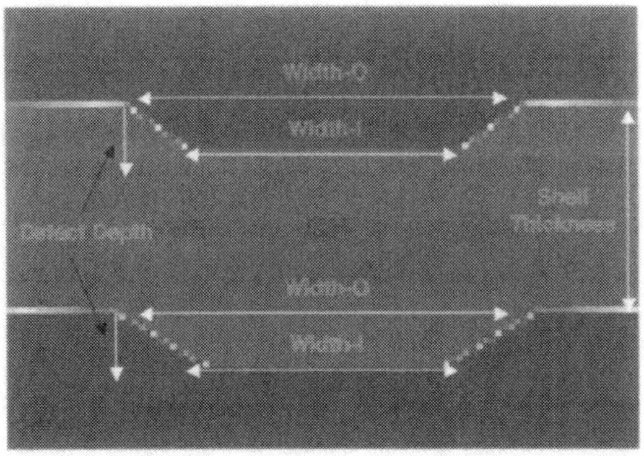

Figure 2. Example of a calculated defect.

Our starting point was 1-D Lasnex calculations. For a fixed absorbed energy, the thermal flux limiter was varied until the experimental peak burn time (bang time) was matched. Using the same absorbed energy and flux limiter, symmetric 2-D lasnex simulations were than done to compare with 1-D and confirm 2-D results. The next step was to do 1-D calculations with the AMR code. Using the same flux limiter and absorbed energy as the Lasnex simulation, the shell source thickness was varied until the density and temperature profiles matched the 1-D Lasnex results. A variance of about 20% in value amplitude and 5% in timing was allowed. The symmetric AMR 2-D simulation was than checked by using the 1-D absorbed energy, flux limiter, and shell source thickness. As the tables illustrate General Atomics provided a detailed metrology of each capsule defect. This information was applied to both the Lasnex and AMR code calculations. A typical defect example is shown in figure 2. Lasnex and AMR defect simulations were done using the constraints found in the 1-D and symmetric 2-D calculations. The results of some of the calculations and the experiment are shown in figure 3. The yields for the perfect thin (8.5 micron-exploding pusher) and thick (15.1micron) shell capsules are plotted (see Tables 1 and 2). The other symmetrically driven capsules gave similar results. The yields for the defect thin (8.5 micron) and thick (15.1 and 15.2 micron) shells are also plotted. These yields give a good representation of the simulations and the data.

The imploded core density and ion temperature of the exploding-pusher-defect-capsule (8.5 microns) is shown in figure 4. This was an AMR calculation with a resolution of 0.25 microns both in the shell and in the DT gas. The time, 0.96 ns, was close to peak burn time. The simulation shows a considerable amount of density and temperature structure. This is a result of a complex series of shocks that occur within the DT gas. The shock breaks through the defect region before the other parts of the shell. An azimuthally directed shock is created and produces an almost cylindrical imposion on axis. This squirts DT plasma along that cylindrical axis while another shock propagates in the opposite direction outward toward the incoming shell. At a slightly later time, multiple shocks make the implosion very complex.

The surface roughness for these capsules was very good. The surface roughness was measured to be 1-9, 20-30 and 50-100 nanometers for high, intermediate and low modes. Since these are very small amplitudes, we expect the modes to be in the nonlinear regime. Using the nonlinear Rayleigh-Taylor growth rate and an Atwood number of one, we estimate the nonlinear growth rate to be 5-10x10^9 for the exploding pusher and about half that for the thick shell capsule.

Both codes greatly over predict the yield of the symmetric implosion. For the exploding pusher (8-8.5 micron shells), the codes predict about a factor of 3-4 decrease in neutron yield over the unperturbed results. This is little larger than we measured experimentally. The yield for the thicker defect targets was predicted to drop by a factor of 10-20 over its unperturbed value. The experiments, however, showed at most a factor of 2 drop in the yield.

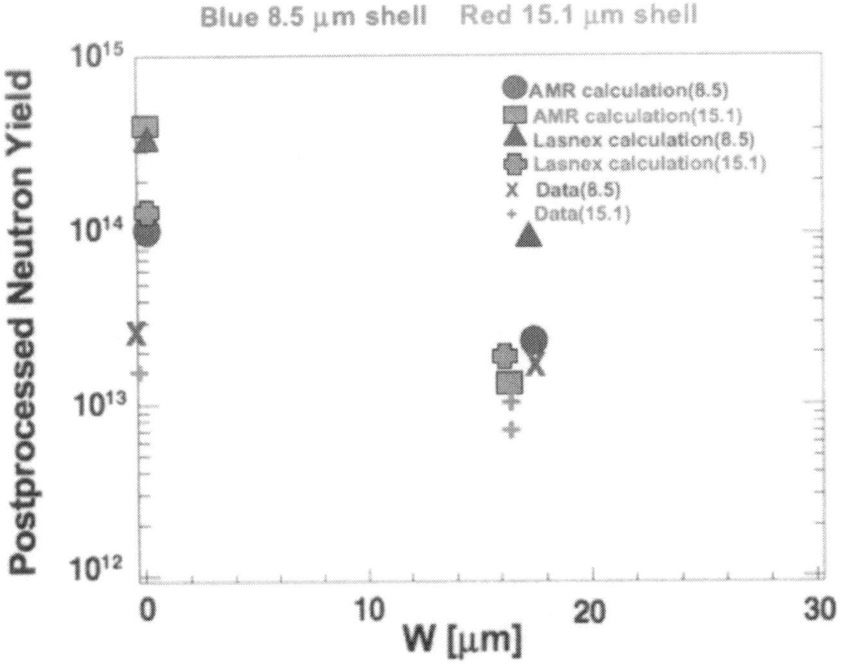

Figure 3. Yield as a function of defect width (**W**).

To address the large over prediction of the unperturbed capsule yield a number of physics additions will be made. A simple constant reduction of 35% in the energy was used in our calculations. LLE has the absorbed time dependent profile for these capsules. Future calculations will use this profile. As discussed, the capsules were mounted on stalks with a glue spot at the capsule-stalk interface. The stalk effect will be added. Lawrence Livermore Laboratory has found some problems with the standard Equation-Of-State (EOS) table. Their new EOS will be used. LLE has a method for estimating the direct drive symmetry on the capsule. This method will be applied to our calculations. Finally, both codes have mix models and these models will be added to our simulations.

4. Summary

Our hypothesis is that our fine scale (feature and mix) radiation-hydrodynamics and burn is not being modeled accurately in our codes. As a result, an experimental program has been started to address this accuracy. The results of direct-drive capsule laser experiments have been presented as our first step. Capsule shell thicknesses were varied to study both the exploding pusher and thick pusher regime. The unperturbed capsules gave experimentally reproducible yields while the capsules with defects gave significantly lower yields. Both AMR and Lasnex calculations greatly over predict the observed yield of the unperturbed capsule. Possible reasons for this discrepancy are outlined and future calculations with the additional physics are planned.

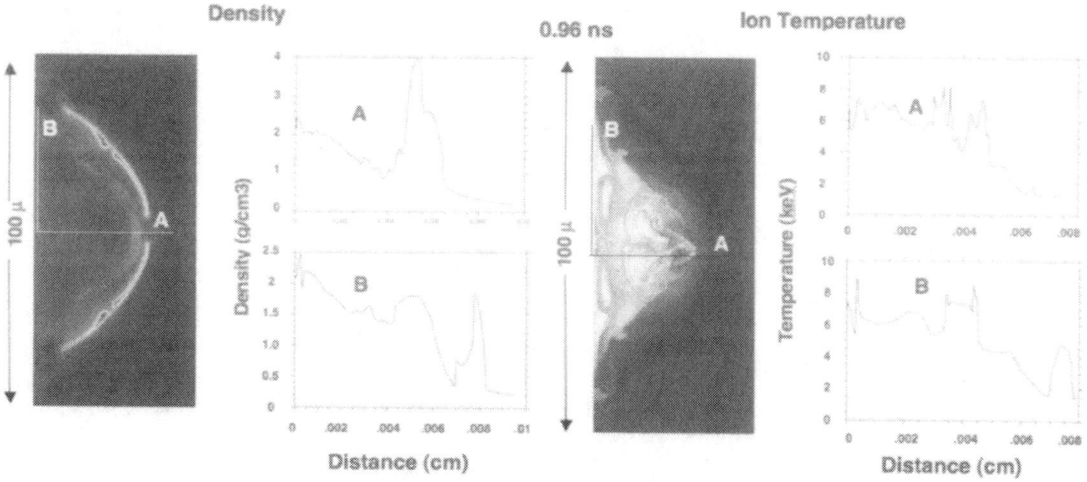

Figure 4. Density and Ion temperature near the peak of burn for the exploding pusher capsule.

5. References

[1] Lanier N E el al. 2006 *Phys. Plasmas* **13** 042703
[2] Fincke J R et al. 2005 LA-UR-05-3718
[3] Kyrala G A et al. 2005 *Laser and Particle Beams* **23** #2 187-92
[4] Edwards John et al. 2005 *Phys. Plasmas* **12** 056318
[5] Bennet G R et al. 2007 *Phys. Rev. Letters* **99** 205003
[6] Langer S H et al. 2007 *High Energy Density Physics* **3** 169-174
[7] LLE private communication

Acknowledgements
We would like to thank the Staffs of GA, LLE and LANL MST-7 for their help and support.

Rayleigh - Taylor instability in elastic-plastic solids

A. R. Piriz, J. J. López Cela

E.T.S.I. Industriales, Universidad de Castilla-La Mancha and Instituto de Investigaciones Energéticas, 13071 Ciudad Real, Spain

N. A. Tahir

GSI Helmholtzzentrum für Schwerionenforschung Darmstadt, Planckstrasse 1, 64291 Darmstadt, Germany

E-mail: roberto.piriz@uclm.es

Abstract. The incompressible Rayleigh - Taylor instability of elastic-plastic solid plates accelerated uniformly by a constant pressure is studied in the linear regime. The instability threshold is found and the boundary for the elastic to plastic transition is also determined. It is demonstrated that transition from the elastic to the plastic regime is a necessary condition for the onset of instability but not a sufficient one. The theory is in excellent quantitative agreement with the results of two-dimensional numerical simulations and reveals the main physical mechanisms that control the instability.

1. Introduction

Rayleigh - Taylor instability (RTI) in solids [1-6] is of fundamental relevance in many experiments on high-energy density physics. The present work is motivated by the requirements of the LAPLAS (Laboratory Planetary Sciences) experiment that we are designing in the framework of the new facility FAIR (Facility for Antiproton and Ion Research) at GSI Darmstadt in Germany [7]. This experiment is dedicated to the study of high-energy density matter generated by using a low entropy cylindrical implosion driven by an intense ion beam. On the other hand, RTI in solids is also important in current experiments where solid material strength is inferred from the direct measurement of the growth rate of perturbations preformed on the surface of an accelerated plate [8, 9]. In the latter context, the stabilizing effects of the solid mechanical properties has been erroneously interpreted as the effect of a very high viscosity [8, 9]. Such an interpretation may have sense only if the shear strain rate is constant [10] but this is not usually the case in the above mentioned experiments. Actually, the more complex constitutive properties of an elastic-plastic solid should be considered.

Here we present an analysis of the linear RTI in solids with elastic-plastic constitutive properties. We come up with a consistent stability criterion that shows that transition from a stable elastic regime to a plastic regime is a necessary condition for instability but not at all a sufficient one. This is done by considering a well defined physical situation so that we can study pure incompressible RTI in a thick plate ($kh >> 1$, $k = 2\pi/\lambda$, h is the plate thickness and λ is the perturbation wave number) under a constant and uniform gravity acceleration g. The analysis is performed by means of an approximate but accurate model based on the Newton's second law [11] which uses a Prandtl-Reuss constitutive model with the von Mises yield criterion

[2]. The study is completed by extensive 2D numerical simulations performed with the finite element code ABAQUS [12].

2. The analytical model

For constructing the model we assume a uniform plate of density ρ which is thick enough ($kh >> 1$) so that finite thickness effects can be neglected. The plate is in the plane (x, z) and it has been accelerated during a very long time until $t = 0$ by a constant and uniform pressure p_0 that represents a low density fluid accelerating the plate with an acceleration $\vec{a} = -g\vec{e_y}$ ($g = p_0/\rho h$). At $t = 0$ a ripple $\delta p = p_0(\xi_0/h)\sin kx$ is superposed to the uniform pressure p_0. This approach is equivalent to the alternative of considering a uniform pressure driving a plate with preformed corrugations of amplitude ξ_0. We have chosen this option because it was easier to realize in our numerical simulations. With the previous considerations the linear evolution of the perturbation amplitude can be well described by the following equation of motion [11]:

$$\frac{\rho}{k}\ddot{\eta} = \rho g(\eta + \eta_0) - S_{yy} \quad , \tag{1}$$

where $\eta(x, y, t) = \xi(t)e^{ky}\sin kx$ because we assume that the perturbed velocity field can be approximated by the one corresponding to an inviscid ideal fluid. $\xi(t)$ it is the instantaneous perturbation amplitude on the interface and S_{yy} represents the force per unit area due to the mechanical properties of the medium. S_{yy} is given by the normal component of the deviatoric part of the stress tensor $\sigma_{ij} = -p\delta_{ij} + S_{ij}$ (p is the thermodynamic pressure and δ_{ij} is the Kronecker tensor) and it can be obtained from the Prandtl-Reuss equations: $S_{yy} = 2kG\eta$, for $\eta \leq \eta_p$; and $S_{yy} = Y/\sqrt{3}\sin kx$, for $\eta \geq \eta_p$; where η_p is the value of η for which the elastic limit is achieved. G and Y are the shear modulus and the yield strength of the solid, respectively. Since the onset of plastic flow occurs first on the interface and then it progresses towards the plate interior its effects on the instability are not felt until it has affected a region with thickness of the order of k^{-1}. Therefore, by evaluating Eq.(1) at $y = y_p \approx -k^{-1}$ we get: $\ddot{\xi} = kg(\xi+\xi_0) - 2k^2G\xi/\rho$, for $\xi \leq \xi_p$; and $\ddot{\xi} = kg(\xi + \xi_0) - \alpha Yk/\rho\sqrt{3}$, for $\xi \geq \xi_p$ ($\xi_p = \alpha Y/2\sqrt{3}kG$). Since $\alpha = e^{k|y_p|}$ it is taken $\alpha = 3$. In addition, since we choose to perturb the driving pressure, it is $\xi(0) = 0$. In the alternative case of perturbing the plate surface it would be $\xi(0) = \xi_0$ and the corresponding equation of motion is modified from Eq.(1) by replacing η by $\eta' = \eta + \eta_0$.

It is convenient to introduce the following dimensionless magnitudes: $z = \xi/\xi_0$, $\tau = t\sqrt{kg}$, $\hat{\lambda} = \rho g\lambda/4\pi G$, $\hat{\xi} = \rho g\xi_0/\sqrt{3}Y$. Then, Eq.(1) reads: $\ddot{z} = z(1 - \hat{\lambda}^{-1}) + 1$, for $z \leq z_p$, and $\ddot{z} = z + 1 - \hat{\xi}^{-1}$, for $z \geq z_p$ ($z_p = \hat{\lambda}/\hat{\xi}$). This equation can be explicitly solved with the initial conditions $z(0) = 0$ and $\dot{z}(0) = 0$ to yield the complete evolution of the perturbation amplitude. More details of the model and complete solutions has been presented elsewhere [13]. Here, for the study of the stability boundaries it is enough to consider the previous equation together with its first integral: $\dot{z}^2 = z^2(1 - \hat{\lambda}^{-1}) + 2z$ ($z \leq z_p$), and $\dot{z}^2 = (z + 1 - \hat{\xi}^{-1})^2 - C^2/\hat{\xi}^2$ ($z \geq z_p$), with $C^2 = (1 - \hat{\xi})^2 - \hat{\lambda}$. In the pure elastic regime the instability threshold reads $\hat{\lambda} > 1$ [11, 14, 15]. For the general elastic-plastic case the interface is also unstable for $\hat{\lambda} > 1$. However, for $\hat{\lambda} < 1$, when the interface is stable in the pure elastic regime, in the elastic-plastic case it can be stable or unstable depending on the value of $\hat{\xi}$. This conclusion is clearly extracted from the complete solution but it can also be obtained directly from the equations for $\ddot{z}$ and $\dot{z}$ by noticing that stability can only exist if at the time τ_m when $\dot{z}(\tau_m) = 0$, also $\ddot{z}(\tau_m) \leq 0$ so that at $\tau = \tau_m$ we have a maximum. Otherwise, if $\ddot{z}(\tau_m) \geq 0$ we have an inflection point and the amplitude will continue growing with no limit. Therefore, the conditions for marginal stability are that $\dot{z}(\tau_m) = 0$ and $\ddot{z}(\tau_m) = 0$. They can only be satisfied if $z(\tau_m) = z_m \geq z_p$, which means that the interface will become unstable once the transition from the stable elastic regime to a plastic regime has already taken place. This plastic regime can be stable provided that $\hat{\xi}$ is sufficiently small. Actually, as it is seen from the exact solutions of the interface equation of motion, this stable plastic regime consists of a plastic phase in which a maximum is achieved and then it is followed by pure elastic oscillations with an amplitude determined by the previous plastic phase. Instead, the amplitude in the unstable plastic solutions for $\hat{\lambda} < 1$ will continue to grow in the plastic regime. In summary, the transition from elastic to plastic regime is a necessary condition for instability but once the transition is achieved the interface may be stable or unstable depending on the value of $\hat{\xi}$. It can be more easily appreciated from the conditions of marginal stability determining the instability threshold:

$$(\rho g \xi_0/\sqrt{3}Y)_{ITH} = 1 - \sqrt{\rho g \lambda/4\pi G} \quad . \tag{2}$$

We can also get the condition for the elastic to plastic transition (EPT) by requiring that the maximum amplitude z_m^e be equal to the amplitude $z_p = \hat{\lambda}/\hat{\xi}$ necessary to reach the elastic limit:

$$(\rho g \xi_0/\sqrt{3}Y)_{EPT} = (1 - \rho g \lambda/4\pi G)/2 \quad . \tag{3}$$

Eqs.(2) and (3) determine the boundaries for the existence of the stability after the elastic limit

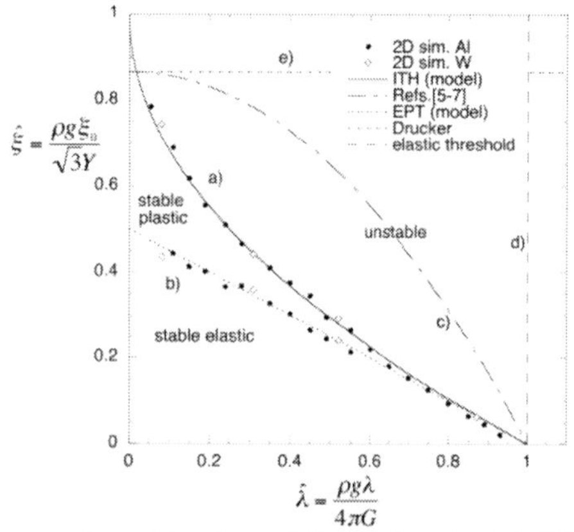

Figure 1. The instability threshold (ITH) and the elastic-plastic transition boundary (EPT).

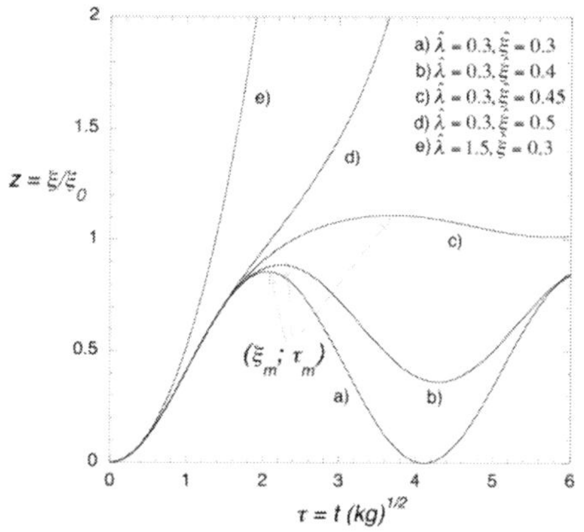

Figure 2. Perturbation amplitude for different dimensionless amplitudes $\hat{\xi}$ and dimensionless wave lengths $\hat{\lambda}$.

has been reached. It may be worth noticing here that the RTI growth is governed by the average motion of the mass contained within a distance of the order of k^{-1} from the interface [11, 14, 15]. Therefore, the boundary for the EPT given by the previous equation is not determined by the onset of the plastic flow at any particular material point of the plate. Rather, it is the achievement of the elastic limit on a region with thickness of the order of k^{-1} what determines the transition from a regime controlled by the elastic properties to one controlled by the plastic behavior. So, the limit for the onset of plastic flow at the interface ($y = 0$) calculated in Ref.[16] by using the elastic theory of Ref.[14] has no special meaning for the stability analysis. For the same reason, such a local condition cannot be compared with the average condition derived in Refs.[4-6] . The latter, on the other side, has the major flaws that the EPT boundary was erroneously identified with the instability threshold and that it was actually derived for thin plates and was then extrapolated to arbitrary thick plates. Nevertheless, it probably predicts the EPT boundary for the case of thin plates correctly but not, of course, the instability threshold.

In Fig. 1 we present the instability threshold [a)] and the EPT boundary [b)] together with the instability threshold derived in Refs.[4-6], what actually was their EPT boundary [c)]. For comparison, we show in Fig.1 the instability threshold for perfectly elastic media [d)] [11, 14, 15].

We also show the Drucker threshold [e)] for the instability of a perfectly rigid plastic solid [3] which is smaller than our upper limit (for $\hat{\lambda} = 0$) by a factor $\sqrt{3}/2$. In Fig.2 we show the different kind of stable and unstable solutions of the interface equation of motion.

3. 2D numerical simulations

For completion of our analysis we have performed extensive 2D numerical simulations with a finite element code ABAQUS [12] in order to compare with the analytic results. We have adopted the Mie-Grüneisen EOS with a Grüneisen coefficient $\Gamma = \rho_0\Gamma_0/\rho$ (Γ_0 is a parameter of the material and ρ_0 is the initial material density). In addition, we have taken the usual linear relationship between the shock velocity u_s and the particle velocity u_p: $u_s = c_0 + su_p$. The constants c_0 and s are parameters characteristic of the material. In the present numerical simulations, however, we have taken $c_0' \sim 10c_0$ in order to ensure incompressibility [6]. Furthermore, we have taken two sets of values, one corresponding to Al parameters: $\rho_0 = 2.7\ g/m^3$, $s = 1.337$, $\Gamma_0 = 2.16$, $c_0 = 5380\ m/s$; and a second one corresponding to W parameters: $\rho_0 = 19.3\ g/m^3$, $s = 1.237$, $\Gamma_0 = 1.54$, $c_0 = 4030\ m/s$. The values of G, Y, ξ_0 and λ were varied in order to span the complete range of values of $\hat{\xi}$ and $\hat{\lambda}$ between 0 and 1 and to meet the conditions for marginal stability.

The results of the ABAQUS simulations for the instability threshold and for elastic plastic transition are also shown in Fig.1. For the latter we have looked at the value of $\hat{\xi}$ for a fixed value of $\hat{\lambda} < 1$ for which the amplitude of the elastic oscillations starts to deviate from the pure elastic case. We can also calculate the perturbation amplitude on the instability threshold. From the equation for $\dot{z}$ we put $C = 0$ and the instantaneous perturbation amplitude turns out: $[\hat{\xi}(z_m + 1)]_{TH} = 1$ or in dimensional form $\left[\rho g(\xi_m + \xi_0)/\sqrt{3}Y\right]_{TH} = 1$. This expression reduces to the Drucker criterion (except by the factor $\sqrt{3}/2$) for $\lambda = 0$, when no perturbation growth occurs ($\xi_m = 0$), and its accuracy is also confirmed by our numerical simulations. This is a very interesting result since generally speaking, it says that in the elastically stable regime ($\hat{\lambda} < 1$), instability will occur if the pressure difference $\rho g(\xi_m + \xi_0)$ across a crest (or a valley) becomes equal to $\sqrt{3}Y$.

References

[1] J. F. Barnes, P. J. Blewet, R. G. McQueen, K. A. Meyer, and D. Venable, J. Appl. Phys. **45**, 727 (1974).
[2] J. W. Swegle and A. C. Robinson, J. Appl. Phys. **66**, 2838, 2859 (1989).
[3] D. C. Drucker, Mechanics Today **5**, 37 (1980); D. C. Drucker, Ing. Arch. **49**, 361 (1980).
[4] P. N. Nizovtzev and P. N. Raevskii, VANT Ser. Teor. Prikl. Fizika **3**, 11 (1991).
[5] A. I. Lebedev, P. N. Nizovtzev, P. N. Raevskii and V. P. Solovev, VANT Ser. Teor. Prikl. Fizika **3**, 11 (1991).
[6] S. M. Bakharakh, O. B. Drennov, N. P. Kovalev, A. I. Lebedev, E. E. Meshkov, A. L. Mikhailov, N. V. Neumerzhitsky, P. N. Nizovtsev, V. A. Rayevsky, G. P. Simonov, V. P. Solovyev, and I. G. Zhidov, Lawrence Livermore National Laboratory Report No. UCRL-CR-126710, 1997 (unpublished).
[7] N. A. Tahir, D. H. H. Hoffmann, A. Kozyreva, A. Tauschwitz, A. Shutov, J. A. Maruhn, P. Spiller, U. Neuner, R. Bock, H. Juranek, and R. Redmer, Phys. Rev. **63**, 016402 (2001).
[8] J. D. Colvin, M. Legrand, B. A. Remington, G. Shurtz, and S. V. Weber, J. Appl. Phys. **93**, 5287 (2003).
[9] H.-S Park, B. A. Remington, R. M. Cavallo, B. R. Maddox, M. J. May, S. M. Pollaine, S. T. Prisbrey, R. E. Rudd, Oral 1.5.4; B. A. Remington, H.-S.Park, S. T. Prisbrey, S. M. Pollaine, R. M. Cavallo, J. A. Hawreliak, A. G. MacPhee, R. E. Rudd, J. S. Wark, E. M. Bringa, M. A. Meyers, Poster 2.10.095. This conference proceeding .
[10] J. P. Poirier, *Earth's Interior*, (Cambridge U. Press,New York, 2000), Chap. 6, p. 174.
[11] A. R. Piriz, J. J. Lopez-Cela, O. D. Cortazar, N. A. Tahir and D. H. H. Hoffmann, Phys. Rev. E **72**, 056313 (2005); A. R. Piriz, O. D. Cortazar, J. J. Lopez-Cela and N. A. Tahir, Am. J. Phys. **74**, 1095(2006).
[12] ABAQUS, Finite Element Code, Version 6.6 (Hibbit, Kalsson and Sorensen Inc. Pawtucket, RI, 2006).
[13] A. R. Piriz, J. J. Lopez-Cela, N. A. Tahir, Phys. Rev. E **80**, 046305 (2009);
[14] B. J. Plohr and D. H. Sharp, Z. angew.Math. Phys. **49**, 786 (1998).
[15] G. Terrones, Phys. Rev. E **71**, 036306 (2005).
[16] G. Terrones, J. Appl. Phys. **102**, 034908 (2007).

3D ray-tracing package dedicated to the irradiation study of a directly driven inertial confinement fusion capsule

M. Temporal[1], R. Ramis[1], and B. Canaud[2]

[1] ETSIA, Universidad Politécnica de Madrid, Spain

[2] CEA/DAM/DIF, Bruyères-le-Châtel, F-91297, Arpajon Cedex, France

E-mail: mauro.temporal@hotmail.com

Abstract. The compression of a directly driven all-DT capsule has been simulated by the hydrodynamic code Multi1D. The capsule absorbs a total energy of about 100 kJ allowing for a maximum density of 650 g/cm^3 and a maximum confinement parameter ρR of 1.6 g/cm^2. An ad-hoc 3D ray-tracing package has been developed to evaluate the uniformity of the deposition energy provided by the direct irradiation of the capsule. Two irradiation configurations with 32 and 48 laser beams characterized by a super-gaussian intensity profile have been considered. A robustness study has been performed as a function of the power imbalance. Neglecting any pointing errors and taking into account only for the power imbalance it is found that half of the shots performed with the two configurations provide the same irradiation uniformity. The 3D ray-tracing package has been also used to estimate the direct irradiation of a re-entrant cone inserted into the capsule. It is found that the direct irradiation of the cone surface reach a maximum intensity of 3 10^{14} W/cm^2.

1. Introduction

In the Direct Driven Inertial Confinement Fusion, a spherical capsule is irradiated by multiple overlapping laser beams. A fraction of the laser energy is deposited into the external side of the shell of the capsule which expands and compresses a small amount of nuclear fuel contained in the capsule. A successful implosion allows for generation of the ignition conditions of the Deuterium-Tritium fuel which can provide energy gain as high as 100. The uniformity of the irradiation is crucial [1-4] to avoid the growth of the Rayleigh-Taylor (RT) [5,6] instability. During the acceleration phase such hydrodynamic instability tends to mixes the low-density fluid of the corona with the high-dense plasma of the compressed shell. Furthermore, during the deceleration phase, the inner surface of the compressed shell becomes RT instable [7]. The uncontrolled growth of the hydrodynamic instability could damage or destroy the capsule before reaching the stagnation thus reducing the energy gain or even avoiding ignition. To prevent the early development of this hydrodynamic instability is crucial to control the uniformity of the energy deposition.

A 3-D ray-tracing code [8] has been developed in order to evaluate the uniformity of the absorbed energy in the capsule. The spherical capsule is irradiated by a given number Nb of beams. The degree of the uniformity is evaluated as the root-mean-square (rms) deviation of the deposited energy into a volume over the sphere surface close to the critical density. The capsule implosion is assumed perfectly spherical and is calculated by mono-dimensional hydrodynamic simulations.

2. 3D Ray-tracing package

A ray-tracing package has been developed to calculate the spatial energy deposition and the absorption of a laser beam. The 1D hydrodynamics simulation performed with the code Multi-1D [9] provides the radial density and temperature profiles of the plasma. From the point of view of the ray tracing package the plasma has spherical symmetry, thus density and temperature vary only along the radial direction. The ray-tracing algorithm considers a laser beam with arbitrary direction and power profile. The beam is divided in beamlets with associated a fraction of the laser power. The directions of the beamlets are initially parallel to the beam axis. The photon trajectories of each beamlet are evaluated by solving the eikonal equation and the energy deposition along the trajectories is calculated via inverse bremsstrahlung. A detailed description of the raytracing methods is given in the Section 3 of reference [10].

2.1. 1D capsule calculations

The capsule is similar to the baseline target designed for the HiPER project [11]. The capsule shown in Fig. 1 is composed by a cryogenic shell of Deuterium-Tritium with an internal radius of 800 μm and thickness of 250 μm thus the total fuel mass is 0.68 mg. The hydrodynamic evolution of the implosion of the capsule is simulated using the Multi-1D code. The incident laser pulse power is shown in Fig. 1, the laser wavelength is 351 nm and the total incident laser energy is around 100 kJ. A maximum fuel density of 650 g/cm^3 is reached in the compressed shell while the maximum areal density $\rho R = 1.6$ g/cm^2 and the maximum in-flight-aspect-ratio (IFAR) is 30. The density and temperature profiles evaluated by the Multi-1D code are used as input for the ray-tracing package which evaluates the laser energy deposition via inverse-bremsstrahlung.

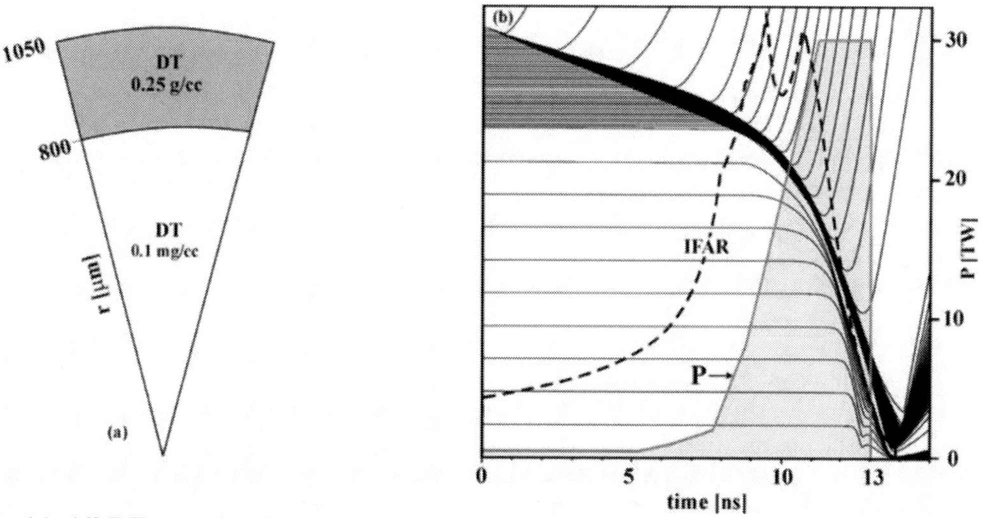

Fig. 1. (a) All-DT capsule. (b) Temporal evolution of some selected lagrangian cells. The IFAR parameter and the incident laser power (P) are also shown as a function of time.

2.2. Irradiation uniformity

Two different laser beam configurations with 32 and 48 laser beams are considered. It is assumed that the directions of all the beams are centered. The laser intensity profile of the beams is supergaussian $I(r) = Io \exp(-r/\Delta)^m$; where Δ is the full width at 1/e and m the supergaussian exponent. A parametric study is performed to estimate the rms deviation σ of the laser energy absorption. The ray-tracing package evaluates the energy deposition into the volume of plasma close to the critical density ρ_c in a volume limited by the sphere where the plasma density is equal to ρ_c and $\rho_c/10$.

Journal of Physics: Conference Series **244** (2010) 022008 doi:10.1088/1742-6596/244/2/022008

The parameters Δ and m that define the laser profiles have been varied between Δ=500 μm and Δ = 1000 μm, while the exponent m varies from 1 to 6. For each couple of values, the rms deviation σ of the energy deposition is calculated. Fig. 2 shows the isocontour of σ evaluated at time 1 ns as a function of Δ and m. It is possible to see that the minima values of σ are near 0.16% for both configurations with 32 and 48 beams. Moreover, the configuration with 48 beams provides two minima of the rms deviation located approximately at Δ = 650 μm, m = 2 and at Δ = 870 μm, m = 4.2. The configuration with 32 beams shown only one minima around Δ = 800 μm and m = 3.5 [8].

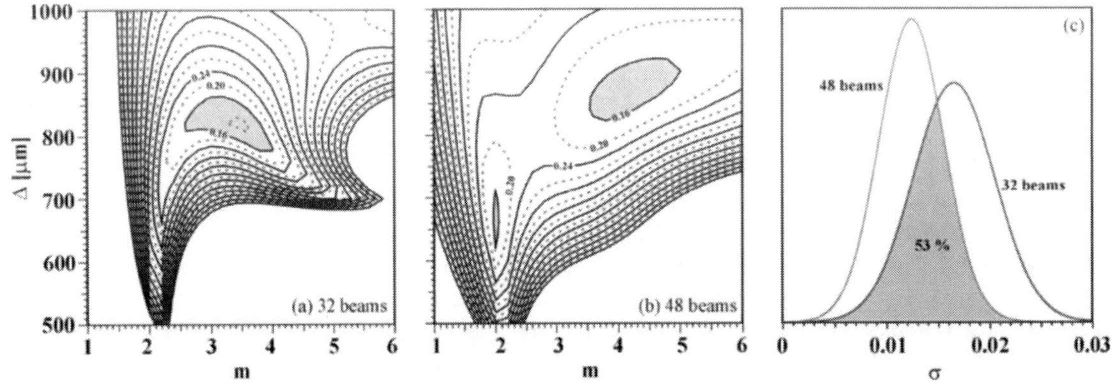

Fig. 2. Rms deviation (σ) of the energy deposited as a function of the parameters Δ and m for the (a) 32 and (b) 48 beams configurations. (c) Densities of probability to obtain a given σ of the uniformity for the 32 and 48 beams configurations and a power imbalance of 5 %.

The uniformity in Fig. 2 is calculated assuming that all the laser beams are perfectly balanced. To perform a robustness study a set of 10000 calculations are performed varying randomly the power of the Nb beams. The power of the beams follows a Gaussian distribution characterized by a standard deviation σ_P = 5 %. The rms deviation of the uniformity provided of the 10000 calculations is fitted by a gaussian law with a mean value and a standard deviation. These Gaussian fits represent the density of probability shown in Fig. 2(c) for the two beam configurations. In this figure it is shown that for a power imbalance of 5 % the average σ produced by the 48 beam configuration is smaller than the one due to the 32 beams configuration. Nevertheless, a quite large overlap (shadowed area) of the two densities of probability (53%) indicates that the two configurations provide the same σ in about half of the shots [8].

3. Irradiation of a conically guided capsule.

The 3D ray-tracing package is also used to evaluate the direct irradiation of the cone of a conically guided ICF capsule. In this case, a cone with a full aperture of 30° is inserted in the capsule irradiated by the 32 beam configurations. The same capsule and laser pulse discussed in Sec. 1 are considered. The ray-tracing package is used to evaluate the direct irradiation of the cone surface I(r,t) [W/cm^2], where r is the distance from the capsule centre and t the time. The intensity I(r,t) has been averaged over the azimuthal angle φ. The total irradiation of the cone surface is shown in Fig. 3 as a function of the position r and the time t. The vertical dashed line indicates the initial position of the external radius of the capsule. It is possible to see that the maximum irradiation is located inside the capsule and reach a maximum of 3 10^{14} W/cm^2 [12].

The intensity I(r,t) has been used as input to the code MULTI2D [13] in order to simulate the behavior of the cone under irradiation. Considering a gold cone, it is found that ablation pressures up to 5 Mbar are generated, providing a strong shock through the cone wall. In Fig. 4, are shown the density into the gold cone as a function of the distance r. The density refers to the time t = 14 ns which is one ns later the end of the laser pulse. At that time the maximum penetration of the shock wave is about 30 μm. Thus, for our specific laser-capsule configuration, a gold cone thicker than 30 μm is needed to avoid the shock breakout [12].

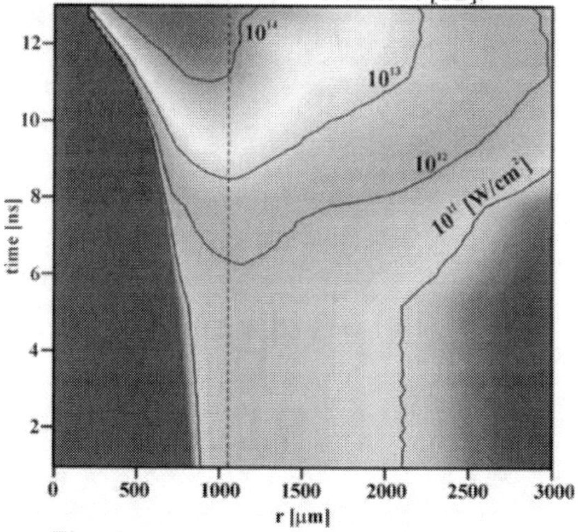

Fig. 3. Laser intensity incident onto the cone surface as a function of the distance from the capsule center r and of the time.

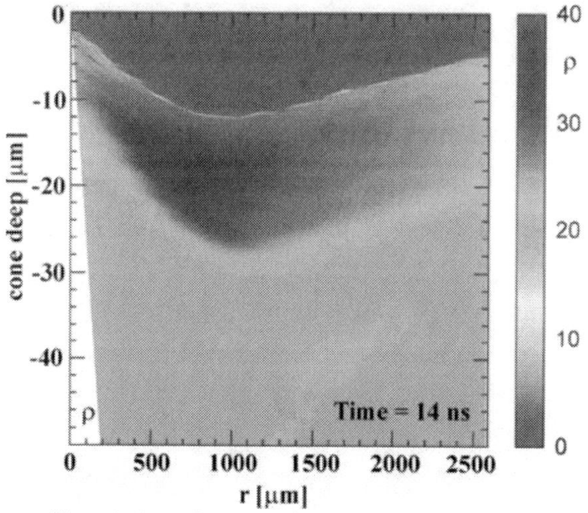

Fig. 4. Density map into the cone wall evaluated at time t = 14 ns.

4. Conclusion

A 3D ray-tracing package dedicated to the direct irradiation of an inertial confinement fusion capsule has been developed [8]. The ray-tracing has been used to evaluate the uniformity of the irradiation provided by a 32 and 48 laser beam configurations. It has been shown that the 48 beams configurations allows for a better uniformity. Nevertheless, a power balance study indicates that in about half of the shots the two configurations provide the same irradiation uniformity. The ray-tracing has been also used to analyze the direct irradiation of the external surface of a cone in a conically guided target. For a specific laser-capsule configuration it has been shown that the cone is irradiated with a peak intensity of 3 10^{14} W/cm^2 [12].

This work was performed in the framework of the HiPER project. MT was supported by the contract Ramon y Cajal-2007-00447 of the Spanish Ministry of Education and Research. RR was supported by the EURATOM/CIEMAT association in the framework of the IFE Keep-in-Touch Activities.

References

[1] Skupsky S and Lee K 1983 *J. Appl. Phys.* **54** 3662
[2] McKenty PW, Verdon CP, Skupsky S, McCrory RL, Bradley DK, Seka W and Jaanimagi PA 1990 *J. Appl. Phys.* **68** 5036
[3] Canaud B, Fortin X, Garaude F, Meyer C, Philippe F, Temporal M, Atzeni S and Schiavi A 2004 *Nucl. Fusion* **10** 1118
[4] Canaud B *et al.* 2007 *Plasma Phys. Control. Fusion* **49** B601
[5] Bodner SE 1974 *Phys. Rev. Lett.* **33** 761
[6] Takabe H, Mima K, Montierth L and Morse RL 1985 *Phys. Fluids* **28** 3676
[7] Temporal M, Jaouen S. and Canaud B 2006 *Phys. Plasmas* **13** 122701
[8] Temporal M and Canaud B 2009 *Eur. Phys. J. D* **55** 139
[9] Ramis R, Schmaltz R and Meyer-ter-Vehn J 1988 *Comput. Phys. Commun.* **49** 475
[10] Temporal M, Jacquemot S, Bonnet L and Decoster A 2001 *Phys. Plasmas* **8** 1363
[11] Dunne M 2006 *Nature Phys.* **2** 2
[12] Temporal M, Ramis R and Canaud B 2009 *Phys. Plasmas* **16** 074503
[13] Ramis R, Meyer-ter-Vehn J and Ramírez J 2009 *Comput. Phys. Commun.* **180** 977

Progress on LMJ targets for ignition

C. Cherfils-Clérouin, C. Boniface, M. Bonnefille, P. Fremerye, D. Galmiche, P. Gauthier, J. Giorla, F. Lambert, S. Laffite, S. Liberatore, P. Loiseau, G. Malinie, L. Masse, P.E. Masson-Laborde, M.C. Monteil, F. Poggi, P. Seytor, F. Wagon, J.L. Willien

CEA, DAM, DIF, F-91297 Arpajon, France

catherine.cherfils@cea.fr

Abstract. Targets designed to produce ignition on the Laser MegaJoule are presented. The LMJ experimental plans include the attempt of ignition and burn of an ICF capsule with 160 laser beams, delivering up to 1.4MJ and 380TW. New targets needing reduced laser energy with only a small decrease in robustness have then been designed for this purpose. Working specifically on the coupling efficiency parameter, i.e. the ratio of the energy absorbed by the capsule to the laser energy, has led to the design of a rugby-shaped cocktail hohlraum. 1D and 2D robustness evaluations of these different targets shed light on critical points for ignition, that can be traded off by tightening some specifications or by preliminary experimental and numerical tuning experiments.

1. Introduction

The Laser Mégajoule (LMJ) facility is located at the Commissariat à l'Energie Atomique (CEA) center CESTA near Bordeaux in France. It is part of the French "Simulation Program" and is devoted to laboratory experiments in a large domain of Plasma Physics [1]. It will deliver up to 1.8 MJ and 550 TW of $\lambda = 0.351$ μm light [2]. The standard approach for achieving ignition on the LMJ relies on the indirect drive scheme. Our nominal point design target A1040 is composed of a graded doped plastic capsule in a cylindrical hohlraum; specifications based on current CEA capabilities give a security margin towards modeling uncertainties and technological dispersions. The experimental plans include the attempt of ignition and burn of an ICF capsule with 160 beams in 2 cones (33.2° and 49°), delivering up to 1.4 MJ and 380 TW at 3ω. New targets needing reduced laser energy with only a small decrease in robustness have been designed for this purpose. They are presented in section 2. Our strategy is to work specifically on the coupling efficiency parameter, i.e. the ratio of the energy absorbed by the capsule to the laser energy. A rugby-shaped hohlraum is proposed, exhibiting a better coupling efficiency and symmetry control of the implosion than a cylindrical hohlraum for the specific energy balance of the 160-beam configuration [3]. Section 3 details 1D and 2D robustness studies.

2. Ignition target design

We have designed 300 eV capsules which need less energy with a small decrease of robustness compared to our baseline capsule A1040 [3]. The capsule designs have been first defined by a 0D-model [4], and then confirmed by capsule-only 1D simulations optimizing the temperature drive. The selected capsules are the capsules A850 (about 110 kJ absorbed) and A943 (about 140 kJ absorbed). Both capsules have the same in flight aspect ratio than the capsule A1040 (about 160 kJ absorbed), which is a signature of the shell break-up risk, while the 1D robustness quantified by the excess in kinetic energy above the ignition threshold is reduced by less than 20%. Figure 1 shows the capsule thicknesses optimized for an uniformly germanium doped plastic ablator, while the LMJ point design is now including gradually doped ablators. As the capsule sensitivity to hydrodynamic instabilities is very dependant on the x-ray spectrum produced by the hohlraum, the ablator thickness and dopant structure need to be optimized for every 2D configuration.

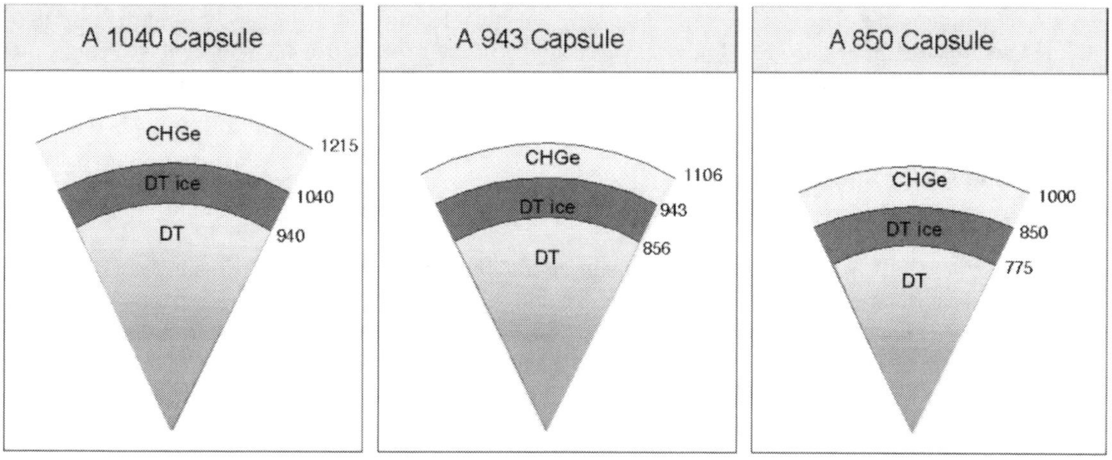

Figure 1 : Pie diagrams of the different capsules studied for the 2-cone configuration (size in μm).

Integrated simulations of these capsules have been performed with our 2D hydro-radiative code FCI2, which includes all the target physics except microscopic laser-plasma interactions and high-mode hydrodynamic instabilities, badly described by the mesh. The scaling is done for a given ratio of the holhraum equatorial radius over the capsule radius, as this parameter governs the flux uniformity around the capsule. In this 2-cone configuration, the radius of the LEH is taken equal to 1.5 mm for cylindrical hohlraums, to compare to 1.75 mm for the 3-cone design. The gas filling, used to limit the wall motion, is a H2-He equimolar mixture, with a density ranging from 0.8 to 1.3 mg/cc, contained by a polyimid window. Controlling the time-dependent symmetry in this specific 1/2-1/2 balance configuration is more difficult than in the 3-cone configuration. The symmetry control can be obtained by increasing the aspect ratio (half-length over equatorial radius) of the cylinder. An alternative way is to break the inner cone in two, which leads back to a configuration close to the 3-cone one.

All A850 integrated simulations, corresponding to small design variations, produce around 10 MJ yield with an energy budget of 850 kJ and a power laser up to 300 TW. Igniting the A943 capsule in a gold cylindrical hohlraum is still in the budget (1.2 MJ, 360 TW) and produces around 20 MJ. Getting more margin to leave some room for design uncertainties, like laser plasma instabilities (LPI) evaluation, such as backscattering processes, is possible by using either cocktail walls or a rugby shape for the hohlraum, or both. Rugby-shaped hohlraums have been shown to have many advantages over cylindrical hohlraums for the 160-beam configuration [5]. First, minimizing the hohlraum wall surface while keeping the equatorial radius constant improves the coupling efficiency, as approximately half the laser energy is used to heat the walls. But the rugby shape has also a significant advantage for the control of the radiation asymmetry on the capsule with a 1/2-1/2 energy balance. Bending the walls toward the center of the hohlraum, helps increasing the outer ring contribution near the level of the inner ring and leads to a better drive of the capsule. The core at ignition shown in Figure 2 presents around 5% hotspot perturbations, illustrating an adequate symmetry. A laser energy of 0.9 MJ and a laser power of 315 TW are required in this case to drive the A943 capsule to ignition. This configuration can also be used to ignite the A1040 capsule with a reduced energy budget, around (1.1 MJ, 320 TW), with a yield of about 30 MJ.

Laser Plasma Instabilities (LPI) may be enhanced in these small targets and their control has to be one of the major requirements of the design process. Two key parameters are the laser intensity and the electronic density inside the hohlraum; these factors can be optimized by playing on the size of the laser focal spot and on the composition and density of the gas filling the hohlraum. Linear growth rates of the Raman (SRS) and Brillouin (SBS) instabilities are evaluated by post-processing the 2D FCI2 simulations with our code PIRANAH. Total amplification is then calculated by integration along the laser beam [6]. The goal is to bring the gain values down to about 10. Work is in progress to lower the SBS by adding a low-Z dopant in the wall, such as boron. Another approach is to work with a lower drive temperature. A 275eV target (BT275 capsule) leading to a larger hohlraum has also been designed with the same laser budget; a key point is then to control its sensitivity to hydrodynamic instabilities (Figure 3).

3. 1D and 2D robustness studies

3.1. 1D robustness

Exhaustive robustness studies [7] have been conducted on the targets to estimate the yield sensitivity to target fabrication, modeling uncertainties and experimental tuning. The various 1D sources of uncertainties have been combined in order to give a net error budget. These sources are small enough that by themselves they do not significantly affect the yield, but a combination of them can be harmful for ignition. The dominant error among the 22 parameters studied was shown to be the total ablator thickness, which provided guidance for target fabrication and led us to tighter specifications. Considerable effort has been done by the CEA target fabrication laboratory on the thickness and doping accuracy control of the ablator shell [8], lowering their contribution to the error budget to acceptable values. The next five main sources of error all come from experimental tuning, and they cannot be improved only by tightening specifications. The Fusion Campaign is elaborated in order to bring them down.

3.2. Controlling hydrodynamic instabilities

The tolerable roughness of the different interfaces of the capsules is determined by the use of different capsule-only simulations.

For sufficiently small initial amplitudes (still large enough to dominate the numerical noise), the evolution of a perturbed interface can be obtained with a simulation-based linear analysis, by combining numerical growth rates with an assumed initial spectrum of perturbations. Haan's saturation model [9] is then used to estimate the final deformations at the end of the acceleration phase. As these simulations cannot tell us by how much the capsule yield is affected, they are complemented with multimode calculations having a realistic power spectrum (modes 12 to 200) and a variable multiplier on the nominal roughness, in order to find the cliff in performance.

The tolerable roughness of the external surface of the plastic shell, leading to 75% of the nominal yield is estimated by these simulations to be between 70 and 150 nm, depending on the capsule. This leaves a reasonable security margin compared to the 20 nm state of the art of the CEA target lab [8]. The tolerable roughness of the internal surface of the solid DT shell (high modes only) is less than 1 μm for all designs.

Complete parametric studies presented on Figure 3 were done with uniformly germanium doped plastic ablator, as its sensitivity to hydrodynamic instabilities can be considered as a lower limit compared to what is obtained with gradually doped ablators. A gradually doped ablator can indeed strongly limit the development of the perturbations, but it has to be finely adapted to the capsule drive, and more specifically to the hohlraum x-ray spectrum, which is not very well known. Work is in progress on a laminated ablator, made of a succession of doped and undoped plastic layers. This multilayered design makes use of the stabilizing effect of anisotropic diffusion on the ablative Rayleigh-Taylor instability, and is now tested experimentally on Omega [10].

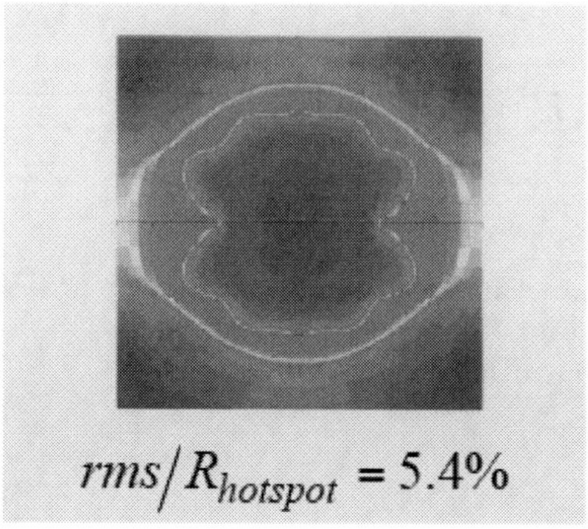

Figure 2: Core at ignition for the A943 capsule in a rugby-shaped hohlraum.

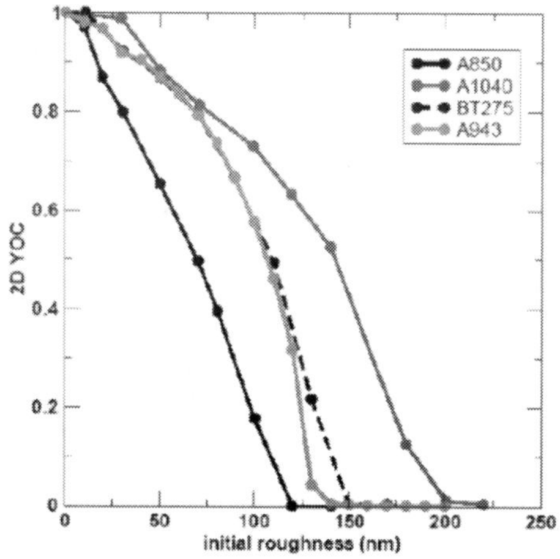

Figure 3: Gain curves of the different ignition capsules versus outer initial roughness

4. Conclusion

Different ICF targets have been designed to achieve ignition on the Laser MegaJoule. The rugby-shaped hohlraum appears as a natural candidate to save energy and achieve ignition with the specific 160-beam configuration. Controlling the laser-plasma instabilities in the hohlraum is now part of the design process, as it stays a major drawback in these small hohlraums. Robustness evaluations of the different targets are compulsory for the definition of mature designs. Integrated simulations of the complete designs are performed to set requirements for the details of the hohlraum. Important effort has gone into designing the capsules in order to optimize their hydrodynamic stability; tolerable roughnesses are then set by 2D capsule-only simulations. Sensitivity studies have shown the importance of the experimental tuning, than can be partially prepared by experiments on other laser facilities and by preliminary experiments on surrogate targets, as THD capsules.

References

[1] Lion C, *J. Phys. : Conf. Ser.*, these proceedings
[2] Ebrardt J and Chaput JM, *J. Phys. : Conf. Ser.*, these proceedings
[3] Holstein P-A *et al.* 2000 *C. R. Acad. Sci. Paris* **t1 IV** 693 ; Holstein P-A *et al.*2008 *CEA Report No.* **R-6182**
[4] Saillard Y 2004 *Laser Part. Beams* **22** 451
[5] Vandenboomgaerde M *et al.*2007 *Phys. Rev. Lett.* **99** 65004
[6] Laffite S, Liberatore S and Loiseau P. 2007 *Bull. Am. Phys. Soc.***52** 16.
[7] Giorla J *et al.* 2009 *Phys. Plasmas* **16** 032704
[8] Legay G *et al.* 2009 *Fusion Sci. Technol.* **55** 438 ; Collier R *et al.*, *J. Phys. : Conf. Ser.*, these proceedings
[9] Haan S 1989 *Phys. Rev. A.* **39** 5812
[10] Masse L 2007 *Phys. Rev. Lett.* **98** 245007 ; Masse L *et al.*, *J. Phys. : Conf. Ser.*, these proceedings

Journal of Physics: Conference Series **244** (2010) 022010 doi:10.1088/1742-6596/244/2/022010

Plastic ablator ignition capsule design for the National Ignition Facility

Daniel S. Clark, Steven W. Haan, Bruce A. Hammel, Jay D. Salmonson, Debra A. Callahan and Richard P. J. Town

Lawrence Livermore National Laboratory, Livermore, California 94550, USA

E-mail: clark90@llnl.gov

Abstract. This paper describes current efforts to develop a plastic ablator capsule design for the first ignition attempt on the National Ignition Facility. The trade-offs in capsule scale and laser energy that must be made to achieve ignition probabilities comparable to those with other candidate ablators, beryllium and high-density carbon, are emphasized. Large numbers of 1-D simulations, meant to assess the statistical behavior of the target design, as well as 2-D simulations to assess the target's susceptibility to Rayleigh-Taylor growth are discussed.

1. Introduction

Three types of ablator material are currently under study for the for the National Ignition Facility (NIF) ignition capsule [1]: beryllium doped with copper, plastic (CH) doped with germanium, and nano-crystalline synthetic diamond doped with tantalum. Each of these materials shows various strengths and weaknesses that recommend it as the first choice ablator. Equally, each of these capsule ablator materials has been intensively investigated both theoretically and experimentally. This paper summarizes the current state of the CH capsule design, as represented by a large data base of simulations run with the radiation hydrodynamics code LASNEX [2]. As these designs constantly evolve in response to improved physics modeling and experimental input, it should be recognized that this paper is necessarily a snapshot of a work in progress.

2. 1-D design considerations

The current CH ablator capsule design is shown in Fig. 1. A layer of cryogenic DT ice lines the inside of a spherical ablator shell. The inner region of the ablator is doped with high-Z germanium to adjust the x-ray opacity, while the outer region, which makes up the majority of the material ablated, is left undoped. The modification of the opacity introduced by the doping is designed to minimize the unstable Atwood number that develops at the fuel-ablator interface at peak velocity. The presence of the high-Z dopant strongly absorbs the incident x-ray flux leaving the undoped layer of the ablator closest to the fuel colder and hence denser than the rest of the ablator. The precise concentration of dopant and width of the doped layer are carefully adjusted taking into account the anticipated hard x-ray flux.

In addition to the capsule dimensions and compositions, the 1-D design is characterized by a tuned radiation pulse shape giving good entropy and implosion velocity. Fig. 2 shows the radiation temperature pulse shape used to drive the capsule. The radiation temperature is extracted from an integrated hohlraum simulation including laser beam propagation and absorption and accounting for expected losses due to laser-plasma interactions. The total laser power used in the integrated hohlraum simulation is also shown. When properly tuned, the four shocks launched by this pulse shape implode the capsule with the low fuel averaged entropy of 0.447 kJ/mg/eV and high implosion velocity of 0.384 mm/ns.

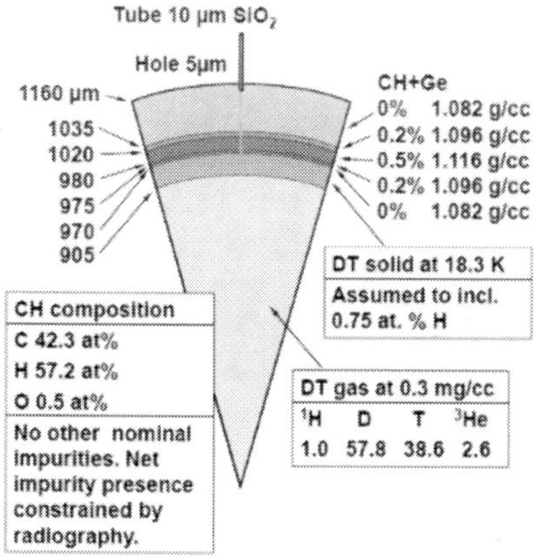

Figure 1. "Pie diagram" for the current CH capsule design.

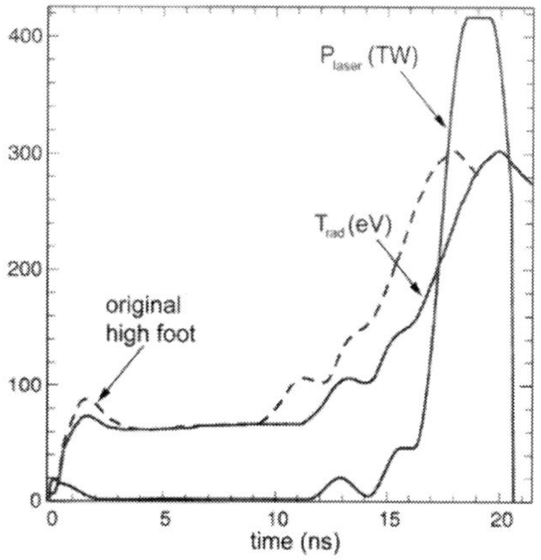

Figure 2. Four-step pulse shape used to drive the CH capsule.

3. Low- and intermediate-mode stability

Once the 1-D considerations of implosion velocity and entropy have been adequately tuned, the 2-D and 3-D considerations of hydrodynamic stability can be addressed. These considerations broadly break down into low mode perturbations which dominate the perturbations around the igniting hot spot and intermediate and high mode perturbations which threaten the shell integrity near peak velocity. Low-mode perturbations are assessed using simulations that include the complete 180° angular extent of the capsule and typically resolve modes up to Legendre mode number $\ell = 30$ with roughness applied to all of the interfaces of the capsule: the inner ice surface, fuel-ablator interface, the internal dopant layers, and the outside of the ablator. Fig. 3 shows the results of such simulations plotted as yield versus multiplier applied to the nominal ice surface roughness. The low-mode yield curve for CH is compared to the analogous yield curve for a beryllium ablator capsule design, showing that the CH design achieves comparable to or slightly higher low-mode robustness.

The effects of intermediate modes are assessed by performing single-mode Lagrangian simulations with infinitesimal perturbations applied to the ablator outer surface and extracting linear growth factors at peak velocity at the ablation front. Fig. 4 shows these ablation front growth factors for modes $\ell = 10 - 160$. The nominal capsule design uses the latest tabular equation of state (EOS) for CH, LEOS 5310, which is more compressible than the earlier table LEOS 5105. The significant increase in peak growth factor with the latest and most accurate EOS compared to the earlier table is apparent.

Also shown in the figure are the growth factors resulting from an earlier version of the design using the latest EOS but a pulse shape with a higher picket on the foot of the pulse (also shown in Fig. 2 as the dashed curve). The stronger first shock with the higher foot reverses the phase of the growth factors in the mode range $\ell = 100 - 150$ relative to the low foot. Since fabricated CH shells show features (surface bumps) in this mode range, the enhanced growth factors for $\ell = 100 - 150$ lead to unacceptable perturbations to the shell and necessitated shifting to the lower foot pulse shape. Simulations of individual isolated bumps, described elsewhere [3, 4], dramatically bear this out.

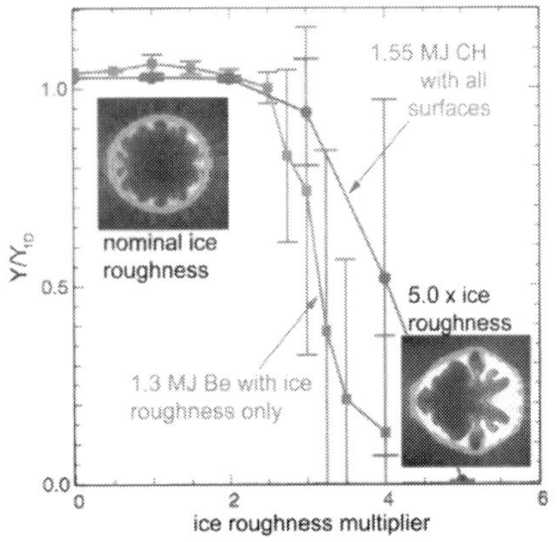

Figure 3. Capsule yield normalized to 1-D versus initial ice roughness for a beryllium ablator design (■) and the CH design (●). Each data point represents an average of seven simulations with different realizations of the random surface roughnesses. The error bars give the variance of the yields. Example hot spots are shown as insets.

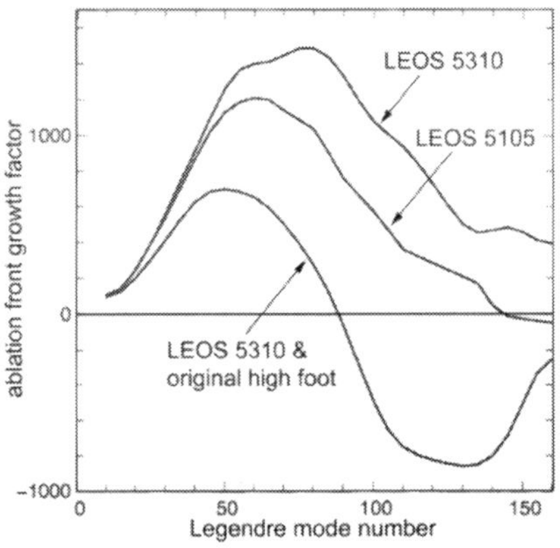

Figure 4. Growth factors for ablator surface perturbations at peak velocity for Legendre mode numbers $\ell = 10 - 160$. Curves for the nominal CH EOS, LEOS 5310, and an earlier, less compressible EOS, LEOS 5105, are shown. Also shown are the negative growth factors that result from using an earlier high-foot pulse shape.

4. Statistical assessments of robustness

In addition to optimizing the capsule design in 1-D and ensuring its stability in 2-D, it is important to recognize that a variety of random defects can occur in any given capsule or in the laser performance from shot to shot. Assessing the effects of these random defects requires a statistical approach. The approach followed here was pioneered and extensively developed by Salmonson for the beryllium ablator design [5]. It consists of tabulating all possible random sources of error to be expected under actual experimental conditions, bracketing those sources of error within specified bounds and with specified statistical distributions, and then generating a data set of random instantiations of the given capsule design with all sources of error. Once the full data set has been run, the simulations may be post-processed to extract, for example, the distribution of expected yields given the assumed uncertainty.

Fig. 5 summarizes the results of such a study on the current CH capsule in the space of capsule yield versus ignition threshold factor (ITF), a measure of capsule performance relative to the threshold of ignition and previously referred to as capsule margin. It can be represented as a product of capsule scale, implosion velocity, fuel averaged entropy, and hot spot perturbation fraction and has been shown to be a reliable predictor of capsule performance [6]. Three data sets are shown in the figure: each of the red dots represents an instantiation of the 1-D capsule design including all 1-D errors with all deviations drawn from nominal distributions, *i.e.*, as expected under experimental conditions. There are 10,000 total instantiations sampled. Each of the blue dots represents a similar 1-D instantiation but with deviations drawn from distributions with twice the width of the nominal distributions. Finally, the green triangles show the results of sub-sampling the 1-D nominal deviations into 2-D by including nominal roughnesses on all interfaces. The total 2-D data set consists of seventy simulations.

What is immediately evident from Fig. 5 is that the CH capsule design is highly robust in 1-D as well as 2-D. With nominal deviations, none of the capsules sampled in 1-D or 2-D failed in the sense of yielding less than 1.0 MJ. Even with the deviations doubled in 1-D, only one instantiation gave a yield less than 1.0 MJ. The utility of the data set with double nominal deviations becomes apparent here since the ITF is normalized by the location of the yield cliff, by definition ITF = 1.0. The location of this cliff is ambiguous with the data set of nominal deviations and only becomes apparent with the double nominal distributions. With this normalization, the nominal 1-D design, without any deviations, is located at ITF = 5.5. With the addition of 1-D nominal deviations, the ITF is degraded to between 2.0 and the original 5.5. With the addition of 1-D double nominal deviations, the distribution of ITF increases further to the edge of the yield cliff. Finally, the cloud of green triangles indicate that with nominal deviations in 2-D an average ITF of approximately 3.0 can be expected.

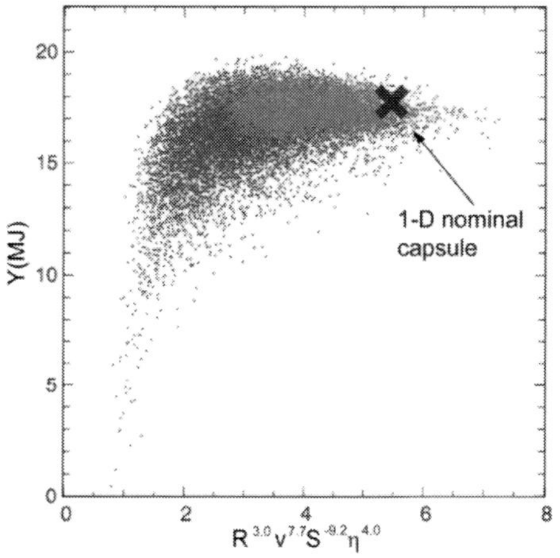

Figure 5. Results of statistical scans over all expected capsule defects summarized as a scatter plot of yield versus ITF. The red ensemble is the result of 10,000 1-D samples with nominal deviations, the blue ensemble is the result of 10,000 1-D samples with double nominal deviations, and the green ensemble results from sub-sampling the 1-D nominal ensemble into 2-D including random roughnesses on all interfaces.

5. Conclusions

This paper summarizes the current state of the plastic ablator ignition capsule design for NIF. Broadly, a balance has been sought between the many competing considerations of 1-D performance optimization, 2-D stability requirements, and 1-D and 2-D statistical reliability in an effort to reach a design most likely to achieve ignition. Further revisions of the design will necessarily continue as physics understanding improves and new experimental data become available.

Acknowledgments

This work was performed under the auspices of the U. S. Department of Energy by Lawrence Livermore National Laboratory under Contract DE-AC52-07NA27344.

References

[1] Lindl J D, Amendt P, Berger R L *et al.* 2004 *Phys. Plasmas* **11** 339
[2] Zimmerman G B and Kruer W L 1975 *Plasma Phys. and Controlled Fusion* **2** 51
[3] Hammel B A, Haan S W, Clark D, Edwards M J, Langer S H, Marink M, Patel M, Salmonson J and Scott H A 2009 *Proc. 2nd Int. Conf. on High Energy Density Physics (Austin, TX)* (Elseveir: New York)
[4] Clark D S, Haan S W, Hammel B A, Salmonson J D, Callahan D A and Town R P J 2010 *in preparation*
[5] Salmonson J D 2008 *Bull. Am. Phys. Soc.* **52** 273
[6] Clark D S, Haan S W and Salmonson J D 2008 *Phys. Plasmas* **15** 056305

Journal of Physics: Conference Series **244** (2010) 022011 doi:10.1088/1742-6596/244/2/022011

Calibrating mix models for NIF tuning

Nelson M Hoffman[1], **Bruce A Hammel**[2], **and Oleg Schilling**[2]

[1]Los Alamos National Laboratory, Los Alamos, NM 87545 USA
[2]Lawrence Livermore National Laboratory, Livermore, CA 94550 USA

E-mail: nmh@lanl.gov

Abstract. Mixing may play a role in early "tuning" implosions at the National Ignition Facility (NIF), intended for optimizing target geometry and laser operating parameters prior to the first ignition shots. Turbulent transport models allow mixing to be represented in integrated capsule-plus-hohlraum simulations, so that its effects can be predicted. Such models typically have many adjustable coefficients, so it is important to calibrate the coefficients systematically, whether by comparison to experiments, theory, or detailed higher fidelity simulations. We have begun by comparing to results from high-resolution 2D multimode simulations of a NIF Rev3 CH(Ge) ignition capsule, using the time-varying angle-averaged radial profiles of ion species composition as the calibration reference. 1D implosion simulations are performed with either of two turbulence models, and the models are calibrated so that the 1D composition profiles roughly match the 2D reference profiles at a sequence of five times near the end of the implosion. Details of the reference profiles, however, such as non-monotonicity resulting from structures in the flow, cannot be represented by these models. Parameter searches, besides identifying optimal sets of model coefficients, also exhibit model sensitivity to parameter variations.

1. Introduction

An imploding ICF capsule is subject to a variety of hydrodynamic instabilities resulting from strong gradients of density, pressure, and velocity in the converging flow [1]. The instabilities may amplify small perturbations in the flow field, leading to turbulence if the amplification proceeds far enough into the nonlinear regime. Nonlinear instability and turbulence in turn may drive "mixing", i.e., the directed or diffusive interpenetration of ion species initially separated by a material interface (contact discontinuity). If instability, turbulence, and mixing arise in an implosion, they are expected to have a variety of deleterious effects on capsule behaviour. For example, in ignition capsules, early fine-scale instability and turbulence may lead to the break-up of the imploding shell. At later times, large-scale instability can deform the hotspot while fine-scale mixing contaminates the fuel. In implosion experiments performed during the tuning phase of NIF (prior to the first ignition tests), mixing can alter the size and shape of the imploded core of symmetry capsules, and change the areal density of the imploding capsule ablator. Thus, inferences about the adequacy of drive symmetry or the amount of unablated mass may be incorrect, if mixing is not accounted for.

---

[1] To whom any correspondence should be addressed.

Given the importance of mixing and the potential severity of its effects, it is crucial to have a capability for representing it in numerical radiation–hydrodynamic simulations used for designing and analyzing future NIF implosions, to permit accounting for and mitigating its effects. The use of turbulent transport ("mix") models to represent mixing in a NIF Rev3 CH(Ge) ignition capsule is described here. In particular, an approach for calibrating the coefficients in the models, the degree to which such models succeed in matching a reference simulation, and the sensitivity of model predictions to variations in the coefficients are discussed.

2. Numerical models of instability, turbulence, and mixing

The most widely accepted technique for credibly simulating the effects of instability, turbulence, and mixing in ICF capsules consists of high-resolution 2D and 3D nonlinear multimode simulations [2]. Such simulations, explicitly resolving spherical harmonic perturbation modes up to $l = 1000$ or higher, are computationally demanding and typically require representing the capsule alone (i.e., omitting the surrounding hohlraum). The effect of the hohlraum in creating a (possibly asymmetric) radiation drive is represented by a temporally and spectrally (and possibly angularly) varying radiation boundary condition on the capsule. This approach requires extracting the multidimensional radiation field information from a separate hohlraum calculation and mapping it to the nonlinear multimode simulation of the "stand-alone" capsule, a process that may introduce inaccuracies owing to phase-space averaging, inconsistent definition of boundaries, and the omission of hydrodynamic interactions between the hohlraum and capsule. While integrated (i.e., capsule plus hohlraum) simulations are certainly routine, to date they have not been used to resolve modes much above $l = 12$. It may be feasible to extend such simulations up to $l = 30$ or so, but high-resolution integrated simulations up to $l \sim 1000$ will not be feasible for the foreseeable future.

Given such limitations, it is worthwhile to consider turbulent transport ("mix") models for representing mixing in integrated simulations. Such models introduce new field variables to describe the turbulent flow statistically, along with associated governing equations requiring simultaneous solution with the usual radiation–hydrodynamic equations. However, mix models typically have many adjustable coefficients and input parameters, so it is important to calibrate the coefficients systematically, with attention to the generality of and uncertainties in the result.

In the present study, a high-resolution 2D nonlinear multimode simulation of the type described in [2] has been used as a reference to calibrate two turbulent transport models: one developed by one of us (OS), which is referred to for brevity as the "Schilling K-ε"or "S$K\varepsilon$" model, and one developed by Zhou, Zimmerman and Burke [3], which is referred to as the "ZZB" model. The S$K\varepsilon$ model uses two equations to describe the evolution of two quantities representing the turbulent field: K, the turbulent kinetic energy density (with units L^2/T^2) and ε, the turbulent kinetic energy dissipation rate (with units L^2/T^3). The modelled turbulent transport equations are

$$\rho\frac{DK}{Dt} = -\frac{\nu_t}{\sigma_\rho\,\rho}\vec{\nabla}\rho\cdot\vec{\nabla}p - \vec{\tau}:\vec{\nabla}u - \rho\varepsilon + \vec{\nabla}\cdot\left(\frac{\mu_t}{\sigma_k}\vec{\nabla}K\right)$$

$$\rho\frac{D\varepsilon}{Dt} = -C_{\varepsilon 0}\frac{\varepsilon}{K}\frac{\nu_t}{\sigma_\rho\,\rho}\vec{\nabla}\rho\cdot\vec{\nabla}p - C_{\varepsilon 1}\frac{\varepsilon}{K}\,\vec{\tau}:\vec{\nabla}u - C_{\varepsilon 2}\frac{\rho\varepsilon^2}{K} + \vec{\nabla}\cdot\left(\frac{\mu_t}{\sigma_\varepsilon}\vec{\nabla}\varepsilon\right)$$

where ρ, p, t, and u denote density, pressure, time, and mean velocity, respectively, and $D/Dt = \partial/\partial t + u\cdot\vec{\nabla}$. These equations are coupled to the mean flow equations. The turbulent stress tensor is denoted by $\vec{\tau}$, the turbulent viscosity by $\nu_t = \mu_t/\rho$, and the coefficients σ_ρ, σ_k, σ_ε (the turbulent Schmidt numbers) and $C_{\varepsilon 0}$, $C_{\varepsilon 1}$, and $C_{\varepsilon 2}$ are to be calibrated. In the current study, a subset of these quantities was selected for calibration: the initial turbulent kinetic energy K_0 and turbulent length scale $l_0 \equiv K_0^{3/2}/\varepsilon_0$, where ε_0 is the initial turbulent dissipation rate; multipliers M_{RT} and M_{RM} on the

Journal of Physics: Conference Series **244** (2010) 022011 doi:10.1088/1742-6596/244/2/022011

Rayleigh–Taylor part and the Richtmyer–Meshkov part, respectively, of the buoyancy production term, though they are usually constrained to be equal; and σ_ρ, σ_k, and σ_ε, where σ_k and σ_ε are always constrained to be equal. The four terms on the right-hand side of each equation represent buoyancy production of turbulence, shear and dilatation production, dissipation, and diffusive transport, respectively. In addition to these equations, there are other equations expressing the transport of each ion species by turbulent diffusion along a gradient.

ZZB is a two-scale turbulence model: it has equations for K and ε similar to those in $SK\varepsilon$, but the buoyancy production term is omitted. Instead, two additional equations are employed to represent the effect of buoyancy. One equation describes the buoyancy-drag velocity V_i

$$\frac{dV_i}{dt} - \beta A g = -C_D \frac{\rho_i}{\rho_1 + \rho_2} \frac{V_i|V_i|}{h_i},$$

where A is the Atwood number, g is acceleration, β is a parameter, C_D is the drag coefficient, and h_i is the mixing layer width in region i, with $i = 1$ or 2 corresponding to light or heavy fluids, respectively. The other equation is for the buoyancy energy E_B:

$$\rho \frac{DE_B}{Dt} = S_B - p_B \vec{\nabla}_L u - \frac{\rho E_B^{3/2}}{L_B} + \vec{\nabla} \cdot (\rho \vec{D}_B \cdot \vec{\nabla} E_B).$$

The dissipation rate of buoyancy energy (the third term on the right-hand side) appears as a source term in the K equation. The two scales are a buoyancy length scale L_B, related to the mixing layer width, and a turbulent length scale $l \equiv K^{3/2}/\varepsilon$. The parameters used for calibration were the initial values h_{10} and h_{20} of the perturbation amplitude in the light and heavy fluids respectively, the time t_{on} at which the model is turned on, and a multiplier S on the turbulent diffusion coefficients.

3. Calibration

3.1. Calibration reference

Reduced-physics model coefficients can be calibrated *via* comparison to experiments, theory, or detailed higher fidelity simulations. As no actual ICF implosions conducted to date resemble pulse-shaped low-adiabat NIF implosions particularly well, a 2D multimode high-resolution simulation of a NIF ignition capsule [2] has been used for calibration, especially because of the detailed temporal and spatial information that can be extracted from it. Here, the time-varying radial distribution of chemical composition in the imploding capsule predicted by the turbulence models is compared to the results from the high-resolution simulation. The simulation focused on a NIF Rev3 CH(Ge) ignition capsule, with outer radius 1.092 mm, a 165-μm-thick ablator with five layers of CH plastic doped with varying amounts of germanium (up to 0.6% by atom), a 75-μm-thick cryogenic DT fuel layer, and an inner DT vapour region with $\rho = 0.322$ mg/cm^3. The simulation was performed in a 9-degree sector, and an initial surface perturbation containing spherical harmonic modes $l = 40$–1000 was imposed on the fuel/ablator interface, with rms roughness 3.84 nm. The implosion was driven by a four-step Haan ignition pulse, input as a frequency-dependent radiation source whose equivalent temperature peaked at 310 eV at approximately 17 ns. Figure 1 (a) and (b) show radial profiles of the angle-averaged mass fraction of CH ablator extracted from this simulation at five times near the end of the implosion as open circles. The curves are labelled with their corresponding time, relative to an arbitrary reference: -600 ps, -400 ps, 0 ps, +400 ps, +800 ps. These curves constitute the calibration reference.

3.2. Calibration procedure and results

Numerous 1D simulations using either ZZB or $SK\varepsilon$ were run, varying the model parameters and coefficients to achieve the closest possible match to the five reference profiles of Figure 1 at the corresponding times. This process was aided by scripts that varied the model inputs over desired ranges and automatically ran the simulations, and by use of a quantitative metric for goodness-of-fit (the average horizontal distance between the reference profile and the 1D model profile at each time).

Figure 1(a) shows a comparison of profiles of ablator mass fraction $f_{m,abl}$ from a 1D simulation using $SK\varepsilon$ to the reference profiles, with model parameters that give a reasonably good fit to the reference. Figure 1(b) shows a similar comparison for a 1D simulation using ZZB.

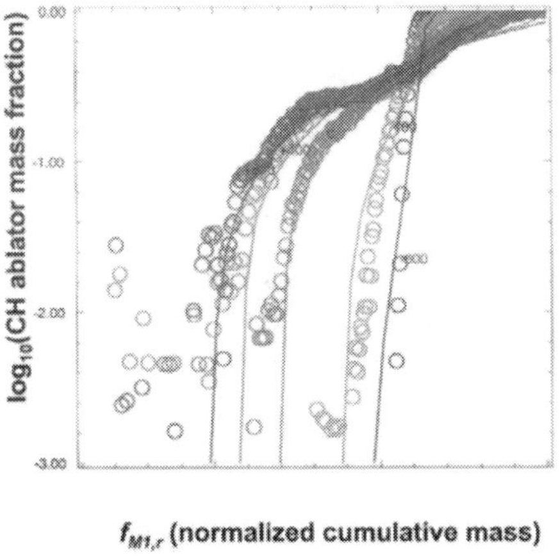

(a)

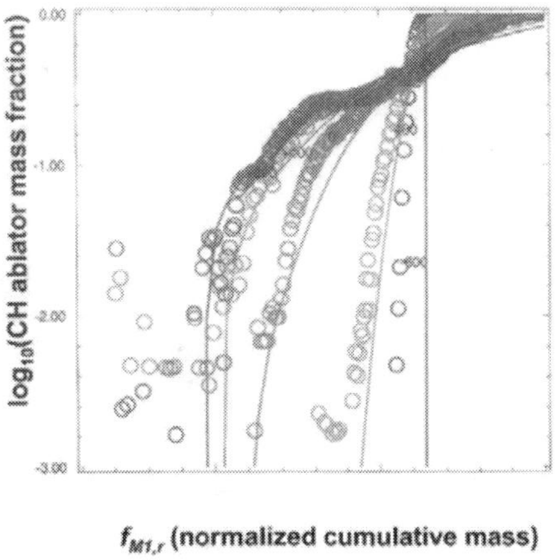

(b)

Figure 1. (a) Radial profiles of ablator mass fraction $f_{m,abl}$ plotted vs Lagrangian cumulative mass variable, at five times near the end of the implosion. Open circles show results from a high-resolution 2D multimode simulation [2]. The mixing front moves progressively inward to the left as time advances. Solid curves show results from a 1D simulation using $SK\varepsilon$ with $K_0 = 10^{-20}$ cm^2/ns^2, $l_0 = 10^{-5}$ cm, $M_{RT} = M_{RM} = 0.47$, $\sigma_\rho = 6.7$, and $\sigma_k = \sigma_\varepsilon = 170$. (b) Same as (a), but solid curves show results from a 1D simulation using ZZB with $h_{10} = h_{20} = 2$ μm, $t_{on} = -530$ ps, and $S = 2$.

Figure 1 shows that both calibrated models agree reasonably well with the reference profiles, at all five times, for each profile's "foot", i.e., $1\% < f_{m,abl} < 20\%$. The agreement is not so good for smaller $f_{m,abl}$ (the "toe", between 0.1% and 1%) or for larger $f_{m,abl}$ (the "plateau", seen at $t = +800$ ps at $f_{m,abl} \sim 25\%$, and in the outward penetration of the mixing layer, at $f_{m,abl} > 30\%$). The discrepancies result to some extent from the definition of the goodness-of-fit metric, which focused on the foot. Other details of the reference profiles, such as the non-monotonicity seen at $t = -400$ ps at $f_{m,abl} \sim 0.2\%$, cannot be represented by the gradient-diffusion approximation for transport.

One result of the calibration exercise is the insight gained regarding the sensitivity of model predictions to variations in the inputs. For example, with the parameter settings used in Figure 1, variations of approximately $\pm 3\%$ in σ_ρ in the $SK\varepsilon$ model lead to variations of approximately $\pm 5\%$ in the penetration of the mixing profile at $t = 800$ ps. The model is less sensitive to variations in σ_k and σ_ε; variations of $\pm 50\%$ in those coefficients give $\pm 5\%$ variation in the mixing penetration.

In future work, the predictive capability of the mix models will be assessed by calibrating against a similar implosion using a different initial perturbation. It is hoped that agreement with a reference can be achieved by varying only initial conditions, and not the model coefficients.

4. References

[1] Hoffman N M 1995 Hydrodynamic Instabilities in Inertial Confinement Fusion *Laser Plasma Interactions 5: Inertial Confinement Fusion (Proceedings of the 45th Scottish Universities Summer School in Physics, St Andrews, UK, August 1994)* ed M B Hooper Institute of Physics Publishing Bristol pp 105–137

[2] Hammel B A, Edwards M J, Haan S W, Marinak M M, Patel M, Robey H and Salmonson J 2008 *J. Phys.: Conf. Ser.* **112** 022007

[3] Zhou Y, Zimmerman G B, and Burke E W 2002 *Phys. Rev. E* **65** 056303

The National Ignition Campaign (NIC) "Blue Team / Red Team" Simulated Campaigns (Sim-Cams)

M.D. Rosen (Sim-Cam referee), J. Edwards (Blue Team leader) and L. J. Suter (Red Team leader)
Lawrence Livermore National Laboratory, Livermore, CA 9450

E-mail: rosen2@llnl.gov

Abstract. The Point design Target of the National Ignition Campaign (NIC) has specifications and tolerances that have been set using a multi variable sensitivity study (MVSS). The sub component interactions and sensitivities that feed the MVSS have been calculated using sophisticated target physics radiation hydro-codes. However, it is cannot be guaranteed that the point design configuration has been specified with enough precision for ignition to occur without first correcting for possible off-sets due to physics uncertainties in the data models used in these studies. For this reason the NIC includes a series of tuning experiments, which have been designed and sequenced to systematically remove potential off-sets in specified target and laser parameters as efficiently and effectively as possible. In order to test the tuning techniques and logic the NIC has been executing Simulated Campaigns (Sim-Cam's). In the Sim-Cam's are a Blue Team and a Red Team. The Blue Team conducts the experiments, making shot-to-shot decisions on how to adjust the laser pulse or target parameters, based on "experimental" data provided by the Red Team who simulate the laser, target and diagnostics performance. To capture a plausible off-set between nature and models the Red Team construct a new virtual reality by adjusting the physics data and models, keeping the changes within their best estimate of uncertainties. The Red Team also include much of the shot-to-shot variability due to small laser and target variations that can be expected in NIF experiments, as well as diagnostic noise. The details of all of this are kept hidden from the Blue Team. The Sim-Cam's also serve to test and improve the infrastructure and processes for executing the NIF experiments. The role of the Simulated Campaigns in preparing for the NIC experimental plan will be described, and some of the results discussed.

1. Introduction

The National Ignition Campaign (NIC) has laid out a strategy to achieve ignition by first doing a series of campaigns that fine-tune the point design specifications for target and laser parameters. This is necessary because there is no guarantee that our data models, as implemented in the hydro-codes that specify the design, are sufficiently accurate to obtain ignition on the first try, without such fine-tuning. It is necessary then, to test this strategy. In particular, we seek to answer several questions: Principally, does the sequence of campaigns converge within a reasonable number of fine tuning attempts, or, for example, does correcting shock timing, and correcting symmetry, end up in an endless do-loop? Does the inevitable shot to shot variability in target and laser parameters compromise the process, or equivalently, will our specifications on the allowed variability prove adequate? As we prepare to enter the real experimental campaigns, are our theoretical data analysis tools in place to do the job? Are our processes properly in place to communicate all critical issues within all aspects of this very complex operation? It is all of these vital questions that we seek to answer by doing simulated campaigns (Sim-Cams).

2. The Sim-Cam methodology

We form two teams. The Blue Team will execute this virtual NIC campaign as if it were real. They will be handed "data" and make decisions, based on that data, of what the next shots will be, by specifying target and laser parameters. They will do so until they reach (virtual) ignition, or give up in frustration, or until some time limit is imposed that stops the game.

The Red Team has two roles and two phases of operation. In phase one the red team will determine a physics/data model that is reasonable yet different from the one currently implemented in the hydro-codes. They will ensure that ignition can be achieved under this new model, via a "re-design" that changes the point design target and laser parameters in acceptable ways. By acceptable, we mean, within the capabilities of target fabrication and of the National Ignition Facility (NIF).

In phase two the Red Team will then act as a virtual facility, simulating the irradiation of the targets "ordered" by the Blue Team, by a laser pulse, also specified by the Blue Team, and using their "Red Team Physics Model". The results of the virtual experiment will be reported out to the Blue Team in terms of virtual diagnostic outputs, e.g. scope voltages vs. time, x-ray pinhole camera images, etc., all convolved with spatial and temporal blurring response functions as well as noise. The Blue team is to be kept in the dark as to exactly what target and laser was used (which vary shot to shot within the NIC specifications). They will also, obviously, not be told how the Red Team Physics Model differs from their own. A referee is put in place to ensure the efficacy of this entire exercise by monitoring the virtual campaign and maintaining separation of Red Team knowledge from the Blue team.

3. The NIC methodology / strategy

A multivariable sensitivity study (MVSS) is done to place the ignition campaign into a unified context. What emerges is an ignition threshold factor (ITF) [1], that if greater than 1, implies a very high probability of reaching ignition. Margin, as defined in references [2,3], is equal to ITF-1. The ITF involves 4 terms. Achieving a low adiabat implosion, a sufficiently high implosion velocity, a sufficiently clean (namely, only partially mixed) fuel, and a sufficiently round, symmetric, implosion.

The adiabat is achieved via tuning of the height and timing of the laser pulse that leads to 4 shocks. The velocity is achieved via tuning the height and duration of the last part of the pulse, where most of the laser power and energy resides, as well as by tuning the capsule ablator thickness. The mix problem is controlled by varying the high Z dopant levels in the ablator, as well as by varying the ablator thickness. Symmetry is achieved by varying the pointing of the laser beams as well as by controlling the time variation of the power balance between the "inner beams" that hit the waist of the hohlraum vs. the "outer beams" that hit the hohlraum wall closer to the laser entrance hole.

After the hohlraum drive is "certified", and laser plasma interaction (LPI) issues under sufficient control, the NIC strategy is to do the shock timing and symmetry campaigns in a somewhat interlaced mode, as we proceed from first through fourth shock. We then do a campaign called THD [4] wherein the D is minimized and replaced by H (T is still needed for beta layering smoothness of the DT ice) in order to cut down on neutron flux, so that it is easier for x-ray diagnostics to inform us of adiabat, mix and shape issues in the final fuel assembly. The THD campaign will be followed by attempts at DT ignition.

4. Let the Games Begin

The Red Team varied (via a one-time throw of the dice) the equations of state and opacities of all capsule and hohlraum materials, within the 1σ (typically 10-20%) uncertainties that exist. Far more uncertainty (~ factor of 2) exists for non-LTE collisional excitation rates and for electron thermal conduction. Those were also varied accordingly. With this new model the point design failed. A re-tuned target/laser combination restored ignition to nominal robustness.

As this was the first such Sim-Cam exercise we simplified it by legislating that the hohlraum drive / LPI issues were near nominal. We also did not explore 3-D issues at this point, nor include mix issues at the ablator DT ice interface. Our main concern was to see if the shock timing / symmetry interlaced campaigns would converge in a reasonable number of virtual shots, and not enter an endless do-loop.

5. The Blue Team succeeds

The first experiments surprised the Blue Team in that the x-ray drive on the first foot of the pulse came in low. They ordered the next shot to have a higher laser power in the foot to establish a slope of drive vs. laser power, and by so doing could then specify the correct laser power to achieve the required first shock intensity. This is the empirical approach of the NIC strategy and it worked quite well. The Blue team attempted to formulate a new model (via an Au hohlraum wall opacity model) though the "reality" was that the discrepancy was much more due to the electron conduction changes in the Red Team model. The main point is that the tuning was done successfully, empirically. (Fig. (1))

Similar discrepancies occurred in the symmetry campaign. There too, a slope of symmetry vs a symmetry tuning parameter (in this case the color separation between inner and outer beams that controls cross beam transfer [5]) was established empirically. It was parallel to the Blue Team's original expectations, but off-set. Again, the main point is the ability to tune empirically via this slope method. (Fig. (2)).

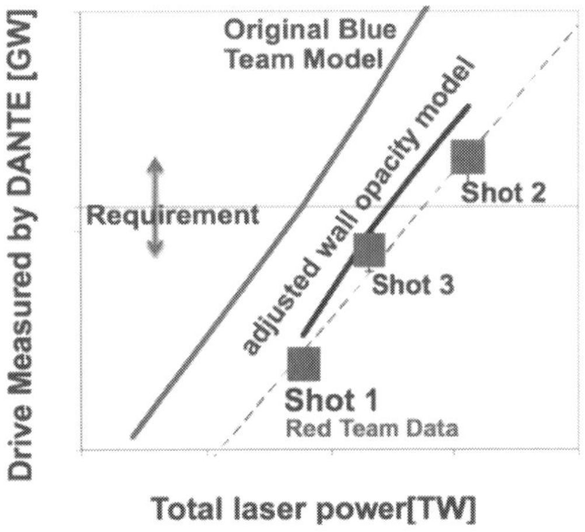

Figure 1: Laser-pulse's foot drive-level tune

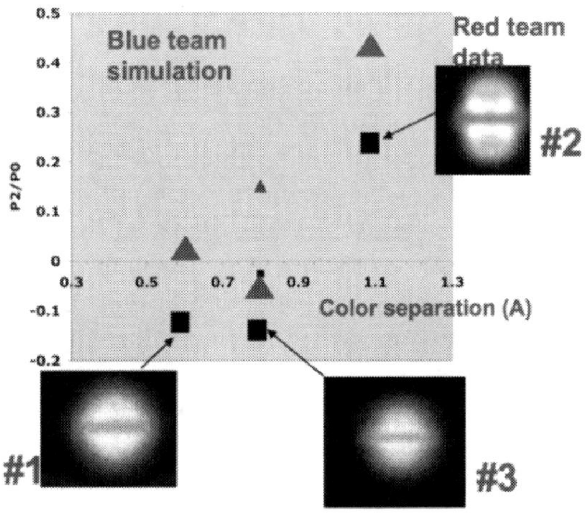

Figure 2: Symmetry tune P2 vs. color separation

After 25 shots the Blue Team was ready to specify an ignition target. Three targets were "shot", first in THD mode, and then, to save time, the very same were then shot in DT mode. In THD mode, the X-ray self emission implied a hot, warm and very hot result, respectively. The back-lit images implied, respectively, a dense, fluffy, and very dense core. (Fig. (3)). In DT mode those 3 targets gave 4, 0.1, and 14 MJ yields respectively, thus achieving ignition on 2 of the 3 shots. In retrospect the middle, failure shot, had target parameters out of the allowed specifications. The corresponding ITFs and margins of these 3 shots are consistent with their performance.

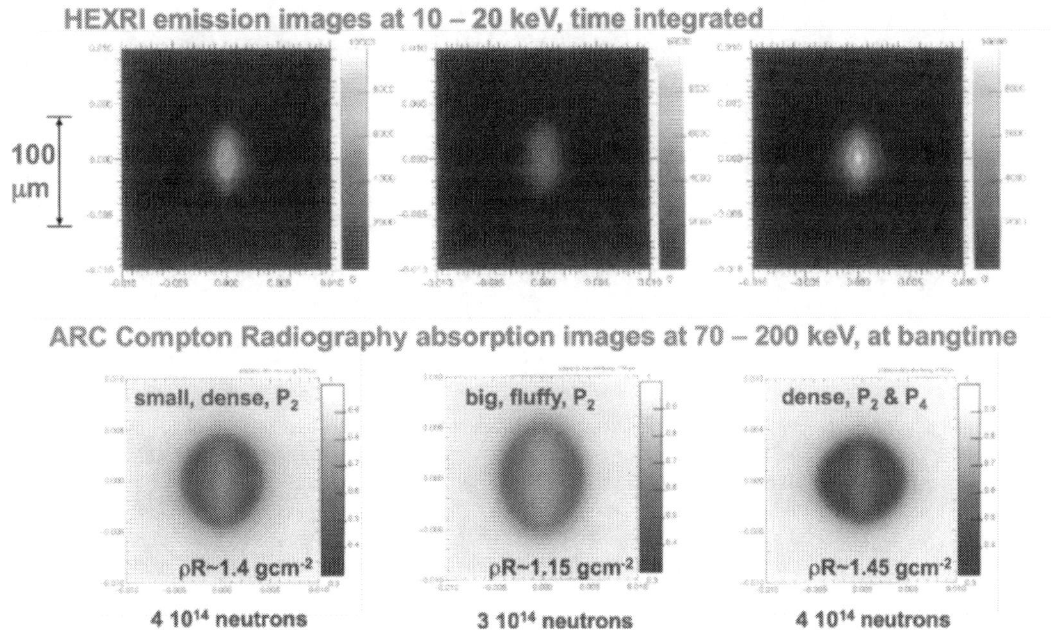

Fig 3: Virtual THD 3 shot attempts. Top row: Self emission images. Bottom row: Back-lit images

6. Other lessons learned and Conclusions

In doing this exercise we learned many operational lessons. The need for a back-up "responsible person" for every principal campaign. The efficiency increase when campaigns are interlaced- to allow more time for analysis between shots of any given campaign. Many practical matters regarding communication and quality control between diverse elements of this vastly complex operation were identified and fixed. In conclusion, this first Sim-Cam was an extremely valuable exercise which confirmed the basic soundness of the NIC strategy for achieving ignition on the NIF.

7. References

[1] S. W. Haan et. al., to be submitted to Phys. Plasmas
[2] S. W. Haan et. al., Phys. Plasmas **12**, 056316 (2005)
[3] D. S. Clark, S. W. Haan, and J. D. Salmonson, Phys. Plasmas **15**, 056305 (2008)
[4] J. Edwards et al, these proceedings
[5] P. Michel et. al., Phys. Rev. Lett. **102**, 025004 (2009)

Acknowledgments

This work was performed under the auspices of the U.S. Department of Energy by Lawrence Livermore National Laboratory under Contract DE-AC52-07NA27344. We are grateful to the 31 members of the blue team from LLNL, LANL, and GA and to the 14 members of the Red Team from LLNL, and SNL for their extremely hard work and dedication throughout this process. We regret that space limitations would not allow their names being listed.

Journal of Physics: Conference Series **244** (2010) 022013 doi:10.1088/1742-6596/244/2/022013

Target design for shock ignition

G Schurtz, X Ribeyre, and M Lafon

CELIA, Université Bordeaux 1,33405 Talence Cedex, France

E-mail: schurtz@celia.u-bordeaux1.fr

Abstract. The conventional approach of laser driven inertial fusion involves the implosion of cryogenic shells of deuterium-tritium ice. At sufficiently high implosion velocities, the fuel ignites by itself from a central hot spot. In order to reduce the risks of hydrodynamic instabilities inherent to large implosion velocities, it was proposed to compress the fuel at low velocity, and ignite the compressed fuel by means of a convergent shock wave driven by an intense spike at the end of the laser pulse. This scheme, known as shock ignition, reduces the risks of shell break-up during the acceleration phase, but it may be impeded by a low coupling efficiency of the laser pulse with plasma at high intensities. This work provides a relationship between the implosion velocity and the laser intensity required to ignite the target by a shock. The operating domain of shock ignition at different energies is described.

1. Introduction

Inertial Confinement Fusion (ICF) is a process to obtain the thermonuclear fusion of small amounts (1 mg or less) of hydrogen isotopes compressed up to several thousands times and heated to the temperature ~ 10 keV. Compression is achieved by means of implosion of a cryogenic solid DT shell, driven by lasers. The central hot spot initiates a thermonuclear combustion wave in the cold, dense fuel around it. A hot spot is obtained at the centre of the target at the end of implosion by conversion of the shell kinetic energy into the internal energy of the innermost parts of the shell. This central ignition scheme demands implosion velocities close to 400 km/s, which present a danger of shell break up during the acceleration stage. Slower implosions, though safer, need an additional source of heating. The principle of shock ignition is to boost up a sub ignition central hot spot by means of a convergent shock wave launched in the target at the end of the compression phase [1, 2]. This shock is formed by a high intensity spike in the laser temporal pulse shape, which may present a risk of excitation of various plasma instabilities. A safe target design for shock ignition results from a trade off between the risks of hydrodynamic and parametric instabilities. The determination of a relation between the spike intensity and the implosion velocity is the goal of this paper.

2. Hot spot pressure

Conditions for hot spot ignition are given by the Rosen-Lindl model for an isobaric fuel assembly [3] and considered in the non-isobaric case in Ref. [4]. The key ignition parameter is the product ρRT of the hot spot areal mass ρR by the ion temperature T. A substitute for the product ρT is the hot spot pressure P_h [2]. The hot spot self heating rate per unit volume is written as:

$$\dot{E}_h \propto \rho_h^2 < \sigma v >, \tag{1}$$

where E_h and ρ_h are the hot spot energy and mass densities, $<\sigma v>$ is the Maxwell-average nuclear cross section. The ignition temperatures are in the range of 6 – 12 keV. In this domain, $<\sigma v>$ scales as T^p, with $2 < p < 3$. So the volumic heating rate and characteristic heating time τ_h scale approximately as:

$$\dot{E}_h = 3\dot{P}_h/2 \propto P_h^2, \qquad \tau_h = P_h / \dot{P}_h \propto P_h^{-1} \tag{2}$$

The hot spot pressure depends on the implosion velocity V_i as $P \propto V_i^a$, the power a depending on the implosion process. An adiabatic compression corresponds to a = 5, whereas in ICF conditions, the Mach scaling a = 3 [5] applies at a given in-flight pressure and entropy parameter. The minimum hot spot pressure P_{ign} required for ignition is related to a minimum implosion velocity V_{ign}. In order to increase the pressure of shells imploded at sub ignition velocities, it was suggested to launch a convergent shock in the target prior to the time of shell stagnation [1, 2]. Such a converging shock encounters several amplification events related to two different hydrodynamic phenomena.

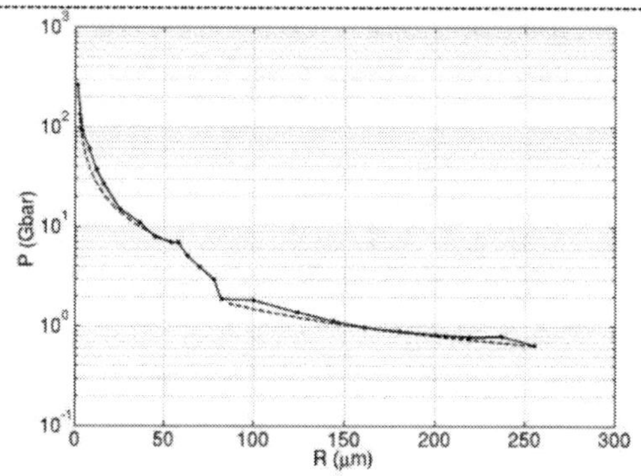

Figure 1. Shock pressure as a function of the shock position. Data from numerical simulations (solid line) is compared to the Guderley model (dashed line)

Following the analysis of Guderley [5], the shock pressure is first amplified from convergence by a factor $C^{0.9}$ where C is the shock convergence ratio. The second amplification stage results from the collision with another shock that is reflected from the center. Figure 1 shows the pressure variation in function of the shock location obtained from a numerical simulation with the code CHIC. The pressure first grows as predicted by the model [5], until the radius of 70 μm, where is collides with the rebound shock. The pressure is amplified 5 times in the collision and then the shock resumes its convergence.

3. Shock collision

Let us assume that the final hot spot pressure is proportional to the pressure arising from the shock collision P_h. Then this post shock pressure must match the criterion for the hot spot self ignition. If P_r is the pressure of the rebound shock and P_s the pressure of the converging shock before the collision, the pressure P_h follows from the Rankine-Hugoniot conditions. The shock amplification factor, $P_h/\max(P_r,P_s)$, is shown in Fig. 2 in function of the ratio P_s/P_r. The maximum of the pressure amplification for a mono-atomic ideal gas is 6 for $P_r = P_s$, and decreases to 3 for $P_s = 5P_r$ or $P_s = P_r/5$. This curve can be approximated by a simple analytical expression:

$$P_h \simeq (P_s^2 + P_r^2 + 10P_sP_r)/(P_s + P_r) \tag{3}$$

It is 20% accurate in the domain studied in Fig. 3. Assuming that P_r scales as V_i^a, and demanding that P_h attains the ignition value P_{ign}, we solve Eq. (3) for the shock pressure in function of the dimensionless implosion velocity $u = V_i/V_{ign}$.

$$P_s = 0.5P_{ign}\left[1 - 10u^a + \left(96u^{2a} - 16u^a + 1\right)^{1/2}\right] \tag{4}$$

This relation in plotted on Fig. 3 for a = 3. The optimal matching condition $P_s = P_r$ corresponds to $P_s = P_{ign}/6$. It is obtained for u = 0.54 (u = 0.65 for a = 4). Above this critical velocity, the shock pressure decreases almost linearly. At u = 0.4, the rebound pressure P_r is only 6.4% of the ignition pressure and the required shock pressure is $P_{ign}/2$. For u = 0.7, the predicted shock pressure ratio is 0.1. The corresponding pressure ratio P_s/P_r is close to 0.3 and our approximation (3) is reasonably accurate. Our model can only be applied for velocities between 0.5 and 0.75 of the self ignition velocity.

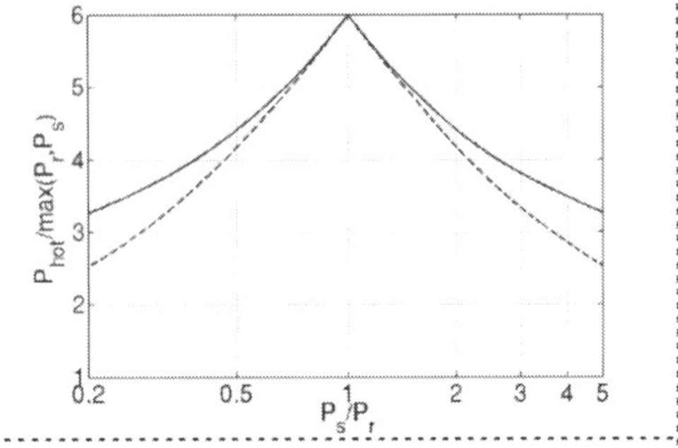

Figure 2. Pressure P_h after shock collision versus the ratio between the shock pressures P_r and P_s. Dashed line presents a simplified relation (3).

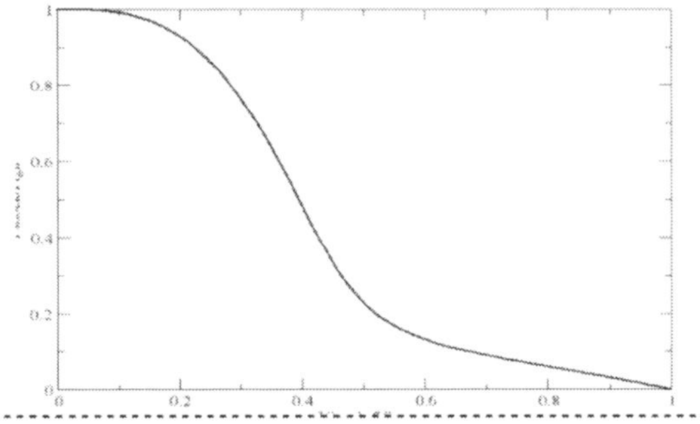

Figure 3. Shock pressure P_s/P_{ign} in function of the implosion velocity $u = V_i/V_{ign}$ for a = 3. A perfect matching condition is achieved at u = 0.55.

4. Laser power for ignition

The shock pressure P_s is proportional to the ablation pressure P_a produced by a intense spike in the laser pulse. The ratio P_s/P_a is given, according [5], by the ratio between the radius of the ablation front and the hot spot radius. The ablation pressure scales with the laser intensity I, as $I \propto P_a^{3/2}$. We considered a shock ignition of the HiPER target [7], imploded at different velocities. The numerical constants of the model are compared with simulations using the CHIC hydrodynamic code in Fig. 4, where the absorbed laser spike power at the ignition threshold is plotted as a function of the implosion velocity. Above 320 km/s, the target is close to self ignition, and the required additional power is comparable to the power for compression. Below 200 km/s, the spike power increases sharply, in excess of 300 TW, which may be inacceptable in terms of laser plasma instabilities. We also considered Euler scaled targets, by multiplying time and dimensions by a factor h, energies by h^3 and power by h^2. Then, the hot spot pressure required to satisfy the ignition criterion $(PR)_{ign}$ is reduced by h, the spike intensity varies as $h^{-3/2}$, and the ignition power grows as $h^{1/2}$. This was not observed in simulations (not shown), where the power for marginal ignition depends very little on the target scale, for velocities in the 200 – 260 km/s range. Since the area of the critical surface varies as h^2, the intensity in the spike decreases as $h^{-3/2}$, according to the model, and h^{-2}, according to simulations.

5. Design issues

The design constraints for shock ignition are three fold. First, the hydrodynamic robustness calls for low implosion velocities, and this is one of the main motivations for shock ignition and fast ignition. Second, one would like to minimize the energy invested in the target. The fuel assembly resulting from the shock collision is non-isobaric and it requires less energy to ignite [2]. According to Fig. 3, an optimal shock configuration is obtained when the target is driven at 55% of its self ignition velocity. Finally, low implosion velocities necessitate high laser powers in the spike, which not only stresses the laser technology, but also presents a risk of low laser plasma coupling efficiency. So the design has to trade off between the dangers of unstable fast implosions, and low laser coupling.
Assuming that the ignitor intensity decreases with the target scale as $h^{-3/2}$, whereas the self ignition velocity decreases as $h^{-1/3}$, it is clear that the trade off will be different according to the target scale. Figure 5 shows the laser spike intensity as a function of the implosion velocity, for different values of the compression energy. The laser intensity is given under the assumption of a 50% absorption efficiency and we accounted for the fact that lowering the velocity at constant energy allows to increase the fuel mass. All targets considered in Fig. 5 are scaled, but not hydrodynamically equivalent. We fixed safety limit at 10^{16} W/cm^2 for the intensity and at 280 km/s for the implosion velocity. There is no sensible shock ignition design using less than 100 kJ of compression energy. At

200 kJ, one can plan implosions in the 230 – 280 km/s range, but with a little margin. The 500 kJ curve offers a much wider range for design, with a working point at 250 km/s, 4×10^{15} W/cm^2. This curve is representative of what can be achieved on the NIF, accounting for a strong reduction of hydrodynamic efficiency due to the polar direct drive irradiation scheme envisioned for this facility.

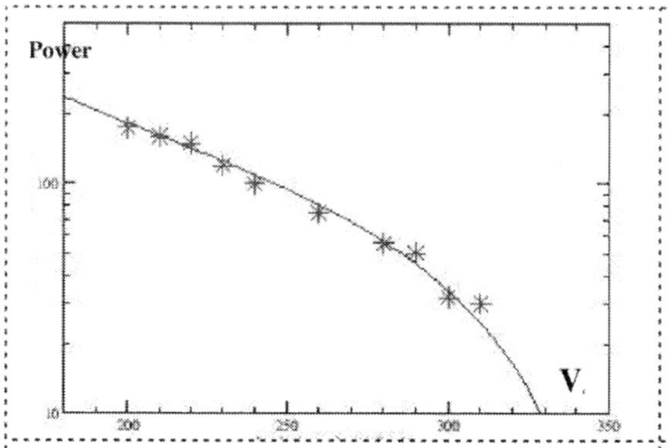

Figure 4. Minimum ignition laser power (TW) as a function of the implosion velocity (km/s). The solid line, Eq. (4), is compared with numerical simulations (stars).

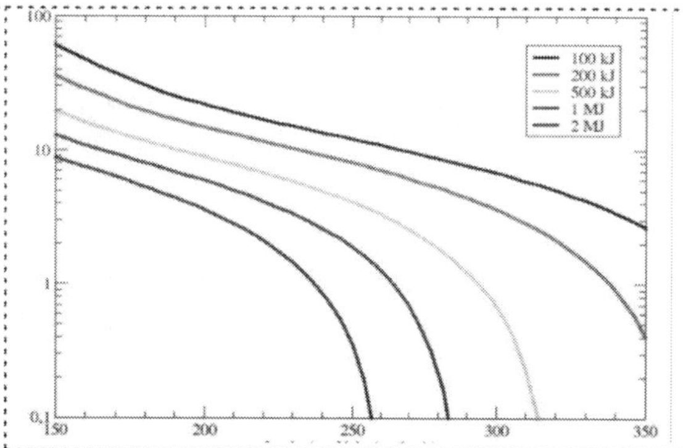

Figure 5. Intensity of the ignition spike (PW/cm^2) as a function of the implosion velocity (km/s) for the compression energies 100, 200, 500 kJ and 1 and 2 MJ.

6. Conclusion

A shock may ignite ICF targets imploded at sub-ignition velocities due to two successive mechanisms of pressure amplification: shock convergence and shock collision. The latter effect offers the pressure amplification factor of 6. Such a shock may be launched from an ablation pressure of several hundreds of Mbar, which necessitates laser intensities larger than 5×10^{15} W/cm^2. However, in this high intensity regime, parametric instabilities may decrease the laser coupling efficiency. Using a simple hydrodynamic model, we derived a relationship between the implosion velocity and the intensity of the ignition laser pulse. Numerical simulations were used to expand the operating domain in the implosion velocity and laser intensity space. The HiPER project envisions the gains close to 80 from a target imploded with 200 – 250 kJ of the laser energy. Our results demonstrate that the implosion velocity should then range between 240 and 280 km/s, with intensities for the shock ignition between 6×10^{15} and 10^{16} W/cm^2. The operating domain at 500 kJ is much wider, allowing implosion velocities as low as 190 km/s. This domain is representative of anticipated polar direct drive shock ignition experiments on the NIF at 750 kJ, with less efficient but more robust targets.

Acknowledgment. This work is partly supported by the Aquitaine Region Council and performed in the framework of the HiPER project (EC FP7 project number 211737).

[1] Ribeyre X *et al.* 2009 *Plasma Phys. Control Fusion* **51** 015013
[2] Betti R 2007 *Phys. Rev. Lett.* **98** 155001
[3] Rosen M D and Lindl J D 1984 *UCRL-50021-83 LLNL* pp 3.5-3.9
[4] Ribeyre X *et al.* 2009 accepted in *Plasma Phys. Control Fusion*
[5] Kemp A. *et al.* 2001 *Phys. Rev. Lett.* **86** 3336
[6] Von Guderley G. 1942 *Luftfart-Forsch.* **9** 302
[7] Atzeni S *et al.* 2009 *Nucl. Fusion* **49** 055008
[8] Zhou C D and Betti R 2008 *Physics Plasmas* **15** 102707
[9] Perkins L J *et al.* 2009 *Phys. Rev. Lett.* **103** 045004

Journal of Physics: Conference Series **244** (2010) 022014 doi:10.1088/1742-6596/244/2/022014

PREDICTION OF IGNITION IMPLOSION PERFORMANCE USING MEASUREMENTS OF LOW-DEUTERIUM SURROGATES

B. K. Spears[1], S. Brandon[1], D. Clark[1],C. Cerjan[1], J. Edwards[1],O. Landen[1], J. Lindl[1], S. Haan[1], S. Hatchett[1], J. Salmonson[1], P. Springer[1], S.V.Weber[1], and D. Wilson[2]

[1] Lawrence Livermore National Laboratory, L-477, 7000 East Ave, Livermore, CA, 94568, USA

[2] Los Alamos National Laboratory, Los Alamos, NM, USA

E-mail: `spears9@llnl.gov`

Abstract. The National Ignition Campaign (NIC) will use non-igniting "THD" capsules with cryogenic ice layers to study and optimize the hydrodynamic assembly of the fuel without burn. These capsules are characterized by the ratios of T:H:D. The species ratios are set with two goals in mind: (1) control T:D in order to adjust the nuclear energy production and (2) preserve the average atomic number of the fuel at 2.5 to maintain hydrodynamic similarity with the DT ignition capsule.

We have developed an experimentally observable ignition threshold factor (ITFX) that uses measurements from THD experiments to predict the performance of DT ignition implosions. It was developed and tested on multiple large databases of 2D radhydro simulations. Each of the thousands of simulations includes twin DT and THD simulations with a variety of physical failure mechanisms – drive asymmetry, capsule roughness, continuum mixing, fabrication errors, among others.

The results of our numerical database and the ITFX metric have allowed us to develop an experimental estimate of the probability of DT ignition based on THD experiments. The analysis accounts for both diagnostic precision and the effects of a finite number of shots. The NIC expects to field a combination of diagnostics and experimental attempts that result in a 15 to 20 percent uncertainty in the experimentally inferred probability of ignition.

This work was completed under the auspices of the U.S. Department of Energy by Lawrence Livermore National Laboratory under Contract DE-AC52-07NA27344.

PACS numbers: 52.57.-z,47.20.Ma,02.50.Sk,47.70.Mc

Submitted to: *Proceedings of IFSA 2009*

Journal of Physics: Conference Series **244** (2010) 022014 doi:10.1088/1742-6596/244/2/022014

1. Introduction

NIF implosion performance can be characterized by a theoretical metric called ITF. ITF is a measure of the probability of ignition and is a function of both 1D and 2D/3D quantities. The probability of ignition is a relatively steep function of ITF, rising rapidly around ITF=1 where the probability is 50 percent. At the nominal value of ITF, 2, associated with the point design the probability of ignition is high. However, when the NIC begins in September, the ITF will be less than 1 because of physics uncertainties that need to be corrected for, empirically, during the ensuing "tuning" campaigns. [1]

We will then field low-deuterium "THD" targets with dudded cryogenic fuel layers. The THD capsules are designed to be hydrodynamic analogues of DT implosions by specifying the T:H:D ratios to be about 74:24:2. Their low deuterium content leaves them unaffected by thermonuclear energy production, yet their radiation and hydrodynamic transport mirror that of DT implosions up to the point where DT implosions become perturbed by alpha production and deposition.

Below we describe a database of simulations that we have used to elucidate the relationship between THD and DT implosions. We develop an experimental version of ITF called ITFX. This metric allows us to diagnose directly the performance of a THD experimental implosion to predict the performance of the analogous DT implosion.

2. The relationship between THD and DT implosions via ITFX

To further investigate the relationship between THD and DT implosions, we have developed a numerical database of simulated implosions. The implosions are located in parameter space according to NIC estimates for target and laser variability (see [2]). We account for variation of many 1D parameters (capsule thickness, peak laser power,etc.) and 2D parameters (capsule surface roughness, radiation asymmetry, continuum mixing,etc.). Our database is composed of 1000 sample points in the parameter space, each point constructed from 4 simulations: a 1D clean DT implosion, a 2D DT implosion, a 1D clean THD implosion, and a 2D THD implosion. For each quadruplet, we model not only the radiation hydrodynamics, but also the output from a variety of x-ray cameras, neutron spectrometers, and other diagnostics. This simulated diagnostic data forms the key link between the experimental observables produced by THD implosions and the related DT implosion performance.

The THD and DT implosions show identical peak mass-averaged fuel velocity and indentical integrated entropy, two key theoretical implosion metrics. The simulated diagnostic data show even more striking trends. Presently, we have considered the simulated nTOF neutron spectrometry data most closely. We find that two features from the THD neutron spectrum are particularly predictive. These are the number of primary DT neutrons (or yield), defined as those neutrons detected with energies between 12 and 17 MeV, and the down scattered fraction, defined as the number of neutrons between 10 and 12 MeV divided by the primary yield. From basic physics

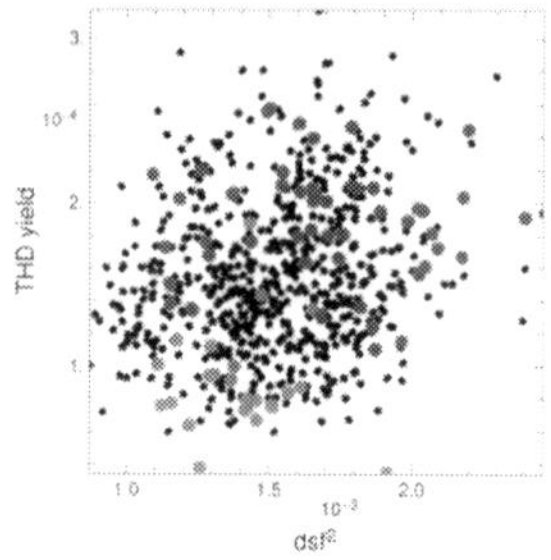

Figure 1. Implosions cluster in (yield,dsf^2)-space. Here, red implosions are highly asymmetric with strong mixing of ablator and fuel. Blue points represent nearly spherical, lightly mixed implosions. The two clusters are nearly separable.

reasoning, the downscattered fraction (dsf) is expected to be proportional to the rhoR of the cold, stagnated fuel. The yield (y), is generally proportional to the rhoR of the hotspot as well as its temperature. Using decision tree and regression analysis, we find that implosions cluster tightly in the space (y,dsf^2) (see figure 2). Here the colored clusters represent implosions filtered by 2D behavior. The red represent asymmetric, highly mixed implosions, whereas the blue are nearly spherical and lightly mixed. This clustering suggests that dsf and y not only encode the 1D implosion behavior implied by our physics intuition, but 2D shape and mix characteristics as well.

Regression analysis shows that the yield of the analogous DT implosion is strongly correlated with an experimental THD metric, $ITFX\ y{*}dsf^{2.3}$ (see figure 2(a)). The general shape of the dependence of DT yield on ITFX reveals the expected threshold behavior typical of ignition phenomena. We have normalized the metric so that ITFX of 1.0 represents marginal implosions. Implosions with ITFX of 1.2 or higher are likely to ignite; implosions with ITFX substantially less than 1 are expected to give very low fusion yields. This ITFX measurement is found to be a power law of the standard theoretical ITF metric ([3]). Further comparison with a threshold metric developed by Betti and co-workers (c.f., [4])), ITF3D, also reveals linear dependence. The detailed physics of the tight correspondence of the ITF, ITFX, and ITF3D continues to be investigated in order to clarify the fundamental physics basis for the strong predictive nature of ITFX.

For a given ITFX, there is a nearly Gaussian probability distribution of associated DT yields. The probability density of DT yield given ITFX is shown in figure 2(a), where the color map represents likelihood. This can be turned directly into a conditional probability function that gives the probability that DT yield will be greater than 1 MJ, that is the probability that the DT capsule will ignite, for a given ITFX value. We call this conditional probability function F.

In addition to F, one needs the probability distribution for ITFX itself to predict the expected probability of ignition. This can be obtained experimentally my measuring the dsf and y for a number of THD shots. The resulting distribution for ITFX values will possess inherent variability from two sources: (1) experimental measurement

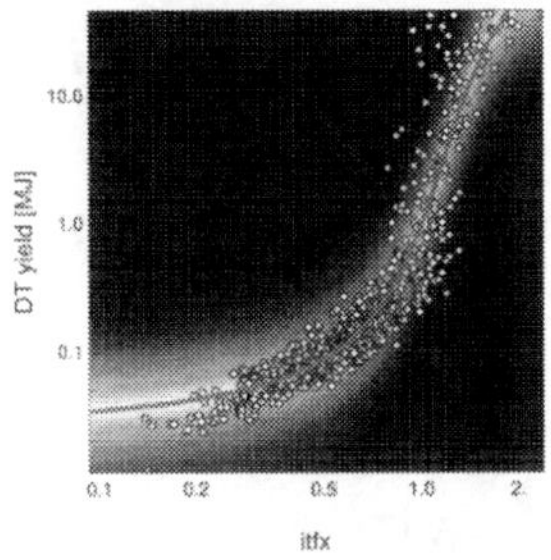

(a)

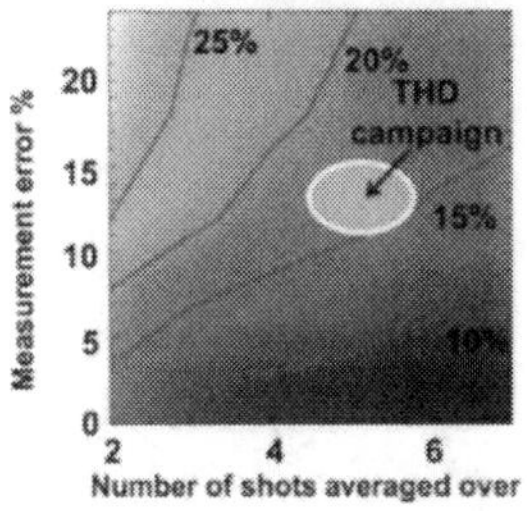

(b)

Figure 2. ITFX and our numerical database can together be used to estimate probability of ignition and our uncertainty in the probability. (a) Conditional probability of DT yield given ITF is estimated from a large database of 2D simulations. (b) Uncertainty in probability of ignition depends on number of shots and measurement error.

uncertainties in dsf and y and (2) uncertainty from finite n-shot sampling of the true underlying ITFX distribution. These two sources of variability combine to set up the precision with which one can know ITFX.

The probability distribution for ignition can be found as

$$POI = \int F(ITFX)P(ITFX)dITFX$$

where POI is probability of ignition and P represents the probability distribution for ITFX. The standard deviation of this distribution is explicitly dependent on the ITFX width, and thus is also dependent on the number of THD shots and the measurement precision. We show in figure 2(b) the uncertainty in probability of igntion versus both number of shots and measurement error.

3. Conclusions

We have produced a large numerical database of radiation hydrodynamics simulations that elucidate the relationship between low yield THD and DT implosions. Analysis of simulated diagnostics signatures shows that key features of the experimental neutron spectrum can be used to predict DT performance and can be used to give the expected probability of ignition based on a set of THD experimental measurements. The uncertainty in this inferred probability is set by both diagnostic precision and the number of shots taken. The NIC expects to operate the THD campaign with enough THD shots to give an uncertainty in ignition probability of 15 to 20 percent.

[1] J. Edwards, B. K. Spears, and J. Lindl, Physics of Plasmas (in preparation) (2010).
[2] S. W. Haan, D. A. Callahan, M. J. Edwards, B. A. Hammel, D. D. Ho, O. S. Jones, J. D. Lindl, B. J. MacGowan, M. M. Marinak, D. H. Munro, S. M. Pollaine, J. D. Salmonson, B. K. Spears, and L. J. Suter, FUSION SCIENCE AND TECHNOLOGY **55**, 227 (2009).
[3] D. S. Clark, S. W. Haan, and J. D. Salmonson, PHYSICS OF PLASMAS **15**, (2008), 49th Meeting of the DPP of the APS, 2007.
[4] C. D. Zhou and R. Betti, Physics of Plasmas **16**, (2009).

The Use of Tritium Rich Capsules with 25-35% Deuterium to Achieve Ignition at the National Ignition Facility

D C Wilson[1], B K Spears[2], S P Hatchett II[2], C J Cerjan[2], P T Springer[2], D S Clark[2], M J Edwards[2], J D Salmonson[2], S V Weber[2], B A Hammel[2], G P Grim[1], H W Herrmann[1] and M D Wilke[1]

[1]Los Alamos National Laboratory, Los Alamos, NM, USA, 87545

[2] Lawrence Livermore National Laboratory, Livermore, CA, USA, 94550

E-mail: dcw@lanl.gov

Abstract. Diagnostics such as neutron yield, ion temperature, image size and shape, and bang time in capsules with >~25 % deuterium fuel show changes due to burn product heating. The comparison of performance between a THD(2%) and THD(35%) can help predict ignition in a TD(50%) capsule. Surrogacy of THD capsules to TD(50%) is incomplete due to variations in fuel molecular vapour pressures. TD(25-35%) capsules might be preferred to study hot spot heating, but at the risk of increased fuel/ablator mixing.

1. Introduction

The path to ignition at the National Ignition Facility (NIF) [1] relies upon using experimental information to tune target conditions. Hohlraum radiation drive, shock timing, symmetry, and mass ablation will all be tuned using surrogate capsules without a cryogenic fuel layer. THD (tritium, hydrogen, deuterium) capsules that include a fuel layer were proposed to perform final adjustments [2], [3]. By maintaining the fuel atomic weight these capsules can be made hydro-dynamically equivalent to T(50%D) capsules, up to the time when burn product heating becomes significant. Capsules with ~ 2% deuterium are now included in ignition tuning. Above ~ 10% alpha particle heating becomes significant. As ignition is approached ion temperature increases, both X-ray and neutron images become larger and more symmetric, and peak yield rate is delayed. Target parameters crucial to hot spot formation, such as central gas density, ice mass, and ice roughness, can be tuned to optimize ignition. Trade-offs that affect burn, such as between un-ablated mass and fuel/ablator mix can be finalized.

A data set of ~ 700 THD capsule implosions with a variety of initial conditions was created based on the 285 eV 1.3MJ Rev 2 beryllium ignition capsule (very similar to the Rev 3 reported in [4]). Spears et al. [2] described how ~ 40 initial parameters affecting the drive were varied in a Latin Hypercube design to represent variation that might occur in a tuned ignition capsule. 2D static hohlraum drive asymmetries up to mode 10 were included. 2D roughness perturbations on all surfaces were taken as multiples of the capsule roughness specification. Mix between the beryllium shell and the layered fuel was simulated. The total variations are more than we expect from a tuned ignition capsule. The complete set of ~700 capsules was simulated with 0.3, 2, 25, 35, and 50% deuterium to simulate THD and TD(50%) implosions. Of these TD implosions only 20% achieve a yield > 1 MJ.

Other smaller sets used intermediate deuterium fractions. Spears et al. [3] reported on results from the 0.3% set. We have since chosen 2% as our nominal deuterium fraction to increase the neutron yield to ~ 2 10^{14} and still be able to use X-ray diagnostics. Here we focus on the higher deuterium fractions.

2. Changes as ignition is approached

One objective in using THD capsules is to be able to predict the yield of the equivalent TD(50%) ignition capsule. Ignition has been defined as a yield of 1.0 MJ, equivalent to 3.5e+17 DT reactions, or ~ 3e+17 neutrons between 12 and 17 MeV. Expected experimental errors in yield measurements are ±10% and in ion temperature ± 0.1 keV. Figures 1, 2 and 3 plot the THD capsule neutron yield on the horizontal axis vs the energy yield (MJ) from the equivalent TD(50%) capsule. In figures 1 and 2 the horizontal span is a factor of 10. In figure 3 it is 1000. As the deuterium fraction is increased from 2% to 25 and 35%, the THD yield increases. At the same time the range of the Doppler broadened ion temperature also increases, allowing better discrimination among capsules whose TD(50%) equivalent would ignite. 2% of the 35% D capsules ignite with yields > 1MJ. These are not shown in figure 6 since their ion temperatures reach up to 25 keV.

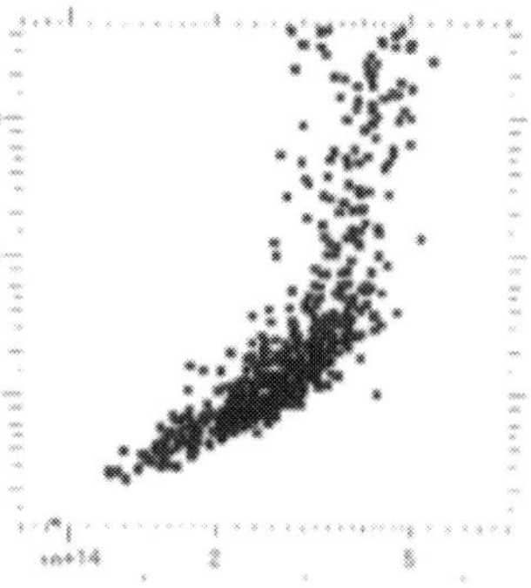

Figure 1. The energy (MJ) produced by a TD(50%) capsule compared to the neutron yield (12-17 MeV) from a THD(2%) capsule.

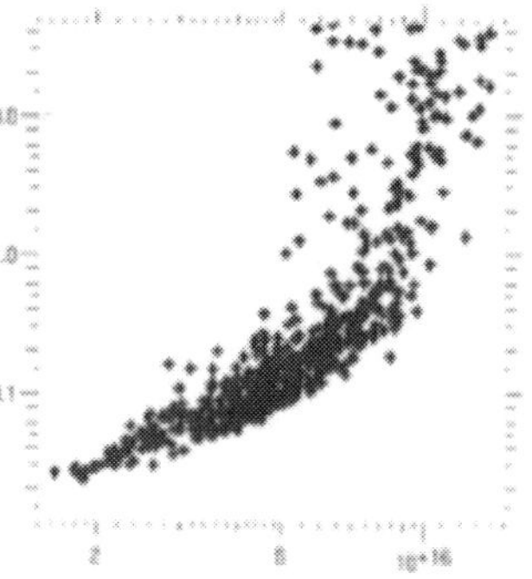

Figure 2. The energy (MJ) produced by a TD(50%) capsule compared to the neutron yield (12-17 MeV) from a THD(25%) capsule..

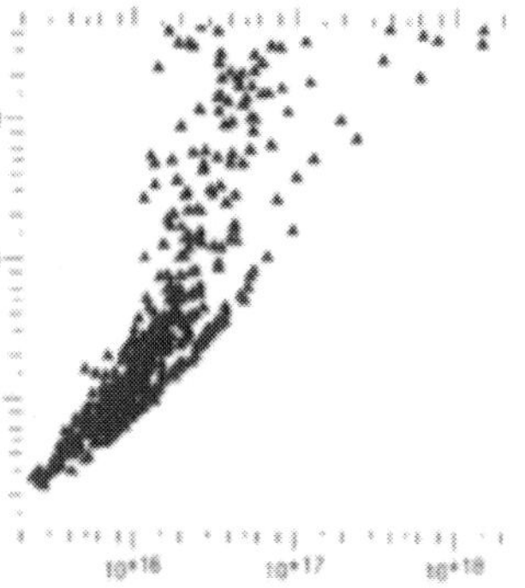

Figure 3. The energy (MJ) produced by a TD(50%) capsule compared to the neutron yield (12-17 MeV) from a THD(35%) capsule.

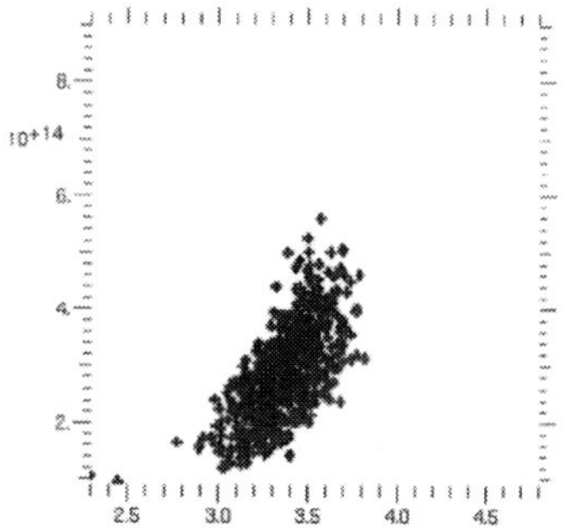

Figure 4. The 12-17 MeV yield vs. the ion temperature (keV) from a THD(2%) capsule.

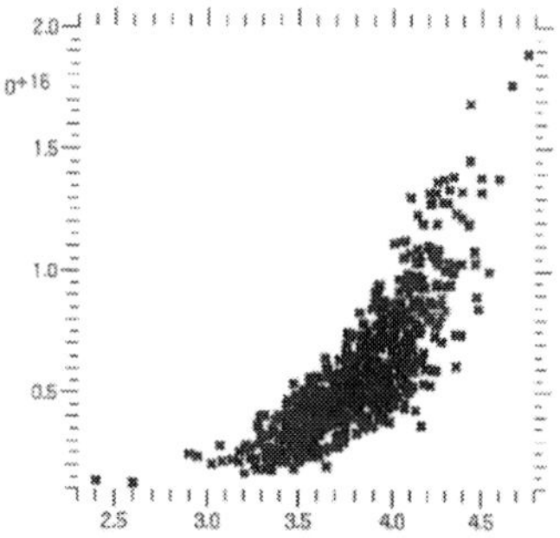

Figure 5. The 12-17 MeV yield vs. the ion temperature (keV) from a THD(25%) capsule.

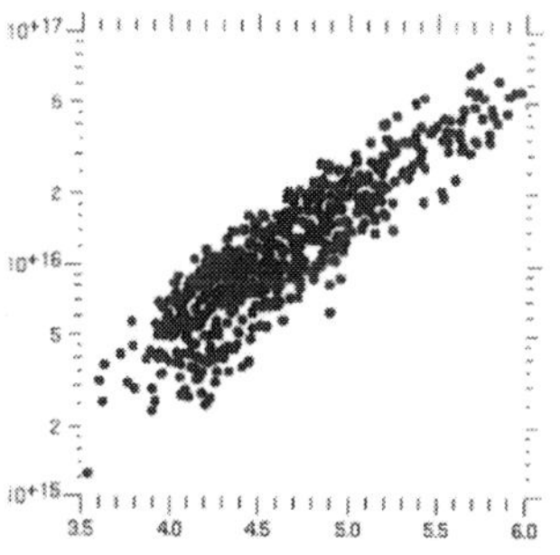

Figure 6. The 12-17 MeV yield vs. the ion temperature (keV) from a THD(35%) capsule.

The TD reaction rate from a capsule is proportional to the product of tritium fraction times deuterium fraction. Figure 7 compares the reaction rates, divided by this product, from 2% (A), 25%(B), 35%(C) and 50%(D). t=0 is the peak in the TD(50%) burn rate. Before ~ -150ps all scaled rates agree. Later when burn product energy deposition becomes important the rate grows, its peak moves later. As a THD capsule approaches ignition conditions and as the fuel temperature increases as shown in figures 5 and 6, the size of the compressed fuel becomes larger than in a capsule with less yield. Figure 8 shows the radius of the down-scattered neutron (10-12 MeV) image and the down-scattered fraction (dsf) (6-12 MeV neutrons / 12-17 MeV). Spears et al. [3] found that the dsf^2*yield(12-17 MeV) of a THD(0.3%) was a good predictor of DT capsule performance. For capsules with > 25% D, the size increases and down-scattered fraction decreases as the yield increases. The dots shown in figure 8 contain 35% D. The other points include the 2 and 5% D capsules. The points with less than dsf < 0.05 have high yields > 1MJ. dsf is proportional to fuel ρr. Thus the target radius ∝ $dsf^{(-1/2)}$, the scaling shown by the solid line.

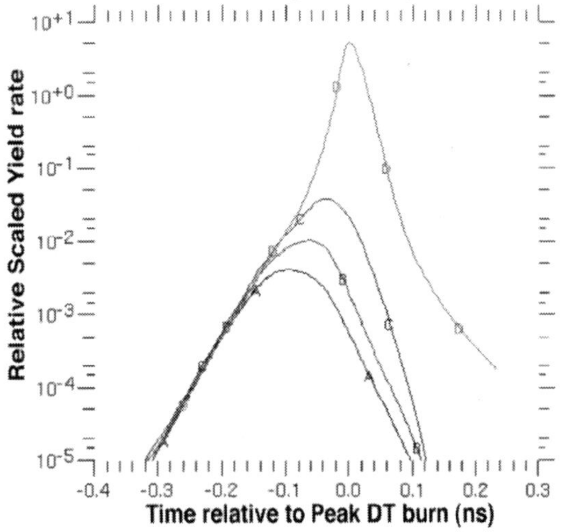

Figure 7. Scaled neutron yield rate from THD(2%), 25%, 35% and 50% relative to the time of peak yield rate of TD(50%).

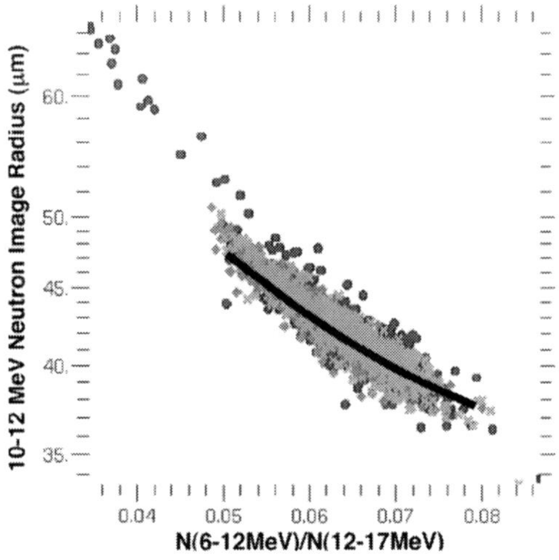

Figure 8. 10-12 MeV neutron image size vs down-scattered fraction.

3. Predicting Ignition

The performance of a capsule in which fusion product heating is significant, relative to a THD(2%) can help predict ignition of a TD(50%) capsule. Figure 9 shows the ratio of neutron yields, and figure 10 the increase in ion temperature. To predict a TD(50%) yield greater than 1 MJ (ignition) the yield ratio should be > 45 and the increase in ion temperature > 1.4 keV

.

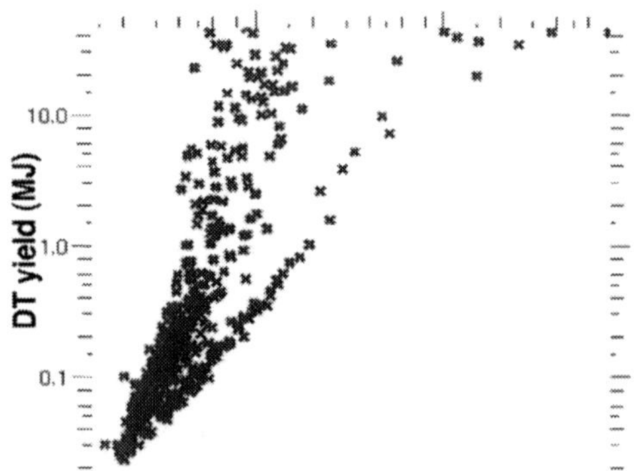

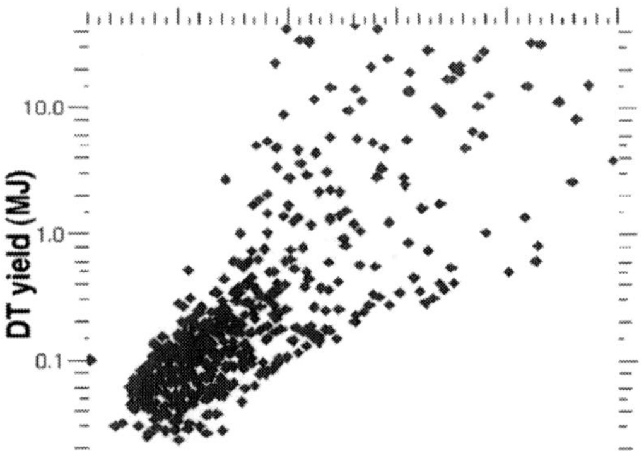

Figure 9. TD(50%) yield (MJ) vs the yield ratio of THD(35%) to THD(2%).

Figure 10. TD(50%) yield (MJ) vs the difference in Tion in THD(35%) and THD(2%).

4. Remaining Surrogacy Errors

THD capsules are not complete hydrodynamic surrogates for TD(50%) capsules. Largely because hydrogen is added, the composition of the central gas differs from the ice due to the different vapor pressures of HH, DD, TT, HD, HT, and DT. In the filling process ice is formed near its triple point, allowed to symmetrize for many hours, then in a few seconds is cooled to produce the desired central gas density. For a TD(50%) capsule at 18.3K , the 3e-4 g/cc central gas is 62% D and 38% T, with DT reactivity (0.62*0.38) essentially the same as the ice. To create a THD ice layer with roughness comparable to the DT layer, the capsule should be fielded at approximately the same 0.9 K below its triple point, or 17.4 K. The central gas then is 79%H 1%D and 20%T and has a molar density 1 ½ times the DT gas. Its reactivity is 14% that of the ice and it contributes very little to the hot spot burn. This halves the DT yield of a THD(2%) capsule, and has an even stronger effect on capsules with more deuterium. If a TD(25%) capsule is fielded at 18.7 K or TD(35%) at 18.5 K we can closely match both the reactivity of the gas and ice, and match both the molar pressure and the density of the gas to the TD(50%) capsule, but the ice density is too high. Hammel *et al.* [5] calculate this will change the aspect ratio at the ablator/ice interface, and cause more instability growth. To study ablator/ice mixing we might use a capsule with hydrogen, but to study hot spot ignition, a capsule without could be better.

References

[1] J Paisner J Boyes S Kumpan W Lowdermilt and M Sorem 1994 *Laser Focus Word* **30**, 75
[2] D C Wilson P Bradley C Cerjan J Salmonson B Spears, S Hatchett II H Herrmann and V Glebov, *Rev. Sci. Instrum.* 2008 **79**, 10E525
[3] B Spears S Brandon D Clark C Cerjan J Edwards O Landen J Lindl S Haan S Hatchett J Salmonson P Springer SWeber and D Wilson 2009 Prediction of ignition performance using measurements of low-deuterium surrogates, *IFSA Conference Proceedings* (to be published)
[4] S Haan et al. 2009 Fusion Science and Technology **55**, 227-32
[5] B Hammel S Haan D Clark M Edwards S Langer M Marinak M Patel J Salmonson and H Scott, High-mode Rayleigh Taylor growth in NIF ignition capsules, 2nd International Conference on High Energy Density Physics (Austin TX) to be published in *High Energy Density Physics*

Acknowledgment

This work was supported by the US Department of Energy and performed at Los Alamos and Lawrence Livermore National Laboratories. LA-UR-09-07214

Transport of hot electron currents in solid targets irradiated by high intensity short laser pulses

P.Antici[1,2,3], P. Audebert[3], M. Borghesi[4], T. Cowan[5,6], L. Gremillet[7], Y. Sentoku[6], J. Fuchs[3]

[1]ILE - Ecole Polytechnique - CNRS –ENSTA – Iogs - UP Sud, Batterie de l'Yvette, 91761 Palaiseau, France

[2]Istituto Nazionale di Fisica Nucleare, Via E. Fermi, 40 - 00044 Frascati, Italy

[3]Laboratoire pour l'Utilisation des Lasers Intenses, UMR 7605 CNRS-CEA-École Polytechnique-Univ. Paris VI, 91128 Palaiseau, France

[4]School of Mathematics and Physics, The Queen's University, Belfast, United Kingdom

[5]Physics Department, MS-220, University of Nevada, Reno, Nevada 89557, USA

[6]Forschungszentrum Dresden Rossendorf, POB 510119 ; 01314 Dresden, Germany

[7]CEA, DAM, DIF, 91297 Arpajon, France

Abstract. We have analyzed the transport of hot electrons generated in the interaction between a short-pulse, ultra-high intensity laser beam (pulse duration τ<ps, intensity $I\lambda^2 > 10^{18}$ W.cm^{-2}.μm^2) and a solid or dense target through the use of proton emission imaging. We used targets of different material (Cu, Al, Au) with a regularly modulated rear target surface in order to compare the electron transport in different conditions. As result, we see that the electron transport depends on the target material and on the interaction conditions.

Hot electron currents obtained in the interaction between a short-pulse, ultra-high intensity laser beam (pulse duration τ<ps, intensity $I\lambda^2 > 10^{18}$ W.cm^{-2}.μm^2) and a solid or dense target are crucial for many applications, such as fast ignition or the generation of secondary sources of particle (ions, X-rays, positrons, neutrons). However, the stability and uniformity of those currents transported through the targets is an important parameter affecting the efficiency of these processes. Several papers have concentrated on analyzing the electron transport within targets ([1] and references therein). One method to get information about the hot electron beam is using proton imaging.

Protons used for the imaging are produced by irradiating a thin foil with a short-pulse, ultra-high intensity laser [2,3,4] and are predominantly accelerated from the rear-surface of the target [5]. The fast electrons propagated through the foil generate a ~TV/m electrostatic sheath field [4] on its rear surface that ionizes the surface atoms and accelerates the resulting ions. The resulting accelerated ion beam is composed mostly of protons originating primarily from contaminant layers of water vapor and hydrocarbons on the target surface [4]. The accelerating sheath field will initially follow the target rear-surface. Since proton acceleration takes place normally at the local electron sheath surface, any deformation of either the electron sheath [6] or the target surface [7] will affect the proton angular distribution.

Recently it has been shown that hot electron beams are more uniform in conductors than in insulators [6], since in insulators there is a lack a sufficient density of conduction band electrons to initially provide the return current that neutralizes the forward electrons. We have also shown that interfaces between insulators and conductors could lead to unstable transport [6]. In this paper, we show that even conducting targets are subject to unstable high-current transport, provided that the current density is large enough or that the target is thin enough. The transport is resolved through the use of proton beam emission imaging.

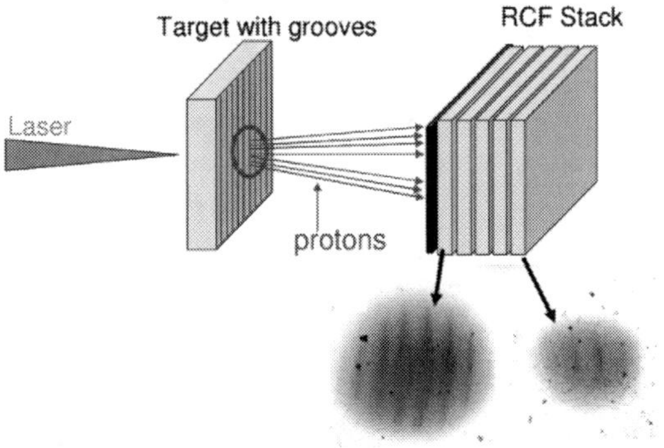

Figure 1: Experimental setup: a short-pulse, high energy laser beam hits a solid target. Protons stemming out from the rear target surface are imaged on RCFs. In some cases, targets with a grooved target rear surface were used, which allows seeing regular modulations on the target rear surface, as illustrated in the figure.

The experiments were performed using the 100 TW short pulse laser at the Laboratoire pour l'Utilisation des Lasers Intenses (LULI). Laser pulses of ~20-30 J of 1 μm light (350 fs) were focused at I_{max}~6 10^{19} W.cm^{-2} onto the front surface of various conductor targets. We used as the main diagnostic proton imaging, i.e. we concentrate on the angular pattern of the accelerated protons that are detected in multiple layers of radiochromic film (RCF) densitometry media [8]. The spatial distribution of the protons in a given RCF layer gives the angular emission pattern at a specific interval of proton energy. We used Al and Cu targets with different thicknesses, some of which had on their rear surface a grooved pattern, i.e. periodic (quasi-sinusoidal) shallow modulations of the target rear-surface, 200 nm deep and spaced 3.6 μm (with a ~20% variation when scanning with an AFM at various locations on the target surface), that allow to imprint regular modulations in the proton beam that thus allow to image the accelerating sheath surface [9]. Figure 1 shows the experimental setup.

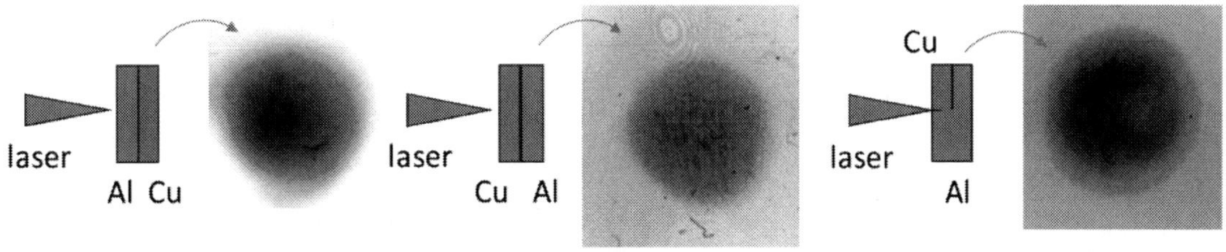

Figure 2: Proton angular distributions at 6 MeV, 3 cm away from the target for different targets (the Al thickness is 17 μm and the Cu is 20 μm), as indicated. The distributions at other proton energies display the same characteristics. In the experimental setup of the figure on the right side, the laser pointing shot-to-shot fluctuation is ~10 μm, similar to the laser FWHM of ~10 μm. This is quite lower than the diameter over which electrons eventually spread (see Fig.4.b). Repeating this shot yielded qualitatively the same result, ensuring the meaningfulness of the result shown despite the laser pointing fluctuation at the interface between the two materials.

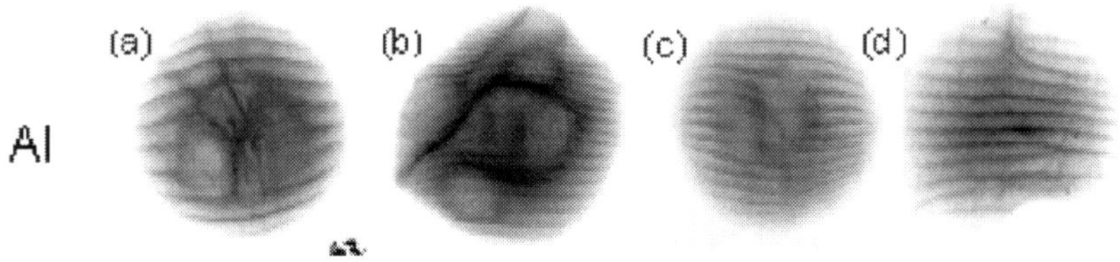

Figure 3: Proton angular distributions at 6 MeV, 3 cm away from the target for various target thicknesses of Al (a) 30 μm, (b) 40 μm, (c) 60 μm and (d) 90 μm.

Firstly, we have analyzed the influence of interfaces between conductors within a compound target. Figure 2 shows that such interfaces do not disrupt transport. In the figure are shown three different experimental setups, combing Al and Cu, i.e. putting Al in the front of the target (Figure 2.a) or at the back (Figure 2.b), and combining half (Figure 2.c). The proton imaging generally shows a smooth profile for the protons, independent of the source being the Al or Cu.

Now, if we use only pure materials but with varying thicknesses, we can observe filamented transport in thin targets even composed of conductors. Figure 3 shows that, for identical laser irradiation conditions, the electron beam reaching the target rear-surface is more and more perturbed when the Al thickness is reduced. In Figure 3 we use Al targets, the rear-surface of which has been imprinted with regular grooves. When going to thinner targets, we observe increasing small-amplitude modulations of the electron sheath that induce local perturbation of the regular modulation of the proton beam. It is important to note that the observed perturbations are local deformations of the sheath and not multiple separated filaments since we are able to follow, even distorted, the regular line pattern through the perturbations. For the thinnest Al target we used (30 μm), the modulations are strong enough to locally induce a strongly reduced proton dose. When going to 60 μm thick Al target there is still a strong central modulation (see Figure 3.c) that eventually disappears for a 90 μm thick target where only a smooth proton beam (as in Figure 3.d) is observed.

Under the same irradiation conditions, the proton beam produced from a 48 μm thick Au targets (i.e. in between the cases of Figure 3.b and c) does not display any perturbation [6, figure 1 therein]. This shows that transport behaves differently depending on the Z and the conductivity of the conductor material. Figure 4.a shows that the maximum proton energy produced from targets of different metal materials but of same thickness is similar, i.e. that the energy density of the hot electrons does not vary appreciably if the target material is changed. This indicates that in Figure 3, it is the small-scale variations within the angular distribution of the hot electrons at the target rear surface that is modified, and not the electron energy density.

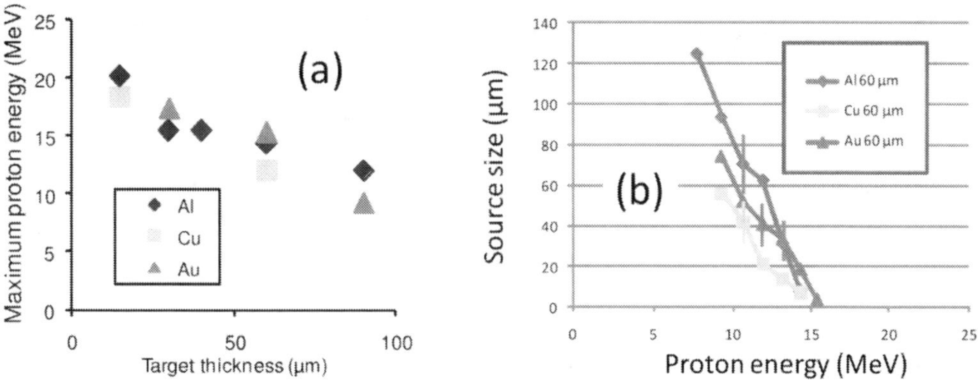

Figure 4: (a) Maximum proton energy as function of the target thickness for Al, Cu and Au. (b) Source size deduced from grooves imprinted on the rear target surface for different materials. It is deduced by counting the number of grooves in each RCF layer. The error bars on the source size, displayed only on a few points for an easier reading of the figure, take into account the uncertainty on the groove spacing.

Figure 4.b shows moreover that the proton source size, taking into account the error bars (not displayed for all points, but ~20 %), can be considered roughly consistent for different materials, i.e. the size of the zone over which the hot electrons spread at the target rear surface, depending on the target material if the thickness does not vary appreciably. This, together with Figure 4.a, hints at the fact that the varied modulations observed after hot electron transport in different materials is indeed related to small-scale modulations of the hot electron front and not to overall different spread or damping of hot electron energy.

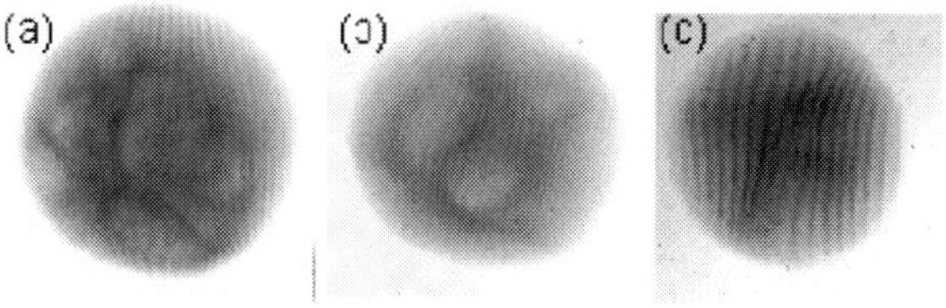

Figure 5: Proton angular distributions at 6 MeV, 3 cm away from the Al 40 µm target for laser intensities (a) I_{max}, (b) $I_{max}/1.8$ and (c) $I_{max}/4$.

The modulations in the sheath are reduced with the laser intensity, as shown in Figure 5. For these shots we reduced the laser intensity by reducing the laser energy. We obtained similar results by enlarging the focal spot diameter or by lengthening the pulse duration – even if the obtained maximum proton energy did change. In summary, proton beams accelerated from conductor targets of a given thickness displays completely different angular patterns depending on the target conductivity and/or Z (e.g. Au or Al), although the electrons have the same energy density (as is inferred from Fig.4). The proton beam from the Al target is perturbed while the one from the Au is not. Complementarily, Fig.5 implies that the appearance of the modulations, i.e. the growth of the observed instability, is very likely linked with high the energy density of the hot electrons transported through the target. There are at least three possible scenarios to justify the observed electron sheath modulations at high intensity and for thin targets:

1-the modulations could be produced at the laser interface and propagating through the target (this statement is supported by experiments and modeling [9])

2-the modulations could be generated during the electron transport, e.g. due to some propagation instabilities as shown in some simulations [10]

3-the modulations could be generated at the rear vacuum interface, due to strong induced magnetic fields that appear on the target rear surface [11]

We are currently performing simulations in order to analyze the propagation of these modulations within the target and to see when and over which range the modulations are washed out, or generated. This will be the object of future investigation.

We acknowledge the expert support of the LULI laser team. This work was supported by grant E1127 from Région Ile-de-France and UNR grant DE-FC08-01NV14050 and the projects ELI and SPARX-FEL.

References

[1] P. Antici *et al.*, *Phys. Rev. Lett.* **101**, 105004 (2008)
[2] S. Hatchett *et al.*, *Phys. Plasmas* **7**, 2076 (2000) ; S. C. Wilks *et al.*, *Phys Plasmas* **8,** 542 (2001)
[3] E. Clark *et al.*, *Phys. Rev. Lett.* **84**, 670 (2000), S. Gaillard *et al.*, *Phys. Rev. Lett.* **96,** 249201 (2006) and E.L. Clark *et al.*, *Phys. Rev. Lett.* **96,** 249202 (2006)
[4] R. A. Snavely *et al.*, *Phys. Rev. Lett.* **85**, 2945 (2000) ; A. Maksimchuk *et al.*, *Phys Rev Lett.* **84,** 4108 (2000)
[5] J. Fuchs *et al.*, *Phys. Rev. Lett.* **94**, 045004 (2005)
[6] J. Fuchs *et al.*, *Phys. Rev. Lett.* **91**, 255002 (2003)
[7] M. Roth *et al.*, *Phys Rev ST-AB* **5**, 061301 (2002)
[8] N. V. Klassen *et al.*, *Med.Phys.* 24, 1924 (1997)
[9] T.E. Cowan et al., *Phys. Rev. Lett.* **92**, 204801, (2004); H. Ruhl et al., *Phys. Plasmas* **11**, L17 (2004); M. Schollmeier *et al.*, *Phys Plasmas* **15**, 053101 (2008)
[10] L. Gremillet *et al.*, *Phys. Plasmas* **9**, 941 (2002)
[11] A. Kemp *et al.*, *Phys. Rev. E* **75**, 056401 (2007)

Simple collision operators for direct Vlasov simulations of laser plasma interaction and transport

T. D. Arber[1] and N. J. Sircombe[2,1]

[1]Centre for Fusion, Space and Astrophysics, Department of Physics, University of Warwick, Coventry, U.K.
[2]AWE Plc, Aldermaston, Reading, Berkshire, U.K.

E-mail: T.D.Arber@warwick.ac.uk

Abstract. Non-local electron transport effects have a direct influence on the compression of cryogenic targets in laser driven ICF and target heating in high energy density experiments. There is a growing need for self-consistent models of laser plasma interactions coupled to non-local transport. We present a direct Vlasov solver that includes multiple species and a simple collision operator. This BGK model operator - which conserves particle density, energy and momentum – is fully implicit. For collisionless plasmas it has been shown that a double layer may be formed in which an accelerated, kinetic ion population satisfies the zero current condition. Here we extend this result to collisionalities of interest to laser driven ignition to assess the validity of nonlocal electron transport models based on fluid ions.

1. Introduction

Recent experimental campaigns on the OMEGA laser[1] have again highlighted the importance of non-local electron transport effects for direct drive of cryogenic ignition targets. There is therefore a growing need for self-consistent models of laser plasma interactions coupled to non-local transport. This is true both for long pulses where flux limiters may be employed and short pulses where fast electron transport may be inhibited. Considerable progress has been made[2,3] using traditional hydro models of the laser plasma interactions coupled to a non-local electron transport model based on a BGK[2], collision operator. However, these models cannot describe the detailed kinetics of fast electron generation and transport. To obtain a fully consistent description of the electron dynamics one must resort to a Fokker-Planck treatment. These have been successful in modelling many aspects of hot electron transport; see for example[5]. To be computationally tractable these full Fokker-Planck treatments rely on an expansion in spherical harmonics, or similar. To accurately model the laser plasma interaction region the Vlasov equation is usually solved on an Eulerian phase space grid. Currently the inclusion of the full Fokker-Planck term in such Eulerian grid based codes is often too computationally expensive. There is therefore a requirement for a simplified approach to the collision operator.

We present a direct Vlasov solver[6] that includes multiple species and a simple collision operator, based on the approach taken in reference [2], in which the collision frequency is energy dependent. This BGK model operator, which conserves particle density, energy and momentum, is fully implicit. While the fully kinetic treatment can only be applied to a smaller domain than was used in fluid models[1] the electric field is solved for self-consistently. In the fluid limit one has to determine the electric field under the assumption of zero net current for the electrons only. For collisionless plasmas

Journal of Physics: Conference Series **244** (2010) 022017 doi:10.1088/1742-6596/244/2/022017

it has been shown[7] that a double layer may be formed in which an accelerated, kinetic ion population satisfies the zero current condition. Here we extend this result to collisionalities of interest to laser driven ignition to assess the validity of nonlocal electron transport models based on fluid ions.

2. BGK Collision Operator

We solve the relativistic Vlasov-Poisson system with one spatial and one momentum dimension. Vlasov's equation for the distribution function f_j is given by:

$$\frac{\partial f_j}{\partial t}+\frac{u}{\gamma}\frac{\partial f_j}{\partial x}+\frac{q_j}{m_j}E\frac{\partial f_j}{\partial u}=C(f_j)$$

where $f=f(\mathbf{x},\mathbf{u})=f(x,u)$, $u=\gamma v$, and $\gamma=\sqrt{1+u^2/c^2}$. Vlasov's equation is solved using the VALIS code[6] which uses a split-Eulerian conservative scheme based on the PPM method[8]. For speed, simplicity and scalability the electric field is updated through Ampere's law rather than Poisson's equation. For the ions the collision term is given simply by $C(f_i)=-\nu_{ii}(f_i-F_i)$ while for the electrons it is given by $C(f_e)=-\nu_{ee}(f_e-F_{e1})-\nu_{ei}(f_e-F_{e2})$. Following the treatment devised by Mannheimer *et al.*[2] we use the high and low speed collision frequencies from the NRL Plasma Formulary[9] analytically matched. For example, for the electron-electron collisions we use $\nu_{ee}=2\nu_e\left(1+0.75\left(m_e v/2k_BT_e\right)^{3/2}\right)^{-1}$ and expressions for $_{ei}$ and $_{ii}$ are given in reference [2]. The Krook operator is treated implicitly so that for the ions, $f_i^{n+1}-f_i^n=\nu_{ii}\Delta t(f_i^{n+1}-F_i)$ giving $f_i^{n+1}-f_i^n=\nu_{ii}\Delta t\left(f_i^n-F_i\right)/(1+\nu_{ii}\Delta t)$. By taking velocity moments of this latter expression it is clear that the Krook operator can be made to conserve energy, mass and momentum by a suitable choice of the distribution functions F_i, F_{e1} and F_{e2}. For the ions the requirement is that F_i satisfies the set of equations defined by $\int \nu_{ii}\left(1+\Delta t\nu_{ii}\right)^{-1}v^n\left(f_i^n-F_i\right)du=0$ where n = 0, 1 or 2. The Krook operator must also be consistent with the relaxation towards a Maxwell-Jüttner distribution. Hence the procedure is to assume $F_i=a\exp(-(1+(u-b)^2)/d)$ and find the set of (a,b,d) which satisfy the conservation of mass, momentum and energy. This only requires knowledge of f_i^n which is of course available at the start of each step so the specification of F_i is complete. This is a simplification over the full Fokker-Planck treatment but has the advantage of being computationally tractable for large grids and mobile kinetic ions.

3. Test Results

An initial tanh profile for the electrons (from 400 eV down to 100eV) was used for testing against previous VFP work. These tests have stationary ions (Z=4) and an electron number density of $2.97\times10^{26}/m^3$ so that there are 100 particles per Debye sphere for the 100 eV electrons. The computed heat flux compared to Spitzer-Harm is plotted in Figure 1. This is after 1000 plasma periods for the hot electrons. The figure shows the computed heat flux from the simulation divided by the free streaming flux. Since for the Spitzer-Harm conductivity this quantity is a function only of mean free path over scale length this Spitzer-Harm value is also plotted as a solid line. This shows heat flux limiting of ~0.1. The computed heat flux is multi-valued when plotted in this way. The lower section of the dashed curve corresponds to the hot side of the temperature profile and exhibits thermal flux limiting. The upper section corresponds to the cold side of the simulation and exceeds the Spitzer-Harm value due to the contribution from hot electrons from the higher temperature region. These results are in qualitative and quantitative agreement with previous simulations based on the full Fokker-Planck collision operator[5].

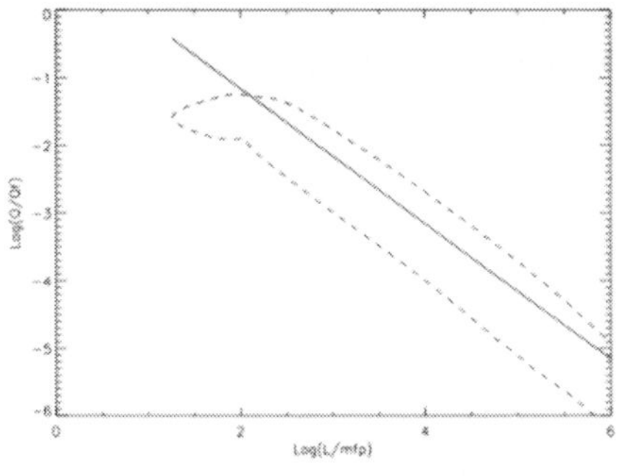

Figure 1: Ratio of calculated electron heat flux (Q) to the free streaming limit (Qf) as a function of scale-length over electron mean free path. Solid line is the Braginskii (Spiter-Harn) predicted heat flux. The dashed line is the computed heat flux

4. Ion Kinetic Effects on Fast Electron Transport

Previous studies of flux limited thermal transport have ignored ion kinetic effects. It is known that hot electron populations in solar coronal flares are capable of triggering thermal fronts[7]. In these the initial flux of hot electrons generates a polarization electric field. This draws a return current of cold electrons until there is no net current. For sufficiently large initial temperature jumps the hot electrons draw a cold electron return current which triggers Ion Acoustic Turbulence (IAT) in the hot electron region. The hot electrons must now overcome the polarization electric field and scattering off IAT to conduct energy from the hot region. It is possible that this suppression of electron current allows an ion current to flow through the polarization field region. The net effect is a zero-current double layer in which electron heat transport is completely suppressed while initially cold ions are accelerated up to hot electron energies.

The theory and simulations[7] of solar coronal thermal fronts are based on collisionless ions and electrons. Here we assess whether a similar process might occur in plasma conditions relevant to laboratory laser plasma experiments by using the BGK model collisions described above. Initial conditions are a 300eV background ion distribution and density of 10^{23} cm^{-3}. These initial tests are for a hydrogen plasma. The electron temperature drops from a hot population down to 300eV. We vary the hot electron temperature from 3keV up to 300 keV and also vary the scale-length of the temperature drop so that a range of mean free path over scale lengths are covered.

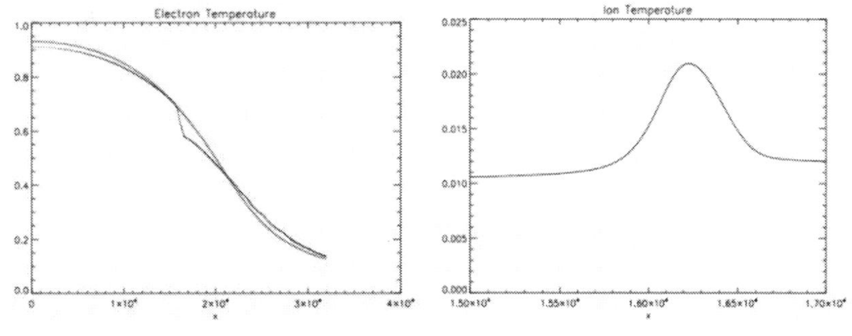

Figure 2: Hot electrons at 3 keV with typical mean free path over temperature scale length of ~0.4. The solid black lines are the simulated electron and ion kinetic energies after 5000 plasma periods and the red line is the same simulation with immobile ions.

When the hot electrons are at 3keV and the ratio of mean free path to scale length is ~0.05 the ions have no effect on the electron temperature profile. At 3keV with a mean free path to scale length ratio of ~0.4 the ion motion is sufficient establish a small temperature jump in the electron profiles. These results are shown in Figure 2. In all figures temperatures are in units of the initial hot electron temperature and lengths are in Debye lengths. At higher hot electron temperatures (300 keV) the cold electron return current is sufficient to trigger IAT despite the presence of collisions. When this happens the electron heat flux becomes almost completely suppressed and the ions begin to be accelerated through the polarization field. This can be seen in Figure 3, which repeats Figure 2, but for hotter electrons. At the thermal front the electron flux is orders of magnitude less than the free streaming limit and ions are accelerated to 40% of the initial hot electron energy.

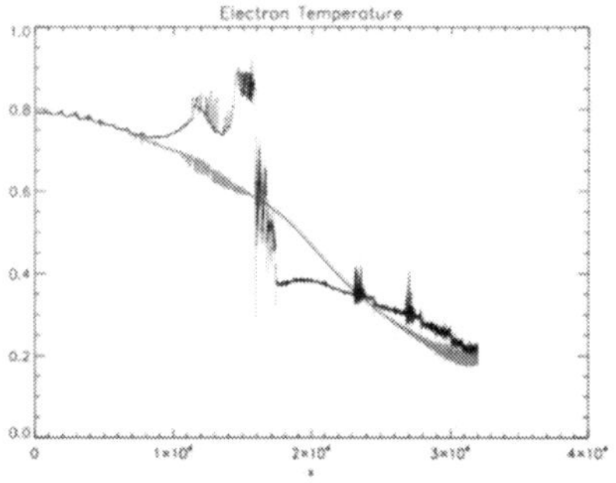

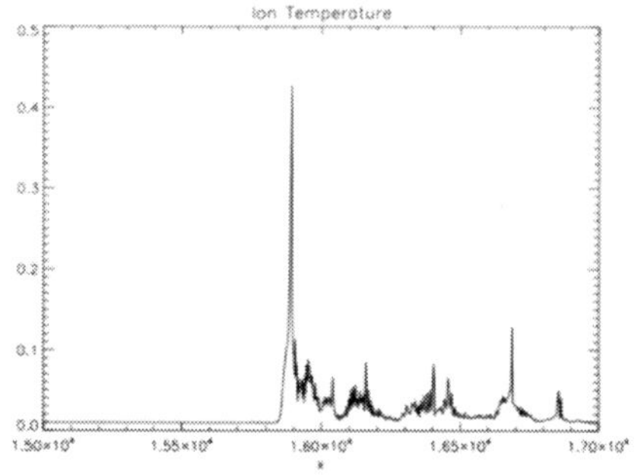

Figure 3: Hot electrons at 300 keV with typical mean free path over temperature scale length of ~1000. The solid black lines are the simulated electron and ion kinetic energies after 5000 plasma periods and the red line is the same simulation with immobile ions.

5. Conclusions

We have solved the 1D1P Vlasov equation with a BGK operator. The Maxwellian to which the BGK operator will relax is determined on each time-step so as to conserve number density, momentum and energy. This is therefore a highly simplified model but it does have relaxation on the correct time-scale; conserves all fluid quantities required for collisions; satisfies the H-theorem; and has been shown to accurately reproduce flux limited thermal conduction calculations from full Fokker-Planck treatments.

Applying this code to the possibility of thermal fronts in laser plasmas showed that for mean free paths of ~0.05 the ion motion was irrelevant. For a mean free paths of ~0.4 ion motion does introduce a step in the electron temperature profile but collisions are still sufficient to suppress a thermal front. With hot electrons of 300 keV and large mean free paths the collisions are no longer sufficient to suppress kinetic effects and the cold electron return current triggers IAT. This in turn inhibits further hot electron conduction and allows ion acceleration through the polarization field. The final state is a current free double layer with completely suppressed electron thermal transport and accelerated ions.

These simulations are of course highly idealized but do suggest that thermal fronts may play a role in short pulse laser heating. To verify if this is indeed true will require modelling laser electron heat input, increased ion mass and charge and possibly two spatial dimensions.

6. References

[1] Goncharov V N et al., *Phys. Plasmas*, **15**, 056310 (2008)
[2] Manheimer W et al., *Phys. Plasmas*, **15**, 083103 (2008)
[3] Colombant D and Manheimer W, *Phys. Plasmas*, **15**, 083104 (2008)
[4] Bhatnagar P L, Gross E P, and Krook M, *Phys. Rev.* **94**, 511 (1954)
[5] Bell A Ret al., *Plasma Phys. Contril. Fusion* **48**, R37 (2006)
[6] Sircombe N J and Arber T D, *J. Comp. Phys.*, **228**, 4773 (2009)
[7] Arber T D and Melnikov V F, *Astrophysical Journal*, **690**, 238 (2008)
[8] Colella P and Woodward P R, *J. Comp. Phys.*, **54**, 174 (1984)
[9] Huba J D 2006 *NRL Plasma Formulary* (Washington D. C. : Naval Research Laboratory)

Effect of gold nano-particle layers on ablative acceleration of plastic foil targets

L J Dhareshwar[1], N K Gupta[1], S Chaurasia[1], P Ayyub[2], N Kulkarni[2], J Badziak[3], T Pisarczyk[3], A Kasperczuk[3], P Parys[3], M Rosiński[3], J Wołowski[3], E Krousky[4], J Krasa[4], K Masek[4], M Pfeifer[4], J Skala[4], J Ullschmied[4], A Velyhan[4], D Margarone[4], A Mezzasalma[5], and P Pisarczyk[6]

[1]Bhabha Atomic Research Centre, Mumbai, India
[2]Tata Institute of Fundamental Research, Mumbai, India,
[3]Institute of Plasma Physics and Laser Microfusion, 01-497 Warsaw, Poland
[4]PALS Research Centre ASCR, 1822 Prague 8, Czech Republic
[5]University of Messina, Italy
[6]Warsaw University of Technology, ICS, 00-661 Warsaw, Poland

e mail: dharesh@barc.gov.in

Abstract: Presence of nano-particles on target surface has been observed to lead to increased laser absorption of laser pulse in plasma. Therefore, a coating of nano-particles on foil targets could lead to an enhanced ablative acceleration. The work presented in this paper concerns this possibility. The results of experiments performed with PALS laser system (125 J, ~250 ps at 1.3 μm) with a focused intensity of about 10^{14} W/cm^2 are presented. 15 μm thick Polyethylene teraphthalate $(C_{10}H_8O_4)_n$ or PET foils show an almost 40% increase in target movement when coated with a layer of gold nano-particles. Comparison between targets with coating of bulk gold and nano-gold shows about 15% higher target movement in gold nano-particle coated PET targets as compared to bulk gold coating. This result is a clear indication of enhanced laser energy absorption in targets with nano-structured surface of gold. We also present evidence to show the effect of nano-particle coating on lateral thermal conduction.

Introduction

Laser plasma interaction in nano-particle targets have been widely studied in the recent past [1-4] with high intensity ultra short laser pulses. Increased soft and hard x-ray emission has been reported in nano-particle coated and surface structured solid targets as compared to plane polished targets, by several authors. There have been studies on emission of highly energetic charged particles from targets with such surface structures. Several physical processes have been attributed to this effect with a consequent indication of higher laser to plasma coupling efficiency [5]. Most of these reports concern the interaction of 100-800 fs pulses with intensity -10^{15} to 10^{16} W/cm^2. High energy,long pulse (sub-nanosecond to nanoseconds) duration lasers are important for experiments on implosion physics of ICF targets. While working with longer pulses of hundreds of picoseconds duration, several issues arise. Does the presence of nano-particle layers on target surface affect absorption for longer pulses also? Even though the nano-particles are ablated away in the very early part of the laser pulse, does their presence make an overall contribution to increase absorption? One of the means of confirming this fact is to record motion of thin foil targets coated with nano-particle layers. An increased velocity of the laser accelerated thin foil targets would be a clear indication of higher absorption. The work presented in this paper concerns this possibility. It also addresses another equally important question as to whether the nano-particle layer on a solid target surface affects axial and lateral thermal conduction.

Journal of Physics: Conference Series **244** (2010) 022018 doi:10.1088/1742-6596/244/2/022018

Experiments

Experiments were performed with the Prague Asterix Laser System (PALS) with 250 J energy per pulse and 250 ps duration at a wavelength of 1.315 μm. The focused intensity on the target was about 4 x 10^{14} W/cm^2 over a spot diameter 400 μm. 15 μm thick Polyethylene teraphthalate $(C_{10}\,H_8\,O_4)_n$ or PET foils were used as base targets. A 2 μm thick layer of gold nano-particles was deposited on PET foils by dc magnetron sputtering (at 325V) from a 50 mm diameter solid gold disk at 10^{-1} torr pressure on a foil kept at 3000^0 K. PET Foils coated with a 2 μm thick layer of bulk gold were also used for comparison. The average particle size was obtained from a measurement of the x-ray diffraction line broadening and was about 15-25 nm. In some experiments, where the lateral thermal conduction was investigated, 100 μm thick gold disks of 10 mm diameter were coated with a 2 μm layer of Au nano-particles. Plasma density and velocity of the target foil were measured with three-frame interferometry/shadowgraphy set up using a 3ω probe beams having 2 ns, 5 ns and 8 ns delay with respect to the main laser beam. A CCD camera with digital data acquisition system supported by requisite software could yield 3-D plasma density profiles as well as shadowgrams of the accelerated foil target.

Results and Discussions

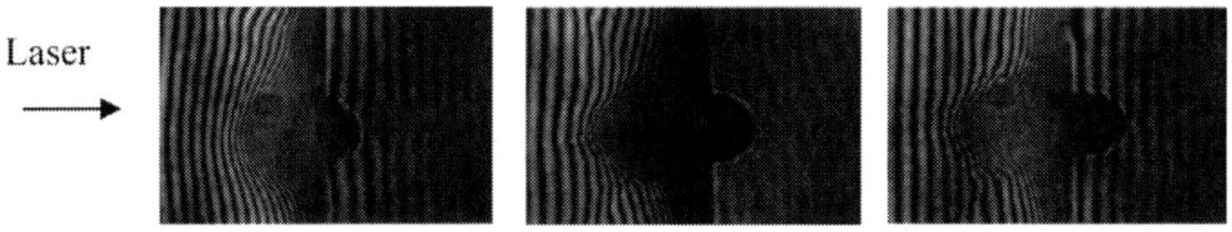

Figure 1. Interferograms of the foils (a) 15 μm PET foil, (b) 15 μm PET foil with 2 μm layer Au in bulk form and (c) 15 μm PET foil with 2 μm layer of Au in nano-particle form.

Figure.1shows the interferograms of the bare foils and foils coated with bulk as well as nano-particles of gold at identical laser energy of 125 J in three different laser shots recorded at probe beam delay of 8 ns with respect to the peak of the irradiating laser pulse. The increased motion of foils coated with bulk gold and nano-particles of gold (b) and (c) is evident as compared to the uncoated PET foil shown in (a). Au-nano-particle coated target shows the maximum movement, 40% higher as compared to uncoated PET. Also, this foil shows a 15% increase in foil motion as compared to PET foil with bulk gold coating. High Z coated plastic targets have been reported to have increased foil motion due to the contribution of X-rays. However, the difference in foil motion for bulk Au and Au-np clearly shows enhanced laser absorption due to presence of nano-particles. Target foil velocity in all cases was measured from a graph of foil motion versus probe time delay and the velocity for all three cases is given in the table below.

Table 1. Target foil velocity measurement for various targets at laser intensity 4 x 10^{14} W/ cm^2.

Target	Velocity (cm/sec)
15μm PET foil	6.25 x10^7
15μm PET foil with 2μm layer of Au bulk	7.5 x10^7
15μm PET foil with 2μm layer of Au -np	8.75 x10^7

Lateral thermal conduction losses play a major role in laser to target coupling efficiency and the hydrodynamic efficiency of laser driven targets in nano-second and even sub-nanosecond duration laser pulses. Lateral thermal conduction in long laser pulses contributes to smoothening of laser non-uniformities, but, it can result in a reduction of laser to target coupling and thus lowering of ablation pressure. The laser was focused to a spot diameter of 200 μm, on 100 μm thick and 10mm diameter gold disk targets. Some of the targets were plane polished whereas the others were coated with a 2 μm thick layer of Au-np for comparison. Figure 2(a) shows the interferograms of the plane Au target without a coating of Au-np, at 2ns and 8ns delay after peak of laser pulse. (b) are the interferograms for Au target with a Au-np coating at 2ns and 8ns delay. Both shots were recorded at laser energy of 135J. It is evident that for Au-np coated target, the plasma diameter is much smaller than for uncoated Au target which indicates a reduction in lateral energy transport for targets having a layer of nano-particles on the surface. It is also interesting to note an intense plasma emission over a large volume within the target seen in 2(b) in the case of Au-np layer which is absent in (a). These experiments thus indicate that the presence of Au-np coatings on a solid target surface affects plasma hydrodynamics.

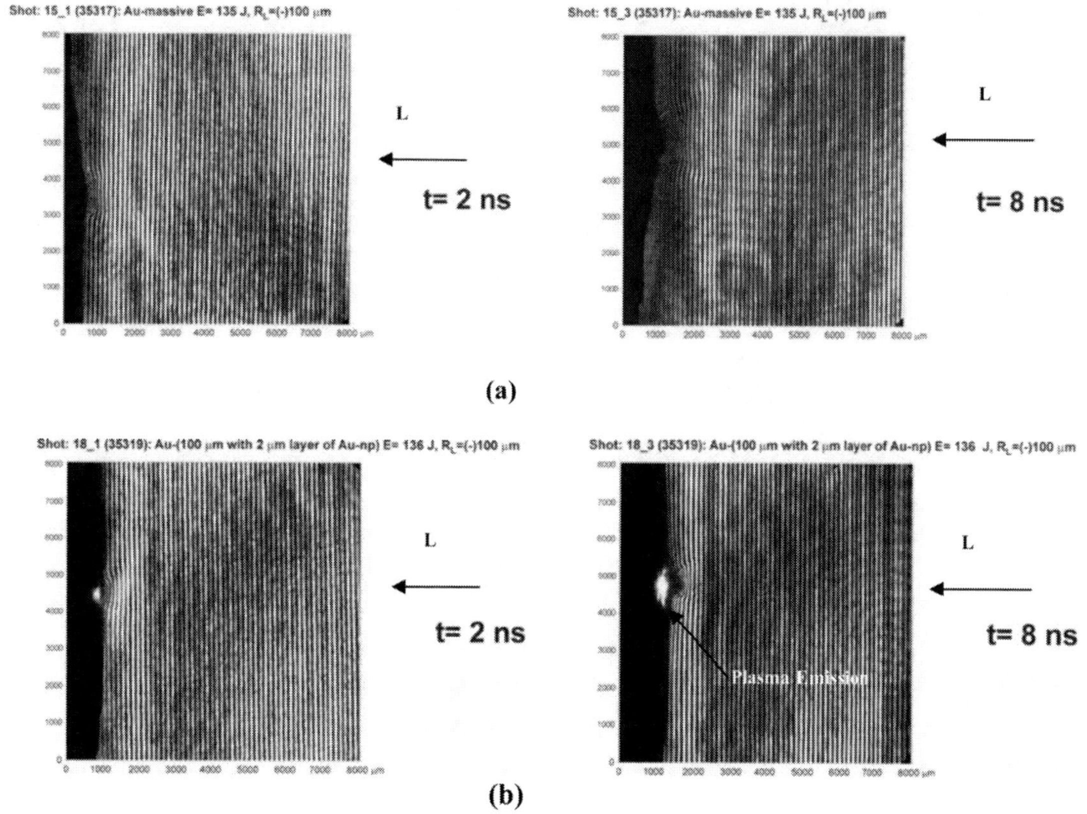

Fig.2. Interferograms of a 10mm diameter, 100μm thick gold disk. (a) Au target without coating,(b) Au target with a 2μm over layer of Au-np coated on the surface.

A simple analytical model is invoked to explain the results. The geometry of the target shown in the Figure 3a has the nano-particle layer with a random distribution of gold particles and void spaces as shown in (b). The gold nano-particles are treated as inclusions and PET is host medium. Such a medium can be characterized by an effective dielectric function.

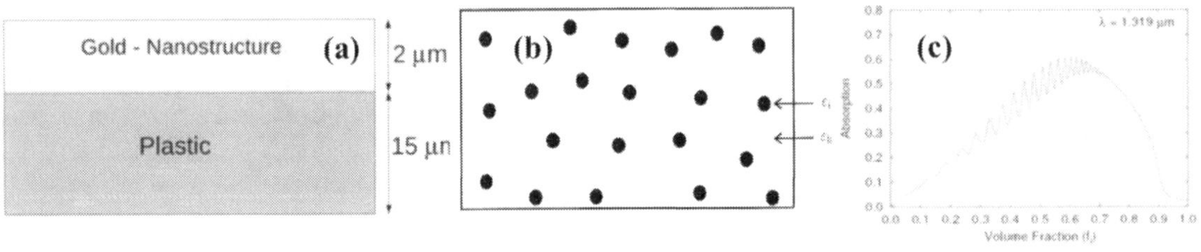

Figure 3. Lateral (a) and top (b) view of the composite target. Dots denote the nano-particles. (c) Shows the fractional laser absorption as a function of volume fraction of solid gold nano particles on solid PET foil.

Laser absorption in the nano-structured targets was simulated using Maxwell-Garnett effective medium approach to calculate the effective dielectric constant of the composite. The effective dielectric constant is then used to calculate the laser absorption using matrix method [6]. For the Maxwell-Garnett model, the effective dielectric function is expressed in the form

$$\varepsilon_{eff}(\omega) = \varepsilon_h(\omega)\frac{1+2\eta(\omega)f_i}{1-\eta(\omega)f_i} \qquad (1)$$

In this equation, f_i is the volume fraction of the inclusions and $\eta(\omega)$ is given by the expression

$$\eta(\omega) = \frac{\varepsilon_i(\omega)-\varepsilon_h(\omega)}{\varepsilon_i(\omega)+2\varepsilon_h(\omega)} \qquad (2)$$

where $\varepsilon_h(\omega)$ is the dielectric function of the host and $\varepsilon_i(\omega)$ is the dielectric function of the inclusions. In these calculations, PET is considered to be a host and gold nano-particles as inclusions. At the laser wavelength used in experiments, the complex refractive index of gold is 0.419 + i 8.42 and refractive index for PET is 1.5 [7].The dielectric constant for PET with gold nano-particles is related to refractive index as, $\varepsilon = \eta^2$.

In Figure 3(c), the variation in fractional laser light absorption by the solid target as a function of volume fraction (f_i) of the gold nano particles is shown. It should be noted that, $f_i = 0$ corresponds to no inclusions of nano-particles and therefore no absorption since PET is a dielectric. On the other hand, when $f_i = 1$, it is equivalent to pure bulk metallic gold target which is expected to have a small absorption 0.023. However, presence of gold nano-particles enhances the laser absorption to more than 60% as compared to bulk gold target. It is also observed that there is an optimum volume fraction of about 0.6 of the nano-particles for which laser absorption peaks. We may note here that in this simple model, only metallic dielectric constants are used and a complete hydrodynamic simulations of the plasma generated have not been performed. This requires a model for space and time dependent dielectric constant as plasma evolves in time and space.

References :

[1] McPherson A, *et al.*, 1994 *Nature (London)* 370, 631.
[2] Ditmire T, *et al.*, 1997 *Nature(London)*, 386, 54.
[3] Kulcsar G, *et al.*, 2000 *Phys. Rev. Lett.* 84, 5149.
[4] Rajeev P P, *et al.*, 2003 *Phys. Rev. Lett.* 90, 115002.
[5] Link S,Wang Z L,El-Sayed M A, 1999, *Jour of Chem.B* **103**,3529
[6] Born M and Wolf E, Principles of Optics, 1970,Oxford,UK, Pergamon press.
[7] Palik E D, Handbook of Optical Properties of Solids, 1985,New York, Academic Press.

Analyses of laser-plasma interactions in NIF ignition emulator designs

D. E. Hinkel, D. A. Callahan, N. B. Meezan, L. J. Suter,
C. H. Still, D. J. Strozzi, E. A. Williams, and A. B. Langdon
Lawrence Livermore National Laboratory, Livermore, CA, USA

hinkel1@llnl.gov

Abstract. The National Ignition Campaign is currently conducting energetics experiments at the National Ignition Facility (NIF). These experiments directly test all aspects of ignition hohlraum performance. An important aspect of good performance is understanding and mitigation of laser-plasma interactions, which enables laser coupling to the target. The hohlraum energetics target and pulse-shape are designed to produce an ignition-hohlraum-like plasma to test laser-plasma interactions. The pre-shot laser-plasma interaction predictions for the first energetics targets (one room temperature target, and one cryogenic target) are presented here. These findings correlate well with the experimental data, namely that the primary concern is stimulated Raman scatter of the inner beams deep in the target. Such scatter is mostly re-absorbed, but modifies the energy deposition profile and thus the symmetry. Stimulated Brillouin scatter is at low levels, both in pre-shot predictions and in the experiments.

1. Introduction

Currently, we are conducting an energetics campaign at the National Ignition Facility (NIF). This campaign is designed to directly test all aspects of ignition performance. The goals of this campaign are to demonstrate: (1) acceptable radiation drive, (2) acceptable backscatter levels, (3) acceptable hot-electron preheat, (4) implosion symmetry tuning, and (5) power flow accounting in the target. Laser-plasma interactions (LPI) directly impact all of these goals, and thus are a main concern.

There are two distinct types of laser-plasma interactions in ignition targets: single-beam LPI, and multiple-beam LPI. Single-beam LPI includes effects such as: (1) backscatter, where the incident laser light resonantly scatters off electron plasma waves (stimulated Raman scatter, or SRS) or ion acoustic waves (stimulated Brillouin scatter, or SBS); (2) filamentation, where intensity modulations create density depressions that can refract and intensify the light; (3) refractive intensification, where the laser light is intensified by beam refraction over the capsule; and (4) re-absorption of scattered light, which modifies the energy deposition profile. These single-beam LPI phenomena will be discussed in this paper. Multiple-beam LPI, such as power transfer from one beam to another via a mutually shared plasma wave also occurs in these targets. Such phenomena are the subject of a paper by P.A. Michel *et al.* [1] in these conference proceedings.

NIF emulator designs[2] are 78% of ignition scale, and use 500-600 kJ of laser energy. The first emulator target that was shot as part of this experimental campaign was the room temperature emulator. It is composed of a gold hohlraum wall lined on the interior with a 60% Au/ 40% B liner about 1 μm in thickness. This liner is used to suppress SBS in the wall plasma blow-off. The capsule is filled with propane, and has a CH ablator. The target gas fill is neopentane. The laser entrance hole (LEH) is 64% of the target radius. For this target we predict (from beam propagation simulations with pF3D[3]) 0.5% SRS generated over the capsule at the end of peak power, which, after cross-beam amplification in the LEH would suggest a reflected power of 1-2%. We find negligible SBS on both inner and outer beam quads, and negligible SRS on the outer beam quads.

The second type of emulator shot at NIF was a cryogenic target. This target is of the same dimensions as the room temperature target, with the same wall composition. The capsule is filled with He, and has a CH ablator. The gas fill is H_4He. For this target, we predict (from beam propagation simulations) ~3-4% SRS generated over the capsule, which, after cross-beam amplification in the LEH would suggest a reflected power of ~ 20-30%. We find negligible SBS on both inner and outer beam quads, and negligible SRS on the outer beam quads for this target as well. There is no evidence of filamentation in these beam propagation simulations. There is refractive intensification of the inner beam quads of the cryogenic target, which results in a larger reflectivity for the cryogenic target. Also in the cryogenic target, re-absorption of SRS as it propagates out of the target somewhat modifies the energy deposition profile.

The remainder of this paper details the analyses that led to these conclusions.

2. LPI assessment at the hydrodynamic scale

We first assess LPI by calculating the gain exponents for stimulated backscatter amplification. The code *newlip*[4], developed by E. A. Williams at Lawrence Livermore National Laboratory, calculates the kinetic, steady state convective gain exponent for SBS and SRS along rays in the radiation-hydrodynamics simulations performed of the emulator targets[2]. This provides a distribution of gains across a given beam, as in the transverse direction the beam is interacting with different materials with transverse as well as longitudinal variations. We formulate the cumulative sum of the gain distribution to find the fraction of power above gain (FOPAG). Here we compare FOPAGS of the room temperature and cryogenic emulators.

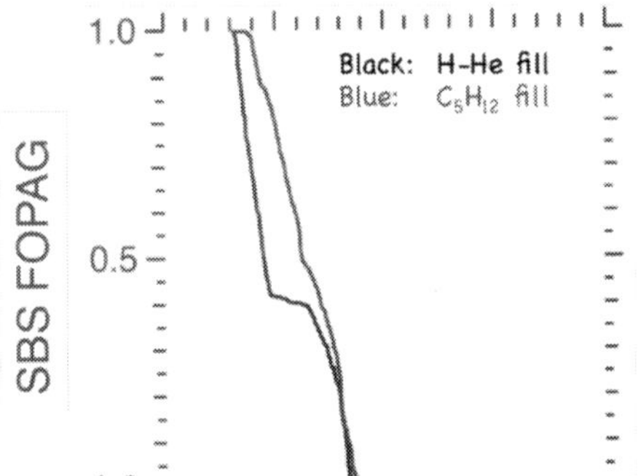

Figure 1. SBS FOPAGs for the 30^0 beams near the end of peak power. The blue curve is the room temperature target FOPAG, and the black is that of the cryogenic target. No ray sees a gain exponent above 9.

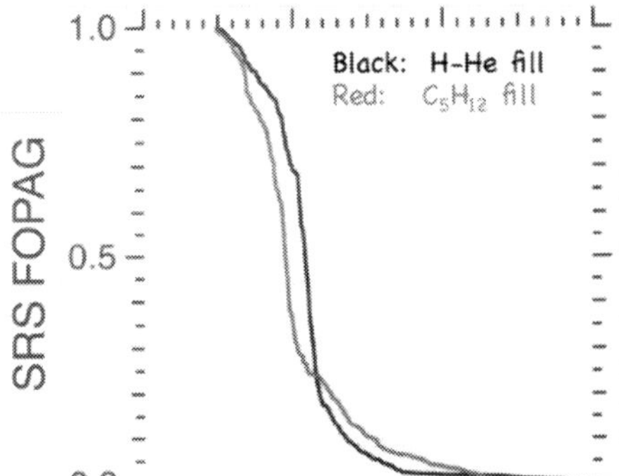

Figure 2. SRS FOPAGs for the 30^0 beams near the end of peak power. The red curve is the room temperature target FOPAG, and the black is that of the cryogenic target. Approximately 50% of the power is in gain exponents > 10.

Figure 1 is a plot of the SBS FOPAG along the 30^0 beams near the end of peak power. The blue curve is the FOPAG for the room temperature target, and the black curve is the FOPAG for the cryogenic target. The gains are quite similar in these targets. The slight difference at lower gains is where the rays propagate through the gas fill, C_5H_{12} for the room temperature target and H_4He for the cryogenic target (more strongly damped). No ray has a gain exponent above 10.

Similarly, Figure 2 is a plot of the SRS FOPAG along the 30^0 beams near the end of peak power. The red curve is the FOPAG for the room temperature target, and the black curve is the FOPAG for the cryogenic target. The gains are approximately equal at this time. Approximately 50% of the power is in gains above 10.

Along the outer beams at the end of peak power, SBS occurs primarily in the wall blow-off, which is composed of Au_6B_4 (B was added to suppress SBS). Here, no ray has a gain exponent above 8. For SRS, no ray has a gain exponent above 5. Thus our primary concern is SRS along the inner beams. To further analyze this, we perform beam propagation simulations with our massively parallel code, pF3D.

3. Beam propagation simulations

pF3D[3] is a laser beam propagation code that couples paraxial wave equations for the incident and reflected waves to a plasma fluid model. These simulations account for effects such as longitudinal and transverse gradients, laser speckle physics, refractive intensification, plasma self-smoothing, and spatial, temporal and polarization beam smoothing, to name a few. The incident light wave is depleted by an SBS coupling term, an SRS coupling term (depending on the reflected light wave and density fluctuation amplitudes), and a refraction term (that depends upon density fluctuations and the incident light wave amplitude). The reflected light waves are driven by the amplitude of their respective density fluctuations coupled to the incident wave. The acoustic and electron plasma wave density fluctuations satisfy respective wave equations driven by the ponderomotive force of the light waves. The background plasma is described by a standard, multi-material nonlinear fluid model that couples to the laser via ponderomotive pressure and inverse bremsstrahlung.

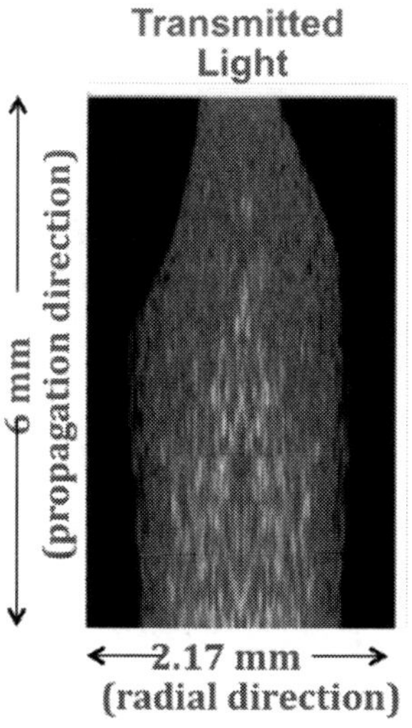

Figure 3. A plot of an azimuthal slice of the forward intensity pattern of a 30^0 quad at peak power for the cryogenic emulator. Laser speckles are evident at higher intensity (white). The propagation path is 6 mm long and the box is 2 mm in cross section. The beam enters the target 1/6 of the way through the box. It is channeled between the ablator and the wall plasma in the latter half of the box.

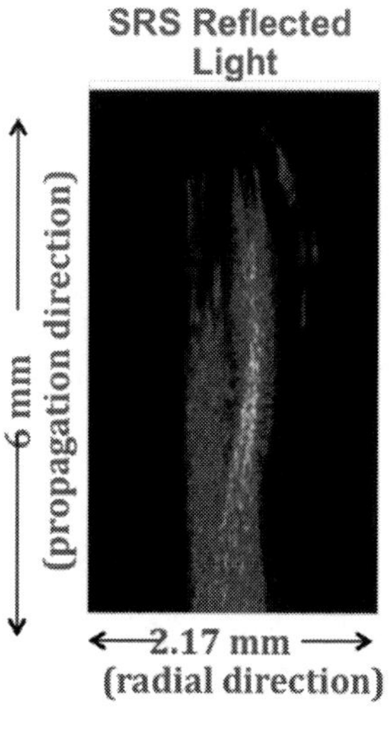

Figure 4. A plot of an azimuthal slice of the SRS intensity pattern of a 30^0 quad at peak power for the cryogenic emulator. SRS is generated near the end of the simulation box, up over the capsule. The longer wave-length SRS refracts differently than the incident light as it propagates out of the target. About 3-4% of the incident light propagates out of the front of the box as SRS light.

Figure 3 is a plot of an azimuthal slice of the forward intensity pattern of a 30^0 quad at peak power for the cryogenic emulator. As the beam enters the target (1/6 of the way through the box), it refracts off the LEH and propagates toward the wall. Halfway through the simulation box the beam begins to get channeled between the ablator plasma (dark plot of plot in top left half) and wall plasma (dark part of plot in top right half). There is evidence of refractive intensification for the cryogenic emulator, and this is what increases the reflectivity in the cryogenic emulator.

Figure 4 is a plot of an azimuthal slice of the SRS intensity pattern of a 30^0 quad at peak power for the cryogenic emulator. SRS is generated over the capsule in the gas fill plasma between the ablator and wall plasma. This longer wavelength light will undergo greater absorption (via inverse bremsstrahlung) as it propagates out of the target than the incident light underwent as it propagated inward (approximately 50% of the incident light is absorbed before it reaches the wall). It will also refract differently than the incident light, as evidenced in Figure 4. We find approximately 3-4% of the incident light is reflected out of the front of the simulation box as SRS light. This does not account for multiple beam effects near the LEH, which will further amplify the SRS reflectivity. We find negligible SBS in this simulation (<< 1%).

4. Summary

We have analyzed a variety of laser-plasma interactions for the first energetics experiments at NIF. These targets are below threshold for filamentation, and show low levels of SBS. This is in agreement with the experiments. We find SRS occurs on the inner beams, and is larger for the cryogenic emulator than for the room temperature emulator. This is also in agreement with the experiments. The SRS seed that is generated in these simulations (~ 0.5% for the room temperature target at peak power, and ~ 3-4% for the cryogenic target at peak power) will be further amplified at the LEH of the target by multiple beam effects, the subject of a paper by P. A. Michel *et al.* [1] in these conference proceedings.

5. Acknowledgments

The authors wish to thank the Multi-programmatic and Institutional Computing (M&IC)) department and the Advanced Simulation and Computing (ASC) program at LLNL for access to high-performance computing platforms. This work was performed under the auspices of the U. S. Department of Energy by the Lawrence Livermore National Laboratory under Contract DE-AC52-07NA27344.

References

[1] Pierre Michel, Laurent Divol, Nathan Meezan, Robert Kirkwood, Ed Williams, David Strozzi, Debbie Callahan, Siegfried Glenzer, and Larry Suter, these conference proceedings.

[2] N. B. Meezan, D. A. Callahan, L. Divol, S. N. Dixit, M. J. Edwards, D. H. Froula, P. A. Michel, J. D. Moody, S. V. Weber, K. Widmann, S. H. Glenzer, and L. J. Suter, these conference proceedings.

[3] R. L. Berger, C. H. Still, E. A. Williams, and A. B. Langdon, *Phys. Plasmas* **5**, 4337 (1998); C. H. Still, R. L. Berger, A. B. Langdon, D. E. Hinkel, L. J. Suter, and E. A. Williams, Phys. Plasmas **7**, 2023 (2000); D. E. Hinkel, D. A. Callahan, A. B. Langdon, S. H. Langer, C. H. Still, and E. A. Williams, *Phys. Plasmas* **15**, 056314 (2008).

[4] D. J. Strozzi, E. A. Williams, D. E. Hinkel, D. H. Froula, R. A. London, and D. A. Callahan, *Phys. Plasmas* **15**, 102703 (2008).

The Sixth International Conference on Inertial Fusion Sciences and Applications IOP Publishing
Journal of Physics: Conference Series **244** (2010) 022020 doi:10.1088/1742-6596/244/2/022020

Assessing the $2\omega_{pe}$ instability and other preheat considerations in ignition-scale hohlraums

W L Kruer[1], Nathan Meezan[2], S P Regan[3], Richard Town[2], David Strozzi[2], Scott Wilks[2], Edward Williams[2], Donald Meeker[2], and L S Suter[2]

[1] University of California, Davis, Davis, CA 95616 USA
[2] Lawrence Livermore National Laboratory, Livermore, CA 94551 USA
[3] Laboratory for Laser Energetics, Rochester, NY 14623 USA

wlkruer@ucdavis.edu

Abstract. In recent experiments[1] Sean Regan, *et. al.* for the first time observed the $2\omega_{pe}$ instability from window plasma in hohlraum targets. This instability can also operate[2] at peak power near the edge of the inner beams in the ablator plasma and near the edge of the outer beams in the liner plasma. Fortunately only a small fraction of the laser energy was estimated to be at risk. A more quantitative assessment of the energy at risk at peak power will here be given. We show that the instability threshold can be significantly reduced for laser beams with an angle of incidence of about 60 degrees due to the swelling of the laser field near its turning point. A simple model is given. It is also shown that for frequently cited plasma conditions, the Raman-scattered light wave can itself drive the $2\omega_{pe}$ instability. This effect is relevant for the nonlinear saturation of stimulated Raman scattering (SRS) and the resulting heated electron generation. Some estimates are given. Finally we conclude with a few remarks about hot electron preheat.

1. Introduction/Summary

The two plasmon decay instability can be operative in ignition scale hohlraums. The intensity threshold for this instability can be quite low and further reduced by overlapped beam effects discussed in direct drive applications[3-4] and by intensity swelling when the laser light is incident onto the quarter critical surface at ~60°. This instability is a particular concern since it can generate hot electrons with an effective temperature of ~70 keV, which are much more effective at preheating than are electrons with an effective temperature of ~30 keV (attributed to the stimulated Raman instability). To implode fuel to high density, preheating by high energy electrons needs to be minimized. These hotter electron distributions have indeed been observed in Omega experiments[1] in which gas-filled hohlraums are irradiated with a shaped laser pulse. Here these electrons were correlated with the measured emission near $3/2\omega_0$, a signature of the $2\omega_{pe}$ instability, and attributed to excitation of this instability near quarter critical density in the window plasma as it blows down to low density. Preheating by high energy electrons generated in the window plasma can be avoided by blowing up the window using lower intensity laser light.

The $2\omega_{pe}$ instability has been predicted[2] to be operative at other times and places; for example in the ablator plasma irradiated by the inner beams. A simple model for the laser energy reaching quarter-critical density with intensity above threshold has been developed and applied to monitor energy at risk to this instability. This energy at risk appears quite tolerable for a recent design for an ignition-scale hohlraum with a Be ablator. Finally, for common plasma conditions, the $2\omega_{pe}$ instability can also be excited by modest levels of Raman-scattered light when this frequency down-shifted light reaches its quarter-critical density. Various consequences are discussed, including increased preheat by a component of electrons heated to higher temperature.

2. Some general features of the $2\omega_{pe}$ instability

The $2\omega_{pe}$ instability[5-8] corresponds to the decay of an incident light wave into two electron plasma waves, a process which occurs near $.25n_{cr}$ (the critical density). The threshold intensity for this instability can be rather low even for .35μm laser light. For normally incident light, the gradient threshold intensity (I_G) is[8]

$$I_G \cong \frac{5.9x10^{15}T_{keV}}{\lambda_\mu L_\mu}\frac{W}{cm^2}, \qquad (1)$$

where T_{keV} is the electron temperature in keV and λ_μ (L_μ) is the laser wavelength (density scale length) in microns. . As an example, consider T_{keV}=2.5, λ_μ=.35 and $L_\mu=10^3$, giving I_G~$4x10^{13}W/cm^2$. The collisional threshold intensity (I_c) can also be rather low in low Z plasmas:

$$I_c \cong \frac{1.4x10^9}{\alpha^2}\left(\frac{Z\ln\Lambda}{T_{keV}^{3/2}\lambda_\mu^2}\right)^2\frac{W}{cm^2}, \qquad (2)$$

where Z is the charge state and lnΛ is the Coulomb logarithm. Here α is a function of how far above gradient threshold one is (α=1 when $I>>I_G$). Again considering a Be plasma with T_{keV}=2.5 and λ_μ=.35 and taking lnΛ=7 and α=1/3, I_c~$5x10^{13}W/cm^2$.

It's interesting to note that for an angle of incidence of ~60° the threshold intensity can be significantly reduced. Obliquely incident light turns at $n=n_{cr}\cos^2(\theta)$. As the light turns, its group velocity slows down and so the intensity locally swells. This swelling of the electric field is readily estimated using the Airy function solution for the propagation of a plane light wave obliquely incident with an angle of incidence θ onto a plasma with a linear density profile:

$$E_{max}^2 \cong E_{FS}^2 3.7\cos^2(\theta)(\omega L/c)^{1/3}. \qquad (3)$$

Here E_{max} (E_{FS}) is the maximum value (free space value) of the electric field of the laser light with frequency ω in a plasma with a linear density scale length L. Hence for θ=60° the electric field intensity swells by about $(\omega L/c)^{1/3}$. Typically $(\omega_0 L/c)^{1/3}>10$, giving rise to a significantly lower threshold intensity. It would be valuable to explore the reduction in threshold due to this swelling in simple experiments using disk targets. It has been found that the swelling of the electric field intensity of normally incident light near the critical density very significantly reduces the threshold for excitation of parametric instabilities by radiofrequency waves in the ionosphere [9].

It's possible that this swelling effect significantly contributes to the observed lower intensity threshold usually attributed to overlapping beam effects. For example, in the Regan experiments one of the three cones struck the hohlraum window at 60°. Likewise, it is suggested that the beam overlap in directly driven targets be scrutinized to see if one can minimize any beams with an angle of incidence near 60°. Perhaps this might be done by adjusting the beam pointing and focal spot sizes.

3. A ray trace model to monitor energy at risk to the $2\omega_{pe}$ instability

A very simple model based on ray tracing is used to estimate the energy at risk to the $2\omega_{pe}$ instability at other times and places in a holraum. When a ray strikes the $.25n_{cr}$ surface, one checks to see if its intensity is above both the gradient and collisional thresholds. If so, the energy remaining in this ray is counted as energy at risk. This model (without a collisional threshold and using overlapped beam intensity) was first used to help understand the window hot electrons and provides a simple early warning system for when the $2\omega_{pe}$ instability can become a concern.

As an example, Figure 1 shows the power at risk versus time for the inner beams (cones with angles of 23.5° and 30°) in a NIF ignition hohlraum design with a Be ablator driven with 1MJ laser energy in a shaped pulse. The energy estimated to be at risk in this design is about 4kJ, principally due to the 23.5° cone of beams. No energy at risk was found for the outer beams with cone angles of 44.5° and 50°. Since we expect[8] that the absorption efficiency of this instability is <10%, the fraction of the laser energy into $2\omega_{pe}$-generated hot electrons is estimated to be <.04% of the laser energy. This would be quite tolerable, since even for T_{hot}~70 keV, the preheat spec is an order of magnitude higher. This tool is continuing to be improved; for example, the thresholds are being refined.

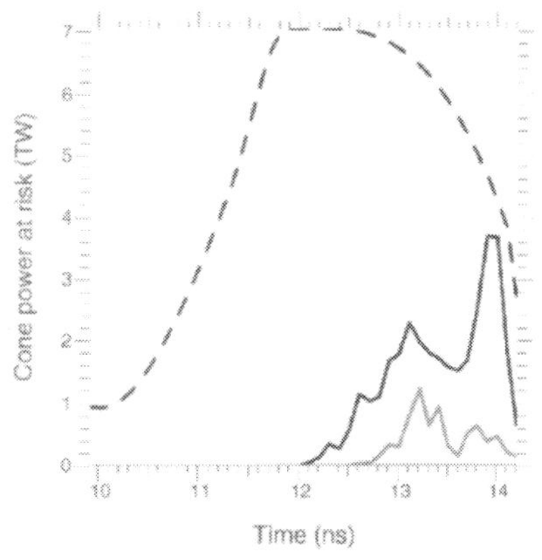

Figure 1: Power at risk for the $2\omega_{pe}$ instability in a NIF ignition design. The dashed curve is the power versus time in an inner quad; the black (red) curve is the power at risk for a 23.5° (30°) inner beam.

4. Excitation of other instabilities by Raman-scattered light

For conditions commonly accessed by the inner beams in an ignition-scale hohlraum, the Raman-scattered light can in turn excite the $2\omega_{pe}$ instability (and/or the Raman instability near $.25n_{cr}$). For example, consider Raman backscatter in a plasma with density of .1 n_{cr} and an electron temperature of 2.5keV. Then $\omega_{pe}=.316\omega_0$ and $\omega_{sc}=.633\omega_0$; i.e., $\omega_{sc}\sim 2\omega_{pe}$! For scattering at higher density $\omega_{sc} < 2\omega_{pe}$, but the scattered light still encounters its quarter-critical density as it propagates to lower density plasma. For even a modest level of scattering, the intensity of the Raman-scattered light (I_{sc}) can exceed the threshold for the $2\omega_{pe}$ instability. For example, assume a scattered intensity of $.2I_0$, where $I_0=5\times10^{14}$W/cm^2. Then $I_{sc}\sim10^{14}$W/cm^2, well above the $2\omega_{pe}$ threshold. If one notes that $\lambda_{sc}\sim.55\mu$m and takes a density scale length of 1mm, the gradient threshold intensity would be ~2.5×10^{13}W/cm^2, and the collisional threshold even less. Perhaps the most important consequence would be the generation of a population of heated electrons with a higher effective temperature (~70keV when the instability is well above threshold). These higher energy electrons are a preheat concern. It is also possible that this generation of other instabilities can help reduce the magnitude of the Raman scattering.

5. Concluding remarks on hot electron preheat

For ignition-scale capsules electrons with an energy greater than about 150-200 keV penetrate the ablator[10]. The fraction of the hot electron population with such energies is quite dependent on T_{hot}. One can tolerate about a factor of about 5 fewer hot electrons with a temperature of 70keV as compared with a temperature of 30 keV. It is clearly important to better understand the magnitude and

Journal of Physics: Conference Series **244** (2010) 022020 doi:10.1088/1742-6596/244/2/022020

dependences of T_{hot} (and the detailed heated energy distributions) as well as techniques for reducing T_{hot}.

Acknowledgements: This work performed under the auspices of the U.S. Department of Energy by Lawrence Livermore National Laboratory under Contract DE-AC52-07NA27344. Part of this work was performed under LDRL tracking number 08-ERD-017.

References

[1] Regan S *et al* (*Phys. Plasmas* in press))
[2] Kruer W L, 2007, *37th Anomalous Absorption Conference* Paper Po6
[3] Stoeckl C et al. 2003 *Phys. Rev. Lett.* **90**, 235002-1
[4]. Seka W *et al.* 2008 *Phys. Plasmas* **15**, 056312
[5] Liu C S and Rosenbluth M N 1976 *Phys. Fluids* **19**, 967
[6]. Simon A, Short R, Williams E A, and Dewandre T 1983 *Phys. Fluids* **26**, 3107
[7]. Afeyan B B and Williams E A 1997 *Phys. Plasmas* **4**, 3827 (ibid **4**, 3845 (1997))
[8.] Langdon, A B, Lasinski B F, and Kruer W L 1979 *Phys. Rev. Lett.* **43**, 133
[9] Wong A Y and Taylor R J, 1971 *Phys. Rev. Lett.* **27**, 644
[10] Meeker D, Suter L, Haan S and Robey H 2008 *Bull. Am. Phys. Soc.* **53**, 227

Journal of Physics: Conference Series **244** (2010) 022021 doi:10.1088/1742-6596/244/2/022021

Laser-plasma interaction physics in multi kilojoule experiments

C Labaune[1], S Depierreux[2], V T Tikhonchuk[3], D T Michel[1, 2], C Stenz[3], N G Borisenko[4], Ph Nicolaï[3], S Hüller[5], D Pesme[5], P Loiseau[2], P-E Masson-Laborde[2], M Casanova[2], M Grech[6], G Riazuelo[2], C Riconda[1], S Weber[3], S Darbon[7], R Wrobel[2], E Alozy[2], A Casner[2], C Meyer[7], P Romary[7], G Thiell[7], W Nazarov[8], J Limpouch[9]

[1] LULI, Ecole Polytechnique, 91128 Palaiseau cedex, France
[2] CEA DAM DIF, F91297 Arpajon, France
[3] CELIA, Université Bordeaux 1, 33405 Talence cedex, France
[4] Lebedev Physical Institute, 53 Leninskii Prospect, Moscow, 119991 Russia
[5] Centre de Physique Théorique, Ecole Polytechnique, 91128 Palaiseau cedex
[6] Max Planck Institute for the Physics of Complex Systems, Dresden, Germany
[7] CEA CESTA, F33114 Le Barp, France
[8] University of St. Andrews, Fife KY16 9ST, Scotland, UK
[9] FNSPE, Czech Technical University in Prague, 115 19 Prague 1, Czech Republic
E-mail: christine.labaune@polytechnique.fr

Abstract. An experiment carried out on the LIL facility is presented. The experiment was designed to investigate the propagation of a multi-kJ laser beam through a plasma produced by the irradiation of a low-density plastic foam. Laser beam self-smoothing has been observed and its consequences on laser imprint and Stimulated Brillouin Scattering (SBS) are discussed.

1. Introduction

Laser-plasma interaction physics (LPI) is one of the most critical issues for the success of laser ignition. After long standing efforts, a significant progress in the understanding of the coupling mechanisms has been obtained and theoretical modeling is now close to a predictive capability for experiments. A large part of this progress has been obtained thanks to numerous experiments done in well-controlled interaction conditions that allowed to test and to improve the physics included in the models. These experiments were carried out on small systems producing a large number of shots and equipped with sophisticated diagnostics having high spatial resolution. These results have been obtained at the cost of a low total energy and, consequently, with small-size and low temperature (Te $\leq$ 1 keV) plasmas. Now, there is a strong need for investigating interaction physics in multi-kilojoule experiments with large, hot, fusion-size plasmas, despite the fact that the number of shots is reduced. Inertial Confinement Fusion at megajoule laser facilities, NIF and LMJ, requires an efficient propagation and coupling of the laser beams to plasmas both for the direct and indirect drive approaches.

2. The LIL experiment

We have designed an experiment based on the multikilojoule system LIL (Ligne d'Integration Laser) aimed at achieving well-defined interaction conditions by using a large focal spot (mm), thus producing hot plasmas ($T_e \geq 2$ keV) with one-dimensional expansion along the laser axis. Among the possible targets (thin exploded foils, thick foils, gas jets, gas bags, hohlraums) we chose low-density (sub-critical average density) foam targets to produce long (mm) underdense plasmas [1]. They have shown favorable conditions for fusion studies (long scale-lengths of several hundreds microns) at high temperatures (keV range) that last for times of the order of the pulse duration [2]. Underdense foams have been used in many interaction experiments related to fundamental material properties [3], production of controlled density fluctuations [4] and interaction physics [5].

The objectives of our experiment were to study the propagation of a high intensity laser beam in a long and warm plasma, to measure the modifications of its coherence properties [6] and their consequences on laser imprint and parametric instabilities. These objectives were met by using a combination of three targets: i) a cylinder of low-density foam, ii) the same cylinder of foam with a 10 μm thick copper foil at the back, iii) a 10 μm thick copper foil. The targets of the first type make it possible to study the propagation of the laser beam through the underdense plasma, produced by the foam irradiation, the analysis being based on the transmitted light features. The targets of the second type allow to study interaction physics both in the foam and in the copper plasmas irradiated by laser light transmitted through the foam plasma. The targets of the third type allow to investigate interaction physics in copper plasma alone. A scheme of the experiment is shown in Fig. 1.

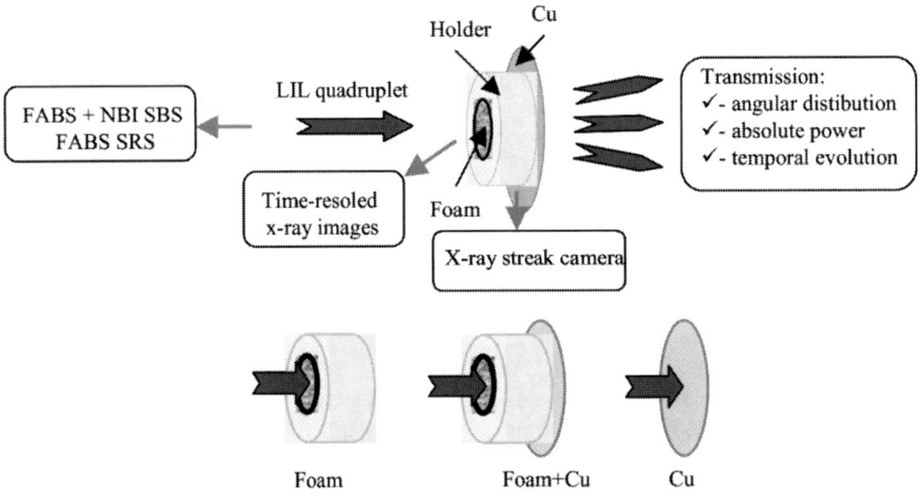

Figure 1. Scheme of the LIL interaction experiment.

The foam densities and lengths were chosen based on numerical simulations using the radiation hydrodynamic code CHIC [7] as well as preliminary experimental studies using the LULI installations. The foams were either TAC ($C_{15}H_{20}O_6$) or TMPTA with densities between 3 and 10 mg/cc which correspond to electron densities of 10 to 35 % of the critical density for 3w light. The scale length of the initial pore structures were in the submicron range and the foam lengths, along the laser axis, were varied between 300 μm and 1 mm. The LIL laser delivered ~ 12 kJ at 351 nm in a 2.7ns square pulse. The main diagnostics were the temporally and spectrally resolved measurements of stimulated Brillouin and Raman scattering, the temporally and angularly resolved analysis of the transmitted light, and spatial and temporal hard x-rays images in the domain of 1 – 5 keV x-rays.

3. Experimental results

With the foam alone, the time-resolved transmitted light and the transverse x-ray streak camera provided consistent information about the laser propagation through the foam. We found that 500 μm long, 10 mg/cc foam was fully ionized in 1.2 ns. The corresponding average speed of the ionization front was 0.5 mm/ns. This value is larger than the acoustic velocity, $c_s \sim 0.32$ mm/ns, which was deduced both from the trajectory of the rarefaction wave front in the x-ray image, as well as from temperature measurements from SBS spectra. So, the first result is that a combination of a 10 mg/cc foam density with a few 10^{14} W/cm^2 irradiation intensity provides a supersonic ionization wave, sustained for a time longer than 2 ns [8].

The second result concerns the modification of the spatial coherence property of the transmitted light, which was demonstrated via the angular broadening of the light transmitted through the foam. Inside the incident cone, as well as outside, we observed broadening of the transmitted light through the foam compared to vacuum. An example of the angular diagram outside the incident cone in the case of a 500 μm-long, 7 mg/cc, foam is shown in Fig. 2. Some light is scattered outside the initial 6° aperture, with a half angle of 16° and a sharp profile. The angular broadening of the transmitted radiation is due to multiple scattering of the laser light on the laser-induced density fluctuations caused by forward stimulated Brillouin scattering (FSBS). FSBS of the RPP beam is responsible for self-induced smoothing. Calculations, using the hydrodynamic plasma parameters, show that the interaction conditions are above the FSBS threshold [8, 9].

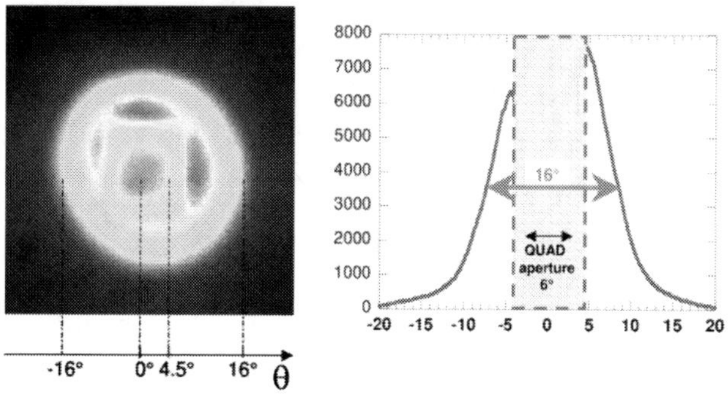

Figure 2. Angular distribution of the transmitted light outside of the incident cone.

The reduction of the laser imprint caused by the laser beam smoothing due to its propagation through the foam was demonstrated by the comparison between the 2D x-ray images, integrated over 200 ps, of the laser imprint on the Cu foil alone and on the Cu foil covered with a 500 μm, 10 mg/cc foam. The comparison was done at the very beginning of the laser impact on the foil, which corresponds to the time t = 0 for the foil alone and to t = 1.2 ns for the foil covered with 500 μm-long 10 mg/cc foam. Large intensity fluctuations with the same spatial scale as in the focal spot distribution in vacuum are observed when the foam is absent (panel a in Fig. 3). The emission of the copper foil covered by the foam plasma (panel b) is much smoother than in the previous case. The inhomogeneities of size ~ 50 μm have been removed and the amplitudes of small-scale fluctuations have been strongly reduced.

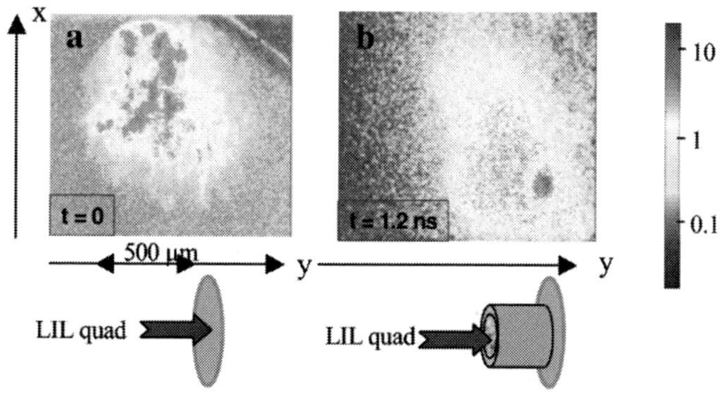

Figure 3. Two-dimensional x-ray images of the Cu foil emission integrated over 200 ps and recorded at the very beginning of the emission: a) Cu alone; b) Cu foil covered with 500 μm of foam.

The last results concern the SBS emission in the backward direction for the three types of targets, as shown in Fig. 4. With the foam alone, the SBS emission occurs at the beginning of the laser pulse and lasts for less than 1 ns. Similar behavior, with even shorter peak, is observed for the Cu foil alone. For the combined target (foam + copper), two distinct temporal components of SBS emission are observed. The first one is similar to the one recorded for the foam alone, while the second one starts 1.2 ns (/~1.9 ns) later for the 500 (/750) μm-long foam. This second signal is clearly associated with the combination foam layer/Cu foil, as it occurs later when the foam length is longer, consistently with the increased delay of the Cu foil irradiation. These results are presently being analyzed by means of our LPI code in which the plasma density and temperatures profiles are imported from the radiation hydrodynamic calculations [10]. Our numerical results show the existence of a dip in the expansion flow spatial profile in the vicinity of the foam/Cu interface, resulting in a high SBS gain factor. The low SBS reflectivity is then being interpreted as due to the laser beam self-induced smoothing.

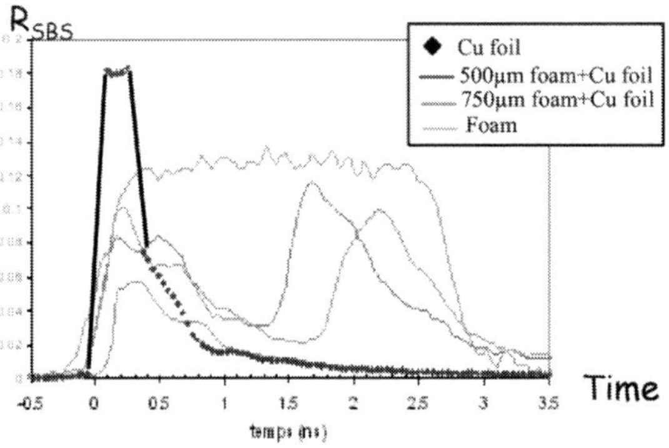

Figure 4. Temporal evolution of the SBS reflectivity inside the laser pulse in backward direction from three types of targets

4. Conclusion

In conclusion, the LIL experiment, in the multikilojoule regime, has demonstrated that the propagation of a laser beam through a thin (~ 500 μm), low density (~ 10 mg/cc), foam located in front of a solid target leads to an efficient beam smoothing and reduces the laser imprint consequently. These results show that it would be possible to design new ablators for direct-drive ICF targets with enhanced absorption and improved smoothing characteristics. The modification of SBS reflectivity from a Cu foil covered by the foam is another consequence of the beam smoothing effect.

Acknowledgments

The authors gratefully acknowledge the efficient participation of the CEA/DCRE group and LIL facility operating teams that made the success of these experiments. This work was partly supported by the ANR contract CORPARIN and by the computing centre IDRIS of the CNRS.

References

[1] Figueroa H *et al. Phys. Fluids* 1984 **27** 1887.
[2] Tanaka K A *et al. Phys. Fluids* 1985 **28** 2910.
[3] Koch J *et al. Phys. Plasmas* 1995 **2** 3820.
[4] Moody J *et al. Phys. Plasmas* 2000 **7** 2114.
[5] Limpouch J *et al. Plasma Phys. Controlled Fusion* 2004 **46** 1831.
[6] Labaune C *et al. C.R. Acad. Sci.* 2000 **4** 727.
[7] Maire P-H, Abgrall R, Breil J and Ovadia J *SIAM J. Sci. Comput.* 2007 **29** 1781.
[8] Depierreux S *et al. Phys. Rev. Lett.* 2009 **102** 195005.
[9] M. Grech et al., Phys. Rev. Lett. 102, 155001 (2009).
[10] Hüller S *et al. Phys. Plasmas* 2006 **13** 022703 ; *J. of Physics: Conf. Series* 112, 022031 (2008).

Saturation of Raman instability in gas jet plasma in LULI 2000 laser experiments

D.T. Michel[1,2], S. Depierreux[2], V.Tassin[2], C.Stenz[3], M.Chateau[1], C. Labaune[1]

1) LULI, Ecole Polytechnique, 91128 Palaiseau cedex, France

2) CEA, DAM, DIF, F_91297Arpajon

3) CELIA, 351 cours de la libération 33405 Talence

E-mail : tomline.michel@polytechnique.edu

Abstract. This paper presents results obtained in an experiment performed at the LULI 2000 laser facility where we measured saturated Stimulated Raman Scattering (SRS) in a supersonic gas jet of ethylene (C_2H_4). Our conditions correspond to the following parameters: $k_{EPW}\lambda_D$ = [0.2-0.3], T_e = [0-1] keV and n_e = [1.2-10] % of critical density at 526 nm. The laser was smoothed with a random phase plate and had a maximal average intensity of 1.2×10^{15} W/cm² in the focal spot.

1. Introduction

The Raman instability is a three wave coupling which appears in laser plasma interaction physics, in particular in the field of inertial confinement fusion [1]. Controlling its linear and nonlinear behaviour is an important issue in this field. In case of high laser intensity, long plasma, high electronic density, or low electronic temperature the SRS instability can be saturated, that is to say that the SRS backscattered intensity no more increases exponentially with the incident laser intensity. Various saturation processes have been identified during experiments or in simulations. A key parameter in these processes is the damping of the electron plasma wave (EPW) [2] which is correlated to the value of $k_{EPW}\lambda_D$, where λ_D represents the electron Debye length and k_{EPW} the wavenumber of the EPW. The experiment was designed to study the saturation of the Raman instability at 526 nm, for moderate intensity ($I_{MAX} \approx 1.2\times10^{15}$ W/cm²) and intermediate $k_{EPW}\lambda_D$ value ($k_{EPW}\lambda_D$ = [0.2-0.3]) with a final objective of comparing the saturation level of the SRS at 526 and 351 nm. This $k_{EPW}\lambda_D$ regime has already been investigated at lower intensity, with a 1.054 µm beam, using exploding foil plasma which exhibits a different electronic density [3, 4]. In our experiment, we used gas jet plasma, which allows a control of the maximum electronic density of the plasma which increases with the jet pressure. The experiment is described, the experimental plasma conditions are given and preliminary results are presented and discussed.

2. Description of the experiment

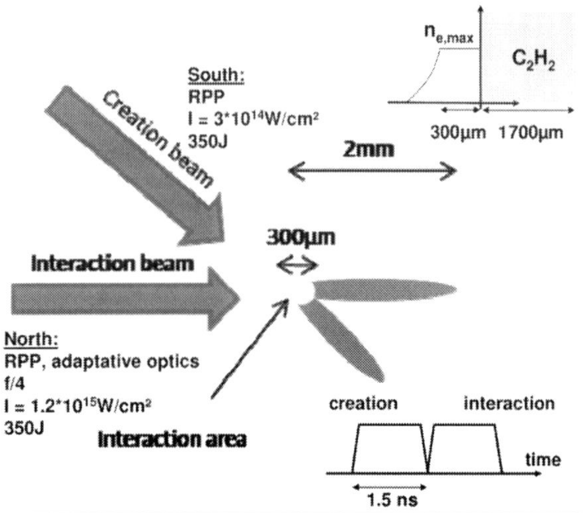

Pressure (bar)	12	22	30	35	50
T_e (eV)	250	350	420	500	600
$(n_e/n_c)_{gas\ jet\ fully\ ionised}$	1.2	3.1	4.2	6.1	10
$k_{EPW,\ gas\ jet\ fully\ ionised}$	0,3	0,27	0,24	0,22	0,22

Figure 1: experimental set-up for the LULI 2000 experiment.

Table 1: electronic temperature (T_e), electronic density divided by critical density calculated at 526 nm (n_e/n_c) assuming that the jet is fully ionized, and corresponding wavenumber of the EPW (k_{EPW}) driven by the SRS times the electron Debye length (λ_D) as a function of the pressure at the entrance of

The experiment was performed using the two kilojoule-beams of the LULI 2000 laser facility. The two 1.053 µm laser beams were separated by an angle of 45°, doubled in frequency and could reach 350 J at 0.526 µm in a 1.5 ns flat topped pulse. The first beam was used to create the plasma whereas the second beam was used to interact with the plasma. The creation beam was focused to a circular focal spot using an f/4 plan-convex lens and a random phase plate (RPP) with 2 mm approximately square elements. The interaction beam was RPP smoothed with 4 mm square elements, corrected with an adaptative optics operated in the laser bay, and focused with an f/4 lens. We calculated for the creation beam a best focus FWHM (full width half maximum) of 260 µm, and for the interaction beam a FWHM of 130 µm. The interaction beam reached an intensity of $1.2\text{x}10^{15}$ W/cm² at maximum energy. The delay between the two beams was 1.5 ns so that at the end of the creation beam pulse, the interaction beam was fired.

A high quality beam splitter was placed in the beam path of the interaction laser to sample both the incident beam energy and the light reflected from laser-plasma instabilities. The backscattered light was collected in the incident beam aperture, was collimated by a mirror and relayed to a series of diagnostics. We used a spectrometer coupled to an optical streak camera with electronic readout to measure the time-resolved spectra from SRS and Stimulated Brillouin Scattering (SBS). For SRS (respectively for SBS), the measurements were done in a range of λ = 500–900 nm (respectively λ = 522–530 nm) and during 5 ns with spectral and temporal resolutions of 10 nm and 200 ps (respectively 0.1 nm and 200 ps). A S1 (respectively S20) photodiode filtered to suppress the SBS and let the SRS pass (respectively suppress the SRS and let the SBS pass) was used to measure the entire SRS (respectively the entire SBS) backscattered energy. Photodiodes were absolutely calibrated and their signals were recorded on fast transient digitizers. Configuration of the experiment and laser characteristics are summarized in Fig. 1.

3. Description of the plasma.

The target was a supersonic gas jet of ethylene (C_2H_4). The jet was circular with 2 mm of diameter. The two laser beams were focused 1mm upon the nozzle and crossed in the front part of the jet so that the interaction beam did not pass through non ionised gas before reaching the plasma as shown in Fig. 1. Due to the angle between the two beams, the length of interaction was 280 µm. We have characterised the atomic density of the gas jet, by using an interferometer technique [6], as a function of the pressure put at the entrance of the nozzle. Fig 2a) and 2b) present the electronic density divided by the critical density calculated at 526 nm (n_e/n_c) of the jet, assuming that the jet is fully ionised, obtained at 1mm above the nozzle for different backing pressure. Fig. 2a) shows the dependence of

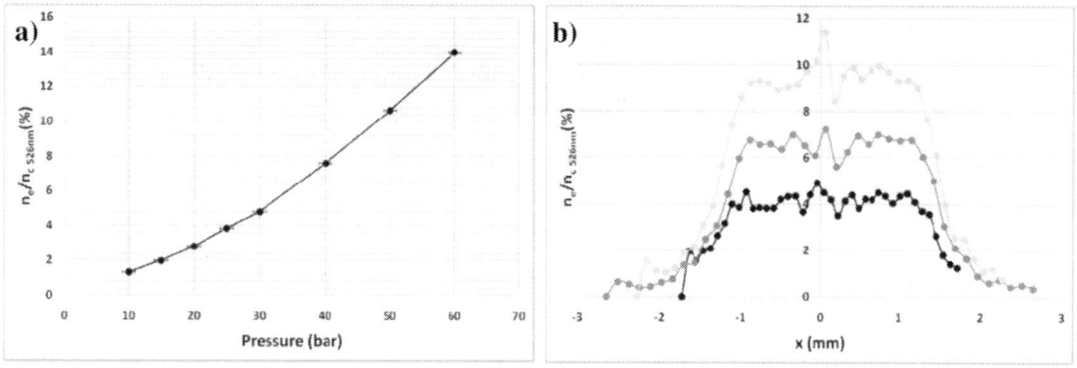

Figure 2: Measurement by interferometry of the electronic density of the gas jet, assuming that the gas is fully ionized: a) electronic density as a function of the pressure at the entrance of the nozzle at 1mm above the nozzle, b) Transverse electronic density profile at 1mm above the nozzle assuming that the jet has a cylinder symmetry for different pressures. The grey curve is calculated at 50 bars, the dark-grey curve is calculated at 40 bars, the black curve is calculated at 30 bars.

n_e/n_c as a function of the pressure. Fig. 2b) shows the transverse profile: we observed that the jet was almost homogeneous over a circular area of about 2.1 mm of diameter. Simulations with the hydrodynamic code FCI2 showed that the electronic temperature (T_e) was approximately constant and did not vary significantly during the interaction pulse. So, assuming that the plasma was stationary, we deduced T_e of the plasma from the SBS spectral shifts. In Table 1, we summarize the plasma characteristics for each pressure used in our experiment. We notice that T_e was higher when n_e/n_c was higher, as expected from the collisional absorption.

4. Experimental results

The Raman backscattered spectra analysis indicates that for each pressure, n_e/n_c which corresponds to the fully ionised gas was actually present in the plasma and corresponded to the maximum n_e/n_c over which the instability occurred. A typical SRS spectrum measured for a backing pressure of 35 bars at the maximal laser intensity of $I = 1.2x10^{15}$ W/cm² is shown in Fig. 3a). The corresponding time integrated spectrum is plotted in Fig. 3b). We notice that the maximal backscattered wavelength appearing in Fig. 3a) and 3b) is 710 nm which corresponds to a value of $n_e/n_c = 6\%$. This value is similar to the electronic density measured by interferometry, assuming that the gas jet is fully ionised. The Raman spectra were relatively narrow, so we deduce that the plasma was fairly homogeneous. The maximal backscattered wavelength in the SRS backscattering spectrum decreases in time. This has been interpreted as due to a decrease of the plasma density because of the plasma expansion. During the interaction, the value of the short wavelength cut-off decreases slightly, indicating that the electron temperature slightly goes down.

For this intermediate pressure condition and from the backscattered wavelengths, we calculate that the SRS occurs for n_e/n_c ranging from 4% to 6%. Assuming that T_e = 450 eV, we calculate that the laser intensity is above the absolute instability threshold intensity for a homogeneous media [1]. At 750 ps, the SRS backscattering drops off. There are two possible explanations for this: (i) the density of the plasma is too small to provide conditions for an absolute instability, (ii) the density profile becomes too sharp and the plasma is not enough homogeneous.

To study the effect of the laser intensity, it was varied from $0.5x10^{14}$ to $1.2x10^{15}$ W/cm² in gas jets at 35 and 50 bars which corresponded to initial densities of 6% and 10% of critical respectively. For both pressures, the SRS backscattering level increased linearly with the laser intensity demonstrating the saturation of the instability. Indeed, if in the growing regime, we could expect an exponential growth instead of linear. We found a saturation level of [10-20] % at 35 bars and [15-30] % at 50 bars. Then we studied the saturation of the SRS as a function of the electronic density of the plasma. The intensity of the laser beam was adjusted to a value around $1.2x10^{15}$ W/cm² and the density of the

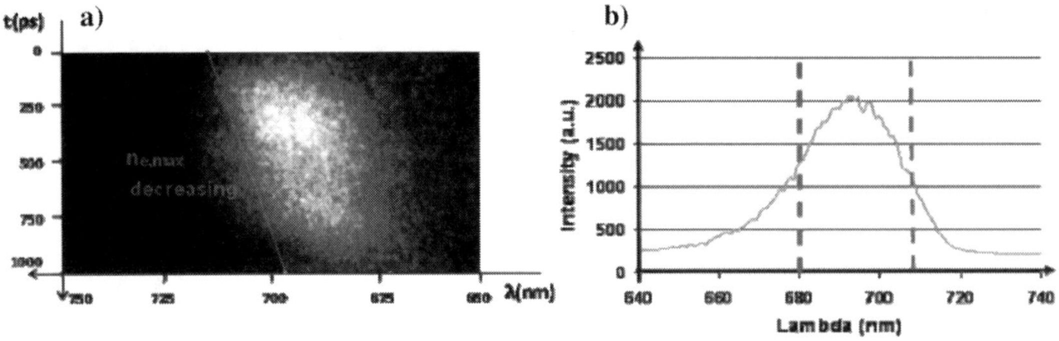

Figure 3: a) Time-resolved SRS spectrum from a 35 bar gas jet at I_{MAX}, b) SRS spectrum for 35 bars at I_{MAX}, integrated in time.

plasma was changed by modifying the backing pressure. We measured with photodiodes SRS and SBS energies for pressure of [12, 22, 30, 35, 50] bars. Results show that the instantaneous SRS intensity at the beginning of the interaction increases with the density approximately linearly. Such behaviour of the saturated SRS reflectivity with plasma density is consistent with saturation due to coupling of the EPWs with ion acoustic wave dynamics

5. Summary

We have presented experiments in gas jets of different pressures, at 0.536 μm. Maximal electron densities of the plasma were in a range of n_e/n_c = [1.2-10], electronic temperature extends from 300 eV to 650 eV and $k_{EPW}\lambda_D$ values extends from 0.2 to 0.3. We have observed the saturation of the SRS at 35 and 50 bars. We have measured a linear dependence of the SRS backscattered maximum reflectivity with the electronic density of the plasma for a laser intensity of $I = 1.2\text{x}10^{15}$ W/cm². A first interpretation of SRS spectra has been proposed. Further investigations are in progress with the hydrodynamic code FCI2 and SRS gain calculation with the PIRANAH post processor.

Acknowledgments
The authors gratefully acknowledge the efficient participation of the LULI 2000 facility operating team that makes the success of this experiment.

References

[1] D. Pesme and C. Labaune, in R. Dautray and J. Watteau, eds., "la Fusion Thermonucléaire Inertielle par Laser", Partie 1, Volume 1, Eyrolles, Paris, 1994
[2] D. A. Russel, D. F. Dubois and Harvey A. Rose Phys. Plasma **6**, 1294 (1999)
[3] S. Depierreux *et al.* Phys. Rev. Let. **84,** 2869 (2000)
[4] S. Depierreux *et al.* Phys. Rev. Let. **89,** 045001 (2002)
[5] B. La Fontaine *et al.* Phys. Rev. Let. **68**, 484 (1992)
[6] R. Benattar *et al.*, J. de Physique C1-234, Tome 39 (1978)

Production of dense laser-driven plasma jets using a cylindrical channel

J. Badziak [1], T. Pisarczyk [1], T. Chodukowski [1], A. Kasperczuk [1], P. Parys [1], M. Rosiński [1], J. Wołowski [1], E. Krousky [2], J. Krasa [2], K. Mašek [2], M. Pfeifer [2], J. Skala [2], J. Ullschmied [2], A. Velyhan [2], Yong-Joo Rhee [3], P. Pisarczyk [4], L. J. Dhareshwar [5], N.K. Gupta [5] and L. Torrisi [6]

[1] Institute of Plasma Physics and Laser Microfusion,01-497 Warsaw, Poland
[2] PALS Research Centre ASCR, 18200 Prague 8, Czech Republic
[3] KAERI, Daejeon, 305-353, Korea
[4] Warsaw University of Technology, ICS, 00-661 Warsaw, Poland
[5] Bhabha Atomic Research Centre, Mumbai, India
[6] University of Messina, Faculty of Physics, Messina 98166, Italy

E-mail: badziak@ifpilm.waw.pl

Abstract: A simple method of production of supersonic plasma jets with parameters relevant to laboratory astrophysics applications is proposed and demonstrated. The method consists in using a cylindrical channel for guiding and collimating the plasma generated from a laser-irradiated thin foil target. In the experiment, a 120 J, 1.315 μm, 0.3 ns laser pulse irradiated a thin (10 or 20 μm) CH foil placed at the entrance of the cylindrical channel (made in the aluminum cylinder) of diameter d_{Ch} = 0.3 mm and length of 2 mm. It was found that when the focal spot diameter (d_L) is well matched to the channel diameter ($d_L \approx (2/3)d_{Ch}$), the channel can form a collimated, high-density (up to 10^{20} cm^{-3}) plasma jet of the Mach number ≥10. The method seems to be flexible in the generation of jets of various hydrodynamic parameters and atomic numbers and does not require high-energy lasers for the jet production.

Generation of dense supersonic plasma jets is a subject of great interest in the laboratory study of astrophysical phenomena [1] and is also investigated in the ICF context [2]. Such plasma jets can be produced by laser irradiation of conical [3], planar [4] or triangular (V-shaped) [5] surface of a solid target. In these cases the jet is emitted in a backward direction (against the laser beam or in some angle to the beam), which can limit practical use of the jet for investigation of its interaction with some specific objects. Moreover, efficient production of plasma jets from planar targets seems to be limited to the high-Z or medium-Z plasmas. For the production of forward-emitted plasma jets a strong shock wave produced by a high-energy (≥1kJ) laser in a massive target with a hole [2, 6] or in the foam filling a short conical channel [7] was used.

In this contribution, we propose and demonstrate the formation of a supersonic forward-emitted plasma jet using a cylindrical channel for guiding and collimating plasma produced from a laser-irradiated thin foil target. This method seems to be flexible in the production of jets of various hydrodynamic parameters and Z numbers and does not necessary need high-energy lasers.

The experiment was performed at the Prague Asterix Laser System (PALS) facility in Prague. A 1.315-μm, 0.3-ns laser pulse of energy about 120 J and intensity up to 4×10^{15} W/cm^2

Journal of Physics: Conference Series **244** (2010) 022023 doi:10.1088/1742-6596/244/2/022023
irradiated a polystyrene (C_8H_8) foil placed at the entrance of the cylindrical channel (made in the aluminium cylinder) of diameter d_{Ch} = 0.3 mm and length L_{Ch} = 2 mm (Fig. 1). The parameters of the plasma flux at the channel output were measured using three-frame interferometry [4] and ion diagnostics (the ion charge collector and the Thomson parabola spectrometer). The effect of the laser focal spot diameter (d_L = 100, 200 or 400 μm) and the foil thickness (l_T = 10 or 20 μm) on the plasma flux parameters were investigated.

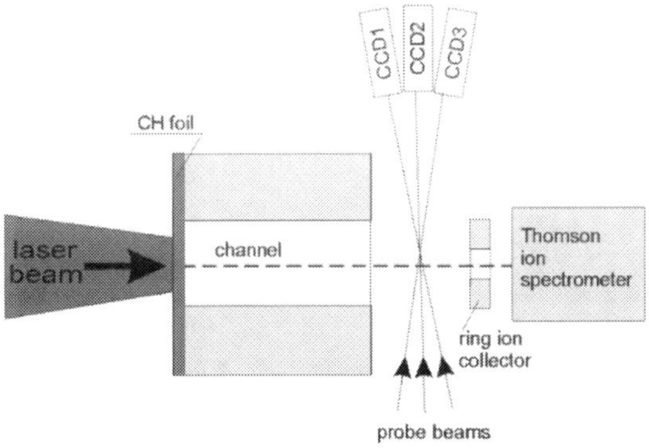

Fig. 1. Simplified scheme of the experimental set-up.

The interferograms and the corresponding electron isodensitograms and spatial distributions of electron density for the plasma leaving the channel of the length L_{Ch} = 2 mm at various instants after the laser pulse (t = 0) are presented in Fig. 2. It can be seen that a collimated plasma jet of electron density increasing in time (up to 10^{20} cm^{-3} at t = 17 ns) is formed at the channel output. Comparing the plasma flux leaving the channel with that produced without the channel we observed that the first one is much more collimated, denser, faster, and the total number of electrons in the flux is higher than that of freely expanding plasma (without the channel).

The parameters of the plasma jet formed in a channel depend on the irradiated foil's thickness l_T. The time evolutions of the plasma jet velocity and the total number of electrons in the jet produced from the CH target of l_T = 10 or 20 μm are shown in Fig. 3. The jet produced from the thinner target is faster, while the number of electrons in the jet is only a little bit higher than that for the thicker target. The spatial distributions of electron density in the jets were similar for both cases.

Another factor determining efficacy of the jet formation is the ratio d_L/d_{Ch}. We observed that at fixed laser energy the most collimated and the fastest jet was produced at d_L = 200 μm ($d_L/d_{Ch} \approx 2/3$). The parameters and collimation of the jet both at d_L = 100 μm and at d_L = 400 μm were not so good, probably due to too large angle of the plasma expansion in the first case and due to too low laser intensity in the second case.

For application of laser-produced plasma jets in the laboratory study of astrophysical phenomena some dimensionless numbers characterizing these jets should be of the same order of magnitude as the ones for astrophysical jets. In particular, this relates to [1, 7]: the Mach number $M = v_j/c_s$, the Peclet number $P_e = v_j r_j/\kappa$, and the Reynolds number $R_e = v_j r_j/\nu$, where v_j and r_j are the jet characteristic velocity and transverse radius, respectively, $c_s = 9.79 \times 10^5 (\gamma Z T_e/A)^{1/2}$ cm/s is the ion sound velocity [8] in the jet, γ is the polytropic index, $\kappa = 2 \times 10^{21} T_e^{5/2}/\Lambda(Z+1)n_e$ cm^2/s is the thermal diffusivity [9], $\nu = 2 \times 10^{19} T_e^{5/2} A^{-1/2} Z^{-3}/\Lambda n_e$ cm^2/s is the kinematic viscosity [9], T_e, Z, A and n_e are the electron temperature, ion charge state, atomic mass number and electron density of plasma in the jet, respectively, and Λ is the Coulomb logarithm. For astrophysical objects like the Herbig-Haro jets M ~ 10, P_e >>1, R_e >>1 [1, 7]. To estimate the values of T_e and Z, which we have to know to determine the above dimensionless numbers for the jets produced in our experiment, we utilized the results of measurements using the ion diagnostics as well as numerical simulations with the one-dimensional HYADES code. The electron temperature of plasma at the source (close to the CH target rear surface) was estimated from the formula [10]: $T_e \approx E_i/4(Z+1)$ where E_i is the mean ion energy of a thermal group of ions emitted forward from the plasma (determined from the ion collector measurements) and Z is the average ion charge state (estimated from the Thomson parabola measurements). For the CH target of the thickness l_T = 10 μm we obtained $T_e \approx 110 \pm 26$ eV.

Journal of Physics: Conference Series **244** (2010) 022023 doi:10.1088/1742-6596/244/2/022023

This value of temperature corresponded fairly well to the one obtained from the numerical simulation which, at the end of laser-target interaction ($t \approx 0.5$ ns), resulted in $T_e \approx 90 - 100$ eV. After the time $t \approx 8$ns, which is roughly equal to that needed to pass through the 2-mm channel by the front part of expanding plasma, the calculated plasma temperature dropped to 12 eV and the plasma front velocity reached about 2×10^7 cm/s. The consistency of these results with our measurements justifies the assumption that the temperature in the plasma jet observed at the 2-mm channel output is roughly equal to $T_{ej} \approx 10$ eV $\approx T_e/10$. Taking the above value of the plasma jet temperature and also: $v_j \approx (1.5 - 4) \times 10^7$ cm/s (Fig. 3), $r_j \approx 0.015$ cm, $n_e \approx 3 \times 10^{19}$ cm^{-3}, $Z = 3$, $A = 12$ we arrive at: $c_s \approx 1.8 \times 10^6$cm/s, $M \approx 8 - 22$, $P_e \approx (1.2 - 3.3) \times 10^2$, and $R_e \approx (2.4 - 6.5) \times 10^4$.

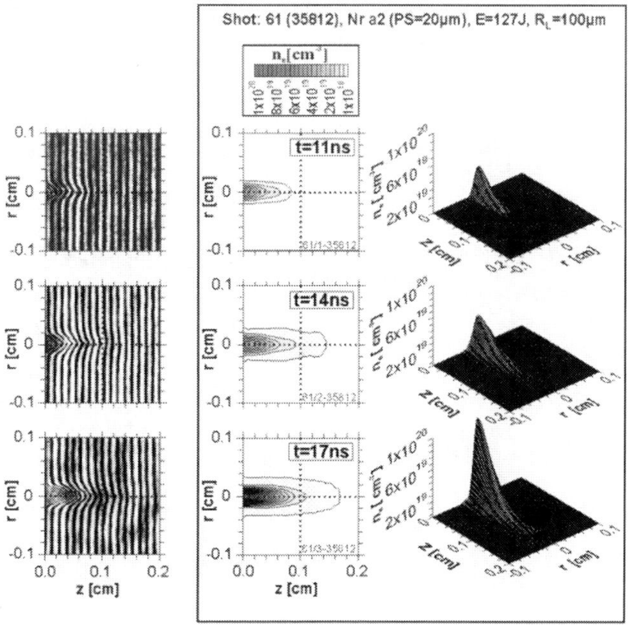

Fig. 2. The interferograms and the corresponding electron isodensitograms and space profiles of electron density distributions for the plasma leaving the channel of the length of 2 mm. $l_T = 20$ μm, $d_L = 200$ μm, $E_L = 127$ J.

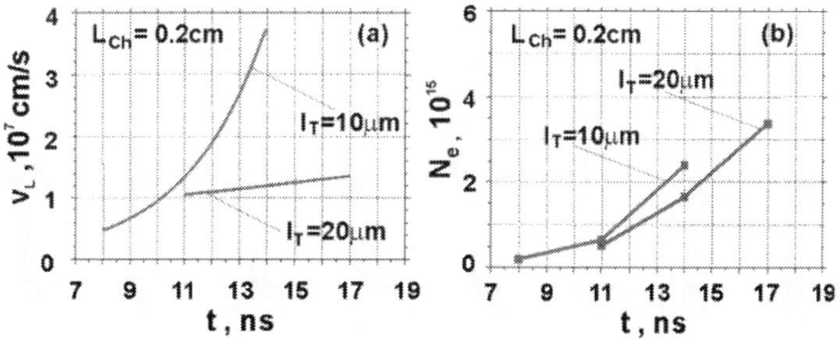

Fig. 3. Evolution of the plasma outflow velocity (a) and the total number of electrons (b) in the plasma flux leaving the 2-mm channel for the CH targets of different thicknesses. $d_L = 200$ μm, $E_L \approx 120 \pm 10$ J.

For the CH target of $l_T = 20$ μm the electron temperature at the source (deduced from the measurements) was lower than that for the thinner target and was estimated to be $T_e \approx 70 \pm 20$ eV. The measured velocity of the plasma jet at the 2-mm channel output was also lower: $v_j \approx 1.3 \times 10^7$ cm/s (Fig. 3) and the values of dimensionless parameters of the jet were as follows: $M \approx 8$, $P_e \approx 32$, and $R_e \approx 6.3 \times 10^4$. We can see that for both $l_T = 10$ and 20 μm the values of M, P_e and R_e correspond fairly well to the ones required for a proper simulation of astrophysical jets.

In conclusion, it has been shown that using a cylindrical channel for guiding and collimating plasma created at its entrance at the interaction of a laser pulse with a thin foil target can be an effective way of production of high-Mach-number, dense ($n_e \sim 10^{19} - 10^{20}$ cm^{-3}), forward-emitted plasma jets of parameters relevant to laboratory astrophysics applications. The production of jets using the channel does not necessary need high laser energy as efficacy of the jet formation is determined mainly by laser intensity I_L and the d_L/d_{Ch} ratio (for a fixed laser pulse duration) and the required values of these parameters can also be attained with low-energy (~J) lasers. It opens a prospect for using compact commercial lasers for the jet production and for some laboratory astrophysics experiments.

Acknowledgement
This work was supported in part by the HiPER project under Grant Agreement No 211737. The experiment was performed within the Access to Research Infrastructure activity in the Seventh Framework Programme of the EU (Contract No 212025, Laserlab Europe-Continuation)

References

[1] B.A. Remington, R.P. Drake, and D.D. Ryutov, Rev. Mod. Phys. **78**, 755 (2006).
[2] B.E. Blue, S.V. Weber, S.G. Glendinning et al., Phys. Rev. Lett. **94**, 095005 (2005).
[3] D.R. Farley, K.G. Estabrook, S.G. Glendinning et al., Phys. Rev. Lett. **83**, 1982 (1999).
[4] A. Kasperczuk, T. Pisarczyk, S. Borodziuk, et al., Phys Plasmas **13**, 062704 (2006).
[5] J. Grava, M.A. Purvis, J. Filevich et al., Phys. Rev. E **78,** 016403 (2008).
[6] J.M. Foster, B.H. Wilde, P.A. Rosen et al., Astrophys. J. **634**, L77 (2005).
[7] B. Loupias, M. Koenig, E. Falize et al., Phys. Rev. Lett. **99**, 265001 (2007).
[8] D. L. Book, NRL Plasma Formulary (Naval Research Laboratory, Washington D.C., 1987)
[9] L. Spitzer, *The Physics of Fully Ionized Gases,* 2nd ed. (Interscience / Wiley, New York, 1962).
[10] J. Badziak, J. Makowski, P. Parys et al., J. Phys. D: Appl. Phys. **34**, 1885 (2001).

Direct spectroscopic observation of ion deceleration accompanying laser plasma–wall interaction

O Renner[1], E Krouský[1], R Liska[2], M Šmíd[2], O Larroche[3], E Dalimier[4,5] and F B Rosmej[4,5]

[1] Institute of Physics of the ASCR, Na Slovance 2, 182 21 Prague, Czech Republic
[2] Czech Technical University in Prague, FNSPE, 115 90 Prague, Czech Republic
[3] CEA DIF, Bruyères le Châtel, 91297 Arpajon Cedex, France
[4] Université Pierre et Marie Curie UPMC, LULI, UMR 7606, Paris, France
[5] École Polytechnique, LULI, PAPD, Palaiseau, France

E-mail: renner@fzu.cz

Abstract. Interactions of plasma jets with solid surfaces are extensively studied in context with development of future fusion devices. In experiments carried out on the iodine laser system PALS, the energetic ions were produced at double–foil Al/Mg targets irradiated by one or two counter–propagating laser beams. The plasma jets from the rear surface of the laser–exploded Al foil streamed towards the Mg target representing the wall preheated by the action of the high–energy photons, particle and/or laser beams. Instead of being trapped by the cold secondary–target material, the forward–accelerated Al ions collided with the counter–propagating matter ejected from the wall. The environmental conditions in near–wall plasmas were analyzed with the high–resolution x–ray spectroscopy and temporally–resolved x–ray imaging. The deceleration of the incident Al ions in the near–wall region was directly observed and quantitatively characterized via Doppler shifts of the J–satellite from the Al Lyα spectral group. The interaction scenario was modelled using the 2D arbitrary Lagrangian Eulerian hydrocode PALE and the multifluid code MULTIF.

1. Introduction

The detailed knowledge of transient phenomena accompanying the interaction of plasma jets with surfaces of solid materials (hereafter plasma–wall interaction, PWI) is of paramount importance for numerous applications, starting from the construction of future fusion reactors and the design of sophisticated targets in indirect drive inertial confinement fusion schemes up to the astrophysics and laboratory experiments modeling various astrophysical situations (see e.g. [1] and references therein).

Due to the complexity of the problems studied, theoretical models of the PWI phenomena based on fluid hydrodynamic simulations, kinetic particle–in–cell approximations or their hybrids [2] provide only qualitative predictions of the shock formation induced by the ion impact and the subsequent plasma evolution. The acquisition of complex information on processes involved in near–wall plasma collisions depends mostly on the realization of well characterized experiments. In particular, laser–produced plasmas with a large span of selectable jet parameters (particle distribution, energy, degree of collimation) facilitate realization of diverse interaction regimes. Here we report precise x-ray spectroscopic measurements characterizing plasma interactions at laser–irradiated double–foil targets.

Journal of Physics: Conference Series **244** (2010) 022024 doi:10.1088/1742-6596/244/2/022024

2. Experiment

The scheme of the experiment performed at the Prague iodine laser system PALS [3] is shown in figure 1. The double–foil targets consisting of two parallel foils of Al (thickness 0.8 μm) and Mg (thickness 2 μm) with a variable spacing were irradiated at normal incidence with one or two counter–propagating laser beams. These beams delivered 5–200 J of frequency–tripled radiation (0.44 μm) in a pulse length of 0.25–0.3 ns. Being focused to a diameter of 80 μm (main beam) or 50 μm (auxiliary beam), they yielded a maximum intensity of 1×10^{16} W/cm^2 on the target. The energetic plasma jets produced at the rear (non–irradiated) surface of the Al foil streamed towards the Mg target representing the wall preheated by the action of the high–energy photons, particle and/or laser beams. Thus instead of being trapped by the cold secondary–target material, the forward–accelerated Al ions collided with the counter–propagating matter ejected from the wall. In addition to the previously published results of the experiments with the single–side irradiated double–foil targets [4], here we concentrate on the evaluation of data obtained at foils separated by a distance of 600 μm and double–side irradiated by the laser energy of 115 J (Al foil) and 6 J (Mg foil).

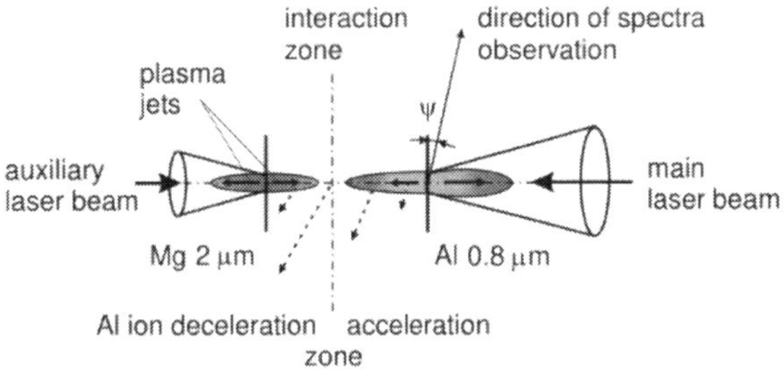

Figure 1. Scheme of the plasma jet formation and spectra observation at laser–irradiated double–foil Al/Mg targets.

The plasma interaction was monitored using a pinhole coupled to a low–magnification x–ray streak camera, optical spectroscopy and several x–ray spectrometers [5]. The time–resolved streak image presented in figure 2a demonstrates the temporally synchronized evolution of the plasma jets at both foils and, after approximately 1 ns, their strong interaction resulting in an enhanced x–ray emission from the plasma region close to the Mg foil. The primary diagnostic data was recorded using a vertical dispersion Johann spectrometer (VJS) fitted with the cylindrically bent crystal of quartz (100). The VJS provides simultaneously two sets of mirror–symmetric spectra [6]. The time–integrated spectrum shown in figure 2b is characterized by the high spectral (at the level of 8000) and 1D spatial resolution (8 μm) along the axis of the plasma jets expansion. By observing the plasma emission at an angle of $\psi = 0.8°$, i.e., in the direction almost parallel to the Al foil surface, the spectra integration over strong plasma gradients perpendicular to the foil surface was avoided.

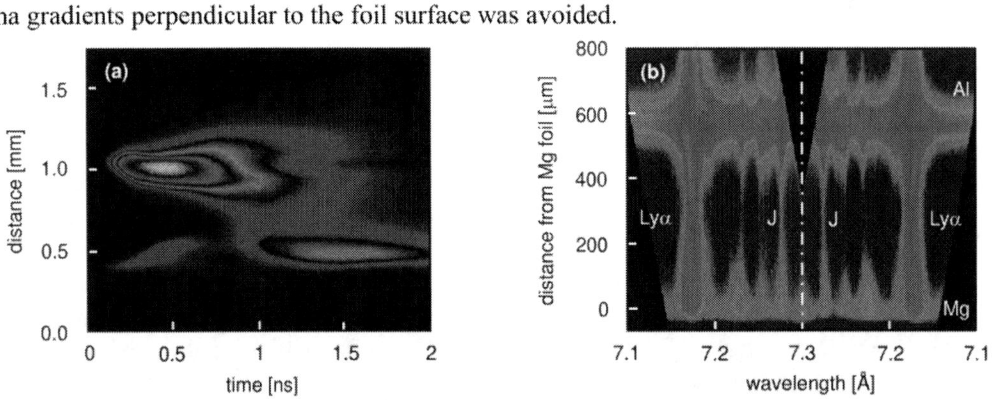

Figure 2. X-ray streak image of the plasma evolution at the two-side irradiated Al/Mg target (a) and the spatially resolved emission spectrum of the Al Lyα spectral group (b). The main laser beam irradiates the Al foil from above, the auxiliary beam strikes the Mg foil from below.

3. Results and discussion

The high–resolution spectra presented in figure 2b were used to derive the environmental conditions in the plasma interaction zone. The outer pair of the dominant spectral lines belongs to Lyα doublet of the hydrogenic Al, the inner pairs of lines are identified as dielectronic satellites $2l2l' \rightarrow 1s2l'$ with the J–satellite closest to the axis of symmetry. The satellite–rich structure observed at the rear surface of the Al foil gradually reduces to the emission of the J–satellite only, the full satellite structure reappears near the Mg foil. This spectra behavior reflects the variable macroscopic parameters of the plasma jets. By comparing the experimental data with synthetic spectra produced by the collisional–radiative code MARIA [7], the effective plasma conditions corresponding to the Al Lyα group emission at the rear surface of the Al foil are characterized by the electron density $n_e = (3\text{–}5)\times10^{21}$ cm^{-3}, temperature $T_e = 300$ eV, and the photon path length $L = 200$ μm. Close to the midplane, the plasma parameters are best fitted by $n_e = 5\times10^{20}$ cm^{-3}, $T_e > 700$ eV, and $L = 350$ μm, whereas in the interpenetration region close the Mg foil surface the plasma conditions are $n_e \sim 3\times10^{21}$ cm^{-3}, $T_e \sim 220$ eV, and $L \sim 500$ μm.

The validity of this interpretation of the observed spectral profiles was confirmed by theoretical modeling. An example of the results obtained by the Prague Arbitrary Lagrangian Eulerian hydrocode PALE [8] is shown in figure 3a. In this approach, after several steps of Lagrangian simulations the deformed moving mesh is reconstructed and the conservative quantities are remapped (Eulerian part) on to a smoother grid. Simulations indicate that already at time $t = -150$ ps before the laser maximum, the upper Al foil burns through and the Al internal energy starts to decrease (mostly being transformed into the kinetic energy of expansion) despite a progressive laser absorption at the border of the hole in the Al foil. The lower Mg foil burns through slightly later at $t = -80$ ps; around this time, the compression of the counter–streaming plasmas attains its maximum and afterwards the plasmas between the foils start to expand in the radial direction r. After $t = -50$ ps, the axial kinetic energy of Al ions starts to decrease, the Al plasma expansion in the direction of the laser axis is gradually slowed down by the counter–propagating Mg plasma.

The interpenetration of the Al and Mg ions was simulated with the multifluid code MULTIF [9]. MULTIF computes the hydrodynamics of an arbitrary number of the ion specii on a common Eulerian grid in the presence of a neutralizing electron background, and takes into account the interactions among all ion and electron fluids due to Coulomb collisions through slowing–down and heating terms. The geometry is 1.5–dimensional, with the main spatial resolution axis along the intertarget distance and a self-similar expansion model in the transverse directions. Each individual ion species is tracked by means of pseudo–Lagrangian markers, whose motion for the case of the Al ions interaction with the cold Mg foil is displayed in figure 3b. The trajectories of the fluid elements evidence regions where the Al and Mg plasmas interpenetrate and/or stagnate with respect to each other.

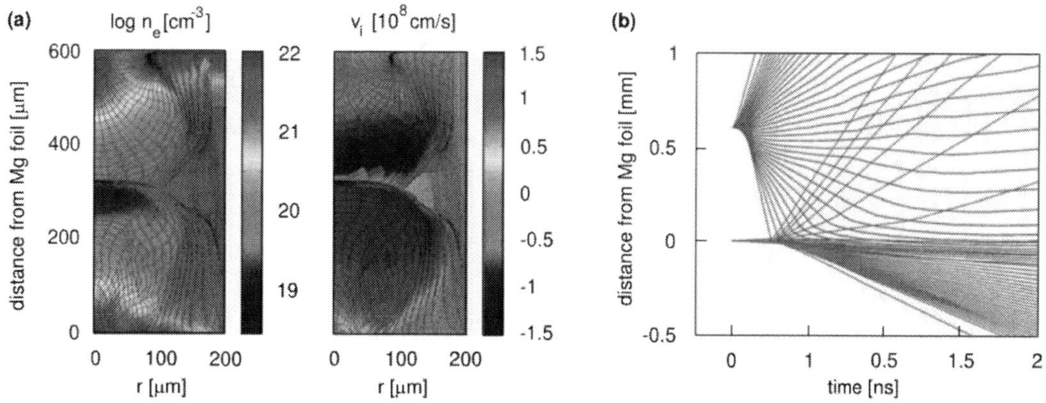

Figure 3. Modeling of the plasma jet interactions. The shown results of the PALE code simulations (a) correspond to the electron density and ion velocity distribution at the laser pulse maximum. The interaction of the Al plasma jet with the cold Mg foil was simulated using the MULTIF code (b).

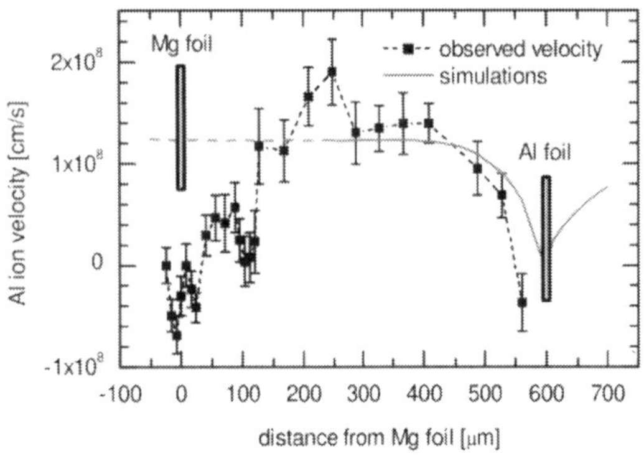

Figure 4. Effective velocities of Al ions (points with error bars) derived from Doppler shifts of the J–satellite emission. Ion deceleration in the interaction zone is compared with theoretical predictions of the ion velocity distribution (solid and dashed line) at free–standing single–side irradiated foil.

The region close to the Mg foil is of particular interest for studying the effects of the ion–wall interaction and plasma interpenetration phenomena. The presence of the Al emission near and below the Mg foil provides a clear evidence of the plasma interpenetration. Moreover, the distinct frequency shifts of spectral lines, which are clearly seen in figure 2b, provide an unique opportunity to measure precisely the ion deceleration via Doppler effect. The analysis of the frequency shifts measurements [1] indicates that the optically thin Al J–satellite is a suitable candidate for the visualization of the Al ion stopping. The previous experiments with single–side irradiated double–foil targets revealed almost monotonically decreasing red shifts of the J–satellite with the decreasing distance from the Mg foil [4]. This J–satellite behavior agrees with the expected stopping of the Al ions at the Mg foil.

The more complicated dependence of the ion deceleration close to the wall was found in plasmas produced at the double–side irradiated Al/Mg targets. The velocity distribution of the Al ions follows from the above described simulations. Starting from the inner foil surface, the Al ions are accelerated until the zone of the intense plasma interaction; this results in the shift of the J–satellite to red. After passing a distance of about 250 μm from the Mg foil, the Al ion deceleration reveals in the decreasing Doppler shifts. The conversion of the observed shifts to ion velocities is shown in figure 4. Assuming the gradual deceleration and simple trapping of the Al ions at the Mg surface, the velocities should monotonically decrease to zero. In contrast, their detailed spatial distribution exhibits oscillations and even negative values which reflect the more complicated scenario of the ion deceleration including the ion back–scattering. The detailed interpretation of this phenomenon based on rigorous 2D description of the interpenetrating plasmas and post-processing of the hydrodynamic data is in progress.

Acknowledgments
This research was supported by the Czech Science Foundation Grant No. P205/10/0814 and the CNRS PICS project No. 4343. The experiments and their simulations were performed under the patronage of the Czech Ministry of Education, Youth, and Sports projects No. LC528 and 6840770022. The authors gratefully acknowledge the assistance of the PALS staff in performing the experiments.

References
[1] Renner O, Liska R and Rosmej F B 2009 *Laser Part. Beams* **26** in print
[2] Evans R G 2006 *High Energy Density Physics* **2** 35–47
[3] Jungwirth K *et al.* 2001 *Phys. Plasmas* **8** 2495–501
[4] Renner O *et al.* 2007 *High Energy Density Physics* **3** 211–7
[5] Rosmej F B *et al.* 2006 *Europhys. Lett.* **76** 815–21
[6] Renner O *et al.* 1997 *Rev. Sci. Instrum.* **68** 2393–403
[7] Rosmej F B 2001 *Europhys. Lett.* **55** 472–8
[8] Liska R., Limpouch J, Kucharik M and Renner O 2008 *J. Phys.: Conf. Series* **112**, 022009
[9] Chenais-Popovics C *et al.* 1997 *Phys. Plasmas* **4** 190–208

Simulating NIF laser-plasma interaction with multiple SRS frequencies

C. H. Still, D. E. Hinkel, A. B. Langdon, J. P. Palastro, and E. A. Williams

Lawrence Livermore National Laboratory, P. O. Box 808, L-472, Livermore, CA 94551-0808

E-mail: bertstill@llnl.gov

Abstract. Understanding the energetics of a NIF ignition hohlraum is important to achieving ignition. Laser-plasma interactions (LPI) can reduce the radiation drive if backscatter occurs, and can also affect the hohlraum energetics by modifying the laser beam energy deposition which in turn can alter the implosion symmetry. The addition of a second SRS frequency to the modeling code pF3d can capture physics which would otherwise have been omitted. In the case of a wide or bi-modal SRS spectrum, this physics can be important. We discuss the modifications to the pF3d computational model, and exhibit its effect in a NIF ignition-relevant LPI simulation.

1. Introduction

The National Ignition Facility (NIF) has been designed to achieve fusion ignition in a laboratory setting by imploding a small DT-filled capsule inside a hohlraum. Laser-plasma interactions (LPI) within the hohlraum can have a significant effect on the energetics, thereby altering the symmetry and drive of the capsule implosion. Speifically, diverse machanisms such as stimulated backscatter, beam spray, beam bending, and beam refraction can all alter the profile of the laser beam's energy deposition, resulting in a modified x-ray conversion. The massively parallel electromagnetic hydrodynamic beam propagation code pF3d ([1, 2, 3]) is used to model LPI within a beam volume inside the hohlraum and calculate the energy deposition along the beam path. This information is then used in hohlraum design simulations to optimize the target design.

For stimulated Brillouin backscatter (SBS), the stimulated light approximately matches the frequency of the incident light. Thus, the light equations can be enveloped around that frequency. For stimulated Raman bacscatter (SRS), the reflected light is at a different frequency, thus a specific value must be chosen to calculate the matched SRS light. In the case where the SRS spectrum has one well-defined peak, which is not overly wide (the numerical half-bandwidth of the algorithm is about 10 nm), the choice is clear. For many ignition relevant designs, this has been the case. But there are cases where the choice for a matching frequency is not so clear (see Figure 1, such as a bimodal SRS spectrum, or one which is fairly broad. In this case, the addition of a second frequency enables the simulation to capture the SRS physics matching more of the spectrum.

In the next section, we discuss the computational SRS model and the addition of a second SRS frequency. In the following section, we apply the model to the ignition design with the SRS

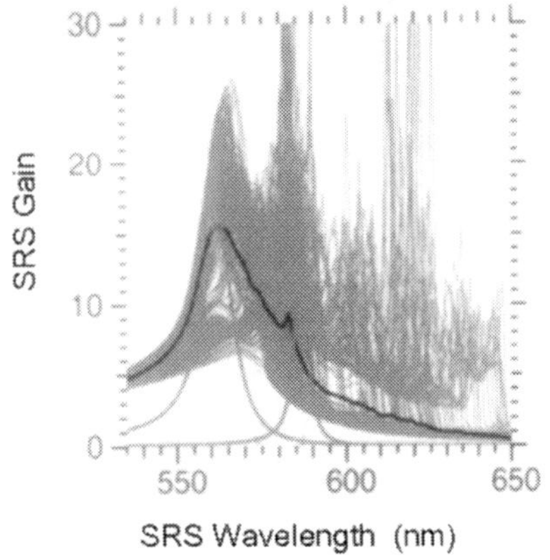

Figure 1. The SRS spectrum for a 285 eV Be-ablator ignition target design. The red curves represent the spectrum for each individual ray used to model the 30° quad of laser beams in the target design, and the black curve is the ray-averaged spectrum. Note this spectrum is not well-matched to a single SRS frequency.

spectrum from Figure 1 and compare the results to a simulation run with only one matching frequency.

2. The computational model

Although stimulated Brillouin scatter (SBS) will be used in the simulations below, its details are not germane to the computational implementation of the SRS. Thus, for simplicity, the SBS model will be excluded here.

The computational model for the light equations follows the paraxial equation. The laser field A_0 is enveloped around the laser frequency and wavenumber ω_0, k_0 $(ck_0(z) = \left({\omega_0}^2 - \bar{\omega}_{pe}^2(z)\right)^{\frac{1}{2}})$ and the SRS field A_R is enveloped around the specified matching frequency ω_R. Adding a second SRS equation primarily consists of adding two more equations to the usual set and modifying the pump wave to add the coupling to the second SRS wave. Using SRS matching frequencies $\omega_{R,j}$for$j = 1, 2$, the resulting equations are

$$\left(\frac{\partial}{\partial t} + v_{g0}\frac{\partial}{\partial z} + \frac{1}{2}v'_{g0} - \frac{ic^2}{2\omega_0}\nabla_\perp^2 + \nu_0 - \frac{i}{2\omega_0}\delta\omega^2\right)\tilde{A}_0 = -\frac{i}{4}\omega_0\tilde{A}_{R,1}\frac{\delta n_{l,1}}{n_c} - \frac{i}{4}\omega_0\tilde{A}_{R,2}\frac{\delta n_{l,2}}{n_c}$$

$$\left(\frac{\partial}{\partial t} - v_{gR,j}\frac{\partial}{\partial z} + \frac{1}{2}v'_{gR,j} - \frac{ic^2}{2\omega_{R,j}}\nabla_\perp^2 + \nu_{R,j} - \frac{i}{2\omega_{R,j}}\delta\omega^2\right)\tilde{A}_{R,j} = -\frac{i}{4}\frac{{\omega_0}^2}{\omega_{R,j}}\tilde{A}_0\frac{\delta n_{l,j}^*}{n_c}$$

where $\delta\omega^2 = {\omega_{pe}}^2 - \bar{\omega}_{pe}^2$. The Langmuir density $\delta n_{l,j}$ is enveloped around $\omega_0 - \omega_{R,j}$,

$$\left(\frac{\partial}{\partial t} - v_{gl,j}\frac{\partial}{\partial z} - \frac{3iv_e^2}{2\omega_{l,j}}\nabla_\perp^2 + \nu_{l,j} + i\Delta_{l,j}\right)\frac{\delta n_{l,j}}{n_c} = -\frac{i{\omega_{pe}}^2k_{l,j}^2}{4\omega_{l,j}{\omega_0}^2}\left(\frac{e\tilde{A}_0}{m_ec}\right)\left(\frac{e\tilde{A}_{R,j}^*}{m_ec}\right)\mathcal{F}_j.$$

where$\Delta_{l,j}$ is the spatially dependent mismatch. The multipliers $\mathcal{F}_j = \mathcal{F}_j(n_e, T_e)$ are chosen to be spatially disjoint (point-wise) $\{x : \mathcal{F}_1(x) \neq 0\} \bigcap \{x : \mathcal{F}_2(x) \neq 0\} = \emptyset$ to prevent a single spatial location from providing a source term to more than one SRS wave. Action is conserved point-wise via

$$\frac{\partial}{\partial t}A_0 = -i\sum_{j\neq 0} c_jA_{R,j} \quad \text{and} \quad \frac{\partial}{\partial t}A_{R,j} = -ic_j^*A_0$$

with

$$A_0 = -2\sqrt{\omega_0}\tilde{A}_0 \qquad c_j = -\frac{\omega_{peR,j}^2}{4\sqrt{\omega_0\omega_{R,j}}}$$
$$A_{R,j} = -2\sqrt{\omega_{R,j}}\tilde{A_{R,j}}$$

which gives

$$|\tilde{A_0}(t)|^2 + |\tilde{A_{R,1}}(t)|^2 + |\tilde{A_{R,2}}(t)|^2 = \text{const}.$$

We have found this approach to work well for well-separated peaks in a Raman spectrum. For closely located peaks, simulations are performed using only one SRS matching frequency.

3. Applying the model to an ignition target

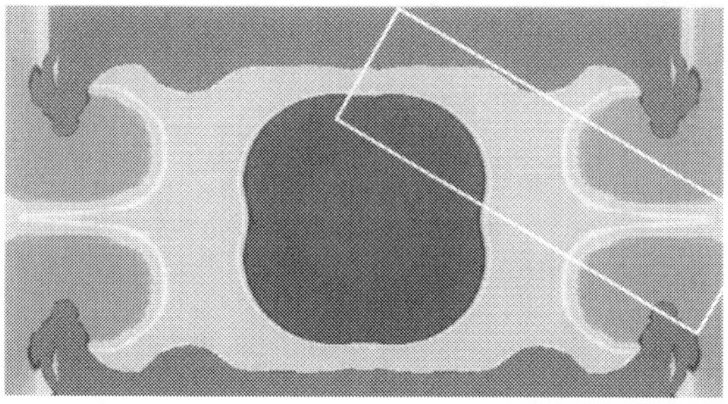

Figure 2. The material plot for a 285 eV Be-ablator ignition target design at peak power. Shown are the pink high-Z hohlraum wall, the blue CH LEH liner, the orange H_4He fill gas and the purple Be ablator. The white rectangle shows the radial projection of the simulation volume.

The target design shown in Figure 2 has significant gain at different densities (c.f. Figure 1), and hence is a good candidate for a two-group simulation. We first performed a near whole beam simulation over the entire beam path length. The full radial extent of the beam was included, but the azimuthal extent was reduced to a subset just large enough to provide adequate beam statistics. Only one matching SRS frequency (at 562 nm) was used. This single SRS group simulation was run on 4,096 (AMD Opteron) processors for approximately ten days before reaching a steady state (~100 ps). Slicing planes aligned with the hohlraum radius through the 3D plasma volume are shown in Figure 3 which have been taken as time averages over 25 ps at steady state. The laser propagation direction is vertical, bottom to top. At the upper left corner

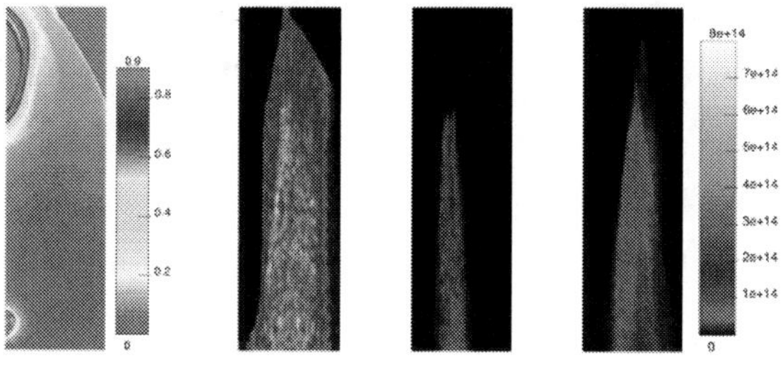

Figure 3. (left to right) Whole beam electron density normalized to 0.8 n_{crit}. Laser intensity, SBS intensity, and SRS intensity normalized to 8×10^{14} W/cm^2.

of the density plot is the capsule ablator plasma, and at the upper right is the hohlraum wall. The feature in the lower left corner is the lip of the laser-entrance-hole (LEH). The incoming laser intensity refracts away from the higher density blowoff from the expanding lip liner as it propagates into the holhraum. The SBS amplifies in the ablator plasma (left-hand side of the plot), and the SRS in the fill gas closer to the hohlraum wall (right-hand side of the plot).

Performing the same near whole beam simulation with the addition of the second SRS matching frequency (at 587 nm) reveals a slightly different story. The two-group simulation took about 20% longer on 4,096 cpus to reach ~100 ps. Averaging over 25 ps at steady state (Figure 4), we find some of the energy which went into SBS in the prior simulation now goes into the alternate SRS wave. This observation is consistent with the reflectivity plots, depicted in Figured 5. The SBS level is higher in the one-group simulations than the two-group calculation. The SRS reflectivity which matches 562 nm is unchanged between the two runs, but the SRS reflectivity of the 587 nm light comes in around 1.5%, which is almost the amount of decrease in the SBS.

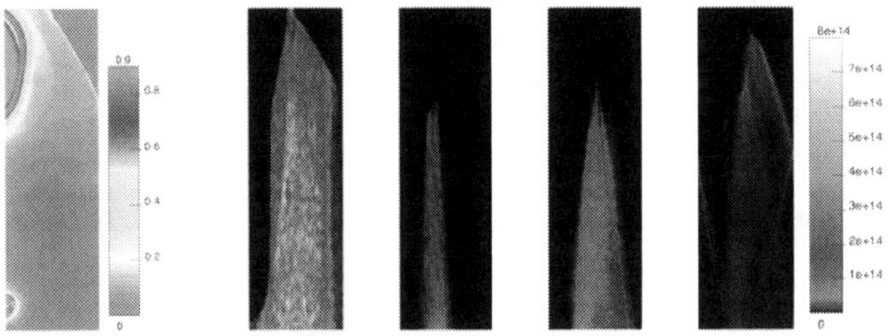

Figure 4. (left to right) Whole beam electron density normalized to 0.8 n_{crit}. Laser intensity SBS intensity, SRS intensity, and alternate SRS intensity normalized to 8×10^{14} W/cm^2.

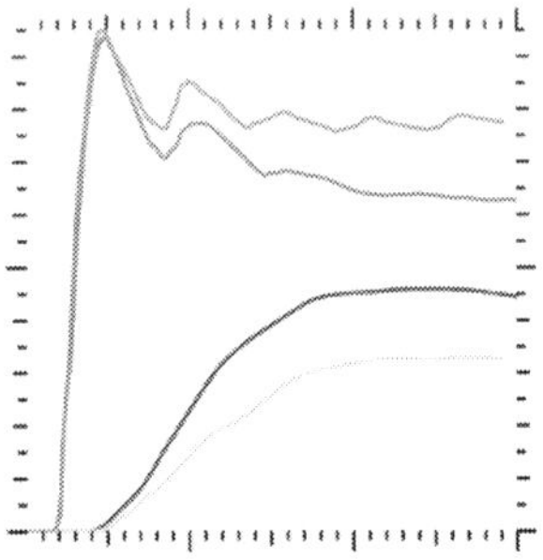

Figure 5. Comparison between the two simulations of reflectivity as a fraction of input intensity versus time [ps]. 1-group: SBS=blue and SRS=red; 2-group: SBS=cyan and SRS=magenta.

4. Conclusions

For target designs in which there is more than one SRS matching region, performing a simulation using a second SRS matching frequency captures physical effects that would have otherwise been overlooked. As a proof-of-principle, we have implemented the two-group model in the current pF3d beam propagation code, and performed comparison LPI simulations of a specific NIF ignition-relevant target. We found the single-group simulation overpredicts the SBS reflectivity compared to the two-group calculation, and the difference in energy goes into the second SRS wave. We have now begun to routinely apply the new code to analyze NIF ignition and ignition-emulator targets.

5. Acknowledgments

The authors would like to thank the Multiprogrammatic and Institutional Computing (M&IC) department and the Advanced Simulation and Computing (ASC) program at LLNL for access to high-performance computing platforms.

This work was performed under the auspices of the U. S.Department of Energy by the Lawrence Livermore National Laboratory under Contract DE-AC52-07NA27344.

This document is identified by the release number LLNL-CONF-418575.

References

[1] D. E. Hinkel, D. A. Callahan, A¿ B. Langdon, S. H. Langer, C. H. Still, and E. A. Williams, *Phys. Plasmas* **15** 056314 (2008).

[2] C. H. Still, R. L. Berger, A. B. Langdon, D. E. Hinkel, L. J. Suter, and E. A. Williams, *Phys. Plasmas* **7**, 2023 (2000).

[3] R. L. Berger, C. H. Still, E. A. Williams, and A. B. Langdon, *Phys. Plasmas* **5**, 4337 (1998).

Journal of Physics: Conference Series **244** (2010) 022026 doi:10.1088/1742-6596/244/2/022026

Hot Electron Generation and Transport Using Kα Emission

K. U. Akli[1], R. B. Stephens[1], M. H. Key[2], T. Bartal[3], F. N. Beg[3], S. Chawla[3], C. D. Chen[1], R. Fedosejevs[4], R. R. Freeman[5], H. Friesen[4], E. Giraldez[1], J. S. Green[6], D. S. Hey[1], D. P. Higginson[3], J. Hund[1], L. C. Jarrott[3], G. E. Kemp[5], J. A. King[3], A. Kryger[5], K. Lancaster[6], S. LePape[2], A. Link[5], T. Ma[3], A. J. Mackinnon[2], A. G. MacPhee[2], H. S. McLean[2], C. Murphy[3], P. A. Norreys[6], V. Ovchinnikov[5], P. K. Patel[2], Y. Ping[2], H. Sawada[3], D. Schumacher[5], W. Theobald[7], Y. Y. Tsui[4], L. D. Van Woerkom[5], M. S. Wei[3], B. Westover[3], T. Yabuuchi[3]

[1]General Atomics, San Diego, CA, 92186, USA

[2]Lawrence Livermore National Laboratory, Livermore, CA 94550, USA

[3]University of California San Diego, San Diego, CA, 92186, USA

[4]University of Alberta, Edmonton, Alberta, Canada T6G 2V4

[5]Ohio State University, Columbus Ohio, 43210, USA

[6]Rutherford Appleton Laboratory, Chilton, Oxon OX11OQX, United Kingdom

[7]Laboratory for Laser Energetics, New York 14623, USA

E-mail: akli1@llnl.gov

Abstract. We have conducted experiments on both the Vulcan and Titan laser facilities to study hot electron generation and transport in the context of fast ignition. Cu wires attached to Al cones were used to investigate the effect on coupling efficiency of plasma surround and the pre-formed plasma inside the cone. We found that with thin cones 15% of laser energy is coupled to the 40μm diameter wire emulating a 40μm fast ignition spot. Thick cone walls, simulating plasma in fast ignition, reduce coupling by x4. An increase of pre-pulse level inside the cone by a factor of 50 reduces coupling by a factor of 3.

1. Introduction

In the fast ignition scheme to inertial confinement fusion, a high intensity short pulse laser is used to generate fast electrons. These energetic electrons propagate to the dense core and deposit their energy in a small spot igniting the fuel. Critical to the success of fast ignition is how efficiently the short pulse laser energy is converted to fast electrons. Substantial work has been carried out regarding conversion efficiency in flat targets [1,2] To date, only Three relatively small "proof of concept" integrated fast ignition experiments have been reported in the literature [3-5] and only one dealt with details of fast electron transport and conversion efficiency [6]. Larger scale integrated fast ignition experiments will be carried out in the near future on OmegaEP, FIREX, and NIF ARC. However, experiments on relatively small-scale lasers can still provide valuable information relevant to fast ignition, provided that the targets are designed carefully. For example, wires attached to cones provide a simplified geometry to study most of the aspect of electron generation and transport. When laser-generated electrons leave the tip of the cone, they enter the wire. Their flux, mean energy, and conversion efficiency can be determined from Kα emission they induce in the wire. Moreover, a wire with 40μm cross-sectional area, the same size as the required fast ignition spot, allows us to study the forward-directed electrons that are more likely to make it to the fast ignition spot.

In this paper, we report on conversion efficiency experiments conducted with cone-wire targets using Kα emission as a diagnostic. We discuss the consequences of neglecting electron refluxing on the interpretation of Kα emission.

2. Experimental setup and diagnostics

Our experiments were conducted on two laser facilities: the Vulcan laser at Rutherford Appleton Laboratory and the Titan laser at Lawrence Livermore National Laboratory. In the Vulcan experiment, the targets were 1 mm long 40 μm thick Cu wires attached to Al cones with 12μm wall thickness [Fig.1-a]. They were irradiated with 350J of laser energy delivered in 1ps and focused to a peak intensity of 2×10^{20} W/cm^2 with an f/3 off-axis parabola. An x-ray imager and x-ray spectrometer, with 30° and 50° viewing angles respectively with respect to the cone-wire axis, were used to collect Kα emission from the wires. In the Titan experiments, the cone wall thickness was varied from 12μm to 160μm and the wires were similar to the ones used in Vulcan. The laser energy was 150J delivered in 0.7ps with 50% of the energy contained in 15μm diameter spot. An improved Kα imager, which uses an optimized assembly of collimators and magnets to increase signal-to-noise ratio, was used get high-resolution spatially resolved information [Fig.1-b]. A newly developed highly ordered pyrolythic graphite spectrometer (DCHOPG) [7] was used to get the number of photons induced by fast electrons propagating through the wires. The viewing angles with respect to the cone-wire axis of the Kα imager and DCHOPG were 43° and 18° respectively.

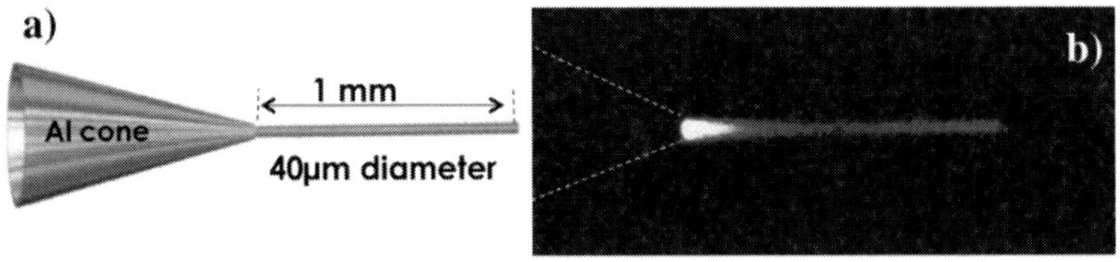

Fig. 1: cone wire target (a); typical Kα image from Titan experiment (b)

3. Results and discussions

In the cone-wire experiments two quantities are measured: the absolute number of Cu Kα photons induced by fast electrons and the slope of Kα emission along the wire. To determine laser-to-electron conversion efficiency and the mean energy of fast electrons, we have used a numerical transport model [8]. This model includes binary collisions, electric fields, and temperature dependent resistivity. We first inject a distribution of electrons with a given total energy and given fast electron temperature. Then a Kα emission profile is numerically generated along the wire. This is a two parameter fit where we seek a combination of electron total energy and temperature that gives a good fit to the experimental data [Fig.2].

The Vulcan data shows that 15% of laser energy is converted to fast electrons [9]. The electron beam temperature is 750 keV, which is much lower than the value predicted by ponderomotive scaling at $2x10^{20}$ W/cm^2. This could be a consequence of density steepening, seen in PIC simulations, at high laser intensities [10]. Our transport model neglects the transport through the cone tip and the effect of magnetic fields. LSP modeling of cone wire targets is underway. The 15% conversion efficiency to 40μm wire, emulating the 40μm fast ignition spot, is encouraging. However, these results are obtained with thin-walled cones. In real fast ignition experiments, the cone is embedded in plasma surround, which provides a path for the electrons that are generated at wide angles to escape. To study this effect we have used Al cones with variable wall thickness (12, 25, and 160μm). The absolute Kα emission, recorded with DCHOPG, is plotted as function of cone wall thickness in Fig 3-a. This data shows that increasing cone wall thickness by a factor of 12 reduces Kα emission in the wire by a factor of 4.

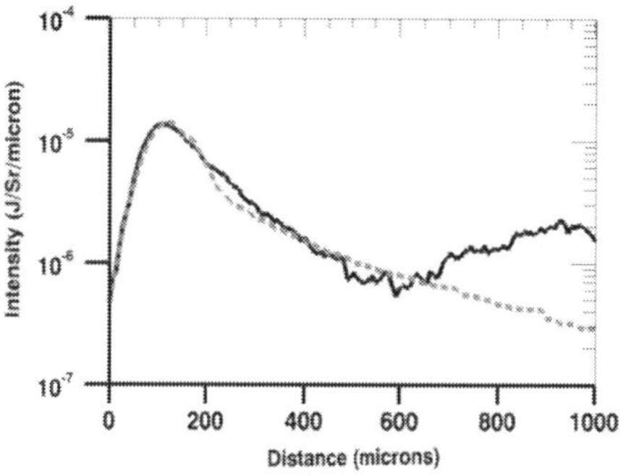

Fig.2: black line-experimental profile; red line-modeled profile

This due to the fact that hot electron density close to the tip of the cone is higher for thin cones compared to the thick ones (Fig 3-b). In the case of thick cones, electrons that are generated with wide angles are lost to the wall and do not make it to the wire. In the case of thin cones, sideways electrons are trapped in the wall due to electrostatic sheaths. Some of them will eventually make it the wire and contribute to Kα emission. This means that electron refluxing in thin cones enhances coupling to the wire and the coupling is not simply due to forward-directed source electrons. Since electron refluxing may not occur in real fast ignition targets with plasma surround, it is very important to take refluxing into account when inferring coupling efficiencies using thin cones. Experiments with 10μm Al/30μm Cu (refluxing flat target) and 10μm Al/30μm Cu/1000μm Al (non-refluxing flat target) show that refluxing enhances Kα emission by a factor of 5.

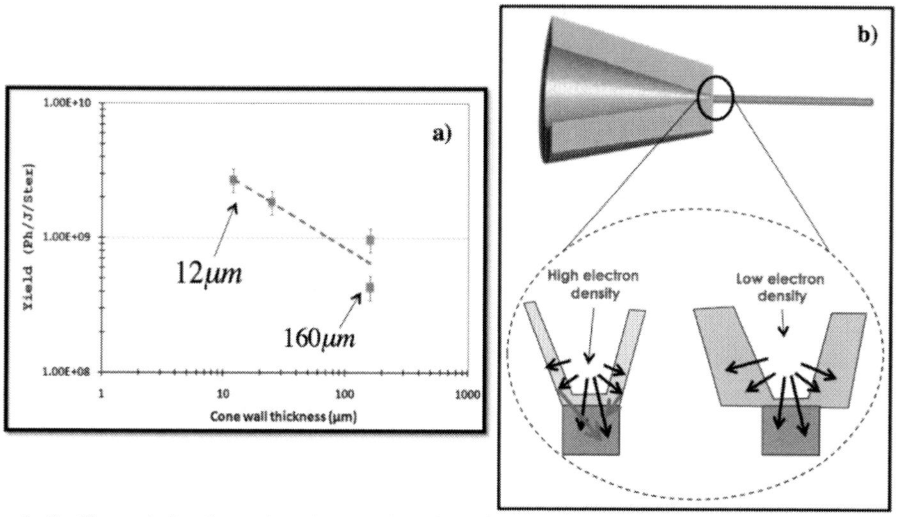

Figure 2: Cu Kα emission from the wire as a function of cone wall thickness (a); cartoon showing electron behaviour inside the cone for the two cases of thin and thick cones (b)

Fast ignition lasers may have hundreds of milli-Joules in pre-pulse and the pre-formed plasma may have a detrimental effect on conversion efficiency. Work by Baton et al. [11] with 1ω vs. 2ω indicate significantly reduced coupling with pre-pulse. We have conducted experiments with non-refluxing thick walled cones attached to wires and we varied the amount of pre-pulse from 8 mJ to 400 mJ. Kα emission data from the wire show that an increase of pre-pulse by a factor of 50 reduces the coupling to the wire by a factor of 3. In term of conversion efficiency, this data indicate that 400mJ of pre-pulse together with surrounding plasma may reduce the coupling to the wire to ~ 1%. Work by MacPhee et al. [12] shows details of electron generation inside the cone. At a pre-pulse level of 100 mJ, hot electrons are generated further away from the cone tip and are accelerated at wide angles toward the walls. Due to pre-formed plasma, the laser undergoes filamentation and self-focusing far away from the tip of the cone.

In conclusion, experiments with cone-wire targets show that electron refluxing in thin cones enhances coupling to the wire and the coupling is not simply due to forward-directed source electrons. Refluxing is unlikely to occur in integrated FI experiments due to the plasma surround. Therefore, the coupling efficiency obtained with thin cones has to be considered an upper limit. It has been shown here that the addition of controlled pre-pulse reduces the coupling efficiency. Further experiments are needed to optimize energy transfer, particularly the trade off between increased energy absorption in the coronal plasma with larger density scalelengths and beam filamentation in the extended region between the critical density and the ablation front. The results presented here indicate that prepulses with energies < 10mJ may be required for full-scale FI lasers. However, these experiments were done with laser parameters (150 J and 0.7 ps) that are different from those required for FI. Extending this study to higher laser energies and longer pulse-length is an additional but necessary step for both experiments and simulations

Acknowledgments:
The authors wish to thank the staff of the Jupiter Laser Facility, LLNL. This work was performed under the auspices of the U.S. DOE under contracts No.DE-FG02-05ER54834, DE-FC0204ER54789 and DE-AC52-07NA27344.

References:
[1] Wharton, KB, et al. *Physical Review Letters* (APS) 81, no. 4 (1998): 822-825.
[2] Chen, C. D, et al. *Physics of Plasmas* 16, no. 8 (Jan 2009): 082705.
[3] Kodama R. et al. *Nature (London)* **412**, no 6849, (2001) 798-802.
[4] Kodama, R, et al. *Nature (London)* 418, no. 6901 (2002): 933-934.
[5] Norreys, PA, et al. *Physics of Plasmas* 11 (2004): 2746.
[6] Key, MH, et al. *Physics of Plasmas* 15 (2008): 022701.
[7] Akli, KU, et al. (under preparation, to be submitted to RSI)
[8] Akli, KU, et al. *Physics of Plasmas* 14 (2007): 023102.
[9] Chrisman, B, Y Sentoku, and AJ Kemp. *Physics of Plasmas* 15 (2008): 056309.
[10] King, J. A, et al. *Physics of Plasmas* 16, no. 2 (Jan 2009): 020701.
[11] Baton, SD, et al. *Physics of Plasmas* 15 (2008): 042706.
[12] MacPhee, AG, et al, (Submitted to *Physical Review Letters*)

The Sixth International Conference on Inertial Fusion Sciences and Applications IOP Publishing
Journal of Physics: Conference Series **244** (2010) 022027 doi:10.1088/1742-6596/244/2/022027

X-ray diagnostics of fast electrons propagation in high density plasmas obtained by cylindrical compression

B Vauzour[1], M Koenig[2], D Batani[3], S Baton[2], F N Beg[4], C Benedetti[5], E Brambrink[2], S Chawla[4], F Dorchies[1], C Fourment[1], M Galimberti[6], L A Gizzi[7], R Heathcote[6], D P Higginson[4], S Hulin[1], R Jafer[3], P Köster[7], L L Labate[7], K L Lancaster[6], A J MacKinnon[8], A G MacPhee[8], W Nazarov[9], Ph Nicolaï[1], J Pasley[10], F Perez[2], X Ribeyre[1], M Richetta[11], J J Santos[1], G Schurtz[1], A Sgattoni[5], C Spindloe[6], L Volpe[3]

1 Centre Lasers Intenses et Applications, Université Bordeaux 1-CNRS-CEA, 33405 Talence, France
2 Laboratoire pour l'Utilisation des Lasers Intenses, 91761 Palaiseau, France
3 Dipartimento di Fisica, Università degli Studi di Milano-Bicocca, 20126 Milan, Italy
4 University of California, San Diego, La Jolla, USA
5 Dipartimento di Fisica, Università di Bologna, Italy
6 Central Laser Facility, Rutherford Appleton Laboratory, Didcot, UK
7 LIL-IPCF, Consiglio Nazionale delle Ricerche, Pisa, Italy
8 Lawrence Livermore National Laboratory, USA
9 University of St Andrews, Fife KY16 9AJ, Scotland
10 Department of Physics, University of York, YO10 5DD, UK
11 Dipartimento di Ingeneria Meccanica, Università di Roma Tor Vergata, Italy

E-mail: `Santos.Joao@celia.u-bordeaux1.fr`

Abstract. We report on X-ray diagnostics results from an experiment on fast electrons propagation in cylindrically compressed targets. It was performed on the VULCAN TAW laser facility at RAL (UK) using four long pulses (1ns, 70 J each at 2ω) to compress a cylindrical polyimide target filled with CH foam at 3 different initial densities. The cylindrical geometry allows us to reach temperatures and densities higher than those obtained in planar geometry compression. 2D hydrodynamic simulations predicted a core density range from 4 to 8 g/cm^3 and a core temperature from 30 eV up to 175 eV at maximum compression. An additional short laser pulse (10 ps, 160 J at ω) was focused on a Ni foil at one of the cylinder edges in order to generate a fast electrons current propagating along the compressed target. A X-ray radiography diagnostic was implemented in order to estimate the core plasma conditions of the compressed cylinder. Moreover two Bragg X-ray spectrometers collected the Kα fluorescence from the target so as to determine the variations of fast electrons population during the compression.

1. Introduction

In the context of the fast ignition scheme, a high energy electrons current produced by an ultra-intense laser beam heats highly compressed DT plasma. The fast electrons, produced near the

critical density, must propagate over a few 100 μm of overdense plasma and deposit their energy ($\approx$ 10 kJ) into the high density fuel [1]. In order to predict and understand the fast electron generation and transport some experiments have been successfully carried out on solid targets. However in this case the achieved densities and temperatures are lower than those expected for the fast ignition [2]. Such differences may involve some modifications in the predicted behaviour of the electrons such as their stopping power or their divergence in the compressed matter.
An alternative geometry consist in radially compress a cylindrical target [3]. It is thus possible to obtain plasmas with densities 10 times greater than the solid density and temperatures of a few hundreds eV. In fast ignition experiments this state of matter well known as warm and dense matter (WDM) is representative of the coronal plasma near the fast electron source and also presents a similar level of degeneracy of the compressed fuel. Using this geometry we have thus performed an experiment on the Vulcan TAW laser facility at RAL. The experiment was split in two phases. A first phase consisted in studying the cylindrical target compression to evaluate the targets' core densities and temperatures profiles at different moments of the implosion. A second phase was devoted to study the fast electrons propagation in such compressed targets.

2. Experimental setup

Four long pulse (LP) laser beams (4×70 J, 1ns, $\lambda = 0.53\,\mu$m) were used to radially compress 200 μm long, 220 μm outer diameter and 20 μm thick polyimide cylinders ($\rho_{poly} = 1.1\,\mathrm{g/cm^3}$). The tube was filled with CH foam at different densities: 0.1, 0.3 and 1 g/cm^3 and closed on both sides by 20 μm thick foils, made of Ni at the front side and Cu at the rear side. An additional short pulse beam (SP) ($5 \times 10^{18}\,\mathrm{Wcm^{-2}}$, 10 ps, $\lambda = 1.06\,\mu$m) was focused on the Ni foil to produce fast electrons. In order to prevent the alteration of the SP beam from the low density plasma corona a tube-shaped plastic coated gold shield was stuck onto the Ni foil (see Figure 1). The fast electrons were produced at various times during compression by introducing a delay ($0 \leq \tau \leq 3.5$ns) between SP and LP beams with an accuracy of $\pm$100 ps (due to the jitter).

The first phase of the experiment was dedicated to the study of the compressed matter's characteristics i.e. the core temperature and density at different stages of the compression. So SP beam was not employed to generate electrons but was used as a backlighting source by focusing it on a Ti foil (25 μm thick) for X-ray radiography diagnostic [4]. The X-ray backlight source placed 10 mm before the target transversally probed the cylinder at different delays during compression. A quartz crystal ($2d = 2.748$Å, $R_c = 380$ mm) situated $\sim$ 200 mm after the target, reflected the Ti Kα radiation (4.5 keV) and imaged the compressed cylinder onto an imaging plate (IP) positioned $\sim$ 2m away with a total magnification of $\sim$ 10.7. The spatial resolution has been calculated with the help of a grid and estimated to $\sim$ 20 μm on the target plane. In order to have a better absorption of the X-rays through the cylinder and thus have a better contrast ratio, CH foam was also doped with 30% Cl in mass (for this phase only). In order to predict the evolution of the cylinders' compression simulations have been performed with the 2D hydrodynamic code CHIC [5, 6]. This code, based on a cell-centered Lagrangian scheme, is mainly devoted to numerical simulations of laser driven ICF.

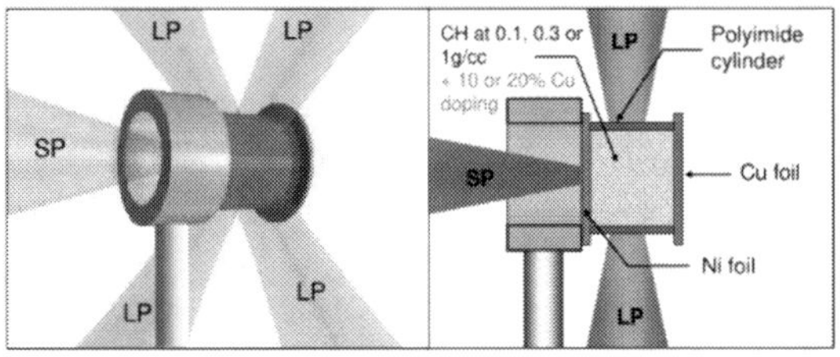

Figure 1. In principle scheme and target.

During the second phase, fast electrons propagation through the compressed cylinder was monitored by the K-shell fluorescence from the Ni and Cu foils. To follow the electrons during their propagation between the two foils CH foam was also doped with 20% Cu in mass (only 10% for 1 g/cm^3 targets). The K-shell fluorescence in the range 7.3 - 9.3 keV was collected by two

Journal of Physics: Conference Series **244** (2010) 022027 doi:10.1088/1742-6596/244/2/022027

X-ray Bragg spectrometers situated at the rear side of the target. The first one was composed by a flat highly oriented pyrolitic graphite (HOPG) [7] crystal with a high reflectivity but low spectral resolution ($\approx 50\,\mathrm{eV}$). The second one consisted of a high spectral resolution ($\leq 5\,\mathrm{eV}$) cylindrical quartz crystal ($2d = 2.024\text{Å}$, $R_c = 100\,\mathrm{mm}$) in Von Hamos configuration.

3. Experimental results and discussion

During the first phase X-ray radiography allowed us to estimate the core temperatures and densities of the compressed targets. This analysis concerned only $1\,\mathrm{g/cm^3}$ targets. A typical image is presented on Figure 2 (a) representing the compressed cylinder at maximum compression i.e. $\tau = 2.2\mathrm{ns}$. It has been possible to simulate the X-ray transmission profile for each delay supposing a laminar flux of X-rays transversally probing a simulated cylinder (2D hydrodynamic simulations).

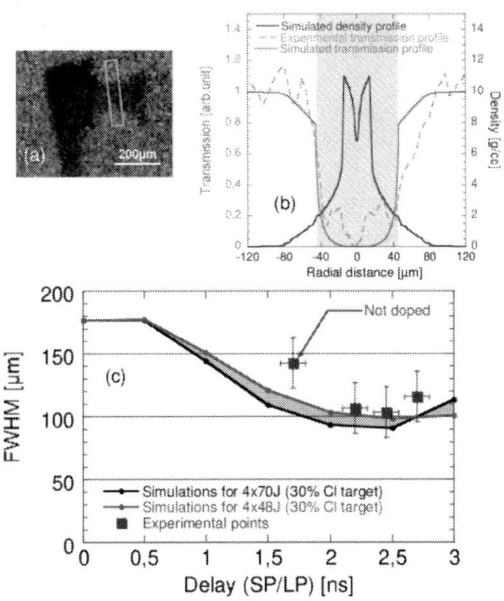

Figure 2. (a) Typical X-ray radiography image obtained for $1\,\mathrm{g/cm^3}$ target at maximum compression (τ =2.2ns). The blue rectangle corresponds to the analyzed area. (b) Experimental vs. simulated transmission profiles of the compressed area. Density is also plotted. The hatched region represents the Cl-doped core (c) Overall results of the X-ray radiography. The grey area corresponds to an uncertainty domain due to energy variations of the LP laser beams.

Figure 2 (b) shows the relatively good agreement between simulated and experimental profile of the transmission obtained for a $1\,\mathrm{g/cm^3}$ target at maximum compression. We can also note that the real radius of the compressed area, given by the density profile, doesn't correspond to the observed radius given by the X-ray transmission profile: $R_{density} \approx 20\,\mu\mathrm{m} \neq R_{Xray} \approx 45\,\mu\mathrm{m}$.

Figure 2 (c) shows the overall results of the X-ray radiography. One can see clearly the compression as a function of the delay and the relatively good reliability of the simulations. It is thus possible to estimate the core temperatures and densities of the $1\,\mathrm{g/cm^3}$ targets (for $4 \times$ 70J LP beams energy) at maximum compression thanks to the simulations:

$$T(2.2\mathrm{ns})_{1\,\mathrm{g/cm^3}} \approx 35\,\mathrm{eV}$$
$$\rho(2.2\mathrm{ns})_{1\,\mathrm{g/cm^3}} \approx 8\,\mathrm{g/cm^3}$$

This good agreement between simulations and experiment for 1g/cc targets allows us to also estimate the core parameters (ρ,T) at maximum compression of $0.1\,\mathrm{g/cm^3}$ and $0.3\,\mathrm{g/cm^3}$ targets with a relatively good confidence:

$$T(1.72\mathrm{ns})_{0.1\,\mathrm{g/cm^3}} \approx 220\,\mathrm{eV}$$
$$\rho(1.72\mathrm{ns})_{0.1\,\mathrm{g/cm^3}} \approx 4\,\mathrm{g/cm^3}$$
and
$$T(1.9\mathrm{ns})_{0.3\,\mathrm{g/cm^3}} \approx 80\,\mathrm{eV}$$
$$\rho(1.9\mathrm{ns})_{0.3\,\mathrm{g/cm^3}} \approx 4\,\mathrm{g/cm^3}$$

During the second phase the two Bragg spectrometers detected the Ni and Cu emission looking at the target's rear side. The Kα and Kβ fluorescence yields from the Ni foil on the front side of the cylinder are interpreted as a signature of the fast electrons source. The Cu-Kα fluorescence yields emitted either from the Cu doped CH foam or from the rear side Cu layer were used as a diagnostic of the fast electrons propagation

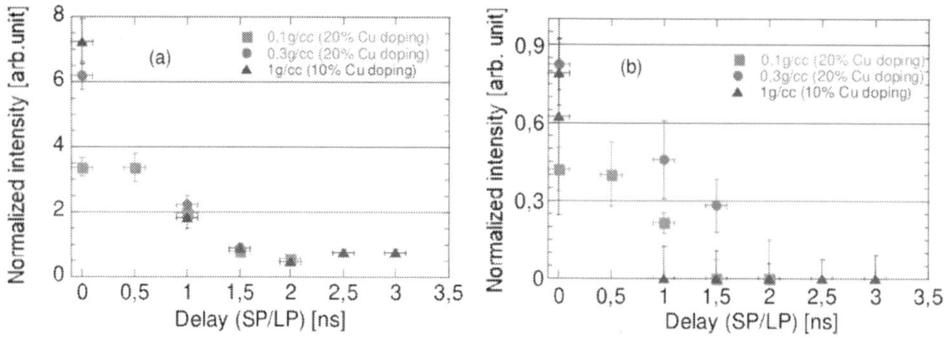

Figure 3. (a) Overall results from the HOPG spectrometer: Cu-Kα/Ni-Kβ ratio as a function of the Delay τ (b) Overall results from the Von-Hamos spectrometer: Cu-Kα/Ni-Kα ratio as a function of the Delay τ . The Cu-Kα/Ni-K$\beta(\alpha)$ ratio corresponds to the relative quantity of electrons that reached the rear side Cu foil divided by the relative quantity of electrons that are generated on the front side Ni foil.

range. A compilation of the results obtained with the HOPG spectrometer is presented in Figure 3 (a). To account for the small shot-to-shot variations of the fast electrons source, the Cu-Kα yields were adjusted to the Ni-Kβ yields for each shot. The Cu-Kα/Ni-Kβ ratio is used as a normalized measure of the fraction of the hot (> 8 keV) electrons reaching the rear surface of the target. We clearly observe a decrease of the ratio against SP/LP delay i.e. against compressed foam density. It means that electrons are slowed down more efficiently when the density grows up. However one can not observe a clear dependence on the initial foam density. Results from the Von Hamos spectrometer confirm these trends (see Figure 3 (b)).

4. Conclusion

The first phase was devoted to the characterization of the compression. With the help of the X-ray radiography diagnostic it has been possible to observe the cylinder's compression as a function of the time. Comparison of the experimental results with the 2D hydrodynamic simulations from CHIC allowed us to estimate the core temperature and density of the compressed target taking into account laser beam energy variations.

In order to study the fast electrons propagation the second phase was monitored by two X-ray spectrometers analysing the K-shell emission from the compressed targets. For all the tested initial target densities, the fraction of the fast electrons population reaching the cylindrical targets rear side decrease during compression. 2D hybrid simulations of the fast electrons transport are still underway in order to explain this behaviour.

5. Acknowledgments

This experiment is a part of the HiPER experimental roadmap and has been fully supported by the HiPER project.

References

[1] Tabak M *et al* , 1994 *Phys. Plasmas* **1** 1626-1634
[2] Atzeni S *et al* , 2008 *Phys. Plasmas* **15**, 056311
[3] Nakamura H *et al* , 2008 *Phys. Rev. Lett.* **100**, 165001
[4] Aglitskiy Y *et al* , 1998 *App. Opt.* **37**, 5253-5261
[5] Maire P H and Breil J, 2008 *Int. J. Nummer. Methods Fluids* **56**, 1417-23
[6] Maire P H *et al* , 2007 *SIAM J. Sci. Comput.* **29**, 1781
[7] Pak A *et al* , 2004 *Rev. Sci. Instr.* **75**, 3747-3749

X-ray Polarization Spectroscopy from Ultra-Intense Interactions

N. Booth[1], R. Clarke[4], P. Gallegos[4], L. Gizzi[3], G. Gregori[2,4], P. Koester[3], L. Labate[3], T. Levato[3], B. Li[2,4], M. Makita[5], J. Pasley[1,4], P.P. Rajeev[4], D. Riley[5], E. Wagenaars[1], J. N. Waugh[1], and N. C, Woolsey[1]

[1]Department of Physics, University of York, Heslington, York, UK, YO10 5DD

[2]Department of Physics, University of Oxford, UK, OX4 1PJ

[3]ILIL-IPCF, Consiglio Nazionale delle Ricerche, and INFN, PISA, Italy

[4]Central Laser Facility, STFC Rutherford Appleton Laboratory, Didcot, UK, OX11 0QN

[5]Department of Physics and Mathematics, Queens University Belfast, Belfast, UK, BT1 4NN

Email nb505@york.ac.uk

Abstract. Detailed knowledge of fast electron energy transport following the interaction of ultrashort intense laser pulses is a key subject for fast ignition. This is a problem relevant to many areas of laser-plasma physics with particular importance to fast ignition and X-ray secondary source development, necessary for the development of large-scale facilities such as HiPER and ELI. Operating two orthogonal crystal spectrometers set at Bragg angles close to 45° determines the X-ray s- and p- polarization ratio. From this ratio, it is possible to infer the velocity distribution function of the fast electron beam within the dense plasma. We report on results of polarization measurements at high density for sulphur and nickel buried layer targets in the high intensity range of $10^{19} - 10^{21}$ Wcm^{-2}. We observe at 45° the Ly-α doublet using two sets of orthogonal highly-orientated pyrolytic graphite (HOPG) crystals set in 1st order for sulphur and 3rd order for nickel.

1. X-ray Polarization Spectroscopy

Knowledge of fast electron energy transport within dense plasmas is a key area for the development of inertial confinement fusion [1], particularly with the current development of large scale facilities such as HiPER [2] and has bearing on secondary source production relevant to the development of ELI [3]. Fast electron generation is the primary method of energy transportation through solid density plasmas beyond the critical density. The propagation of fast electrons through the solid density plasma is mostly collision free. In order to sustain a high fast electron current the propagation of the relativistic electrons through the plasma is compensated by a cold electron return current [4] which collisionally ionizes the target plasma to the K-shell.

Magnetic sub-levels (with quantum numbers M_J) within an atomic energy level are preferentially populated into $M_J=\pm\frac{1}{2}$ through excitation by the electron beam. Emission from the magnetic sub-levels is polarized ($\Delta M_J = 0$ for π- polarization, and $\Delta M_J = \pm 1$ for σ- polarization) [5], and only some M_J's are populated, which produces a net polarization effect in the emission from the plasma. Polarized X-ray emission is associated with the strong anisotropy of the fast electron velocity distribution function [6]. This anisotropy is related to the mechanism responsible for the generation of fast electrons in intense laser-plasma interactions. In a study by Keiffer et al [7] the degree of polarization is shown to be given by $P = (I_\sigma - I_\pi)/(I_\sigma + I_\pi)$, and a number of further studies have been performed to observe polarization measurements of inner shell X-ray emission. These measurements have demonstrated the technique as a probe of hot electron velocity distributions at incident laser intensities up to 2×10^{19} Wcm^{-2} [8]. Through the development of inertial confinement fusion and secondary source developments, there is a real need to expand these measurements into the highly relativistic regime at intensities above 10^{19} Wcm^{-2}.

This paper presents results of an experiment to perform polarization spectroscopy measurements in the relativistic regime at intensities from 8.5×10^{18} Wcm^{-2} to 1.0×10^{21} Wcm^{-2}, to analyze the Ly-α doublet, $1s_{1/2}$ - $2p_{3/2}$ (Ly-α1) and $1s_{1/2}$ - $2p_{1/2}$ (Ly-α2), X-ray emission from hydrogen-like sulphur and nickel and to determine if spectroscopic polarization measurements are possible at these high intensities and in a single shot. Ionizing Ni to the K-shell demands incident laser intensities of approximately 10^{21} Wcm^{-2}, and requires return current heating to high temperatures [9].

In order to observe both the p- and s- polarizations (with respect to the plane of incidence at the crystal) from the plasma in a single shot we demonstrate the use of a pair of Bragg reflecting crystals positioned at a Bragg angle of approximately 45°, where the intensity of the diffracted X-rays is given as $I_p \sim R(\theta)\cos^2(2\theta_B)$. By operating the crystals at an angle of 45° the p- polarized component is absent and only the s- component of the polarized X-rays is reflected. By positioning the HOPG crystals orthogonally to one another at 45° to the target, the crystals act as ideal polarizers and monitor the X-ray p- and s- polarized emissions. Percival and Seaton [10] have shown that the emission from the Ly-α2 line is unpolarized and as such act as a calibration between the two spectrometers. The ability to observe both the p- and s- polarizations in a single shot is critical as it removes the problems of analyzing data with inherent shot-to-shot differences in the plasma.

2. Experiment

The experiment was performed in the Vulcan Petawatt target area at the Central Laser Facility, UK. The Petawatt laser operates at a wavelength of λ = 1054 nm, has a pulse duration of 600 fs and contained a maximum of 320 J of energy on target. The beam was focused to a best spot size of 5×6 μm^2 using an f/3 off-axis parabola and incident at 40°, which achieved peak intensities on target of 1.0×10^{21} Wcm^{-2}. The beam was defocused at the target position to a spot size of approximately 50×50 μm^2 to reduce the intensity on target to 8.0×10^{18} Wcm^{-2}.

The principal diagnostic used in this experiment was a pair of highly oriented pyrolytic graphite (HOPG) (002) crystals which had a mosaic spread of 0.4 ± 0.1°. The two crystals were positioned above the target to view X-ray emission perpendicular to the fast electron quantization axis. Both crystals were positioned 200 mm from the target and both with Bragg angles of 45°. The high integrated reflectivity of HOPG crystals is essential in this experiment in order to record weak X-ray emissions from within a high level of noise. This will then enable accurate single shot polarization measurements to be performed.

By using the spectrometers at $\theta_B = 45°$, the number of transitions which are able to be observed is severely limited. Here we demonstrate the use of two crystal spectrometers with an identical set-up, which can be used to observe the emission from hydrogen-like sulphur viewed in first order and hydrogen-like nickel which is viewed in third order. The integrated reflectivity of HOPG crystals in the s- polarization is very high, whilst the p- polarization integrated reflectivity is significantly reduced, which allows the HOPG crystals to act as ideal polarizers.

Thin foil targets of polysulphane (H_2S_n, $n > 2$) and nickel of 25 μm and 10 μm thicknesses cut to 100 μm × 100 μm squares were mounted on copper stalks and placed in a rotational mount to enable fresh targets to be moved into position without the need to break the vacuum. Some targets were tamped on the front surface and/or the rear surface of the foils by 0.2 or 0.5 μm thicknesses of an aluminum diagnostic layer.

Further diagnostics were also used in order to observe additional characteristics of the target plasma. An electron spectrometer [11] was used to observe the velocity distribution of the electrons, a single hit camera with a pinhole array [12] to observe the energy of absorbed photons in single events, and a high dispersion toroidal spectrometer to observe the K-α and K-β emission from the surface coatings of aluminum.

3. Results

Figure 1 shows the sulphur X-ray emission of both the p- and s- polarizations from a polysulphane plasma recorded from two HOPG crystals operating orthogonally to each other and observing emission from the same shot. The target was polysulphane with no aluminum surface tamping, and was irradiated by the laser at an intensity of 4.3×10^{20} Wcm^{-2}. The Ly-α1 and Ly-α2 are indicated in the figure, and demonstrates that it is possible to resolve the Ly-α doublet emission using two orthogonal spectrometers in a single shot.

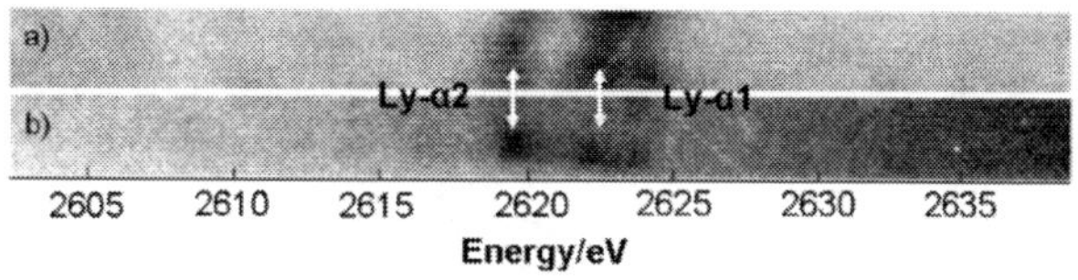

Figure 1. Sulphur Ly-α Emission spectra of a) the p- polarization and b) the s- polarization from the HOPG crystal spectrometers observing emission from a thin foil polysulphane target. The Ly-a doublet can be seen to be resolved in the centre of each spectrum. The background signal is different in each due to the different positions of the spectrometers and the shielding issues.

In order to process the recorded spectra from the HOPG crystals, lineouts are taken across the central region of the spectrum. The data is then processed to remove the wavelength dependent transmission of the spectrometer filtering [13]. In this case the filtering was 50 μm and 25 μm thicknesses of beryllium for the p- and s- polarizations respectively plus a 6 μm thick CH foil. Digitisation noise and the background signal are also removed. Figure 2 shows the lineouts of the spectra shown in Figure 1, clearly resolving the two peaks of the Ly-α doublet of sulphur, with Ly-α1 at 2622.6 eV and the Ly-α2 at 2619.6 eV.

Lineouts of the emission from a thin (10 μm) foil Ni target also with no aluminum tamping layer is shown in Figure 3. The target was irradiated by the laser at best focus and full energy at an intensity of 6.0×10^{20} Wcm^{-2}. The resolved Ly-α doublet is again clearly shown, with the Ly-α1 emission line at 8101.4 eV and the Ly-α2 emission line at 8072.8 eV. Further processing to calculate the integrated line intensities contained within the peaks will allow the calculation of the degree of polarization produced in the laser-target interaction from the formula given earlier.

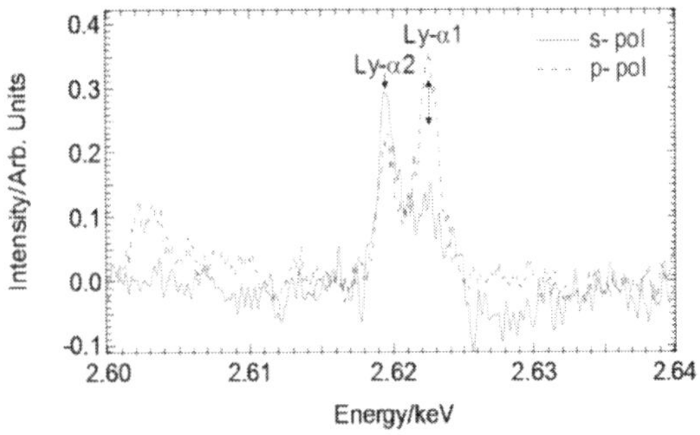

Figure 2. Lineouts of the sulphur X-ray emission from a thin polysulphane target (as shown in figure 1). of the p- and s- polarization X-ray emissions recorded at the HOPG crystal spectrometers

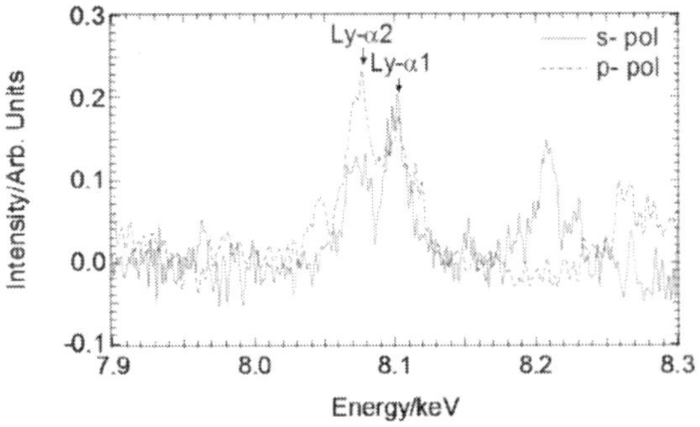

Figure 3. Lineouts of the emission from a thin nickel target of the p- and s- polarization nickel X-ray emissions recorded at the HOPG crystal spectrometers.

4. Conclusions

This paper presents measurements of the degree of polarization from inner shell X-rays produced during high intensity laser-plasma interactions between 10^{19} and 10^{21} Wcm^{-2}. We have demonstrated that it is possible to observe both the p- and s- polarizations of the X-ray emission in a single shot by the use of two HOPG crystals positioned orthogonally to one another at a Bragg angle of 45°. Due to the high integrated reflectivity of the s- component at the crystal plane, the crystals act as ideal polarizers. The results presented here show the resolved Ly-α doublet X-ray emission from thin foil targets of sulphur and nickel, and further analysis of the relative peak intensities of the Ly-α1 emission will allow the degree of polarization of the plasma X-rays to be calculated.

This technique allows polarization spectroscopy from an X-ray emission line to be observed from a single shot at high laser intensities and provides new avenues for direct measurement of electron transport processes within a high density plasma, critical to inertial fusion energy and X-ray secondary source developments.

References

[1] M. Tabak *et al*, *Phys. Plasmas*, **1** 1626 (1994)
[2] The HiPER project: http://www.hiper-laser.org/
[3] The Extreme Light Infrastructure European project: http://www.extreme-light-infrastructure.eu/
[4] R. Kodama *et al*, *Nature*, **412** 798 (2001)
Y. Senoku *et al*, *Phys. Rev. Letts.*, **90** 155001 (2003)
[5] T. Fujimoto and S.A. Kazantsev, *Plasma. Phys. Control. Fusion*, **39** 1267 (1997)
[6] L. Labate *et al*, *App. Phys. B*, **86** 229 (2007)
[7] J.C. Keiffer *et al*, *Phys. Rev. Letts.*, **68** 480 (1992)
ibid., Phys. Rev. E, **48** 4648 (1993)
[8] T. Kawamura *et al*, *Phys. Rev. Letts.*, **99** 115003 (2007)
Y. Inubushi *et al*, *Phys. Rev. E*, **75** 026401 (2007)
Y. Inubushi et al, J. Quant. Spect. Radiat. Trans., **99** 305 (2006)
[9] K.U. Akli *et al*, *Phys. Rev. Letts.*, **100** 165002 (2008)
[10] I.C. Percival and M.J. Seaton, *Philos. Trans. R. Soc. London, Ser. A*, **251** 113 (1958)
[11] S.P.D. Mangles *et al*, *Phil. Trans. R. Soc. A*, **364** 663 (2006)
[12] L. Labate *et al*, *Rev. Sci. Instr.*, **78** 103506 (2007)
[13] X-ray solid transmission: http://henke.lbl.gov/optical_constants/filter2.html

The Sixth International Conference on Inertial Fusion Sciences and Applications IOP Publishing
Journal of Physics: Conference Series **244** (2010) 022029 doi:10.1088/1742-6596/244/2/022029

Swarm of ultra-high intensity attosecond pulses from laser-plasma interaction.

S. S. Bulanov[1,2], V. Yu. Bychenkov[2,3], K. Krushelnick[1], A. Maksimchuk[1], K. I. Popov[3], W. Rozmus[3]

[1] FOCUS Center and Center for Ultrafast Optical Science, University of Michigan, Ann Arbor, Michigan 48109, USA
[2] Institute of Theoretical and Experimental Physics, Moscow 117218, Russia
[3] Theoretical Physics Institute, University of Alberta, Edmonton T6G 2J1, Alberta, Canada
[4] P. N. Lebedev Physics Institute, Russian Academy of Sciences, Moscow 119991, Russia

Abstract. We report on the realistic scheme of intense X-rays and γ-radiation generation in a laser interaction with thin foils. It is based on the relativistic mirror concept, *i.e.*, a flying thin plasma slab interacts with a counterpropagating laser pulse, reflecting part of it in the form of an intense ultra-short electromagnetic pulse having an up-shifted frequency. A series of relativistic mirrors is generated in the interaction of the intense laser with a thin foil target as the pulse tears off and accelerates thin electron layers. A counterpropagating pulse is reflected by these flying layers in the form of a swarm of ultra-short pulses resulting in a significant energy gain of the reflected radiation due to the momentum transfer from flying layers.

The development of sources of intense ultra-short electromagnetic (EM) pulses, X-rays, and even γ-rays is an important potential application of intense laser matter interactions [1]. One of the most promising ways to generate such sources is the use of a reflection of EM radiation from a flying relativistic mirror. This was first studied by Einstein in [2] as an example of Lorentz transformations. The radiation frequency up-shift is proportional to the square of the mirror Lorentz factor, making the scheme very attractive for the generation of high frequency pulses.

The principal idea of the relativistic plasma mirror has existed for a long time [3]. Recently several ways to create such mirrors have been proposed. One way is to use the plasma waves in the wakefield of a high intensity pulse as it travels through low density plasma in the wave breaking regime [4, 5]. Another potential method is the interaction of intense linearly polarized electromagnetic pulses with solid density plasma, where either sliding [6] or oscillating mirrors [3, 7, 8, 9] can be formed. The relativistic mirrors can also be formed in the regimes of laser-thin foil interaction previously considered in regard with the ion acceleration [10, 11, 12, 13].

In this letter we propose a more realistic mechanism for generation of ultra-short EM pulses in laser-solid density target interaction. In the proposed scheme such pulses are created in the course of a counterpropagating laser pulse interaction with a series of flying electron layers, *i.e.* the Relativistic Multilayer Reflection (RMR) mechanism. Such layers are produced when an intense laser pulse interacts with a thin solid density target and extracts and accelerates thin electron layers [14, 15, 16, 17]. The high density and relativistic velocity of these electron layers make it possible that such structures will reflect the incoming radiation, acting as flying mirrors.

Let us first estimate the properties of the flying electron layers using a simple 1D model. Serial layer production occurs at the instants when the laser field is high enough to extract electrons from the foil, which attracts them back by the Coulomb force. Such values of the field are reached at consecutive maxima and minima of laser electric field. The number of electrons per flying layer can be obtained from the requirement that the Coulomb attraction force should be compensated by the Lorentz force exerted by the EM field on the electrons and the fact that each electron layer, escaping the attraction of the ion core, increases the charge separation field that should be compensated for by the laser pulse field [17]. Let us approximate the field of the laser as $a = a_0 \exp[-t^2/\tau^2]\cos[2\pi t/T]$, where τ is the half of the duration of the pulse and T is the period of the EM wave. The maxima and minima of such a wave are at $t_j = Tj/2$, $j = 0, \pm1, \pm2, \ldots$. Then for some field maximum a_j the charge separation field that already exists is determined by a_{j+1}. Then the number of electrons evacuated is $\Delta N_j^e = \lambda R^2 n_{cr}(a_j - a_{j+1})$, here R is the radius of the irradiated area. The total number of evacuated electrons will be $\Delta N^e = \sum_j \Delta N_j^e = \lambda R^2 n_{cr} a_0$, *i.e.* is determined by the maximum of the vector potential only. The duration of the bunch can be found by solving the equation $a(t_j + \xi_j) = a_{j+1}$:

$$\frac{\xi_j}{T} = \frac{1}{2\pi}\arccos\left\{\exp\left[-\frac{T^2}{\tau^2}\left(\frac{j}{2}+\frac{1}{4}\right)\right]\right\}. \qquad (1)$$

For a laser pulse with $\tau = 5T$ and $T = 3$ fs the electron bunch has an attosecond duration: $\xi_0 = 60$ as, which is in good agreement with the results of 2D PIC simulations presented below.

In order to estimate the reflection coefficient of the flying electron layer we perform Lorentz transformation to the reference frame co-moving with the electron layer and use the results of Ref.[8]: reflection $\rho = \epsilon_j/(i + \epsilon_j))$ and transmission ($\tau = i/(i + \epsilon_j)$) coefficients. Here $\epsilon_j = \epsilon_0^j/(1+\beta)\gamma = \epsilon_0^j/2\gamma$ (since $\beta \sim 1$) is the parameter governing the transparency of the foil in the co-moving with the foil frame and $\epsilon_0^j = \pi \Delta n_j^e \xi_j / n_{cr} T$ is the transparency parameter of the electron layer in the laboratory frame [8]. For the incident laser intensity, I_0, the reflected pulse in the laboratory frame will have the up-shifted frequency by a factor of $4\gamma^2$ and the increased intensity and energy

$$I = \frac{16\epsilon_0^{j\,2}\gamma^4}{4\gamma^2 + \epsilon_0^{j\,2}} I_0, \quad \mathcal{E}_r = \frac{4\epsilon_0^{j\,2}\gamma^2}{4\gamma^2 + \epsilon_0^{j\,2}} \mathcal{E}_0. \qquad (2)$$

For $\gamma \ll \epsilon_0^j$ $\rho \to 1$ and reflected intensity and energy are determined by the Lorentz factor alone. In the second case, $\gamma \gg \epsilon_0^j$, the foil moves so fast that it becomes increasingly transparent for incoming radiation and the energy of the reflected pulse is limited by the transparency parameter $\mathcal{E}_r \sim \epsilon_0^{j\,2}$. Thus the efficiency of the light intensification is determined by $\gamma^2 \times \min\{\gamma^2, \epsilon_0^{j\,2}\}$ for intensity and $\min\{\gamma^2, \epsilon_0^{j\,2}\}$ for energy.

Below we present the results of 2D PIC simulation using code REMP [20]. The targets are composed of fully ionized carbon C^{+6} with an electron density of $400n_{cr}$. The grid mesh size is $\lambda/200$, space and time scales are given in units of λ and $2\pi/\omega$, respectively, the simulation box size is $15\lambda \times 12.5\lambda$, where λ and ω are high-intensity laser wavelength and frequency respectively. The number of particles per cell is 225. The 1.6 PW laser pulse which generates flying electron layers is introduced at the left boundary and propagating along x axis from left to right. The pulse is linearly polarized along the y axis (P-polarization), tightly focused ($f/D = 1$), and has Gaussian transverse and longitudinal profiles. The counterpropagating laser is introduced at the right boundary and propagates from right to left along the x axis. It is polarized along the z axis (S-polarization) in order to distinguish between the radiation generated by the accelerating pulse and the reflected one. It has $a_0 = 1$ and $\lambda_0 = 4\lambda$.

Below we present the results of 2D PIC simulations for the cases of a mass limited target and a thin foil (see Fig. 1). We consider the mass limited target to show the formation of flying electron layers without a transverse flow of electrons towards the irradiated spot. In the first case the counterpropagating laser pulse is reflected from the flying electron layers accelerated

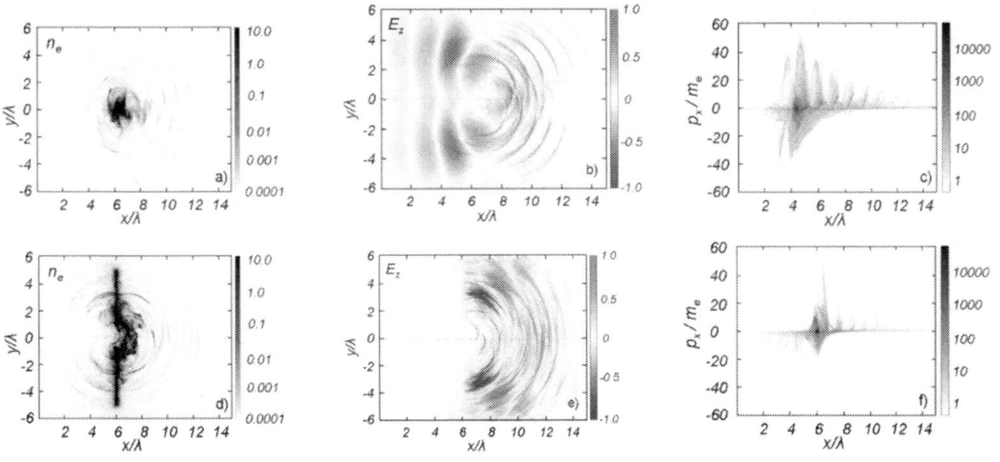

Figure 1. (color on-line)The reflection of laser pulse by accelerated electron layers in the case of a mass limited target (upper row: a, b, and c) and a thin foil (lower row: d, e, and f). The electron density distribution a) at $t = 20$ and d) at $t = 18$; the distribution of counterpropagating pulse electric field after reflection, b) at $t = 20$ and e) at $t = 18$, the field is measured in units of $m_e c\omega/e$; the distribution of electrons in (p_x, x) phase plane at $t = 12$, c) and f).

from a disk with diameter 1λ and thickness 0.1λ placed at $x = 6.0\lambda$, *i.e.* before the focus of the accelerating pulse, which is focused at $x = 8.0\lambda$ (Figs. 1a-c). In the second case the pulse interacts with a 0.1λ thick foil placed at the focus ($x = 6\lambda$) of the accelerating pulse. The electron density distributions at $t = 20$ are shown for both cases in Figs. 1a and 1d. The thin electron layers that act as flying relativistic mirrors can clearly be seen. The density of these layers vary from 5 to $15n_{cr}$ for the mass limited targets and from 10 to $20n_{cr}$ for the thin foil. The duration of these bunches is about 70 as. The distance between the layers is equal to λ in the half-spaces $y > 0$ and $y < 0$. At the same time the layers in $y > 0$ and $y < 0$ half-spaces are shifted by $\lambda/2$ in the x direction with respect to each other. The generation of swarms of short EM pulses through the RMR mechanism is demonstrated in Figs 1b and 1e. In order to determine the properties of flying electron layers and determine Lorentz factors of mirrors the distributions of electrons in (p_x, x) phase plane are shown in Figs. 1c and 1f for $t = 12$. The formation of flying relativistic mirrors with $\gamma \sim 5$ can clearly be seen.

Further intensification can possibly be achieved if we are able to focus the reflected pulse into a diffraction limited spot. Let us estimate the resulting intensity and possible applications to the study of QED effects in such fields. For a single pulse it will lead to

$$I_f \simeq (D/\lambda_r)^2 I = \begin{cases} 256\gamma^8 (D/\lambda_0)^2 I_0, & \gamma \ll \epsilon_0^j \\ 64\gamma^6 \epsilon_0^{j\,2} (D/\lambda_0)^2 I_0, & \gamma \gg \epsilon_0^j \end{cases} \tag{3}$$

where D is the reflected pulse width before focusing and $\lambda_r = \lambda_0/4\gamma^2$. Then for $\epsilon_0^j \sim 10^2$, $\gamma \sim 10$, $D = 3\lambda_0$ and $I_0 \sim 10^{18}$ W/cm^2 the resulting intensity will be of the order of the intensity characteristic for the effects of nonlinear QED, *i.e.* Schwinger intensity, $I_S \sim 10^{29}$ W/cm^2, [18]. At this intensity the probability of one of the most profound processes of nonlinear QED, the e^+e^- pair production in vacuum by strong EM field, becomes optimal. We should note here that the plane EM wave does not produce pairs in vacuum [18], because in this case both field invariants, $\mathcal{F} = \mathbf{E}^2 - \mathbf{H}^2$, $\mathcal{G} = \mathbf{EH}$, are equal to zero, which is not the case for the focused pulse [19].

Let us estimate the threshold field (intensity) needed to produce one electron-positron pair by such a field in vacuum. Since the pairs are produced primarily near the focus, we take $\pi R^2 c\tau$ for the spatial volume of the pulse and estimate the number of pairs produced, according to [18], as

$$N \approx \frac{e^2 E_S^2}{4\pi^2 \hbar^2 c} \pi R^2 c\tau^2 \bar{\epsilon}\,\bar{\eta} \coth\frac{\pi\bar{\eta}}{\bar{\epsilon}} \exp\left(-\frac{\pi}{\bar{\epsilon}}\right), \quad (4)$$

here $\bar{\epsilon}$ and $\bar{\eta}$ are the averaged over time values of dimensionless field invariants $\epsilon = \sqrt{(\mathcal{F}^2+\mathcal{G}^2)^{1/2}+\mathcal{F}}/E_S$ and $\eta = \sqrt{(\mathcal{F}^2+\mathcal{G}^2)^{1/2}-\mathcal{F}}/E_S$ in the focus and have the meaning of electric and magnetic fields in the reference frame where they are parallel, see, e.g., Ref. [18]. Here $\mathcal{F}$ and $\mathcal{G}$, are the invariant of the electromagnetic field. In order to find $\bar{\epsilon}$ and $\bar{\eta}$ we utilize the results of Ref. [19], where a 3D model of electromagnetic field in focus [21] was used. Then $\mathcal{F}_e(0,0) = 4\Delta^2 E_0^2 g^2\left(\frac{\varphi}{\tau}\right)$, $\mathcal{G}_e(0,0) = 0$, $\epsilon = 2\sqrt{2}\Delta g E_0$. Here $\Delta \equiv c/\omega R = \lambda/2\pi R$ and R is the radius of the focal spot. The intensity in this model is $I = G\frac{c}{4\pi}E_0^2$, $G = \sqrt{\pi/32}$ for $g = \exp(-4(t-z)^2/\tau^2)$. Then $\bar{\epsilon} = (128\pi)^{1/4}(I/I_S)^{1/2}\Delta$, $\quad I = (\bar{\epsilon}^2 I_S)/(8\sqrt{2\pi}\Delta^2)$. The average threshold field, i.e. the field that is needed to produce one electron-positron pair per pulse, is given by $\bar{\epsilon}_{th} = (\pi/\Lambda)(1-2/\Lambda\log(\pi/\Lambda))$, $\Lambda = 2\log\left[(c\tau\lambda)/(4\pi^2 l_c^2\Delta)\right]$, here l_c is the electron Compton wavelength. Since the focus pulse an up-shifted frequency by a factor of $4\gamma^2$ and is compressed by the same factor, $\Lambda = 18$ for $\lambda_r = \lambda/4\gamma^2$, $\tau_r = \tau/4\gamma^2$, $\Delta = 0.1$ and $\gamma = 10$. Then $\bar{\epsilon}_{th} = 0.2$ and the intensity is $I_{th} = 9.7\times 10^{28}\mathrm{W/cm^2}$.

As a conclusion, in this letter we considered a new way to generate ultra bright high intensity X-rays and γ-rays by reflecting EM pulse from the relativistic mirror. In the proposed scheme the role of the flying mirror is taken by laser accelerated electron layers, which are formed in the process of the intense laser pulse interaction with thin solid density targets. The reflected pulses have an up-shifted frequency and increased intensity. The reflected pulse intensification is determined by $\gamma^2 \times \min\{\gamma^2, \epsilon_0^{j\,2}\}$. Further intensification of the reflected light can be achieved by its focusing into a diffraction limited spot that will bring the resulting peak intensity well into the domain of nonlinear QED with laser systems, which are presently available.

This work was supported by the National Science Foundation through the Frontiers in Optical and Coherent Ultrafast Science Center at the University of Michigan and Russian Foundation for Basic Research.

[1] Krausz F and Ivanov M 2009 *Rev. Mod. Phys.* **81** 163.
[2] Einstein A 1905 *Ann. Phys. (Leipzig)* **17** 891.
[3] Bulanov S V, Naumova N M, and Pegoraro F 1994 *Phys. Plasmas* **1** 745.
[4] Bulanov S V, Esirkepov T Zh, and Tajima T 2003 *Phys. Rev. Lett.* **91** 085001.
[5] Kando M, *et al.* 2007 *Phys. Rev. Lett.* **99** 135001; Pirozhkov A S, *et al.* 2007 *Phys. Plasmas* **14** 123106.
[6] Pirozhkov A S, *et al.* 2006 *Phys. Lett. A* **349** 256; 2006 *Phys. Plasmas* **13** 013107.
[7] Lichters R, Meyer-ter-Vehn J, and Pukhov A 1996 *Phys. Plasmas* **3** 3425.
[8] Vshivkov V A, *et al.* 1998 *Phys. Plasmas* **5** 2727.
[9] Zepf M, *et al.* 2007 *Phys. Rev. Lett.* **98** 103902.
[10] Meyer-ter-Vehn J and Wu H -C 2009 *Eur. Phys. J D* **55** 433; Kulagin V V, *et al.* 2007 *Phys. Rev. Lett.* **99** 124801.
[11] Habs D, *et al.* 2008 *Appl. Phys. B* **93** 349.
[12] Esirkepov T Zh, *et al.* 2009 *Phys. Rev. Lett.* **103** 025062.
[13] Qiao B, *et al.* 2009 *New J. Phys.* **11** 103042.
[14] Naumova N, *et al.* 2004 *Phys. Rev. Lett.* **92**063902; *Phys. Rev. Lett.* **93** 195003.
[15] Ma Y -Y, *et al.* 2006 *Phys. Plasmas* **13** 110702.
[16] Tian Y, *et al.* 2008 *Phys. Plasmas* **15** 053105.
[17] Popov K I, *et al.* 2009 *Phys. Plasmas* **16** 053106.
[18] Schwinger J 1951 *Phys. Rev.* **82** 664.
[19] Narozhny N B, *et al.* 2004 *Phys. Lett. A* **330** 1; Bulanov S S, *et al.* 2006 *JETP* **129** 14.
[20] Esirkepov T Zh 2001 *Comput. Phys. Comm.* **135** 144.
[21] Narozhny N B, Fofanov M S 2000 *JETP* **117**, 867.

Collimation of relativistic laser-generated high energy electron beams via double cone target in fast ignition scheme

Hong-bo Cai[1,2], Kunioki Mima[2,3], Tomoyuki Jozaki[2], Hideo Nagatomo[2], Atsushi Sunahara[4], Wei-min Zhou[5], Rodney J. Mason[6]

[1]Institute of Applied Physics and Computational Mathematics, PO Box 8009, Beijing 100088, China
[2]Institute of Laser Engineering, Osaka University, 2-6 Yamadaoka, Suita, Osaka 565-0871, Japan
[3]The Graduate School for the Creation of New Photonics Industries, 1955-1, Kurematsu, Nishiku, Hamamatsu, Sizuoka 431-1202, Japan
[4]Institute for Laser Technology, 2-6 Yamadaoka, Suita, Osaka 565-0871, Japan
[5]Laser Fusion Research Center, China Academy of Engineering Physics, Mianyang, Sichuan, 621900, China
[6]Research Applications Corporation, Los Alamos, New Mexico 87544, USA

E-mail: hongbo-cai@ile.osaka-u.ac.jp

Abstract. The success of fast ignitor scheme will ultimately relies on the efficient generation and propagation of enormous number of high-energy charged particles. Particle-in-cell simulations aimed at improving the coupling efficiency of input laser energy deposited to a compressed core by using a double cone are described. Quasistatic magnetic fields generated at the vacuum layer inside the double cones are found to play an important role in confining the high energy electrons. The double cones result in the confinement and focusing of about 15% of the input energy for deposition in the compressed core.

1. Introduction

In the Fast Ignitor [1], a relativistic electron beam is considered to be the most suitable source for igniting a hot spot much smaller than the dense compressed DT core. In order to improve the efficiency of the coupling and transport of the energy into dense plasma, cone targets[2,3] have been used in the fast ignition scheme. The merits of cone targets have been shown both in experiments [2] and simulations [4,5].

However, it is found that electrons accelerated by the laser field can escape freely from cone sides to the surrounding coronal plasma, resulting in the decrease of the energy flux through the cone tip. Nakamura et al.[4] have shown that the double cone confines the electrons for hundreds of femtoseconds with an immobile ion background. In their simulations the electrons are blocked by the sheath electric field inside the vacuum gap. However, the plasma expansion could be very significant. Over several hundreds of femtoseconds the sheath electric field inside the gap could be reduced to a very low level. Therefore, important questions arising are: 1) whether the double cone is still effective in confining the high energy electrons after one picosecond, and 2) how much energy can be confined and focused to the compressed core by using the double cone? These are the principal subjects of this Letter.

2. Numerical study of double cone target

Figure 1(a) is a sketch of the geometry of the simulations. The width of our inner cone wing is $5\lambda_0$ and the width of the gap is $3\lambda_0$. The plasma consists of three species: electrons, protons with $m_p/m_e = 1836$ outside the cone, and heavy ions (gold ion with an assumed charge state $Z_i = 40$) and $m_i/m_e Z_i = 195.4/40$. The gold cone, whose edge is drawn with the dashed lines, is surrounded by a hydrogen plasma. Both the plasma density of the gold cone and that of the hydrogen plasma are $40n_c$. The p-polarized laser pulse at $\lambda_0 = 1.06\mu m$ wave length, and $1.2 \times 10^{19} W/cm^2$ intensity irradiates the target from the left boundary. The intensity profile is Gaussian in the y-direction with a spot size of $5.0\mu m$(FWHM). The laser rises in $20T_0$, where T_0 is the laser period, after which the laser amplitude is kept constant. A typical simulation duration is $450T_0$, which corresponds to about 1.5ps for $\lambda_0 = 1.06\mu m$.

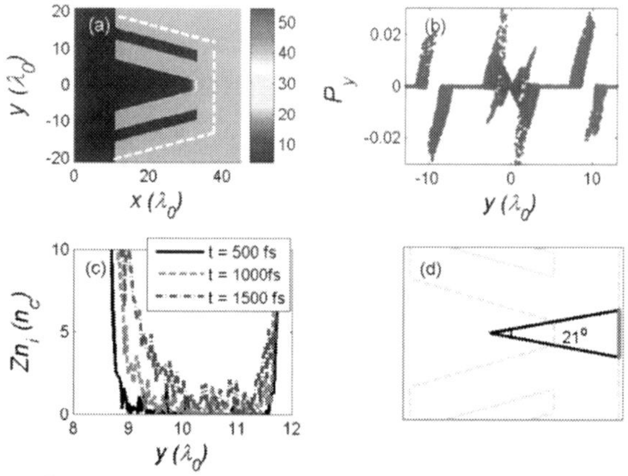

Figure 1. (a) Initial density profile of the double cone target. (b) Phase space of high Z ions in the region $x \in (25, 30)\lambda_0$ at $t = 500fs$. (c) A slice of the ion density profile at position $x = 25\lambda_0$ inside the upper gap of the double cone target. (d) Sketch of collection cone (dark) for the double cone target (green).

Here, we use a grid size of $\Delta x = \Delta y = \lambda_0/64$ with 2800×2688 grid cells. The time step used is $0.01T_0$. Fifty particles are used in one mesh, and the total number of particles is about 2.55×10^8. Both the fields and particles boundary conditions are absorbing boundary conditions, either in the x and y direction. In order to eliminate the unphysical sheath field at the boundaries, we set cooling buffers in our simulation. Furthermore, in order to reduce the restrictions on the grid size compared with the Debye length, we used a fourth order interpolation scheme to evaluate fields and currents [6].

To determine the plasma expansion inside the gap, we plot the phase space of high Z ions in the region $x \in (25, 30)\lambda_0$ at time $t = 500fs$ in Fig. 1(b). Clearly, after $1ps$ of interaction the vacuum gap is filled with plasma a density that can be as high as several critical densities. Thus, we may conclude that we can not expect the sheath electric fields to be still effective in confining high energy electrons after $t > 1ps$.

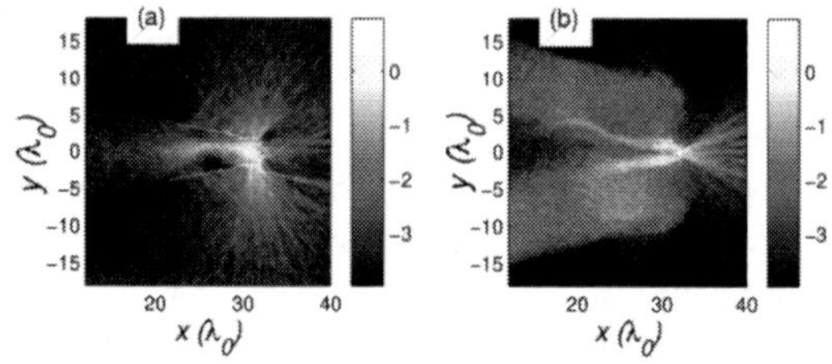

Figure 2. The natural logarithm of the electron energy density for single cone (a) and double cone (b) at time $t = 1000fs$.

In Figure 2 the energy density distributions of electrons with energy between $0.5 \leq E[MeV] \leq 2.0$ are plotted. It is clearly seen that the high energy electrons are mainly accelerated at the cone tip and cone side wall. In the single cone case, some of high energy electrons move freely into the surrounding corona plasma and then the energy flux decreases through the cone tip. Alternatively, in the double cone case, few electrons can "leak" out into the surrounding corona plasma even after one picosecond.

Figure 3(a) shows that the sheath electric field peaks at time $t = 250fs$, and eventually decreases to very low level after time $t = 600fs$. In comparison, the quasistatic magnetic field still keeps growing after it grows up at time $t = 330fs$. It is important to stress that the gyro-radius of a typical $1MeV$ hot electron is less than $\lambda_0/2$ under the quasistatic magnetic fields which are the order of $100MG$. Therefore, the electrons escaping into the vacuum gap can be effectively reflected back by the magnetic fields.

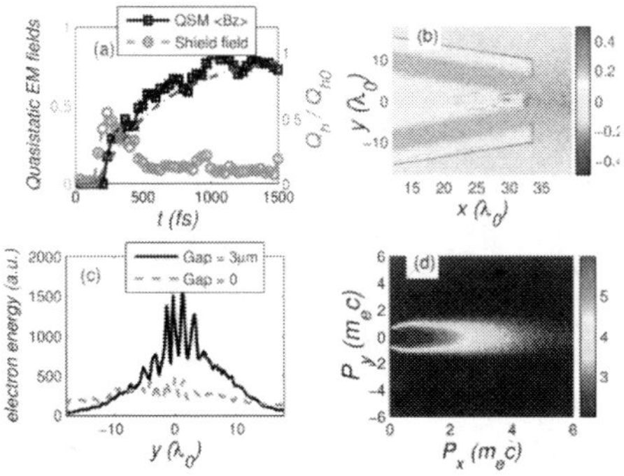

Figure 3. (a) The maximum values of the quasistatic magnetic field (solid line) and the sheath electric field (dashed line) inside the gap. The dash-dotted line is the plot of $Q_h(t)$ in Eq. (2). (b) The time averaged current j_x at time $1500fs$. (c) The transverse distribution of the energy of the escaped electrons from the right boundary. (d) The natural logarithm of the momentum distribution of the collected high energy electrons.

3. Theoretical analysis of the generation of magnetic fields

In order to evaluate the total energy of the recirculating high energy electrons Q_h, we employ

$$\frac{dQ_h}{dt} = G - Q_h \frac{cS}{3V}(1-\beta) - Q_h \frac{cS_0}{3V}, \quad (1)$$

where G is the input laser energy, the second term is the energy flux escaping to the side, β is the reflectivity of high energy electrons by the electromagnetic fields, S is the side area, V is the total volume of the system, and the third term is the energy flux emitted from the cone tip. Also, S_0 is the area of the cone tip. We assume $|dln\beta/dt| \ll \Omega \equiv cS/3V$, then integrate Eq. (1) to obtain:

$$Q_h = \frac{G}{\Omega[(1-\beta)+S_0/S]}\{1 - e^{[-\Omega(\int_0^t (1-\beta(t))dt + tS_0/S)]}\}. \quad (2)$$

Eq. (2) indicates that the total energy of high energy electrons increases with time. As a numerical example, we evaluate Eq. (2) for the present double cone case. According to Fig. 3(b), the cone tip area S_0 is about $10\mu m$ and the side area S is about $40\mu m$. On the other hand, simulation shows the flux escaping from the side wall is about 1/4 of the flux escaping from the tip. Those facts indicate that $(1-\beta)$ is about 1/16. Since $V \simeq 500\mu m^2$, $\Omega \equiv cS/3V = 8 \times 10^{12}s^{-1}$. So that the electron accumulation time, $\tau_a = [dln(Q_h)/dt]^{-1}$ can be estimated as $\tau_a = [\Omega(1-\beta) + S_0/S]^{-1} \sim 0.5ps$. Then $Q_h(t)$ of Eq. (2) is plotted in Fig. 4(a). Note that the quasistatic magnetic field $\langle B_z(t) \rangle$ is proportional to $Q_h(t)$.

The escaped high energy electrons are observed at the boundaries. In the single cone target case only 47.6% of the total absorbed laser energy is contained in the forward energy flux by high energy electrons. When using a double cone, this fraction can be as high as 79.4%. In fact, a significant amount of high energy electrons escape from the side walls - about 19% percent of the total input laser energy - is saved for the double cone case compared to that in the single cone case.

In Fig. 3(c) we plot the time-integrated electron energy observed at $x = 40\lambda_0$ as a function of y-coordinate. To determine the energy flux of electrons injected into the core, we measure the high energy electrons across a series of planes within the cone; see Fig. 1(d). We found about 14.8% of input laser energy can deposit in the core. In Fig. 3(d) the momentum distribution of this group of collected electrons with energy larger than $100keV$ shows that the double cone target works effectively in collimating and focusing high energy electrons.

4. Wider cone tip size and larger pre-plasma inside the cone

In FIREX-I project in Osaka University, the ignition laser pulse is energy of 10kJ and pulse duration of 1ps to 10ps which is avoidably accompanied by a ns ASE-pedestal. The ASE pulse preheats the target and generates a large pre-plasma. For FIREX-I heating laser, hydrodynamic simulations [7] show that under the irradiance of 1ns ASE pulse with intensity $I = 10^{11}W/cm^2$, the pre-plasma has three segments. In the vicinity of solid density, the scale length is λ_0 and the lower end of the density slop terminates at $3n_c$. In over-dense region, the scale length is $7\lambda_0$ and the lower end of the density slop terminates at $0.3n_c$. While in the underdense region, the scale length is $17\lambda_0$ and the lower end of the density slop terminates at $0.08n_c$; see Fig. 4(a). Here, the cone tip size is $6\mu m$. The Gaussian laser pulse rises in $1.6ps$ and its FWHM duration is $2ps$. The laser intensity is $1.2 \times 10^{19}W/cm^2$ and its spot size is $10\mu m$ (FWHM). It is found the pre-plasma plays an important role in the generation of high energy electrons. Fig. 4(b) shows the high energy density at time $2ps$. one can see that the high energy electron beam is easily deviated with the laser axis, which introduces some uncertainty for FI. In order to avoid this deviation, double cone target is needed. Computations in this last case are currently in progress, and will be the object of a future publication.

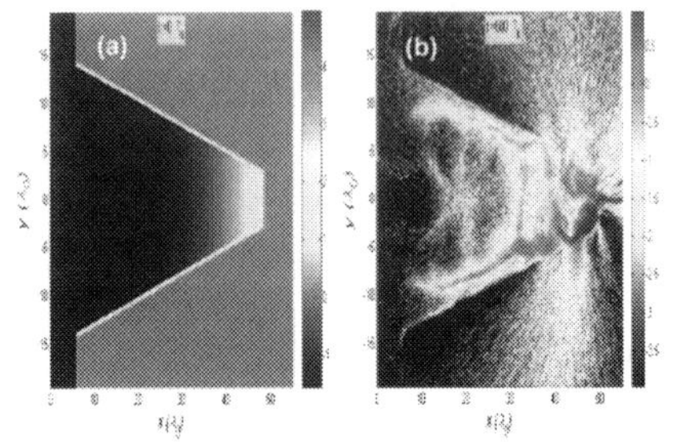

Figure 4. (a) Initial density profile of the cone target with a large pre-plasma. (b) The natural logarithm of the electron energy density at time $t = 2ps$.

5. Conclusion

In summary, we performed 2D3V-PIC simulations to study the merits of double cone in fast ignition. Our simulation results indicate that the energy flux through the cone tip in the double cone is much larger and more tightly focused than that in the single cone. It has been shown that in the single cone case, the high energy electrons, about 30% of the total input laser energy, escape into the surrounding corona plasma from the side walls. In comparison, in the double cone case the high energy electron flux, about 11% percent of the total input laser energy, escapes from the side walls.

This work was supported by MEXT, a Grant-in-Aid for Creative Scientific Research (15GS0214).

References

[1] M. Tabak *et al.*, Phys. Plasmas **1**, 1626 (1994).
[2] R. Kodama *et al.*, Nature **418**, 933 (2002); *ibid.* **412**, 798 (2001).
[3] R. J. Mason, Phys. Rev. Lett. **96**, 035001 (2006); R. J. Mason, Bull. Am. Phys. Soc. **50**, 139 (2005).
[4] T. Nakamura *et al.*, Phys. Plasmas **14**, 103105 (2007).
[5] Y. Sentoku *et al.*, Phys. Plasmas **11**, 3083 (2004).
[6] Y. Sentoku *et al.*, J. Comput. Physics **227**, 6848 (2008).
[7] A. Sunahara et al., submitted to IFSA 2009.

Hot electron transport modelling in fast ignition relevant targets with non-Spitzer resistivity

D A Chapman[1], S J Hughes[2], D J Hoarty[1], and D J R Swatton[1]
[1] Plasma Physics Department, AWE, Aldermaston, Reading, Berkshire, RG7 4PR, UK
[2] Computational Physics Group, AWE, Aldermaston, Reading, Berkshire, RG7 4PR, UK
E-mail: david.chapman@awe.co.uk

Abstract. The simple Lee–More model for electrical resistivity is implemented in the hybrid fast electron transport code THOR. The model is shown to reproduce experimental data across a wide range of temperatures using a small number of parameters. The effect of this model on the heating of simple Al targets by a short–pulse laser is studied and compared to the predictions of the classical Spitzer–Härm resistivity. The model is then used in simulations of hot electron transport experiments using buried layer targets.

1. Introduction

Fast ignition (FI) Inertial Confinement Fusion schemes, and many laser–driven High Energy Density Physics experiments, rely on the efficient propagation of an energetic electron beam through high density, partially ionised (and potentially degenerate) matter [1]. The transport of the beam energy is strongly affected by the resistivity of the target, which dictates the heating due to the return current via Ohm's law, $\mathbf{E} \approx -\eta(T)\mathbf{J}$. In this work the resistivity model of Lee and More [2] is used in simulations of fast electron transport in FI relevant targets using the relativistic explicit hybrid transport code THOR. The code is based on the work of Davies *et al.* [3], i.e. the transport of the hot electron population is modelled through interaction with a cold fluid background of electrons and ions on an Eulerian grid. The code is also computationally inexpensive, scales well on parallel systems compared to many PIC or VFP schemes, and can treat both 2D and 3D Cartesian and axisymmetric geometries.

2. Lee–More resistivity model

The classical Spitzer scaling of electrical resistivity [4] is known to break down in the low temperature/high density conditions produced during short–pulse laser–matter interactions. In low temperature regions of the target the resistivity is then significantly overestimated (see Figure 1a), leading to far larger return current heating and magnetic fields than may be expected. Consequently, the Spitzer resistivity is often capped, $\eta_{\max} = \eta_{\rm SH}(100\ {\rm eV})$, to prevent unphysical behaviour in transport calculations. However, it may be naive to assume that such an approach is sufficient for describing the evolution of solid density targets initially at room temperature. The resistivity model of Lee and More (L–M) is a practical alternative to rigorous calculations [5] which adequately captures the effects of strong coupling and degeneracy, is material dependent and easily implemented. The resistivity is calculated from an effective collision frequency,

$\eta_{\rm LM} = \nu_{\rm eff}/A(\beta\mu)\,\varepsilon_0\omega_{pe}^2$, appropriate to the temperature regime. A summary of the L–M model used in this work is given below

$$\nu_{\rm eff} = \min\left(\nu_{\rm max}, \left(\nu_{\rm hot}^{-1} + \nu_{\rm cold}^{-1}\right)^{-1}\right), \tag{1}$$

$$A(\beta\mu) = \frac{4}{3}\frac{F_2(-\beta\mu)}{(1+e^{-\beta\mu})F_{1/2}^2(-\beta\mu)}, \tag{2}$$

$$\nu_{\rm hot} = f(Z^*)\frac{Z^* e^4 n_e \ln(\Lambda)}{3(2\pi)^{3/2} m_e^2 \varepsilon_0^2 v_T^3}, \tag{3}$$

$$\nu_{\rm max} = \frac{1}{r_0}\left(v_F^4 + v_T^4\right)^{1/4}, \tag{4}$$

$$\nu_{\rm cold} = \frac{\nu_{\rm max}}{50}\frac{T}{T_m}, \tag{5}$$

where $f(Z^*) \approx (1 + 3.3/Z^*)$ is an empirical factor to account for electron–electron collisions [6], $v_{T,F}$ is the thermal/Fermi speed, r_0 is the mean ion sphere radius, T_m is the melting temperature of the material, $\beta\mu$ is the normalised chemical potential and F_j is the Fermi–Dirac integral of order j. A good fit to experimental data (blue curve) [7] can be obtained using the L–M model (red curve) with a set of adjustable parameters α_j, that scale the collision frequency in the cold, saturation, and plasma regimes respectively. It is clear that in the low temperature regime the capped Spitzer model (dashed black curve) is inadequate to describe the resistivity.

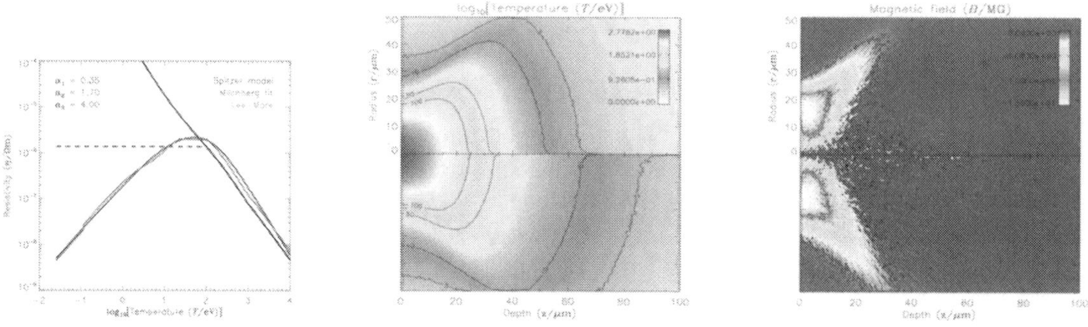

(a) (b) (c)

Figure 1: (Colour online) (a): temperature scaling of resistivity models. (b–c): comparisons of the background temperature (b), and magnetic field (c) for the optimised Lee–More model (top half) and capped Spitzer (bottom half) models in initially cold ($T_0 = 0.025$ eV) material.

To gauge the effect of non-Spitzer resistivity on the heating and field structure a number of simulations of simple Al blocks have been performed using THOR. The electron source is created near the surface with Gaussian spatial and temporal profiles (to represent the laser pulse), and a Maxwell–Boltzmann distribution of energies. The distribution is characterised by a temperature, given by a suitable scaling on the laser irradiance, and by an angular spread, which is also Gaussian with a width of $\theta_{1/2} = 15°$. The energy of the source is 10 J, which is determined by the absorbed intensity $f_{\rm abs}I = 0.2 \times 10^{19}$ Wcm^{-2}, spot size $\sigma \approx 20$ μm, and pulse length $\tau = 0.5$ ps. The resulting heating profiles at $t = 2$ ps through the simulations are similar for both models near the surface, but begin to diverge in the low temperature region. The average temperature for $z \gtrsim 25$ μm is then greater by a factor of ~ 2 for the capped Spitzer model (Figure 1b). Consequently, the structure of the B–field, which is primarily generated near the surface, is largely unaffected, although the lobe extends further into the target for the Spitzer model (Figure 1c). However, this difference is too small to make a noticeable impact on the collimation of the electron beam for these conditions. Furthermore, changing the laser wavelength from 1ω ($\lambda = 1.054$ μm for Nd:Glass laser) to 2ω ($\lambda = 527$ μm), and thereby decreasing the hot electron temperature, gives greater heating near the surface due to increased field stopping and drag whilst the heating at large depths is not affected. As expected, if the initial temperature is high ($T_0 \sim 100$ eV) then the heating and field growth converge to the predictions of the Spitzer model.

Whilst this result suggests that the Spitzer model is not unreasonable for hot electron transport modelling, it should be noted that the seemingly good agreement with the L–M model seen here is most likely due to the relatively high energy and deposition rate of the source; the target is rapidly heated to the temperature regime in which the models give similar results. Greater differences may therefore be expected if the energy of the source were substantially reduced, i.e. for ultra–short–pulse laser or X–ray driven targets.

3. Modelling fast electron heating in buried layer targets

Recent AWE experiments [8] have measured the heating of thin (~ 0.2 μm) Al foils buried in plastic targets using 1ω and 2ω p–polarised light from the CPA arm ($\tau = 0.5$ ps) of the HELEN laser (see Figure 2a). These experiments are of interest to fast ignition as an understanding of the deposition rate of the electron beam energy in dense materials is critical to meeting ignition criteria and modelling isochoric heating experiments. A number of THOR simulations using L–M resistivity have been performed to model the range of buried layer depths and laser intensities used in the experiment; $z_{\rm Al} = 2 - 20$ μm and $I = 6 \times 10^{17}$ Wcm^{-2} (2ω), 6×10^{18} Wcm^{-2} (1ω), 1×10^{19} Wcm^{-2} (2ω). In the experiments the laser pre–pulse was mitigated to ensure isochoric heating by using a plasma mirror for the 1ω shots, and from the conversion crystal for the 2ω shots. This is modelled in the simulations by reducing the energy in the laser pulse from 50 J (a conservative estimate for HELEN) to 40 J for 1ω and 25 J for 2ω. For Gaussian spatial and temporal intensity profiles the characteristic radius of the laser spot is then given by $\sigma = 5.47(E_{\rm J}/I_{18}\tau_{\rm ps})^{1/2}$ μm. Beg's law, $T_{\rm hot} = 100(I_{17}\lambda_\mu^2)^{1/3}$ keV, is used for the hot electron temperature (since $a_0 \leq 2.2$ for all intensities considered) and estimates of the absorption and divergence are made from experimental data [9, 10].

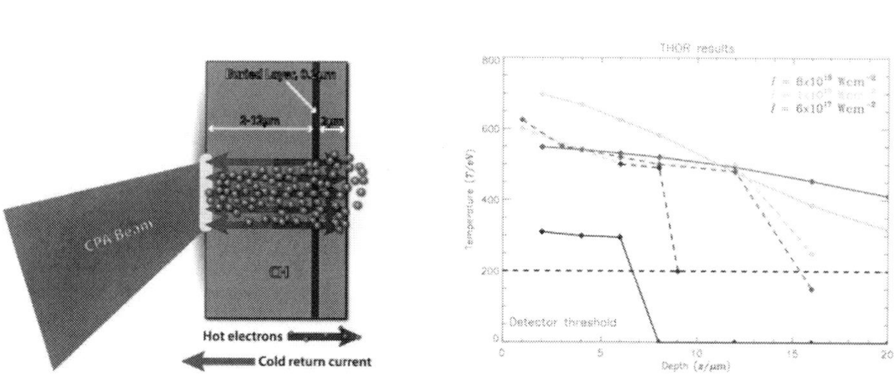

(a) (b)

Figure 2: (Colour online) (a): Schematic representation of the experimental set up. (b): comparisons of the experimental data (dashed curves) and THOR calculations (solid curves).

Figure 2b shows the calculated temperatures in the buried Al layers at the end of the laser pulse (solid curves) and the experimental data (dashed curves) for comparison. The average temperature has been calculated by integrating over the thickness of the buried layer and out to a cutoff radius corresponding to the point at which the temperature reaches the detector threshold ($T \approx 200$ eV). The shape of the heating profiles from the high intensity runs (red and green curves) are broadly consistent with the trends observed in the data up to $z \approx 12$ μm, showing a gradual decrease in temperature. However the rapid drop–off in temperature beyond this point is not observed. The high intensity 2ω run (green curve) also produces greater heating near the surface and reduced heating at depth compared to the 1ω run. Whilst this is consistent with simple estimates of the Ohmic heating rate; $\dot{T} \propto (I/\lambda)^{4/3}$, where the current density is estimated as $J_{\rm hot} = ef_{\rm abs}I/k_BT_{\rm hot}$, the data does not support this. A sharp drop–off in temperature is seen for the low intensity 2ω run (blue curve), although this occurs slightly earlier in the target and the temperatures are significantly lower than observed. This could suggest that the absorption of the low intensity pulse may have been considerably larger ($\sim 60\%$) than has been considered in this work. Improved agreement with the experiment may be achieved via a semi–integrated approach to modelling the electron source, e.g. taking the distributions of electron energy and angular spread from PIC or direct Vlasov simulations, and by better modelling the detector.

4. Conclusions

The Lee–More resistivity model has been implemented in the hot electron transport code THOR, and the subsequent effects on hot–electron heating have been studied with a number of simulations of simple Al targets. It is shown that a simple capped Spitzer model overpredicts the heating at large depths ($50 - 100$ μm) by a factor of ~ 2 compared to the L–M model, although the structure and peak value of the resulting magnetic field are not strongly affected. This result is largely independent of the hot electron temperature, as well as the initial temperature of the simulations. Initial simulations of buried layer targets heated by short–pulse produced hot electron beams have also been performed using this improved model. Whilst the results show that the heating profiles calculated by THOR broadly agree with experimental observations, the characteristic temperature drop–off for depths of $z \approx 15$ μm is not well reproduced, particularly for runs with high intensity. Future work in this area will focus on refining the resistivity model and using integrated modelling including radiation–hydrodynamics and full kinetic simulations.

References

[1] Tabak M *et al.* 1994 *Physics of Plasmas* **1** 1626–1634
[2] Lee Y T and More R M 1984 *Physics of Fluids* **27** 1273–1286
[3] Davies J R 1997 *Phyical Review E* **56**
[4] Cohen R S *et al.* 1950 *Physical Review* **80** 230–238
[5] Ichimaru S 2004 *Statistical Plasma Physics Volume II: Condensed Plasmas* (Westview Press)
[6] Atzeni S and Meyer-Ter-Vehn J 2004 *The Physics of Inertial Confinement Fusion* (Oxford Press)
[7] Milchberg H M *et al.* 1988 *Physical Review Letters* **61** 2364–2369
[8] Hoarty D J *et al.* 2009 *IFSA Plenary 3.0.2*
[9] Ping Y *et al.* 2009 *Physical Review Letters*
[10] Green J S *et al.* 2008 *Physics Review Letters* **100** 015003–1–4

The Sixth International Conference on Inertial Fusion Sciences and Applications IOP Publishing
Journal of Physics: Conference Series **244** (2010) 022032 doi:10.1088/1742-6596/244/2/022032

Integrated simulations of ignition scale fusion targets for the HiPER project

A Debayle[1], J J Honrubia[1], E D'Humières[2], V T Tikhonchuk[2], S Micheau[3] and M Geissler[3]

[1] ETSI Aeronáuticos, Universidad Politécnica de Madrid, Spain
[2] CELIA, Université Bordeaux 1, CNRS, CEA, Talence, France
[3] CPP, Queen's University of Belfast, Belfast, United Kingdom

E-mail: arnaud.debayle@upm.es

Abstract. Integrated simulations of fusion targets with a re-entrant cone are presented. Fuel assembly is modeled with one- and two-dimensional (2D) radiation hydrodynamics. Fast electron acceleration in the cone is simulated with 2D planar particle-in-cell (PIC) codes. The fast electron transport and the fuel ignition are described self-consistently with a 2D cylindrical electron hybrid code coupled to hydrodynamics. It is found that at laser intensities $\sim 10^{20}$ W cm^{-2}, the front rippling at the cone tip generates strong magnetostatic fields, producing a large electron beam divergence. Fuel ignition at acceptable values of the ultra high intensity (UHI) laser energy requires a substantial reduction of the electron beam divergence.

The method of Fast Ignition (FI) of inertial confinement fusion (ICF) targets assumes a separation of compression and heating phases of the entire process [1]. The use of cone targets provides a possibility to generate a relativistic electron beam near the dense core. It is supposed to carry a significant fraction of the laser energy to a small hot spot where the fusion reactions are triggered. While integrated experiments have demonstrated a high efficiency of coupling the UHI beam to the core [2], it is not clear whether these results apply to intensities and energies required for ignition-scale targets [3]. In this paper, we estimate the UHI laser beam requirements, to ignite a target with a gold double cone inserted, as proposed in Ref. [4], within the context of the HiPER project [5]. The entire modeling is conducted in three steps. First, we design the fuel assembly with special attention to the integrity of cone tip during the compression phase. Simulations are performed with the 1D/2D radiation hydrodynamic code SARA [6]. Second, fast electron acceleration by a UHI laser pulse in a double cone is studied with 2D PIC simulations [7] without collisions. The pre-plasma generated by the ASE laser pulse was neglected as we are concerned by a reference calculation in the ideal case. Third, the fast electron transport and energy deposition are studied with a hybrid code coupled to hydrodynamics [8].

The re-entrant cone can be seriously damaged during the deuterium-tritium (DT) shell implosion and compression due to i) direct irradiation by nanosecond laser beams, ii) shock wave generated at the shell-cone interface, and iii) shock wave and plasma jet produced when the shell collapses at the center [9]. For simplicity, we have considered a gold cone without a plastic coating in the preliminary calculations shown here. Radiation transport has not been included in the simulations to avoid the ablation of the gold cone surface and the tip by the X rays generated in the shell, as reported in Ref. [9]. In order to decrease the jet velocity and the shock pressure at stagnation, we have designed a target with an implosion velocity of

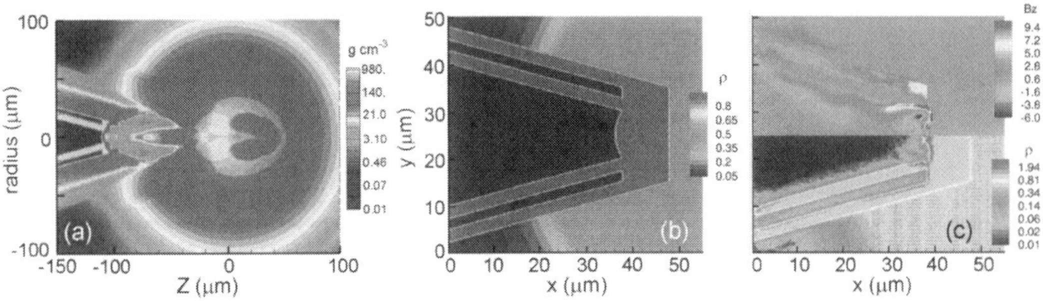

Figure 1. a) Density profile at the time of peak ρR with asymmetric implosion velocity (highest velocity near the cone wall). The doble cone has a half-angle of 15°, with an inner radius of 10 μm, an outer cone inner radius of 20 μm, and a distance between the initial shell center (z = 0) and the cone tip of 75 μm. The thickness of the inner cone wall and the void region between inner and outer cones is 5 μm. **b)** Initial density profile normalized to $80n_c$ used in the PIC simulation. (The cell size is $\lambda_0/56$ in both directions, with 40 electrons and 1 ion per cell. The size of the cone and the laser beam has been scaled-down by a factor of 2.) **c)** Density profile normalized to $80n_c$ (bottom) and magnetic field normalized to $m_e\omega_0/e$ (top) near the end of simulation, $t \sim 1.2$ ps.

2×10^7 cm s^{-1}, significantly lower than that of the HiPER reference target (2.8×10^7 cm s^{-1}) [10]. Our capsule has a total radius of 800 μm and is composed of a shell of 30 μm of CH and 220 μm of DT ice wrapping 550 μm of DT gas at a density of 0.04 mg cm^{-3}. The 1D parameters of the capsule are as follows: $\sim$ 140 kJ of absorbed energy at the wavelength of 0.35 μm, the peak $\rho R \sim 1.6$ g cm^{-2}, and the peak entropy parameter during implosion $\alpha \sim 0.77$. The absorbed power pulse is designed such as all shocks are weak and arrive at the inner shell surface equally spaced in time. The maximum power is 40 TW. Due to the lack of an ALE (Arbitrary Lagrangian Eulerian) capability in the available radiation-hydrodynamics codes, we have studied the target implosion by 1D Lagrangian hydrodynamics simulations and the fuel compression including the cone by means of 2D Eulerian hydrodynamics. The 1D simulation is carried out until the peak implosion kinetic energy is attained. Then, the density, internal energy and velocity profiles are remapped onto a 2D cylindrical mesh with the gold double cone inserted and the hydrodynamic evolution continues until the peak ρR is achieved. We use an asymmetric fuel implosion to mitigate the cone tip destruction and to optimize the mass distribution of the compressed core. This is modeled by setting a polar angle dependent implosion velocity at the remapping time (which can be obtained by a polar variation of the initial shell thickness). Assuming a P_1 asymmetry of the implosion velocity, the most favourable configuration of the compressed core is obtained when the maximum velocity is 10% over the average at the cone walls. In this case, a small plasma jet preceded by a shock wave moves towards the cone tip (see figure 1(a)), while the main jet produced at the shell collapse time is directed just in the opposite direction. The thickness of the outer cone and the tip is 30 μm in order to avoid any plasma filling of the inner cone and the void between the cone double walls. Gold or heavier materials are the most appropriate for the cone fabrication because lighter materials would lead to a faster destruction of the cone tip due to their higher shock velocity. It is worthwhile noticing that in the compressed core configuration of figure 1(a): i) the distance between the inner surface of gold cone tip and the high density core is only $\sim$ 80 μm, ii) the ρR along the axis (r = 0) has increased to $\rho R = \int_{core} \rho dz/2 \sim 2.3$ g cm^{-2} due to the jet-like flow, and iii) a dense plasma jet spreads between the cone tip and the core. It will act as a wire to guide the fast electrons.

At the time of peak ρR, we assume that an UHI laser beam is launched inside the double cone generating a fast electron beam, whose characteristics have been estimated quantitatively

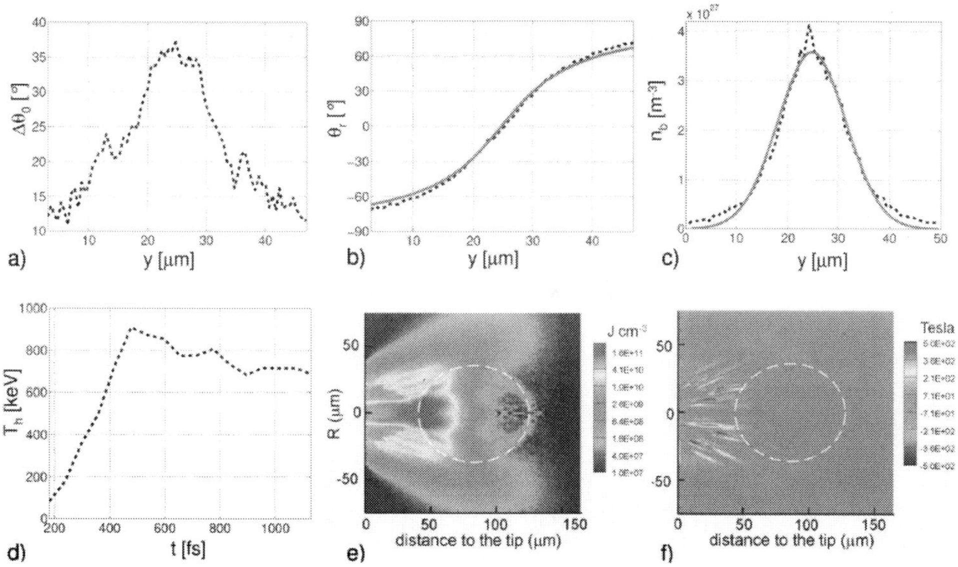

Figure 2. **a)** Dispersion angle $\Delta\theta_0$ versus radial coordinate y at time $t \sim 650$ fs for $41\,\mu$m$< x < 43\,\mu$m. **b)** Fast electron mean propagation angle θ_r versus radial coordinate y at the same time and position. **c)** Beam density versus transverse coordinate y at the same time. **d)** Hot electron temperature, in keV, depending on time for electrons with $0.25 \leq \mathrm{E} \leq 2.5\,$MeV. **e)** Energy density deposited. Fast electrons are injected at the left surface of the simulation box. **f)** Azimuthal magnetic field at the end of the pulse.

by PIC simulations. The laser maximum intensity is $I_0 = 2 \times 10^{20}\,\mathrm{W\ cm^{-2}}$, p-polarization, the wavelength is $\lambda_0 = 1\,\mu$m, the pulse duration is 1 ps, and the cone density is 80 n_c, where $n_c = m_e\epsilon_0\omega_0^2/e^2$ is the critical density. The laser beam has a temporal and spatial Gaussian shape with a radius of $10\,\mu$m at half maximum (HWHM). A short laser pulse length and a low cone density have been chosen in order to obtain hole boring conditions comparable with the case of a realistic laser pulse length and cone density. Figure 1 (b) shows the initial plasma density profile. According to figure 1 (c), the plasma density is non-uniformly modulated by the laser pulse due to the natural front rippling enhanced by the interference fringes that appear at the cone tip inner surface. The hole boring velocity, proportional to $\sqrt{I}$, is two times higher in the fringes of maximum intensity, $I_{\max} \simeq 4\ I_0$. A strong front corrugation, coupled with the transverse ponderomotive force, generates strong defocusing magnetic fields [3], that increase the electron beam divergence. The local electron angular distribution $f_l(\theta, y, t)$, with $\theta = \arctan(p_y/p_x)$, can be approximated by a Gaussian function $f(\theta) \propto \exp[-(\theta - \theta_r)^2/\Delta\theta_0^2]$, where $\theta_r(y,t)$ is the mean propagation angle and $\Delta\theta_0(y,t)$ the angular dispersion. Figures 2 (a) and (b) show $\Delta\theta_0$ and θ_r versus the radial position for electrons with energies higher than 250 keV behind the acceleration zone. The dispersion angle is maximum at the cone tip inner surface due to the strong magnetic field. However, its value is small, $\sim 15 - 30°$ at HWHM. The high electron divergence comes from the radial velocity, given by θ_r. For instance, electrons at the distance $y \sim 8\,\mu$m are ejected in the direction of $\sim 35°$. Consequently, the full electron angular distribution function, $f(\theta) = \int f_l(\theta, y)dy$, presents a high dispersion angle with a HWHM of $\Delta\theta \sim 55°$. Both parameters, $\Delta\theta_0$ and θ_r, are constant in time. However, the fast electron beam radius increases with time, and it attains the value of the outer cone inner radius because of the double cone confinement. This effect increases the full divergence with time, since more electrons have a high radial velocity. The local fast electron energy spectrum is also estimated versus transverse direction and time, and it is characterized by a Gaussian function for electrons with energy between 250 keV and 2.5 MeV. The fast electron temperature does not depend on radial direction due to the electron mixing in strong magnetic fields. Moreover, as seen in figure 2(d), the electron temperature saturates below 1 MeV due to the profile steepening effect [11]. The laser-to-electron conversion efficiency is about 35 % for electrons with energy $E \geq 100$ keV.

Assuming a cylindrical geometry for the initial radial velocity and the dispersion angle, the parameters of the fast electron source obtained by PIC simulations were used in the electron transport code. It is important to point out that the damaged cone tip has been artificially removed from the simulation box to avoid the scattering of fast electrons due to Coulomb collisions [12] and the magnetic defocusing caused by resistivity gradients at the Au/DT interface. Special attention has been payed to the electron initial divergence. If the initial radial current is neglected and electrons are injected symmetrically around the propagation axis, the beam is strongly collimated by the self-generated resistive magnetic field. On the contrary, if electrons are injected with the initial radial current observed in PIC simulations, the beam diverges just after injection, the magnetic field is much weaker and the beam collimation almost disappears. This is a central result of our analysis. Even with all the approximations assumed, our results show that the target of figure 1(a) can be ignited only with an unacceptably high electron beam energy (≥ 50 kJ) for the divergence $\Delta\theta = 55°$ found in PIC simulations. A reduction of $\Delta\theta$ to 35° lowers the electron beam ignition energy to the level of 40 kJ (delivered in 20 ps, FWHM), which could be achieved with the laser energies around 100 kJ envisioned in the HiPER project [5]. Figures 2(e) and (f) show the energy deposition and self-generated magnetic field in this case. It is worth remarking that the density jet that spreads from the compressed core to the cone tip contributes to the electron beam guiding. This can be explained the resistive magnetic field, $j_b \times \nabla\eta$, generated at the edges of the jet due to its lower Ohmic heating, lower temperature and higher resistivity when compared with the surrounding low density plasma.

In summary, we have performed integrated simulations relevant for FI. 2D hydrodynamic simulations show that an asymmetric shell implosion and sufficiently thick gold cone walls and tip prevent its destruction before peak compression. PIC simulations show unacceptable high electron beam divergence $\Delta\theta \sim 55°$. Strong radial currents of accelerated electrons are due to the laser ponderomotive force and strong density modulations that generate defocusing magnetic fields. Fuel ignition at the electron beam energy level of 40 kJ can be achieved if the initial beam divergence can be reduced to at least 35° by means of optimized designs of the cone and the UHI laser beam temporal and spatial profiles.

Acknowledgments

This work has been partially supported by the HiPER project and the research grants ENE2006-06339 and CAC2007-13 from the Spanish Ministry of Education and Research. Calculations have been performed at the CESVIMA, CINES and CCRT supercomputing centers.

References

[1] Tabak M *et al.* 1994 *Phys. Plasmas* **1** 1626
[2] Kodama R *et al.* 2004 *Nature* **432** 1005
[3] Baton S D *et al.* 2008 *Phys. Plasmas* **15** 042706
[4] Nakamura T *et al.* 2007 *Phys. Plasmas* **14** 103105, Cai H-B *et al.* 2009 *Phys. Rev. Lett.* **102** 245001
[5] Dunne M 2006 *Nat. Phys.* **2** 2
[6] Honrubia J J. 1993 *J. Quant. Spectrosc. Radiat. Transf.* **49** 491
[7] Sentoku Y and Kemp A J 2008 *J. Comput. Phys.* **227** 6846, Geissler M *et al.* 2006 *New J. Phys.* **8** 186
[8] Honrubia J J and Meyer-ter-Vehn J *Plasma Phys. Control. Fusion* **51** 014008
[9] Nagatomo H *et al.* 2007 *Phys. Plasmas* **14** 056303, Atzeni S *et al.* 2008 *Phys. Plasmas* **15** 056311
[10] Atzeni S *et al.* 2009 *Nucl. Fusion* **49** 055008
[11] Chrisman B *et al.* 2008 *Phys. Plasmas* **15** 056309
[12] Johzaki T *et al.* 2009 *Phys. Plasmas* **16** 062706

Omega EP, Laser Scalings and the 60 MeV Barrier: First Observations of Ion Acceleration Performance in the 10 Picosecond Kilojoule Short-Pulse Regime

Kirk Flippo[1], Teresa Bartal[2], Farhat Beg[2], Sugreev Chawla[2], Jim Cobble[1], Sandrine Gaillard[1,5], Daniel Hey[3], Andrew MacKinnon[3], Andrew MacPhee[3], Phillip Nilson[4], Dustin Offermann[1], Sebastien Le Pape[3], Mark J. Schmitt[1]

[1] Los Alamos National Laboratory, PO Box 1663, Los Alamos, NM, USA
[2] University of California, San Diego, Center for Energy Research, 9500 Gilman Dr., La Jolla, CA 92093, USA
[3]Lawerence Livermore National Laboratory, 7000 East Avenue, Livermore, CA 94550
[4]University of Rochester-Laboratory for Laser Energetics, 250 E. River Rd., Rochester NY 14623-1299
[5] Forschungszentrum Dresden-Rossendorf, Rossendorf, Germany

kflippo@lanl.gov

Abstract. Omega EP is capable of producing 1000 J in 10 ps and is currently the most energetic short-pulse laser in the world. The performance of EP in terms of proton beam energies is compared with other laser systems worldwide at similar intensities. Omega EP results are discussed in the context of these lasers and the empirical ~ 60 MeV barrier, which has existed since the discovery of forward laser-accelerated protons in 2000 [1-2].

1. Introduction

Since the discovery of short-pulse laser forward-accelerated ions in 2000 [1-2] due to the Target Normal Sheath Acceleration (TNSA) mechanism [3] the maximum observed energy has been stuck for almost 10 years at below 60 MeV, starting with data from the Nova Petawatt with nearly 500 J producing 58 MeV [1], and lately in 2009 from Trident with 90 J producing 58.5 MeV (reported here and [4]). Though the proton energies obtained from flat foil targets have not increased past this empirical barrier over the last decade, the amount of energy needed to achieve the same acceleration has dropped, first by a small amount [5] and then by a factor of 6 (see [4, 6], and Figure 4); and as recently reported in this issue, micro-cone targets were able this year to finally break this once seemingly impenetrable barrier [4, 7]. In addition theoretical schemes have been published that promise to go beyond the TNSA mechanism at higher intensities [8]. PIC simulations and theory do not yet show any obvious reason for which this barrier should exist; however, to date, no flat foil has been reported to produce more than 60 MeV. In this paper, data from Omega EP, the world's current highest energy (1000 J) short-pulse (10 ps) system, are presented in relationship to other systems and the 60 MeV barrier.

2. Omega EP proton beam observations

The Omega EP laser is currently capable of producing a 1000 J of 1 μm light in 10 ps focused to a 40 μm spot. This present Omega EP data set was collected over 3 shot days at a nominal laser energy of 1000 J during 2009. Two target materials were used, i.e. a dielectric (CVD diamond), and a conductor (Cu). The primary diagnostic was Radiochromic Film (RCF) in the LLNL designed and built Proton Film Pack II (PFPII), shown in Figure 1, and a LANL modified version with a hole in the stack for access to a Thomson Parabola (TP) Twelve Inch Manipulator (TIM) based diagnostic. The PFPII is simple a 7x7 cm holder for a ~2 cm thick sandwiched stack of RCF and aluminum filters for Radiochromic Film Imaging Spectroscopy (RIS) [9].

Figure 1: Diagram for the PFPII (grey and blue) in the Omega EP chamber showing the sidelighter (red) and backlighter (pink) EP beams focusing on a target (not shown, center). The PFPII is 7 cm away from the target, which is normal to either the sidelighter or backlighter beam. Only one beam was used at a time. The LANL PFPII modification has a hole in the center to allow for ions to stream through the PFPII to a TIM based Thomson Parabola on the sidelighter axis.

Figure 2 compares the beam quality and spectrum from a 15 μm thick CVD target (right) and Cu (left) target. In comparison to the proton beam from the CVD target, the proton beam from the Cu target is a larger, more defuse beam, with a darker inner beam, and a characteristic proton beam temperature of 3.79 MeV. The proton beam from the CVD target has a well defined outer edge and a higher temperature spectrum, i.e. 5.58 MeV, than the copper. Overall, the Cu target beam contains more protons in the 4-15 MeV range, but fewer protons in the higher energy end of the spectrum, i.e. in the 15-36 MeV range. This results in the fact that the Cu target's beam contains about 1.8% of the laser energy in protons, while the CVD target's beam contains 2%.

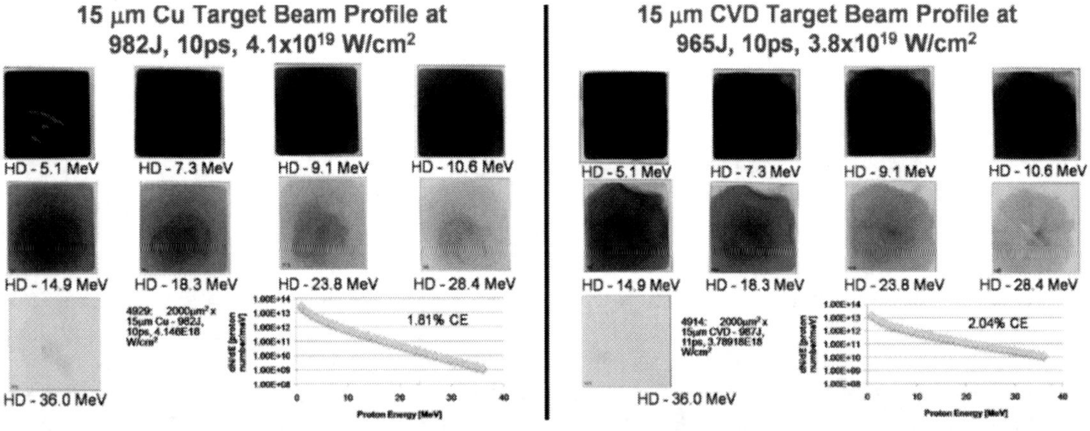

Figure 2: Comparison of 15 μm copper target (left panel) with 15 μm CVD diamond target (right panel). The panels show PFPII data from the RCF film stack the energy associated to each layer, and a calculated convolved proton spectrum fitting the data in the lower right of each panel.

In the modified PFPII stacks, i.e. those with an added central hole on axis, a Thomson Parabola (TP) was used. Figure 3 shows the proton trace from a 35 μm thick, 0.8 mm diameter CVC hemispherical shell target at 1000 J and 10 ps. The protons were recorded on an Image Plate (IP) from Fujifilm, which was placed behind a piece of CR-39 plastic nuclear track detector. CR-39 is used to detect heavier ions, and in this case, the IP is used to measure the protons which travel through the ~1 mm thick CR-39, i.e. with energies above ~ 9.4 MeV. For ion and proton detection, IPs are not as sensitive as CR-39, but they can detect protons over a much larger energy range. In the case illustrated in Figure 3, unlike the RCF stack, the TP did not record the highest energy portion of the beam , i.e. 36 MeV, because the TP only samples a very small portion of the beam, at a much greater resolution, while separating the different ions species; the high energy portion of the beam did not make it through the hole and therefore was not measured by the TP.

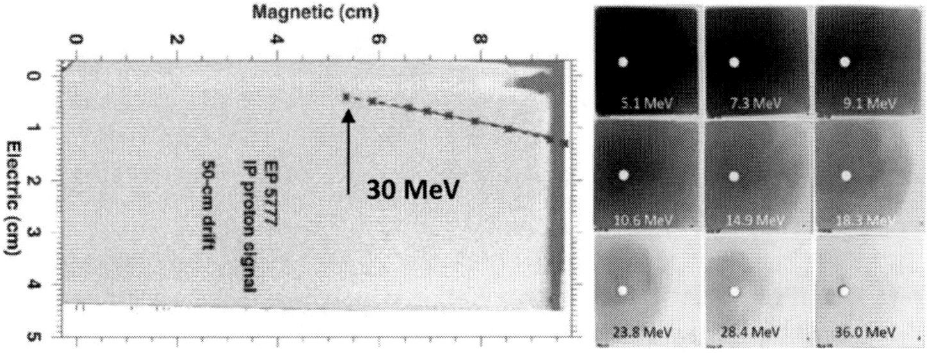

Figure 3: (left) TP trace of the protons on an IP in false color from a 35 μm thick CVD target. A theoretical energy curve (blue *'s) is overlaid on the ion trace, which extends slightly further then the plotted end of 30 MeV. (right) The PFPII RCF stack showing the beam of 36 MeV. The last frame is contrast enhanced for clarity. The observed TP cutoff of ~30 MeV is consistent with the fact that the highest energy portion of the beam (on the last RCF) did not intersect the hole.

3. The 60 MeV Barrier and Omega EP

For just about 10 years now, the field of short-pulse laser accelerated protons has been limited to less than 60 MeV using flat-foil targets. Novel conical target geometries have recently been shown to be able to access the region above 60 MeV for the first time [4, 7], but complex target designs obviously increase the risk of implementing applications requiring high rep-rates, such as hadron therapy. Thus, increasing the energy from a simple flat foil would go a long way to making such applications a reality. Currently EP has demonstrated 1000 J, but will be able to deliver 2600 J in 10 ps within a year (circa 2010), bringing the potential to produce ions above the barrier. Currently, even at a modest intensity of ~5×10^{18} W/cm^2, the laser is able to produce protons of almost 50 MeV.

Figure 4 shows how Omega EP currently compares to other laser systems in the world with relatively similar contrasts (namely ~10^{-6} to >10^{-8}). Plotted are data from these lasers ranging from ~ 30 fs to > 1 ps in pulse duration and from 10^{17} to nearly 10^{21} W/cm^2 in intensity. One can see that most of the lasers fall along a line, fitted with a power law with an exponent of nearly 0.7. The EP data (blue ■'s) is shifted quite a bit to the left of most of the data and when fitted with another power law, yields an exponent of ~0.44, close to the ponderomotive scaling of 0.5 [3]. This indicates that the EP laser is out-performing all other lasers in the world in terms of proton energy at a given intensity, not surprising given the fact that it has 2 to 100 times the energy of most of those lasers. However, EP (at 100 TW) has yet to produce a beam above 50 MeV, unlike the Nova PW, the Vulcan PW, or the Trident 200 TW (0.2 PW) lasers, albeit at 10 times the intensity or more. Thus, until more energy or a smaller spot is available, the 60 MeV barrier remains to be tested.

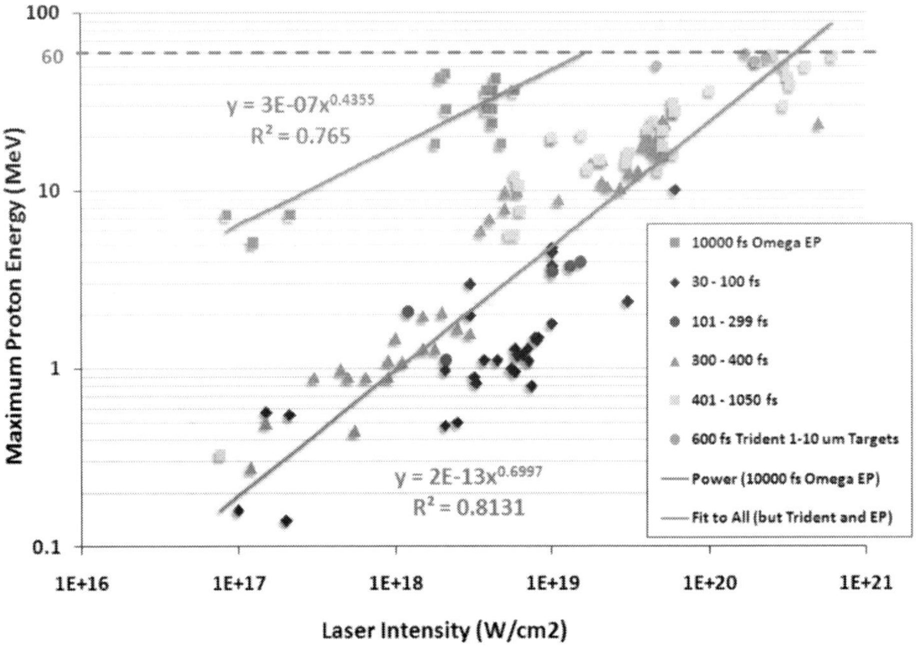

Figure 4: OMEGA EP Proton scaling as a function of laser intensity plotted with data from other laser systems around the world. Data plotted is taken from [1, 3-4], and [10]. The data is broken down into pulse duration groups, 10000 fs, Omega EP (light blue ■'s); 30-100 fs (dark blue ♦'s); 101-299 fs (purple ●'s); 300-400 fs (green ▲'s); 401 -1050 fs (yellow ■'s); and 600 fs, Trident (orange ●'s). Two fits are shown, one for all the data excluding the EP data, and one for the EP data.

This work supported by US DOE/NNSA, performed at LANL, operated by LANS LLC under Contract DE-AC52-06NA25396.

References

[1] R.A. Snavely, M.H. Key, S.P. Hatchett, *et al.*, *Phys. Rev. Lett.*, 85, 2945 (2000).
[2] A. Maksimchuk, S. Gu, K. Flippo, *et al.*, *Phys. Rev. Lett.*, 84, 4108-4111 (2000); E.L. Clark, K. Krushelnick, J.R. Davies, *et al.*, *Phys. Rev. Lett.*, 84, 670 (2000).
[3] S.C. Wilks, A.B. Langdon, T.E. Cowan, *et al.*, *Phys. Plasmas*, 8, 542 (2001).
[4] S.A. Gaillard, K.A. Flippo, M. Lowenstern, *et al.*, *J. of Phys.: Conf. Series*, this volume, (2010).
[5] L. Robson, P.T. Simpson, R.J. Clarke, *et al.*, *Nat. Phys.*, 3, 58-62 (2007).
[6] K.A. Flippo, J. Workman, D.C. Gautier, *et al.*, *Rev. Sci. Inst.*, 10E534 (2008).
[7] S.A. Gaillard, K.A. Flippo, B. Gall, *et al.*, *Nat. Phys.*, (in preparation).
[8] L. Yin, B.J. Albright, B.M. Hegelich, *et al.*, *Physics of Plasmas*, 14, 056706-8 (2007); K. Flippo, B.M. Hegelich, B.J. Albright, *et al.*, *Laser and Part. Beams*, 25, 3-8 (2007); A. Macchi, F. Cattani, T.V. Liseykina, *et al.*, *Phys. Rev. Lett.*, 94, 165003/1-4 (2005); A. Macchi, S. Veghini, and F. Pegoraro, *Phys. Rev. Lett.*, 103, (2009).
[9] F. Nurnberg, M. Schollmeier, E. Brambrink, *et al.*, *Rev. Sci. Inst.*, 80, (2009).
[10] M. Zepf, E.L. Clark, K. Krushelnick, *et al.*, *Phys. Plasmas*, 8, 2323 (2001); A.J. Mackinnon, M. Borghesi, S. Hatchett, *et al.*, *Phys. Rev. Lett.*, 86, 1769 (2001); J. Badziak, P. Parys, A.B. Vankov, *et al.*, *Phys. Rev. Lett.*, 87, 215001 (2001); T.E. Cowan, M. Roth, J. Johnson, *et al.*, *Nuc. Inst. & Meth. Phys. Res. Sec. A*, 455, 130-139 (2000); P.K. Patel, M.H. Key, A.J. Mackinnon, *et al.*, *Plasma Phys. Cont. Fus.*, 47, B833-B840 (2005); K.W.D. Ledingham, P. McKenna, T. McCanny, *et al.*, *J. Phys. D. App. Phys.*, 37, 2341-2345 (2004); K. Krushelnick, E.L. Clark, M. Zepf, *et al.*, *Phys. of Plasmas*, 7, 2055 (2000).

Proton acceleration from ultrahigh-intensity short-pulse laser-matter interactions with Cu micro-cone targets at an intrinsic ~10^{-8} contrast

S. A. Gaillard[1,2], K. A. Flippo[2], M. E. Lowenstern[3], J. E. Mucino[3], J. M. Rassuchine[1], D. C. Gautier[2], J. Workman[2] and T. E. Cowan[1]

1 Forschungszentrum Dresden-Rossendorf, P.O. Box 510119, Dresden, Germany
2 Los Alamos National Laboratory, P-24, MSE526, Los Alamos, NM, USA
3 Atmospheric, Oceanic & Space Sciences, University of Michigan, Ann Arbor, USA

Corresponding author: sandrine.gaillard@gmail.com

Abstract. In this paper, we report on experiments comparing various geometries of conical Cu targets to Cu flat foils, which were performed on the 200 TW Trident laser (~80 J, 600 fs, ~7 μm spot size, S-polarization and ~1.5×10^{20} W/cm^2) at an intrinsic (to the system's regenerative amplifier) ASE contrast of 10^{-8}. The current work builds on previous results obtained on Trident (~20 J, ~14 μm spot size, P-polarization, ~10^{19} W/cm^2, also at the intrinsic contrast of 10^{-8}) demonstrating enhanced proton energies and laser-proton conversion efficiencies (η) using Flat Top Cone (FTC) targets [1]. An electron spectrometer and a Cu Kα imaging diagnostic were added to respectively assess the electron population, and determine the characteristics of laser absorption in FTCs. Results indicate a linear correlation between electron temperatures and proton energies, as well as laser absorption taking place in a preplasma filling the cone, preventing the previously observed enhancement in proton energies.

1. Motivations and introduction

In previous experiments performed on the Trident laser at ~20 J and at the intrinsic 10^{-8} contrast, we found that flat-top cone (FTC) targets yielded an increase in proton energy from 19 MeV to >30 MeV and in laser-proton conversion efficiency (η) from 0.5 % to 2.5 %, as compared to flat-foil targets [1]. These results could be due to improved laser guiding toward the cone tip, which would yield higher laser intensities and hotter electrons. Improved electron production and transport can lead to an increase in the electron density and temperature at the flat-top. In addition, a longer electron confinement on the flat-top can lead to reduced mass target (RMT) effects such as resistive/confining edge fields and enhanced target/top charge up. Finally, when the laser is misaligned and cannot reach the cone tip, or is absorbed farther from the flat-top due to an excess in preplasma, the proton acceleration is neither as efficient nor as energetic.

The experiment described here was performed after the Trident laser energy upgrade from ~20 J to ~80 J to verify whether the enhanced proton energy would scale with laser energy and whether the predictions (from simulations) concerning the effect on the preplasma level in the FTC neck were correct [1]. To diagnose the laser absorption zone inside the FTC, Cu targets were used, as well as Cu-Kα 2D imaging. Indeed, if the lower energy end of the hot-electron population deposits its energy in the FTC neck, then Kα-x-ray emission is direct indication of preplasma filling the cone tip [2]. In this paper, we show that, with this laser energy enhancement, the spot size decrease (due to the addition of a deformable mirror) from ~14 μm down to ~7 μm FWHM (with 47 % of the energy in the spot), and the resulting intensity increase from ~10^{19} W/cm^2 to ~10^{20} W/cm^2, at the intrinsic contrast level, the laser is able to propagate to the top of both the FTC and the Funnel Cone (FC) targets, i.e. Cu Kα emission is observed from the cone tip. However, most of the emission originates at the cone walls (i.e. cone wall emission or CWE is predominant), due to a relatively large preplasma. This preplasma prevents the laser from being efficiently absorbed closer to the cone tip [2, 3]. The hot electron population is generated in the preplasma region, and not close to the flat-top. This negatively impacts the proton acceleration, especially in the case of thin FTC necks [1]. Electron spectroscopy indicates that the temperature of the escaping electrons correlates linearly well with proton energy.

A more recent experiment performed on Trident with an enhanced contrast (>10^{-10}) shows that proton energies are increased to 67.5 MeV at only ~80 J of laser energy [4], versus previous petawatt laser records of 58 MeV at ~400 J. It also shows that not only is it crucial to obtain laser absorption at the tip (note that tip heating is dependent on laser contrast and intensity [2]), but it is even more important to find the optimum balance [4] between CWE wall versus top emission (TE).

2. Description of the targets and of the experimental set-up

The experimental set-up and diagnostics are pictured in Figure 1 (left). The various cone targets shown in Figure 1 (right), i.e. funnel cones (FC) (a-c) and an FTC (d), as well an RMT that were the ~10 μm thick, 300 μm diameter broken-off tops of FTCs (d-2), were all shot at normal (0°) incidence. Regular large flat foils (2×7 mm^2 for all target thicknesses, except the 1 μm case which was ~1×1 mm^2) were shot at 22.5° incidence to avoid back-reflections and damage to the focusing parabola.

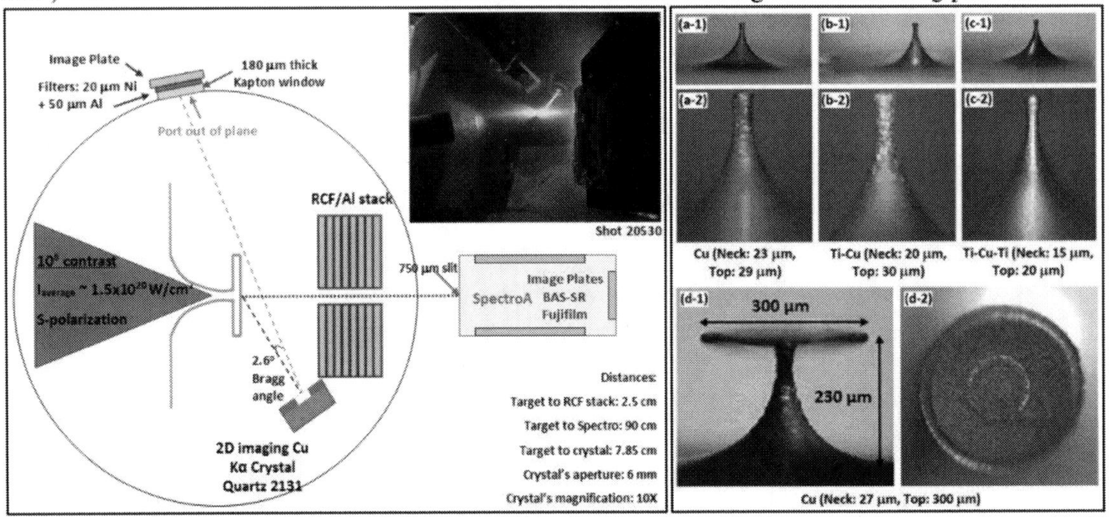

Figure 1: (left) Set-up of the Aug. 2008 Trident Cone beamtime; (left inset) actual image of an FTC target shot *(Photo credit: J. S. Cowan)*; (right) Various geometries of cone targets shot, i.e. 3 types of FCs (a-c): (a) pure Cu (b) 500 nm of Ti above the Cu, (c) 500 nm of Ti above and below the Cu; and FTCs (d). RMTs were also shot, which were the broken-off tops of the FTC (d-2).

3. Proton energies, conversion efficiencies η, and Cu Kα emission of the various targets

Table 1 shows proton energy ranges and η (using RIS [5] on a MicroD) for various target geometries. Note the ~5 MeV variation in the maximum proton energy from 10 μm thick flat foils between shots with similar laser energies; and the much greater deviation (i.e. ~10 MeV) for the RMTs with fairly constant laser conditions. The proton energies reported in Table 1 are the energies of the proton beams accelerated from the target via the target sheath normal acceleration (TNSA) mechanism, observed with RadioChromic Film (RCF) stacks, and in the case of cone targets, accelerated from the top of the cone. This *top beam*, emitted normally from the top of the FTC or the FC, is almost always round and uniform; while the *side beam*, emitted from the cone walls, tends to emit as an annular non-uniform spray and is ignored, given that the electrons involved do not reach the spectrometer.

Target type		Thickness	Proton energy variation	range J, shot number; or e.g.	Respective η
Flat	Large	1 μm – 150 μm	58.5 MeV [6] – ~25 MeV	90 J, 20549 – 97 J, 20542	3.73 % – 1.32 %
	Large	10 μm	~50 MeV to ~55 MeV	92 J, 20534 to 94 J, 20537	1.35 % & 1.6 %
	RMT	10 μm	~32 MeV to ~44 MeV	88 J, 20552 to 82 J, 20508	2.15 % & 2.12 %
Cone	FC	10 μm	~12 MeV to ~47 MeV	e.g. 26.7 MeV at 76 J, 20518	0.79 %
	FTC	10 μm	~5 MeV to ~30 MeV	e.g. 27.8 MeV at 86 J, 20532	1.62 %

Table 1: Proton energy ranges and conversion efficiencies η for flat-foil targets and cone targets.

Usually, when the laser is misaligned (Figure 2 (left)), the *side beam* is more energetic than the *top beam*, but never greater than that from a flat foil (i.e. in the best case, see Figure 3(b), >47.2 MeV compared to 55 MeV). *Top beams* are always accelerated from the tip of FCs, unless the neck diameter is too small (<~ 20 μm outer diameter neck), and/or the neck is too long, such that, even when the laser is well aligned, no proton beam can form off the tip and there only is all the proton emission comes from the walls (Figure 2 (right)). As the neck becomes larger (~25 μm), the *top beams* are accelerated to higher proton energies, and the ratio of the *top beam / side beam* proton maximum energies increases.

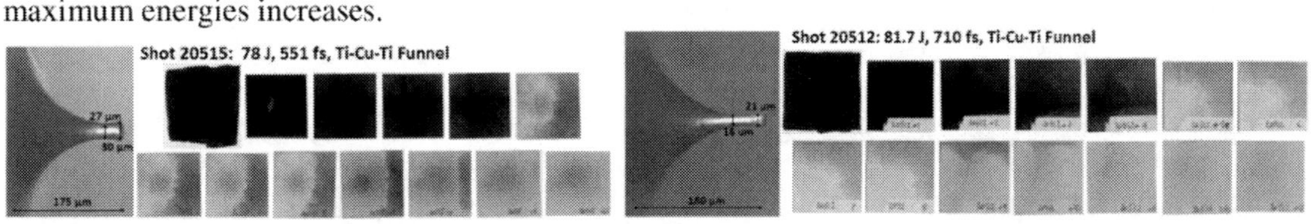

Figure 2: (left) Shot 20515, FC, laser misaligned, E_p *side beam* > E_p *top beam*; (right) Shot 20512, FC, laser aligned, neck too small such that there is no normal proton beam emission at high energy.

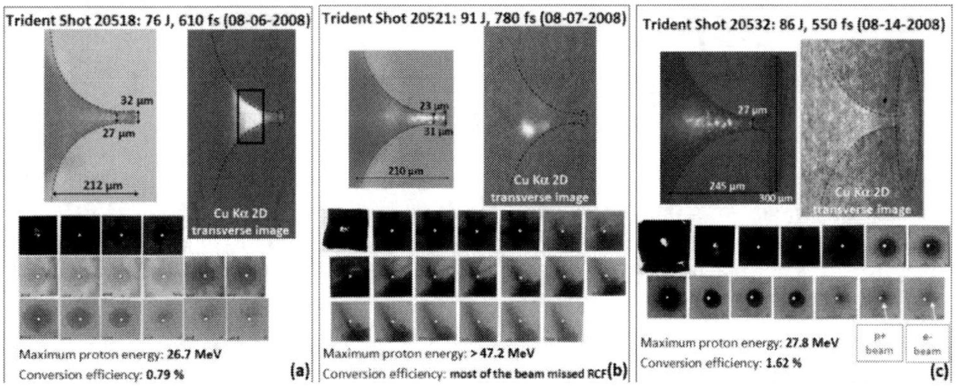

Figure 3: Pictures of the targets, Cu Kα images and RCF stacks for: FCs (a) shot 20518 (aligned) and (b) shot 20521 (misaligned); FTC (c) shot 20530 (aligned). The conversion efficiency is higher for the FTC (1.62%) than it is for the FC (0.79%) since the protons can be accelerated from a larger top area.

Figure 3 compares Cu Kα images and RCF stacks obtained from two FCs and an FTC. In the aligned cases (Figure 3(a)&(c)), Cu Kα imaging confirms that the cone is filled with preplasma, and thus the laser penetrates poorly to the tip. However the protons are accelerated in nice, uniform *top beams*, but with no proton energy increase. Note in (c) that the fairly strong electron beam emitted from the flat top in the normal direction does show the presence of hot electrons, mainly in the same direction as the protons. Also, even if the laser is aligned and reaches the tip, yielding Cu Kα TE, it does not mean that the proton energy will be enhanced. This suggests there has to be the right balance of CWE and TE [4], which is unachievable in the intrinsic contrast case, where CWE >> TE. In Figure 3(b), Cu Kα imaging confirms that the laser misalignment is responsible for the proton *side beam*. The *top beam* is hard to distinguish (2nd to last detector on the 2nd row of RCFs) and has a relatively low energy (27.8 MeV), while the side beam is very energetic (> 47.2 MeV). Lastly, the very energetic oblique (upper left to lower right) cone wall spray from the FC indicates that the electrons born on the cone wall are emitted normal to all its length.

4. Correlation between electron temperature and proton energy

Figure 5 (left) shows the electron population from the electron spectrometer [7] as well as how the lower (T_1) and higher (T_2) electron temperatures are obtained. In the case of the cones (Figure 5 (right)), the average of the lower and higher temperatures $(T_1+T_2)/2$ correlates better linearly with proton energies (of the top, not the side, beam) rather than T_1 alone. For the flat foils (various thicknesses), the proton energies are higher, but there is a much weaker linear correlation with electron temperature (not shown here). In all cases, the linear correlation between proton energy and electron temperature is in agreement with the Wilks scaling [8].

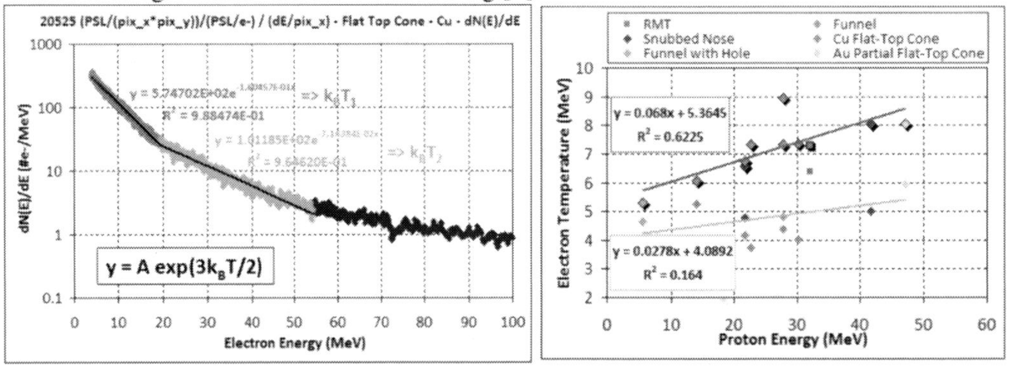

Figure 5: Electron temperatures as a function of proton energy: (left) example of an electron spectrum: T_1 is obtained using the red data points, T_2 using the orange ones; (right) T_1 and $(T_1+T_2)/2$ temperatures of the various cones. The T_1 values are represented by the smaller data points (lighter colored trendline). The $(T_1+T_2)/2$ values are represented by the larger shadowed data points (darker purple trendline).

5. Summary

The four-fold increase in laser energy from ~ 20 J to ~ 80 J at the intrinsic 10^{-8} contrast seems to considerably increase the plasma pre-fill inside the FTC, as indicated by the Cu Kα x-ray absorption images. This prevents further proton energy enhancement even with higher laser energy and FTC targets. In terms of proton energy, flat foils actually perform better than aligned or misaligned FTCs. To obtain an increase in proton energy at ~ 80 J, a higher contrast must be used [4]. The FTC proton energies appear to correlate well with the electron temperatures, suggesting that the electron acceleration in a plasma-filled cone is not of sufficient temperature to support higher energy ions.

6. Acknowledgments

The authors thank Nanolabz for targets; the Trident crew: F. L. Archuleta, R. P. Gonzales, T. R. Hurry, R. P. Johnson, S.-M. L. Reid, T. Shimada, S. A. Letzring, and D. S. Montgomery for laser support; J. S. Cowan for scanning RCF; and the LANL LDRD program for supporting this project.

References

[1] K. A. Flippo *et al.*, Phys. Plasmas 15 056709 (2008).
[2] J. Rassuchine *et al.*, Phys. Rev. E 79 0364408 (2009); J. Rassuchine *et al.*, J. Phys.: Conf. Series 112 022050 (2008).
[3] S. D. Baton *et al.*, Phys. Plasmas 15 042706 (2008).
[4] S. A. Gaillard, T. Kluge, K. A. Flippo, B. Gall *et al.*, in preparation for Nat. Phys. (2010).
[5] F. Nürnberg *et al.*, Rev. Sci. Inst. 80 033301 (2007).
[6] K. A. Flippo, T. Bartal, F. Beg *et al.*, this conference (2009).
[7] T. E. Cowan *et al.*, Proc. of the 8th Workshop on Adv. Acc. Concepts (1998); T. E. Cowan *et al.*, Laser and Part. Beams 17 773 (1999).
[8] S. Wilks *et al.*, Phys. Plasmas 8 542 (2001); S. Wilks *et al.*, Phys. Rev. Lett. 69 1383 (1992).

Journal of Physics: Conference Series **244** (2010) 022035 doi:10.1088/1742-6596/244/2/022035

Stable single channel formation in long scale plasma for fast ignition

H Habara[1,2], M Hatakeyama[1,2], T Tanimoto[1,2], N Kageiwa[1,2], S Hino[1,2], A.L Lei[3], K.A Tanaka[1,2]

[1]Institute of Laser Engineering, Osaka University, 2-6, Yamada-oka, Suita, Osaka, 565-0871, Japan

[2]Graduate School of Engineering, Osaka University, 2-1, Yamada-oka, Suita, Osaka, 565-0871, Japan

[3]Shanghai Institute of Optics and Fine Mechanics, Chinese Academy of Sciences, Shanghai 201800, China

Email: habara@ile.osaka-u.ac.jp

Abstract. We observe a single plasma channel in long scale plasma created on the plane target. Smaller electron divergence and enhancement on moderate energy electrons are obtained than those from plane target. The experimental results also show that slight laser energy is used for the channel formation and the conversion efficiency to electrons is comparable to the plane target. These results imply efficient heating at "Super-Penetration" fast ignition.

1. Introduction

Since the Fast Ignition (FI) scheme is proposed in 1994 [1], interactions of ultra-intense laser beam with plasmas have been intensively investigated and the fruitful results significantly contribute for expansion of scientific understanding and technological applications such as high energy density science. In FI scheme, the required dense fuel is formed via conventional laser driven implosion, but the ignition temperature is achieved via quick heating by a separate ultra-intense laser beam, resulting in significantly reduction of total laser energy required for the ignition. However, the intense laser should propagate into mm scale plasma where the many laser instabilities such as SRS or filamentation can be driven before reaching the critical density. The cone-in-shell geometry, which can reduce such instabilities, was successfully demonstrated about 100 times of heating of the core compared with no intense beam injection [2]. However, this geometry breaks the implosion symmetry and therefore makes difficult to achieve high-gain target designs.

An alternative approach is to inject ultra-intense laser pulses directly into the coronal plasma, so called "Super-Penetration" fast ignition [3]. When the laser power exceeds the critical power of relativistic self-focusing, the laser beam can propagate in the overdense plasma due to relativistic induced transparency and laser hole boring. In our previous works, a long plasma channel was observed penetrating into overdense plasma when the intense laser pulse irradiated onto a long scale plasma with 80μm overdense region [4]. In this experiment, the transmitted laser energy is also detected and we found that the transmittance depends on the focus position of the intense laser beam in the long scale plasma [5].

Journal of Physics: Conference Series **244** (2010) 022035 doi:10.1088/1742-6596/244/2/022035

From the view point of FI, it is important to measure the fast electrons under these conditions. In order to observe energy distribution of the accelerated electrons, and to estimate the conversion efficiency from laser energy to fast electrons, we measured energy distribution of fast electrons generated by an intense laser interacting with a long scale plasma.

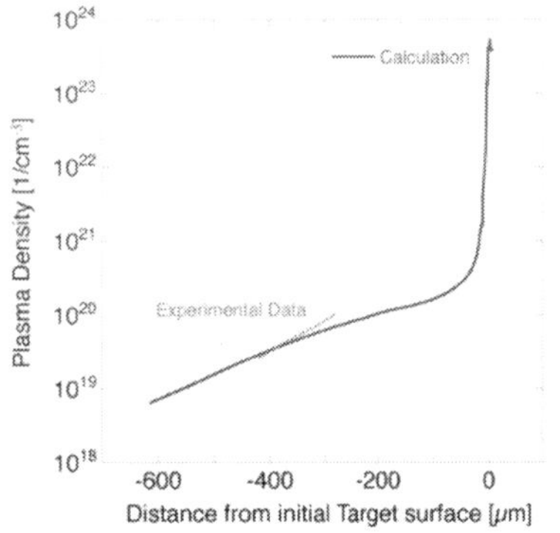

Figure 1. Initial electron density distribution by experiment (dash) and calculation (solid).

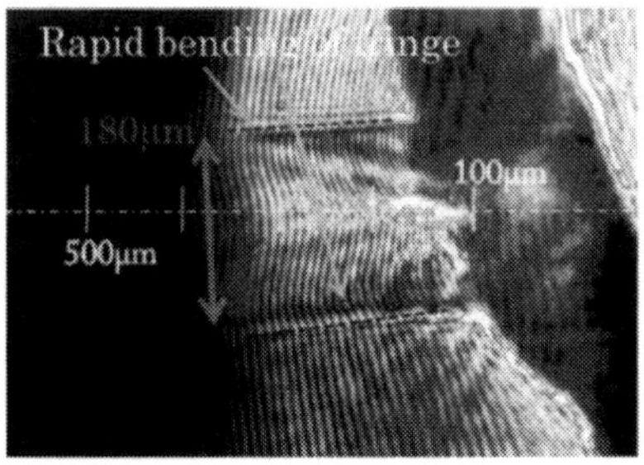

Figure 2. Interferometer image when the short pulse beam is irradiated on 200µm in front of initial target surface.

2. Experimental setup

The experiment is performed at GEKKO MII 60TW system in Osaka University. This system has three separate beams: one is the short pulse beam operated in 1µm wavelength. The output energy is 30J with 500fs pulse duration. The laser is focused onto a thin plastic target via F3.3 off-axis parabolic mirror to be around 20~30µm diameter focal spot, resulting in up to 10^{19}W/cm^2 of peak intensity. The long pulse beam (/30J/0.5ns) is irradiated onto the target from 25° away of the short pulse beam to create the pre-created long scale plasma. The low energy probe beam (2 /5ps) is injected parallel to the target surface to observe the plasma density gradient via optical interferometer. We put several diagnostics such as electron spectrometer (ESM) on the laser axis and the side to observe the electron energy distributions, X-ray Pin-hole camera (XPHC) for X-ray image on the interaction point, and K photon counting detector to estimate the number of electrons inside the target.

Figure 1 shows the electron density scale of plasma created by the long pulse beam focused with 200µm dia. (~10^{14}W/cm^2). A 1D-radiation MHD calculation, represented by solid line, well reproduces the density scale from the experiment by the dashed one [6]. According to the calculation, the scale length to n_c is about 30µm and there is a long under dense plasma from 10^{19} to 10^{20}cm^{-3} about 400µm length. In this experiment, the focusing position of intense laser pulse is changed from 100 to 500µm from the initial target surface in order to find an optimal laser penetration.

3. Channel formation inside plasma

When the laser is focused around 200~500µm from the initial target surface, we observed a long plasma channel inside the underdense region. Figure 2 shows an interferometer image taken at -200µm focusing conditions (~ 1×10^{20}cm^{-3} density plasma). In the image, short pulse laser came from left boundary and created a straight plasma channel indicated by black dashed lines where the fringe lines are rapidly bended. Observed channel position is from 100 to 400µm where the plasma density is relatively constant (0.5-1.0×10^{20}cm^{-3}). Although similar channel images are observed for other focusing conditions such as -500µm, for 200µm focusing condition a strongest 2 self-emission is detected exactly corresponding to the critical surface position (light blue spot on the axis in the figure). In addition, we observed a strong X-ray emission only for this condition whereas no signal is measured for other focusing positions. From these results, we assume that most of laser energy is propagated around critical surface at the 200µm focusing condition.

Analyzing the fringe shifts, the electron densities of channel wall and center are about 2 and 1/10 of background density (immediately before short pulse injection), respectively. Channel expansion speed is estimated from the probe beam timing and the length of channel, resulting in 90μm/80ps = 1.1×10^{8}cm/s. This ion sound speed is comparable to the previous channel formation experiments, ex. 5×10^{8}cm/s [7]. In order to estimate the energy to create the channel, we adapt the similarity solution assuming blast wave expansion in cylindrical symmetry. The rate of expansion can be given by [8,9]

$$R(t) = (E/\rho_0)^{1/4}\sqrt{t} \qquad (1)$$

Here *R(t)* is the channel expansion length in given *t*, *E* the deposited energy per channel length, $_0$ the plasma density and *t* the time of observation. Substituting *R*=90μm, $_0$=8.27×10^{-4}g/cm^3, *t*=80ps, and *L*=300μm, we obtain the energy to create plasma channel is 1.6J. In this experiment, the laser energy in the focal spot is about 12J, so that the energy for channel creation is only 13% of laser energy.

4. Conversion efficiency

Electron energy distribution is measured with two different methods: one is the magnet spectrometer (ESM) for escaping electrons from the target and other is K photon counting from added Cu layer at the target rear. Figure 3 represents typical ESM electron spectra from -200μm (green) and -500μm (red) focusing conditions with the plane target (blue) result. Comparing the electron spectra from plane target and plasma cases, two significant difference can be clearly appeared: (a) electron spectrum from the side for plane target has a higher temperature than the preplasma cases whereas the almost same for on-axis spectra, and (b) moderate energy (up to 10MeV) electrons at on-axis for 200μm focus condition are almost 20 times enhanced compared with the plane target. These results, the reduction of beam divergence angle and enhancement of moderate energy electrons, have already presented in the previous papers [3], which implies relativistic self-focusing of intense laser beam in long scale plasma resulting in laser penetration into higher density plasma.

It is known that ESM electrons detected with some distance in vacuum suffer a strong sheath potential created at the target rear side [10], so that the detected spectra might be modified in particular for lower energy electrons. For this reason, we also measure Cu-K emission intensity generated through scattering of electrons passing through the Cu layer. We also added a Ti layer between plastic and Cu (10μm) to reduce the electron energy reaching to Cu layer. By changing the Ti thickness, we can estimate electron energy distribution according to the energy reduction at the Ti layer. The stopping power and ranges in various Ti thickness are calculated with Electron Gamma Shower (EGS) code. Table 1 shows the K photon intensities changing the Ti layer thickness. Clearly the photon counts decrease with increasing of Ti thickness.

Figure 4 shows the electron energy distributions from K data, converting from X-ray intensity to the electron number according to K fluence cross section, as shown by solid dots, and also several ESM spectra (dashed lines). In the result, the electron energy distribution seems two temperature Maxwellian (K T_e: 165keV and ESM T_e:5MeV). Assuming the electron spectrum inside the target is the two temperature Maxwellian, the total energy of electrons is 3.1J (Emission angle is assumed to be 30° for ESM data [3]). In the result, conversion efficiency from laser energy to fast electrons is 26%. Surprisingly this efficiency is comparable to the efficiency for plane target [11,12].

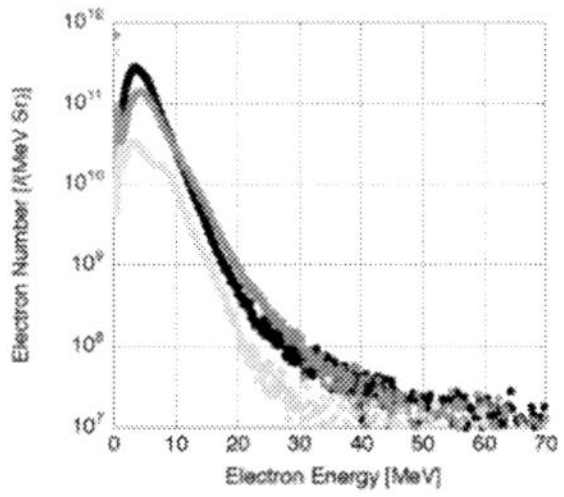

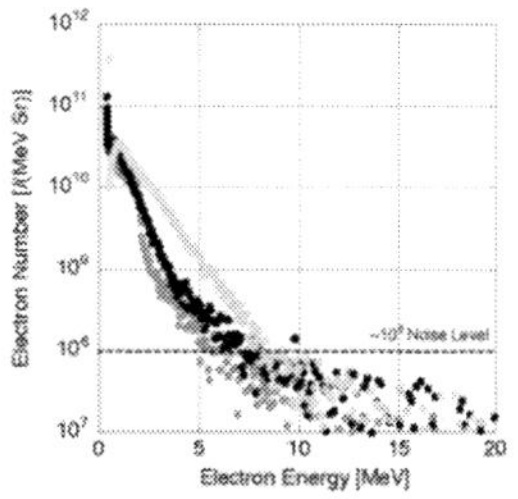

Figure 3. Electron energy spectra for laser axis (left) and side (right). Black, Dark gray, and Light gray dots represent the spectra from 200, 500μm focusing position and plane target.

Journal of Physics: Conference Series **244** (2010) 022035 doi:10.1088/1742-6596/244/2/022035

Ti thickness (μm)	0	50	100	300
K counts ($\times 10^{11}$)	2.91	0.97	0.60	0.25

Table 1. K photon intensities changing the Ti layer thickness.

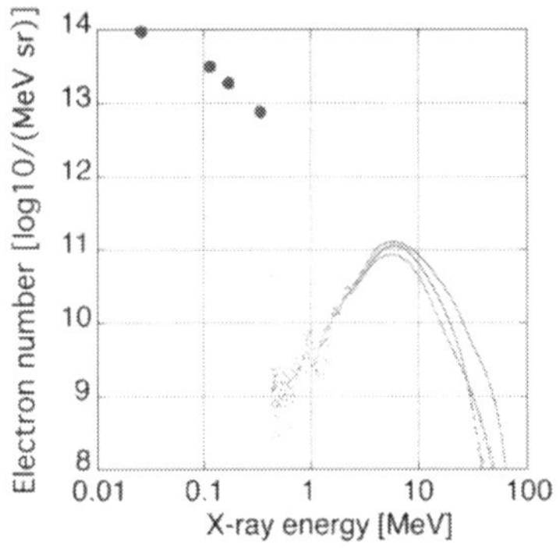

Figure 4. Electron energy distribution from K emission (dotted circles) and 3 ESM spectra (taken in different shots with same laser and plasma conditions)

5. Summary

In the summary, a stable single channel is created in a long scale plasma when the laser is focused around 200~500μm from the initial target surface. From the analysis of optical interferometer, the channel formation uses 13% of laser energy. Under this condition, electron emission angle becomes smaller and moderate energy electrons up to 10MeV are enhanced compared with those from plane target. ESM spectra and K photon counting indicates that the conversion efficiency from laser to electrons is about 26% where is comparable to the plane target. These results imply that most of laser energy can be transported around critical surface without significant consumption by optimizing focusing position to the density scale gradient, and can be expected effective heating to the core.

It is interesting to note that Atzeni and co-workers show required electron energy for ignition to NIF size plasma, E_{ig}=22kJ×(1-exp(-1.45MeV/ E))$^{-1}$, where E is the temperature of accelerated electrons [13]. In our case, it is required 252kJ short pulse from the conversion efficiency with the electron spectra in Fig. 4. On the other hand, Hatchett et al. indicate 60% conversion efficiency in the plane target when the laser intensity increases to 10^{20}W/cm^2, resulting that only 109kJ laser energy is needed. If we can obtain such high conversion efficiency under long scale plasma for high intensity laser beam, "Super-Penetration" will be feasible and attractive way to real ignition.

References

[1] M. Tabak *et al.*, Phys. Plasmas **1** (1994) 1626.
[2] R. Kodama *et al.*, Nature **418** (2002) 933.
[3] A.L. Lei *et al.*, Phys. Plasmas **16** (2009) 056307.
[4] A.L. Lei *et al.*, Phys. Rev. E **76** (2007) 066403.
[5] T. Matsuoka *et al.*, Plasma Phys. Control. Fusion **50** (2008) 105011.
[6] K. Adumi *et al.*, Phys. Plasmas **11** (2004) 3721.
[7] J. Fuchs *et al.*, Phys. Rev. Lett. **80** (1998) 1658.
[8] A.J. Mackinnon, *et al.*, Phys. Plasmas **6** (1999) 2185.
[9] Ya.B. Zel'dovich, *et al.*, "Physics of shock waves and high temperature hydorodynamic phenomena" (Academic New York, 1966).
[10] T. Yabuuchi, *et al.*, Phys. Plasmas **14** (2007) 040706.
[11] M. Nakatsutsumi, *et al.*, New J. Physics **10** (2008) 043046.
[12] R. Kodama, *et al.*, Phys. Plasmas 8 (2001) 2268.
[13] S. Atzeni *et al.*, Phys. Plasmas **15** (2008) 056311.
[14] S. Hatchett *et al.*, Phy. Plasmas **7** (2000) 2076; K. Yasuike *et al.*, Rev. Sci. Instrum. **72** (2001) 1236.

Journal of Physics: Conference Series **244** (2010) 022036 doi:10.1088/1742-6596/244/2/022036

Beam collimation and transport of laser-accelerated protons by a solenoid field

K. Harres[1], I. Alber[1], A. Tauschwitz[2], V. Bagnoud[2], H. Daido[3], M. Günther[1], F. Nürnberg[1], A. Otten[1], M. Schollmeier[4], J. Schütrumpf[1], M. Tampo[3] and M. Roth[1]

[1]Technische Universität Darmstadt, Institut für Kernphysik, Schlossgartenstraße 9, 64289 Darmstadt, Germany
[2]GSI - Hemholtzzentrum für Schwerionenforschung GmbH, Plasmaphysik and PHELIX, Planckstrasse 1, 64291 Darmstadt, Germany
[3]Photo Medical Research Center, JAEA, 8-1 Umemidai, Kizugawa-city, Kyoto, 619-0215, JAPAN
[4]Sandia National Laboratories, Albuquerque NM 87185, USA

E-mail: k.harres@gsi.de

Abstract. A pulsed high field solenoid was used in a laser-proton acceleration experiment to collimate and transport the proton beam that was generated at the irradiation of a flat foil by a high intensity laser pulse. 10^{12} particles at an energy of 2.3 MeV could be caught and transported over a distance of more than 240 mm. Strong space charge effects occur, induced by the high field of the solenoid that forces all co-moving electrons down the the solenoid's axis, building up a strong negative space charge that interacts with the proton beam. This leads to an aggregation of the proton beam around the solenoid's axis and therefore to a stronger focusing effect. The collimation and transport of laser-accelerated protons is the first step to provide these unique beams for further applications like post-acceleration by conventional accelerator structures.

1. Introduction

Laser-accelerated ion beams have triggered an extensive discussion about their possible applications due to their unique beam parameters. Ultra-low emittances [1] and highest beam fluxes [2] were experimentally proven. The main problems that are still under strong investigations are the exponential energy spectrum and the high envelope divergence of the beams [3]. Many attempts were done to shape the energy distribution to a more mono-energetic one. Special targets with proton or carbon monolayers [4, 5] on the target surface were used as well as deuterated droplets [6]. Additionally, the beam divergence could be reduced by different techniques like the laser-triggered micro lens [7] or with hemi-spherical targets [8]. In another experiment a quadrupole doublet was used to guide the proton beam [9]. The drawbacks of these systems were either the complexity of the set-up or the high particle losses caused by the collimation device. To avoid this we have established an alternative to control the transport of laser-accelerated protons that uses a pulsed high field solenoid to collimate the beam directly behind the target foil. The coil of the solenoid consists of a brass helix with a length of 75 mm and its open aperture is 44 mm in diameter. This allows to catch nearly all protons at a particle

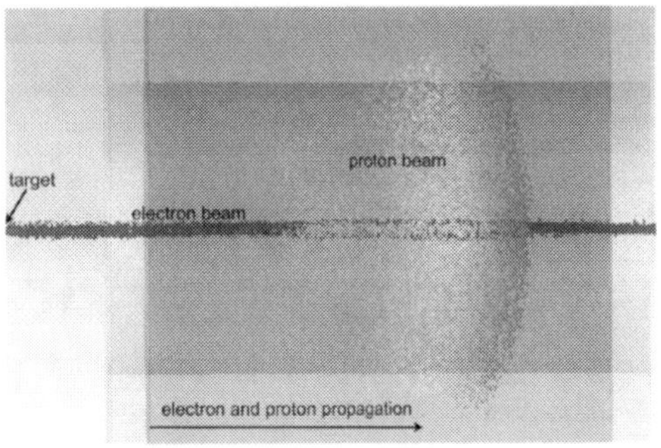

Figure 1. Electron (blue) and proton (cyan to red) flux through the solenoid. The beam has entered the solenoid from the left side of the picture. The electrons are circulating around the solenoid's axis at their gyro radius, leading to a strong negative space charge that results in a proton beam aggregation around the axis.

energy of 2.5 MeV. In the experiment the solenoid ran at 8.6 T. At these strong fields the magnetic pressure that causes high magnetic stress inside the coil is a main concern. The solenoid could successfully be tested up to 15 T which equals 90 MPa magnetic pressure.

2. Simulations

Several simulations were done to investigate the propagation of a proton beam through the solenoid. The simulation program was CST Particle Studio, which is based on the finite volume method used to solve the Maxwell equations. It includes a particle tracker as well as a PIC (particle-in-cell) solver.

During the laser-plasma interaction not only protons but also electrons are accelerated in forward direction. Especially the co-moving electrons, i.e. electrons that travel at the same velocities as the proton beam, have to be considered in the simulations. Due to their much lower energies these electrons get forced down to the solenoid's axis directly at the target position (the magnetic field strength at the target in a distance of 17 mm to the solenoid is around 0.95 T). This effect has strong influence on the proton propagation. The normally existing quasi neutrality of the beam is not preserved. Hence, the protons get attracted by the negative space charge of the electron beam, leading to a proton aggregation around the solenoid's axis. Figure 1 shows the proton and electron beam inside the solenoid 2.75 ns after the laser has hit the target. Electrons can only be found in a very small radius around the solenoids axis (blue dots) and the strong proton aggregation can clearly be seen (green dots - lowest proton energy (1 MeV) - to red dots - highest proton energy (5 MeV)). This effect has strong influence on the proton beam transport and changes the beam profile. Three time steps of the propagation of the protons are shown in figure 2. No electrons are shown in this figure so that the proton aggregation around the axis gets much clearer. The energies of the particles are the same as in figure 1, (1 - 5) MeV. At the latest time step the aggregation starts to smear out. The electrons are not circulating around the axis anymore, because behind the solenoid the magnetic field lines diverge. This leads to a break up of the collimated electron beam and therefore to a broadening of the proton beam aggregation.

The total number of particles in the simulations were 49530. Each particle species was represented by a bunch with a Gaussian distribution with a total charge of 10^{-7} C which equals $6\text{x}10^{11}$ electrons and protons, respectively. The mesh cells were 100 μm in all directions and the time step width was 407 fs.

3. Experimental results

The experimental campaign was done at the PHELIX (Petawatt High Energy Laser for Ion eXperiments) facility at GSI - Helmholtzzentrum für Schwerionenforschung GmbH in Darmstadt, Germany. PHELIX delivers 700 fs short laser pulses with energies up to 150 J on target. This

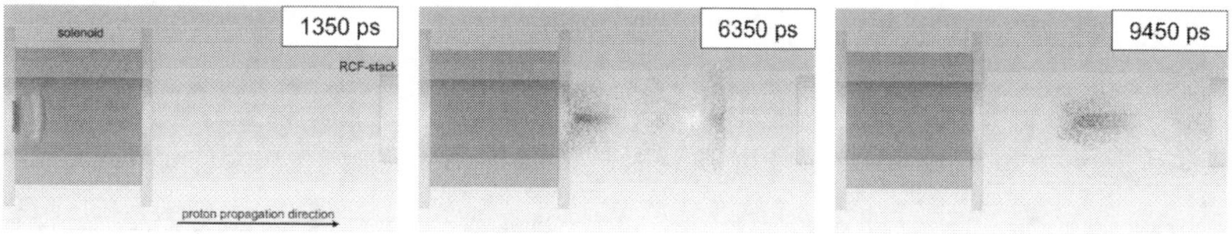

Figure 2. Propagation of the proton beam through the solenoid (left side) to the RCF detector (right side) at different time steps. The strong aggregation of the proton beam due to the electron's space charge is clearly visible. The color of the particles indicates their energy from the lowest value (blue) to the highest one (red).

campaign was the first run of the laser system working at relativistic intensities and during the first shots laser-accelerated protons with energies exceeding 30 MeV could be observed.
The main diagnostic used in the experiment were radiochromic films (RCF). These are dosimetry films sensitive to ionizing radiation. During irradiation the films turn from transparent to blue. The obtained optical density of the films gives information about the deposited dose or rather the proton numbers. An energy resolved measurement was done by using the film in a stack configuration [10].
The solenoid was placed 17 mm behind the target. As mentioned above the magnetic field strength was 8.6 T which let to a collimation of 2.5 MeV protons. Figure 3 shows the first six layers of three different RCF stacks, respectively. The energies (2.3 MeV - 7.6 MeV) belong to protons that are stopped in the specific RCF layer. The first six films on the left, figure 3 a), show the proton beam measured 40 mm behind the target without the solenoid in place. A well defined almost round beam could be observed. Figure 3 b) shows the reference shot at a distance of 130 mm behind the solenoid which equals 241 mm to the target. The solenoid was in place but turned off. Due to the strong divergence of the beam (25 degrees half opening angle) only a few particles are detected by the RCF. At energies higher than 6.0 MeV only a weak proton background is visible due to the fact that the number of protons for these energies is below the RCF detection threshold of 10^8. In comparison figure 3 c) shows RCF films from a shot where the solenoid was in use. A strong signal could be measured for a proton energy of 2.3 MeV as expected from the simulations. The particle numbers exceed 10^{12} as well as for the first layer

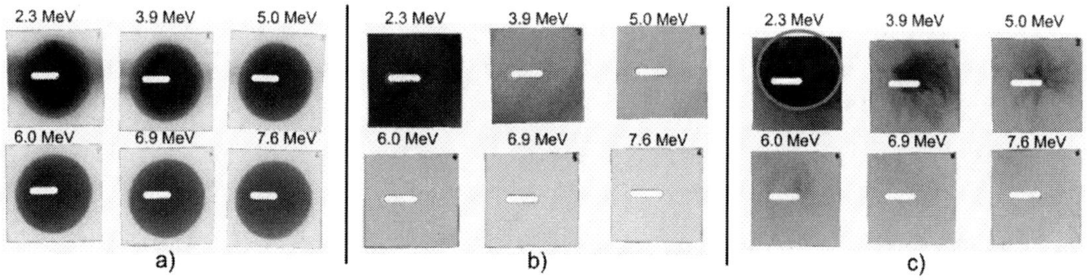

Figure 3. The proton beam measured by RCF at a distance of 40 mm to the target a) and at a distance of 241 mm b). c) The proton beam measured behind the solenoid at the same position as b). Due to the collimation a strong proton signal was detected at 2.3 MeV particle energy. Even for higher energies a clear proton signal could be observed due to the additional focusing effect by the co-moving electrons.

of figure 3 a). The beam size is 39 mm in vertical and 42 mm in horizontal direction, marked by the red circle. A diameter of 40 mm was calculated by the simulations. Not only in the first layer but also in layer two to four a clear proton signal was detected. This happened due to the electron attraction on the proton beam. Normally, only a few particles are expected to be measured at energies higher than 2.5 MeV due to the strong divergence of the beam. Even particles up to 6 MeV are affected by the electron electric field leading to the strong signals in the RCF.
Additionally, strong beam modulations could be observed in the RCF layers when using the solenoid in comparison to a standard laser-proton acceleration without any collimation device. The reason is that eddy currents were induced in the target by the solenoid's magnetic field (0.95 T at the target position) which let to a strong bending of the foil. Figure 3 shows a picture of the laser-plasma interaction zone. The front and rear plasma expansion can easily be identified. The induced eddy currents caused a bending of the target by 18 degrees. Simulations that were done with a 18 degrees bent target show that nearly 50 % of the protons get lost inside the solenoid and the the intensity maximum of the beam image at the RCF position is moved out of the center by 15 mm which can be seen as well in figure 3 c). For future experiments the bending will be suppressed by using non conductive targets.

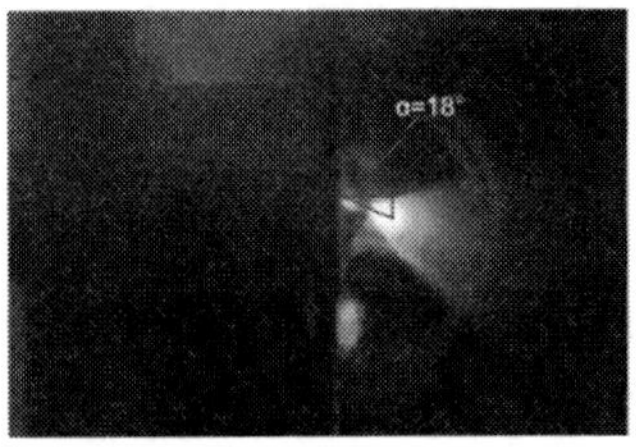

Figure 4. Image of the laser-plasma interaction. The picture shows a shot where the target got bent by the strong magnetic field of the solenoid inducing eddy currents in the target.

4. Conclusion

In this article we have shown that the collimation of a laser-accelerated proton beam by a pulsed high field solenoid is possible and leads to great results in terms of collimation efficiencies. 10^{12} protons could be transported over a distance of 241 mm. Inside the solenoid strong space charge effects occurred due to the co-moving electrons leading to a proton beam aggregation around the axis. This resulted in a stronger focusing of the beam so that even higher energies than the expected 2.3 MeV could be observed by the RCF.

5. Acknowledgements

The authors would like to thank the divisions of PHELIX and plasma physics at GSI for their excellent support. This work was supported by Bundesministerium für Bildung und Forschung (BMBF), support code 06 DA 122 I.

References

[1] T.E. Cowan *et al.* 2004 *Phys. Rev. Lett.* **92** 204801
[2] M.A. Stoyer *et al.* 2001 *Rev. Sci. Instrum.* **72** 767
[3] M. Zepf *et al.* 2003 *Phys. Rev. Lett.* **90** 064801
[4] H. Schwoerer *et al.* 2006 *Nature (London)* **439** 445
[5] B.M. Hegelich *et al.* 2006 *Nature (London)* **439** 441
[6] S. Ter-Avetisyan *et al.* 2006 *Phys. Rev. Lett.* **96** 035005
[7] T. Toncian *et al.* 2006 *Science* **312** 410
[8] P.K. Patel *et. al* 2003 *Phys. Rev. Lett.* **91** 125004
[9] M. Schollmeier *et al.* 2008 *Phys. Rev. Lett.* **101** 055044
[10] F. Nürnberg *et al.* 2009 *Rev. Sci. Instrum.* **80** 033301

Effects of Preformed Plasma of CH Foam on Fast Electron Generation

M. Hata[1], H. Sakagami[2], A. Sunahara[3], T. Johzaki[4], and H. Nagatomo[4]

[1]Department of Physics, Nagoya University, Nagoya, Aichi 464-8602, Japan

[2]Department of Simulation Science, National Institute for Fusion Science, Toki, Gifu, 509-5292, Japan

[3]Institute for Laser Technology, Suita, Osaka, 565-0871, Japan

[4]Institute of Laser Engineering, Osaka University, Suita, Osaka, 565-0871, Japan

hata.masayasu@nifs.ac.jp

Abstract. In fast ignition, the intensity of an unavoidable pre-pulse of a heating laser is still high to generate plasmas even if the contrast ratio is high. So, a preformed plasma (pre-plasma) is created before a main pulse of the heating laser irradiates a target. The recent research suggests that coating an inner surface of the cone of the cone-guided target with the low-density foam enhances fast electron generation. But the pre-plasma and the foam structure, which consists of dense CH plasmas and voids, are ignored in that research. In this paper, effects of the pre-plasma and the foam structure on the fast electron generation are investigated with the use of an one-dimensional Particle-In-Cell code. It is found that effects of the pre-plasma on the fast electron generation are small and the foam structure weakens the generation of fast electrons with the moderate energy.

1. Introduction

The cone-guided target, of which an inner surface of the cone is coated with the low-density foam, is proposed for fast ignition experiments in FIREX-I, because it is reported that the low-density foam enhances fast electron generation [1]. In fast ignition, an ultrahigh intense laser is used to heat the core. A pre-pulse of the heating laser is unavoidable and its intensity is still high to generate plasmas even if the contrast ratio is high. Therefore a preformed plasma (pre-plasma) is created before a main pulse of the heating laser irradiates the target. An importance of the pre-plasma on the fast electron generation is suggested in recent researches [2,3]. The pre-plasma is ignored in previous research [1], therefore pre-plasma effects should be investigated. The foam is actually constructed with dense CH plasmas and voids, but the foam structure is also ignored and the foam density is assumed to be uniform in that research. Thus, effects of the foam structure should be also investigated.

We perform simulations for both cases with and without the pre-plasma using an one-dimensional Particle-In-Cell code and investigate the effects of the pre-plasma on the fast electron generation. The foam structure is realized to put dense CH plasma and vacuum gap in turn due to one dimensionality. So, we perform simulations with the different gap lengths and also investigate effects of the foam structure on the fast electron generation.

2. Pre-plasma and Foam Structure

In simulations, the target is constructed from CH foam pre-plasma, CH foam, Au and CD plasmas. The CH foam pre-plasma consists of three parts. Each part has an exponential profile with the scale length (L_{pre}) of 56.9 ($0.10 \sim 0.22n_{cr}$), 18.0 ($0.22 \sim 0.80n_{cr}$) and 0.869[μm] ($0.80n_{cr} \sim n_{foam}$), where n_{foam} is the CH foam density. Supposing full ionization, the ionization degree (Z) and the mass number (A) of the pre-plasma are assumed to be 3.5 and 6.5 as averaged values of C and H, respectively. The density profile of the pre-plasma is determined by simulations performed with STAR1D, an one-dimensional Lagrangian radiation-hydrodynamic code [4]. STAR1D simulates pre-plasma formation by the pre-pulse which is set to $I_L = 10^{11}$[W/cm^2] and $t_{flat} = 1$[ns]. The ionization degree and the mass number of the CH foam plasma, which has the flat profile with 20[μm] thickness and n_{foam}, are the same as the CH foam pre-plasma. The profile of the Au (CD) plasma is also flat with Z = 8, A = 197, 10[μm] thickness and $100n_{cr}$ (Z = 3.5, A = 7, 35[μm] thickness and $100n_{cr}$). The main pulse of the heating laser is set to $I_L = 10^{20}$[W/cm^2], $t_{rise/fall} = 375$[fs] and $t_{flat} = 1.5$[ps]. To investigate the difference of the fast electron generation by the foam density, we set n_{foam} to 10, 20, 40 and $80n_{cr}$.

2.1. Pre-plasma

To evaluate fast electron characteristics, we observe fast electrons in the CD plasma, 10[μm] behind the Au-CD boundary. Figure 1 shows time evolutions of fast electron beam intensity (a) with pre-plasma and (b) without pre-plasma. The time when the fast electron beam intensity rises for the first time in the figure 1 (a) is 200[fs] later than that in the figure 1 (b). The time is defined as $t = 0$ when the heating laser starts to interact with a leading-edge of the pre-plasma, so fast electrons take more time to arrive at the observation point in the case with pre-plasma than that without pre-plasma. As the length of the pre-plasma is about 70[μm] and fast electrons propagate with almost light speed, the time lag is estimated at 233[fs] and it is a good agreement with the observed delay. In the cases of 40 and $80n_{cr}$, the electron beam intensity with pre-plasma increases more and is maintained with a higher level than that without pre-plasma. But in the cases of 10 and $20n_{cr}$, the electron beam intensities with and without pre-plasma are almost same.

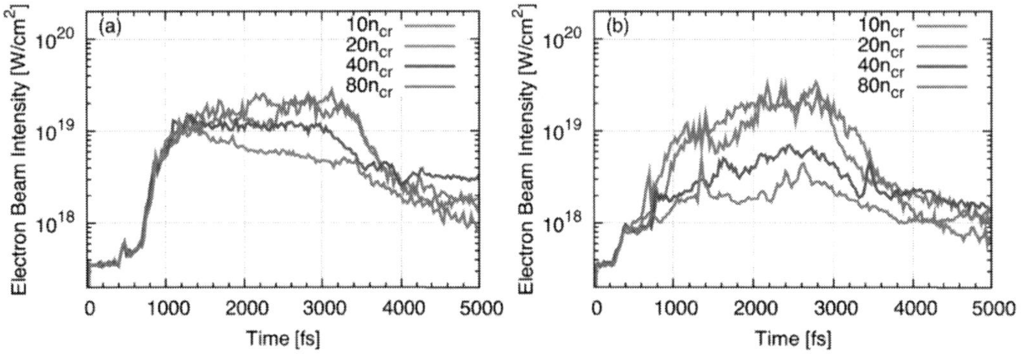

Figure 1. Time evolutions of fast electron beam intensity (a) with pre-plasma and (b) without pre-plasma. Red, green, blue and purple lines indicate n_{foam} = 10, 20, 40 and $80n_{cr}$, respectively.

Figure 2 shows time-integrated energy spectra of fast electrons (a) with pre-plasma and (b) without pre-plasma. In the cases of 10 and $20n_{cr}$, the energy spectrum with pre-plasma is very similar to that without pre-plasma in whole range of electron energy. However, in the cases of 40 and $80n_{cr}$, fast electrons with pre-plasma increases more than those without pre-plasma of a low-energy (1~5 [MeV]) component. Even though fast electrons of a high-energy (>5 [MeV]) component are not generated at all without pre-plasma, those are generated in the case with pre-plasma.

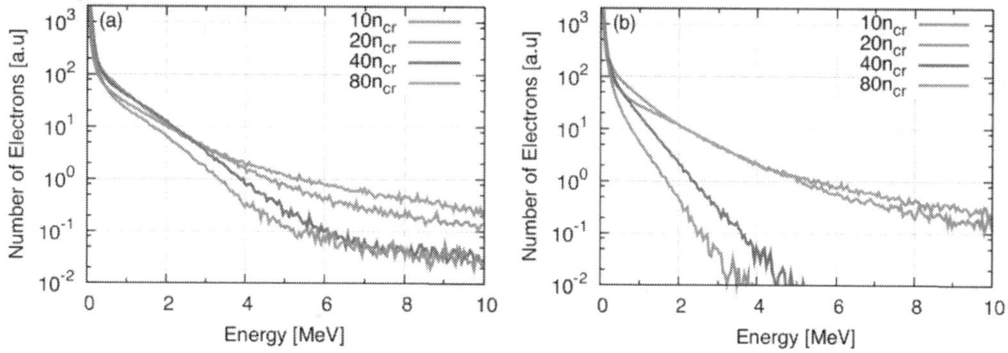

Figure 2. Time-integrated energy spectra of fast electrons (a) with pre-plasma and (b) without pre-plasma. Red, green, blue and purple lines indicate n_{foam} = 10, 20, 40 and $80n_{cr}$, respectively.

Figure 3 shows electron density profiles with and without pre-plasma in the case of (a) $n_{foam}=20n_{cr}$ and (b) $n_{foam}=40n_{cr}$. We compare results at difference time, because fast electrons with pre-plasma reach the observation point 200[fs] later than those without pre-plasma as stated above. In the case of $20n_{cr}$ without pre-plasma, the low-density plasma is generated in front of the CH foam plasma and expected to interact with the heating laser just same as the pre-plasma. As a result, the spectrum without pre-plasma is same as that with pre-plasma. In the case of $40n_{cr}$ without pre-plasma, no low-density plasma is found in front of the CH foam plasma. Consequently, fast electrons of the high-energy component are not generated. On the other hand, the electron density at the laser front in the steepened density profile without pre-plasma is found to be $\sim 100n_{cr}$ and it is higher than that ($\sim 80n_{cr}$) with pre-plasma. As a result, in the case of $40n_{cr}$ without pre-plasma, the laser directly interacts with the higher density plasma and fast electrons of the low-energy component are generated less than those with pre-plasma [1].

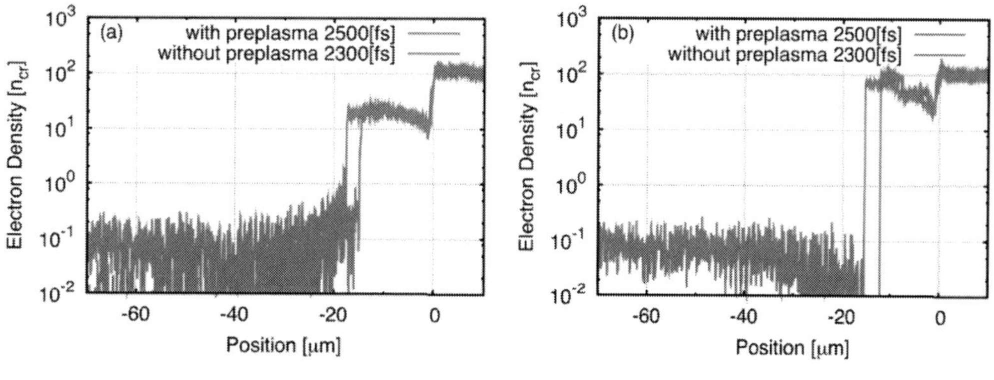

Figure 3. Electron density profiles with and without pre-plasma for (a) $n_{foam}=20n_{cr}$ and (b) $n_{foam}=40n_{cr}$. Red and green lines indicate the cases with pre-plasma and without pre-plasma, respectively.

2.2. Foam Structure

The foam is actually constructed with a solid CH and voids. After laser irradiation, the solid CH becomes plasmas, which are quickly diffused into voids. Thus the density of the foam is defined as an averaged density and the foam density is assumed to be uniform in previous research [1]. However, the foam structure should affect the fast electron generation if the size of voids is enough large not to fill them with plasmas instantaneously. So, we investigate effects of the foam structure on fast electron generation. In one-dimensional simulations, the foam structure can be imitated by dense CH plasmas and vacuum gaps in turn. Fast electrons with energy of ~1 [MeV] can heat the core most efficiently [2], and they are generated most in previous simulations when the foam density is $20n_{cr}$. So we choose the foam with $n_{foam}=20n_{cr}$ for foam structure investigations, and set the density of the dense CH plasma to $100n_{cr}$ and the length of it four times longer than that of the vacuum gap. Simulations are performed with the different gap lengths, namely 2, 4 and 8[μm], and other parameters are same as that of previous simulations.

Figure 4 shows time evolutions of electron beam intensity and time-integrated energy spectra of fast electrons. If the width of the vacuum gap is wider, the rising of electron beam intensity occurs at later time. Looking at spectra, as the width of the gap is wider, less electrons of the low-energy component are generated and more electrons of the high-energy component are generated. It takes longer time to fill the vacuum gap by plasma diffusion when the length of the gap is longer, and then an electrostatic sheath field is maintained for longer time. While the electrostatic sheath field is preserved, fast electrons of the low-energy component are reflected by this field and cannot pass through the vacuum gap. These reflected electrons of the low-energy component are propagated to opposite direction of laser irradiation, accelerated by the reflected laser. Then they are reflected again by the electrostatic sheath field in front of the pre-plasma edge. As a result, if the width of the gap is wider, the electrostatic sheath field is maintained for longer time and more fast electrons of the high-energy component are generated.

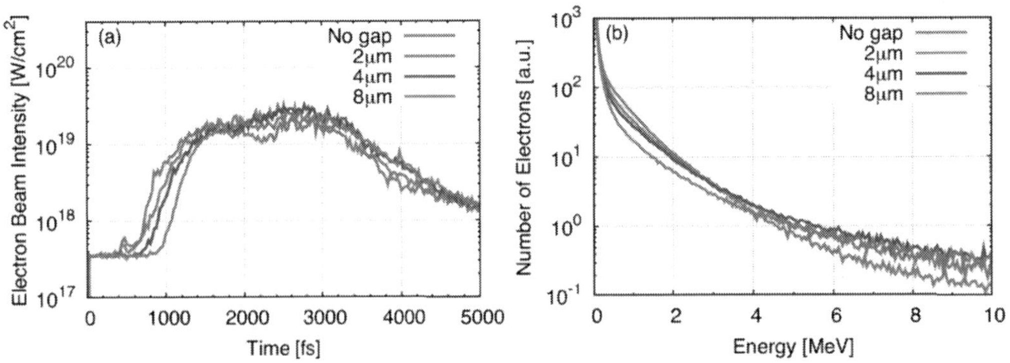

Figure 4. (a) Time evolutions of fast electron beam intensity. (b) Time-integrated energy spectra of fast electrons. In two figures, red, green, blue and purple lines indicate that no gap, the gap size of 2, 4 and 8[μm], respectively.

3. Summary

We have investigated effects of the pre-plasma on the fast electron generation for low-density foam coated targets in fast ignition. Fast electron generation in the case of $n_{foam}=20n_{cr}$ with pre-plasma is same as that without pre-plasma. However, in the case of $n_{foam}=40n_{cr}$ with pre-plasma, fast electrons are generated more than those in the case without pre-plasma. Even such difference, we can conclude that effects of the pre-plasma are not significant for the fast electron generation. We have also investigated effects of the foam structure and found that the number of fast electrons of the low-energy component decreases, but these of the high-energy component increases when the length of the vacuum gap is 8 [μm]. Even so, we can also conclude that the foam structure does not significantly affect on the fast electron generation.

References

[1] Sakagami H, Johzaki T, Nagatomo H and Mima K 2009 *Nucl. Fusion* **49** 075026
[2] Johzaki T, Sakagami H, Nagatomo H and Mima K 2007 *Laser Part. Beams* **25** 621-9
[3] Kemp A J, Sentoku Y and Tabak M 2008 *Phys. Rev. Lett.* **101** 075004
[4] Sunahara A, Nishihara K and Sasaki A 2008 *Plasma Fusion Res.* **3** 043

Journal of Physics: Conference Series **244** (2010) 022038 doi:10.1088/1742-6596/244/2/022038

Fast ignition by laser-driven carbon beams

J.J. Honrubia,[1] J.C. Fernández,[2] M. Temporal,[1] B.M. Hegelich,[2,3] and J. Meyer-ter-Vehn[4]

[1] ETSI Aeronáuticos, Universidad Politécnica de Madrid, Madrid-28040, Spain
[2] Los Alamos National Laboratory, Los Alamos, NM-87544, USA
[3] Fakultät für Physik, LMU München, D-85748 Garching, Germany
[4] Max-Planck-Institut für Quantenoptik, D-85748 Garching, Germany

E-mail: javier.honrubia@upm.es

Abstract. Two-dimensional simulations of ion beam driven fast ignition are presented. Ignition energies of protons with Maxwellian spectrum and carbon ions with quasi-monoenergetic and Maxwellian energy distributions are evaluated. It is found that quasi-monoenergetic ions have better coupling with the compressed Deuterium-Tritium and substantially lower ignition energies. These energies are similar to those found for relativistic electrons, provided that a laser to quasi-monoenergetic carbon ion conversion efficiency around 10% can be achieved.

Fast ignition by laser-driven ion beams [1] offers the advantages of the classical interaction of ion beams with compressed fuels, their much more localized energy deposition, and the stiffer ion transport with the possibility of beam focusing. Recent particle-in-cell (PIC) simulations have shown that circularly polarized pulses with irradiances around 10^{22} W/cm^2 can accelerate ions up to hundreds of MeV by means of radiation pressure. These ions have quasi-monoenergetic energy distributions, small divergence angles [2–5] and laser-to-ion conversion efficiencies around 10% [3]. The laser break-out afterburner scheme has been also proposed to generate almost perfectly collimated quasi-monoenergetic ions with conversion efficiencies of a few percent [6] at laser intensities about 10^{21} W/cm^2. These new schemes are suitable to accelerate ions heavier than protons to the energies required for fast ignition applications. Fernández et al. [7] suggested the use of quasi-monoenergetic carbon ions with kinetic energies of a few hundreds of MeV to ignite pre-compressed fusion targets. Use of heavier ions has been studied by Shmatov in in Ref. [8]. One advantage of quasi-monoenergetic ion beams is the possibility to place the source far from the compressed core avoiding the use of re-entrant cones. However, the feasibility of the carbon ion fast ignition scheme (CFI) relies on the demonstration of conversion efficiencies comparable to those found for protons. Here, we assume a laser to quasi-monoenergetic carbon ions conversion efficiencies of about 10% and compare the potential of the proton fast ignition (PFI) and CFI schemes.

We consider the ideal case of perfectly collimated cylindrical ion beams of 30 μm diameter impinging on a DT blob with a peak density around 500 g/cm^3. Scattering of ions by the background DT plasma has been neglected because it is small for temperatures lower than the ignition temperature (≈10 keV). The two different configurations of the compressed DT shown in figure 1 are assumed. The first one is a simplification of the configuration obtained from the 2D implosion calculations of cone-targets. The second one is similar to that obtained by Clark and Tabak [9] from 1D implosion calculations of a direct-drive target to get an almost isochoric configuration of the compressed fuel.

This target has 0.9 mg of DT that are compressed by a laser pulse of 485 kJ delivered in 34 ns [9]. The calculations shown below have been performed with the 2D radiation-hydrodynamics code SARA, which includes flux limited electron conduction, multigroup radiation transport, ion energy deposition, DT fusion reactions and α-particle transport [10].

The simulation box is shown in figure 1. Ions come from the left and propagate towards the blob through low density plasma. We assume that ions are generated instantaneously with a Maxwellian energy distribution for protons ($N(\varepsilon) = 2N_0\varepsilon^{1/2}\exp[-\varepsilon/kT]/\pi^{1/2}(kT)^{3/2}$, where N_0 is the total number of ions and T their temperature) and a Gaussian energy distribution for carbon ions ($N(\varepsilon) = N_0\alpha^{1/2}\exp\{-\alpha[(\varepsilon-\varepsilon_0)/\delta\varepsilon]^2\}/\pi^{1/2}\delta\varepsilon$, where $\alpha = 4\ \ln(2)$, ε_0 is the mean energy and $\delta\varepsilon$ the full width at half-maximum, FWHM). If the energy spread is $\delta\varepsilon/\varepsilon_0 = 0.1$, we refer to this last distribution as "quasi-monoenergetic". We assume Instantaneous emission of the beam ions because the spread of the time of flight from the source to the target ($\approx$ 10 ps) is much longer than the typical ion acceleration time ($\approx$ 1 ps). We used analytical formulas [11] to compute the kinetic energy and the beam power on target of ions accelerated instantaneously at a distance *d*. Beam power and ion kinetic energy of 10 kJ Maxwellian proton beam generated at *d* = 0.5 mm and quasi-monoenergetic carbon ion beams generated at *d* =1.35 cm are shown in figure 2. These two distances have been chosen in order to have the same peak power and approximately the same pulse duration on target ($\approx$ 8 ps) with both beams. It is worth noting that the almost negligible time spread of ions with quasi-monoenergetic energy distributions allows to place the source at much higher distances *d* than Maxwellian ions. Because beam focalization over distances of a few centimeters may be difficult, several focusing techniques have been proposed such as: i) ballistic transport [12], ii) focusing by fields self-generated in hollow microcylinders by intense subpicosecond laser pulses [13] and iii) focusing by magnetic lenses [14]. We assume that, ideally, ion beams are perfectly collimated without any divergence at the source and during the transport towards the compressed fuel.

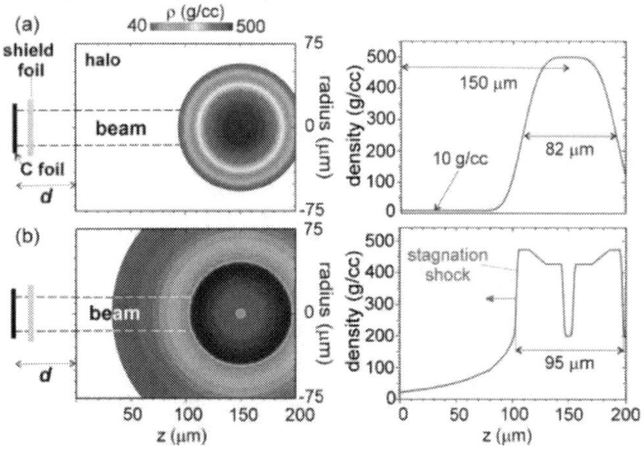

Figure 1. Density isocontours and radial profiles of the targets considered. (a) Super-Gaussian density blob and (b) density distribution taken from Fig. 8 of Ref. [9].

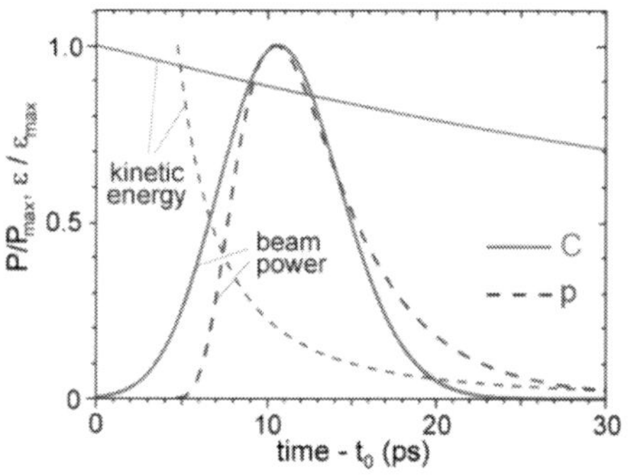

Figure 2. Beam power and ion kinetic energy at the left surface of the simulation box as a function of time. The dashed line corresponds to a Maxwellian proton beam with $T_p = 4$ MeV, d=0.5 mm, t_0=0, ε_{max}= 58 MeV and P_{max} = 1.12 PW. The solid line corresponds to a Gaussian carbon ion beam with $\delta\varepsilon/\varepsilon_0$=0.1, d=1.35 cm, t_0 = 158 ps, ε_{max} = 458 MeV and P_{max}=1.12 PW.

Range lengthening effect is important when plasma electron velocities become comparable to the ion velocity. In the case of Maxwellian ions generated far from the fuel, this effect may be beneficial for ignition because the decrease of ion kinetic energy with time (see figure 2) is balanced by their range lengthening as the DT is heated up, keeping the ion range almost constant with time [11]. On the contrary, range lengthening may be a disadvantage for quasi-monoenergetic ions because their kinetic energy on target has only a small variation with time when compared with Maxwellian ions (see figure 2). However, it is worthwhile noting that range lengthening is less pronounced for ions

Journal of Physics: Conference Series **244** (2010) 022038 doi:10.1088/1742-6596/244/2/022038

with high kinetic energies per nucleon. For instance, the range of 400MeV carbon ions increases by a factor of ≈3 for plasma temperatures from hundreds of eV to 10 keV, while this factor is around 20 for 5 MeV protons [15].

The ignition energies E_{ig} as a function of proton temperature and kinetic energy of carbon ions are shown in figures 3(a) and (b), respectively. They have been obtained as the minimum beam energy for which the thermonuclear fusion power has an exponential or higher growth sustained in time. We found an optimal proton temperature T_p around 4 MeV, for which E_{ig} = 12.7 kJ. This energy is higher than the ≈ 9 kJ found in Ref. [16] for an ideal isochoric fuel configuration with a DT density of 500 g/cm^3 and the same source – target distance d = 0.5 mm. This difference accounts for the energy deposited in the plasma surrounding the dense fuel. For the same ideal isochoric configuration used in the reference, our simulations give E_{ig} = 9.5 kJ, in good agreement with [16]. This result evidences the importance of the plasma surrounding the dense fuel, which increases substantially ignition energies. The details are given in Ref. [15].

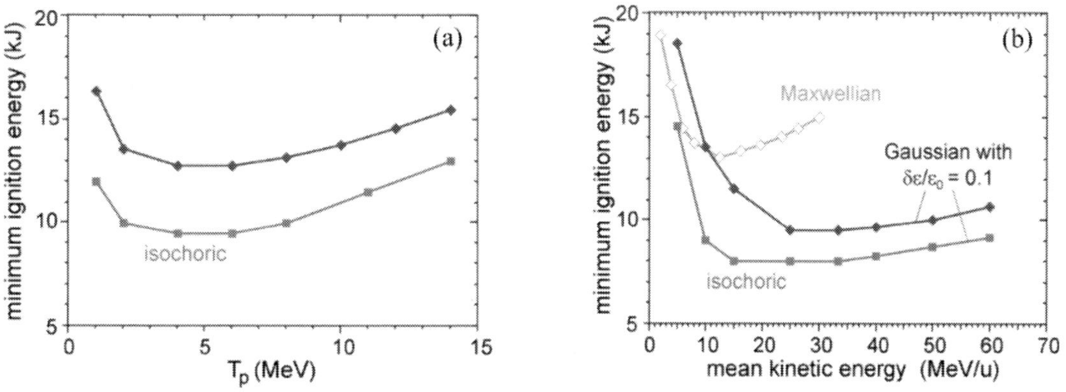

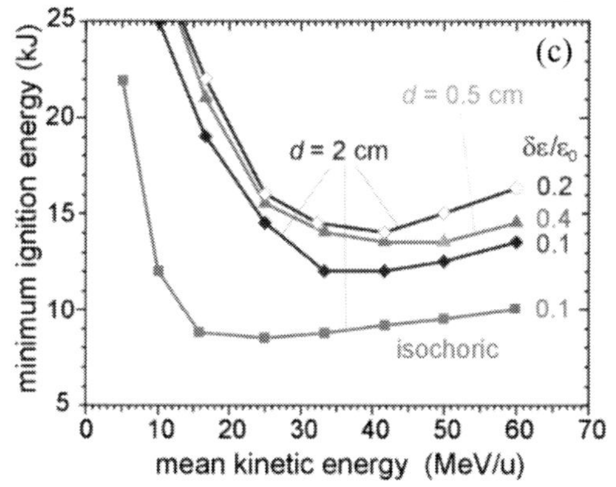

Figure 3. Minimum ignition energies of the targets shown in Fig. 1. (a) target with the super-Gaussian density distribution heated by Maxwellian proton beams with temperature T_p and a source-target distance d = 0.5 mm. (b) target with the super-Gaussian density distribution heated by carbon ion beams with Gaussian and Maxwellian energy distributions as a function of the mean kinetic energy per nucleon. d = 1.35 cm for the beams with Gaussian energy distributions and d = 0.5 mm for the Maxwellian beam. (c) target of Fig. 1(b) heated by carbon ion beams with Gaussian energy distribution and energy spread $\delta\varepsilon/\varepsilon_0$. Curves are labeled by the source-target distance. The ignition energies of an ideal isochoric blob at 500g/cm^3 are shown in all plots.

Quasi-monoenergetic carbon ions ($\delta\varepsilon/\varepsilon_0$ = 0.1) have a better coupling with the dense fuel. Figure 3(b) shows that ignition energies are around 9.5 kJ for the optimal energy range from 25 to 40 MeV/u, which are about 25% lower than those obtained for Maxwellian protons. Similarly to PFI, the ignition energies can be reduced further by removing the coronal plasma surrounding the dense core. For "ideal" isochoric targets, E_{ig} is reduced to 8 kJ for 15 – 32 MeV/u ions. It is interesting to note that the effect of the surrounding plasma on E_{ig} is more pronounced for ion energies lower than 25 MeV/u. Maxwellian carbon ions have ignition energies much higher than quasi-monoenergetic ions and comparable to those found for Maxwellian protons.

Ignition energies of quasi-monoenergetic carbon ions with different energy spreads $\delta\varepsilon/\varepsilon_0$ heating the imploded fuel configuration shown in figure 1(b) are plotted in figure 3(c). It is remarkable that these energies are at least 25% higher than those obtained for the target with the super-Gaussian density distribution of figure 1(a). This is mainly due to the energy deposition in the plasma surrounding the core, as can be evidenced by comparing the ignition energies with those obtained for the ideal isochoric configuration. Note that the effect of the surrounding plasma on E_{ig} is particularly important for relatively low ion kinetic energies. Thus, the optimal energy range of quasi-monoenergetic carbon ions is shifted from the 25 – 40 MeV/u obtained for the super-Gaussian density distribution to the 33 – 50 MeV/u shown by the curve labeled with $\delta\varepsilon/\varepsilon_0 = 0.1$ in figure 3(c).

Ignition energies are very sensitive to the ion energy spread, as can be seen by comparing the curves labeled by $\delta\varepsilon/\varepsilon_0 = 0.1$ and 0.2 in figure 3(c). Energy spreads higher than 0.2 lead to high ignition energies >20 kJ for $\delta\varepsilon/\varepsilon_0 = 0.4$ and $d = 2$ cm. These energies can be reduced by placing the ion source closer to the target. For instance, the ignition energy of 50 MeV/u carbon ions with $\delta\varepsilon/\varepsilon_0 = 0.4$ generated at a distance $d = 0.5$ cm turns out to be $E_{ig} = 13.5$ kJ, slightly higher than the 12.5 kJ obtained for ions with the same kinetic energy, $\delta\varepsilon/\varepsilon_0 = 0.1$ and $d = 2$ cm. Yet, a source - core distance $d = 0.5$ cm is still high enough to use target designs without a re-entrant cone inserted. Therefore, we can conclude that the requirement of quasi-monoenergetic spectra can be relaxed in such a manner that ion beams with broad energy distributions generated relatively far from the compressed core can ignite realistic imploded fuel configurations with moderate beam energies. If the plasma surrounding the dense core is minimized or removed, these ignition energies could be reduced further.

In summary, the use of quasi-monoenergetic ions heavier than protons improves the beam coupling efficiency and reduces substantially the ignition energies. In addition, quasi-monoenergetic ions allow placing the ion source far away from the fuel capsule, simplifying target design and fuel implosion and compression, provided that ions can be focused onto a spot of about 30 μm over distances of a few centimeters. This conclusion is still valid for ion energy spreads as high as 40%. In this case, the increase of the ignition energy can be compensated by reducing the source – capsule distance, which is still high enough to avoid the use of re-entrant cones. Prior to the application of the fast ignition by quasi-monoenergetic ions scheme, laser to ion conversion efficiencies of the order of 10% have to be demonstrated experimentally. If these conversion efficiencies are achieved, our calculations show that the laser energy requirements are similar to those found for fast ignition driven by relativistic electrons. In this case, the advantages of the ion fast ignition scheme are attractive enough to envision ion fast ignition as a candidate to demonstrate fast ignition in future facilities such as HiPER [17].

Acknowledgements

One of the authors (J.J.H.) would like to thank the fruitful discussions and the hospitality of the P-24 group of LANL. This work was partially supported by the Research Grants ENE2006-06339 and CAC-2007-013 of the Spanish Ministry of Education, and by the IFE Keep-in-touch Activities of EURATOM. M.T. has been supported by a contract of the Spanish Ministry of Education (Ramón y Cajal 2007-0447).

References

[1] M. Roth et al., Phys. Rev. Lett. **86**, 436 (2001).
[2] A. P. L. Robinson et al., New J. Phys. **10**, 013021 (2008).
[3] O. Klimo et al., Phys. Rev. ST Accel. Beams **11**, 031301 (2008).
[4] N. Naumova, et al., Phys. Rev. Lett. **102**, 025002 (2009).
[5] M. Chen, A. Pukhov, T. P. Yu, and Z. M. Sheng, Phys. Rev. Lett. **103**, 024801 (2009).
[6] L. Yin et al., Phys. Plasmas **14**, 056706 (2007).
[7] J. C. Fernández et al., Nucl. Fusion **49**, 065004 (2009).
[8] M. L. Shmatov, J. Phys.: Conf. Ser. **112**, 022061 (2008).
[9] D. S. Clark and M. Tabak, Nucl. Fusion **47**, 1147 (2007).
[10] J. J. Honrubia, J. Quant. Spectrosc. Radiat. Transf. **49**, 491 (1993).
[11] M. Temporal, J. J. Honrubia, and S. Atzeni, Phys. Plasmas **9**, 3098 (2002).
[12] M. H. Key, Phys. Plasmas **14**, 055502 (2007).
[13] T. Toncian et al.,Science **312**, 410 (2006).
[14] M. Schollmeier et al., Phys. Rev. Lett. **101**, 055004 (2008).
[15] J.J. Honrubia et al., Phys. Plasmas **16**, 102701 (2009).
[16] M. Temporal, J. J. Honrubia, and S. Atzeni, Phys. Plasmas **15**, 052702 (2008).
[17] M. Dunne, Nat. Phys. **2**, 2 (2006).

PLASMA PHOTONIC DEVICES WITH COMPLEX REFRACTIVE INDEX IN EUV REGION

Y. Inubushi, T. Tanaka, A. Higashiya[1)], J. Ishida[2)], T. Ishikawa[1)], H. Kimura[1,3)], T. Kumagai, S. Morimoto, M. Nagasono[1)], H. Ohashi[1,3)], M. Yabashi[1)], Y. Yamaguchi[2)], H. Yoneda [2)], and R. Kodama

Graduate School of Engineering, Osaka University, Suita, Osaka, Japan

1) RIKEN XFEL Project Head Office, Sayo, Hyogo, Japan

2) Institute for Laser Science, University of Electro-Communications, Chofu, Tokyo, Japan

3) Japan Synchrotron Radiation Research Institute, Sayo, Hyogo 679-5148, Japan

yinubusi@eie.eng.osaka-u.ac.jp

Abstract. We report two types of plasma photonic devices using complex refractive index around an absorption edge. Cherenkov radiation using laser-produced electron beam having a Maxellian distribution is evaluated by theoretical calculation. In the calculation result, this Cherenkov radiation is monochromatic in spite of energy distribution. This monochromaticity is due to characteristic of real part of complex refractive index around an absorption edge. Nonlinear transmission in metal using EUV light pulse is also reported. Transmission becomes nonlinearly higher with increase in energy density of EUV light from 6 J/cm^2. This nonlinear phenomenon is caused by change of absorption coefficient, that is, imaginary part of complex refractive index.

1. Introduction

Novel plasma photonic devices [1] have been developed to generate and control high energy density particles and radiation. These high energy density beams are expected for applications such as inertial fusion, x-ray generation and laboratory astrophysics. By using characteristics of complex refractive index, novel plasma photonic devices to generate and control x rays and/or EUV light can be realized. Complex refractive index N is described by

$$N = n + i \cdot k. \tag{1}$$

Here, n and k are real part and imaginary part of complex refractive index. In EUV region, real part n only around absorption edge exceeds 1. Imaginary part k has large difference between shorter wavelength side and longer wavelength side of absorption edge, and absorption coefficient also has large difference.

In this paper, we report extensive study of plasma photonic devices using complex refractive index in EUV region. Two types of plasma photonic device are reported. One is the monochromatic EUV radiation source based on Cherenkov radiation using ultra-high intensity laser-produced relativistic electron beam. For this radiation source, real part of refractive index n exceeding 1 around absorption edge was used. Previously Cherenkov radiation in this region was generated by using monochromatic electron beam provided from an accelerator [2]. We show possibility of use of laser-produced electron beam having an energy distribution. The other is nonlinear transmission in metal using focused EUV-FEL (Free Electron Laser) pulse. Observed nonlinear phenomenon indicates that imaginary part of complex refractive index k around absorption edge is drastically changed. This phenomenon can be applied to novel optics in EUV region such as auto-correlator and x-ray pulse slicer.

2. Monochromatic EUV radiation source

Generally, in x-ray and EUV region, real part of complex refractive index n is smaller than 1. However, n exceeds 1 around the absorption edge for some matters. In this area, Cherenkov radiation can be emitted. This radiation is called "X-ray Cherenkov Radiation (XCR)" hereafter. XCR is monochromatic because refractive index exceeding 1 is in a very narrow region. Moreover, XCR has adjustability of wavelength by changing material of emitter.

In the case that an electron propagates linearly in a matter having a real part of complex refractive index n, emission angle θ and photon number N taking into account relativistic effects and absorption in the matter [3] are shown in

$$\cos\theta = \frac{\gamma}{n\sqrt{\gamma^2 - 1}} \quad (1)$$

$$\frac{dN}{d\varepsilon} = \frac{2\pi\alpha}{hc}\left\{1 - \frac{\left(1+\frac{E}{E_0}\right)^2}{n^2\left[\left(1+\frac{E}{E_0}\right)^2 - 1\right]}\right\}\int_0^d \exp\left(-\frac{1}{l_{abs}}\frac{z}{\cos\theta}\right)dz \quad (2)$$

where γ is Lorentz factor, c is light speed in vacuum. ε and α are photon energy of XCR and fine structure constant of the matter, respectively. z is propagation length of electron in the matter. E is electron energy and E_0=0.511 MeV, $l_{abs}=-hc/4\pi nk\varepsilon$ is absorption length. Emitter thickness d of 10 µm was used, because absorption length of this XCR is much shorter than 10 µm. High energy density electron beam produced by ultra-high intensity laser has energy distribution. In the calculation, electron slope temperature of 2 MeV, which was derived by using a plasma photonic device [1], is assumed. Divergence angle of electron beam is assumed to be 5°. This divergence angle is realized by

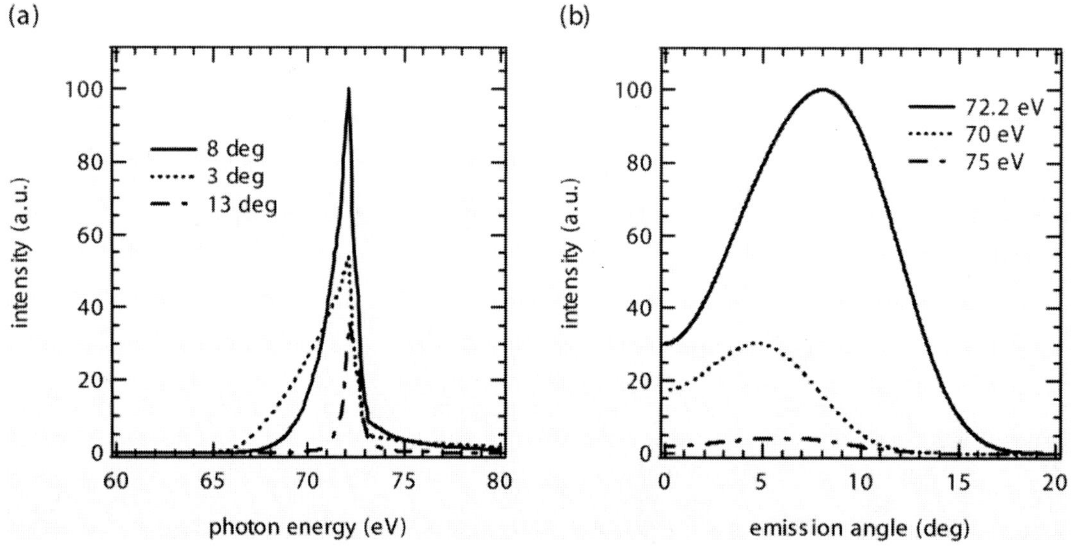

Figure 1. Calculation results of XCR emitted from Al. (a) Spectrum. (b) Angular distribution. This XCR has emission peak at 8.0° and 72.2 eV.

using a plasma photonic devices [1].

Figures 1 show calculation result of XCR emitted from aluminium (Al) emitter. This XCR has intensity peak at 72.2 eV and 8.0°. Although photon energy of peak of *n* at Al L absorption edge is 72.7 eV [4], emission peak shifts to 72.2 eV due to absorption in Al emitter. Figure 1(a) shows spectra dependent on emission angles. In spite of energy distribution of electron beam, XCR is highly monochromatic. Spectral width at peak of angular distribution (8.0°) is 1.2 eV, this corresponds to 1.7% bandwidth. Figure 1(b) shows angular distribution. This XCR has cone-like angular distribution dependent on photon energy.

Luminescence and efficiency of XCR was evaluated. Laser energy of 1 J is assumed, conversion efficiency from laser to electron beam assumed to be 10%.Calculated luminescence is $5x10^{22}$ photons/mm^2 mrad^2 sec 0.1%b.w. Efficiency from laser energy to XCR is about $6x10^{-4}$. This luminescence is higher than the case with electron beam produced by an accelerator. This XCR can be a compact monochromatic EUV radiation source, because generation of electron slope temperature of 2 MeV is possible with table-top laser system.

3. Nonlinear transmission using focused FEL

Development of free electron lasers (FEL) producing high energy photons are expected to open many new possibilities for science and engineering. One of the interesting studies with these FEL is nonlinear effect in this wavelength region. In the previous study, nonlinear transmission in Al was investigated with 13.5 nm-EUV light pulse [5], and focal spot was evaluated by ex situ observation of irradiated PMMA target [6]. We reported nonlinear transmission in tin. In the experiment we used a new method a using fluorescent material for in situ observation of EUV-FEL focal spot and transmission. This method is very convenient due to use of visible fluorescence.

EUV-FEL provided from XFEL prototype accelerator SCSS in SPring-8 [7] are used. Wavelength was 51 nm with bandwidth of < 1%. Pulse duration is <100 fs. Typical energy at entrance of experimental chamber was 10 μJ/pulse. This EUV pulse was focused by using two same cylindrical mirrors as a Kirkpatrick-Baez (K-B) optics [8]. Tin (Sn) was used as a target material for the saturable absorber. According to reference data [4], Sn has an N-shell edge at 52 nm. A three-layered target was used. The front side of Sn is coated by a thin Au film (thickness: 5 nm) in order to prevent a

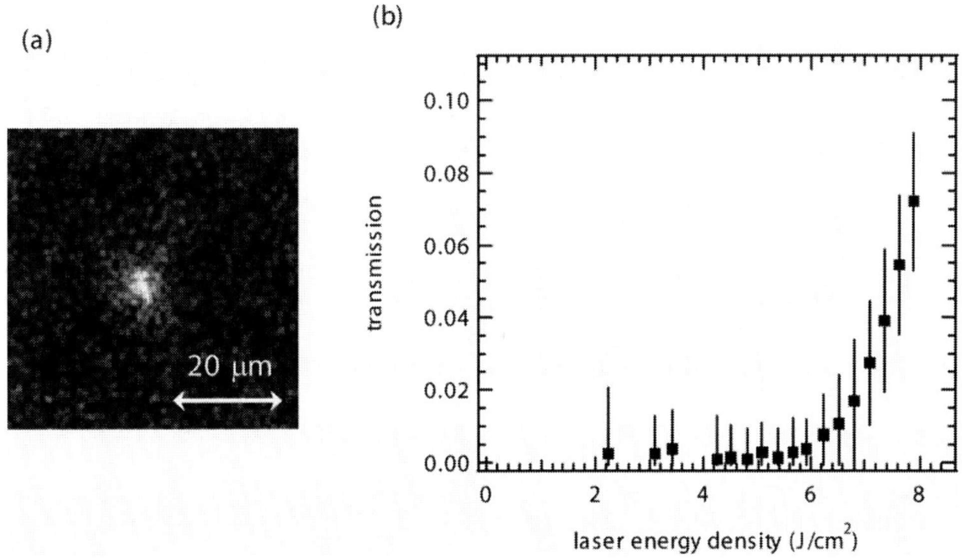

Figure 2. (a) Focal spot of EUV-FEL observed with cooled CCD camera. Spot size is 5.6 μm x 4.9 μm. (b) Measured transmission of EUV-FEL in Sn layer (thickness: 80 nm).

contamination oxidation layer. The rear side is attached to a fluorescent material. A fused silica glass (SiO_2) was used as a fluorescent material. The band gap energy of the SiO_2 is 9.6eV so that in the wavelength range of this study, single photon absorption fluorescence occurs. Intensity and spot image of visible fluorescence were observed with a photomultiplier and a cooled CCD camera, respectively. Incident laser energy was monitored by surface scattered light with an electron multiplier calibrated with pyroelectric detector. The target was moved so that the EUV pulse always hits a fresh surface of the target.

Figure 2(a) shows focal spot of EUV-FEL observed with cooled CCD camera. Spot size of 5.6 μm x 4.9 μm was measured by in situ observation of fluorescent image. This focal spot size was kept during the experiment. Incident laser energy density was changed by using a gas attenuator. Before measurement of transmission in Sn layer, an experiment with two-layered target (Au / SiO_2) was performed, and it was confirmed that intensity of transmitted VUV light in Au layer is linearly proportional to incident laser energy density. Transmission in Sn layer (thickness: 80 nm) was derived from comparison with fluorescent intensity of two-layered target and that of three-layered target. Figure 3 shows transmission of EUV-FEL in Sn layer dependent on incident laser energy density. When the laser energy density is low, EUV light almost does not transmit in Sn layer. However, transmission becomes nonlinearly higher with increase in laser energy density from 6 J/cm^2. Measured transmission is much lower than reference data [4] in which transmission of 51-nm EUV light is 0.15 and that of longer wavelength region is around 0.5. This reason is seemed to be oxidation contamination in target fabrication. Measured transmission at $> 6\ J/cm^2$ is 100 times higher than that at $< 6\ J/cm^2$. This increase is more drastic than that in reference data. Additional investigation such as pump-probe experiments and calculation of atomic process will clarify details of this nonlinear effect.

4. Summary

We report two types of plasma photonic devices using complex refractive index around an absorption edge. X-ray Cherenkov radiation is an attractive radiation source. Especially, in spite of laser-produced electron having energy distribution, this radiation is monochromatic. Nonlinear transmission in Sn layer was realized using focused EUV-FEL. Fluorescent material was used for in situ measurements. This phenomenon can be applied for auto-correlator and x-ray pulse slicer and so on.

Acknowledgements

This work was partially supported by grants for the Global COE Program, "Center for Electronic Devices Innovation", and the X-ray Free Electron Laser Utilization Research Project, from the Ministry of Education, Culture, Sports, Science and Technology of Japan and Kakenhi (21740398).

References

[1] Kodama R, Sentoku Y, Chen Z L, Kumar G R, Hatchett S P, Toyama Y, Cowan T E, Freeman R R, Fuchs J, Izawa Y, et al, 2004 Nature **432** 1005.

[2] Knulst W, Luiten O J, van der Wiel M J, and Verhoeven J, 2001 Appl. Phys. Lett. **79** 2999.

[3] Bazylev V A, Glebov V I, Denisov E I, Zhevago N K, Kumakhov MA, Khlebnikov A S, and Tsinoev VG, 1981 Sov. Phys. JETP **54** 884.

[4] Henke B L, Gullikson E M, and Davis J C, 1993 Atomic Data and Nuclear Data Tables **54-2** 181.

[5] Nagler B, Zastrau U, Fäustlin R R, Vinko S M, Whitcher T, Nelson A J, Sobierajski R, Krzywinski J, Chalupsky J, Abreu E, at al, 2008 Nature Photonics **2** 555.

[6] Chalupsky J, Juha L, Kuba J, Cihelka J, Hájková V, Koptyaev S, Krása J, Velyhan A, Bergh M, Caleman C, at al, 2007 Opt. Exp. **15** 6036.

[7] Shintake T, Tanaka H, Hara T, Tanaka T, Togawa K, Yabashi M, Otake Y, Asano Y, Bizen T, Fukui T, at al, 2008 Nature Photonics **2** 555.

[8] Kirkpatrick P, and Baez A V, 1948 J. Opt. Soc. Am. **38** 766.

Integrated simulations of core heating in cone-guiding fast ignition, FIREX-I

T. Johzaki[1], H. Nagatomo[1], A. Sunahara[2], H.-B. Cai[1], H. Sakagami[3], K. Mima[1]

[1]Institute of Laser Engineering, Osaka University, 2-6 Yamadaoka, Suita, Osaka 565-0871, Japan
[2]Institute for Laser Technology, 206 Yamadaoka, Suita, Osaka 565-0871, Japan
[3]Department of Simulation Science, National Institute for Fusion Science, Toki, Gifu 509-5292, Japan

E-mail: tjohzaki@ile.osaka-u.ac.jp

Abstract. The core heating properties of a Au cone-attached CD shell target in FIREX-I were investigated by the integrated simulations with FI^3 code systems. The importance of fat electron transport in the deformed cone tip was shown. In addition to the collisional scattering and drag in the Au cone, the strong resistive magnetic field is generated at the top of the cone tip when the tip shape is deformed. Due to this field, the fast electrons are strongly scattered, and then the core heating efficiency becomes smaller compared with the case neglecting the magnetic field.

1. Introduction

In the cone-guiding fast ignition, an imploded core is heated by the energy transport of fast electrons generated at the cone inner surface. In FIREX-I [1], our goal is the demonstration of efficient core heating ($T_i \sim 5$keV) using a newly developed 10kJ- LFEX laser.

We evaluated the core heating in FIREX-I by the 1D PIC and Fokker-Planck (FP) simulations [2], and also 2D FP simulations [3]. In the high-Z cone case (*e.g.*, Au), the rapid density steepening of the interaction surface and the strong scattering due to highly ionized atoms occur, which reduce the conversion efficiency of heating laser to fast electrons. In addition, since the Au atoms in cone tip reaches highly ionized state rapidly, the fast electron beam quality deteriorates due to the collisional and resistive drags and scattering by Au ions in the cone tip. As the results, the core heating rate becomes lower with time even for the fast electron beam with the constant intensity. Thus, we proposed a low-Z material (*e.g.*, CH) as an alternative of cone tip material to reduce the collisional effects. We also found in the 2D simulations assuming a spherical core that the magnetic field generated around the core edge due to the thermoelectric current weakens the self-guiding effect [3, 4, 5] due to the resistive magnetic field, which reduces the core heating efficiency.

In the above simulations [2, 3], we did not consider the implosion dynamics. In the implosion of cone-attached shell targets, the core shape is far from the spherical one and the cone tip is deformed by the jet-like plasma flow from the imploded core, which may affect the core heating properties. In the present paper, on the basis of integrated simulation using FI^3 code system [6], we evaluate the core heating properties in FIREX-I; *e.g.*, the effects of core and cone tip deformation.

2. Implosion simulation for cone-guiding CD shell target

First, we carried out an implosion simulation with a 2D radiation-hydro code "PINOCO" [7] for a CD shell target (8 μm thickness, 250 μm inner radius and 10^{-4}g/cm^3 gas in the shell). An Au cone (30 degree opening angle, 50μm offset from the shell center, 10μm tip inner radius, 7μm tip thickness and 5μm wall thickness near the tip) is attached to the shell. The shell is uniformly irradiated by 2.5kJ Gaussian-pulse-shaped green laser (λ_L= 0.53 μm). The compressed core profiles are shown in **Fig.1**. The maximum compression is obtained at t =2.02ns. The most part of the shell was highly compressed although the imploding shell contacting to the cone wall rolled up and was dragged by the Kelvin-Helmholtz instability. The core density reaches higher than 120 g/cm^3 at the maximum. Because of the pressure imbalance due to the existence of the cone, the hot spot is moved toward the cone tip. This flow hits the cone tip 50ps before the maximum compression. At the maximum compression, the cone tip considerably collapsed and the low density plasma penetrates into the vacuum region inside the cone, which is not preferable for core heating. Thus, in the following core heating analysis, we use the imploded core profile at t = 1.97ps. At this moment, the cone tip is already deformed, but the vacuum region inside the cone is still kept clean. The optical size of the core is $\int \rho dz$ ~ 0.14g/cm^2.

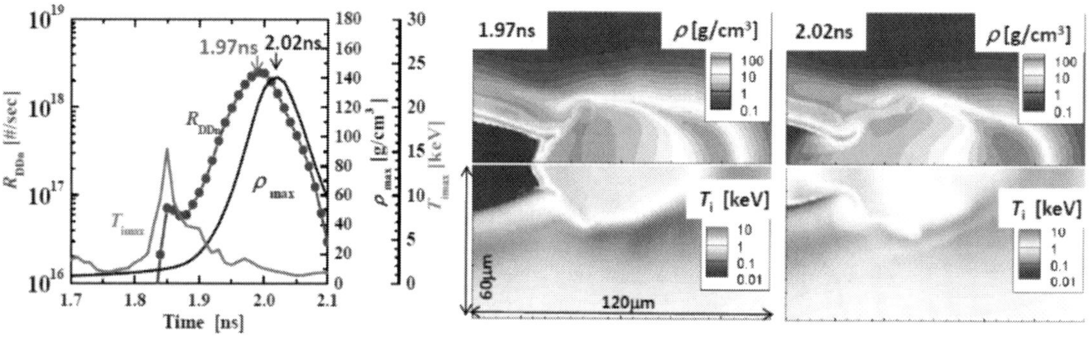

Figure 1. Imploded core profiles obtained from a PINOCO simulation; (a) temporal profile of maximum density (ρ_{max}), maximum ion temperature (T_{imax}) and DD fusion reaction rate (R_{DD}), (b) spatial profiles of density and ion temperature at 1.97ns and (c) spatial profiles at 2.02ns.

3. Core heating by fast electron

Using the imploded core profile at t = 1.97ns (Fig.1 (b)) as the initial bulk plasma profile, we carried out the core heating simulations, where the bulk plasma is treated with radiation-hydro model and the fast electron transport is treated with Fokker-Planck transport model [3, 8]. The fast electron beam having the slop temperature of T_{fe}= 1.0MeV, the supper Gaussian radial profile with r_{FWHM} = 15μm, the angular spread of 20 degree full angle and the duration of τ_{fe} = 5ps (constant intensity during τ_{fe}) is injected inside the Au cone tip. The injected beam energy E_{fe} is 5kJ, which corresponds to the energy coupling of $\eta_{L\to fe}$ = 50% from the 10kJ LFEX laser to the fast electron beam. To evaluate the magnetic field effects, we carried out the simulations assuming two different field conditions (with and without the magnetic-field).

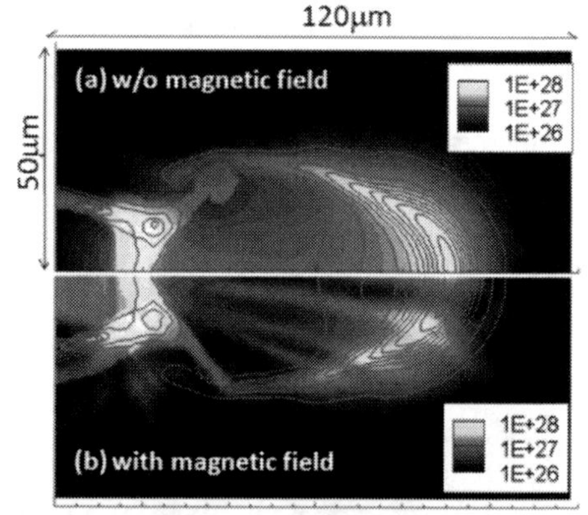

Figure 2. spatial profiles of fast electron heating rates [W/m^3] at t = 3ps for the cases (a) without and (b) with the magnetic field

In **Fig.2**, the spatial profiles of fast electron heating rate at t = 3ps are shown for the cases without and with the magnetic field. As was pointed out in our previous work [3], when a high-Z material such as Au is used as the cone material, the fast electrons lose their kinetic energy in the cone tip due to the collisional and resistive drag. Thus in Fig.2, high heating rates in the cone tip are observed in the both cases in Fig.2. As for the heating rate in the dense core, the uniform heating is observed in the case without magnetic-field. Contrary to this, in the case with the magnetic field, the multi-spots of heating rate are observed in the core region. In addition, the heating by the transverse fast electron flow is observed in the plasma surrounding the cone side wall. These are caused by the magnetic-filed around the cone tip.

The spatial profile of azimuthal magnetic field observed at t = 3ps is shown in **Fig.3**. Together with the result of the present simulation including magnetic field, the magnetic field profile obtained at t = 3ps in the previous work [3] is shown, where the "clean" cone tip (not deformed) and spherical CD core were assumed and the fast beam condition was the same as that in the present simulation. Even in the "clean" cone tip case (Fig.3(b)), the strong magnetic field generated around the cone side wall, which is due to the resistivity jump at the material contact surface between Au cone and the imploded CD plasma [9, 10]. Compared with the outer coronal CD plasma, the inner Au cone has a larger resistivity because of the larger Z, so that the resistive magnetic field ($\nabla \eta \times \boldsymbol{j}_f$, where η is the plasma resistivity and $\boldsymbol{j}_f$ is the fast electron current density) is generated at the contact surface. However, at the top of the cone tip, the directions of $\nabla \eta$ and $\boldsymbol{j}_f$, are nearly parallel in the "clean" cone tip case, so that the magnetic field is not strong. On the other hand, in the case using the imploded core profile (Fig.3(a)), the cone tip is deformed by the attack of the plasma flow from the imploded core. At the top of the cone tip, hence, $\nabla \eta$ and $\boldsymbol{j}_f$, are not parallel, which results in generating the strong magnetic field. The fast electrons are hence scattered by this magnetic field. In addition to the collisional scattering in the Au cone, because of the scattering by the magnetic field, the beam divergence becomes larger and the self-guiding magnetic field, which is observed at the core heating simulations neglecting the fast electron transport in the cone tip [3, 4, 5], is not clearly formed. As the result, the fast electron beam breaks into several filamentous beams, which results in the multi-spot heating of the core.

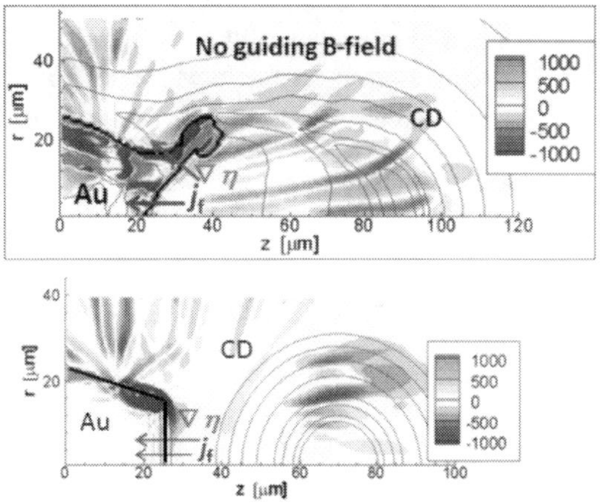

Figure 3. spatial profile of azimuthal magnetic field [T] observed at t = 3ps obtained for the cases for (a) using the imploded core profile and (b) assuming "clean" cone tip and spherical core. The density contours are presented by red thin lines.

The temporal evolution of $\langle T_e \rangle_{DD}$ and $\langle T_i \rangle_{DD}$ are shown in **Fig.4**, where $\langle T_e \rangle_{DD}$ and $\langle T_i \rangle_{DD}$ are the spatially-averaged bulk electron and ion temperatures weighted by DD reaction rates. The initial peak of $\langle T_e \rangle_{DD}$ observed around t = 0.6ps results from the heating in the low density region between the cone tip and the dense core. At the initial phase, the fusion reactions mainly occur at this low-density and relatively-high-temperature region, since the temperature in the dense core region is so low (<0.2keV) (shown in Fig.1). With time passing, the temperature in dense core region rises due to the fast electron heating, and the fusion reactions mainly take place in the dense core region. So, the $\langle T_e \rangle_{DD}$ after t = 1.5ps represents the mean

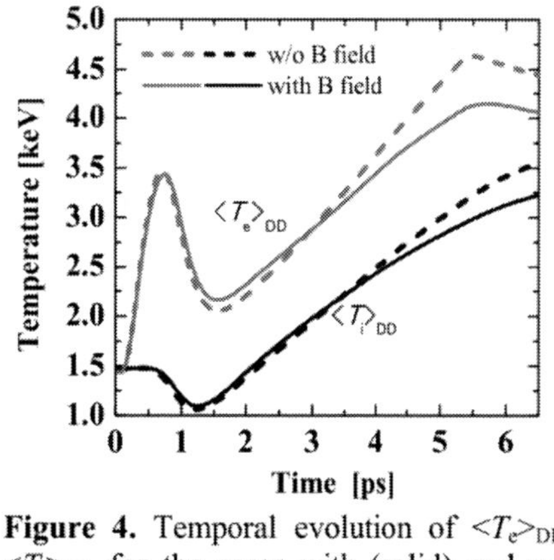

Figure 4. Temporal evolution of $\langle T_e \rangle_{DD}$ and $\langle T_i \rangle_{DD}$ for the cases with (solid) and without (broken) magnetic field

temperature of the dense core region. The difference between $<T_e>_{DD}$ and $<T_i>_{DD}$ comes from the slower temperature relaxation compared with the bulk electron heating by the fast electrons. Between the two cases (with and without magnetic field), the difference in $<Te>_{DD}$ is small in the early phase ($t < 3.0$ps). However, in the later phase, $<Te>_{DD}$ for the case with the magnetic field becomes smaller compared with the case without the magnetic field since the scattering by the magnetic field becomes significant. This result is contrary to the simulations without the fast electron transport in the cone tip, where the core heating rate and the resultant core temperature are enhanced due to the self-guiding resistive fields. The present simulation result indicates the importance of inclusion of the fast electron transport in the cone for evaluating the core heating. The core heating properties are summarized in **Table I**. The energy coupling of the fast electron to the core and the maximum values of $<T_e>_{DD}$ and $<T_i>_{DD}$ are $\eta_{fe\text{->}core}$ = 20.8%, $<T_e>_{DD,max}$ = 4.15keV and $<T_i>_{DD,max}$ = 3.06keV for the case including the magnetic field, and the values are smaller than those for the case neglecting the magnetic field.

Table I Magnetic field effect on core heating properties

B-fields	$E_{dep,core}$ ($\eta_{fe\text{->}core}$)	$<T_e>_{DD,max}$	$<T_i>_{DD,max}$
w/o	1.83kJ (36.6%)	4.63keV	3.70 keV
with	1.04kJ (20.8%)	4.15 keV	3.24 keV

* $E_{dep,core}$ and $\eta_{fe\text{->}core}$ are the fast electron deposited energy in the core ($\rho > 5g/cm^3$) and the energy coupling of fast electron to the core.

4. Summary

The core heating properties of a Au cone-attached CD shell target in FIREX-I were investigated by the integrated simulations with FI^3 code systems [6] and the importance of the transport in the deformed cone tip was shown. In addition to the collisional scattering and drag in the Au cone, the strong resistive magnetic field is generated at the top of the cone tip if the tip shape is deformed. The fast electrons are strongly scatted by this magnetic field and then the core heating efficiency becomes smaller compared with the case neglecting the magnetic field. Under such a situation, the self-guiding due to the resistive field can no longer be expected.

In the present study, we assumed an ideal fast electron beam profile. Further investigation including fast electron generation is required for overall understanding of the core heating dynamics.

ACKNOWLEDGMENTS

This work was supported by MEXT, Grant-in-Aid for Scientific Research (C) (20540482).

References

[1] H. Azechi and the FIREX Project, Plasma Phys. Control. Fusion **48**, B267 (2006).
[2] T. Johzaki, et al., Plasma Phys. Control. Fusion **51**, 014002 (2009).
[3] T. Johzaki, Y. Nakao, and K. Mima, Phys. Plasmas **16**, 062706 (2009).
[4] J. J. Honrubia and J. Meyer-ter-Vehn, Nucl. Fusion **46**, L25 (2006); Plasma Phys. Control. Fusion **51**, 014008 (2009).
[5] A. A. Solodov, et al., Phys. Plasmas **15**, 223803 (2008).
[6] H. Sakagami and K. Mima, Laser Part. Beams **22**, 41 (2004), H. Sakagami, et al., Laser Part. Beams **24**, 191 (2006), T. Johzaki, et al., Laser Part. Beams **25**, 621 (2007), H. Sakagami, et al., Nucl. Fusion **49**, 075006 (2009), H. Nagatomo, et al., Nucl. Fusion **49**, 075028 (2009).
[7] H. Nagatomo, et al., Proc. of 2nd IFSA (Kyoto, 2001), Elsevier, 140 (2001).
[8] T. Yokota, et al., Phys. Plasmas **13**, 022702 (2006).
[9] R. L. Robinson, and M. Sherlock, Phys. Plasmas **14**, 083105 (2007).
[10] T. Johzaki, et al., J. Phys. Conf. Series **112**, 022091 (2008).

Journal of Physics: Conference Series **244** (2010) 022041 doi:10.1088/1742-6596/244/2/022041

Magnetic Collimation Of Petawatt Driven Fast Electron Beam For Prospective Fast Ignition Studies

S. Kar, D. Adams, M. Borghesi, K. Markey, B. Ramakrishna, M. Zepf

School of Mathematics and Physics, Queen's University, Belfast, BT7 1NN, U.K.

K. Lancaster, P. Norreys, A.P.L. Robinson

Central Laser Facility, Rutherford Appleton Laboratory, Chilton, OX11 0QX, U.K.

D.C. Carroll, P. McKenna, M. Quinn, X. Yuan

Department of Physics, University of Strathclyde, Glasgow, G4 0NG, U.K.

C. Bellei, J. Schreiber

Blackett Laboratory, Imperial College London, London SW7 2BZ, U.K.

E-mail: s.kar@qub.ac.uk

Abstract. Collimated transport of fast electron beam through solid density matter is one of the key issues behind the success of the fast ignition scheme by means of which the required amount of ignition energy can be delivered to the hot spot region of the compressed fuel. Here we report on a hot electron beam collimation scheme in solids by tactfully using the strong magnetic fields generated by an electrical resistivity gradient according to Faraday's law. This was accomplished by appropriately fabricating the targets in such a way that the electron beam is directed to flow in a metal which is embedded in a much lower resistivity and atomic number metal. Experimental results showed guided transport of hot electron beam over hundreds of microns length inside solid density plasma, which were obtained from two experiments examining the scheme for petawatt laser driven hot electron beam while employing various target configurations.

1. Introduction

Achieving collimated transport of fast electrons through solid density plasmas is a hot topic because of its relevance for fast ignition (FI) [1, 2], high energy density physics [3], laser induced radiation sources [4] and particle acceleration [5]. Experimental studies, via laser and target parameter scans, on hot electron transport in solids show minimum spray angle of 30-40 degrees [6, 7, 8] - which limits the efficiency with which energy can be coupled to the hot spot in FI. Therefore various schemes are being investigated in order to achieve less divergent or even collimated transport. In addition to the ideas such as the cone [9] and the wire [10] guiding, recent theoretical studies proposed a concept based on self-generated magnetic fields at resistivity boundaries [11]. Combining Ohm's law with Faraday's law one obtains $\frac{\partial \mathbf{B}}{\partial t} = \eta \nabla \times \mathbf{J}_f + (\nabla \eta) \times \mathbf{J}_f$,

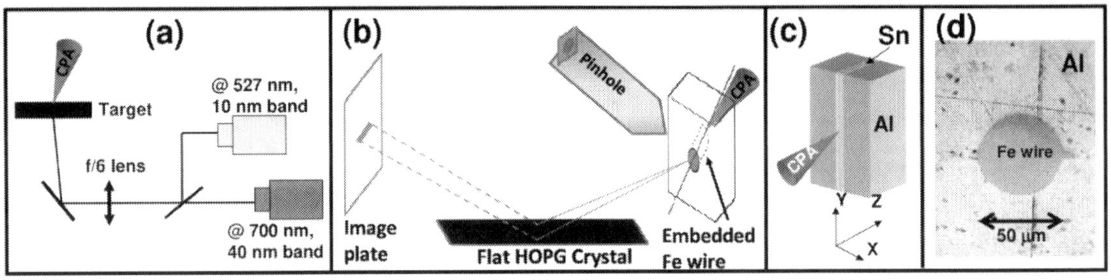

Figure 1. **(a)** and **(b)** show schematics of the setups in the experiments, EXP-1 and EXP-2 respectively. **(c)** Schematic of the target with one dimensional guiding geometry.**(d)** shows the image of the transverse cross-section of the target of two dimensional guiding geometry taken by an optical microscope.

where η is the resistivity($\propto Z/T_e^{1.5}$ in the Spitzer limit) and $\mathbf{J}_f$ is the density of the fast electron current flowing into the target. The second term on the right hand side of this equation tells about the generation of magnetic field due to a resistivity gradient which pushes fast electrons towards regions of higher resistivity. Once established, this field is further enhanced by magnetic field due to the current density gradient. Simulations have shown that for the parameters of the Vulcan PW, the field strength is sufficient to collimate an MeV electron beam [11]. One key point for experiments, where the targets are initially cold, is to maintain the sign of the resistivity gradient across the Z-boundary as the target heats up to hundreds of eV during the interaction. This can simply be achieved by using a core material (in which the electron beam is intended to be confined) with a higher Z and higher initial core resistivity than the surrounding medium.

The experimental proof of principle of the magnetic collimation scheme was carried out at Vulcan Petawatt laser facility employing one dimensional structured target [12], where the role of strong magnetic field at the resistivity gradient is proved by the help of computer simulations. In a follow-on experiment at the same facility and employing hard X-ray diagnostics, we extended our studies further with the same one dimensional as well as two dimensional guiding configurations - targets exhibiting a core/cladding structure analogous to optical waveguides, where fast electrons were forced to propagate inside a thin ($\sim$ 50 μm diameter) Iron wire of higher electrically resistivity and atomic number than the aluminum in which it is embedded.

2. Experimental setup

Both of the experiments (let us call the first proof-of-principle experiment as 'EXP-1' and the follow-on experiment as 'EXP-2' - schematics of the setups are shown in the figure. 1(a) and (b) respectively) were performed at Rutherford Appleton Laboratory employing Vulcan Petawatt laser system. After reflection from a plasma mirror, which was employed in order to reduce the prepulse and amplified spontaneous emission intensities on the target, the laser pulse delivered 150 J of energy on target in fwhm duration of 1 ps. Laser spot size on the target was 20 μm, reaching peak intensity of $\sim 10^{20}$ W/cm^2 on target at near normal incidence. A schematic of the one dimensional electron guiding target fielded in the experiments is shown in the figure 1(c). The CPA was targeted to irradiate the embedded 12 μm thick Tin (Sn) layer sandwiched between two Aluminium (Al) slabs. The target thickness was 200 μm. In case of the target for cylindrical guiding fielded in EXP-2, Fe wire of 50 μm diameter was embedded in Al, and the target thickness was 250 μm. An optical microscope image of transverse cross-section of the target is shown in figure 1(d). The materials were chosen based on the contrast of their

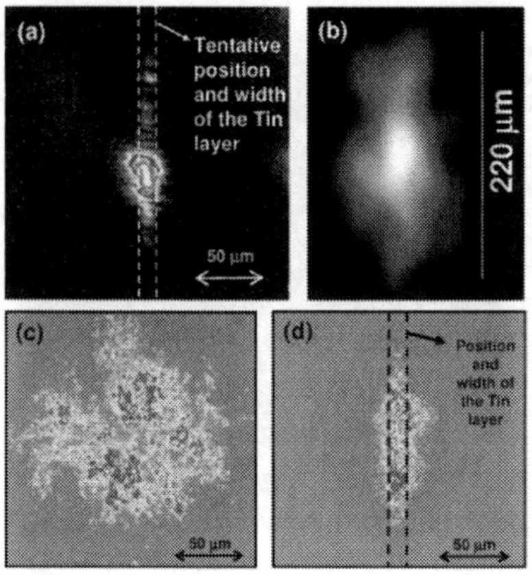

Figure 2. **(a)** and **(b)** are respectively the time integrated image of optical transition radiation and X-ray pinhole camera image of the Al-Sn-Al one dimensional guiding target (shown in Fig 1(a)) rear surface obtained in EXP-1 and EXP-2, respectively. **(c)** and **(d)** shows 2D spatial distribution of square of the hot electron density at the rear surface a 200 μm thick Al foil target and the guiding target, as employed in the experiment, obtained from 3D hybrid code described in the text.

atomic weights and electrical resistivities, as well from the point of view of fabrication. In the EXP-1, the target rear surface was lapped to a roughness of 50 nm rms and was gold coated (of sub micron thickness) and in EXP-2, 15 μm Cu foil was glued to the target rear surface. In the EXP-1, time integrated images of the optical transition radiation (OTR) at 527±5 nm and Plankian emission at 700±20 nm wavelengths (using bandpass optical filters) from the target rear surface were simultaneously recorded by CCD cameras. The spatial resolution of the images at 527 nm and 700 nm were 4 μm and 5.5 μm respectively. In the EXP-2, the electron beam size at the target rear surface was diagnosed by Xray diagnostics such as HOPG (highly oriented pyrolytic graphite) spectrometer [13] and pinhole camera, mainly looking at 8.05 KeV X-rays from the Cu tracer layer at the target rear surface due to K-α transitions in Cu by the hot electrons passing through it.

3. Results

According to the Faraday's law, the magnetic field is created across the resistivity gradient, which is not limited to any specific symmetry. Therefore, strong resistivity gradient along the Al-Sn boundary will lead to a rapid magnetic field growth and consequently guiding of the fast electrons should still occur in the target with one dimensional guiding geometry (as shown in figure 1(a)), albeit the fast electrons will only be strongly confined in one dimension normal to the Sn slab. Indeed, we obtained elucidating results showing the effect of the strong collimating resistive magnetic field while employing one dimensional guiding targets in both experiments. Highly elliptical shape of the hot electron beam distribution across the one dimensional guiding target rear surface is inferred from the OTR and X-ray pinhole images obtained from the EXP-1 and EXP-2 respectively, shown in the figure 2 (a) and (b) respectively. It is striking to see the oblong shape, width and the orientation of the image in the experimental data which correlates very well with the sandwiched Tin layer of the guiding target and contrasts with the typical circular patterns obtained from flat foil targets [6, 7]. Weaker magnetic confinement parallel to the Sn slab, as compared along normal to the tin layer, should result in an elongated electron beam profile at the target rear surface.

Simulations ware carried out to study the hot electron beam transport in the presence of resistive boundaries by modeling the 1D guiding geometry [12]. 3D hybrid code, called ZEPHYROS [11], was employed which uses similar approximations and methods to the hybrid code of Davies [14], i.e. a particle-based hybrid code with static background plasma. As expected, the simulation for the pure Al target produced a fairly uniform electron distribution of 100 μm fwhm at the rear surface (i.e. across the z =200 μm plane) as shown in the figure 2(c). On the other hand, the asymmetrical electron distribution obtained for the case of sandwich

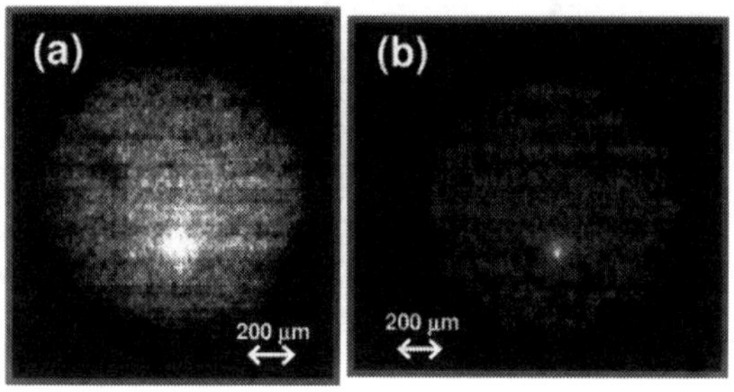

Figure 3. (a) and (b) are respectively the Xray pinhole camera images of the rear surface of 250 μm thick Fe foil and Al-Fe-Al two dimensional guiding target (shown in Fig 1(b)) obtained in EXP-2.

target can be seen in figure 2(d). Due to the resistivity gradient across the interfaces, the strong magnetic walls generated at the interfaces confine electrons within it, whereas some of the electrons managed to escape before the strength of the magnetic field could be sufficient to confine them. The simulation also reproduces the ratio between the number of electrons escaped to the number of electrons confined along the X axis (the collimation axis), thus establish the underlying dynamics of the magnetic field. The experimentally obtained heating profile of the target rear surface provided further evidence of collimated electron flow in the target [12]. While employing the 2D collimator target in EXP-2, we found a clear reduction in the electron beam size at the target rear surface as compared to the reference Fe target of same thickness (see figure 3).

4. Conclusion

It is experimentally demonstrated that the divergence of hot electrons inside a solid density target can be controlled by targets with suitable resistivity gradients around the beam. The resistivity gradients at the interfaces are practically maintained, for the relevant time scale, by choosing central metal of higher atomic number as well as higher electrical resistivity than the surrounding one. This scheme may have substantial positive impact on the Fast Igniter approach to fusion energy production, by enhancing coupling efficiencies of the relativistic electron beam to the fusion fuel and by relaxing some geometrical constraints on the target design.

5. Acknowledgements

The authors would like acknowledge funding support from (Engineering and Physical Science Research Council (EPSRC) and the Royal Society, UK and support for target fabrication from the QUB mechanical workshop.

References

[1] M. Tabak *et al.*, Phys. Plasmas, **12**, 057305 (2005).
[2] S. Nakai and K. Mima, Rep. Prog. Phys., **67**, 321 (2004).
[3] E.M. Campbell, Nucl. Instrum. methods A, **415**, 27 (1998).
[4] H.S. Park *et al.*, Phys. Plasmas, **13**, 056309 (2006).
[5] M. Borghesi *et al.*, Fusion Sci. Tech., **49**, 412 (2006).
[6] R.B. Stephens *et. al.*, Phys. Rev.E, **69**, 066414 (2004).
[7] J.J. Santos *et. al.*, Phys. Rev. Lett. **89**, 025001 (2002).
[8] K. Lancaster *et al.*, Phys. Rev. Letts., **98**, 125002 (2007).
[9] R. Kodama *et al.*, Nature, **432**, 1005 (2004).
[10] J.S. Green *et al.*, Nature Phys., **3**, 853 (2007).
[11] A.P.L. Robinson *et al.*, Phys. Plasmas, **14**, 083105 (2007).
[12] S. Kar *et. al.*, Phys. Rev. Letts., **102**, 055001 (2009)
[13] A.W. Moore, *Highly oriented pyrolytic graphite, Chemistry and Physics of Carbon*, Vol. 11, Marcel Dekker, New York, 1973.
[14] J. Davies, Phys. Rev. E, **65**, 026407 (2002).
[15] S.C. Wilks *et al.*, Phys. Rev. Letts., **69**, 1383 (1992).

Vlasov-Fokker-Planck simulations of fast-electron transport with hydrodynamic plasma response

R J Kingham, M Sherlock, C P Ridgers, and R G Evans
Plasma Physics Group, Imperial College London, London SW7 2AZ, UK
E-mail: rj.kingham@imperial.ac.uk

Abstract. We report on kinetic simulations of the transport of laser-produced relativistic electron beams (REB) through solid-density plasma, including the hydrodynamic response of the plasma. We consider REBs with parameters relevant to fast-ignition of compressed inertial confinement fusion capsules. We show that over the 10–20ps timescales required for fast-ignition, thermal pressure (from Ohmic heating) can significantly modify the density which in turn strongly affects the propagation of injected fast-electrons; it allows them to re-collimate into a narrow, intense beam under conditions where they initially undergo beam-hollowing. Similar static-density calculations do not show re-collimation. The re-collimation effect is attributed to PdV cooling in the pressure-induced density-channel, which in turn suppresses defocusing magnetic fields generated by resistivity gradients. These simulations have been carried out using the new 2D-3V Vlasov-Fokker-Planck (VFP) code FIDO running in hybrid mode.

1. Introduction

The generation of energetic electron-beams during the interaction of ultra-intense laser pulses with solid targets, and the subsequent propagation of these beams into the target has received a lot of interest, both experimental and theoretical, in the last 15 years. This interest stems in a large part from Tabak's proposed "fast-ignition" concept for inertial confinement fusion (ICF) [1]. The HiPER project, currently in the preparatory phase, is a prime example of a laser-facility designed to achieve inertial fusion energy via fast ignition. To date, all experimental and most theoretical work on fast-electron transport has considered timescales of up to about 1ps. This is too short for solid-density plasma to move – ions are essentially frozen – and much shorter than the 10–20ps duration of a viable fast-ignition laser. On these 0.1–1ps timescales, signatures of collimated propagation [2] and propagation in an annular ring [3] have been observed under some conditions. A criterion for "resistive collimation" of an electron beam was proposed in [4] and an explanation of beam-hollowing (e.g. annular ring) can be found in [5]. Both used kinetic simulations to help tease out the underlying mechanisms. The basic description of these resistive phenomena is that the collisionless fast-electron current $\mathbf{j}_f$ is balanced almost exactly by a return current of cold, collisional background electrons, $\mathbf{j}_c$. Using a basic Ohm's law, the E-field is $\mathbf{E} = \eta\mathbf{j}_c \approx -\eta\mathbf{j}_f$ (where η is resistivity) which yields a B-field induction equation $\partial\mathbf{B}/\partial t = \eta(\nabla \times \mathbf{j}_f) + \nabla\eta \times \mathbf{j}_f$, with the first & second terms giving rise to collimating and beam-hollowing type B-fields, respectively. Results from 30ps duration, hybrid simulations of fast-electrons propagating into an imploded DT fuel core have recently been reported [6]. They include hydrodynamic response and show that expansion of the core is not important during energy deposition, but do not consider the gold cone-tip.

Back of the envelope calculations hint that significant cavitation of the plasma - through which the REB propagates - could occur over the 10–20ps beam duration required for ignition. Previously, the effect of *static* density-modulations of the plasma has been considered [7]. In this paper, we consider the effect of plasma density evolution on REB propagation. We show, for the first time, that density changes in the solid target, caused by the REB, greatly affect its propagation on timescales exceeding $\sim$5ps.

The requirements for fast-ignition [8] are currently thought to be an electron-beam energy of at least 25kJ, duration 10–20ps, radius $\sim 20\mu$m and electron kinetic energy of $T_f \approx 0.3$–$2\,$MeV. A suitable laser needs an intensity $I \approx 2\times10^{20}\,$W/cm^2 at wavelength $\lambda_o = 0.53\mu$m and power $P_l \approx 3\,$PW assuming a conversion efficiency (to fast electrons) of $\eta_{abs} \approx 0.3$. Ohmic heating of a solid-density target by the high current of fast electrons under these conditions can easily raise the temperature of the background electrons by several keV in a picosecond, in the vicinity of beam. We estimate a temperature evolution of $T_e \approx 12(t/\mathrm{ps})^{2/5}$ keV based on [4] (eqn. on pg. 2) at solid density $n_e = 10^{24}\mathrm{cm}^{-3}$, $Z \ln \Lambda = 40$ for a beam with $P_{abs} = 1\,\mathrm{PW}$, $T_f = 1\,\mathrm{MeV}$ and radius 20μm. The strong pressure gradients in turn cause density modulations to develop on the timescale of the sound-speed transit time across the heated region, typically a few picoseconds. Linearising the fluid continuity equation and momentum equation, subject to a prescribed, fixed, temperature perturbation ($T_e = T_o + \delta T \cos(2\pi x/\lambda)$) and assuming an initially uniform density, it is straightforward to estimate the density perturbation driven up by the pressure modulation: $\delta\rho/\rho_o = 2\pi^2 \left(\Delta C_s t/\lambda\right)^2 \approx 3.3(\overline{Z}/A)\delta T_{keV} t_{ps}^2/\lambda_{\mu m}^2$ where ΔC_s is the sound-speed associated with the overheat δT. Hence the time for 10% modulation is estimated to be 1.1ps for a gold plasma (assuming $\overline{Z} = 30$) with $\delta T = 4$keV at $\lambda = 5\mu$m, for example.

2. Simulation details

Calculations were carried out using FIDO which solves the relativistic Vlasov-Fokker-Planck equation for electrons, together with Maxwell's equations on a 2D Cartesian spatial grid. The momentum space for the electron distribution function (EDF) is discretized on a spherical polar grid, as in [9], so $f_e(p, p_\theta, p_\phi, x, y, t)$ and the fields E_x, E_y and B_z are included. FIDO predominantly uses explicit finite-difference methods for advancing the EDF but e-i and e-e collisions are implicit. FIDO can run in either full-kinetic or hybrid mode. The results reported here used hybrid mode where only the injected fast-electrons are treated kinetically. Magneto-hydrodynamic motion of the ion-fluid (background-plasma fluid) in full-kinetic (hybrid) mode in response to thermal & magnetic pressure-gradients is implemented using a method adapted from Ziegler [10]. FIDO incorporates a suite of EM-field solvers; an implicit moment method was used here. Spitzer resistivity and ideal-gas electronic heat-capacity are used in hybrid mode.

The "base" simulation consists of an initially uniform plasma – $T_{eo} = 100$eV and $n_{eo} = 1.2 \times 10^{24}\mathrm{cm}^{-3}$ – with $Z \ln \Lambda = 40$ and $m_i = 6m_p$. A laser of peak intensity $I = 5 \times 10^{19}\mathrm{W/cm}^2$ and Gaussian transverse profile of $1/e$ width $R_s = 3\mu$m ($P_l \approx 14$ TW) is incident onto this plasma. Fast-electrons are injected along the bottom edge of the domain, of size 30×30 microns, in the +ve (positive) $y-$direction with an energy distribution characterised by $T_f = 1.8$MeV and a Gaussian angular distribution with $1/e$ half-cone angle $\theta_{inj} = 28°$. The spatial profile of the injected energy flux mimics the laser radial profile and an absorption efficiency of $\eta_{abs} = 0.3$ is used. (T_f is kept the same across the injection region.) The spatial grid resolution is $\Delta x = \Delta y = 0.2\mu$m and open boundary conditions are used. The momentum grid uses $np = 50$ ($p_{max} = 22\, m_e c$) $n\theta = 1(-5)$ and $n\phi = 96$. (Typically $n\theta = 1$ is sufficient to give good results since B_z rotates particles in ϕ.)

3. Results

Initially the injected electrons ($\theta_{inj} = 28°$) collimate within 100fs, under the action of the resistive $\eta(\nabla \times \mathbf{j}_f)$ B-field source. This is consistent with the self-collimation criterion, $\Gamma > 1$, given in [4], where $\Gamma = R_s/(r_g \theta_{inj}^2)$ and r_g is the electron gyro-radius in the self-generated B-field. (A full expression for Γ in terms of density, Z, electron energy, etc., is given in [4].) Under our conditions $\Gamma \geq 1$ for $t \geq 40$fs. By 1ps, the beam hollows beyond 10μm into the target, consistent with eqn. (13) in [5] which suggests the onset of beam-hollowing in about 0.5ps for a $5 \times 10^{19}\mathrm{W/cm}^2$ laser and the conditions considered here. With hydrodynamic background plasma response included, an intense, narrow beam of fast-electrons has already re-formed on axis by 5ps as shown in figure 1(a). By 10ps, virtually all the injected electrons are funnelled down this central filament [fig. 1(b)]. Formation of this central filament does not occur when the background density is static (no hydro); instead a hollowed beam persists, as shown in figure 1(c). At 5ps the peak plasma temperature is $T_e \approx 2.2$keV and fig. 1(d) shows that pressure gradients under the REB have already significantly cavitated the background plasma density by up to 40%.

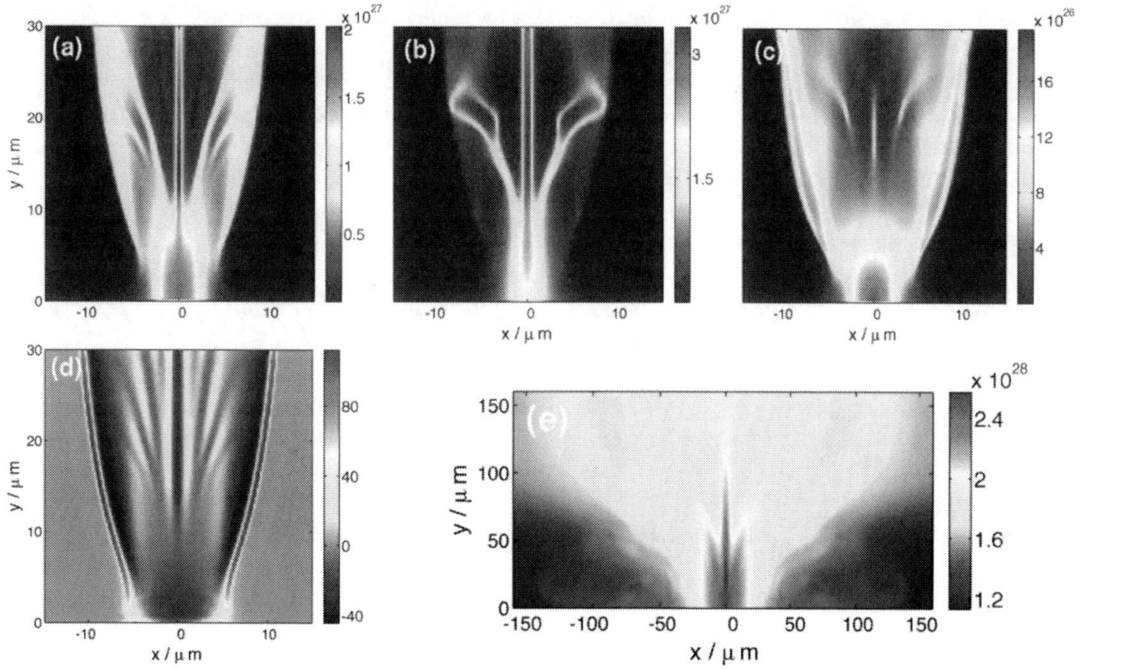

Figure 1. Re-emergence of central REB – beam density n_f $[m^{-3}]$ (a)–(c) & (e) ; % plasma density change $\delta\rho/\rho_o$ (d). 14TW, $R_s = 3\mu$m laser: (a) 5ps, (b) 10ps, (c) 10ps without hydro, (d) 5ps. 3PW, $R_s = 20\mu$m + gold: (e) 20ps. [Hydro is used unless otherwise indicated.]

This phenomenon - formation of a strong, central filament when hydrodynamic density changes are included - has been seen under other conditions, including tighter injection ($\theta_{inj} = 7°$) with lower background charge state ($Z \ln \Lambda = 4$). The possibility that the central filament is purely a numerical phenomenon has been ruled out by tilting the direction of the injected electrons by 20°. The central filament still occurs and only when hydro is included. The emergence of the tightly collimated REB is also seen under conditions needed for fast-ignition. Figure 1(e) shows the fast-electron density at 20ps with hydrodynamic background-plasma response included. There is virtually no beam without hydro. The conditions here are $R_s = 20\mu$m, $Z \ln \Lambda = 70$, $T_{eo} = 200$eV, $n_{eo} = 2.1 \times 10^{24}$cm^{-3}, $I = 2 \times 10^{20}$W/cm^2, $\theta_{inj} = 22°$ and $\Delta x = \Delta y = 1\mu$m.

4. Explanation of refocusing mechanism

The emergence of the central filament once density cavitation develops is linked with PdV cooling of the expanding plasma in the cavity. This in turn permits suppression of the resistivity-gradient driven decollimating/hollowing B-field source $\nabla\eta \times \mathbf{j}_f$ and re-establishment of the dominance of the focusing $\eta(\nabla \times \mathbf{j}_f)$ resistive source. With a static density profile and no thermal conduction, a rigid beam model of resistively driven B-field generation in an Ohmically-heated plasma predicts that the two sources largely cancel each other out leaving a small net defocusing $\dot{B}$ in the beam body. This is seen in our hybrid-VFP "base" simulations with static density [fig. 2(a)]. With hydro included, this delicate balance is disrupted. The PdV term in the electron-fluid energy equation contributes to evolving the temperature and therefore influences the resistivity profile (since $\eta \propto T_e^{-3/2}$ assuming Spitzer). In the limit where PdV effects dominate, the temperature evolves adiabatically (ignoring thermal conduction) so that $n_e/T_e^{3/2} =$ const. which acts to reduce the temperature under the REB, with the effect initially scaling as $\delta T_{\rm PdV}/T \propto \delta n/n \propto -t^2$. Figure 2(b) shows the modulated density & temperature profiles across the plasma at $y = 2.5\mu$m, t = 5ps which clearly shows the lowering & flattening of $T_e(x)$ by PdV effects. The flatter temperature profile in turn decreases $\nabla\eta$ so that $\partial_x\eta$ is reduced from 0.03 to 0.018Ω at x = 3.5μm. This ensures that with hydro, the net B-field source generates a collimating B-field across the body of the REB. As shown in fig. 2(c), the beam-hollowing field [-ve (negative) B_z] is completely removed by 5ps, replaced with a ~100 T collimating field in the region $0 \leq x \leq 3\mu$m. It is thought that this field focuses the bulk of the injected electrons into the tight beam seen further into the plasma. FIDO currently neglects thermal conduction in the background in hybrid mode. We estimate that thermal conduction has less effect than PdV effects for transverse features of wavelength exceeding about 10μm. Hence thermal conduction should not greatly affect the temperature dynamics in the crucial beam-focusing region just beyond the injection region. If present, it should further flatten the temperature profile thereby helping to inhibit beam-hollowing.

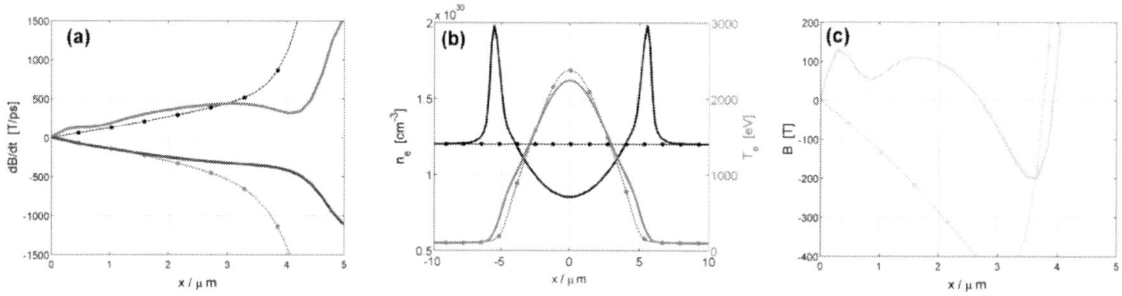

Figure 2. B-field sources + PdV effect for 14TW laser – (a) $\nabla\eta \times \mathbf{j}_f$ (-ve) and $\nabla \times \mathbf{j}_f$ (+ve) sources, (b) plasma density & elec. temp. , and (c) B-field. All at $y = 2.5\mu$m, t = 5ps. Solid (dashed with • markers) curves are with (without) hydro.

5. Conclusions

We have directly seen the influence of hydrodynamic density evolution on the propagation of intense 10–20ps duration REBs in solid targets. It permits the reversal of beam-hollowing and re-collimation of the beam after some picoseconds. Pressure gradients in the Ohmically heated background plasma significantly modify the density within ~ 5ps. Lowering of the plasma density just beyond the fast-electron injection point is accompanied by a PdV cooling effect which is sufficient to inhibit and reverse the formation of beam-hollowing fields. Whether or not this re-collimation mechanism persists into the highly compressed core of a fast-ignition target needs to be addressed. The results indicate that hydrodynamic motion must be included when addressing fast-electron transport through solid-density regions of a fast-ignition target.

Acknowledgments

This work has been carried out within WP9 of the HiPER project. Use of computer resources from the KITSI project, headed by G. Schurtz (CELIA, Bordeaux), is gratefully acknowledged.

[1] Tabak M *et al* 1994 *Phys. Plasmas* **1** 1626
[2] Tatarakis M *et al* 1998 *Phys. Rev. Lett.* **81** 999
[3] Storm M *et al* 2009 *Phys. Rev. Lett.* **102** 235004
[4] Bell A R and Kingham R J 2003 *Phys. Rev. Lett.* **93** 035004
[5] Davies J R, Green J S and Norreys P A 2006 *Plasma Phys. Control. Fusion* **48** 1181
[6] Honrubia J J and Meyer-ter-Vehn J 2009 *Plasma Phys. Control. Fusion* **51** 014008
[7] Robinson A P L, Kingham R J, Ridgers C P and Sherlock M 2008 *Plasma Phys. Control. Fusion* **50** 065019
[8] Atzeni S *et al* 2008 *Phys. Plasmas* **15** 056311
[9] Yokota T, Nakao Y, Johzaki T and Mima K 2006 *Phys. Plasmas* **13** 022702
[10] Ziegler U 2004 *J. Comput. Phys.* **196** 393

Implosion hydrodynamics and heating synchronization measurement using X-ray framing cameras

Mayuko Koga, Keisuke Shigemori, Hiroyuki Shiraga, Hiroshi Azechi

Institute of Laser Engineering, Osaka University,
2-6 Yamada-Oka, Suita, Osaka 565-0871, Japan

koga-m@ile.osaka-u.ac.jp

Abstract. In fast ignition laser fusion experiments, it is important to know the implosion hydrodynamics and the heating mechanism of imploded core. However, it is difficult to measure the imploded core and heating laser injection at the same time because of their large spectral differences. In this paper, we propose a simultaneous measurement of an implosion process and an injection time of heating laser by using X-ray framing camera (XFC).

1. Introduction

Fast ignition [1-4] is one of the proposed ways to generate fusion plasmas. This scheme separates the compression and heating phases. Nano second pulse lasers are irradiated to a fuel target and generated imploded core plasma. Then, pico second short pulse laser is injected into the tip of the gold cone. When the laser interacts with the gold, energetic electrons are produced. These electrons deposit their energy and raise the fuel to fusion temperatures. Fast ignition scheme is considered as a powerful method for generating laser fusion plasma because the laser energy required for implosion in fast ignition scheme is expected to be much smaller than that in the central ignition scheme.

In Fast ignition fusion experiment, it is important to know the implosion hydrodynamics and the heating mechanism of the imploded core plasma. Especially, to measure a heating laser injection time and observe a core plasma heating process is necessary for efficient heating of the imploded core plasma, However, the large spectral difference between thermal X-ray from the imploded core plasma and high energy X-ray from hot electrons generated by the heating laser causes difficulty in measuring both X-rays at the same time. In this study, we report a simultaneous measurement of both X-rays and discuss the possibility of heating laser injection time measurement with an X-ray framing camera (XFC).

2. Experiment

X-ray framing cameras have been widely used for diagnosing laser-driven implosions [5-8]. Figure 1 (a) shows the schematic diagram of our X-ray framing camera system made by Hamamatsu photonics, Model C5896. The X-ray framing camera has two micro channel plates (MCP) which size is 40 mm × 50 mm. The first MCP is for gating and the second MCP is for amplification of a signal. Four Au striplines are deposited on the first MCP. Four electric pulses propagate through each stripline with arbitrary delay. The first MCP becomes sensitive only when electric pulse exist, therefore time evolution of 2D X-ray images are recorded by using a pinhole imager array with very

Journal of Physics: Conference Series **244** (2010) 022043 doi:10.1088/1742-6596/244/2/022043

high-speed time resolution (about 80 ps). Electrons through tandem MCP hit the fluorescence plate and the visible fluorescent lights are recorded with a CCD camera.

The experiments were performed using Gekko XII laser system and LFEX laser system at the Institute of Laser Engineering, Osaka University. 9 beams of Gekko XII irradiated the plastic shell target to create the imploded core plasma and after that LFEX laser were injected to heat the core plasma. The pulse width of LFEX laser is about 4 ps. The targets were about 500 μm in diameter and 7 μm in thickness with gold cones, which had a 30° opening angle. Observation direction of XFC is shown in Fig. 1 (b).

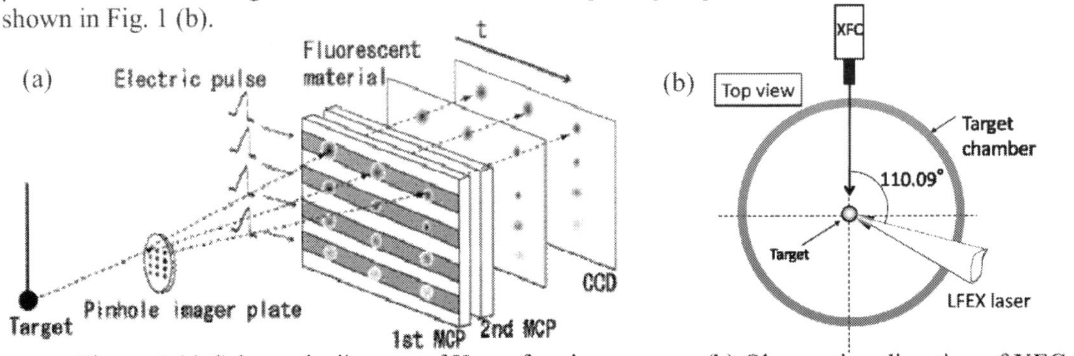

Figure 1 (a) Schematic diagram of X-ray framing camera, (b) Observation direction of XFC.

3. Results and discussion

Figure 2 shows the time- and space-resolved images obtained by X-ray framing camera. As shown in Fig. 2, bright zones are observed in third stripline. The bright zones were observed only when additional heating laser was injected. In fast ignition experiments, hot electrons generated by the heating laser and MeV class high energy electron emit sub MeV ~ MeV class X-ray through bremsstrahlung. These high energy X-rays pass through pinhole array base materials and form a bright zone on the image data. Thus, it is considered that the peak position of these bright zones become indicators of the injection time of additional heating laser.

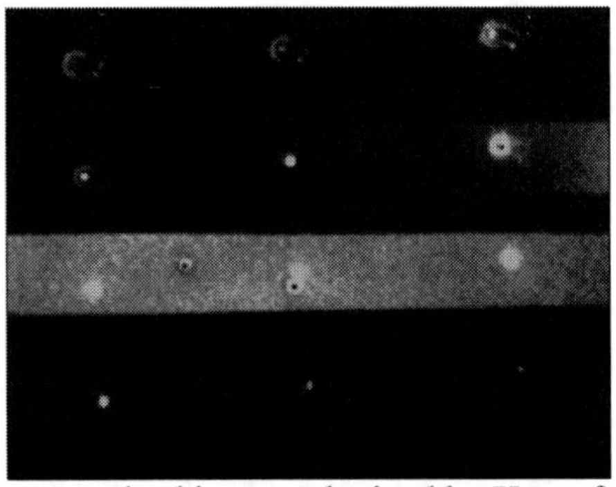

Figure 2. The time- and space-resolved images obtained by X-ray framing camera. The target was deuterated plastic shell with 30 degree gold cone. GXII laser energy was 133J/beam and LFEX laser energy was 179J.

The temporal evolution of X-ray intensity profile reconstructed from the X-ray image is shown in Fig. 3. Two experimental data are shown as a black line and a gray line. The X-ray intensity peak is consistent with the above mentioned most bright zone in the X-ray image. Although this peak does not show the heating laser injection time directly, but as we said before, it is related to the heating laser injection time. For example, in the experiment shown as a gray line, the injection time of LFEX laser was set to be delayed 160 ps compared to the experiment shown as a black line. As shown in Fig. 3, it is found that the X-ray peak position of gray line is about 1000 ps, which is delayed for 250 ps

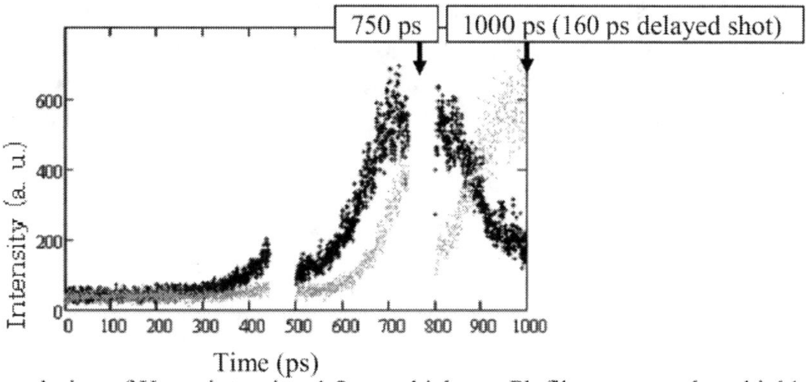

Figure 3. The temporal evolution of X-ray intensity. 1.2 mm thickness Pb filter was used to shield high energy X-ray under the detectable intensity. The LFEX laser injection time is different in the two experiment. The other experimental conditions were same. The target was plastic shell with 45 degree gold cone. GXII laser energy was about 200J/beam and LFEX laser energy was about 550J.

compared with that of black line, about 750 ps. The laser injection time is considered to have an error bar less than 100 ps due to jitter. Therefore, we consider this result is reasonable and we can estimate the heating laser injection time from the peak position of high energy X-ray intensity profile.

The FWHM of MCP gain is considered to be about 80ps in our X-ray framing camera system. However, as shown in Fig. 3, the intensity of high energy X-ray shows broad profile over 200 ps. This broadening cannot be explained by the simple dynode model used in usual MCP gain calculation. Therefore, we are trying to build the new model which explains this broadening. The concept of dynode model is as follows. One dynode is defined by electrons travelling one side wall to the other side wall of the channel. One dynode gain δ is defined as follows [9];

$$\delta = \left(V_z / V_c\right) \cdot k \qquad (1)$$

Here, Vz is the voltage between one dynode, Vc is the first crossover potential (i.e., the minimum potential for unity secondary emission ratio) and k is a secondary emission curvature coefficient assumed based on experimental measurements. The total MCP gain G of n dynodes is the multiplication of all dynode's gain.

$$G = \delta_1 \cdot \delta^{k(n-1)} \qquad (2)$$

Here δ1 is gain of the first input electron.

In thermal X-ray measurement, n is typically fixed at about 9.8 and δ is considered to be proportional to the applied electric pulse voltage. Therefore, FWHM of MCP gain becomes about 105 ps when FWHM of electric pulses is 330 ps. This nonlinear gain response makes the high temporal resolution measurement possible. However, in this experiment, high energy X-ray may penetrate both Pt pinhole imager plate and MCP and generate electrons throughout the MCP channel. Electrons generated in mid MCP channel have smaller n compared to that of electrons generated on MCP channel surface. This decrease of n may broaden FWHM of MCP gain. In previous paper [10,11], we confirmed this concept by calculation. We calculated gain profiles by using various n values and found that the FWHM of the gain in the case of small n value is wider than that in the case of large n value. This means that the gain profile of electrons generated in mid MCP channel are broadened compared with that of electrons generated on MCP surface. Therefore, in the case of high X-ray incidence, total gain profile is sum of these various n gain profiles and it is broadened compared with the case of thermal X-ray incidence. This theory is considered to explain the broad high energy X-ray profile qualitatively. Thus, it is considered that we can estimate the heating laser injection time from the peak position of X-ray intensity by using this model calculation. For more precise modelling, it is necessary to know the incident energy of X-rays.

Therefore, we tried to estimate the incident X-ray energy by using metal filter. First, we found that the aluminium filter base material (2 mm thickness) has no effect to shield the incident X-ray. The transmission rate of 2 mm thickness aluminium is 0.9 for 80 keV X-ray. Therefore, the incident X-ray energy is considered to be above 80 keV. Second, we used Pb filter with 1.2mm thickness. It is found that the incident X-ray intensity was decayed into about 10%~40%, although the decay rates are slightly changed shot by shot. Figure 4 shows the transmission rate of 1.2mm thickness Pb filter. 10%~40% transmission rate correspond to the case of 150～250 keV X-ray. From these results, it is considered that the incident X-ray energy is about 150～250 keV. We are now improving the fitting model considering this result.

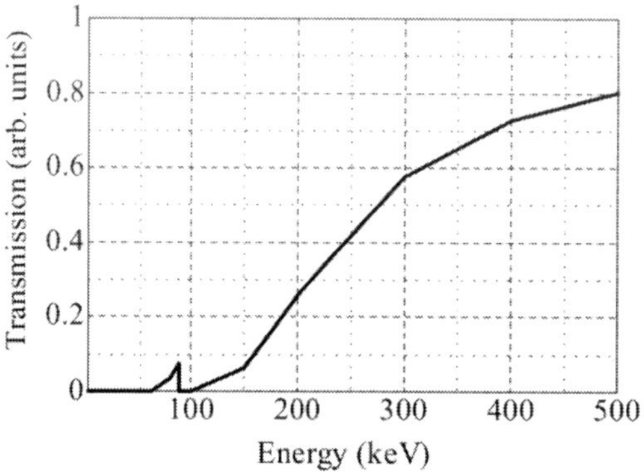

Figure 4. The transmission rate of 1.2mm thickness Pb filter.

4. Summary

We succeeded in measuring heating laser injection time and the imploded core plasma simultaneously by using X-ray framing camera. LFEX laser injection time was observed as bright zones on striplines. The peak position is found to delay corresponding to the delayed setting of LFEX injection time. MCP gain of high energy X-ray calculated by using dynode model explains the broadening of high energy X-ray intensity profile qualitatively. The incident X-ray energy is measured by metal filters and found to be 150 keV～200keV. The more precise model based on this result will be build in near future and XFC will become powerful tools in fast ignition experiments to observe the implosion hydrodynamics and the heating synchronization simultaneously.

References

[1] N. G. Basov, S. Yu. Gus'kov, and L. P. Feoktistov, J. Soviet Laser Research **13**, 396 (1992).
[2] M. Tabak, J. Hammer, M. E. Glinsky, W. L. Kruer, S. C. Wilks, J. Woodworth, E. M. Campbell, and M. D. Perry, Phys. Plasmas **1**, 1626 (1994).
[3] R. Kodama, P. A. Norreys, K. Mima, A. E. Dangor, R. G. Evans, H. Fujita, Y. Kitagawa, K. Krushelnick, T. Miyakoshi, N. Miyanaga et al., NATURE **412,** 798 (2001).
[4] R. Kodama, H. Shiraga, K. Shigemori, Y. Toyama, S. Fujioka, H. Azechi, H. Fujita, H. Habara, T. Hall, Y. Izawa et al., NATURE **418**, 933 (2002).
[5] D. K. Bradley, P. M. Bell, J. D. Kilkenny, R. Hanks, O. Landen, P. A. Jaanimagi, P. W. McKenty, and C. P. Verdon, Rev. Sci. Instrum. **63**, 4813 (1992).
[6] J. D. Kilkenny, Rev. Sci. Instrum. **63**, 4688 (1992).
[7] D. K. Bradley, P. M. Bell, O. L. Landen, J. D. Kilkenny, and J. Oertel, Rev. Sci. Instrum. **66:** 716 (1995).
[8] M. Katayama, M. Nakai, T. Yamanaka, Y. Izawa, and S. Nakai, Rev. Sci. Instrum. **62**, 124 (1991).
[9] Edward H. Eberhardt, Applied Optics, **18**, 1418 (1979).
[10] M. Koga, T. Fujiwara, T. Sakaiya, M. Lee, K. Shigemori, H. Shiraga, H. Azechi, Rev. Sci. Instrum. **79**, 10E909 (2008).
[11] M. Koga, T. Fujiwara, T. Sakaiya, M. Lee, K. Shigemori, H. Shiraga, H. Azechi, T. Ozaki, J. Plasma Fusion Sci. Series.**8,**1218 (2009).

Real-time ultrafast dynamics of dense, hot matter measured by pump-probe Doppler spectrometry

Amit D Lad[1], S Mondal[1], V Narayanan[1], Saima Ahmed[1], P P Rajeev[2], A. P. L. Robinson[2], J Pasley[3,2] and G Ravindra Kumar[1]
[1] Tata Institute of Fundamental Research, 1 Homi Bhabha Road, Colaba, Mumbai - 400005, India
[2] Central Laser Facility, Rutherford-Appleton Laboratory, Chilton, Oxfordshire, U. K.
[3] Department of Physics, University of York, Heslington, York, U. K.

E-mail: amitlad@tifr.res.in

Abstract. A detailed understanding of the critical surface motion of high intensity laser produced plasma is very crucial for understanding the interaction. We employ the two colour pump-probe technique to report the first ever femtosecond scale ultrafast dynamics measurement of the critical surface of a solid plasma produced by a relativistically intense, femtosecond pump laser beam (10^{18} W/cm^2, 30 fs, 800 nm) on an aluminium target. We observe the Doppler shift of a time delayed probe laser beam (10^{12} W/cm^2, 80 fs, 400 nm) up to delays of 30 ps. Such unravelling of dynamics has not been possible in earlier measurements, which typically used the self reflection of a powerful pump pulse. We observe time dependent red and blue shifts and measure their magnitudes to infer plasma expansion velocity and acceleration and thereby the plasma profile. Our results are very well reproduced by 1D hydrodynamic simulation (HYADES code).

1. Introduction

Intense, femtosecond laser pulses explosively ionize solids and produce some of the highest energy densities achievable in the laboratory [1]. The resulting plasma not only offers great opportunities for studying extreme states of matter, but can also provide a source for high brilliance, high energy, tuneable, ultrashort sources of x-rays, electrons, ions and positrons [1]. The physics of such plasmas has deep implications for the fast ignition scheme of laser fusion [2]. At ultrahigh intensities the physics is challenging, given the variety of absorption mechanisms that can operate, the relativistic nature of the particles and the complex nature of their transport [1]. The instantaneous profile of the plasma density is a crucial parameter that controls much of the light coupling physics [3], generation of plasma instabilities [4] etc. The light coupling in steep and high density plasmas occurs essentially at the critical surface where for example, strong plasma oscillations can be excited (resonance absorption) [5]. In the four decades of investigation of laser produced plasmas, the dynamics of coupling of long (nanosecond) pulses have been investigated very thoroughly in experiments, analytical theory and simulations [6], but the actual motion of the plasma near to the critical surface was integrated out during the long pulse durations. In the ultrashort regime, there have been far fewer studies of the dynamics at extremely high intensities, though there are studies at moderate intensities

[1] To whom any correspondence should be addressed.

[7,8]. Liu and Umstadter [8] reported that the ponderomotive pressure at even as low an intensity as 10^{15} W/cm^2 is enough for a 400 fs laser pulse to steepen the electron density profile. Kalashnikov *et al.* [9], reported red and blue shifts as a function of the laser pulse shape for a 2 ps pulse at 10^{17} W/cm^2, measuring the self reflection of the incident pulse. Sauerbrey *et al.* [10] reported detailed and interesting measurements of reflectivity, x-ray generation and spectral shifts at 10^{17} W/cm^2 for 350 fs laser pulses. They reported red and blue shifting and dependence of the spectral shifts on intensity and chirp of the input laser pulse. Sauerbrey [11] also presented a detailed model of the laser–plasma interaction and demonstrated the conditions of the laser pulse and target parameters that could produce either red-shift or blue-shift. In many of these studies [7-13], the effects that were considered were essentially 'instantaneous' as the measurements relied on reflection of the irradiating pulse itself.

Here we create plasma on a metal target at 10^{18} W/cm^2 with an 800 nm, 30 fs laser pulse and interrogate the plasma at different later 'instants' with a 400 nm, 80 fs laser pulse, tracking the plasma all the way up to 30 ps. We present clear evidence for the initial inward motion of the critical surface which bounces back at later time. We are able to derive 'instantaneous' velocity and acceleration profiles of the initially kicked plasma. Our results are well reproduced by 1D hydrodynamic simulation (HYADES code) [14]. We believe our results are important for studies that seek to understand energy and mass transport in the plasma at relativistic intensities.

2. Experimental

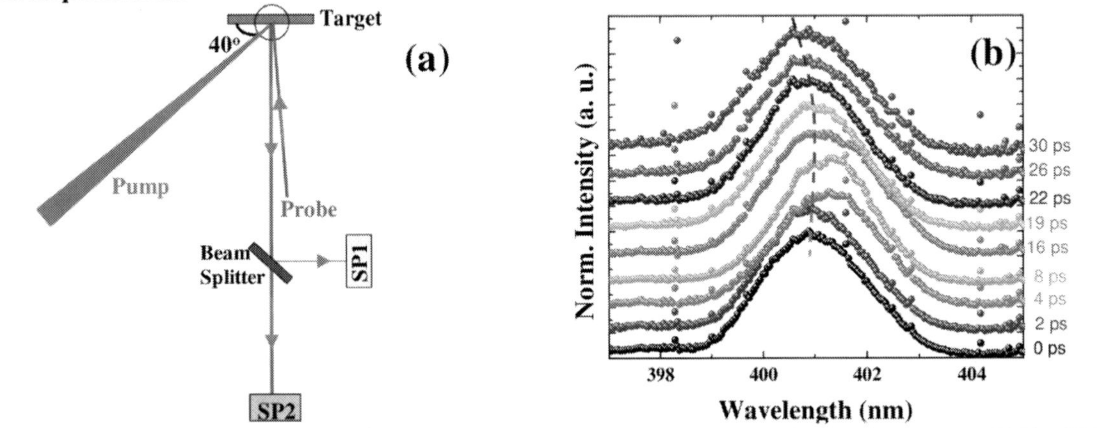

Figure 1. (a) Schematic of the experimental setup: Avantes (SP1) and Ocean Optics HR-2000 (SP2) UV-visible spectrometers; (b) Normalized reflected spectra of the probe beam from the plasma surface at several time delays.

We use the two colour pump-probe technique for measuring the dynamics of the critical layer. (figure. 1(a)). The pulses are derived from a 20 TW, 30 fs Ti-Sapphire chirped pulse amplification laser. A *p*-polarized pump pulse at 800 nm (having 37 nm band-width), pulse at repetition rate of 10 Hz, is incident at 40° to the target normal. Its nanosecond prepulse contrast is 3×10^{-6}. A small portion of the pump pulse is extracted and frequency doubled in a 1 mm thick BBO crystal to generate a probe pulse at 400 nm (having 3 nm band-width), which has duration of 80 fs. The probe pulse is delayed; with a resolution of 1 μm using a PLX retroreflector mounted on a linear translation stage and is passed through a BG-39 filter to remove residual 800 nm light. The pump pulse is focused with an f/4 gold coated off axis parabolic mirror to a diameter of 20 μm on optically polished aluminium target. The probe is focused to 60 μm with an f/30 lens so as to be collimated on the plasma created by the pump.
Typical pump intensities generated are about 5 x 10^{18} W/cm^2, while the probe is at ~10^{12} W/cm^2. The pump-probe simultaneity at the target (i.e. zero time delay) is obtained by monitoring the reflectivity of the probe beam as a function of the delay; at zero time, there is a sharp transition in reflectivity from a high value to a lower value. The probe pulse reflected from the target is divided by a beam splitter and fed to two different spectrometers (SP1 and SP2) (figure 1). An 'Avantes' UV-visible spectrometer (SP1) operates in the range of 245 to 505 nm having spectral resolution of 1.3 Å, while

another (SP2, 'Ocean Optics HR-2000') works in the range of 350 to 445 nm, having a spectral resolution of 0.5 Å. Here we present only the spectra measured by SP2 because of its higher resolution. 50 spectra are collected and averaged at each time delay. The narrow spectrum of probe pulse facilitates the observation of small wavelength changes, as evident in our data.

3. Results and Discussion

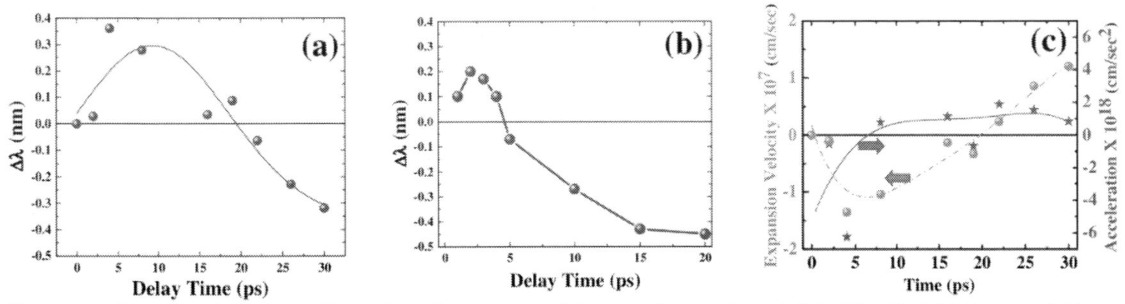

Figure 2. Doppler shift in reflected probe spectra (a) experimental and (b) 1D HYADES simulations. Calculated instantaneous (c) expansion velocity and acceleration from Doppler shift.

Figure 1(b) illustrates the normalized reflected spectra of the probe beam from the plasma surface. The black curve is the probe spectrum at zero time delay. The figure 2(a) shows that for aluminium, the reflected probe pulse is red-shifted at early times (up to 19 ps) while the shift reverses and becomes increasingly blue for larger delays. The normally incident 400 nm probe pulse reflects from its critical surface located at electron density of 6.3×10^{21} cm^{-3}. Our data indicate that critical surface initially recedes from the probe, pushing into target. At later times, this surface moves back towards vacuum, getting progressively higher in speed. Our data clearly indicate a 'turn around' time of around 19 ps.

The temporal resolution in our experiment enables a deduction of 'instantaneous' velocity and acceleration of the critical surface, shown in figure 2(c). The shifts are consistent with velocities of the order of 10^7 cm/s expected from simple analytical estimates based on $0.5c\,(\Delta\lambda/\lambda)\,(\cos\theta)$ [8], while the acceleration is found to be of the order of 10^{18} cm/sec^2 [11]. The velocities are similar to those found by Kalashnikov *et al.* [9] which were derived from the 'instantaneous' second harmonic reflection from the plasma.

The modelling of the target hydrodynamics under the influence of the laser was performed using the 1D radiation hydrodynamics code HYADES [14]. HYADES is a 1D Lagrangian radiation hydrodynamics simulation code with a flux limited diffusion model for electron conduction and a multi-group diffusion model of thermal radiation flow within the target. The results of HYADES modelling as pertains to the interaction of the main pulse with the target cannot be expected to accurately reflect the interaction since a PIC or PIC-Hybrid code would be necessary for this purpose. It should be borne in mind, therefore, that the location and quantity of energy deposited during the main pulse will not be accurately represented by HYADES. The prepulse interaction should, however, be well represented by the models incorporated in the code.

In the modelling, a 500 μm thick target is divided into 100 Lagrangian cells, which are feathered toward the driver facing surface. Two different laser sources are incorporated, one at 800 nm which represents the pump laser, and the other the probe at 400 nm. The probe is modelled so as to determine whether the interaction of the probe with the plasma is affecting the result of the experiment. A flux limiter of 0.06 is employed, which is appropriate for the prepulse interaction of the pump beam. The calculations described in the previous subsection were analyzed by estimating the leading order wavelength shift that the moving critical surface would induce, i.e. $\Delta\lambda = (2\,\chi_{crit}\,\lambda_{probe})/c$, where, the velocity of the critical surface was approximated to the fluid velocity at the critical surface. Post-processing the above calculation yields the wavelength shift as a function of probe time delay, and this is shown in figure 2(b). Although there is not exact quantitative agreement, the qualitative behaviour of the wavelength shift with probe delay is reproduced and the absolute wavelength shifts are within a factor of 2 of the experimental measurements (figure 2(a)). This discrepancy is likely explained by the fact that HYADES poorly represent the energy deposition of the main pulse.

By examining the simulation output in light of figure 2(b), the experimental measurements can now by physically interpreted as follows. The pump laser launches a compression wave into the front surface plasma which has been formed by the laser prepulse. At early times the position of the critical surface for the shorter wavelength probe beam tracks this compression wave as it moves into the target. Thus at early times a red shift is measured. At later times the compression wave has propagated into a region of overdense plasma, and so the critical surface for the probe now sits in a region that is undergoing rarefaction and thus the critical surface is now moving into the vacuum and towards the laser. Therefore a blue shift is measured.

From the analysis of the simulation results the qualitative behaviour of the spectral shift can be understood. The simulations are within a factor of two of an accurate quantitative description with an experiment. This means that the experimental approach used in here can provide detailed insights into the plasma dynamics in the crucial energy deposition region, with sufficient time resolution that it can be readily compared to a variety of different simulation types. This can consequently allow for a better theoretical understanding of the interaction of the main pulse with the solid target.

4. Conclusions

We report the first ever pump-probe dynamics of the critical surface of solid density plasma produced by high intensity, femtosecond laser. Time-resolved ultrafast Doppler spectrometry is a powerful technique to infer fast dynamics in dense plasma systems. The variation in the peak shift and modified spectral-width of the probe beam is observed as a function of probe delay. At intensities we use in the experiment here, the plasma motion is mostly driven by hydrodynamics and hence occurs in picoseconds timescales. Ponderomotive pressure becomes more significant at higher intensities, and critical surface oscillations can be as fast as the speed of light. The oscillating surfaces can up-shift the driving pulse to high harmonic attosecond pulses. Doppler spectrometry with few-cycle pulses can provide invaluable information regarding these otherwise obscure ultrafast dynamics.

5. References

[1] Mourou G A, Tajima T and Bulanov S V 2006 *Rev. Mod. Phys.* **78**, 309
[2] Tabak M, Hammer J, Glinsky M E, Kruer W L, Wilks S C, Woodworth J, Campbell E M, Perry M D and Mason R J 1994 *Phys. Plasmas* **1**, 1626
[3] Gibbon P and Forster E 1996 *Plasma Phys. Contr. Fusion* **38**, 369
[4] Eliezer S 2002 *The Interaction of High Power Lasers with Plasmas* (IOP Publsihing, Bristol)
[5] Sandhu A S, Kumar G R, Sengupta S, Das A and Kaw P K 2005 *Phys. Rev. Lett.* **95**, 025005
[6] Atzeni S and Meyer-ter-Vehn J 2004 *The Physics of Inertial Fusion - Beam Plasma Interaction, Hydrodynamics and Hot Dense Matter* (Clarendon Press, Oxford)
[7] Michberg H M and Freeman R 1990 *Phys. Rev.* A **41**, 2211
[8] Liu X and Umstadter D 1992 *Phys. Rev. Lett.* **69**, 1935
[9] Kalashnikov M P, Nickles P V, Schlegel Th, Schnuerer M, Billhardt F, Will I and Sandner W, 1994 *Phy. Rev. Lett.* **73**, 260
[10] Sauerbrey R, Fure J, Le Blanc S P, van Wonterghem B, Teubner U and Schafer F P 1994 *Phys. Plasmas* **1**, 1635
[11] Sauerbrey R 1996 Phys. Plasmas **3**, 4712
[12] Veres G, Kocsis G, Racz E and Szatmari S 2004 *Appl. Phys.* B **78**, 635
[13] Tarvin J A and Schroeder R J 1981 *Phys. Rev. Lett.* **47**, 341
[14] HYADES is a commercial product of Cascade Applied Sciences, email Larsen@casinc.com

Journal of Physics: Conference Series **244** (2010) 022045 doi:10.1088/1742-6596/244/2/022045

Lateral propagation of fast electrons at the laser-irradiated target surfaces

Y. T. Li[1,3], X. X. Lin[1], [a]F. Liu[1], B. C. Liu[1], F. Du[1], S. J. Wang[1], C. Li[1], M. L. Zhou[1], L. Zhang[1], X. Liu[1], J. Wang[1], [b]F. Liu[1], X. L. Liu[1], L. M. Chen[1], Z. H. Wang[1], J. L. Ma[1], Z. Y. Wei[1], J. Zhang[1,2,3]

[1] Beijing National Laboratory for Condensed Matter Physics, Institute of Physics, Chinese Academy of Sciences, Beijing 100080

[2]Shanghai Jiao Tong University, Shanghai 200240, China

Email: ytli@aphy.iphy.ac.cn or jzhang@aphy.iphy.ac.cn

Abstract. Lateral propagation of fast electrons at the target surfaces irradiated by femtosecond intense laser pulses is measured by k_α x-ray imaging technique when a preplasma is presented. An annular halo surrounding a bright spot is observed in the x-ray images when the scale length of the electron density is large. For an incidence angle of 70° the x-ray images show a non-symmetrical distribution peaked to the laser propagation direction. The x-ray photons in the halo are mainly excited by the fast electrons that flow in the preplasma when their paths intersect the high density regions near the target surface.

1. Introduction

Lateral propagation of fast electrons at the target surfaces irradiated by intense laser pulses is important to understand the physics in cones, with which Kodama *et. al.* have demonstrated the enhancement of neutron yield in the fast ignition (FI) experiments [1]. Evidence of the lateral transport has been found by x-ray [2,3,4] and ion emission [5]. Fast electron jets emitted into vacuum close to target surface have been also observed with simple plane targets [6,7,8, 9]. The lateral propagation of fast electrons also enlarges the x-ray source size. This has to be controlled for the imaging applications because large-size sources reduce the spatial resolution.

Recently two-dimensional x-ray imaging measurements show that a preplasma in the cone may reduce the fast electron propagation and energy deposition [10,11]. For the real conditions of fast ignition the preplasma effect on the electron propagation must be considered because the intensity of the heating short pulse is much higher. In this paper, the electron propagation at the front target surface when a preplasma is formed before the main laser pulse is studied by k_α x-ray imaging technique. The origins of the k_α x-ray photons measured are identified. The dependence of the x-ray images on the laser incidence angles and the preplasma scale length are presented.

2. Experimental setup

[3] To whom any correspondence should be addressed.

Journal of Physics: Conference Series **244** (2010) 022045 doi:10.1088/1742-6596/244/2/022045

The experiments were carried out using the Xtreme Light II (XL-II) Ti: sapphire laser system at the Institute of Physics, Chinese Academy of Sciences. The laser system can deliver a linearly-polarized pulse with energy up to 0.6 J in 60 fs at 800 nm. The laser pulse was focused onto a target using an f/3.5 off-axis parabolic mirror. The diameter of the focal spot was ~ 4.5 μm. This gave an intensity up to 5×10^{18} W/cm^2. The contrast ratio was ~ 10^{-5} measured by a third order cross-correlator (Sequoia) at 10 ps. A prepulse with a duration of 60 fs, split from the main pulse, was used to generate a preplasma. The main diagnostic was K_α x-ray imaging technique. A spherically bent quartz crystal was used to image the Cu K_α x-ray emission at 8.048 keV onto a 16 bit charge-coupled device (CCD) camera with a magnification of 8x. The energy resolution was about 11eV. The astigmatism-limited spatial resolution was ~ 20 μm. A Be filter was set in front of the CCD camera to remove visible light and reduce background x-ray noise.

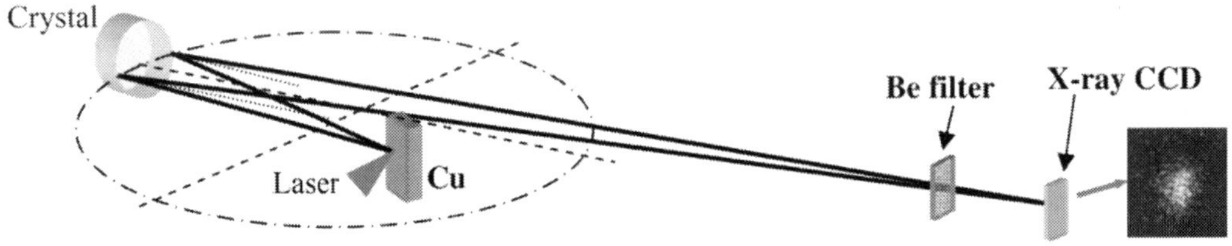

Figure 1. Schematic view of the experimental setup.

3. Results and discussions

Figure 2 shows the x-ray images taken for three laser prepulse conditions. The laser incidence angle is 45°. For all the images a halo structure surrounding a bright spot is observed. The size of the halo increases with the preplasma scale, while the x-ray photon number decreases. Another interesting phenomenon is that a ring-like halo is presented in Fig. 2(b).

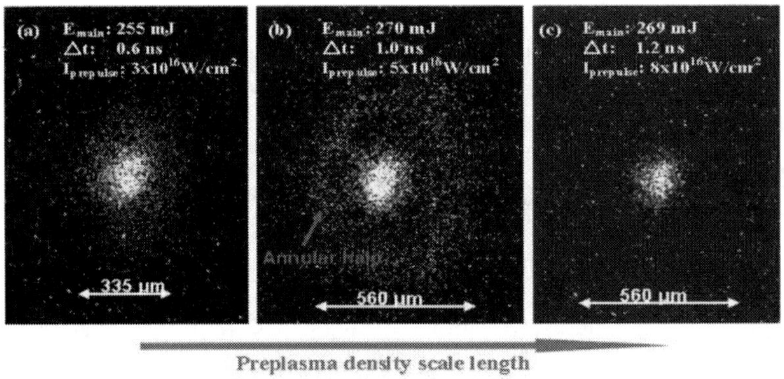

Figure 2. K_α x-ray images for different prepulse conditions. The energy of the main pulse (E_{main}), delay time between the prepulse and the main pulse (△t), and intensity of the prepulse ($I_{prepulse}$) are listed on each image.

Figure 3 shows the images taken for the incidence angle 45° and 70°. The laser pulse comes in from the right. For 45° the halo is basically symmetrical, while for 70° the halo becomes non-symmetrical and more x-ray photons are excited along the laser propagation direction.

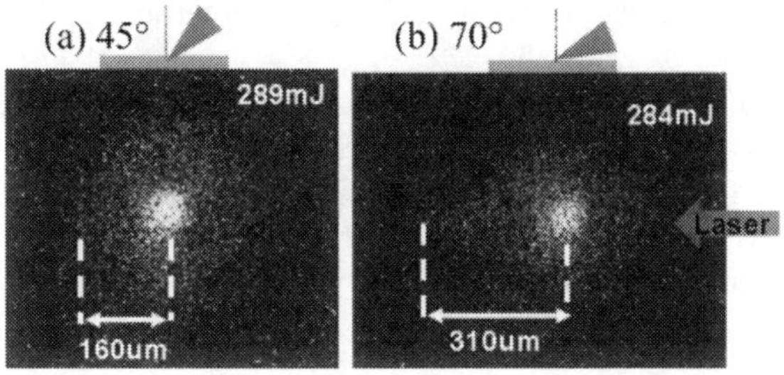

Figure 3. K_α x-ray images for the incidence angle of 45° (a) and 70° (b).

The k_α x-ray photons measured can be possibly excited by three groups of fast electrons, as shown in Fig. 4(a). The first group transports longitudinally into the target bulk and excites x-ray photons at the deep regions (not so deep that x-ray still can escape into vacuum and seen by the imaging crystal). The second one transports laterally as a current along the initial target surface due to the confinement of the surface magnetic and electrostatic fields [6,12]. However, this current is significant only when the electron density profile is steep. The third one is the electrons that diffuse in the preplasma. K_α x-ray photons are excited when the path of such electrons intersect the high density regions near the target surface. We used a step-like target to distinguish the three mechanisms. Figure 4 (b) shows the target configuration and optical image taken by a microscope. An 18 μm thick copper foil is coated on a CH plastic substrate partially. The laser pulse shines on the CH substrate near the copper coating. In this case the Cu k_α x-ray photons are excited only by the electrons above the CH substrate. Figure 4(c) and (d) shows the comparison of the step-like target with the standard plane copper target. One can see that the signals at the halo regions are similar to each other. This indicates that the longitudinal and the lateral fast electron current under the initial surface do not contribute much to the x-ray signal measured. The measured halo mainly originates from the third group of fast electrons.

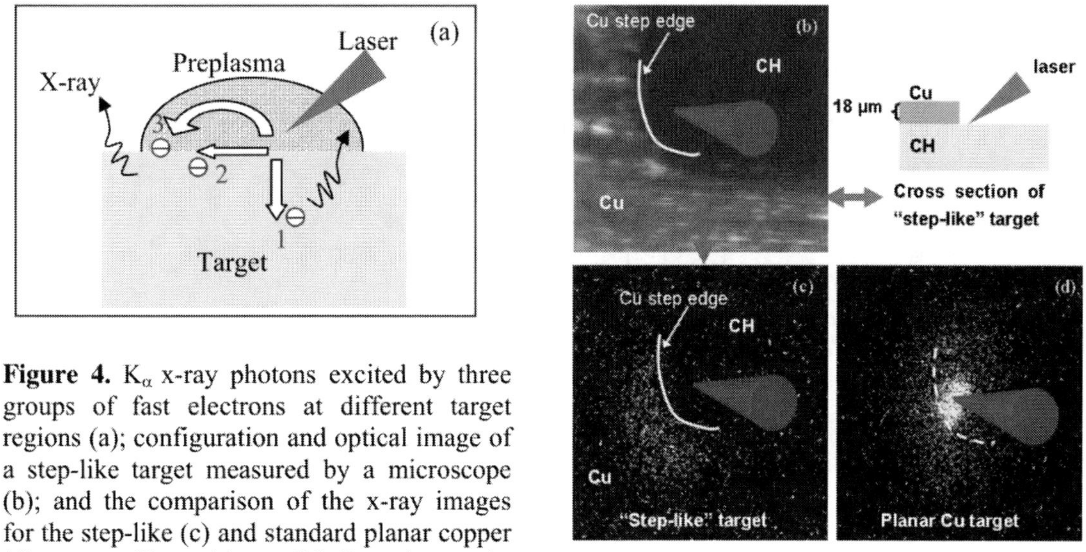

Figure 4. K_α x-ray photons excited by three groups of fast electrons at different target regions (a); configuration and optical image of a step-like target measured by a microscope (b); and the comparison of the x-ray images for the step-like (c) and standard planar copper (d) target. The white solid line shows the position of the boundary between the copper coating and the CH substrate.

When the intensity of the prepulse and the delay between the prepulse and main pulse are increased, the preplasma scale length is increased accordingly. The third group of fast electrons can go further away with the preplasma from the laser focus laterally. Therefore, the size of the x-ray halo observed is increased with the laser prepulse. When the laser incidence angle is as large as 70°, the light pressure of the main pulse will lead to a non-symmetric distribution of the plasma. This can explain the tail-like distribution in the laser propagation direction shown in Fig 3(b).

The ring structure in Fig. 2(b) is probably caused by the self-generated magnetic field in the preplasma. It is well know that an azimuthal magnetic field can be generated due to the nonlinearity of the electron density gradient and the electron temperature gradient [13]. Recent measurements with proton deflectometry show that the strong magnetic fields can be generated at the edge of a plasma [14,15]. Such fields can modify the fast electron trajectories and result in the observed x-ray ring.

4. Conclusions

Lateral propagation of fast electrons at the front target surfaces irradiated by intense laser pulses has been studied by k_α x-ray imaging technique. It is found that a halo structure is formed around the laser focus. The halo size increases with the preplasma scale length and a ring structure is presented. A non-symmetrical halo distribution is also observed for the laser incidence angle 70°. However, this does not directly correlated with the fast electron current at the initial target surface because the observed x-ray photons in the halo region are mainly excited by the fast electrons flowing and diffusing in the preplasma. One should be careful to distinguish the possible x-ray origins when using a crystal imaging technique to study fast electron transport.

This work is supported by the National Nature Science Foundation of China (Grant Nos. 10925421, 10675164, 60621063, 10734130), National Basic Research Program of China (973 Program) (Grant No.2007CB815101) and the National High-Tech ICF program.

References

[1] R. Kodama *et al*, Nature **41**, 2798 (2001).
[2] R.B.Stephens *et al.*, Phys. Rev. E **69**, 66414 (2004).
[3] T. Feurer *et al.*, Phys. Rev. E **56**, 4608 (1997).
[4] H. Langhoff *et al.*, Phys. Plasmas **16**, 072702 (2009).
[5] P. McKenna *et al.*, Phys. Rev. Lett. **98**, 145001 (2007).
[6] Y. T. Li *et al.*, Phys. Rev. Lett. **96**, 165003 (2006).
[7] H. Habara *et al.*, Phys. Rev. Lett. **97**, 095004 (2006).
[8] X. H. Yan *et al.*, Phys. Plasmas **15**, 013106 (2008).
[9] Y. Ping *et al.*, Phys. Rev. Lett. **100**, 085004 (2008).
[10] L. Van Woerkom *et al.*, Phys. Plasmas **15**, 056304 (2008).
[11] S. D. Baton *et al.*, Phys. Plasmas **15**, 042706 (2008).
[12] T. Nakamura *et al.*, Phys. Rev. Lett. **93**, 265002 (2004).
[13] J. A. Stamper *et al.*, Phys. Rev. Lett. **40**, 1177 (1978).
[14] C. K. Li *et al.*, Phys. Rev. Lett. **99**, 055001 (2007).
[15] C. K. Li *et al.*, Phys. Rev.E **80**, 016407 (2009).

Computer simulation of laser proton acceleration for hadron therapy

V A Lykov and G V Baidin

Russian Federal Nuclear Center – VNIITF named after academician E I Zababakhin
Vasilyeva str 13, Snezhinsk, Chelyabinsk reg., 456770, Russia

E-mail: v.a.lykov@vniitf.ru

Abstract. The *LegoLPI* code was used for simulation of laser-driven proton acceleration with the purpose to find an optimal conditions for the generation of proton beams with parameters required for hadron therapy. The 2D PIC simulations were done for various type of targets (double-layer foils, polyethylene foils, hydrogen jets) and target irradiation conditions (different intensities and polarizations, laser pulse wavelengths and durations, focal spot sizes). It is shown that efficiency of proton generation would be high enough for the optimal condition of target irradiation, but the spectrum of accelerated protons is rather broad for both linearly and circularly polarized laser pulses because of the two-dimensional effects arising when the proton acceleration length is comparable with the focal spot. As follows from the 2D *LegoLPI* calculations, the efficiency of formation of proton spectrum in energy range of 200-250 MeV and normalized to laser energy may reach values $(0.5\text{-}1)\cdot10^{7}$ protons$\cdot$(MeV$\cdot$J)$^{-1}$ for the considered types of laser - targets systems. The performed calculations specify an opportunity of generation proton beams whose parameters meet proton therapy requirements with the use of a laser that would have a peak power of about 1 PW and average power up to 1 kW. The powerful ultra-short pulse lasers, which are under construction in some countries around the world, are very close to the facilities necessary for this purpose.

Introduction

The experimental and theoretical studies of laser-driven ion beams attract considerable attention due to their potential applications in medicine, science and technology [1-4]. The practical applications (proton radiography and therapy, laser ion sources for linear accelerators) require quasi-monoenergetic and even monoenergetic beams. Following from results of numerical simulations, it was proposed that ion beams with quasi-monoenergetic spectra can be generated if use a high-Z thin foil with hydrogen or water vapor present on its surface [3]. Circularly polarized laser piulses may be of special interest to the generation of monoenergetic proton beams. This problem is discussed in a number of inetersting papers [5-7], however, feasibility of implementing the advantages of circularly polarized pulses for the generation of quasi-monoenergetic ions from thin foils is still to be studied out. Laser-driven acceleration of protons to high energies from thin foils irradiated by high-intensity laser pulses imposes rigid requirements for laser pulse contrast. Thus it is necessary more practicable laser-target systems for proton acceleration to high energies. The gas-jet targets seem quire realistic for this purpose. These targets have already been studed by multidimensional PIC-codes [8-9] and it used in experiments on laser-driven ion acceleration to high energies [10].

The 3D parallel relativistic PIC-code *LegoLPI* to simulate the high-intensity laser – plasma interaction, the electron and ion acceleration was developed by G V Baidin in RFNC-VNIITF [11].

Journal of Physics: Conference Series **244** (2010) 022046 doi:10.1088/1742-6596/244/2/022046

The paper presents results of the 2D *LegoLPI* calculations performed with the purpose to find an optimal conditions for the generation of proton beams with parameters required for hadron therapy: the proton energy must be about 200 - 250 MeV with monochromism 1-2%, and proton beam intensity - above 10^{10} protons•s^{-1} [2]. The calculations were done for different targets (double-layer foils of aluminum and polyethylene, polyethylene foils and hydrogen jets of different densities) and pulse parameters (intensity, duration and wavelength, linear and circular polarizations, focal spot size). The main criterion in their comparison was attaining the highest possible efficiency in the laser acceleration of protons to 200-300 MeV.

1. The simulation of protons acceleration at laser irradiation of double-layer foils

The target was the aluminum foil with thickness of 100 nm and the polyethylene disk with diameter of 6 µm and thickness 50 nm that was placed at the rear surface of Al-foil. Matter was assumed fully ionized, i.e. two plasma layers were initially present: the first layer had consisted of electrons and Al^{13+} - ions with number densities n_e=7.8•10^{23} cm^{-3} and n_{Al}= 6.0•10^{22} cm^{-3}; the second one - electrons, protons and $^{6+}$ - ions with number densities n_e=3.4•10^{23} cm^{-3}, n_p=8.5•10^{22} cm^{-3}, and n_C=4.25•10^{22} cm^{-3}, respectively. The laser pulse was characterized by Gaussian profiles in time and transverse space direction. It had p- polarization, wavelength =1 µm, focal spot size d_{FWHM} = 3 µm. The laser pulse had duration $_{FWHM}$= 30 fs and maximum of intensity: $\langle I\rangle_{max}$= 1.4•10^{22} W•cm^{-2} ($\langle I\rangle_{\max} = c(a\cdot a_0)^2\cdot(8\pi)^{-1}$, $a = 100, a_0 = 2\pi m_e c^2 (q_e \lambda)^{-1}$). The maximal intensity $\langle I\rangle_{max}$ was reached at the moment t_{max} = 2• $_{FWHM}$. The computational domain was 40x24 (µmx µm). It was divided into square cells with side length of 0.01 µm. For the 2D case, the total laser pulse energy was E_{2D} = 14.3 J•μm^{-1}.

The protons were accelerated mainly during time 100-150 fs, achieving for this time about 70% of energy that they obtained at last moment of this run (see figure 1).

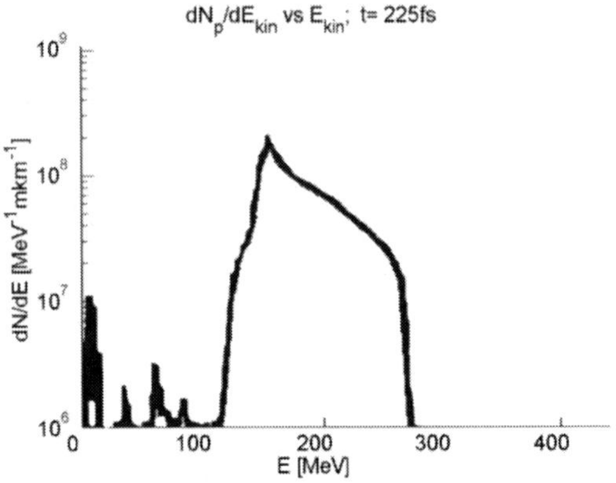

Figure 1. The proton spectrum for a moment of time 225 fs in calculation of double-layer foil.

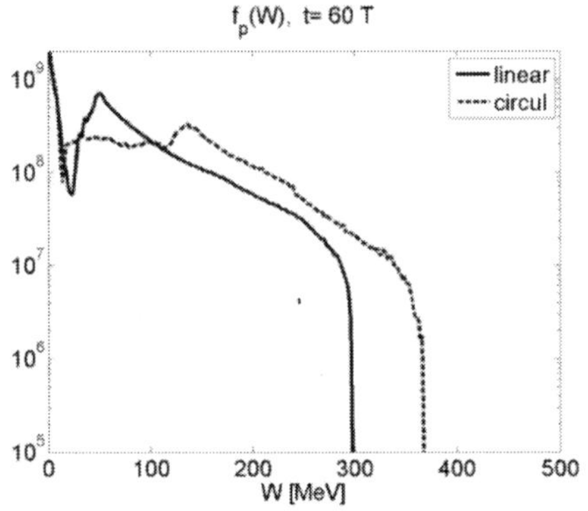

Figure 2. The proton spectrum for a moment of 160 fs in calculation of CH_2 - foil irradiated by laser pulse with a linear (continuous line) and a circular (dashed line) polarization.

Our calculations show that the efficiency of laser energy conversion to protons with energies above 100 MeV is about 2-4%, the spectrum of accelerated protons is broad and the number of protons accelerated to energy of 200-250 MeV per unit of a proton energy and normalized to laser energy may reach (2-5)•10^6 protons•(MeV•J)$^{-1}$.

2. The simulation of proton acceleration to 200-400 MeV from polyethylene foil irradiated by linearly and circularly polarized laser pulses

The figure 2 shows results of the 2D *LegoLPI* calculations performed for polyethylene foils with thickness of 100 nm that irradiated by laser pulses with linear and circular polarization.

The laser pulse was Gaussian in time and super-Gaussian (n=4) over the focal spot. It had follows parameters: wavelength =0.8 µm, focal spot size d_{FWHM} = 4 µm. The maximum intensity $<I>_{max}$ was equal to 6.8•10^{21} W•cm^{-2} and reached at t_{max} = 30 fs that was equal to the total pulse duration at half maximum (t_{max} = $_{FWHM}$) the total laser pulse energy was E_{2D} = 8.5 J•$µm^{-1}$.

The target was assumed to contain fully ionized polyethylene with number densities of electrons, protons and $^{6+}$- ions, respectively: n_e=3.6•10^{23} cm^{-3}, n_p=9•10^{22} cm^{-3}, and n_C=4.5•10^{22} cm^{-3}

The computational domain was 50*24 µm*µm in size, grid was square with side length of 0.01 µm. The number of particles of each species in one cell was 100.

For the above mentioned parameters of laser pulse, the foil thickness was varied in the range of 50-200 nm to obtain the highest possible efficiency of laser energy conversion to 200-400 MeV protons. The 2D calculations performed by *LegoLPI* code for the polyethylene foils irradiated by laser pulses with linear and circular polarization shown that the spectra of accelerated protons are rather broad for both polarizations. This is due to the fact that the foil significantly bends during acceleration: protons obtain the main part of their energy while passing distance of 2-4 d_{FWHM}. The efficiency of laser proton acceleration to energy of 200-250 MeV may reach value of (0.5-1)•10^7 protons•$(MeV•J)^{-1}$, as it follows from the 2D -calculations performed for the polyethylene foils irradiated by the 30 fs laser pulses at intensity of 7•10^{21} W•cm^{-2}. It should be noted that the circularly polarized pulses are more effective than the linearly polarized ones in 2-3-times (see figure 2).

3 The simulation of proton acceleration to 200-400 MeV from low-density hydrogen jets irradiated by laser pulses with wavelength of 3.3 µm

Since laser proton acceleration to high energies from thin foils irradiated by short laser pulses imposes rigid requirements for laser pulse contrast, it is necessary to seek for more practicable laser-target systems. Gas jet targets seem quite realistic for laser-driven ion acceleration.

Here we present results of 2D calculations performed by the *LegoLPI* code for hydrogen jets irradiated by pulses from a PW-laser with wavelength of 3.3 µm. The diod-pumped laser LUCIA (=3.3 µm, 400 TW, 10 Hz) being built on the OPCPA scheme [12] can be taken as a prototype.

The laser pulse was set to be Gaussian in time and super-Gaussian (n=4) over the focal spot. Its parameters were as follows: wavelength =3.3 µm and the focal spot size d_{FWHM} = 24 µm. The laser intensity maximum was equal to 3•10^{20} W•cm^{-2} at time t_{max} = 200 fs (18 laser periods) that was equal to pulse duration $_{FWHM}$.

The target was assumed to be a 60-µm-thick layer of fully ionized hydrogen. The initial number densities of electrons and protons were varied in the range (0.25-10)•n_{cr}, where n_{cr} = 10^{20} cm^{-3}.

The computational domain had sizes 256*96 µm*µm and was divided into square cells with side of 0.125 µm. The number of particles of each species in one cell was 25.

The figure 3 and figure 4 show results of the 2D *LegoLPI* code calculations performed for above mentioned parameters of laser pulse and initial number density equal to 3•10^{20} cm^{-3} (n_e =3•n_{cr}). The efficiency of laser-to-proton energy conversion was about 20% and the efficiency of proton spectrum formation in energy region of 200-250 MeV was about (5-8)•10^6 protons•$(MeV*J)^{-1}$ in this simulation. It is significant that these values close enough to results obtained in the 2D simulations of the double-layer foils and the polyethylene foils irradiated by linearly polarized pulses.

The efficiency of proton spectrum formation by laser pulses with different wavelengths is conserved under concurrently changing: laser intensity $I \sim {}^{-2}$, pulse duration $t \sim$, target and focal spot sizes L, $d \sim$, electron number density in target $n_e \sim {}^{-2}$. The 2D *LegoLPI* code calculations confirmed these simple scaling laws that conserve all parameters of laser accelerated proton beams.

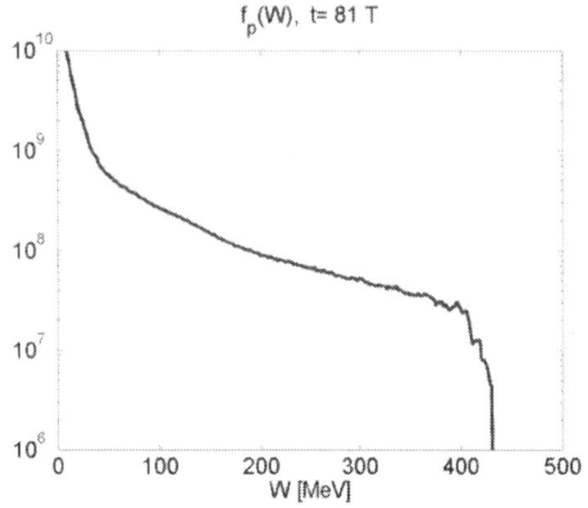

Figure 3. The proton spectrum at moment 890 fs according to the 2D *Lego-LPI* code calculation performed for the PW laser with wavelength of 3.3 μm and hydrogen jet with initial electron number density of $3\cdot10^{20}$ cm^{-3}

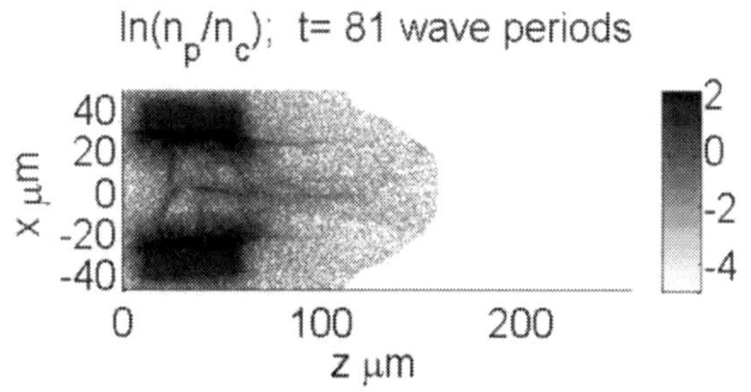

Figure 4. The spatial profile of proton number density at moment 890 fs according to the 2D *Lego-LPI* code calculation performed for the PW laser with wavelength of 3.3 μm and hydrogen jet with initial electron number density of $3\cdot10^{20}$ cm^{-3}

It should be noticed that the maximum of proton energy in these calculations is about two times more then energy needed for hadron therapy. So we hope that main results of the 2D *LegoLPI* code simulations of laser proton acceleration for hadron therapy remain true and after the 3D-calculations performing.

A beam of protons with energies of 200-250 MeV and monochromism about 1-2% that required for proton therapy can be formed from broad spectrum of laser-accelerated protons with the help of electromagnetic systems and filters with high efficiency. It is known also that for proton therapy, the intensity of a proton beams to be about 10^{10} protons per second on average. So, performed calculations specify an opportunity of generation proton beams whose parameters meet proton therapy requirements with the use of lasers that would have a peak power of about 1 PW and average power up to 1 kW. The powerful ultra-short pulse lasers that are under construction in some countries are close to facilities necessary for this purpose.

The work was supported by the ISTC under Project 2289.

References

[1] Nemoto K et al. 2000 *Appl. Phys. Lett.* **78** 595
[2] Bulanov S V and Horoshkov V S 2002 *Plasma Physics Report* **28** 453
[3] Esirkepov T Zh et al. 2002 *Phys. Rev. Lett.* **89** 175003
[4] Bulanov S S et al. 2008 *Med. Phys.* **35** No 5 1170
[5] Yan X Q et al. 2008 *Phys. Rev. Lett.* **100** 135003
[6] Robinson A P L et al. 2008 *New Journal of Physics* **10** 013021
[7] Chen M et al. 2009 *Phys. Rev. Lett.* **103** 024801
[8] Bulanov S V et al. 2000 *JETP Letters* **71** No 10 407
[9] Bulanov S S et al. 2008 *Bull. Am. Phys. Soc., Series II* **53** No 14 125
[10] Willingale L et al 2006 *Phys. Rev. Lett* **96** 245002 .
[11] Lykov V and Baydin G V 2008 *Bull. Am. Phys. Soc., Series II* **53** No 14 218
[12] Chanteloup J–C and Bahbah S 2006 *"LUCIA short pulses applications"*, Report at 3rd International Workshop on High Energy Class - Diod Pumped Solid State Lasers, LLNL, USA, May 17-19, 2006 (www.laser.llnl.gov/programs/psa/pdfs/hec-dpssl06/07lucia_short_pulse.pdf)

ePLAS MODELING OF HOT ELECTRON TRANSPORT IN NAIL-WIRE TARGETS*

R. J. Mason*[1], R. Faehl[1] and R. Kirkpatrick[1], T. Ma[2], M. S. Wei[2], F. N. Beg[2], M. H. Key[3], and R. B. Stephens[4]

[1]Research Applications Corporation, Los Alamos, NM 87544, USA
[2]University of California, San Diego, Ca 92093, USA
[3]Lawrence Livermore National Laboratory, Livermore, CA 94550, USA
[4]General Atomics, San Diego, CA 92186, USA

E-mail: rodmason01@msn.com

Abstract. Relativistic electron transport in short-pulse laser illuminated nail-wire and foil targets has been studied with the e-PLAS implicit/hybrid plasma simulation code in a cylindrical geometry. The intensities are typical of the Vulcan and Titan petawatt lasers (1.06 μm, 1.7x10^{20} W/cm^2) in 10 μm diameter Gaussian spots, producing hot electrons at a relativistic $\gamma = 3.4$ according to Beg scaling, and with 20% absorption assumed. The targets are 200 μm long nail-headed copper wires 20 μm in diameter, and copper foils 120 μm thick. The code dumps energy at constant intensity for a picosecond at the critical surface into an isotropic relativistic Maxwellian particle-in-cell hot electron distribution. The emitted hot particles draw a cold electron return current scattering off the background ions taken as ionized at a fixed Z = 15. Transport of the hot electrons is impeded by the electric field from a Spitzer resistivity acting on the cold electron return current. Assumed emission angle is shown to seriously affect hot e^- penetration in the wires.

1. Introduction

A number of experiments have recently been done on the Livermore Titan and Rutherford Vulcan [1] laser systems to investigate the transport short pulse laser-driven relativistic electrons in copper wire targets. Lateral access in wires can give K_α diagnostic access to show hot electron penetration depth. The lasers typically deliver 1.7 x 10^{20} W/cm^2, 1.06 μm pulses in 10 μm FWHR Gaussian spatial pulses of picosecond duration to 300 μm long, 20 μm diameter copper wires with nail-like heads. We have explored the dynamics of the resultant hot electron transport with the ePLAS code [2]. It is assumed that 20% of the laser energy is absorbed at critical with Beg thermal scaling [3]. We examine here crucial elements of the transport modelled in cylindrical geometry. First, we look at idealized behavior of axial-directed electron beams in the wire showing reflections off the centerline and wire surface. Then, the net hot electron transport as a function of the assumed emission angle at the source is investigated. We find that 10° emission gives deep penetration, and notable surface transport, while 90° emission can be largely trapped near in the nail head. The predicted cold background temperatures are shown to track such penetration consistently. Finally, we categorize the hot penetration history for a typical run, in which the *E*- and *B*-fields, cold electron scatter, and hot electron drag determine the background cold electron and ion heating with surface transport over 1 ps.

2. Calculational approach

The ePLAS code calculates 2-D self-consistent *E*- and *B*-fields using the Implicit Moment Method. Here, the modelling is cylindrical (r,z) geometry, following E_r, E_z and B_ϕ components The light is tracked by ray-tracing to a depth below the critical surface, and dumped there, converting cold fluid to hot particles in a conically-directed, relativisitic Maxwellian momentum distribution. The spot is a Gaussian with a 5 μm HWHM deposited over 1 ps. The code follows the laser-generated hot electrons as mass-weighted relativistic PIC particles running though background fluid cold electrons and ions. Cylindrical particle updates use a variant of the Boris CYLRAD coding, 1970. The particle hot electrons undergo scatter of the ions and drag on the cold electrons. Eulerian, 2nd order Van Leer modelled fluids are used, permitting a large dynamic range in the densities. The returning current of cold fluid electrons undergoes Joule heating, drag deposition from the hots, and PdV compressive heating. Full ionization to Z=15 is assumed. The implicit E-field treatment facilitates modelling at the high electron target densities ~ 10^{25} electrons/cm^3 with 1.5 μm wide cells and 0.5 fs time steps - with finite-grid instability numerical heating avoided. The retention of the cold electron and ion inertia permits solutions in vacuum regions neighboring the wire target without the introduction of an artificial vacuum resistivity. Particle hot electrons help to capture kinetic effects arising in the transport, such as resistive Weibel instability. The electron scatter ν_c is done at Spitzer rates, modified for metals by a floor (typically 100 eV) on the background cold temperature. Hot electron scatter ν_h and drag α rates are relativistically corrected to match Jackson's and Mosher's analysis. The cold electron and ion temperatures are mutually coupled at Spitzer rates, such that for high enough densities, cold heating is directly mirrored in the ions. Typical runs, on a 50 x 200 mesh with 50 particles emitted/time-step and covering 1 ps, complete in 1000 cycles, and require less than 15 minutes on a 2.1 GHz PC processor. IDL is used for the graphics.

3. Computational results and conclusion

Initially we have performed test simulations of a beam directed toward the axis to test both cylindrical axial reflections, and those off the wire surface. Hot scatter, drag and cold current resistivity were "off". As evident in Fig. 1, the beam can reflect multiple times on passing down the wire. The frame on the upper left shows contours of the full nail/wire with an enlarged head, and a selected point for laser deposition and beam creation. The remaining frames show e$^-$ beam transit from 286 to 547 fs.

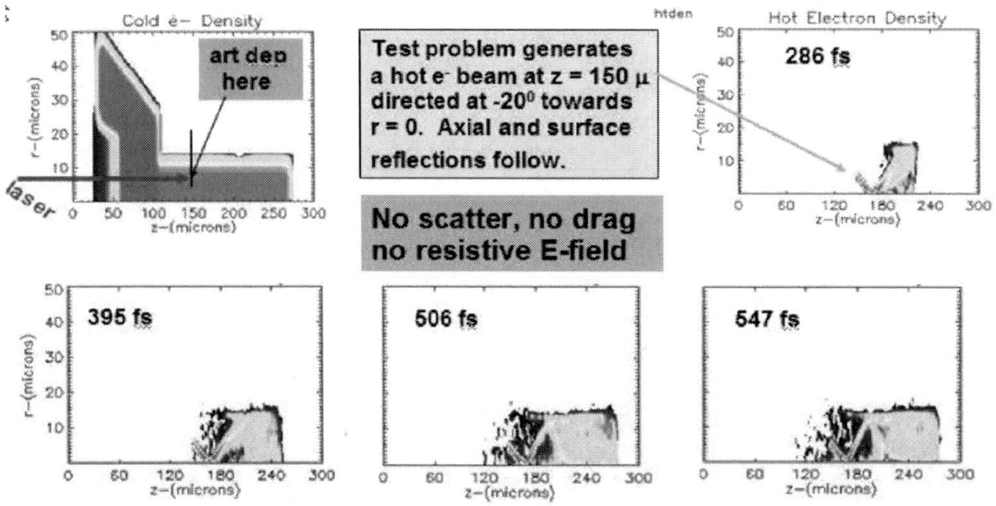

Figure 1. Internal reflections at the wire surfaces can slow hot electron penetration.

Journal of Physics: Conference Series **244** (2010) 022047 doi:10.1088/1742-6596/244/2/022047

Next with all the physics "on", full simulations with varied conical emission angles were run, showing reduced penetration as the assumed angle was increased - a 100-fold decay shifting from a 100 μm range for 10° emission, down to just 50 μm for 90° emission.

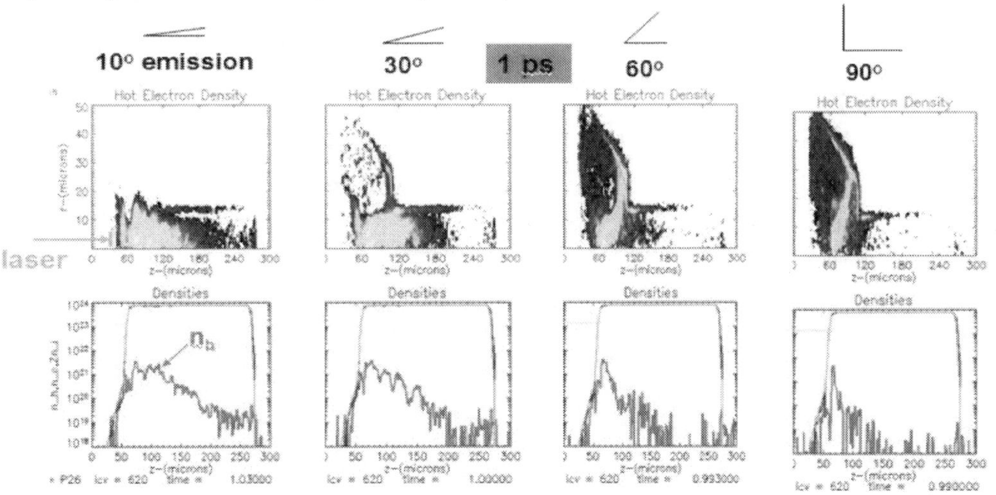

Figure 2. Nail/wire hot e- penetration is sensitive to assumed electron emission angle from the laser source.

From experimental experience choosing 30° for the emission angle, we acquire, finally, the results of Fig. 3. The upper left frame shows hot electron density contours for 1 ps, calibrated by the axial rough (red) curve below - peaking near 10^{22} cm^{-3} at z = 70 μm and dropping to 10^{19} by 220 μm. Surface transport associated with spontaneous B-fields is evident here. The Gaussian laser profile is contoured in the upper right frame of Fig. 3 (intensity surface-plot below). Temperatures at 1 ps are collected in the mid-lower frame. The hot electrons (jagged-red) are at nearly 1 MeV. Next, the cold electrons (blue) peak at near 3 keV, and the ions (brown) within the rod are nowhere hotter than 400 eV. Additionally near the surfaces, elevation of cold electrons to hot electrons is employed by the model.

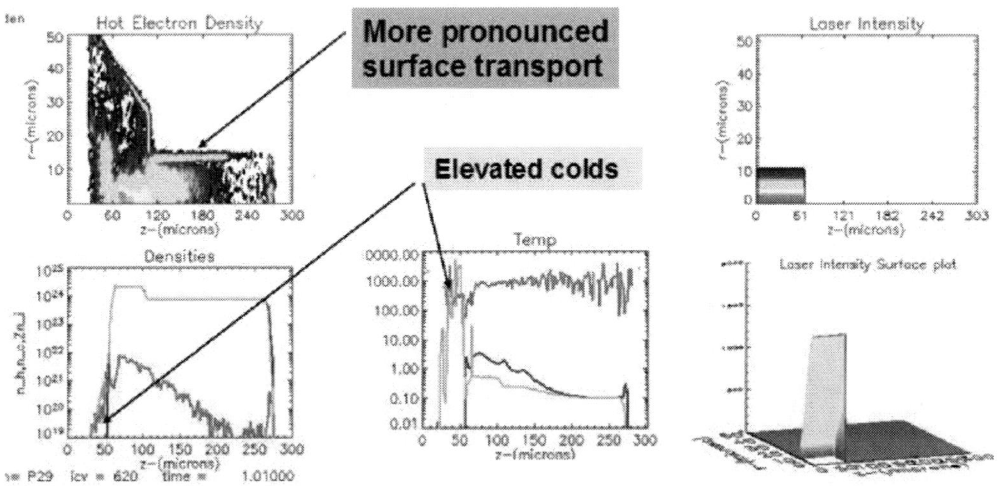

Figure 3. Calculated densities, temperatures and intensities at 1 picosecond.

Acknowledgments
The authors are grateful to D. Barnes, W. Atchison, D. Henderson and Y.C.F. Thio for helpful discussions.

References

[1] T. Ma, et al., "Transport of Energy by Ultra-Intense Laser-Generated Electrons in Nail-Wire Targets," Phys. of Plasmas - to be published, Nov. 2009.
[2] R. J. Mason, J. Comp. Phys. **71**, (1987) 429.
[3] M. G. Haines et al., Phys. Rev. Lett. **102,** 045008 (2009).

Hot electron generation at a steep interface in super intense laser-matter interaction

R. Mishra[1], **Y. Sentoku**[1], **A. J. Kemp**[2]

[1]Department of Physics, University of Nevada, Reno, Nevada 89557, USA
[2]Lawrence Livermore National Laboratory, California 94551, USA

E-mail: `rohini@physics.unr.edu`

Abstract. Super intense laser light ($> 10^{20}$W/cm^2) is able to sweep preplasma over short times and compress the preplasma density gradient typically generated by the prepulse of today's high-intensity, high energy laser systems. Hot electron generation at steep plasma density gradients has been studied in a previous paper [Kemp et al., Phys. Rev. Lett. (2008)], which identified a mode of hot electron acceleration that is characterized by the formation of low-density shelf in front of the target. In this paper, we study the direct interaction mode, which appears after the shelf plasma interaction. We propose an analytical model to describe the absorption, explain electron energy spectra, and identify the parameter regime where the direct interaction mode becomes important.

1. Introduction

Current advanced laser technology is capable of producing high contrast ratio (10^{10}) laser pulses with relativistic intensity ($> 10^{18}$ W/cm^2) and ultra short duration (few tens of fs). Such high contrast short laser pulses form only a small pre-plasma thus can directly interact with steep interface of targets, which is important because of various potential applications e.g. isochoric heating of dense plasmas [1], inertial confinement fusion, bright hard and soft x-ray sources [2]. Although strong bulk heating beyond 100 eV is expected in this direct heating mode, the underlying heating mechanism is not fully understood.

In an intense laser solid interaction, there are basically three different stages in a one-dimensional interaction scenario. The first important acceleration is the ponderamotive acceleration, near critical density, n_c, according to Wilks's scaling [3]. Then a long and intense laser pulse can sweep off the preformed plasma, and form a shelf plasma with the density $\sim n_c$ in front of the steepened interface [4]. Finally the shelf plasma density is being reduced through the interaction, and drops much lower than n_c. Following that, the laser light interacts with dense plasma directly at the steepened interface. In this paper we study this direct interaction mode.

2. Numerical study of LPI at steep density gradient

In this section we demonstrate a one-dimensional simulations, for laser-matter interaction which focus on direct interaction mode, using our PICLS code [5]. Hot electrons produced in this mode are important for energy transport inside a solid target. A slab target with fully ionized deuterium plasma extends from 20 - 40 λ in a 40 λ simulation box. We used a computational resolution of $\Delta x = 1\mu m/78$ and $\Delta t = \tau/78$ to resolve the highest density, $\sim 400n_c$. Each mesh (Δx) contains 300 deuterium ions and 300 electrons. A linearly polarized laser pulse ($\lambda = 1\mu$m) with semi infinite envelope and ($a = 10$), rising up in 15 τ with a gaussian profile, incidents normally on slab targets of width 20λ. Absorbing boundaries are used for particles and electromagnetic fields. To account plasma density dependence on hot electron generation and their energy scaling we performed a set of simulations with different slab targets of densities ranging between 40 to $400n_c$. Simulations were carried out for approximately 500 fs (150 τ). Figure 1 shows time evolved electron phase plots at the irradiation surface during one laser oscillation period. Electrons are trapped in electrostatic potential, at target surface, and form a blob structure of thickness $\sim 3l_s = 3c/\omega_p$. Peak longitudinal momentum of this blob depends on electrostatic potential, which is balanced by laser photon pressure [4]. Note here that the total number of electrons trapped in this blob are almost constant during the interaction. Leaving electron jets are balanced by return current which supplies electrons to the blob. In next sections we will introduce analytical models for the electron acceleration in the blob. In Fig. 1, (b), (c) and (d) we see the continued oscillation of trapped electrons and 2ω electrons extraction from outer blob region by $\mathbf{J}\times\mathbf{B}$ mechanism [6] in one laser period and then injected back into the target. Except 2ω jets, additional 5-6 electron jets with lower energy are observed emerging from the rear end of blob. We call them *DC-ponderomotive (DCP) electrons*, as they are accelerated via DC field of laser as discussed later. The energy of the DCP electrons depends on the peak momentum of the blob electrons. Note here that the electrons in the central region are always trapped and never come out from the blob, so the electrons forming 2ω and DCP electrons are originally located deeper in the target and they are introduced to the outer most orbit of the oscillating blob by means of the return current.

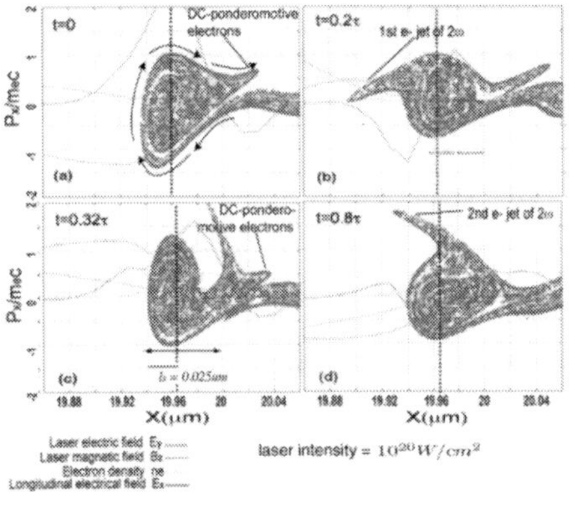

Figure 1. Electron phase evolution, $x - p_x$. (a) Electrons emerging, arrows show electron flow (b) 1st jet of 2ω extraction (c) 2ω and DCP electrons at later time (d) 2nd jet of 2ω. Arrows point in the direction of rotating blob in phase space. Dashed line shows the position of peak electron density.

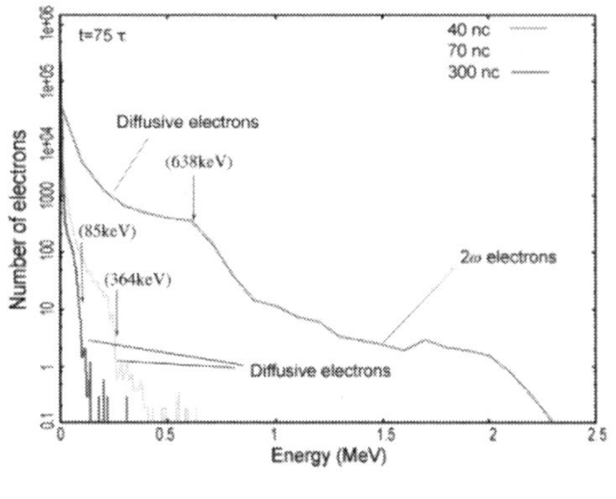

Figure 2. Electron energy spectrum obtained in various density targets. Arrows indicate the energy of oscillating electrons in the trapped region, energies in brackets are estimated peak energy of DCP electrons by Eq. (1), introduced in the later section.

For $200n_c$ target, we observed a similar blob structure of thickness $\sim 3l_s$ and DCP electrons jets, but we do not see 2ω jet. However the height of blob is reduced by a factor of $\sqrt{1/n_e} \sim \sqrt{40/200} = 0.44$. Again, the maximum number of DCP electrons group is five to six in one laser period, similar to $40n_c$ density case. The number of DCP electron jets has a weak density dependence [7]. In other density cases such as, $70n_c$, $100n_c$, $300n_c$ and $400n_c$, except $70n_c$ which has an insignificant 2ω jets, we do not observe 2ω electrons. Electron energy spectra obtained for different target densities are plotted together in Fig. 2. This picture gives a clear evidence for disappearance of high energy tail in spectrum due to the 2ω jet suppression.

3. Model for direct interaction: DC-ponderomotive (DCP) electrons

A normally incident laser on target exerts a pressure P_L, $P_L = (1+\eta)I/c$ where η is reflectivity and I is laser intensity. This pressure, which results due to the non oscillating component of ponderomotive force, pushes electrons more than the ions, hence a charge separation is established and an electrostatic potential ϕ is excited at the target surface [7]. This potential gap can be estimated by conserving the momentum flux of mass flow (plasma pressure) with the light pressure, $(1+\eta)I/c = n_e e\phi$, at the reflection surface [4],[8]. If electrons accelerate in this potential well up to energy $e\phi$ and gain longitudinal momentum P_x, then their energy is,

$$\epsilon = m_e c^2 \left(\sqrt{1 + P_x^2/m_e^2 c^2} - 1 \right) = (1+\eta)I/n_e c, \tag{1}$$

here P_y is neglected, m_e, c, n_e are electron mass, light velocity and electron density, respectively. In simulations we observed that the DCP electrons escaping from the oscillation region have peak energy which is the average energy of the oscillations, then the energy of DCP electrons is given by $\langle\epsilon\rangle = m_e c^2 a^2 (1+\eta) n_c / 4n_e$, here, time averaging in one laser period gives $< P_x^2 > \equiv P_x^2/2$ and using relations $I/c = m_e c^2 a^2 n_c/2$ and a=$eE_L/m_e c\omega$. The peak energy of oscillations in the blob for $40n_c$, $70n_c$ and $300n_c$ density cases are calculated by Eq. (1) and indicated with arrows in Fig. 2, which are consistent with the observed spectrum. For example, in 40 n_c case the arrow clearly shows the transition point between DCP electrons and $\mathbf{J} \times \mathbf{B}$ electrons energy in the spectrum.

Fig. 1 shows that the electrons are trapped at the target surface and form a blob like structure of thickness $\sim 3l_s$. Note here that area of blob in phase space is conserved, meaning no net accumulation of charge or energy, particle conservation gives a condition; $v_d n_d \simeq v_r n_e$, here v_d, n_d and v_r, n_e are the speed and density of DCP and return current electrons, respectively. Energy flux conservation gives us the condition $I_{in} - I_{rf} = \chi I_{in} \simeq \epsilon_d v_d n_d$, here χ is absorption coefficient and ϵ_d is energy of the DCP electrons. The energy of electrons in return current, which scales with $(n_d/n_e)^2$, is negligible compared to that of the DCP electrons. Plugging in the DCP electron velocity v_d in energy flux conservation condition calculated from the expression which gives average energy of DCP electrons.

$$\chi \simeq (2/a)(n_d/n_c) \cdot (1 - 1/\gamma_d) \sqrt{n_c/n_e}, \tag{2}$$

here $\gamma_d = \sqrt{1 + < P_x^2 > /m_e c^2} \sim 1 + a^2 (n_c/n_e)$, and $(1+\eta)/2 \sim 1$. The factor $(1 - 1/\gamma_d)$ in Eq.(2) is very small for $a^2 \cdot (n_c/n_e) \ll 1$. To estimate maximum absorption we solve $\partial\chi/\partial a = 0$ and find $\chi_{max} \simeq 0.6 \cdot (n_d/n_e)$ at $a \simeq 1.27\sqrt{n_e/n_c}$, which indicates that the absorption by DCP electrons becomes efficient for $a > 10$ and $n_e > 100n_c$. A rough estimation of DCP electron density, is $n_d \sim n_e/20$ for $400n_c$ and $\sim n_e/10$ for $100n_c$. Though this is a small fraction of input energy but DCP electrons deposit energy effectively inside the target in a small volume, which is important to create keV energy - solid density plasmas.

In Fig. 3 we plot Eq. (1) (DCP electrons), and $\mathbf{J} \times \mathbf{B}\,(2\omega)$ scaling obtained from Helmholz boundary soultion [9]. Electron momentum observed in simulations agrees well with 2ω scaling below $100n_c$, while with the DCP energy scaling above $100n_c$. A transition occurs around $100n_c$, in dominating mechanism to produce the maximum energy electrons. 2ω jets are disappeared in higher density targets (>$100n_c$) since they do not have sufficient energy to overcome potential gap so remain trapped, also their suppression reduces the total absorption. Since a circularly polarized laser pulse does not produce 2ω electrons, then the absorption will occur mainly because of DCP electron acceleration. To see laser polarization effect on DCP electron we

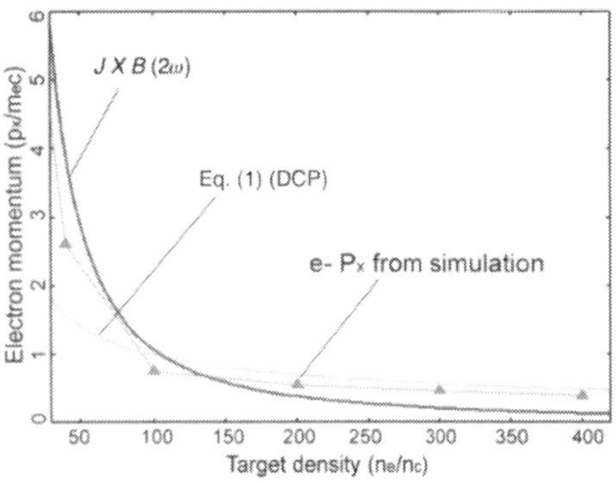

Figure 3. Analytical scaling for longitudinal P_x in Eq.(1) (DCP electrons) and $\mathbf{J} \times \mathbf{B}$ with the observed maximum electron momentum in simulation for various densities.

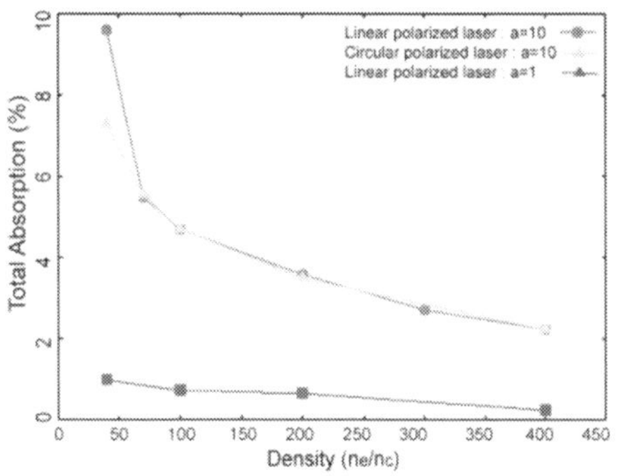

Figure 4. Total absorption measured in various density targets for linearly (red) and circularly polarized (green) with a=10, and linearly polarized laser (blue) with a=1.

performed simulations with circularly polarized laser light while keeping other laser and target parameters similar. Note here that in simulations we have found that the ion contribution to total absorption is extremely small, such as $< 1\%$ for a=10, and it drops to $\sim 0.001\%$ for a=1. Hence the absorption is mainly by the electrons. The ion absorption rates are consistent with the Denavit [10]. We measured the total absorption for the period of 150 $\tau(\sim 500fs)$. In Fig. 4, we can see that the absorption is low $\sim$ 1% and almost constant with $a = 1$, since the DCP electrons are inefficient. Also, the absorption for circularly polarized laser light is identical to the absorption of linearly polarized light, above $100n_c$. We can conclude here that for density $> 100n_c$, DCP remains the only absorption mechanism.

4. Summary

We studied electron acceleration at overcritical plasma surface and demonstrated the production of DC-ponderomotive (DCP) electrons by non-oscillating laser photon pressure. We have also proposed an analytical model of '*ponderomotive*' electron acceleration and demonstrated that the injection frequency and energy of DCP electrons are independent of laser polarization. More importantly, the DCP mechanism for electron acceleration dominates in high density targets, such as $n_e \geq 100n_c$. Absorption by DCP electrons does not depend on laser polarization above $n_e > 100n_c$. These DCP electrons, with short stopping ranges and relatively high density, $\sim$ 10% of the target density, have a great potential to heat the target in a limited volume. We derived optimal conditions for the absorption via DCP electrons, which suggest, i.e., in targets of density 100 n_c or above the contribution from DCP electrons to the total laser absorption will be maximum for a pulse with $a \sim 10$.

References

[1] P. M. Nilson, W. Theobald, J. F. Myatt *et al.*, Phys. Rev. E **79** 016406 (2009).
[2] M. Tabak *et al.*, Phys. Plasmas **1**, 1626 (1994).
[3] S. C. Wilks, W. L. Kruer, M. Tabak, and A. B. Langdon , Phys. Rev. Lett., 69, 1383 (1992).
[4] A. J. Kemp, Y. Sentoku, and M. Tabak, Phys. Rev. Lett. **101**, 075004 (2008).
[5] Y. Sentoku, and A. J. Kemp, J. Comput. Phys. **227**, 6846 (2008).
[6] W. L. Kruer, and K. Estrabook, *Phys. Fluids*, **28**, 430-432 (1985).
[7] R. Mishra, Y. Sentoku, and A. J. Kemp, Phys. Plasmas **16**, 112704 (2009).
[8] Y. Sentoku, W. Kruer, M. Matsuoka, and A. Pukhov, Fusion Science and Technology **49**, 278 (2006).
[9] P. Gibbon, Short Pulse Laser Interaction with Matter, Imperial College Press, Chapter-5, section(5.2.2) p.139.
[10] J. Denavit, Phys. Rev. Lett., **69**, 3052, (1992).

Journal of Physics: Conference Series **244** (2010) 022049 doi:10.1088/1742-6596/244/2/022049

MEASUREMENT OF HOT ELECTRON TRANSPORT IN OVERDENSE PLASMA VIA SELF INDUCED GIANT MAGNETIC PULSES

S Mondal[1], V Narayanan[1], Amit D Lad[1], Saima Ahmed[1], S Sengupta[2], A Das[2], Z M Sheng[3], P K Kaw[2], G Ravindra Kumar[1]

[1]Tata Institute of Fundamental Research,1-Homi Bhabha Road, Colaba,Mumbai-400005,India
[2]Institute of Plasma Research, Bhat, Gandhinagar - 382428, India
[3]Institute of Physics, CAS and SJT University, P. R. China

E-mail: msudipta@tifr.res.in

Abstract. Spatial and temporal resolved ultrashort(8ps) multimegagauss(65 MG) magnetic field has been measured in plasma produced on Al-coated BK-7 glass by the interaction of a relativististic intensity laser($4\text{x}10^{18} W/cm^2$, 30 fs) using pump-probe polarimetry. The 2D profile of magnetic field is captured using a CCD camera. Mapping of this magnetic field maps the transport of relativistic electrons in the plasma. The magnetic field profiles indicate filamentary behavior (Weibel-like instability). Particle in cell simulation are used to explain the result obtained.

1. Introduction

The largest terrestrially available magnetic fields are generated by interaction of solid targets with ultrashort super intense laser pulses($> 10^{18} W/cm^2$)[2, 3, 4]. Since the origin of the magnetic field is directly related to the electron transport, the study of this magnetic field has attracted considerable attention as it can map electron transport properties inside the plasma. In recent years lasers capable of producing intensity$10^{19} W/cm^2$ in many laboratories and are driving electrons in relativistic regime [6]. There is increasing interest in development of laser driven high energy particle accelerators, short pulse x-ray sources, neutron sources and fast ignition scheme for inertial fusion. Understanding electron transport is important for all these studies. There are many simulations [7, 8] on electron transport and giant magnetic fields produced inside plasma but the experiments are fewer and in particular spatial resolved dynamics of the magnetic field inside the plasma has not been studied.

In this experiment we monitor electron transport in the plasma by measuring self generated magnetic field. Light from an ultra-short super intense laser pulse(I$\sim 5\times10^{18} W/cm^2$, $\tau = 30fs$) is used to create a dense, hot plasma and launch the electron jets and the polarization change of a weak, time delayed probe beam is used to monitor the magnetic field.

2. Experimental setup

Experiments are performed using a table-top terawatt Ti:sapphire CPA laser, which can deliver 30fs, 10 Hz pulses of 600mJ each at 800 nm central wavelength. The schematic of the experimental setup is shown in the fig. 1. The p-polarized pump beam is focused at an angle of 40° with respect to the target normal with a f/4 off-axis parabola on an Al coated BK-7 glass target. The target was placed inside a vacuum chamber of pressure 10^{-5} torr and was mounted on a computer controlled xyz-θ precision stage so that every laser pulse irradiates a fresh spot on the target. The pump spot size on target is 20 μm. The pulse energy on target is 120mJ, giving a peak intensity $\sim 5 \times 10^{18} W/cm^2$.

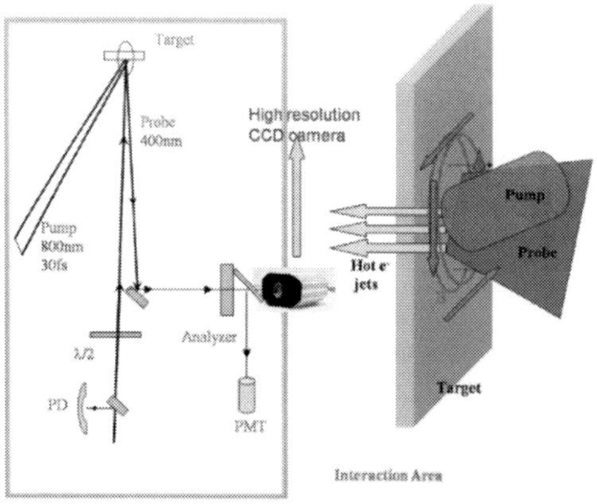

Figure 1. Schematics of experimental setup

The thickness of the Al-coating on the BK-7 glass is kept much larger than the skin depth ($\delta_s \sim c/\omega_p$) at 800nm($\sim$ 100nm).So the incident pump laser interacts only with the Al coating and creates plasma on Al layer and the generated hot electron can ionize the glass and propagate through the resulting hot medium. We extracted 10 percent of the main beam and converted to second harmonic(2ω) and used as probe so that it can penetrate and sample the over dense region (4 times the critical density n_c) for 800nm of the plasma. At critical density most of the laser light is absorbed and hot electrons are expected to be generated. One more advantage of using 2ω probe is noise from pump can be filtered. We kept probe intensity very low $10^{12} W/cm^2$ and weakly focused on the target so that probe does not modify the plasma and can sample larger area. The probe sense magnetic field in the plasma and its polarization state changes, acquiring ellipticity (Cotton-Mouton effect[1]). The probe is delayed by high precision translation stage(Akribis). The spatial overlapping of two pulses is monitored by a CCD camera coupled to high zoom video microscope lens. The temporal matching of the two pulse iss done by looking at the reflectivity of probe as a function of time delay.

The reflected probe is divided into two parts for simultaneous measurements by two detectors. One arm has a photo-multiplier tube to measure space integrated ellipticity and other a CCD camera to measure spatially resolved ellipticity. We measure the reflected probe signal from plasma with parallel and crossed position of the analyzer at different time delays and derive time dependent ellipticity.

3. Result and Discussions

From the ellipticity we can calculate magnetic field. We can write $\beta(t) = (e^2/m_e c^3 \omega n_c) \int n_e(l,t) B^2(l,t) dl$ to derive B-field from ellipticity [2, 4]. From the spatially integrated ellipticity data the magnetic field is deduced assuming a spatially uniform B-field over the entire crosssection and an exponential density gradient $n_e(x) = 4n_c^{400}\exp(-x/L(t))$, where L(t) is the plasma slab length, $4n_c^{400}$ corresponds to the turning point of 400nm probe at normal incident. Where $L(t) = L_0 + v_{exp}t$, where L_0 is 1 μm, t is time delay and v_{exp} is the expansion velocity of the plasma critical surface estimated to $5 \times 10^6 cm/sec$ from Doppler shift measurement of the reflected probe from plasma. For the spatially resolved data (from the CCD), we get the ellipticity for each pixel by doing the above procedure for that pixel and get 2D spatial map of magnetic field at critical surface of 400nm probe.

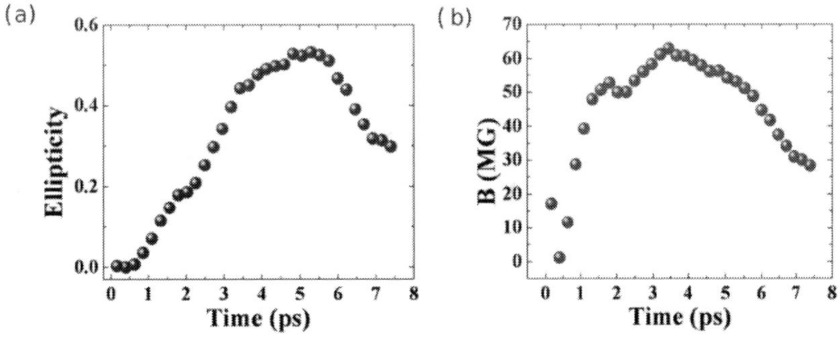

Figure 2. Space integrated time resolved (a) ellipticity and (b) magnetic field

Fig.2(a) shows the space integrated ellipticity of as captured in photomultiplier tube and fig.2(b) shows the magnetic field derived from this ellipticity. Integrated magnetic pulses is 7 ps long in time and maximum magnetic field is around 65 megagauss at the critical surface of the probe (400nm).

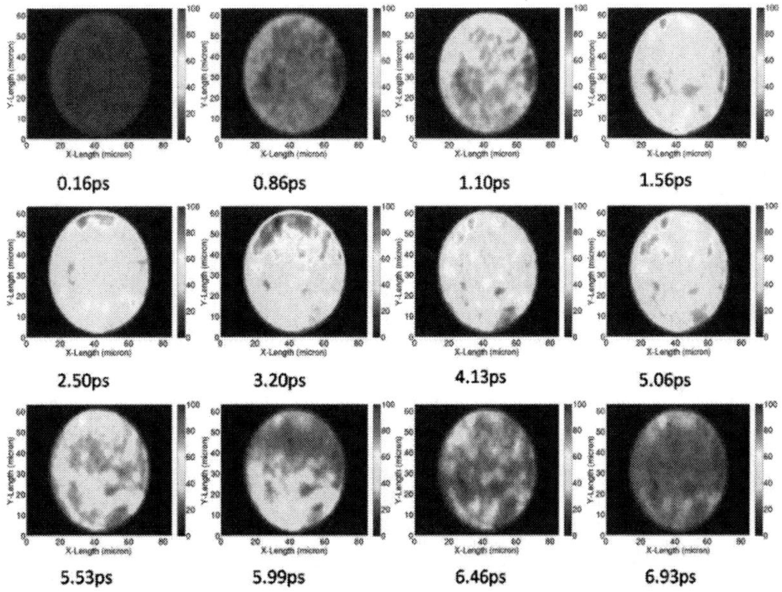

Figure 3. Space and time resolved magnetic field

Fig.3 shows the x-y profile of the B-field as captured by the CCD camera, These are the first ever measurements of the simultaneous space and time resolved magnetic field.

The 2D profile of the B-field is not cylindrically symmetric but it has structure on it which indicates the current distribution which is responsible for this magnetic field has structure, ie the current distribution is fillametary. This filamentary behavior of the current distribution was known in 2D and 3D PIC simulations in much earlier time, but experimental evidence of dynamics of this current fillamentation, which is known as Weibel instability, is observed first time as it happens.

We have run 2D PIC simulation for our experimental condition for long time. The peak magnetic field is about the same magnitude of the experimental value. We are improving the simulation to understand out results better.

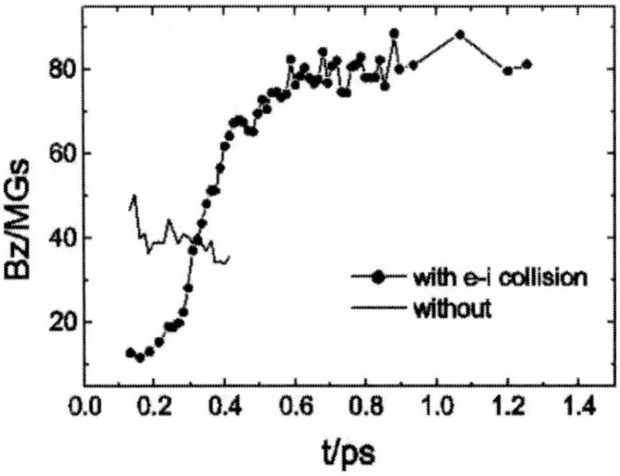

Figure 4. 2D PIC simulation of the magnetic field

References

[1] Tatarakis M Watts I Beg F N Clark E L Dangor A E Gopal A Haines M G Norreys P A Wagner U Wei M S Zepf M Krushelnick K 2002 *Nature* **415** 280

[2] Sandhu A S Dharmadhikari A K Rajeev P P Kumar G R Sengupta S Das A and Kaw P K 2002 *Phys. Rev. Lett.* **89** 225002

[3] Wagner U Tatarakis M Gopal A Beg F N Clark E L Dangor A E Evans R G Hanies M G Mangles S P D Norreys P A Wei M S Zepf M and Krushelnick K 2002*Phys. Rev. E* **70** 026401

[4] Kahaly S Mondal S Sengupta S Das A and Kaw P K 2009 *Phys. Plasma* **16** 1

[5] Segre S E 1998*Plasma Phys. Control. Fusion* **41** R57

[6] Drake R P 2006 *High Energy Density Physics-Fundamentals, Inertial Fusion and Experimental Astrophysics* (Springer-Verlag, Heidelberg)

[7] Sentoku Y Sheng Z M Kaw P Nishihara K Nishikawa K 2002 *Phys. Plasma* **65** 046408

[8] Honda M Meyer-ter-Vehn J and Pukhov A 2000 *Phys. Plasma* **7** 1302

Journal of Physics: Conference Series **244** (2010) 022050 doi:10.1088/1742-6596/244/2/022050

Controlling dynamics of imploded core plasma for fast ignition

H. Nagatomo[1], T. Johzaki[1], A. Sunahara[2], H. Shiraga[1], H. Sakagami[3], H. Cai[1], K. Mima[1]

[1]Institute of Laser Engineering, Osaka University, 2-6 Yamada-oka, Suita, Osaka, 565-0871, Japan
[2]Institute for Laser Technology, 2-6 Yamada-oka, Suita,Osaka, 565-0871, Japan
[3]National Institute for Fusion Science, Oroshi, Toki, 509-5292, Japan

naga@ile.osaka-u.ac.jp

Abstract. In the Fast ignition, formation of highly compressed core plasma is one of critical issue. In this work, the effect hydrodynamic instability in cone-guided shell implosion is studied. Two-dimensional radiation hydrodynamic simulations are carried out where realistic seeds of Rayleigh-Taylor instability are imposed. Preliminary results suggest that the instability reduces implosion performance, such as implosion velocity, areal density, and maximum density. In perturbed target implosion, the break-up time of the tip of the cone is earlier than that of ideal unperturbed target implosion case. This is crucial matter for the Fast ignition because the pass for the heating laser is filled with plasma before the shot of heating laser. A sophisticated implosion design of stable and low in-flight aspect ratio is necessary for cone-guided shell implosion.

1. Introduction

Fast ignition is an attractive scheme in laser fusion [1]. At the first phase of Fast ignition scheme, highly compressed fuel core plasma is formed by nano-second-order implosion laser, and then, just around the maximum density time, it is heated by peta watt laser to achieve a fusion burning condition. Mostly, a shell targets for Fast ignition is fitted with reentrant gold cone target to preserve a path for heating laser which is ultra-intense short pulse. Therefore, understanding of the cone guided implosion dynamics is one of significant problem. We have been studied the problems, especially, on the formation of high density, and breakdown of the tip using two-dimensional radiation hydrodynamics simulation code, PINOCO [2]. In the result, we have proposed an advanced target for the scheme [3]. However, these studies are ideal cases, and realistic conditions must be taken account into simulation. In general, a high temperature hot-spot in imploded core is not required for Fast ignition before heating timing. But this leads to misunderstanding that the hydrodynamic instability of imploding shell target is considered less serious. Therefore, in this preliminary study, effect of Rayleigh-Taylor instability on formation of high density core and timing of tip breakdown is investigated numerically. A target and an implosion scales are relevant to FIREX-I (Fast Ignition REalization Experiment, Phase-I) at ILE Osaka University.

2. Simulation of coned-guided implosion

Two-dimensional radiation hydrodynamic simulation is performed. A configuration of a typical cone-guided shell target which is used in Gekko-XII experiment at ILE Osaka University is shown in figure 1. The gold cone with an opening angle of 30 degree is attached to a spherical shell of CD (8 μm thick). The tip position from center of the shell and thickness of the tip of the cone are 50 μm and 8 μm respectively. In order to investigate the effect of hydrodynamic instability, target surface is perturbed initially. The amplitude of initial target surface perturbation is shown in figure 2. Red line ("measured" line) indicates a typical target roughness which was measured at ILE. The measured imprint is modeled [4] and imposed on the target surface roughness, of which amplitude is indicated as blue line ("imprint model" line). And the sum of these amplitudes is indicated as black line ("total" line). From the linear theory $l \approx 200$ must be taken account. However, due to the limitation of computational resources, the mode numbers $l \leq 80$ are considered in this study. The shell target is irradiated by uniform laser of which wavelength, energy and pulse duration are λ= 0.53 μm, 3.0 kJ, and 1.5 ns respectively.

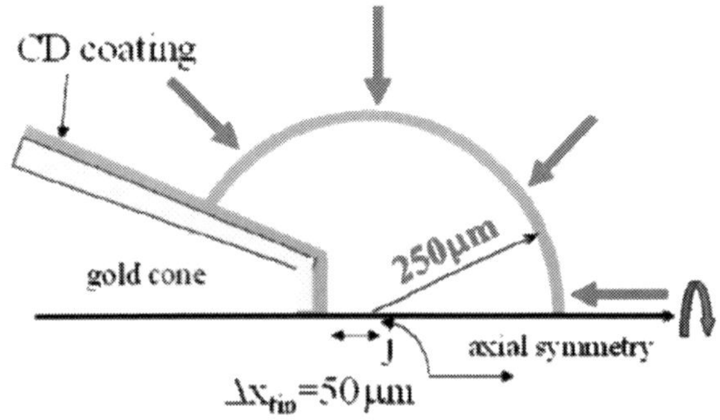

Figure 1. Configuration of target, which is typical size and materials for GXII experiment. Thickness of the tip is 8 μm, and opening angle of the cone is 30 degree.

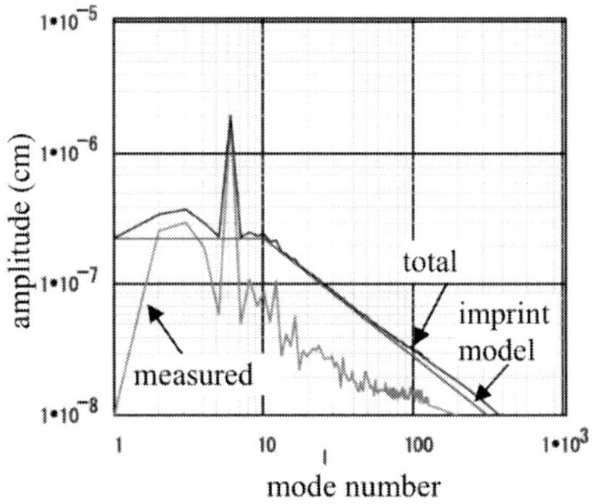

Figure 2. Amplitude of initial target surface perturbation. Red line ("measured" line) indicates a typical target roughness. The measured imprint is modeled and imposed on the target surface roughness, of which amplitude is indicated as blue line ("imprint model" line). The sum of these amplitudes is indicated as black line ("total" line).

For comparison, two simulations with and without initial perturbations are executed. Figure 3 shows mass density and electron temperature contours at t=1.90 [ns] and at the maximum density time (t=2.07 [ns] and t=2.08 [ns]). In the perturbed target case, at t=1.90 [ns] during the target acceleration phase, Rayleigh-Taylor instabilities grow rapidly and reach non-linear phase. Some plasma expanded from broken shell has filled inside the shell already, where mass density is nearly solid density (figure 3(a)). This plasma inside the shell is compressed on the axis, and forms the massive jet flow which strikes toward the tip of the cone. Similar jet flow was observed in GXII experiment [5]. At the maximum compression time, the tip is already blown away (figure 3(b)). The distance between the critical density in the cone and the center of the imploded core is more than 150 μm which is too far to achieve the high heating efficiency. On the contrary, the tip of the cone survives, and the inside of the cone is kept clean in unperturbed target case (figure 3(d)). The maximum implosion velocities of perturbed target and unperturbed target are 2.7×10^7 [cm/s], and 2.9×10^7 [cm/s] respectively. In perturbed target case, the deceleration rate is higher than that of unperturbed target case, due to the plasma inside the shell. The radius of the compressed core is nearly 40 μm, and the density of the core is less than 40 [g/cm^3], which is insufficient to absorb hot electrons produced by heating laser.

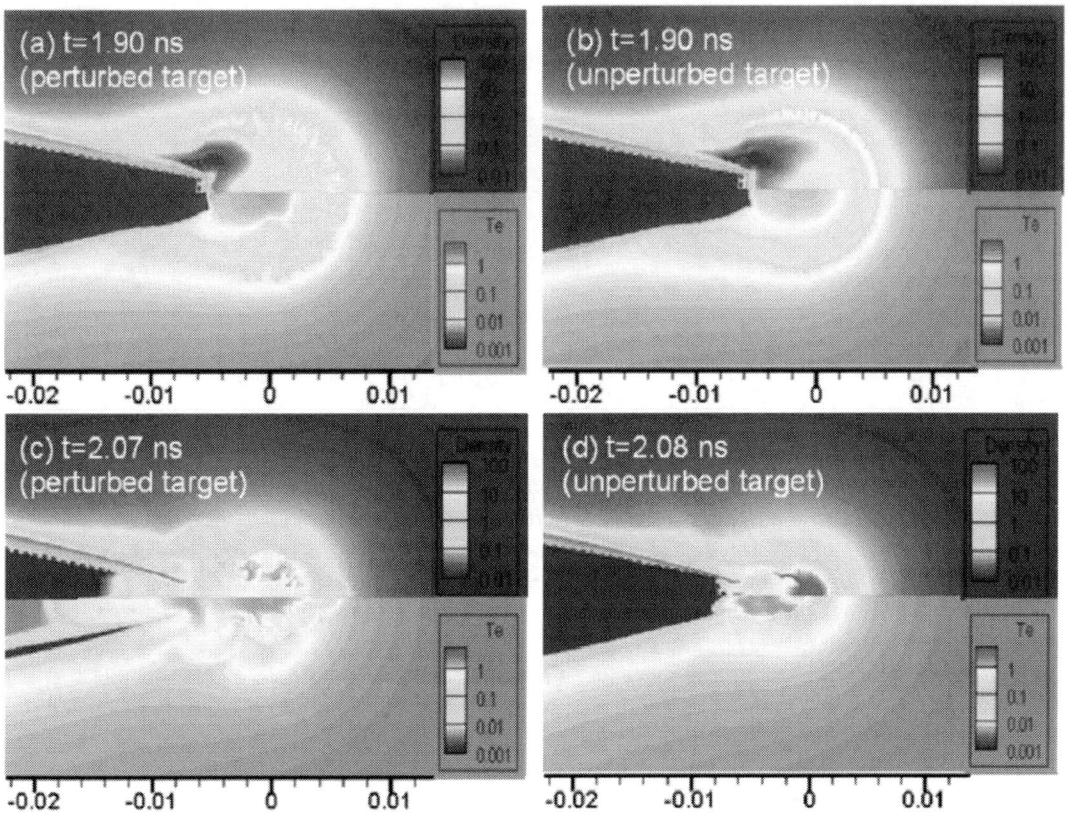

Figure 3. In each figure, (top) mass density contour [g/cm^3] and (bottom) electron temperature contours [keV] are shown. (a) perturbed target at t=1.90 [ns], (b) unperturbed target at t=1.90 [ns], (c) perturbed target at t=2.07 [ns], (d) unperturbed target at t=2.08 [ns] are shown respectively.

Figure 4 shows the time history of angler averaged areal density. In perturbed target case, the maximum areal density is reached only 0.15 [g/cm^2], which is about one fourth of the unperturbed target case. Stagnation cannot be observed clearly on the axis, and the sharp peak of density (or areal density) is not appeared, momentum energy remain

From the linear perturbation theory, initial perturbation of mode number $l < 200$ is necessary. in this kind of simulations. However, due to the limitation of computational resources, it is imposed to be $l < 80$. That is to say, severer condition is predicted in experiment. Robust target against hydrodynamic instability must be designed. A slow implosion method [6] is one of the attractive methods, where thick and massive shell is imploded with a low implosion velocity to achieve high areal density.

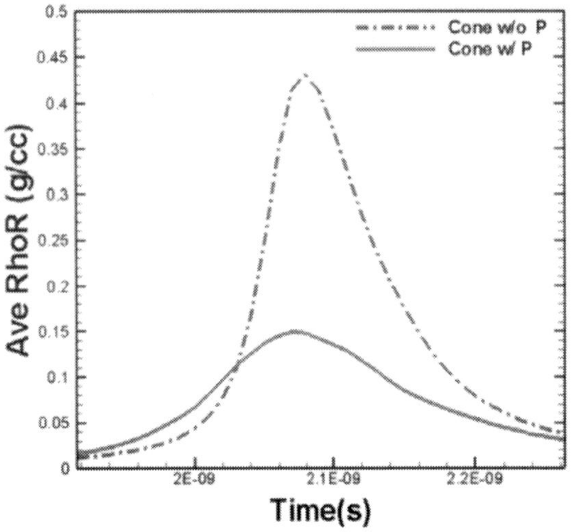

Figure 4. Time history of averaged areal density (ρR [g/cm^2]). An unperturbed initial target case (Dot-dash line), and perturbed initial target case (solid line).

3. Conclusion and Summary

A preliminary study of the effect of hydrodynamic instability is presented. The simulation results suggest that in cone-guided implosion with perturbed target shell, the tip of the cone is destroyed earlier than the ideal unperturbed case. The maximum areal density is one fourth of unperturbed case, and the maximum compression time comes earlier than that of unperturbed case.

The implosion design of Fast ignition is different from the conventional central hot-spot type implosion. For the future, specialized target design for Fast ignition is necessary. A slow implosion method which can achieve high density and high areal density is effective way. In this method, in-flight target thickness is thick enough to prevent the crucial target break-down caused by the hydrodynamic instability. Controlling the formation of the jet flow from core plasma toward the tip of the cone must be considered carefully. They will be considered in our next advanced target design.

The material and thickness of the tip are very significant factors not only from the viewpoint of implosion, but also from the viewpoints of laser plasma interaction and hot electron transport [7]. However, they are not discussed in this paper. Fully integrated simulation where implosion, LPI, and hot electron transport code are linked will be carried out in near future for our next advanced target design.

Acknowledgments
These simulations were executed at Cyber Media Center, Osaka University, and ILE Osaka University. The authors would like to appreciate the technical staffs of supercomputer room at CMC and ILE Osaka University.

References

[1] Tabak, M., et. al., *Phys. Plasmas* **1**, 1626-1634, (1994).
[2] Nagatomo, H., *et al.*, *Phys. Plasmas* **14** 056303 (2007).
[3] Nagatomo, H., *et al.*, *Nucl. Fusion* **49** 075028, (2009).
[4] Azechi, H. *et al.*, *Phys. Plasmas* **4**, 4079 (1997).
[5] Shiraga, H., *et al.*, in Fusion Energy 2008 (Proc. 22nd Int. Conf. Geneva, 2008) .
[6] Betti, R. and Zhou, C., *Phys. Plasmas* **12** 110702 (2005).
[7] Johzaki, T., *et al.*, *Plasma Phys. Control. Fusion,* **51** 014002 (2009).

Journal of Physics: Conference Series **244** (2010) 022051 doi:10.1088/1742-6596/244/2/022051

Toward Laser Driven Proton Medical Accelerator

M. Nishiuchi[1], H. Sakaki[1], T. Hori[1], K. Ogura[1], A. Yogo1, A. S. Pirozhkov[1], A. Sagisaka[1], S. Orimo[1], M. Mori[1], H. Sugiyama[1], H. Kiriyama[1], M. Tampo[1], I. Daito[1], H. Okada[1], S. Kanazawa[1], M. Tanoue[1], T. Shimomura[1], Y. Nakai[1], H. Sasao[1], D. Wakai[1], F. Sasao[1], P. R. Bolton[1], H. Daido[1], S. Kawanishi[1], Y. Iseki[2], T. Nagafuchi[2], K. Maeda[2], K. Hanawa[2], T. Yoshiyuki[2], H. Souda[3], and A. Noda[3]

[1]Japan Atomic Energy Agency, [2]Toshiba Corporation, [3]Kyoto University

nishiuchi.mamiko@jaea.go.jp

Abstract. Towards our final goal, such as to establish the laser-driven proton accelerator for the medical application, one of the most important things to establish is to develop the proton transport system. In this continuous work, we demonstrate the focusing system of the laser-driven proton beam with permanent magnet qadrupoles (PMQs), with which a 2.4 MeV laser-driven proton beam, having a divergence angle of ~10 degrees at birth, is focused to a spot whose size is 3x8mm^2 at 640mm downstream from the target.

1. Introduction

Laser-driven proton beam is generated from intense laser plasma interactions with solid targets[1]. In this setting, protons are accelerated from the rear surface of the solid target by a quasi-static electric field with micrometer scale lengths and gradients of order MeV/μm. In contrast to conventional accelerators the combination of very high peak currents >10^{13} protons within bunch duration of a few ps with extremely low transverse emittance [2] (attributed to small source size and low transverse temperature) makes laser-driven proton beams unique with the promise for compact accelerator development.

This recent progress in laser-driven proton acceleration makes us aware of the realization of the compact laser-driven proton accelerator for the medical application. Photo Medical Research Center (PMRC) is a JAEA program fostered by KPSI and funded by both PMRC corporate partners and by

JAEA through *"The Special Coordination Fund (SCF) for Promoting Science and Technology commissioned by the Ministry of Education, Culture, Sports, Science and Technology (MEXT) of Japan"*. Under this program, our goal is to realize the compact laser-driven proton accelerator for the medical application. To establish such an accelerator, we have to develop, 1) compact, stable, and high-repetitive laser system, 2) stable, high repetitive target system which can supply us laser-driven proton beam with sufficient amount of number and kinetic energy, and 3) transport and irradiation system of the proton beam. Here we report on our progress on transport system of laser-driven proton beam.

Adding to the novel characteristics already mentioned above, laser-driven proton bunches are typically characterized by a large divergence (half angles >10°) and energy spread (near 100%) with an exception of a few exotic target designs [3]. In the development of the beam transport system, we should establish the beam focusing system as well as proton mono-energetized system. We report here the focusing and spectral enhancement of the laser-driven multi-MeV proton beam using permanent quadrupole magnets[4].

2. Experimental Results

The 100 TW class Ti:sapphire laser system called J-KAREN laser[5], which belongs to the Kansai Photon Science Institute of the Japan Atomic Energy Agency, is used for the experiment. In these experiments an incident laser pulse whose energy, central wavelength and diameter are 0.7J, 800 nm and 50 mm, respectively. The pulse width is 30 fs (full width at half maximum). A 12.5-μm-thick polyimide tape target ($–[C_{22}H_{10}O_4N_2]_n–$) is irradiated with the P-polarized beam from the incidence angle of 45 degrees. Motion of the tape provided a fresh target surface for 1Hz repetition rate. Peak irradiances of up to 10^{20}Wcm^{-2} are achieved by focusing with an off-axis parabola (152.4 mm focal length) to spot sizes of 6x3 μm^2(FWHM).

From this setup proton beam with maximum energy of ~3MeV is stably produced by 1Hz repetition rate. A typical proton energy spectrum monitored by the absolutely calibrated TOF spectrometer [6] is shown in Figure. The beam pattern, 80mm downstream from the target is detected with the CR-39 nuclear track detector covered with 40 μm Al-energy range filter is also shown in Figure. From the beam divergence measurement, the half divergence angle is determined to be (10 ± 1) degrees. The number of the protons within the energy interval of (2.4 ± 0.1)MeV is (1.3 ± 0.3)x10^7 protons/shot.

Figure also shows the experimental setup. In our focusing system, two permanent magnet quadrupole (PMQs) are used as is shown in Figure. The magnetic field strengths are 55 and 60 T/m for the first and second quadrupoles, respectively. The first one is placed at 92 ± 1 mm from the target while the second is placed at 195 ± 1 mm from the downstream end of the first one.

The proton beam pattern passing through the focusing system is measured with the CR-39 nuclear track detector for a single shot exposure at some distances away from the target as is shown in Fig.3.

The energy range of the protons detected on the CR-39 is 2.2 MeV<E<3.1MeV. The focal spot size is measured to be 3 mm (horizontally) and 8 mm (vertically). The measured value is also much larger than that expected from its small transverse emittance, which mainly comes from the chromatic aberrations in the quadrupole pair. The asymmetry of the spot shape can explain from the uncertainty of the setting position of the CR-39 in the experiment ($\pm$1mm).

The Monte Carlo simulation is conducted using 2×10^6 protons emitted simultaneously and uniformly spread across a divergent half angle of 10 degrees. The full spectrum is chosen to match the observed one. The PMQ focusing field gradients are applied. The simulation is conducted under the absence of space charge effects, therefore, the simulation result is not influenced by the number of particle used in the simulation. The simulated beam patterns in the proton energy range of 2.2 MeV<E<3.1MeV at successive positions are also shown in Figure. The simulated results are in reasonable agreement with the experimental results.

The TOF spectra measured at 1930 mm from the target with and without PMQs and the simulated results with PMQs are shown in Figure. Simulated spectrum is uniformly scaled by a factor determined as the ratio between the total number of the protons measured in the experiment and the number of particles used in the simulation. The focusing action of the PMQ pair significantly alters the transmitted spectrum such that it appears quasi-mono-energetic spectrum. The simulated spectrum at the TOF position, which includes the response of the TOF detector is in good agreement the measured one. The proton bunch width is approximate 5 ns at the TOF detector.

We, therefore, know from the simulation the proton beam spectra at the focus position, as is shown in Figure. The spectrum is accumulated from the circular region around the focal spot with diameter of 3 mm. According to the simulation the resultant quasi-monochromatic proton beam in the energy interval of 2.4$\pm$0.1 MeV at the focal plane includes ~5×10^4 protons (i.e., which corresponds to ~3% of the original spectrum), representing a pulse duration of ~1 ns. The simulated transmittance of 2.4$\pm$0.1 MeV protons through the PMQ pair is about 30%.

3. Summary

We have developed the focusing system of the laser-driven proton beam using PMQs, with which we demonstrated repetition-rated and stable focusing of 2.4 MeV laser-driven proton beam. These continued developments and optimizations of proton beam optics in the laser-driven case is sure the first step to realize the novel Laser-driven proton accelerator for the medical treatment.

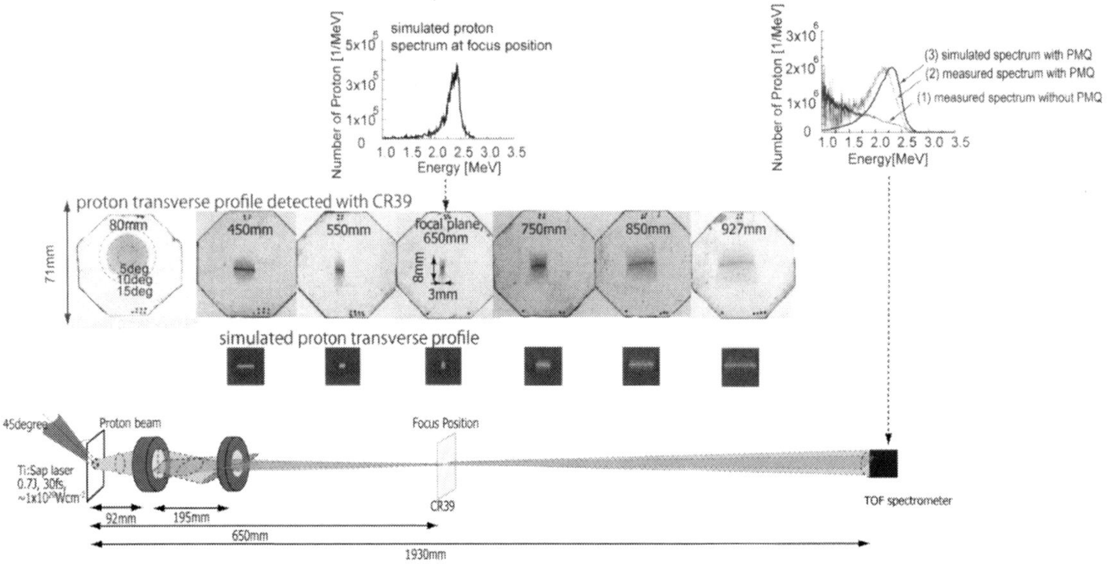

Figure Setup of the experiment as well as the proton transverse profiles measured with CR-39 nuclear track detectors at successive positions and the simulated ones. The Time of Flight spectrum of the proton beam measured at TOF positions in the cases of with and without PMQs, and simulated one at TOF position with PMQs are shown at upper-right corner. The simulated proton spectrum at focal plane is also shown.

References

[1] E. L. Clark et al. Phys. Rev. Lett. **84**, 670 (2000); Hatchett et al. Phys. Plasmas. **5**, 2076 (2000); Maksimchuk et al. Phys. Rev. Lett. **84**, 4108 (2000); Snavely et al. Phys. Rev. Lett. **85**, 2945 (2000)

[2] T.E. Cowan et al. Phys. Rev. Lett. **92**, 204801 (2004)

[3] Hegelich et al. Nature **439**, 441 (2006); Schwoerer et al. Nature **439**, 445 (2006); Yin et al. Laser Part. Beams **24**, 291 (2006); Yin et al. Phys. Plasmas **14**, 056706 (2007).

[4] M. Schollmeier et al. Phys. Rev. Lett. **101**,055004 (2008); S. Ter-Avetisyan et al., Laser Part. Beams **26**, 637 (2008) ; M. Nishiuchi et al. Appl. Phys. Lett. 94, 061107 (2009)

[5] H. Kiriyama et al. Opt.Lett. **33**, 645 (2008)

[6] S. Nakamra et al. Jpn. J. Appl. Phys., Part 2 **45**, L913 (2006)

Warp simulations for capture and control of laser-accelerated proton beams

Frank Nürnberg1, 2, A. Friedman2, 3, D.P. Grote2, 3, K. Harres1, B.G. Logan2, M. Schollmeier4 and M.Roth1

[1] Institut für Kernphysik, Technische Universität Darmstadt, 64289 Darmstadt, Germany
[2] Lawrence Berkeley National Laboratory, Berkeley CA 94720, USA
[3] Lawrence Livermore National Laboratory, Livermore CA 94550, USA
[4] Sandia National Laboratories, Albuquerque NM 87195, USA

E-mail: f.nuernberg@gsi.de

Abstract. The capture of laser-accelerated proton beams accompanied by co-moving electrons via a solenoid field has been studied with particle-in-cell simulations. The main advantages of the Warp simulation suite that we have used, relative to envelope or tracking codes, are the possibility of including all source parameters energy resolved, adding electrons as second species and considering the non-negligible space-charge forces and electrostatic self-fields. It was observed that the influence of the electrons is of vital importance. The magnetic effect on the electrons outbalances the space-charge force. Hence, the electrons are forced onto the beam axis and attract protons. Beside the energy dependent proton density increase on axis, the change in the particle spectrum is also important for future applications. Protons are accelerated/decelerated slightly, electrons highly. 2/3 of all electrons get lost directly at the source and 27% of all protons hit the inner wall of the solenoid.

1. Introduction

Ion acceleration from high-intensity, short-pulse laser irradiated thin foils [1, 2] has attracted much attention during the past decade. The emitted ion and, in particular, proton pulses contain large particle numbers (exceeding 10^{12} particles) with energies in the multi-MeV range and are tightly confined in time (<ps) and space (source radius a few micrometers). The generation of these high-current beams is a promising new area of research and has motivated pursuit of applications such as tabletop proton sources or preaccelerators.

Requirements for an injector are controllability, reproducibility and a narrow (quasi-monoenergetic) energy. However, the source provides a divergent beam with an exponential energy spectrum that exhibits a sharp cutoff at its maximum energy [3]. The laser and plasma physics group of the TU Darmstadt, in collaboration with GSI Helmholtzzentrum für Schwerionenforschung and LBNL/LLNL, is studying possibilities for transport and RF capture in conventional accelerator structures. The first experiments were carried out at the PHELIX laser system [4].

The results indicate the need for theoretical characterization. The effect of external magnetic fields on the expanding proton beam, current neutralized by co-moving electrons, is not investigated up to now. Emerging space charge issues due to charge separation and the self-fields of the proton beam have a significant influence on the proton beam transport, because the intense beam behaves as a non-neutral plasma. Therefore, any analysis of beam dynamics needs to include the electrostatic self-fields of the beam. Such an analysis is an ideal application for the particle-in-cell (PIC) simulation technique which has been heavily used for neutral plasma simulations. It became evident that the Warp simulation code, developed by the LLNL/LBNL Heavy Ion Fusion group, is the perfect tool to analyze this beam behavior. WARP [5] is used to optimize the experimental setup as well as the ion optics to transport the proton beam. These controllable beams will simplify isochoric heating of samples. They will also enable fundamental studies in FI research, using a proton ignitor beam that comes to be separated spatially from the plasma that created by the initial laser-matter interaction.

2. Simulation setup

Warp suite of simulation codes was developed to study high current ion beams for application in heavy-ion driven inertial confinement fusion. Therefor, high current beams are necessary. Warp implies the dominant space-charge forces and the electrostatic self-fields of the beam. The code combines the PIC technique (uses the Lorentz equation of motion to advance in time simulation particles) with a description of the accelerator "lattice" of elements. The effects of the space-charge is included by a global solution of Poisson's equation, giving the electrostatic potential, at each time step. The macro-particles are advanced in time using a combination of the "leap frog" and "isochronous leap frog" methods.

2.1. Proton beam parameters

Warp compared to other simulation codes can import user-specific particle source parameters. Hence, proton beam parameters, e.g. energy distribution, envelope opening angle, source size and transverse emittance, could be included energy resolved. The reference beam was recorded at the Phelix laser system, delivering 130 J in 700 fs at intensities about 4.5×10^{19} W/cm^2. The protons follow an exponential energy distribution from 2.3 MeV to a cut-off energy at 28.7 MeV and have an integrated particle number of 8×10^{12}. The opening angle decreases from 54° full angle to 10° for increasing proton energy. The proton source size diameter follows a parabolic behavior with its maximum at 480 μm. And the transverse RMS emittance decreases from 1.6 to 1.0 mm-mrad for increasing proton energy.

2.2. Particle source and external field

The protons are injected before the first time step with the "inverse transform sampling" method. Limited by the calculating capacity and the length of a simulation, the simulation particle number is set to 10^6. Additionally, co-moving electrons ($v_e = v_p \rightarrow 1.2$ keV $< E_e <$ 15.7 keV) are injected as separate species but with the same beam parameters. For the first simulations the electrons with energies > 15.7 keV will be neglected. By setting up a volume source, the plasma frequency can be resolved with an acceptable simulation time step of 0.1 ps. At later time during expansion, the density is low enough to go to 1 ps time steps. As external field a 8.6 T solenoid is included. Its field is obtained from the analytic field profile of a cylindrical current sheet. The field off axis is given by multipole expansion. The solenoid dimensions are 72 mm in length with a front and a back 6 mm insulator flange. The inner radius of the coil is 22 mm and the distance to the source is 17 mm.

2.3. Simulation grid

The simulation grid for the field solver is defined by resolving the Debye length. A rough initial estimate yields to 0.5 mm per grid cell in radial direction and 1 mm per grid cell in beam direction (cylinder symmetry). The boundary condition for the particles is set to absorption and the potential vanishes.

3. Simulation results

3.1. Beam neutrality

Before studying the influence of the solenoid on the proton beam, Warp has to guarantee the beam neutrality. For this case, a simulation without solenoid field was done. During the first 400 ps the co-moving electrons neutralize the proton beam. Shortly after, a potential in the order of some mV appears due to numerical heating. At 800 ps the potential builds up to 1 V and when the beam hits the solenoid inner wall it adds up to 100 V. Finally, the neutrality at the beginning is enough for the simulation setup, because the existing magnetic field affects the beam neutrality already during the first time steps.

3.2. Proton and electron beam expansion

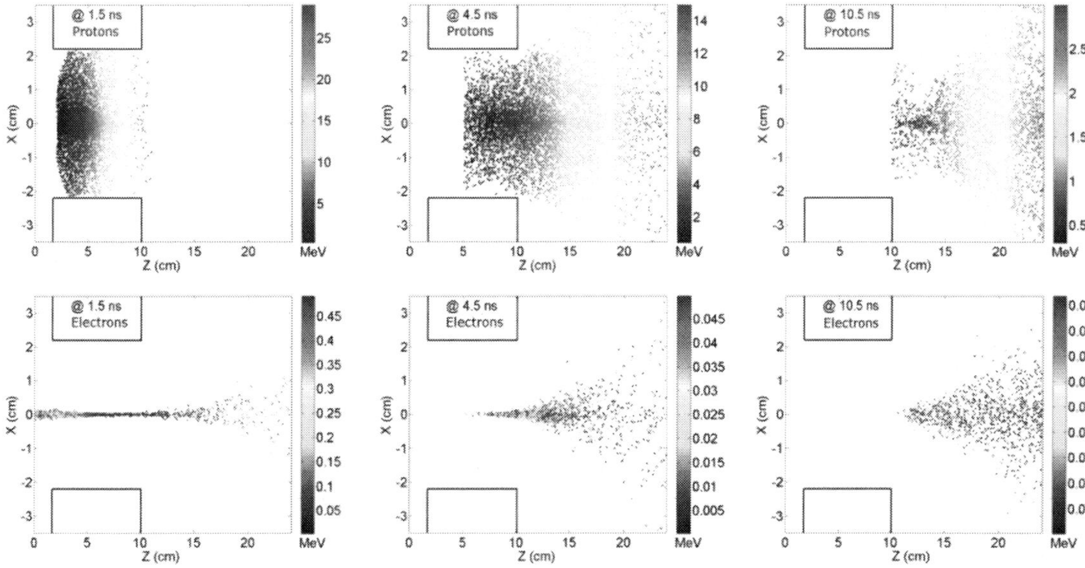

Figure 1. Snapshots of the proton (top) and electron (bottom) expansion for 3 different time steps: 1.5, 4.5 and 10.5 ns. Changing color scale for the particle energy because of better illustration, $E_{p,max} = 29.5$ MeV and $E_{e,max} = 1.25$ MeV.

Due to the higher charge to mass ratio of the electrons compared to the protons, the electrons rotate on a much smaller radius in the magnetic field. The Larmor radius is given by: $r_L = \sqrt{2E_{kin}m_e}/eB$ and is < 0.5 mm at the source position (B = 1.4 T) for the energy range of the co-moving electrons. Hence, most of the electrons get immediately bent down to the beam axis, the electron density increases and the expansion is not neutral anymore. Electrons are decelerated and accelerated (up to 1.25 MeV), not only forwards also backwards, because of the negative potential growth on axis. The magnetic effect outbalances the space-charge force. So, the electrons are not pulled back by the positive proton potential, the protons are attracted to the beam axis. This proton density increase can be observed for a wide range of energies, see figure 1. When the beam leaves the solenoid and the field attenuates (0.3 T 5 cm behind the solenoid), the electrons start to expand and follow the protons. Additionally, a proton beam waist and an over-focusing can be seen (simulation time of 10.5 ns in figure 1).

3.3. Proton and electron energy distribution

Figure 2 shows the proton and electron energy distribution for three different times: the initial distribution, the lost particles and the spectrum behind the solenoid. The acceleration/deceleration for the electrons is explicitly pointed out by the spectrum description. But the maximum electron energy given above of 1.25 MeV does not appear in the spectrum, because this kinetic energy was generated and reduced during the expansion. A very important observation is the loss of 66% of all electrons already during the first time steps. They are decelerated, pushed backwards and absorbed at z=0, the boundary of the simulation box. Thus it appears that the guaranteed beam neutrality is enough for the first picoseconds and finally the expansion is dominated by a positive potential. But still 2/3 of all electrons are sufficient to generate a high enough negative potential on axis to attract the protons.

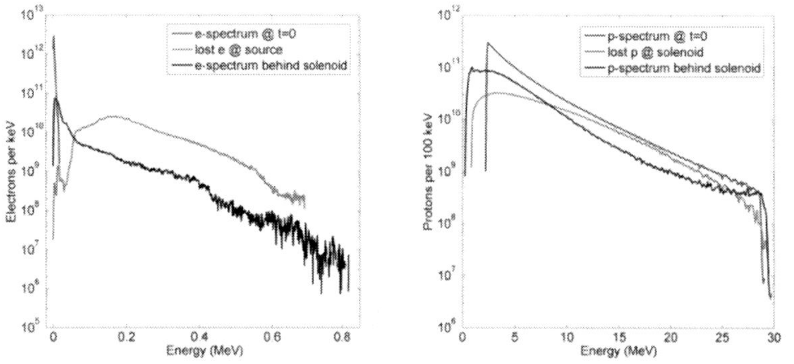

Figure 2. Electron (left) and proton (right) spectra: initial distribution, lost particles and spectrum behind solenoid.

As well as the electrons, the protons get slightly accelerated and decelerated. But there are no proton return currents. All lost protons are absorbed at the inner wall of the solenoid illustrated by the dip in the proton transmission spectrum. An overall proton transmission of 63% could be observed. But it is not possible to calculate an energy dependent transmission because of the acceleration/deceleration, the protons per energy interval jump.

3.4. Proton beam collimation

A first review shows only a proton density increase on axis, not really a collimation. The proton beam waist 1) could be an evidence for beam collimation at an energy of around 0.8 MeV (envelope simulations without space charge effects predict 2.5 MeV). By checking the RMS or edge radius of each proton energy, no constant radius can be observed. Additionally, the radial velocity component only decreases slightly. But for collimation it has to be close to 0. That is also the reason, why the beam waist increases for later times. The recent simulations data are still subject to further analysis and results will be published elsewhere.

[1] Snavely R *et al* 2000 *Phys. Rev. Lett.* **85** 2945
[2] Wilks S *et al* 2001 *Phys. Plasmas* **8** 542
[3] Nürnberg F *et al* 2009 *Rev. Sci. Instrum.* **80** 033301
[4] Harres K *et al* 2009, Beam collimation and transport of laser-accelerated protons by a solenoid field, *Proc. Int. Conf. on Inertial Fusion Sciences and Applications (San Francisco)*
[5] Grote D P *et al.* 2005 *AIP Conf. Proc.* vol 749 issue 1 pp 55-58

Journal of Physics: Conference Series **244** (2010) 022053 doi:10.1088/1742-6596/244/2/022053

Carbon ion beam focusing using laser irradiated, heated diamond hemispherical shells

D.T. Offermann[1], K.A. Flippo[1], S.A. Gaillard[1,2], D.C. Gautier[1], S. Letzring[1], J.C. Cobble[1], G. Wurden[1], R.P. Johnson[1], T. Shimada[1], D.S. Montgomery[1], R.P. Gonzales[1], T. Hurry[1], F. Archuleta[1], M.J. Schmitt[1], S-M. Reid[1], T. Bartal[3,4], M.S. Wei[3], D.P. Higginson[3], F.N. Beg[3], M. Geissel[5] and M. Schollmeier[5]

[1] Los Alamos National Laboratory, P.O. Box 1663, Los Alamos, New Mexico, 87545.
[2] Forschungszentrum Dresden-Rossendorf, Institut fuer Strahlenphysik, PF 51 01 19, 01314 Dresden, Germany.
[3] Mechanical and Aerospace Engineering Department, Univ. of California San Diego, La Jolla, California.
[4] Lawrence Livermore National Laboratory, Livermore, California.
[5] Sandia National Laboratory, Albuquerque, New Mexico.

E-mail: `offerman@lanl.gov`

Abstract. Experiments preformed at the Los Alamos National Laboratory's Trident Laser Facility were conducted to observe the acceleration and focusing of carbon ions via the Target Normal Sheath Acceleration (TNSA) mechanism using hemispherical diamond targets. Trident is a $200\,TW$ class laser system with $80\,J$ of $1\,\mu m$, short-pulse light delivered in $0.5\,ps$, with a peak intensity of $2 \times 10^{20}\,W/cm^2$. Targets where Chemical Vapor Deposition (CVD) diamonds formed into hemispheres with a radius of curvature of $400\,\mu m$ and a thickness of $5\,\mu m$. The accelerated ions from a hemisphere were diagnosed by imaging the shadow of a witness copper mesh, located $2.4\,mm$ behind the target, onto a film pack located $5\,cm$ behind the target. Ray tracing was used to determine the location of the ion focal spot. The TNSA mechanism favorably accelerates hydrogen found on the targets. To make the carbon beam detectable, targets were first heated to several hundred degrees Celsius using a CW, $532\,nm$, $8\,W$ laser. Imaging of the carbon beam was accomplished via an auto-radiograph of a nuclear activated lithium fluoride window in the first layer of the film pack. The focus of the carbon ion beam was determined to be located $630 \pm 110\,\mu m$ from the vertex of the hemisphere. LA-UR 09-06918

1. Introduction

The goal of this work was to observe the focused carbon ion beam produced by the TNSA [1, 2] mechanism. The focusing of a high energy carbon ion beam into a micron scale spot size has applications ranging from Inertial Confinement Fusion (ICF) Fast Ignition (FI) to cancer radiotherapy [3]. Carbon ions have a narrower Bragg peak than protons, and thus deposit more energy in a smaller volume. It has also been proposed that a carbon beam could be used to perturb the surface of an imploding ICF capsule. Ion beam acceleration from micron thick, solid targets irradiated by ultra-intense, ultra-short pulse lasers has been studied at great length [4, and references therein]. Experiments have demonstrated that curved foil targets can be used to focus the beam [5]. Most of the work that has been done has concentrated on diagnosing the proton beams. This is primarily because of the difficulty in detecting other ion species with detectors like RCF packs. Spectrometers, which can identify different charge to mass species such as Thomson parabola spectrometers, have provided information of the ion beam in small solid angles. The full beam profile has been typically measured using stacks of radiochromic film [6, 7] or nuclear activation [8]. The problem with most of these techniques for detecting ions heavier than hydrogen is that they cannot discriminate between species. Placing filters in front of the detector to block the hydrogen is not possible because protons have a greater range in the filter material than heavier ions.

In this work, Chemical Vapor Deposited (CVD) synthetic diamond targets were irradiated with laser light from the Trident short-pulse beam at Los Alamos National Laboratory. Targets were pre-heated using a CW laser in order to remove water and hydrocarbon contaminants. The abundance of each ion species emitted is highly dependent upon the target surface composition [9, 10, 11]. The CVD targets were hemispheres with a radius of $400\,\mu m$ and a shell thickness of $5\,\mu m$. The spatial profile of the resulting carbon ion beam was measured using a novel nuclear activation technique, whereby a $1\,mm$ thick lithium fluoride plate in the ion beam path reacted with the carbon beam to produce radioactive nuclides. The number of nuclides per unit area is proportional to the incident ion flux, thus the beam profile could be measured via an auto-radiograph of the activated LiF plate. Nitrogen-13 produced in the $C^{12}(Li^7, He^6)N^{13}$ reaction provided a means of further distinguishing carbon ions from any remaining protons. Using a witness copper mesh and ray tracing, the focus of the carbon beam was determined to be at $630 \pm 110\,\mu m$ from the vertex of the hemisphere.

2. Experimental set-up

The experiment was conducted at the Trident laser facility at Los Alamos National Laboratory using the short-pulse beam. The laser is a $200\,TW$ class system which delivers $80\,J$ on target with a FWHM pulse length of $0.5\,ps$ with a peak intensity of $2 \times 10^{20}\,W/cm^2$. The wavelength is $1053\,nm$. The target was positioned so that the focal plane of the $f/3$ off-axis parabolic mirror was centered on the outer surface of the CVD diamond shell at normal incidence. A copper witness mesh with a grid spacing of $127\,\mu m$ was located $2.4\,mm$ downstream of the laser focal plane. A second laser beam was used to heat the target to several hundred degrees prior to the Trident beam firing in order to remove the contaminants. The laser was a Coherent Verdi V18 CW laser with an output of up to $18\,W$ at $532\,nm$. The beam was focused to a spot size of about $1\,mm$ and was typically run between $4\,W$ and $8\,W$ for several minutes prior to firing the Trident short-pulse. The heating laser was stopped after the shot was completed.

The principal diagnostic was a film pack which consisted of stacked layers of Gafchromic HD-810 radiochromic film, a $1\,mm$ thick LiF window and filters made of Kapton polymide film or aluminum foil. The film pack was placed $5\,cm$ behind the target and had diameter of $5\,cm$. The layers from front to back (front facing the target) were $7\,\mu m$ of Kapton, $1\,mm$ of LiF, $90\,\mu m$ of Kapton, $13\,\mu m$ of aluminum, 1 piece of HD-810 film, $1\,mm$ of aluminum and 1 piece of HD-810. The HD-810 film is multi-layered with a $0.75\,\mu m$ gelatin layer in the front, a $7.5\,\mu m$ active layer and a $95\,\mu m$ polyester base. With this configuration, protons had to have a minimum incident energy of $14\,MeV$ to reach the first HD-810 film layer and needed $0.5\,MeV$ to reach the LiF. Carbon ions required $300\,MeV$ and $10\,MeV$ for the same. A Thomson parabola spectrometer was located behind the target and could be used when the LiF/RCF pack was not used. The spectrometer was used to measure the relative abundance of protons in the beam and the effectiveness of heating.

After a shot was complete, the LiF/RCF pack was disassembled so that the LiF layer could be scanned. The LiF was placed in direct contact with a Fujifilm BAS-SR imaging plate in order to auto-radiograph the activation. This method has been used to measure proton beam profiles using copper as the activated plate [8]. The initial time for scanning was 10 minutes after irradiation and integrated for about 20 minutes. This time window was chosen because the signature nuclide, due to the carbon beam, comes from the $C^{12}(Li^7, He^6)N^{13}$ reaction. The N^{13} is a positron emitter with a half-life of 9.8 minutes. After 10 minutes, all nuclides with half-lives of a minute or less decayed to background levels and longer lived nuclides did not have enough activity to contribute significantly with the exception of C^{11} and F^{18}. The production of C^{11} is unlikely. The data could have been confused, however, by the production of F^{18} (half-life of 110 minutes) which can be produced protons. In one scenario, Ne^{18} would decay rapidly into F^{18} after being produced by the $(p, 2n)$ reaction with F^{19} in the LiF. The F^{18} nuclide could also be produced directly by (p, d) and $(p, p + n)$ reactions. The lowest threshold for these reactions is $8.6 MeV$. Thus, the requirement for heating the targets was that the peak proton energy fall below this threshold. The nuclide abundance during the scan window was determined by simultaneously placing the LiF window and imaging plate between two scintillators and making a coincidence counting measurement of the 511 keV gamma-ray produced by the positron decay. The relative abundance was inferred from the average half-life of the LiF decay.

3. Results and discussion

For targets which were not heated by the CW laser, the proton energies were above $40\,MeV$ and made profiling the carbon beam unfeasible. Data from the coincidence counter revealed a typical average half-life of about 1 hour, likely from the decay of F^{18} and N^{13} combined. However, when the target was heated, the average half-life dropped to 10 minutes, consistent with a nearly pure N^{13} decay (figure 1). Thomson parabola data confirmed that when the target was heated the peak proton energy went from the usual $40\,MeV$ to $3\,MeV$. This is still energetic enough for the (p, n) reaction with Li^7. However, this produces Be^7 which has a 53 day half-life and a thousandth of the activity of N^{13}. Reasonable fits to the coincidence counting data were achieved with either 97% N^{13} and 3% Fe^{18} or 98% N^{13} and 2% Be^7.

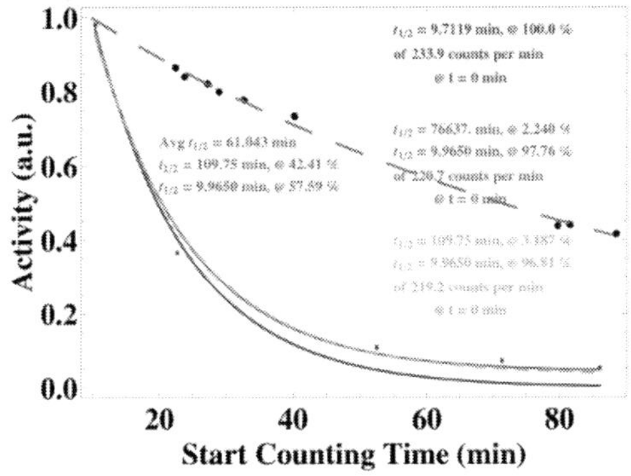

Figure 1. Half-life Fits: Heated target shown in gray, blue and dashed green, Unheated shown in dashed red.

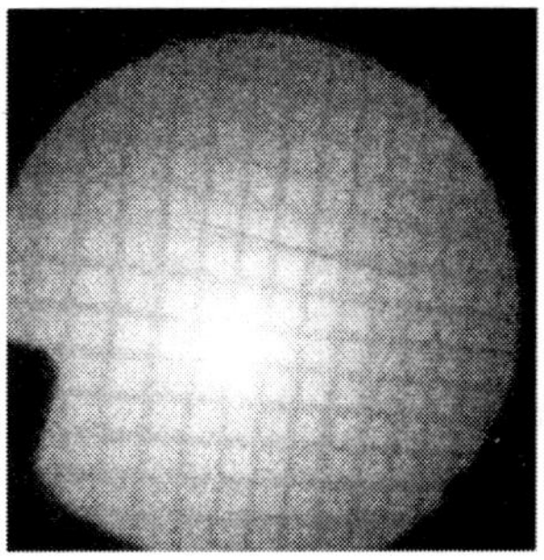

Figure 2. Auto-radiograph of the activated LiF from a heated target.

The auto-radiograph from the heated target showed a clear picture of the shadow cast by the copper witness mesh (figure 2). Ray tracing techniques [7] were used to locate the position of the ion beam focus. Figure 3 shows the image collected on the imaging plate and the ray tracing, which matched points on the mesh to points on the image. The point where the rays cross nearest to one another (the solution to the linear least squares problem) was identified as the focus. For a hemisphere with a $400\,\mu m$ radius, the focus was found to be $630 \pm 110\,\mu m$ from the vertex of the hemisphere (figure 3). The uncertainty in the position is dominated by the uncertainty in the relative spacing between the target and the witness mesh; assumed to be within $100\,\mu m$. This focal distance is consistent with the $1.6 \times R$ proton focusing observed in past works [5].

Figure 3. Ray tracing from a heated CVD diamond hemisphere. Figure is zoomed so that the target and mesh positions can be seen relative to the ion focus. Purple disks are the plus minus range of the focal position, the orange disk diameter is the maximum radius made by a ray, and the yellow oval is the RMS radius of $35\,\mu m$.

The edge sharpness of the mesh lines in figure 2 suggests a source size that is between $60\,\mu m$ and $100\,\mu m$ in diameter, with the caveat that the source is much larger than the mesh bar size ($40\,\mu m$). The RMS distance of the rays from the focal point was $35 \pm 20\,\mu m$. The uncertainty is the standard deviation of radii about the focal point.

4. Acknowledgments

This work would also like to acknowledge Applied Diamond, Inc. for the quality of the targets they made. This work supported by US DOE/NNSA, performed at LANL, operated by LANS LLC under Contract DE-AC52-06NA25396.

References

[1] Hatchett S P, Brown C G, Cowan T E, Henry E A, Johnson J S, Key M H, Koch J A, Langdon A B, Lasinski B F, Lee R W, Mackinnon A J, Pennington D M, Perry M D, Phillips T W, Roth M, Sangster T C, Singh M S, Snavely R A, Stoyer M A, Wilks S C and Yasuikec K 2000 *Phys. Plasmas* **7** 2076

[2] Wilks S, Langdon A, Cowan T, Roth M, Singh M, Hatchett S, Key M, Pennington D, MacKinnon A and Snavely R 2001 *Phys. Plasmas* **8** 542

[3] L Yin BJ Albright B H and Fernandez J 2006 *Laser and Particle Beams* **24** 291–298

[4] Fuchs J, Antici P, d'Humieres E, Lefebvre E, Borghesi M, Brambrink E, Cecchetti C, Kaluza M, Malka V, Manclossi M, Meyroneinc S, Mora P, Schreiber J, Toncian T, Pepin H and Audebert P 2006 *Nature* **2** 48

[5] Patel P, Mackinnon A, Key M, Cowan T, Foord M, Allen M, Price D, Ruhl H, Springer P and Stephens R 2003 *Phys. Rev. Lett.* **91** 125004

[6] Hey D S, Key M H, Mackinnon A J, MacPhee A G, Patel P K, Freeman R R, Woerkom L D V and Castaneda C M 2008 *Rev. Sci. Instrum.* **79** 053501

[7] Nürnberg F, Schollmeier M, Brambrink E, Blažević A, Carroll D C, Flippo K, Gautier D C, el M G, Harres K, Hegelich B M, Lundh O, Markey K, McKenna P, Neely D, Schreiber J and Roth M 2009 *Rev. Sci. Instrum.* **80** 033301

[8] McKenna P, Ledingham K W D, McCanny T, Singhal R P, Spencer I, Santala M I K, Beg F N, Krushelnick K, Tatarakis M, Wei M S, Clark E L, Clarke R J, Lancaster K L, Norreys P A, Spohr K, Chapman R and Zepf M 2003 *Phys. Rev. Lett.* **91** 075006

[9] Offermann D T, Freeman R R, Woerkom L D V, Foord M E, Hey D, Key M H, Mackinnon A J, MacPhee A G, Patel P K, Ping Y, Sanchez J J, Shen N, Bartal T, Beg F N, Espada L and Chen C D 2009 *Phys. Plasmas* **16** 093113

[10] Foord M, Mackinnon A, Patel P, MacPhee A, Ping Y, Tabak M and Town R 2008 *J. Appl. Phys.* **103** 056106

[11] Hegelich M, Albright B, Audebert P, Karsch S, Blazevic A, Brambrink E, Cobble J, Cowan T, Fuchs J, Gauthier J, Gautier C, Geissel M, Habs D, Johnson R, Karsch S, Kemp A, Letzring S, Roth M, Schramm U, Schreiber J, Witte K and Fernandez J 2005 *Phys. Plamsas* **12** 056314

Isochoric heating of low-Z, reduced-mass targets with high intensity laser pulse

S. Ohshima[1,2], H. Nishimura[1], D. Batani[3] T. Desai[3], M. Veltcheval[3], H. Nakamura[1], M. Tanabe[1], T. Fujiwara[1], T. Kawamura[4], F. Koike[5], R. Mishra[6], Y. Sentoku[6], R. Mancini[6], P. Hakel[6], , and K. Mima[1]

[1] Intitute of Laser Engineering, Osaka University, Suita, Osaka, Japan
[2] Pioneering Research Unit for Next Generation, Kyoto University, Uji, Kyoto, Japan
[3] University of Milano Bicocca, Milan, Italy
[4] Tokyo Institute of Technology, Yokohama, Kanagawa, Japan
[5] Physics Laboratory, School of Medicine, Kitasato University, Sagamihara, Kanagawa, Japan
[6] Department of Physics, University of Nevada, Reno
E-mail: shinsuke.ohshima@kupru.iae.kyoto-u.ac.jp

Abstract. Energy deposition by hot electrons in ultra-short, high intensity laser produced dense plasma was investigated by X-ray spectroscopy. Kα lines from partially ionized chlorine embedded in a triple layered, reduced mass target was observed to derive electron temperatures. Fast electrons generated by intense laser were confined in the smaller mass target, showing efficient increase of temperature. Heating efficiency by electron refluxing was one percent of laser energy and surface heating is significant at the laser intensity of ~ 1×10^{18} W/cm^2.

1. Introduction+

Recent advances of ultra-short, high intensity laser have expanded widely in various applications after the technique of chirped pulse amplification was discovered. The extremely intense field realized by the laser causes many interesting phenomena, such as generation of X-ray, γ-ray and energetic particles. Fusion in the fast ignition scheme is one of the most important applications, where a large energy, high intensity laser pulse heats and ignites hot core plasma compressed by multiple laser beams[1-3]. In this scheme and in other applications, understanding hot electron transport and laser energy deposition are one of essential issues.

Upon a classical model, laser-generated electrons penetrate into a material of solid density and dissipate their energy [4-9]. The conversion efficiency from laser to the electrons is typically 10-40%, depending on laser intensities [3,10,11]. Initially, return current of bulk electrons, needed for charge neutrality, heat the target in the penetration process. Then, the fast electrons start to recirculate after reaching rear surface in the case of a target with a finite size in the longitudinal direction because they are trapped and rebounded in plasma sheath created both on the front and the rear surfaces [12,13]. The process is called as "electron refluxing", and a simple calculation shows that 99% electrons reflux in target with the range of laser intensity from 10^{18}-10^{19} W/cm^2[14].

In this experiment, we aimed to realize isochoric heating of a matter using a mass-reduced target. If the target is substantially smaller than the attenuation length of hot electrons, the electrons can recirculate many times in the target and can heat it uniformly. Such an uniformly heated plasma can allow us to evaluate the temperature using spectroscopic diagnostics without considering temperature and density gradient, and enable us to investigate the physics of the hot dense matter. Several experimental researches have been done using a monolithic or multi layered target consisting of metallic materials [14-19]. However, it is well known that there exists a steep gradient appearing particularly at the target front side, this may result in difficulty in temperature derivation by X-ray spectroscopic method.

In this work, we made an experiment to assess energy transport and deposition to target using triple layered, all plastic, mass reduced target. X-ray spectroscopic measurement of K shell lines was employed in this study [20,21].

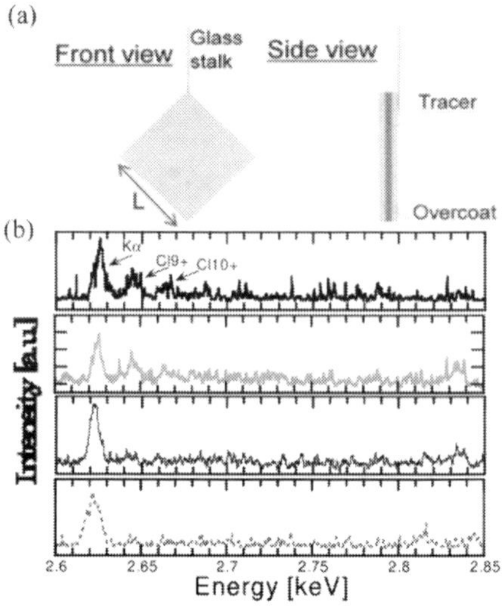

Fig. 1(a)Front and side view of three layered, reduced mass target. Four kind of target with lateral size L=50, 100, 300, 1000 μm were used. (b)X-ray spectra observed for L=50, 100, 300, 1000 μm targets.

2. Experimental Setup

The Experiments was performed using Gekko MII (GMII) laser system. The energy delivered from the system was E_L ~ 10 J per pulse of typically 500 fs in duration, and the wavelength was 1 μm. The laser was focused to target surface with a spot size of ~ 20-35 μm at the incidence angle of 20 degrees with respect to the target normal. Typical laser intensity was ~ 5×10^{17} - 1.5×10^{18} W/cm^2.

Four targets having different sizes were employed. All the targets consisted of 5-μm-thick C_8H_8(Parylene-N)/5-m-thick C_2H_3Cl(Polyvinyl chloride)/5-μm-thick C_8H_8, as shown in Fig. 1(a). The chlorine embedded in the middle layer was a tracer and its K-shell lines were investigated as a function of target total mass. The outer shape of the target was square and the lengths of each side, L, were 50, 100, 300 and 1000 μm. They are called type A, B, C and D. These conditions provide mass per incident energy of 4.6×10^{-9}~1.6×10^{-6} g/J, which are comparable to those of past feasibility experiment (2.8×10^{-9}~2.6×10^{-8} g/J)[2]. In addition, a polyvinyl chloride film of 5-μm thick and 300×300 μm in lateral size without overcoating was used to measure their target surface temperature.

An X-ray Spectrometer used in this experiment, consisted of a conically bent crystal (α-quartz (10-11)) and a back illuminated CCD camera (Princeton instruments inc.). This spectrometer covered spectral range from Cl-Kα (2.62 keV) to Heα (2.79 keV).

3. Results and Discussion

X-ray spectra for type A, B, C and D targets are shown in Fig. 1(b). The most intense line is Cl-Kα emissions around energy of 2.62 keV for all the targets. Besides, energy shifted Kα lines emitted from partially ionized Cl ions were observed around 2.64 and 2.66 keV particularly for type A and B targets.

Blue shift in the peak position and change in overall shape of near-neutral Kα lines are obtained, as shown in Fig. 2(a) of expanded view around 2.62 keV. From comparison of spectra between these for type A and D, energy of the Kα line is shifted

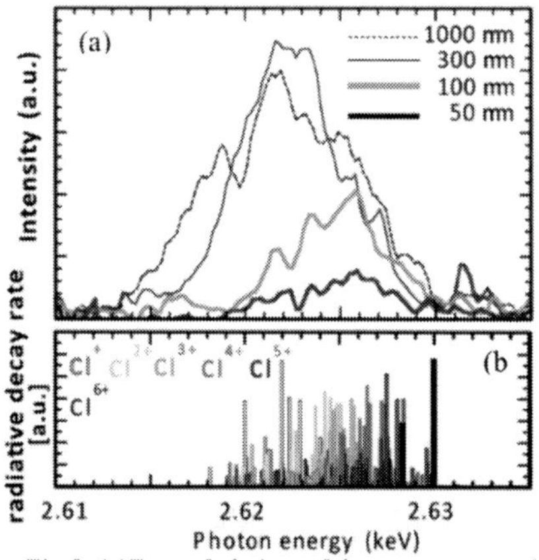

Fig.2 (a)Expanded view of the spectra around 2.62 keV. The vertical axis is normalized to laser energy and the values of intensity can be compared. (b) Radiative decay rate for Cl^{+}-Cl^{6+} ions obtained by calculation of grasp92 code.

about 5 eV owing to the change of ion population to highly ionized stage.

Figure 2(b) shows radiative decay rates of Kα lines from each ionization states of Cl ions predicted with a multi configuration Dirac-Fock atomic code GRASP 92[22] and RATIP[23]. Here, we considered Kα transitions for Cl^{1+}–Cl^{6+}, assuming their outermost bound electrons are not excited, namely, Cl^{N+}: $1s2s^2 2p^6 3s^2 3p^{(6-N)+}$ -> Cl^{N+}:$1s^2 2s^2 2p^5 3s^2 3p^{(6-N)+}$(N=1-6). In the calculation result, line emissions from Cl^{+} to Cl^{6+} overlap each other and the energy shift caused by ionization of M-shell electrons is very small. Nevertheless, general trend of energy shift in blue side is seen for higher ionization stages. The Kα lines observed experimentally are identified in the range of about 2610-2630 eV, which is corresponding to the emission from Cl^{+}-Cl^{6+} of the calculation. Emissions from Cl^{7+} and Cl^{8+} ions are not calculated in the calculation, while emission from Cl^{9+}– Cl^{10+} can be identified in the experimental data. The spectroscopic results imply that a small temperature gradient may exist in lateral direction of the target.

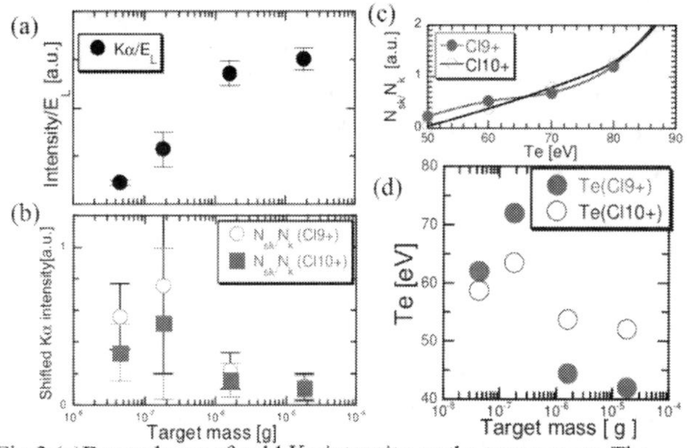

Fig.3 (a)Dependence of cold Kα intensity on the target mass. The intensities are normalized by laser energy. (b)Dependence of shifted Kα intensities(Cl^{9+}, Cl^{10+}) normalized by that of cold Kα intensities. (c)Line intensity ratios of the Kα line group to cold Kα to neutral Kα given as a function of electron temperature derived by FLYCHK, and (d)Electron temperature derived from experiments using the result shown in (c)

Figure 3(a) shows dependence of cold Kα (Cl^{+}-Cl^{8+}) intensity N_k on the target mass, where the values of intensity were obtained by integrating the peak around 2.62 keV and normalized by laser energy. The intensities of Kα lines close to the neutral Kα line decrease with decrease in the target size. This result suggests that Kα emission from highly ionized Cl ions become more dominant over that from ions close to neutral atoms.

Furthermore, Fig.3(b) shows that dependence of shifted Kα line intensities N_{sk} normalized by N_k on the target mass. Here, Kα yield is described as $N_k = Ne\int_0^{\infty} dE_0 f(E_0)\int_{E0}^{0} dE\,\omega_k n_{Cl}\sigma_k (dE/ds)^{-1}$, where Ne, E_0, ω_k, σ_k, and n_{Cl} are respectively the total number of hot electrons, initial energy of hot electrons, fluorescence yield, collisional cross section between hot and bulk electrons, and the number of chlorine atoms in the target [6]. Simplifying the above equation, we can express cold and shifted Kα yields as $N_k \sim n_{neutral} \cdot f(E_L)$ and $N_{sk} \sim n_{ion} \cdot f(E_L)$ and then can obtain $N_{sk}/N_k \propto n_{ion}/n_{neutral}$, which is an indication of relative number of highly ionized ions to that of ions close to neutral atom (Cl^{+} - Cl^{8+}). In Fig. 3(b), the normalized intensity for Cl^{9+} and Cl^{10+} ions has tendency to increase with decrease in the target mass, showing progress of ionization stages in plasma. These results successfully demonstrated the effectiveness of reduced-mass target for efficient heating. Note that for type A target the laser energy was eventually lower by 1/3 (E_L ~4.7 J) than those of another shots (E_L ~10-12 J) and the intensity of the emission was not intensive. Nevertheless, spectral intensity is comparable to that of the case for target B within error bar, showing the effective heating.

Using an atomic kinetic code FLYCHK [24], we tried to determine the temperature from X-ray spectra. Assuming the electron density ne ~ 1 x 10^{23} cm^{-3}, $n_{hot}/n_e \sim 0.01$ and T_{hot}= 500 eV, we obtained a relationship between plasma temperature and the line ratios for N_{sk}/N_k as shown in Fig.3(c). Dependence of electron temperature derived from comparison between the experiment and FLYCHK calculation on the target mass is shown in Fig. 3(d). The highest temperature of 70 eV is obtained for the mass reduced target.

The total energy deposited in the volume of tracer layer of solid density at 70 eV and the average Z value of 7 is only ~1 % of incident laser energy. The energy needed for ionization is about ~0.02 J and for heating is ~0.03 J. On the contrary, surface heating by laser-matter interaction is very significant in laser energy deposition. Electron temperature derived from the X-ray spectrum for the polyvinyl chloride foil without overcoating was around 1 keV, that is also reproduced in a collisional 1D PIC simulation as shown in Fig.4. The details of the code were described in reference [25]. This result is also consistent with previous experimental result using a monolithic Cu target [8-11]. Despite existence of hot electron refluxing, local heating of target was experimentally found. This might be attributed to the energy distribution of hot electrons[25]. Hot electrons involved in the lower energy side, components lower than the slope temperature, dominate over the overall energy carried by hot electrons. They travel in a shorter distance and deposit their energy in a plasma cose to the target surface. Further experiments and simulations will be needed for different experimental conditions, for example, investigating the dependence of heating efficiency on laser intensity, for which fast electrons have different energy distribution, is important and interesting.

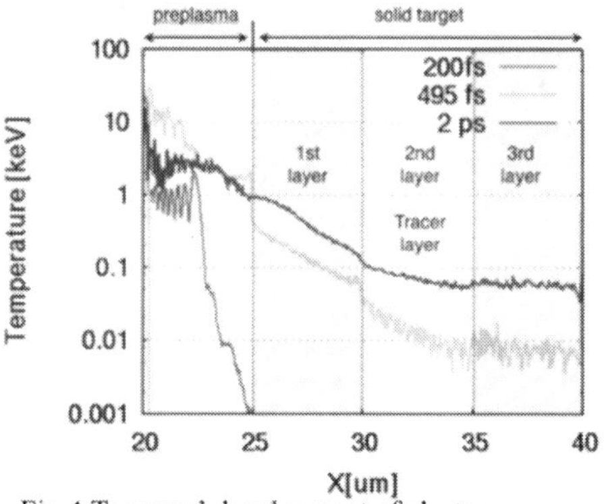

Fig.4 Temporal development of electron temperature calculated by 1D PIC simulation at 200 fs, 495 fs and 2ps.

4. Summary

Energy deposition by hot electrons into mass-reduced target has been studied by X-ray spectroscopic measurement, and analyzed with simulations. The experimental results support the hot electron confined in a small target by the sheath self-created in the target surface. However, there still remains a target temperature gradient along the depth of the target. The PIC simulation shows the tracer layer sandwiched with a surface tamper and a rear substance is moderately uniform in terms of temperature and density so that the isochoric heating seems to be realized in the localized region. Near future, we will operate an experiment using a chlorinated and deutrinated target to measure electron and ion temperature simultaneously. This will clarify physics of energy relaxation from hot electrons to bulk ions comprehensively.

[1] M. Tabak *el al.*, Phys. Plasmas 1, 1624(1994).
[2] R. Kodama *el al.*, Nature 418, 933(2002).
[3] M.H.Key, Phys. Plasmas 14, 055502 (2007).
[4] H. Teng *el al.*, Phys. Rev. E, 67 026108 (2003).
[5] M. Borghesi *el al.*, Phys. Rev. Lett., 83 4309 (1999).
[6] J. Fuchs *el al.*, Phys. Rev. Lett., 91, 255002 (2003).
[7] R.B. Stephens *el al.* Phys. Rev. E, 69, 066414 (2004).
[8] A.R. Bell *el al.*, Plasma Phys. Control. Fusion 39, 653 (1997).
[9] R.R. Freeman *el al*, Fusion Sci. Tech. 49 297 (2006).
[10] S.P.Hatchett *el al.*, Phys. Plasmas 7, 2076 (2000).
[11] M. H. Key *el al.*, Phys. Plasmas 5, 1966 (1998).
[12] H.S.Park *el al.*, Phys. Plasmas 13, 056309 (2006).
[13] R. P. J. Town *el al.*, Nucl. Instrum. Methods Phys. Res. A 544, 61 (2005).
[14] J. Myatt *el al.*, Phys. Plasmas 14, 056301(2007).
[15] P. M. Nilson *el al.*, Phys.Plasmas, 15, 056308 (2008).
[16] K.U.Akli *el al.*, Phys. Plasmas 14 023102 (2007).
[17] W. Theobald *el al.*, Phys. Plasmas 13, 043102(2006).
[18] S. D. Baton *el al.*, HEDP, 3, 358(2007).
[19] J. Rassuchine *et al.*, Phys. Rev. E 79, 036408 (2009)
[20] T. Kawamura *el al.*, Phys. Rev. E, 66, 016402 (2002).
[21] H. Nishimura *el al.*, J. Quant. Spectrosc. Radiat. Transf. 81, 327 (2003).
[22] F. A. Parpia *el al.*, Comput. Phys. Commun. 94, 249(1996).
[23] F.Fritzsche *el al.*, Phys. Scr., T100, 37, (2002).
[24] H.K. Chung *el al.*, J. Quant. Spectrosc. Radiat. Transf. 81, 107 (2003).
[25] A.J.Kemp *el al.*, Phys. Rev Lett., 97, 235001 (2006).

Journal of Physics: Conference Series **244** (2010) 022055 doi:10.1088/1742-6596/244/2/022055

Propagation dynamics of an ultrashort, high energy laser pulse via self-modulation in gas medium with atmospheric pressure for laser compression

Takamitsu Otsuka[1], Masashi Kudo[1], Shohei Sakai[1], Takeshi Higashiguchi[1,2,3], Noboru Yugami[1,2,3], Toyohiko Yatagai[2], and Ryosuke Kodama[3,4]

[1]Department of Advanced Interdisciplinary Sciences, Utsunomiya University, Yoto 7-1-2, Utsunomiya, Tochigi 3218585, Japan

[2] Center for Optical Research & Education (CORE), Utsunomiya University, Yoto 7-1-2, Utsunomiya, Tochigi 3218585, Japan

[3] Japan Science and Technology Agency, CREST, 4-1-8 Honcho, Kanagawa, Saitama 3320012, Japan

[4] Graduate School of Engineering, Osaka University, 2-6 Yamada-oka, Suita, Osaka 5650871, Japan

E-mail: higashi@cc.utsunomiya-u.ac.jp

Abstract. Self-compression of 20-mJ laser pulses has been demonstrated in a free-space argon-filled cell. A 130-fs pulse was compressed to less than 60 fs (full width at half maximum) with an output energy of 17 mJ at an argon gas pressure of 25 kPa, corresponding to an input peak power of 3.65 times the self-focusing critical power through a single filament in a 10-mJ energy region.

1. Introduction

Ultrashort laser pulses are used for x-ray lasers [1], high-order harmonic generation [2], and charged-particle acceleration [3]. Laser systems with an energy of about 10 mJ at a pulse width on the order of 10 fs are expensive because of their complex optical layout. To obtain ultrashort, intense laser pulses, compression by self-phase modulation (SPM) and dispersion compensation has been demonstrated [4]. The spectral broadening produced by the SPM through nonlinear propagation has been studied. It results from a temporal change in the refractive index of $n = n_0 + n_2 I$, where n_0, n_2, and I are the linear refractive index, the nonlinear refractive index, and the laser intensity, respectively. The change in refractive index due to the Kerr effect thus depends on the laser intensity. The resulting change in the optical phase produces new wavelength components in the spectrum. The effective SPM is induced both by the optical Kerr effect and by long filamentation produced when the laser power exceeds the self-focusing critical value of $P_{\text{crit}} = \lambda^2/2\pi n_0 n_2$ in the propagation of the laser pulse (of wavelength λ) in the medium. When the laser intensity is strong enough, field ionization also occurs and a plasma is produced, which acts as a negative lens to defocus the laser light. The balance between the self-focusing and the plasma defocusing results in stable filamentation, called self-channeling, which keeps the laser intensity at around 10^{13} W/cm^2 over the Rayleigh length.

To induce efficient SPM, a pulse compression method in a gas-filled hollow fiber has been demonstrated to produce femtosecond pulses [5]. A hollow optical fiber with a pressure gradient has also been used as a nonlinear medium to broaden the spectrum, resulting in a compressed pulse width of 8.4 fs with an energy of 5 mJ [6]. The incident pulse energy, however, is limited to a few mJ due to the low damage threshold of the fiber wall. When the input power is higher than a critical value in a gas-filled hollow fiber, the mode coupling efficiency between the laser beam and fiber can be low due to self-focusing, resulting in reduced transmission. To overcome the limitation to low input energies for a hollow fiber [7, 8], one can take advantage of self-channeling for high energy density physics (HEDP) applications. To achieve effective self-compression in 10-mJ energy regions, it is important to study the dynamics of ultrashort laser pulses in the filamentation.

In the present paper, we report self-compression at laser energies of 20 mJ with a stable single filament. We observe 60-fs output with 17-mJ energy for 130-fs input at an energy of 20 mJ. We also observe splitting of the pulses for low laser energy and argon gas pressure, which is attributed to multiple filamentation at the focal point.

2. Experiments and discussion

We used a femtosecond laser pulse from a chirped-pulse-amplification Ti:sapphire laser system, operating at a repetition rate of 10 Hz. The pulse width of the laser was 130 fs (full width at half maximum, or FWHM) with a maximum output energy of 40 mJ and a central wavelength of 800 nm with a FWHM bandwidth of 7 nm. The laser pulse was focused into a 1-m-long pressurized Ar gas cell using a planoconvex lens having a focal length of 1.5 m. The focused laser intensity must be higher than 1.1×10^{13} W/cm^2 in the absence of self-focusing. To measure the pulse width, the output beam was sent into a single-shot second-harmonic autocorrelator. The spectra were measured using a multichannel spectrometer (Ocean Optics HR-4000). The beam diameter and profile at the focal point and at the exit of the gas cell were measured using a CCD camera.

High intensity of the laser pulse is maintained over an extended distance in the filaments, leading to substantial spectral broadening through the nonlinear interaction with the gas. The Kerr effect depends on the laser intensity and the gas pressure in the cell. Figure 1 plots the dependence of the output pulse duration on the critical-power-normalized input power. The circles corresponded to the pulse duration of the compressed pulse through the gas cell. The rectangles also showed the uncompressed pulse duration when the dual pulse shape was observed on the autocorrelator. The onset of self-compression due to the Kerr effect was observed at $P_{\mathrm{in}} \approx 3.65 P_{\mathrm{crit}}$, where $P_{\mathrm{crit}} \approx 42$ GW for an input energy of 20 mJ, pulse width of 130 fs, and Ar gas pressure of 25 kPa. The shortest measured pulse duration was about 60 fs (FWHM). When P_{in} was less than $3P_{\mathrm{crit}}$, pulse splitting was observed due to multiple filamentation. The duration of the split pulse was nearly the same of that of the input laser pulse. Self-compression was not observed for $P_{\mathrm{in}} \geq 8P_{\mathrm{crit}}$, because of beam diffraction and a shortened filamentation length due to plasma generation in the gas cell.

The focused laser beam profile was measured to study its splitting. Figure 2 shows radial profiles at the focal point 165 cm away from a lens inside the cell for input powers of $P_{\mathrm{in}} \approx 0.12P_{\mathrm{crit}}$, $P_{\mathrm{in}} \approx 3.07P_{\mathrm{crit}}$, and $P_{\mathrm{in}} \approx 10.23P_{\mathrm{crit}}$. Several hot spots are observed, indicating multiple filamentation at $P_{\mathrm{in}} \approx 0.12P_{\mathrm{crit}}$ and at $P_{\mathrm{in}} \approx 10.23P_{\mathrm{crit}}$. On the other hand, we only observe a single filament at $P_{\mathrm{in}} \approx 3.07P_{\mathrm{crit}}$.

Spectral broadening results in a longer interaction length, as shown in Fig. 3. This figure plots the broadening at $P_{\mathrm{in}} \approx 3.65P_{\mathrm{crit}}$ with a FWHM bandwidth of 18 nm due to SPM in a single filament, compared to the input laser bandwidth of 7 nm. The incident laser pulse overlaps longer wavelengths of the output pulse. In contrast, self-compression occurs at shorter wavelengths in the spectrum. This phase difference results from dispersion in the plasma.

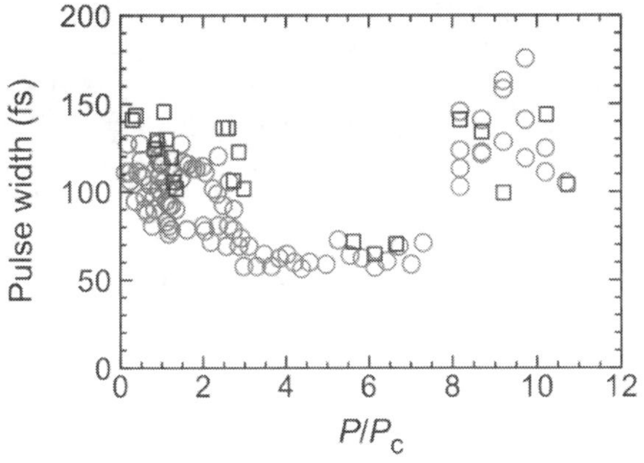

Figure 1. Dependence of the pulse duration on the input power normalized to the self-focusing critical power. The circles and rectangles correspond to compressed and uncompressed pulses, respectively.

Pulse splitting at $P_{in} \geq 8P_{crit}$ arises due to the different group velocities in the multiple filaments. The main pulse has reduced group velocity dispersion due to a strong nonlinearity [9]. The remnant pulse (for weak nonlinearity) separates in the filamentary region, leading to the pulse splitting. The weak pulse then propagates with a pulse width comparable to the input laser width.

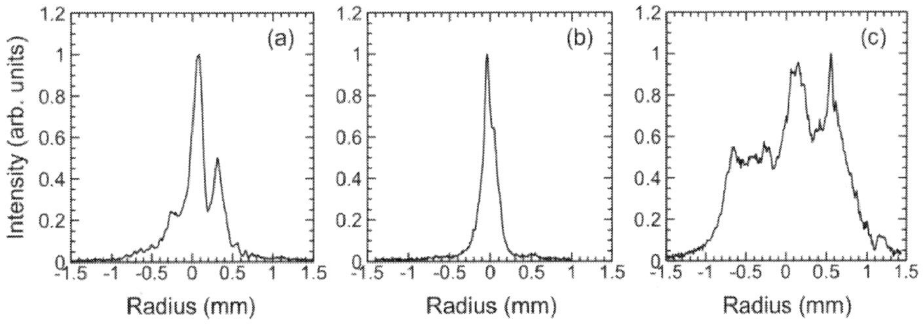

Figure 2. Radial profiles of the laser beam at the focal point at values of P_{in} of (a) $0.12P_{crit}$, (b) $3.07P_{crit}$, and (c) $10.23P_{crit}$.

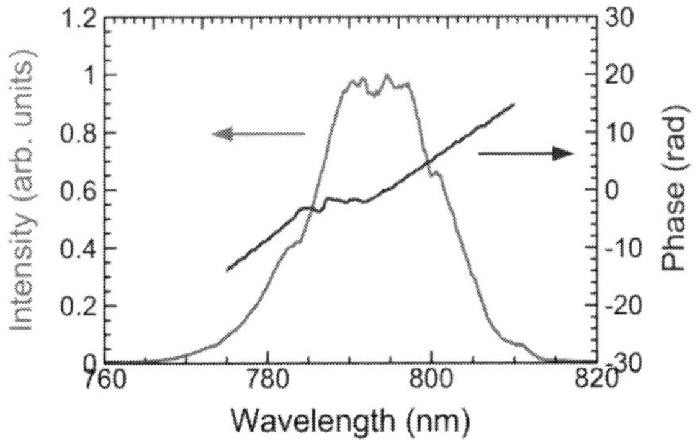

Figure 3. Time-integrated spectrum of a compressed output pulse for an input laser energy of 20 mJ and Ar gas pressure of 25 kPa, corresponding to $P_{in} \approx 3.65P_{crit}$.

The pulse duration decreases as the input laser energy increases. For low input energy, the duration of the output pulse is about the same as that of the input pulse. But at an input pulse energy of 20 mJ, a self-compressed pulse is observed with a FWHM duration of less than 60 fs and an output energy of 17 mJ at an argon gas pressure of 25 kPa, corresponding to $P_{in} \approx 3.65P_{crit}$, as plotted in Fig. 4. The shortening of the pulse width results from dispersion compensation, because the 60-fs output for a 130-fs input is 1.5 times the Fourier transform limit of 40 fs according to the spectrum of the output beam. The transmission was observed to be 85%, due to the input and output windows of the gas cell. The transmitted coefficient was experimentally measured to be unity through the gas medium without the laser window transmission without the gas cell. Therefore, the laser energy loss, which was attributed to the multiphoton ionization process to form the filament, was very small. On the other hand, the transmission (throughput) is limited to 50 – 70% when using a hollow fiber. For further self-compression at input laser energies exceeding 20 mJ, one could use a cascading gas cell having a long effective interaction length that suppresses the multiple filamentation and diffraction due to plasma production.

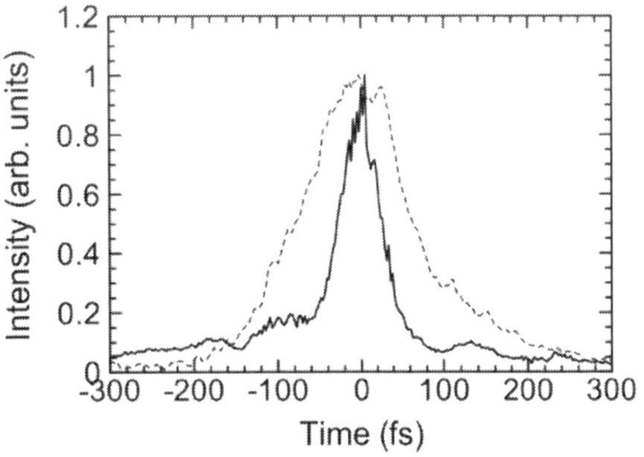

Figure 4. Temporal waveforms of the input laser pulse (dashed line) and the self-compressed output pulse (solid line) at an input laser energy of 20 mJ and Ar gas pressure of 25 kPa, corresponding to $P_{\text{in}} \approx 3.65P_{\text{crit}}$.

3. Summary

In summary, we produced filamentation in an argon-gas-filled cell without a hollow fiber by irradiating it with a high-intensity laser to initiate SPM. Self-compression resulted in a pulse width as short as 60 fs with an output energy of 17 mJ for an input laser energy of 20 mJ. This self-compression method is simple and it can compress the input pulse by as much as 30% in width using dispersion compensation.

Acknowledgements

This work was supported by CREST (Core Research for Evolutionary Science and Technology) of the Japanese Science and Technology Agency (JST). Part of this work was performed under the auspices of MEXT (Ministry of Education, Culture, Science, and Technology of Japan). One of the authors (T.H) received support from the Takahashi Industrial and Economic Research Foundation and Research Foundation for Opto-Science and Technology.

References

[1] Nagata Y, Midorikawa K, Kubodera S, Obara M, Tashiro H, and Toyoda K 1993 *Phys. Rev. Lett.* **71** 3774
[2] Takahashi J E, Kanai T, Ishikawa L K, Nabekawa Y, and Midorikawa K 2008 *Phys. Rev. Lett.* **101** 253901
[3] Leemans P W, Nagler B, Gonsalves J A, Toth Cs, Nakamura K, Geddes R G C, Esarey E, Schroeder B C, and Hooker M S 2006 *Nature Phys.* **2** 696
[4] Shank V C, Fork L R, Yen R, and Stolen H R 1982 *Appl. Phys. Lett.* **40** 761
[5] Nisoli M, De Silvestri S, and Svelto O 1996 *Appl. Phys. Lett.* **68** 2793
[6] Suda A, Hatayama M, Nagasaka K, and Midorikawa K 2005 *Appl. Phys. Lett.* **86** 111116
[7] Skupin S, Stibenz G, Bergé L, Lederer F, Sokollik T, Schnürer M, Zhavoronkov N, and Steinmeyer G 2006 *Phys. Rev.* E **74** 056604
[8] Nurhuda M, Suda A, Hatayama M, Nagasaka K, and Midorikawa K 2002 *Phys. Rev.* A **66** 023811
[9] Hatayama M, Suda A, Nurhuda M, Nagasaka K, and Midorikawa K 2003 *J. Opt. Soc. Am.* B **20** 603

Development of the compact electron spectrometer for the FIREX-I Project in Gekko XII

T Ozaki, M Koga[1], H Shiraga[1], H Azechi[1], H Sakagami and FIREX Group[1]
National Institute for Fusion Science, 322-6, Oroshi, Toki 509-5292, Japan
[1]Institute of Laser Eng., Osaka Univ., 2-6, Yamada-oka, Suita, 565-0871, Japan

E-mail: ozaki@nifs.ac.jp

Abstract. The high energetic electron measurement is one of the most important issues to research the ignition mechanism in the Fast Ignition Realization EXperiment Project. It is also important for the energy spectra with angular distribution because the electron spread is different by the target design. Therefore we have been developed the compact Electron SpectroMeter so as to be installed on different angular potions. We have performed the calibration using L-band LINAC in the Institute of Scientific and Industrial Research, Osaka University. The analyzer has been tested to measure energetic electrons from the aluminum and gold plain targets irradiated by LFEX laser (maximum energy of 10 kJ) up to 800 J. The electron measurement has been performed at the integrated experiments using CD-shell with Au cone. The non-Maxwellian spectrum can be observed when the effective core heating by the electron is occurred.

1. Introduction

In the laser fusion, the implosion and the central ignition of the DT shell have been the most popular method to realize the inertial confinement fusion [1]. However in order to achieve the break-even, the compression over 1000 times of the solid density requires huge laser energy over 1 MJ. In Institute of Laser Engineering, Osaka University, the fast ignition concept which is the imploded core heating using the high energetic electron beam produced by the pulse-compressed laser, has been proposed [2]. The high-energy electron is generated by a strong electric field based on the interaction between the high intensity laser and the pre-plasma produced by the laser pre-pulse [3]. The electron beam energy spectrum is strongly depended on the pre-plasma scale length and the main pulse intensity [4]. To obtain suitable electron energy less than several MeV for fast ignition, the pre-plasma scale length should be suppressed enough by choosing target materials, the pre-pulse control and the target area density.

In the Fast Ignition Realization EXperimental (FIREX) Project [5], the electron generated from the guided gold-cone irradiated by the heating laser, is utilized to heat the imploded core to initiate the fusion burning. The high energetic electron measurement is one of the most important issues to research the ignition mechanism. It is also important for the energy spectra with angular distribution because the electron spread is different from the target configuration. Therefore we have been developed the compact Electron SpectroMeter (ESM) so as to be set on different angular potions.

The electron spectra obtained by ESM is not reflected directly to the electron profile, which is contributed to the core heating, because the electron is decelerated by electric field due to the virtual cathode which is produced by intensive electron emission. Therefore low energy part less than 0.5

MeV cannot be observed due to the virtual cathode potential. However the electron spectrum included the low energy part can be obtained from the high -energy tail of the electron spectrum if we assume the Maxwellian distribution. Although the result should be compare with the result from the Cherenkov radiation [6], ESM is still an important tool for investigating the heating mechanism in core region.

2. ESM design

Number and size of viewing ports are limited in Gekko XII [7] target chamber I. We use the most popular 6-inches port. The triangle shape of the magnet is chosen due to the compactness and the wide energy range. We can save space since the magnet is hidden in the target chamber. The electron beam enters obliquely into the analyzer in order to obtain the wide observable energy range. A neodymium alloy is used as the permanent magnet. The magnet circuit is determined as to minimize the leakage of the magnetic field. Therefore small leakage of the magnetic field still remains near the top of the triangle. The two-dimensional magnetic field has been measured every 5 mm. The beam orbit has been calculated by using the observed magnetic field. The typical magnetic field strength and the magnet gap are 0.7 T and 8 mm, respectively. Figure 1(a) shows the photograph of the ESM analyzer.

The inexpensive and commercial based imaging plate (IP, Durr Dental Co.) [8], is used as the beam detector. This has merits about no electrical noise and the wide dynamic range for the electron intensity. Two IP folders are installed for measurements in high and low energy regions in order to extend observable energy ranges. The X-ray shield of 10-20 cm thickness covers the IP holder to avoid strong X-ray irradiation from the target. The IP is also shielded from the lights by the shutter in the holder. The holders with the light shield are brought to IP reader (Vista Scan, Durr Dental Co., 12.5 micron/step) after the electron irradiation. The intensity data have 16-bit resolution. The electron irradiates within the 8 mm width on IP. Therefore real signal can be obtained by subtracting signals within 8 mm width from the background in the other area of IP.

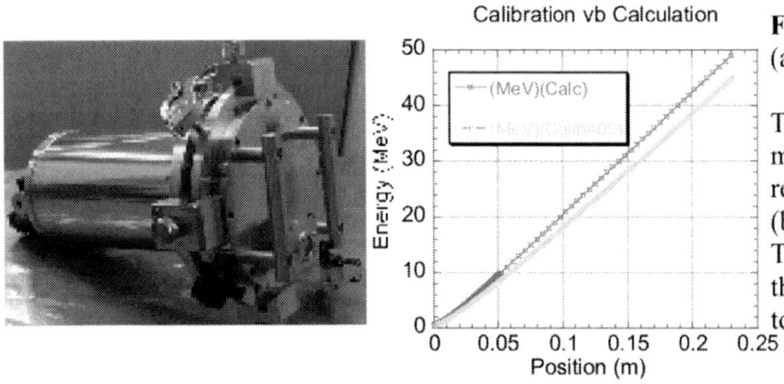

Figure 1.
(a) The photograph of the compact ESM. [9] Total length and diameter of the main part are 30 cm and 15 cm, respectively.
(b) The energy calibration. The difference may come from the leakage magnetic field at the top of the magnet.

3. ESM calibration

The purposes of the calibration are as follows;

(a) comparison of the beam orbit with the calculation, (b) the beam intensity calibration.

We have performed the calibration of the analyzer using L-band LINAC [10] in the Institute of Scientific and Industrial Research, Osaka University. L-and LINAC has an ability of strong and ultra short pulse electron generation of maximum 91 nC and 20 ps. The calibration has been performed using single pulse at two different energy of 9.5 MeV, 27.1 MeV. Main energies and charges in the calibration were 9.5 MeV, 27.1 MeV and from 0.1-10 pC. Electron beam from LINAC passes among air. Own vacuum chamber is prepared because the beam scattering by air should be minimized. The beam size is 5 mm at the exit of the beam line and 10 mm at 13 cm from the exit in air. The energy spreads are 0.2 MeV at 9.5 MeV and 0.3 MeV at 27.1 MeV, respectively.

The energy calibration is different from that in calculation. The main reason may be due to the magnetic field leakage near the top of the triangle. We can obtain the calibration curve from those results as shown in Fig 1(b). In this fitting, a magnetic field strength added parabola-like modification is assumed. The calibration of the intensity was also obtained by comparison between the incident electrons and the counts on IP. The real signal is obtained by the elimination of the background from the irradiated signal. The intensity calibration factor is found to be 1.35×10^{-8} pC/counts.

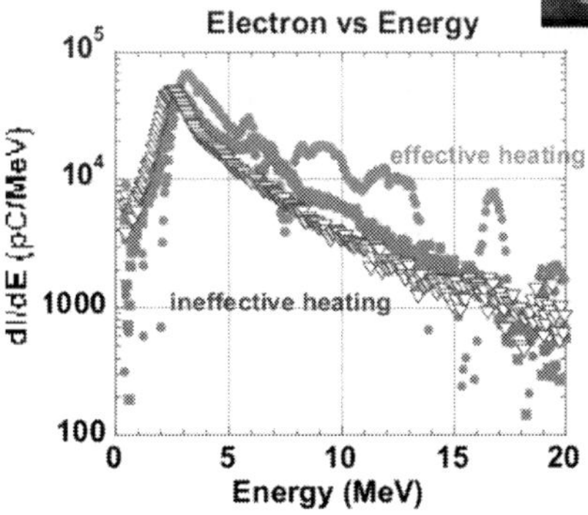

Figure 2.
(a) IP irradiated by the electron.
Electron intensity is recorded as 16 bit digital data.
(b) Typical energy spectra in effective and non-effective heating cases.
The spectrum is non-Maxellian when the effective heating is occurred.

4. Target irradiation by LFEX laser and the implosion-heating experiments using CD-shells with Au-cone

The analyzer is tested to measure energetic electrons from the aluminum plain target with 10 μm thickness irradiated by LFEX laser (maximum energy 10 kJ, the wave length of 1.05 mm, 4 beamlets) [11] with the pulse duration of 4 ps. The analyzer is installed on the Gekko XII Target chamber I at 20.9 degrees against the laser injection direction where is at the rear side of the target. This shot has been done by compression of one beam let LFEX laser. The maximum electron energy of 3 MeV can be observed when the LFEX laser is collimated up to 75 x 110 mm and the laser intensity of 3.5×10^{17} W/cm^2. The electron spectrometer only detects the escaped electrons over 0.5 MeV into and the amount of the electrons observed are strongly limited by the high electrostatic potential formation by the electrons.

LFEX laser irradiates the CD shells (deuterium polystyrene) with Au cone imploded by the Gekko XII laser. CD-shells specification is 500 μm diameter and 7 μm thickness with gold cone of 30-60 degrees. Mainly 9 laser beams of Gekko XII with 200–400 J/beam are used to compress the CD-shell. The injection timing of LFEX laser is adjusted so as to irradiate the imploded core of CD-shell using the optical delay system between LFEX and Gekko XII. The imploding time is determined by the shell specification and Gekko XII laser intensity. Until now the laser timing does not match each other accurately, we are finding the best timing for heating in the combination of the shell and laser energy.

Figure 2(a) and (b) show the electron exposure signal on IP and the typical energy spectra in three different cases of irradiated CD-Au shells. The real signal can be obtained by the subtracting the x-ray noise because much X-ray still remains on IP in spite of thick metal shielding. The calibration data in Sec. 2 are used to obtain the absolute spectra. In the figure, the uniform spatial distribution is assumed. The spectra in the effective heating cases are different from that in ineffective heating case.

Figure 3(a) shows the LFEX laser energy dependence of the electron flux. The electron flux on Au target is larger than on Al target because the there are more electrons in Au. The electron flux in the simple Au-cone is compared with that in CD-shell with Au cone. In CD-shell with Au cone, the LFEX laser should be injected at the final phase of the shell implosion ideally. However the implosion duration and the life-time of the imploded core are nano-seconds and several tens pico-seconds, respectively, which are larger than the LFEX laser pulse duration. Therefore the shell implosion process is independent on the energetic electron production mechanism by LFEX laser if the Au cone is not destroyed during the implosion. The electron flux in the Au cone should be equal to that in the CD-shell with Au cone. However the flux in CD-shell is obviously larger than that in Au cone. Two reasons are considerable. One is that the Au cone may be broken during implosion. Another possibility is that the ablation plasma makes short circuit so as to prevent the virtual cathode production. Figure 3(b) shows the injection timing dependence of the electron flux. At 200 ps in advance of the imploding, the Au cone is expected to remain without destroy. However the electron flux on CD-shell with Au cone is ten times larger than that on the simple Au cone. This means that the electron flux is enhanced by the ablation plasma.

In the electron measurement using ESM, we can find whether the heating is successful or not by the spectrum shape rather than the electron flux. At the successful heating, the low energy part in the spectrum is disappeared due to the electron absorption in the imploded core. For example, the intensity of the non-effective heating decreases at 2.3 MeV. However the intensity of the effective heating decreases at 3 MeV as shown in Fig.2(b). The spectra in the effective heating is the non-Maxwellian distribution because the electron may be absorbed by the imploded core. We find the spectrum has strong angular dependence from another detector, which is positioned just in front of the LFEX laser direction. [12]

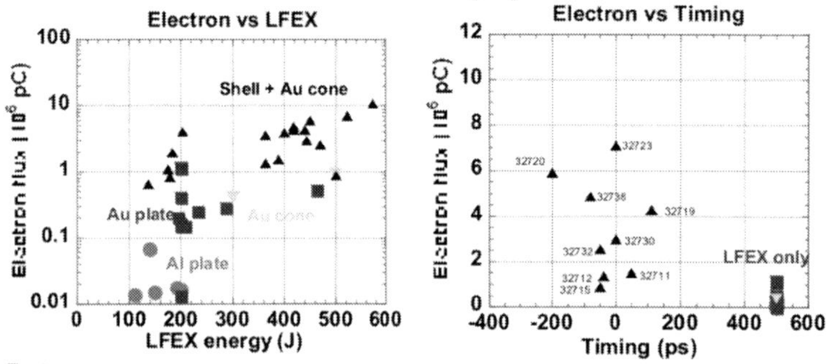

Figure 3.
(a) LFEX energy dependence of the electron fluxes for four different targets..
(b) Electron intensity dependence of the timing between the LFEX laser and the core implosion.

5. Summary

The high energetic electron measurement is one of the most important issues to research the ignition mechanism. We have been developed the compact electron spectrometer so as to be set on different angular potions. ESM is calibrated by using L-band LINAC in the Institute of Scientific and Industrial Research, Osaka University. The analyzer was tested to measure energetic electrons from the plain target and Au cone irradiated by LFEX laser. The combination experiments between the Gekko XII and LFEX laser starts using CD-shell with Au cone. We can obtain the energy spectra when the hot electron produced by the LFEX laser heats the imploded core of CD-shell produced by the Gekko XII.

6. References

[1] Azechi H, et al., 1991 *Laser Part. Beams* **9** 193-207
[2] Kodama R, et al., 2001 *Nature* **412** 789-802
[3] Key M, et al., 1998 *Phys. Plasmas* **5** 1966-1972
[4] Tajima T and Dawson J M 1979 *Phys. Rev. Lett.* **43** 267 - 270
[5] Azechi H 2006 *Plasma Phys. Control Fusion* **48** 267—B275
[6] Zheng J, et al., 2005 *Phys. Plasmas* **12**, 093105
[7] Yamanaka C, et al., 1985 *proc. 10th Int. Conf. London* **3** 3
[8] Ozaki T, et al., *NIFS report* (to be published)
[9] http://www.duerr.de/
[10] http://www.sanken.osaka-u.ac.jp/labs/rl/facility.html
[11] Miyanaga N, et al., 2003 *in Proc. of IFSA'03*, ThF1.1
[12] Habara H, et al, private communication.

Lateral hot electron transport and ion acceleration in femtosecond laser pulse interaction with thin foils

Jan Psikal[1,2], Vladimir Tikhonchuk[2], Jiri Limpouch[1], Ondrej Klimo[1]

[1] Faculty of Nuclear Sciences and Physical Engineering, Czech Technical University in Prague, Brehova 7, 11519 Praha 1, Czech Republic

[2] Centre Lasers Intenses et Applications, CEA - CNRS - University Bordeaux 1, 351 Cours de la Liberation, 33405 Talence, France

E-mail: psikajan@fjfi.cvut.cz

Abstract. In this paper, the lateral electron transport in a thin foil, limited in transverse sizes, is studied by numerical PIC simulations for two linear polarizations (p, s) of femtosecond laser pulse incident on a foil at various angles. The transport is due to hot electron recirculation forth and back and electron guiding on the foil surface by quasi-static magnetic and electric fields. It is demonstrated that the second mechanism takes place for larger incidence angles, although the recirculation is still important. There, ions accelerated from a lateral foil edge, which is out of the laser focal spot, can have higher energies than the ions from the rear foil side.

1. Introduction

Lateral electron transport in thin foils is a result of two phenomena. The first one is a fast electron current propagating along the target surface revealed in recent theoretical analysis, numerical simulations, and experiments [1], [2]. This current is generated by a short laser pulse of a relativistic intensity ($I\lambda^2 \approx 10^{19}$ Wcm$^{-2}\mu$m^2) when it is incident on a plane target at a large angle ($\theta \approx 70°$). Generated strong quasi-static magnetic and electric fields confine electrons in a potential well along the target surface and the electrons are resonantly accelerated by laser electric field inside the potential well [3]. This could result in electron energies exceeding ponderomotive potential and in transport of those electrons along the target surface far beyond interaction region.

The second mechanism of the lateral electron transport is due to hot electron recirculation. In this case, most of accelerated electrons pass through the target and are reflected in the expanding Debye sheath on the rear surface. This results in reversing of the normal component of electron velocity while the transverse velocity is largely unaltered. Hot electrons reflux in the foil many times and propagate towards target edges. The resulting ion emission from the edges of thin target foil or from moving electron sheath towards the foil edge, was investigated in Ref. [4] and Ref. [5], respectively.

In this paper, we investigate the influence of these two mechanisms on fast electron transport towards foil edges and on resulting ion acceleration. We performed 2D PIC simulations of the interaction of femtosecond laser pulse with an ionized thin foil target and studied the dependence of the electron lateral transport on the laser incidence angle. The latter is also quantified by analysis of characteristics of fast protons emitted from foil edges.

2. Simulation method and parameters

Our relativistic colisionless particle-in-cell (PIC) code in two spatial directions and with three velocity components is described in Ref. [6]. To investigate the lateral electron transport itself and its effect on proton acceleration, several simulation setups were chosen. The incidence angles of 30°, 45°, 60°, and 75° were considered for both p and s polarizations of the laser beam. Other simulation parameters are similar for all runs.

A fully ionized foil of the size $77\lambda \times 2\lambda$, consists of electrons and protons of the initial density 20 n_c (where n_c is the electron critical density). The transverse foil size is set to be two times larger than the laser focal spot for the largest incidence angle, i.e., $(10\lambda/\cos 75°) \times 2 = 77\lambda$, in order to observe the lateral transport and dissipation of surface electron current outside the interaction zone. A laser beam at the wavelength $\lambda = 1.0$ μm has a super-Gaussian profile (n=5) in the perpendicular plane with the width 10λ at half maximum. The temporal laser pulse profile has a trapezoidal shape with a constant maximum intensity 3.4×10^{19} W/cm^2 (the dimensionless amplitude $a_0 = 5.0$) of duration 20τ (in laser periods, 67 fs), and two linear ramps of duration 5τ (17 fs) at the beginning and at the end of the pulse.

3. Results and discussion

In the case of p-polarization, our simulations with large laser pulse incidence angles (75° and 60°) show that some electrons are pulled out from the target front by the electric field of the laser wave and their trajectories are bent parallel to the surface by the magnetic field of the wave. A part of those ejected electrons is pulled back into the target, but some of them rest close to the target front and form a bunch which moves along the surface in the direction of incident laser wave vector projection onto the foil. The fast electrons are moving outside the target while the return current is flowing in a dense plasma at the target surface. The bunches are distributed regularly with $\lambda/\sin(\theta)$ interval. That indicates that the Brunel absorption mechanism dominates in this case.

In the case of s-polarization, the oscillating electric field is parallel to the foil surface and perpendicular to the simulation plane, the electrons cannot be pulled out from plasma directly by the electric component of the laser wave, but they can be ejected by $j \times B$ heating. Therefore, the bunches of electrons are ejected twice per laser period and are distributed regularly with $\lambda/\sin(2\theta)$ interval. Thus, the period is twice shorter and the maximal current is smaller compared with the p-polarization case.

Fig. 1 shows averaged densities of the most energetic electrons (their kinetic energy is higher than 3 MeV) on the front foil surface for both laser beam polarizations and for incidence angle of 75°. The densities are calculated in two surface regions of sizes $4\lambda \times 1.5\lambda$ - "region 1" is located at the laser focal spot and "region 2" outside the spot region. In the region 1, one can see electron bunches moving along the foil surface with the velocity close to c as it was discussed above. Outside the spot region (region 2), accelerated bunches of electrons are gradually broadened into continuous electron current and dissipated, more rapidly in the case of p-polarization.

We also analyzed electron energy spectra on the foil front surface in both regions. For larger incidence angles (75° and 60°), the spectra contain a high energy tail. The energy of electrons in this tail is much higher than the value of ponderomotive energy $\varepsilon_{eh} \approx m_e c^2(\sqrt{1 + a_0^2/2} - 1) \approx$ 1.5 MeV. Such spectra are in agreement with simulations in Ref. [3]. That confirms that the surface acceleration takes place in the laser irradiated zone. The physical mechanism of electron surface acceleration is rather similar to the betatron acceleration in laser channels [7], [8].

In order to assess the influence of lateral hot electron transport on ion acceleration, we compare fast ion characteristics in simulations with four laser beam incidence angles (30°, 45°, 60°, 75°). Dependencies of maximum proton energy ε_{imax} and ion energy fluence W_{itot} (that is, the total energy of ions accelerated from a unit surface) on the incidence angle are shown in Fig. 2. These values are calculated from spectra "measured" in the strip of width 1.75λ,

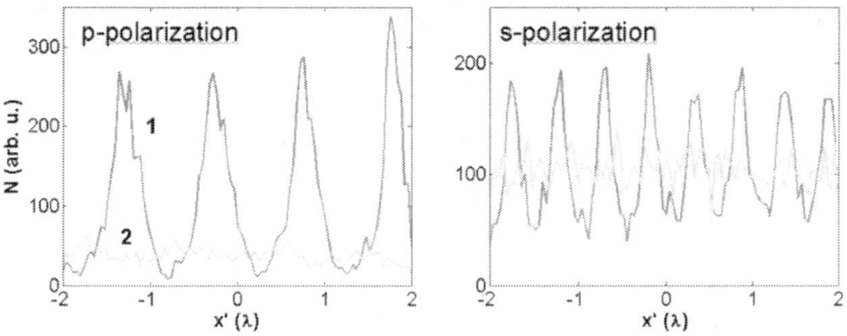

Figure 1. Density of hot electrons along the target front in two time instants in surface regions 1 and 2 (incidence angle of 75°). Only electrons with energy higher than 3 MeV are taken into account. The first region is located in the center of interaction zone. The second region is shifted along the front surface 30λ to the lateral right side and recorded about 30τ later than the first, which corresponds to the shortest time needed for electrons to be moved from the first to the second region. The regions are located from $y = -1.0\lambda$ to $y = 0.5\lambda$, initial plasma-vacuum interface at $y = 0$.

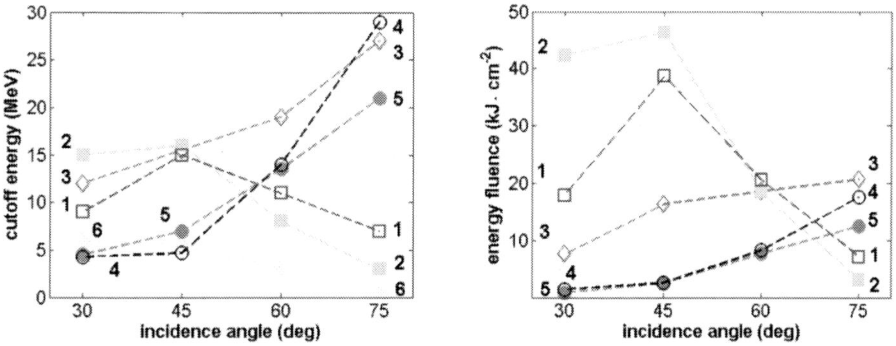

Figure 2. Left panel - Dependence of maximum energies of protons on the position of the emission zone on the target and the incidence angle: 1 - middle front, 2 - middle rear, 3 - lateral right, 4 - lateral right, *s*-polarization, 5 - lateral right r.b.c., 6 - lateral left; "r.b.c." denotes the simulation cases where the fast electrons escaping to vacuum are replaced by thermal electrons of Maxwellian distribution with initial temperature about 1 keV; all the lines except for line 4 are related to *p*-polarized laser pulse.
Right panel - Energy fluences of protons emitted from different target regions for incidence angles of 30°, 45°, 60°, and 75°. Only protons with kinetic energy higher than 1 MeV are taken into account. Line numbering has the same meaning.

perpendicular to the surface in the center of foil on its front, rear sides, and on lateral sides at the time of 250 fs after the interaction of the laser pulse with the target, when the ion acceleration process is terminated. The acceleration from left side is relatively weak and decreases with increasing incidence angle as the hot electrons are mostly accelerated in the direction of the laser wave vector projection onto the foil surface towards lateral right side. There, a strong proton acceleration takes place for large angles.

The maximum proton energy on lateral right side is 27 MeV and 29 MeV and the corresponding ion energy fluence about 21 kJ/cm^2 for *p*-polarization and 16 kJ/cm^2 for *s*-

polarization, respectively, in comparison with the energy fluence of laser pulse about 2.8 MJ/cm^2. The number of accelerated ions is higher for the p-polarization in spite of a stronger confinement of absorbed laser energy in the case of s-polarized pulse. This fact can be explained by the electron energy distribution which is not favorable for an efficient proton acceleration in the case of s-polarization, because it is dominated by a relatively small number of fast electrons. These most energetic electrons overcome the potential barrier created by the ions and by themselves, and are lost from the system (they are accumulated on the boundaries of the simulation box), whereas less energetic "bound" electrons participate in the acceleration, according to a theoretical model of Ref. [9].

In the case of p-polarization, additional simulations were performed to demonstrate explicitly the effect of electron surface guiding on proton acceleration. The runs where hot electron recirculation is artificially suppressed shows that the recirculating electrons (which are mostly observed in the p-polarization case) contribute significantly to proton acceleration, even for the largest incidence angle. A significant difference between a standard simulation and the run with the substitution of hot by thermal electrons on the rear foil side can be observed, both, in the cutoff energy and in the energy fluence related to the lateral right foil side. The difference is more pronounced in the proton energy fluence as the most energetic electrons guided along the surface (which are presented in all simulations) enhance mainly the cutoff energy, whereas recirculating electrons (less energetic, but more numerous) are contributing into the enhancement of the fluence (as it depends more strongly on hot electron density).

4. Conclusion

Lateral electron transport in a thin foil due to hot electron guiding along the foil front surface is investigated by 2D PIC simulations. While a femtosecond laser pulse is incident on the foil with a large angle (more than 60°), a part of electrons is confined on the foil front. They are accelerated to very high energies exceeding the ponderomotive energy, and transported towards an edge of the foil in the direction of laser wave vector projection onto the foil front surface. The ions emitted from the foil edge (lateral side), which is out of the interaction region, by the electric field created by these electrons can reach even several times higher maximum energy than the ions accelerated from the rear foil surface in the target center, although their total number is rather low. These guiding and acceleration effects are not observed for smaller incidence angles. In this case, the lateral transport is mainly due to multiple recirculation of electrons through the foil and the energies of transported electrons are much lower.

To clearly observe the effect of the electron guiding along foil front surface on proton acceleration, an artificial boundary condition cooling down recirculated electrons is applied on the rear foil surface. It is shown that hot electron recirculation forth and back still plays an important role even for very large incidence angles, which are considered for the cone targets used for fast ignition. Therefore, the schemes that reduce looses of recirculating electrons along the cone wall would increase the electron transport efficiency.

References

[1] Nakamura T *et al.* 2004 *Phys. Rev. Lett.* **93** 265002
[2] Li Y T *et al.* 2006 *Phys. Rev. Lett.* **96** 165003
[3] Nakamura T *et al.* 2007 *Phys. Plasmas* **14** 053112
[4] McKenna P *et al.* 2007 *Phys. Rev. Lett.* **98** 145001
[5] Nakamura T *et al.* 2008 *Phys. Rev. E* **77** 036407
[6] Psikal J *et al.* 2006 *Czech. J. Phys.* **56** B515
[7] Pukhov A *et al.* 1999 *Phys. Plasmas* **6** 2847
[8] Psikal J *et al.* 2009 Lateral hot electron transport and ion acceleration in femtosecond laser pulse interaction with thin foils *Phys. Plasmas* submitted
[9] Passoni M and Lontano M 2008 *Phys. Rev. Lett.* **101** 115001

Journal of Physics: Conference Series **244** (2010) 022058 doi:10.1088/1742-6596/244/2/022058

Interaction of high intensity laser pulses with pre-formed plasma in fast ignition targets

M G Ramsay[1], **N J Sircombe**[1,2] **and C S Brady**[2]

[1] Plasma Physics Department, AWE Plc., Aldermaston, Reading, Berkshire, RG7 4PR, UK
[2] Centre for Fusion, Space and Astrophysics, University of Warwick, Coventry, CV4 7AL, UK

E-mail: martin.ramsay@awe.co.uk

Abstract. Two-dimensional PIC simulations of the interaction of a high intensity, short pulse laser with planar foil targets preceded by a pre-formed density ramp have been performed. We observe a trend towards hot electron sources with higher characteristic energies and larger divergence angles in targets with longer pre-plasma scale lengths. We also present the results of simulations of cone geometry targets containing pre-formed plasma, and examine the case of a misfire involving the incident laser being offset from the central cone axis. The resultant hot electron populations produced in the presence of both short and long scale length pre-plasmas suggest that beam accuracy may be critical when scaling to a full fast ignition facility.

1. Introduction

The fast ignition approach to inertial confinement fusion (ICF) [1] commonly uses a short pulse laser to accelerate electrons, which deliver their energy to the core of a compressed fusion capsule. The requirements for ignition of the fuel capsule place constraints on the divergence angle, energy and duration of the hot electron beam. Recent studies [2, 3] of cone-guided fast ignition targets [4] suggest that plasma filling of the cone tip due to a pre-pulse will have a significant impact on the production of hot electrons. Some advanced fast ignition target designs include a low density foam lining in the cone [5] to create a controlled pre-plasma in an attempt to maximise the absorption of incident light and the production of energetic electrons.

Using the particle-in-cell code, EPOCH[1], we have studied the effect of varying the pre-plasma scale length on hot electron production for the case of a high density foil with a long scale length, low density pre-plasma. We have considered foils at normal and oblique incidence, in order to represent the cone tip and walls. We also present the results for 2D cone geometry targets. All simulations were performed without collisions.

2. Planar foil simulations

The simulated targets consisted of a 25 μm thick, high density ($n_e = 5 \times 10^{28}\,\mathrm{m}^{-3}$), high Z plasma with a front surface pre-plasma of exponential density scale length 5 or 50 μm. Initial temperatures of 1 keV in the pre-plasma and 100 eV in the foil were used. The density and temperature profiles were based on 1D simulations of a 1 ns pre-pulse with an intensity contrast

[1] EPOCH has been developed under EPSRC S&I funding at the University of Warwick, and its continuing development is funded under an EPSRC CCP at three UK universities.

ratio of 10^6 performed using AWE's hydro-code, `NYM` [6]. Targets were simulated at 0 and 70° (equivalent to a 20° half-angle cone) to the laser axis respectively, at both pre-plasma scale lengths. The ions were assigned mass and charge values of $40m_p$ and $20e$ respectively. The incident short pulse laser was modelled as a 0.5 ps FWHM top-hat pulse with a short rise time, focused to a 10 μm diameter spot and peak intensity of $I\lambda^2 = 10^{20}$ Wcm$^{-2}\mu$m^2 for $\lambda = 1054$ nm.

Hot electrons produced by the laser interaction with the pre-plasma region were recorded by particle probes placed at a plane 10 μm into the high density foil region. The energy spectra of the electrons recorded by the particle probes are shown in figure 1a. By fitting an exponential to the high energy tails of the electron distributions, it was possible to estimate the characteristic hot electron energy, T_h[2]. The values of T_h, fraction of laser energy absorbed into electrons detected by the probes, and their mean direction of motion (relative to the laser axis) for each simulated case are given in table 1.

Table 1: Table of the values of T_h for each scale length (L) and angle of incidence (θ) case. Also tabulated are the fraction of laser energy absorbed into hot electrons detected by the probe plane, and their mean direction of motion ($\langle\phi\rangle$). Note, the ideal value of T_h is 1–5 MeV [1, 8].

L (μm)	θ (°)	T_h (MeV)	Abs. frac.	$\langle\phi\rangle$
5	0	5.3	18%	− 1.2°
5	70	7.6	19%	50.6°
50	0	25.4	35%	− 4.6°
50	70	37.0	36%	58.3°

The presence of a long scale length pre-plasma resulted in a higher maximum electron energy, and an increase in T_h. At oblique incidence the laser effectively encounters a longer scale length pre-plasma. Thus moving away from normal incidence continued the trend towards higher peak electron energies, higher T_h and more energy absorbed from the laser overall. The increase in T_h with pre-plasma scale length may be linked to the production of small numbers of very high energy electrons in the low density channel ($n_e < 10^{26}$ m^{-3}) formed by the laser pulse, analogous to the laser wakefield accelerator [9, 10].

The hot electron divergence angles (defined as the standard deviation in their direction of motion) showed a noticeable increase in the normal incidence, long scale length results, as shown by figure 1b. This may have been due to the onset of plasma instabilities such as laser filamentation [11]. Any filaments produced in the oblique incidence case were refracted away from the foil, and thus similar divergence angles to the short scale length cases were recorded by the probes.

3. Cone target simulations

The interaction of the laser pulse with the cone tip was predicted to yield similar results to the normal incidence, planar targets. However, the laser missing the cone tip and striking the side of the cone may result in further interaction events occuring elsewhere within the cone. Therefore, to investigate the potentially detrimental impact of such a misfire in the presence of a pre-plasma, the central axis of the laser pulse was deliberately offset from the cone axis by 25 μm. The 2D cone geometry targets were simulated as having a thickness of 25 μm, inner tip diameter of 30 μm and a 20° half-angle. Exponential density profiles, of scale lengths 5 and

[2] Note that we avoid comparison with the ponderomotive energy (3.9 MeV) due to ambiguities over factors such as the range of energies over which to calculate T_h [7].

(a) Electron energy spectra for planar foils.

(b) Divergence half-angle as a function of energy.

Figure 1: (Colour online) Data recorded by the particle probes for 5 (red) and 50 (blue) micron scale length pre-plasmas at 0 (solid) and 70 (dashed) degrees to the incident laser.

50 μm, were added to the front (inside) of the cones. The laser parameters were unchanged from the planar foil simulations.

Multiple interactions were observed in the presence of a short scale length pre-plasma. The reflection of the laser pulse from the upper wall, cone tip and lower wall is visible in the magnetic field (B_z) plot, figure 2a. This behaviour produced multiple hot electron sources, each directed away from the expected position of the assembled fuel (see table 2). In the long scale length case, however, the laser pulse was refracted in the cone tip (as shown in figure 2b). The refraction of the laser resulted in a larger and higher energy, single hot electron source with a greater divergence angle than any of the individual short scale length sources. The mean directions of motion and divergence angles recorded by the particle probes are given in table 2. Using the combined energy and angular distribution data, the fraction of electron energy which could potentially be deposited in the assembled fuel ($E \leq 100$ MeV and $-45° \leq \phi \leq 45°$) for the short and long scale length cases was calculated to be 56% and 42% of the total forward travelling electron energy respectively.

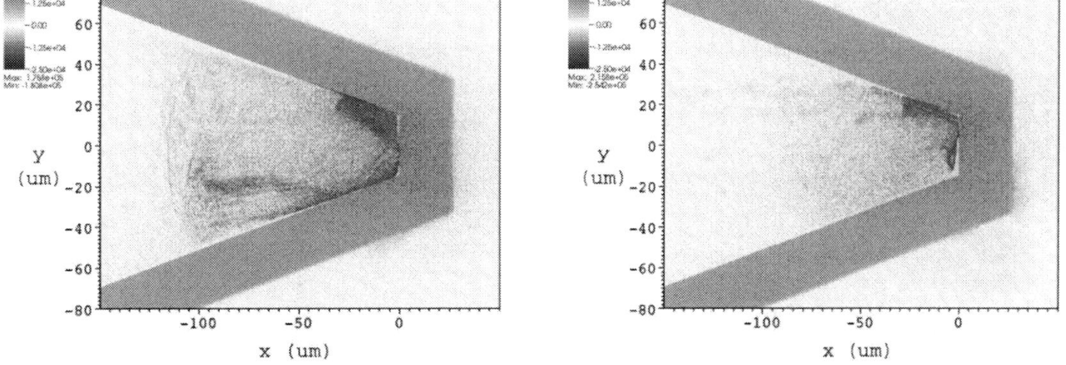

(a) 5 micron scale length pre-plasma.

(b) 50 micron scale length pre-plasma.

Figure 2: (Colour online) Z-component of the magnetic field (B_z), taken at $t = 1.25$ ps, from simulations of a laser offset 25 μm from the cone tip.

Table 2: Table of the mean angle of motion ($\langle\phi\rangle$) of the hot electrons passing through each of the cone walls for the short and long scale length cases, relative to the laser axis. The divergence half-angles are given in parentheses.

Probe position	$\langle\phi\rangle$ for $L = 5\,\mu$m	$\langle\phi\rangle$ for $L = 50\,\mu$m
Upper wall	47.8° (42.7°)	40.8° (28.1°)
Cone tip	−22.3° (25.3°)	−23.2° (22.8°)
Lower wall	−93.9° (27.7°)	−80.5° (36.7°)

4. Conclusion

For the purposes of hot electron generation in the cone-guided fast ignition approach to ICF, the presence of a pre-plasma within the cone appears to be undesirable. In the case of a long scale length pre-plasma, the laser pulse must propagate through a large region of low density plasma. The interaction of a short pulse laser with underdense ($n_e < n_c$) plasma is known to produce highly energetic electrons [9, 10] which have a much greater stopping distance than is suitable for fast ignition. A long scale length pre-plasma allows more energy to be coupled to the hot electrons overall. However, reducing the extent of the pre-plasma (possibly via pre-pulse mitigation techniques) is more likely to yield hot electron populations with energy distributions more suitable for fast ignition purposes. The presence of a shorter scale length pre-plasma also appears to reduce the divergence of the hot electrons. This may be due to the conditions being less suitable for the growth of instabilities such as laser filamentation [11] and Raman scattering [12].

When the full cone geometry is taken into consideration (in 2D), a cone wall interaction has a detrimental effect on the hot electron population. A short scale length pre-plasma leads to multiple interactions as the laser is reflected from the cone surfaces, producing multiple hot electron sources with few electrons directed towards the assembled fuel (56% of the forward travelling electron energy). Increasing the pre-plasma scale length, however, causes the laser pulse to refract in the cone tip, producing a single, large, highly divergent source of hot electrons (with 42% of the forward travelling electron energy directed towards the assembled fuel).

Further PIC simulations at a range of beam offsets and angles are required to better characterise the potential impact of a misfire. Full 3D runs will also be required to determine if these effects remain the same in a full 3D geometry.

We have found that neither of the cone geometry cases considered are desirable for fast ignition. Thus we conclude that ensuring a consistently high degree of beam accuracy may prove to be a critical factor in the design of a high repetition rate fast ignition facility.

References

[1] Tabak M *et al.* 1994 *Phys. Plasmas* **1** 1626
[2] Baton S D *et al.* 2008 *Phys. Plasmas* **15** 042706
[3] Van Woerkom L *et al.* 2008 *Phys. Plasmas* **15** 056304
[4] Kodama R *et al.* 2001 *Nature* **412** 798
[5] Lei A L *et al.* 2006 *Phys. Rev. Lett.* **96** 255006
[6] Roberts P D 1980 *J. Phys. D* **13** 1957
[7] Sherlock M 2009 *Phys. Plasmas* **16** 103101
[8] Deutsch C *et al.* 1996 *Phys. Rev. Lett.* **77** 2483
[9] Tajima T and Dawson J M 1979 *Phys. Rev. Lett.* **43** 267
[10] Bingham R *et al.* 2004 *Plasma Phys. Control. Fusion* **46** R1
[11] Max C E *et al.* 1974 *Phys. Rev. Lett.* **33** 209
[12] Drake J F *et al.* 1974 *Phys. Fluids* **17** 778

The influence of the irradiation symmetry and the foam absorber on the target compression characteristics for the HiPER

V. Rozanov[1], G. Vergunova[1], S. Guskov[1], N. Demchenko[1], I. Doskoch[1], R. Stepanov[1], D. Barishpoltsev[1], R. Yakhin[1], E. Aristova[2], N. Zmitrenko

[1] P.N. Lebedev Physics Institute, RAS, Moscow, Russia
[2] Institute of Mathematics Modeling, RAS, Moscow, Russia

E-mail: rozanov@sci.lebedev.ru

Abstract. The calculations and analysis of the target compression state are reported for the HiPER facility under deviation of the baseline target irradiation conditions from the completely symmetric conditions considered earlier. The irradiation asymmetry readily arises at a small number of irradiating laser beams (or the directions of irradiation, for example, two or 8-20 directions) if the targets are used in a thermonuclear reactor where the number of beams is limited.
The calculations have been taken by the scheme proposed earlier, which includes plotting of the irradiation maps and histograms, 1D- and 2D simulations of the target irradiation and compression. The calculations of the foam absorber targets have been compared to the experiments on the plane target irradiation at the LIL facility, where the absorber size and the irradiation conditions are in agreement with the conditions necessary for the HiPER facility.

1. Introduction

An experimental verification of the main prediction in ICF – the achievement of a thermonuclear yield with the gain coefficient > 10, which becomes possible with the help of the NIF facility, makes topical several questions related to the scheme of fast ignition. In a symmetrical version, for the fast ignition targets the more high gain values are predicted as compared to the hot spot. However, it still remains unclear how the deviation from symmetrical irradiation, which may occur under chamber-reactor conditions, could influence the thermonuclear yield. The worsening of the irradiation symmetry in the reactor, as compared to the experimental symmetrical facility, is connected with a smaller number of laser beams and impossibility to use several directions of irradiation due to the reactor design. The present paper considers the influence of the irradiation symmetry on the compression parameters of the base target for the HiPER facility [1] using 2, 8, 12, 20 and 48 beams. In the simulations a standard target R_0=1044 μm [1,2] is used. The beam FWHM is taken 870 μm. It is planned to make use of 48 laser beams in the experimental research with the HiPER facility. The simulations were made by the scheme developed and proposed earlier [3,4]. Briefly present the executive sequence of this scheme: the target illumination map is plotted, the data on the focusing system is used, the intensity distribution in the beam, and the arrangement of the beams on the sphere. Then a histogram of the target surface distribution over different levels of illumination is plotted. A histogram in the form of δ-function corresponds to an ideal symmetry of illumination. For the parts with different illumination the 1D simulations are performed, which form the basis for 2D simulations. The same 2D map may be compared with many 2D simulations, which correspond to a large number of spherical target cross-sections along the equator or meridians (under appropriate choice of the point corresponding to the "north pole"). From 1D simulations we get the information (the distributions of density, temperature, velocity, energy sources), which will be used further to define the initial conditions for 2D simulation. For the given cross-section a new illumination map is obtained by the turn of the primary map. For the 2D calculations are taken the variants, which are the most close to the axis-symmetrical ones. For the beam arrangement on the sphere corresponding to the Platonic one (the beam number 6, 8, 12, 20 …), the most close to the axis-symmetric are the variants when at the "north pole" of the sphere is located either a beam or a point, which is the most distant from the near beams. A rather complicated scheme of simulations is necessary to replace the 3D simulations adequate to the present task. 3D simulations are difficult not only due to the computational reasons, but to many unclear physical reasons as well, for example, the account of beam polarization.

The HiPER facility is presently under projection and analysis. That is why in this paper the authors consider a possibility to perform the experiments on the physics of targets using the LIL facility for the conditions close to the ones expected for the future HiPER facility. The data on the irradiation parameters and the target size possible for LIL are taken from the data on the experiments on the foam target irradiation, and the analysis of these experiments is presented in [5]. The laser beam energy in the LIL facility is 10kJ), the pulse duration ~ 3ns makes 1/12 and 1/4 of the expected HiPER values, and, so, the verification of the behavior of 1/12 of the target surface is quite possible.

2. Irradiation of baseline HiPER by N=2, 8, 12, 20 and 48 beams

The variant N=2 is expectedly bad [3], and it is not considered. Figure 1a,b,c,d shows the illumination maps for N=8, 12, 20, 48. The calculations are performed for the radiation total energy of 110 kJ, the energy geometrical losses make ~ 13%.

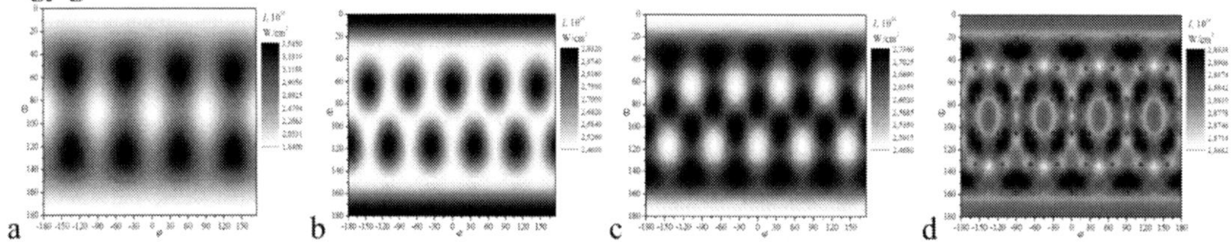

Fig.1. Maps for irradiation of the HiPER target by different number of laser beams.

Figure 2 a,b,c,d presents the histograms of these variants. The maximal intensity ranges important for these variants are: a – N=8, I [10^{14}W/cm^2]=1.7÷3.6; b – N=12, I=2.5÷3; c – N=20, I=2.5÷2.8; d – N=48, I=2.70÷2.73. The intensity maps are plotted for the time moment of the laser pulse maximum. At the increase of the laser beam number the intensities on the map tend to the average value on the target surface. If the number of the beams is decreased (the pulse energy being conserved), the intensity maxima grow and tend to the intensity in the centre of the focusing spot. Te smallest intensity value is associated with the energy distribution in the beam, and its analytical evaluation presents difficulties. One can see that with an increase of the beam number the histogram nears the δ-function, which corresponds to homogeneous distribution.

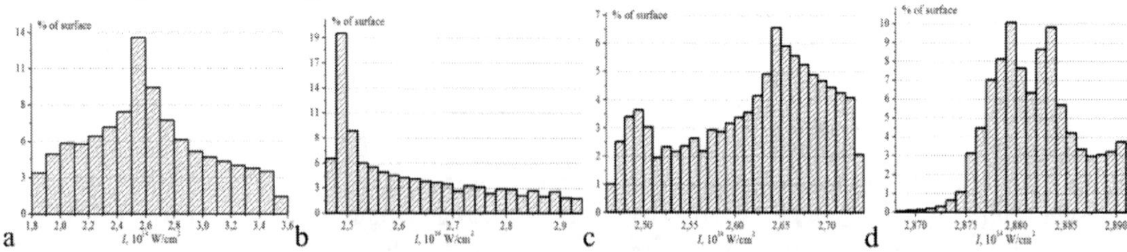

Fig.2. The histograms for irradiation of the HiPER target by different number of laser beams.

One can use different values to characterize the quality of target compression. Figure 3 shows the average for different direction of value ρR.

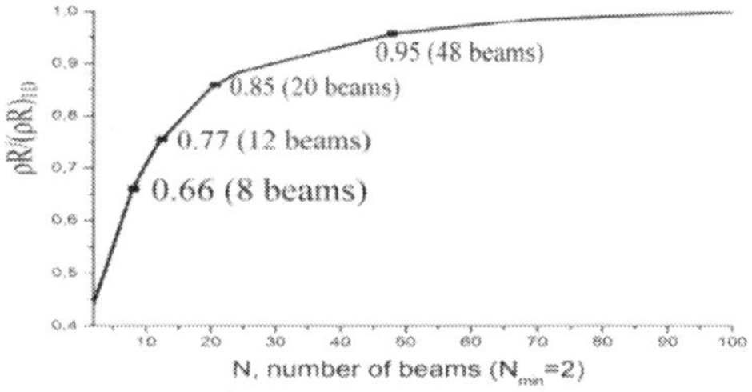

Fig.3. Dependence ρR from number of laser beams N.

3. About a possibility to imitate of HiPER target irradiation and shell acceleration on a LIL facility

Within the framework of the plane target experiments one can verify a behavior of some values, which play an important role at spherical compression. During the experiments on a LIL facility [5]. It was performed irradiation of a foam-like TMPTA ($C_{15}H_{20}O_6$) target having the density 6.5-10 mg/cm^3, pore dimension 1-2 μm, and target thickness 600-1000 μm. The laser pulse parameters were as follows: λ=0.351 μm, duration 3ns, intensity $4 \cdot 10^{14}$ W/cm^2, radius of an irradiated region R=500 μm, and energy 10 kJ. It is found [5] that a laser pulse is absorbed in a target, the reflection does not exceed several per cent, and the last half of the pulse has transmitted the foam practically without absorption due to the average foam density, which is below critical (integrally, for t>1.5 ns the absorption is not higher than 20%). The target for the HiPER facility has a DT-ice shell of 0.25 g/cm^3 density at thickness 211 μm. We proposed that the DT-layer could be replaced by TMPTA foam of 0.25 g/cm^3 density. The EOS had been measured experimentally for that type of foam [6]. For the shock wave velocity, D_f=21.7 км/s in TMPTA, the pressure P_f=0.76 Mbar, and velocity of the matter behind front u_f=16.7 км/s proved to be close to the analogous values for the DT-ice, i.e. D_f=21.7 км/s, P_f=0.81 Mbar, and u_f=15 км/s. Replacement of the ice at density 0.2-0.25 g/cm^3 allows one to perform experiments without using cryogenic technologies, which makes the experiments much easier. We report the results of the two-dimensional LATRANT code calculations on acceleration of a shell made of the "ice" imitator and ablator. That is a TMPTA foam at density p=0.25 g/cm^3 being under the action of a conventional LIL pulse on a 900 μm-thickness absorber also made of TMPTA, but at density 6.5 mg/cm^3. The results are presented in Fig.4, where 4a is the scheme of the target and 4b is the pulse shape; 4c is the density ϱ, parameter $\alpha=p/p_F$, p_F – Fermi pressure, pressure p and velocity V in respect of the z-coordinate on the beam axis for t=4.1 ns; 4d is the two-dimensional map of temperature distribution for t=4.1 ns.

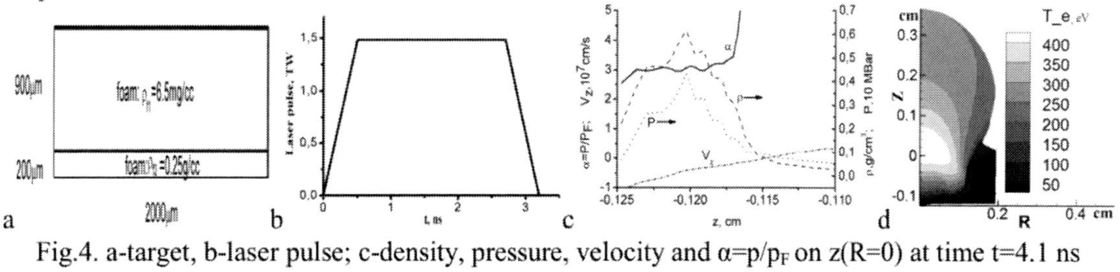

Fig.4. a-target, b-laser pulse; c-density, pressure, velocity and $\alpha=p/p_F$ on z(R=0) at time t=4.1 ns (Elas=4kJ), d-2D field of temperature at the same time.

From the represented data it is seen that the calculations reproduce satisfactorily the parameters of an accelerated shell at an earlier stage of acceleration: p~1-10 Mb, α~3-4, V=100-200 км/s. They are within the range corresponding to the HiPER conditions. However, for the targets with the foam absorber it is needed to optimize the pulse shape anew because the standard pulse provides a lesser value of the compression parameter ρR. The analogous experiments intended to imitate the baseline

HiPER targets (without foam) could also be made if the LIL facility would provide the pulse shape that is adequate to a low-adiabatic compression of the targets [1].

4. Conclusion

For the targets of the HiPER facility we have developed the irradiation patterns (maps) and the histograms of the target surface distribution over the irradiation levels. The 2D compression calculations have been made for the target irradiation under conditions of a reactor with a small number of beams, N=8,12,20. For N=48, the irradiation corresponds to the spherically symmetric conditions. It is shown that the parameter of the target compression ρR for HiPER is reduced by factor 1.5 if the number of irradiating laser beams is decreased to the range of 8-20 as compared to the symmetric calculation (48 orders). Using the foam absorber at plane target irradiation one can see that the parameters of the accelerated shell prove to be close to the HiPER target parameters at an earlier acceleration stage. A possibility is shown of an experimental verification of the LIL target acceleration on the HiPER facility. The EOS parameters of the TMPTA and DT-ice shells were found to be close. That makes it possible to verify experimentally the regimes of the shell acceleration without using the cryogenic technologies.

The present work has been supported by the RFBR grant #08-02-01083-a, the grant of the RAS Presidium program 12P, joint grant from the RFBR and Japan Society for the Promotion of Science #08-02-91202-JF.

References

[1] www.hiperlaser.org
[2] S.Atzeni, C.Belli, A.Schiavi. Phys.Plasmas v.14, 052702 (2007).
[3] V.Rozanov et al. XXX ECLIM 31.08.2008. Darmstadt, Germany, Report MO.0701. Proceedings of ECLIM.
[4] V.Rozanov et al. IFSA 2007, Kobe, Japan Journal of Physics. Conference Series v.112 (2008), 022010.
[5] V.Rozanov et al. XXX ECLIM, 31.08.-05.2008. Darmstadt, Germany, Report MO.0702. Proceedings of ECLIM.
[6] M.Koenig et al. Astrophysical Journal Suppl.Series 127: 385-388, 2000, April.

The Sixth International Conference on Inertial Fusion Sciences and Applications IOP Publishing
Journal of Physics: Conference Series **244** (2010) 022060 doi:10.1088/1742-6596/244/2/022060

Fast electron propagation in high-density plasmas created by 1D shock wave compression: experiments and simulations

J J Santos[1], D Batani[2], P McKenna[3], S D Baton[4], F Dorchies[1], A Dubrouil[1], C Fourment[1], S Hulin[1], E d'Humières[1], Ph Nicolaï[1], L Gremillet[5], A Debayle[6], J J Honrubia[6], P Carpeggiani[2], M Veltcheva[2], M N Quinn[3], E Brambrink[4] and V Tikhonchuk[1]

[1] Université Bordeaux 1, CELIA, Talence, France
[2] Dipartimento di Fisica "G. Occhialini", Univ. degli Studi di Milano-Bicocca, Milan, Italy
[3] SUPA, Department of Physics, University of Strathclyde, Glasgow, UK
[4] LULI, Ecole Polytechnique - CNRS - CEA, Palaiseau, France
[5] CEA-DPTA, Bruyères-le-Châtel, France
[6] GIFI, Universidad Politécnica, Madrid, Spain

E-mail: `Santos.Joao@celia.u-bordeaux1.fr`

Abstract. We present results from an experimental characterization of fast electron transport in high density plasmas created by 1D shock wave compression. The Kα fluorescence from a Cu layer embedded in Al or CH foil targets is measured. We use long laser pulses (LP) with 180 J, 1.5 ns, 0.53 μm to compress the foils by shock wave propagation to 2-3 times their solid density and heat them to $\sim$ 4 eV (close to the Fermi temperature). A counter-propagating high-intensity short laser pulse (SP), with 40 J, 1 ps, 5×10^{19} Wcm^{-2}, generates intense currents of fast electrons which propagate through the deep regions of the target just before shock breakthrough. The results are compared to the uncompressed, solid density case (without the LP beam). The complete set of measurements is compared to numerical results, including a 2D hydrodynamic description of the compression and pre-pulse effects, 2D PIC simulations of the SP beam interaction and both hybrid and PIC simulations of the electron transport in the target depth and sheaths. In the case of the non-compressed targets we need to take fast electron refluxing into account to reproduce the experimental results. By exploring the domain of warm temperatures, we identify a regime for the incident fast electron current density, $10^{10} < j_h < 10^{12}$ Acm^{-2}, for which the collective mechanisms of electron transport differs appreciably between solid density and compressed matter.

1. Introduction

In the fast ignition scheme for inertial confinement fusion, an energetic electron current, produced by an ultra-intense laser, heats a highly compressed DT plasma core. The fast electrons must transport a few 10s kJ of energy from their source, near critical density, over a few 100 μm of overdense plasma, to the high density fuel (at some 1000 times the solid density). Experiments have been successfully carried out in initially cold, solid targets to understand the mechanisms of fast electron transport in dense matter. However, in those experiments the initial target conditions are far from those expected in a fast ignition target, and fast electron propagation

Journal of Physics: Conference Series **244** (2010) 022060 doi:10.1088/1742-6596/244/2/022060

will therefore probably be different under such conditions. Whilst not yet being able to reproduce the fusion target core conditions, it is important to study fast electrons transport in warm and dense plasmas [1, 2], representative either of the degeneracy of the compressed fuel or of the density and temperature levels near the fast electron source.

Planar 1D compression with a counter-propagating electron flux geometry is adopted in this work. This has the advantage of ensuring plasma homogeneity over a large region around the fast electron propagation axis, but does not enable ρr to be varied, and therefore the effects of compression on the electron collisional stopping power to be measured. This configuration is adapted to identify compression-induced changes in the collective fast electron transport mechanisms. The so-called anomalous stopping power can be a non negligible source of energy loss for the injected fast ignition electron beam, mostly in regions close to the fast electron source (after the cone tip for a cone-based target or at the edge of a density channel hole bored in the coronal plasma), where the injected electrons current density is very high. The warm and dense plasma conditions we obtained by shock propagation in foil targets, reaching densities of $\sim 5\,\mathrm{g/cc}$ and temperatures $\sim 4\,\mathrm{eV}$, are representative of the fusion target coronal plasma after the fast electron source. Most of the dimensioning studies for fast ignition targets ignore this stopping power. Even if the fast electron energy deposition in the highly compressed nuclear fuel core is dominated by collisions, it is important to quantify the resistive stopping power in regions closer to the electron source, as this is likely to determine the maximum energy transported by the fast electrons.

2. Experimental results

Our experiment was conducted using the PICO 2000 laser system at LULI. We used long laser pulses (LP) with 180 J, 1.5 ns, $0.53\,\mu\mathrm{m}$, $500\,\mu\mathrm{m}$ diameter flat-top focal spot at 22.5° angle of incidence (with respect to the target normal axis) to compress the foils by shock-propagation. A counter-propagating high-intensity short laser pulse (SP), with 40 J, 1 ps, $1.06\,\mu\mathrm{m}$, giving a focused intensity $\approx 5\times10^{19}\,\mathrm{Wcm^{-2}}$ at 45° angle of incidence, generated the intense currents of fast electrons which propagated deep into the target. We used foil targets with propagation layers of different thickness (between 10 and $40\,\mu\mathrm{m}$) of either aluminium (Al) or plastic (CH). All the targets had a $10\,\mu\mathrm{m}$ CH as a generation layer (SP side of the target) and a $10\,\mu\mathrm{m}$ Cu tracer after the propagation layer, followed by $10\,\mu\mathrm{m}$ Al $+10\,\mu\mathrm{m}$ CH double layer on the LP side. The results for compressed targets were compared to the uncompressed, solid density targets (interactions without the LP beam). For each target we had previously studied the shock wave generated by the LP beam, in particular its velocity and the time needed to breakout on the SP beam side of the target. This was done by streaked-optical-pyrometry (SOP) of the self-emission from the SP-irradiated side of the target to achieve the correct relative timing of the LP and SP beams such that the compressed length of the target propagation layer was maximised whilst maintaining a cold, solid surface for SP beam interaction [3].

The results of the measurements of Kα emission from the Cu tracer are summarized in Figure 1. We imaged the Cu-Kα emission with a spherical Bragg crystal. From the radius of the signal as a function of the overdense thickness of the target (Fig. 1-a), we deduce a divergent fast electron propagation with a half-angle of $36\pm4^\circ$. The abscissa corresponds to the length between the relativistically corrected critical density and the middle of the Cu layer [3]. We do not measure a difference in the divergence as a function of the propagation-layer material, Al or CH, solid or compressed.

The Cu-Kα emission yields are plotted in Fig. 1-b) as a function of the target areal density (again corresponding to the distance from the coordinate of the critical density to the middle of the Cu layer). These yields correspond to the integration of the emission spectral lines measured with a conical Bragg crystal. We see that: i) in solid targets the Cu-Kα signal is stronger for an Al compared to a CH propagation layer; ii) in compressed targets the signal yields for Al and

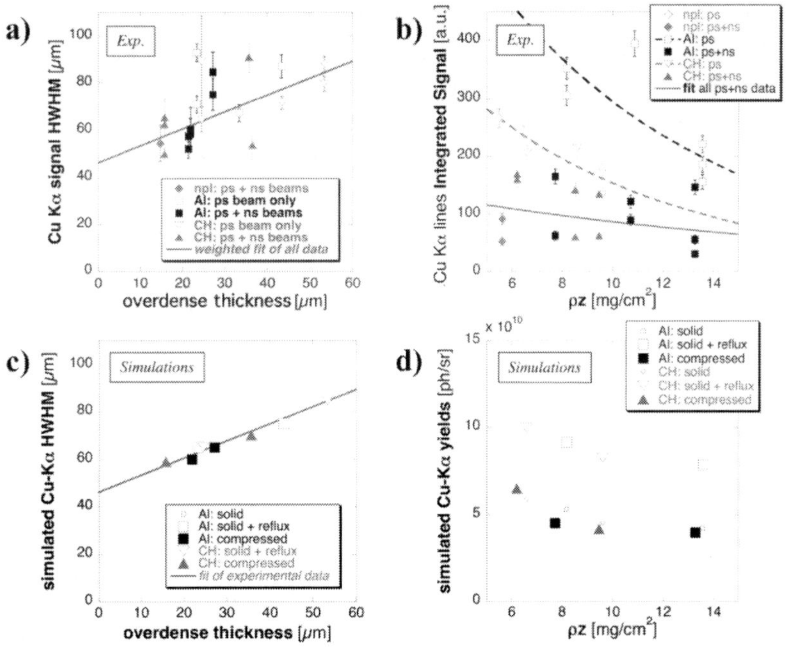

Figure 1. **a)** Experimental radius of the Cu-Kα emission as a function of the targets' overdense thickness. **b)** Experimental Cu-Kα yields as a function of the targets' areal density (the exponential curves fitting the data are plotted only to guide the eye). **c)** and **d)** Numerical equivalents of **a)** and **b)**, respectively.

CH are comparable; iii) signal yields are lower in compressed than in solid targets for both Al and CH propagation layers, the difference being larger in the case of Al.

3. Discussion

Point i) is explained by the electric field inhibition of the fast electron propagation in cold CH layers, as evidenced in previous work [4].

Point ii) is also well understood as both compressed Al and compressed CH propagation layers are partially ionised and behave like conducting plasmas. In fact, 2D hydro-radiative simulations using the code CHIC, accounting for the effects of the LP-induced shock and of the SP ASE-pedestal, showed that both the compressed Al and CH were heated up to $\sim 4\,\mathrm{eV}$ and had comparable densities of free electrons and comparable degrees of ionization.

Hybrid simulations of the electron transport help to understand point iii). The fast electrons source is injected as dictated by a CALDER-PIC simulation of the SP laser beam interaction with the density gradient created by the ASE-pedestal (the pre-plasma density profile was evaluated for $10^{19} < n_e < 10^{20}\,\mathrm{cm}^{-3}$ by side-on interferometry 300 ps before the intense part of the pulse and cross-checked by 2D CHIC hydro calculations [3]). The results from the hybrid transport simulations, including a Kα emission package, are summarized in Figures 1-c) and 1-d). From this first set of simulations, we can appreciate in Fig. 1-c) that the calculated radial width of the Kα emission from the Cu fluorescent layer perfectly reproduces the linear fit of the experimental measurements. In agreement with the experiment, there is no divergence dependence on the propagation-layer material. In Fig. 1-d) the calculated Kα emission yields qualitatively reproduce the experimental trends for the compressed targets (full symbols) but this is not the case for the solid targets (small open symbols).

The higher fluorescence yields measured for the solid targets can only be reproduced if fast electron re-injection into the target at the unperturbed rear surface is included in the simulations (big open symbols in Fig. 1-d). There is a high probability that these refluxing fast electrons propagate again across the Cu layer, in the opposite direction, raising (approximately doubling) the time-integrated fluorescence yields. This cannot happen in the compressed case, as the LP beam creates a several-100s μm-long and underdense coronal plasma behind the ablation front. We investigated the effect of the refluxing electrons in the hybrid simulations using ideal elastic scattering at the target rear surface. This edge condition is completely artificial, but is justified by PICLS simulations (without collisions) of the fast electron transport including large regions beyond the targets initial limits, thus simulating the electrons trajectories in the front and rear side sheaths. The simulations reveal that after 2 ps, a factor of ~ 2 times more electrons having crossed the Cu fluorescent layer compared to the case without refluxing. This supports the use of the artificial edge-reflection conditions in the hybrid simulations.

Even with electron refluxing we cannot, however, explain the measured differences between $K\alpha$ yields from solid CH and solid Al targets. We tested the effect of changing the electron source radius and injection angular distribution in the simulations, within the limits for which the resulting $K\alpha$ signals compare quite well to the experimental data and their error bars. By altering these two parameters, it was found that if the fast electron current density j_h is kept between 10^{10} and $10^{12}\,\mathrm{Acm^{-2}}$ through $15\,\mu$m of material, the electron beam energy loss rate due to collective mechanisms is 10-20% higher in compressed Al than in solid Al at equal depth. This is mostly attributed to the resistivity increase of the compressed Al as it is heated to a temperature close to the Fermi temperature. The difference is relatively small because the resistivity differences exist only for the first 0.8 - 1 ps. This is the time needed for the target bulk to be heated by the fast electron beam to temperatures $> 10\,\mathrm{eV}$, attaining the Spitzer transport regime in either the initial solid or warm-compressed Al, and where no differences are expected between the two cases. For $j_h < 10^{10}\,\mathrm{Acm^{-2}}$, resistive effects are negligible compared to the collisional energy losses. For $j_h > 10^{12}\,\mathrm{Acm^{-2}}$, the rate of collective energy losses increases, but the head of the fast electron beam heats the background material so rapidly that we can practically consider the Spitzer regime as the initial state for both solid and compressed cases.

For the CH targets, the yields of the $K\alpha$ signal are also weaker for the compressed case compared to the solid case, but the difference is less than for Al - the effects of electron refluxing in the non-compressed case compensates for the effects of the higher resistivity of cold plastic. For our intense interaction regime, producing high current densities, the initial higher resistivity of CH compared to Al is very rapidly cancelled by the fast electrons energy deposition and target heating. This is different to what has been reported in previous work performed in a non relativistic regime [5].

4. Conclusions

We have identified a regime for the fast electron current density, $10^{10} < j_h < 10^{12}\,\mathrm{Acm^{-2}}$, for which differences in the collective transport in solid Al compared to compressed Al are measurable for our experimental conditions. The complete simulated picture of the experiment is ongoing and involves correct modelling of the CH conductivity as a function of temperature and the fast electron source radius as a function of electron energy and time. Nevertheless, our results show the importance of electric field effects, changes in plasma conductivity and insulator-to-conductor phase transitions on fast electron transport in dense matter.

References

[1] Nakamura H *et al.* 2008 *Phys. Rev. Lett.* **100** 165001
[2] Perez F *et al.* 2009 *Plasma Phys. Control. Fusion* **51** 124035
[3] Santos J J *et al.* 2009 *Plasma Phys. Control. Fusion* **51** 014005
[4] Pisani F *et al.* 2000 *Phys. Rev. E* **62** R5927
[5] Hall T *et al.* 1998 *Phys. Rev. Lett.* **81**, 5, 1003

Journal of Physics: Conference Series **244** (2010) 022061 doi:10.1088/1742-6596/244/2/022061

Design Considerations for a Cone in a Fast Ignition Capsule

Henry D. Shay[1], Daniel S. Clark, Peter A. Amendt, Max Tabak, Michael H. Key, Michael M. Marinak, and Mehul V. Patel

Lawrence Livermore National Laboratory, Livermore, CA 94551

hdshay@llnl.gov

Abstract. An alternative to inertial fusion with central ignition is "fast ignition", in which one laser compresses the DT fuel adiabatically and a second laser with a short, very intense pulse heats the compressed core with super-thermal electrons[1]. One approach to fast ignition entails the introduction of the second laser beam via a hollow cone that pierces the side of the capsule. Critical considerations for the design of the cone in such an experiment include:

- Perturbation of the implosion by the cone
- Minimization of the column density of material between the critical density surface for the ignitor beam and the converged high density region
- Positioning, alignment, and shape of the cone to minimize deleterious hydrodynamic effects
- Effect of radiation gradients around the cone on the symmetry of the implosion.

This study entails the 2D and 3D simulations of a fast-ignitor experiment having a cryogenic deuterium-tritium capsule imploded within a high-*Z* hohlraum heated by about 650kJ of 3ω laser beams on the NIF.

1. Introduction

Inasmuch as the relativistic electrons generated by the high-intensity, short-pulse ignitor beam will probably have a broad angular distribution[2], absorption at the critical density surface of a capsule will result in few of the electrons traversing the compressed DT core and, consequently, only a small fraction of the ignitor beam energy absorbed in the compressed core. Using a hollow reentrant cone to provide an unobstructed path to the compressed core might be a solution for this problem, but introduces a number of its own difficulties: (1) asymmetry of implosion, (2) cone metal impurity mixing with the DT fuel, (3) metal spray from shocking the inner surface of the cone and subsequently occluding the ignitor beam, (4) the tip of the cone being crushed and/or penetrated by the high pressures generated in the compressed DT core. This study seeks a design which successfully addresses these problems.

Our method is to use a single capsule design[3] and to conduct 2D HYDRA[4] simulations to find an optimal cone design for that capsule. The capsule design has Be and Be/Cu ablator layers and DT ice. It is driven by x-radiation with a peak equivalent radiation temperature of 210eV. The time of peak ρR in the compressed DT fuel is about 32.5ns. Calculations reported here have used a time and frequency dependent x-radiation source generated by 1D radiation-hydrodynamic hohlraum calculations. After

[1] To whom any correspondence should be addressed.

completing this 2D study, 3D HYDRA calculations that model the radiation fluxes in the cylindrical hohlraum with the cone protruding from the hohlraum wall at the center plane and with NIF laser illumination of about 650kJ will be conducted.

Several design parameters have been investigated:

- inner and outer angles of the conical surfaces with respect to the axis of the cone
- thickness of the nose of the cone on axis
- distance of the nose of the cone from the center of the capsule
- distance from the center of the capsule to the intersection of the projection of the outer conical surface with the axis of the cone
- coefficient of the P1 asymmetry in the x-ray flux incident upon the ablation surface.

Key results from the simulations are summarized by several quantities as noted in Figure 1:

- peak ρR in the DT, evaluated perpendicular to the axis
- X, distance from the critical density surface to the 100g/cc contour in the compressed core at the time of peak ρR
- ρX, the column density along the distance X
- location of the shock front on the inner surface of the cone near the tip at the time of the ignitor beam

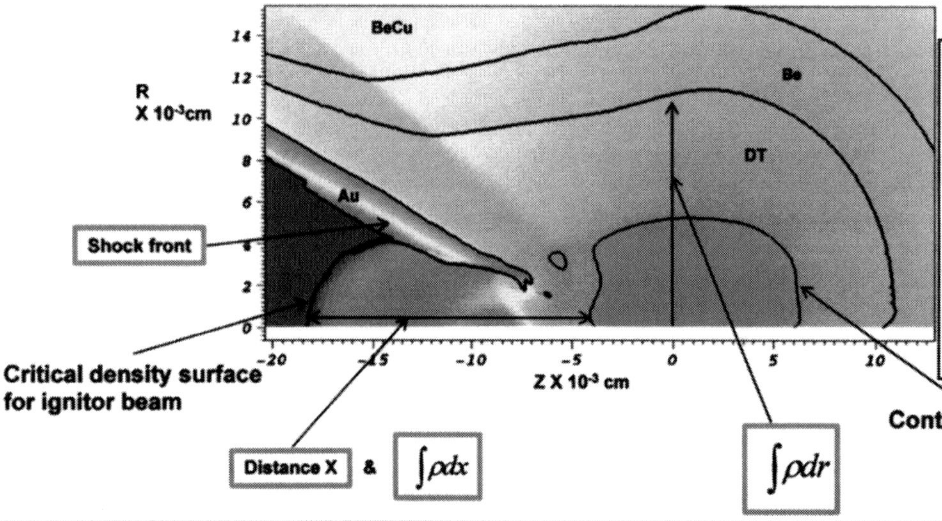

Figure 1. This picture shows the region boundaries and a pseudocolor plot of log ρ. Four key figures of merit that result from a 2D HYDRA simulation, taken at 32.5ns, roughly the time of peak ρR are shown in red boxes.

As illustrated in Figure 1, the pressure generated in assembling the DT core is so great, typically exceeding 50Gbars, that the DT punches through the tip of the cone. According to these calculations, then, the ignitor beam would be incident on the DT jet, with only a thin wisp of cone metal (gold in these simulations) remaining.

Since the fusion efficiency of the DT is proportional to $\rho R/(\rho R + 6g/cm^2)$, it is clearly advantageous to maximize ρR. Because the energetic electrons generated by the ignitor beam have such a broad angular distribution, the distance X should be less than or comparable to the size of the compressed core; its 100g/cc contour typically has a radius of about 60μm. To reduce energy loss of these electrons in traversing the remaining metal of the cone tip and the low density DT jet, the column density over X should be small, well less than $1g/cm^2$. If a shock breaks out of the inside surface of the cone near the tip before the illumination of the ignitor beam, the path of the laser beam may well be occluded.

2. HYDRA calculations

These HYDRA calculations were conducted with 6 user blocks and 40 domains on 40 processors. The angular zoning in the capsule was about 1°. The finest zone on the inner surface of the DT ice was 225nm, as determined by a convergence study. The finest zone in the Be ablation region was 75nm. The finest zone on the outer surface of the cone was 175nm. These calculations employed flux-limited x-radiation diffusion approximation for radiation transport.

These calculations display extreme shear between the motions of the imploding DT and capsule shells and the relatively immobile cone surface. Several Arbitrary Lagrange Eulerian (ALE) options were exercised in order to maintain fairly regular zoning. One technique was to permit the nodes to "relax" as various criteria were satisfied and, additionally, to schedule more extensive remapping at specified intervals. A second technique was to map the zoning from the unperturbed side of the implosion onto the side with the cone perturbation during much of the implosion. Thereafter, techniques similar to the first ALE methods were employed. Aggressive ALE methods, used to reduce mesh distortion, had the deleterious effect of coarsening the zoning on the outer surface of the cone and so artificially reducing any ablation of the cone resulting from short wavelength Au M- and L-band radiation (2-5keV and 8-12keV, respectively) penetrating through the ablators shells.

As shown in Figure 2, the implosion along the cone lags the implosion elsewhere in the capsule. This effect does not stem from the influence of viscosity. Near the cone a lower ablation pressure is achieved since a shock is launched into the cone, thus relieving some of the ablation pressure. Moreover, outside of the capsule the cone is strongly ablated sending a large plume of ablated gold that subsequently shadows the capsule near the cone.

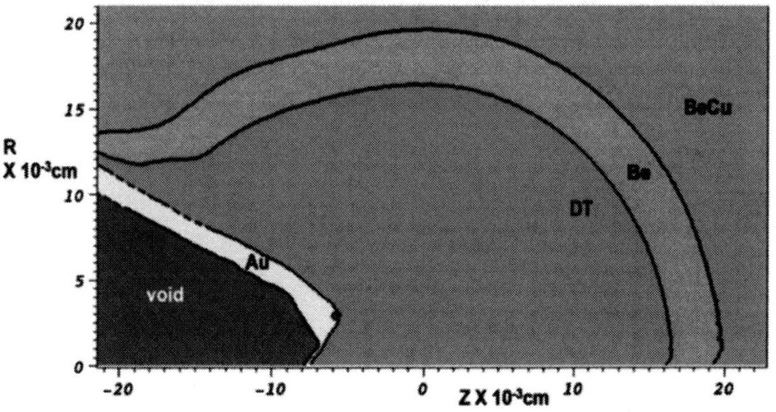

Figure 2.
This illustration, taken at 32ns, shows material boundaries. The implosion is clearly lagging on the side with the cone.

The implosion along the cone lags so much that, by the time of peak ρR (see Figure 1), some of the DT has been trapped next to the cone. This implies that gold ablated from the surface of the cone would not be entrained and carried down to contaminate the compressed core. The coarsening of the mesh by ALE relaxation might have resulted in underestimating the ablation of the gold and might have caused this entrainment to be neglected. Test 1D calculations with finer zoning in the gold and with the gold zoning kept Lagrangian strongly suggest that toward the end of the implosion and near the tip of the cone short wavelength M- and L-band photons originating in the areas of laser absorption will generate ablated plumes of gold extending as much as a few tens of microns. Test 1D calculations indicate that much of that ablation can be adequately suppressed with low-*Z* coatings (eg., Be, CH, or diamond).

Another consequence of the implosion lagging along the cone is that the shock from the unperturbed side strikes the nose of the cone before the arrival of shock from the side with the cone. Consequently, we have explored calculations with a first-order Legendre perturbation (P1) to advance the implosion on the cone side. In order to perform calculations using flux-limited diffusion, we inhibited transverse (not radial) diffusion outside of the ablation surface. To represent adequately the ablation of the cone at radii greater than that of the ablation surface, we permitted transverse diffusion closer to the cone. This suppression factor varied smoothly by eighteen orders of magnitude from angles near the cone to those angles on the side without the cone. Because the implosion had a relatively low Mach number (~3) as the implosion "launched" off the tip of the cone, it tended to fan out rather than proceeding directly ahead as it would with a much higher Mach number. This factor tended to make the implosion near the tip of the cone cylindrical and to exacerbate tip penetration by the DT jet.

3. Conclusions

These simulations attained peak ρR of 1.3 to 1.9g/cm^2, distances X of 100 to 140μm, column densities along X of 0.15 to 0.30g/cm^2, and shock breakout times preceding the time of peak ρR by 100 to 500ps. As the ρR reaches its peak, it tends to plateau so that illuminating with an ignitor beam earlier would not greatly decrease the burn efficiency. This expedient would decrease X and retard the timing of the shock break-out time until after the ignitor beam. Thickening the cone near the tip by larger values of projection of the outer conical surface past the center would also shift the shock break-out times later. We are continuing this tuning and are making progress in ensuring cone tip survival.

While the column density between the critical density surface and the compressed core is small enough to be negligible. The distance X is larger than needed to ensure the absorption of much of the energy of the electrons, and we are continuing the tuning to shorten this distance. Some recent papers[5] suggest that self-generated magnetic fields might constrain the angular divergence of the electrons though it is unclear whether this effect is present in the parameter regime of interest.

We are also conducting HYDRA calculations with implicit Monte Carlo transport of x-radiation photons in order to compute more accurately the variation of the ablation pressure with polar angle.

This work was performed under the auspices of the U.S. Department of Energy by Lawrence Livermore National Laboratory under Contract DE-AC52-07NA27344.

References:

[1] Max Tabak, *et al.*, Phys. Plasmas **1**, 1626 (5).
[2] R. B. Stephens, *et al.*, Phys. Rev E **69**, 066414 (2004).
Y. Sentoku, *et al.*, Phys. Plasmas **6**, 2855 (1999).
J. Green, *et al.*, Phys. Rev. Letters **100**, 015003 (2008).
K. Lancaster, et al., Phys. Rev. Letters **98**, 125002 (2007).
[3] D. S. Clark and M. Tabak, Nuclear Fusion **47**, 1147 (2007).
[4] M.M. Marinak, *et al.*,Phys. Plasmas **8** , 2275 (2001).
[5] A. P. L. Robinson and M. Sherlock, Phys. Plasmas **14** , 083105 (2007).
S. Kar, Phys. Rev. Letters **102**, 055001 (2009).
M. Storm, *et al.*, Phys. Rev. Letters **102**, 235004 (2009).

VALIS: A direct Vlasov solver for modelling laser plasma interaction

N. J. Sircombe[1,2] and T. D. Arber[2]

[1]AWE Plc, Aldermaston, Reading, Berkshire, U.K.
[2]Centre for Fusion, Space and Astrophysics, Department of Physics, University of Warwick, Coventry, U.K.

nathan.sircombe@awe.co.uk

Abstract. An accurate treatment of the relativistic Vlasov-Maxwell system is of fundamental importance to a broad range of plasma physics topics of direct relevance to inertial confinement fusion and fast ignition. With recent advances in high performance computing, the direct solution of Vlasov's equation is becoming increasingly viable as an additional modelling tool, alongside established particle in cell methods. We introduce VALIS, an algorithm for the numerical solution of the Vlasov-Maxwell system in two spatial dimensions with two momentum dimensions, and demonstrate its application to the problem of modelling short-pulse laser absorption in short-scale length, over-dense targets with system parameters comparable to those achieved on the HELEN laser system at AWE.

1. Introduction

With high intensity, short pulse, laser systems we are able to heat matter to extreme temperatures creating exciting opportunities for high energy density physics experiments and exotic inertial fusion schemes, such as Fast Ignition[1]. The mechanisms by which the short-pulse beam delivers its energy, how this energy is distributed and how it is transported into the target are complex and interdependent. A detailed understanding of these problems is essential for designing and interpreting experiments on future laser systems, such as ORION and VULCAN 10PW, and will be of paramount importance to the success of the Fast Ignition path to Inertial Confinement Fusion (ICF).

Here we consider the absorption of incident laser energy in a short scale-length density ramp using a recently developed direct Vlasov solver, VALIS[2]. Kinetic models, such as VALIS, derive from a statistical treatment of the plasma retaining information about the distribution of particle momenta absent from hydro models. In addition to the conventional spatial dimensions, a kinetic model may have up to three velocity (or equivalently momentum) dimensions. Together, these describe a phase space of up to six dimensions. The complexity of such models allows a detailed consideration of non-linear kinetic phenomena, but renders problems analytically and computationally challenging. These models can be solved, together with the appropriate equations of electrodynamics, either directly or via particle in cell (PIC) methods). The PIC approach is widely adopted owing to its flexibility, stability and underlying simplicity but the noise free solution offered by direct Vlasov solvers is often desirable[3].

2. The relativistic 2D Vlasov Maxwell system

We adopt a relativistic Vlasov-Maxwell system with two spatial (x,y) and two momentum (ux, uy) dimensions – a '2D2P' system. Vlasov's equation for the electron distribution function f is given by:

$$\frac{\partial f}{\partial t}+\frac{\mathbf{u}}{\gamma}\cdot\nabla_{\mathbf{x}}f-\frac{e}{m_e}\left(\mathbf{E}+\frac{\mathbf{u}}{\gamma}\times\mathbf{B}\right)\cdot\nabla_{\mathbf{u}}f=0 \qquad (1)$$

where: $f=f(\mathbf{x},\mathbf{u})=f(x,y,u_x,u_y)$; $\mathbf{u}=\gamma\mathbf{v}=\gamma(v_x,v_y,0)$; $\gamma=\sqrt{1+|\mathbf{u}|^2/c^2}$; $\mathbf{E}=(E_x,E_y,0)$ and $\mathbf{B}=(0,0,B_z)$.

Maxwell's equations are updated on a staggered grid via a second order finite difference scheme. In these simulations, the ions are treated as an immobile, initially neutralising, background but VALIS has recently been enhanced to allow for the consideration of an arbitrary number of species, on different momentum grids with arbitrary charge and mass.

The most common numerical approach to Vlasov's equation is to adopt some form of time-splitting scheme[4,5]. Conservative[6,7] and semi-Lagrangian[8] approaches have been most popular, more recently some attempts have been made at adaptive Vlasov codes[3]. In a comparison of conservative schemes[7], the Piecewise Parabolic Method (PPM)[9] has been shown to be both robust and accurate across a range of tests. The PPM method is conservative and formally third order away from strong limiters. PPM also ensures monotonicity so that false oscillations, potential sources of kinetic instabilities, cannot be introduced. While for some tests the semi-Lagrangian scheme may be more accurate than high-order conservative schemes (such as PPM), these differences are small[10]. In addition semi-Lagrangian schemes require extra steps to enforce positivity, these additional steps have to date only been implemented in dimensionally split form. Unfortunately the semi-Lagrangian technique must be applied in full multidimensional, i.e. unsplit, form for relativistic problems to ensure global conservation[11]. VALIS is parallelised via domain decomposition - this is a necessity given the need to advance quantities on a 4D grid. The simulations discussed here were run on grids of at least (nx, ny, nux, nuy) = (640, 192, 210,210) distributed over between 256 and 640 processing elements (PEs). VALIS has been tested on up to 1024 PEs on AWE's CRAY XT3 system and demonstrates near-ideal scaling (a speed up of ~1.994 for each doubling in the number of PEs).

3. Short pulse absorption in short scale-length plasmas

We consider the interaction of a short laser pulse with a steep, exponential density ramp. VALIS adopts a system of normalised, dimensionless, units in which velocities are normalised to the speed of light in vacuum and time is normalised to the inverse of the electron plasma frequency. Key system parameters for the simulations discussed here are given in Table 1. We consider three density scale lengths: 'short', 'medium' and 'long' - although all are effectively short when compared with long-pulse laser-plasma interactions. The density ramp is angled at between 0 and 30 degrees to the P-polarised incident beam.

Figure 1 shows the transverse electric field and electron density profiles for one case. The interaction of the incident beam with the density ramp is clearly visible as is the reflection of the beam at the critical surface. In all cases, populations of hot electrons are generated which move into the target (see Fig. 2). It is common for this hot population to be characterised by a single temperature T_{hot}. However, it is often inadequate to describe the complex phase space structures generated in the absorption region with a single 'temperature'. Often the distribution function contains beams and plateaux, this makes estimating T_{hot} a problematic and subjective undertaking. We define T_{hot} in terms of the normalised directed energy calculated from the electron distribution function

$$T_{hot}\approx\frac{\int E_{n_x}(\mathbf{u})K(\mathbf{u})f_e\,d\mathbf{u}}{\int K(\mathbf{u})f_e\,d\mathbf{u}} \qquad (4)$$

where E_{nx} is the directed energy in x and K is the kinetic energy, both as a function of momentum. Since the aim, ultimately, is to model how the deposited energy is absorbed and transported into the dense material, it is also useful to compare the change in the energy of the electron population at the top of the ramp - in the dense material but away from the absorption region.

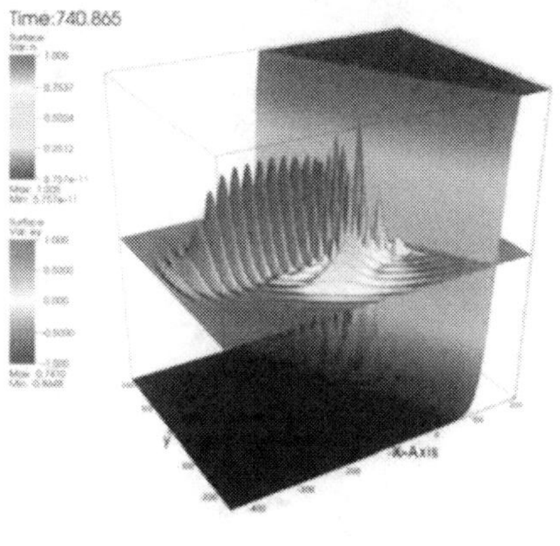

Figure 1: Density profile and transverse field for the medium scale length short-pulse laser plasma interaction simulation, angled at 30 degrees, at t = 740, equivalent to 105fs.

Parameter	Simulation Parameters	VALIS units
Temperature	5keV	0.02
Number density	1.58e28 m^{-3}	1.0
Scale length, 'short'	0.05 μm	1.25
'medium'	0.5 μm	12.5
'long'	3.4 μm	80.0
Target angles	0°, 10°, 30°	0, 10, 30
Laser wavelength, λ	1.06 μm	8π
Laser pulse length	78.9 fs	560
Laser spot size	5 μm	120
$I\lambda^2$	5.5e18 $Wcm^{-2}\mu m^2$	157.9
Distance	0.042 μm	1

Table 1: System parameters for short-pulse LPI demonstration.

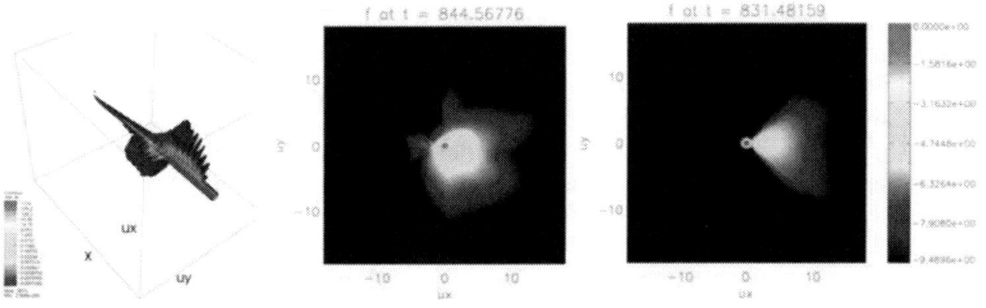

Figure 2: Left: Electron (x, u_x, u_y) phase space showing hot electrons produced by absorption of laser pulse into an overdense ($16n_c$) plasma ramp with a scale-length of ~0.05 laser wavelengths. Centre and right: Electron (u_x, u_y) phase space spatially integrated over the top of the plasma ramp; short scale length angled at 30 degrees (centre); long scale length angled at 0 degrees (right).

Table 2 summarises the results for various target conditions. The long scale length case absorbs much more of the incident energy but the result is to heat the foot of the ramp. This generates a hot, but not particularly numerous, population with an effective hot electron temperature in excess of 400keV (see Fig. 4), this population is not particularly well suited to carrying energy into the dense material, as can be seen in Table 2. In the short scale length case the laser interacts directly with the dense material, and despite utilising a smaller fraction of the available EM energy, the laser is able to heat the dense material more effectively. The medium scale length case, at normal incidence, is broadly similar to the short case. The average hot electron energy and absorption fraction are almost identical, however with the absorption region farther away from the top of the density ramp, we find that the heating in the dense material is reduced. The medium scale length case is improved significantly when the target is angled. The mean energy of the hot population in this case is significant and the absorption climbs to over 50%, although the total energy change at the top of the ramp remains smaller than all three short scale length cases.

Changing the angle of the short scale length target also produces an increase in the mean hot electron energy (as we would expect since the laser's electric field is now has a component into the density ramp) and improves absorption slightly. However, it proves less effective at raising the temperature at the top of the ramp.

The long scale length case was generally unaffected by a change in target angle from 0 to 10 degrees. At higher angles the refraction of the beam should be clearly apparent.

Run	f_{abs}	T_{hot}	ΔE_{dense}
Short, 0°	0.30	100 keV	35 keV
Short, 10°	0.31	110 keV	34 keV
Short, 30°	0.37	200 keV	22 keV
Med., 0°	0.28	145 keV	12 keV
Med., 30°	0.52	300 keV	20 keV
Long, 0°	0.88	400 keV	8 keV
Long, 10°	0.90	440 keV	10 keV

Table 2: Absorption fraction, T_{hot} and average kinetic energy change of electrons the top of the ramp, in each case.

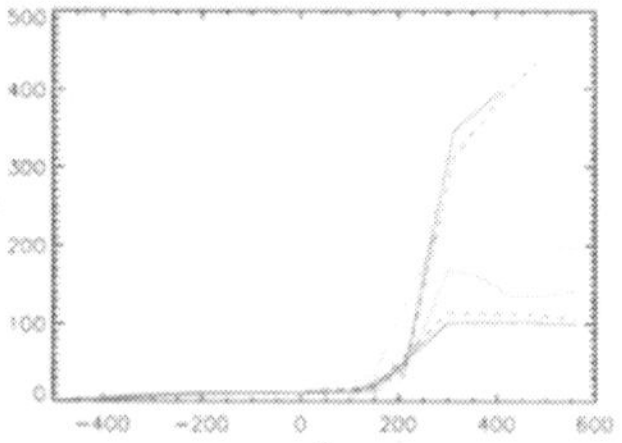

Figure 4: Estimate of T_{hot} at the top of the ramp for the long (red), medium (green) and short (blue) scale length cases at normal (solid line), 10° (dashed line) and 30° (dotted line) incidence. Time is given in VALIS units, relative to the instant when the laser reaches the front of the target.

4. Conclusions

Collimated, energetic beams will be able to heat material to higher temperatures and at greater depths, so the properties of the hot electron population produced during short pulse laser-matter interaction is an important consideration for high energy density physics experiments[12] and Fast Ignition schemes. These results indicate that the longer scale-length system absorbs more incident energy and produces the hottest population of electrons in the absorption. However there are other considerations, such as how much of this is delivered to the dense material of the target – in these simulations the average energy of the electrons, at the end of the simulation, is higher in the medium scale length case angled at 30 degrees to the laser normal. But the temperature change of the dense material is greatest in the shortest scale lengths, where the laser is interacting with the material directly. The impact of return current heating and collisonal effects will dominate transport beyond the absorption region, so a self consistent model for absorption and transport is needed to fully asses these issues. Also, it is not possible, in practice, to control the scale length of the pre-plasma so tightly. Even if measures are taken to mitigate the pre-pulse (as was the case for experiments on the HELEN laser at similar intensities[12]) there will still be shot-to-shot variation in the plasma scale length. Future work will study the properties of the hot electrons produced in similar, but more extensive systems, with mobile ions and over a wider range of initial parameters – particularly the irradiance. This will allow us to better understand the mechanisms which produce hot electrons and how these, along with the properties of the hot electron population, are influenced by the laser and plasma conditions.

5. References

[1] M Tabak et al.Phys. Plasmas, 1, 1626 (1994)
[2] N J Sircombe and T D Arber. J. Comp Phys., 228, 4773 (2009)
[3] N. Besse *et al.* J. Comp. Phys. 227, 7889 (2008).
[4] C.Z. Cheng and G. Knorr. J. Comp. Phys., 22, 330 (1976).
[5] A. Mangeney, F. Califano, C. Cavazzoni and P. Tavnicek. J. Comp. Phys., 179, 495 (2002).
[6] F. Filbet, E. Sonnendrücker and P. Bertrand. J. Comp. Phys., 172, 166 (2001).
[7] T.D. Arber and R.G.L. Vann. J. Comp. Phys., 180, 339 (2002).
[8] E. Sonnendrücker, J. Roche, P. Bertrand and A. Ghizzo. J. Comp. Phys.149, 201 (1999).
[9] P. Colella and P.R. Woodward. J. Comp. Phys. 54, 174 (1984).
[10] F. Filbet and E. Sonnendrucker. J. Comp. Phys. 150, 247 (2003).
[11] F. Huot *et al.* J. Comp. Phys. 185, 512 (2003).
[12] D. J. Hoarty et al."PPD Annual Report 2007", available from www.awe.co.uk.

Journal of Physics: Conference Series **244** (2010) 022063 doi:10.1088/1742-6596/244/2/022063

Simulations of electron-beam transport in solid-density targets and the role of magnetic collimation

A. A. Solodov, M. Storm, J. F. Myatt, R. Betti, D. D. Meyerhofer, P. M. Nilson, W. Theobald, and C. Stoeckl

Fusion Science Center and Laboratory for Laser Energetics, University of Rochester, Rochester, NY 14618

E-mail: asol@lle.rochester.edu

Abstract. Three-dimensional simulations of solid-target electron-transport experiments have been performed, using the hybrid-PIC code LSP. The experimentally observed fast-electron divergence half-angle of 16° in the target was reproduced assuming an initial divergence half-angle of ~56°, close to the value expected from the simple ponderomotive acceleration formula: $\theta_{1/2} = \tan^{-1}\left[\sqrt{2/(\gamma-1)}\right]$, where γ is the electron relativistic factor. The simulations accurately reproduce the details of the electron transport observed in the experiment. The electron beam propagates as an expanding annulus that breaks into filaments due to the resistive filamentation instability. The electron-beam partial collimation and annular propagation is due to the resistive azimuthal magnetic field generated at the outer edge of the electron beam.

1. Introduction

Powerful lasers focused onto the surface of a solid, produce large numbers of energetic electrons. Such energetic electrons can be used in the fast ignition approach to inertial confinement fusion (ICF). In fast ignition ICF [1] a laser generated beam of energetic electrons deposits its energy inside a precompressed deuterium-tritium fuel pellet, heating a small hot spot fast enough before the fuel expands hydrodynamically. Fast ignition requires that the energetic electrons remain collimated as they propagate through the target. Early experiments using multiterawatt laser systems observed a collimated fast electron flow in low Z targets [2]. Later experiments using imaging of K_α emission from high Z buried layer targets (see Ref. [3] and references therein) measured a divergent electron flow with half-angle divergence increasing with the laser intensity: from $\theta_{1/2}$≈12° at I≈3×10^{18} W/cm^2 to $\theta_{1/2}$≈27° at I≈5×10^{20} W/cm^2 according to Fig. 2 of Ref. [3]. Hot-electron collimation in the experiments of Ref. [2] was explained by the presence of self-generated resistive magnetic fields [4]. Resent two-dimensional hybrid-particle-in-cell (PIC) simulations [5] showed that the effectiveness of magnetic collimation decreases with the laser intensity and that the experiments of Ref. [3] can be explained by a partial (not-complete) collimation of hot electrons. It has been, however, noticed [5] that more complete three-dimensional simulations including the details of the resistivities for the target materials are required for a better quantitative agreement of the electron divergence half-angle in the experiments and simulations.

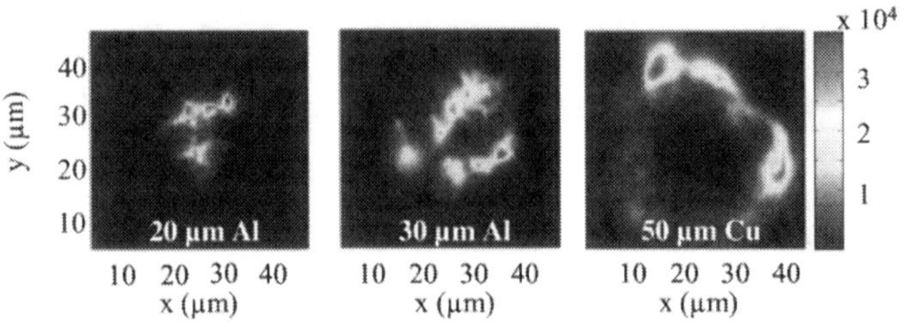

FIG. 1 Targets's rear-surface emission (in arbitrary units of intensity).

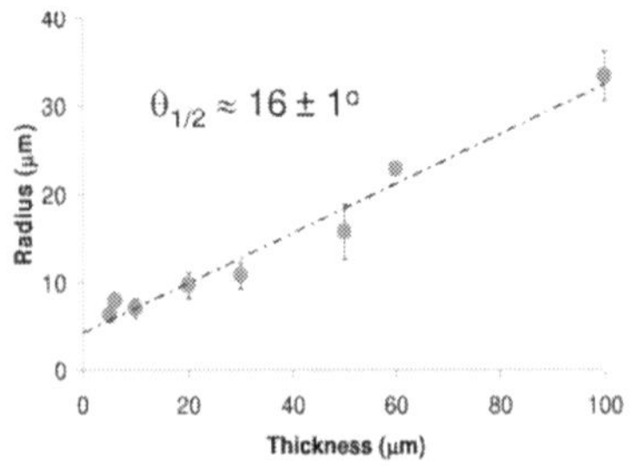

FIG. 2 Transverse size of the rear-surface emission versus target thickness.

Here we report the results from three-dimensional (3D) hybrid-PIC simulations of the electron-transport experiments recently performed at the Laboratory for Laser Energetics [6]. In this work, the effects of the target resistivity, including different target ionization levels, have been studied in detail. This detailed analysis provides more accurate modeling of resistive magnetic fields and electron transport than in Ref. [6] where electron collimation depended on the degree of target ionization (assumed to be fixed). The simulations confirm that electron transport can be explained by partial collimation of hot electrons by resistive magnetic fields. The initial divergence half-angle of the hot electrons in the target is found to be close to the value predicted by the ponderomotive scaling.

2. Summary of experiments

The experiments were conducted on the Multi-Terawatt (MTW) Laser Facility at the University of Rochester's Laboratory for Laser Energetics. A laser pulse of wavelength λ_L=1.053 μm, with an energy of ~5 J and a duration of Δt_L~650 fs, was focused with normal incidence to a 4-μm-radius spot, producing an intensity of ~10^{19} W/cm^2. The Al, Cu, Sn, and Au foil targets had transverse dimensions of 500 μm and thicknesses ranging from 5 to 100 μm. A coherent transition radiation (CTR) diagnostic was fielded to acquire images of the rear-side optical emission with a spatial resolution of ~1.4 μm.

Figure 1 shows three images of the rear-side emission plotted in arbitrary units of intensity. From left to right, the targets are 20-μm-thick Al, 30-μm-thick Al, and 50-μm-thick Cu. The emission contains small-scale structures, with a mean diameter of ~4.0 μm, superimposed on a larger annular feature whose diameter increases with target thickness.

Figure 2 shows how the size of the rear-surface emission region grows with target thickness. No dependence on the target material was observed. The half-angle electron-beam divergence was inferred to be $\theta_{1/2}$=16° using a least-squares linear fit.

3. The simulation model and parameters

The three-dimensional hybrid-PIC code LSP [7] has been used to model the transport of hot electrons in solid targets. The collisional model in LSP has been recently modified to include relativistic effects for hot electrons [5]. Separate species for hot electrons in different energy ranges are used to insure the correct scattering and slowing-down rates for electrons at different energy levels. The collisional model was tested to reproduce the correct ranges, blooming, and straggling of hot electrons, the correct resistive electric and magnetic fields, and to conserve energy. The simulations shown in this paper use the Thomas-Fermi ionization model [8] and equation of state [9] to calculate changes in the ionization state and specific heat capacity of the background electrons with the target temperature. The simulations also use the Lee and More resistivity model [10].

Hot electrons are promoted from the background of plasma electrons at the left-hand-side target boundary, having an exponential energy distribution, $\sim \exp(-E/\langle E_h \rangle)$. The mean energy is given by the maximum of the ponderomotive [11] and Beg [12] scalings: $\langle E_h \rangle[MeV] = \max\left\{0.511\left[\left(1 + I\lambda_0^2/2.8\times10^{18}\right)^{1/2} - 1\right], 0.1\left(I\lambda_0^2/10^{17}\right)^{1/3}\right\}$. Here I is the intensity of a

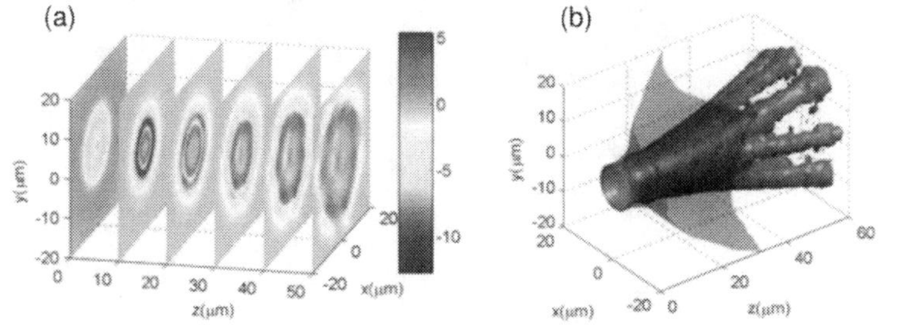

FIG. 3 (a) Cross sections through the azimuthal magnetic field (in units of MG) for the 60-μm target, 350 fs after the laser-pulse peak. (b) The fast-electron density isosurface at 50 % of the peak density in each (x, y) plane (red-solid) and the equivalent isosurface with the magnetic field artificially suppressed (blue-transparent).

Gaussian in space and time laser pulse having the maximum intensity of 1.44×10^{19} W/cm^2, full width at half-maximum (FWHM) of 5.54 μm, and duration τ=650 fs. The energy conversion efficiency to hot electrons is 20%, in agreement with LSP simulations of the experiments on MTW [13]. Similar conversion efficiencies for the present intensities were reported by others [14]. Electrons with energy $E_h = (\gamma - 1)mc^2$ are randomly injected in a cone with a half-angle $\theta_{1/2} = \alpha \tan^{-1}\left[\sqrt{2/(\gamma-1)}\right]$ which for $\alpha = 1$ describes the angle at which electrons are ejected from a focused laser beam by the ponderomotive force [15]. Simulations have been performed for different parameters α and Al targets with thicknesses 20-100 μm and transversal dimensions of 120 μm.

4. Simulation results

Figure 3 (a) shows cross sections through the azimuthal magnetic field 350 fs after the peak of the laser pulse in the simulation for $\alpha = 1$ and 60-μm thick target. Figure 3 (b) shows the location of the hot-electron-density isosurface (red-solid) at 50 % of the peak density in each transverse plane at the same moment of time; the blue semi-transparent surface corresponds to the case with the magnetic field artificially suppressed. The hot-electron beam is partially collimated by the resistive magnetic field generated at the outer edge of the electron beam. This field is most intense in the first 20 μm in the target. The magnetic field reduces the initial beam divergence half-angle of $\theta_{1/2} \approx 56^0$ (averaged within the FWHM of the beam spatial and temporal distribution) to $\theta_{1/2} \approx 16^0$, close to the beam divergence half-angle in the experiment. The variation of the beam density distribution with the propagation distance resembles an expanding annulus which breaks into filaments due to the resistive filamentation instability. Electron beam develops an annular shape because electrons propagating at larger angles to the axis are deflected by the edge magnetic field more than electrons propagating at smaller angles. The electron angular distribution peaks at about 16° to the axis and the density distribution develops a peak at the beam edge. The annular beam distribution is reinforced by magnetic fields generated at the inner and outer sides of the annulus, having opposite signs and collimating electrons into the annular region.

The collimating effect of magnetic field is significant, as it is seen from comparison of the beam density isosurfaces with and without magnetic field in Fig. 3 (b). The magnetic field, however, is not sufficient to completely collimate the beam because of its high initial divergence. Figures 4 and 5 compare the results of simulations for different α. The divergence half-angle increases (Fig. 4) from 0° (complete collimation) for $\alpha \le 0.7$ to more than 16° for $\alpha > 1$. Figure 5 shows the hot-electron-density distributions (for electrons with E>0.25 MeV) on the back of a 50-μm thick target, averaged in time using $\int n(\mathbf{r},t)dt / 1.065\tau$, for three values of α: $\alpha = 0.7$, 1, and 1.2. The annular density distribution in Fig. 5 (b) for $\alpha = 1$ is similar to the CTR distributions in Fig. 1. Notice that CTR is thought to be produced by the high-energy part of the hot-electron distribution. We studied the spatial distributions of electrons with energies between 1.5 MeV and 2 MeV and above 2 MeV (representing a relatively small fraction of electrons) and found similar annular distributions as for the bulk of hot electrons in Fig. 5 (b).

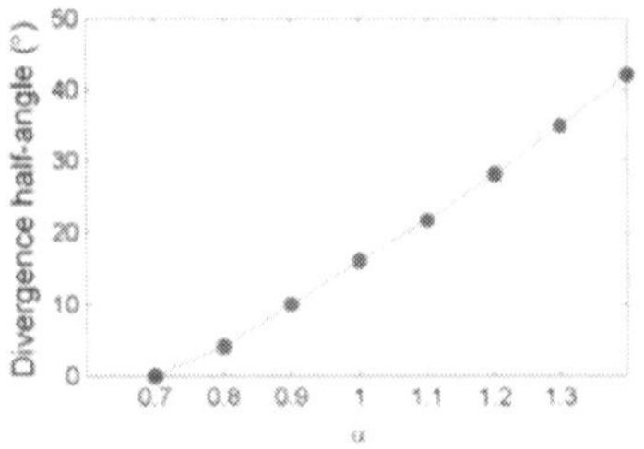

FIG. 4 Divergence half-angle versus α.

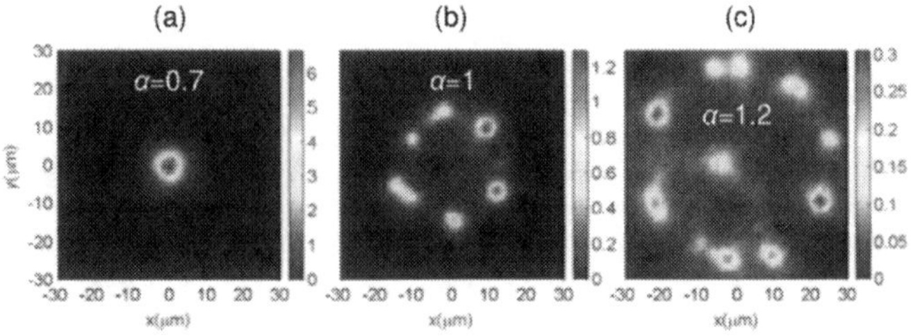

FIG. 5 Rear-surface transverse-density distributions of hot electrons for the 50-μm thick target.

5. Summary

3D hybrid-PIC LSP simulations are shown to reproduce the details of electron transport in the experiments on MTW. The hot electrons are partially collimated by the self-generated resistive magnetic field. The electron beam propagates as an expanding annulus that breaks into filaments due to the resistive filamentation instability. The experimentally observed fast-electron divergence half-angle of 16° has been reproduced for Al targets assuming an initial divergence half-angle in the target of ~56°, close to the value expected from the simple ponderomotive acceleration formula. Similar results were obtained in the simulations using approximate models for the plasma resistivity and specific heat capacity, when a constant degree of ionization of 8 for Al plasma was used [6]. Simulations for higher-Z targets are in progress. Preliminary results show more-diffused (less-spiky) electron-beam density distributions due to enhanced electron scattering in the target. The overall divergence of hot electrons, however, is not increased for sufficiently thin targets for which the CTR signal in the experiment was distinguishable [6], generally thinner than the hot-electron mean-free pass. These simulations will be reported elsewhere.

6. Acknowledgments

This work was supported by the U.S. Department of Energy under Cooperative Agreement DE-FC02-04ER54789 (Fusion Science Center, Office of Fusion Energy Science) and DE-FC52-08NA28302 (Office of Inertial Confinement Fusion), the University of Rochester, and the New York State Energy Research and Development Authority. The support of DOE does not constitute an endorsement by DOE of the views expressed in this article.

References

[1] M. Tabak *et al.*, Phys. Plasmas **1**, 1626 (1994).
[2] M. Tatarakis *et al.*, Phys. Rev. Lett. **81**, 999 (1998); M. Borghesi *et al.*, Phys. Rev. Lett. **83**, 4309 (1999); L. Gremillet *et al.*, Phys. Rev. Lett. **83**, 5015 (1999).
[3] J. S. Green *et al.*, Phys. Rev. Lett. **100**, 015003 (2008).
[4] A. R. Bell and R. J. Kingham, Phys. Rev. Lett. **91**, 035003 (2003).
[5] A. A. Solodov, K. S. Anderson, R. Betti, *et al.*, Phys. Plasmas **16**, 056309 (2009).
[6] M. Storm, A. A. Solodov, J. F. Myatt, *et al.*, Phys. Rev. Lett. **102**, 235004 (2009).
[7] D. R. Welch *et al.*, Phys. Plasmas **13**, 063105 (2006).
[8] R. M. More, Adv. At. Mol. Phys. **21**, 305 (1985).
[9] A. R. Bell, Rutherford Appleton Laboratory Report RL-80-091, 1980 (unpublished).
[10] Y. T. Lee and R. M. More, Phys. Fluids **27**, 1273 (1984).
[11] S. C. Wilks and W.L. Kruer, IEEE J. Quantum Electron. **33**, 1954 (1997).
[12] F. N. Beg *et al.*, Phys. Plasmas **4**, 447 (1997).
[13] P. M. Nilson, W. Theobald, J. F. Myatt, *et al.*, Phys. Rev. E **79**, 016406 (2009).
[14] K. Yasuike *et al.*, Rev. Sci. Instrum. **72**, 1236 (2001).
[15] C. I. Moore, J. P. Knauer, and D. D. Meyerhofer, Phys. Rev. Lett. **74**, 2439 (1995).

Journal of Physics: Conference Series **244** (2010) 022064 doi:10.1088/1742-6596/244/2/022064

Divergence of laser-generated hot electrons generated in a cone geometry

R.B. Stephens[1], K.U. Akli[1], T. Bartal[2], F.N.Beg[2], S. Chawla[2], C.D. Chen[4], L. Divol[4], R. Fedosejevs[5], R.R. Freeman[3], H. Friesen[5], E. Giraldez[1], D.S. Hey[4], D.P. Higginson[2], C. Jarrot[2], G.E. Kemp[3], M.H. Key[4], A. Krygier[3], D. Larson[4], S. Le Pape[4], A. Link[3], T.Y. Ma[2], A.J. MacKinnon[4], H.S. MacLean[4], A.G. MacPhee[4], C. Murphy[2], V. Ovchinnikov[3], P.K. Patel[4], Y. Ping[4], H. Sawada[2], D. Schumacher[3], Y. Tsui[5], M.S. Wei[2], L.D. Van Woerkom[3], B. Westover[2], S.C. Wilks[4], T. Yabuuchi[2]

[1]*General Atomics, San Diego, California 92186, USA*
[2]*Department of Mechanical and Aerospace Engineering, University of California San Diego, La Jolla, California 92093, USA*
[3]*The Ohio State University, Columbus, Ohio 43210, USA*
[4]*Lawrence Livermore National Laboratory, Livermore, California 94550, USA*
[5]*Department of Electrical and Computer Engineering, University of Alberta, Alberta T6G 2R3, Canada*

Email: rich.stephens@gat.com

Abstract. Short-pulse, ultra-intense lasers generate hot electrons at the cone tip in a Fast Ignition target. Core heating and cone-wire experiments find that about 20% of the incident laser energy is coupled into a target, but do not characterize electron propagation direction, a critical parameter for ignition. Previous studies using flat foils suggest they propagate forward, diverging by ~40°. Buried cone targets—conical cavities in multilayer metal foils—were developed to allow divergence measurements in an FI relevant geometry. Preliminary results show increased electron divergence in a 30 μm diameter cone tip which disappears for 90 μm diameter tips. Implications of the experiment are discussed.

1. Introduction

A defining feature of the fast ignition concept for inertial fusion energy is ignition of the fuel, during the short interval of maximum density, by energy injected using a separate short-pulse laser. [1] A reentrant cone was added to the originally proposed configuration to reduce the required energy transport distance. [2] The tip of that cone must be narrow to avoid interfering with the fuel compression. So the electrons must be launched from a confined volume. Forward-going electrons generated in this way have been extracted from a cone tip with attached Cu wire. [3] Laser to electron energy coupling has been shown to be ~15% for the largest diameter wire (40 μm). This is very similar to ~20% determined from integrated core-heating experiments. [4, 5] But in both cases, the spread of the electrons as they leave the cone tip, critical for the heating of a small ignition volume, could not be determined; the former because the electrons are artificially confined to the cone-wire assembly by electrostatic fields at its outside surfaces, and the latter because of lack of a spatially resolving electron diagnostic.

Divergence of laser generated electrons has been characterized using a flat interface, aluminum as the plasma, and the *K*-edge fluorescence from a buried Cu layer to determine their number and spread (Fig. 1a). [6] The typical electron divergence angle is 40°, apparently increasing with laser intensity. [7] In principle, this data should also apply to electrons generated inside a cone; a typical fwhm laser spot is ~10 μm diameter and, at the Titan laser where our experiments have been performed, anecdotal evidence suggests a pointing accuracy of ~5 μm; these dimensions are small enough to routinely land the laser energy on the 30 μm diameter flat cone tip. Recent measurements showing a reduction of

Journal of Physics: Conference Series **244** (2010) 022064 doi:10.1088/1742-6596/244/2/022064

~5X in coupling from cone tip to wire as the cone wall thickness increases, [8] as well as PSC PIC modeling showing strong perturbations caused by cone wall plasma from only nominal prepulse energy, [9] strongly suggested that electron divergence from cones might be very different than from flats.

In this paper we test that thesis with new buried cone targets that allow electron propagation characterization while approximating fast ignition conditions—a plasma-free conical space embedded in the blow-off from a compressed shell.

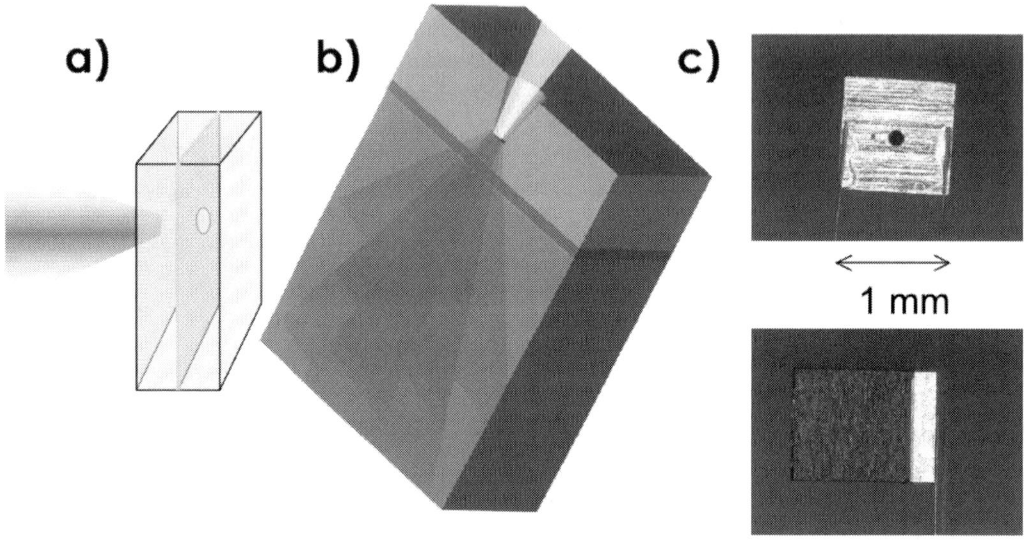

Figure 1: Targets used for measurement of electron divergence. a) Flat Al foil containing a buried Cu layer, b) Conical cavity in an Al foil. The incident laser beam and resulting electron flux are overlain in yellow and red, respectively. The far side of the Al is first coated with a 25 μm thick Cu layer, and then a C block added. c) Top and side view of a buried cone target

2. Experiment

The targets were 200 μm thick Al foils plated on one side with 25 μm thick Cu to which 1 mm thick C was glue, and a cone-shaped cavity cut into the other side (Fig. 1b). The cavity walls had a 15° half-angle opening and tip diameters of 30 μm (standard for cone experiments) and 90 μm. The cavities were cut either 100 μm or 190 μm deep so that their tips were 100 μm or 10 μm from the Cu layer. They were shot at the Titan laser facility at Lawrence Livermore National Laboratory (LLNL) using ~0.7 ps, ~150 J pulses; the f/3 beam was focused to the flat tip of the cone. A water-filter-protected fast diode monitored the prepulse and an equivalent plane setup monitored the focus on every shot. [10]

3. Results

The Cu *K* fluorescence was measured using an HOPG spectrometer, [11] for total emission, a spherically bent Bragg mirror imaging 8.03 keV radiation on an x-ray ccd camera [12] with a spatial resolution ~10 μm. Only the Bragg imager data has been analyzed to date. The fluorescence images show a spot diameter very similar to those from flat foils for the cones with 90 μm diameter tips, and with larger divergence for those with 30 μm diameter [Fig. 2]. Analysis of the fluorescence partition between peak and diffuse background, between shallow and deeply buried fluor layers is underway.

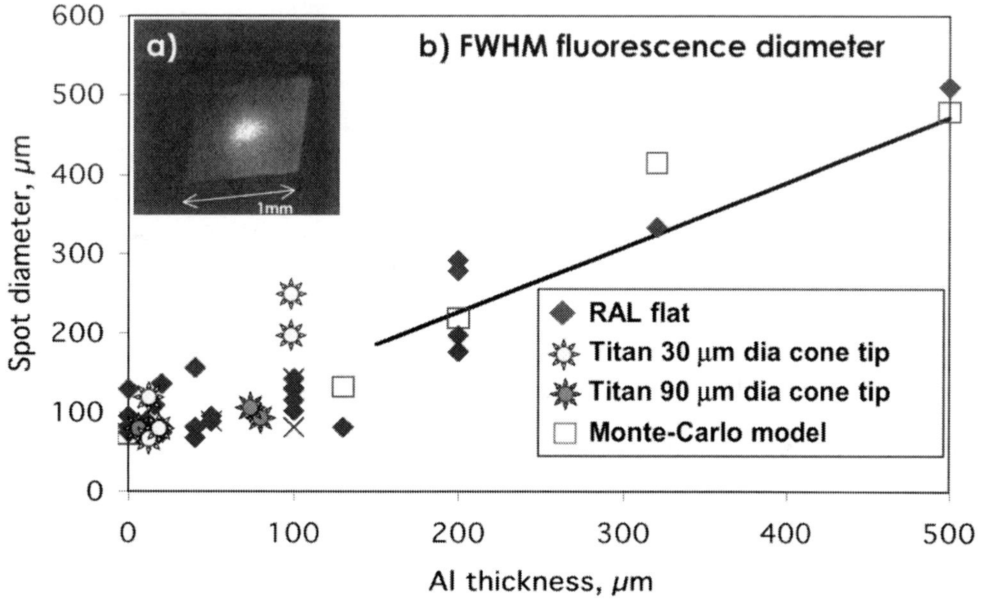

Figure 2: a) 8 keV fluorescence image from a buried cone target shown in Fig. 1 b,c. b) FWHM diameter of fluorescence from all those targets (yellow and red stars) overlaid on flat foil data (the blue diamonds) and a Monte-Carlo model (the open squares) from Ref. 6.

4. Discussion

One can see from the data in Fig. 2 that the cone walls, for the standard 30 μm diameter cone tip, clearly increased the electron dispersion; the data from ~100 μm deep fluor is substantially outside the scatter of data from the flat foils. More surprisingly, they did **not** change the spot diameter for a shallowly buried fluor layer. Aside from one large diameter outlier caused by a bad laser focus (as determined by the equivalent plane imager), they are virtually identical to each other and to the earlier flat foil results, and all ~3X larger than the 30 μm cone tip diameter only 5 μm from the fluor layer. The ~100 μm fwhm spots had been previously justified as the result of combined temperature and density gradients generating quasi-static fields [13] allowing electrons to flow out along the surface. [14, 15] That image is hard to justify when the surface is only 30 μm diameter. The flat foil data was explained in Ref. 5 with a model (the open squares in Fig. 2) in which the electrons dispersed widely enough to cause a large spot in the shallow buried fluor layer are of such low energy that they do not reach a deeper fluor. Alternatively plasma build-up inside the cone could cause electrons to be generated some considerable distance up the cone away from the tip; the PSC PIC simulation in Ref. 9 shows that even for a nominal 7.5 mJ prepulse, electron may be generated as much as 50 μm up the cone away from its tip. Analysis of the relative strengths of these spots combined with detailed modeling will be necessary to evaluate these alternatives.

5. Acknowledgements

The authors wish to thank the staff of the Jupiter Laser Facility, LLNL. This work was performed under the auspices of the U.S. DOE under contracts No.DE-FG02-05ER54834, DE-FC0204ER54789 and DE-AC52-07NA27344.

References

[1] M. Tabak, et al., *Phys. Plasmas* **1** 1626-1634 (1994).
[2] S. Hatchett et al., *Bull. Am. Phys. Soc.* **46**, 47 (2001).
[3] J.A. King et al., *Phys. Plasmas* **16**, 020701 (2009).
[4] R. Kodama et al., *Nature* **418**, 933 (2002).
[5] M.H. Key et al., *Phys. Plasmas* **15**, 022701 (2008).
[6] R.B. Stephens et al., *Phys. Rev. E* **69**, 066414 (2004).
[7] J.H. Green et al., *Phys. Rev. Lett.* **100**, 015003 (2008).
[8] K.U. Akli et al., "Hot electron generation and transport using Kα emission," elsewhere in this issue.
[9] A.G. MacPhee et al., "Limitation on pre-pulse level for cone-guided fast-ignition ICF," submitted to *Phys. Rev. Lett.* See Fig. 3a.
[10] S. LePape et al., *Opt. Lett.* **34**, 2997-2999 (2009).
[11] K.U. Akli et al., under preparation, to be submitted to *Rev. Sci. Instrum.*

[12] Y. Aglitskiy et al.," *Rev. Sci. Instrum.* **70**, 530-535 (1999).
[13] D.W. Forslund and J.U. Brackbill, *Phys. Rev. Lett.* **48** 1614-1617 (1982).
[14] J.M. Wallace, *Phys. Rev. Lett.* **55**, 707-710 (1985).
[15] R.R. Freeman et al., *J. Quant. Spect. & Rad. Transf.* **81**, 183-190 (2003).

Fast ignition transport simulations for NIF

D. J. Strozzi, D. P. Grote, M. Tabak, B. I. Cohen, R. P. J. Town, A. J. Kemp

Lawrence Livermore National Laboratory, 7000 East Avenue, Livermore, CA 94550, USA

E-mail: strozzi2@llnl.gov

Abstract. This paper shows work at Lawrence Livermore National Lab (LLNL) devoted to modeling the propagation of, and heating by, a relativistic electron beam in a idealized dense fuel assembly for fast ignition [1]. The implicit particle-in-cell (PIC) code LSP is used. Experiments planned on the National Ignition Facility (NIF) in the next few years using the Advanced Radiography Capability (ARC) short-pulse laser motivate this work. We demonstrate significant improvement in the heating of dense fuel due to magnetic forces, increased beam collimation, and insertion of a finite-radius carbon region between the beam excitation and fuel regions.

1. Simulation approach

A basic dilemma in fast ignition (FI) transport simulation is capturing the physics of the relativistic electron beam, while affordably describing the dense, cold ($\lesssim$ 1 keV) background plasma. Resolving high-frequency Langmuir waves is prohibitive, and not essential since they are strongly damped by high collisionality. Light waves also need not be resolved, since the plasma is far above critical density for laser light. To solve this problem we use the implicit particle-in-cell (PIC) method, which numerically damps unresolved modes ($\omega\Delta t \gg 1$) while fully resolving the physics for $\omega\Delta t \lesssim 1$ subject to some numerical modification of the dielectric shielding [2]. Unlike some hybrid approaches which use simplified Maxwell equations or a resistive Ohm's law, this method can simulate the complete physics at low density and *smoothly* transition to an effectively reduced description at high density.

We use the PIC code LSP [3] in 2D cylindrical r-z geometry for our simulations. We employ an electromagnetic direct-implicit (as opposed to implicit-moment) PIC algorithm related to that of Ref. [4]. However, the "implicit magnetization" term is neglected. In the numerical field equations (see Eq. (97a) in [2]), the term containing the implicit magnetization is smaller than the term containing the electrostatic "implicit susceptibility" (included in LSP) by roughly the factor $v_e\Delta t/2L$, where v_e is the electron drift speed and L is a length scale. For the extreme case of $v_e = c$ and $L = \Delta x$, this term is $< 1/2$ as long as the light Courant condition is respected (which we always do). We run with the fully-damped D1 scheme [5]. We do not simulate the short-pulse LPI, but instead excite an electron beam with a specified distribution from the background plasma. The E and B fields are found with an iterative ADI solver.

To reduce the cost of simulating the background plasma, we utilize LSP "fluid particles." They each carry mass, position, momentum (representing the drift of a thermal Maxwellian), and a temperature governed by an energy equation. Their evolution equations contain terms for fluid effects like heat conduction and pressure-gradient forces. These are automatically included in standard PIC with collisions, but must be explicitly added for fluid particles, whose momenta do not reflect any thermal spread. The fluid-fluid collision rate is the classical one for a hot plasma, without taking Fermi degeneracy or electron-electron collisions into account; this gives a resistivity scaling like $T_e^{-3/2}$ where T_e is electron temperature. This may be inadequate for our conditions, and is being improved. We initially have nine fluid particles per cell per species, much less than would be needed by conventional PIC to resolve a thermal Maxwellian.

We extensively modified the collisions of the beam electrons off other species. We found LSP's original algorithm generated significant energy error. Our method scatters a test species off a field species, but not itself. This suffices for beam number densities much less than the background, which applies here. We follow a spatially grid-based method similar to Manheimer [6] and Lemons [7] (we adopt Lemons' spherical-coordinate algebra). We find cell-averaged field-species moments (e.g., density and drift velocity), and increment each test particle's momentum $\vec{u} = \gamma\vec{\beta}$. On average, this yields a deterministic energy loss, and stochastic heating and angular spread. In the field species' drift frame, $\vec{u}$ changes from $\vec{u}_0$ to $\vec{u}_0 + \Delta\vec{u}$ with

$$\Delta u = -\nu_{ud,j}\Delta t + [\nu_{us,j}\Delta t]^{1/2} N_u \tag{1}$$

$$\Delta\theta = [\nu_{\theta,j}\Delta t]^{1/2} N_\theta, \qquad \Delta\phi = 2\pi U_\phi \tag{2}$$

$u = |\vec{u}|$, (θ, ϕ) are the (polar, azimuthal) angles with respect to $\vec{u}_0$, U_ϕ is a uniform deviate on $[0, 1)$, and N_u, N_θ are normal deviates with zero mean and unit variance. $j = (e, i)$ for field (electron, ion) species. The inverse process, namely collisions of the field off test species, is done by energy and momentum conservation in each cell.

Our exact formulas are those of Davies [8], which essentially match Solodov's [9].

$$\nu_{ud,e} = \nu_{0,e}\beta^{-2}L_{ud}, \qquad \nu_{ud,i} = \nu_{us,j} = 0, \qquad \nu_{\theta,j} = 2\nu_{0,j}\gamma^{-2}\beta^{-3}L_{\theta,j} \tag{3}$$

$$L_{ud} = \ln\left[\frac{m_e c^2}{\hbar\omega_{pe}}\frac{\beta(\gamma-1)^{1/2}}{2^{1/2}}\right] + 9/16 + \gamma^{-1}(\ln 2 + 1/8)(-1 + 1/(2\gamma)) \tag{4}$$

$$L_{\theta,e} = \ln\Lambda - (1/2)(1 + \ln[2\gamma + 6]), \qquad L_{\theta,i} = \ln\Lambda - (1/2)(1 + \beta^2) \tag{5}$$

$\nu_{0,j} = 4\pi c r_e^2 n_j Z_j^2$, $r_e = e^2/(4\pi\epsilon_0 m_e c^2) \approx 2.82$ fm, $\Lambda = 2\lambda_{De} m_e c\gamma\beta/\hbar$, and λ_{De} is the background electron Debye length. The energy loss is appropriate for a fast electron scattering off a cold, dense background, such that all electrons are effectively free (Fermi's density effect dominates any ionization-energy effects [8]). Inclusion of nonzero field species temperature would give $\nu_{us,j} \neq 0$ (stochastic heating).

2. Calculations on an idealized NIF-ARC coupling target

We present LSP transport calculations for a compressed target appropriate to coupling (as opposed to ignition) experiments on NIF-ARC. ARC entails converting four NIF long-pulse beams (one quad) to a short-pulse laser. This should deliver $\approx$ 9 kJ of laser energy in a 5 ps pulse, at a wavelength of 1054 nm, into a spot with intensity half-width at half-max of ≈ 25 μm. We do not simulate the short-pulse laser, but instead excite an electron beam inside the simulation domain with a prescribed distribution function. In particular, we excite a beam with 4.8 kJ of kinetic energy (assuming a feasible laser to beam conversion efficiency of 53%) in a 5 ps flattop pulse, with linear ramps of 0.1 ps.

The beam distribution is $f/f_0 = n(\vec{r}, t) f_E(E) f_\theta(\theta)$ where $\int dEd\theta\, f = n$, n is the beam density, f_0 is a scale factor, and θ is the polar angle of $\vec{u}$ with respect to the u_z axis. $f_E(E) = \exp[-E/(0.3\,\mathrm{MeV})] + 0.05\exp[-E/(3\,\mathrm{MeV})]$ for kinetic energy E between 0.2 and 10 MeV, and $f_E = 0$ otherwise. We commonly see such two-temperature distributions in explicit-PIC simulations of the short-pulse LPI. In angle, we use $f_\theta(\theta) = \sin\theta\exp[-(\theta/\Delta\theta)^2]$; the $\sin\theta$ factor is absent if one considers number per *solid* angle. The density profile is

Figure 1. (a) Initial plasma conditions for base case. Solid black: mass density ρ, solid red: temperature (same for all species), dash black: total pressure (initial peak value: 74 Gbar). (b) Increase in thermal energy in ignition region, as a fraction of total added beam energy. Solid black: base case; solid red: base case but material is pure D instead of CD; solid green: base case without magnetic forces; dashed black: base case but $\Delta\theta = 60°$; dashed red: pure D with $\Delta\theta = 60°$; dashed blue: D with C wire, $\Delta\theta = 60°$.

$n = n_0(t)\exp[-(r/25\mu\text{m})^4]$, with $n_0(t)$ giving the temporal pulse. We do not vary f_E or f_θ in space or time. For this beam, the peak intensity and n_0 are $6.4\cdot10^{19}$ W/cm^2 and $1.0\cdot10^{22}/\text{cm}^3$, or about 10 times the critical density for ARC. The beam electrons were excited from the background electrons in the region $(r,z) = (0,25)-(35,30)\,\mu$m.

Figure 1(a) displays the initial profiles, which are spherically symmetric about $z_0 = 140\,\mu$m. The initial plasma conditions represent an optimistic fast-ignition fuel assembly, in which no high-pressure "jet" coming from the core to the cone pushes the critical surface far from the fuel. All ions are fully ionized, which is reasonable given that we only use C and D ions and the initial temperature is >100 eV. The domain extends from $(r,z) = (0,0)-(100,240)\,\mu$m, the grid spacing was 0.5 μm except for an expansion of the r spacing to 1 μm over the first 5 μm to reduce spurious B fields near axis. The time step was 0.7 fs, and the run took 200 minutes of wall time on 38 Opteron cpu's to reach 7 ps of run time.

Figure 1(b) shows for several runs the fuel heating, or increase in background thermal energy in the "ignition region" of $(r,z) = (0,80)-(30,120)\,\mu$m. Our base case (solid black curve) has a C-D (equal parts atomic) plasma everywhere, and beam divergence $\Delta\theta = 35°$. About 30% of the beam energy gets deposited in the ignition region. The final pressure, and background T_e at several times, are shown for this run in Fig. 2. Turning off magnetic forces (solid green) on all species reduces the coupling efficiency to 13%. Replacing C-D with pure D and keeping the same mass density (and thus electron density) slightly increases the coupling to 34%. One might expect a decrease in resistive B fields with pure D due to the lower Z, and thus a reduction in magnetic collimation and finally coupling. A possible reason this does not happen is that although the average Z and resistivity are lower, the increased ion number density means the heat capacity of pure D is increased. Thus, the plasma stays cold and more resistive longer.

The CD and D cases were rerun with magnetic forces and a larger angular spread of $\Delta\theta = 60°$, giving the dashed black and red curves. They are close to each other, and the no magnetic force run. To mitigate the reduced coupling, we inserted into the pure-D case a pure-C "wire" from $(r,z) = (0,0)-(30,70)\,\mu$m, and smoothly varied the composition from C to D over 5 μm. A pinching azimuthal B is generated due to the radial resistivity gradient at the C-D interface (see Fig. 3). The resulting coupling (dashed blue curve) increases to about 21%. This shows the potential benefits of tailored material profiles to compensate for a poorly-collimated beam.

Journal of Physics: Conference Series **244** (2010) 022065 doi:10.1088/1742-6596/244/2/022065

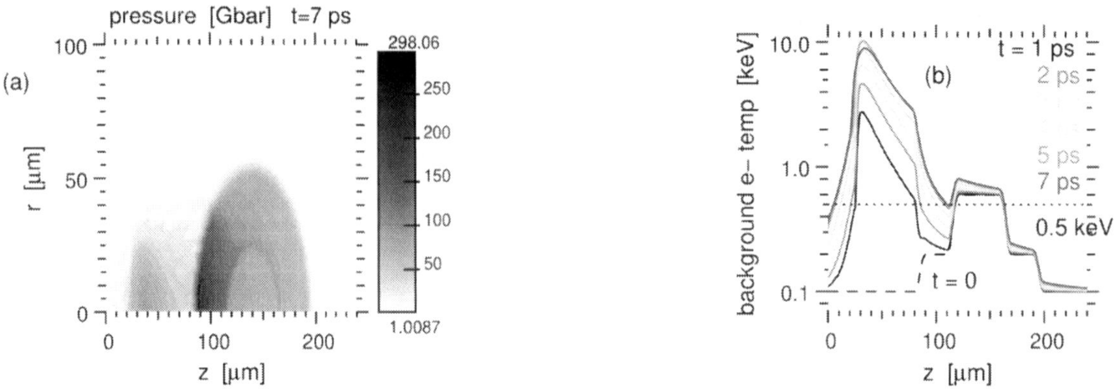

Figure 2. (a) Final (t=7 ps) background pressure profile for base case. (b) background electron temperature, averaged from $r = 0 - 10\,\mu$m, for several times for base case.

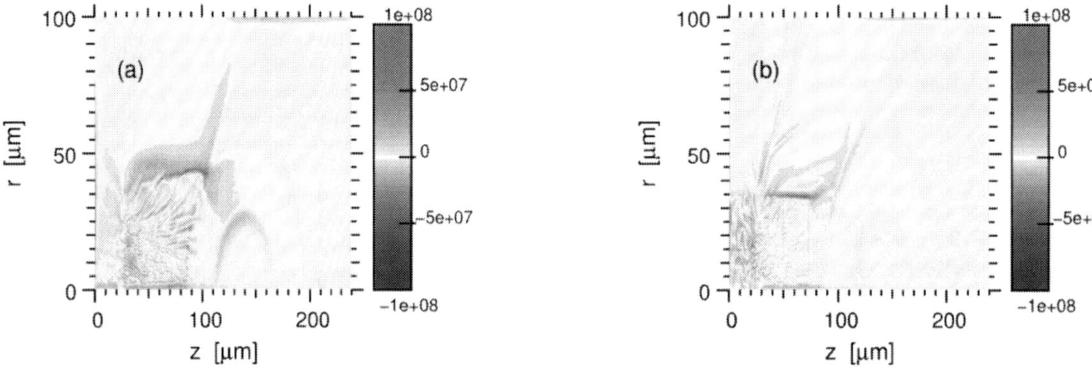

Figure 3. Azimuthal magnetic field B_θ at $t = 5$ ps for (a) base case and (b) D plasma, $\Delta\theta = 60°$, with C wire (dashed blue curve from Fig. 1(a)). $B_\theta < 0$ gives an inward radial force on electrons.

These results suggest the energy coupling for plausible electron beams and targets can be 20-30%. We are extending this work to more realistic configurations, and the full ignition problem.

Acknowledgments

We thank L. I. Divol, A. B. Langdon, and D. R. Welch for fruitful discussions. This work prepared by LLNL under Contract DE-AC52-07NA27344, LDRD tracking number 07-SI-001.

References

[1] Tabak M, Hammer J, Glinsky M E, Kruer W L, Wilks S C, Woodworth J, Campbell E M, Perry M D and Mason R J 1994 *Phys. Plasmas* **1** 1626–1634
[2] Langdon A B and Barnes D C 1985 *Multiple Time Scales* ed Brackbill J U and Cohen B I (Orlando, FL: Academic Press, Inc) pp 335–375
[3] Welch D R, Rose D V, Cuneo M E, Campbell R B and Mehlhorn T A 2006 *Phys. Plasmas* **13** 063105
[4] Hewett D W and Langdon A B 1987 *J. Comput. Phys.* **72** 121–155
[5] Cohen B I, Langdon A B and Friedman A 1982 *J. Comput. Phys.* **46** 15
[6] Manheimer W, Lampe M and Joyce G 1997 *J. Comput. Phys.* **138** 563–584 ISSN 0021-9991
[7] Lemons D S, Winske D, Daughton W and Albright B 2009 *J. Comput. Phys.* **228** 1391–1403
[8] Atzeni S, Schiavi A and Davies J R 2009 *Plasma Phys. Controlled Fusion* **51** ISSN 0741-3335
[9] Solodov A A and Betti R 2008 *Physics. Plasmas* **15** 042707

Features of a point design for Fast Ignition

M.Tabak,D.Clark,R.P.J.Town,M.H.Key,P.Amendt,D.Ho,D.J.Meeker,H.D.Shay,B.F.Lasinski,A.Kemp,L.Divol,A.J.Mackinnon,P.Patel,D.Strozzi,D.P.Grote

Lawrence Livermore National Laboratory P.O.Box 808,Livermore,CA,USA 94550

Tabak1@lln.gov

Abstract. Fast Ignition is an inertial fusion scheme in which fuel is first assembled and then heated to the ignition temperature with an external heating source. In this note we consider cone and shell implosions where the energy supplied by short pulse lasers is transported to the fuel by electrons. We describe possible failure modes for this scheme and how to overcome them. In particular, we describe two sources of cone tip failure, an axis jet driven from the compressed fuel mass and hard photon preheat leaking through the implosion shell, and laser prepulse that can change the position of laser absorption and the angular distribution of the emitted electrons.

1. Introduction

The energy required to ignite fuel in the Fast Ignition scheme[1] depends on both the assembled fuel density and the coupling efficiency. We require that the ignition region has column density about 0.6 g/cm^2 in order trap alpha particles during the heating phase and that this ignition region is embedded in a fuel with total column density at least 1.5 gm/cm^2 so that the fuel has time to climb in temperature from the 10 keV of ignition to the 30+ keV of robust burn. For fixed column density and ignition temperature, the ignition energy scales like ρ^{-2}. We can produce this high fuel density by imploding the fuel.

This leaves us with the problem of how to efficiently couple the short pulse laser energy to the fuel. We use a cone embedded in the implosion system[2]to provide open access for the short pulse laser to reach the fuel. The required implosion can have deleterious side effects on the coupling efficiency. First, the stagnated fuel itself can produce a high pressure jet, that pushes back the cone tip and the critical surface facing the short pulse laser. This effect is discussed in section 2. The interaction of the laser driving the implosion either with the hohlraum wall or with the ablator material can produce hard photons that penetrate the imploding shell and ablate some of the cone material. This cone material can mix with fuel and inhibit ignition or impede the stagnation. This xray preheat problem is discussed in section 3. Another preheat problem is caused by the prepulse of the ignition laser. Lasers using OPCPA typically have 10^{-5}-10^{-4} of the incident energy appearing 1-3 nanoseconds before the main pulse. The effect of this prepulse in the cone is discussed in section 4. Section 5 summarizes the work.

2. Cone tip survival and electron transport efficiency

The transport efficiency is a strong function of the distance between the electron source and the ignition region. LSP[3] was used to model the electron transport between sources and ignition regions for two electron density distributions abstracted from hydrodynamic simulations. The electron source was abstracted from explicit PIC simulations driven at $I\lambda^2 = 5\ 10^{19}$ W cm^{-2}µm^2. The simulations resulted in an electron distribution with the median energy 4.4 MeV and half-width-half-max of the angular distribution 50°. Table 1 summarizes the results for the nominal electron distributions and those with half the opening angle. The simulations based on the smaller source to fuel region distance had substantially greater coupling efficiency.

Coupled energy	30 micron separation	110-130 micron separation
Nominal angle	20%	4%
Half angle	50%	18%

Table I. Coupling efficiency as function source to fuel separation and source opening angle

What causes the large standoff between the fuel and the electron source in most simulations when we can initialize the cone-center of implosion distance arbitrarily? Simulations of this system show an axis jet that flows from the converged fuel to the cone tip. The pressure in this jet is comparable to the stagnation pressure of the compressed fuel and adequate to send a strong shock through the cone tip. When the shock breaks out from the back of the cone, the ensuing rarefaction can drive the critical surface hundreds of microns away from the compressed fuel. This jet is generated because the stagnation pressure of the fuel is balanced by the ram pressure of the incoming flows in all directions except in the cone direction.

Can we abate this jet by interposing additional material between the compressed core and the cone tip? We use a simple system to give insight into the physics. The initial state is a 1D slab composed of 60 microns of 500 g/cm^{-3} hydrogen touching 70 microns of 10 g/cm^{-3} hydrogen. Naively, we would expect that the side with the tamping would show slower shock propagation than the untamped side. The simulation showed that there was little difference between the tamped and untamped sides. This occurs because the adiabatic expansion of the dense plasma is limited by its internal sound speed, not the relatively low-density tamper.

The jet has time to impact the cone tip because the fuel comes to high pressure before enough column density has assembled. A possible solution is to assemble the part of fuel near the cone early so that when the shock has propagated through this fuel the entire fuel has assembled. That is, we can win the race with the reflected shock. Figure 1 shows an example of this strategy. Two capsules were imploded around an immovable cone. One capsule was symmetric, while the other had the cone side leading the opposite by 100 microns when the average radius was 300 microns. The density distributions of both capsules at stagnation are similar. However, the symmetric capsule generates a high pressure as its flow stagnates on the cone. No such feature exists for the asymmetric one. The asymmetric design generates pressure more than an order of magnitude smaller at this location. That pressure is due to the Prandtl-Meyer flow splashing on the symmetry axis—not the jet from the core.

3. Hard photon preheat and cone tip expansion

The long pulse laser driving the fuel assembly produces electron temperatures of several keV in the blowoff of the hohlraum wall for indirect drive or in the capsule corona for direct drive. These heated electrons can produce multi-keV photons by bremsstrahlung, free-bound radiation or line radiation. These hard photons can, in turn, penetrate the capsule and heat the high-Z cone. The high-Z material can then expand, impede the implosion and possibly mix with the fuel. When the mass fraction of Au in DT reaches 0.006, the required ignition energy deposited in the fuel triples. For our indirectly driven designs,we estimate the amount of cone blowoff by using the photon spectra calculated in integrated hohlraum+capsules simulations, determining the spectra that leak through the capsule shell, then driving a gold sphere with these leakage photons and determining how far the surface of the sphere expanded under various conditions. There are two ways we can reduce the cone blowoff: introduce seeding in the ablator in order to attenuate these photons. For example we graded the doping of iodine(maximum 1 per 15 $CH_{1.3}$) over the nominal payload for a capsule driven with a peak radiation temperature of 200 eV. The doping must be graded as to avoid density gradients that can increase the growth of the Rayleigh-Taylor instability. This effectively doubles the payload mass, and lowers the peak compression. Or we can cover the cone with a low Z(carbon or beryllium) tamper; here 10 microns of density 3.47 carbon. Table II summarizes the results.

shield	Tamper(micron)	Au excursion(micron)
yes	10	10
yes	0	130
no	10	33
no	0	180

Table II. Au expansion as function of tamper and shield

Other implosion scenarios are possible. The design above used a conventional pulse shape, where 4 shocks are timed to coalesce just across the gas/inner payload interface. These pulseshapes come to peak power before the shell has moved far. They efficiently couple the radiation to the capsule because the capsule still has large surface area at this time. This is also the time of peak laser intensity and hard photon production Unfortunately, this is before the shell has thickened because of convergence effects. The attenuation of the preheat photons is the same as it is for plane slab. We have also considered another pulse shape that was originally designed to produce a completely isochoric fuel assembly[4]. In that pulseshape, peak intensity is delayed until the capsule converged more. The efficiency of radiation absorption is reduced, but the column density of the dopant has increased. Hence a given mass of dopant attenuates the hard photons an order of magnitude more efficiently than the conventional design and the mass that was taken by the doped ablator can be returned to the ablator or real payload. This scheme has been used in integrated calculations[5]. Another possibility is to replace the high-Z cone with a high density carbon one. The opacity of carbon for these hard photons is sufficiently small that it is not heated above the boiling point. There will be increased hydrodynamic losses to the cone, however.

Directly driven capsules can also suffer from this cone ablation. A capsule driven by 1 MJ of 354 nm light at peak intensity 1.4 10^{15}W/cm^2 drove a gold expansion of 65 microns. This expansion wasn't seen in high gain capsules driven at 3 10^{14} W/cm^2. An integrated 2-D simulation of recent Omega experiments saw the cone expand a few tens of microns.

4. Reducing the prepulse from the shortpulse laser.

Shortpulse lasers often have prepulses arriving 1-3 nanoseconds before the main pulse and carrying 10^{-3}-10^{-5} of the total energy. These 10's of milllijoule prepulses produce plasmas extending over 100 microns. There is anecdotal evidence from existing experiments and simulations using cone targets[6] that these preplasmas significantly move the location of the laser absorption upstream from the cone tip and broaden the angular distribution. Calculations without prepulse show the source of hot electrons at the cone tip with a 40° to 50° half angle[7]. Baton et.al., inferred a reduction in forward going electrons of more than an order of magnitude when an approximately 10 mJ prepulse was injected into a cone compared a frequency doubled pulse without prepulse. Simulations suggested that the electrons were driven sideways into the cone walls, not forward. Our understanding of this situation is incomplete, but the trend is discouraging. Full scale Fast Ignition may require short pulse laser energies of order 100kJ, leading to prepulses in the 1-10J region. Developing techniques to reduce these prepulses is a sensible precaution.

Laser based schemes that can reduce this prepulse include double OPCPA[8] and intensity dependent polarization rotation[9]. Implementing such schemes would be the preferred way to solve this problem. However, such implementations may be difficult for a given laser. As an alternative we looked for plasma based solutions. Plasma mirrors in which light with intensity above 10^{15} forms a plasma that then specularly reflects light have been demonstrated in short pulse laser experiments. The low intensity prepulse does not form a plasma and therefore does not reflect the light. Unfortunately, the reflection efficiency at high intensity is only 50% and after about 1 ps the surface roughens and the reflection is no longer specular.

We suggest using inverse Bremsstrahlung absorption to selectively absorb the prepulse while allowing the main pulse through. For example, consider a 3 ns 1 J prepulse of 1 micron light incident on a 4 eV hydrogen plasma of density 10^{-5}g/cm^3 1.5 cm long followed by a 10 ps 100kJ pulse. The plasma is preheated so that there are sufficient electrons for efficient inverse Bremsstrahlung. The prepulse intensity was reduced by at least 5 orders of magnitude(the code cutoff), while only 4.5% of the high intensity pulse was absorbed because this plasma was sufficiently heated(1keV-100 keV) to turn off the inverse Bremsstrahlung. At an electron density of 5 10^{18}/cm^3 and power 10^{16}W, this system is well above the threshold for relativistic self-focusing. Further work is required on the focusing and laser-plasma instability properties of this system.

5. Summary

We have described three detrimental issues for Fast Ignition and novel approaches to deal with them. Jets launched from the compressed fuel can destroy the cone tip and further separate the critical surface from the fuel. Asymmetric implosions can delay this jet formation. Hard photons made in the hot coronal plasmas can preheat the high-Z cone used to guide the implosion. The high-Z blowoff can mix with the fuel or retard the implosion. Mitigation measures are adding dopants to the ablator, tamping the cone with low-Z, changing the drive pulse shape or using a diamond cone. Laser prepulse can dramatically alter the shortpulse laser coupling to the fuel. Using a low density hydrogen is one way to eliminate this prepulse. This work performed under the auspices of the U.S. Department of Energy by Lawrence Livermore National Laboratory under Contract DE-AC52-07NA27344.

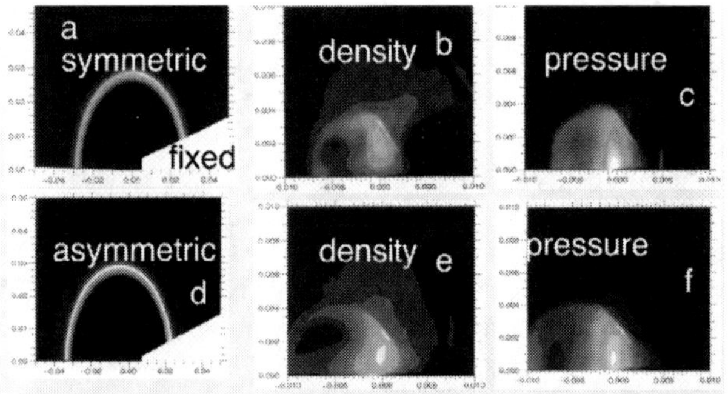

Figure 1. Density(a,b,d,e) and pressure(c,f) for symmetric(a-c) and asymmetric(d-f) implosions. a,d show the initial states. b,c,e,f show stagnation

References

[1] M.Tabak,et.al.,Phys. Plasmas **1**,5(1994)1626.
[2] S.Hatchett and M.Tabak,30th Anomalous Absorption Conference 2000;S.Hatchett et.al.,Fusion Science and Technology 49,327(2006),M.Tabak,et.al.,LLNL document IL8826B(1997).
[3] D.R.Welch,et.al.,Physics of Plasmas **13**(2006)063105.
[4] D.S.Clark and M.Tabak,Nuclear Fusion,47(2007)1147-1156.
[5] H.D.Shay,et.al.,these proceedings.
[6] S.Baton,et.al.,Phys.Plasmas **15**(2008)042706, K.Akli, et.al., these proceedings, A.Kemp, et.al., these proceedings., L.Van Woerkem, et.al. ,Phys. Plasmas **15**,5(2008)056304.
[7] B.F.Lasinski,et.al.Phys.Plasmas **16**,1(2009)012705.
[8] C.Siders,private communication.
[9] M.Zepf, private communication.

Measurement of Fast Electrons Spectra Generated by Interaction between Solid Target and Peta Watt Laser

T. Tanimoto[1], H. Habara[1], K. A. Tanaka[1], R. Kodama[1], M. Nakatsutsumi [2], K. L. Lancaster [3], J. S. Green [3], R. H. H. Scott [3,4], M. Sherlock [3,4], P. A. Norreys [3,4], R. G. Evans [4], M. G. Haines [4], S. Kar [5], M. Zepf [5], J. King [6], T. Ma [6], M. S. Wei [6], T. Yabuuchi [6], F. N. Beg [6], M. H. Key [7], P. Nilson [8], R. B. Stephens [9], H. Azechi [10], K. Nagai [10], T. Norimatsu [10], K. Takeda [10], J. Valente [11], and J. R. Davies [11]

[1] Graduate School of Engineering, Osaka University, 2-6 Yamada-oka, Suita, Osaka 565-0871, Japan

[2] Laboratoire pour l'Utilisation des Laser Intenses (LULI), CNRS, Ecole Polytechnique, Route de Saclay, 91128 Palaiseau, Cedex F 91128, France

[3] STFC Rutherford Appleton Laboratory, Harwell Science and Innovation Campus, Didcot, Oxfordshire OX11 0QX, United Kingdom

[4] Department of Physics, Imperial College London, Blackett Laboratory, Prince Consort Road, London SW7 2BZ, United Kingdom

[5] Centre for Plasma Physics, School of Mathematics and Physics, Queens University Belfast, University Road, Belfast BT7 1NN, United Kingdom

[6] Department of Mechanical and Aerospace Engineering, UC San Diego, 9500 Gillman Drive 0411, La Jolla, California 92093-0411, United States of America

[7] Lawrence Livermore National Laboratory, P.O.Box 808, Livermore, California 94550, United States of America

[8] Laboratory for Laser Energetics, University of Rochester, 250 East River Road, Rochester, New York 14623, United States of America

[9] General Atomics Corp. PO Box 86508, San Diego, California 92186-5608, United States of America

[10] Institute of Laser Engineering, Osaka University, 2-6 Yamada-oka, Suita, Osaka 565-0871, Japan

[11] Groupo Lasers e Plasmas, Centro de Fisica dos Plasma, Instituto Sperior Tecnico, Av. Rovisco Pais, 1, 1049-001 Lisbon, Portugal

E-mail tanimoto@ile.osaka-u.ac.jp

Abstract. Fast electron energy spectra have been measured for a range of intensities between 10^{18} Wcm^{-2} and 10^{21} Wcm^{-2} and for different target materials using electron spectrometers. Several experimental campaigns were conducted on peta watt laser facilities at the Rutherford Appleton Laboratory and Osaka University. In these experimental campaigns, the pulse duration was varied from 0.5 ps to 5 ps. The laser incident angle was also changed from normal incidence to 40° in p-polarized. The results show a reduction from the ponderomotive scaling on fast electrons over 10^{20} Wcm^{-2}.

1. INTRODUCTION

The fast ignition (FI) scheme for inertial confinement fusion was first proposed by M. Tabak [1] in 1994. In FI scheme, when an ultra intense laser (UIL) pulse is irradiated into an imploded plasma, fast electrons are generated from laser-plasma interactions and are used to heat the fuel core. Recently, LFEX laser at Osaka University and OMEGA-EP laser at University of Rochester are constructed for FI integral experiments. These laser intensities exceed 10^{20} Wcm^{-2}. Moreover, pulse duration becomes about 10ps and longer than previous experiment. Therefore, in these conditions, the characteristics of fast electrons should be studies.

We deduced slope temperatures from the measured fast electron spectra and compared those with pondero motive scaling in the range 10^{18} Wcm^{-2} to 10^{21} Wcm^{-2}.

2. EXPERIMENT

The experiments were performed using both the VULCAN peta watt (VULCAN PW) [2] and Gekko XII peta watt (GXII PW) laser systems [3]. The VULCAN PW has a measured intensity contrast ratio of 4×10^{-8} and delivered pulses in the range 0.5 – 5 ps. Laser pulses with up to 300 J were delivered on target. The wavelength was 1.054 μm. The laser was focused onto target using an f/3 parabola and the spot size was 7μm Full Width at Half Maximum (FWHM). 20% of the laser energy was contained within that focal region [4]. The laser intensity is changed from 10^{18} Wcm^{-2} to 10^{21} Wcm^{-2} by varying the energy delivered and the pulse duration. The laser was incident at both normal (0°) and p-polarized 40°. The GXII PW has a measured intensity contrast ratio of 1.5×10^{-8}. The pulse duration was between from 0.6 to 0.7 ps. Up to 100 J was delivered on target. The wavelength is 1.053μm. The laser was focused onto target using an f/7.6 parabola and the spot size was 15 μm FWHM. The laser intensity was from 10^{18} Wcm^{-2} to 10^{19} Wcm^{-2}. The laser was incident at an angle of 26° to the target normal [5].

These experiments used a variety of target materials in both single and multilayer structures. The interaction surfaces of target were Cu, Au, Al and Ti and the target thickness ranged from 5 μm to 100 μm. Two different electron spectrometers (ESM) were used for the fast electron spectrum characterization. The ESM was placed behind the target and measured the fast electron energy spectra. Fuji film imaging plates were used as the detectors. In spite of such broad range of conditions, only the laser intensity showed a clear dependence on the temperature.

Electron spectra were measured along the laser axis. The measured energy spectra were fitted with a relativistic Maxwellian distribution with the electron temperature T of the form [6,7], $N(E) = N_0E^2\exp(-E/T)$. Here N_0, E and T are the electron number, the electron energy and the electron temperature, respectively.

Figure 1 shows the measured fast electron slope temperature versus the incident laser intensity. The experimental results are compared with the Wilks' ponderomotive scaling of the form [8]

Journal of Physics: Conference Series **244** (2010) 022067 doi:10.1088/1742-6596/244/2/022067

$T[MeV] = 0.511\left(\sqrt{1 + I_{18}\lambda_{\mu m}^2/1.37} - 1\right)$, where T, I_{18} and $\lambda_{\mu m}$ are the electron temperature, the laser intensity in units of 10^{18} Wcm^{-2} and the wavelength in microns, respectively. The dotted line is the calculation of the electron temperature from this ponderomotive scaling. The broken line is Haines' relativistic model. The solid line shows a least square fitting on the data.

In Fig. 1 the data shows a clear departure from the Wilks' scaling. The fast electron temperature becomes lower than ponderomotive scaling. In a most recent study of fast electrons measured both within targets and at ESMs show that the spectra are very much alike [9]. It is of great importance to point out here that this scaling is in favor for fast ignition up-coming integral experiments; rather high irradiation laser intensity can be used to generate fast electrons within a modest slope temperature. Those experiments will use 1-10 ps ultra-intense laser pulse to fast heat a highly compressed fuel to achieve several keV core temperature [10,11].

A least squares fit to the experimental points provides almost the $I^{1/3}$ fit, as shown in Figure 1, but some consideration is needed in the interpretation of this scaling. In the absorption region, fast electron energy distribution is of particular concern since a number of recent computational studies have shown the fast electron spectrum softens with time. Those simulations indicate that this softening results from the preformed plasma in the corona being swept away by the ponderomotive force early in the interaction, even with pico second duration pulses. Thereafter, the excursion distance of electrons pulled from the skin layer is much smaller than the full laser wavelength, resulting in the reduced mean energy [12,13]. We believe that the main part of the acceleration is performed by the v x B force even though we believe that our experiments suffer little effect of pre-plasmas. Especially the relatively long laser pulse may sweep up any plasma in front of the solid surface and then the laser starts facing its effective critical density plasma whose charge states are determined by the laser heated temperature at the area. If this situation is realized, the fast electron temperature may drop further [12,13]. As discussed in several works, the pre-pulse may alter the characteristics of ultra-intense laser matter interactions drastically [14,15]. Pre-pulse level of this experiment can be seen in Fig. 10 in Ref. 16 which shows that the interactions should occur at the effective critical density.

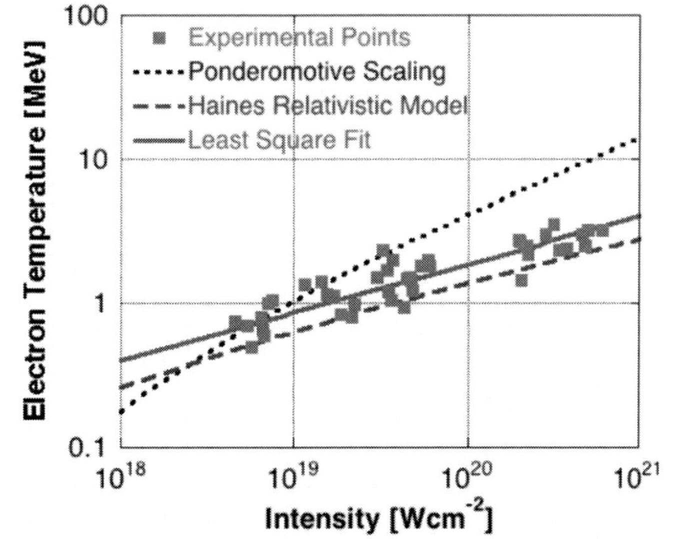

Figure 1. Plot of the measured electron temperature as a function of intensity on target. Also shown is the ponderomotive scaling as dotted line [8] and Haines' relativistic model as broken line [19]. A least square fit to the data as solid line shows that the fast electron temperature scales as T [MeV]= $0.4\,(I_{18}\lambda_{\mu m}^{2})^{1/3}$.

A novel explanation is briefly reviewed for the $I^{1/3}$ intensity scaling in Fig. 1, one that reproduces previous results up to 10^{19} Wcm^{-2} [17] as well as more recent Bremsstrahlung radiation measurements of electron entering the target up to 10^{21} Wcm^{-2} [18]. The laser-plasma interaction region is treated as a one-dimensional 'black box', the thickness of which is a few collision less skin-depths. Relativistic conservation equations are then applied to this region, rather in the same way as in a shock transition. This very simple model leads to Haines' relativistic model. This model is discussed in detail in Ref. 19. The scaling based on the relativistic model shows very good consistency with the experiment within the laser intensity range between $I = 10^{18}$ and $I = 10^{21}$ Wcm^{-2} in Fig. 1.

3. CONCLUSION

In conclusion, the electron temperature of fast electron escaping to an electron spectrometer located on the target chamber wall has been measured for a variety of different laser conditions using two different peta watt laser facilities. These new results are consistent with both Bremsstrahlung radiation measurements generated from electrons entering the overdense plasma [18] as well as analytic models that include both energy and momentum conservation [19]. It has been found that the departure from the ponderomotive scaling of the fast electron temperature as a function of intensity on target is robust and occurs under a range of different irradiation conditions. The obtained scaling gives a wider range of laser intensity window for modest temperature of fast electrons that could be utilized for designing up-coming integral fast ignition experiments compared the one based on the ponderomotive scaling.

4. ACKNOWLEDGEMENT

We are grateful to the support from the project: global center of excellence (GCOE) at the Department of Electrical, Electric, and Information, the Graduate School of Engineering, Osaka University, Osaka Japan and Japan-UK collaboration program of the international collaboration for high energy density science (ICHEDS) of Japan society for the promotion of science (JSPS) at Osaka University, Osaka, Japan. A part of the research is supported by Grant-in-Aid for Scientific Research of JSPS (Contract No. 19206099) and by Japan/U.S. Fusion Cooperation Program (Contract No. 2036) and by Kansai Research Foundation for technology promotion program (Contact No. 2009F008).

References

[1] M. Tabak et al., Phys. Plasma, **1**, 1626 (1994).
[2] C. N. Danson et al., Nuclear Fusion **44**, S239-S246 (2004).
[3] Y. Kitagawa et al., IEEE J. Quantum Electronics, **40**, 281 (2004).
[4] P. K. Patel et al., Plasma Phys. Controlled Fusion **47**, B833 (2005).
[5] S. P. D. Mangles et al., Nature (London) **431**, 535 (2004).
[6] T. Yabuuchi et al., Phys. Plasmas, **14**, 040706 (2007).
[7] T. W. Phillips et al., Rev. Sci. Instrum., **70,** 1213 (1999).
[8] S. C. Wilks et al., Phys. Rev. Lett. **69**, 1383 (1992).
[9] H. Habara, private communications (2008).
[10] K. Mima et al., Eur. Phys. J.D., **44**, 259(2007).
[11] D. N. Maywar et al., J. Phys. Conf. Ser., **112**, 032007 (2007).
[12] B. Chrisman et al., Phys. Plasmas **15**, 056309 (2008).
[13] A. J. Kemp et al., Phys. Rev. Lett. **101,** 075004 (2008).
[14] G. Malka et al., Physc. Rev. E **66**, 066402(2002).
[15] G. Malka et al., J. Opt. Soc. Amer. B **14**, 2091 (1997).
[16] W. Theobald et al., Phys. Plasmas **13**, 043102 (2006).
[17] F. N. Beg et al., Phys. Plasmas **4**, 447 (1997).
[18] H. Chen et al., Phys. Plasmas **16**, 020705 (2009).
[19] M. G. Haines et al., Phys. Rev. Lett. **102**, 045008 (2009).

Journal of Physics: Conference Series **244** (2010) 022068 doi:10.1088/1742-6596/244/2/022068

Plasma parameters of a capillary discharge-produced plasma channel to guide an intense laser pulse

Hiromitsu Terauchi[1], Jin-xiang Bai[1], Takeshi Higashiguchi[1,2,3], Noboru Yugami[1,2,3], Toyohiko Yatagai[2], and Ryosuke Kodama[3,4]

[1]Department of Advanced Interdisciplinary Sciences, Utsunomiya University, Yoto 7-1-2, Utsunomiya, Tochigi 3218585, Japan

[2] Center for Optical Research & Education (CORE), Utsunomiya University, Yoto 7-1-2, Utsunomiya, Tochigi 3218585, Japan

[3] Japan Science and Technology Agency, CREST, 4-1-8 Honcho, Kanagawa, Saitama 3320012, Japan

[4] Graduate School of Engineering, Osaka University, 2-6 Yamada-oka, Suita, Osaka 5650871, Japan

E-mail: yugami@cc.utsunomiya-u.ac.jp

Abstract. We demonstrated the production of an optical waveguide in a capillary discharge-produced plasma. Plasma parameters of its waveguide were characterized by use of both a laser interferometer and a hydrogen plasma line spectrum. A maximum temperature of 3.3 eV with electron densities of the order of 10^{17} cm^{-3} was observed at a discharge time of 150 ns and a maximum discharge current of 400 A. An ultrashort, intense laser pulse was guided over long vacuum Rayleigh lengths using this plasma channel.

1. Introduction

The guiding of intense laser pulse in a plasma has a lot of applications which include high-order harmonic radiation [1], x-ray laser [2], and laser-plasma-based electron acceleration [3]. These applications require intense laser pulse propagating, generally above the ionization threshold of atoms. For femtosecond laser technology generated by a chirped-pulse-amplification scheme, the high laser intensity and the high optical field are achieved in a tightly focused beam. For a focused Gaussian beam, the interaction length is approximately a Rayleigh length of less than 1 mm, which is attributed to the tightly focused beam. To overcome this short interaction length, it is necessary to appropriate optical waveguides that are efficiently coupled with an intense laser pulse. The choice of optical waveguides may thus become as important as that of high intensity laser. Intense laser pulse may be guided in a plasma if the refractive index along the optical pulse propagation direction can be increased sufficiently to balance diffraction.

There are several different kinds of optical plasma waveguides used in high-intensity laser-plasma interaction experiments, such as gas/cluster jet [4], hollow fiber [5], and discharge-produced plasma waveguide [6, 7, 8, 9, 10]. Among these, capillary-self-breakdown-ablated channel in vacuum [7, 8] and hydrogen gas-filled channel [9, 10] have been actively studied as possible efficient plasma channel to guide the intense laser pulse. The hydrogen gas-filled capillary plasma waveguide forms a fully ionized plasma. The plasma channel is formed by Ohmic heating to the capillary wall, which causes the electron density to be decreased, near the capillary centre. In order to understand the dynamics and channel formation of its waveguide, it is important to understand the temporal evolution of the electron temperature in the capillary discharge-produced plasma channel. A very limited number of investigations, however, have been reported to characterize the temporal evolution of the plasma parameters in the hydrogen plasma channel.

In this paper, we characterize the temporal evolution of the plasma parameter of the hydrogen plasma channel. The hydrodynamics of the plasma channel resulted in a maximum electron temperature of 3 eV with the parabolic electron densities of the order of 10^{17} cm^{-3}. Intense laser pulse with a spot diameter of 36 μm (full-width at half maximum: FWHM) was guided in the plasma channel through a 1.5-cm long plasma channel waveguide.

2. Experiments and discussions

A Pyrex-glass capillary (Japan Cell Co., Ltd.) was placed in a vacuum chamber. A 1.5-cm long pipe with a diameter of 500 μm as the capillary is made in a rectangular solid Pyrex-glass block. The hydrogen gas was flowed from the gas port of the structure, as shown in Fig. 1(a) and was filled in a capillary. The maximum discharge voltage was 30 kV by use of a pulsed power supply with a condenser capacity of 2 nF. The pulse width of the discharge current is 100 – 150 ns (FWHM). The maximum discharge current was about 400 A at the discharge time around 150 ns. To measure the electron density, an 8 ns (FWHM) laser pulse at 532 nm is used as a probe-pulse. The probe beam passes through the capillary plasma and is relayed to an imager by a lens with a focal length of 15 cm. In the case of a Mach-Zehnder interferometer, the probe beam was strongly refracted by the capillary wall. The visibility less than 0.5 in the interferogram, which was attributed to the low energy balance between probe and reference beams, seemed too low for high spatial resolution measurements. To overcome the low contrast, a Nomarski interferometer is inserted in front of the imager [11]. The contrast of the interferogram is achieved to be 0.8. The electron density is determined from the phase shifts of the interferogram. The spatial distribution of the electron density is converted by use of the inverse Abel transformation. The spectra in visible spectral region from the plasma channel were measured. A spectrometer with 2400 grooves/mm grating is positioned at 45° with respect to the optical axis in the capillary. Time-resolved image-intensified CCD (ICCD) camera coupled with the spectrometer is used to detect the visible emission from the plasma. We observed the temporal history of the spectral lines at the different delays with respect to the onset of the discharge. By using the intensity ratio of Hydrogen H_α ($\lambda = 656.3$ nm) to H_β ($\lambda = 486.1$ nm) lines, the electron temperature is determined.

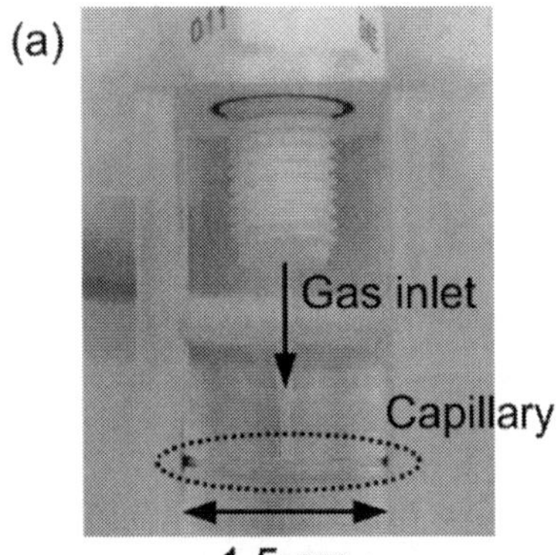

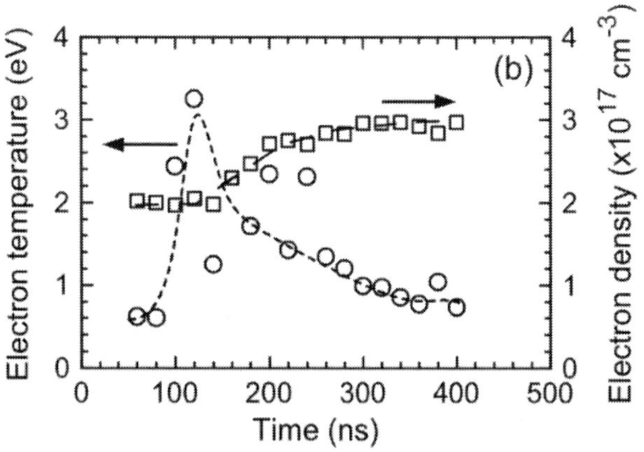

Figure 1. (a) Photograph of the capillary. (b) Electron temperature (circles) and its density (rectangles) with the dashed lines.

Journal of Physics: Conference Series **244** (2010) 022068 doi:10.1088/1742-6596/244/2/022068

The temporal evolutions of the electron temperature and its density on the axis are shown in Fig. 1(b). The electron temperature increases when the time and current of the discharge increase, and reaches its maximum of 3.3 eV at the discharge time around 150 ns. Later, the electron temperature decreases and reaches at the value of 0.5 eV, which is corresponded the time-integrated electron temperature, when the discharge current completely vanishes. The temporal evolution of the axial electron density is almost constant in the range of $(2-3) \times 10^{17}$ cm^{-3} during the discharge time between 60 and 400 ns.

To evaluate the electron temperature and channel formation dynamics, we compare with the previous work and simple numerical simulation. Broks *et al.* have described the nonlocal thermal equilibrium (LTE) model to analyze a slow pulsed capillary discharge-produced plasma channel waveguide [12]. The model of the plasma channel was found to give similar values for electron temperature due to the capillary diameter, peak discharge current, and pulse duration. The model, therefore, could reproduce a clear insight of the plasma heating and the channel formation, and may be useful for design of the future experiments. In order to understand the mechanisms and the effects of the electron temperature by which the plasma channel is produced, it is important to compare the experimental results and numerical simulations. The spatiotemporal calculation of the electron temperature was evaluated to be 6 eV by Bobrova *et al.* [13]. The numerical calculation became 2 times as large as that observed by our measurements. The electron temperature is expected to maximize on the axis, which is attributed to a result of the fact that the Joule heating is balanced by the thermal conduction to the cold capillary wall. As a result, the parabolic density profile is formed in the axial minimum due to the hydrogen is fully ionized in the capillary.

The electron density profile is given by $n_e(r) = n_{e0}(r=0) + \Delta n_e (r/r_0)^2$, where $n_{e0}(r=0)$ is the axial electron density and Δn_e is the difference at radius $r = r_0$. In the case of a Gaussian laser profile, the matched spot size of the plasma channel is evaluated to be $w_0 = [r_0^2/(\pi r_e \Delta n_e)]^{1/4}$, where r_e is the classical electron radius. The matched spot diameter of the plasma channel is largely independent of the shape of the electron density profile [10]. Figure 2(a) shows the radius distributions of the electron density in the uniform central section of a 1.5-cm long, 500-μm diameter capillary at discharge times of 80, 120, and 160 ns. These profiles are produced the rather parabolic profiles and almost kept to be constant in the discharge times of 80 – 160 ns. The axial electron density and the difference at the discharge time of 160 ns were observed to be $n_{e0} \approx (1-2) \times 10^{17}$ cm^{-3} and $\Delta n_e \approx 4 \times 10^{17}$ cm^{-3} with $r_0 \approx 250$ μm. The correspond matched spot diameter was evaluated to be $w_0 = 49.4$ μm. Figure 2(b) shows the longitudinal electron density on the axis at the discharge time of 150 ns, and the electron density was $(1-3) \times 10^{17}$ cm^{-3}, and its density is almost uniform along the capillary.

To confirm the guiding effect of the plasma channel, we observed the output beam profile of an intense femtosecond laser ($\lambda_L = 800$ nm, $\tau_L = 130$ fs, $\varepsilon_L = 45$ mJ) at the exit of the capillary. The laser pulse is focused onto the entrance of the capillary with a focal diameter of 30 μm (FWHM) in vacuum using a lens with a focal length of 17 cm. The focused laser intensity is 5×10^{16} W/cm^2. Figure 3 shows photographs of the laser beam profile through the 1.5-cm long capillary with and without the plasma channel at the discharge time of 150 ns. A comparison of these photographs shows evidence of the optical guiding by use of the plasma channel. Without the plasma channel, the output beam through the capillary could not be propagated, as a result, the transmittance coefficient was less than 1%. In contrast, it was found that the beam diameter of the transmitted laser pulse with the plasma channel is about 36 μm (FWHM). This is smaller than that of the theoretical expected value. The transmittance coefficient of the laser pulse through the plasma channel is up to 50%. Note that the intensities of the output beam cannot be compared with these photographs due to the different scale of the CCD camera sensitivity.

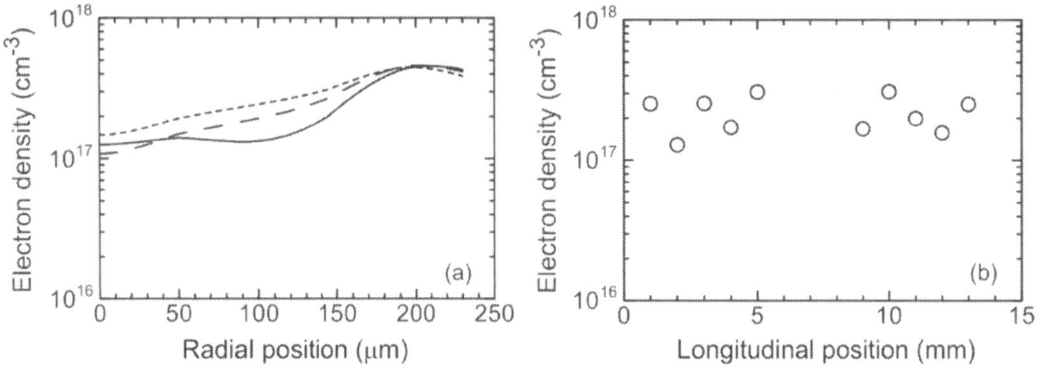

Figure 2. (a) The radial distributions of the electron density at the discharge time of 80 (dotted), 120 (dashed), and 160 ns (solid). (b) The longitudinal electron density on the axis at the discharge time delay of 150 – 160 ns.

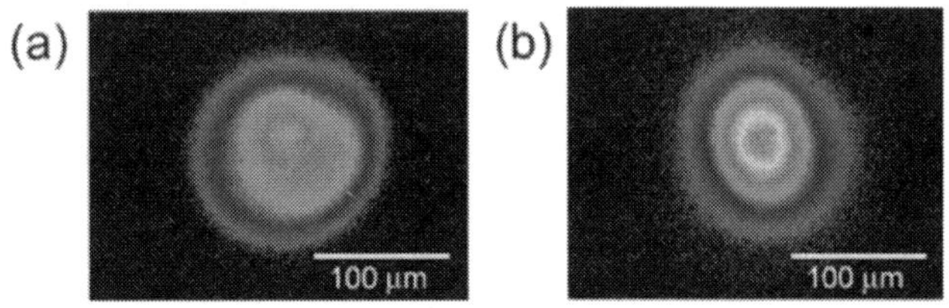

Figure 3. Photographs of the transmitted laser profile at the capillary exit without (a) and with (b) the plasma channel, respectively.

3. Summary

In summary, we characterized the time-resolved plasma parameters of the capillary discharge-produced plasma channel with the electron density of the order of 10^{17} cm^{-3}. It was found that the electron temperature is 3.3 eV at the peak discharge current. As a result, a plasma channel was produced with the axial density of $(1 - 3) \times 10^{17}$ cm^{-3}. This low density plasma channel made the optical waveguide possible and realized the laser guiding with long vacuum Rayleigh lengths.

References

[1] Gaudiosi M D, Reagan B, Popmintchev T, Grisham M, Berrill M, Cohen O, Walker C B, Murnane M M, Kapteyn C H, and Rocca J J 2006 *Phys. Rev. Lett.* **96** 203001
[2] Butler A, Gonsalves J A, McKenna M C, Spence J D, Hooker M S, Sebban S, Mocek T, Bettaibi I, and Cros B 2003 *Phys. Rev. Lett.* **91** 205001
[3] Leemans P W, Nagler B, Gonsalves J A, Toth Cs, Nakamura K, Geddes R G C, Esarey E, Schroeder B C, and Hooker M S 2006 *Nature Phys.* **2** 696
[4] Durffe G C, Lynch J, and Milchberg M H 1994 *Opt. Lett.* **19** 1937
[5] Dorchies F, Marquès R J, Cros B, Matthieussent G, Courtois C, Vélikoroussov T, Audebert P, Geindre P J, Rebibo S, Hamoniaux G, and Amiranoff F 1999 *Phys. Rev. Lett.* **82** 4655
[6] Hosokai T, Kando M, Dewa H, Kotaki H, Kondo S, Hasegawa N, Nakajima K, and Horioka K 2000 *Opt. Lett.* **25** 10
[7] Kaganovich D, Zigler A, Hubbard F R, Sprangle P, and Ting A 2001 *Appl. Phys. Lett.* **78** 3175
[8] Palchan T, Kaganovich D, Sasorov P, Sprangle P, Ting C, and Zigler A 2007 *Appl. Phys. Lett.* **90** 061501
[9] Butler A, Spence J D, and Hooker M S 2002 *Phys. Rev. Lett.* **89** 185003
[10] Gonsalves J A, Rowlands-Rees P T, Broks P H B, M van der Mullen A J J, and Hooker M S 2007 *Phys. Rev. Lett.* **98** 025002
[11] Tao Y, Tillack S M, Harilal S S, Sequoia L K, and Najmabadi F 2007 *J. Appl. Phys.* **101** 023305
[12] Broks P H B, Garloff K, and M. van der Mullen M J J 2005 *Phys. Rev.* E **71** 016401
[13] Bobrova A N, Esaulov A A, Sakai -I J, Sasorov V P, Spence J D, Butler A, Hooker M S, and Bulanov V S 2001 *Phys. Rev.* E **65** 016407

Journal of Physics: Conference Series **244** (2010) 022069 doi:10.1088/1742-6596/244/2/022069

Relativistic hole boring and fast ion ignition with ultra-intense laser pulses

V T Tikhonchuk[1], T Schlegel[2], N Naumova[3], I V Sokolov[4], C Regan[1], M Temporal[5], J-L Feugeas[1], Ph Nicolaï[1], X Ribeyre[1], C Labaune[6], G Mourou[7]

[1] Centre Lasers Intenses et Applications, Université Bordeaux 1, Talence, France
[2] GSI Helmholtzzentrum für Schwerionenforschung GmbH, Darmstadt, Germany
[3] Laboratoire d'Optique Appliquée, ENSTA - École Polytechnique - CNRS, Palaiseau, France
[4] Space Physics Research Laboratory, University of Michigan, Ann Arbor, MI 48109, USA
[5] ETSIA, Universidad Politecnica de Madrid, Spain
[6] Laboratoire d'Utilisation Lasers Intenses, École Polytechnique, Palaiseau, France
[7] Institut de la Lumière Extrême, ENSTA - École Polytechnique - CNRS, Palaiseau, France

E-mail: tikhonchuk@celia.u-bordeaux1.fr

Abstract. An analytical model and numerical simulations demonstrate that pulses with intensities exceeding 10^{22} W/cm^2 may penetrate deeply into the plasma and accelerate efficiently ions in the forward direction. We propose a new scheme for fast ignition of precompressed DT fusion targets by using two laser pulses. The first pulse (or a sequence of several pulses) creates a channel with diameter $\sim 30\,\mu$m through the plasma corona up to the fuel density ~ 1 g/cm^3. The second pulse with a higher intensity accelerates the deuterium and tritium ions at the head of this channel. The overall ignition energy is ~ 100 kJ.

1. Introduction

The laser ponderomotive force may provide an efficient ion acceleration in bulk dense targets and evacuate a channel enabling further laser beam propagation. A quasi-stationary model of a laser piston predicts the general parameters of the acceleration process in one-dimensional geometry. The particle-in-cell (PIC) simulations confirm the estimated characteristics in a wide range of laser intensities and ion densities and show advantages of circularly polarized laser pulses compared to linear polarization. The characteristics of channel formation and the angular distribution of accelerated ions are found from two-dimensional PIC simulations. This model and integrated numerical simulations are used for design of a new scheme of ion fast ignition.

2. Quasi-stationary model of ponderomotive ion acceleration

A quasi-stationary model of a laser piston predicts general parameters of the acceleration process [1]. An expression for the instantaneous piston velocity follows from momentum conservation of the photon flux coming from the laser side and of the opposite plasma flow coming from the upstream region of the piston. Assuming a circular polarization of the laser light and neglecting the electron heating, this velocity, $v_f = \beta_f c$, is determined by the relation $\beta_f = B/(1+B)$, where the dimensionless parameter $B = (I_{\rm las}/\rho c^3)^{1/2}$. The ions are accelerated in the thin double layer and they are propagating upstream with the velocity $v_i = 2\beta_f c/(1+\beta_f^2)$ and with the energy $\epsilon_i = 2m_i c^2 B^2/(1+2B)$. In particular in the limit of a non relativistic piston, $B \ll 1$, $v_i \simeq 2v_f$ and the ion energy obeys the asymptotic solution $\epsilon_i = 2m_i I_{\rm las}/\rho c$.

Knowing the density dependence of the piston velocity and of the ion energy, one can describe also the acceleration process in an inhomogeneous plasma. For an exponential density increase in laser propagation direction with the scale length L, we obtain analytical results for the ion fluence F_i, the laser pulse duration $\tau_{\rm las}$ and the mean energy of accelerated ions:

$$F_i \simeq \frac{2I_{\rm las}L}{c}\ln\frac{\rho_{\rm max}}{\rho_{\rm min}}, \quad \tau_{\rm las} = 2L\sqrt{\frac{\rho_{\rm max}c}{I_{\rm las}}}\left(1-\sqrt{\frac{\rho_{\rm min}}{\rho_{\rm max}}}\right), \quad \langle\epsilon_i\rangle \simeq \frac{2I_{\rm las}m_i}{(\rho_{\rm max}-\rho_{\rm min})c}\ln\frac{\rho_{\rm max}}{\rho_{\rm min}}. \quad (1)$$

3. Numerical simulations of ion acceleration

One-dimensional PIC simulations [2] confirm the analytical predictions of the piston model. Two-dimensional [1, 3] and three-dimensional [4] PIC simulations of the ponderomotive ion acceleration with intense laser pulses provide an additional information about the angular divergence of the accelerated ions and the stability of the hole boring process. The mean divergence of the ion bunch depends on the laser beam intensity distribution, and for a flat top laser intensity profile accelerated ions are confined in a cone with the opening angle $\Delta\theta \approx 6°$. The plasma channel is formed with the velocity predicted by the piston model and the laser

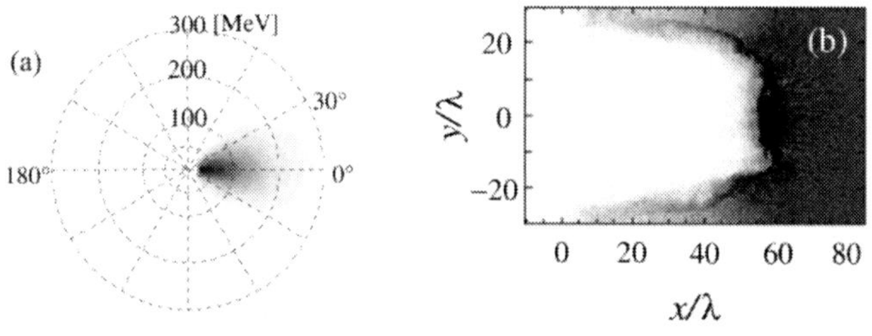

Figure 1. Ion angular distribution (a) and channel formation (b) in the laser piston regime. The laser intensity is 4×10^{22} W/cm^2, the wavelength is 0.8 μm, the beam width 16 μm, and the plasma density increases from 5 to 100 n_c over the length of 50 μm.

propagation in the channel is stable because almost no plasma has left behind the piston as it is shown in Fig. 1b. It was found from numerical simulations that the radiation losses from relativistic electrons entering the vacuum laser field behind the piston have a strong positive effect by maintaining escaped electrons in a close vicinity of the piston. This effect of electron cooling need to be accounted for laser intensities exceeding 10^{22} W/cm^2.

4. Fast ignition with ponderomotively accelerated ions

The proposed ignition process proceeds in two stages. Similarly to the scheme of the electron fast ignition [5], the first laser pulse creates a channel through the plasma corona and the second pulse accelerates the deuterium and tritium ions in the bottom of this channel. The hole boring process can be also divided in two steps. For the hole boring of an underdense plasma one can utilize a laser pulse of an intensity $\sim 10^{21}$ W/cm^2 and, according to the scaling proposed in Ref. [6], a 1 mm hole can be produced in 3 ps with the laser power of 3.5 PW. The hole boring in the overdense plasma requires the intensities at least 3×10^{21} W/cm^2 and it may take more than 10 ps to create a channel to the densities about 1 g/cm^3.

More intense laser pulses are needed for ignition. According to [7], the following scaling laws define the ignition energy $E_{\rm ig}$, the hot spot radius $r_{\rm ig}$ and the energy deposition time, $\tau_{\rm ig}$, as functions of the fuel density ρ:

$$E_{\rm ig} \approx 18(\rho_{\rm ref}/\rho)^{1.85}\,{\rm kJ}, \quad r_{\rm ig} \approx 20(\rho_{\rm ref}/\rho)^{0.97}\,\mu{\rm m}, \quad \tau_{\rm ig} \approx 21(\rho_{\rm ref}/\rho)^{0.85}\,{\rm ps}$$

where $\rho_{\rm ref} = 300$ g/cm^3. However these scalings were defined for a homogeneous sphere of fuel and for particles with a fixed range of 1.2 g/cm^2.

Knowing the ignition energy, $E_{\rm ig}$, and the scale length of the plasma density profile, L, from Eqs. (1) we define the required laser power $P_{\rm las} = \pi r_{\rm acc}^2 I_{\rm las}$ and the length of the acceleration zone $l_{\rm acc} = L\ln(\rho_{\rm max}/\rho_{\rm min})$, assuming that the density ratio is chosen, $\rho_{\rm max}/\rho_{\rm min} \sim 4-6$. The radius of the acceleration zone, $r_{\rm acc} = r_{ig} - \Delta\theta D$ is defined by the mean beam divergence $\Delta\theta$ and the distance D between the acceleration zone and the dense fuel. The latter is found from the condition that the position of the acceleration zone is such that the range of accelerated ions is approximately 1 g/cm^2. This corresponds to the average energy of deuterium and tritium ions ~ 10 MeV. This qualitative estimates provide a starting point for more detailed integrated numerical simulations. Three examples of fusion targets are explored: one of them is a relatively

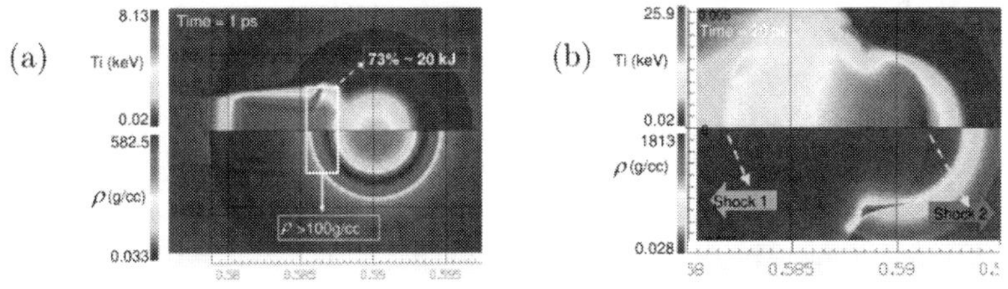

Figure 2. Temperature (upper part) and density (lower part) distributions in the HiPER target. Color codes for each panel show the maximum and minimum values. Panel (a) shows the profiles at the time of 1 ps, when ions deposited their energy. Panel (b) shows the profiles at the time of 20 ps, when the shock waves produced by the expansion of the heated parts of the shell collide inside the central hot spot and ignite the fusion reactions that propagate radially. The cell size in the figures is 50 μm.

small all-DT target designed for the HiPER project [8], other two targets of a larger size [9, 10] are more adapted for a future fusion energy plant. The integrated numerical simulations of the ignition and combustion processes were conducted with the radiation hydrodynamic codes CHIC [11] and DUED [12] combined with the modules of ballistic ion transport.

The baseline HIPER target [8] at the stagnation consists of a 25 μm thick shell of the mean density of 400 g/cm^3 with the mean radius of 50 μm. This target requires 200 kJ laser energy for compression. It was ignited with the laser pulse of intensity $I_{\rm las} = 9.3 \times 10^{21}$ W/cm^2 focused in the spot of the radius of $r_{\rm acc} = 23\,\mu$m at the density of 1 g/cm^3. The acceleration zone of a length 30 μm is located at the distance $D = 60\,\mu$m from the shell. In this example shown in Fig. 2, 20 kJ (73%) of the ion energy is dumped to the dense shell and 23% is deposited along the ion path between the acceleration zone and the dense shell. The laser pulse power is 150 PW and a corresponding laser energy is 350 kJ. The released fusion energy is 8.5 MJ, corresponding to the energy gain of about 15. The ignition process is not optimized because the shell is too thin and too much energy is lost for expansion of the heated region. According to Fig. 3a, the fusion reactions are stagnated at a low level first 20 ps after the ion energy deposition, and the ignition occurs when the exploded part of the shell collides with the opposite, cold part. Targets with a thicker core and smaller hot spot need to be designed for fast ignition scheme. Two thicker

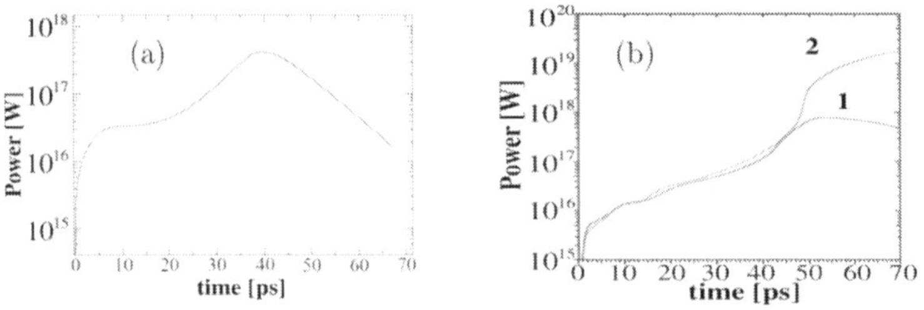

Figure 3. Output fusion power released in the simulations with the HiPER target [8] (a) and (b) with two capsules analyzed in Refs. [9] (1) and [10] (2).

targets described in Refs. [9, 10] are more suited for the ion ignition. As it is shown in Fig. 3b, the fusion energy power grows continuously immediately after the energy deposition. Therefore, the reactions are ignited directly in the energy deposition zone. However, these targets require more laser energy for compression: 500 kJ for [10] and 1 MJ for [9]. The target of Ref. [10] requires 17 kJ of the ion energy for ignition and releases $\sim$ 90 MJ, and the target [9] requires 12.7 kJ of the ion energy and released $\approx$ 200 MJ. According to the piston model, these targets require the ignition laser energy of 100 kJ and the power of 30 PW.

5. Conclusions

Laser ion acceleration by the ponderomotive pressure is an efficient process at high intensities, although a high contrast and circular polarization are needed. The model of a laser piston that predicts the ion energy spectrum in function of the laser and plasma parameters is confirmed with numerical PIC simulations. The process of channel formation is stabilized by the effect of electron radiation losses. The actual baseline HiPER target has a very thin shell at the stagnation time and it requires too much energy of the ignitor laser pulse, three times more than the energy estimated for the electron fast ignition. Such thin shell targets are not adapted for fast ignition. Thicker targets allow to reduce the ion ignition energy to the level of 100 kJ, and to obtain the energy gain more than 200. Future optimization of the ion fast ignition could be related to acceleration of ions with higher charge, such a carbon or oxygen. An experimental demonstration of the hole boring process is desirable for evaluation of reliability of this scheme.

Acknowledgements. The authors are thankful to G. Schurtz for helpful comments. The work is supported by the European program MC IRSES, and by the HiPER and ELI projects. M.T. is supported by the research contract (Ramon y Cajal-2007-00447) of the Spanish Ministry of Education and Research.

References

[1] Naumova N *et al.* 2009 *Phys. Rev. Lett.* **102** 025002
[2] Schlegel T *et al.* 2009 Phys. Plasmas **16** 083103
[3] Klimo O, Psikal J, Limpouch J and Tikhonchuk V T 2008 *Phys. Rev. ST Accel. Beams* **11** 031301
[4] Liseykina T V, Borghesi M, Macchi A and Tuveri S 2008 *Plasma Phys. Control. Fusion* **50** 124033
[5] Tabak M *et al.* 1994 *Phys. Plasmas* **1** 1626
[6] Li G *et al.* 2008 *Phys. Rev. Lett.* **100** 125002
[7] Atzeni S 1999 *Phys. Plasmas* **6** 3316
[8] Atzeni S, Schiavi A and Bellei C 2007 *Phys. Plasmas* **14** 052702
[9] Temporal M, Honrubia J J and Atzeni S 2002 *Phys. Plasmas* **9** 3098
[10] Clark D S and Tabak M 2007 *Nucl. Fusion* **47** 1147
[11] Maire P-H, Abgrall R, Breil J and Ovadia J 2007 *SIAM J. Sci. Comput.* **29** 1781
[12] Atzeni S 1986 *Comput. Phys. Comm.* **43** 107

Journal of Physics: Conference Series **244** (2010) 022070 doi:10.1088/1742-6596/244/2/022070

Electron energy deposition to the fusion target core for fast ignition

W M Wang[1,4*], Z M Sheng[1,2,‡], P A Norreys[3], M Sherlock[3], R Trines[3], A P L Robinson[3], Y T Li[1], B Hao[1] and J Zhang[1,2]

[1]Beijing National Laboratory of Condensed Matter Physics, Institute of Physics, CAS, Beijing 100190, China

[2]Department of Physics, Shanghai Jiao Tong University, Shanghai 200240, China

[3]Central Laser Facility, CCLRC Rutherford Appleton Laboratory, Chilton, Didcot, Oxfordshire, OX11 0QX, United Kingdom

[4]Graduate School of Engineering, Utsunomiya University, 7-1-2 Yohtoh, Utsunomiya 321-8585, Japan

E-mail: *hbwwm1@aphy.iphy.ac.cn; ‡zmsheng@sjtu.edu.cn

Abstract. Heating of the target core for fast ignition by electron beams is investigated by two-dimensional collisional particle-in-cell simulations. It is found that the electron beams emitted from the core surface with the initial energy of $1.4MeV$, $2.4MeV$, and $4.2MeV$ can heat most efficiently the core with $\rho r = 0.75\text{g/cm}^2$, 1.5g/cm^2, and 3g/cm^2, respectively, when taking $\rho = 300\text{g/cm}^3$, where ρ and r are the mass density and radius of the core, respectively.

1. Introduction

As an alternative route to realize laser fusion energy, the fast ignition (FI) scheme has attracted significant attention since it was proposed by Tabak *et al.* some 15 years ago [1]. The first FI integrated experiments were conducted by Kodama *et al.* in 2001 [2]. In their experiments the 'cone-guiding' technique was used to reduce the propagation distance of electron beams to the core and consequently $20-30\%$ coupling from the ignition laser to the core was obtained. Subsequently many relevant studies have been reported, referring to hot electron production and optimization [3, 4, 5, 6], the electron beam transport in coronal plasmas [7, 8, 9, 10], and the core heating mechanism [11, 12, 13], as well as integrated studies of these processes [14, 15, 16, 17]. In particular, Honrubia *et al.* investigated the process of the transport of electron beams produced at the cone tip to the fusion target core with hybrid particle-in-cell (PIC) simulations [16]; The studies by Chrisman *et al.* include all of the relevant physical processes except the laser propagation in the cone using collisional PIC simulations [17]. Both of these results show that collisions dominate core heating.

Previous studies for the core heating usually follow the sequential procedure according to the actually processes, i.e. from the hot electron generation to the core heating. Here, we follow a different procedure in order to find the optimized laser and target parameters for the FI realization. In this procedure, first, assuming electron beams have reached the core surface, one studies the electron beam energy requirement to heat the core most efficiently, where collisional stopping is most significant; secondly, assuming electron beams have been produced at the cone tip, one studies how to get the optimum beam energy required in the previous step when the

Journal of Physics: Conference Series **244** (2010) 022070 doi:10.1088/1742-6596/244/2/022070

beams have transported through the coronal plasmas of about $100\mu m$, where collective stopping is significant; thirdly, one studies the parameters for the ignition lasers and targets, which may lead to more optimal electron beams. Every step can be performed numerically using the same plasma density and size as the actual experiments, which provides realistic reference for integration simulations and experiments. The current paper presents simulation results aiming at the first step mentioned above.

2. Simulation Results

Since the core heating in this step is dominated by collisions between particles and collective effects can be ignored, we use a collisional PIC code without solving the Maxwell's equations, setting the electromagnetic fields vanished in the extremely high density core. The collision of weighted macro-particles is treated according to Refs. [18, 19, 20].

In the two-dimensional PIC simulations, we take the core as uniform deuterium tritium plasma with the density of $5.4 \times 10^4 n_c$ (corresponding to $\rho = 300g/cm^3$) and distributing within the circle with the radius $r = 25\mu m$, where $n_c = 1.1 \times 10^{21} cm^{-3}$. The plasma density outside the core decreases exponentially (following $exp(-r/r_0)$ with $r_0 = 9.2\mu m$) and the coronal plasma on the right side of the core is neglected, which enables us to calculate the energy deposition only inside the core more exactly. The plasma electron and ion temperatures are all at $300eV$. The test beams have the density of n_c, include 10000 electrons, and distribute within $1\mu m \times 1\mu m$. The beams are initially located on the left surface of the core from $x = 199\mu m$ to $x = 200\mu m$. The beam electrons have the same energy moving along the +x-direction only. For convenience we take the period T_0 of $1\mu m$ wavelength lasers as the time unit (i.e. $3.3fs$).

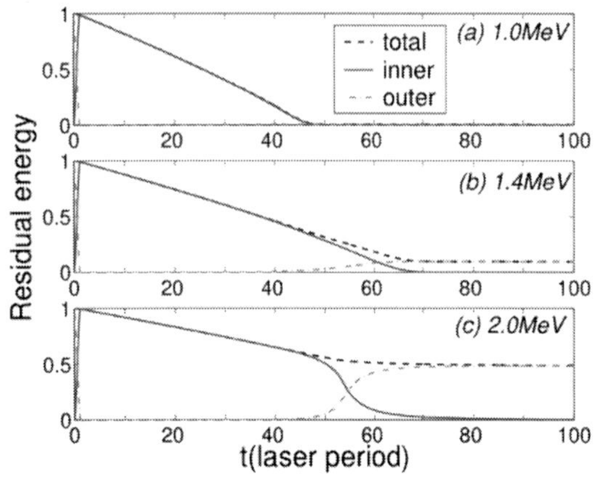

Figure 1. Temporal evolution of energy deposition of beams, where the black broken line, blue solid line, and red dotted line represent the total beam energy, the beam energy inside the core, and the beam energy outside the core, respectively. The initial energy of beams is $1.0MeV$, $1.4MeV$, and $2.0MeV$ in (a), (b), and (c), respectively.

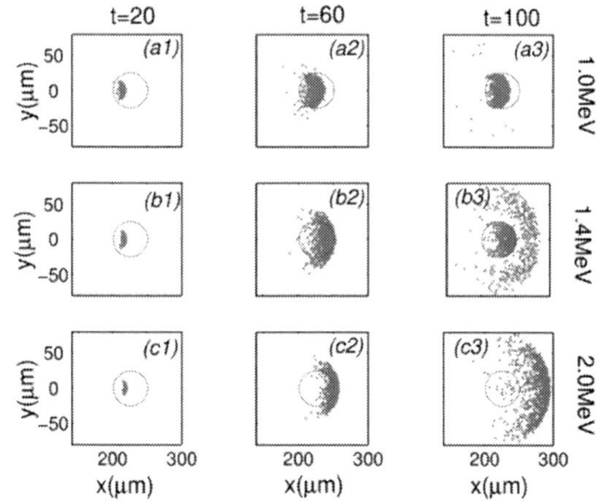

Figure 2. Spatial distributions of beam electrons, where the blue circle is the core boundary. The three rows represent initial beam energy of $1.0MeV$, $1.4MeV$, and $2.0MeV$, respectively. The three columns correspond to the time of 20, 60, and 100 laser periods, respectively.

The temporal evolution of energy deposition of beams is displayed in Fig. 1. One can see that the strong collision between the beam electrons and the core particles leads to the gradual energy deposition of beams and thus the residual energy of electrons inside the core reduces with time. For low energy beam, e.g. $1MeV$, the two curves for the total beam energy and the beam energy inside the core nearly coincide, which implies that the beam electrons deposit almost all energy to the core and they cannot flee from the core area [also see Fig. 2(a)]. The corresponding energy deposition ratio, which is defined as the deposited energy inside the core region divided by the total beam energy, is 99% obtained from Fig. 1(a). Increasing the beam energy to $1.4MeV$, some of electrons can escape from the core, as seen from Fig. 2(b) clearly. Furthermore, one can see that electrons flee away mainly in the transverse direction due to their large transverse momenta accumulated during collisions. The energy deposition ratio in the core area is decreased to 90% obtained from Fig. 1(b). When the beam electrons have higher individual energy, e.g. $2MeV$, most electrons are left with high residual energy after they propagate through the core, mainly in the longitudinal direction [see Figs. 1(c)], and the energy deposition ratio in the core area drops considerably to 51%, as obtained from Fig. 2(c). It should be noticed that the results above have some difference from the reference [21] because different models are taken, which will be studied further.

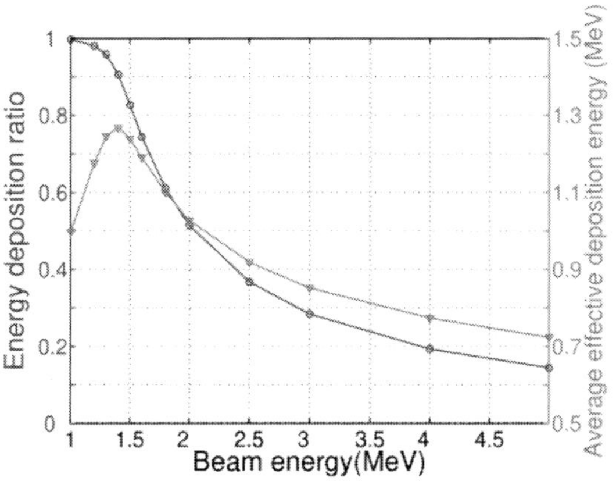

Figure 3. The energy deposition ratio (circles) and the average effective deposition energy (triangles) of electrons as a function of the initial beam energy.

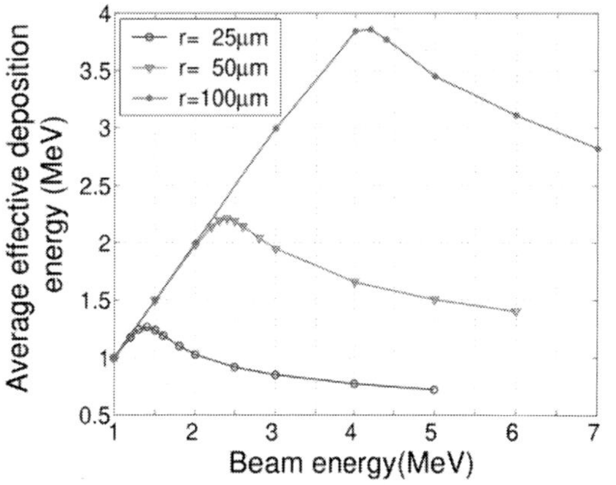

Figure 4. The average effective deposition energy of electrons as a function of the initial beam energy for cores with the radii of $25\mu m$, $50\mu m$, and $100\mu m$, respectively.

The blue line with circles in Fig. 3 shows the energy deposition ratio as a function of the initial beam energy. One can see that the energy deposition ratio reduces gradually from $1MeV$ to $1.4MeV$. However, from $1.4MeV$ to $2MeV$ the energy deposition is diminished rapidly because more electrons with higher residual energy go through the core. For the beams with energy higher than $2MeV$, the deposition goes down continuously and it becomes only 14% at $5MeV$. This is because when electrons have higher energy, their collision frequencies become smaller.

We obtain the average effective deposition energy of electrons, which is a product of the energy deposition ratio and the corresponding initial beam energy, as shown by red line with triangles in Fig. 3. One can see that the maximum average effective deposition energy is about $1.3MeV$, which appears at the initial beam energy of $1.4MeV$. This value goes down when increasing the beam energy. Thus for the size and density of the core with $\rho = 300g/cm^3$ and $\rho r = 0.75g/cm^2$, the optimum beam electron energy is $1.4MeV$ for energy deposition. This optimum beam energy is about the value with which some beam electrons start to go through the core region along the longitudinal direction. Obviously, this indicates that the ignition laser intensity should be controlled at certain magnitude so that the average electron energy is around the optimum value for efficient core heating.

In the same way, we obtain the optimum beam energy for the cores with the radii of $50\mu m$ ($\rho r = 1.5g/cm^2$) and $100\mu m$ ($\rho r = 3g/cm^2$), respectively, which are presented in Fig. 4. One can see that the optimum initial beam energy is $2.4MeV$ and $4.2MeV$ for the cores with the radii of $50\mu m$ and $100\mu m$, respectively. The corresponding average effective deposition energy is about $2.2MeV$ and $3.9MeV$, respectively. Also, these optimum beam electron energy is about the values with which some beam electrons start to go through the cores in the longitudinal direction.

It should been pointed that the optimum beam energy presented above can provide a reference for the energy level of electrons for effective absorption at the given target size. For larger size, the optimal energy shall be larger. Since electrons produced by intense lasers usually have broad energy spectra, electrons with different energy will deposit their energy in different positions. Our simulation may help to estimate where the target achieves the highest temperature by fast electron heating.

3. Conclusion

In summary, preliminary investigation on the energy deposition of electron beams incident from the core surface to the fusion core has been conducted by use of collisional particle-in-cell simulation. It is found that the low energy beams (i.e. $1MeV$) can transfer almost all of their energy to the core. When increasing the beam energy, the energy deposition will decrease gradually. When the beam energy rises to the value with which the beam can just go through the core, a slight growth in the beam energy can cause a rapid decrease of the energy deposition ratio. The optimum beam energy for the core heating is about this value. For the three typical cores with $\rho r = 0.75\text{g/cm}^2$, 1.5g/cm^2, and 3g/cm^2, the optimum beam energy is $1.4MeV$, $2.4MeV$, and $4.2MeV$, respectively.

Acknowledgments

This work was performed partially during WMW's stay at the STFC Rutherford Appleton Laboratory, U.K. This work was supported in part by the CAS-STFC collaboration program, the NSFC (grants No. 10674175, 10734130, 60621063), National High-Tech ICF Committee of China, and National Basic Research Program of China (grants No. 2007CB815101, 2007CB815105).

References

[1] Tabak M *et al* 1994 *Phys. Plasmas* **1** 1626
[2] Kodama R *et al* 2001 *Nature (London)* **412** 798
[3] Chen Z L *et al* 2005 *Phys. Rev. E* **71** 036403
[4] Lei A L *et al* 2006 *Phys. Rev. Lett.* **96** 255006
[5] Sheng Z M *et al* 2000 *Phys. Rev. Lett.* **85** 5340
[6] Li Y T *et al* 2006 *Phys. Rev. Lett.* **96** 165003; Li Y T *et al* 2001 *Phys. Rev. E* **64** 046407
[7] Santos J J *et al* 2002 *Phys. Rev. Lett.* **89** 025001
[8] Matsumoto T, Taguchi T, and Mima K 2006 *Phys. Plasmas* **13** 052701
[9] Li Y T *et al* 2008 *Phys. Rev. E* **77** 016406
[10] Zhou C Y, He X T, and Yu M Y 2008 *Appl. Phys. Lett.* **92** 151502
[11] Sentoku Y, Mima K, Kaw P, and Nishikawa K 2003 *Phys. Rev. Lett.* **90** 155001
[12] Kemp A J, Sentoku Y, Sotnikov V, and Wilks S C 2006 *Phys. Rev. Lett.* **97** 235001
[13] Mason R J 2006 *Phys. Rev. Lett.* **96** 035001
[14] Norreys P A *et al* 2004 *Phys. Plasmas* **11** 2746
[15] Ren C *et al* 2004 *Phys. Rev. Lett.* **93** 185004; Ren C *et al* 2006 *Phys. Plasmas* **13** 056308
[16] Honrubia J and Meyer-ter-Vehn J 2006 *Nucl. Fusion* **46** L25
[17] Chrisman B , Sentoku Y, and Kemp A J 2008 *Phys. Plasmas* **15** 056309
[18] Takizuka T and Abe H 1977 *J. Comput. Phys.* **25** 205
[19] Nanbu K 1997 *Phys. Rev. E* **55**, 4642.
[20] Sentoku Y and Kemp A J 2008 *J. Comp. Phys.* **227** 6846.
[21] Tabak M *et al* 2005 *Phys. Plasmas* **12** 057305

Journal of Physics: Conference Series **244** (2010) 022071 doi:10.1088/1742-6596/244/2/022071

Experimental investigation to demonstrate Impact Fast Ignition scheme

Watari T[1], Azechi H[1], Nakai M[1], Hironaka Y[1], Sakaiya T[1], Nakamura H[1], Shiraga H[1], Shigemori K[1], Hosoda H[1], Arikawa Y[1], Homma H[1], Norimatsu T[1], Murakami M[1], Jhozaki T[1], TMS group[1], GOD group[1], Karasik M[2], Gardner J[2], Bates J[2], Colombant D[2], Weber J[2], Aglitsky Y[2], Obenschain S[2]

[1] Institute of Laser Engineering, Osaka univ., Japan
[2] Naval Research Laboratory, Washington DC, USA

E-mail: twatari@ile. osaka-u. ac. jp

Abstract. We have proposed a new ignition scheme of Fast Ignition, called "Impact Fast Ignition (IFI)", in which a compressed fuel is ignited by impact collision of a fragment of separately imploded fuel. We performed integrated experiments on impact ignition, in which a portion of a deuterated polystyrene (CD) shell was accelerated to about 600 km/s and was collided with precompressed CD fuel. The shell was imploded using 9 beams and the impactor was accelerated using 3 beams of the GEKKO XII laser system. The laser energy was 350 J per beam. The kinetic energy of the impactor was efficiently converted into thermal energy generating a temperature of about 1.8 keV., Observed maximum neutron yield was 2×10^6. This yield was 80 times as large as that without impactor. We will present the experimental details and results, demonstrating the high potential of impact ignition for fusion energy production.

1. Introduction

In our institute, the integrated experiments of Fast ignition were performed [1]. The results of these experiments have shown that substantial fraction of the incident Peta watt laser is converted to the thermal energy of the compressed core plasma. However the fast ignition approach requires the understanding of physics of hot electron generation and transport. Present physical understanding of the physics is insufficient to make a quantitative prediction of the ignition. Recently, Impact Fast ignition (IFI) [2] was proposed as a new ignition scheme of Fast ignition that totally eliminates these complex problems while keeping the advantage of the compactness of the fast ignition. In the IFI method, we accelerated a small portion of the fuel (called "impactor") to a super-high velocity to collide with a pre-compressed main fuel. Figure 1 shows the schematic view of the initial target structure of the impact ignition overlapped with the compressed fuel image at maximum compression. Since the kinetic energy of the impactor is converted to the thermal energy of both ions and electrons, its velocity needs to be higher than 1000 km/s to get the ignition temperature of 5 keV [via $m_{DT}v_{\rm imp}^2/2=2(3/2)T$, where m_{DT}, $v_{\rm imp}$, T are the average mass of the DT ion, the velocity of the impactor and T is the required temperature]. For example, if the impactor energy $E_{\rm imp}$ is 10 kJ under typical ignition conditions (T = 5–10 keV, ρR = 0.4 g/cm^2), the impactor density requirement before shock compression will be ρ_{imp} = 31–44 g/cm^3. In order to study an effectiveness of IFI method, we had performed fundamental experiments to create a high speed impactor, In this experiment, we conducted experiments on ablatively accelerated planar foil targets accelerated to velocities approaching 1000 km/s [3,4]. In the recent follow-up experiment performed at Naval research laboratory, the impactor velocity of 1000 km/s was achieved. In this paper, we report the results of the latest integrated experiment on IFI, in which a portion of a deuterated polystyrene (CD) shell was accelerated to about 600 km/s and was collided with precompressed CD fuel.

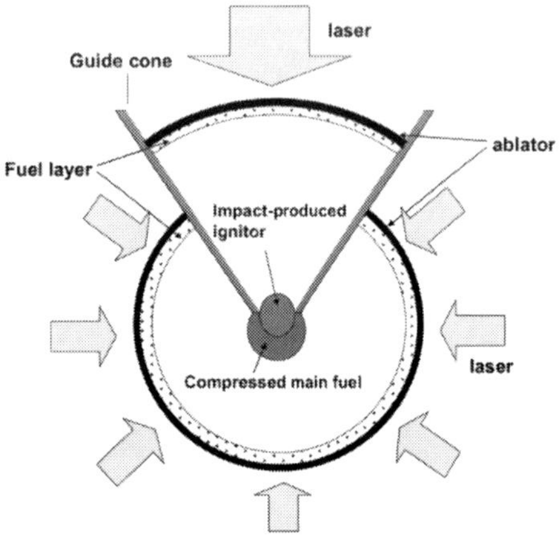

Figure 1. The schematic view of the initial target structure of the impact ignition overlapped with the compressed fuel image at maximum compression.

2. Integrated experiment

In the integrated experiment, we used a CD shell target and CD impactor attached on an Au cone. Figure 2 shows the target design and picture used in this experiment. The main body of fuel was a CD shell that served as a fusion fuel surrogate (500-μm diameter, 7-μm thickness). A gold cone with an apex angle of 90° was inserted into the shell of the main body of fuel. The cone had an open apex with a diameter of 250 μm to allow the impactor to transmit freely through the apex. The impactor was a CD shell (900-μm diameter, 10-μm thickness) attached to the entrance of the cone. A CD impactor was accelerated by the high power laser and collides to a compressed CD shell. The shell was imploded using 9 beams and the impactor was accelerated using 3 beams of the GEKKO XII laser system. The laser energy was 350 J per beam and the laser wavelength was 0.53 μm. The laser pulse shape was gaussian with FWHM of 1.3 ns. The timing of the impactor laser pulse relative to the implosion laser pulse was changed at 100 ps or 50 ps intervals.

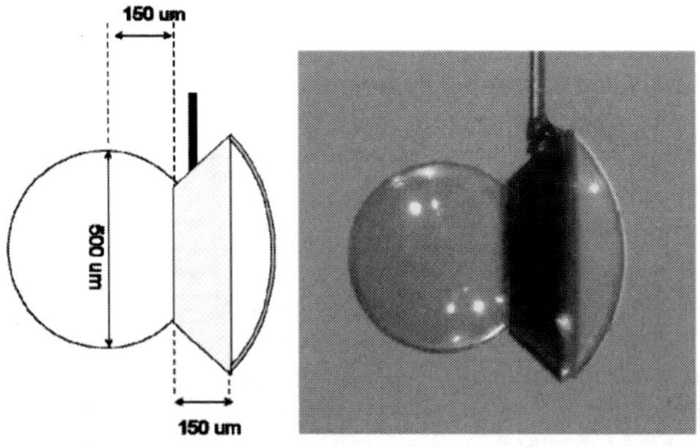

Figure 2. The target design and picture used in this experiment

Figure 3 shows the collision of the impactor with the precompressed main body of fuel measured from the temporal evolution of the streaked x-ray images at energies of 1-3 keV. When the timing between the implosion laser and the impactor laser is optimized, the impactor can be expected to collide with the main body of fuel near the initial center of the main body of fuel. These data enables us to determine the collision timing by comparing x-ray pulses from the impactor with those from the main body of fuel.

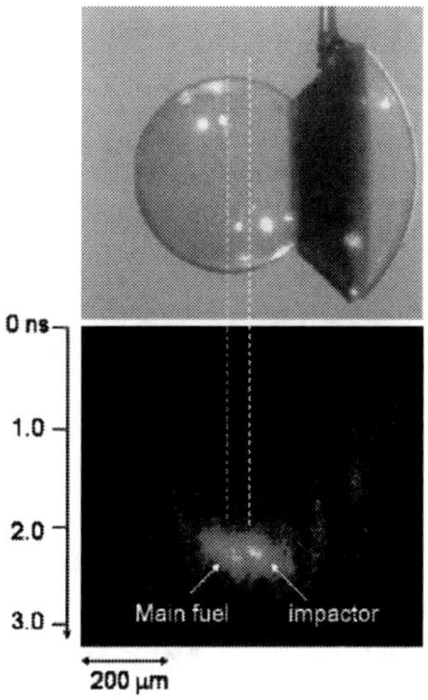

Figure 3. Picture of the target and streaked x-ray image showing the collision of the impactor with the precompressed main body of fuel..

We observed the yield and isotropy of the neutron emission with four neutron detectors, which were operated in current mode. These detectors consisted of fast plastic scintillators (Bicron BC422 with a 1% benzophenone quencher) with either 18 cm (diameter) × 2.5 cm (thickness) or 10 cm (diameter) × 5 cm (thickness) coupled with fast photomultiplier tubes (Hamamatsu R2083). A 5-mm lead shield was placed in front of each detector. Two detectors were located in the nearly forward and backward directions with respect to the acceleration direction of the impactor, while the other two detectors were in nearly orthogonal directions. The distances between the target and the detectors were 47, 320, 350, and 410 cm. The observed values of neutron yields of five detectors were comparable to each other and the significant energy shift of the neutron spectrum was not observed, so the neutrons observed in this experiment were concluded to be generated by the thermonuclear fusion reactions. We measured neutron yields varying the time lag between the incidence time of laser for impactor and laser for main fuel to research the timing dependence of neutron yields (Fig. 4). The horizontal axis of Fig.4 means the collision timing. The collision timing for the peak neutron yield was determined by comparing x-ray emission peak from the impactor with those from the main body of fuel on a streaked x-ray image (Fig. 3). In figure 3, a time of the emission peak from the impactor appeared 10 ps before those from the main fuel (collision timing: -10 ps). Those for the other data were obtained from the timing between the implosion laser pulse and the impactor laser pulse. In these shots, the density and mass of the impactor just before the collision were essentially the same because the target specification and the laser condition were identical in every shot. It was shown that the generated neutron yields depended on the timing for the impactor to collide with the imploded main fuel. The dramatic increase of the neutron yield in about 100 psec demonstrates a sharp increase of the fuel density by stagnation. When a measured neutron yield was maximum (2×10^6), a density of a main fuel was considered to be the highest in this experiment as was reproduced by a computer simulation. The solid line in Figure 4 shows the temporal evolution of peak density at main fuel, which was calculated from the one-dimensional hydrodynamic simulation code "ILESTA-1D" [5,6]. This code includes an equation of state based on the Thomas-Fermi model [7] and the electron energy transport based on the Fokker-Plank (FP) equation [8]. The origin of time is set to be the onset of the maximum compression time.

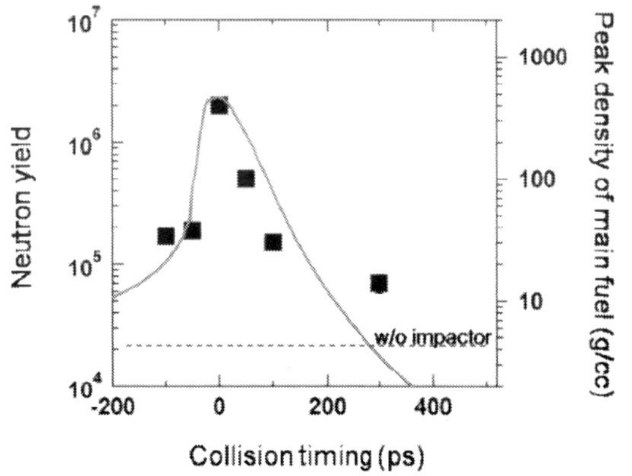

Figure 4. Experimental results of the integrated experiment and simulated peak density of main fuel.

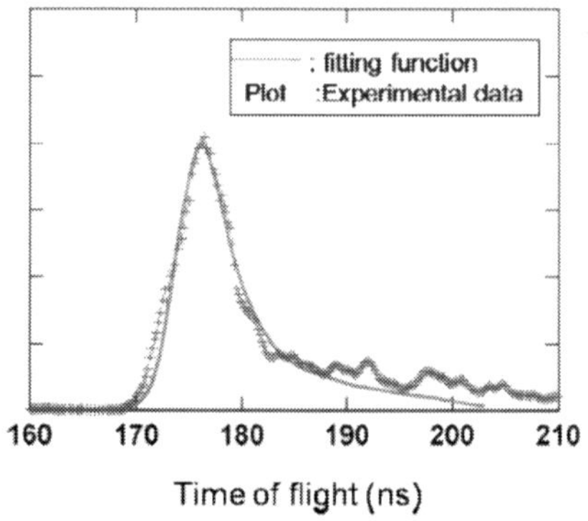

Figure 5. DD neutron TOF spectrum observed by current mode detector and fitting line.

In figure 4, collision timing dependence of the neutron yield and simulated peak density time evolution of main fuel shows a similar shape. It is shown that the difference of generated neutron yields came from the difference of density of the main fuel to collide. The temperature of collided impactor increases as the density of compressed main fuel increases. Figure 5 shows the neutron time-of-flight (TOF) spectrum observed by current mode detector at the shot when we observed maximum neutron yield. Ion temperature of heated plasma was obtained from the Doppler broadening of this neutron spectrum as $1.81^{+0.41}_{-0.58}$ keV.

Additionally, we also imploded the target without an impactor in a reference experiment. As is shown by the broken line in the Fig. 4, we observed 2×10^4 neutrons in this reference experiment. This result means that the neutron yield had increased by a factor of 100 at the maximum by impact heating.

3. Conclusion

We found that the generated neutron yields depend on the timing for the impactor to collide with the imploded main fuel. It obviously means that the generated neutron yields depend on the density on pre-compressed main fuel. 2.0×10^6 neutrons were observed at the maximum value. In this shot, the impactor was heated to 1.8 keV by the collision. And the impact heating increased the neutron yield by a factor of 100.

References.

[1] Kodama R *et al.*, 2001 Nature **412** 798
[2] Murakami M, and Ngatomo H, 2005 Nucl. Instrum. Meth. Phys. Res. A **544** 67
[3] Azechi *et al.*, Phs. Rev. Let. **102** 235002 (2009)
[4] Watari T *et al.*, J. Phys.: Conf. Ser. **112**, 022065 (2008)
[5] Takabe H *et al.*, Phys. Fluids **31**, 2884 (1988)
[6] Sunahara A *et al.*, Phys. Rev. Lett. **91**, 095003 (2003)
[7] M. More R *et al*, Phys. Fluids **31**, 3059 (1988)
[8] Sunahara A, in *Annual Progress Report of Institute of Laser Engineering*, (Osaka University Press, Osaka, 1999) p.295

Journal of Physics: Conference Series **244** (2010) 022072 doi:10.1088/1742-6596/244/2/022072

Plasma currents and inverse bremsstrahlung absorption under strong dc/ac electric fields

Su-Ming Weng[1,3], Zheng-Ming Sheng[2,1,‡], and Jie Zhang[2,1]

[1]Beijing National Laboratory of Condensed Matter Physics, Institute of Physics, CAS, Beijing 100190, China

[2]Department of Physics, Shanghai Jiao Tong University, Shanghai 200240, China

[3]Theoretical Quantum Electronics (TQE), Technische Universität Darmstadt, D-64289 Darmstadt, Germany

E-mail: [‡]zmsheng@sjtu.edu.cn

Abstract. Plasma current induced by a strong dc electric field is studied by Fokker-Planck (FP) simulation. Based upon this, a hydrodynamic-like equation, similar to Spitzer's but without the weak-field limit, is given for calculating the current. This equation is suitable for application in hybrid-PIC simulations relevant to the fast ignition (FI) scheme. Furthermore inverse bremsstrahlung (IB) absorption is studied in a wide laser intensity range. In particular, we introduce an IB operator which is similar to Langdon's, but without the low laser intensity limit. This operator enables one to treat IB absorption properly in the inertial confinement fusion (ICF) schemes with the NIF/LMJ-scale laser intensity beyond 10^{15} W/cm^2.

1. Introduction

As is well known, a dc electric field will produce a current in plasma, while an ac electric field (usually laser field) induces IB absorption. Plasma current produced by a weak dc electric field can be well described by Spitzer's theory [1], and IB absorption in a low intensity laser can be treated consistently with the evolution of the electron distribution function (EDF) by Langdon's IB operator [2]. However, both Spitzer's theory and Langdon's operator are based on perturbation theory, assuming small anisotropy of EDF. Therefore, Spitzer's theory is valid only for $E_{dc} \ll E_c = m_e v_e \nu_{ei}/e$ and Langdon's operator for $E_{ac} \ll m_e v_e \omega/e$. Consequently, the observed electric conductivities are usually much lower than that predicted by Spitzer's theory in the non-weak dc electric fields [3], and Langdon's IB operator overestimates the absorption rates in an intense laser [4]. In the following sections, plasma dynamics in dc/ac electric fields are studied by a FP code [5], with a new equation introduced for calculating plasma currents and a modified IB operator for calculating absorption rate with high dc/ac electric field strength.

2. Plasma current in dc electric fields

Since the EDF is crucial to plasma properties, in Fig. 1 we draw the EDFs of plasma that obtained from FP code [5, 6] with ionic charge state $Z_i = 1$ under dc electric fields with different strengths. In the weak field, almost no shift of the EDF from $v = 0$ is seen. In the strong field, the EDF is almost collectively drifted away from $v = 0$. While the EDF in the moderate field is a hybrid of a stationary and drifting Maxwellian, as a bridge between the EDFs in weak and

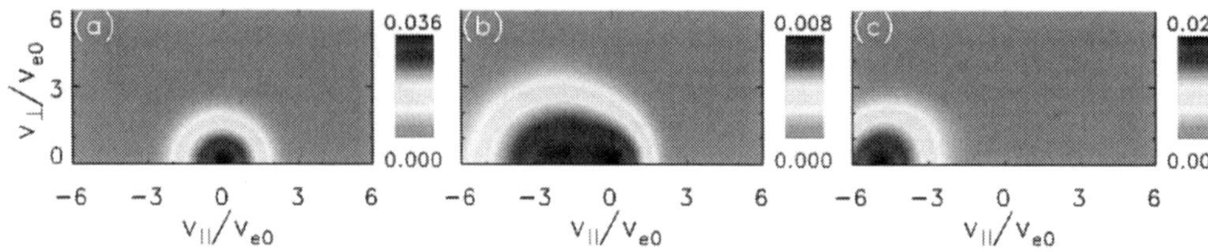

Figure 1. EDFs under different dc electric fields: (a) $0.01E_c$ after $500\tau_{ei}$, (b) $0.1E_c$ after $50\tau_{ei}$, and (c) $1.0E_c$ after $5\tau_{ei}$, where $\tau_{ei} = 1/\nu_{ei} = v_{e0}^3/(Z_i\Gamma^{e|e})$ and the EDF in unit of n_e/v_{e0}^3.

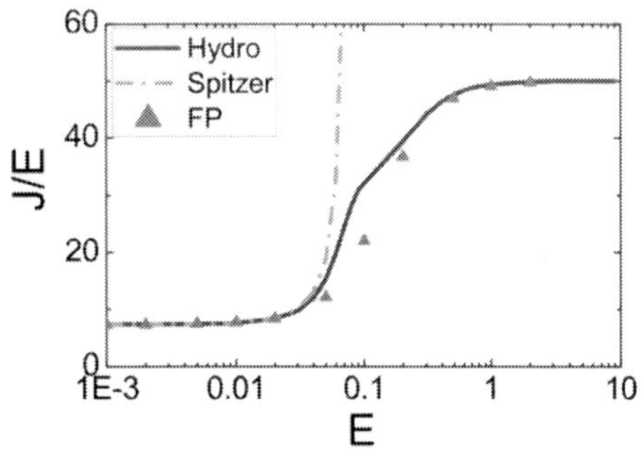

Figure 2. Plasma conductivities J/E as functions of dc electric field, obtained by Spitzer's model, by Eq. (2), and by FP code, at $t = 50\tau_{ei}$ after these electric fields are inducted into the plasmas with same initial parameters, where J/E in unit of $n_e e^2\tau_{ei}/m_e$ and E in unit of E_c.

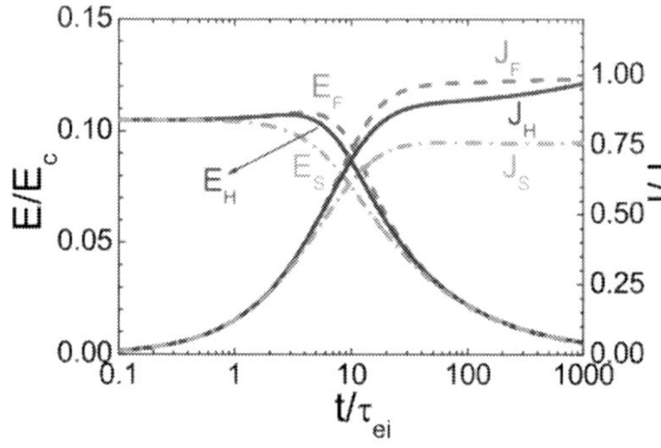

Figure 3. Electric field updated by Spitzer's (E_S), by hydrodynamic-like Eq. (2) (E_H), and by FP code (E_F), as well as the produced currents of these fields as a current of fast electrons $J_f = 3.5$GA with a radius of $20\mu m$ transports into a uniform plasma with initial temperature of 500eV and density of 5gcm^{-3}.

strong field. Therefore, the EDFs in arbitrary dc electric fields can be expressed generally as

$$f(\mathbf{v}) = \delta\frac{n_e}{(2\pi v_{te1}^2)^{3/2}}\exp\left(-\frac{v^2}{2v_{te1}^2}\right) + (1-\delta)\frac{n_e}{(2\pi v_{te2}^2)^{3/2}}\exp\left[-\frac{(\mathbf{v}-\mathbf{v_d})^2}{2v_{te2}^2}\right], \quad (1)$$

where δ is the proportion of the stationary Maxwellian component, the first and second terms of RHS are the stationary and drifting Maxwellian components, respectively.

If we define parallel temperature $T_\parallel = m_e\int f(\mathbf{v})v_\parallel^2 d\mathbf{v}$ and perpendicular temperature $T_\perp = m_e\int f(\mathbf{v})v_\perp^2 d\mathbf{v}/2$, it will be a good approximation to calculate plasma current as [6]

$$J = \sigma_0 E\left[1-\exp(-\frac{t}{\tau_r})\right]\exp\left(\frac{T_\perp - T_\parallel}{T_\perp}\right)\left(\frac{T_\perp}{T_0}\right)^{3/2} + n_e e\left[1-\exp(\frac{T_\perp - T_\parallel}{T_\perp})\right]\left(\frac{T_\parallel - T_\perp}{m_e}\right)^{1/2}, \quad (2)$$

where σ_0 is Spitzer's conductivity [1] for initial temperature T_0, τ_r is the response time defined as the time when the current is $1-1/e$ (here e is Euler's number) of the final steady value, $T_\parallel$ and $T_\perp$ can be updated by $dT_\parallel/dt = 2JE - 2\nu_{ei}(v_{eff})(T_\parallel - T_\perp)$ and $dT_\perp/dt = \nu_{ei}(v_{eff})(T_\parallel - T_\perp)$, where $\nu_{ei}(v_{eff}) = Z_i\Gamma^{e|e}/v_{eff}^3$ is the effective e-i collision frequency with $v_{eff} = \sqrt{(T_\parallel^2 + 2T_\perp^2)/m_e}$. From Eq. (2) one can calculate plasma conductivity under a dc electric field of arbitrary strength. And it agrees well with FP code, while Spitzer's theory obviously overestimates the plasma conductivity in non-weak dc electric field as shown in Fig. 2.

As shown in Fig. 3, we simulate the generation of return current (RC) during the transport of a fast electron beam $J_f = 3.5$GA in a plasma with $T_e = 500$eV and density of 5gcm^{-3}. Since Spitzer model overestimates the conductivity, it underestimates the electric field and produces a RC only about $0.75J_f$. While our hydrodynamic-like equations give a good estimation of the electric field to produce the RC as large as J_f. Therefore, it is more suitable for treating the RC and the relevant joule heating during the fast electron transport in the FI targets [7].

3. Inverse bremsstrahlung absorption in ac electric fields

Since the EDF oscillates with $\mathbf{u} = -u_0 \sin(\omega t)\mathbf{e}_z$ ($u_0 = eE/m_e\omega$ is the peak electron oscillating velocity) in an intense laser, it is more convenient to rewrite FP equation in the oscillating system (v', θ') with transformation $\mathbf{v}' = \mathbf{v} - \mathbf{u}$ as $\partial f'/\partial t = C'_{ei}(f') + C'_{ee}(f')$, where $C'_{ee}(f')$ is identical to $C_{ee}(f)$, and $C'_{ei}(f')$ is given by

$$
\begin{aligned}
C'_{ei}(f') &= \frac{\nu^*_{ei}}{2}\left[\frac{u^2 + 2uv'\cos\theta' + u^2\cos^2\theta'}{v'}\frac{\partial f'}{\partial v'} + \frac{2u\sin\theta'(v' + u\cos\theta')}{v'}\frac{\partial^2 f'}{\partial v'\partial\theta'}\right. \\
&+ \left.\frac{v'^2 + 2uv'\cos\theta' + u^2\cos 2\theta'}{v'^2\tan\theta'}\frac{\partial f'}{\partial\theta'} + (u\sin\theta')^2\frac{\partial^2 f'}{\partial v'^2} + \left(\frac{v' + u\cos\theta'}{v'}\right)^2\frac{\partial^2 f'}{\partial\theta'^2}\right], \quad (3)
\end{aligned}
$$

with $\nu^*_{ei} = Z_i\Gamma^{e|e}/(u^2 + 2uv'\cos\theta' + v'^2)^{3/2}$. Using Legendre expansion $f' \simeq f'_0(v') + f'_1(v')\cos\theta'$ and assuming $\partial f'_1/\partial t \simeq -i\omega f'_1$ for $\omega \gg \nu_{ei} \equiv Z_i\Gamma^{e|e}/v_e^3$ (which is usually fulfilled), we get

$$\frac{\partial f'_0}{\partial t} \simeq \frac{u^2}{3v'^2}\frac{\partial}{\partial v'}\left[v'^2\nu'_{ei}g(\nu'_{ei})\frac{\partial f_0}{\partial v'}\right] + C'_0, \quad (4)$$

where C'_0 is the e-e collision term only relevant to f'_0 and $g(\nu'_{ei}) = 1 - b(\nu'_{ei}/\omega)^2/[1 + b^2(\nu'_{ei}/\omega)^2]$ with $\nu'_{ei} = Z_i\Gamma^{e|e}/(u^2 + v'^2)^{3/2}$ and $b = (2u^2 + 5v'^2)/5v'^2$. However, it is still difficult to integrate Eq. (4) over time for arbitrary ratios u_0/v_e. By numerical simulation, we find that [8]

$$\frac{\partial f'_0}{\partial t} \simeq \frac{u_0^2}{6v'^2}\frac{\partial}{\partial v'}\left[v'^2\nu'_{eff}g(\nu'_{eff})\frac{\partial f_0}{\partial v'}\right] + C'_0, \quad (5)$$

which can generate the absorption rate in good quantitative agreement with that obtained from our FP code for $0.2 \le u_0/v_e \le 65$ as shown in Fig. 4, where $\nu'_{eff} = Z_i\Gamma^{e|e}/(v'^2 + u_0^2/\zeta)^{3/2}$ with $\zeta = 3.84 + (142.59 - 65.48u_0/v_e)/[27.3u_0/v_e + (u_0/v_e)^2]$.

In Fig. 4, we compare the absorption rates R calculated from different models. At low intensity R obtained from all methods agrees well with each other and increase linearly with the intensity as predicted [4]. However, Langdon's IB operator has already overestimated R when $I = 10^{14}$ W/cm^2 with $u_0 \simeq 2v_e$, and this deviation grows dramatically to several orders of magnitude with increasing intensity, while R calculated from our IB operator decreases slowly with the intensity and still shows a very good quantitative agreement with FP simulation and molecular dynamic simulation [9] at $I > 10^{14}$ W/cm^2. This illustrates that our IB operator can produce the proper absorption rate at high laser intensity and can be used as the generalized version of Langdon's IB operator for a variety of practical applications [10].

As an example, we simulate the IB heating of Ref. [11], which is relevant to indirect drive ICF scheme. For $T_e = 284\ eV$, which is the peak temperature in the non-magnetic case at 440 ps in Ref. [11], u_0/v_e is about 0.96, and it will be larger for other lower temperatures. Therefore, Langdon's IB operator, which is valid for $u_0/v_e \ll 1$, is no longer suitable for treating IB absorption in this case. In Fig. 5, we compare the absorption rates and temperatures obtained from different models. It is found that our IB operator accurately estimates the absorption rate. However, Langdon's operator results in an overestimated absorption rate and an overheated plasma, which may affect the heat transport and subsequent processes in the ICF scheme [11].

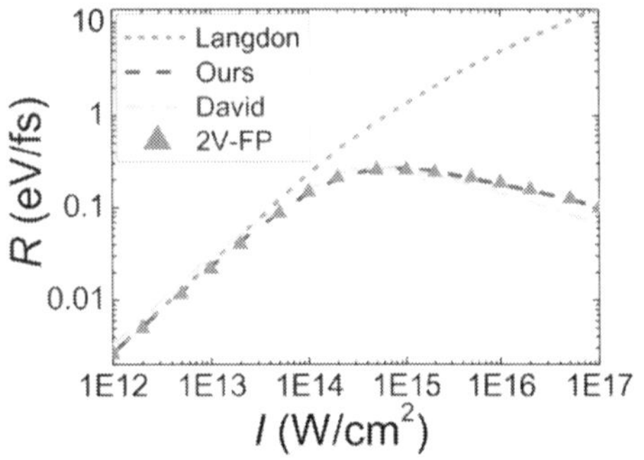

Figure 4. The absorption rates R VS laser intensity obtained from FP code, Langdon's IB operator, our IB operator, and David's fitted formula (6) in Ref. [9] from molecular dynamic method. The parameters are: $n_e = 10^{20}$ cm^{-3}, $T_{e0} =$ 10 eV, $Z_i = 1$, and laser wavelength $\lambda = 1.06\mu$m.

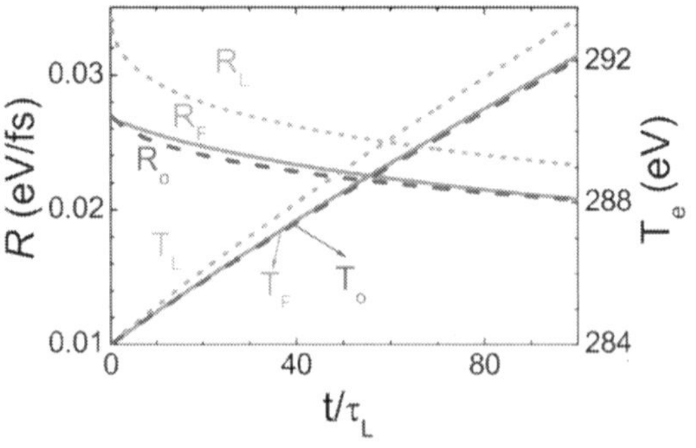

Figure 5. The absorption rates R and plasma temperatures T_e updated by FP code (R_F and T_F), Langdon's IB operator (R_L and T_L), and our IB operator (R_o and T_o). The parameters are: electron density of 1.5×10^{19} cm^{-3}, initial temperature of 284 eV, ionization state of $Z_i = 7$, laser wavelength of 1.054μm, and intensity of 6.3×10^{14} W/cm^2.

4. Summary

We have found that the EDF in a dc electric field can be presented as a hybrid of a stationary and drifting Maxwellian. According to the form of EDF, we derive the hydrodynamic-like Eq. (2), which can be used as Spitzer's model but without the weak-field limit. For fast electron transport in the FI targets, it is found that the RC obtained with Eq. (2) can compensate the beam current almost completely, whereas the one obtained with Spitzer's model cannot. Furthermore we have studied the IB absorption in a wide range of laser intensity. It is found that if $u_0 > v_e$ the absorption will be inhibited with the increasing intensity. Based upon the simulation, we have introduced an IB operator without the low laser intensity limit. A simulation example relevant to the indirect drive ICF scheme shows that our IB operator provides a better evaluation of the absorption rate than Langdon's at high intensity over a few times of 10^{14}W/cm^2.

Acknowledgments

This work was supported by the NSFC (grants No. 10674175, 60621063), National High-Tech ICF Committee of China, and National Basic Research Program of China (grants No. 2007CB815101, 2007CB815105). One of the authors (S.M.W.) wants to thank Professor P. Mulser of the Technische Universität Darmstadt for fruitful discussions and suggestions and acknowledges support from the Alexander von Humboldt Foundation.

References

[1] Cohen *et al.* 1950 *Phys. Rev.* **80** 230, Spitzer L *et al.* 1953 *Phys. Rev.* **89** 977.
[2] Langdon A B 1980 *Phys. Rev. Lett.* **44** 575.
[3] Benage J F *et al.* 1999 *Phys. Rev. Lett.* **83** 2953; Lencioni D E, *Phys. Fluids* **14** 566.
[4] Pert G J 1995 *Phys. Rev. E* **51** 4778.
[5] Weng S M, Sheng Z M, and He M Q *et al.* 2006 *Phys. Plasmas* **13** 113302.
[6] Weng S M, Sheng Z M, and He M Q *et al.* 2008 *Phys. Rev. Lett.* **100** 185001.
[7] Mason R J, 2006 *Phys. Rev. Lett.* **96** 035001; Hauer A, and Mason R J, 1983 *Phys. Rev. Lett.* **51** 459.
[8] Weng S M, Sheng Z M, and Zhang J, 2009 *Phys. Rev. E* **80** 056406.
[9] David N, Spence D J, and Hooker S M 2004 *Phys. Rev. E* **70** 056411.
[10] Bell A R, and Kingham R J 2003 *Phys. Rev. Lett.* **91** 035003 and reference therein.
[11] Ridgers C P, Kingham R J, and Thomas A G R 2008 *Phys. Rev. Lett.* **100** 075003

Journal of Physics: Conference Series **244** (2010) 022073 doi:10.1088/1742-6596/244/2/022073

SIMULATION OF THE HOHLRAUM FOR A LASER FACILITY OF MEGAJOULE SCALE

M.N. Chizhkov, M.Yu. Kozmanov, S.N. Lebedev, V.A. Lykov, V.V. Rykovanova, V.N. Seleznev, K. I. Selezneva, O.V. Stryakhnina, A.A. Shestakov, A.V. Vronskiy

Russian Federal Nuclear Center – VNIITF
Vasilieva str. 13, Snezhinsk, Chelyabinsk reg., Russia, 456770

E-mail: M.N.Chizhkov@VNIITF.ru

Abstract. 2D calculations of the promising laser hohlraums were performed with using of the Sinara computer code. These hohlraums are intended for achievement of indirectly-driven thermonuclear ignition at laser energy above 1 MJ. Two calculation variants of the laser assembly with the form close to a rugby ball were carried out: with laser entrance hole shields and without shields. Time dependent hohlraum radiation temperature and x-ray flux asymmetry on a target were obtained.

Introduction

The 2D Sinara computer code was developed in VNIITF for simulation of accidental conditions in fast neutron reactors [1]. It was used to simulate the destruction of the reactor core casing and consequences from the destruction of structural components of a nuclear power plant, etc. Sinara code was also used for thermal analysis of the experimental facility ATLAS [2]. Latter its capabilities were extended to allow radiation transport simulations in multigroup kinetic approximation by the finite-difference DSn-method [3]. Then the Sinara code has been adapted for calculations of laser hohlraums, specifically, the equation of energy for matter now considers electronic heat conductivity with flux limitation, a 3D model of laser light absorption is implemented.

Results of first calculations of a laser hohlraum by the Sinara code are presented in the report. For laser energy saving and improving of x-ray flux uniformity on the target the configuration of hohlraum in the form of a rugby ball [4] was used. The calculation variant with the laser entrance hole shields [5] was also performed. Time dependent hohlraum radiation temperature and x-ray flux asymmetry on a target were calculated.

1. Calculation setup

The chosen design of the laser hohlraum is presented in Fig 1. Length of the system is 1.1 cm, equatorial radius – 3.5 mm, radius of laser entrance holes (LEH) – 1.8 mm, LEH are covered by a CH film with thickness of 1 μm. The internal cavity of the hohlraum is filled by the equimolar mixture of hydrogen and helium with the density of 0.8 mg/cc. In the system centre there is a spherical pellet with the external radius of 1.2 mm.

The variant of calculation with the additional shields covering LEH is also performed. LEH shields are golden disks 0.5 mm in radius and 20 μm in thickness. Their hole-facing surfaces are covered with a CH layer 10 μm thick. The shields are 1.2 mm distant from the outer surface of the target. The position and size of shields were optimized using the view factor code [6] in fixed geometry.

Journal of Physics: Conference Series **244** (2010) 022073 doi:10.1088/1742-6596/244/2/022073

Laser radiation transfer and absorption were simulated with the 3D model. Refraction and reflection of laser rays was not modeled (fixed rays trajectories). Irradiation scheme is shown in Fig.2. Laser radiation of a wavelength $\lambda = 0.35$ μm was directed into the hohlraum in six cones, three through each LEH. Total laser power versus time was defined in accord with French work [7] (Fig.3). The distribution of laser power between the cones changes with time, the powers of the middle and outer cones are identical. The time dependent energy fraction in the inner cone shown in Fig.4 was defined so that to reduce the second harmonic of x-ray flux on the target to less than 3%.

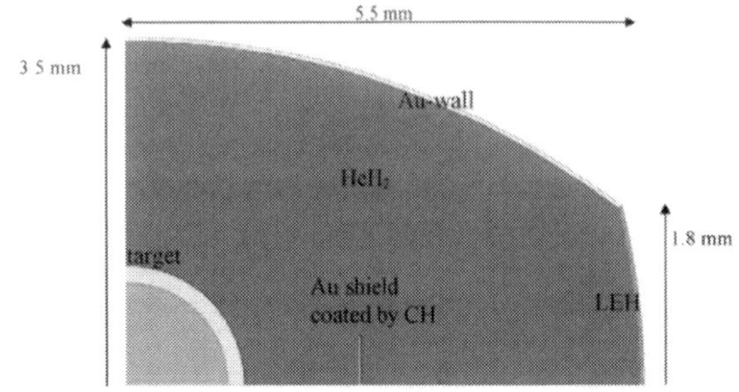

Fig.1 – Hohlram geometry

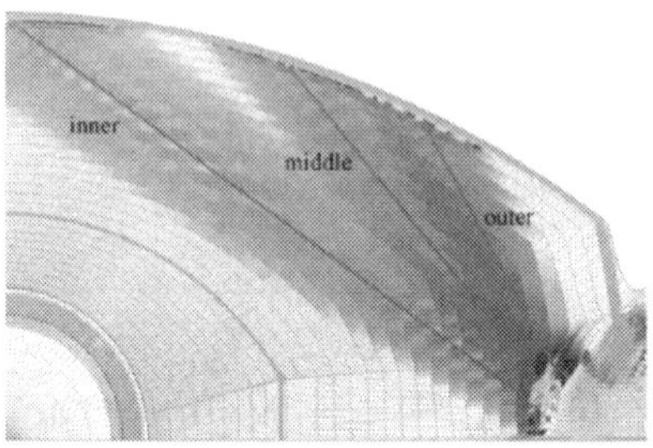

Fig.2 – Spatial grid and irradiation scheme

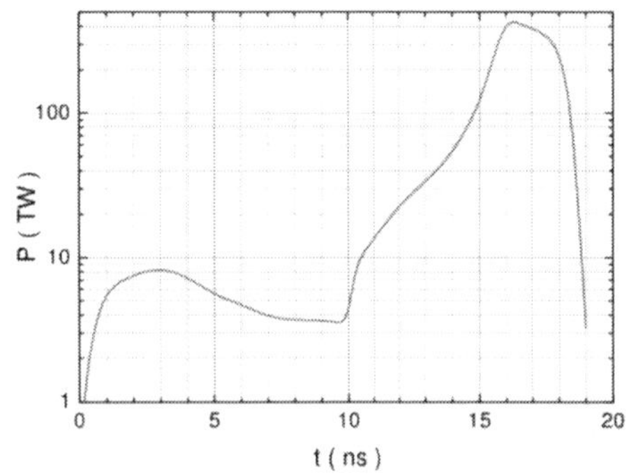

Fig.3 – Total laser power [7]

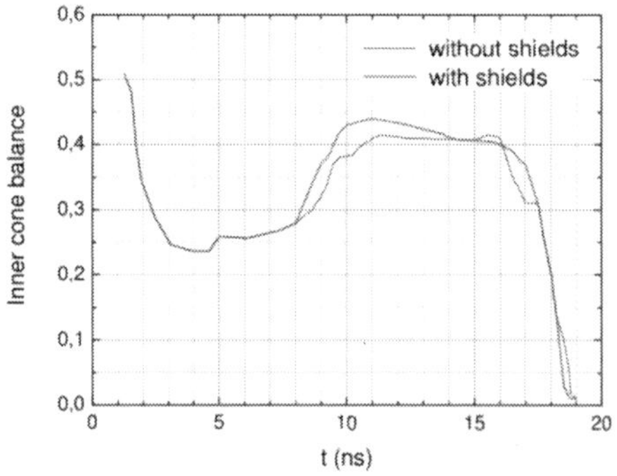

Fig.4 – Energy fraction in the inner cone

The spatial grid is essentially not orthogonal and consists from about 60000 cells (Fig.2). Quantity of photons energy groups is 10, number of photons flight directions – 144. Calculations were done with interpolated equations of state in the form proposed in [8] and opacities calculated with the average atom model [9].

2. Calculations Results

X-ray radiation temperature in the hohlraum is one of the most important characteristics for the indirect drive ignition. Fig.5 shows results of the Sinara code calculations for the hohlraum with and without LEH shields. Using of shields has allowed increasing the peak x-radiation temperature on 2% from 310 eV to 317 eV.

For the quantitative analysis of irradiation asymmetry the radiation flux on the target was expanded in terms of Legendre polynomials. Figures 6 and 7 show the amplitudes of radiation flux non-uniformity for Legendre modes l = 2 and 4.

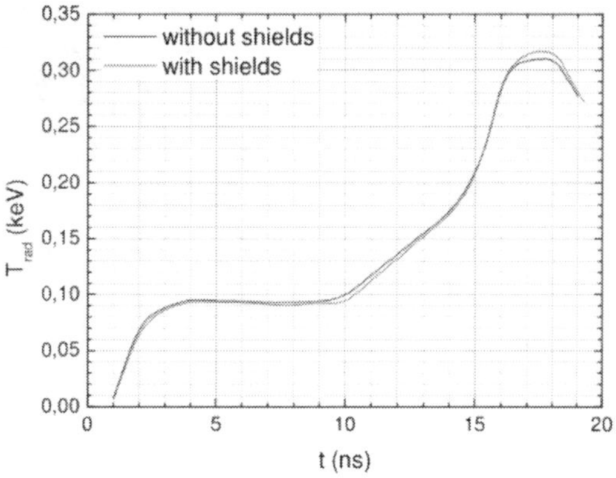

Fig.5 – Time dependent x-ray temperature on the equator of the target

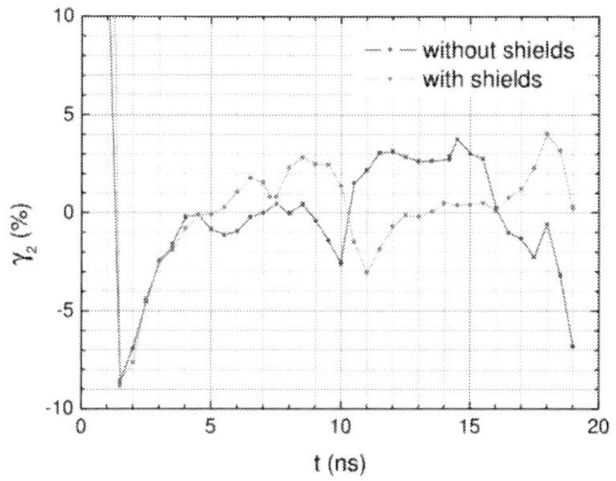

Fig.6 – Second harmonic versus time

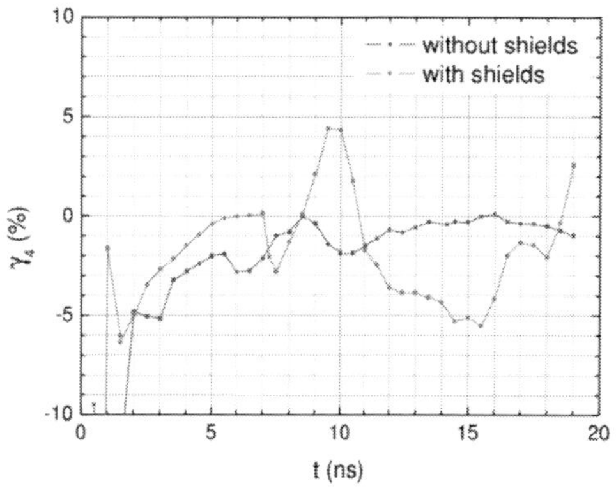

Fig.7 – Fourth harmonic versus time

The maximum value of amplitude of the second and fourth harmonics most the time does not exceed 5%. Time averaged value of the second harmonic has made 0.29 % (root mean square $\sigma = 2.1$ %) in hohlraum without shields, and in hohlraum with shields – 0.59 % ($\sigma = 2$ %). Average value of the fourth harmonic has made -1.5 % ($\sigma = 2.3$ %) and -1.2 % ($\sigma = 2.4$ %) for assemblies with and without shields, accordingly. For the symmetry improvement the further optimization is required.

Figures 8-11 show state of the system at the time of the maximum laser power.

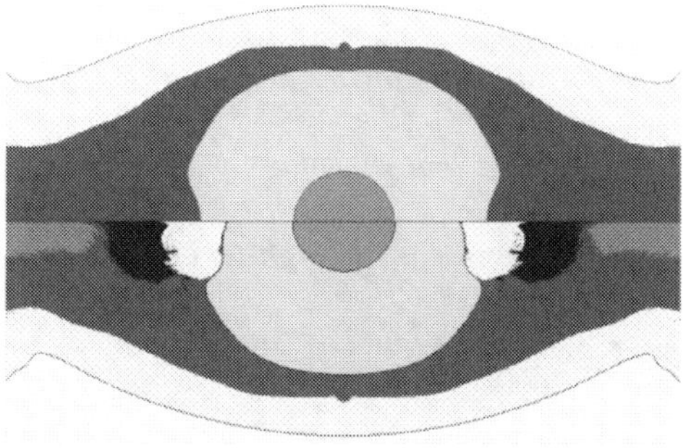

Fig.8 – Material distribution

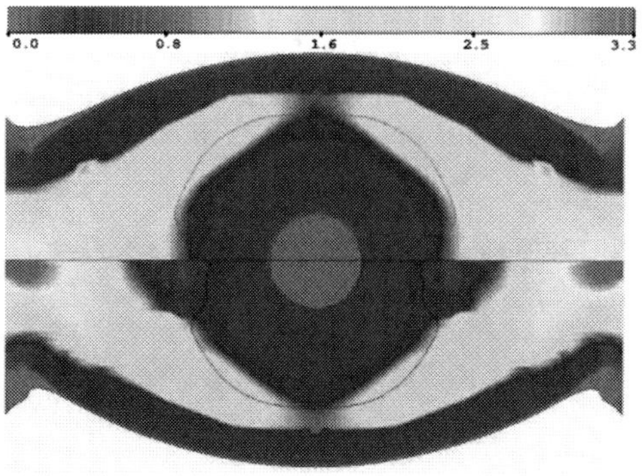

Fig.9 – Electron temperature (keV)

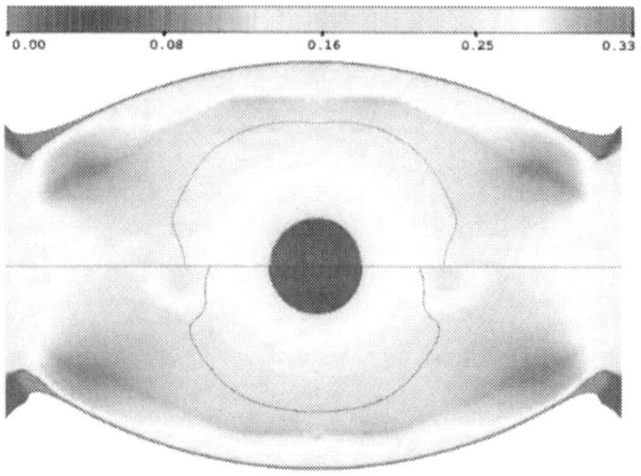

Fig.10 – Photon temperature (keV)

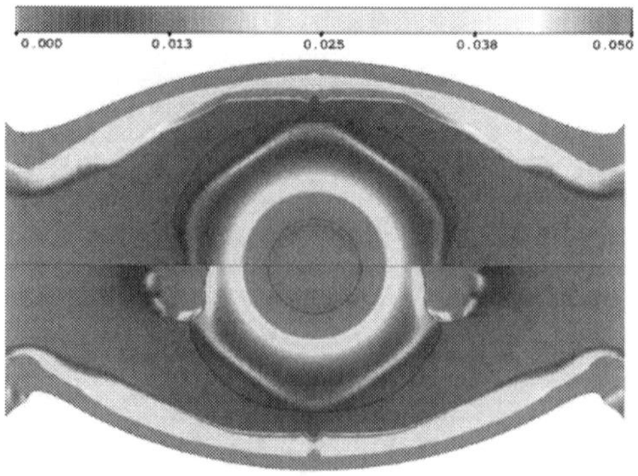

Fig.11 – Density (g/cc)

Conclusion

Performed two-dimensional calculation has confirmed the possibility of laser hohlraums modeling with the Sinara code. The calculated symmetry of the target irradiation could be improved in both considered system. However it would be excess of calculation accuracy because physical models included in the Sinara code require the further development. Firstly, the implemented laser absorption model does not consider the refraction and reflection of laser rays. Its consideration may lead to some redistribution of laser radiation absorption in a hohlraum. Secondly, Sinara calculations did not allow for the influence of the nonequilibrium ionic structure on absorption coefficients and radiation emission in golden walls. Development of the Sinara code will be continued. It could be used for design of indirectly driven experiments on the Iskra-6 facility which is planned to be constructed in Russia [10].

References

[1] A.D.Gadzhiev, V.V.Gadzhieva, S.Y.Kuzmin et al. *SINARA code package for modeling dynamics of abnormal processes in fast reactors.* VANT **3**, pp. 25-35, 2000 (in Russian).

[2] A.A.Lapaksin, S.N.Lebedev, T.V.Lebedeva et al. *Modeling of convective heat-mass transfer in ATLAS facility.* Proceedings of 6th international seminar of ISTC "Science and Computer Technology", vol. 1, pp. 39-55, Moscow, 2003 (in Russian)`.

[3] A.D.Gadzhiev, V.N.Seleznev, A.A.Shestakov. *DSn-method with artificial dissipation and VDM-method of iteration acceleration for the numerical solution of 2D kinetic radiation transfer equations.* VANT **4**, pp. 33-46, 2003 (in Russian).

[4] P.Amendt, C. Cerjan, A. Hamza et al. Phys. Plasmas 14, 056312, 2007.
M. Vandenboomgaerde, C. Cerjan, A. Hamza et al. PRL, 065004, 2007.

[5] D. A. Callahan, P. A. Amendt, E. L. Dewald et al. Phys. Plasmas 13, 056307, 2006.

[6] N.I. Chentsov, G.I. Dumkina, N.S. Shoydina. *Calculation of view factors in axially symmetric geometry.* Engineering Physics Journal **34**(2), pp. 306-312, 1978 (in Russian).

[7] S. Laffite, M. Bonnefille, P. Cargo et al. *Numerical study for the LMJ target 2D design.* In IFSA-2001. Elsevier, pp. -230, 2002.

[8] G.M. Eliseev, G.E. Klinishov. *Equation of states for solids and its spline approximation.* IPM Preprint #173, Moscow, 1982 (in Russian).

[9] A.F. Nikiforov, V.G. Novikov, V.B. Uvarov. *Quantum-statistical high-temperature plasma models. Fizmatlit,* Moscow, 2000 (in Russian).

[10] G.A. Kirillov, G.G. Kochemasov, A.V. Bessarab et al. *Status of laser fusion activity at VNIIEF (Arzamas-16).* Laser Part. Beams **18**, pp. 219-228, 2000.

First hot electron measurements in near-ignition scale hohlraums on the National Ignition Facility

E.L. Dewald[1], L.J. Suter[1], C. Thomas[1], S. Hunter[1], D. Meeker[1], N. Meezan[1], S.H. Glenzer[1], E. Bond[1], J. Kline[2], S. Dixit[1], R.L. Kauffman[1], J. Kilkenny[3] and O.L. Landen[1]
[1]Lawrence Livermore National Laboratory, 7000 East Avenue, Livermore, CA 94550.
[2]Los Alamos National Laboratory, Los Alamos, NM 87545 USA.
[3]General Atomics, P.O. Box 85608, San Diego, CA 92186.
E-mail: dewald3@llnl.gov

On the National Ignition Facility (NIF), the hot electrons generated in laser heated hohlraums are inferred from the >20 keV bremsstrahlung emission measured with the FFLEX broadband spectrometer. New high energy (>200 keV) time resolved channels were added to meet requirements for ignition and to infer the generated >170 keV hot electrons that can cause ignition capsule preheat. First hot electron measurements in near ignition scaled hohlraums heated by 96-192 NIF laser beams are presented.

1 Introduction

The National Ignition Campaign (NIC) that has begun recently on the National Ignition Facility (NIF) [1] will attempt to achieve fusion ignition and energy gain using mega joule laser-driven hohlraums [2]. During the laser beams propagation inside the hohlraums hot electrons can be generated by laser-plasma instabilities [3]. For the current ignition design [4] the hot electrons with energies >170 keV can penetrate through the capsule ablator causing preheat of the deuterium-tritium (DT) fuel [5]. The preheat results in entropy increase and hence less fuel compression, resulting in less PdV work on the hotspot to initiate fusion. In the indirect drive ignition design on the NIF the high-Z hohlraums are heated by 192 laser beams with a shaped pulse shown in Figure 1a.

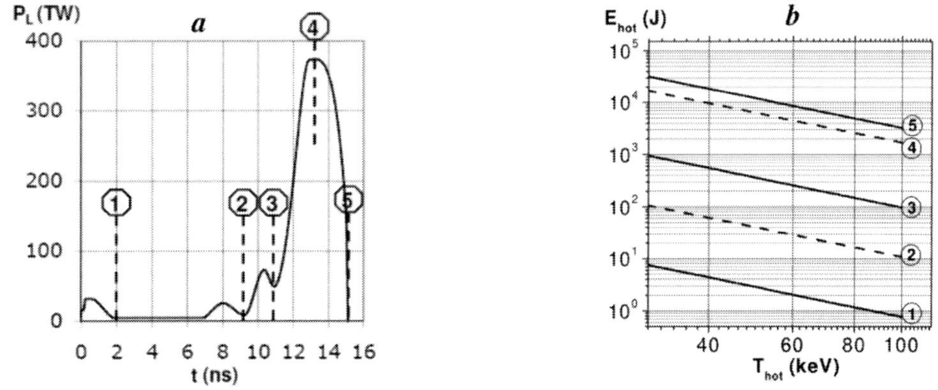

Figure 1 (a) Laser power for NIF ignition design and (b) hot electron thresholds for ignition as a function of T_{hot} integrated up to different times during the ignition pulse

Calculations show that in order to maintain fuel entropy to levels acceptable for ignition the hot electron thresholds are much more stringent at early times than at late times during the ignition pulse [5]. Since only the >170 keV hot electrons can cause fuel preheat the calculated hot electron thresholds for ignition $E_{hot} \sim 1/T_{hot}^{1.8}$ assuming a maxwellian hot electron distribution. Figure 1b shows the calculated thresholds for ignition [6] up to different times during the laser drive for an ignition design using a DT filled, Cu doped Be capsule (2.35 mm diameter) placed in a U hohlraum (5.9 mm diameter, 10 mm long) [4]. These thresholds assume isotropic hot electron generation inside the hohlraum. For T_{hot}=30 keV the allowable fraction of hot electrons out of the total laser energy ranges from 0.02% (E_{hot}=7.6 J) for the first 2 ns to 2.5% (E_{hot}=30 kJ) at the end of the ignition laser pulse. Since most of the hot electrons are stopped inside the hohlraum, we infer them from the hard x-ray bremmstrahlung spectra they generate in the hohlraum walls. The thick target bremsstrahlung spectra generated by a Maxwellian hot electron distribution is approximated by [3]:

$$I[keV/keV.sr] = \frac{5}{4\pi} \times 10^{11} \cdot \frac{Z}{79} \cdot E_{hot}[J] \cdot e^{-h\nu/kT_{hot}[keV]} \quad [1]$$

On the NIF we measure the time integrated absolute >20 keV hard x-ray bremsstrahlung emission from laser heated targets using the Filter-Fluorescer diagnostic (FFLEX) [7].

2. The FFLEX diagnostic

2.1 Calibration and performance qualification

The FFLEX detectors consist of NaI(Tl) scintillators coupled to XP2008 photomultiplier tubes (PMT) that are connected to charge sensitive amplifiers (CSA) [8]. An early version of FFLEX with 8 filter-fluorescer channels in the 20-150 keV spectral range was activated in 2004 during the first hohlraum experiments during NIF Early Light (NEL) Campaign using the first four NIF beams [8,9]. Recently, absolute response and PMT gain measurements were performed in single photon counting mode using an Am241 radioactive source in preparation for the NIC. The measured PMT voltage gain exponents, in the 6.7-7.5 range, were consistent with 2004 calibrations and the scintillator-PMT absolute responses decreased by 5% in average. To qualify the performance of FFLEX we reproduced one of the NEL experiments [9] using 1.6 mm diameter, 1.5 mm long vacuum halfraum heated by a NIF quad with a 2 ns square laser pulse and 13 kJ total energy. Figure 2 shows a comparison between the FFLEX spectra measured for the similar experiments during NEL and during NIC.

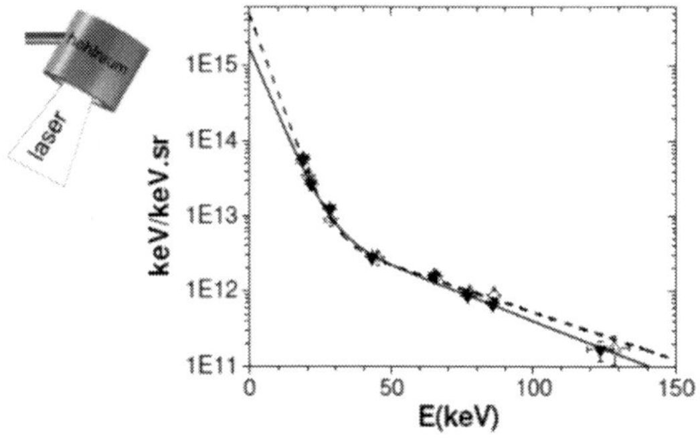

Figure 2 Experiment layout and corresponding FFLEX spectra measured on NEL [8,9] (empty points, dashed fit) and on a similar 2009 NIC experiment (solid points and fit)

The spectral points in the 8 FFLEX filter-fluorescer channels in 2004 and 2009 experiments are consistent with each other within 15% errors in the channel spectral and absolute responses, validating the diagnostic. The measured spectra are fitted by two-temperature distributions with a thermal electron component with T_l=4-5 keV temperatures characteristic for the laser spot plasma generated at the hohlraum wall and a hot electron component with T_{hot}=30 keV characteristic for hot electrons generated by stimulated Raman scattering (SRS) and $f_{hot} \leq 1\%$ (E_{hot}=100 J).

2.2 FFLEX upgrade with high energy time resolved channels

For NIC, FFLEX must be able to measure minimum Ehot of ~1 J which was demonstrated during NEL [8,9]. Furthermore, early and late time hot electron thresholds (Fig.1b) that will be measured separately on experiments using truncated ignition laser pulses (Fig.1a) require a shot-to-shot dynamic range of 10^5 which is satisfied by the broad gain range of the used PMT's and CSA's. The 8-channel FFLEX, however, cannot distinguish hard x-rays generated by >170 keV preheat hot electrons, nor can it measure a ~100 keV superhot electron component that would adversely preheat the fuel. Since at >170 keV photon energies no element absorption edges exist, the potential new high energy channels are high band-pass filter only ones [10]. With the estimated 15% calibration uncertainties per channel, calculations show that two new channels with typical filter only responses with >200 keV and >350 keV spectral band-passes are required to resolve two temperatures 30+100 keV hot-superhot electron distributions. They will also distinguish preheat hot electron signatures and, since experiments with a reduced number of truncated laser drives will be performed, the new channels are required to have <1 ns time resolution. Recently, in addition to the existing 8 time integrated filter-fluorescer channels, the required two high energy, time resolved channels were added to FFLEX. Figure 3 shows the layout and design of the upgraded FFLEX.

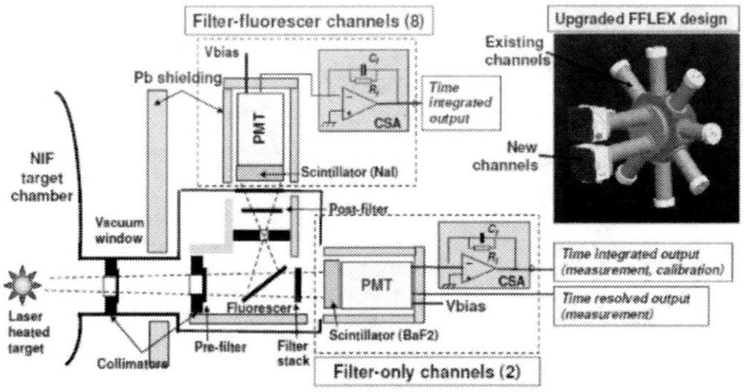

Figure 3 Layout and design of FFLEX upgraded with the two high energy, filter only channels

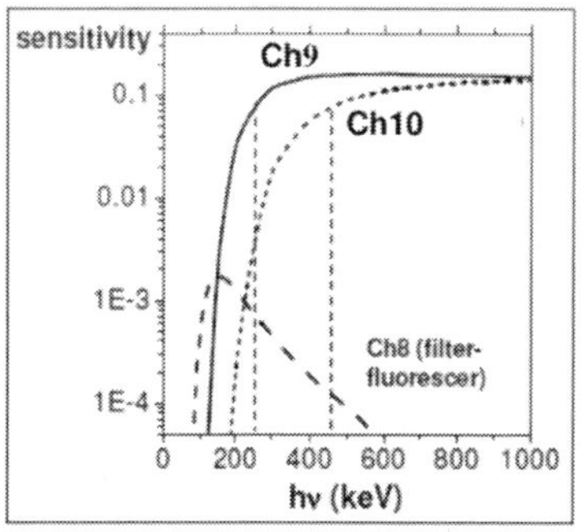

Figure 4 Spectral response of the new high energy channels

For the new channels fast Hamamatsu R5320 photomultipliers with a rise time of 0.7 ns are used, coupled to BaF_2 scintillators. The time response of the system that includes 50 m long signal cables measured with a Co60 source in single photon count mode revealed a 0.7 ns rise time and ~2 ns FWHM impulse response that depends slightly on the PMT voltage. Time resolved and time integrated signals via CSA are acquired by tapping PMT signals off both anode and last dynode.
Figure 4 shows the spectral responses for the new >200 keV and >350 keV channels, obtained by using filter stacks consisting of Au/Pt/Cu/Al (0.5/0.8/5/3 mm thick) coupled to a 10 mm thick scintillator and Pt/Cu/Al (4/5/3 mm thick) coupled to a 15 mm thick scintillator, respectively. Cu/Al filter combinations are used to suppress the K shell fluorescence of the high-Z filters. The absolute calibration of the high energy channels is ongoing in parallel to NIC experiments.

3. First hot electron measurements in multi-beam NIF hohlraum experiments

First 96-192 multi-beam hohlraum experiments were performed in scale 0.7-0.9 hohlraums of the full ignition targets to validate the hohlraum x-ray drive for ignition. Scale 0.7 (3.6 mm diameter x 6.5 mm long) vacuum hohlraums were heated by 96 laser beams with 2 ns square laser pulse with 150-310 kJ total energy. At corresponding average laser intensities at the wall of 4, 6 and $8\cdot10^{14}$ W/cm^2 we measured 2-T distributions with f_{hot}/T_{hot} of 0.01%/17 keV, 0.02%/20 keV and 0.2%/20 keV respectively. The results are consistent with previous experiments [8,9,11] which show that hot electron fractions scale with the laser intensity and start to be significant (~>1%) at ~10^{15} W/cm^2.
Scale 0.9 (4.6 mm diameter x 8.4 mm long) vacuum and gas filled hohlraums were heated by 192 laser beams and Figure 5 shows a summary of measured FFLEX spectra. Vacuum hohlraums heated by 600 kJ laser energy with 2 ns square laser pulses yielded 2-T spectra with T_1~10 keV and f_{hot}=0.2% at T_{hot}=30 keV. At similar $8\cdot10^{14}$ W/cm^2 laser intensity, this result is consistent with that measured the scale 0.7 NIF hohlraums and during NEL [8,9]. The result is lower than f_{hot}~>1% measured at 10^{15} W/cm^2 intensity (Fig. 2). In conclusion, for similar plasma filling conditions [8], the f_{hot} scaling with laser intensity in NIF multi-beam ignition scale vacuum hohlraum experiments is consistent with previous data measured in reduced scale hohlraums heated by >40x less total laser energy.
Gas filled (CH_4) hohlraums at room temperature were heated with ignition-like laser pulses (Fig. 1) with pulse lengths of 11 ns and ~500 kJ total laser energy. For a laser intensity of $5\cdot10^{14}$ W/cm^2, i.e.

lower than in vacuum hohlraums (Fig 5), the measured FFLEX spectra are also fitted by 2-T distributions with T_l~10 keV and T_{hot}~30 keV, but with 10x less spectral flux, yielding lower f_{hot}=0.02% as expected.

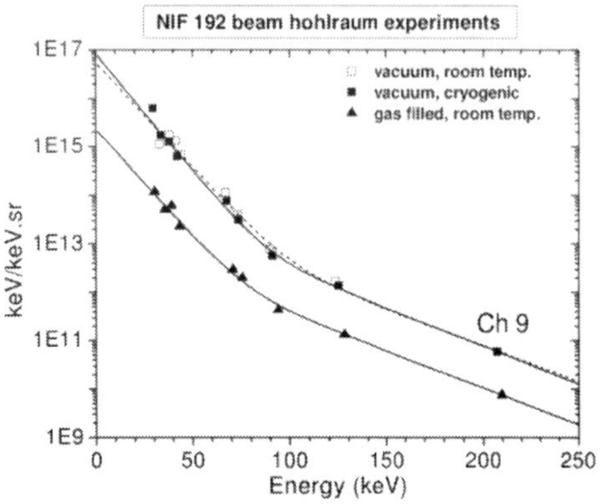

Figure 5 FFLEX spectra in room temperature and cryogenic vacuum hohlraums and in gas filled hohlraums at room temperature

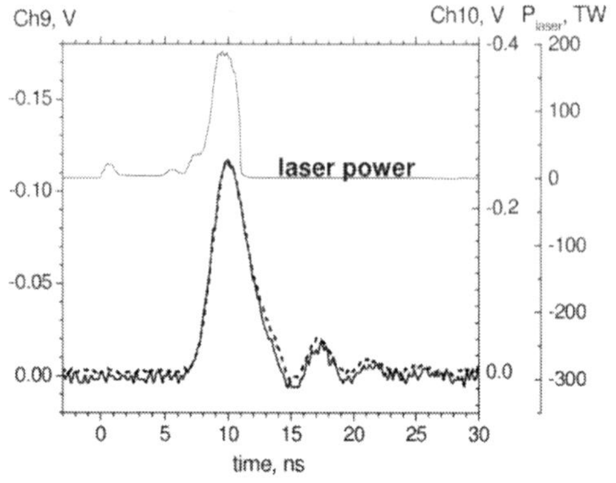

Figure 6 Total laser power, Ch. 9 (dashed line) and Ch. 10 (solid line) hard x-ray signals

The new >200 keV Channel 9 gives spectral points consistent with the first 8 channels (Fig. 5), showing that there is no high energy cutoff of the hot electron distribution at hν~4kT_{hot}, as suggested by some physics models [3]. Channels 9 and 10 provide also the time history of the hard x-ray signals corresponding to hot electrons. For vacuum hohlraums, hard x-ray signals with 6 ns FWHM were measured for 2 ns square laser pulses. The longer duration of the x-ray signals compared to the laser pulses is attributed to the high-Z wall plasma stagnation on the hohlraum axis. Figure 6 shows the total laser power and the time resolved x-ray signals measured in channels 9 and 10 for scale 0.9 gas filled hohlraums at room temperature. X-ray signals were generated only during the peak of the laser power, with no measurable early time emission within the single shot dynamic range of the scopes of 200. Furthermore, since the gas fill suppresses the wall plasma stagnation, their FWHM of 3 ns is comparable to the main laser pulse. In the ongoing NIC FFLEX will be employed in cryogenic gas filled (He-H mix, pure He) hohlraum experiments, similar to ignition hohlraums [2,4]. Furthermore, Channel 10 will provide data that will probe the existence of a superhot electron component [3].

4 References

[1] G.H. Miller, E.I. Moses and C.R. Wuest, *Nucl. Fusion* **44**, 228 (2004).
[2] J.D. Lindl, P. Amendt, R.L. Berger, S.G. Glendinning, S.H. Glenzer, S.W. Haan, R.L. Kauffman, O.L. Landen, *Phys. Plasmas* **11**, 339 (2004).
[3] W. Kruer, *Physics of Laser Plasma Interactions,* Addison-Wesley Publishing Co. (1988).
[4] S.W. Haan, D.A. Callahan, M.J. Edwards, B.A. Hammel, D.D. Ho, O.S. Jones, J.D. Lindl, B.J. MacGowan, M.M. Marinak, D.H. Munro, S.M. Pollaine, J.D. Samolson, B.K. Spears, L.J. Suter, *Fusion Sci. Technol.* **55**, 227 (2009).
[5] D. Meeker, N. Meezan, L.J. Suter, S. Haan, *Bull. Am. Phys. Soc.* (CO1.002) (2004).
[6] S.W. Haan, LLNL, private communication 2008.
[7] C.L. Wang, *Rev. Sci. Instrum.* **52**, 1317 (1981).
[8] J.W. McDonald et. al., *Phys. Plasmas* **13**, 1 (2006).
[9] E.L. Dewald, O.L. Landen, L.J. Suter et al, *Phys. of Plasmas* **13**, 1, (2006).
[10] C. Stoeckl et al *Rev. Sci. Instrum.* **72**, 1197 (2001).
[11] M.D. Cable, H.N. Kornblum, R.L. Kauffmann, LLNL Internal Annual Report (1985).

The Sixth International Conference on Inertial Fusion Sciences and Applications IOP Publishing
Journal of Physics: Conference Series **244** (2010) 022075 doi:10.1088/1742-6596/244/2/022075

Modeling Heat Conduction and Radiation Transport with the Diffusion Equation in NIF ALE-AMR

A.C. Fisher[1], D.S. Bailey[1], T.B. Kaiser[1], B.T.N. Gunney[1], N.D. Masters[1], A.E. Koniges[2], D.C. Eder[1], R.W. Anderson[1]

1: Lawrence Livermore National Laboratory, P.O.Box 808, Livermore, CA 94551, USA
2: Lawrence Berkeley National Laboratory, 1 Cyclotron Rd. Berkeley, CA 94720, USA

E-mail: `fisher47@llnl.gov`

Abstract. The ALE-AMR code developed for NIF is a multi-material hydro-code that models target assembly fragmentation in the aftermath of a shot. The combination of ALE (Arbitrary Lagrangian Eulerian) hydro with AMR (Adaptive Mesh Refinement) allows the code to model a wide range of physical conditions and spatial scales. The large range of temperatures encountered in the NIF target chamber can lead to significant fluxes of energy due to thermal conduction and radiative transport. These physical effects can be modeled approximately with the aid of the diffusion equation. We present a novel method for the solution of the diffusion equation on a composite mesh in order to capture these physical effects.

1. Introduction

This work is focused on improving the modeling capability for NIF debris and shrapnel simulations. Such simulations are being used by the NIF program to predict fragment sizes and velocities in the aftermath of a NIF target shot. This capability allows us to detect fragments that may damage NIF optics and diagnostics beforehand and redesign targets in order to reduce such risks before NIF target shots are taken.

In this paper we present recent progress in adding heat conduction and radiation transport effects to the ALE-AMR code. In order to add these effects an AMR (Adaptive Mesh Refinement) capable diffusion solver is required. We will describe our diffusion solver and the testing we used to show that it is *2nd* order accurate. We will also describe the work required to introduce heat conduction and radiation transport modules into ALE-AMR. Finally, we will present a demonstration simulation using these new capabilities.

2. AMR Capable Finite Element Diffusion Solver

To work with ALE-AMR our diffusion solver must be capable of operating on the multi-level, multi-processor, block structured SAMRAI (Structured Adaptive Mesh Refinement Application Infrastructure) data representing the ALE-AMR variables. The Finite Element Method (FEM), however, requires data in a single level composite mesh format. Also after the solution is obtained in the composite mesh format it must be used to update the ALE-AMR data in the multi-level format. This necessitates a mapping capable of translating between the SAMRAI representation and the composite mesh representation.

To begin forming this mapping we detect which nodes in the SAMRAI representation are at the finest level for that location and thus need to be nodes in the composite mesh. For each finest node that is detected a unique, non-contiguous fine ID number can be computed using the index of the node and vice versa as follows

$$\begin{aligned} F = i + jN_i + kN_iN_j \\ i = F\%N_i,\ j = (F/N_i)\%N_j,\ k = F/(N_iN_j) \end{aligned} \tag{1}$$

where (i, j, k) is the SAMRAI index location of the node assuming it was refined to the finest level and (N_i, N_j, N_k) is the number of nodes in each direction assuming the entire block was refined to the finest level, and the % is the modulo operator. Every unique node is then counted in sequence to yield a contiguous global ID numbering for the nodes. A map between this contiguous global ID number and the non-contiguous fine ID number is constructed and stored for later use. When more than one CPU is used each CPU counts the nodes it owns and the numbers are shifted by the number of nodes owned by the lower numbered CPUs. For nodes that are shared by more than one CPU in the SAMRAI representation, a tie breaker is used to determine which CPU owns the node and all the other CPUs sharing the node store the global ID in another map for later use. These maps and (1) allow straightforward translation between indices in the SAMRAI representation and global ID numbers for the composite mesh representation.

In addition to this translation between representations, a family of finite elements is required to handle all of the cases found in the composite mesh. For elements that are not at a coarse-fine boundary, standard bilinear quads in $2D$ and trilinear hexes in $3D$ are used. However, there are many cases of elements at the coarse-fine boundary with each case having a different combination of element faces refined. For these elements we use an approach similar to the transition elements found in [1]. The extra nodes on the faces due to the transition have hat basis functions along that face with the value reaching 1 at that node and 0 at the other nodes on the face. In the dimension not on the face the basis function is simply linear. The corner basis functions in our transition elements are the standard linear functions with fractions of the new transition basis functions subtracted out in order to ensure that the corner basis function is 0 at all the transition nodes. This method of construction yields a set of basis functions that satisfies the interpolation property and also enforces continuity across all the element faces.

Using the composite mesh mapping and the family of transition elements it is now possible to apply the FEM within the framework of ALE-AMR. We now turn our attention to the solution of the following diffusion equation.

$$\nabla \cdot \delta \nabla u + \sigma u = f \tag{2}$$

Applying the standard Galerkin approach yields the following linear system approximation

$$\begin{aligned} A\mathbf{u} + \mathbf{b} = \mathbf{f} \\ A = M_\sigma - K_\delta \\ (M_\alpha)_{ij} = \int_\Omega \alpha \phi_i \phi_j d\Omega \\ (K_\alpha)_{ij} = \int_\Omega \alpha \nabla \phi_i \cdot \nabla \phi_j d\Omega \\ \mathbf{b} = 0 \end{aligned} \tag{3}$$

where M is the mass matrix, K is the stiffness matrix, and an insulating boundary means $\mathbf{b} = 0$. A set of quadrature rules is needed to approximate the integrals and construct the matrices. For the standard elements we use a basic mass lumping integration rule with quadrature points located at the corners of the element. Using this quadrature rule generates an A matrix that has the M-matrix property and thus is inverse positive. Inverse positivity is an important property for physical models such as heat conduction since temperatures are expected to stay above

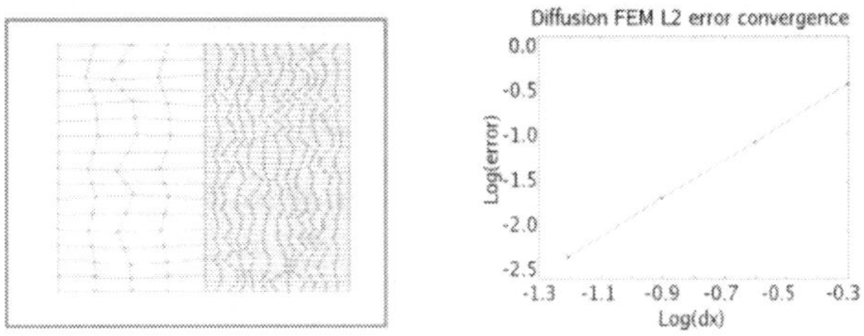

Figure 1. Convergence test on a mesh with transition elements reveals a slope of 2 in the log-log plot. This indicates 2nd order covergence.

absolute zero. For the transition elements a set of mass lumping quadrature rules is constructed by evaluating points at every node location in the element including the corner nodes and the nodes on the transition faces. Unfortunately the gradient of the transition basis functions is undefined at the transition node locations. We step around this problem by taking the limits of the gradient from each distinct region touching the point (2 in $2D$ and 4 in $3D$) and averaging those limits. This allows the formation of lumped mass and stiffness matrices found in (3). That linear system is solved by using the HYPRE library providing an approximate solution to the diffusion equation.

We measure the accuracy of this solution using a standard L_2 error convergence test. A randomized mesh is generated with the right side refined using a ratio of 3:1 in order to test the transition elements Figure 1. The entire random mesh is then refined 3 more times yielding a total of 4 meshes. Each of these meshes is used to approximate the solution to a simple Possion problem with non-zero Dirichlet boundary. These approximations are then compared to the analytical result to obtain the L_2 norm of the error. The slope of 2 the error norms shown in Figure 1 indicates that the method has 2nd order convergence.

3. Heat Conduction and Radiation Transport Modeling

Now that we have a diffusion equation solver, both heat conduction and radiation transport can be modeled with relative ease. For heat conduction the equation can be time evolved implicitly by using the solver at each time step yielding

$$\begin{gathered} C_v \frac{T^{n+1}-T^n}{\Delta t} = \nabla \cdot D^n \nabla T^{n+1} - \alpha T^{n+1} \\ \delta = D^n,\ \sigma = -\alpha - \frac{C_v}{\Delta t} T^n,\ f = -\frac{C_v}{\Delta t} T^n \end{gathered} \tag{4}$$

where C_v is the specific heat, T is temperature represented at the nodes, D is the heat conductivity, and α is the absorptivity. The variables δ, σ, and f are the diffusion equation parameters from (2). Similarly the diffusion approximation to radiation transport can be implicitly time evolved yielding

$$\begin{aligned} \frac{E_R^{n+1}-E_R^n}{\Delta t} &= \nabla \cdot \lambda(\tfrac{c}{\kappa_r}) \nabla E_R^{n+1} + \tilde{\kappa}_p (B^n - c E_R^{n+1}) \\ C_v \frac{T^{n+1}-T^n}{\Delta t} &= -\tilde{\kappa}_p (B^n - c E_R^{n+1}) \\ \delta = \lambda(\tfrac{c}{\kappa_r}),\ \sigma &= -\tilde{\kappa}_p c - \tfrac{1}{\Delta t},\ f = -\tfrac{1}{\Delta t} - \tilde{\kappa}_p B^n \end{aligned} \tag{5}$$

where E_R is the radiation energy represented at the nodes, λ is a function used to impose flux limiting on the diffusion approximation, c is the speed of light, κ_r is the Rosseland opacity, $\tilde{\kappa}_P$ is a modification to Planck opacity which is used to linearize the equation as in [2], and B is the blackbody intensity. After the nodal E_R and T are evolved through the above equations, the cell based material temperatures and energies must be updated to reflect the changes. In order

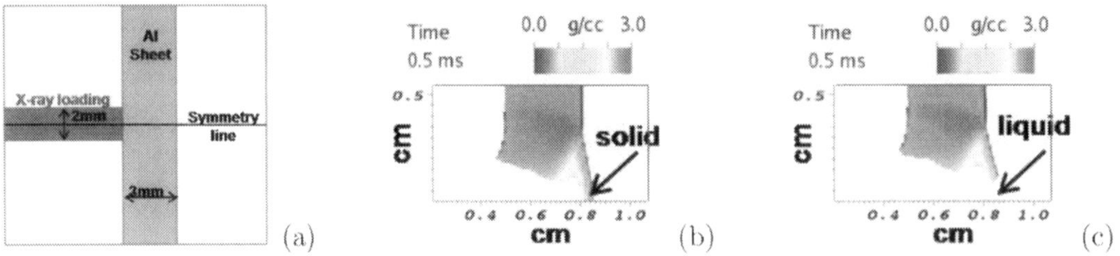

Figure 2. (a) Setup for an Al X-Ray interaction simulation. A $10^5 J/cm^2$ source of X-Rays was loaded onto the surface of an Al foil. (b,c) Density of remaining solid Al after 0.5 μs without and with radiation transport effects respectively.

to handle this we use the mapping developed by [3] in which changes in variables are mapped between nodes and cells. These mappings make it possible to transfer energy between nodes and cells without introducing large amounts of artificial diffusion.

4. Simulation Results

To demonstrate this new capability in ALE-AMR we chose a simulation related to the NIF debris and shrapnel work. In these $2D$ simulations a $1ns$ $10^5 J/cm^2$ x-ray pulse was loaded onto $2mm$ of the surface of a $3mm$ thick Al foil. The simulations were allowed to run to $0.5\mu s$ both with and without the new radiation transport module active. At the end of the simulations we captured density plots of the Al, and thresholded those densities by the material state. All of the Al that had a temperature above the melting point was thresholded to white while all of the Al that was still solid was drawn with the color scale (see Figure 2). While the differences between the simulations appear small they are still significant. In the simulation without radiation transport the foil appears ready to spall off solid material from the back side. In the simulation with radiation transport the material coming off the back side has melted which means it will be more likely to break into smaller droplets in transit to any optics or diagnostics. The droplets are a lower debris risk which is exactly what we are tasked to measure for NIF targets. This highlights the importance of modeling radiation transport to the NIF debris and shrapnel project.

5. Conclusions

The ALE-AMR code is now capable of modeling heat conduction and radiation transport effects. An AMR capable diffusion solver was developed in order to model these effects. This diffusion solver was put through an L_2 error convergence test to show that it is $2nd$ order accurate. A technique for mapping variables from cells to nodes and vice versa was developed to tie the node based solver into the cell based ALE-AMR variables. Also an X-Ray Al foil interaction was simulated with and without radiation transport effects, highlighting the importance of radiation transport to the NIF debris and shrapnel work.

Acknowledgments

This work performed under the auspices of the U.S. Department of Energy by Lawrence Livermore National Laboratory under Contract DE-AC52-07NA27344

References

[1] Gupta A 1978 *International Journal for Numerical Methods in Engineering* **12** 35–45
[2] Shestakov A, Harte J and Kershaw D 1988 *Journal of Computational Physics* **76** 385–413
[3] Shestakov A, Milovich J and Prasad M 2001 *Journal of Computational Physics* **170** 81–111

Journal of Physics: Conference Series **244** (2010) 022076 doi:10.1088/1742-6596/244/2/022076

On a statistical scattering model to explain capsule implosion asymmetry in vacuum hohlraums with radiation temperatures of order 100eV

S. R. Goldman, G. A. Kyrala, E. S. Dodd, J. P. Grondalski, N. M. Hoffman, H. A. Rose, A. Seifter

Los Alamos National Laboratory

srg@lanl.gov

Abstract. We apply a statistical scattering model of laser ray propagation to obtain improved agreement with measurements for capsule emission symmetry in one of three sets of experiments. Linearized post-processing of the simulations foramplification of a variety of plasma instabilities, coupled with an accounting for intensity seed perturbations in the laser beams suggests that thermal filamentation can cause the density fluctuations. Although this mechanism does not appear to be significant in the remaining two series, its identification adds to the knowledge base in the use of symmetry capsules to measure hohlraum drive asymmetry.

1. Introduction

Researchers from Los Alamos National Laboratory have conducted shot sequences on symmetry capsules in hohlraums at the LLE facility of the University of Rochester to probe contributions to symmetry behavior of NIF capsules due to the first of four pulses in the NIF ignition profile[1]. The concept of symmetry capsules has been elucidated in an earlier paper [2]. The capsule and hohlraum dimensions were scaled to 0.7 of NIF scale 1 size. The hohlraum temperatures were designed to be about 100eV, approximately the temperature for the NIF hohlraum during the first pulse.

The characteristic geometry for the individual shots is shown in Figure 1; for simplicity we consider one half of the total hohlraum, and assume reflection symmetry around the z=0 plane. Up to three cones of beams were incident through the laser entry hole (LEH) at each end of the hohlraum. All shot series were primarily concerned with the predictability of capsule symmetry at peak emission.

A limited number of random phase plates were available for the first set of shots; since it appeared likely that it would be hardest to control the cone 1 beams, which penetrated furthest into the hohlraum, the phase plates were put on these beams. For the latter two shot sets, random phase plates were available for all cones. Other differences between the original and the latter shots were pulse-width (3ns FWHM for the original series vs. 1ns FWHM for the latter series) and hohlraum temperature (lower for the original series because the available beam energies were approximately 270 J/beam instead of 500 J/beam).

In the absence of azimuthal variation around the hohlraum symmetry (z) axis, the capsule shape for any emission intensity at the time of peak total emission can be expressed as $a_0 + \sum_{n=1}^{\infty} a_{2n} P_{2n}(\cos\theta)$, where P_{2n} is the Legendre polynomial of order 2n and θ is the angle between the line from the capsule origin to the contour point and the symmetry axis. For $a_2/a_0 < 0$, the capsule is oblate; for $a_2/a_0 > 0$, the capsule is prolate. The differences in a_2/a_0 between experimental and simulational results were initially significantly greater for the original series.

In this paper we present and provide justification for a model of statistical scattering of laser rays, which brings the simulational results for the original shot series into substantially better agreement with experiment. We then assess the impact of this model on the latter two shot series.

2. Comparison of experiments and simulations for the original shot series

The targets for this series were CH shells, nominally 15μm thick and 1400μm in diameter. Figure 2 summarizes the results for a pointing scan consisting of three shots with the intersection of the central ray of the 42° cone (2) with the z-axis successively at -500μm, -250μm, and +500μm with respect to the intersection of the z-axis and the LEH plane. (The geometry for -250μm displacement is shown in Figure 1; cones 1 and 3 make angles of 21° and 58° with the negative z-axis.) Values for a_2/a_0 and a_4/a_0 for the image 30% contour are plotted vs. z-axis – cone2 intersection location in microns.

In addition to calculations with standard Lasnex modeling, simulations were also done with a statistical laser ray scattering model (also available in Lasnex) based on the assumption of a scattering length due to unresolvable small-scale fluctuations in the electron density n_e. The scattering length has the form: $L_s=L_0(n_e/n_{crit})$, $n_e \geq n_{cut\text{-}off}$, $L_s=0$, otherwise. The value for L_0 (= 10^{-7}cm) provides a reasonably close match to the experimental data. $n_{cut\text{-}off}$(= 10^{19}cm^{-3}) was selected to avoid spurious low electron density interactions; for our case n_{crit} is the critical electron density for 3ω light.

The usage of $L_0=10^{-7}$cm instead of the more plausible $L_0=10^{-4}$cm is only significant for electron densities between 10^{19} and 10^{20} cm^{-3} (where the resulting values for L_s are more physically reasonable), but this has the effect of greatly increasing the size of the region of significant ray scattering. Higher densities are spatially close to the region of significant absorption, so the effect of L_0 variation on the energy deposition location due to the scattering from them is small.

Experimental values are connected by the black lines. Results for standard calculations, are linked by green lines, and results including statistical scattering are joined by magenta lines. Solid lines connect values for a_2/a_0, and dashed lines are appropriate to a_4/a_0 . Experimental emission images are in color, and simulated emission images with statistical scattering are shaded in grey.

Considering the plots for a_2/a_0 and a_4/a_0 separately, it is clear that the agreement between simulation and experiment is significantly improved by the addition of the statistical scattering model to the simulations.

3. Physical basis for the application of the statistical scattering theory

Plasma instabilities can possibly amplify seed perturbations due to intensity fluctuations in the incident laser beams to produce plasma density microstructure for beam scattering. Since the beams in cones 2 and 3 do not pass through phase plates, they are more likely to contribute to instability growth than smoothed beams with speckle structure. The NEWLIP code [3] has been used to obtain linear calculations of the integrated gain exponent G for various instabilities along an averaged laser ray trajectory for the beams in each cone. Spatial gains obtained by Kruer [4] were used for the filamentation processes. A set of results for the 42 degree cone with pointing at -250μm is shown in Figures 3a-3c. Instability growth within the hohlraum, e.g. within the gold plasma ablated from the hohlraum walls, appears to be dominated by thermal filamentation.

For purposes of an estimate, the size of seed perturbations in intensity can be found from laser intensity plots, which are provided on the University of Rochester LLE web-site [5]. Figure 4 is the plot for a beam with pulse shape, energy, and defocus at the hohlraum side wall, similar to those in the experiment. Using IDL to Fourier transform the beam in two-dimensions and then taking a diagonal cut of the resulting image, one readily obtains Figure 5 as a representation for the variation of intensity fluctuations as a function of spatial wave number (normalized to k_0, the wave number for 3ω light). Normalizing to the amplitude of the uniform (k=0) mode ($\approx 10^{4.33}$), one finds that $3.e\text{-}5 \leq \delta I/I_0 \leq 3.e\text{-}2$, where δI is the intensity seed amplitude and I_0 is the mean intensity.

From Figure 3a, we note that the estimated values of amplified perturbations ($\eta = e^G \delta I/I_0$) are readily greater than unity; i.e. for G=15 and $\delta I/I_0 = 3.e\text{-}5$, $\eta \approx 10$. Hence it is plausible for this set of shots, that thermal filamentation instability can produce substantial fluctuations for beam scattering. It is also possible that structuring of cone1 beams could derive from cone2, because, as shown in Figure 1, the deposition regions for the two cones tend to overlap on the hohlraum side wall.

For thermal filamentation the fractional electron density fluctuations, $\delta n/n$, and intensity fluctuations, $\delta I/I$, are related by [4]: $\delta n/n = -(1/7)(\delta I/I)(I/2n_{crit}T_e c)(\nu_{ei}/kv_e)^2$, with T_e, ν_{ei}, and v_e respectively the electron temperature, electron-ion collision frequency and electron thermal velocity. For $I = 10^{14}$W/cm^2, and $T_e = 600$ev, with λ in microns, this becomes [6]: $\delta n/n \approx 2.75x10^{-4}\lambda^2(\delta I/I)(n/10^{21})^2$, from which at $k/k_0 = 0.01$, and n=10^{21}cm^{-3}, $\delta n/n \approx 0.3\,\delta I/I$.

4. The effect of statistical scattering through thermal filamentation on capsule emission symmetry for the latter two shot series

The laser intensities at the hohlraum side wall in these two series are within about a factor of 2 of those in the first series. With phase plates, a number of factors should act to decrease the magnitude of density fluctuations: 1. transfer of beam energy from larger to smaller wavelengths, which have lower intrinsic instability growth rates, 2. weaker linear coupling of the resulting intensity fluctuations to density fluctuations, 3. creation of speckles in the direction of light propagation to limit coherent filamentation growth lengths. Hence it is not surprising that simulation results are closer to experiment for these series.

For the second series, which did not use cone 1 beams, the calculated a_2/a_0 values for the pointing scan agreed with experiment to within 0.02. For the third series, which included beams at all three cone angles, for CH capsules with oblate observed shape ($a_2/a_0 < 0$), the calculated values were typically high by ≈ 0.25, still a relatively small fraction of the discrepancy with standard Lasnex modeling for the original series.

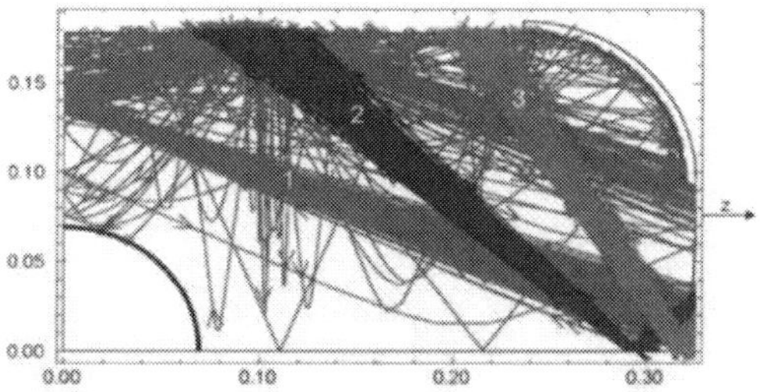

Fig. 1 Original shot series: region boundaries and beam cones at t=0.

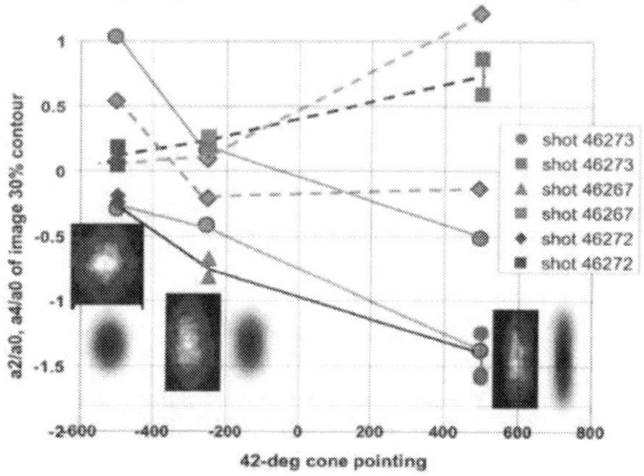

Fig. 2 Graphical summary of pointing scan results for the initial shot series.

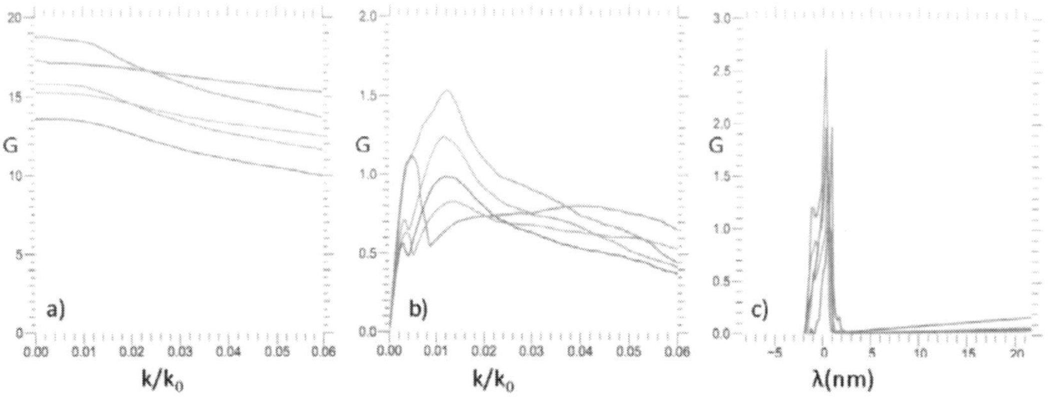

Figure 3 Total gain exponent G for instabilities at selected times after the start of the laser pulse: red-0.7ns, green-1.2ns, blue-1.7ns, cyan-2.2ns, magenta-2.7ns. Abscissae are either wave number k normalized to k_0, the wave number for 3ω light or wavelength λ. Instabilities are: a) thermal filamentation, b) ponderomotive filamentation, c) backward stimulated Brillouin scattering. The peak G value for backward stimulated Raman scattering (not included) is 0.6.

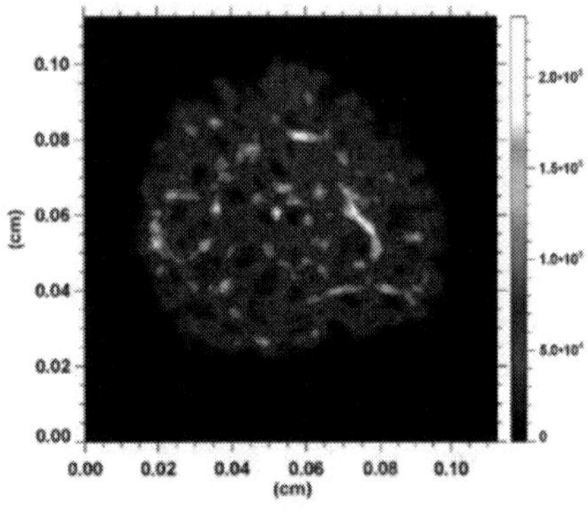

Figure 4. Relative intensity plot for a beam similar to the beams in cones 2 and 3.

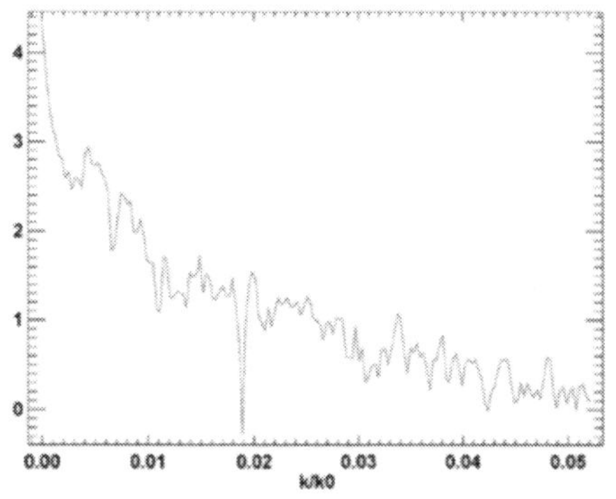

Figure 5. Plot of $\log_{10}[F(k)]$ vs k/k_0, for the beam of Fig. 4. Here F(k) denotes the Fourier amplitude of the intensity perturbation at magnitude k for wave vectors in the diagonal direction.

References

[1] A. Seifter *et al., Laser and Particle Beams* 27, 123 (2009)
[2] N. M. Hoffman *et al., Journal of Physics: Conference Series* 112 022075 (2008)
[3] See National Technical Information Service Document No. DE200110420 (E. A. Williams, "Platform Independent Laser-Plasma Interaction Post-processor for Lasnex," in Lawrence Livermore National Laboratory ICF Annual Report UCRL-L-105820-98, 13,1998). Copies may be ordered from the National Technical Information Service, Springfield, Va 22161.
[4] W. L. Kruer, *Comments Plasma Phys. Controlled Fusion* 9, 63 (1985)
[5] University of Rochester LLE Website: www.lle.rochester.edu
[6] NRL Plasma Formulary 2007, J.D. Huba, Ed. Naval Research Laboratory, p 32

3D Simulations of the NIF Indirect Drive Ignition Target Design

O.S. Jones, J.L. Milovich, D.A. Callahan, M.J. Edwards, O.L. Landen, J.D. Salmonson, D.S. Clark, S.W. Haan, M.M. Marinak, M.V. Patel, S.H. Langer, S.M. Sepke, P.T. Springer, S.P. Hatchett

Lawrence Livermore National Laboratory, P.O. Box 808, Livermore, CA 94551, USA

Email: jones96@llnl.gov

Abstract. The radiation hydrodynamics code Hydra is used to quantify the sensitivity of different NIF ignition point designs to several 3D effects. Each of the 48 NIF quads is included in the calculations and is allowed to have different power. With this model we studied the effect on imploded core symmetry of discrete laser spots (as opposed to idealized azimuthally-averaged rings) and random variations in laser power.

1. Introduction

Historically, most of the design work to specify the combination of hohlraum dimensions, materials, and laser power and pointing that delivers the desired radiation drive for a given ICF capsule has been done using axi-symmetric calculations. Calculations that assume axial symmetry have generally been found to give good agreement with many hohlraum experiments done on laser facilities such as Omega. However, the very tight tolerances on implosion symmetry that are required for an ignition target mean that nonideal 3D effects need to be included in more detailed calculations in order to assess the sensitivity of a particular design to these effects.

To do these calculations, we used the radiation hydrodynamics code Hydra [1]. Approximately 30 million census photons were used for the implicit Monte Carlo radiation transport. The capsule and hohlraum materials were divided up into about 2 million computational zones. The capsule angular zoning is 2.5 degrees in the polar direction and 4.5 degrees in the azimuthal direction. This type of zoning is adequate for resolving low modes (up to mode 8) that characterize the radiation drive asymmetry. Effects of higher mode perturbations due to capsule and fuel surface roughness are not included in this type of calculation. The thin plastic window across the laser entrance hole (LEH) that holds in the hohlraum fill gas is included, as are the different layers of wall material (typically gold, gold/boron, and uranium). The wall material in the calculations is 50 microns thick at the barrel and 100 microns thick at the LEH insert. This is thicker than the actual thicknesses of the hohlraum (30 microns at the barrel and 50 microns at the LEH insert). The thicker wall was needed to accommodate the zoning strategy employed near the LEH, which required that the outer boundary of the wall to be fixed

We define ignition time for an igniting capsule as the time at which the central temperature exceeds 12 keV. At that time, we define the local hotspot radius as the radius at which the density along any ray emanating from the center equals half of the maximum density along that ray. Since the hotspot surface can be expressed in terms of the spherical harmonics, Y_{lm}, then the rms deviation of the hotspot from its average value can be expressed in terms of the coefficients of the spherical harmonics, a_{lm}. The expression for the hotspot rms is given by

$$\xi = \sqrt{\frac{\iint (r - R_{avg}) d\Omega}{4\pi} \frac{1}{R_{avg}}} = \sqrt{\sum\sum\left(\frac{a_{lm}}{a_{00}}\right)^2} . \quad (1)$$

The m=0 spherical harmonic are the axisymmetric modes, and the Y_{l0} functions are proportional to the more familiar Legendre polynomials (P_2, P_4, etc.).

2. Calculating the intrinsic 3D radiation asymmetry

The NIF laser consists of 192 laser beams grouped into 48 quads of beams. The quads are grouped into 4 cones per side at polar angles of 23.5, 30, 44.5, and 50 degrees. The beams are pointed such that the 23.5 and 30 degree cones ("inner cones") illuminate the hohlraum near the midplane and the

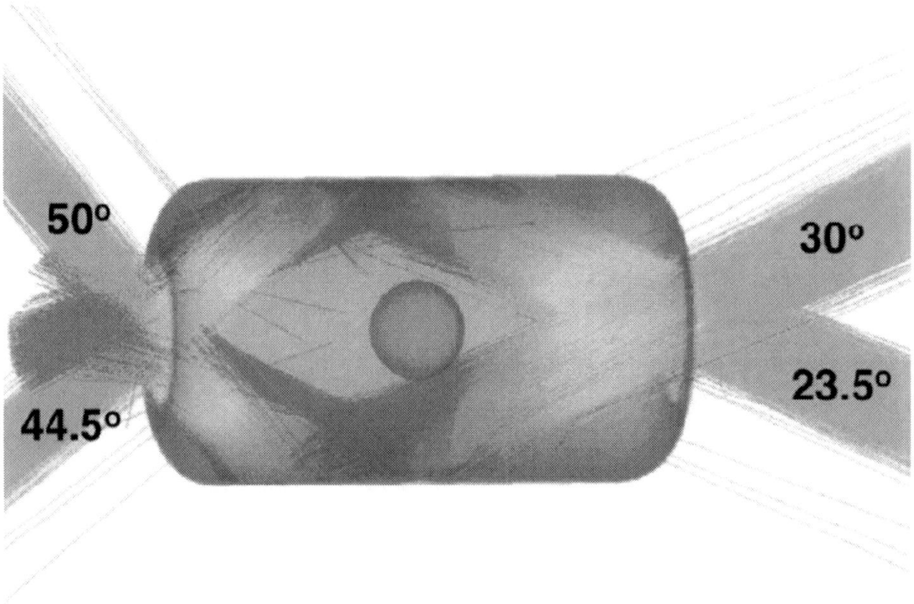

Figure 1. Material boundary plot from 3D calculation showing one quad at each polar angle.

44.5 and 50 degree cones ("outer cones") illuminate the hohlraum wall farther out toward the LEH, as shown in Figure 1. There are 4 quads per side for the 23.5 and 30 degree cones and 8 quads per side for the 44.5 and 50 degree cones. To control the lowest even Legendre mode asymmetry (P_2), a different laser power versus time is used for the inner and outer cones. The P_4 asymmetry is controlled by varying the hohlraum length to change the polar angle of the outer ring of beams relative to the capsule. Each NIF quad is treated as a single f/8 beam.

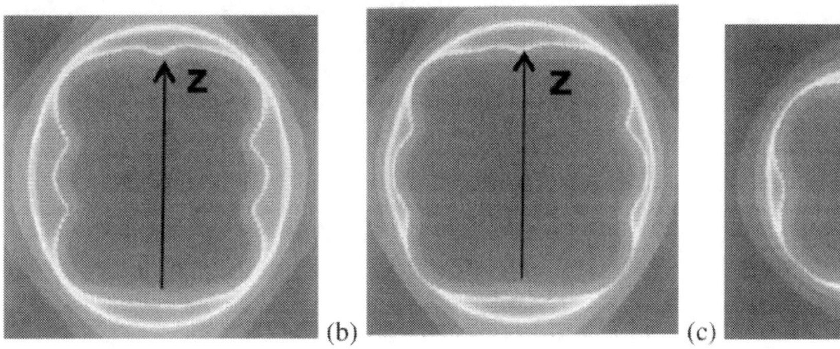

Figure 2. Density slices at maximum compression, (a) ϕ=0 , (b) ϕ=90, (c) z=0

The amount of intrinsic 3D radiation drive asymmetry due to the 3D illumination pattern was investigated for a 1180 micron radius doped beryllium ablator capsule driven with a peak radiation drive of 285 eV [2]. The pointing and inner/outer cone balance were adjusted so that an axisymmetric calculation gave a symmetric implosion. Then the same parameters were used for an ideal (no random errors in pointing or power) 3D calculation. Figure 2 shows contours of density at the time of maximum compression. Slices in the ϕ=0 and ϕ=90 degree planes show the Y40 asymmetry, and a slice across the midplane clearly shows an m=4 asymmetry. The m=4 is caused by the variation in brightness between 23.5 and 30 degree beams. This core has a total hotspot rms of 9.6%. There is 6.4% from a_{20}, 3.7% from a_{40}, 2.8% from a_{44}, and 1.4% from a_{54}. So the m=4 modes contribute about 3% to the total rms asymmetry.

3. Effect of laser power balance on core shape

The sensitivity to laser power balance was assessed for a 1000 micron radius Be capsule driven with a peak radiation drive temperature of 300 eV [3]. The power balance was estimated from a sequence of 16 full power single beam shots on NIF. The shot-to-shot differences in laser power were due to variations in the power and energy injected into the main laser, differences in amplifier gain, and differences in crystal detuning angle. Figure 3 shows the fractional difference in the laser power for each shot relative to the power of the average shot in the 16 shot sequence. Also shown is the rms variation of the 2-ns-averaged pulses as compared to the NIF specification.

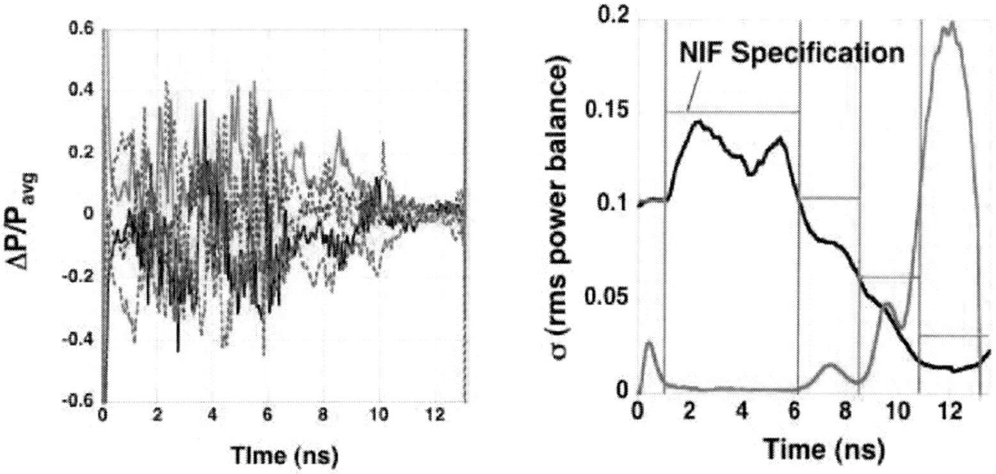

Figure 3. Fractional deviation in power from average for 16 sequential single beam NIF shots

Each measured power difference was multipled by two small random numbers to create 32 more statistically equivalent traces. The difference in power was then mapped onto a laser pulse shape tuned numerically for the 285 eV Be ablator design. The same 48 power traces were used for all runs, with different random realizations accomplished by assigning a particular quad a different power trace each run. Figure 4 shows the contribution of laser power balance to the hotspot rms as a function of a power balance multiplier. The net contribution due to power imbalance is obtained for each case by subtracting the hotspot rms for a baseline case with perfect power balance. We concluded that the measured power imbalance contributed about 0.1 to the hotspot rms.

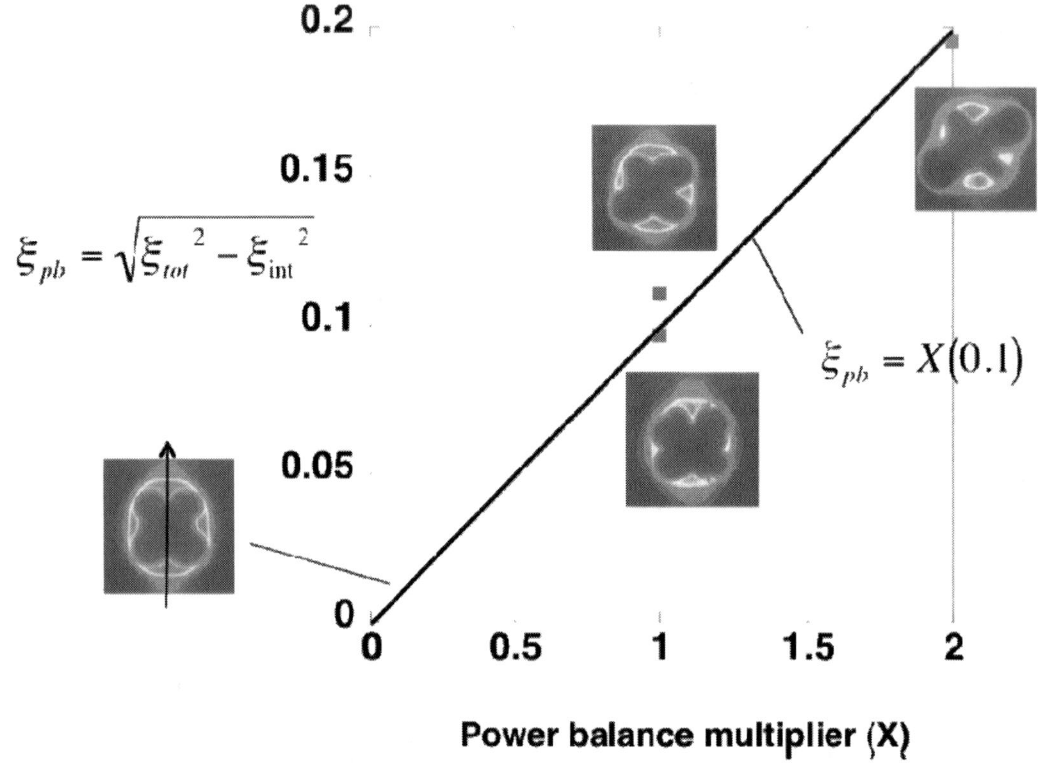

Figure 4. Sensitivity of hot spot rms to power balance multiplier for 285 eV Be ablator design

4. Conclusions

The Hydra radiation code was used to determine the sensitivity of NIF point designs to 3D effects. The radiation asymmetry due to the intrinsic 3D illumination pattern contributed about 0.03 to the final imploded core hotspot rms. The amount of laser power imbalance inferred from NIF measurenents contributes 0.1 to the hotspot rms.

References

[1] M. M. Marinak, et al., *Phys. of Plasmas*, **8**, 5, 2275 (2001)
[2] J. D. Salmonson, et al., Bull. Am. Phys. Soc., **52**, 273 (2007)
[3] S. W. Haan, 300eV Be design ref

The Sixth International Conference on Inertial Fusion Sciences and Applications IOP Publishing
Journal of Physics: Conference Series **244** (2010) 022078 doi:10.1088/1742-6596/244/2/022078

A new attempt for radiation hydrodynamics simulation with anisotropic and non-equilibrium distribution

N Ohnishi, K Sugai and Y Ogino

Department of Aerospace Engineering, Tohoku University, 6-6-01 Aramaki-Aza-Aoba, Aoba-ku, Sendai 980-8579, Japan

E-mail: ohnishi@rhd.mech.tohoku.ac.jp

Abstract. We present a computational method for solving time-dependent radiation moment equations with an Eddington tensor which represents an anisotropic field of radiation. For estimating the Eddington tensor, we propose a conical-ray method covering the whole solid angle with which the steady-state radiative transfer equation is solved. Computed radiation flux shows an excellent agreement with an analytical solution using the proposed conical-ray method even if a light source is located at a distant place from the computed point.

1. Introduction

A flow dynamics in high-temperature gases or plasmas is often influenced by energy transport of radiation emitted from flowfield itself. When we analyze such a flowfield numerically, it is necessary to couple radiative transfer equation with hydrodynamic equations in a certain form. If the density of the flowing matter is low enough, i.e., for optically thin case, we can conduct a formulation with a steady-state assumption of radiation field since electromagnetic waves spread more quickly than the characteristic speed of the flowfield. On the other hand, a diffusion approximation can be applied for solving the radiative transfer equation because of an isotropic radiation field for optically thick case in which the matter density is relatively high. However, for a flowfield rapidly changing with space and time such as laser ablation plasma, supernova explosion (neutrino transfer), and so on, radiative transfer property depends on the dynamics of the field. In such a case, a phenomenon cannot be described appropriately with one of two above-mentioned limits, but a method of connecting two limits smoothly is needed.

For moment equations of radiative transfer, an Eddington tensor defined by radiation energy density and stress tensor is responsible for an anisotropic property of radiation field. In the present paper, a new computational technique which takes the anisotropic radiation suitable for radiation hydrodynamics simulation into consideration by nonlocally determining the Eddington tensor is examined as following the works of Stone et al. [1] and Hayes & Norman [2]. We introduce a conical-ray method for estimating a precise Eddington tensor even with a point source which is difficult to be resolved with the conventional ray method. Test problems have been performed under a simple condition, and the proposed method is validated as compared with the conventional one.

2. Numerical modeling

2.1. Estimation of Eddington tensor with ray-tracing technique

The value of the Eddington tensor highly depends on optical thickness of the matter considered as mentioned in the previous section. In the optically thick case, since the radiation is easy to be absorbed, and influence of it, therefore, cannot reach a distant place, the Eddington tensor can be described as a function of locally obtained physical quantities. In this case, the transport process of radiation becomes in diffusion, then it is approximated by $\mathsf{f}_\nu = 1/3\mathsf{I}$ with this limit, where I is the unit tensor. However, when not optically thick, or when optical thickness is nonuniform, it is necessary to estimate the Eddington tensor from nonlocal physical quantities because the radiation field receives the influence of the matter in a distant place. In this paper, in order to take such a situation into consideration, a steady-state transfer equation is solved only for estimating the Eddington tensor including the anisotropic information of radiation, and then the time-dependent moment equations are solved with it. This procedure was first proposed by Stone et al. [1].

In the previous our work [3], radiation rays were set to be at the solid angles corresponding to the Gauss points of the discrete ordinate (so-called S_N) method [4]. If we obtain a specific radiation intensity $I_{\nu n}$ by a ray-tracing manner with a n-th ray, the Eddington tensor can be approximated by the following discrete integration,

$$\mathsf{f}_\nu = \mathsf{P}_\nu / E_\nu = \frac{\displaystyle\int_0^\infty I_\nu \boldsymbol{\Omega}\boldsymbol{\Omega} d\boldsymbol{\Omega}}{\displaystyle\int_0^\infty I_\nu d\boldsymbol{\Omega}} \simeq \frac{\displaystyle\sum_n w_n I_{\nu n} \boldsymbol{\Omega}_n \boldsymbol{\Omega}_n}{\displaystyle\sum_n w_n I_{\nu n}}, \tag{1}$$

where P_ν, E_ν, and $\boldsymbol{\Omega}$ are the radiation stress tensor, the radiation energy density, and the unit vector of propagation direction, respectively. Subscripts ν and n denotes the frequency and the ray index. The variable w_n is a weight of the n-th Gauss point for the numerical integration.

If the rays are discretely arranged in a solid angle as treated in the conventional work, it is impossible to take in the information on the cell which does not exist along the path of the rays and may become important in the optically thin media. Although increasing a number of rays simply improves this fault, a huge number of rays and, of course, a tremendous computational cost are required for touching all the cells. Then, in order to include the information other than the ray passing cells, we consider a conical ray (see Fig. 1) which has a certain margin around the original ray and do the ray-tracing with the physical quantities volume-averaged over the cells within this margin. Other procedures such as the original ray setting and the numerical integration are performed in the same manner as the previous study.

Figure 2 shows a sample of the arrangement of conical rays in a 1/8 sphere. The original rays are set to the angles corresponding to the S_8 method. Three types of conical ray can be considered in this case. There may be many candidates of the method to determine the solid angles (i.e., the radii r_a, r_b, and r_c in Fig. 2) covered by each cone. In this study, however, we set them to let the following value I be minimum; $I = \sum_i^{npairs} (r_i + r_j - L_i)^2$ for $r_i + r_j - L_i > 0$, where j expresses an adjoining cone and L_i is an interval with the adjacent cone. The sum is performed for all the types of pair connecting with the red lines in Fig. 2.

2.2. Numerical procedure for solving radiation moment equations

The definition position of each physical quantity follows the manner of Ref. [2] with staggered grids. Therefore, E_ν and the diagonal element of P_ν are the cell-centered quantities, the radiation energy flux $\mathbf{F}_\nu$ is defined at the cell interface, and the non-diagonal element of P_ν is defined at the cell edge. Each element of the Eddington tensor is similarly defined as P_ν.

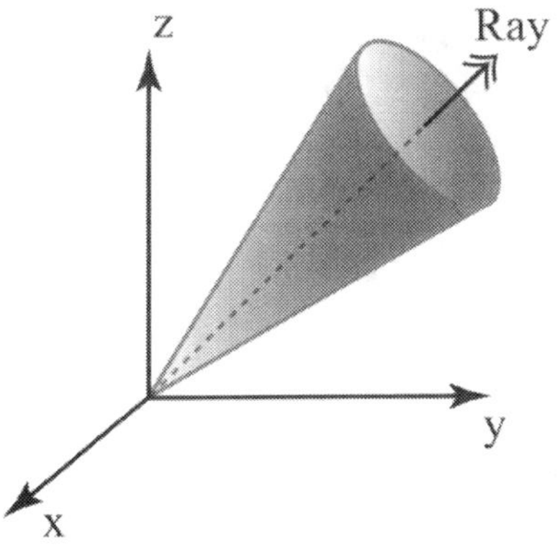

Figure 1. Schematic view of a conical ray.

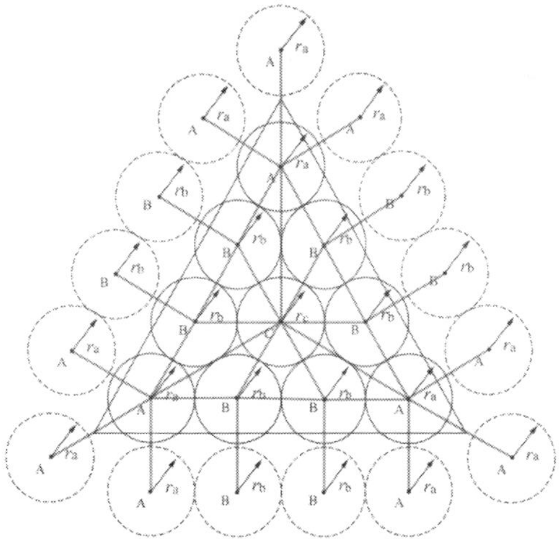

Figure 2. An example of conical rays and their radii in a 1/8 sphere.

With the definition given in this way, the 0th- and 1st-order moment equations can be written in the following forms with a partially implicit method,

$$E_{\nu i,j}^{n+1} = E_{\nu i,j}^{n} + \Delta t \left[4\pi\eta_{\nu i,j}^{n} - c\chi_{\nu i,j}^{n} E_{\nu i,j}^{n+1} + \left(\nabla \cdot \mathbf{F}_{\nu}^{n+1}\right)_{i,j} \right], \tag{2}$$

$$\mathbf{F}_{\nu i,j}^{n+1} = \frac{\mathbf{F}_{\nu i,j}^{n}}{1 + c\chi_{\nu i,j}^{n}\Delta t} - \frac{c^2 \Delta t}{1 + c\chi_{\nu i,j}^{n}\Delta t} \nabla \cdot \left(\mathsf{f}_{\nu}^{n} E_{\nu}^{n+1}\right)_{i,j}, \tag{3}$$

where η_ν and χ_ν are the emissivity and the opacity, respectively. These forms are also almost same as proposed by Hayes & Norman [2]. Note that our formulation is developed for two-dimensional simulation while this paper only presents the one-dimensional results. We solve the system of equations for $E_{\nu i,j}^{n+1}$ using a family of conjugate gradient method.

3. Test problem

A simple problem for validating our proposed method is presented. A constant heat flux F_0 from surface of a small sphere is given as a boundary condition for a spherically symmetric simulation. Propagation of the radiation to a medium with a uniform opacity was calculated without fluid motion. For further simplicity, it is supposed that the opacity does not have dependency for frequency, and the emissivity from the medium is set to be zero. Considering an optically thin situation in which the solution is more sensitive for anisotropic radiation, we performed simulations with opacity $\chi = 0.2$ over the region where the optical depth is comparable to or larger than unity. In Figs. 3 and 4, the horizontal axis shows the optical depth from the surface of the central sphere. Red lines denote plots of radiation energy flux F_r at each $\Delta t = 7.5(\chi/c)$.

Figure 3 shows a computational result with the conventional ray method. One can see the reasonable agreement with an analytical solution by first two plots. This feature indicates the advantage using the Eddington tensor estimated by ray-tracing method while the P_1 method (almost diffusion approximation) must deviate from the analytical solution even at this early stage [3]. However, after the next plot, the obtained solution suddenly depart from the analytical one. Here, a blue line in the figure represents the diagonal element of the calculated Eddington tensor. Note there is a discontinuity at optical depth $\sim$ 0.7. This discontinuity actually results from that the all arranged rays miss the central light source. This is an unavoidable

Journal of Physics: Conference Series **244** (2010) 022078 doi:10.1088/1742-6596/244/2/022078

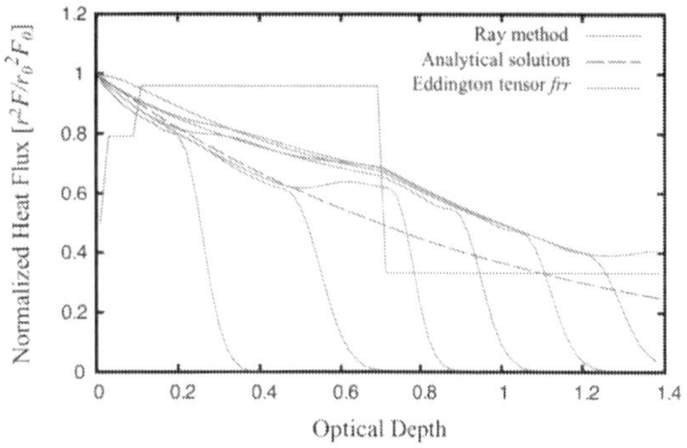

Figure 3. Radiative heat flux ($\chi = 0.2$) with the conventional ray method.

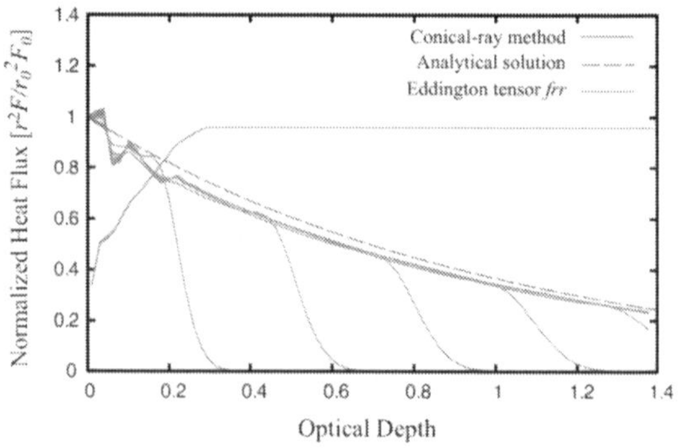

Figure 4. Radiative heat flux ($\chi = 0.2$) with the proposed conical-ray method.

shortcoming of the conventional ray method. After missing the light source, the computed flux is always overestimated, and the behavior of the solution is similar to the P_1 method because the Eddington tensor in that regime is close to the isotropic one $(1/3\,\mathrm{I})$.

Figure 4 shows the result using the proposed conical-ray method. With the conical-ray method, the radiation flux shows an excellent agreement with the analytical solution even if the light source is located at a distant place from the computed point. Moreover, the computed Eddington tensor smoothly approaches to unity which is the exact value in the case of free-streaming. However, the small deviation is found at the vicinity of the light source. This discrepancy comes from a relatively coarse resolution rather than the conventional method due to the averaging over the overlapping cones. An inaccurate solution near the light source may lead to an issue for radiation hydrodynamics simulations. Further investigation is needed to overcome this unwelcome feature.

4. Conclusion

We have developed a new method for solving the radiation transfer equation aiming to be coupled with a radiative flowfield. Conical rays are introduced in order to estimate the Eddington tensor accurately by solving the steady-state transfer equation even with a distant light source. The computed results shows that the proposed method reasonably reproduces an analytical radiation energy flux with a light source in an optically thin situation while the conventional ray method cannot at the distant place from the source. Since the present conical-ray method is slightly expensive for coupling computation with a flowfield, we should explore an effective computation within a feasible cost, e.g., utilizing the GPU technology.

Acknowledgement

This study was supported by Program for Improvement of Research Environment for Young Researchers from Special Coordination Funds for Promoting Science and Technology (SCF) commissioned by the Ministry of Education, Culture, Sports, Science and Technology (MEXT) of Japan.

References

[1] Stone J M, Mihalas D and Norman M L 1992 *Astrophys. J. Suppl.* **80** 819
[2] Hayes J C and Norman M L 2003 *Astrophys. J. Suppl.* **147** 197
[3] Sugai K, Ohnishi N, Ogino Y and Sawada K 2007 *Proceedings of the third Asian-Pacific Congress on Computational Mechanics (Kyoto)* GS16-2
[4] Modest M F 2003 *Radiative Heat Transfer 2nd Ed.* (San Diego: Academic Press)

Journal of Physics: Conference Series **244** (2010) 022079 doi:10.1088/1742-6596/244/2/022079

EFFECTS OF PRE-FORMED PLASMA INSIDE A GUIDING CONE IN FAST IGNITION SCHEME

Atsushi Sunahara*, **Hongbo Cai**[1], **Tomoyuki Johzaki**[1], **Hideo Nagatomo**[1], **and Kunioki Mima**[1]

Institute for Laser Technology Japan*,
Institute of Laser Engineering Osaka University [1]
2-6 Yamadaoka Suita Osaka Japan 565-0871* [1]

E-mail: suna@ile.osaka-u.ac.jp

Abstract. We simulated the plasma expansion of the cone wall due to the pre-pulse irradiation in fast-ignition experiments. Our radiation hydrodynamic simulations show the temporal evolution of the pre-formed plasma inside the cone. At the onset of the pre-pulse, collision within the expanding plasma near the cone axis generate a plasma jet. At the later stage of the pre-pulse the large density shelf above the critical density of 1.06 μm wavelength laser is formed near the tip inside the cone, and subsequently this critical density surface moves away from the inner surface of the cone tip. Our simulation gives roughly $80\mu m$ shift of the critical density surface under the typical pre-pulse condition of LFEX laser at Osaka University for 45 deg open angle cone. This shift of the critical density can reduce the energy coupling between the heating laser and the imploded core.

1. Introduction

Fast ignition is one of the attractive schemes of the inertial confinement fusion, proposed by Tabak[1], and its principle was experimentally verified by Kodama[2]. The aim of this scheme is: an implosion laser generates an imploded high dense plasma core, and before the core expands, it is heated by fast electrons created by an intense short pulse laser. In oder to heat the core efficiently, fast electrons should be generated in the space close to the imploded core. For this purpose, we use the cone guiding target so that the main heating laser should penetrate deeply without interacting with the long scale plasma. However, the space inside the cone is not completely plasma-free, since all short intense pulse lasers have pre-pulse before the main pulse and this pre-pulse generates plasma. LFEX laser at ILE Osaka University is the OPCPA laser system, in which the ratio of main pulse/pre-pulse is estimated to be 10^8. This intensity is enough to generate the plasma inside the cone.

In the pioneering work, Baton [3] reported that the existence of the pre-formed plasma could scatter the fast electrons, and thus, reducing the component of the fast electron that contributes to core plasma heating. H-B. Cai [4] also reported that the energy distribution of fast electrons was changed by the pre-plasma. The relatively low energy component of the fast electrons, that contributes to the heating of the core, was reduced. Inversely high energy component which is not contributing to the core heating, was increased. Therefore, understanding of the effect of the pre-plasma inside the cone is important in the fast ignition scheme, and the pre-plasma profile should be used for the initial condition of the Particle in Cell simulation in modeling the generation of fast electrons by the intense laser. In this paper, we carried out 2D radiation hydrodynamic simulations of the pre-plasma generation inside the cone target in the typical pre-pulse condition found in LFEX laser. Our results show that collisions within the expanding plasma near the cone axis create a plasma jet at the onset of the pre-pulse, and in the later stage of the pulse, ablated plasma fills inside the cone, especially near the tip and the critical density surface moves outwards. This spatial move reduces the energy coupling between the heating laser and the imploded core.

2. Simulation

In simulation of the plasma generation inside the cone, we used the 1D and 2D radiation hydrodynamic simulation code (STAR). This code was originally developed for the extreme ultra-violet (EUV) research for the semi-conductor lithography [5]. For the EUV research, we simulated the laser-produced tin plasma (Sn, Z=50) for the laser of intensity ranging from 10^{10} to 10^{12} W/cm^2. and 1.06 μm wavelength, and the results of the simulation were verified by many experimental observations. This confirms that the essential data used in the code, such as the equation of state (EOS), and opacity for tin, are correct [6]. Therefore, we assume a tin cone, using the tin EOS and opacity data used in EUV work, although the cone is usually made of gold (Au, Z=79). To check the validity of using tin, we conducted the 1D simulations with various atomic numbers at the laser of intensity $1 \times 10^{11} W/cm^2$ and $1.06\mu m$ wavelength and investigated the dependence of the pre-plasma generation on atomic number Z with the opacity based on the screened hydrogenic averaged ion model. In Fig.1, the plate target of materials of various Z values are irradiated by the laser at the normal incidence, and at the time of 1ns irradiation, contour of the electron number density n_e is plotted. We see that as the atomic number Z increases, the plasma density scale length decreases because of the high heat capacity that leads to a lower sound velocity. For Au, the density scale length is $7\mu m$ at $n_e = 10^{21} cm^{-3}$, which is nearly the same as that of tin. Also the profiles of tin and Au are similar. From this comparison, we conjecture that results of simulations using tin cone will show the essential feature of simulated results of Au cone.

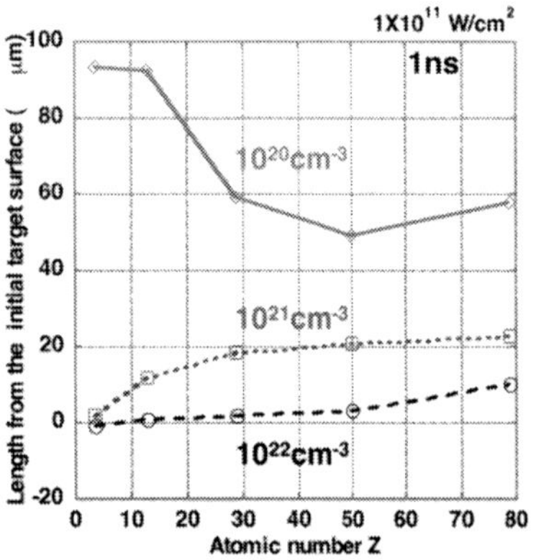

Figure 1. The atomic number Z dependence on the expanded plasma profiles with laser intensity $1 \times 10^{11} W/cm^2$ and 1ns square pulse. Each curve shows the constant electron density line.

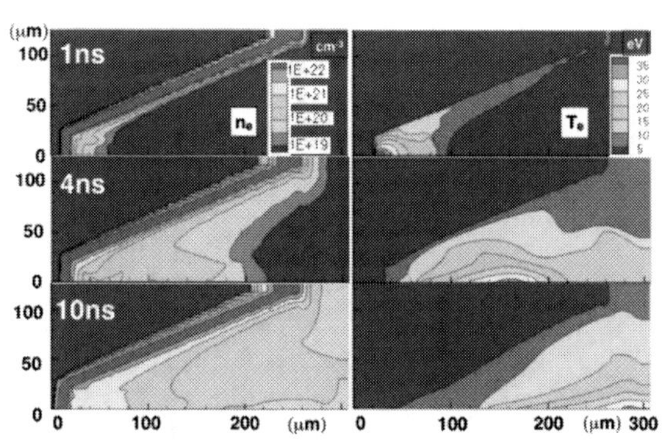

Figure 2. Temporal evolution of the expended plasma tin plasma inside the cone. Electron density (left) and Electron temperature (right).

Next we show the temporal evolution of the pre-formed plasma obtained from the 2D simulation. We used the Star2D radiation hydrodynamic code with Euler coordinate and cylindrical geometry, where initial vacuum region is not exactly zero density, and it is set to be $3 \times 10^{15}\ cm^{-3}$ ion density. The cone material is solid tin and thickness of the cone is 10 μm at the cone tip. The cone open angle is 45 deg. The diameter of the cone tip is 40 μm (the typical diameter in experiments). The laser wavelength is 1.06 μm and its intensity is $1.3\times 10^{11} W/cm^2$ and pulse duration is $10ns$ with the gaussian profile in space and square pulse in time. These parameters are derived from the currently used LFEX laser. The laser propagation is calculated using the straight ray approximation without ray refraction. For electron conduction, we used the flux-limited diffusion model with the flux-limiter of 0.1. The radiation transport is calculated by the multi-group diffusion approximation with the photon energy group of 40 with the range from 0 to 1.5keV. Laser absorption is treated as the inverse-bremsstrahlung process. The detailed numerical scheme of our 2D simulation code (Star-2D) was already described in [?]. In Fig. 2 we show the temporal evolution of the plasma inside of the cone with 45deg open angle. At the beginning of the pulse, the cone tip is ablated and generated plasma is outwardly blown off. As the time elapses, collisions within the ablated plasma near the cone axis generate the plasma jet. The maximum velocity of plasma jet is $4 \times 10^6 cm/s$. Eventually the plasma from the side wall of the cone fills inside the cone tip and the density shelf over the critical density of $1.06\mu m$ is formed. In Fig. 3, we show the density profiles along the cone axis at time = 10ns, with the angle of 30 deg and 45 deg angle cone, respectively. We can see that a large amount of plasma comes from the side wall of the cone and we see the density plateau over the critical density of $1.06\mu m$ wavelength laser, near the cone tip. We did not see this long density plateau for the laser irradiation on the plate target. This is the characteristic property of cone targets. Also, the scale length of the density plateau is significantly longer than that given in Baton paper [3]. At time = 10ns, as the cone angle decreases from 45deg to 30deg, the density plateau becomes longer and the position of the critical density of $1.06\mu m$ wavelength laser shifts away from the cone tip. From this comparison, we conclude that a smaller angle cone gives longer scale length of the pre-plasma and also a longer gap between the critical density and the target center that is position of the high density core. With the 45 deg cone angle case, our 2D simulation gives the position of the critical density at $80\mu m$ away from the cone tip. In our first-integrated fast ignition experiment conducted on July 2009, we had a x-ray emission ring at the distance of $\sim 100\mu m$ from the cone tip when a heating short pulse laser was irradiated. We conjecture that this x-ray emission ring can be an indication of the shift of the critical density position from cone tip. This pre-plasma generation can decrease the coupling between the main heating laser

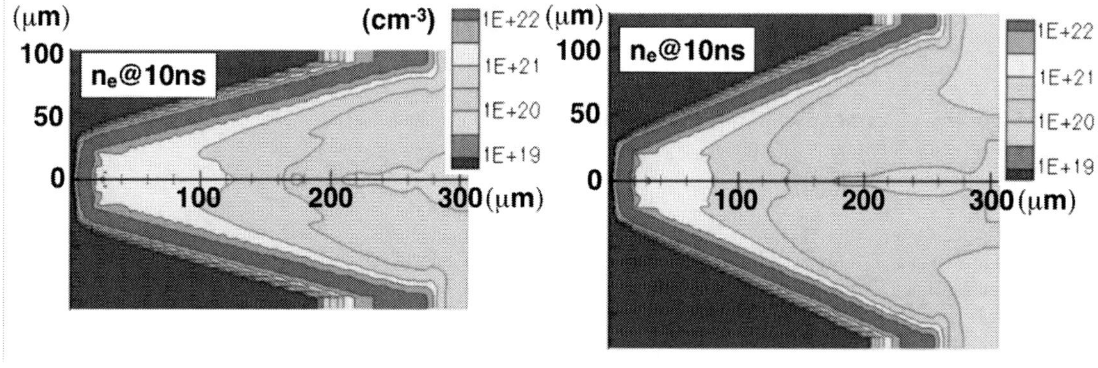

Figure 3. Electron density profiles of pre-formed plasma at 10ns for 30 deg open angle cone (left) and for 45 deg angle cone (right), respectively.

and the imploded core. If we assume some angular spread of fast electrons generated near the critical density, the energy coupling between main heating laser and the imploded core decreases in proportional to the inverse square of the length between core center and critical density. In our fast ignition experiment conducted on July 2009, we set the distance from core center to cone tip to be $50\mu m$. If the position of the critical density is shifted outwardly by $80\mu m$, the gap between core center to the critical density becomes $130\mu m$, and the coupling of the energy decreases by the factor of $(50/130)^2$. In order to suppress this effect, one possible way is to increase the intensity of the main heating laser pulse. For the heating pulse laser, the quiver velocity of the electrons is close to speed of light. In such a condition, the critical density n_{cr} is relativistically increased as;

$$n_{cr} = \frac{\gamma m_e \omega^2}{4\pi e^2} \tag{1}$$

Here, γ is relativistic index, m_e is electron mass, ω is oscillation frequency of the laser and e is electron charge, respectively. We show the γ in function of the laser intensity for linearly and circularly polarized laser in Fig.4. As the laser intensity increase, γ increases. At the laser intensity of $2 \times 10^{19} W/cm^2$, we expect n_{cr} to be three times larger than the critical density in non-relativistic limit. In Fig.3, the position of n_e of $3 \times 10^{21} cm^{-3}$ is very close to the cone tip. We think that increase of the heating pulse can reduce shifting of the position of the critical density caused by the preformed plasma.

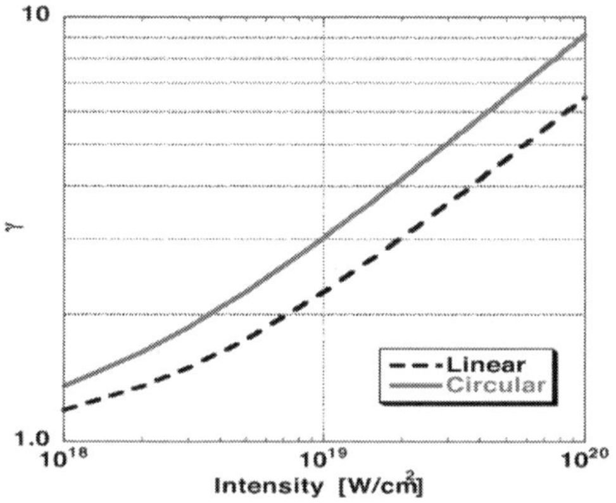

Figure 4. Relativistic gamma in function of the laser intensity

3. Conclusion

We have studied the pre-plasma generation inside the cone caused by the pre-pulse before the arrival of the main pulse by the 1D and 2D radiation hydrodynamic simulations. In a cone target, plasma produced from the side wall is a significant part of the plasma filling near the tip inside cone. Consequently, the density plateau region appears, and the critical density surface moves away from the cone tip, reducing the energy coupling. To suppress this effect of the pre-plasma, it might be necessary to increase the intensity of the main heating laser.

Acknowledgments
A part of this work has been supported by JSPS.

References

[1] Tabak, M. et al., *Phys. Plasmas* **1** (1994) 1626.
[2] Kodama, R. et al.,*Nature* **412** (2001) 798. ; Kodama, R. et al.,*Nature* **418** (2002) 933.
[3] Baton, S. D. et al.,*Phys. Plasmas* **15** (2008) 042706.
[4] Cai, H-B. et al., *submitted to Phys. Plasmas*
[5] Sunahara, A. et al., *Journal of Physics: Conference Series* **112** (2008)042048-1-4.
[6] Nishihara, K. et al., *Phys. Plasmas* **15** (2008) 056708-1-11.

Journal of Physics: Conference Series **244** (2010) 032001 doi:10.1088/1742-6596/244/3/032001

Assessment and Mitigation of Electromagnetic Pulse (EMP) Impacts at Short-pulse Laser Facilities

C G Brown Jr., E Bond, T Clancy, S Dangi, D C Eder, W Ferguson, J Kimbrough, A Throop

Lawrence Livermore National Laboratory, 7000 East Ave., Livermore, CA 94550, USA

E-mail: brown207@llnl.gov

Abstract. The National Ignition Facility (NIF) will be impacted by electromagnetic pulse (EMP) during normal long-pulse operation, but the largest impacts are expected during short-pulse operation utilizing the Advanced Radiographic Capability (ARC). Without mitigation these impacts could range from data corruption to hardware damage. We describe our EMP measurement systems on Titan and NIF and present some preliminary results and thoughts on mitigation.

1. Introduction

Electromagnetic pulse (EMP) is a known issue for short-pulse laser facilities such as Vulcan, Titan and Omega-EP [1]. We expect the National Ignition Facility (NIF) will be impacted by EMP in both long-pulse, which is the normal NIF operation mode, and short-pulse operation. However, we expect the largest impacts during short-pulse operation, such as with the short-pulse based Advanced Radiographic Capability (ARC). Therefore, understanding and mitigating EMP is an important part of the effective use of this and other laser facilities. Short-pulse ($\sim$ few ps) lasers can produce very energetic (MeV) electrons that can be used to produce short-pulse, high-energy, x-ray backlighting as in ARC experiments. In general, the high energy of these electrons causes more to escape the target than the lower-energy "hot electrons" ($<$100 keV) often produced by laser-plasma interactions in long-pulse ($\sim$ few ns) laser-plasma interactions.

Even in the case of short-pulse operation, the number of electrons that escape ($\sim 10^{12}$) is a small fraction of the total number produced in the target, and the associated charge is a small fraction of a Coulomb. However, the short duration causes the escaping electrons to produce very large transient currents and EMP. In addition, the short duration translates to a very broad spectrum with the potential of high-frequency EMP. In both cases (short- and long-pulse) escaping electrons can be a significant source of EMP.

For effective shielding it is critical that the spectrum of the EMP, as well as the time-domain behavior, is known. Both measurements and simulations are necessary to understand EMP in short- and long-pulse lasers. We have fielded a large suite of diagnostics, including B-dot (loop) and D-dot (monopole) sensors, inside and outside the Titan short-pulse laser target chamber, which is a part of the Jupiter Laser Facility at Lawrence Livermore National Laboratory (LLNL) [2], in order to characterize both the EMP and the escaping electrons. We have used our Titan experience to field an EMP measurement system with B- and D-dot sensors [3] inside and outside of NIF's target chamber. In order to establish predictive capabilities and improve mitigation techniques, we have extended the 3D electromagnetic simulation code EMSolve [4] to model EMP generation. A specified number and spatial distribution of escaping electrons are used as input to EMSolve simulations. In this paper, we describe our measurement systems and signal processing techniques employed on Titan and NIF. We also give some preliminary results and thoughts on mitigation. Refer to [5] for information on how we model EMP in EMSolve.

2. Measurement systems

We fielded a wide variety of diagnostics at the Titan short-pulse laser facility in order to understand EMP and measure escaping electrons in such facilities. Our diagnostics include multiple B- and D-dot sensors, electron spectrometers, TLD's (Thermo-Luminescent Detectors), UTLD's (Ultra-Thin TLD's), image plates, and a Faraday cup. Our B- and D-dot measurement system is well shielded, consisting of differential sensors for common-mode noise rejection; shielded conduit to attenuate fields impinging on baluns and measurement cables; and a shielded box for the oscilloscopes. We have B-dot sensors both inside and outside the laser chamber and a D-dot sensor inside to measure the levels of electromagnetic fields. It is important to measure both the time-varying magnetic and electric fields because the ratio of those fields inside resonant chambers generally is not the intrinsic impedance of free space (377 Ω). Figure 1 is a photograph of the interior of the Titan chamber. The labeled sensors are a Prodyn RB-270 B-dot sensor (mid frequency); two RB-230 B-dot sensors (high frequency); and a Prodyn RB-50 sensor (low frequency). The remaining interior sensor in the picture, a Prodyn AD-55 D-dot, is located in the upper, right portion of the photograph.

Our NIF EMP measurement system is similar. The present configuration in NIF is a temporary, early deployment for initial NIF operation that does not provide the full Faraday-cage shielding that was used on Titan. However, measurements with sensors covered with electromagnetic shielding indicate little, if any, corruption due to extraneous EMP noise in the present configuration. A well-shielded measurement system, similar to the Titan system, is being installed for longer-term operation.

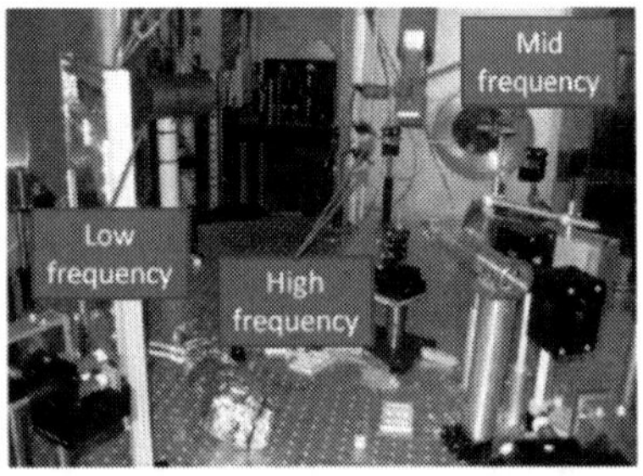

Figure 1. Inside of Titan chamber during our measurement campaign.

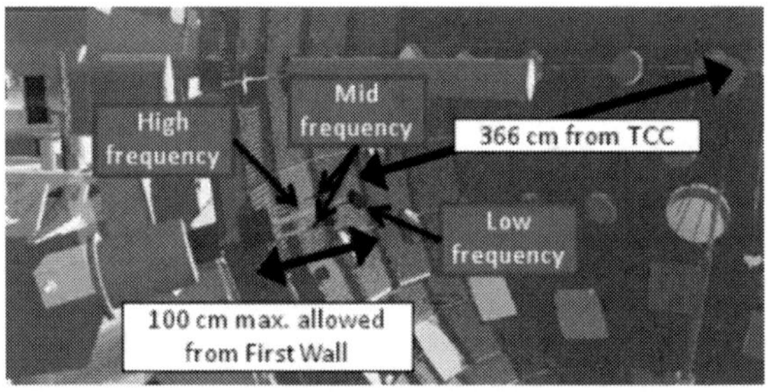

Figure 2. CAD image of EMP diagnostic in target chamber.

3. Signal processing techniques

In order to remove the effects of the measurement system, we perform channel compensation, or deconvolution. We construct calibration files for each component of the measurement system, consisting of sampled transfer functions in the frequency domain. The measurement system transfer function is then simply the product of all of the frequency-domain transfer functions of the individual components. A straight-forward method of performing the channel compensation would be to transform the data from the oscilloscopes into the frequency domain; apply the inverse of the system transfer function; and then transform back into the time domain. However, this method tends to magnify noise in the frequency bands where the transfer function is small, producing unrealistic signals. In order to reduce this noise magnification, we use Tikhonov regularization with L-curve analysis to determine the optimal regularization parameter [6]. Additionally, we low-pass filter the channel-compensated signal at the frequency of the analog low-pass filter in the measurement system.

4. Preliminary results and thoughts on mitigation

We applied the signal processing techniques above to the measured data from Titan and NIF. First, our preliminary results from short-pulse Titan (~ ps) indicate large EMP, on the order of tens to hundreds of kV/m, that has a duration (before decaying into the noise) of up to about a few hundred ns. For example see Fig. 3, which is processed data from one of our RB-230 (high-frequency) sensors located about 32 cm from target chamber center (TCC). The resulting magnetic flux density values (B field) have been converted to electric field strength (E field) by multiplying by c, the speed of light in vacuum: $E \approx cB$. Note that this conversion is only approximate, since the relationship does not hold in general for resonant cavities. The peak magnitude of the resulting E-field signal is 167 kV/m. The transient duration before decaying into the noise is ~ 400 ns. A smoothed estimate of the one-sided energy spectral density (ESD) using just the transient portion is displayed in Fig. 4. The ESD indicates significant EMP out to greater than ~ 5 GHz. We cannot observe much higher than ~ 5 GHz because of the oscilloscope analog bandwidth (~ 6 GHz) and a 5 GHz low-pass filter in the measurement system. Further, we have observed a clear trend with target size in the Titan data (Fig. 5). Note that the peak magnitude of the E field tends to increase with the target size, a possible explanation being that a larger target reduces the electrostatic field allowing more electrons to escape. The data in Fig. 5 is for square silver targets 12 μm thick. The laser pulse width is 2 ps with laser energies ranging from 123 to 155 J (~ 75 TW average power).

Our preliminary results from long-pulse (~ ns) NIF indicate EMP with peak magnitudes up to ~ 10 kV/m. For example see Figs. 6, 7, and 8, which are processed data from our RB-230 (high-frequency), RB-270 (mid-frequency), and RB-50 (low-frequency) sensors, respectively, located about 4 m from TCC. The shot is N090808 (233 kJ, 2 ns, 116 TW, vacuum hohlraum). Note that if we assume $1/r^2$ scaling, the fields could be ~ 16× higher 1 m from TCC. Again, the measured B-field values have been converted to E field using $E \approx cB$. A smoothed estimate of the one-sided ESD using just the transient portion of the signals is displayed in Fig. 9. The ESD is plotted for the different sensors only over the frequency range where they are "valid", which is judged by comparison with the noise floor. The ESD indicates EMP with a maximum at ~ 1 GHz with significant EMP out to greater than ~ 5 GHz. As before we cannot observe much higher than ~ 5 GHz because of the oscilloscope analog bandwidth and a 5 GHz low-pass filter in the measurement system. Also, we measured volt-level noise, likely due to EMP, on terminated signal cables on a NIF diagnostic instrument manipulator (DIM) inside the target chamber. We have not yet measured any EMP above the noise outside of the target chamber.

Based on these preliminary findings, we are reassessing diagnostic shielding inside the target chamber, starting the process of preparation for ARC operation, which we estimate could produce roughly an order of magnitude higher fields. The present focus of mitigation efforts is on placing high-frequency, Faraday-cage shielding around cables and diagnostics inside the target chamber. The amount of shielding required will vary depending on the incident EMP, coupling parameters, noise level requirements, and the susceptibility of the diagnostic's electronics. Accordingly, it is difficult to make a general statement about the level of EMP fields that would affect the diagnostics. However, in view of our current measured data we estimate that for normal NIF operation at least 60 dB of shielding is required for the DIM signal cabling, and flexible shielded conduits and other shielding methods are being evaluated for this purpose. Similarly, we are providing Faraday-cage shielding for airboxes and diagnostics mounted on the

DIMs. Additionally, fingerstock contacts on the DIMs are being installed. Ferrites for noise current suppression and electromagnetic interference windows for shielding apertures are also being evaluated.

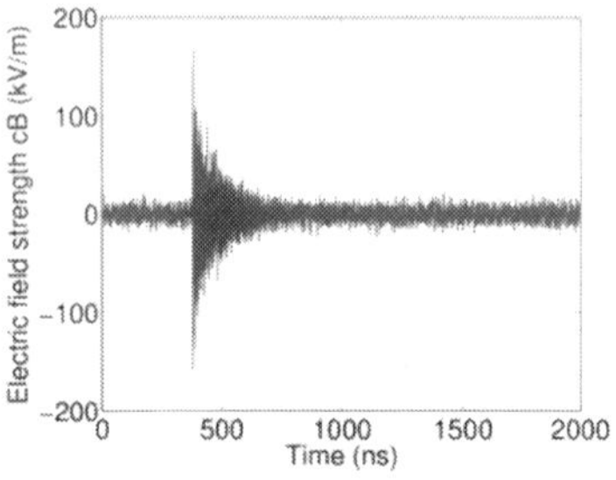

Figure 3. Processed Titan data from high-frequency B-dot sensor (with $E \approx cB$).

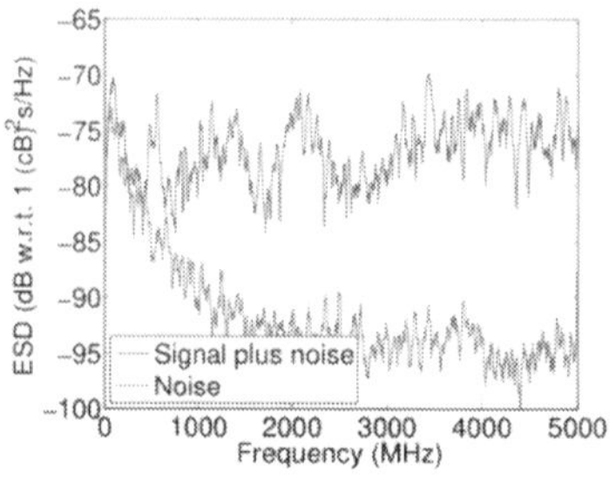

Figure 4. An estimate of the ESD of the transient portion of Fig. 3.

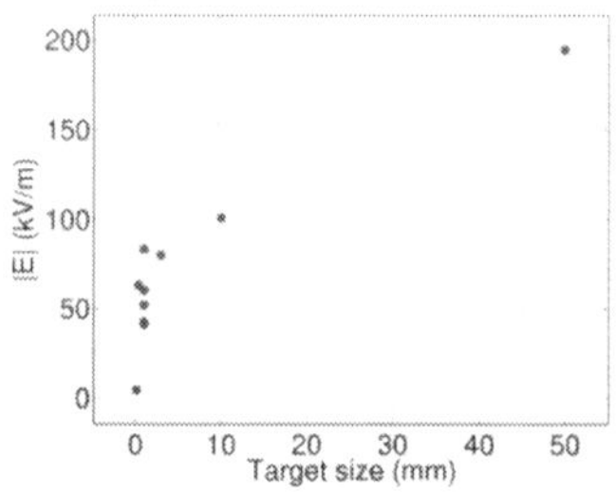

Figure 5. Peak magnitude of electric field in Titan versus target size.

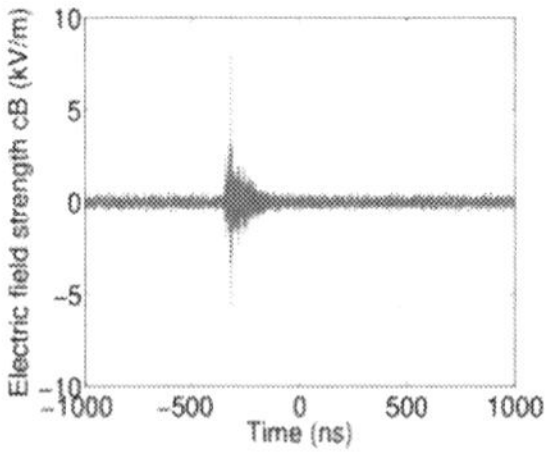

Figure 6. Processed NIF data from high-frequency B-dot sensor (with $E \approx cB$).

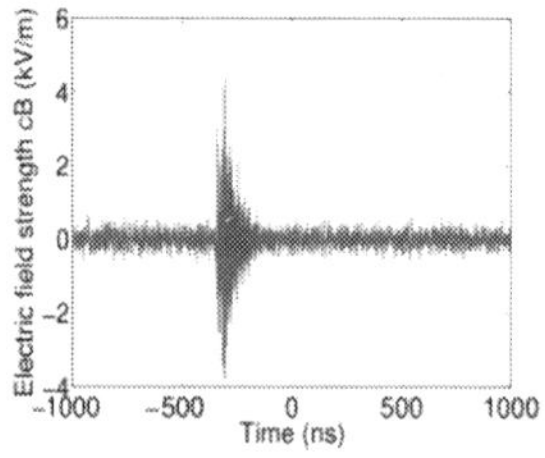

Figure 7. Processed NIF data from mid-frequency B-dot sensor (with $E \approx cB$).

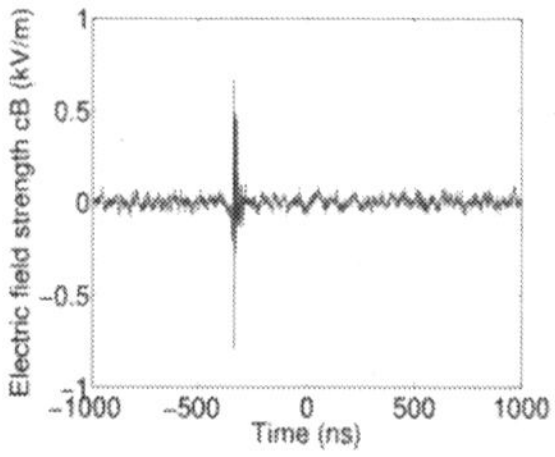

Figure 8. Processed NIF data from low-frequency B-dot sensor (with $E \approx cB$).

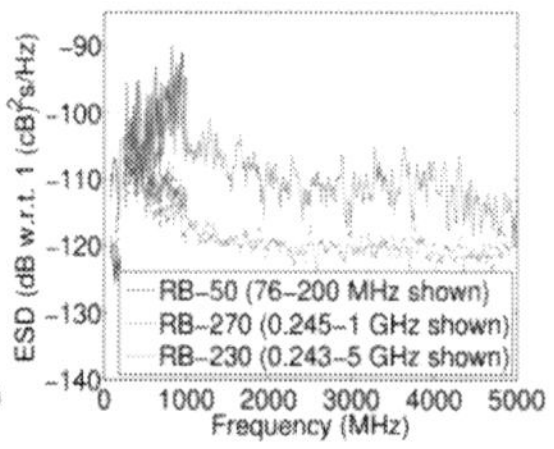

Figure 9. Composite spectrum of processed NIF data and noise floor from Figs. 6 to 8.

5. Summary and conclusions

We have deployed measurement systems on both Titan and NIF. Our preliminary data from Titan indicate high-level (on the order of tens to hundreds of kV/m), broad-spectrum EMP. We have also observed a clear trend in EMP level with target size. Our preliminary data from long-pulse NIF indicates EMP up to ~ 10 kV/m, which we estimate to be higher (~ 100 kV/m) 1 m from TCC. We postulate that these fields are generating the volt-level signals that we are observing on the diagnostic cables. Based on these preliminary findings, shielding for target chamber diagnostics is being reassessed, starting the process of preparation for ARC operation.

Acknowledgments

Prepared by LLNL under Contract DE-AC52-07NA27344. We are grateful to the Jupiter Laser Facility staff, including J Bonlie, R V Maren, D Price, etc. Many others from both National Security Technologies, LLC (NSTec), Z Ali, C Brooksby, J Fornes, L MacNeil, K Moy, V Tran, and J Yuhas, and LLNL, B DeHope, also contributed. Finally, we are indebted to the NIF team.

References

[1] Mead M J, Neely D, Gauoin J, Heathcote R and Patel P 2004 *Rev. Sci. Instrum.* **75** 4225
[2] http://jlf.llnl.gov (Accessed September 2009)
[3] Edgel W R Prodyn Application Note 192 Prodyn Technologies
[4] White D and Stowell M 2004 *IEEE T. Microw. Theory* **52**(5) 1404
[5] Stowell M L 2008 Tech. Rep. LLNL-CONF-409029 Lawrence Livermore National Laboratory
[6] Aster R, Borchers B and Thurber C 2005 *Parameter Estimation and Inverse Problems* (Academic Press)

Journal of Physics: Conference Series **244** (2010) 032002 doi:10.1088/1742-6596/244/3/032002

Pulse compression below 40fs at 1μm: the first step towards a short-pulse, high-energy beam line at LULI

Xiaowei Chen[1, 2, 3,*], Jiping Zou[1,†], Luc Martin[1], Francois Simon[1], Rodrigo Lopez-Martens[3], Patrick Audebert[1]

[1] LULI, Ecole Polytechnique, CNRS, CEA, UPMC, 91128 Palaiseau, France

[2] ILE, ENSTS, Chemin de la Hunière, 91761 Palaiseau, France

[3] LOA, ENSTA, ParisTech, CNRS, Ecole Polytechnique, 91761 Palaiseau, France

[*] xiaowei.chen@ensta.fr

[†] ji-ping.zou@polytechnique.edu

Abstract. We present the upgrading project ELFIE (Equipement Laser de Forte Intensité et Energie) based on the "100TW" mixed Nd:glass CPA laser system at 1μm at LULI, which includes an energy enhancement and the development of a short-pulse, high-energy, good temporal contrast beam line (50fs/5J). We report the first experimental step towards the short-pulse, high-energy beam line: spectral broadening above 60nm from 7nm and temporal pulse compression below 40fs from 300fs at 1μm through a Krypton-filled hollow fiber compressor.

1. ELFIE project

The goal of the ELFIE project is to upgrade the 100TW laser facility [1] at LULI which consists of a Ti:sa and Nd:mixed-glass CPA laser system. In order to support different plasma experiments, this laser facility provides a main laser beam of 30J/300fs optically synchronized with a 60J/500ps uncompressed (chirped) beam, an auxiliary 10J/300fs beam and a 100 mJ/few ps tuneable probe beam. High laser operation efficiency has been demonstrated since its commissioning in 1997, and the laser quality and performance have been improved constantly. With the dynamic wavefront control, the laser system can fire a full energy shot every 20 minutes while keeping excellent beam focusing ability on target.

The ELFIE project was initiated in order to increase the laser intensity available on target and to provide a more versatile laser tool for plasma physics research. Fig.1 shows a schematic of the ELFIE system design that mainly includes a boost in energy of the original mixed Nd:glass laser chain based on CPA technique, and the development of a new short-pulse, high-energy, good temporal contrast beam line based on OPCPA technique. The total amplified pulse energy delivered by the upgraded mixed Nd:glass laser system will be up to 200J, providing the synchronized pump source for the last OPCPA stage of the short-pulse beam line, a 50J/500ps laser beam for plasma creation and two 30J/300fs compressed laser beams. The high-energy short-pulse good contrast beam line aims at achieving 50fs laser pulses with the energy of 5J to be used as probe beam with a good temporal resolution in the plasma physics experiments. This versatile laser facility will enable the use of a variety of secondary sources of particles or radiation in order to study high-energy density matter (e.g. fusion or astrophysical plasmas) or fast evolving plasmas.

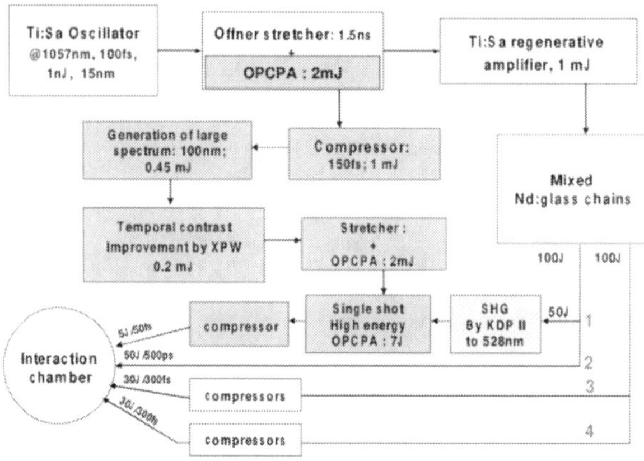

Fig.1 Schematic layout of the ELFIE project.

2. 50fs/5J beam line

The development of the short-pulse high-energy good temporal contrast beam line will consist of five stages, as highlighted by the orange parts in Fig.1:

1) An OPCPA pre-amplifier boosting the stretched nJ pulses up to several mJ at 10Hz without spectral narrowing, achieving 1mJ/100fs pulses after compression.
2) Spectral broadening from ~15nm to above 60nm by a noble gas-filled hollow fiber followed by temporal compression below 50fs using chirped mirrors, with an overall energy efficiency of ~50%.
3) Temporal contrast enhancement of the >0.4mJ/<50fs pulses by a nonlinear cross-polarized wave generation (XPWG) filter [2, 3] for 3~4 orders, with a typical throughput >20% for Gaussian input beams using a two-crystal arrangement. XPWG also has the advantage of broadening and smoothing the incident laser spectrum. We expect to seed the booster OPCPA stages with high temporal quality broadband pulses [4].
4) Boosting of the temporally filtered pulses from the μJ level up to 7J by two broadband high-energy OPCPA stages while preserving the 60nm spectral bandwidth.
5) Pulse compression using a new high-efficiency broadband grating compressor enabling the delivery of 50fs/5J pulses on target.

This beam line will provide new opportunities for high-field plasma physics research.

3. Spectral broadening and pulse compression through gas-filled hollow fiber technique

3.1. Pulse compression in hollow fibers

Because the laser source we plan to use is in 100fs level, spectral broadening is necessary in order to achieve pulse duration below 50fs. Nonlinear propagation through a noble gas-filled hollow fiber is a widely used technique to obtain efficient spectral broadening in the VIS-IR spectral range [5, 6]. Moreover, the confinement of the dielectric waveguide ensures spatially uniform broadening and an excellent output beam profile. However, as far as we know, few experiments have been carried out in mJ level in 1μm range till now. In order to demonstrate the affectivity of this technique in 1 μm region, we tested it on the front-end regenerative amplifier of the current LULI 100TW laser system. Here, we report the high spatial quality spectral broadening from 7nm to >60nm and temporal pulse compression down to 38fs by using a Krypton-filled hollow fiber compressor. This work is a key step in development of our 50fs/5J beam line.

3.2. Experimental setup

The laser source consists of a Ti:sa oscillator operating at 1057nm, an Offner-type stretcher and a Ti:sa regenerative amplifier working at 10Hz. The bandwidth of the seed pulses is decreased from

Journal of Physics: Conference Series **244** (2010) 032002 doi:10.1088/1742-6596/244/3/032002

15nm to 7nm due to the spectral gain narrowing effect after ~100 roundtrips inside the regenerative amplifier. After amplification >1mJ pulses are sent to a temporary grating compressor yielding 0.7mJ/300fs compressed pulses. These pulses are then coupled into a hollow fiber (Femtolasers GmbH) by an f=80cm focusing lens. The hollow fiber is 1-m-long with an inner diameter of 250μm and rests inside a glass tube filled with Krypton. The experimental setup is shown in Fig.2. The dispersion of the output pulses from the hollow fiber is then carefully compensated using six commercial broadband chirped mirrors with $-50fs^2$/bounce (Layertech) and a pair of fused silica wedges to get the shortest compressed pulses. The compressed pulses are finally characterized by a fiber spectrometer (Ocean Optics), a CCD camera, a home-made autocorrelator, and an energy meter.

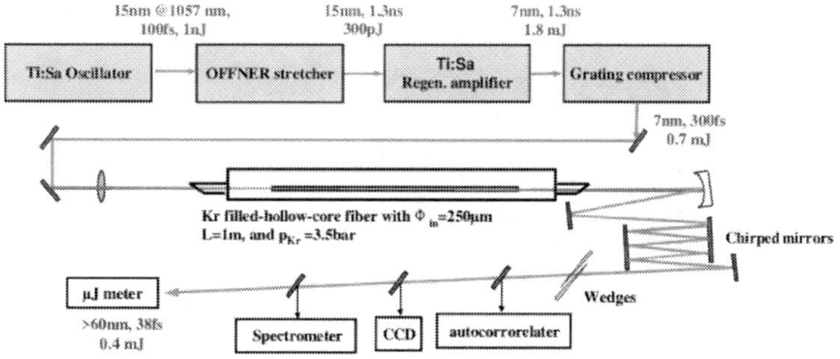

Fig.2 Krypton-filled hollow fiber pulse post-compressor setup.

3.3. Experimental results and analysis

In the experiment, we gradually increase the Krypton pressure while monitoring the measured output. Spectra at different gas pressures are shown in Fig.3a. We observe the characteristic spectral broadening due to self-phase modulation (SPM) because of the higher nonlinear refractive index of the gas. The spectral clipping at the gas pressures of 3bar and 3.5bar comes from the bandwidth limitation of the spectrometer. Comprehensive numerical simulations of nonlinear pulse propagation inside the fiber have also been done, in which a Gaussian pulse was assumed and all the pulse, hollow fiber and gas parameters were set according to the experimental condition. The 2D+1 simulation code is based on the nonlinear envelope equation (NEE) and typical nonlinear effects such as Kerr effect, self-steepening and gas ionization were considered. Fig.3b shows a calculated spectral profile and spectral phase of the output pulses from the fibre at a Krypton pressure of 3.5bar. Except the fast modulations observed around the central wavelength, the simulation shows qualitative agreement with the experimental measurement on the spectral broadening. The simulation results confirm that SPM is the dominant broadening process in our experimental conditions. In Fig.3b, the deep spectral modulations and the ripple shape of the spectral phase both stem from typical SPM effects. This ripple-shape spectral phase includes high-order phase terms which are difficult to compensate simply by using chirped mirrors. However, the frequency chirp is relatively linear across the whole bandwidth, ensuring efficient pulse compression. At a Krypton pressure of ~3.5bar, the pulse duration can be compressed below 40fs after fine compensation of spectral chirp. A typical measured autocorrelation trace of the compressed pulses is shown in Fig.4b, corresponding to a pulse duration of 38fs (FWHM). For comparison, the measured autocorrelation trace of the input laser pulses is also shown in Fig.4a. Thanks to the mode confinement of the hollow fiber, the generated 38fs laser beam has an excellent spatial profile, which is measured by a CCD camera in the experiment, as shown by the colourful insert figure in Fig.4b. In addition, the energy stability of the compressed pulses has been preserved quite well, only slightly decreased from ~3% to ~5% due to the nonlinear propagation in the hollow fiber.

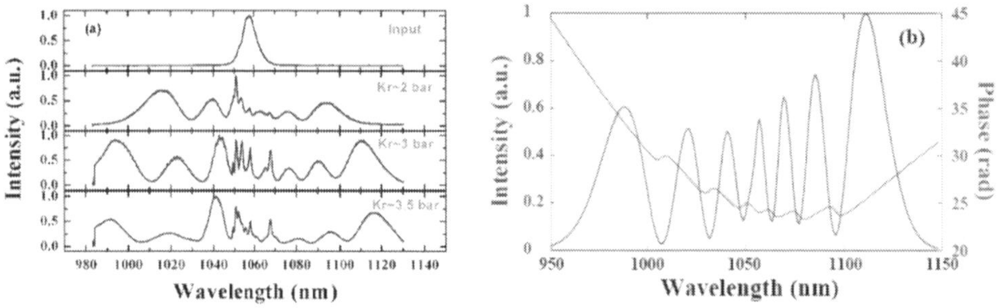

Fig.3 (a) Experimental spectrum evolution of the output pulses from the hollow fiber at different Krypton pressures. (b) Simulated spectral profile and phase of the output pulses from the hollow fiber at the gas pressure of 3.5bar, in which the input pulses are assumed as 0.7mJ/300fs.

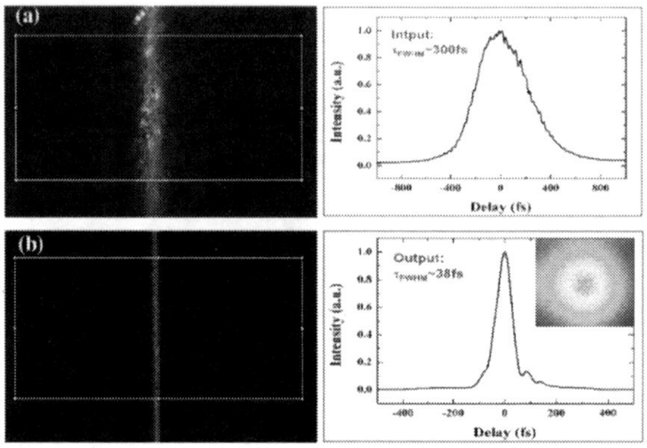

Fig.4 Measured autocorrelation trace of the compressed 38fs pulses (b), compared with that of the 300fs input laser pulses (a). The insert shows the measured spatial profile of the compressed laser beam.

4. Conclusion and prospects

In conclusion, we demonstrate pulse compression by a factor of 7.5 at 1 μm wavelength with an overall energy efficiency close to 60% using a Krypton-filled hollow fiber compressor. The mode confinement of the waveguide ensures spatially uniform broadening and an excellent output beam profile. These pulses can then be efficiently temporally filtered by using XPWG technique before seeding the high-energy OPCPA stages. This work is currently underway as a significant step towards completion of 50fs/5J beam line for ELFIE project.

Since the residual phase of the laser pulses affects a lot the XPWG energy efficiency as well as the pulse spectral profile, fine dispersion compensation is one of the important issues for the following XPW stage in the development of 50fs/5J beam line. Broader and less modulated spectrum can be expected by seeding the hollow fiber with shorter input pulses with higher energy, which is feasible under the broadband OPCPA front-end configuration.

References

[1] J. P. Zou et al., 3rd IFSA, 7-12 June 1998, Monterey, CA, USA, SPIE **3492**, 94 (1998).
[2] A. Jullien et al., Opt. Lett. 30, 920 (2005).
[3] A. Cotel et al., Applied Phys. B **83**, 7 (2006).
[4] V.Chvykov et al.,Opt. Lett. **31**, 1456 (2006).
[5] M. Nisoli etal., Appl. Phys. Lett. **68**, 2793 (1996).
[6] X. Chen et al., Opt. Lett. **34**, 1588 (2009).

Progress on Converting a NIF Quad to Eight, Petawatt Beams for Advanced Radiography

J. K. Crane, G. Tietbohl, P. Arnold, E. S. Bliss, C. Boley, G. Britten, G. Brunton, W. Clark, J. W. Dawson, S. Fochs, R. Hackel, C. Haefner, J. Halpin, J. Heebner, M. Henesian, M. Hermann, J. Hernandez, V. Kanz, B. McHale, J. B. McLeod, H. Nguyen, H. Phan, M. Rushford, B. Shaw, M. Shverdin, R. Sigurdsson, R. Speck, C. Stolz, D. Trummer, J. Wolfe, J. N. Wong, G. C. Siders, C. P. J. Barty

Lawrence Livermore National Laboratory, 7000 East Ave., Livermore, CA 94550, USA

Email: crane1@llnl.gov

Abstract. We are converting a quad of NIF beamlines into eight, short-pulse (1-50 ps), petawatt-class beams for advanced radiography and fast ignition experiments. This paper describes progress toward completing this project.

1. Introduction

To diagnose the high-density core of an imploding fusion capsule on a time scale that is relevant for successfully optimizing target and driver parameters, ~100 keV, ps x-ray backlighters are required. We are converting four (one quad) of the NIF laser system's 192 long-pulse beamlines to eight, petawatt-class, high-intensity beams. Each of these eight beams can be independently timed, pointed and focused onto high-Z, x-ray-backlighters in the vicinity of the cryogenic ignition target, producing a multi-frame, short-pulse x-ray movie of the cryogenic fusion target during the compression and ignition phases. We call this x-ray generation system the Advanced Radiographic Capability (ARC)[1]. In addition to generating short-pulse x-rays to support advanced radiography for fusion and other high energy density science experiments, the eight beams may be combined into a single pulse on target to explore fast ignition physics in the 10 kJ, 5 ps regime.

2. Description

We will convert a quad of long pulse (ns-scale) beamlines of the NIF laser system [2] into ps-scale, high intensity beamlines using conventional chirp-pulse amplification techniques [3], adapted to the overall architecture of the NIF laser system. Figure 1 shows the architecture for converting a NIF quad to ARC. ARC hardware through the front end has been built and tested; the rest is in various stages of design and fabrication.

2.1. ARC front end

The seed pulse for ARC is produced by a Yb:doped fiber mode-locked oscillator [4]. We stretch the pulse to several ns using a chirped-fiber-Bragg-grating (CFBG) that is supplied by the University of Southampton. We split the output of the CFBG to make two separate fiber beamlines. The positive group delay produced by the CFBG is balanced by the fixed negative group delay in the large final compressors and small, adjustable compressors, or pulse width controllers (PWCs) that are rack-mounted in the NIF master oscillator room [4]. We amplify the output energy in the fiber front end up to 1 nJ before transporting the pulse to the laser bay.

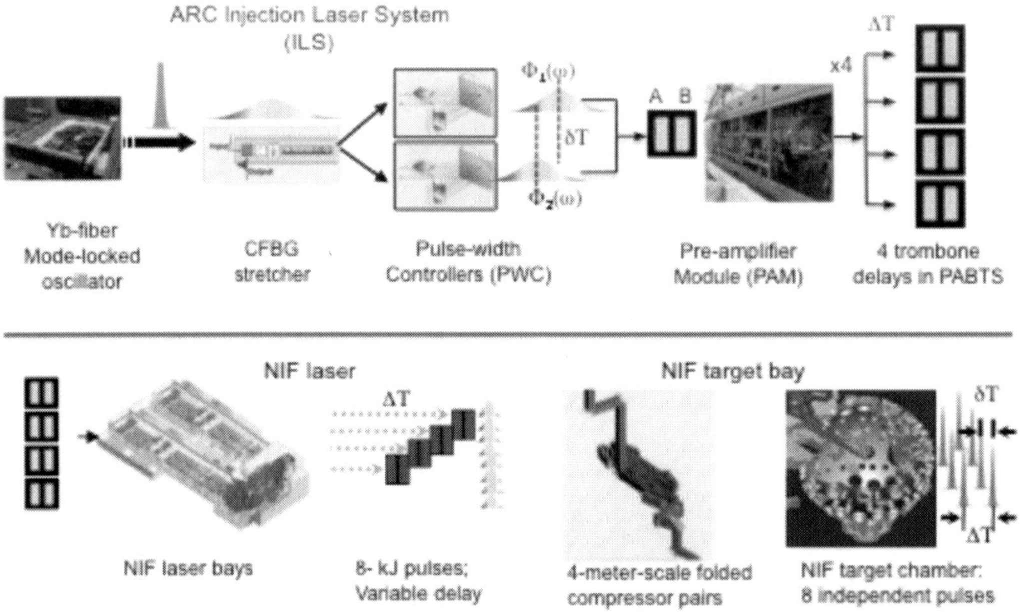

Figure 1. Layout shows the principal subsystems for the conversion of a NIF quad to ARC. Arrows indicate propagation direction: pulse starts at upper left, ends at lower right.

In the laser bay the two, nJ fiber outputs from the ARC fiber front end seed identical regenerative amplifiers (regens) that include intracavity birefringent filters to tailor the pulse spectrum. The two regen outputs are spatially shaped and combined in the Preamplifier Module (PAM) to form a split beam pair (A and B) whose outer perimeter is the same size as a NIF beam. The ARC split beam pair will be amplified in a modified NIF chain up to the kJ level.

2.2. *Pulse width, timing and dispersion balance*

The two fiber beamlines are split after the CFBG and go to two different PWCs. We can separately adjust the group delay and timing, δT (see Fig. 1), between the stretched pulses in the two fiber systems by adjusting the grating slant distance, roof mirror positions or by adding fiber jumpers for larger delays. The two fiber outputs are combined after the regens into a single, split-beam aperture to be amplified through the NIF chain (see top right in Fig. 1). The amplified, split-beam at the PAM output will be split 4 ways in the Preamplifier Beam Transport System (PABTS)[2] before being injected into the four beamlines of the NIF quad (upper right image in Fig. 1). The ARC PABTS will have two-meter long trombone sections with stepper motor controls to adjust the delays, ΔT, among the beamlines in the quad. Finally each of the eight, amplified ARC split beams is sent to a four grating, folded compressor in one of the large vessels in the target bay. To generate eight identical pulses at the ARC output, dispersion must be carefully balanced in each beamline. We developed a group delay diagnostic that measures dispersion in the separate dispersive systems, e.g. CFBG, PWC, transport fiber, grating compressors [5].

2.3. *Target bay systems*

The eight, kJ split-beam outputs from the laser bay are picked off in the switchyard by insertable mirrors on rails. Additional mirrors direct the split beams into four, folded, diffraction grating compressors [1], housed in two large vacuum vessels. The lower right portion of Fig. 1 shows the layout for one of the split beam pairs (A, B). The pulse compressors have four pairs of side-by-side, multi-layer dielectric (MLD) gratings [6]. One grating in the pair disperses split beam A, the other beam B. The grating pairs have slightly different groove densities to prevent crosstalk on the dispersed beams on grating pairs two and three.

After compression, four large mirrors direct each split beam to a separate target for generating ps x-rays to backlight experiments at the center of the NIF target chamber. The first of the four output turning mirrors leaks a fraction of the light to a diagnostics system that measures the spatial and temporal characteristics of the compressed ARC pulse. The third output mirror is an off-axis parabola with a ~ 9 meter focal length to focus ARC at the target chamber center. After the parabola a pair of mirrors with independent adjustment directs the pair of focusing pulses (A and B) to separate targets near target chamber center. Thin, disposable debris shields protect the final optics from target debris.

3. Requirements and performance

The top level requirements for ARC, shown in Table 1, are determined by user needs and laser and optical system limitations. These top-level requirements are flowed back to the front end to determine the requirements at each subsystem interface. For example, the peak irradiance is driven by the need to generate enough 100 keV x-rays to produce spatially resolved images of the imploding capsule in the vicinity of "bang time". However this peak value is limited by damage on the final grating, parabola, and final turning mirror at the specified pulse width. The damage values are determined from measurements of coating witness samples and deterministic models [7] for damage growth. To estimate encircled energy at the ARC focus we use NIF and CEA propagation codes [8,9]. Detailed interferograms from the manufactured ARC gratings, wavefront files for large mirrors, and data from the NIF output sensor package [2] at ARC energy levels provide inputs for the propagation codes.

Table 1. Top level requirements for ARC.

Description	Value	Basis
Peak irradiance	$>10^{18}$ W/cm^2 @ 2 ps	sufficient 100 keV x-rays for radiography missions
Encircled energy	>80% in 300 μm diameter circle at TCC	
Pre-pulse contrast	<-70 dB in power <10^{-5} in energy	maintain target integrity
Pulse width range	1-50 ps	satisfy different HEDS missions
Pointing stability	< 75 μm rms	consistent with current NIF requirements
Timing jitter relative to NIF average	< 30 ps rms	x-ray movie range for HEDS experiments
Timing adjustment relative to NIF	0-75 ns	
Back-scatter protection	>10% of output energy	Protect NIF laser system

4. Recent progress

We have made some major modifications in the past year to better protect the ARC laser system from target back-reflections. We replaced the two final fiber amplifiers in the original design with two, high-bandwidth, regenerative amplifiers, boosting the energy of the ARC injection laser system from 100 mJ to 5 J. At the same time we will reduce the stored energy in the main NIF amplifier chain used for ARC by reducing the number of slabs in the main amplifiers from eleven to eight. To further protect the front end from backward propagating energy from the target, we plan to replace one of the amplifier slabs with a polarizer and triple-pulse the plasma electrode Pockels cell (PEPC)[2] to further reduce backward propagating 1053 nm light. These changes should protect the laser against a return pulse with up to 10% of the total forward propagating energy.

We built a four grating, folded compressor at the output of the ARC Injection Laser Test facility, shown in Figure 2. We use this test facility to measure performance and diagnose systems issues in the ARC front end: dispersion balance, temporal fidelity, amplified spontaneous emission, focal spot quality and ILS energetics. We also test the temporal diagnostics that will be used at the ARC output. The compressor gratings are 100 mm X 200 mm MLD gratings that are identical in line spacing to the ARC A gratings. We double pass the compressor to reduce the footprint while duplicating the group delay of the ARC compressor. Figure 3 shows a 1 TW pulse measured at the output of the ARC test facility using single-shot, second-harmonic-generation, frequency-resolved optical gating (SHG FROG). The small bumps at the front of the pulse are due to imperfect balance of the third-order dispersion in the front end plus group delay ripple from the CFBG.

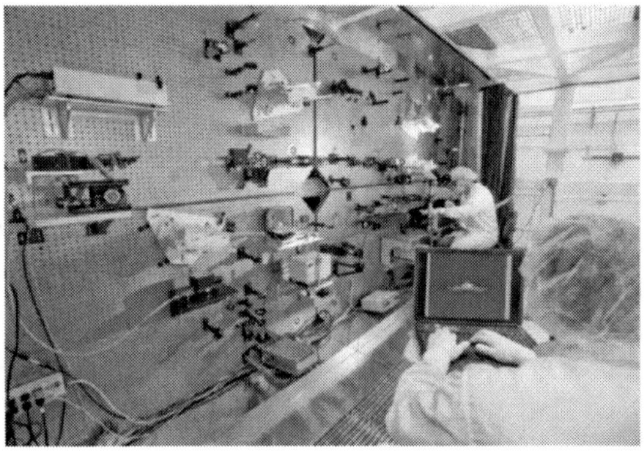

Figure 2. ARC Injection Laser Test Facility.

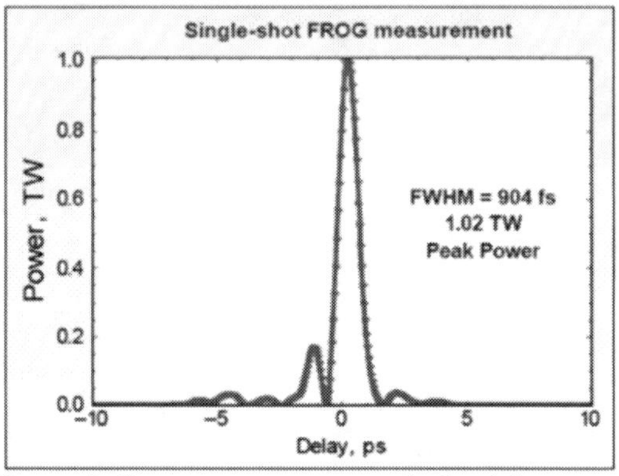

Figure 3. Power measurement derived from energy and temporal diagnostic at the ARC Injection Laser Test Facility output.

Many of the major systems for ARC are complete, including the ARC ILS, the triple-pulsed PEPC, most of the large optics (including the gratings and parabolas), and the large vacuum vessels for the compressors and parabolas.

Acknowledgments

This work was performed under the auspices of the U.S. Department of Energy by Lawrence Livermore National Laboratory under contract DE-AC52-07NA27344.

References

[1] C. P. J. Barty, et.al., 2004 *Nucl. Fusion* **44** S266
[2] C. A. Haynam, et. al. 2007 *Appl. Optics* **46** 3276
[3] D. Strickland and G. Mourou 1985 *Opt. Commun.* **56** 219
[4] J. W. Dawson, et. al. 2009 *IEEE J. Select Topics in Quantum Electr.* **15** 207
[5] C. Haefner, et. al. 2009 Dispersion balancing of complex CPA-systems usings the phase-shifting technique *Proc. of Conf. on Lasers and Electro-Optics* (Baltimore, MD
[6] J. A. Britten, et. al., 2005 Advanced dielectric grating technology for high-energy petawatt lasers *Quant. Electr. and Laser Science Conf* JFB5
[7] B. Stuart, M. Feit, A. Rubenchik, B. Shore, M. Perry 1995 *Phys. Rev. Lett.* **74** 2248
[8] R. A. Sacks, et. al., 2008, *Journal of Physics: Conf. Ser.* **112** 032024
[9] O. Morice 2003 *Opt. Eng.* **42** 1530

Plasma mirror effect with a short-pulse KrF laser

I B Földes[1], A Barna[1], D Csáti[2], F L Szűcs[1], S Szatmári[2]

[1] KFKI Research Institute for Particle and Nuclear Physics, Association EURATOM HAS, H-1525 Budapest, P.O.B. 49. Hungary
[2] University of Szeged, Department of Experimental Physics, H-6720 Szeged, Dóm tér 9. Hungary

E-mail: foldes@rmki.kfki.hu

Abstract. Plasma mirror effect is demonstrated for short-pulse KrF lasers. It is shown that the reflectivity of the plasma starts to increase at the plasma threshold of 10^{12}W/cm^2 and it saturates at 10^{14}W/cm^2. The maximum reflectivity was ~33% for 45° angle of incidence, and ~50% was reached for 12.4°. This strong reflection allows even its direct application for short-pulse 248 nm systems whereas using it before the last amplifier may even clean the pulse from spatial inhomogeneities.

1. Introduction

Plasma mirrors or self-induced plasma shuttering[1] are at present the final tools to clean laser pulses from prepulses. Ultrashort laser pulses are generally preceded by prepulses originating either from the incomplete compression of the chirped pulses or from the long ASE of the amplifier as in our case which will be called pedestal throughout this paper. Pedestals can be detrimental for the investigations of high intensity laser-matter interactions, because they may generate preplasmas even if the intensity of the pedestal is several orders of magnitude less than that of the main pulse. In this case the main laser pulse interacts not with the solid target but with the preformed plasma. The principle of plasma mirror is that in case the intensity of the laser pulse which falling onto a transparent solid material is chosen so that only the leading edge of the main ultrashort pulse is above the threshold for plasma production, the pedestal of lower intensity will be transmitted. Thus the part of the pulse reflected from the plasma (which does not have time to expand during the short pulse duration) can be separated resulting in a contrast improvement up to several orders of magnitude. Several successful experiments were carried out, cleaning the pulses of titanium-sapphire and OPA laser systems by plasma mirrors [2,3]. Such a scheme was applied to the Vulcan facility allowing high-harmonics generation from solid surfaces to the multi keV level[4].

The short pulse KrF laser in our laboratory is based on direct amplification of 500-600fs pulses without using the chirped pulse amplification (CPA) scheme. Therefore the pedestal originates only from the ASE of the KrF amplifiers allowing a contrast as high as 10^{10}. However, in the case of KrF laser pulses of ~5eV photon energy free ions are generated by photoablation and photoionization below 10^7W/cm^2 intensity as reported earlier [5]. It means that for intensities above 10^{17}W/cm^2 free ions will be in the focal volume by the time the main pulse arrives. According to the results of mass spectroscopy [5] it is desirable to keep the prepulse level below 2×10^6W/cm^2.

Herewith we report about experiments in which we investigate the possibilities for applications of plasma mirrors for short–pulse KrF systems. It is shown that the plasma mirror effect can be observed at this wavelength with a similar behavior than in the infrared. Although the shorter wavelength of the KrF laser results in a lower reflectivity than for Nd or Ti-sapphire systems, it is shown that plasma mirrors may be applied for short-pulse KrF systems, too. The optimum intensity range with the corresponding plasma temperature is estimated below, and possible schemes for plasma mirrors in KrF systems are given.

2. Experimental arrangement

Our short-pulse KrF laser system is based on a double discharge EMG150 Lambda-Physik excimer laser [6]. The XeCl laser pumps a dye-laser chain which - after frequency conversion – provides the seed-pulse for the KrF amplifier. A half-wave plate was inserted after the frequency conversion in order to change the polarization of the beam. In order to reach the highest reflectivity, s-polarized beam was used for the reflection experiments. The beam has two off-axis passes in the preamplifier, then - after a vacuum pinhole spatial filter – two additional off-axis passes through the main amplifier. This system gives 70mJ pulses of 620±10fs duration in a nearly diffraction-limited beam. The beam can be focused with an off-axis parabola mirror to a spot of 2μm diameter[7] reaching ~10^{18}W/cm^2 intensity.

In the present investigations the laser beam was focused by an F/10 lens onto an antireflexion-coated glass-plate. Reflection experiments were carried out both using 45° and 12.4° angle of incidence. The target was moved by stepping motors shot-by-shot. The laser energy was kept constant, the intensity was varied by shifting the lens relative to the target, which offered more than 4 orders of magnitude intensity range, from less than 10^{12}W/cm^2 up to above 10^{16}W/cm^2. The energy was monitored for each shot and the reflected radiation was measured by calibrated photodiodes. The diode signal was integrated by a peak-hold detector and digitized, controlled by a microprocessor. Fiber-based communication provided the isolation from electronic noises.

In addition, VUV spectroscopy was carried out using a Jobin-Yvon toroidal grating. The toroidal grating of 550 lines/mm imaged the target onto an MCP detector equipped with a phosphor screen. The visible light of the screen was imaged to a CCD camera. For the spectral investigations the glass target plates were coated by a low-Z material, (i.e. by LiF), spectra were obtained in the 10-20nm spectral range with~0.1nm resolution.

3. Results and discussion

Figure 1. illustrates the intensity dependence of the reflectivity for 45° angle of incidence. The reflectivity is less than 1% from the AR-coated target below the plasma threshold of ~10^{12}W/cm^2. Above the plasma threshold the increase is nearly logarithmic up to 30-35% at ~10^{14}W/cm^2 while the scatter of the data increases significantly with increasing intensity. For higher intensities (above 10^{15}W/cm^2) the reflectivity even decreases with strong shot-to-shot variations. The saturation behavior is similar to the observations of Ziener et al.[2] for the infrared radiation but the maximum observed reflection is significantly lower. Thus, the experiment clearly demonstrates that the plasma mirror effect is present for the KrF radiation as well.

Even higher reflectivity was reached when the angle of incidence was nearly perpendicular, i.e. 12.4° as shown in Fig. 2. Here again we can observe the logarithmic increase of reflectivity at low intensities, and a saturation near 10^{14}W/cm^2 laser intensity. The saturation behavior is similar to the observations with infrared radiation [2] but the maximum reflectivity is below the one observed therein. It was found to be a little above 30% for 45° angle of incidence whereas it exceeds 40% (in some shots even 50%) for 12.4°. Thus it is still well below the 70-80% obtainable reflectivity for the infrared radiation and it is also less than that obtained with KrF laser pulses of shorter duration [8].

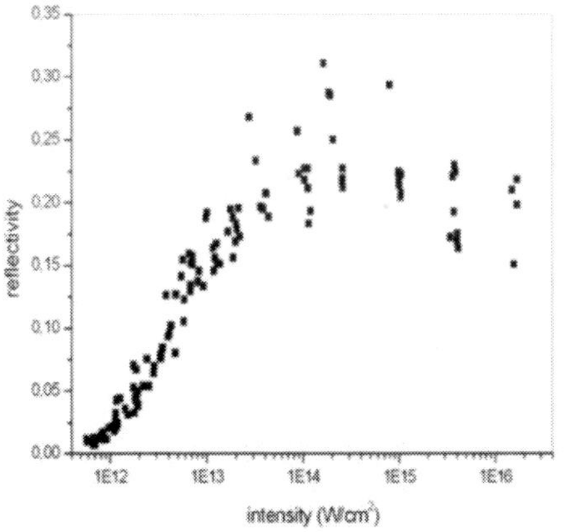

Figure 1. Intensity dependence of the reflectivity for 45° angle of incidence.

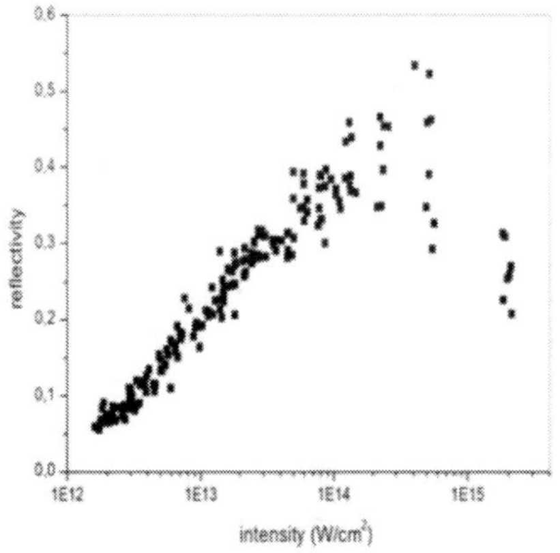

Figure 2. Intensity dependence of the reflectivity for 12.4° angle of incidence.

In order to estimate the optimum plasma parameters spectroscopic investigations were carried out in the VUV. The spectra at 10^{14}W/cm^2 intensity show strong H-like and He-like radiation from Li, whereas the B- and Be-like F-lines appear between 14 and 18 nm. Even Li-like F-lines can be observed at wavelengths shorter than that of the Lyman-α radiation of the Li ions, referring to temperatures of 20-25eV [9].

The observation that the reflectivity was lower in our case with 620fs laser pulses than with the earlier ones using 250fs duration is not surprising. The larger plasma in the present case simply leads to the enhancement of absorption. The decrease of reflectivity above the optimum conditions is a probable consequence of the initiation of nonlinear phenomena as high harmonics generation [10]. These effects reduce the absorption and they also contribute to the larger shot-to-shot variation of the measured reflectivity, which deteriorate the quality of the reflected beam, too.

4. Conclusions

Plasma mirror effect was demonstrated for ultrashort KrF laser pulses. The reflectivity is shown to increase above the plasma threshold logarithmically and it saturates below 10^{14}W/cm^2 when the plasma temperature is ~25eV. The measured maximum reflectivity was found to be ~33% for 45°angle of incidence, and it was nearly 50% for 12.4°. Although this reflectivity is lower than that for 800 nm radiation, these results show that plasma mirrors can be applied even directly for increasing the contrast of short-pulse high-power KrF systems. For a pulse of ~100fs [6] higher reflectivity is expected.

There is however another possibility for its application using the properties of short pulse amplification of KrF systems. We can use the fact that KrF amplifiers work in saturation regime. Consequently, if one applies the plasma mirror before the final amplifier the output energy will not be significantly reduced, whereas only the pedestal of the ASE from the final amplifier will remain. This ASE does not give a significant contribution in the focus because - due to its larger divergence - it will remain practically unfocused in the focal plane.

A further advantage of this setup can be utilized if the plasma mirror is positioned in the focal plane. This not only relaxes the energy requirements but due to the diffraction limited property of the KrF laser beam - the plasma mirror will not deteriorate the beam quality. There is no structure within the diffraction limited focal spot, and thus the plasma mirror acts as a secondary source of radiation operating as a spatial filter [11]. Such a scheme is capable for cleaning the laser beam from the Fourier-components with higher spatial frequency, thus from the spatial inhomogeneities of the main pulse.

Acknowledgments
This work, supported by the European Communities under the contract of Association between EURATOM and the Hungarian Academy of Sciences, was carried out within the framework of the European Fusion Development Agreement. The views and opinions expressed herein do not necessarily reflect those of the European Commission. It was also supported by the Hungarian OTKA foundation Nr.60531 and by the IAEA CRP on Pathways to Inertial Fusion – An Integrated Approach, Research Contract No. HUN 13759. It was also a part of a COST activity, MP0601.

References

[1] Kapteyn H, Murnane M, Szoke A and Falcone R 1991. *Opt. Lett.* **16** 490-492
[2] Ziener Ch, Foster P S, Divall E J, Hooker C J, Hutchinson M H R, Langley A J and Neely D 2003 *J. Appl. Phys.* **93** 768
[3] Doumy G, Quéré F, Gobert O, Pendrix M, Martin P, Audebert P, Gauthier J C, Geindre J P and Wittmann T 2004 *Phys. Rev. E* **69** 026402
[4] Dromey B, Kar S, Bellei C, Carroll D C, Clarke R J, Green J S, Kneip S, Markey K, Nagel S R, Simpson P T, Willingale L, McKenna P, Neely D, Najmudin Z, Krushelnick K, Norreys P and Zepf M 2006 *Phys. Rev. Lett.* **99** 085001
[5] Földes I.B, Bakos J S, Gál K, Juhász Z, Kedves M Á, Kocsis G, Szatmári S and Veres G 2000 *Laser Physics* **10** 264-269
[6] Szatmári S 1994 *Appl. Phys. B* **58** 211-223
[7] Rácz E, Földes I B, Kocsis G, Veres G, Eidmann K, Szatmári S 2006 *Appl. Phys. B* **82** 13-18
[8] Fedosejevs R, Ottmann R, Sigel R, Kühnle G, Szatmári S and Schäfer F P 1990 *Appl. Phys B* **50** 79-99
[9] Salzmann D 1998 *Atomic Physics in Hot Plasma* Oxford University Press Oxford UK ISBN 0-19-510930-9
[10] Földes I B, Bakos J S, Veres G, Bakonyi Z, Nagy T, Szatmári S 1996 *IEEE J.Selected Topics in Quantum Electronics* **2** 776-781
[11] Szatmári S, Bakonyi Z, Simon P 1997 *Opt. Commun.* **134** 199-204

The Sixth International Conference on Inertial Fusion Sciences and Applications IOP Publishing
Journal of Physics: Conference Series **244** (2010) 032005 doi:10.1088/1742-6596/244/3/032005

PERFORMANCE MEASUREMENTS OF THE INJECTION LASER SYSTEM CONFIGURED FOR PICOSECOND SCALE ADVANCED RADIOGRAPHIC CAPABILITY

C. Haefner, J. E. Heebner, J. Dawson, S. Fochs, M. Shverdin, J.K. Crane, K. V. Kanz, J. Halpin, H. Phan, R. Sigurdsson, W. Brewer, J. Britten, G. Brunton, B. Clark, M. J. Messerly, J. D. Nissen, B. Shaw, R. Hackel, M. Hermann, G. Tietbohl, C. W. Siders and C.P.J. Barty

Lawrence Livermore National Laboratory
7000 East Avenue, Livermore, CA 94550, USA

Email: haefner2@llnl.gov

Abstract. We have characterized the Advanced Radiographic Capability injection laser system and demonstrated that it meets performance requirements for upcoming National Ignition Facility fusion experiments. Pulse compression was achieved with a scaled down replica of the meter-scale grating ARC compressor and sub-ps pulse duration was demonstrated at the Joule-level.

1. Introduction

Dynamic x-ray radiography of Inertial Confinement Fusion (ICF) implosions is a core requirement for diagnosing key experimental parameters in the National Ignition Campaign. The Advanced Radiographic Capability (ARC) upgrade project at the National Ignition Facility (NIF) is designed to deliver short-pulse kilo-Joule laser pulses for X-ray backlighting of NIF fusion capsules. When complete, eight Petawatt-class high-intensity beamlines with controllable delays will add the capability to record the dynamics of fuel compression and fusion ignition in a multi-frame X-ray "motion-picture"[1, 2]. ARC will also support a variety of other high energy density experiments and fast igniter studies on NIF. Full scale fast ignition on NIF will require high-energy (tens of kJ), short pulse duration (~2-10 ps), and small focal spot size (tens of microns).

2. ARC architecture

ARC is a Petawatt-class, chirped pulse amplification (CPA) system implementing eight, meter-scale four-grating compressors to deliver compressed pulses just before focusing into the NIF target chamber. ARC will operate in a split beam configuration, dividing each NIF aperture into two subapertures that fit on pairs of side-by-side, 91 cm wide, compressor gratings. This split-beam architecture produces eight, high-intensity, short pulses from a quad of NIF beamlines. The short pulse is generated in a mode-locked fiber oscillator, ensuring stability and robust operation in a compact geometry[3]. Pulses are stretched for CPA in a chirped-fiber-Bragg grating (CFBG), specially designed as the conjugate of the calculated dispersion of the overall laser system. After further amplification the beam is split into two, and launched into two small bulk grating pre-compressors, functioning as pulse width controllers (PWC). By adjusting the group velocity dispersion of each pulse, these PWCs will control the temporal pulse width at the final ARC compressor output from the sub-ps transform limit to 50 ps. The pulses are then routed by a transport fiber and amplified by a Yb-doped fiber up to 1 microJoule just prior to injection into bulk Nd:Glass regenerative amplifiers. To compensate for gain-narrowing, we added a birefringent, gain-flattening filter to the regen cavity, extending the bandwidth from 2.2 nm to 4.5 nm at the output. The two regen outputs are spatially shaped, apodized and combined as two ARC sub-apertures whose outer perimeter matches a NIF beam at the input of the 4-pass rod amplifier. The shaping is designed to pre-compensate the spatial gain profiles in the NIF rod and slab amplifiers to produce a pair of rectangular, flat-top, near-field profiles at the NIF amplifier chain output. The regen outputs are then combined and shaped into a pair of subapertures in a Split Beam Injection (SBI) module. The combined shaped beam is then further amplified in a 4-pass Nd:Glass rod amplifier, producing up to 6 Joules in a spatial format tailored for direct injection into the NIF main amplifiers. The main amplifiers will amplify each individual subaperture to ~1 kJ (dependent on pulse duration). Two vacuum vessels containing 4 compressors each compress the pulses; the output pulse duration is dependent on the PWC setting. Space constraints in the NIF target bay required a compact folded compressor design. There are actually four pairs of 4 grating compressors: one grating in the pair is illuminated by one ARC sub-aperture and the second grating of the pair is illuminated by the other sub-aperture. Each grating in the pair has slightly different groove densities to reduce crosstalk between the subaperture beams. Each of the compressors employs four different 91cm x 45 cm multi-layer dielectric diffraction gratings, made by LLNL[4].

Four major contributions to group velocity dispersion need to be carefully balanced: Material dispersion of several tens of meters of glass, CFBG, PWC, and the compressor (Fig. 1 left). Since the system does not allow adjustment of the dispersion among the four different NIF apertures, alignment of the 8 ARC compressors has to be performed such that the overall residual group delays are less than 1 ps across all subapertures to meet specifications. A radio-frequency (RF) phase-shifting technique will be used to precisely measure the group delay of each individual subsystem and to match the dispersion characteristics of each individual ARC compressor before shots are fired.

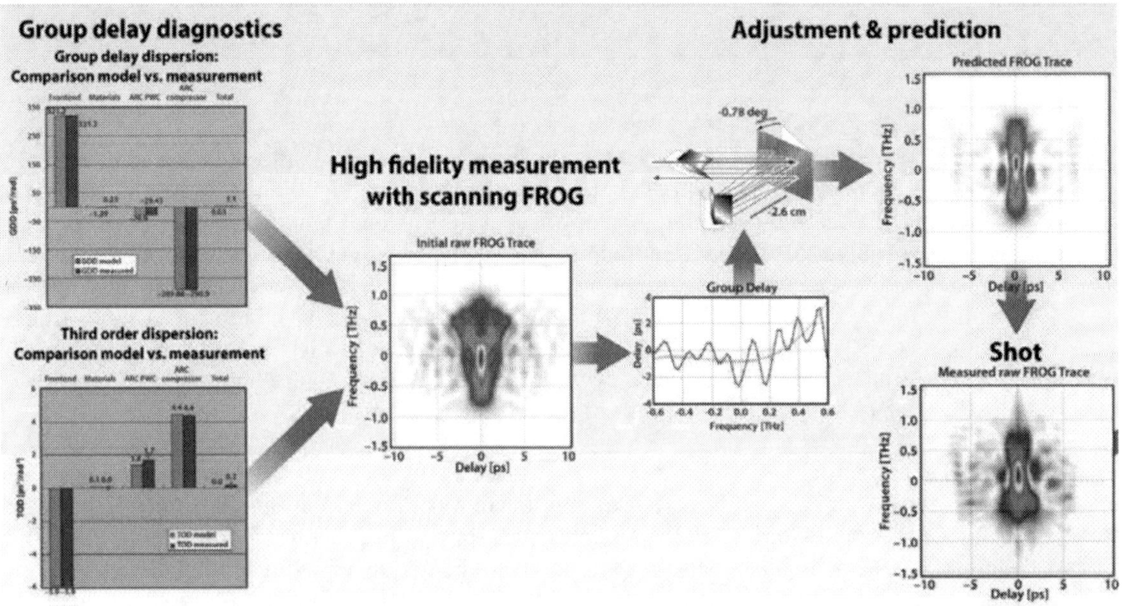

Figure 1: Strategy for balancing dispersion on the ARC testbed: offline GD-measurements are compared with the model, and dispersion in the full system is compensated to better than 1 ps. A high fidelity scanning FROG reveals residual GD, and by minimizing RMS GD deviations adjustments are made to the PWC accordingly.

3. ARC frontend test facility

The performance of ARC relies critically on the front-end performance and the dispersion pre-compensation in the system. We have built and qualified the ARC front-end prototype (up to the 4-pass Joule level amplifier) in an offline facility. We use this test facility to demonstrate dispersion balance, measure temporal fidelity, and characterize the near- and far-field spatial profiles for the compressed ARC front end. We also use this facility to provide real data for our propagation and energetics models that predict overall performance of the ARC system. In order to test compressed pulses, we constructed a scaled down compressor designed with the same calculated dispersion values to demonstrate our dispersion management scheme. After installation, we measured the group delay (GD) dispersion of the compressor using the phase shifting technique described in [5, 6]. Earlier measured GD-data of the fiber front was used to design the PWC. We also test various temporal diagnostics on this test facility that will be used for recording shot data at the ARC output. The temporal diagnostics include an ultra low-energy (picoJoule), scanning cross-correlator, a single-shot FROG (frequency resolved optical gating) device allowing detection of pulses up to 5 ps in duration, and a homebuilt, multi-shot scanning FROG with ultra-wide temporal window (several tens of ps and true ~1:400 dynamic range). We also record for each shot a high resolution spectrum to validate our results.

4. Results

Using the birefringent intracavity filter in the regen, the spectral width of the laser output was optimized. Gain narrowing in the pre-amplifier module using a LHG-8 rod is compensated by broadening and shifting the regen spectrum to the blue. The spectrum was centered at 1052.6 nm and

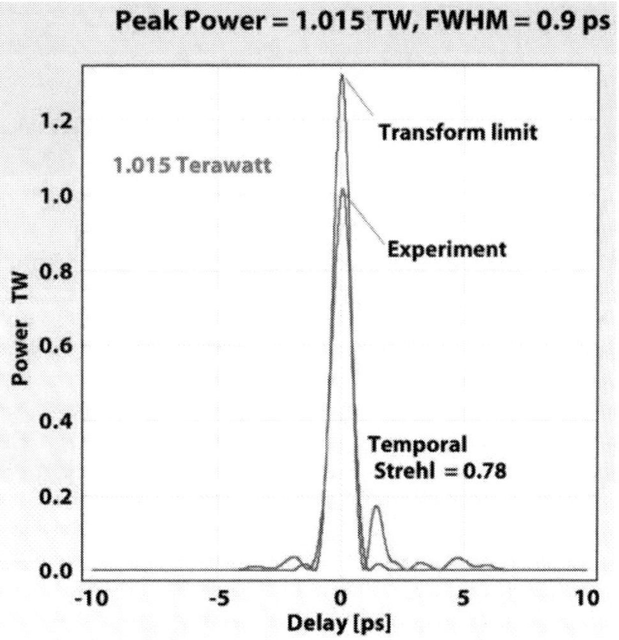

Figure 2: Power measurement derived from energy and single shot FROG diagnostics at the compressor output in the ARC Injection Laser Test Facility.

broadened up to 4.5 nm with super-gaussian shape. Figure 1 shows our strategy for balancing dispersion in the ARC injection laser test bed: We set the PWC grating angle of incidence (AOI), and slant distance to the nominal values predicted by our group delay diagnostic measurements and dispersion balance model. Small adjustments to the PWC and compressor alignment are made to be in good agreement with modeled dispersion values. The total residual group delay derived from the group delay measurements is less than 1 ps. Then we measured the compressor output pulse width with a single-shot autocorrelator while optimizing the PWC slant distance for the shortest autocorrelation. A change of ~2mm in slant distance from the model prediction, equivalent to ~ 0.4 ps^2/rad, was necessary to produce the shortest pulse on the autocorrelator. Next we made a high dynamic range, multishot FROG measurement, and performed a high resolution, precision FROG analysis based on principal components generalized projections[7] yielding $E(t, \lambda)$. We mimimized the RMS deviations of the GD across the spectrum using the angle of incidence and slant distance of an analytic Treacy formalism[8] as parameters (see Fig.1). The obtained values were applied to the PWC and the measurement repeated. Fig. 1 shows also the predicted FROG trace with the change in dispersion based on the Treacy model for the PWC. After changing angle of incidence and slant distance of the PWC according to the analysis a nearly transform limited pulse duration of 814 fs was measured with the scanning FROG. Then the system was configured for Joule-level amplification and the diagnostics configured for single shot. We fired a shot at 1.3 J and recorded the temporal pulse characteristics with single shot FROG. The trace is shown on the right side in Fig. 1. Figure 2 shows the pulse measurement derived from energy and temporal diagnostics, yielding 1.02 TW in peak power. The slight increase in pulse width from 814 fs to 898 fs is expected and attributed to gain narrowing in the 4-pass Nd:Glass rod amplifier. The remaining small pre-pulse ~1.3 ps ahead of the main pulse is attributed to the group delay ripples in the CFBG and does not affect the overall performance of ARC.

Acknowledgments
This work was performed under the auspices of the U.S. Department of Energy by Lawrence Livermore National Laboratory under contract DE-AC52-07NA27344.

References

[1] C. P. J. Barty *et al.*, Nuclear Fusion **44**, S266 (2004)
[2] C. A. Haynam *et al.*, Applied Optics **46**, 3276 (2007)
[3] J. W. Dawson *et al.*, IEEE Journal of Selected Topics in Quantum Electronics **15**, 207 (2009)
[4] J. A. Britten *et al.*, 2005 Quantum Electronics and Laser Science Conference (QELS) Baltimore, (2005)
[5] J. K. Crane *et al.*, in Conference on Lasers and Electro-Optics (CLEO), San Jose, (2008)
[6] C. Haefner *et al.*, in Conference on Lasers and Electro-Optics (CLEO), Baltimore, (2009)
[7] D. J. Kane, Journal of the Optical Society of America B-Optical Physics **25**, A120 (2008)
[8] E. Treacy, IEEE Journal of Quantum Electronics **QE-5**, 454 (1969)

The Vulcan 10 PW Project

C Hernandez-Gomez , S P Blake, O Chekhlov, R J Clarke, A M Dunne, M Galimberti, S Hancock, R Heathcote, P Holligan, A Lyachev, P Matousek, I O Musgrave, D Neely, P A Norreys, I Ross, Y Tang, T B Winstone, B E Wyborn and J Collier

Central Laser Facility, Science and Technology Facilities Council, Chilton, Didcot Oxon, OX11 0QX, United Kingdom

Cristina.hernandez-gomez@stfc.ac.uk

Abstract. The aim of this project is to establish a 10 PW facility on the Vulcan laser system capable of being focussed to intensities of at least 10^{23} Wcm^{-2} and integrate this into a flexible and unique user facility This paper will present progress made in Phase one developing the 10PW Front End as well as the concept for the new Vulcan 10 PW facility. The new facility will be configured in a unique way to maximise the scientific opportunities presented through a combination with the existing capabilities already established on Vulcan. This ground breaking development will open up a range of new scientific opportunities.

1. Introduction

The Vulcan 10 PW project will upgrade the Vulcan laser, located at the Central Laser Facility (CLF) to beyond the 10 PW power level and provide focussed intensities of greater than 10^{23} Wcm^{-2} to its international user community. This will be achieved by generating pulses with energies of 300J and with durations less than 30fs using the Optical Parametric Chirped Pulse Amplification (OPCPA)[1] technique. The generation of a high energy and ultra short laser pulses using the OPCPA concept offers high broadband gain with good contrast. This technique opens an affordable way of harnessing the high energy produced by laser systems such as Vulcan into multi-Petawatt ultra-short pulse systems.

Over the last decade the CLF through a number of research projects has demonstrated the scaling of the OPCPA technique from the table top to the Petawatt regime [2] to the large aperture demonstration [3] of pulse amplification up to 35J, with 70 nm bandwidth and 80 fs duration.

The Vulcan 10 PW design is based on a triple OPCPA amplification scheme with initial amplification occurring in the picosecond domain and subsequent amplification occurring in the nanosecond domain to guarantee a high contrast pulse. We divided the project into two Phases. The objective of phase one was to develop a novel Front End - sub 30 fs pulses at 910 nm with 1 J of energy that will seed the subsequent amplification stages to reach the required output energy. In addition another important aspect of phase one was to develop the concept design of the 10 PW facility and identify the technology that would need to be developed in advance to minimise any risks during phase 2, such as grating technology.

2. Methodology

The proposed scheme for the Vulcan 10 PW facility is shown in figure 1. The Front End comprises of the mJ broadband seed, a stretcher and an OPCPA J level amplifier based on LBO crystals. The sub-Joule output from the Front End system will enter an OPCPA amplification system, based on two large aperture deuterated Potassium Di-Hydrogen Phosphate (DKDP) crystals for amplification to the 500 J level. These two crystals will be pumped by long pulse beams from Vulcan. To achieve this level of energy two Vulcan high energy beam lines will be upgraded through the installation of additional 208 mm diameter amplifier chains. The amplified output beam will then be compressed in time and directed to different areas to be focussed between 10^{22} Wcm^{-2} and above 10^{23} Wcm^{-2}.

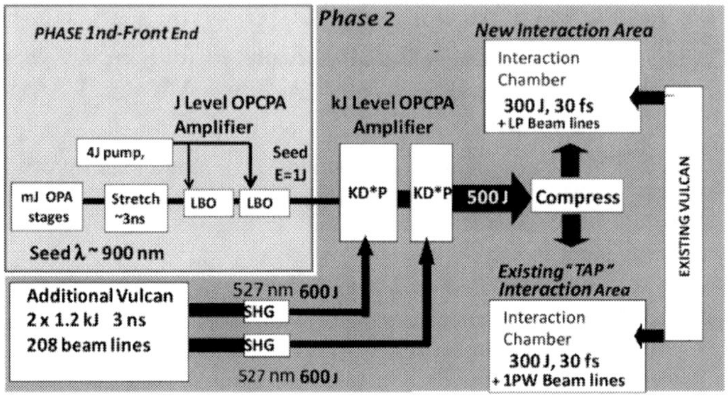

Figure 1. Schematic of the Vulcan 10 PW Facility

3. The 10 PW Front End

A novel Front End has been developed during phase one to seed the remainder of the system that will be built as part of phase 2. It currently delivers ultra short pulses at 910 nm, with energies of up to 0.4 Joule and with sufficient bandwidth to support a pulse with a duration of 30fs. The development of a broadband seed at 900nm [4] with a 150nm bandwidth specification has been key element of this system. The theoretical calculations show that the OPCPA scheme described above can support more than 150nm. Using this bandwidth is advantageous as the seed needs to be stretched to 3 ns pulse for amplification in the large aperture OPCPA stages.

The 910 nm seed currently generated with a chirped scheme has a bandwidth of 165 nm and energy of 40 μJ per pulse. This seed is then stretched to 1.87 ns FWHM on a test stretcher and amplified in two LBO crystals pumped by a commercial temporally shaped pump laser. The design of the Joule stage in terms of contrast, gains and energy were based on a millijoule of seed energy at the input. As the input seed energy is currently limited to 1~ μJ due to stretcher inefficiencies, additional gain over that originally anticipated has been provided in the first stage by placing two crystals consecutively. In this way the un-depleted pump energy from the first crystal is recycled to extract gain from the second one In using this scheme the output energy achieved so far is >15 mJ for stages 1 and 2 and 0.4 J for stage 3. We have demonstrated an amplified bandwidth of 100 nm at the output. There is a spectral loss of the input seed due to the limitation by the gratings size of the stretcher currently limiting the output bandwidth. Further loss in spectra of the amplified pulse is caused by limitation of currently used broadband mirrors .The evolution of the amplified pulse spectrum on the J level OPCPA stages is shown in figure 2.

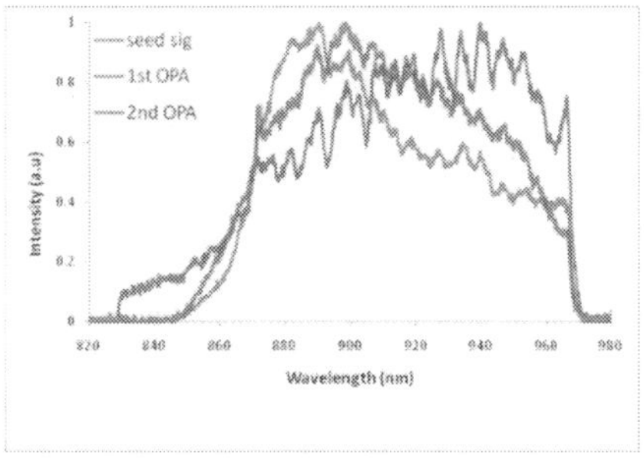

Figure 2. Spectra of the Front End system at different points of amplification

The output beam from the Joule level OPA stages was directed to a test compressor and diagnostics suite for characterisation. We were able to measure the pulse length of the compressed pulse down to shorter than 30 fs using a commercial device.

To achieve the contrast specification of the Front End it is necessary to increase the output pulse energy of the mJ-stage OPA on ps time scale significantly from its current ~40μJ to ~1mJ. This will allow the first stage to operate with a single crystal as originally designed, work is currently on going to increase the energy of the broadband seed to the mJ level and to characterise the contrast.

4. Large aperture OPCPA amplifiers

The high energy laser chain used for the pump generation and OPCPA stages have been designed and modelled to a level that ensures that the design meets the technical specifications required to deliver 300J in 30 fs. The laser chain required to pump the OPCPA stages will be a combination of the existing Vulcan Nd: glass amplifier chain and a new set of amplifiers.

The additional laser amplifier chain has been designed using a minimum number of amplifiers in a double pass configuration using angular multiplexing. A long pulse shaping system is required in order to ensure that the correct temporal shape of the pump pulse is achieved. The current Vulcan laser is composed of a series of small aperture rod amplifiers and large aperture circular disc amplifiers. To provide an optimised geometry for the OPCPA amplification and particularly pulse compression, square pump beams are required. The design requires the final 208 amplifiers with square slabs (as opposed to the current elliptical ones) which is a new type of technology to the one currently being used in Vulcan. As part the first phase of the project we have developed the technology to build this type of amplifiers.

There are two large aperture OPCPA stages to amplify the pulse from 1 J to 500J. We have performed 2D simulations on the amplification of the large aperture DKDP stages. An input signal of 1J centered at 910 nm with $\Delta\lambda$=150nm stretched pulse duration of 3ns is amplified in 2 DKPD crystals 220mm square pumped by pump pulses generated by frequency doubling the Vulcan beams to 600J at λ=527nm ,Δt=3ns flat top. The theoretical model shows output energy prior to compressor of 592J over the entire bandwidth.

5. Compressor and grating technology

The compressor design is based on a single pass four grating scheme. The compressor gratings need a line density of between 900 l/mm and 1200 l/mm. Calculations performed indicate that the optimum line density for the gratings of the large aperture compressor for the bandwidth we intend to use is 900 l/mm. The grating should be efficient supporting 150 nm required pulse spectral bandwidth centred around 910 nm .These low line density gratings are not readily available at the large aperture required from any manufacturer. We theoretically modelled the performance of the gratings under different configurations to establish the optimum geometrical configuration to use in the compressor. The calculations of expected diffraction efficiency from 900 l/mm gold gratings at S-polarisation for the required spectral region show very good efficiency of over 95 % within our pulse spectrum.

In addition working with the Plymouth Grating Laboratory (PGL) to investigate the theoretical performance over a large bandwidth of gold and dielectric gratings at 900-1100 l/mm, and to establish the feasibility to have these manufactured at large diameters apertures. Following these studies a second contract was placed with PGL to actually manufacture small grating samples in order for us to test their performance. We have received two sets of grating samples from the PGL over last year. First, gold gratings of binary type with different line densities with groove densities 900 l/mm, 1000 l/mm and 1100 l/mm. The second set were gold and silver gratings with 900 l/mm groove densities of Sinusoidal and Binary types. The tests showed that both silver and gold gratings of the binary type operating at Littrow out of plane configuration have quite uniform diffraction efficiency above 90%.

6. Facility

As part of phase one of this project an overall design for the new facility that will be built during phase two was developed. A significant extension to the Vulcan building infrastructure is required to accommodate the large amount of new equipment. The Vulcan laser facility comprises of laser hall, front end room, control rooms, capacitor bank and three target areas. The requirements of the new facility are to house a new Front end, new laser amplification chain and compressor and extend two of the existing target areas with the appropriate shielding.

The proposed new building requires the existing building to be extended on the South and West sides and a second floor built in above the two of the current Target Areas, Target Area West (TAW) and Target Area East (TAE). The upper floor would then house the additional laser chain, OPCPA amplifiers and compressor. TAE, traditionally used for low intensity experiments, would be transformed to become a dedicated *High Intensity* 10 PW area (HIA). The existing Target Area Petawatt (TAP) would be enhanced to become a combined 1+10 PW area (TAP10). Both areas HIA and TAP10 will be shielded appropriately to provide adequate radiation protection facility.

References

[1] I. N. Ross et al, Opt. Commun. 144, 125–133 (1997)
[2] J.L.Collier et al, IFSA 2003, American Nuclear Society, Pub: Elsevier, 603-607, (2004)
[3] J Collier *et al,* Applied Optics, 38, No. 36, (1999) 7486
[4] Y Tang et al, Optics Letters, 33, No20, 2386-2388 (2008)

Interferometric adaptive optics for high-power laser beam correction in fast ignition experiments

D. Homoelle, K. L. Baker, P. K. Patel, E. Utterback, M. C. Rushford, C. W. Siders, C. P. J. Barty

Lawrence Livermore National Laboratory, 7000 East Avenue, Livermore, CA 94550, USA

E-mail: homoelle1@llnl.gov

Abstract. We present the design for a high-speed adaptive optics system that will be used to achieve the necessary laser pointing and beam-quality performance for initial fast-ignition coupling experiments. This design makes use of a 32x32 pixellated MEMS device as the adaptive optic and a two-channel interferometer as the wave-front sensor. We present results from a system testbed that demonstrates improvement of the Strehl ratio from 0.09 to 0.61 and stabilization of beam pointing from ~75μrad to <2μrad.

1. Introduction

With the recent completion of the National Ignition Facility (NIF) at Lawrence Livermore National Laboratory, an experimental facility now exists which is capable of conducting the first full-scale fast-ignition (FI) experiments. As the first step toward this goal, a near-term experimental campaign is being planned to quantify the total fast-ignition drive-energy requirements.

This experiment will make us of the Advanced Radiographic Capability (ARC) diagnostic system which is currently under construction at the NIF. This system uses one quad of the NIF amplifier chain to generate high-peak-power pulses, with pulse widths ranging from 1 to 50 picoseconds, and total quad energies ranging from ~6 to ~20 kJ, limited by the damage threshold of the final optics. As shown in Fig. 1(a), each beam in the quad is split vertically in two to produce four split-pairs. The gap in each beam allows for the use of eight independently-timed laser beams for the ARC diagnostic. Each half of the beam uses separate compressor optics.

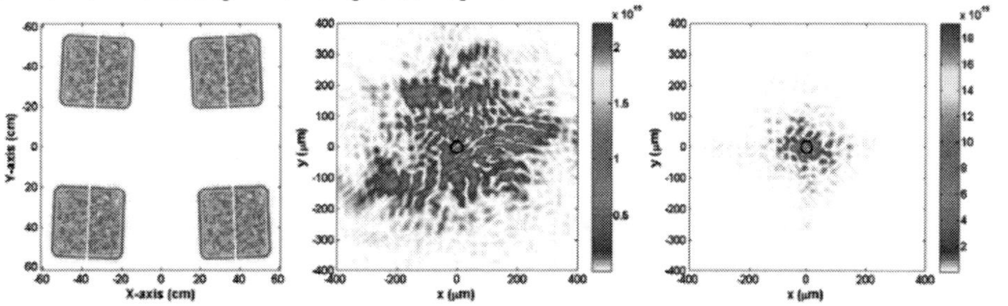

(a) (b) (c)

Figure 1. (a) Near-field intensity of ARC quad, (b) simulated instance of the far-field intensity for an uncorrected ARC quad, and (c) simulated far-field intensity of the ARC beam with proposed correction. The black circle in (b) and (c) indicates the 40-μm cone tip.

The goal of the first sub-ignition-scale experiments is to measure the laser-to-electron energy coupling. Currently this conversion efficiency has a large uncertainty, with the value expected to be anywhere from 10-40% [1]. In order to plan for the full-scale FI experiments, this parameter must be known to greater precision. Hydrodynamic and particle-in-cell simulations have indicated that for these initial FI coupling experiments, the ARC quad of beams will need to deliver 3-4 kJ of total energy to within a 40-50μm diameter spot at the end of a FI cone target. The laser parameters for the current ARC configuration and those necessary to achieve the coupling experiment goals are listed in Table 1.

Requirement	Expected ARC Performance	Campaign I FI Electron Coupling Demo
Energy in 40-50μm diameter	< 300J	3-4 kJ
Intensity (W/cm²)	~1.0E+18	1.0E+20
Pointing Stability	±75 μm	±10 μm
Pulse Length	1-50 ps	5 ps

Table 1. Laser requirements for the fast ignition campaign.

High-speed measurements taken at 1.3 kHz on one of the NIF beamlines indicated that the RMS spot-displacement from target chamber center (TCC) was <17μm over a 12-second period, and that peaks in the vibration spectrum occurred between 20 and 60 Hz [2]. Because of the additional optics needed in the ARC beamline for compression, the pointing RMS is expected to be $\geq$ 75μm, and the Strehl ratio of a single beam of the quad is expected to be anywhere from 0.09 to 0.36. To achieve the laser requirements for the fast ignition experiments a system must be in place to improve laser pointing and beam quality. We have proposed using an interferometric adaptive optics system to achieve the laser pointing and beam-quality requirements on the ARC beamline [3].

2. Adaptive Optics System Design

Because the capability to do fast-ignition experiments was not included as part of the original design for NIF and ARC, the architecture of these existing systems limited the design space for the adaptive optics (AO) upgrade. We chose to use an interferometric adaptive optics system based on a previously demonstrated system at LLNL[4] because it can be incorporated into the existing system architecture. The interferometric adaptive optics system consists of a wavefront sensor (WFS) camera, a MEMS device which consists of a 32x32 array of piston-only, 300-μm-pitch actuators, and a single-longitudinal-mode laser which serves as both the reference in the interferometer arm and the point source at target chamber center. The system operates as follows:

1. Just before the shot, a ~10nJ, nanosecond, narrow-linewidth laser propagates from an angle-cleaved fiber optic placed across the fast-ignition cone opening.
2. The laser back-propagates through all eight of the ARC split-beam beamlines to four separate phase conjugation engines (PCE's) at Relay Plane 10 (RP10) on NIF.
3. At the PCE, shot-noise-limited coherent-detection is used by interfering the returning laser pulse with a local oscillator, which is the same illumination laser used in step 1. The shot-noise limit is important because of the high-losses in the NIF amplifier chain (~60dB) that are necessary for reducing laser backscatter.
4. The two-channel interferogram (a sine and cosine channel is measured) is processed to yield the wavefront (tip/tilt and higher-order wavefront distortion) characterizing the beam path.
5. The wavefront correction & electronic offset (e.g. pointing shift) is applied to the MEMS device and tip-tilt mirror.
6. The fast ignition laser pulse samples the MEMS and is pre-corrected for RP10-to-Target aberrations.

Because we plan to use a pixellated MEMS device, phase-unwrapping will not be necessary, which minimizes the calculation requirements and allows for higher-speed operation. Furthermore, because the point source will be attached to the target, any small motions in the target will be accurately tracked by the system. This is important for hitting the target to within 10μm (given the focal length of the ARC beamline, this translates to a ~1μrad pointing requirement).

A scaled testbed has been assembled to test the performance of the proposed adaptive optics system before implementation on the ARC beamline. The goal of the testbed is to measure the system performance as characterized by a handful of parameters: pointing accuracy, pointing correction bandwidth, residual-wavefront-error as a function turbulence-strength and spatial- and temporal-frequencies, and the ability to phase the two halves of a split beam.

The degree to which the wavefront must be corrected has been estimated with the aid of Monte Carlo simulations. We stochastically varied the beam pointing, intrabeam phasing, and the phase for each realization. Each random phase realization had fixed r.m.s amplitude and a structure function similar to that of a measured NIF phase aberration. These simulations suggest that with a pointing error of 10μm and λ/10 phasing between shared beam apertures, the single-beam residual r.m.s wave-front error must be < λ/10 in order to have a significant probability of achieving >3kJ in a 50-μm bucket. This corresponds to a single-beam Strehl ratio *S*~0.65. These results are shown in Fig. 2.

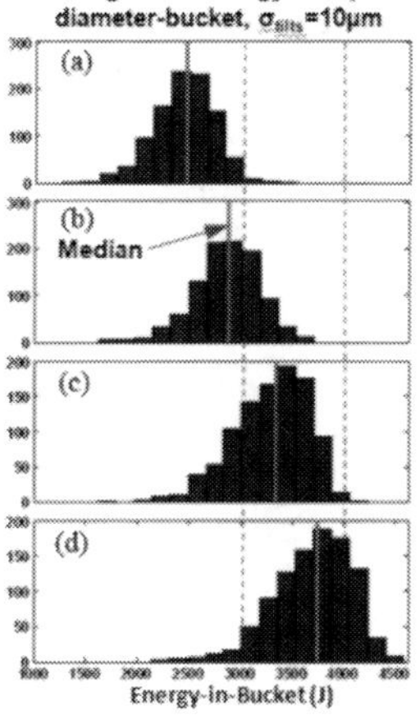

Figure 2. Histogram of energy in a 50μm bucket with a pointing uncertainty of ±10μm for a r.m.s. residual phase of (a) ~λ/7, (b) ~λ/8, (c)~λ/10, and (d) ~λ/12. The 3-kJ and 4-kJ points are indicated by the dotted line, and the median value of the distribution is indicated with the solid red line.

3. Testbed Results

Because significant improvements to the ARC performance can be made by pointing-only improvements (no wavefront correction), we first tested the ability of the adaptive optics system to correct for pointing-only errors. We did this by using two fast-steering mirrors, each placed in a relay plane of the final focusing objective, one to induce a randomly-generated pointing jitter on the beam, and the second to correct for this disturbance. Because the wavefront sensing system is sensitive to waves of tilt across the aperture and not absolute pointing angle in the far-field, achieving pointing stability to less than 9.3μrad in the lab corresponds to pointing stability of less than 1μrad on NIF. Similarly, 75μm of tilt at NIF target chamber center corresponds to an angle-error of σ_{ARC} =8.4 μrad on the ARC beamline which corresponds to σ_{tilts} ~78 μrad in the lab. The timing of the lab has also been scaled. It is expected that dedicated hardware will be required to run the system on NIF at ~1000 Hz, but for flexibility the lab demonstration is operated through software which limits the repetition rate of the system to ~100 Hz.

We compared system performance for five different tilt-calculation algorithms: one used the Goldstein algorithm [5] to unwrap the phase and extracts the tilts by a least-squares fit of a plane to the phase, one uses a simple technique of averaging the tilts between neighbouring pixels in the phase calculation, and the remaining three determine the far-field focal spot position by taking the FFT of the measured phase for three different resolutions. The results for one particular test are shown in Fig.

3. In this test the measurements were taken and the corrections were applied at 100 Hz, and the bandwidth of the applied jitter was 0.2Hz (a) and 2.0Hz (c). Three of the five algorithms achieved the fast ignition requirement that $\sigma_{tilts} < 9.3$ μrad. It is not yet known what the jitter spectrum will be on the ARC beamline, but these results indicate that a 1-kHz system on the ARC beamline could sufficiently correct for a 10-Hz vibration with a 8.4-μrad amplitude.

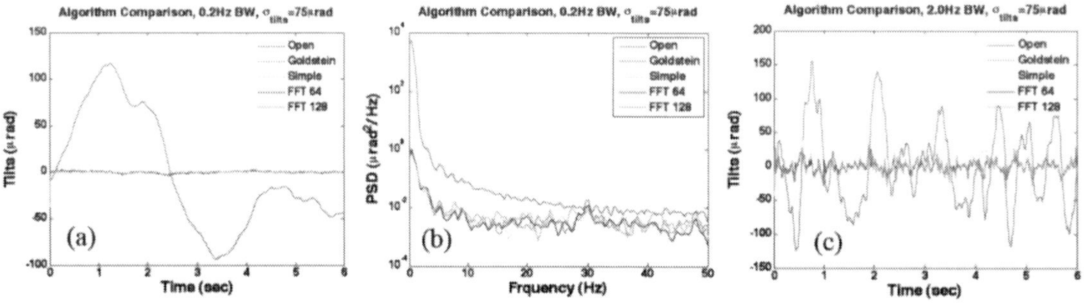

Figure 3. (a) Residual tilts for 5 algorithms and 0.2Hz bandwidth, 75μrad disturbance, compared to open loop, (b) PSD of measured tilts, and (c) residual tilts for 2.0Hz bandwidth disturbance.

Preliminary tests have also been performed using the 32x32 MEMS device to correct for a static phase disturbance. In these tests a static phase plate was placed in the beam, the two-channel interferogram was measured, and the wrapped phase correction was applied to the beam. Figure 5(a) shows the strength of the phase disturbance, and the RMS of the phase is σ_{phase}=4.1 radians. Figure 5(b) shows the far-field spot when no phase correction is applied to the MEMS device, and the resulting Strehl ratio is 0.09. Figure 5(c) shows the far-field spot with the MEMS correction applied and results in a Strehl ratio of 0.61. Even though this level of phase disturbance is greater than we expect to encounter on the ARC beamline, we still achieve a Strehl ratio in line with what is required to meet the fast ignition requirements. We are currently transitioning to a high-speed correction system and will at that point measure the dynamic properties of the phase correction system.

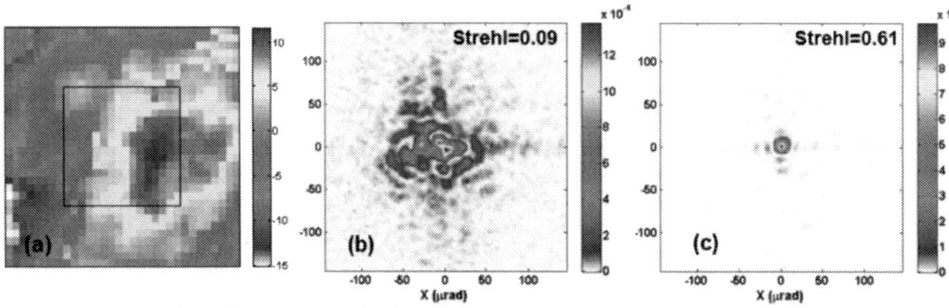

Figure 4. (a) Measured and unwrapped phase disturbance. The black box indicates the area of the MEMS that was used for this measurement, which has been restricted due to manufacturing turned-down-edge of the substrate. (b) Far-field of the uncorrected beam, (c) far-field of corrected beam.

This work performed under the auspices of the U.S. Department of Energy by Lawrence Livermore National Laboratory under Contract DE-AC52-07NA27344.

References

[1] Zuegel, J. D. *et al., Fusion Science and Technology*, **49** 453 (2006).
[2] Zacharias, R. A. *et al.*, Opt. Eng. **43**, 2873 (2004).
[3] Baker, K. L. *et al., Proc. SPIE*, **7209** 720907 (2009).
[4] Baker, K. L. *et al.*, Appl. Opt., **43**, 5585 (2004).
[5] Ghiglia, D. C., Pritt, M. D. *Two-Dimensional Phase Unwrapping*, John Wiley & Sons (1998).

Geometrical optimization of an ellipsoidal plasma mirror toward tight focusing of ultra-intense laser pulse

A.Kon[1], M. Nakatsutsumi[1,2], S. Buffechoux[2], Z. L. Chen[1,4], J. Fuchs[2], Z.Jin[3], R. Kodama[1,3,4]

[1] Graduate School of Engineering, Osaka University, 2-1 Yamada-oka, Suita, Osaka 565-0871, Japan

[2] Laboratoire pour l'Utilisation des Lasers Intenses, UMR 7605 CNRS-Ecole Polytechnique-CEA-Univ. Paris VI Palaiseau, France

[3] Photon Pioneers Center, 2-1 Yamada-Oka, Suita, Osaka 565-0871, Japan

[4] Japan Science and Technology CREST

Email; konakira@ile.osaka-u.ac.jp

ABSTRACT We developed for the first time, very compact (<1 cm^3) extremely low f-number (f/# = 0.4) confocal ellipsoid focusing systems. Direct measurement of the laser focal spot using a low-energy laser beam indicates 1/5 reduction of the spot size compared to standard focusing (using a f/2.7 optics). Such mirror is thus able to achieve significant enhancement of the focused laser intensity without modifying the laser system itself. The mirror is then used under plasma mirror regime which enables us to compactify the size, to liberate us from the anxiety of protecting the optics from target debris after shots, and to enhance the temporal contrast. In this paper, we focus our attention to designing and optimizing the geometry of such innovative plasma optics.

1. Introduction and motivation

With the rapid advances in laser technology, laser beams are now available that can be routinely focused to intensities approaching >10^{21} W/cm^2. There are already ongoing projects worldwide aiming to reach and even exceed the PetaWatt power level. A host of potential applications are proposed ranging from fast ignition of compressed inertial fusion targets, ion and electron acceleration, and diagnostics for high energy density plasma science. At very high intensity levels, such as $I_{Laser} > 10^{22}$ W/cm^2, physical processes are presently poorly understood but a wide area of potential applications is proposed. There are several methods to bring the intensity of light to such levels, for example, by increasing the input energy or shortening the pulse duration. An alternative method is to obtain a tight focus which is more effective because the intensity is proportional to the square of the inverse of the focal spot size. Recently, 0.8 μm focal spot size was achieved using small f-number off axis parabola focusing optic (f/0.6) [1]. However, the distance between the target and the focusing optics decreases with decreasing the f/#, so care must be taken to protect the mirror surface from debris induced by the exploded target flow after laser irradiation. One way to alternate the concern of damage only the optics is to use plasma-based optics, there so called "plasma mirror (PM)" [2] were proposed and investigated in detail [3].Such plasma-based mirror is single-use, thus concerns about debris will naturally eliminated. In addition, as plasmas are already ionized, the PM can be used under much higher laser energy fluence than solid optics where fluence is restricted by their damage threshold (~10^{11} W/cm^2), i.e. under at least 100-times higher fluence compare to the damage threshold of solid optics (typically 0.1-1 J.cm^{-2} for ~1ps laser pulse [4]). Thus, the optics size can be reduced by at least a factor of 10. Thus, the PM is more like laser experiment targets, rather than permanent mirrors. This change in thinking frees us from the anxiety of protecting the focusing optics from solid target debris after shots, thus enabling us to shorten dramatically the distance between the mirror surface and the target. In addition, a plasma mirror can enhance the temporal and spatial contrast ratio of the laser [5] [6] which improve the light pulse quality for many applications, such as the generation of high-order harmonics, directed electron beams, relativistic ions. Here, we developed for the first time, a plasma-based extremely small f-number focusing mirror to enhance the laser intensity through reduction of the spot size. In previous study, plasma mirrors are applied to only enhancement of temporal contrast of the laser (i.e., an improvement of the light pulse quality that is absolutely crucial for solid-interaction applications, such as the generation of high-order harmonics [6], ion acceleration [7], etc.). In this paper, we concentrate on the design of such extremely low f/# plasma-focusing mirror and demonstration of focal spot reduction by using low energy laser. Around 10-fold enhancement of the laser intensity was obtained under plasma-mirror condition in high-energy laser experiment.

2. Design of the mirror, experiment and discussion

2.1. Design of the ellipsoid focusing mirror

Our aim was to design a plasma-based extremely small low-f/# focusing mirror. We chose an ellipsoid (ellipsoid of revolution) for our focusing mirror, rather than parabolic or hyperboloid, because the ellipsoid was point-to-point imaging optics, and strict alignment accuracy was not required to create the image free from astigmatism. In addition, as there are two foci, the ellipsoid mirror can be simply placed after the focal point of the OAP, with one of the foci of the ellipsoid coinciding with the OAP focus. We optimized its shape for the purpose of enhancing the laser intensity through reduction of the focal spot size. The ellipsoid can be expressed as $x^2/a^2+y^2/b^2+z^2/b^2=1$ (a>b). Here a and b represent the major- and minor-axis, respectively. As expressed in Fig.1(a), ρ represents the distance between the first focus of the ellipsoid (here, the focal point of the OAP) and the cross point between the chief ray and the ellipse, and r' is the distance between the cross point and the secondary focus (the other focus).

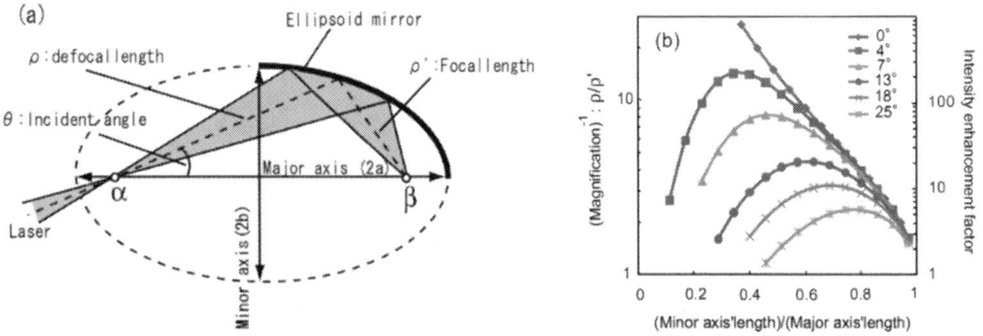

Fig.1 (a) Schematic drawing of an ellipsoid mirror to focus a laser to a tight spot. The incoming laser is focused by conventional parabola mirror on the focus point α. (b) Magnification from the first focus (α) to the second focus point (β) as a function of the ratio b/a (minor-/major- axis). The right vertical axis indicates an estimated enhancement factor of the focused laser intensity through reduction of the focal spot. The calculation was done for various incident angle of the laser respect to the major axis of the ellipsoid.

The magnification from the first focus to the second one can be simply estimated as $\rho'/\rho = [(1+\varepsilon^2)-2\varepsilon\cos\theta] / (1-\varepsilon^2)$ [8]. Here, ε is its eccentricity and $\varepsilon = (1-b^2/a^2)^{1/2}$. The magnification is calculated based on this expression for various incident angle of the laser respect to the major axis of the ellipsoid (θ) and is summarized in Fig. 1(b). The horizontal axis indicates the ratio between the minor- and the major-axis (b/a). The equivalent intensity enhancement factor is simply estimated as $(\rho'/\rho)^{-2}$. We found that optimized geometry (thus, smaller magnification) varies with θ, and smaller magnification is expected for smaller incident angles. However, if the incident angle is too small, the incoming laser will hit the target (which is placed at the focus point after the ellipsoid) before reflecting from the ellipsoid surface. Finally, the incident angle is decided by taken into account the f-number of incoming beam (determined by the OAP). In our experiment, such f-number was 2.7 (10.5° in collecting half angle), thus we chose 13 $^{\circ}$ as the incident angle. Figure 1(b) suggests that the optimum ratio of the minor-axis to the major-axis at this angle is 0.6. The size of the mirror is determined by taking with account the laser energy fluence on the ellipsoid surface, as this ellipsoid mirror will be used under the PM condition. The working fluence for the PM is between 10 and 1000 J/cm^2 [3] [9]. Finally, we decided our ellipsoid mirror geometry to be a=3.5 mm, b=2.0 mm. In this case, the estimated average fluence on the surface of the EPM is 100-300 J/cm^2 ($3\text{-}8\times10^{14}$ W/cm^2) for 3-11 J laser energy. The Resulting f-number after the EPM becomes 0.4. To our knowledge, such f/# is the smallest value even achieved for a high intensity laser system. Note that the ratio of the f/# (0.4/2.7~0.13) is not same as the magnification because the paraxial approximation is not valid for such low f/#. The expected magnification of spot size and corresponding intensity enhancement are 0.24 and 19, respectively. The material of the mirror was chosen to glass (K-PBK40).

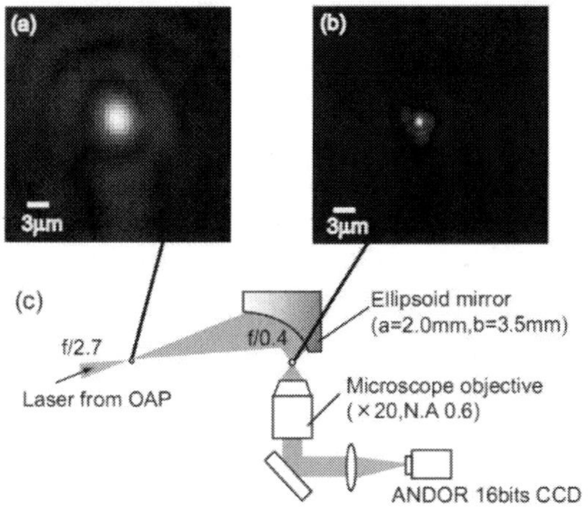

Fig.2 focusing image (a) after the f/2.7 off-axis parabola mirror (OAP), 4.4 μm FWHM (b) after the f/0.4 EPM, 0.9 μm FWHM (c) Set up for the focal spot measurement,

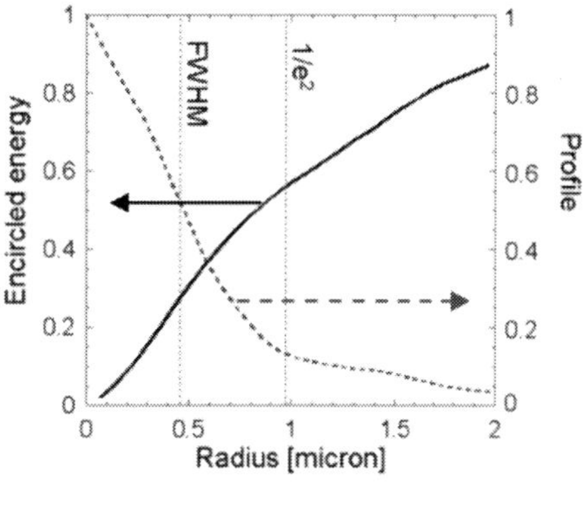

Fig. 3 azimuthally averaged radial profile of the measured focal spot after reflecting from the ellipsoid mirror (blue dotted line corresponding to Fig2. (b)). The FWHM size is ~0.9 μm. Encircled energy for the same focal spot (solid line).

2.2. Spot size measurement after reflecting from the ellipsoid mirror

To confirm our design and set-up, we directly measured the focal spot of the laser after reflecting from the ellipsoid mirror. This measurement was performed at the LULI 100TW laser facility. As shown in fig.2, to cover the large aperture after the ellipsoid mirror, we placed a infinity-corrected large aperture microscope objective (Edmond optics, ×20, N.A.0.6) at around 1.3 cm after the focal spot of the ellipsoid mirror. The laser beam was then collimated and was re-focused on an ANDOR 16 bits CCD camera by an achromatic lens (400 mm focal length). The total magnification of the system was 40. To avoid a damage of the objective, the laser energy was reduced at the μJ level (10 Hz operation). In this case, the ellipsoid mirror was not turned into plasmas but act as conventional solid-mirror. After doing rough positioning using a mock-up alignment system outside the chamber, we did put the ellipsoid mirror in its position determined by to the focal spot position of the OAP. Final alignment was done using the 10 Hz laser (i.e. it in the same laser as the one used for high power shots, but without amplification), by monitoring the laser spot at the focal point after the ellipsoid mirror. The laser was frequency doubled (528 nm). After doing precise positioning of the elliptical mirror, we finally obtained a 0.9 μm full width at half maximum (FWHM) spot after the ellipsoid mirror. This is around 5-times a smaller spot size compare to the one obtained at the focal point of the f/2.7 OAP (4.4 μm, Fig.2 (a)). The obtained magnification of 0.2 shows that good agreement with the geometrically expected value, as indicated in Fig. 1(b). Analysis of the focal spot image (shown in Fig. 3) shows the encircled energy for the two configurations is very similar, i.e. 27 % at FWHM and 55 % at $1/e^2$.Then the reflectivity of the PM (30±10%) is measured by using a calibrated calorimeter. So we indeed expect the magnification is 5-10 [10].

3. Conclusion

We proposed an ellipsoidal focusing mirror which is able to focus ultra-intense laser pulses into much reduced size compare to conventional focusing. The geometry was optimised by taking into account the magnification as well as laser energy fluence on the ellipsoid mirror surface. We demonstrated 5-fold reduction of the focal spot size (down to 0.9 μm FWHM) by using this ellipsoid mirror compared to standard focusing (f/2.7 OAP). The measured magnification (~ 0.2) matches the geometrically expected value. This mirror has then been used under plasma-mirror condition (>50 J.cm^{-2}, 2-10 J). We clearly obtained a significant enhancement of laser intensity (by a factor of ~10) by using the ellipsoid plasma-mirror. As a result, the proton maximum energy produced by irradiating a thin (few μm) metal foil with laser pulses was remarkably enhanced by a factor of 3-9 with the ellipsoid plasma-mirror as well as laser intensity. The detailed of this experimental result is described in Ref [10]. We believe such compact, extremely low f/# plasma-optics can contribute to the offspring of the next-generation of high power lasers, opening the door to access high-energy density science and applications.

Acknowledgment

This study was supported by Japan Science and Technology CREST, Global COE program in Ministry of Education, Culture, Sports, Science and Technology (Electronic Devices Innovation) and from the grant E1127 from Région Ile-de-France. We thank to the expert support of the LULI laser team. M.N. acknowledges the support of JSPS Postdoctoral Fellowship for Research Abroad.

References

[1] M.Tabak, et.al. Phys. Plasmas, 1, 1626-1634.
[2] S.-W. Bahk, et. al. Opt Lett 29,2837(2004)
[3] G. Doumy, et. al. Phys. Rev. E, 69, 026402 (2004) and references theirin
[4] M. D. Perry, et . al. J. Appl. Phys., 85, 6803-6810
[5] R Hörlein, et. al. New J.phys.10 (2008) 083002
[6] C. Thaury, et. al. Nature Phys. 3, (2007) 424
[7] D. Neely, Appl. Phys. Lett., 89. 021502 (2006).
[8] T. Wittmann, et. al. Rev. Sci. Inst. 77,083109 (2009)
[9] O.N.Stavroudis , et. al. J. Opt. Soc. Am. A 9, 2083-2088 (1992)
[10] M.Nakatsutsumi, A.Kon, .. et. al., submitted to Nature Photonics.

2.4-Joule chirped pulse operation by a laser-diode-pumped slab laser for pumping non-collinear OPCPA

T Kurita[1], K Sueda[2], T Sekine[1], T Kawashima[1] and N Miyanaga[2]

[1] Hamamatsu Photonics K. K. 5000, Hirakuchi, Hamakita-ku, Hamamatsu, Shizuoka 434-8601, JAPAN

[2] Institute of Laser Engineering, Osaka University, 2-6, Yamadaoka, Suita, Osaka 565-0871, JAPAN

t-kurita@crl.hpk.co.jp

Abstract. We present a diode-pumped chirped pulse amplification (CPA) system that generates multi-joule energy at a central wavelength of 1054 nm at a repetition rate of 10 Hz. The purpose of this laser is to serve as a pump source for non-collinear optical parametric chirped pulse amplification (NOPCPA). A Nd-doped glass slab with zigzag optical path was used as the gain medium of the main amplifier in this system to obtain multi-joule output with repeatable operation. The Nd:glass zigzag slab amplifier system consists of four-pass pre-amplification and four-pass power amplification. The seed pulse that is fed to the main amplifier was generated by a mode-locked fiber oscillator emitting at a 1053 nm central wavelength. The oscillator output was pulse-stretched to 2.7 ns duration with a 4.5 nm spectral bandwidth and amplified to 100 μJ by optical parametric amplification by use of type-I BBO crystals. After the main amplification, 2.4 J of energy in 3.7 nm of spectral bandwidth at 1 Hz repetition rate was obtained. This spectral bandwidth corresponds to a transform-limited pulse duration of 440 fs. This result indicates that our CPA laser is capable of delivering multi-joule pump light after pulse compression and frequency doubling for 30-TW NOPCPA system.

1. Introduction

In recent years, few-cycle ultrashort pulses have attracted much attention for new fields of research such as the production of isolated attosecond pulses [1], monochromatic particle acceleration [2], high-order harmonic generation [3] and probing molecular dynamics over a broad spectral range [4]. Since the first demonstration of non-collinear optical parametric amplification (NOPCPA) that makes ultra broadband amplification possible [5], considerable effort has been directed toward developing shorter pulse duration [6-8] as well as higher energy output [9, 10]. More recently, 130 mJ pulses with a duration of 7.9 fs were obtained using NOPCPA [11].

However, power scaling of few-cycle pulses is challenging because of the lack of pump sources that deliver sufficiently short, high-energy pulses. Repetitive output is of great importance for practical applications. We have been working on the development of a diode-pumped Nd:glass zigzag slab laser as a test bed for an inertial fusion energy (IFE) driver [12, 13]. The key components are a diode-pumped, multipass arrangement and a water-cooled slab amplifier that is capable of delivering output of tens of joules at 10 Hz repetition rate. Our scheme is to utilize this slab laser as pump source for an NOPCPA system using a carrier envelope phase stabilized Ti:Sapphire oscillator and BBO crystals to deliver a peak intensity of 30 TW (150 mJ in 5 fs) after pulse compression. In this paper, we describe a Nd:glass zigzag slab laser based on a chirped pulse amplification (CPA) scheme [14], which can deliver 2.4 J of energy with a pulse duration of 2.5 ns at a spectral bandwidth of 3.7 nm with a repetition rate of 1 Hz before pulse compression.

2. Multi-joule pump laser system for NOPCPA

The schematic of our pump laser system based on CPA is shown in Figure 1. A commercial mode-locked Er doped fiber laser (FemtoFiber®, TOPTICA Photonics) using photonic crystal fiber was used to generate a white-light continuum. A seed laser at the central wavelength of 1053 nm was extracted from the white-light continuum, delivering 200-mW output with 300-fs pulse duration at 97.5-MHz repetition rate. The spectral bandwidth was 34.6 nm (FWHM). The femtosecond seed pulse was stretched to a few nanoseconds by means of an Offner stretcher in order to achieve efficient amplification. The amplifier scheme of the front end was an optical parametric amplification (OPA) where two BBO crystals were pumped by a commercial Nd:YAG laser (Surelite™, Continuum) with 5 ns pulse duration at 10 Hz repetition rate. The output energy of 300 mJ from the Nd:YAG laser was divided to pump both crystals whose sizes were 5 × 5 × 18 mm³ and 5 × 5 × 15 mm³, respectively.

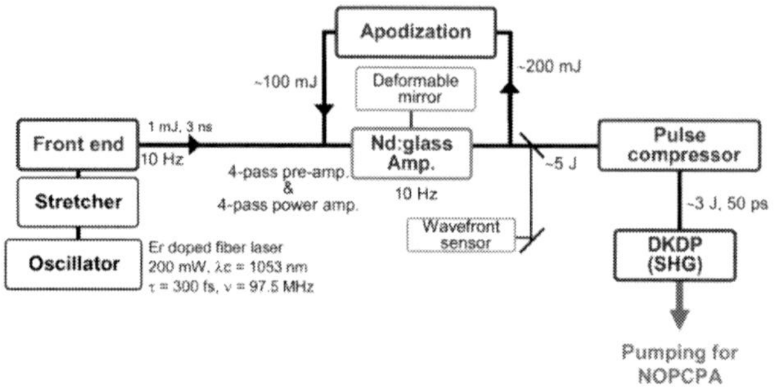

Figure 1. Schematic layout of the zigzag slab CPA laser.

The seed pulse from the front end was fed to a Nd:glass zigzag slab laser amplifier by a relay imaging telescope. The gain medium was Nd-doped phosphate glass (1.0 weight%, Nd_2O_3:HAP-4, HOYA) that was 10 mm thick, 22 mm high, and 335 mm long. Both sides of the slab were pumped by a total of 2,400 diode bars. A quasi-cw, 803 nm, AlGaAs diode bar produced 100 W peak power with 200 μs pulse duration at a repetition rate of 10 Hz. Our CPA laser system consists of four-pass pre-amplification and four-pass power amplification. The output of the pre-amplifier with the beam diameter of 8 mm was extracted by a combination of a quarter waveplate and a polarizer. After pre-amplification, the beam was expanded and apodized with a serrated aperture that mitigates the diffraction effect. The beam size was enlarged to a rectangular shape 20 mm high and 7.8 mm wide to match the cross section of the slab glass. The beam was image-relayed again to the same amplifier for four-pass power amplification. A different incident angle to the slab glass was chosen to realize pre- and power amplification in the same amplifier. Figure 2 depicts the optical path layout of the zigzag slab laser glass.

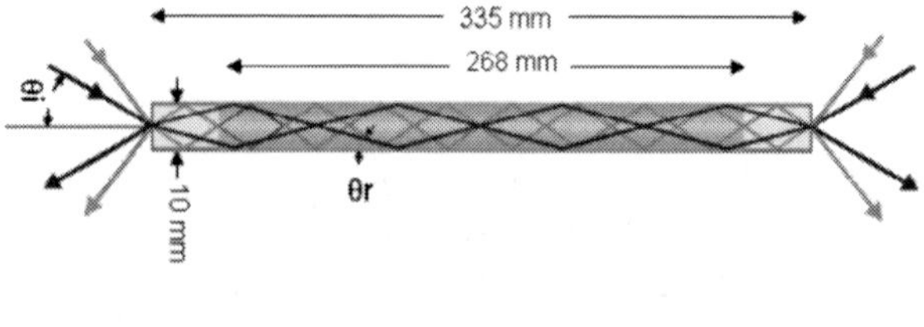

Figure 2. Optical path layout of the zigzag slab laser glass.

The incident angle and the total internal reflection angle for the pre-amplification were $\theta_i = 13.6$ deg. and $\theta_r = 6.8$ deg., respectively. As for the power amplification, $\theta_i = 29.3$ deg. and $\theta_r = 16.6$ deg. were used. This geometry allows us to utilize the whole gain volume, minimizing the non-overlapping area. A deformable mirror (Night N (Opt) Ltd.) and a wavefront sensor (SID-4, Phasics) were used to compensate for the wavefront distortion caused by thermo-optic effects of the power amplification. We used a pair of gratings to shorten the pulse duration to as short as 50 ps, which had not been achieved yet. A DKDP crystal (Type-II, 75 mm diameter, 7 mm thickness) was prepared for frequency doubling.

3. CPA output from a diode-pumped Nd:glass zigzag slab laser

In order to characterize CPA operation, we calculated the effect of spectral narrowing with respect to output energy. The model is based on Frantz-Nodvik analysis [15] that allows for a spectrally broadened input pulse. Figure 3 shows the change of spectral shape after 8-pass amplification when third-order Gaussian input with a spectral bandwidth of 6 nm was applied. As it passed through the amplifier, the bandwidth was narrowed from 6.0 nm to 4.6 nm. In this way, the bandwidth change was calculated for input with 10 nm and 20 nm bandwidths, as shown in Figure 4. The results show that Nd:glass (HAP-4) keeps the bandwidth at around 5 nm, enabling pulse compression to 300 fs.

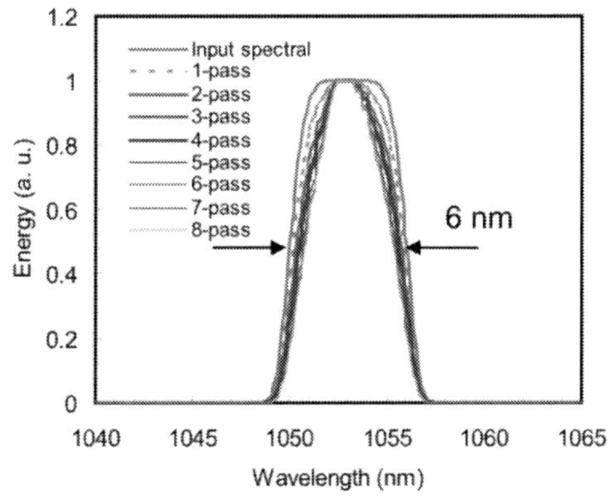

Figure 3. Calculated spectral shape change of input with 6-nm bandwidth.

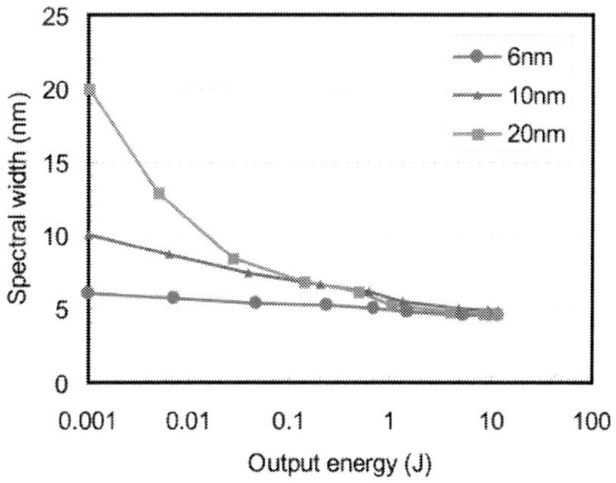

Figure 4. Calculated spectral bandwidth change with respect to output energy after 8-pass CPA.

The CPA experiment was conducted at the pump repetition rate of 1 Hz. The seed pulse for the slab laser was pulse-stretched to 2.7 ns and amplified to 100 μJ at the front end. The spectral bandwidth of the oscillator was truncated by the concave mirror in an Offner stretcher to 4.5 nm. The output characteristics of 8-pass CPA are shown in Figure 5. An output energy of 2.4 J with a pulse duration of 2.5 ns was obtained. Figure 6 indicates the output spectral and the calculated spectral by using the model we described above. The output spectral was narrowed to 3.7 nm, which corresponds to a transform-limited pulse duration of 440 fs. The obtained spectrum is in good agreement with the calculated one. The central wavelength of the input pulse was 1055 nm. The timing-jitter of temporal overlap between the signal and the pump in OPA could be attributed to this gap. The central wavelength of the amplified output was shifted to 1054 nm, which is the maximum gain of the Nd:glass medium.

As mentioned previously, the major limitation of power scaling of few-cycle pulses was the lack of a high-energy pump source. In this study, we achieved multi-joule output, although the pulse compression and frequency conversion still remain to be tested. Since our slab laser design is based on power scaling by enlarging the height of the slab medium [12], this should open the way to increasing the pump energy to the level of tens of joules.

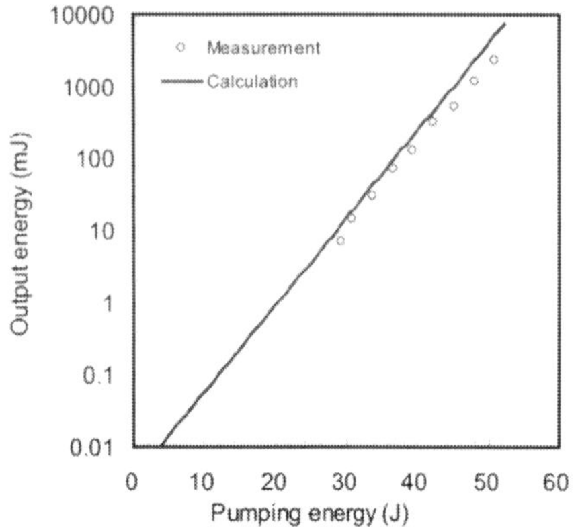

Figure 5. Laser performance of CPA output energy with respect to pumping energy.

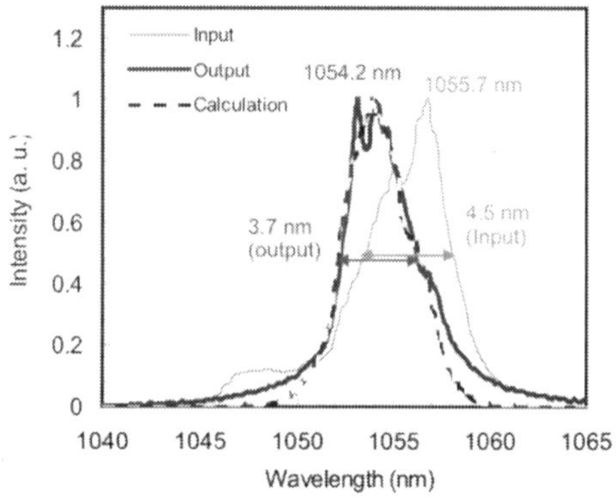

Figure 6. CPA spectra before (thin curve) and after (thick curve) 8-pass CPA and the calculated one.

4. Summary

In this paper, we have described the achievement of a pulse with an energy of 2.4 J for a pulse duration of 2.5 ns and a bandwidth of 3.7 nm at a central wavelength of 1054 nm and a repetition rate of 1 Hz, generated from a diode-pumped Nd:glass zig-zag slab laser. In the future, we plan to amplify the seed pulse from a Ti:Sapphire oscillator by combining the pump laser in order to increase the few-cycle pulse energy to over 100 mJ.

This research was supported by Japan Science and Technology Agency, CREST.

References

[1] Hentschel M, Kienberger R, Spielmann Ch, Reider G A, Milosevic N, Brabec T, Corkum P, Heinzmann U, Drescher M and Krausz F 2001 *Nature* **414** 509–13
[2] Tajima T and Dawson J M 1979 *Phys. Rev. Lett.* **43** 267-70
[3] Corkum P B 1993 *Phys. Rev. Lett.* **71** 1994-97
[4] Brida D, Bonora S, Manzoni C, Marangonil M, Villoresi P, Silvestri S D and Cerullo G 2009 *Opt. Express* **17** 12510-15
[5] Dubietis A, Jonusauskas G and Piskarskas A 1992 *Opt. Commun.* **88** 437-40
[6] Armstrong M R, Cerullo G, Silvestri D, Gallmann L, Matuschek N, Steinmeyer G, Keller U, Angelow G, Scheuer V and Tschudi T 2001 *Opt. Lett.* **26** 1155-7.
[7] Baltuska A, Fuji T and Kobayashi T 2002 *Opt. Lett.* **27** 306-8
[8] Adachi S, Ishi N, Kanai T, Kosuge A, Itatani J, Kobayashi Y, Yoshitomi D, Torizuka K and Watanabe S 2008 *Opt. Express* **16** 14341-52
[9] Witte S, Zinkstok R Th Hogervorst W, and Eikema K S E 2005 *Opt. Express* **13** 4903-8
[10] Tavella F, Nomura Y, Veisz L, Pervak V, Marcinkevičius A and Krausz F 2007 *Opt. Lett.* **32** 2227-9
[11] Herrmann D, Veisz L, Tautz R, Tavella F, Schmid K, Pervak V and Krausz F 2009 *Opt. Lett.* **34** 2459-61
[12] Kawashima T *et al.* 2001 *Jpn. J. Appl. Phys.* **40** 6415-25
[13] Yasuhara R *et al.* 2008 *Opt. Lett.* **33** 1711-3
[14] Strickland D and mourou G 1985 *Opt. Commun.* **56** 219-21
[15] Frantz L M, Nodvik J S 1963 J. Appl. Phys. **34**, 2346-9

Journal of Physics: Conference Series **244** (2010) 032010 doi:10.1088/1742-6596/244/3/032010

Performance of and Initial Results from the OMEGA EP Laser System

D D Meyerhofer[1,2,3], J Bromage[1], C Dorrer[1], J H Kelly[1], B E Kruschwitz[1], S J Loucks[1], R L McCrory[1,2], S F B Morse[1], J F Myatt[1], P M Nilson[1,3], J Qiao[1], T C Sangster[1], C Stoeckl[1], L J Waxer[1] and J D Zuegel[1]

[1] Laboratory for Laser Energetics, University of Rochester, 250 E. River Road, Rochester, NY 14623, USA
[2] also Departments of Mechanical Engineering and Physics & Astronomy, University of Rochester
[3] also Fusion Science Center for Fast Ignition and Extreme States of Matter, University of Rochester

E-mail: ddm@lle.rochester.edu

Abstract. The OMEGA EP Laser System was completed in April 2008. It consists of four NIF-like beamlines that will each produce 6.5 kJ per beam at 351 nm. Two of the beamlines can be configured as high-energy petawatt beamlines that will each produce 2.6 kJ in a 10-ps laser pulse. This paper describes the current status of the OMEGA EP Laser System and some initial experimental results.

1. Introduction

The Laboratory for Laser Energetics completed its new long-pulse ultraviolet and high-energy petawatt (HEPW) laser system, OMEGA EP, in April 2008 [1–3]. It consists of four NIF-like beamlines [4] that will each produce 6.5 kJ per beam at a 351-nm wavelength in a 10-ns pulse. Two of them can be configured as high-energy petawatt (HEPW) beamlines that will each produce 2.6 kJ in a 10-ps laser pulse at a wavelength of 1.053 μm. The current UV laser performance is limited by the installed optics. Higher-damage-threshold UV optics have been ordered to achieve full UV capability in FY11.

The HEPW energy has been ramped to 2.1 kJ in a 12-ps pulse. The energy is limited by the intensity modulations in the beam near field and lower optics vacuum damage thresholds than expected. Significant progress has been made in developing OMEGA EP as a fully equipped user facility. Many diagnostics have been installed and a number of exciting physics results have been obtained.

Section 2 describes the current performance of the OMEGA EP HEPW Laser System. Section 3 describes some of the initial experimental results. Section 4 concludes this article.

2. Performance of the OMEGA EP Laser System

A significant effort has been devoted to increasing the on-target HEPW laser energy. In August–September 2009, the IR energy was ramped to 2.1 kJ in a 12-ps laser pulse. Damage was observed on the final grating and a few downstream optics. The damage appears to be related to near-field modulations on the amplified beam and lower than expected optics damage thresholds in vacuum. The damaged optics have been replaced and routine operation with 1.5-kJ, 10-ps laser pulses is anticipated by mid-FY10. Work will continue in FY10 to improve the beam performance and understanding vacuum optics damage thresholds leading to the generation of full system energy (2.6 kJ in 10 ps) in FY11.

Laser-diagnostic capabilities on OMEGA EP are improving in parallel with system performance. The focal spot is determined from the measured beam-wavefront and phase errors in the optical path to the target chamber [5,6]. This technique has been experimentally validated with direct focal-spot measurements of low-energy laser pulses. Figure 1 shows the encircled energy calculated for a 1-kJ, 10-ps laser pulse, showing a 22.7-μm radius encloses 80% of the beam's energy. The inset shows the calculated intensity distribution in the target plane. Ongoing improvements to the wavefront quality of the OMEGA EP beamlines will reduce the focal-spot size.

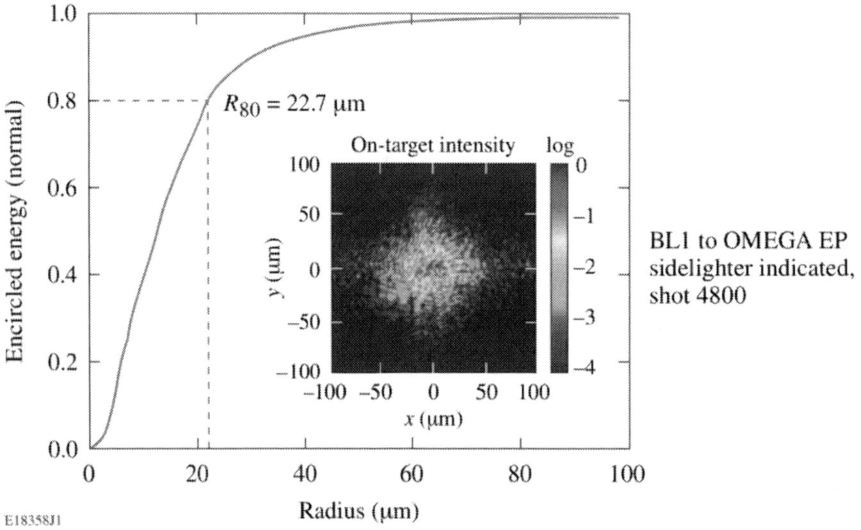

Figure 1. Encircled energy of an OMEGA EP beam at best focus. The inset shows the focal-spot image on a logarithmic scale. The IR pulse energy was 1 kJ in a 10-ps laser pulse.

The optical contrast of high-intensity laser pulses can affect the laser–target interaction physics. LLE is developing two diagnostics to measure the on-shot contrast. Both have been shown to measure the pulse contrast to 80 dB or higher. The optical contrast within 0.5 ns of the peak of the pulse will be measured with a single-shot cross-correlator based on pulse replication [7]. The optical contrast up to 0.5 ns before the peak of the pulse will be measured with a fast photodiode and oscilloscope. This device was installed for a series of 1.5-kJ, 10-ps laser pulse target experiments. Figure 2 shows the optical contrast in this time window for three shots. The pedestal is reproducible at the –60-dB level. This pedestal contains approximately 0.01% of the pulse energy (~150 mJ). Note that if the pulse were compressed to 1-ps duration, the energy in the pedestal would be unchanged and the intensity contrast would be ~70 dB.

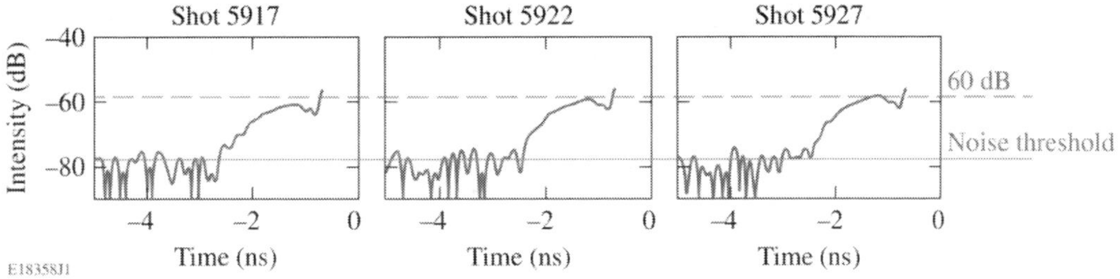

Figure 2. Measurements of the optical contrast of the OMEGA EP HEPW beamline up to 0.5 ns before the main pulse. The laser pulse energy was 1.5 kJ in a 10-ps laser pulse for all shots.

3. Initial results from the OMEGA EP Laser System

When an intense laser pulse interacts with a solid target, a significant fraction of the laser beam's energy is converted into energetic electrons [8,9]. As the electrons reach the rear surface of the target some escape, setting up a sheath field that confines the remaining electrons to the target [10–12]. The electrons collide with the target atoms heating the target and producing K-shell radiation. The amount of K-shell radiation is proportional to the total energy in the energetic electrons [13]. As the target is heated, the atoms are ionized, reducing the population in the levels that produce the K_β radiation so that the ratio of the K_β-to-K_α signals decreases [12,14]. As a result, K-shell spectroscopy provides two ways of determining the laser-to-energetic-electron conversion efficiency. The amount of target heating should be consistent with the conversion efficiency determined from the K_α yield.

A series of experiments were carried out on the MTW Laser System [15] (1 ps, 1 to 10 J) at the Laboratory for Laser Energetics, where the absolute K_α yield and the K_β-to-K_α ratio was measured [16,17]. The K_β-to-K_α ratio data is shown in Fig. 3 (black dots) as a function of laser energy divided by target volume (target energy density). Both the yield and the ratio are consistent with a laser-to-energetic-electron conversion efficiency of 20±10% (20% shown in the solid black line). The right-hand axis shows the temperature of the target determined from this ratio that is consistent with the predicted energetic-electron conversion efficiency. Solid density conditions with temperatures in excess of 200 eV are inferred.

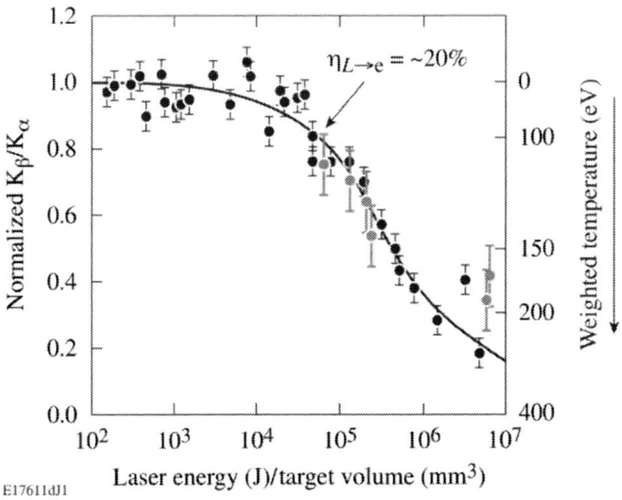

Figure 3: The ratio of the K_β-to-K_α signals from high-intensity laser interactions with small mass targets as a function of the energy density (laser energy/target volume). The black points have a laser energy of 1 J in 1 ps, while the blue points are from OMEGA EP with energies up to 1.3 kJ in 10 ps. The black line is a prediction of line ratio as a function of energy density of a laser-to-electron conversion efficiency of 20%. The right-hand axis shows the solid density target temperature inferred in these experiments.

These experiments were extended to the OMEGA EP Laser System with energies up to 1.3 kJ in a 10-ps pulse [18]. The K_β-to-K_α ratios are shown as the blue points in Fig. 3, again as a function of target energy density. The OMEGA EP data are consistent with the MTW data, suggesting that the conversion efficiency is unchanged from 1 J, 1 ps to 1300 J, 10 ps.

4. Summary

The OMEGA EP Laser System was completed in FY08 and began operation for users in FY09. The laser-system performance and its diagnosis continues to improve. In its first year OMEGA EP conducted 350 target shots for 30 principal investigators. In one series of experiments, the conversion efficiency from high-intensity laser energy to electrons was found to be ~20%, independent of laser energy between 1 and 1300 J and pulse duration between 1 and 10 ps.

5. Acknowledgements

This work was supported by the U.S. Department of Energy Office of Inertial Confinement Fusion under Cooperative Agreement Nos. DE-FC52-08NA28302 and DE-FC02-04ER54789, the University of Rochester, and the New York State Energy Research and Development Authority. The support of DOE does not constitute an endorsement by DOE of the views expressed in this article.

References

[1] Maywar D N *et al.* 2008 *J. Phys., Conf. Ser.* **112** 032007
[2] Waxer L J *et al.* The OMEGA EP High-Energy, Short-Pulse Laser System presented at CLEO/QELS 2008 (San Jose, CA, 4–9 May 2008) paper JThB1
[3] Waxer L J *et al.* 2005 *Opt. Photonics News* **16** 30–6
[4] Moses E I 2008 *Fusion Sci. Technol.* **54** 361–6
[5] Bromage J, Bahk S-W, Irwin D, Kwiatkowski J, Pruyne A, Millecchia M, Moore M and Zuegel J D 2008 *Opt. Express* **21** 16,561–72
[6] Zuegel J D *et al.* 2009 *Rev. Laser Eng.* **37** 437–42
[7] Dorrer C, Bromage J and Zuegel J D 2008 *Opt. Express* **16** 13,534–44
[8] Chaker M, Kieffer J C, Matte J P, Pépin H, Audebert P, Maine P, Strickland D, Bado P and Mourou G 1991 *Phys. Fluids B* **3** 167–75
[9] Kieffer J C *et al.* 1989 *Phys. Rev. Lett.* **62** 760–3
[10] Hatchett S P *et al.* 2000 *Phys. Plasmas* **7** 2076–82
[11] Snavely R A *et al.* 2000 *Phys. Rev. Lett.* **85** 2945–48
[12] Myatt J, Theobald W, Delettrez J A, Stoeckl C, Storm M, Sangster T C, Maximov A V and Short R W 2007 *Phys. Plasmas* **14** 056301
[13] Theobald W *et al.* 2006 *Phys. Plasmas* **13** 043102
[14] Gregori G *et al.* 2005 *Contrib. Plasma Phys.* **45** 284–92
[15] Bagnoud V, Begishev I A, Guardalben M J, Puth J and Zuegel J D 2005 *Opt. Lett.* **30** 1843–45
[16] Nilson P M, Theobald W, Myatt J, Stoeckl C, Storm M, Gotchev O V, Zuegel J D, Betti R, Meyerhofer D D and Sangster T C 2008 *Phys. Plasmas* **15** 056308
[17] Nilson P M, Theobald W, Myatt J F, Stoeckl C, Storm M, Zuegel J D, Betti R, Meyerhofer D D and Sangster T C 2009 *Phys. Rev. E.* **79** 016406
[18] Nilson P M, Theobald W, Myatt J, Stoeckl C, Storm M, Gotchev O V, Zuegel J D, Betti R, Meyerhofer D D and Sangster T C 2009 Effect of Laser Energy and Pulse Duration on Energetic-Electron Production in Intense Laser–Solid Interactions submitted to *Phys. Rev. Lett.*

Journal of Physics: Conference Series **244** (2010) 032011 doi:10.1088/1742-6596/244/3/032011

Single Shot Temporal Contrast Measurement of Subpicosecond Pulses

S. Montant[1*], D. Villate[2], C. Rouyer[2]
1 CELIA, Université Bordeaux 1, CEA, CNRS, 43 rue de Noailles, Domaine du Haut Carré, Université Bordeaux 1 33405 Talence, France
2 CEA CESTA, BP2, 33114 Le Barp Cedex, France

* E-mail: montant@celia.u-bordeaux1.fr

Abstract. We propose a new single-shot third order correlator to measure the temporal contrast of pulses in the picosecond range of high power chain lasers. We use time-to-angular coding to transfer the temporal contrast information to a 3ω probe pulse. We determine the temporal window with the resolution obtained from this technique and predict via simulations the time-to-space coding as well as the time-to-frequency coding. We demonstrate the concept experimentally. The image acquisition is performed with a simple 16 bit CCD camera. An amplitude mask can easily improve the dynamics.

1.Introduction

In general, the ultra short laser pulses from a Chirped Pulse Amplification (CPA) laser suffer from amplified spontaneous emission (ASE) background and pre- and post-pulses associated with the main laser pulse. The pre-pulses combined with ASE will modify the target and thus significantly alter the physics of the laser-matter interaction. In Ultra High Intensity (UHI) physics [1] the contrast ratio between the main pulse, the pre-pulse and ASE must be between 10^7 and 100 femtoseconds up to the nanoseconds scale. In the context of the French Aquitaine Petawatt project [2], the ranged contrast is 10^{10}. Another characteristic of UHI laser is their low repetition rate. So to obtain the contrast ratio of a UHI laser, we need to measure the contrast ratio with a single shot measurement.

Direct measurement of the contrast ratio with an electro optics device utilizing a fast integrator is typically limited to 100 picoseconds resolution. Below this resolution, an optical method is used. To acquire the temporal profile in a single shot configuration, the time must be coded in another dimensionality includes space, wavelength and angular dispersion. Space, wavelength and angular domain can be used. Since the final detector is a CCD spatial detector, the wavelength and angular domain must be converted into spatial domain using a spectrometer and a lens respectively. In 1998, Sirutkaitis et al [3] demonstrate the first single shot third order correlator using time-to-space coding. The third-order correlation was registered over a 3 ps temporal range. A dynamic range of 10^3 was demonstrated and limited only by the photodiode array detector. Collier et al [4] presented a simple linear design. In 2001, Geindre at al. [5] demonstrated a method for obtaining the time-resolved measurement. Based on spectral interferometry, the time-to-wavelength technique allows for single-shot measurements and keeps the temporal resolution consistent with the full bandwidth of the probe pulse. Surville in his thesis [6] demonstrated a time-to-wavelength coding method to measure the spatio temporal evolution of a smoothed laser pulse.

We demonstrate a single-shot third order cross-correlator based on time-to-angular coding [7-8]. Upon presentation of an analytical study of this coding technique, we compare the three different coding methods. In a third part we demonstrate experimentally the time-to-angular coding.

2.Comparison between the different time coding.

To enable us to compare the time-to-angular coding to another encoding method, we performed an analytical study of vectorial coding. An acousto-optic device is typically used for this purpose. However this technique is too slow to enable picosecond resolution. A diffraction grating (N grooves/mm) creates an angular dispersion of the spectral components of the input pulse. In this way the temporal stretch is linked to angular dispersion [7]. To overcome this limitation we used a two step coding method: in a first step we create a time-to-wavelength coding using a simple grating stretcher (F stretching factor). In a second step, a diffraction grating creates a wavelength-to-angular coding. Hence the temporal stretch and the angular dispersion are freely adjusted.

The addressable number of points in single shot diagnostic correspond to independent points that can be acquired. In a time-to-angular coding cross correlator the addressable number of points is the ratio between the dispersion, θ_D , induced by the grating and the natural full angle diffraction, θ_d , of the beam : $N_{points}^{dispersion}=\theta_D/\theta_d$. The dispersion is $\theta_D=N\,\Delta\lambda/\cos\alpha_0$, the divergence is $\theta_d=2\ln2\,\lambda/(d\,\pi)$, λ is the central wavelength, $\Delta\lambda$ is the spectral bandwidth of the pulse, d is the Full Width Half Maximum size of the beam on the diffraction grating and α_0 is the grating diffraction angle. The same results are obtained using a temporal approach and taking into account the ratio between $\tau_R=d\,N\,\lambda/(c\cos\alpha_0)$ the transverse time delay induced by the grating and τ_0 the duration of the transform-limited pulse. Howerver, the input pulse on the diffraction grating is not Fourier Transform limited. When the input chirp pulse, $\tau=F\tau_0$, is longer than the transverse delay time , the number of points is limited by : $N_{points}^{chirp}=F\tau_0/\tau_R$

The optimum corresponds to the maximum of the two numbers of addressable points. The maximum number of addressable points is $N_{points}^{max}=\sqrt{F}$ when $\frac{d}{\sqrt{F}}=\frac{2\ln2}{\pi}\frac{\lambda}{\Delta\lambda}\frac{\cos\alpha_0}{N}$. In this case the resolution is equal to the delay time induced by the dispersion grating: $\tau_{resolution}=\tau_R=\sqrt{F}\,\tau_0$

Table 1. Comparison between different coding methods.

Coding methods :	**Spatial**	**Spectral [6]**	**Angular**
Resolution	τ_0	$\frac{F\tau_0}{2\sqrt{2}}$	$\frac{\sqrt{F}\tau_0}{2\ln2}$
Temporal domain width	$\frac{n\tan\phi/2}{2c}d$	$F\tau_0$	$F\tau_0$
Number of points addressable	$\frac{2c\tau_0}{nd\tan\phi/2}$	$2\sqrt{2}$	$\sqrt{2\ln2}\sqrt{F}$

[a] With τ_0 is the FWHM duration of the fundamental input pulse, n the refractive index and ϕ the angle between the two beams.

The table 1 summarizes the descriptors for each coding method in this investigation. One of them is the very low number of points addressable in the time-to-wavelength coding. This result can be understood because, firstly, the spectral bandwidth of the second harmonic corresponds to the total temporal width of coding and secondly the spectral bandwidth of the second harmonic is roughly equal to the spectral bandwidth of the fundamental beam. Since the number of addressable points is the ratio between the spectral bandwidth of the the third harmonic and the spectral bandwidth of the fundamental beam, the number of points is very low. To perform a high number of usable points, the spectral bandwidth of two beams must be very different. The time-to-space coding has a good resolution but a temporal probe domain width limited by the transverse spatial size of the beam or the third harmonic crystal. The time-to-angular coding has a limited resolution but a freely adjustable temporal domain width.

A more complete analytical evaluation of the electric field demonstrates that the number of addressable points is :

$$N_{points}^{analytical}=\left\{\left[1+2\left(\frac{\tau_R}{\tau_0}\right)^2\right]\left[1+2\left(\frac{\tau_R}{\tau}\right)^2\right]\Big/\left[1+4\left(\frac{\tau_R}{\tau}\right)^2+4\left(\frac{\tau_R^2}{\tau_0\tau}\right)^2\right]\right\}^{1/2}$$

This expression is in good agreement (figure 1 – full line) with the previous approach and with a complete simulation perform under Miro [9] (red dot). Moreover, the limitation induced by the diffraction grating for small beam size (dot line) and by the chirp are plotted (dash line).

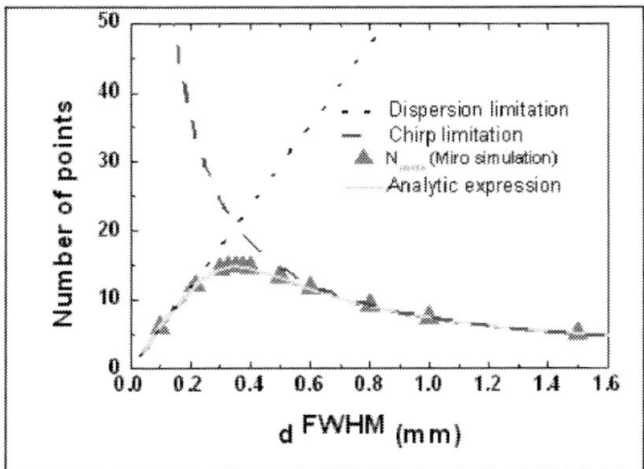

Figure 1. Number of addressable points induced by diffraction grating (dot line), chirp of pulse (dash line), calculated with a complete analytic approach (line) and with simulation under Miro code (red dots).

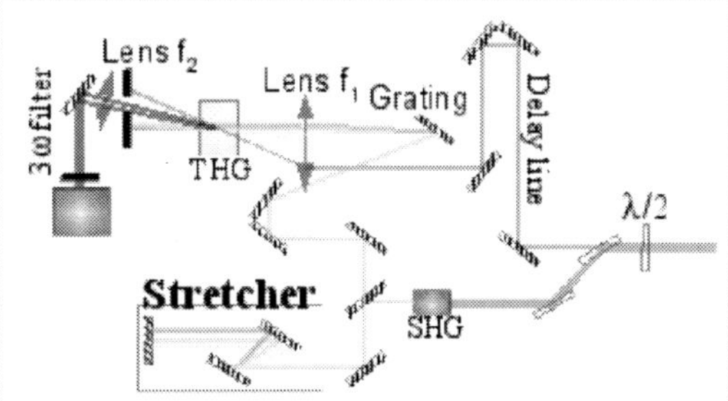

Figure 2. Experimental setup. SHG : second harmonic generation, THG : third harmonic generation.

3.Experimental results

3.1.Experimental setup

This diagnostic is a single-shot third order cross correlator based on a time-to-angular coding. The third order correlation is done between a fraction of the input pulse and the major part second harmonic converted input pulse. After the second harmonic conversion, the time-to-angular coding is realized in two steps: a first one creates a time-to-wavelength coding with a double pass grating compressor. A second step creates a wavelength-to-angular coding using a simple diffraction grating.

The experimental setup is sketched on Figure 2. The input pulse is split into two arms with a half wave plate and a polarizer. Less than 10 % of the energy input is sent through the first arm for the pulse being tested. The ratio between the two arms is slightly adjusted to perform the best Third Harmonic Generation (THG) efficiency. The pulse in the second arm is frequency doubled with a type I BBO crystal. The SHG pulse goes across a double pass compressor. This chirped pulse is sent to 2400 lines/mm diffraction grating with an angle of incidence angle close to Littrow angle. Beams of two arms are mixed in the THG BBO type I crystal with a lens of focal length f_1 located a distance $2f_1$ away from the diffraction grating. The angular coding is transformed in spatial coding on a 16 bit

CCD camera with a lens of focal length f_2 located a distance f_2 from the THG crystal. Spatial and 3ω filters are located in front of the camera to remove the fundamental and second harmonic beams.

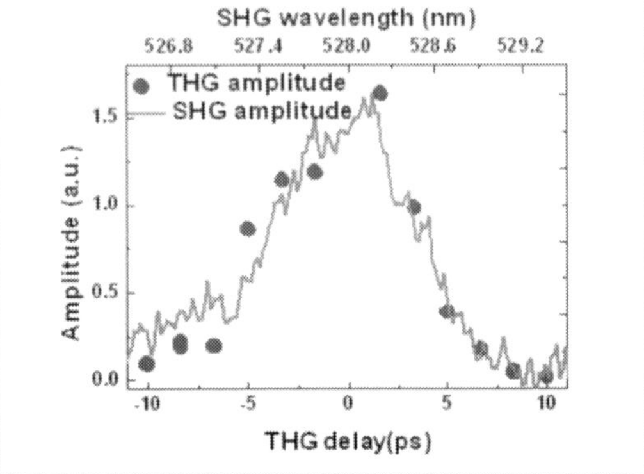

Figure 3. Correlation between the 2ω spectrum and the THG along the temporal interval available.

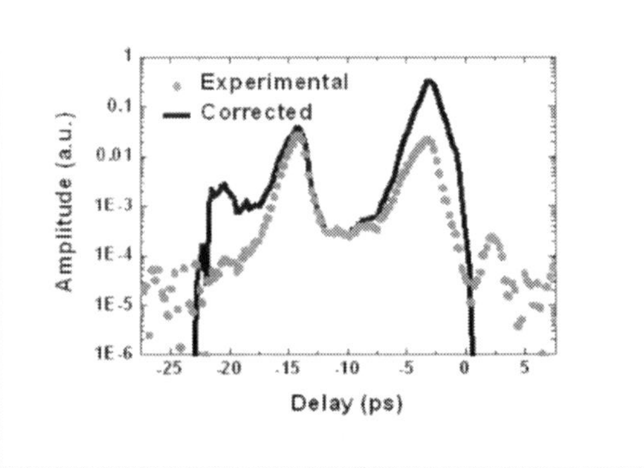

Figure 4. Single shot contrast measurement without amplitude correction (dot) and with an amplitude correction (line).

3.2.Experimental results and discussion

Optical pulses from CPA Titanium: Sapphire regenerative amplifier were used to demonstrate the cross correlator. The amplifier delivers a 200 fs, 500 μJ pulse at 1053 nm at 10 Hz. To validate the cross correlator we insert before the diagnostic a Michelson interferometer with an unbalanced energy and delay arms. Therefore, we can adjust the energy and the delay of the second pulse. We have verified the linearity of the temporal calibration for the CCD camera at 200 fs/pixel with resolution of 600 fs. The Figure 3 plots the variation of THG main pulse along the temporal window moving the delay line. This figure demonstrates the correlation between the 2ω spectrum profile of the probe pulse and THG amplitude signal along the time interval. Therefore, we can define a Gaussian profile amplitude correction using the 23 picoseconds time interval. Using this amplitude correction on a measured single shot contrast measurement (dots on Figure 4) we can calculated the corrected single shot amplitude (line on Figure 4). Only the central part of the two curves are superposed where the corrections is very small, while on the edge of the time interval correction the difference between the two curves is very important. Thus demonstrate the very low contrast quality of the input laser pulse.

4.Conclusion.

We have demonstrated for the first time a single-shot third order cross correlator based on time-to-angular coding. We have made comparison between the different coding methods. The amplitude dynamic is limited by the very low contrast of the laser used. Nevertheless the dynamic of the 16 bit CCD camera can be improved using a spatial amplitude mask.

References

[1] Perry M D and Mourou G 1994 *Science* **264** 917
[2] Blanchot N et al 2008 *Plasma Phys. Control. Fusion.* **50** 124045
[3] Sirutkaitis V, et al 1998 *Lithuanian Physics Journal* **38** 74,
[4] Collier J, et al 2001 *Laser and Particle Beams* **19** 231
[5] Geindre J P, Audebert P, Rebibo S and Gauthier J C 2001 .*Optics Letters* **26** 1612
[6] Surville J 2005 “Developpement et mise en oeuvre ...” *PhD thesis of University Bordeaux 1*
[7] Oba K. Sun P.-S., Mazurenko Y. T. and Fainman Y. 1999 Applied Optics **38** 3810
[8] Jovanovic I. Et al 2006 CLEO Proceedings JThD137
[9] Morice O 2003 *Opt. Eng.* **42** 1530-41

Journal of Physics: Conference Series **244** (2010) 032012 doi:10.1088/1742-6596/244/3/032012

Optimization of contrast ratio and focusable intensity in 700TW femtosecond Ti:sapphire laser facility

Zhao-Hua Wang1, Zhi-Yi Wei1,*, Cheng Liu1, Feng Liu1, Jing-Long Ma1, Hao Teng1 and Jie Zhang1,2

[1] Institute of Physics, Chinese Academy of Sciences, Beijing 100190, China

[2] Department of Physics, Shanghai Jiaotong University, Shanghai, 200240, China

E-mail: * zywei@aphy.iphy.ac.cn

Abstract. In this presentation, we report an improvement technique in our 700TW Ti:sapphire laser facility. In order to improve the contrast ratio, we designed a long cavity ring pre-amplifier. With careful alignment in each dimension of grating, we got a good focal spot after compressor. The size of focal spot was about 25 μ m with an f/10 off-axis-parabolic mirror (OAP). Using an f/3 OAP to focus the laser beam in experiment, we expect the laser intensity of 10^{21}W/cm^2 could be realized on the target.

1. Introduction

In recent years remarkable progresses have been made on the ultrahigh intensity laser with peak power of multi-100TW[1,2], a new facility of ELI toward 200PW is being planned in Europe. In those facilities, the contrast ratio and beam quality will be the most important specifications for the laser matter interaction. In general, the contrast ratio was mainly related with the pre-amplifier, the conventional CPA scheme is only capable of $10^{5\text{-}6}$, which will lead to a considerable pre-plasma and shield the process of the physics phenomena. Although high contrast ratio as 10^{11} was demonstrated in the front stage with new technique, still it is a challenge work to realize such a contrast ratio in PW scale laser. In other side, the focus ability of laser beam is tightly related with the wave-front, laser beam without phase distortion can be well focused into a diffractive limit and support a high focusable intensity. In this presentation, we will report the optimized experiments on the contrast ratio and beam quality in our 700TW Ti:sapphire laser facility (eXtreme Light III, XL-III). By using a filter based on cross-polarized wave (XPW) generation in BaF_2 crystal, the contrast ratio in front stage was greatly improved. Observation on laser beam also showed the focus ability with compressor alignment.

2. System description

Our laser system is made of femtosecond laser oscillator, pulse stretcher, energy amplifier and pulse compressor. A Ti:sapphire oscillator was pumped with 5W 532nm laser, which generated stable mode-locking laser of sub-20fs at repetition rate of 80MHz. To reduce the gain narrow effect, we set an acoustic-optical modulator (DAZZLER™) to pre-shape the laser spectrum, and the bandwidth of spectrum was broadened from 28nm to 36nm. After stretching the laser to about 600ps with an Öffner stretcher, the chirped laser pulse was firstly amplified by a regenerative amplifier which was pumped by a Nd:YAG laser at repetition rate of 10Hz. The second stage was a multi-pass amplifier which was pumped by a doubled frequency Nd:YAG laser at 1Hz repetition rate, which is capable of single pulse energy of 3J at 532nm. The pump laser was customized design with dual beams output, energy of each beam was 50J at 527nm. Considered the total energy can reach 100J, we further enlarge the beam diameter to 60mm in third stage amplifier. In this stage, the chirped laser pulse was boosted to energy of 43.5J. Finally, the amplified laser pulse was enlarged to 120mm and injected into the vacuum compressor. The compressor was consisted of four gold coated holographic gratings (Jobin-Yvon Inc.) with groove of 1480 lines/mm, the size of grating 1 and 4 was 230x180x30mm, one of grating 2 and grating 3 was 460x210x50mm. Pulse duration as short as 31fs was obtained with energy of about 22.5J, corresponding to the peak power more than 700TW.

3. Contrast enhanced design

In the original design, the pre-amplifier was a traditional regen-amplifier. The contrast of ASE from the cavity was just $10^{5\text{-}6}$. In order to improve the contrast to 10^9, a double-CPA (DCPA) with XPW design was expected in our laser system[3,4]. Fig.1 is the layout of the XPW technique used in our laser system. The energy of laser pulses from the 1kHz Ti:sapphire system was about 450 μ J. It passed through two pieces of cross-polarized Glan-Taylor prism, and was focused on the BaF_2 crystal with a 800mm focal lens. The length of BaF_2 crystal was 2mm. Using two pieces of BaF_2 crystals, the clean laser pulse with 40 μ J was obtained from XPW device. Then it was injected into second stage CPA system. In the second stage, we designed a long cavity ring regen-amplifier, which cavity was about 12m, using two pieces of crystals in the cavity (Shown as in Fig.2). The Ti:sapphire crystals were pumped by a frequency doubled Nd:YAG green laser operating at 10Hz. Two concave mirrors with curvature radius of 4-m were positioned in the cavity to satisfy the stable-cavity criterion. We used two focal lens to focus the pump laser to the spot size of 2mm on the crystals. Two Ti:sapphire crystals were used to reduce the heat effect of a single one. All optical components were positioned far from the cavity beam waist which existed between the two concave mirrors. One of the crystals was near the concave mirror in order to allow the beam to double pass the gain medium. The two pockels cells were used in the cavity for the seed injection and amplified laser pulses ejection. With the pump energy of 160mJ, the energy of 20mJ was obtained.

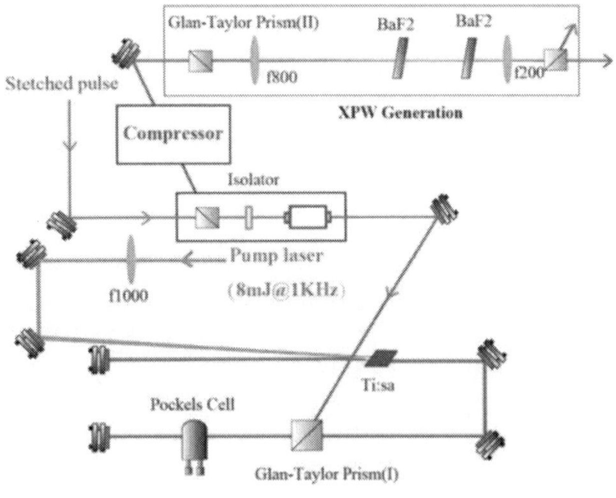

Fig.1 Schematic diagram of XPW device

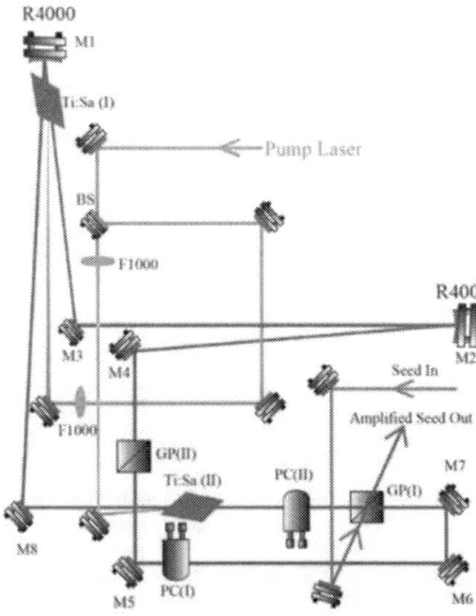

Fig.2 Schematic diagram of the long ring regenerative amplifier. M1,M2, 4m ROC mirrors; M3-M8, flat mirrors; GP, Glan-Taylor Prism; PC, Pockels Cells

4. The alignment of compressor gratings

We performed simulation using commercial optical design software ZEMAX to study the influence of relative misalignment angles between grating pair on the output beam far-field pattern. Fig. 4 shows a sketch of a single pass four-grating compressor which was equivalent to a double pass grating pair compressor with a retro-reflecting mirror. The size of grating 1 and 4 was 230 mm × 180 mm × 30mm, and the size of grating 2 and 3 was 460mm × 210mm × 50mm. The grating groove densities of the four grating are 1480 lines/mm. A laser beam with diameter of 120 mm and incident angle of 20.5° onto grating 1 propagates through the compressor. Seven wavelengths range from 750 nm to 850 nm were used in the simulation. The output beam was focused by a lens with focal length of 1000 mm. The spatial profiles of the laser beam in the focal plane (in x-y plane) are investigated. The focus spots when grating 4 is added by a small angle deviation from its perfectly aligned orientation (shown as in fig.4). The rotation of the grating surface around x-axis (=1 mrad) results in focal spot elongation along y-axis in the diffraction plane as shown in fig 3 (a). The rotation of the grating surface around z-axis (=1 mrad) results in focal spot elongation along x-axis in the plane perpendicular to the diffraction plane in fig 3(b). The effect of rotation of the grating groove around y-axis is similar. When the grating surface has deviation around x-axis and z-axis simultaneously (=1 mrad, =1 mrad), a tilt and elongated focal spot will be observed in fig 3(c). When grating groove rotation around y-axis is added further (=1 mrad; =1 mrad; =1 mrad), the different spectrum components will no longer be in a line but curved as presented in fig 3(d). It is worthy to mention that this phenomenon can be used as a criterion of whether there are both alignment errors of grating surface rotation around z-axis and groove rotation around y-axis, which can not be determined easily by other devices because of only one dimensional diagnostics [5, 6] or two wavelengths used.

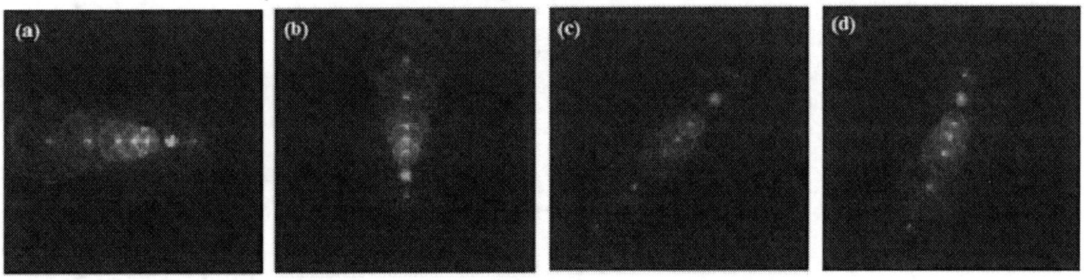

Fig. 3 Focal plane view when gratings 4 have a small angle deviation from its perfectly aligned orientation: (a) α=1 mrad, (b) γ=1 mrad, (c) α=1 mrad, and γ=1 mrad, (d) α=1 mrad; β=1 mrad; γ=1 mrad. The image size is 500 μm × 500μm. The spots corresponding to different wavelengths are due to spectral decomposition in the focal plane.

The gratings were installed on the geometrical position and pre-aligned using the usual method. The oscillator output pulses with beam size being expanded to about 120 mm diameter are injected into the compressor and with the designed incident angle of 20.5° onto the first grating. After propagation through the compressor, the pulses are focused onto a CCD camera to view the far-field pattern. Grating 2 is assumed to have a misalignment error in the dispersion plane with a slight angle around x-axis. The laser beam with incident angle $_i$=20.5° into compressor will have angular chirp when pass through the unparallel grating pair. The output angular difference between wavelength $_1$ and $_2$ is given by = (cos $_1$- cos $_2$)/cos $_i$, where $_1$ and $_2$ is the diffraction angle corresponding to wavelength $_1$ and $_2$ respectively [5]. The double pass of the beam through the grating pair will double the angle difference. When the output beam is focused, the angular chirp will result in a decomposition of different spectral component in the focal plane as shown in fig. 4(c). The focal spots displacement is given by y= f 2 , where f =500 mm is the focal length of the lens. The displacement should be less than half of the focal spot diameter of D=20 μm as shown in fig. 4(d). So we obtain f 2 < 1/2 D, which results in < D cos $_i$ /(4f (cos $_1$- cos $_2$)). For the laser beam from our oscillator with spectrum range from 750nm to 850 nm, we estimate the alignment error to be < 40 μrad. Using the formulas given in [5], the alignment error in other two orientations are estimated to be < 67 μrad; and < 38 μrad respectively.

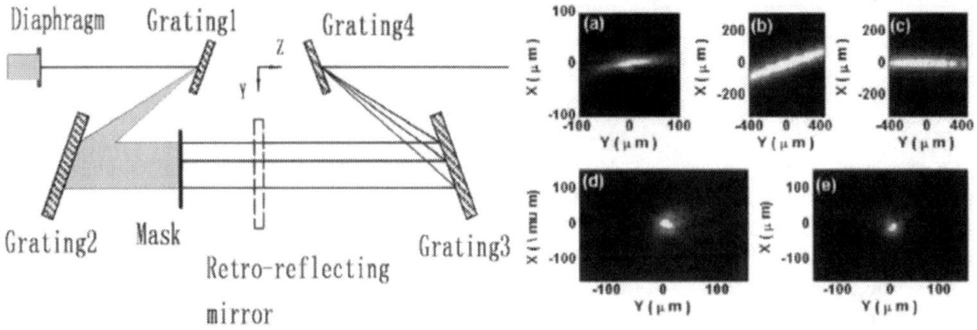

Fig. 4 Example of compression grating parallelism alignment procedure by far-field monitoring: (a) focal spot obtained after the gratings have been aligned by the usual method. (b) adjust the groove rotation around y-axis until the focal spots formed by different wavelength components are in a line instead of a curve. (c) adjust the surface rotation around z-axis until the focal spots are in a horizontal line. (d) the output far-field of beam double passing well aligned 2-grating pair and (e) the output far-field of beam single passing well aligned 4-grating compressor.

5. Summary

In this presentation, we report an improvement technique in our 700TW Ti:sapphire laser facility. With a DCPA design, the contrast ratio of laser was improved from the original 10^5 to 10^8 (shown as in fig.5). With careful optimization, we want to realize 700TW laser output with contrast of 10^9. With careful alignment in each dimension of grating, we got a good focal spot after compressor. The size of focal spot was about 25 μ m with a f/10 OAP(fig.5). Using an f/3 OAP to focus the laser beam in experiment, we expect the laser intensity of 10^{21}W/cm^2 could be realized on the target.

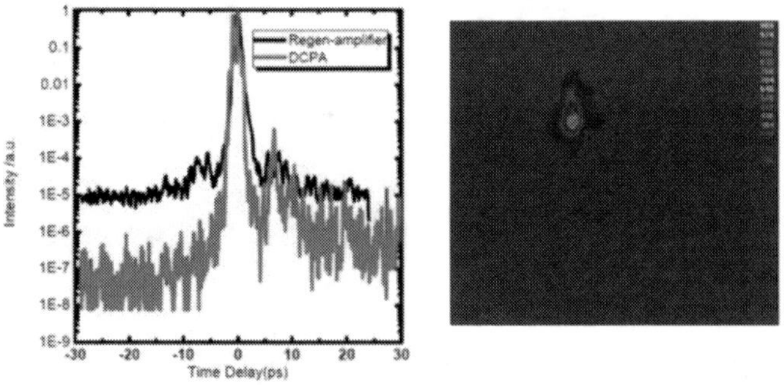

Fig.5 The contrast ratio of laser with DCPA design and the focal spot with f/10 OAP.

6. Acknowledgments

This work is supported by the Instruments Research Program by CAS and National Basic Research Program of China (973 Program) (No. 2007CB815104) as well as NSFC (No. 60621063, 60490280,10804128).

References:

[1] M. Aoyama and K. Yamakawa et al., Opt Lett. Vol.28, 1594(2003)
[2] X.Y. LiANG et al., Opt and Laser in Engineering, Vol. 44, 130(2006)
[3] M. P. Kalashnikov et al., Optics Expres, Vol. 12, 5088(2004)
[4] A. Jullien et al., Opt Lett, Vol. 30, 920(2005)
[5] G. Pretzler, A. Kasper, and K. J. Witte, Appl. Phys. B, **70**,1 (2000).
[6] Z. Sacks, G. Mourou, and R. Danielius, Opt. Lett. **26**,462 (2001).

Development of the 50 TW laser for joint experiments with 1 MA z-pinches.

P P Wiewior, V V Ivanov, O. Chalyy

University of Nevada Reno, Physics Department, Reno, NV 89557

E-mail: pwiewior@unr.edu

Abstract. A 50 TW high-intensity laser (aka "Leopard" laser) was developed for experiments with the 1 MA z-pinch generator at the University of Nevada, Reno. The laser produces short pulses of 0.35 ps; energy is 15 J. Long pulses are 1 ns; energy is 30 J. The output beam diameter is 80 mm. The Leopard laser applies chirped pulse amplification technology. The laser is based on the 130 fs Ti:Sapphire oscillator, Öffner-type stretcher, Ti:Sapphire regenerative amplifier, mixed Nd:glass rod and disk amplifiers, and vacuum grating compressor. An adaptive optics system ameliorates focusing ability and augments the repetition rate. Two beam terminals are available for experiments: in the vacuum chamber of the z-pinch generator (aka "Zebra"), and a laser-only vacuum chamber (aka "Phoenix" chamber). The Leopard laser coupled to the Zebra z-pinch generator is a powerful diagnostic tool for dense z-pinch plasma. We outline the status, design, architecture and parameters of the Leopard laser, and its coupling to Zebra. We present the methods of laser-based z-pinch plasma diagnostics, which are under development at the University of Nevada, Reno.

1. Introduction

The Leopard laser was designed and built in the Nevada Terawatt Facility (NTF) at UNR to increase a capability of the 1 MA Zebra z-pinch generator. The coupling of the high-intensity Leopard laser and the high-current z-pinch opened many new research opportunities for high energy density physics, studied both theoretically and experimentally. Experiments historically performed on Zebra can benefit tremendously from new diagnostics driven by the intense laser (x-ray backlighting, Thompson scattering, proton radiography, x-ray plasma absorption spectroscopy, and others). In addition, the Leopard laser will be used for the study of the interaction of intense laser light with matter in exotic states and environments, such as strongly magnetized plasma, intense radiation sources, plasma jets, and other conditions in hot dense plasma. A more detailed understanding of these phenomena can assist in progress toward fast ignition fusion, as well as laser-driven plasma diagnostics. Last but certainly not least, this invaluable experimental tool will help fulfill the NTF's primary mission to train students and young scientists in high energy density physics.

2. Description of the Leopard laser system

The Leopard laser is a hybrid Ti:Sapphire/Nd:glass laser system suitable to generate 15 J sub-picosecond pulses, with high spatial and temporal beam quality. The laser is composed of several major subsystem: a femtosecond front-end (oscillator, stretcher, and regenerative amplifier), flashlamp-pumped rod and disk amplifiers, a grating compressor, beam transport lines, diagnostics, and control system. Strong capability for joint experiments with Zebra motivated the design of the Leopard laser. The laser is optimized for high energy and intensity delivered on a target, including a high repetition rate front-end for convenient alignment throughout the entire system, ample bandwidth for sub-picosecond pulse length, minimal B-integral, and commercially available diffraction gratings. All together, these features produce long-lasting laser parameters, with easy and affordable maintenance.

2.1. Front-end

The Leopard front-end uses a femtosecond oscillator (Tsunami, SpectraPhysics) pumped by a 9.2 W cw frequency-doubled DPSS Nd:YVO_3 laser (Millenia Pro X, SpectraPhysics). The ultra short pulses from the oscillator are stretched in time from their original duration of 130 fs to almost 1.3 ns in the Öffner-type home-build stretcher. After the pulse picker, the stretched pulses are seeded to a Ti:Sapphire regenerative amplifier (custom made, Coherent) where they are amplified to 1 mJ. The regenerative amplifier repetition rate is 500 Hz. To improve contrast ratio, a fast Pockels cell (UPC, Leysop Ltd.) with 250 ps rise time is located after the regen. The Pockels cell also reduces the repetition rate to 100 Hz. The pulse selected for further amplification passes the serrated aperture where its spatial profile changes to "top hat" geometry; this enables better energy extraction from Nd:glass rod and disk amplifiers. The serrated aperture concludes the front-end of the Leopard laser.

2.2. Amplifiers

The amplifier section includes Nd:glass rod and disk amplifiers, Faraday isolators, Pockels cells, and a spatial vacuum filter/image relay system. Rod amplifiers in the chain are: 6 mm diameter phosphate glass (double pass), 19 mm diameter silicate glass (double pass), second 19 mm phosphate glass (single pass), and 45 mm diameter phosphate glass (single pass). The disk amplifier is 94 mm phosphate/silicate mixed glass (single pass). The laser beam is spatially expanded, spatially filtered and relay imaged between sequential amplification stages. The Pockels cells and Faraday isolators improve contrast ratio and protect the amplifiers from damage due to back-reflections. The 6 mm amplifier is an old Quantel device refurbished at the NTF. The 19 mm and 45 mm amplifiers are units decommissioned from the petawatt laser system at Lawrence Livermore National Laboratory (LLNL), also refurbished at the NTF. The 94 mm disk amplifier is a refurbished unit from LLNL NOVA laser. The disk amplifier is cooled by the flow of nitrogen delivered from liquid nitrogen container; the rod amplifiers are water-cooled. The beam leaving the amplifier chain is spatially filtered and relay imaged to a vacuum pulse compressor.

2.3. Compressor and beam transport

The amplified pulse is compressed in a home-build vacuum compressor. The compressor operates in a double pass, dual diffraction gratings configuration. The gratings used are gold-coated compression gratings manufactured by Jobin-Yvon/Horiba. Each has 1740 grooves per millimeter, delivers 90% of absolute efficiency, with more than $\lambda/8$ surface figure at 1 μm. All crucial optical elements are mounted on remote-controlled rotational and/or translational stages, and the alignment process can be done under the vacuum. The overall compressor efficiency is approximately 60%.

The vacuum beam transport lines deliver the high-energy compressed beam using 15 cm diameter tubes, interconnected with turning boxes. The turning boxes contain large-diameter dielectric mirrors, on remote-controlled mounts. The beam is focused in either the laser-only Phoenix target chamber, where experiments with laser plasma are conducted, or in the Zebra target chamber, where the

Leopard-Zebra coupled experiments are conducted. The mirror in the first turning box after compressor can steer the laser beam to either terminal. An off-axis parabolic mirror focuses the laser beam on the target in the Phoenix target chamber. The Zebra target chamber is presently equipped only with a preliminary lens focusing system, although the final setup will also employ an off-axis parabolic mirror. The entire beam transport system is routinely maintained 10^{-5} Torr vacuum, and is separated from the target chambers by large diameter gate valves.

2.4. Diagnostics and adaptive optics system

The diagnostics and control system are necessary to optimize and monitor the essential parameters of the Leopard laser. Two separate diagnostics sub-systems have been developed: one measures parameters at selected locations in the laser chain, the other measures parameters of the output beam at full energy. Diagnostics inside the Leopard laser include oscillator characteristics (average power, spectrum, pulse duration, stability of the pulse train), regen characteristics (average power, pointing and direction, spectrum, pulse duration), and other selected locations in the beam where the energy, direction, and spatial profiles are monitored. In addition, several photodiodes monitoring the temporal position of amplifier fluorescence, with respect to the amplified pulse, are installed in the amplifier chain. A computer with LabView software records the signals from these photodiodes, as well as from cameras measuring the spatial beam profiles.

Output beam diagnostics necessarily use indirect measurement techniques. A low-energy sample beam is picked off inside the vacuum compressor, and delivered to an adjacent Faraday cage. The sample beam diameter is approximately 80 mm, with maximum energy of 150 mJ. The beam diameter is reduced using a down-collimating telescope; beam intensity is also reduced. We produce four different measurements: an energy measurement (using calibrated energy meter), laser pulse duration time (using FROG Grenouille in single shot mode), focal spot size (using the long distance microscope equipped with 16-bits CCD camera), and a beam wavefront quality (using a Shack-Hartmann sensor). The Shack-Hartmann sensor is a part of an adaptive optics system implemented on the Leopard laser. The system is manufactured by Imagine Optic (France), and is tailored to the particular requirements of the laser. The system was installed and commissioned by Imagine Optic together with NTF personnel. In addition to the Shack-Hartmann sensor, the system consists of a deformable mirror, control electronics, and software with a correction algorithm. The adaptive optics system allows significant correction of the laser beam wavefront aberrations, and leads to greatly improved intensity and focus on target. The corrected wavefront exhibits approximately 50 nm rms variation. The Strehl ratio of the focal spot is approximately 0.8. The measured focal diameter is about eight times smaller than without the adaptive compensation, and intensity in the focus is increased by a factor of 60. The measured fwhm of the focal spot produced by an f/10 lens is approximately 20 microns. Besides the dramatic improvement of the wavefront quality, and the resulting increase in intensity on target, the repetition rate is also greatly improved. Before adaptive optics, cooling time (mainly the disk amplifier) was two hours minimum. We were usually limited to four shots per day. With adaptive optics, the laser can be fired four to five times per hour (10-15 minutes is required for flashlamp cooling). The repetition rate in not longer given by disk amplifier cooling, but by changes to the experimental setup.

3. Development of Leopard-based plasma diagnostics methods.

The Leopard laser is a powerful tool for high energy density physic studies, using the Phoenix target chamber, and through joint experiments with the Zebra. Leopard delivers sub-picosecond or nanosecond pulses to the Zebra vacuum chamber through the vacuum beam path. Zebra is triggered from the clock of the Leopard laser, does not have laser triggering, and the timing setup of laser and pulsed power generator has jitter of ±20 ns. This timing allows an investigation of main stages in wire array implosion.

The Leopard laser is used for astrophysical experiments with a 1 MA z-pinch generator as a source of the 1 MG magnetic field [1]. Several Leopard-based diagnostics are under development for dense z-pinch plasmas. First, the Leopard laser can be used as a backlighter for z-pinch plasma. Experimental methods for the laser backlighting of the z-pinch were developed and are presented, for example in [2, 3]. Second, the absorption spectroscopy has been developed for z-pinches at the NTF. Investigation of the electron temperature T_e and density n_e profiles in the z-pinch plasma and their dynamics is crucial for understanding of the physical mechanisms of plasma heating and x-ray generation. The x-ray spectroscopy is the main method to derive the plasma electron temperature and density in z-pinches at the radiative stage. The absorption spectroscopy can derive the electron temperature and density of the wire-array plasma at the non-radiative ablation and implosion stages. The absorption spectroscopy was applied to laser plasma [4], but still was not used for z-pinches. The Leopard laser provides a broadband x-ray backlighting with Sm and Ge targets. The target will be installed in the Zebra chamber for probing z-pinches at the ablation stage. The x-ray emission of the main z-pinch will be blocked to protect a time-integrated spectrometer.

Next, we suggest investigation of parametric conversion of laser radiation with intensity 10^{14}-10^{15} W/cm^2 to $3/2\omega$ and $1/2\omega$ harmonics in z-pinch plasma. Parametric conversion resulting from a two-plasmon decay can be used to derive a temperature of plasma with the electron density near the 25% of critical density, because a spectral separation of "blue" and "red" satellites in harmonics depends on the electron temperature. For radiation of wavelength ~1 μm the 25% of critical density is 2.5×10^{20} cm^{-3}. This density was measured in z-pinches on the Zebra generator with the x-ray spectroscopy. Parametric conversion is a non-spectroscopic method for estimation of the electron temperature and plasma density in z-pinch plasma.

Acknowledgments

The authors acknowledge the assistance of the team of the Nevada Terawatt Facility at the University of Nevada, Reno (A. Astanovitskiy, S. Batie, B. LeGalloudec, D. Macaulay, V. Nalajala, R. Presura) in the execution of this work. We thank R. Mancini, P. Hakel, and A.P. Shevelko for their input in x-ray absorption spectroscopy. We also thank UNR's students for their input during this project. This work was supported by the U.S. Department of Energy under Cooperative Agreement no. DE-FC52-06NA27616.

References

[1] Plechaty C, Presura R, Wright S, Neff S and Haboub A 2009 *Astrophys. Space Sci.* **322** 195.
[2] Sinars D B, Cuneo M E, Bennett G R et. al. 2003 *Rev. Sci. Instrum.* **74** 2202
[3] Sinars D B, Cuneo M E, Yu E P, at al. 2004 *Phys. Rev. Lett.* **93** 145002
[4] Schott R, Philippe F, Angelo P et al. 2003 *J. Quant. Spectr. Rad. Transport* **81** 441

Journal of Physics: Conference Series **244** (2010) 032014 doi:10.1088/1742-6596/244/3/032014

Multiterawatt Ti:Sapphire/KrF laser GARPUN-MTW as a test bench facility for verification of combined amplification of nanosecond and subpicosecond pulses

V D Zvorykin, A A Ionin, A O Levchenko, L V Seleznev, D V Sinitsyn, and N N Ustinovskii

P N Lebedev Physical Institute of RAS, Leninsky prospect 53, Moscow, 119991 Russia

zvorykin@sci.lebedev.ru

Abstract. The possibility of the same large-aperture KrF laser driver to amplify simultaneously both nanosecond pulses for thermonuclear target implosion and picosecond ones for fuel ignition is discussed relative to KrF-based Fusion Test Facility. In this way experiments were performed at hybrid Ti:Sapphire/KrF GARPUN-MTW facility on amplification of subpicosecond pulses. 2-TW, 330-fs pulses were produced with beam divergence 20 μrad in direct double-pass amplification scheme. Peak power as high as 30–40 TW can be achieved in 50-fs pulses, being combined with long pulses (of a few ns to 100 ns) of ~1 GW power.

1. KrF laser as a prospective driver for the Inertial Fusion Energy

A single-shot 1.8-MJ Nd glass National Ignition Facility has been successfully completed and Thermo Nuclear ignition experiments were started at the LLNL, USA [1] whilst the other facility Laser Mega Joule is under construction in France, being scheduled for operation in 2014 [2]. Against this background the research program to produce Inertial Fusion Energy is underway considering Diode Pumped Solid State Laser and e-beam-pumped Krypton Fluoride laser as candidates for rep-rate, highly-efficient and durable IFE drivers [3]. The ongoing research with KrF lasers aims at obtaining $\geq 3\times10^8$ shots in continuous operation with $\geq$6-7 % overall efficiency at a laser module scalable to 28-kJ energy that is the bench mark for Fusion Test Facility design [4].

To obtain economically attractive IFE power plant two advanced approaches are considered, which allows one to increase TN target gain and to reduce laser driver energy. In a Fast Ignition scheme [5, 6] a powerful tens-ps pulse with intensity I ~ 10^{19} W/cm^2 generates electrons or protons, which off-center ignite the TN fuel preliminary compressed by nanosecond pulse. Simultaneous amplification in the same large-aperture KrF amplifiers stacks of angular-multiplexed nanosecond and picosecond pulses was proposed earlier [7] which seems to be a promising way for the FI IFE. Short pulses of 1 to 20 ps duration can be focused into the same point or distributed over the compressed target [8, 9]. According to the scaling law $E_e = \left[\frac{I\times(\lambda/1.05\mu\text{m})^2}{10^{19}\ \text{W/cm}^2}\right]^{1/2}$ MeV [10] for ponderomotive acceleration of electrons up to energy E_e=1 MeV, which is optimal for electron coupling to the compressed target core

[11], higher intensity $I \sim 2\times10^{20}$ W/cm^2 is required for KrF laser wavelength λ =248 nm. Fortunately, it is easy to achieve due to better beam focusability ($\sim\lambda^2$).

In Shock Ignition scheme [12, 13] hundred-ps final spike with intensity ~ 10^{16} W/cm^2, ten times higher than the main part of the pulse, generates a convergent shock wave, which ignites compressed TN fuel in the central spot. Appropriate target design [14] and laser system analysis [15] have shown that, for a short KrF laser wavelength, target gain as high as 140 may be achieved with modest 0.5-MJ driver energy. A required SI pulse shape can be formed in a course of quasi-steady amplification of angularly multiplexed train of input pulses, which should be specially precompensated for heavily saturated amplifier during high-intensity spike [16]. Alternatively, the idea of simultaneous amplification of short & long pulses [7, 8] can be used to combine a complex SI pulse form directly on a target. On the path to its realization we have investigated amplification of subpicosecond pulses in multistage GARPUN-MTW KrF laser facility [8].

2. Amplification of short and long pulses in KrF laser

Particular features of KrF lasers operating at B→X bound-free transitions (see [8] and references therein) are short lifetime of the excited state of the molecule (radiative $\tau_r \sim 6$ ns; with account for collisional quenching it is $\tau_c \sim 2$ ns), large induced emission cross section $\sigma = 2.5\cdot10^{-16}$ cm^2, small value of saturation energy fluence $Q_s=h\nu/\sigma=2$ mJ/cm^2 (where $h\nu = 5$ eV is the energy of radiation quantum), and significant nonsaturable loss in a gain medium (typical ratio of small-signal gain to absorption coefficient g_0/α=10–20). Because of rapid recovery time of the population inversion in the gain medium ($\tau_c \sim 2$ ns), each short pulse (with duration $\tau_{sh}\leq\tau_c$) does not affect the subsequent pulse if those follow with time interval $\Delta t\geq\tau_c$. Also short pulses will be effectively amplified after termination of a stack of long pulses (with $\tau_{long} \geq \tau_c$, typically $\tau_{long} \sim 5$ ns), which are amplified in a quasi-steady manner. On the other hand, during quasi-steady amplification at optimal intensity (corresponding to maximum efficiency) $I_{opt}=I[(g_0/a)^{1/2}-1]$ the energy stored in the gain medium is not completely extracted. The rest can be used for simultaneous amplification of short pulses, though with less gain $g=(g_0\alpha)^{1/2}$. This also might be useful to reduce Amplified Spontaneous Emission, which decreases a contrast of the short pulse.

3. Experimental setup

Hybrid Ti:Sapphire/KrF multiterawatt laser facility GARPUN-MTW combines the previous version of e-beam-pumped multistage GARPUN KrF laser with recently constructed Ti:Sapphire front-end "START-248 M" [8]. The final large-aperture GARPUN amplifier with a gain volume of 12×18×100 cm^3 is pumped by two counter-propagating 350-keV, 60-kA (50 A/cm^2), 100-ns e-beams guided by magnetic field of ~0.08 T. When operating in free-running oscillation mode with $Ar/Kr/F_2$ gas mixture at 1.4-atm pressure and with specific pumping power W_b = 0.7–0.8 MW/cm^3, it provides up to 100 J in 100-ns pulse. Another 8×8×110-cm^3 BERDYSH module pumped by a single-side magnetic field-guided 350-keV, 50-kA (50 A/cm^2), 100-ns e-beam with W_b = 0.6–0.7 MW/cm^3 produces up to 25 J at 1.8-atm gas mixture. Both amplifiers are synchronized with KrF master oscillator (Lambda Physik EMG TMSC 150 model) producing 200-mJ, 20-ns pulses, which fire laser-triggered switches of HV pulsed power supply of e-beam guns of both Berdysh and GARPUN amplifiers. Frequency tripled Ti:Sapphire front-end "START-248 M" (Avesta Project Ltd.) was designed to generate 10-Hz train of 0.5-mJ, 60-fs pulses at wavelength λ=248 nm matched with KrF gain band.

The layout of experiments on amplification of short pulses is shown in figure 1 with appropriate equipment for measuring pulse energy, duration, spectral and spatial distribution. A single pulse was cut out of a continuous Ti:Sapphire front-end train and synchronized with KrF amplifiers within accuracy of ±5 ns at the top of 100-ns pumping pulse. Double-pass amplification was used for both preamplifier and final amplifier with a spatial filter placed between them. Initial 8-mm front-end beam diameter was successively expanded by convex and concave mirrors to fit amplifier apertures. The only transmissive optics was CaF_2 amplifier windows inclined to prevent parasitic oscillations.

Journal of Physics: Conference Series **244** (2010) 032014 doi:10.1088/1742-6596/244/3/032014

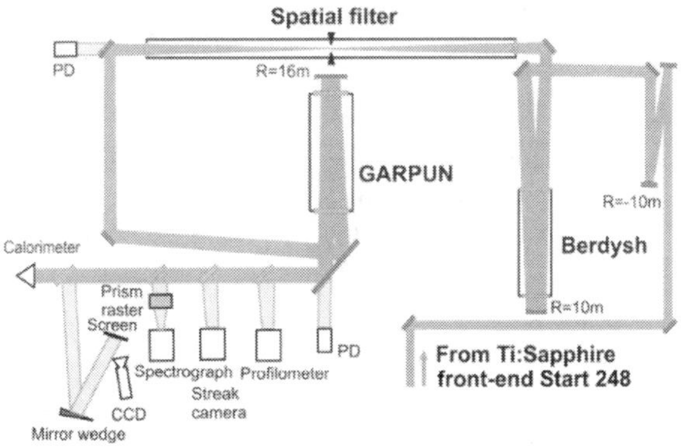

Figure 1. Layout of experiments.

4. Experimental results and discussion

Figures 2, 3 present experimental dependences of output energy density Q_{out} on input one Q_{in} for double-pass amplification in both BERDYSH preamplifier and GARPUN amplifier compared with numerical simulation for various small-signal gain coefficients g_0 and nonsaturable absorption α (shown in the figure inserts). Depletion of the gain by the ASE in folded amplifiers was taken into account with corresponding gain profiles along the amplifiers [8].

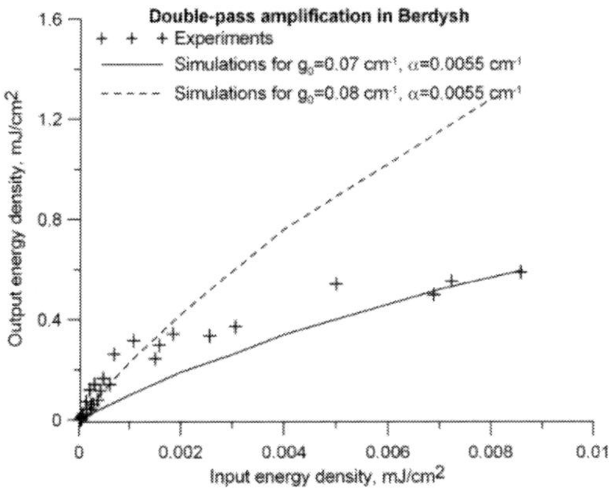

Figure 2. Output energy density Q_{out} vs input Q_{in} for BERDYSH preamplifier.

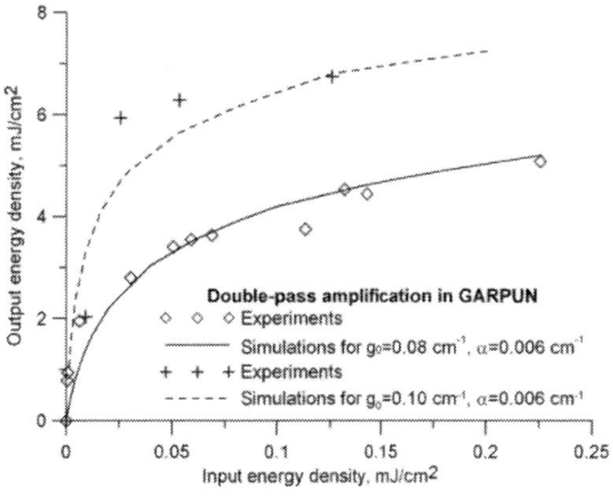

Figure 3. Output energy density Q_{out} vs input Q_{in} for GARPUN amplifier.

The preamplifier operating with maximal Q_{out} = 0.6 mJ/cm^2 is far from saturation (Q_s=2 mJ/cm^2) and has a stage gain G = 70 (Q_{out}/Q_{in}). A deviation of experimental dots from calculated curves was probably due to nonlinear absorption of laser radiation in preamplifier windows, which should increase at higher intensities. Total output energy over laser beam cross-section area S_b = 38.5 cm^2 (it is averaged along the amplifier) was E_{out} = 23 mJ with a fill factor f = 0.6 (a fraction of S_b relative to the whole amplifier aperture). The final amplifier is saturated by Q_{out} = 6.7 mJ/cm^2, which gives total E_{out} = 620 mJ over S_b = 92.5 cm^2 (f = 0.43). Contribution of the ASE in a solid angle of calorimeter (~ 2×10^{-5} sr) was ~ 3%. Higher energy up to 1.5-2.0 J in a single short pulse is expected for double-pass amplification in both Berdysh and GARPUN amplifiers after optimization. A train of short pulses can be also amplified or combined with long pulses of few-ns to 100-ns duration [8].

For 0.1-ps input pulse in present experiments, output pulse was time-broadened by group velocity dispersion (GVD) in amplifier windows. Its duration is estimated by using two methods: by interference of two beams being split by a mirror wedge and by a streak camera. In fact a coherence length is measured using the former approach, which gives a lower limit $\tau_{1/2}$~0.33 ps. The latter measurement is limited by a temporal resolution of the camera and gives upper limit $\tau_{1/2}\leq$1 ps.

Spatial distribution of radiation (see figures 4, 5) was affected by: (i) block-like structure of CaF_2 amplifier windows (it is the best material in respect of nonlinear absorption) that introduces phase modulation at block boundaries and (ii) beam filamentation, as pulse power P~2 TW>> $P_{cr}=3.8\lambda^2/8\pi n_0 n_2 \approx$ 100 MW, critical power for filamentation of UV (λ=248 nm) light [17]. For measured ASE contribution and divergence of a short-pulse beam $\Theta_{1/2}$=20 μrad (figure 5) the contrast ratios of the short pulse to the ASE can be estimated in the future target irradiation experiments: $CR_Q \sim 10^6$ for energy densities and $CR_I \sim 10^{11}$ for intensities (short pulse of ~ 1 ps was assumed).

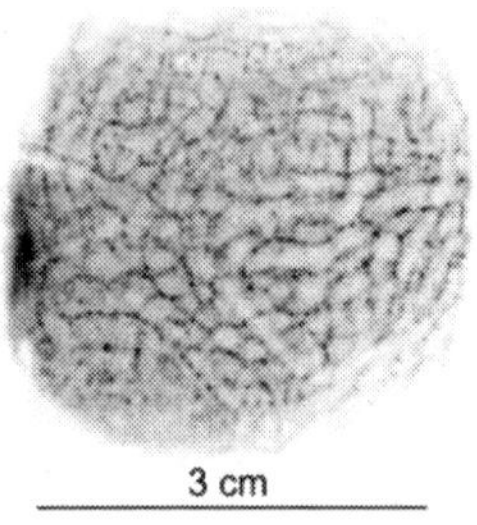

Figure 4. Near-field distribution in laser beam.

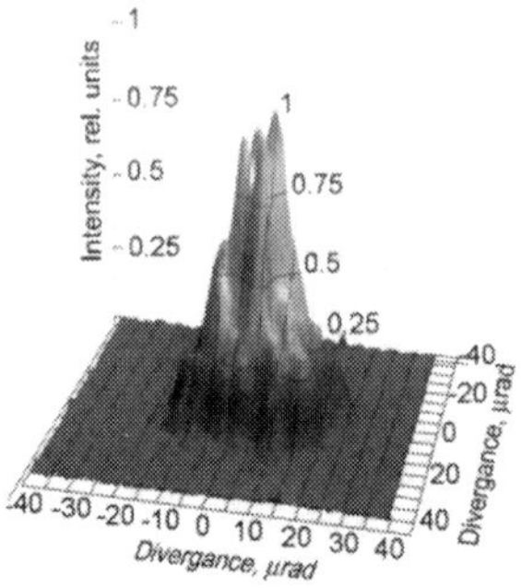

Figure 5. Far-field distribution in laser beam.

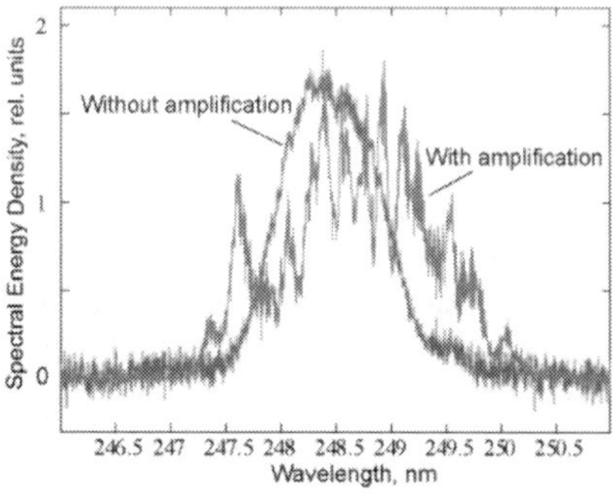

Figure 6. Spectral distribution of short-pulse radiation.

A spectral width of amplified short pulse was measured to be $\Delta_{1/2}$~ 2.3 nm (figure 6), which enables obtaining pulse duration ~50 fs if GVD in amplifier windows will be precompensated by initially introduced negative frequency chirp. Then peak power as high as 30–40 TW can be achieved and with improved beam quality the short pulse can be focused into intensities of up to 10^{20} W/cm^2.

This research was supported by RAS programs of fundamental research and by RFBR grants 08-02-01331, 09-02-12018 OFI-M and 09-07-13593-OFI-TS.

References

[1] Moses E I IFSA 2009 Program Sep. 6-11, 2009, San Francisco, USA, p.175
[2] Lion C and Ebrardt IFSA 2009 Program, Sep. 6-11, 2009, San Francisco, USA, p.177
[3] Sangster T C, *et al.* 2007 *Nucl. Fusion* **47** S686
[4] Obenschain S P, Colombant D G , Schmitt A J, Sethian J D and McGeoch M W 2006 *Phys. Plasmas* **13**, 056320
[5] Basov N G, Gus'kov S Yu and Feoktistov L P 1992 *J. Soviet Laser Res.* **13** 396
[6] Tabak M, Hammer J, Glinsky M E, Kruer W L, Wilks S C, Woodworth J, Campbekk E M, Perry M D and Mason R J 1994 *Phys. Plasmas* **1**, 1626
[7] Zvorykin V D, Lebo I G and Rozanov V B 1997 *Bull. of Lebedev Phys. Inst.*, No. 9-10, 20
[8] Zvorykin V D, *et al.* 2007 *Laser & Particle Beams* **25**, 435
[9] Földes I B and Szatmári S 2008 *Laser & Particle Beams* **26**, 575
[10] Wilks S C, Kruer W L, Tabak M and Langdon A B 1992 *Phys. Rev. Lett.* **69** 1383
[11] Atzeni S 1999 *Phys. Plasmas* **6**, 3316
[12] Sherbakov V A 1983 *Sov. J. Plasma Physics* **9** 240
[13] Betti R, Zhou C D, Anderson K S, Perkins L J, Theobald W and Solodov A A 2007 *Phys. Rev. Lett.* **98** 155001
[14] Schmitt A J, Bates J W, Obenschain S P, Zalesak S T, Fyfe D E and Betti R 2009 *Fusion Sci. Tech.* **56** 377
[15] Obenschain S P, Sethian J D and Schmitt A J 2009 *Fusion Sci. Tech.* **56** 594
[16] Lehmberg R H, Giuliani J L and Schmitt A J 2009 *J. Appl. Phys.* **106** 023103
[17] Khan N, Mariun N, Aris I and Yeak J 2002 *New J. Phys.* **4**, 61.1

Journal of Physics: Conference Series **244** (2010) 032015 doi:10.1088/1742-6596/244/3/032015

Current status of the LUCIA laser system

D. Albach[1,2], M. Arzakantsyan[1], G. Bourdet[1], J-C. Chanteloup[1], Ph. Hollander[1] and B. Vincent[1]

[1]LULI, Ecole Polytechnique, CEA, UPMC Univ. Paris 06; 91128 Palaiseau, France

[2]ISL, 5 rue du Général Cassagnou, 68300 Saint-Louis, France

E-mail: daniel.albach@polytechnique.edu

Abstract. The current status of the LUCIA laser program is discussed. While aiming at 100J, 10Hz, 10ns, a first milestone is set at 10 Joules with a repetition rate of 1-3 Hz. 7ns long, sub-mJ pulses generated by a cavity-dumped oscillator are first preamplified at the sub-J level. Thermal effects limit amplification and repetition rate at this stage. These pulses will be injected into the main amplifier, where amplification is limited by amplified spontaneous emission. It is expected that these pulses reach energy level of ~10J.

1. Introduction

The LUCIA laser system [1] is designed to deliver pulses with energies in the order of 100J with a pulse duration of 10 ns combined with a repetition rate of 10 Hz in its final configuration. The first milestone for the project is set at 10 J with a repetition rate of 1-3 Hz. Main issues to be addressed are Amplified Spontaneous Emission (ASE) and thermal effects management. The impact on the successive stages is somewhat different. Early amplification stages are limited by thermal effects whereas the main amplifier section is limited by ASE [2]. Experimental feedback collected from this program are of great interest for further scaling requirements imposed by the next generation of kJ DPSSL laser beam lines, e.g. for the HiPER project [3,4].

2. System status

Currently, the LUCIA beam line consists of a cavity dumped oscillator, a PreAmplifying Stage (PAS), followed by a beam shaping section and completed by a Main Power Amplifier (MPA).

The oscillator uses Yb^{3+}:YAG as active medium and delivers 7 ns long pulses at 1030 nm with 350 μJ at repetition rates up to 10 Hz. The amplifying medium is not actively cooled and therefore we experience a ~10 % drop in output energy at a repetition rate of 10 Hz compared to the single-shot operation of the oscillator.

These pulses are successively amplified in a pre-amplifier system to the 100 mJ level. A first pre-amplification stage (PAS) section relies on a crystal pumped by one laser diode stack whose emission is concentrated on a 3.5 x 4.8 mm² large surface. The intensity reaches 17kW/cm² in average. The average gain per pass is ~4.5 at the maximum driving current for the laser diode stack of 150A and a pump duration of 1 ms. Amplification is performed through four passes using angular multiplexing and relay imaging. The gain medium is a 30 mm diameter Yb^{3+}:YAG crystal doped at 5 at.% with a thickness of 3.2 mm. As the pump surface is relatively small, we see no evidence of transverse ASE or parasitic oscillations (the large un-pumped periphery acting like an absorbing cladding for 1030 nm radiation). The injected beam has a Gaussian shape with a $1/e^2$ diameter of 3 mm, therefore one can expect a strong impact of the thermal behavior of the active medium on the phase, through focalization and aberrations like astigmatism and coma. As already mentioned, beam propagation is performed using image relay, conserving the intensity distribution in the relay planes. Nevertheless, phase distortions accumulate, increasing the intensity on the imaging lenses. Figure 1a shows the result of a phase measurement at the exit of the PAS for the static phase. A polynomial decomposition on the Zernike basis reveals that the main distortion is due to defocus. 0° astigmatism is important too, as the pumped zone is rectangular. The necessity of a compensation of the phase distortion is currently under observation. Energy limitations are, so far, set by the Laser Induced Damage Threshold (LIDT) of the transport optics used after the PAS. (figure 1b). Lenses designed for a higher LIDT have recently been installed.

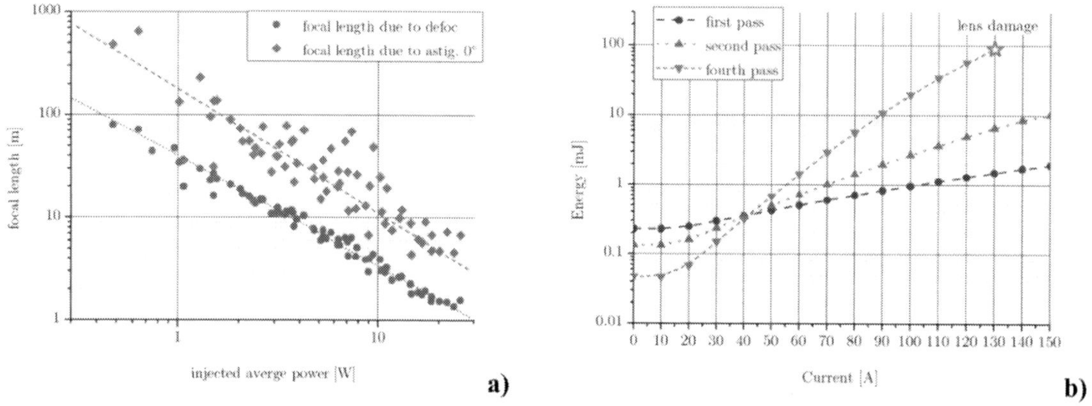

Figure 1: left : Focal length evolution (log vertical scale in meters) versus injected pump power for defocus (circles) and astigmatism (diamonds). Right : Output energy (mJ) versus laser diode stack driving intensity (Amperes) after 1,2 and 4 passes for Lucia PreAmplifer Stage (PAS). Energy build-up was halted at 130 A due to LIDT failure of the lens use for energy monitoring.

After pre-amplification, the beam is transported to the main power amplifier (MPA) using relay imaging combined with a magnifying telescope assembly. The last imaging telescope also filters the high spatial frequencies introduced by a serrated aperture [5] placed in its object plane. The serration pattern is a circular distribution of linear teeth of 1 mm height and a width of 300 µm. The inner diameter of the circular serration pattern is 20 mm.

The MPA is pumped by a partially filled laser diode array, delivering ≈120 J within 1 ms over a ~10 cm² area, resulting in an intensity of up to 12 kW/cm². The active medium is a 60 mm diameter circular Yb^{3+}:YAG crystal doped at 2 at.% with a thickness of 7 mm. Without the negative impact of transverse ASE, considerably high values for the small signal gain (up to $3cm^{-1}$) are theoretically possible. But gain measurements on the MPA show a strong limitation introduced by ASE (see figure 2), as the transverse gain is significantly larger than the gain in extraction direction. The MPA was designed to amplify a rectangular beam. Its pumped area is with its 26 x 38 mm², large enough favouring ASE and parasitic oscillations. Reducing the pumped area to a 30 mm diameter circular aperture by introducing a simple mask, the gain in the amplifier increases significantly from a maximum of ~2 per pass to ~2.8 (green curve). Reflecting the wasted pump energy, with Fast-Axis (FA) concentration mirrors, into the effectively useful pump area, the intensity increases from

~12 kW/cm² to ~ 16 kW/cm², raising the gain again to ~3.2 per pass in the MPA (blue curve). Gain being still limited by the parasitic oscillations within the active medium itself, an index matching liquid was injected in a peripheral cavity located at the outer border of the crystal. A gain of ~3.4 per pass was achieved (pink curve), limited by a back reflection from the crystal mount leading to the parasitic oscillations. Pump durations longer than ~700 µs are not reasonable as they will lead to pure heating and not show any positive contribution to the achievable gain.

Besides the ASE limitations due to the transverse oscillations, accumulated phase resulting in focal chnges on the one hand and depolarization on the other. These were monitored and used as a criterion to qualify the laser head. We studied two different designs for the main amplifier gain medium geometry. The first relied on a 2 at.% doped rectangular crystal with a size of 40 x 36mm² and a thickness of 8.4 mm. Both crystal faces were AR coated at 940 nm, whereas only one face was also AR coated at 633 nm, with the other one HR coated at 633 nm. This offers the possibility to measure both surfaces deformation independently by flipping the crystal in its mount. The second crystal is a cylinder (60 mm diameter, 7 mm thick, 2 at.%) AR coated for both pump and probe (at 1064nm in this case) on one side and HR for both on the other. This second design demonstrates a significant improvement compared to the rectangular one - almost one order of magnitude in the introduced static thermal lens, as shown in figure 3a.

The stress induced birefringence of the 60 mm diameter crystal was measured using a set of crossed polarizers as described in [6]. Minimizing the birefringence is achieved by orienting the crystallographic [111]-axis perpendicular to the pumped surface and parallel to the extraction [7]. As the gain medium exhibits a natural birefringence, the average depolarization (square dots) within a 25 mm diameter beam shows a minimum at about 200 W of injected average power on a 30 mm diameter surface. The depolarization in the centre (circle dots) however rises almost linearly with the injected average power. This can be understood, as the intrinsic birefringence depends on the growth, orientation, mounting and coating of the crystal, whereas the introduced birefringence is linked to the added thermal stress, exhibiting a classical four-foil shape with a local maximum in the centre. The overall average depolarization observed never exceeds 1°.

The working point (for 10J extraction level) for the MPA is in the order of ~100-300 W average pump power in the current pumping configuration. We therefore expect a single pass gain of approximately 3 (including transport losses) and negligible defocusing as well as a minor depolarization in the 4-pass design for the MPA.

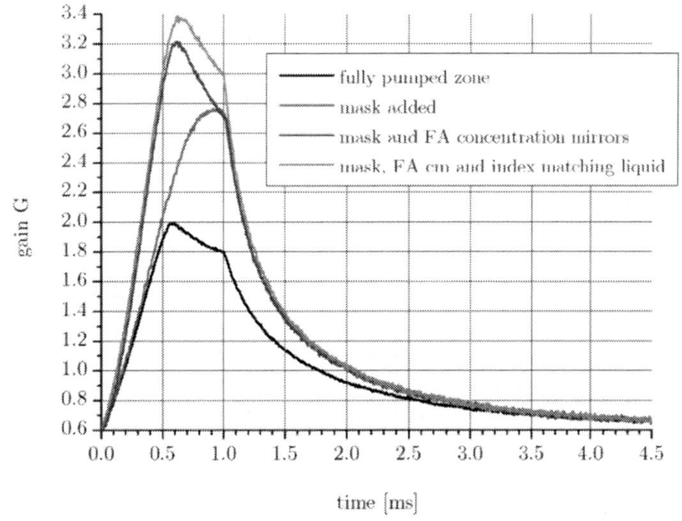

Figure 2:single pass gain evolution with time (ms) during and after a 1 ms pump pulse (see exponential decays starting at the end of the pump pulse for all curves). Each curve is associated to a different transverse ASE control strategy put in place according to inserted legend and full body text explanation.

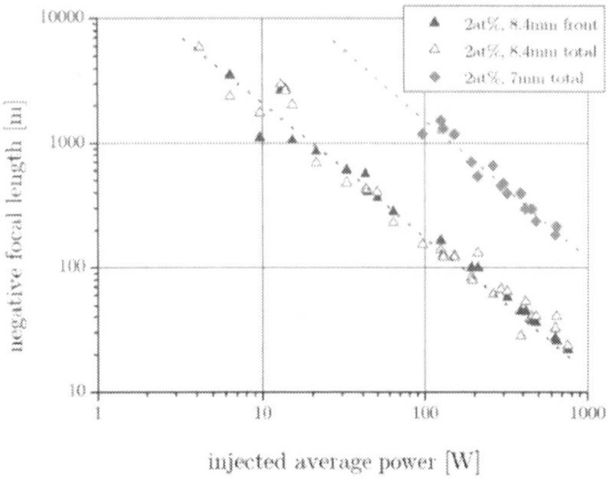

a)

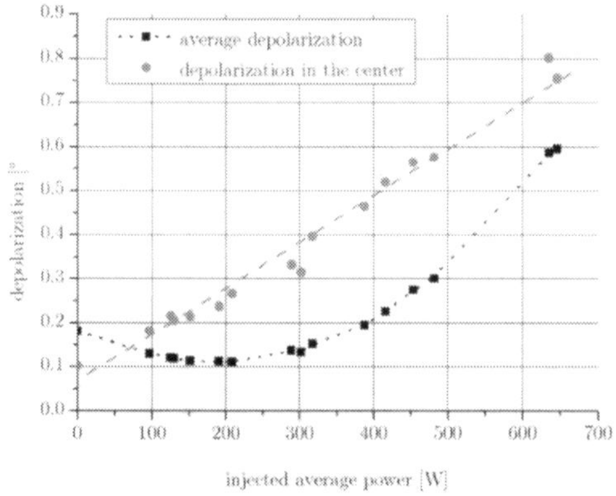

b)

Figure 3: Left : Focal length evolution (log vertical scale in meters) versus injected pump power for 2 crystal geometries (rectangular, triangle points and cylindrical, diamond points). One order of magnitude can be observed between crystals. Right : Evolution of depolarization (in degrees) versus injected pump power for the 60 mm diameter crystal.

3. Conclusion

The influences of thermal effects as well as ASE are studied in the context of the experimental realization of the first energy milestone (10J) for the LUCIA laser project. While ASE plays no important role for the oscillator as well as for the PreAmplifying Stage (PAS), it is important for the Main Power Amplifier stage (MPA). Thermally introduced static deformation is measured for the PAS as well as for the MPA and an evaluation for the introduced birefringence was performed for the MPA as well.

4. References

[1] Chanteloup J.-C., Yu H. , Bourdet G., Dambrine C., Ferré S., Fülöp A., Le Moal S., Pichot A., Le Touzé G. and Zhao Z., *Proc. of SPIE*, **5707** 105-116 (2005)

[2] Albach D., Chanteloup J.-C. and Le Touzé G., "Influence of ASE on the gain distributionin large size, high gain Yb3+:YAG slabs", *Optics Express* **17** (5), 3792-3801 (2009)

[3] Dunne M., *Nature Physics* **2**, 2 - 5 (2006)

[4] Chanteloup J.-C., Albach D., Lucianetti A., Ertel K., Banerjee S., Mason P., Hernandez-Gomez C., Collier J., Hein J., Wolf M., Körner J. and Le Garrec B. "Multi kJ Level Concepts for HiPER Facility", Sixth International Conference on Inertial Fusion Sciences and Applications (IFSA 2009), 6-11 september 2009, San Francisco, USA

[5] Auerbach J. and Karpenko P., "Serrated-aperture apodizers for high energy laser systems", Applied Optics, **33**, 3179-3183, (1994)

[6] Hodgson N and Weber H, "Laser Resonators and Beam Propagation – Fundamentals, Advanced Concepts and Applications", 2nd edition, Springer (2005), chapter 24

[7] Chen Y, Chen B, Patel M, Bass M, "Calculation of thermal-gradient-induced stress birefringence in slab Lasers-I", IEEE Journal of Quantum Electronics, **40**, 909-916 (2004)

The Sixth International Conference on Inertial Fusion Sciences and Applications IOP Publishing
Journal of Physics: Conference Series **244** (2010) 032016 doi:10.1088/1742-6596/244/3/032016

A Laser Technology Test Facility for Laser Inertial Fusion Energy (LIFE)

A.J. Bayramian, R.W. Campbell, C.A. Ebbers, B.L. Freitas, J. Latkowski, W.A. Molander, S.B. Sutton, S. Telford, and J. A. Caird
Lawrence Livermore National Laboratory, P.O. Box 808, Livermore, CA 94550, USA

bayramian1@llnl.gov

Abstract. A LIFE laser driver needs to be designed and operated which meets the rigorous requirements of the NIF laser system while operating at high average power, and operate for a lifetime of >30 years. Ignition on NIF will serve to demonstrate laser driver functionality, operation of the Mercury laser system at LLNL demonstrates the ability of a diode-pumped solid-state laser to run at high average power, but the operational lifetime >30 yrs remains to be proven. A Laser Technology test Facility (LTF) has been designed to specifically address this issue. The LTF is a 100-Hz diode-pumped solid-state laser system intended for accelerated testing of the diodes, gain media, optics, frequency converters and final optics, providing system statistics for billion shot class tests. These statistics will be utilized for material and technology development as well as economic and reliability models for LIFE laser drivers.

1. Introduction

The search for a source of green, carbon-emission-free power source is perhaps one of our most important activities in the twenty first century. Of the many technologies under consideration, the way to harness the most energy from the least amount of fuel is to use nuclear power. In response to this logic, nuclear fission power plants dot the globe, yet they have not entirely taken over the power industry due to issues with nuclear waste storage and weapons proliferation. The "other" nuclear power is nuclear fusion, the same power that fuels our sun and all the stars we see in the sky. Understanding of the efficiency and cleanliness of this power source has generated over fifty years of research and development as we try to harness this process on earth.

For the first time, there are several projects underway worldwide which promise to demonstrate fusion ignition and gain > 1. The first of these facilities to be completed is the National Ignition Facility (NIF) at Lawrence Livermore National Laboratory (LLNL). In 2010, the team of scientists and engineers operating the NIF will start a campaign to demonstrate fusion ignition, which could serve as the basis for a national program to develop Laser Inertial Fusion Energy (LIFE). A LIFE laser driver needs to be designed and operated which meets the rigorous requirements of the NIF laser system. In addition, the LIFE laser will have new requirements of operating at high average power and having a system lifetime of >30 years (10 billion shots). No laser has ever simultaneously met these requirements, yet a credible design must be based on relevant experimental demonstrations. Ignition on NIF will serve to demonstrate laser driver functionality. Previous operations of the Mercury laser system at LLNL have demonstrated the ability of a diode-pumped solid-state laser to run at high average power. However, the final system lifetime requirement of >10 billion shots remains unproven.

One of the first steps toward demonstrating laser viability over a 30 year period is an accelerated test of the basic components and technologies. Proven components and technologies could then be assembled in a full scale laser demonstrator where the engineering and durability would have a high probability of success (expense higher and risk necessarily lower). To accomplish this task, we propose the construction of a Laser Technology Test Facility (LTF), a diode pumped solid state laser system which would operate at 100 Hz, which would enable billion shot tests in <6 months.

2. A 100-Hz Laser System

The LTF laser builds on laser technology and expertise developed at LLNL in both the NIF laser and the Mercury laser system [1]. The basic architecture is that of a master oscillator power amplifier (MOPA) with a fiber based oscillator that enables precision pulse shaping seeding larger aperture bulk amplifiers to produce the required energy.

The seed laser begins with a 15 mW continuous single mode oscillator. Since the gain peak of the Nd:YAG based amplifier is temperature dependent[2], a tunable source is required to maintain the oscillator at the effective gain peak of the diode-pumped power-amplifiers. Temporal shaping is performed using an integrated $LiNbO_3$ electro-optic Mach-Zehnder modulator controlled by a Highland Technologies arbitrary waveform generator. At the output of this system, the pulse energy is in the pico-Joule regime. To meet the micro-Joule energy input requirements of the pre-amplifier system, the pre-amplifier needs micro-Joule class energy, which is achieved with a three-stage fiber-amplifier system. Each stage utilizes polarization maintaining single mode ytterbium doped fiber amplifiers which provide 20 dB gain.

The pre amplifier chain begins with a collimated fiber launch from the final fiber amplifier providing a 6 mm beam with Gaussian spatial profile. The Gaussian is converted to a flat-top spatial beam profile using a refractive shaper (MolTech). To minimize diffractive ripple in the propagating beam, an apodizer is used to generate a 20th order superGaussian. An adaptive optic provides wavefront correction based on a wavefront sensor signal which is located in an amplifier diagnostic at the output of the system. After magnification, the beam passes through two stages of Faraday isolation and is imaged in the first amplifier head. The 10 mm aperture diode-pumped Nd:YAG amplifiers (Northrup Grumman) are capable of Joule class operation at > 100 Hz. The beam is image relayed to a second identical amplifier situated in a 4-pass angular-multiplexed, image-relayed cavity. When activated at 100 Hz, the amplifier heads have a thermal focal length which must be compensated to maintain collimated image relay. By decreasing the length of the relay telescopes, the transmitted beam is divergent and can exactly compensate the thermal lens providing a collimated beam. After passing through the amplifier, the beam is reflected by a mirror along a second multiplexing angle, enabling the beam to pass by the injection mirror, and hit a retro reflection-mirror. This reflected beam retraces the entire beam path back to the first amplifier, where a Faraday rotator ejects the pulse to the targeting telescopes.

The power amplifier is the heart of the LTF laser system, generating valuable diode array and diode driver lifetime data in addition to the 10-J, 100-Hz operation for testing potential LIFE optics. To achieve 1 kW average power, the power amplifier required custom design of the diode array, diode light delivery, and amplifier head to accommodate the increased thermal load.

The diode array technology builds on the SIMM [3] technology developed at LLNL for the Solid State Heat Capacity Laser (SSHCL). This diode packaging offers high average power and brightness in a simple, rugged, scalable architecture that is suitable for large two-dimensional arrays. While the diode bar and packaging technologies meet the needs of the LTF laser system (and perhaps LIFE), the diode power supplies are lacking. Current power supplies are limited to ~20 Hz, the cost per unit is excessive (~ $1/W), and the pulsers are inefficient due to operational parameter adjustability, diagnostics, and the need for long cables between pulser and diode array. Of the many tasks facing the LIFE laser system, one of them is to reduce the cost and increase efficiency of diode pulsers. A new kind of pulser, named a "smart tile," has been designed which is a miniaturized and simplified version of the current pulsers. The central design is based on the following key elements: remove all processing and adjustability, place the pulser in close contact to the diodes it is driving, and utilize the diode array backplane as the heatsink for the pulser. Using a circuit design program (SPICE, RD

Research), our "smart tile" model indicates these devices should meet performance requirements with a baseline component cost which is 40X lower in cost and 20X lower in size while being 15% higher in efficiency. This kind of pulser will be utilized on the LTF both to meet the demanding average power requirements and as a stepping stone towards LIFE technology. Just like the optics being tested in the output portion of the laser, the diode array and diode pulsers will also be tested in an integrated fashion for billions of shots.

The diode light delivery system for the amplifier represents a somewhat different than the duct approach previously used on Mercury [1], the mix of ducts and refractive optics in LUCIA[4], or the direct side pumping techniques used in HALNA [5] and other systems. In the fast axis direction (after microlensing < 1° degree divergence), the diode array is subdivided into individual columns of diode tiles (10 bar diode stacks). A prism is placed in front of each column and is used to direct the diode light toward the amplifier. Rather than focus the light, the diode emission is allowed to diverge to fill the amplifier aperture. The overlay of multiple diode array columns creates a high intensity pump profile without the need for ducting or focusing. Along the slow axis, where the divergence is close to 10°, a pair of cylindrical lenses is used to image the diode array onto the amplifier. Immediately preceding the amplifier, a very short refractive duct is used to clean up the edges of the pump profile to create a sharp edge to the beam thereby improving extraction efficiency and modefill (area of the extraction beam relative to the pump area). A FRED (Photon Engineering) ray-trace model was used to engineer an optimum solution for the array characteristics and amplifier size. The ray trace results include net absorption for the amplifier and indicate >95% energy delivered to the 2.5x2.5 cm^2 aperture with peak-to-valley gain profile of 2:1. Since the diode brightness is conserved, the almost perfectly flat pump beam observed at the end of the duct propagates and diverges, which has an effect on the uniformity and edges of the beam. Still, considering the amplifier thickness is greater than the aperture size, the gain profile is surprisingly flat and sufficient for the LTF laser requirements.

The power amplifier for this high average power laser system requires careful thermal management and design to avoid beam modulation, depolarization, and thermally induced fracture of the amplifier slabs. At 100 Hz, the amplifier slabs must dissipate more than 660 W of thermal power or 105 W/cm^2. As on Mercury, LUCIA, and many other slab amplifiers worldwide, an effective method of minimizing wavefront distortion is to remove the heat in a geometry which creates a quasi-one-dimensional thermal gradient along the same direction as the laser propagation. This geometry ensures that all parts of the propagating laser beam see the same set of thermal gradients and therefore the same index profile as the beam propagates through the system. To first order, a system designed in this way is athermal. For the LTF amplifier, the method we have chosen is to use sapphire, a tough transparent heatsink, since sapphire has a close refractive index and thermal expansion match to Nd:YAG. To mitigate losses due to amplified spontaneous emission (ASE), an edge cladding of SF14 (Schott Glass) doped with copper is bonded to the Nd:YAG perimeter. Using this geometry as the baseline design, a complete 3-D thermal model of the amplifier head was developed using code that was benchmarked on the Mercury laser [1]. The modeled thermal wavefront is approximately 33 waves of power and a residual wavefront that is nominally a 4th order stress induced edge effect (~5.4 waves). The power can be compensated with a small change in the relay telescope length and the higher order wavefront corrected by our adaptive optic in the pre-amplifier or even a static wavefront corrector as was demonstrated on the Mercury laser [1]. Although we have attempted to create a 1-D thermal gradient, the effect is not perfect since sapphire's thermal conductivity is far from infinite at room temperature (~ 30 W/m K). However, considering the average power of this Nd:YAG system, the temperature and thermal wavefront are reasonable and tractable.

The architecture of the LTF borrows from the experience gain on the Mercury laser and the NIF. The 10 J is extracted in four passes, passively switched by taking advantage of angular multiplexing. Utilizing a relatively large multiplexing angle (~1°), the 100 mJ injected beam is spatially separated from the other three passes through the system. After passing through the amplifier twice, the beam exits the telescope where a knife edge mirror sends the beam through a Pockel's cell for parasitic isolation. A second knife edge mirror re-injects the pulse for a 3rd and 4th pass through the system.

After the 4th pass, the pulse exits the system. The entire architecture comfortably fits on a 5'x16' table. For other applications, such as laser peening or pump source for optical parametric amplification (OPA), or Ti:sapphire based chirped pulse amplification systems, this architecture can be easily folded to a much smaller footprint.

3. Component Technology Testing

The central purpose behind the LTF is the accelerated testing of laser systems to shot counts relevant to the lifetime of a LIFE laser driver. At 13 Hz, a LIFE plant would accumulate 12.3 billion shots in 30 years. Development of the materials for LIFE will require all of the quality control, materials purification, fabrication, polishing, and coating developments that have occurred as part of the NIF development and more. To get the most data and statistics from a test, our strategy is to expose many multiple samples of many different kinds of optics simultaneously. By placing optics that have a higher probability of survival upstream from optics that are more questionable, downstream modulation can be minimized. To maintain the flat-top beam and further minimize downstream modulation, image relay telescopes are used to reimage the beam between different types of optics. The image relay minimizes nearfield modulation, allows beam cleanup utilizing pinholes in the relay telescopes, and adds the freedom to change the beam diameter / testing fluence on different kinds of optics. Testing results will be incorporated into optical lifetime models, system availability models, and economic models. Just like material developments for the NIF, the results will also be used in iterative development loops aimed at improving fabrication and finishing processes for the optics as well as testing mitigation techniques for passivating optical defects or repairing optics. There are two baseline testing strategies: cm^2 small area testing for 6 months will generate 1.6 billion shots which is a significant fraction of LIFE plant lifetime, and full aperture raster scanning using the cm^2 beam which in 6 months would generate millions of shots.

In addition to the optical testing, the diode array technology will also be tested including: diode bars, diode packaging, diode pulsers, and the integrated array unit. Finally, the long term tests indirectly test the computer controls, data acquisition, diagnostics, and support equipment. Our intent is that the LTF serve as a scale model of a LIFE driver beamline, containing many of the control points, optics, and diodes tested at LIFE driver relevant levels. This work was performed under the auspices of the U.S. Department of Energy by Lawrence Livermore National Laboratory under Contract DE-AC52-07NA27344.

References

[1] Bayramian A, Armstrong P, Ault E, Beach R, Bibeau C, Caird J, Campbell R, Chai B, Dawson J, Ebbers C, Erlandson A, Fei Y, Freitas B, Kent R, Liao Z, Ladran T, Menapace J, Molander B, Payne S, Peterson N, Randles M, Schaffers K, Sutton S, Tassano J, Telford S and Utterback E, 2007 The Mercury Project: a high average power,gas-cooled laser for inertial fusion energy development *Fus. Sci. & Tech.* **52** 383-7

[2] Kimmelma O, Tittonen I and Buchter SC 2008 Thermal tuning of laser pulse parameters in passively Q-switched Nd:YAG lasers *Appl. Opt.* **47** 4262-6

[3] Skidmore JA, Freitas BL, Crawford J, Satariano J, Utterback E, DiMercurio L, Cutter K and S. Sutton 2000 Silicon monolithic microchannel-cooled laser diode array *Appl.Phys. Lett.* **77** 10-12

[4] Chanteloup JC, Yu H, Bourdet G, Dambrine C, Ferré S, Fülöp A, Le Moal S, Pichot A, Le Touzé G and Zhao Z 2005 Overview of the Lucia laser program: towards 100 Joules, nanosecond pulses, kW averaged power, based on Ytterbium Diode Pumped Solid State Laser *Solid State Lasers XIV: Technology and Devices, Proc. SPIE 5707, edited by Hanna J. Hoffman, Ramesh K. Shori,* (SPIE, Bellingham, WA, 2005) 105-16

[5] Kawashima T, Kanabe T, Matsumoto O, Yasuhara R, Furukawa H, Miyamoto M, Sekine T, Kurita T, Kan H, Yamanaka M, Nakatsuka M and Izawa Y 2006 Development of Diode-Pumped Solid-State Laser HALNA for Fusion Reactor Driver *Elec. Eng. in Jap.* **55** 27-35

Journal of Physics: Conference Series **244** (2010) 032017 doi:10.1088/1742-6596/244/3/032017

LMJ on its way to fusion

J. Ebrardt, JM. Chaput

Commissariat à l'Energie Atomique, Centre Île de France – 91297 Arpajon, FRANCE

jacques.ebrardt@cea.fr

jeanmichel.chaput@cea.fr

Abstract. The Laser Mégajoule (LMJ), is under construction by the French Commissariat à l'Energie Atomique (CEA) at CESTA laboratory near Bordeaux. The LMJ is an important part of the French "Simulation Program". The LMJ is dedicated to high energy density physics experiments, and it is designed to obtain ignition and fusion of a DT fuel capsule. The laser and target systems have been designed to meet the specifications needed for these experiments, in an overall optimization including the fusion target. The current status of the LMJ project is presented.

1. Introduction

LMJ's laser system is designed to deliver up to 1.8 MJ and up to 550 TW of 0.35 μm UV light on target. The pulse duration can be tuned between 0.2 and 25 ns and can be adjusted to the desired pulse shape.

The performances of a quadruplet of beam lines have been demonstrated on the Ligne d'Intégration Laser (LIL), LMJ's prototype [1]. It has been used to test technological improvements until 2008, all laser components and systems being now qualified. LIL is also used for plasma experiments since 2004 [2].

The target system includes different sub-systems around the target chamber, among them the positioning equipment for the diagnostics and the targets (non cryogenic or cryogenic for fusion experiments), the diagnostics themselves, the alignment system… These sub-systems have also make use of the feedback of LIL operation and specific prototyping, for example for cryogenic equipment.

2. Building construction

The main building has been commissioned in December 2008 after five years of work (figure 1). The utilities building is in operation since 2007, and the laser and target bays are air conditioned. The laser bays are in clean room condition.

3. Control system

LMJ control system is based on a layered architecture, with four levels: equipment control, subsystem control, system supervisory including shot planning and operation, and high level supervisory; this last upper level includes maintenance management, shot archiving and analysis and network administration, and is linked to personnel safety. The first two levels are included in the equipment contracts; the two supervisory levels have been contracted to software companies and have passed the advanced design review.

4. Laser system

The laser bundles amplification sections are under assembly, the first laser bay being completed (figure 2). Work is now in progress in the three other laser bays. All laser components and sub-systems have been qualified, taking into account the feedback from experiments and tests on LIL that has been used to make improvements on some of the designs. All these components are in, or ready for, serial production. Operation of the laser lines is based on the concept of "In Line Replaceable Units" (LRU); every type of LRU is now in production.

Figure 1: LMJ building

Figure 2: Laser bay 1

LMJ is using a pair of gratings as focusing system, one on each side of the two frequency conversion crystals. The second grating is focusing the tripled wavelength at 0.35 μm and also rejects the undesired wavelengths, mostly the fundamental at 1.05 μm; they are absorbed outside the target chamber. Combined with 0.5 nm bandwidth and a phase plate, the gratings also provide longitudinal smoothing by spectral dispersion.

A first prototype of the frequency conversion and focusing system (SCF) in operation on LIL since 2002 gave good results. Its operation feedback led to optimisation, to a final LMJ version with a more robust design that will allow simpler and faster operation. This LMJ SCF has been fully tested and is now qualified (figure 3). All the requirements have been fulfilled, for example: superimposing the four focal spots of the individual beams of the quadruplet has proved to be very easy, with a FWHM of 150 μm (figure 4); this quadruplet focal spot stays constant when changing the pointing of the SCF inside the specified area.

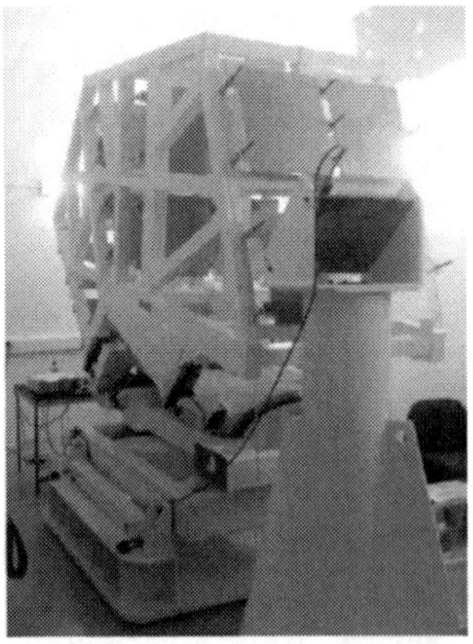

Figure 3: LMJ SCF prototype

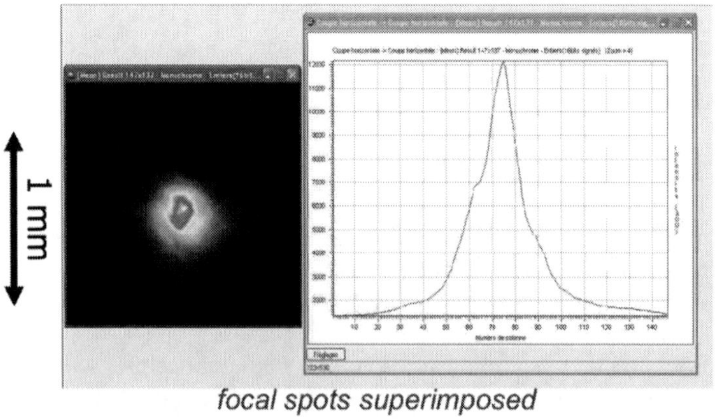

Figure 4: Focal spot of the four beams of a quad with LMJ SCF

5. Target bay

The steel and aluminum frameworks are being installed on the concrete structures around the target chamber (figure 5). Process equipment mounting will start in 2010. CAD mock-up of the target bay is complete and includes all process and utilities parts; it is used for mounting and its control and to prepare operation and maintenance (figure 6).

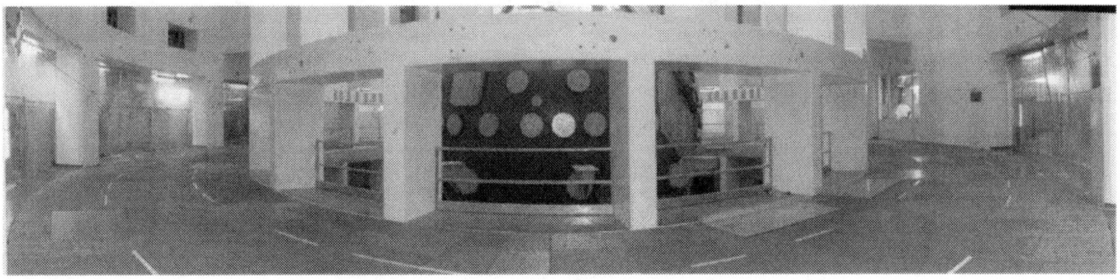

Figure 5: The target chamber seen from the completed mid "equator" level

Figure 6: CAD view of the main process equipment of the target bay when completed

The detailed engineering of the plasma diagnostics inserters (SID) has been completed. Their operation and maintenance have been studied with CAD and real mock-up, taking into account nuclear constraints.

The plasma diagnostics designs beneficiate from the feedback of the LIL diagnostics routinely used for plasma experiments since 2004, while being adapted to LMJ experiments (for example spatial resolution and field of view); their design is also taking into account the hardening necessary to withstand the harsh constraints of hard X-rays and debris. The first plasma diagnostics are ready for fabrication, beginning in 2010.

Prototypes of the specific equipment are built and tested. Two examples are the chamber maintenance robot and the alignment system. The main purpose of the chamber robot is to install and remove protection panels inside the chamber; the robot is undergoing its final tests (figure 7). The alignment system is under tests with both "classic" and cryogenic fusion look-alike targets.

Figure 7: Chamber maintenance robot and CAD view of the protection panels inside the chamber

LMJ cryogenic targets are very sophisticated objects [3]. They have to be handled from the fabrication unit to LMJ's target holder while maintaining cryogenic temperature. Scale 1 prototypes of the different necessary cryogenic units have been made and tested, demonstrating the feasibility of the whole process (figure 8). A thermal shroud is needed to insulate the target from the ambient temperature and must be removed a fraction of a second before the shot, without inducing perturbation of the target temperature nor vibration. Prototypes of the shroud and its fast removal system have been built and have been successfully tested: the requirement of thermal stability of ± 0.2 mK during the gripping and the removal of the shroud has been fulfilled.

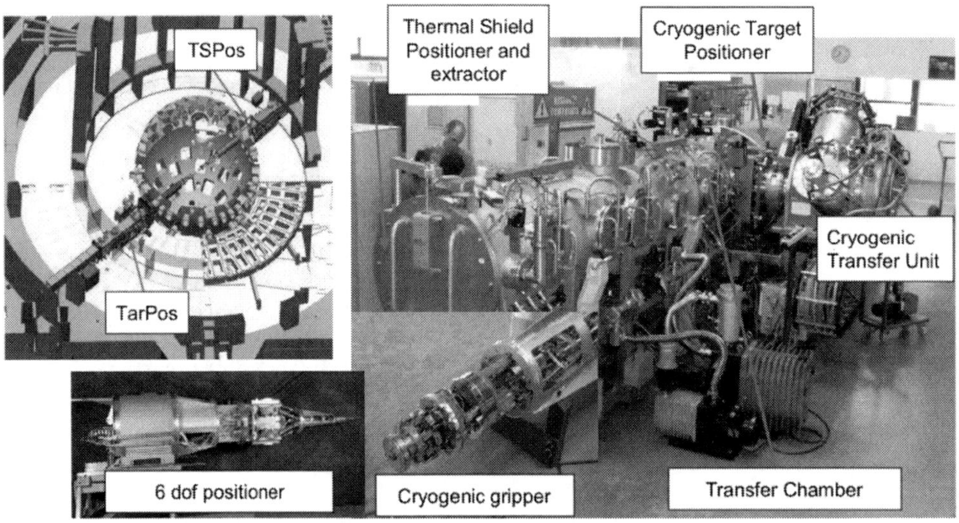

Figure 8: Prototypes of the cryogenic target positioner and transfer unit

6. Conclusion

LMJ building has been commissioned, and laser bundles amplification section assembly is in progress; the first laser bay is complete. The mounting of equipment in the target bay has started, with process equipment mounting beginning in 2010. All laser components and systems have been qualified and meet the specifications; target chamber equipment prototypes under test give expected results.

References

[1] JM. Di-Nicola et al, Journal de Physique, IV, **vol 133**, 595 (2006)
[2] A. Casner et al, J. Phys, : Conf Ser., these proceedings

[3] R. Collier et al, J. Phys, : Conf Ser., these proceedings

Assessment and Mitigation of Radiation, EMP, Debris & Shrapnel Impacts at Megajoule-Class Laser Facilities

D. C. Eder[1], R. W. Anderson[1], D. S. Bailey[1], P. Bell[1], D. J. Benson[3], A. L. Bertozzi[4], W. Bittle[5], D, Bradley[1], C. G. Brown[1], T. J. Clancy[1], H. Chen[1], J. M. Chevalier[6], P. Combis[7], L. Dauffy[1], C. S. Debonnel[7], M. J. Eckart[1], A. C. Fisher[1], A. Geille[6], V. Y. Glebov[5], J. Holder[1], J. P. Jadaud[7], O. Jones[1], T. B. Kaiser[1], D. Kalantar[1], H. Khater[1], J. Kimbrough[1], A. E. Koniges[2], O. L. Landen[1], B. J. MacGowan[1], N. D. Masters[1], A. MacPhee[1], B. R. Maddox[1], M. Meyers[3], S. Osher[4], R. Prasad[1], D. Raffestin[6], J. Raimbourg[7], V. Rekow[1], C. Sangster[5], P. Song[1], C. Stoeckl[5], M. L. Stowell[1], J. M. Teran[4], A. Throop[1], R. Tommasini[1], J. Vierne[7], D. White[1], P. Whitman[1]

[1]LLNL, [2]LBL, [3]UCSD, [4]UCLA, [5]LLE, USA, [6]CEA, DAM, CESTA, France, [7]CEA, DAM, DIF, France

Email: deder@llnl.gov

Abstract. The generation of neutron/gamma radiation, electromagnetic pulses (EMP), debris and shrapnel at mega-Joule class laser facilities (NIF and LMJ) impacts experiments conducted at these facilities. The complex 3D numerical codes used to assess these impacts range from an established code that required minor modifications (MCNP - calculates neutron and gamma radiation levels in complex geometries), through a code that required significant modifications to treat new phenomena (EMSolve - calculates EMP from electrons escaping from laser targets), to a new code, ALE-AMR, that is being developed through a joint collaboration between LLNL, CEA, and UC (UCSD, UCLA, and LBL) for debris and shrapnel modelling.

1. Introduction

Extending the energy of large laser facilities from the 10's of kJ's (Omega) to the MJ range (NIF & LMJ) increases the impact of operation on the facility. In this paper, we discuss the impacts of radiation, EMP, and debris/shrapnel and ways to mitigate these impacts. On the path to ignition, a series of THD (tritium, hydrogen, deuterium) shots are planned with the mix of isotopes chosen to produce relatively low yield/radiation levels for conditions that would produce ignition and high yield for a pure DT capsule [1]. However, even for these lower yields (~ 10^{15} neutrons) appropriate collimation, shielding, and beam dumps must be designed for diagnostics. The majority of this design work is done with the 3D Monte Carlo code, MCNP [2], which we also use for our calculations of worker doses for activities in the NIF facility. It is important that all structures (diagnostics, auxiliary shielding, etc.) be designed with both background levels for diagnostics and radiation levels for workers in mind.

For all experiments on NIF and LMJ, laser-matter interactions will produce energetic electrons. The number of these electrons that escape the target primarily depends on their energy, size of target, and duration of laser pulse. Escaping electrons produce an electromagnetic pulse (EMP) that can impacts electronics in diagnostics and other facility systems. The energy of the electrons depends on the intensity of the laser and the corresponding laser-mater interaction physics. For laser intensities in the 10^{12}-10^{16} W/cm^2 range, laser-plasma interactions (LPI) generally produce "hot electrons" with energies in the 10's of keV range with higher energies tails extending out to a few 100's of keV's. In contrast, very high intensities (10^{18}-10^{20} W/cm^2) can produce very energetic (MeV) electrons. At NIF, long (ns) pulse operation is used to heat hohlraums and other targets with LPI producing hot electrons. Electrons with energies extending out to 10's of MeV will be generated at NIF when we convert 1-4 beams to short (ps) operation as part of the Advanced Radiographic Capability (ARC). We expect large EMP with electric fields in the few MV/m range for ARC operation. We have studied EMP associated with MeV electrons on the Titan laser facility with measured electric fields up to 400 kV/m that are very broadband extending from 50 MHz to 5 GHz or beyond. Good agreement with EMP data has been obtained on Titan with the 3D electromagnetic code, EMSolve [3], using measured spatial distributions of the escaping electrons. Significant modifications of the EMSolve code were required to model EMP in laser chambers. A suite of EMP probes has been installed on NIF at locations inside and outside the chamber. The inside location is ~4 m from target chamber center (TCC).

Debris and shrapnel associated with NIF and LMJ experiments must be assessed and appropriate mitigation steps must be taken to protect optics and diagnostics. Both facilities use relatively thin (1-3 mm) borofloat glass as the final optic. On NIF they are called disposable debris shield (DDS's) and on LMJ they are EPA's. Targets including shields, etc. must be designed so that no large shrapnel fragments with velocities sufficient to penetrate the relatively thin optics are launched towards the optics. Diagnostics components, e.g., pinhole arrays, can become a source of shrapnel that can damage other components, e.g., microchannel plates. Modelling of debris/shrapnel generation is particularly challenging because the simulation domain can includes hot radiating plasma and cold fragmenting solids. A new 3D code, ALE-AMR, has been developed for this application [4]. The code combines arbitrary Lagrangian Eulerian (ALE) hydrodynamics with adaptive mesh refinement (AMR) to allow the code to dynamically add zones at interfaces and in regions with steep gradients. Validation experiments have been done at LLNL's Janus laser and at CEA's LIL laser. The code is used to model NIF experiments and will be used by CEA researchers to model LMJ experiments.

2. Neutron and gamma backgrounds

In order to achieve acceptable performance from ignition diagnostics, it is necessary to reduce neutron and gamma backgrounds by the use of well designed collimators, shields, and beam dumps. The design of these components depends on the spectrum of the neutrons (DD, THD, or DT shots), the neutron yield, and the time range of the instrument. The high-energy x-ray imager (HEXRI) is a time-gated instrument that is sensitive to the neutron and gamma radiation for 40 ns after the laser pulse. The calculated gamma fluence using MCNP at 40 ns is shown in Figure 1 [5]. For this time period, the dominant sources of gammas are n-γ reactions in components near TCC, e.g., the cryogenic target positioner. (For instruments integrating over all time, e.g., image plates, the n-γ reactions in the target chamber wall are the dominant sources of gammas.) We have designed a tungsten collimator that reduces the HEXRI background by more than an order of magnitude [5].

3. Electromagnetic pulses

EMP probes are sensitive to changing electric and magnetic fields, called Ddot and Bdot, respectively. To cover the wide frequency range of interest (~10 MHz to ~5 GHz) a suite of probes is required. We have installed EMP probes inside the NIF chamber as shown in Figure 2. They are ~1 m away from the stainless steel first wall panels and ~4 m from TCC. We have measured EMP for a wide range of target shots with electric fields generally in the 1-10 kV/m range. Given the expected $1/r^2$ scaling for the initial portion of the EMP pulse, fields ~15 times larger are expected for diagnostics located ~1 m from TCC. A significant fraction of the EMP is at relatively high (> 1 GHz) frequencies [6]. EMSolve simulations show that such fields can be produced by ~10^{12} electrons escaping in a few ns.

For ARC operation on NIF, we expect significantly larger EMP with fields at 1 m from TCC to exceed 1 MV/m and frequencies extending out to 5 GHz and beyond. This is based on field measurements done at the Titan and Omega-EP laser facilities. In general, fields of order 1 MV/m must be reduced by ~ 80 dB (4 orders of magnitude) to avoid data disruption and electronic damage. We are testing improved EMP mitigation techniques on NIF. This work is being applied to diagnostic cables where we have observed volt level signal on well shielded, terminated coaxial cables.

For yield operation, the gammas produced from n-γ reactions in target bay components can produce Compton electrons in cables resulting in system generated EMP (SGEMP). This type of EMP requires significant radiation shielding and becomes more important as we go to high yield operation. Our approach is to place essentially all diagnostics behind 2 m of concrete for high yield operation.

4. Debris and shrapnel generation

In general, the primary target, e.g., hohlraum, is completely vaporized in NIF and LMJ experiments. This vaporized material along with debris produced from unconverted light striking the first wall panels in NIF (no unconverted light enters the LMJ chamber), and x-ray ablation from first walls and other components inside the target chamber coats the NIF DDS's and the LMJ EPA's. At both facilities, these thin optics are located in cassettes that can hold up to 10 optics. The total mass of debris is controlled (< 1g/shot) to obtain acceptable replacement rates for the cassettes.

The issue that generally requires mitigation is the generation of molten or solid fragments from auxiliary target components or diagnostics, such as supports, shields, backlighters, pinhole arrays, etc.. For ignition targets on NIF, the high-Z hohlraum is surrounded by an Al thermal-mechanical package (TMP), which has an average thickness of ~500 microns. For equatorial diagnostics located near TCC, the dominant loading is from a debris wind produced by the vaporized TMP. (For polar diagnostics, x-ray loading is important.) The Al deposited on the front of the gated x-ray imager (GXD) in shown in Figure 3. This debris wind was observed to be strong enough to bend and nearly completely shear a Ta pinhole array located behind a 500-micron thick Ta collimator as shown in Figure 4. (This shot was for a debris test with no camera being used.) In following shots, the Al wedge was replaced by a back 500-micron thick Ta collimator and the GXD was able to obtain images of imploding capsules.

It was determined that the flat Si support arms used for cryogenic targets reflect too much 1ω light into other beam lines. One short-term proposal was to replace the flat Si supports with curved Al rods until a long-term approach could be developed. We modeled this proposed configuration with ALE-AMR and found that it was acceptable for shots with ~1 MJ of laser energy but not acceptable for shots with ~1/3 MJ of energy. We show results for this lower energy case in Figure 5. The non-vaporized portions of one of the Al rods are shown at 6 different times. In the last frame, we see that there is a large fragment moving at ~1 km/s in towards an optic. A fragment with this size and velocity would penetrate even a 3-mm thick DDS as shown in Figure 7. For the 1 MJ case, a longer length of rod is vaporized and the remaining portions are directed back along the rod and not towards optics. Given these results, the Al rod idea was dropped and the long-term design using thin waffled Al layers covering the flat Si was used for all shots.

5. Summary/Conclusion

Advanced 3D simulations using a range of codes is essential for the assessment and mitigation of facility impacts associated with radiation, EMP, and debris/shrapnel. A collimator, shielding, and a beam dump have has designed for the HEXRI diagnostic. Measurements of EMP on NIF are being used to test mitigation techniques to prepare the facility for ARC operation. Experiments at Janus and LIL facilities provide additional validation data for the new ALE-AMR code used for debris and shrapnel modeling. CEA plans to use the ALE-AMR code along with their ESTER and HESIONE codes to model debris and shrapnel on LMJ. Experiments have demonstrated the effects of a debris wind on diagnostics located in the equatorial position. Simulations have shown that a proposed configuration to use Al rods to support cryogenic targets is not acceptable with respect to optics damage. A configuration using waffled Al layers over Si supports has been shown to be acceptable.

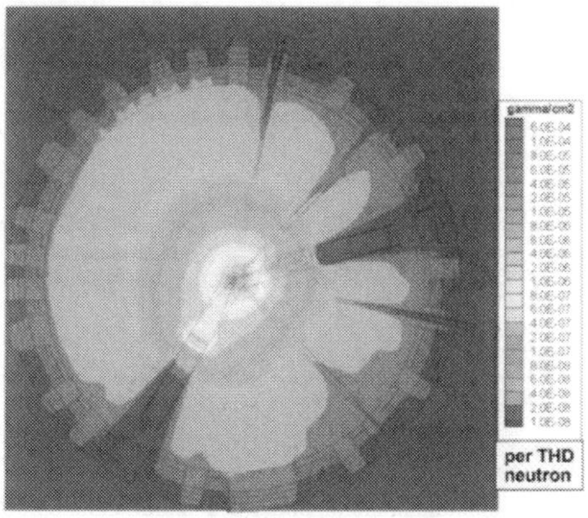

Figure 1 Gamma radiation fluence for a THD shot with 2% D 40 ns after the laser shot.

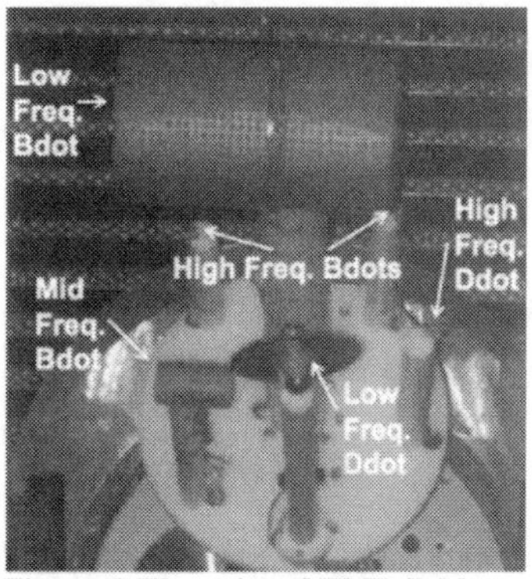

Figure 2 The suite of EMP field probes located ~4 m from TCC.

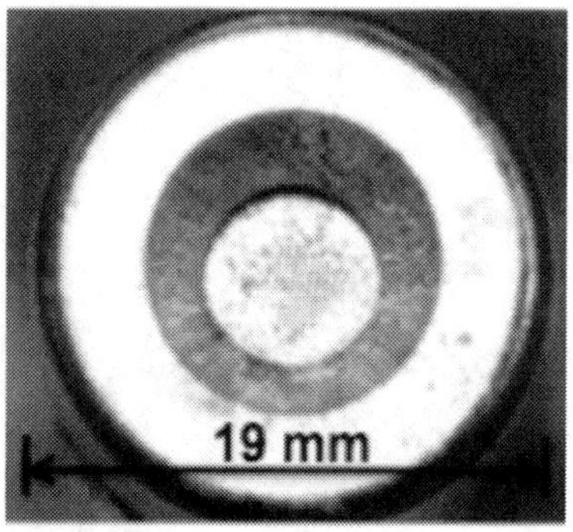

Figure 3 Front of GXD nose cone 8 cm from TCC coated with Al from TMP.

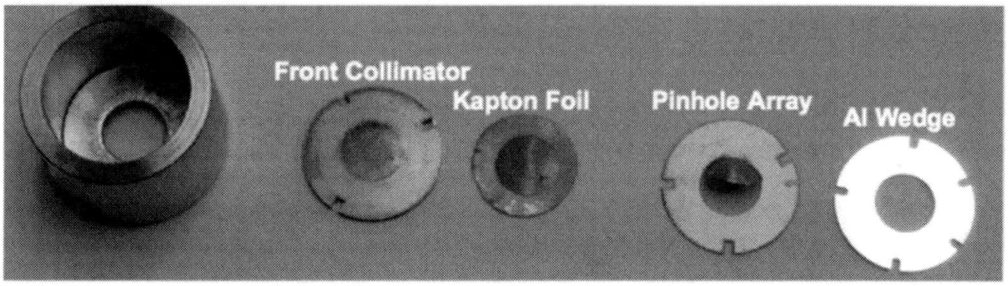

Figure 4 The front collimator, Kapton foil, pinhole array, and Al wedge placed inside GXD nose cone for a shot with nose cone 8 cm from TCC located on an equatorial position exposed to debris wind from target.

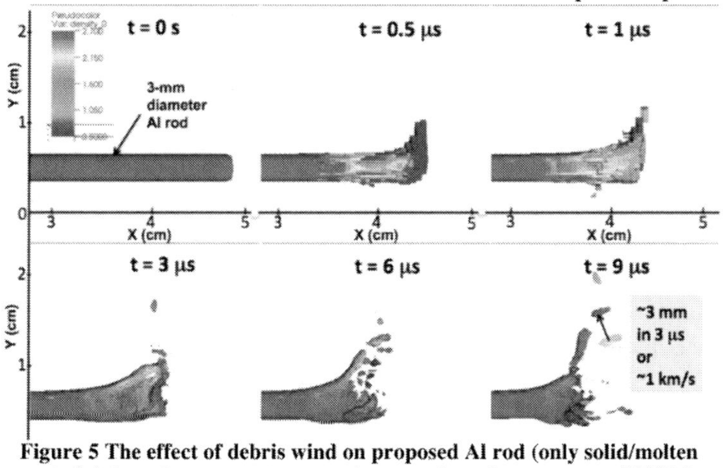

Figure 5 The effect of debris wind on proposed Al rod (only solid/molten material shown) to support cryogenic target for a laser energy of 333 kJ.

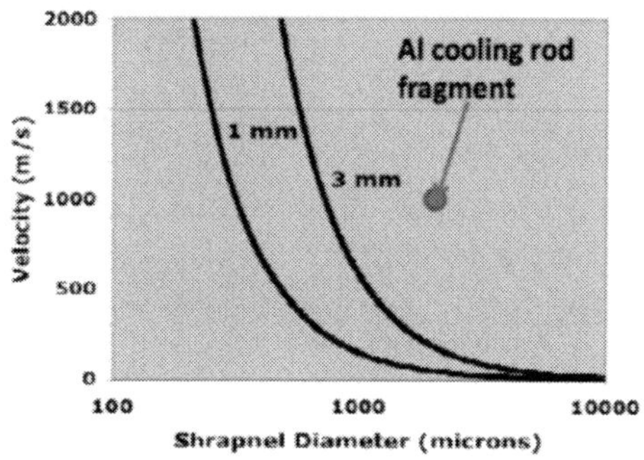

Figure 6 Penetration curves for 1 and 3 mm thick DDS's along with the calculated size and velocity of the fragment shown in last frame of figure 5.

References

[1] Wilson D C, Bradley P A, Cerjan C J, Salmonson J D, Spears B K, Hatchett S P, Herrmann H W, and Glebov V Yu, 2008 *Rev. Sci. Instrum.* **79**, 10E525

[2] Brown F B, Martin W R, and Mosteller R D, 2008 "Monte Carlo - Advances and Challenges," ANS *Reactor Physics Division Topical Meeting (PHYSOR-08), (LA-UR-08-05891)*

[3] White D and Stowell M 2004 *IEEE T. Microw. Theory* **52**(5) 1404

[4] See papers by Koniges A, et al., Masters N, et al., and Fisher A, et al., 2009 in this series

[5] See paper by Khater H, et al., 2009 in this series

[6] See paper by Brown C, et al., 2009 in this series

This work performed under the auspices of US DOE by LLNL under contract DE-AC52-07NA27344.

Journal of Physics: Conference Series **244** (2010) 032019 doi:10.1088/1742-6596/244/3/032019

ALE-AMR: A New 3D Multi-Physics Code for Modeling Laser/Target Effects

A. E. Koniges[1], N. D. Masters[2], A. C. Fisher[2], R. W. Anderson[2], D. C. Eder[2], T. B. Kaiser[2], D. S. Bailey[2], B. Gunney[2], P. Wang[2], B. Brown[2], K. Fisher[2], F. Hansen[2], B. R. Maddox[2], D. J. Benson[3], M. Meyers[3], A. Geille[4]

[1]Berkeley Lab, USA, [2]LLNL, USA, [3]UCSD, USA, [4]CEA, DAM, CESTA, France

Email: aekoniges@lbl.gov

Abstract. We have developed a new 3D multi-physics multi-material code, ALE-AMR, for modeling laser/target effects including debris/shrapnel generation. The code combines Arbitrary Lagrangian Eulerian (ALE) hydrodynamics with Adaptive Mesh Refinement (AMR) to connect the continuum to microstructural regimes. The code is unique in its ability to model hot radiating plasmas and cold fragmenting solids. New numerical techniques were developed for many of the physics packages to work efficiency on a dynamically moving and adapting mesh. A flexible strength/failure framework allows for pluggable material models. Material history arrays are used to store persistent data required by the material models, for instance, the level of accumulated damage or the evolving yield stress in J2 plasticity models. We model ductile metals as well as brittle materials such as Si, Be, and B4C. We use interface reconstruction based on volume fractions of the material components within mixed zones and reconstruct interfaces as needed. This interface reconstruction model is also used for void coalescence and fragmentation. The AMR framework allows for hierarchical material modeling (HMM) with different material models at different levels of refinement. Laser rays are propagated through a virtual composite mesh consisting of the finest resolution representation of the modeled space. A new 2nd order accurate diffusion solver has been implemented for the thermal conduction and radiation transport packages. The code is validated using laser and x-ray driven spall experiments in the US and France. We present an overview of the code and simulation results.

1. Introduction

The extreme range of physical conditions (temperature, density, pressures, etc.) and spatial scales (microns to meters) associated with modeling debris/shrapnel generation at large laser facilities (Omega, NIF, Orion, LMJ, etc.) motivated a new numerical approach to simulations. This approach based on combining Arbitrary Lagrangian Eulerian (ALE) hydrodynamics with Adaptive Mesh Refinement (AMR) provides great flexibility but introduces many algorithmic challenges. The new code, ALE-AMR, builds off the early work in incorporating ALE and AMR for gas-dynamics [1] using the Structured Adaptive Mesh Refinement Application Interface (SAMRAI) [2]. The addition of multi-material and the corresponding interface reconstruction capability provides the flexibility to relax the need to force material boundaries to correspond to zone boundaries. This allows one to place a complex configuration on a mesh and allow the AMR to add zones to refine the interfaces [3]. The AMR capability is also used dynamically to provide higher resolution in regions with steep gradients during simulations. The implementation of an anisotropic stress tensor and support for numerous strength/failure models provides the flexibility to model a wide range of materials using anything from analytical flow-stress models to detailed single-crystal plasticity models [4]. The code models both ductile metals used for shields, diagnostic filters, etc., as well as brittle materials such as Si, Be, and B4C used for cryogenic cooling/support and diagnostic filters.

The initial deposition of energy is generally by 3ω laser light with 1ω and 2ω laser light being important for some structures. Laser rays at the different frequencies are propagated through a virtual composite mesh consisting of the finest resolution representation of the modeled space with inverse bremsstrahlung absorption [5]. The energy deposited by the laser can create hot plasmas with subsequent energy transfer occurring by thermal conduction or radiation. A new 2nd order accurate diffusion solver has been implemented for the thermal conduction and radiation transport packages [6]. As an example of the current capability, we can model a target component heated by a laser that becomes a hot plasma with the resulting radiation striking another component producing ablation and the launch of a shock through the component. This shock can reflect at a free boundary and produce a tensile wave and subsequent spall. The code can calculate the spall including size of spall fragments and velocities. This information is used to determine if optics or diagnostics are at risk. If the shock or radiation causes the spall material to be molten, a surface tension model is required to predict droplet sizes of spall material. We are currently studying different surface tension models with collaborators at UCLA to determine which models are appropriate for the ALE-AMR code.

2. Overview of the ALE-AMR code

The code is based on Lagrangian hydrodynamics with position and velocity being nodal variables and density, internal energy, temperature, pressure, strain, and stress being zonal (cell centered) variables. (The thermal conduction and radiation transport equations are solved using the diffusion approximation with temperature and radiation energy being additional nodal variables.) The hydrodynamic equations are:

$\frac{\partial \rho}{\partial t} + \nabla \bullet (\rho \vec{v}) = 0$	Continuity Equation
$\rho \frac{\partial \vec{v}}{\partial t} = \nabla p + \nabla \bullet \Sigma' + \rho \vec{b}$	Momentum Equation (with deviatoric stress tensor shown)
$\rho \frac{\partial e}{\partial t} + p \nabla \bullet \vec{v} = 0$	Energy Equation
$\sum^{n+1} = f(\Sigma^{n}, \rho, e, \vec{v}, p, T, \vec{h})$	Material Stress Update

The equations are closed through the use of the equation of state (EOS), which is often tabulated. The EOS specifies the thermodynamic relation between state variables, such as internal energy, density (volume), pressure and temperature. Cells can have multiple materials with the volume fraction of each material in a mixed cell being tracked. If needed, e.g., for mesh refinement, interfaces are constructed using the volume fractions of neighboring cells. In mixed cells, the zonal hydrodynamic variables are determined for each material but averaged quantities are calculated and used for the hydrodynamic step. If the mesh becomes unacceptably distorted, the mesh is relaxed and all variables are mapped/advected to the new mesh. The different physics packages are included in an operator splitting approach. Parallel domain decomposition and the corresponding message passing is primarily treated within SAMRAI allowing the code to run efficiently on parallel computing architectures.

3. Code Validation

The code is also being validated with microscopic laser and x-ray induced spall experiments in the US and France. Figure 1 and 2 show results of a laser-induced spall experiment. The properties of the spall material, i.e., size, number, and velocity of fragments, obtained from ALE-AMR simulations are compared to observed data from in-flight images and aerogel samples. Similar experiments have been conducted in France using x-rays from a laser heat Au foil to induce spall in a range of different foils.

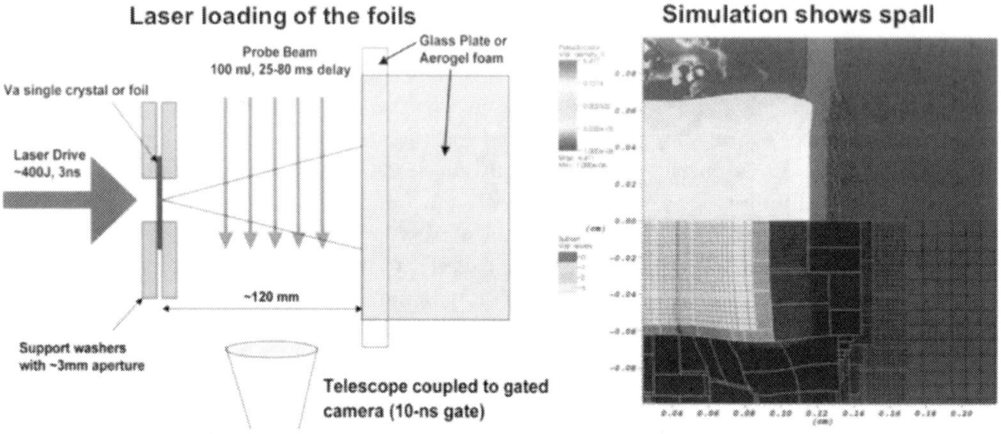

Figure 1 Laser induced spall experiment at LLNL with fragment capture and imaging. ALE-AMR simulation on right with mesh refinement shown in lower half.

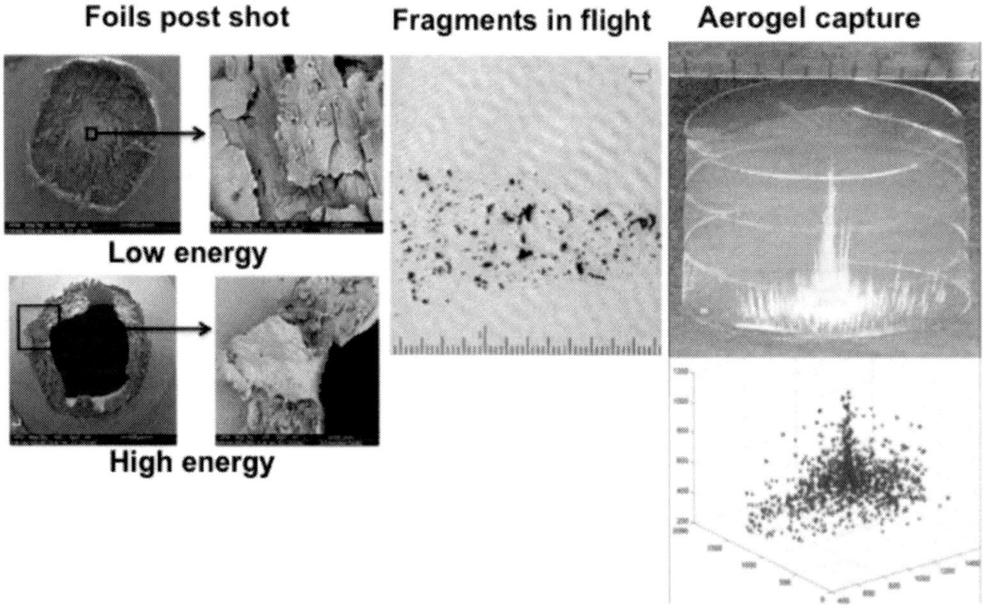

Figure 2 Foils for low-laser energy show spall while high energies leave hole in foil. Fragments observed in flight. Capture in areogel with size and depth of fragments used to calculate velocity.

4. NIF Simulations

A schematic of the NIF chamber is shown in Figure 3. A cryogenic target is inserted from the right and the 3D simulation on the right is for a cryogenic cooling ring. Coming in from the left side is a diagnostic snout used for x-ray imaging. A close up of the snout is given in Figure 4. Inside the snout is a pinhole array with front and rear collimators. We show results of ALE-AMR simulations on a single pinhole for 3 levels of x-ray loading. A wide range of NIF experiments have been modeled.

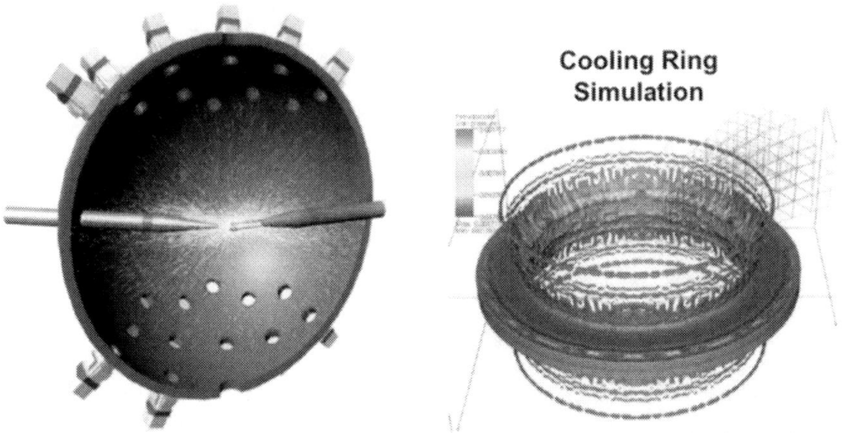

Figure 3 NIF chamber with target and diagnostic inserted. ALE-AMR simulation on right.

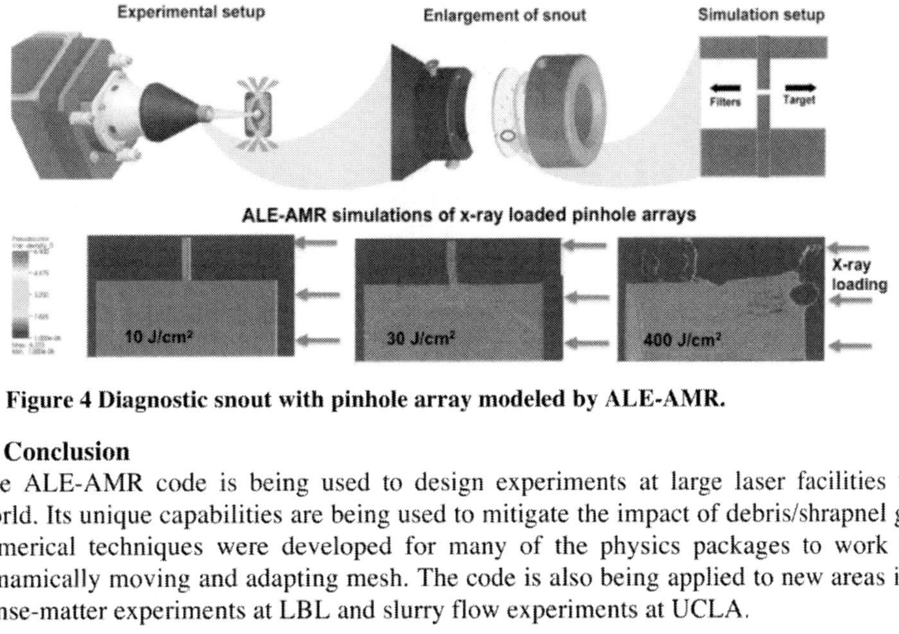

Figure 4 Diagnostic snout with pinhole array modeled by ALE-AMR.

5. Conclusion

The ALE-AMR code is being used to design experiments at large laser facilities through out the world. Its unique capabilities are being used to mitigate the impact of debris/shrapnel generation. New numerical techniques were developed for many of the physics packages to work efficiency on a dynamically moving and adapting mesh. The code is also being applied to new areas including warm-dense-matter experiments at LBL and slurry flow experiments at UCLA.

References

[1] Anderson R W, Elliott N S and Pember R B 2004 *Journal of Computational Physics* **199** 598
[2] Wissink A M, Hornung R D, Kohn S R, Smith S S and Elliott N S 2001 *SC01 Proceedings* (Denver, CO) Also available as LLNL technical report UCRL-JC-144755.
[3] Masters N, Anderson R, Elliott N, Fisher A, Gunney B and Koniges A 2008 *Journal of Physics: Conference Series* **112** 022017
[4] Fisher A C, Masters N D, Dixit P , Benson D J, Koniges A E, Anderson R W, Gunney B T N, Wang P and Becker R 2008 *Journal of Physics: Conference Series* **112** 022027
[5] Masters N D, Kaiser T B, Anderson R W, Eder D C, Fisher A C, and Koniges A E, *in this series*
[6] Fisher A C, Bailey D S, Kaiser T B, Gunney B T N, Masters N D, Koniges A E, Eder D C, and Anderson R W *in this series*

Journal of Physics: Conference Series **244** (2010) 032020 doi:10.1088/1742-6596/244/3/032020

HiPER Laser Architecture Principles

B. J. Le Garrec[1], C. Hernandez-Gomez[2], T. Winstone[2] and J. Collier[2]

[1]CEA-CESTA, 15 avenue des Sablières, BP2, 33114 LE BARP, FRANCE
[2]Central Laser Facility, STFC Rutherford Appleton Laboratory, Didcot, OX11 0QX, United Kingdom

E-mail: bruno.le-garrec@cea.fr

Abstract. The HiPER project [1] is in a preliminary phase and a risk assessment is being conducted for a generic HiPER beamline concept – this is based on an anticipated requirement for a single beamline (or focal spot unit) drawn from the best available knowledge we have to date of the fusion physics requirement. For the moment the general architecture is to first order independent of the specific beamline implementation (i.e compression phase or ignition phase) as the fundamental requirements are very similar. In this paper we are describing the laser beamline as a bundle of unit beams (or beamlets) and we are explaining that optical zooming is quite possible especially in the case of shock ignition.

1. Introduction

The HiPER project [1] intends to demonstrate fast ignition of laser-driven targets by means of a multi-beam, multi-ns laser pulse of about 250-300 kJ for implosion and a short pulse of about 100 kJ delivered in 15 ps for ignition [2]. In this type of direct-drive inertial confinement fusion, the "ignitor" shock can also be launched by a 60 kJ power spike at the end of the laser pulse [3]. At this stage, even proton driven ignition must be considered [4].

On the "laser" side, our objective is to identify the least expensive and most useful driver but not to preclude any alternative laser design that can make the facility more flexible and help achieving fusion.

2. Baseline concepts

The Laser Baseline stems from several key factors that are related to the facility requirements.

2.1. Facility requirements

The conventional direct-drive scheme involves both compression and ignition beams. There is a base-line target [5, 6] but there is of course still great uncertainty as to what this baseline will be.
Compression beams are to be focused as 48 focal spots onto the capsule: 250 to 300 kJ and 50 to 75 TW in a 10 ns shaped pulse at 3ω. Ignition beams will be either shock ignition beams or fast ignition beams. Shock ignition requires a final 400 ps long spike at 3ω: 60 to 80 kJ and 150 to 200 TW. High symmetry is not required but smaller focal spots will enhance the laser absorption. Fast ignition will require 100 kJ of 15ps short pulses certainly at 2ω and the only scheme is the gold conically guided capsule [7]. At this stage, it is not clear whether the ignition beams should be coherently phased or not,

and a small angle set of 20 to 30 beams overlapping somewhere at the cone entrance is quite tricky to design.

The primary HiPER facility will be operated on a "high" repetition rate basis (typically 10 Hz). Again, this repetition rate is not known precisely because repetition rate, laser efficiency and target gain are related to the cost of electricity that one would expect from the power plant design [8, 9].

This "high" repetition rate basis essentially means (in technological terms) that substantial new laser technology development will be required. This stems from the simple fact that the existing "single shot" technology as used by the National Ignition Facility or the Laser MégaJoule is, in general, not viable for high repetition rate requirements, although certain component technologies or techniques could readily be adapted.

Current "high repetition rate" laser technology based on flashlamps could not be scaled in any feasible or credible manner to the levels of efficiency required by HiPER. Our working assumption is that this beamline will be based on Diode Pumped Solid State Laser (DPSSL) technology [10]. This diode pumped technology is very promising technologically but relatively immature in its development and certainly prohibitive in its cost at today's prices when considering the requirements of the HiPER facility.

The availability, operation and performance of large scale/aperture components and component technologies at high average power are an unknown but essential entity. The level of industrial technological maturity with respect to the laser technology needs of HiPER is still someway off, even if in specific areas the industrial potential is evident.

2.2. Stacking of laser cells and bundle design.

Because roughly 380 kJ is to be focused on the capsule in 48 focal spots and assuming 50 % frequency conversion efficiency, this means that 16 kJ is required at the fundamental wavelength per focal spot.

Our assumption is that one laser beam is a bundle of beam lines (or beamlets) and one bundle is focused as a single focal spot onto the target. This assumption is also based on the fact that the unit cell will have a moderate size: the size is related to the affordability of the optical components that can withstand high average power (diode pumped amplifiers, frequency conversion crystals etc…). There is a trade-off between laser gain, amplified spontaneous emission and non linear effects (SRS, SBS). Let us call "beamlet" one full "unit cell" beam line.

The last generation of fusion lasers has shown that high fluence and low gain were compatible and would not preclude high extracted energy. Nevertheless, high saturation fluence should be considered carefully because the maximum fluence at the amplifier output must be lower than the laser damage fluence of the weakest optical component of the laser. As an example, let us consider 10J/cm^2 damage fluence and a beam whose near field modulation depth equals 1.5; this means that the average fluence in the beam shall not exceed 6.7 J/cm^2 and that 1kJ energy can be extracted for a 150 cm^2 aperture.

At that point, there is one reason why the size of the beam cannot be increased. In fact, the amplified spontaneous emission (ASE) gain in the transverse direction of the beam scales as the diameter of the beam (most of the time, one must consider the diameter or the diagonal size of the component). In a typical 1 cm long amplifier whose transverse size is 10 cm (diagonal size D is approximately 15 cm), the ratio of the ASE gain to the laser gain is $e^{15} \approx 3.10^6$. In other words, small signal gain (g) is limited by ASE as soon as the Fresnel loss at the edges (R) cannot compensate for the transverse gain:

$$g \leq \frac{-\ln(R)}{D} \tag{1}$$

When the refraction index is in the range 1.8 to 1.9 (typical values for garnets and sesquioxides), gain cannot exceed 0.12 cm^{-1}. This value is comparable to typical values commonly available in the disk geometry [11, 12]. It is clear that there is a trade-off between gain (limited by ASE effects) and clear aperture of the beam (diameter or diagonal size).

Journal of Physics: Conference Series **244** (2010) 032020 doi:10.1088/1742-6596/244/3/032020

According to the above values, one beam line could be made of 16 beamlets, each of whom would be able to amplify the fundamental wavelength from the front-end level (tens of millijoules) up to the kilojoule's level. The addition of different pulse shapes is possible while stacking the laser cells as described above. This bundle principle has been successfully tested both on the NIF [13] and the LIL facility [14] with the quadruplet design.

Let us consider that each beamlet has its own phase plate and that the bundle is focused with a single lens (as shown on top of figure 1). This technique is called beam conditioning because a speckle pattern is created inside the focal spot and the focal spot energy distribution can be adjusted with different phase plates: different sizes and different shapes are possible.

In the case of a bundle of beam lines, all the beamlet focal spots will overlap and the speckle patterns from each individual beamlet will mix so that the contrast will decrease, leading to a "smoothed" profile of the focal spot (as shown at the bottom of figure 1).

2.3. Optical zooming

Zooming has been proposed by the Naval Research Laboratory for KrF laser fusion [15] and is widely used in their target designs [16, 17]. Optical zooming means reducing the focal spot continuously or step by step during the pulse. This can be done when every single beam line is associated with an arbitrary waveform generator and a phase plate, leading to a time dependant focal spot as shown figure 2.Our bundle principle allows building different complex pulse shapes and overlapping different focal spots. It is based on time-delayed pulse shapes associated with different beamlets and can be considered as optical zooming [18]. Different phase plates and/or different focal lengths can be used.

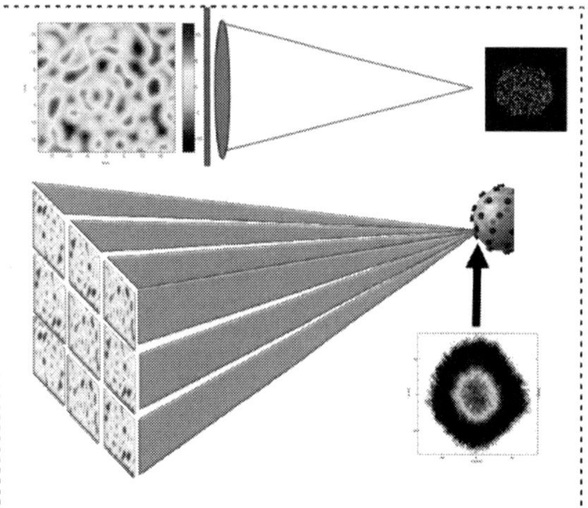

Figure 1. Top: each beamlet has a phase plate to create a speckle pattern inside the focal spot. Bottom: many beamlets overlap to smooth the focal spot.

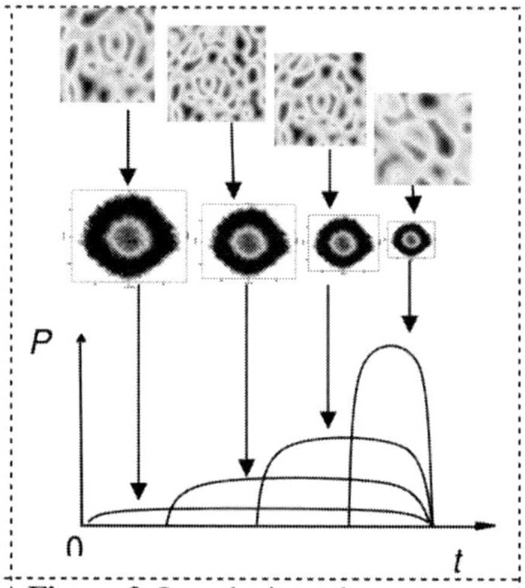

Figure 2. Introducing adequate timing between beamlets leads to optical zooming: spot size is moving during the pulse.

Our bundle principle is very convenient in the case of the direct-drive scheme involving both compression beams and shock ignition beams and this has been pointed out recently [19].

Of course and depending on the system requirements, both coherent and incoherent additions have to be made to ensure enabling pulse synchronization, smoothing and coherent propagation of compression and ignition beams down to the target.

3. Conclusion

We have presented the HiPER laser baseline that is to first order independent of the specific beamline implementation (i.e compression phase or ignition phase). The laser architecture is based on the target symmetry requirement leading to 48 focal spots and roughly 8 kJ at the third harmonic per focal spot.

Assuming Diode Pumped Solid State Laser (DPSSL) technology for the amplifiers, each of the beamline is made of a bundle of unit laser beams (that we call beamlets). Because there is a trade-off between gain (limited by ASE effects) and clear aperture of the beam (diameter or diagonal size), it seems reasonable not to exceed the kilojoule energy level per beamlet. So far, one bundle will be made of approximately 16 beamlets. The good thing is that this bundle principle allows performing optical zooming as soon as it is possible to associate one arbitrary waveform generator and one phase plate per beamlet.

Acknowledgments

The authors gratefully acknowledge the support of the Funding Agencies in undertaking this work (EC FP7 project number 211737): EC, European Commission, MŠMT, Ministry of Education, Youth and Sports of the Czech Republic and STFC, Science and Technology Facility Council of the United Kingdom.

References

[1] Dunne M 2006 *Nature Physics* **2**, 2; also*: http//www.hiper-laser.org*
[2] Atzeni S et al. 2008 *J. of Physics: Conference Series* **112** 022062
[3] Betti R et al. 2008 *J. of Physics: Conference Series* **112** 022024
[4] Roth M et al. 2001 *Phys. Rev. Lett.* **86** 436
[5] S. Atzeni et al. 2007 *Phys. Plasmas* **14** 052702
[6] Ribeyre X at al. 2008 *Plasma Phys. Controlled Fusion* **50** 027007
[7] Kodama R et al. 2001 *Nature (London)* **412**, 798
[8] Lindl J 1994 *Inertial Confinement Fusion* (AIP Press, Springer), chapter 13, p 184
[9] Meier W R 2006 *Proceedings 29th ECLIM ISBN:84-690-2624-0*; 2006 *LLNL report UCRL-CONF*-221936
[10] Chanteloup J-C et al. , *these journal proceedings.*
[11] Bourdet G L and Casagrande O 2005 *Optics Communications* **244** 327
[12] Casagrande O *PhD Thesis* (Ecole Polytechnique, Paris), *in French*
[13] Haynam C A at al. 2008 *J. of Physics: Conference Series* **112** 032004
[14] Di-Nicola J-M et al. 2006 *J. Phys. IV France* **133** 595
[15] Lehmberg R H and Goldhar J 1987 *Fusion Technol.* **11** 532
[16] Bodner S E et al. 2002 *Fusion Eng. Des.* **60** 93
[17] Schmitt A J et al. 2004 *Phys. Plasmas* **11** 2716
[18] Canaud B and Garaude F 2005 *Nucl. Fusion* **45** L43
[19] Schmitt A J et al. 2009 *Fusion Science and Technology* **56** 377

Journal of Physics: Conference Series **244** (2010) 032021 doi:10.1088/1742-6596/244/3/032021

Performance of a high-resolution deformable mirror for wavefront correction on the LIL/LMJ and PETAL baselines

S. Mainguy, C. Grosset-Grange, E. Bordenave

Commissariat à l'Energie Atomique (CEA)
Direction des Applications Militaires (DAM)
Centre d'Etudes Scientifiques et Techniques d'Aquitaine (CESTA)
BP 2; 33114 Le Barp, France

E-mail : stephane.mainguy@cea.fr

Abstract. The performance of a high-resolution deformable mirror ("HRDM") is analyzed for wavefront correction on the CEA high-power laser baselines in France : the Laser Mégajoule (LMJ), the Laser Integration Line (LIL) and the Petawatt Aquitaine Laser (PETAL). This study is achieved using monochromatic numerical simulations with the CEA "MIRÓ" laser propagation code. When compared to the LMJ deformable mirror used alone, the HRDM located at the L-turn is shown to give equivalent performance at filtering pinholes whereas the HRDM located at injection is less efficient. The use of an HRDM improves the impulse response of the LIL/LMJ main amplifier section. As a consequence, on the PETAL 1ω focal spot, it gives a reduction of factor 2.7 of the radius at 80% of encircled energy and an increase up to 40 % of the peak intensity. Finally, maximum mechanical amplitude of 10 nm is numerically evaluated for sinusoidal phase technological defects in the case of the HRDM located at the L-turn.

1. Introduction

The Laser Mégajoule (LMJ) project [1], in France, is part of the CEA-DAM nuclear weapon simulation program. It is dedicated to high energy physics experiments including thermo-nuclear ignition and burn by inertial confinement. The laser system is composed of 240 neodymium glass laser lines arranged in bundles of 8 beams. It is designed to deliver pulses up to 1.8 MJ energy at 351nm-wavelength (3ω) on target, after frequency conversion. Peak power is up to 550 TW depending on pulse shape and pulse duration which can be tuned between 0.3 ns and 25 ns.

The Laser Integration Line (LIL), the prototype of LMJ, is a 4 beam facility. Each beam delivers between 1 kJ and 9.5 kJ energy with a pulse duration ranging from 0.3 ns to 25 ns at 351 nm on target. The LIL was commissioned in 2004 [2] and is currently used to test technological improvements and for plasma experiments.

The Petawatt Aquitaine Laser (PETAL) is an additional line in the LIL facility. It is designed to deliver 500 fs pulses with a 3.5 kJ-beam energy and a 7.2 PW-peak power. It will enable the study of new regimes of the high intensity physics. This laser line uses a chirped pulse amplification (CPA) technique at 1.053 μm (1ω) and a synthetic aperture compression scheme [3]. The line is under construction. Wavefront correction on PETAL is one current challenge to obtain the laser performances at target chamber center including requested focal spot size without any spatial pedestal, peak power and eventually peak intensity.

In this paper, the performance of a high-resolution deformable mirror (HRDM) is evaluated in comparison with a LMJ deformable mirror for wavefront correction on the LIL/LMJ and PETAL baselines. This study is achieved by using the CEA "MIRÓ" laser propagation code. First, the principle and architecture of the considered deformable mirror systems are described as well as the calculation process with the MIRÓ code. Then the performance obtained with the HRDM is presented in terms of the laser spot dimension in the spatial filter pinholes, the residual phase at the main amplifier output and the dimensions of the focal spot at 1ω (PETAL) and at 3ω (LIL/LMJ). Finally, an amplitude specification on the HRDM technological defects is given for the HRDM located in the L-turn and for sinusoidal phase defects.

2. Wavefront correction configurations and calculation method

2.1 Principle and architecture of deformable mirror systems

The main amplifier section (MAS) of a LIL/LMJ single beamline and of the PETAL beamline is shown on figure 1. In order to minimise the spot size, the design includes control loops which are devoted to aberration compensation. Aberrations are either static (due to optical component manufacturing and positioning) or dynamic during system shots (due to the thermal stress inside the large flash-lamp-pumped amplifier slabs).

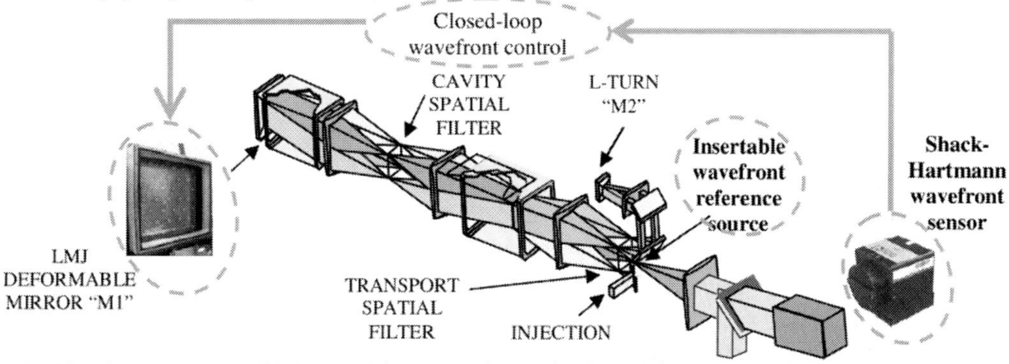

Figure 1. The four-pass amplifying architecture, for a single LIL/LMJ beamline and for the PETAL beamline.

The LMJ deformable mirror "M1" is located in the rear end of the MAS where the beam size is about 400 × 400 mm². The design of this mirror was optimized with 39 mechanical actuators. It has been tested and gave the expected results on the LIL facility [1,4].

The HRDM has 18 × 18 actuators. It is located at injection or at the L-turn : its size is about 40 mm or 80 mm, which is respectively about ten to five times smaller than the LMJ deformable mirror. We assume that an additional compensation of aberration is obtained by adjusting the position of a spatial filter lens, therefore the HRDM can be used alone. Thus five configurations of deformable mirror systems are examined : M1 alone, HRDM at injection, HRDM at L-turn ("M2"), HRDM at injection combined to M1 and HRDM at L-turn combined to M1.

The case of a deformable mirror located at the rear end of the MAS, that would have the same spatial resolution as the HRDM, has not been studied yet.

2.2 Calculation method using the MIRÓ propagation code

The phase correction to apply to the deformable mirror (M1 alone or HRDM alone) is obtained by minimising the residual phase at the MAS ouput. This optimisation is achieved using three-loop optical propagation calculations with the CEA MIRÓ laser propagation code [5]. In the case of a combined use of the M1 and the HRDM, a first three-loop wavefront correction calculation with the

Journal of Physics: Conference Series **244** (2010) 032021 doi:10.1088/1742-6596/244/3/032021

M1 gives a phase correction that is used in a second three-loop wavefront correction calculation with HRDM.

Then for each of the five wavefront solution configurations, a monochromatic MIRÓ calculation which takes into account phase corrections applied on mirrors gives the laser performance.

3. Performance of HRDM on LIL/LMJ and PETAL baselines

3.1 Laser spot performance in the spatial filters

Considering the dimension of spatial filter focal spots, the maximum amplitude required for the HRDM actuators are evaluated in all cases and are compatible with the current range of previous fabricated HRDM.

Laser spot angular radii at 99% of energy in the pinholes planes of the spatial filters are calculated in order to be compared to the pinholes dimensions. These radii are equal to 20 µrad to 50 µrad for the HRDM used alone at L-turn and for the HRDM at L-turn or injection combined to the M1. When the HRDM is used alone at injection, the spot radii are all less then 70 µrad except in the first pinhole where it is equal to 140 µrad, that is 7 times the diameter obtained with the LMJ M1 mirror. Indeed, when the HRDM is used alone at injection, phase pre-compensation before the first pinhole is very high. Thus, considering the 60 µrad LMJ first pinhole nominal dimension, HRDM used alone at injection is not an acceptable wavefront correction solution.

It results that using the HRDM at injection is an acceptable wavefront correction solution only if it is coupled to the M1. The HRDM in the L-turn, alone or combined to M1, gives laser spot performance in the spatial filters equivalent to the M1 alone.

3.2 Residual phase at the main amplifier output

The use of the HRDM (with or without the M1) improves the impulse response of the LMJ laser amplifier section when compared to the M1 used alone. The beam wavefront peak-to-valley amplitude is reduced of 20% (figure 2) and the beam residual wavefront slopes are reduced of 25% for spatial periods greater than 10 mm in the main amplifier section.

3.3 Dimensions of the focal spot at 1ω (PETAL) and at 3ω (LIL/LMJ)

A simplified PETAL baseline is modeled at 1ω with MIRÓ by using a perfect thin focusing lens with a focusing distance of 3.5 m. The calculation is monochromatic and B-integral effects on the focal spot are not studied. The use of the HRDM at L-turn (with or without the M1) in comparison to the M1 used alone, reduces of a factor 2.7 the 1ω focal spot radius at 80% of encircled energy, down to 17 µm (figure 3) and increases up to 40 % the focal spot peak intensity.

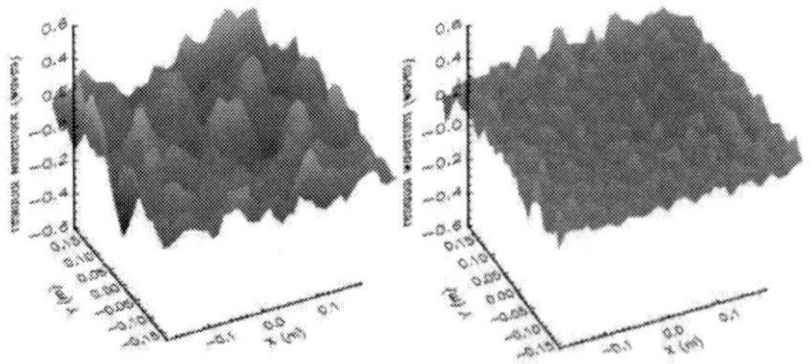

LMJ deformable mirror HRDM at the L-turn

Figure 2. Comparison of 1ω residual wavefronts at the MAS output for wavefront corrections with the HRDM at L-turn (right) and with the LMJ deformable mirror (left).

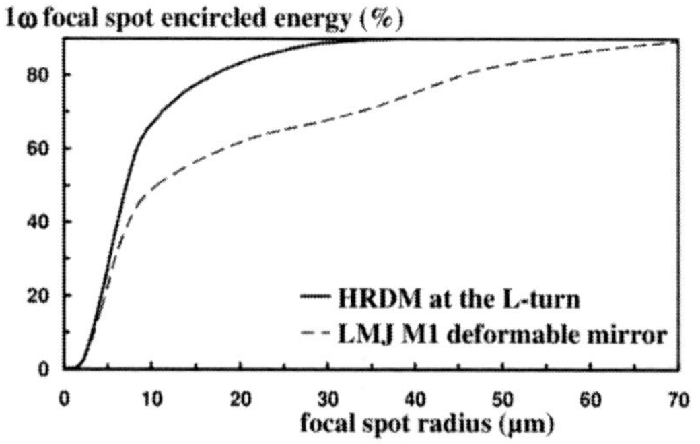

Figure 3. Comparison of 1ω focal spot encircled energies for wavefront corrections with the HRDM at L-turn and with the LMJ M1 deformable mirror.

MIRÓ simulations of the LMJ baseline are achieved for a 5 ns square pulse. Nominal filtering parameters are applied : 100 µrad angular radius at each pinhole except on the first pinhole where it is reduced to 60 µrad to avoid parasitic retro-reflections towards the front end. The injected energy is adjusted in order to obtain 18 kJ at the main amplifier output (1ω). It results that when the HRDM is used instead of the M1, a 12 % decrease of the 3ω focal spot diameter is obtained in the first transverse direction with a 4% decrease in the other direction, corresponding to the pump-induced aberration asymmetry.

4. HRDM specifications on technological defects

Technological defects due, for example, to inter-actuators effects, are modeled by two-dimensional sinusoidal phase defects at the HRDM surface. The case of a L-turn located HRDM with defect periods from 1 mm to 4 mm is considered through MIRÓ monochromatic propagation simulations. It results that for a peak-to-valley mechanical amplitude of the defects less than 10 nm, no critical amplitude modulation or energy loss is observed at the ouput of the MAS and the 1ω focal spot energy and peak intensity are stable up to 0.5%.

5. Conclusion

Wavefront correction performances of the HRDM on the LIL/LMJ and PETAL baselines, is compared to the LIL/LMJ M1 deformable mirror performances with the CEA MIRÓ laser propagation code. The HRDM located at the L-turn gives equivalent performances at filtering pinholes whereas the HRDM located at injection is less efficient. The HRDM improves the impulse response of the LIL/LMJ main amplifier section for spatial periods greater than 10 mm. As a consequence, on the PETAL 1ω focal spot, the HRDM reduces of a factor 2.7 the diameter at 80% of encircled energy and increases the peak intensity up to 40%. Also a maximum mechanical amplitude of 10 nm is numerically evaluated for phase sinusoidal technological defects in the case of an HRDM located at the L-turn.

On the PETAL baseline, the next wavefront correction tasks are the following ones : broad spectrum beam simulations of the PETAL main amplifier section including a compensation system for longitudinal chromatism, optimisation of HRDM actuator number, experimental validation with a LIL/LMJ M1 deformable mirror then with an HRDM in the L-turn (M2), coupled with a second deformable mirror located before the compressor ouput.

6. References

[1] J. Ebrardt, JM. Chapput, "LMJ project status", *The fifth International Conference on Inertial Fusion Sciences and Applications (IFSA2007)*, Journal of Physics IV: Conference Series **112** (2008) 032005

[2] J.M. Di Nicola, N. Fleurot, T. Lonjaret, X. Julien, E. Bordenave, B. Le Garrec, M. Mangeant, G. Behar, T . Chiès, C. Féral, H. Graillot, M. Luttmann, F. Jéquier, E. Journot, O. Lutz and G. Thiell, "The LIL facility Quadruplet Commisionning", *The fourth International Conference on Inertial Fusion Sciences and Applications (IFSA2005),* Journal of Physics IV **133** (2006) 595-600

[3] N. Blanchot, G. Behar, T. Berthier, E. Bignon, F. Boubault, C. Chappuis, H. Coïc, C. Damiens-Dupont, J. Ebrardt, Y. Gautheron, P. Gibert, O. Hartmann, E. Hugonnot, F. Laborde, D. Lebeaux, J. Luce, S. Montant, S. Noailles, J. Neauport, D. Raffestin, B. Remy, A. Roques, F. Sautarel, M. Sautet, C. Sauteret and C. Rouyer, "Overview of PETAL, the multi-Petawatt project on the LIL facility," Plasma Phys. Control. Fusion **50**, 124045-124055 (2008).

[4] C. Grosset-Grange, J. N. Barnier, C. Chappuis, H. Cortey, "Design principle and first results obtained on the LMJ deformable mirror prototype", *Adaptative Optics for Laser Systems and Other Applications*, edited by G. Chériaux, Chris. J. Hooker, Michal Stupka, Proc. of SPIE Vol.6584, 658403, (2007)

[5] O. Morice, "Miró : complete modeling and software for pulse amplification and propagation in high-power laser systems", *Opt. Eng.* **42** (6) 1530-1541, June 2003.

Journal of Physics: Conference Series **244** (2010) 032022 doi:10.1088/1742-6596/244/3/032022

Laser Ray Tracing in a Parallel Arbitrary Lagrangian-Eulerian Adaptive Mesh Refinement Hydrocode

N D Masters[1], T B Kaiser[1], R W Anderson[1], D C Eder[1], A C Fisher[1], A E Koniges[2]

[1]Lawrence Livermore National Laboratory, Livermore, CA U.S.A.
[2]National Energy Research Scientific Computing Center, Lawrence Berkeley National Laboratory, Berkeley, CA U.S.A

E-mail: `masters6@llnl.gov`

Abstract.
ALE-AMR is a new hydrocode that we are developing as a predictive modeling tool for debris and shrapnel formation in high-energy laser experiments. In this paper we present our approach to implementing laser ray tracing in ALE-AMR. We present the basic concepts of laser ray tracing and our approach to efficiently traverse the adaptive mesh hierarchy.

1. Introduction
Shrapnel and debris with the potential of damaging expensive optics, diagnostics, or fixturing, may be generated by the extreme conditions present in high energy laser facilities such as the National Ignition Facility (NIF), Laser Mégajoule (LMJ) and Omega. The ALE-AMR multi-physics hydrocode has been developed as a predictive modeling tool to identify and assess sources of shrapnel and debris and the hazards they pose so these may be mitigated [1–6]. Such simulations generally require large computational domains in which to track fragment formation and trajectories. ALE-AMR implements arbitrary Lagrangian-Eulerian (ALE) hydrodynamics within an adaptive mesh refinement (AMR) framework. This allows the mesh to be adapted in response to the current state: applying additional mesh resolution in regions with rapidly changing dynamics while a coarser mesh may be used in regions that are currently less interesting—enhancing computational efficiency without sacrificing accuracy or resolution. In evaluating debris and shrapnel formation we need to be concerned not only with the energy deposited by the 3 ω (UV) light—critical to target design—but also the unconverted 1 ω and 2 ω (red and green) that may irradiate support and diagnostic structures. Laser ray tracing is an efficient means for simulating laser propagation and energy deposition and provides a realistic energy source for hydrodynamic simulations.

2. ALE-AMR
The Structured Adaptive Mesh Refinement Application Infrastructure (SAMRAI) [7, 8] framework on which ALE-AMR is implemented provides a hierarchy of logically rectangular patches on which the problem physics are advanced. The simulation resolution can be improved in areas of interest, e.g., large gradients or fragmentation, by overlaying refined meshes that represent subsets of the domain

Journal of Physics: Conference Series **244** (2010) 032022 doi:10.1088/1742-6596/244/3/032022

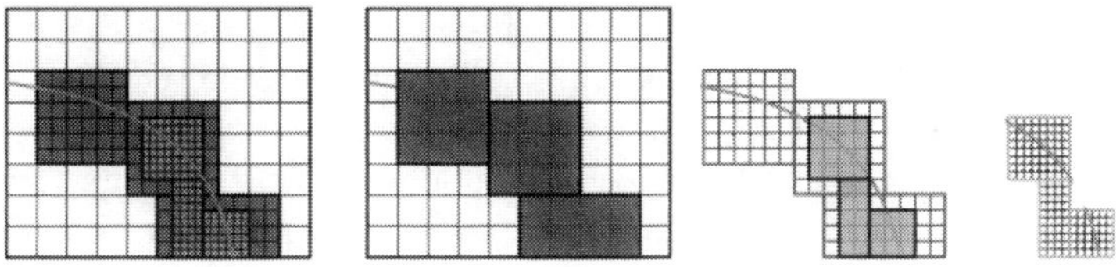

(a) AMR Mesh (b) Level 0 (c) Level 1 (d) Level 2

Figure 1. Rays traverse a virtual composite mesh. Rays are passed to the appropriate patch which may be at a higher, lower, or the same level of refinement

(see Figure 1). The hydrodynamics are evolved using a Lagrange-plus-remap algorithm. After problem initialization (mesh generation and initial conditions) the problem is advanced by repeating the following steps:

Lagrange Step The mesh is deformed by evaluating the forces (pressures and stresses) acting on the nodes and the resulting accelerations and displacements.

Remap Step Periodically the current Lagrange solution is remapped to an new mesh to avoid tangling, this may be either the original mesh (Eulerian) or some arbitrary mesh (ALE), usually a relaxed mesh intermediate to Lagrange and Eulerian configurations.

Adaptive Mesh Refinement Zones are tagged for coarsening or refinement by evaluating the current solution with respect to user defined criteria. A new mesh hierarchy is then created (regridding) onto which the current solution in mapped through coarsening or refining the solution variables.

Additional Physics If additional physics (heat conduction, radiation diffusion, or in this case laser ray tracing) are required they are performed at this point and the cumulative result is used for the next Lagrange step.

The code is designed to run on large parallel machines, therefore each processor is responsible for only a small subset of the problem, with communication limited to the information necessary to correctly synchronize adjacent or underlying patches through patch ghost cells or coarsened and refined solutions.

3. Laser Ray Tracing

Ray tracing is a powerful technique applied to high-energy laser simulations by Friedman [9] and Kaiser [10]. The key assumption is that the length scale of variations in the medium is larger than the laser wavelength over most of the computational domain. This allows the wave equations to be simplified as an equation of motion for rays and tracked through each zone. The electron density gradient is considered to be linear within each zone resulting in parabolic ray trajectories. Discontinuities at zone faces are treated via Snells law and energy deposition is computed as an inverse bremsstrahlung process. For a detailed description of the application of laser ray tracing to high energy laser simulations the reader is referred to the papers cited above.

Lagrangian or ALE modes result in deformed meshes consisting, in general, of quadrilateral elements in 2D and hexahedral elements in 3D (bounded by lines and doubly ruled surfaces, respectively). The intersections of the parabolic ray trajectories with these zones results in quadratic (2D) and quartic (3D) equations that must be solved for each face of the current zone. Although several solutions to these equations may exist valid solutions must fall within the boundaries of the current face. The exit face is identified as the one with the shortest exit time. Ambiguities may arise if a ray passes close to edges and/or corners. Incorrect selection of the exit face will result in a lost ray in the next zone as no valid exit points will be found. Such ambiguities are removed by reevaluating the ray traversal in the current zone after slightly perturbing the entry point within the entry face.

4. Logical Patches and AMR

We implemented laser ray tracing in ALE-AMR by considering the AMR hierarchy as a virtual composite mesh, i.e., rays are only propagated through the finest representation of the physical space as shown in Figure 1. The composite mesh is never explicitly constructed, rather at the end of each zone traversal an exit point and apparent destination zone are identified. The destination zone may be either: (a) on the same patch, but a finer representation exists , (b) on a different patch possibly at a different level (coarser or finer) , or (c) on the same level and patch . As a ray enters a new zone (from a neighboring zone, a different patch, or when introduced to the computational domain) we determine if the zone is the finest representation of the physical space by querying the `is_finest` variable. This variable is updated each time the structure of the AMR grid is modified as follows: (1) `is_finest` patches on Level 0 and Level 1 are filled with ones and zeros, respectively, (2) the finer level is coarsened to fill any underlying coarse regions with zeros, and (3) the process is repeated with each successively finer pair of coarse-fine levels (e.g., Levels 1 and 2). Therefore if `is_finest` = 0 for a given zone, then a finer representation exists to which the ray should be passed. In ALE-AMR, fine nodes in faces that lie on coarse-fine boundaries are constrained to be equally distributed within the parametric space of the coarse face. The fine destination index is found by determining the location of the ray entry point in the parametric space of the entry face (line in 2D, doubly ruled surface in 3D) and identifying the correct fine entry face (and associated fine zone) that covers that fraction of the parametric space. The destination patch is found by querying the SAMRAI `findOverlapIndices` function to identify the patch that overlaps the destination zone and then requesting the processor that holds this patch with the SAMRAI `getMappingForPatch` function (see SAMRAI reference documents supplied with the SAMRAI source [8]). Once identified (assuming that the destination processor is not the current processor) then the ray is queued to be passed to the processor when the current batch of rays have been processed.

For rays leaving the current patch (see (b)) the destination patch is found by again calling `findOverlapIndices` and `getMappingForPatch` for the apparent destination. If found, the rays are queued to be passed to their destinations (note: `is_finest` will be queried when they arrive—which may lead to subsequent refinement). If `findOverlapIndices` returns an empty overlap list then no patch at the current—or finer—level contains the destination zone and the destination must be coarsened. The index is coarsening by dividing the fine destination index by the refinement ratio and flooring the result, e.g., $\text{coarsen}((27,19),\texttt{ratio}=(3,3)) = (9,6)$. `findOverlapIndices` is called again for the coarse index to identify the coarse destination patch. Coarsening and refinement is repeated until the finest representation of the physical destination is found.

Unless either of the above two events ((a) or (b)) happen, rays continue to propagate zone-to-zone on a given patch and deposit energy as detailed in the aforementioned references until the ray power is depleted. Energy deposited by all rays is accumulated in the zones traversed by rays and then coarsened down to underlying patches at the end of the ray tracing step. Coarsening of deposited energy is accomplished by taking a mass weighted average of the energy (or change in energy) in the fine zones, e.g., $e_c = (\sum \rho_i e_i V_i) / (\sum \rho_i V_i)$ where ρ_i, e_i, and V_i are the zonal densities, internal energies, and volumes of the fine zones associated with an underlying coarse zone (the coarse internal energy assigned the value e_c). The deposited energy may then be directly applied to the zonal internal energies or used as a source term in a subsequent diffusion step.

Direct MPI calls are used to first communicate which processors will send rays and the intended recipients, then the queued rays are sent and received and added to the list of rays that have arrived from other processors or are being introduced to that processor at domain boundaries. Laser ray tracing ends when all rays on all processors have terminated.

In the first timestep of ray tracing ray entry points are found by testing for ray crossings of each face touching a domain boundary on the coarsest level. In subsequent timesteps we search in the neighborhood of the last valid entry zone. This local search may fail if the domain boundaries are allowed to move, in which case the search region may be expanded or an exhaustive search again applied. Searching the coarsest level reduces the number of face crossing evaluations and avoids needing to determine which patches touch domain boundaries prior to ray insertion as only the coarsest level is guaranteed to exist at all points in the simulation domain. If the insertion zone is not the finest representation we repeat the face crossing search to accurately determine the entry point on the fine zones that replace the coarse zone.

One additional difficulty arises when rays insertion passes close to patch boundaries during start up. If a ray passes to close to an edge (in 3D, or a corner in 2D) we remove any ambiguity by applying a small random perturbation to the ray trajectory. At startup, such rays will receive independent random perturbations on different processors—possibly leading to multiple copies of the same ray. We avoid this by identifying potential duplicates (any that enter zones along patch boundaries) and compare this to the list of suspects on all other processors and remove the duplicates (preferring those from lower rank processors). Once an initial entry point (and associated search neighborhood) has been found, duplicate rays become very rare.

Current load balancing involves processing rays in batches that cascade to neighboring patches/processors. This evens the load on the processors that will participate in ray tracing. However, some processors may hold patches or levels that are traversed by few if any rays and may remain idle during a given laser ray tracing step. Future work will seek to address load balancing issues.

To illustrate the AMR laser ray tracing algorithm we will follow a ray through the mesh in Figure 1(a). The ray is introduced on the left side of the mesh into a zone on Level 0. After traversing one zone a finer representation of the destination is found Figure 1(b). The ray is passed to the appropriate Level 1 patch and ray propagation continues until a patch boundary is encountered (see Figure 1(c)). The ray is then passed to the neighboring Level 1 patch. After one zone traversal a finer representation is again indicated and the ray passed to Level 2. After traversing this Level 2 patch the ray leaves and no patch exists at the current level to receive it (see Figure 1(d)), the destination is coarsened and the process continued until the ray leaves the domain (or is fully depleted)

5. Conclusions and Future Work

We have implemented laser ray tracing in the ALE-AMR hydrocode to provides an efficient means for modeling laser energy deposition. This provides the means to deposit energy in a realistic manner as we seek to mitigate potential damage through predictive modeling of debris and shrapnel. Future work will explore load balancing and present applications of the ALE-AMR laser ray tracing package in simulating debris formation in high energy laser experiments.

Acknowledgments

Work performed under the auspices of the U.S. Department of Energy by Lawrence Livermore National Laboratory under Contract DE-AC52-07NA27344

References

[1] Masters N, Anderson R, Elliott N, Fisher A, Gunney B and Koniges A 2008 *Journal of Physics: Conference Series* **112** 022017
[2] Fisher A C, Masters N D, Dixit P, Benson D J, Koniges A E, Anderson R W, Gunney B T N, Wang P and Becker R 2008 *Journal of Physics: Conference Series* **112** 022027
[3] Eder D, Koniges A, Landen O, Masters N, Fisher A, Jones O, Suratwala T and Suter L 2008 *Journal of Physics: Conference Series* **112** 032023
[4] Koniges A E, Anderson R W, Wang P, Gunney B T N, Becker R, Eder D C, MacGowan B J and Schneider M B 2006 *Journal de Physique IV* **133** 587–593
[5] Anderson R W, Pember R B and Elliott N S 2001 *40th American Institute for Aeronautics and Astronautics Aerospace Science Meeting and Exhibit* (Reno, Nevada) AIAA 2002-0738
[6] Anderson R W, Elliott N S and Pember R B 2004 *Journal of Computational Physics* **199** 598–617
[7] Wissink A M, Hornung R D, Kohn S R, Smith S S and Elliott N S 2001 *SC01 Proceedings* (Denver, CO)
[8] SAMRAI website `https://computation.llnl.gov/casc/SAMRAI/`
[9] Friedman A 1983 *Bulletin of the American Physical Society* **28** 1124
[10] Kaiser T B 2000 *Physical Review E* **61**(1) 895–905

Journal of Physics: Conference Series **244** (2010) 032023 doi:10.1088/1742-6596/244/3/032023

Temperature dependence of laser-induced damage threshold in silica glass

K. Mikami[1], S. Motokoshi[2], M. Fujita[2], T. Jitsuno[1], J. Kawanaka[1], R. Yasuhara[3]

[1]Institute of Laser Engineering, Osaka University, Suita, Osaka, Japan
Tel: +81-6-6879-8762, Fax: +81-6-6877-0900
[2]Institute for Laser Technology, Osaka, Japan
[3]Central Research Lab., Hamamatsu Photonics K. K., Hamamatsu, Shizuoka Japan
E-mail; kmikami@ile.osaka-u.ac.jp

Abstract. The laser-induced damage thresholds in silica glasses at different temperature conditions (123 K – 473 K) by Nd:YAG laser fundamental (wavelength 1064 nm) and third harmonic (wavelength 355 nm) 4 ns of pulses were measured. In the results, the damage thresholds increased at low temperature. At 1064 nm, the temperature dependence became strong by the concentration of impurities. However, at 355 nm, the temperature dependences of almost sample were almost the same for different concentration of impurities.

1. Introduction

A new high-power laser system with cryogenically cooled Yb:YAG ceramics is expected as inertial fusion driver. The laser system has several advantages in thermal properties such as thermal conductivity and thermal expansion [1]. Laser-induced damage threshold (LIDT) of optical elements is one of important properties because output power of laser systems is limited by LIDT. However, influence of temperature on LIDT is not known well [2]. We have studied the temperature dependences of LIDT in silica glasses by Nd:YAG laser at fundamental (wavelength 1064 nm) and third harmonic (wavelength 355 nm) to clear the temperature dependence of LIDT and damage mechanism.

2. Samples and apparatus

Five grades of silica glasses of different concentrations of impurities were prepared. The concentration of OH and metallic impurities in silica samples are shown in Table 1. The sample size is 30 x 30 x 30 mm^3 and the incidence and the opposite surfaces were optically polished (surface flatness λ, surface roughness <3 nm).

Figure 1 shows the experimental layout. Incident energy of Nd:YAG laser pulse was adjusted by a half-wave plate and a polarizer, and monitored by peak voltages of calibrated biplanar phototube. The third harmonic pulse was converted by two KDP crystals. Each sample and focusing lens (f=30 mm) were set in a vacuum chamber (~50 mTorr) to prevent condensation at low temperature. The focused laser beam had a smooth Gaussian-shape with a spot diameter at $1/e^2$ peak intensity of 70 μm (1064 nm) and 30 μm (355 nm).

In this experiment, we used "N-on-1" damage test method. A fixed site was irradiated, with the fluence increased until damage was observed. Figure 2 shows the shot lapse of irradiation. The damage was detected by three methods; plasma emission, transmitted laser energy, and transmitted co-arraignment He-Ne laser pattern as images shown in Fig. 2.

Table 1 Impurities of samples used in the present study [3].

	Grade	Al	Ca	Cu	Fe	Na
Synthetic Silica	ED-A	<0.01	<0.01	<0.01	<0.01	<0.01
	ED-B	<0.01	<0.01	<0.01	<0.01	<0.01
	ES	0.1	0.1	0.01	0.05	0.05
Fused Silica	S	0.7	<0.01	<0.01	0.05	0.3
	N	8	0.6	<0.01	0.2	0.6
		K	Li	Mg	OH	
Synthetic Silica	ED-A	<0.01	<0.01	<0.01	<200	
	ED-B	<0.01	<0.01	<0.01	<10	
	ES	0.05	0.05	<0.01	1000	
Fused Silica	S	<0.01	<0.01	<0.01	160	
	N	0.1	<0.01	0.04	200	

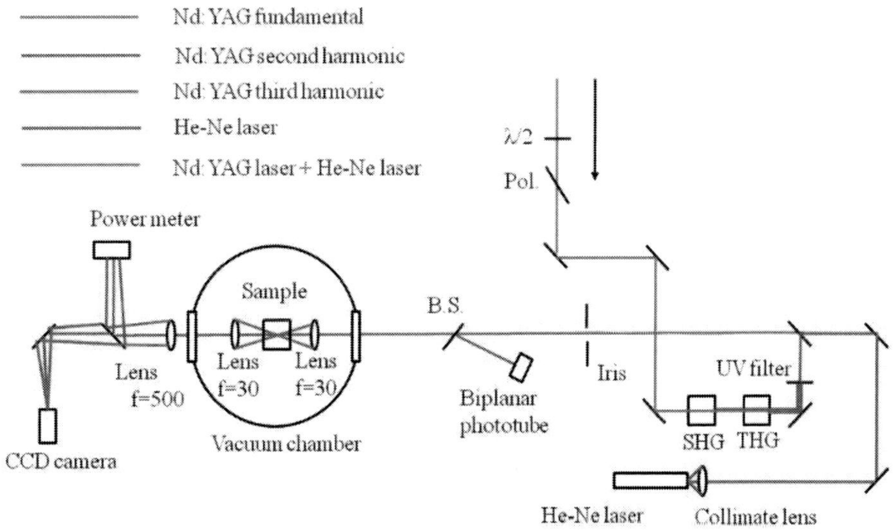

Fig.1 Experimental layout of LIDT measurement.

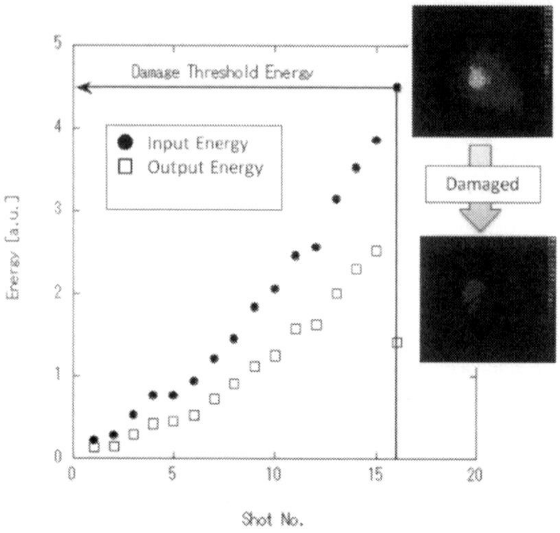

Fig.2 Damage test lapse and the damage detection by transmitted images.

3. Results and discussions

Figure 3 shows LIDTs of different samples at different temperature. The LIDT increased linearly with cooling for all samples in spite of the laser wavelength. The change rates of LIDT to temperature (⊿D. T.), that is the temperature dependences of LIDT, were also different. On the other hand, at 355 nm, the temperature dependences were constant for the almost samples except S-grade. It was suggested the effects of temperature on the damage mechanisms was different with irradiated laser wavelengths.

The relations between the temperature dependence and the concentration of impurities are shown in Table 1. Figure 4 shows the dependence to the concentration of each impurity. Red and blue circles indicate data at 1064 and 355 nm, respectively. The dependences at 1064 nm depend on the contaminant concentration. For the OH concentration particularly, the dependences at 1064 nm is different with the synthetic silica and the fused silica samples. As the results, it could be supposed that the OH or metallic impurities were one of dominant factors on the damage mechanisms influenced by temperature for 1064-nm laser pulse.

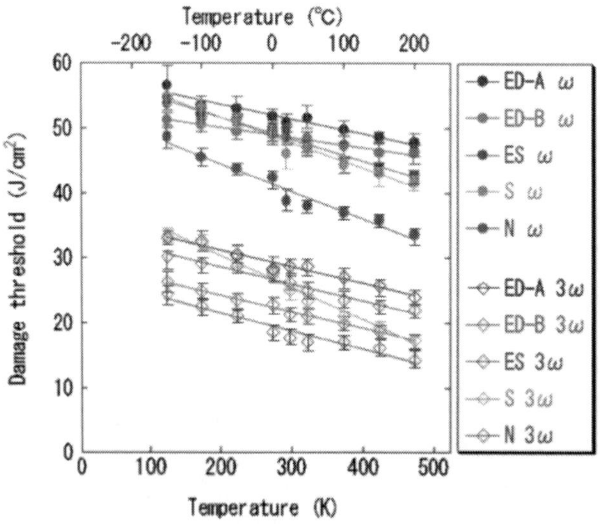

Fig. 3 Temperature dependence of LIDT.

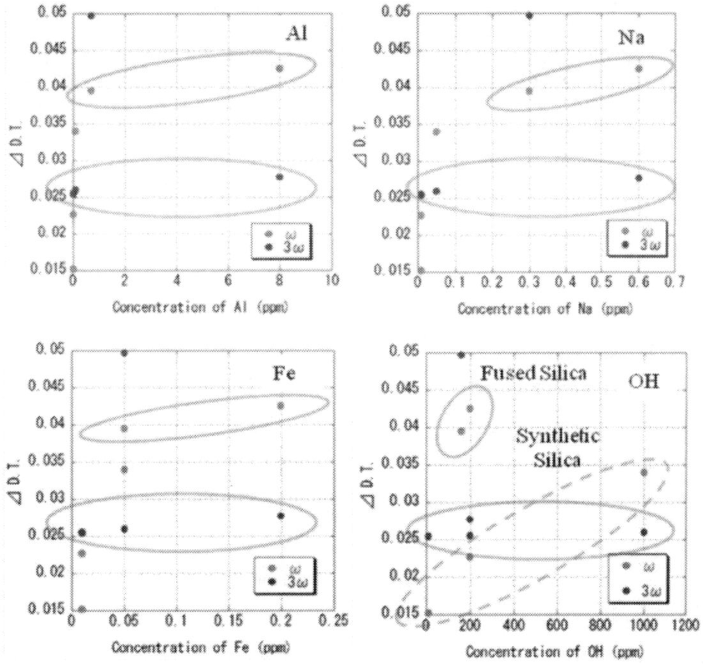

Fig. 4 Dependence of change rate of damage threshold on concentration of impurity materials.

4. Summary

LIDTs of bulk silica glasses at different temperature were evaluated by a fundamental and third harmonic pulse of Nd:YAG laser. As the results, LIDTs were increased by cooling. The relationship between temperature dependence and concentration of impurities at 1064 nm was different with that at 355 nm. Therefore, effects of impurities to temperature dependence will be different in 1064 nm and 355 nm.

References

[1] J. Kawanaka, S. Tokita, H. Nishioka, M. Fujita, K. Yamanaka, K. Ueda, and Y. Izawa, Laser Physics 15,1306 (2005)
[2] L. D. Merkle and D. Kitriotis, Phys. Rev. B, 38 1473 (1988)
[3] TOSOH Quarts catalog

Non-sinusoidal phase modulations for improved performance of high power lasers

Denis PENNINCKX, Steve HOCQUET, Jean-François GLEYZE, Claude GOUÉDARD, and Yves JAOUËN[(*)]

CEA, CESTA, BP n°2, 33114 Le Barp Cedex
[(*)]Institut Telecom / Telecom ParisTech, 46 rue Barrault 75634 Paris

Denis.Penninckx@cea.fr

Abstract. Spectral broadening is required on high power lasers to avoid Stimulated Brillouin Scattering in the laser chains and to smooth the focal spot. Up to now, spectral broadening has been obtained by sinusoidal phase modulations. However, it leads to non-homogeneous spectra, which is not optimal. In this paper we show that, thanks to non-sinusoidal phase modulations, performance of high power lasers may be enhanced. Furthermore, we demonstrate theoretically and experimentally that adjusting non-sinusoidal phase modulations may be almost as simple as adjusting sinusoidal ones.

1. Anti-Brillouin and smoothing functions for high power lasers

Spectral broadening is required on high power lasers such as National Ignition Facility or the Laser MégaJoule (LMJ) to avoid Stimulated Brillouin Scattering (SBS) in the laser beam lines [1] and to smooth the focal spot [2]. Up to now, spectral broadening has been obtained by a sinusoidal phase modulation. A sinusoidal modulation is defined by its frequency f_m and its amplitude m_l (modulation depth). Our calculations show that for LMJ avoiding SBS at 3ω requires that the maximum of the spectral power distribution (SPD) be below around 6%, with 100% corresponding to a not-broadened spectrum. Current values for LMJ are thus f_m=2GHz and m_l=21. Smoothing consists of rapidly moving the speckle pattern of the focal spot. It is characterized by the asymptotical contrast that is equal to the RMS value of the spectrum intensity. The smoothing modulation frequency is set by the smoothing grating to 14.25GHz and the modulation depth m_l is set to 15 at 3ω to reach an asymptotical contrast of 22%.

However, sinusoidal modulations lead to non-homogeneous spectra, as can be seen in the left part of Figure 1. Hence, to reach a given performance, a higher broadening is required. The bandpass of a laser beam line is theoretically limited by the spectral acceptance of the frequency conversion in the final optics assembly and in practice also by other imperfections. A high spectral broadening slightly reduces the total energy and leads to temporal shape distortions through so-called FM to AM conversion [3]. Thus, tailoring the spectrum would be welcome [4, 5], for instance to make it flat, to reach the same optical performance with a narrower spectral bandwidth, as shown in the right part of Figure 1.

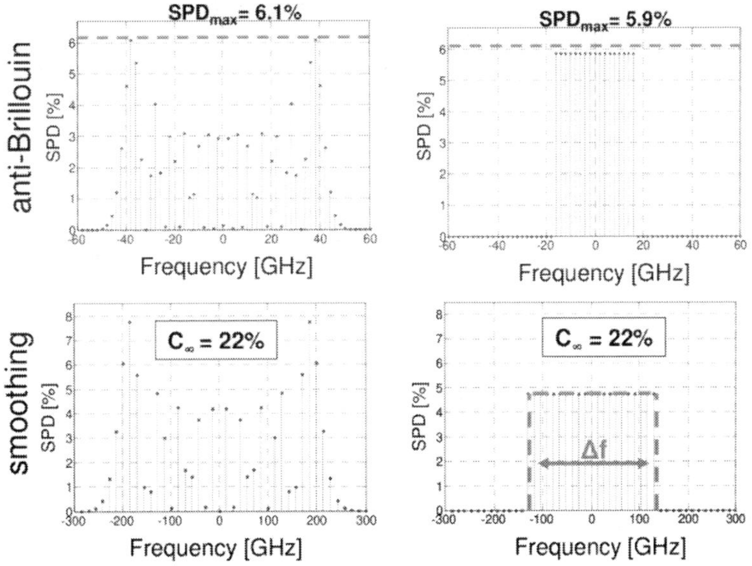

Figure 1: Left: current anti-Brillouin and smoothing spectra on LIL and LMJ facilities.
Right: ideal spectra with the same anti-Brillouin (above) and smoothing (below) efficiencies.

2. Non-sinusoidal phase modulations

It can be shown that such ideal spectra cannot be perfectly tailored with a pure phase modulation [6]. However, they can be approximated by increasing the number of degrees of freedom as compared to a sinusoidal modulation. Since the spectra are composed of Dirac peaks, the phase modulations φ(t) can be developed in Fourier series:

$$\varphi(t) = \exp\left(i\sum_{n=1}^{N} m_n \sin(2\pi n f_m t + \varphi_n) \right)$$

where N is the number of harmonics of f_m, m_n the modulation depth of the nth harmonic and φ_n its relative phase (Figure 2).

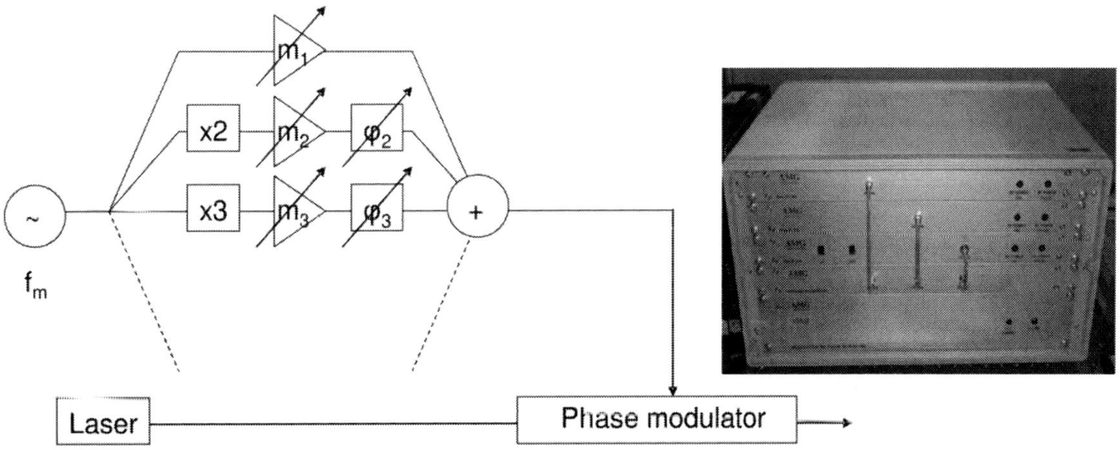

Figure 2: Non-sinusoidal modulation scheme
and picture of one electrical 3-harmonic generator (f_m=10GHz, N=3). The different harmonic amplitudes m_i and phases φ_i can be adjusted separately

Journal of Physics: Conference Series **244** (2010) 032024 doi:10.1088/1742-6596/244/3/032024

Since φ_1 can be used as phase reference (φ_1=0) without lack of generality, the number of degrees of freedom is 2N-1. We choose N=3, which is a good trade-off between complexity and flatness performance (see Figure 3). Note that N=1 corresponds to the sinusoidal case. We have designed two prototypes of electrical drivers with N=3 to generate the spectra, one with f_m=2GHz and the other with f_m=10GHz (Figure 2).

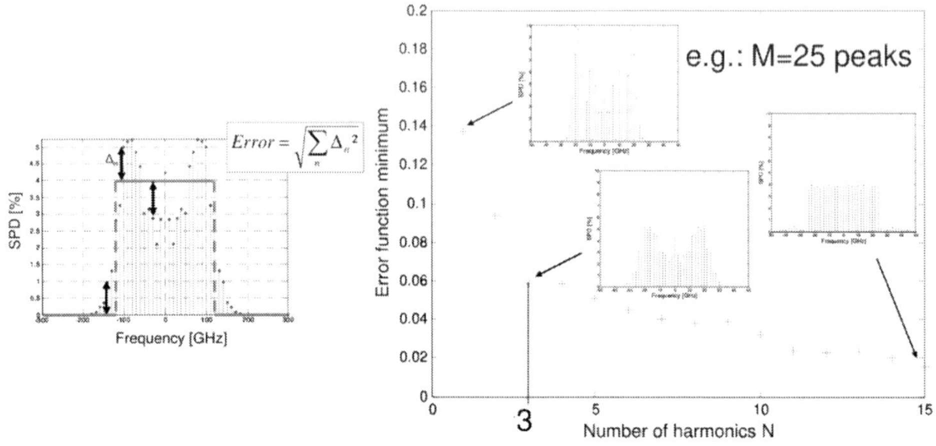

Figure 3: Error between the spectrum obtained with a non-sinusoidal phase modulation and the ideal spectrum versus the number of harmonics N for an ideal spectrum composed of M=25 Dirac peaks.

To avoid shifting the laser wavelength when switching the phase modulation on and off, a symmetrical spectrum is required. This condition leads to φ_2=90° and φ_3=180° [6]. Then, optimizing m_1, m_2 and m_3 to obtain a flat spectrum yields almost constant values of m_1 and m_3 whatever the spectrum width. Hence, as for sinusoidal phase modulations, a single parameter (m_2) is sufficient to control the spectral width whatever the number of Dirak peaks M, as shown in Figure 4 [5].

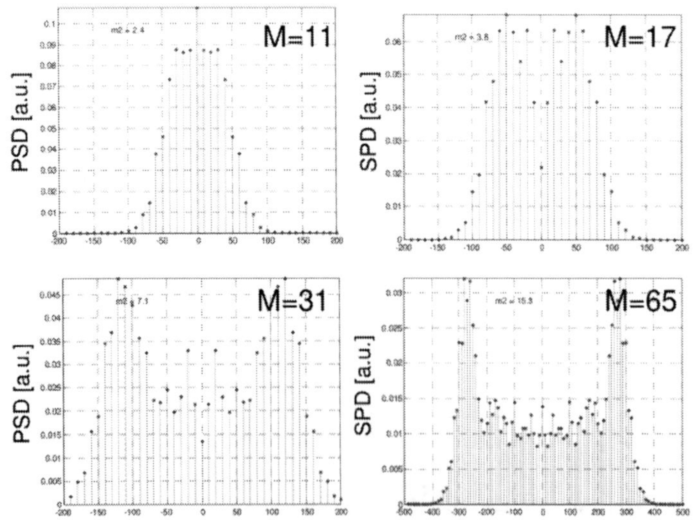

Figure 4: Spectra of various widths, given by the number of Dirac peaks M, obtained with non-sinusoidal phase modulations with N=3.
In every case, m_1=1.4, φ_1=0, φ_2=90°, m_3=0.4, and φ_3=180°.
m_2=2.4 for M=11 ;
m_2=3.8 for M=17 ;
m_2=7.1 for M=31 ;
m_2=15.3 for M=65.

3. Experiments and application to LMJ

We have used the 3-harmonic electrical drivers to generate non-sinusoidal phase modulations for the anti-Brillouin and the smoothing functions. An example of measured spectra for the anti-Brillouin function (f_m=2GHz) is given in Figure 5. Thanks to non-sinusoidal phase modulations, the broadening can be reduced while obtaining the same anti-Brillouin and smoothing performance. For instance, we have shown both numerically and experimentally that spectra (b) and (c) in Figure 5 have the same Brillouin threshold. Similar results were also obtained for smoothing. The broadening reduction allows an increase of the frequency conversion efficiency and a decrease of the FM-AM conversion.

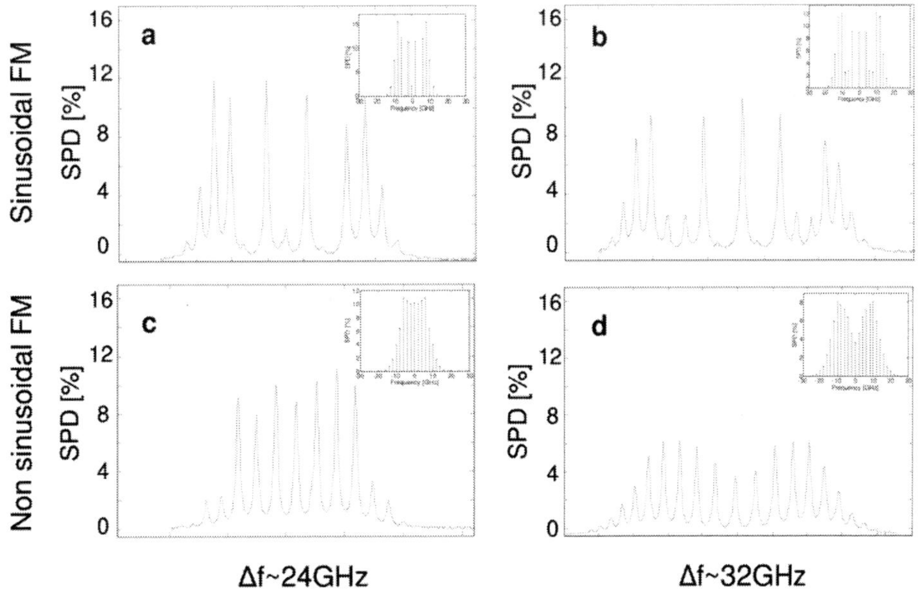

Figure 5: Measured spectra with sinusoidal (or N=1, above) and non-sinusoidal (N=3, below) for two spectrum bandwidth (left 24GHz and right 32GHz). Insets show simulation results for comparison. Note that the maximum SPD is almost equal for spectra (b) and (c) but spectrum (c) is much narrower.

Our simulations show that for LMJ an increase of 5% energy and a decrease of 1/3 of FM to AM conversion is possible thanks to non-sinusoidal modulations while keeping the same asymptotical contrast of smoothing.

References

[1] J.R. Murray et al, J. Opt. Soc. Am. B, Vol.6, No. 12, pp2402-2411 (1989).
[2] J. Garnier et al., J. Opt. Soc. Am. A 14, 1928–1937 (1997).
[3] S. Hocquet et al. Applied Optics, Vol. 47, No. 18, 06/2008 and Vol. 48, No. 13, 05/2009.
[4] Y. Liu et al., Chinese Opt. Lett., 01/2009.
[5] Patent application FR 0858954.
[6] Steve Hocquet, PhD Thesis, Telecom ParisTech, 11/2009.

Journal of Physics: Conference Series **244** (2010) 032025 doi:10.1088/1742-6596/244/3/032025

Neutron Activation of NIF Final Optics Assemblies

S Sitaraman, L Dauffy, H Khater, and S Brereton

Lawrence Livermore National Laboratory, 7000 East Avenue, Livermore, CA 94550, USA

sitaraman1@llnl.gov

Abstract. Analyses were performed to characterize the radiation field in the vicinity of the Final Optics Assemblies (FOAs) at the National Ignition Facility (NIF) due to neutron activation following Deuterium-Deuterium (DD), Tritium-Hydrogen-Deuterium (THD), and Deuterium-Tritium (DT) shots associated with different phases of the NIF operations. The activation of the structural components of the FOAs produces one of the larger sources of gamma radiation and is a key factor in determining the stay out time between shots to ensure worker protection. This study provides estimates of effective dose rates in the vicinity of a single FOA and concludes that the DD and THD targets produce acceptable dose rates within10 minutes following a shot while about 6-days of stay out time is suggested following DT shots. Studies are ongoing to determine the combined effects of multiple FOAs and other components present in the Target Bay on stay-out time and worker dose.

1. Introduction

Construction of the National Ignition Facility (NIF) has been completed and early phase experiments leading to eventual controlled self-sustaining nuclear fusion energy gain have commenced at the facility. The neutrons generated by these experiments activate various components and structures present in the facility. A good understanding of the level of residual radiation present at various locations in the facility is important in determining the stay out time between shots to ensure worker protection.

Final Optics Assemblies (FOAs) provide beam conditioning, frequency conversion, focusing, and debris protection in a compact module. There are four optics modules per FOA with forty eight FOAs attached to the target chamber wall. They focus the 192 laser beams to the target chamber center. Since they are in direct line of sight of the target, the FOAs experience some of the highest neutron fluences outside the target chamber. This study will focus on the activation of the large structural components in the FOA and estimate the effective dose rates in the vicinity of a single FOA. An earlier study [1] was performed with a simpler model of the FOA. While, a more recent study [2] used the same detailed computational model described here to examine the contact dose rates associated with optical elements in the FOA, this study focuses on dose contributions of the massive structural components of the FOA.

2. Computational Model

The radiation transport code, MCNP [3], was used to obtain the neutron flux spectra in various structural components of the FOA. These fluxes were then used to activate these components using the activation code, ALARA [4]. The gamma sources from ALARA for the various structural components of the FOA were then used in a second MCNP calculation to obtain effective contact and 1-ft dose

Journal of Physics: Conference Series **244** (2010) 032025 doi:10.1088/1742-6596/244/3/032025

rates from each of the FOA components. ICRP-74 [5] flux-to-dose rate conversion factors were used to obtain the dose rate estimates.

Targets representing different phases of NIF operation were used in the models. For the Deuterium-Deuterium (DD) shots, a yield of $1x10^{13}$ neutrons/shot was used with a shot schedule of 4 shots per day for 50 days. The DD source spectrum used for the neutron calculations was 99% 2.45 MeV and 1% 14.1 MeV. The Tritium-Hydrogen-Deuterium (THD) shot was assumed to yield $2x10^{14}$ neutrons/shot with a shot schedule of 2 shots per day for 50 days. A full THD neutron source spectrum based on 2% deuterium content was used for the calculations. The Deuterium-Tritium (DT) shot was assumed to yield $7.45x10^{18}$ neutrons per shot with a shot schedule of 60 shots per year with a six day interval between shots. A full DT source spectrum with a 14.1 MeV peak was used in the calculations.

A detailed model of the FOA (Figure 1) was used in the MCNP neutron flux spectrum calculations. All 48 FOAs were included in this model as well as details of the target chamber walls, the floors and structural columns inside the Target Bay (TB) (Figure 2). The MCNP gamma calculation used a simplified model of the FOA that was voided out and divided into eight axial regions each with its own gamma source. Thus, having no self shielding, this model yielded conservative estimates of the dose rates.

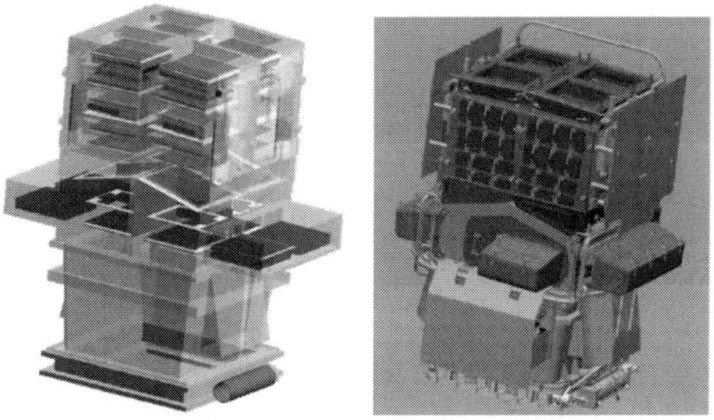

Figure 1. Final Optics Assembly: MCNP model (left), CAD model (right)

Figure 2. MCNP model of the Target Bay (FOAs shown in blue)

3. Effective Dose Rates from an FOA

This section contains a summary of the calculated dose rates for the three target types at various decay times following the last shot in each shot sequence. This will allow for the accounting of isotopes that may accumulate or decay over the time span of the shot schedule. The tables presented for each target type contain both the contact dose rate and the dose rate at 1-ft from the region of the FOA that produced the highest value. It must be noted that in the cases of the THD and DT shots, different regions of the FOA were bounding at contact and 1-ft away.

3.1. Dose Rates following DD Shots

The main contribution to the dose rate in the short term comes from ^{28}Al with the longer term contribution from ^{56}Mn and ^{24}Na. The contribution to the dose by ^{24}Na is not very significant at earlier decay times since the predominant DD neutron energy is well below the threshold of the production cross section of the isotope via an (n, α) reaction in ^{27}Al. These dose rates are negligible at 10 minutes after the last shot. The limiting dose rate in the case of the DD shot was from the part of the FOA containing the disposable debris shield. Table 1 presents the results from DD shots.

Table 1. Effective Dose Rates following DD Shots

Time after last shot (minutes)	Contact Dose Rate (mrem/h)	1-ft Dose Rate (mrem/h)
10	2.39e-2	1.95e-2
30	2.06e-3	1.77e-3
60	1.52e-3	1.23e-3
360	6.24e-4	6.03e-4

3.2. Dose Rates following THD Shots

The main contribution to the dose rate in the short term comes from ^{28}Al, ^{24}Na, and ^{27}Mg. The longer term contributions come mainly from ^{24}Na and ^{56}Mn. The THD neutron spectrum has a low energy tail that promotes the production of both ^{56}Mn and ^{27}Mg via (n,γ) reactions in ^{55}Mn and ^{26}Mg, respectively. In addition, the high energy portion of the THD neutron spectrum produces ^{24}Na from ^{27}Al. Effective dose rates at 10 minutes after the last shot are at acceptable levels of approximately 1 mrem/h. In the case of the THD shot, the limiting component of the FOA was the vacuum/valve assembly that is in contact with the target chamber. Table 2 presents the results from the THD shots.

Table 2. Effective Dose Rates following THD Shots

Time after last shot (minutes)	Contact Dose Rate (mrem/h)	1-ft Dose Rate (mrem/h)
10	1.38	6.21e-1
30	3.43e-1	1.31e-1
60	1.74e-1	6.64e-2
720	8.56e-2	3.03e-2

3.3. Dose Rates following DT Shots

The DT yield shots correspond to 20 MJ of energy and produce the highest amount of activation. The dose rate at 6 days after the last shot on contact with the FOA is approximately 6 mrem/h and over 2 mrem/h at a distance of 1-ft. Thus a minimum 6-day stay-out time is recommended for these shots. This is consistent with the cooling time required for handling individual optics components [2]. In the case of the optical components, the highest dose rates were also caused by activation of the aluminium frames. The limiting component in the case of the DT shot was the vacuum/valve assembly that is in contact with the target chamber, which is predominantly made of aluminium but also has stainless steel components. The main contributions to the dose rates come from ^{24}Na, ^{54}Mn, and ^{58}Co. The DT neutron spectrum with a large component of high energy neutrons lends itself to the production of both ^{54}Mn (via (n,p) reaction in ^{54}Fe) and ^{58}Co (via (n,2n) reaction in ^{59}Co) which contribute to the dose rate at the long decay time of approximately 6 days. Table 3 presents the results at decay times following the last of the 60 shots.

Table 3. Effective Dose Rates following DT Shots

Time after last shot	Contact Dose Rate (mrem/h)	1-ft Dose Rate (mrem/h)
1 hour	4.96e3	1.99e3
12 hour	2.19e3	8.03e2
1 day	1.23e3	4.51e2
6 day	5.82	2.35
10 day	1.1	6.02e-1

4. Conclusions

Structural components of the FOAs produce dose rates that are low enough to allow entry into the TB at 10 minutes after DD and THD shots. However, more realistic entry times (for other reasons) will be closer to an hour after the shots making the radiation levels even lower. For the full yield DT shots, a stay-out time of days will allow radiation levels to reach acceptable values that will minimize radiation exposure to maintenance personnel. This study evaluated the radiation field in the vicinity of a single FOA after shots with different targets. In reality, the radiation levels in the TB will be the cumulative value from multiple FOAs and other structures present at the same floor level. Studies are being conducted to estimate these cumulative radiation levels at various locations within the TB while taking into account the reduction of dose rates associated with most maintenance tasks which take place at larger distances from highly activated components. Preliminary results indicate that for the most part the FOAs would contribute the largest fraction to the dose rate on the floors where they are present.

Acknowledgement

This work was performed under the auspices of the U.S. Department of Energy by Lawrence Livermore National Laboratory under Contract DE-AC52-07NA27344.

References

[1] J.F.Latkowski, "*Neutron Activation of the NIF Final Optics Assemblies and their Effect upon Occupational Doses,*" Fusion Technology 34 (1998)

[2] L.S.Dauffy, H.Y.Khater, S.Sitaraman, S. Brereton, "*Activation Analysis of the Final Optics Assemblies at the National Ignition Facility*", Fusion Science and Technology 56 (2) 736 (2009)

[3] X-5 Monte Carlo Team, "*MCNP - A General Monte Carlo N-Particle Transport Code, Version 5,*" Los Alamos National Laboratory, LA-UR-03-1987 (2005)

[4] P. Wilson and D. Henderson, "*ALARA: Analytic and Laplacian Adaptive Radioactivity Analysis code,*" University of Wisconsin, UWFDM-1070 (1998)

[5] International Commission on Radiological Protection, "*ICRP Publication 74: Conversion Coefficients for Use in Radiological Protection Against External Radiation,*" Pergamon press (1997)

Journal of Physics: Conference Series **244** (2010) 032026 doi:10.1088/1742-6596/244/3/032026

Phase control of SBS PCM seeding by optical interference pattern clarified: Direct applicability for IFE laser driver

Ondrej Slezak and Milan Kalal

Faculty of Nuclear Sciences and Physical Engineering, Czech Technical University in Prague, Brehova 7, 115 19 Prague 1, Czech Republic

E-mail: slezaond@fjfi.cvut.cz

Hong Jin Kong

Department of Physics, Korea Advanced Institute of Science and Technology,Yuseong-gu, Daejeon, Korea

Abstract. Theoretical background of stimulated Brillouin scattering (SBS) seeded by a spatially localized optical interference pattern was investigated. Performed analysis revealed that the SBS may have its origin in the acoustic standing wave produced as a transient state between the initial optical interference field and the resulting stationary density modulation. Acquired knowledge can be found valuable in optimization of laser beam combination scheme considered e.g. for IFE applications.

1. Introduction

Construction of high energy high repetition laser drivers belongs to important IFE challenges. One particular approach to this goal considers a *combination* of less energetic laser beams into more powerful ones using stimulated Brillouin scattering (SBS) phase conjugating mirrors (PCM). The main obstacle of this approach is a *randomness* of the phase of the reflected Stokes field exhibited by individual laser beams before their combination [1, 2]. Hence it became necessary to find some mechanisms which could control the phase of the SBS PCM reflected waves. Over the last twenty years many techniques dealing with this problem were suggested [3]–[9]. Probably the most suitable among them was the technique adopted by Kong et al [10]. In this approach the feedback concave mirrors placed behind the SBS cells are used for a suitable *phase locking*. This technique was found both *efficient* as well as relatively *insensitive* to an optical *alignment*. It makes possible, in principle, to lock the phase of an arbitrary number of simultaneously working SBS PCMs.

2. Phase locking using feedback mirror

To get the SBS reflected Stokes field *phase* under control it is necessary to drive properly the process of an arising acoustic Brillouin wave. In particular, the *initial time* and *position* of such an acoustic wave have to be *set* the right way. It was proven experimentally that the phase difference between two or more SBS PCM reflected Stokes fields may be controlled with a high degree of accuracy by the back-seeding *concave* mirror [10]. Theoretical description of mechanisms playing role in this phase locking technique was presented recently [11]. In that paper the formula was found describing how the Brillouin acoustic wave *initial time* depends on the gradual laser energy *increase* (starting the integration time from the pump pulse leading edge). This way the *initial time* was firmly established. However, the second condition, the exact *spatial location* of this wave, was still awaiting for its correct *mathematical description* as well as *proper understanding*.

When the *concave mirror* is placed behind the SBS cell (through which the parallel laser beam is traveling) the corresponding *interference pattern* is produced. In fact, this pattern is an *optical standing wave*, oscillating with the *optical* frequency. These oscillations are far too fast for any immediate response by SBS medium molecules. Therefore, the *electrostriction* source term coming from the time-averaged optical intensity modulation is employed to drive the acoustic wave [12]. However, such a *stationary* optical intensity field *cannot* become the *source* term for the *moving* density grating.

Let us analyze the SBS PCMs system shown in the Fig. 1.

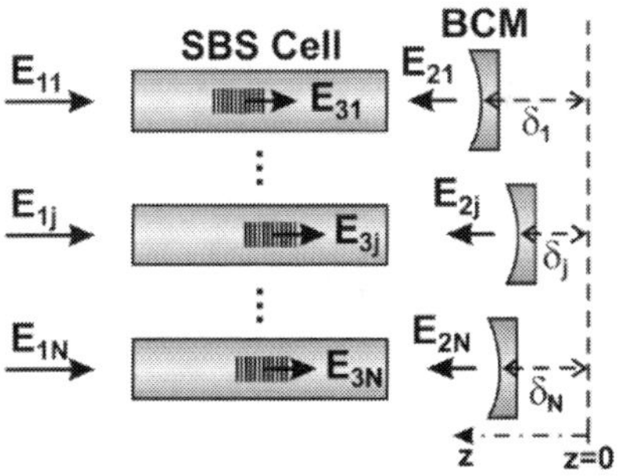

Figure 1. N simultaneously working SBS PCMs with back-seeding concave mirrors (BCM). The incident pump beam E_{1j} for the j-th SBS PCM is going through the SBS cell *unfocused.* It is *reflected* by the j-th concave mirror and *focused* into the SBS cell as E_{2j}. The SBS reflected Stokes wave is denoted as E_{3j}. The z-position of the j-th BCM is given by δ_j.

In order to comprehend the basic *mechanism* of the phenomenon as such, let us analyze the *simplest* model possible. All the waves (E_{ij}, $i = 1\ldots3, j = 1\ldots N$) are considered to be the corresponding components of *linearly polarized monochromatic plane waves* which are *switched on* at the time $t = 0$ with *constant* amplitudes. The fact, that the beam is focused by the concave mirror and so E_{2j} becomes the *Gaussian* beam, allows for further simplification considering just the Rayleigh region close to the focal plane of the mirror. In this area the electromagnetic wave reflected by the mirror may be approximated by the plane wave provided the pump beam cross section radius r_p is significantly smaller then the concave mirror curvature radius R_m ($r_p \ll R_m$). The reflected wave amplitude is then multiplied by the factor μ which represents the amplitude increase caused by focusing.

Under these conditions any optical wave in the system under consideration may be expressed as

$$E_{ij} = \frac{1}{2}A_{ij}\exp\left[i\left(k_{ij}z - \omega_{ij}t + \phi_{ij}\right)\right] + \text{c.c.}, \tag{1}$$

where A_{ij}, k_{ij}, ω_{ij}, and ϕ_{ij} denote the amplitude, wave number, frequency, and phase of the ij-th wave, respectively. The waves reflected by the mirror are counter-propagating and so $k_{2j} = -k_{1j} \equiv k_j$. The mirror reflection does not change the waves *frequency* $\omega_{2j} = \omega_{1j} \equiv \omega_j$. The *phase* after the reflection is given by $\phi_{2j} = \phi_{1j} + \pi + (k_{1j} - k_{2j})\,\delta_j$. Substitution for k_{2j} provides $\phi_{2j} = \phi_{1j}+\pi-2k_j\delta_j$. The *amplitude* after the reflection is given as $A_{2j} = \mu_j A_{1j} \equiv \mu_j A_j$.

By using these formulae it is possible to derive the *intensity interference pattern* in the SBS PCM proximity in the form

$$\left\langle E_j^2(z,t)\right\rangle = (E_{1j}+E_{2j})(E_{1j}+E_{2j})^* =$$
$$= \frac{1}{4}|A_j|^2\left[\frac{1}{2}\left(1+\mu_j^2\right)-\mu_j\exp\left[i2k_j(z-\delta_j)\right]\right]+\text{c.c.}. \tag{2}$$

It should be noted that the intensity interference pattern *nodal point* position *does not depend* on the pump beam phase ϕ_{1j}. It is fully determined by the actual mirror position δ_j.

The acoustic wave equation right-hand side driven by the electrostriction is given as [12]

$$g_j(z,t) = -\frac{\gamma}{8\pi}\frac{\partial^2}{\partial z^2}\left\langle E_j^2(z,t)\right\rangle, \tag{3}$$

where the electrostriction coupling constant γ was introduced. Substitution of (2) into (3) provides

$$g_j(z,t) = -\frac{\gamma\mu_j}{4\pi}|A_j|^2 k_j^2\exp\left[i2k_j(z-\delta_j)\right]+\text{c.c.} \tag{4}$$

The acoustic waves $\varrho_j(z,t)$ in SBS medium are solutions of the acoustic wave equation (for the sake of simplicity *no damping* was considered)

$$\frac{\partial^2\varrho_j}{\partial t^2}-v^2\frac{\partial^2\varrho_j}{\partial z^2} = -\frac{\gamma\mu_j}{4\pi}|A_j|^2 k_j^2\exp\left[i2k_j(z-\delta_j)\right], \tag{5}$$

where v stands for the sound velocity in SBS medium.

The *initial conditions* may be determined as the acoustic noise waves with *random* phases. Considering only the SBS *phase matched* waves and denoting $q_j \equiv 2k_j$ and $\Omega_j \equiv q_j v$ these initial conditions can be expressed the following way

$$\varrho_j(z,0) = \frac{1}{2}S_j\left\{\exp\left[i\left(q_jz+\varphi_j^-\right)\right]+\exp\left[i\left(q_jz+\varphi_j^+\right)\right]\right\}+\text{c.c.},$$
$$\frac{\partial\varrho_j}{\partial t}(z,0) = \frac{1}{2}iS_j\Omega_j\left\{-\exp\left[i\left(q_jz+\varphi_j^-\right)\right]+\exp\left[i\left(q_jz+\varphi_j^+\right)\right]\right\}+\text{c.c.}, \tag{6}$$

where S_j is the acoustic noise amplitude dependent on geometry, temperature, material parameters, and frequency [1] and $\varphi_j^\pm$ denotes the thermal noise random phases.

The *general* solution of the problem (5)+(6) can be found in the form

$$\varrho(z,t) = \varrho_0-\left\{\frac{\gamma\mu_j}{32\pi v^2}|A_j|^2\exp\left[iq(z-\delta_j)\right]+\right.$$
$$+\frac{1}{2}\left(\frac{\gamma\mu_j}{32\pi v^2}|A_j|^2\exp\left[-iq_j\delta_j\right]+S_j\exp\left[i\varphi_j^-\right]\right)\exp\left[i\left(q_jz-\Omega_jt\right)\right]+$$
$$\left.+\frac{1}{2}\left(\frac{\gamma\mu_j}{32\pi v^2}|A_j|^2\exp\left[-iq_j\delta_j\right]+S_j\exp\left[i\varphi_j^+\right]\right)\exp\left[i\left(q_jz+\Omega_jt\right)\right]\right\}+\text{c.c.}. \tag{7}$$

This solution consists of *three* distinct components. The *first* one (ϱ_0) represents the *mean density value* of the medium. The *second* component (remaining part of the first line) stands for the *stationary density modulation.* The terms on the second and the third line express the *standing acoustic wave* written in the form of a superposition of two counter-propagating acoustic waves. One of these waves exactly matches the SBS wave.

Moreover, the acoustic waves from thermal noise background with their phase close to $\varphi_j^\pm = -q_j\delta_j$ undergo the constructive interference with the interference field drive wave and become dominant in the acoustic noise background. Such waves have the highest probability to become the SBS seed. However, in the case of $\frac{\gamma\mu_j}{32\pi v^2}|A_j|^2 \gg S_j$ it is possible to *neglect* the thermal noise background compared to the standing wave. It should be apparent from (7) that the relative phase difference between any SBS cells may be tuned by the change of δ_j parameter.

3. Conclusions

A simple model of the *phase locking* controlled by SBS PCM scheme using a back-seeding concave mirror was analyzed. It was found that the SBS process can be efficiently seeded by a standing acoustic wave produced by an optical interference pattern in SBS PCM proximity. Compared to the thermal noise *random* phase wave seeding considered so far this is a new important result. Formulae were found to determine under which conditions one of these seeding processes would prevail. The corresponding acoustic grating induced by the interference field is spatially localized uniquely with respect to the concave mirror position. Using this approach it is possible to control the relative phase of the backscattered Stokes field for an arbitrary number of SBS cells simultaneously employed thus offering many interesting applications. This standing acoustic wave is in fact a transient phenomenon as for the equation with viscosity included this wave would become gradually extinct and only the stationary density modulation would remain. The solution found is an important contribution to proper understanding of this already observed and in practice employed mechanism of the phase locking.

4. Acknowledgments

This research was supported by the Ministry of Schools, Youth and Sports of the Czech Republic (project No. LC528 and grant KONTAKT ME933), by the IAEA Research Contract No. 13781, and No. 13758/R0, by the Nuclear Research & Development Program of the Korea Science and Engineering Foundation (KOSEF) grant funded by the Korean government (MEST) (grant code: M20090078160), directly by CTU in Prague, Czech Republic, and also by KAIST, Daejeon, Republic of Korea.

References

[1] Boyd R W, Rzazewski K, and Narum P 1990 Noise initiation of stimulated Brillouin scattering *Phys. Review A* **42** 5514
[2] Basov N G, Zubarev I G, Mironov A B, Mikhailov S I, and Okulov A Yu 1980 Phase fluctuations of the Stokes wave produced as a result of stimulated scattering of light *Zh. Eksp. Teor. Fiz.* **31** 645
[3] Rockwell D A and Giuliano C R 1986 Coherent coupling of laser gain media using phase conjugation *Opt. Lett.* **11** 147
[4] Loree T R, Watkins D E, Johnson T M, Kurnit N A, and Fisher R A 1987 Phase locking two beams by means of seeded Brillouin scattering *Opt. Lett.* **12** 178
[5] Moyer R H, Valley M, and Cimolino M C 1988 Beam combination through stimulated Brillouin scattering *J. Opt. Soc. Am. B* **5** 2473
[6] Andreev N F, Khazanov E A, Kuznetsov S V, Pasmanik G A, Shklovsky E I, and Sidorin V S 1991 Locked phase conjugation for two-beam coupling of pulse repetition rate solid-state lasers *IEEE J. Quant. Elect.* **27** 135
[7] Ridley K D and Scott A M 1996 Phase-locked phase conjugation using a Brillouin loop scheme to eliminate phase fluctuations *J. Opt. Soc. Am. B* **13** 900
[8] Kong H J, Lee J Y, Shin Y S, Byun J O, Park H S, and Kim H 1997 Beam recombination characteristics in array laser amplification using stimulated Brillouin scattering phase conjugation *Opt. Review* **4** 277
[9] Bowers M W and Boyd R W 1998 Phase locking via Brillouin-enhanced four-wave-mixing phase conjugation *IEEE J. Quant. Elect.* **34** 634
[10] Kong H J, Lee S K, Yoon J W, and Beak D H 2006 Beam combination using stimulated Brillouin scattering for the ultimate high power-energy laser system operating at high repetition rate over 10Hz for laser fusion driver *Opt. Review* **13** 119
[11] Ostermeyer M, Kong H J, et al. 2008 Trends in stimulated Brillouin scattering and optical phase conjugation *Las. Part. Beams* **26** 297
[12] Boyd R W 1992 *Nonlinear Optics* San Diego: Academic press Inc.

Journal of Physics: Conference Series **244** (2010) 032027 doi:10.1088/1742-6596/244/3/032027

Simulations for experimental study of warm dense matter and inertial fusion energy applications on NDCX-II

J. J. Barnard[1], J. Armijo[2], F. M. Bieniosek[2], A. Friedman[1], M. Hay[2], E. Henestroza[2], B. G. Logan[2], R. M. More[2], P. A. Ni[2], L. J. Perkins[1], S-F. Ng[2,4], J.S. Wurtele[2], S.S. Yu[2,4], A. B. Zylstra[2]

[1]*Lawrence Livermore National Laboratory, Livermore, CA, USA*
[2]*Lawrence Berkeley National Laboratory, Berkeley, CA, USA*
[3]*Tech-X Corporation, Boulder, CO, USA*
[4]*Chinese University, Hong Kong, China*

E-mail: jjbarnard@llnl.gov

Abstract. The Neutralized Drift Compression Experiment II (NDCX II) is an induction accelerator planned for initial commissioning in 2012. The final design calls for a ~3 MeV, ~30 A Li^+ ion beam, delivered in a bunch with characteristic pulse duration of 1 ns, and transverse dimension of order 1 mm. The purpose of NDCX II is to carry out experimental studies of material in the warm dense matter regime, and ion beam/hydrodynamic coupling experiments relevant to heavy ion based inertial fusion energy. In preparation for this new machine, we have carried out hydrodynamic simulations of ion-beam-heated, metallic solid targets, connecting quantities related to observables, such as brightness temperature and expansion velocity at the critical frequency, with the simulated fluid density, temperature, and velocity. We examine how these quantities depend on two commonly used equations of state.

*This work was performed under the auspices of the U.S Department of Energy by Lawrence Livermore National Laboratory under Contract DE AC52 07NA27344, and by the University of California, Lawrence Berkeley National Laboratory under Contract DE AC03 76SF00098.

1. Introduction

Ion beams have a number of advantages for heating materials to the Warm Dense Matter (WDM) state [1 - 4]. Included among these are capabilities for spatially uniform and volumetric energy deposition over relatively large and diagnosable material volumes. The Neutralized Drift Compression Experiment (NDCX II) is now being constructed at Lawrence Berkeley National Laboratory to study Warm Dense Matter questions and to investigate target-beam coupling relevant to heavy-ion driven inertial fusion energy. NDCX II is the successor to the experiment NDCX I that demonstrated the technique of injecting plasma into the final beam path to greatly reduce the effects of space charge and therefore achieve short pulses needed for WDM and IFE studies [5]. The physics design of the NDCX II accelerator is described elsewhere [6]. In section 2, we review the target configurations considered for NDCX II. In section 3, we give an example of simulations of a solid Aluminum foil target. In sections 4 and 5, we examine the expected temperature and velocity at the critical surface.

2. Target configurations for NDCX II

A number of target configurations have been considered for NDCX II. These include spherical [7] and cylindrical bubbles [8] (i.e. spherical or cylindrical voids to create enhanced regions of higher pressure and temperature after the ion heating collapses the voids) and planar solid and planar foam [9] targets. Pulse formats include single pulses of fixed ion energy (but with an energy spread at the target) for WDM studies, and double pulses with varied energy (or single pulses with ion energy that changes over the pulse) to investigate ion-coupling efficiency[10]. Recent heavy ion driven "direct drive" target simulations for inertial fusion energy have shown promisingly high fusion gain [11], by increasing the ion deposition length (the range) over the course of the pulse. This increasing range can be accomplished by increasing the ion energy over the course of the pulse. Experiments that demonstrate increased coupling efficiency, by increasing the range over the course of the pulse, have been simulated[10]. In this paper, however, we focus on the WDM mission for NDCX II.

3. Simulation of beam-heated solid aluminum target

As described in section 2, NDCX II is being designed to heat both solid and porous planar metallic foils, among other options. For the simulations described in this paper we used the radiation hydrodynamics code HYDRA [12]. We assumed the target material was a 3.5 μm thick solid Aluminum, and that the equation of state was either QEOS [13] or LEOS, both of which are accessible by the HYDRA code. LEOS was run with and without the Maxwell construction. (The Maxwell construction is an equilibrium version of the EOS that permits coexistence of liquid and vapor phases, removing the dynamically unstable region of the density-temperature phase space.) The ion intensity varied with time as a parabola of full width duration 1 ns. The ion beam used was a 2.8 MeV Li^+ beam and the simulation used HYDRA's Bethe-Bloch ion deposition algorithm. The intensity was adjusted to yield approximately 20 kJ/g integrated over the pulse. The HYDRA simulations were 2D with an assumed 0.5 mm beam radius; the results here simply describe the 1D evolution of the target along the longitudinal axis (parallel to the beam direction). Figure 1 describes the evolution of the density, temperature, velocity, and charge state Z* of the target during the 1 ns of heating of the pulse.

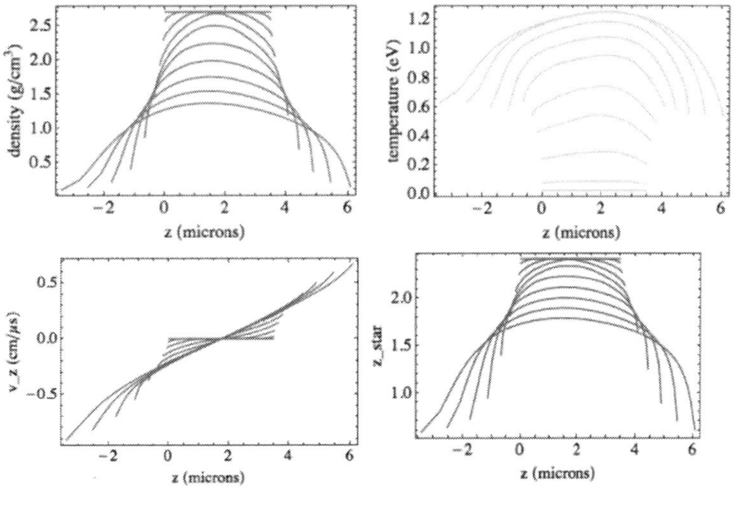

Figure 1. Longitudinal variation of target at radial center (r=0) for ten different times equally spaced between 0 and 1 ns. Ion beam heating ends at 1 ns. Target initially lies between 0 and 3.5 microns. Upper left: Density vs. longitudinal distance z; Upper right: Temperature kT vs. z. Lower left: Longitudinal velocity v_z vs. z; Lower right: Effective charge state of target Z^* vs. z.

4. Predicted brightness temperature evolution for beam heated Aluminum target

For the target described in section 3, with finite initial temperature non-uniformity, using either QEOS or LEOS as the equation of state, and an evolving Z*, the evolution will differ from that described in section 4. As a way of testing the ability of pyrometer measurements [14] to discriminate equations of state in an actual experiment, we estimated the temperature as a function of time as measured by a pyrometer at three widely separated wavelengths, using the ion pulse parameters and two EOS's described in section 3. For the brightness temperature, we assume the following simple model for the brightness temperature T_b: $T_b = T_{max}$ if $\nu > \nu_{critmax}$; $T_b = T(\nu_{crit})$ if $\nu_{critmax} > \nu > \nu_{critmin}$, and 0 if $\nu_{critmin} > \nu$. Here T_{max} is the maximum material temperature within the foil (generally found at the foil center), $\nu_{critmax}$ is the maximum critical frequency in the foil (also usually at the center) and $\nu_{critmin}$ is the minimum critical frequency in the simulation (usually at the outermost zone, numerically limited by the finite density of the lagrangian fluid element). Here ν_{crit} is the local plasma frequency divided by 2π. The results for the NDCX II simulations are found in fig. 2.

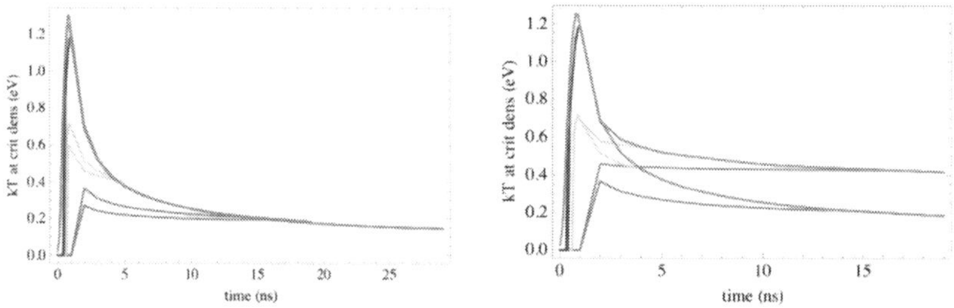

Figure 2. Left: Evolution of brightness temperature (defined in section 4) for three observation photon energies (red: 1500 nm; green 450 nm, and blue: 150 nm) and two equations of state (upper: LEOS; lower: QEOS). Both EOS are without Maxwell construction. The magenta curve is the evolution of T_{max}. Right: Same as left figure, except that the upper curves are LEOS without Maxwell construction, and the lower curves are with the Maxwell construction.

5. Predicted velocity evolution for beam heated Aluminum target

We may similarly calculate the velocity that would be inferred by a reflected laser pulse if the reflection occurs predominantly at the critical frequency. These results are shown in figure 3.

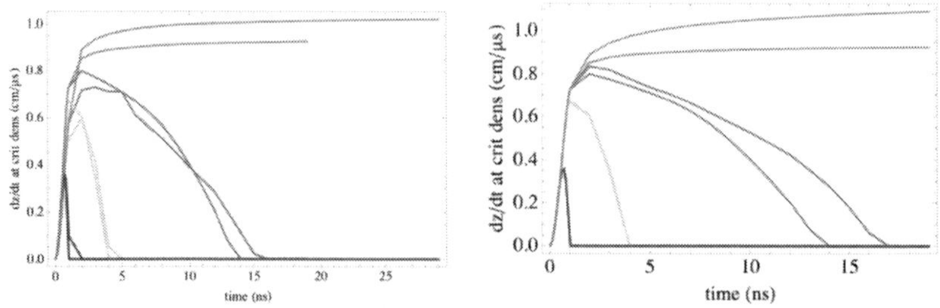

Figure 3. Left: Evolution of velocity at the critical density for three observation photon energies (red: 1500 nm; green 450 nm, and blue: 150 nm) and two equations of state (upper: LEOS; lower: QEOS). Both EOS are without Maxwell construction. The magenta curve is the evolution of v_{max}, the velocity of the outermost zone. Right: Same as left figure, except that the upper curves are LEOS without Maxwell construction, and the lower curves are with the Maxwell construction.

6. Discussion and conclusion

It is evident from figures 2 and 3 that widely spaced wavelength pyrometry measurements of expanding foils will differ at the 15 to 25% level, between the two equations of state (QEOS and LEOS). The brightness temperature is lower at lower frequencies and the pulse duration is longer. It is also worth noting that the choice of whether to use the Maxwell construction in hydrodynamic simulations also makes significant and measurable differences to predicted pyrometry measurements, particularly in the IR for these parameters. The choice of which construction to use is non-trivial. Maxwell construction implies equilibrium has occurred over the length scale of the simulation zone, even in the dynamically changing situation of a rapidly expanding foil. Simulation without the Maxwell construction implies that droplets and bubbles are well resolved, which is unlikely to be true at all scales [15]. Differences in brightness temperature of 15 to 25% should be easily detectable; however, measurements at multiple wavelengths will be needed to separate differences arising from inital beam intensity from differences arising from EOS. For the velocity measurements, the differences between EOS (and between simulations with and without Maxwell construction) are most clearly seen in the longer wavelength results; the differences are on the 10% level for the two candidate EOS's.

We should point out that these calculations have neglected any absorption through the medium and any differences between the propagation of the two polarization states, and so are really only a first look at the ability of NDCX II to clarify EOS questions. We have also used a solid target a more rigorous test (relative to a porous target), as the hydrodynamic time scale for this case is the shortest relative to the pulse duration of NDCX II. Including these additional physics and a wide range of materials is currently under study by the authors. Finally we should note that these calculations can also be applied to (and were partially motivated by) laser heated targets if there is initial temperature equilibration followed by hydrodynamic expansion [16].

7. References

1. B.G. Logan, et al., IFSA 2007, J. of Phys., Conference Series 112 (2008) 032029.
2. N.A. Tahir, et al, Nucl. Inst. and Meth. in Phys. Research B 245 (2006) 85–93.
3. J.J. Barnard, et al., Proc. 2005 Particle Accelerator Conference, p. 2568 (2005).
4. F.M. Bieniosek, et al, Nucl. Inst. Meth. A 606 (2009) 146-151.
5. P.A. Seidl, et al, Nucl. Inst. Meth. A 606 (2009) 75-82.
6. A. Friedman *et al.*, Nucl. Instr. and Meth. A **606**, 6 (2009).
7. S. F. Ng, J. J. Barnard, P. T. Leung, B.G. Logan, and S. S. Yu, in preparation (2009).
8. J. J. Barnard, et al, Nucl. Inst. and Meth. in Phys. Research A, **606**, 134-138, (2009)
9. A. B. Zylstra, J.J. Barnard, and R. M. More, High Energy Density Physics, in press (2009).
10. S. F. Ng, S. Veitzer, J. J. Barnard, P. T. Leung, S. S. Yu, in preparation (2009).
11. B. G. Logan, L. J. Perkins, and J. J. Barnard, Phys. Plasmas **15**, 072701 (2008).
12. J. J. Barnard, et al, Nucl. Inst. and Meth. in Phys. Research A, **577** (2007) 275-283.
13. R. More, K.H. Warren, D.A. Young, G. B. Zimmerman, Phys. Fluids, **31**, 3062 (1988).
14. P.A. Ni, et al, Laser and Particle Beams, December 2008.
15. J. Armijo and J.J. Barnard, in preparation (2009).
16. P. A. Ni, F. M. Bieniosek, M. Leitner, C. Weber, W. L. Waldron, Nuclear Instruments and Methods in Physics Research A, **606**, (2009), 169-171.

The Sixth International Conference on Inertial Fusion Sciences and Applications IOP Publishing
Journal of Physics: Conference Series **244** (2010) 032028 doi:10.1088/1742-6596/244/3/032028

Ion-beam-driven warm dense matter experiments

F.M. Bieniosek[1], J. J. Barnard[2], A. Friedman[2], E. Henestroza[1], J. Y. Jung[1], M. A. Leitner[1], S. Lidia[1], B. G. Logan[1], R. M. More[1], P. A. Ni[1], P. K. Roy[1], P. A. Seidl[1], W. L. Waldron[1]

[1]***Lawrence Berkeley National Laboratory, Berkeley, CA, USA***

[2]***Lawrence Livermore National Laboratory, Livermore, CA, USA***

E-mail: fmbieniosek@lbl.gov

Abstract. As a technique for heating matter to high energy density, intense beams of heavy ions are capable of delivering precise and uniform beam energy deposition to a relatively large sample. The US heavy ion fusion science program has developed techniques for heating and diagnosing warm dense matter (WDM) targets. We have developed a WDM target chamber and a suite of target diagnostics including a fast multi-channel optical pyrometer, optical streak camera, VISAR, and high-speed gated cameras. Initial WDM experiments heat targets by both the compressed and uncompressed parts of the NDCX-I beam, and explore measurement of temperature, droplet formation and other target parameters. Continued improvements in beam tuning, bunch compression, and other upgrades are expected to yield higher temperature and pressure in the WDM targets. Future experiments are planned in areas such as dense electronegative targets, porous target homogenization and two-phase equation of state.

*This work was performed under the auspices of the U.S Department of Energy by Lawrence Livermore National Laboratory under Contract DE AC52 07NA27344, and by the University of California, Lawrence Berkeley National Laboratory under Contract DE AC02-05CH11231.

1. Introduction

The US heavy ion fusion science program is developing techniques for studying ion-beam-driven warm dense matter (WDM) targets [1-3]. Intense ion beams have several attractive features as a technique for generating WDM. These features include: precise control of local beam energy deposition dE/dx, nearly uniform throughout a given volume, and not strongly affected by target temperature, large sample sizes (about 1 micron thick by 1 mm diameter), and the ability to heat any target material, for example, foams, powders, conductors, insulators, solid, gas, etc. The WDM conditions are achieved by space-charge neutralized longitudinal drift compression and transverse focusing of the ion beam to provide a hot spot on the target with a beam spot size of about 1 mm, and compressed pulse length about 2 ns. Initial experiments use a 0.3 MeV, 30-mA K^+ beam (below the Bragg peak) from the NDCX-I accelerator to heat foil targets. The NDCX-1 beam contains an uncompressed pulse up to >10 μs of fluence ≥200 kW/cm^2, and a compressed pulse of fluence ~10 mJ/cm^2. We have recently improved the NDCX beamline and its capabilities for WDM studies. The improvements derive from the replacement of the original Induction Bunching Module (IBM) with a new one that can generate nearly twice the volt-seconds, and from a longer neutralized drift section that allows the longer duration energy modulated pulses to fully compress at the target plane. The IBM waveform has been optimized for several operating beam energies. The waveform fidelity is such that the relative error in energy is less than 1% over ~400-500 ns pulse duration.

2. Target positioner and diagnostics

The target holder has two axes of motor drives to provide two degrees of freedom of remote positioning of the target. These include 2 axes on the target positioner and 2 axes on the light collection optics table. This configuration allows us to remotely position the target and target assembly without breaking vacuum to move the target to a fresh spot. Precision target positioning equipment allows rapid re-positioning of the target foil between shots. Improvements to the target assembly include the installation of a simple final focus cone. The cone enhances beam intensity on target by reflecting additional beam ions that are initially radially outside the target area. Target diagnostics [4] include a fast optical pyrometer, an optical streak-spectrometer, VISAR, and two high-speed gated image intensified cameras. We have installed a new current monitor downstream of the target to measure beam current transmitted through the target as a function of time. The beam current transmitted should be sensitive to the formation of droplets in the target foil.

3. Model of target dynamical response

As the ion beam deposits its energy volumetrically on the solid target, the target goes through a series of phase changes depending on the total beam energy fluence and power. For the beam power density levels available in NDCX-1, in the 100 kW/cm^2 range for the longitudinally uncompressed beam (several μs), and in the >> 1 MW/cm^2 range for the longitudinally compressed beam (2 ns) the target will go through a melting process and vaporize a fraction of its mass reaching temperatures near 0.5 eV (~ 6000 K).

The diameter of the beam spot on target is 1 mm. The thermal diffusivity of the thin-foil targets is of order 1 cm^2/s or less, which means that even on a 10-μs time scale the heat diffuses transversely on a 30 micron scale. Therefore we can model the target dynamics using only time and the spatial dimension along the beam direction. In the model, beam heating is balanced by energy flux from the surface of the foil due to evaporation, radiation, and thermionic (Richardson) emission. Using this model we calculate the temperature at which the cooling mechanisms equilibrate the energy flux from the ion beam (Fig. 1). The thickness of target foils referred to in Fig. 1 are: Au: 150 nm, Al: 350 nm, Pt: 120 nm, Si: 400 nm, and C: 400 nm.

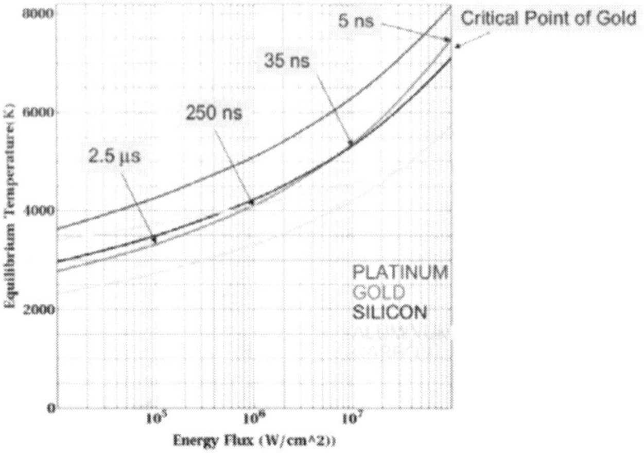

Fig. 1. Equilibrium temperature as a function of beam energy flux. Times indicated refer to time needed to reach equilibrium temperature.

One of the key areas of investigation in warm dense matter is the dynamic transition between liquid and vapor of superheated metals. In particular, as the solid is heated above the melting temperature, expansion will occur, and droplets form due to surface tension effects. A complete theory of precisely how these droplets form is under development. Experimental evidence for droplet formation is available from optical and current transformer data. We are beginning to compare detailed theory that describes the emission spectrum from droplets. The calculation is done by extending the textbook Mie scattering theory to a general dielectric function and adapting it to calculate absorption and emission coefficients for the droplets. This work is in progress.

4. Experimental beam-target data

The optical target emission spectrum is measured by a high dynamic range Hamamatsu streak camera coupled with a spectrometer. Reconstructed temperature, T, is obtained from non-linear least square fit of experimental spectra, I(λ,T) to a radiation model. The model is the Planck formula multiplied by emissivity, ε(λ,T), which is depending on situation, has either linear or square dependence on wavelength [4]. The streak-spectrometer data indicates that the target temperature in platinum reaches 4500 K, in approximate agreement with model predictions. Optical self-emission from the expanding shower of hot debris (liquid droplets) 500 μs after a platinum target shot shows the presence of hundreds of hot droplets, supporting the characterization of the target as forming droplets early in the beam pulse.

The evolution of the Au target during and after the ion-beam-heating can be characterized by 3 distinct phases. The thermal radiation reaches a maximum after about 1.5 μs relative to the head of the beam (Fig. 2, Phase 1).

Phase 2: Au I (atomic vapour) lines appear in spectral records slightly after the radiation peak. Simultaneously the total thermal radiation signal drops. We attribute the early peak in radiation pattern to the disassembly of the foil into droplets. Note that, while overall radiation decreases, the reconstructed temperature gives either a constant temperature or a slight increase (beam still deposits energy into the target).

Phase 3: The atomic lines turn off with the end of the beam pulse and the total radiated signal continues to fall. Since the atomic lines are ion-beam pumped, the moment the lines appear is related to droplet formation, i.e. ions pass through the vacated regions between droplets and excite the Au vapor downstream of the target.

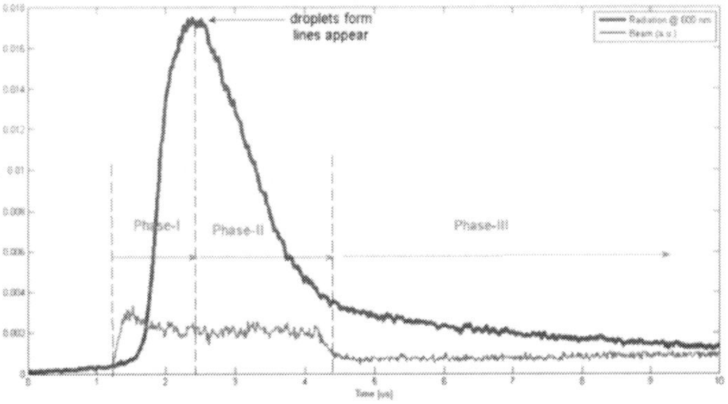

Fig. 2: Thermal radiation at 600 nm and heating-ion-beam intensity.

5. Conclusions

Rapid bulk heating of target foils to temperatures up to ~4500 K (0.4 eV) has been achieved using NDCX-I beams. The mechanism of target heating is described by a simple model of the equilibrium between energy input from the beam and energy loss from the surface of the target due to mechanisms such as vaporization of the target material. Development of techniques for heating and diagnosing targets allows bulk heating of WDM targets in the laboratory using ion beam heating. The NDCX-I environment is conducive to multiple repetitive target experiments for detailed study of target behavior under various conditions and using multiple diagnostics.

Evidence for the formation of liquid metal droplets on the μs time scale is presented, and compared with predictions. Our experiments are expected to shed light on droplet formation in metal targets under WDM conditions and on the properties of the subsequent debris shower. These results could find wide application in areas such as simulating volumetric neutron heating in inertial confinement fusion facilities, and other applications of liquid metal droplets.

Other future areas of study include the equation of state near the solid-liquid and liquid-vapor phase boundaries, and material properties such as the evaporation rate, surface tension, electrical conductivity; identification of critical points and phase boundaries; porous targets; positive ion/negative ion experiments in high electron affinity targets; etc.

The next-step facility, NDCX II, is under construction and will offer additional capabilities [5]. The NDCX I work is essential to benchmarking our beam manipulation, targetry, and diagnostic techniques for NDCX II, which will be able to probe the WDM regime at higher temperatures and pressures. Improvements to the beamline, targetry and diagnostic components will continue to be made, and will improve the facility for a variety of WDM experiments.

6. References

1. B.G. Logan, et al., IFSA 2007, Journal of Physics, Conference Series 112 (2008) 032029.
2. P.A. Seidl, et al., Instrum. Meth. A 606 (2009) 75-82.
3. F.M. Bieniosek, et al., Nucl. Instrum. Meth. A 606 (2009) 146-151.
4. P.A. Ni, et al., Laser and Particle Beams, 27, December 2008.
5. Friedman, A; et al., Nucl. Instrum. Meth. A, 606, 6-10, (2009).

Journal of Physics: Conference Series **244** (2010) 032029 doi:10.1088/1742-6596/244/3/032029

Design of a scaled recirculator for Heavy Ion Inertial Fusion

K. Fiuza, B. Beaudoin, S. Bernal, I. Haber, R. A. Kishek, P. G. O'Shea, C. Papadopoulos, D. Sutter and C. Wu

Institute for Research in Electronics & Applied Physics, University of Maryland, College Park, MD 20742, U.S.A.

kfiuza@umd.edu

Abstract. An alternative concept for Heavy Ion Inertial Fusion (HIF) is the use of a recirculator to accelerate ion beams to energies in the range of 50-100 GeV [1]. The physics of an ion recirculator can be explored by means of scaled experiments in a compact machine like the existing University of Maryland Electron Ring (UMER). UMER has been successfully used for the study of the fundamental physics of space-charge-dominated transport using a 10 keV electron beam with up to 100 mA of current (or 10 nC per a 100 ns pulse) [2]. Due to the low energy and high perveance, the UMER beam accesses the same range of intensities as an HIF driver. In this paper we report on a computational study for the design of an acceleration stage for UMER using an induction cell. Using the two-dimensional transverse slice model in the particle-in-cell code WARP we show that it is possible to accelerate the UMER beam up to 20 keV without major modifications to the machine. Such acceleration enables future experiments on transverse resonance crossing and studies on longitudinal pulse behavior.

Introduction

The physics of the space-charge-dominated electron beam transport is applicable, on a scaled basis, to a large class of other beam systems [3]. In order to accelerate an ion beam to energies in the range of 50-100 GeV, an attractive concept is the use of a recirculator, as is proposed for Heavy Ion Inertial Fusion (HIF) applications [1]. Because the recirculator concept entails the repeated passage of the beam through the same lattice, emittance growth, halo formation and possible beam losses and activation of the vacuum pipe are major concerns. In addition, the crossing of betatron resonances as the beam accelerates is not sufficiently understood for space-charge dominated beam transport.

Employing a scaled experiment using a nonrelativistic electron beam at the University of Maryland Electron Ring (UMER), we are able to address the physics of an ion recirculator. The UMER lattice includes three induction gaps equally spaced in the ring, although at the present moment just one induction gap is in operation, for longitudinal focusing and acceleration [4]. The UMER lattice comprises 36 FODO (focusing-defocusing) cells in an 11.52 m circumference. The beam parameters are: 0.5-100 mA beam current at 10 keV nominal energy; 1.6 - 13 μm, normalized, 4 x rms emittance; < 10 mm average beam radius. The operating tunes are in the range $\nu_{0X,Y}$=5.0 - 8.0. With these parameters, tune depressions in the range 0.14 - 0.9 are possible, which cover those envisioned in a future HIF driver [2].

We report on a computational study using the particle-in-cell code WARP [5] for the design of an acceleration schedule to double the nominal energy of 10 keV, turning UMER into a recirculator. We address the transverse beam dynamics issues involved in accelerating the UMER beam to the maximum possible energy achievable without ramp the bending dipole magnets or implementing any other major change in the machine operation. A future paper will look at accelerating to higher energies under more general operating conditions.

WARP simulations

WARP is a multi-dimensional particle-in-cell and accelerator code [5]. Using the two-dimensional transverse slice model of the WARP code we explored the consequences of the acceleration on UMER. The two transverse dimensions are x and y in the lab frame, in a right handed Cartesian system with x on the horizontal and y on the vertical. Using these coordinates, the unnormalized 4 x rms emittance is $\varepsilon = 4\left[\left\langle\left(x-\langle x\rangle\right)^2\right\rangle\left\langle\left(x'-\langle x'\rangle\right)^2\right\rangle-\left\langle\left(x-\langle x\rangle\right)\left(x'-\langle x'\rangle\right)\right\rangle^2\right]^{1/2}$, where brackets $\langle\ \rangle$ indicate average over a certain distribution. The horizontal coordinate of the beam centroid is $\langle x\rangle$, defined as the center of the charge particle distribution, and the effective beam x-semi axis (envelope x-size) is $2\left[\left\langle\left(x-\langle x\rangle\right)^2\right\rangle\right]^{1/2}$. The WARP simulations with initial kinetic energy at 10 keV, using one induction gap to accelerate, are conducted to investigate a realistic acceleration schedule for UMER. The goal is to find the best feasible design, which ideally keeps a reasonable beam quality, minimizing particle losses and emittance growth, with no engineering changes. In order to do so we tested different parameter combinations such as beam energy, beam current, centroid displacements and envelope matching. A matched beam envelope has the same periodicity of the focusing lattice as the solution of the Kapchinskij-Vladimirskij (KV) envelope equations. We also tested numerical parameters, e.g. number of particles, resolution and time step. The initial particle distribution is semi-Gaussian in all simulations; this distribution corresponds to a uniform density in configuration space, but a Gaussian profile in velocity space.

Since the planned experiments for acceleration in UMER will include longitudinal beam confinement [4], the simulations incorporate the main effect. In the simulations this is taken in consideration by changing the weight of particles as the current increases due to the compression of the beam longitudinally. Analytical calculations show that the current increases from 99.4 mA at 10 keV to 122.04 mA at 20 keV. The paper concentrates, however, on the transverse dynamics as the longitudinal dynamics play a secondary role.

After choosing an acceleration schedule, we explore the limits imposed by such acceleration using different values of bare tune using different sets of current values.

Preliminary results show that one gap of 50 V, i.e., increasing the beam energy by 50 V per turn is a viable option for initial experiments on acceleration in UMER. A simulation with an idealized quadrupole/dipole lattice (no strength or mechanical errors) using a 23.5 mA beam is shown in Figure 1. The graphs illustrate the evolution, over some 100 turns, of (a) kinetic energy, (b) centroid oscillation and (c) 4 x rms unnormalized emittance. The centroid motion remains stable, despite the growing amplitude of the oscillations caused by the constant bending dipole fields (i.e., no ramping).

Figure 2 shows results from a simulation using the same 50 V gap as before but over 250 turns, and including errors. To test the robustness of the design in the simulations, we introduce 1% random errors in the quadrupole strengths, using a Gaussian distribution, as well as initial rms mismatch (envelope) and injection (centroid) errors. The results in Fig. 2 correspond to the worst-case scenario where acceleration was possible up to 20 keV even with a 15% mismatched beam envelope, which drives oscillation of a breathing envelope mode [6]. From left to right: (a) kinetic energy increases gradually to 20 keV; (b) x and y centroid coordinates decouple from each other, with the x component oscillating and growing up to 24 mm, causing the centroid to hit the pipe and cause complete beam loss. The y component doesn't oscillate, as expected for bending on the horizontal plane exclusively; and (c) envelope oscillations that are initially coupled and slightly decrease with energy growth. Also shown is the instant when the x component reaches the pipe and complete beam loss occurs. This occurs at the 235th turn, or 2,707 m. For UMER this is a remarkable result since our previous recent mark was 100 turns using 0.6 mA beam [4].

The simulation results shown in Figure 3 refer to a lower beam current of 7 mA, but a faster one-gap acceleration schedule of 1.0 kV per turn over 5 turns. The initial energy is 10 keV, and the nominal operating tune is close to 6.4. From (a), (b), and (c) we see that faster acceleration leads to faster instabilities, most probably caused by the mismatch between the beam energy and the focusing strength of the lattice. The maximum amplitude of the centroid oscillations is already 10 mm by the second turn, which brings the beam close to the vacuum pipe wall and can lead to additional instabilities, as for example through image forces [7]. The centroid oscillations (a) also point out to transverse resonance crossings, which are expected since the zero-current (bare) tune changes as the energy increases, from 7.6 at 10 keV to 5.09 at 15 keV. However resonance phenomena in this context must be carefully studied. The decreasing envelope pattern seen after some 2.5 turns in Fig. 3 (b) is due to particle losses as well as the breathing envelope oscillations from mismatch [6]. It is seen from (c) that the number of particles decays very quickly due to all these effects together.

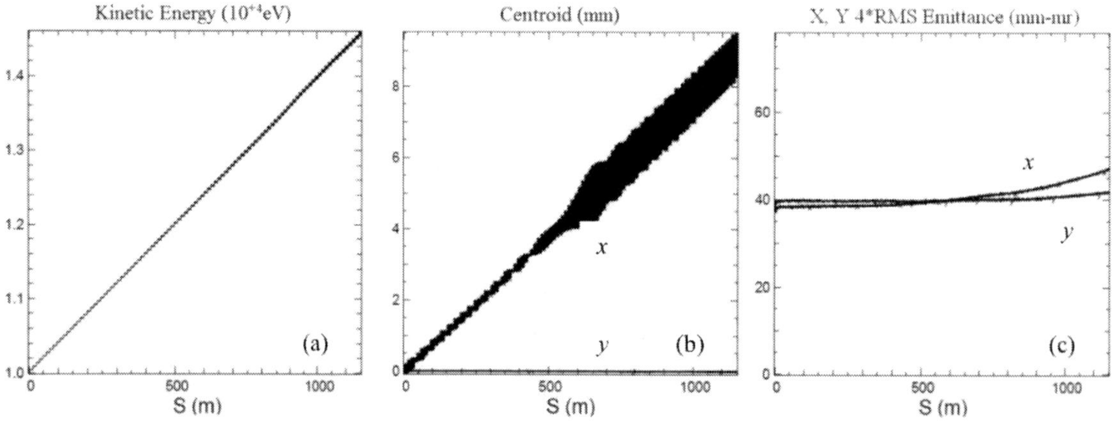

Figure 1- Ideal simulation using a 50 V/turn induction gap to accelerate a 23 mA electron beam from 10 keV to 15 keV over 100 turns. (a) kinetic energy change, (b) the x-component of the centroid oscillates relative to the pipe axis due to the lack of dipole ramping, and (c) the 4 x rms emittance growth seen is modest.

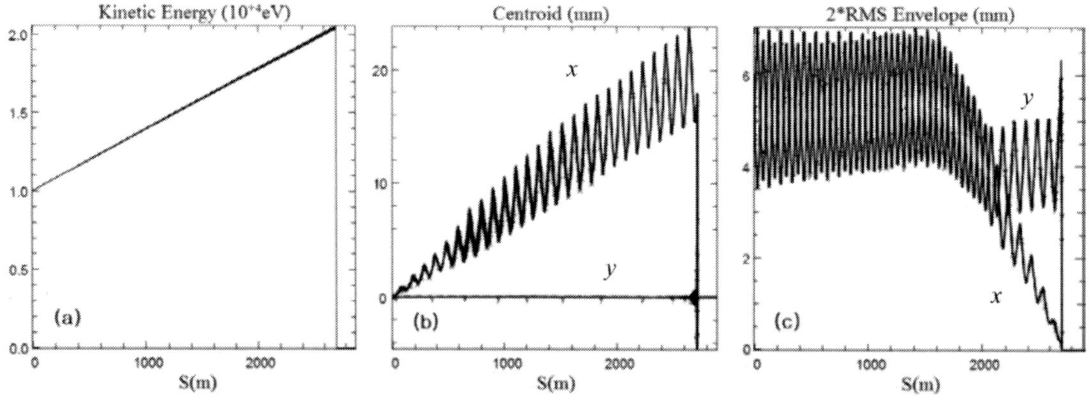

Figure 2- Worst case scenario of acceleration of 23 mA electron beam with 50 V/turn induction gap. The simulation implements a combination of errors: 1% quadrupole strength errors and 15% mismatched envelope. (a) Kinetic energy increases to 20 keV, (b) x-component of the centroid oscillates and grows until reaching the wall, and (c) mismatched 2 x rms envelope oscillations until beam disappearance.

Journal of Physics: Conference Series **244** (2010) 032029 doi:10.1088/1742-6596/244/3/032029

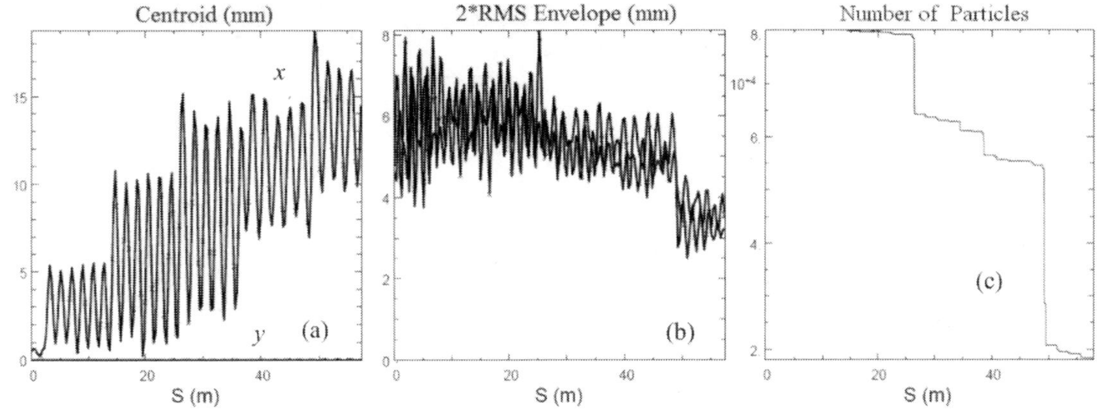

Figure 3 - Testing acceleration limit using a 1 kV/turn induction gap with a 7 mA electron beam. (a) Large amplitude of the centroid oscillations, (b) mismatched 2 x rms envelope, and (c) number of particles decreasing fast, with 75% particle losses after only 5 turns.

Conclusion

The existing induction module in UMER can in principle be used to accelerate electron beams with different acceleration schedules with no dipole ramping. Simulations with the WARP code show that, as expected, acceleration schedules that use lower increments in energy gain per turn help to maintain better beam stability. The simulations demonstrate the possibility to accelerate a 23 mA electron beam in UMER from 10 keV to 20 keV in some 200 turns. However, without ramping dipoles, better control of errors including envelope mismatch is necessary to keep centroid and envelope oscillations from leading to beam losses. Acceleration of a 7 mA beam to 15 keV seems much more feasible at this stage in the UMER project. It opens new opportunities for research in space-charge-dominated beams, especially in resonance crossing [8], an area relevant to HIF and other advanced accelerator concepts.

The results showed in this work are preliminary but are pointing towards the feasibility of acceleration in UMER without major changes to the machine. Currently, we are also studying the possibility to start beams at lower energies in order to have a bigger range for acceleration. The next stage of this research will involve detailed longitudinal dynamics studies using three-dimensional simulations in WARP.

The authors would like to acknowledge the US Dept. of Energy Offices of Fusion Energy Sciences and High Energy Physics, and the Office of Naval Research for funding this work.

References

[1] J. J Barnard, et al., in Phys. Fluids B: Plasma Phys. 5 (1993) 2698.
[2] R.A. Kishek, et al., Nucl. Instr. and Meth., in Phys. Res. A 544 (2005) 179.
[3] I. Haber, et al., Nucl. Instr. and Meth., in Phys. Res. A 606 (2009) 64.
[4] B. Beaudoin, et al., Proceedings of the 2009 Particle Accelerator Conference, Vancouver, CA, to be published.
[5] D.P. Grote, et al., Nucl. Instr. and Meth., in Phys. Res. A 415 (1998) 428.
[6] R.L. Gluckstern, Phys. Rev. Lett. 73, N. 9, (1994) 1247.
[7] K. Fiuza, R. Pakter, F.B. Rizzato, Physics of Plasmas, v. 13 (2006) 023101.
[8] A. Friedmann, et al., Nucl. Instr. and Meth., in Phys. Res. A, 415 (1998) 455.

X-ray pulse shaping in experiments with planar wire arrays at the 1.6 MA Zebra generator

V L Kantsyrev[1], A S Safronova[1], A A Esaulov[1], K M Williamson[1], I Shrestha[1], G C Osborne[1], M E Weller[1], M F Yilmaz[1], N D Ouart[1], V V Shlyaptseva[1], A L Velikovich[2] and L I Rudakov[3]

[1] *Physics Department, University of Nevada, Reno NV 89557*
[2] *Naval Research Laboratory, Washington, DC 20375*
[3] *Icarus Research Inc., Bethesda, MD 20824*

E-mail: victor@physics.unr.edu

Abstract. The shaping of the x-ray radiation pulse is very important in both radiation physics research and Inertial Confinement Fusion studies. The novel planar wire array (PWA) was found to be the effective radiator tested at the university-scale 1.6 MA, 100 ns Zebra generator. The single PWA consists of a single row of wires that are parallel to each other, while the double planar wire array (DPWA) and triple planar wire array (TPWA) include two or three parallel plane wire rows, respectively. All multi-planar geometries resulted in a cascade-type array implosion with a complicated multi-step precursor formation before plasma stagnation. The PWAs (without additional core foam target) feature a dynamic precursor evolution that is a powerful tool for x-ray pulse shaping. The shape and timing of the x-ray pulse from different PWAs were theoretically predicted and experimentally analyzed for a variety of planar wire arrays.

1. Introduction.

The shaping of the x-ray radiation pulse from wire array implosions is needed for both radiation physics and Inertial Confinement Fusion (ICF) studies [1]. It is as important as the efficiency magnetic energy conversion to x-ray radiation, x-ray peak power and the source size. Ideal Z-pinch soft x-ray (in sub-keV spectral region) power history for three-step pulse ICF scheme includes the first prepulse 20-25 ns before x-ray peak, the foot prepulse (step) on the front of the main x-ray burst, and the main burst. The energy in prepulses should be 15-20 % from the main burst radiated output [1]. X-ray pulse shaping experiments were performed early with nested cylindrical wire arrays with a load core foam target [2]. Implosions of novel planar wire array (PWA) have been investigated since 2005 [3] with single- or multi-planar wire rows placed in the center of the z-pinch chamber between anode and cathode. It has shown to be a compact x-ray source with appropriate radiation pulse shaping. The PWA's radiation yields exceed that from kinetic energy conversion E_K (E_K < 4-6 kJ) in 3-5 times, and mechanism of energy coupling is enhanced resistivity of the strongly inhomogeneous plasma (Hall MHD physics) [4]. Double Planar Wire Array (DPWA) with straight wires is currently the best radiator at 0.8-1.4 MA current. It is characterized by the combination of highest yield, power, mm-scale size, and low opacity effects [5, 6]. At 1 MA current, PWAs show comparable yield and power with another loads, compact cylindrical wire array (CCWA) [7]. However, DPWAs demonstrate substantial advantages over cylindrical wire arrays, for example, they have near quadratic load peak current scaling for yields and powers while CCWA has shown only lower than linear scaling (in 1-5 MA peak current range) [6, 8].

All multi-planar array geometries (without additional core foam target) resulted in a cascade-type wire array implosion with a complicated multi-step precursor formation before stagnation [4, 5]. The sub-keV prepulses generated during precursor formation can be applied for radiation pulse shaping [5, 6]. Another possibility for radiation pulse shaping in multi-planar arrays is based on the fact that plane wire rows imploded independently generating sub-keV prepulses in different times (especially when rows have different initial mass or were made from different materials) [5, 6].

Precursor plasma in multi-planar arrays is formed in multi-steps with some locations not coinciding with the stagnating plasma column position [5]. Typically, primary precursor appeared as early as 40-50 ns before x-ray peak, but at this step emitted radiation is not strong enough to form a prepulse [5]. Secondary precursor formation occurred during the final moments of independent implosion of wire rows and appeared as double plasma columns 10-15 ns before the main x-ray peak is clearly seen even in 1 keV time-gated images [5, 6] (Fig. 1). To observe such images as well as 1 keV precursor images in experiments with different cylindrical wire arrays [3, 9], temperature Te of precursor must be more than 200-250 eV. Recent experiments have proved it: Te > 400 eV and Ne = 10^{19} cm^{-3} were found in cylindrical low-wire-number Cu precursor plasmas on Zebra generator [10]. Up to 20 % of L-shell emissions were associated with the precursor. This means that precursor plasma emits strong x-rays to form prepulse and can be applied for radiation pulse shaping from z-pinch plasma.

2. Variation of sub-keV prepulse shape with changing PWA geometry and elemental composition

Experiments were performed on the 1.6 MA, 100 ns Zebra generator at UNR. The diagnostic includes filtered PCD (keV region) and XRD (sub-keV region) detectors, x-ray time-gated and time-integrated imaging, and time-integrated spatially-resolved and time-gated spatially-integrated x-ray spectrometers [3, 6].

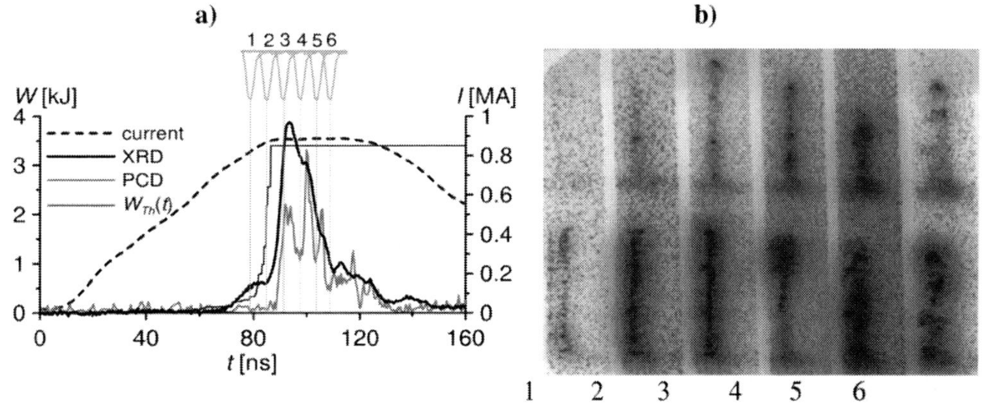

Figure 1. Experimental results from asymmetric W/Al DPWA with an inter-row gap Δ=3 mm: a) the data curves are signals from the current, XRD, and PCD detectors. W_{th} is the total radiation yield predicted by WADM and the total experimental yield W_{exp}= 18.6 kJ. At the top are positions of frames of time –gated x-ray pinhole camera. b) The time-gated x-ray plasma images obtained with microchannel plate (MCP) detector (1 keV is at the top, 3keV is at the bottom, anode is at the bottom in each row).

Journal of Physics: Conference Series **244** (2010) 032030 doi:10.1088/1742-6596/244/3/032030

The promising results were obtained from asymmetric DPWA with straight wires where plane rows were made from different materials, for example, one row made from Al, and another made from W wires. It was seen that sub-keV prepulse was generated during the secondary precursor formation (double column in frame 1, Fig. 1, b). Experimental results were compared with Wire Ablation Dynamic Model (WADM) simulation [11]. Difference between the predicted W_{th} and observed W_{exp} yields can be explained by another energy coupling mechanism that turns on at stagnation stage because WADM takes in account only transformation of kinetic energy into radiation.

DPWA with skewed wires is another load where sub-keV prepulses are generated during precursor formation. Such z-pinch load is a way to mitigate RT instabilities and to provide more control of x-ray generation by creation an axial magnetic field between wire rows [6]. In skewed DPWA the electron temperature Te is higher than in a standard DPWA [6]. Reverse skew angles for opposite wire planes create axial solenoid-type B_z field (Fig. 2, a).

a) **b)** **c)**

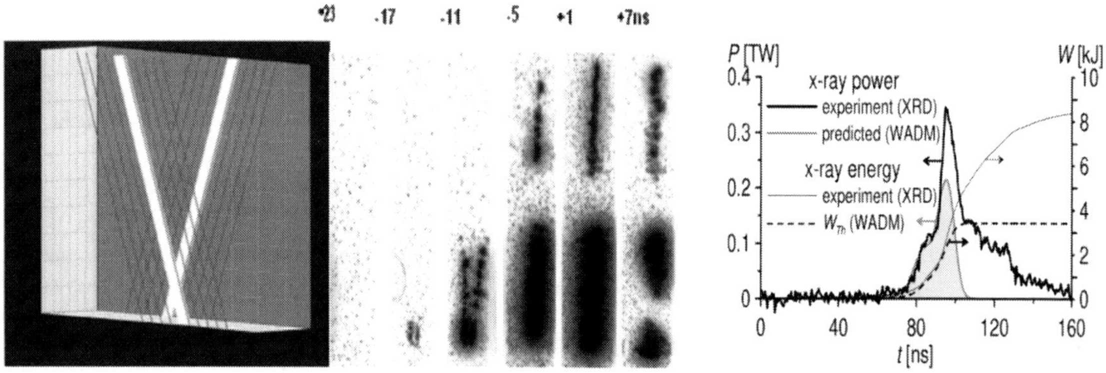

Figure 2. Implosion characteristics of Al DPWA with skewed wires (inter-row gap Δ=6mm and angle between wires is 28^0): a) WADM prediction of positions of secondary precursor plasma columns (in white) before stagnation (original wires positions shown as black dotted lines). Direction of observation is under 45^0 to wire row plane. Azimuthal field $B_{||}$ = 25 T (outside) and axial field B_z = 8 T (inside load) when current reached I = 0.5 MA. b) X-ray time-gated images: at the top >3 keV, at the bottom >1keV. Frame duration is 3ns and interframe interval is 6 ns. Observation is under 45^0 to wire row plane. The time is in ns before x-ray peak. c) WADM simulation vs experiment. W_{th} is the total radiation yield predicted by WADM and W_{exp}.> 8 kJ is experimental sub-keV radiation yield.

After appearance of secondary precursor columns (Fig. 2, b) they start to merge firstly from the center of the anode-cathode gap to form an x-ray burst. A central part of the skewed DPWA implodes several ns earlier than the rest of the array, forming the x-ray prepulse. WADM simulation of the x-ray pulse shape vs experiment with skewed DPWA has shown good correlation of prepulse shape and amplitude. Difference between P_{pred} and P_{exp} at the x-ray peak can again be explained by another energy coupling mechanism that turns on at stagnation stage.

Another perspective load that provides a possibility of radiation pulse shaping is a triple planar wire array. The prediction of radiation pulse shape for Al triple planar wire array was made with WADM model (see Fig. 3).

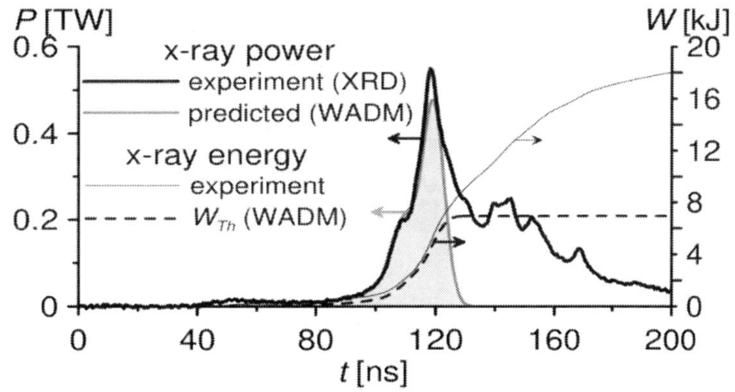

Figure 3. Implosion characteristics of Al (5056) TPWA with distributed mass. Number of wires 8 x 8 x 8. Mass distribution: 2:1:2 (wires diameter: 15 μm / 10 μm / 15 μm.) W_{th} is the total radiation yield predicted by WADM. W_{exp}> 17 kJ is the experimental total radiation yield.

Planar wire rows in triple planar wire arrays are imploded independently generating sub-keV prepulses in different times to form a joint foot prepulse in front of the main x-ray burst. Predicted foot prepulse (step) in front of the main x-ray burst was found in good agreement with simulation.

3. Conclusion.

All multi-planar wire array geometries resulted in a cascade-type implosion with a complicated multi-step precursor formation before stagnation. The PWAs feature a dynamic precursor evolution that may be a powerful tool for x-ray pulse shaping. The PWA has demonstrated possibility of radiation pulse shaping. Experiments with multi-planar wire arrays such as double planar array with straight and skewed wires and triple planar array, have shown results that are in qualitative agreement with theoretical simulations.

Acknowledgement.
Authors thank the UNR/NTF teams for technical support of the experiments on Zebra and J. Kindel, the director of NTF, for programmatic support of the experiments on Zebra. Work was supported by DOE/NNSA under Cooperative Agreements DE-FC52-06NA27586, DE-FC52-06NA27588, and in part by DE-FC52 06NA27616.

References

[1] Vesey R A, Herrmann M C, Lemke R W et al., 2007 *Physics of. Plasmas* **14** 056302
[2] Cuneo M E, Vesey R A, Sinars D B et al., 2005 *PRL* **95** 185001
[3] Kantsyrev V, Rudakov L, Safronova A et al., 2006 *IEEE Trans. Plasma Sci.* **34** 194 and 2295
[4] Chuvatin A S, Rudakov L I, Velikovich A L, 2006 *AIP Conf. Proc.* **808** 343
[5] Kantsyrev V L, Rudakov L I, Safronova A S et al., 2008 *Phys. Plasmas* **15** 030704
[6] Kantsyrev V L, Safronova A S, Esaulov A A et al, 2009 *HEDP* **5** 115
[7] Kantsyrev V L, Rudakov L I, Safronova A S et al., 2007 *HEDP* **3** 136
[8] Jones B, Cuneo M E, Ampleford D J et al., *Sandia National Laboratories Report SAND2008-6166*, Office of Scientific and Technical Information (OSTI) of the U.S. Department of Energy. Document no. 941403, September 2008.
[9] Kantsyrev V L, Fedin D A, Esaulov A A et al., 2006 *IEEE Trans. Plasma Sci.* **34** 2288
[10] Coverdale C A, Safronova A S, Kantsyrev V L et al., 2009 *PRL* **102** 155006
[11] Esaulov A A, Kantsyrev V L, Safronova A S et al., 2009 *HEDP* **5** 166

Analysis of X-ray Iron and Nickel Radiation and Jets from Planar Wire arrays and X-pinches

A S Safronova, V L Kantsyrev, A A Esaulov, N D Ouart, V Shlyaptseva, K M Williamson, I Shrestha, G C Osborne and M E Weller

University of Nevada, Reno, NV 89557 USA

E-mail: alla@unr.edu

Abstract. University-scale Z-pinch devices are able to produce plasmas with a broad range of sizes, temperatures, densities, their gradients, and opacity properties. Radiative properties of such plasmas depend on material, mass, and configuration of the wire array loads. Experiments with two different types of loads, double planar wire arrays (DPWA) and X-pinches, performed on the 1 MA Zebra generator at UNR are analyzed. X-pinches are made from Stainless Steel (69% Fe, 20% Cr, and 9% Ni) wires. Combined DPWAs consist of one plane from SS wires and another plane from Alumel (95% Ni, 2% Al, 2% Si) wires. The main focus of this work is on the analysis of plasma jets at the early phase of plasma formation and the K- and L-shell radiation generation at the implosion and stagnation phases in experiments with the two aforementioned wire loads. The relevant theoretical tools that guide the data analysis include non-LTE collisional-radiative and wire ablation dynamics models. The astrophysical relevance of the plasma jets as well as of spectroscopic and imaging studies are demonstrated.

1. Introduction

It was shown that planar wire arrays (PWA), a new type of the wire array load, produced significant radiation yields on a ns-scale, particularly DPWAs (up to $E_t \sim 25$ kJ and $P_{peak} \sim 1$ TW) and generated bright spots of sub-mm size [1, 2]. X-pinches are also very good sources of x-rays and can be used for studying radiative properties of plasmas with even higher electron temperatures (Te > 2 keV) and densities (Ne > 10^{22} cm^{-3}) than that from PWAs on scales of a few μm to several mm in size [1, 3-4]. They yield short (few ns) x-ray bursts from one or several bright plasma spots near the wire cross point and produce strong electron beams. Recently, the first applications of X-pinches to astrophysics were proposed and described in [5]. Here we present the results of the new experiments with X-pinches focused on the simultaneous study of "hot" Heα and "cold" Kα lines of Fe and Ni. In general, Lyα lines of Fe XXVI and associated satellite lines (and also Ni XXVIII) are the highest temperature lines observed in solar flares [6]. Heα lines of Fe XXV and Ni XXVII (~6.7 keV and ~7.8 keV, respectively) are also indicative of high temperatures (>1.5 keV). These are many similarities between solar flare spectra where the "hot" Fe lines were observed and studied and the spectra of certain other astrophysical sources, such as stellar coronae and supernova remnants. Then, the study of these lines has broader applications than just for solar flares. In addition, the "cold" Kα 6.4 keV iron line provides information on cool material in the vicinity of hard x-ray sources as well as on the characteristics of the x-ray sources themselves and was discovered in the x-ray spectra from the solar flares, x-ray binaries and active galactic nuclei. The understanding of mechanism of origination of "hot" and "cold" Fe lines in various astrophysical objects is very important (see, for example, [7]) and can be improved by studying these lines in laboratory plasma under well controlled experimental conditions.

2. Experimental results and discussion

Experiments with two different types of loads, double planar wire arrays (DPWA) and X-pinches, were performed on the 1 MA (100 ns current rise time) Zebra generator at UNR. X-pinches are made from Stainless Steel (SS 304: 69% Fe, 20% Cr, and 9% Ni) wires. Combined DPWAs consist of one plane from SS wires and another plane from Alumel (95% Ni, 2% Al, 2% Si) wires. The experimental diagnostic set includes filtered PCD and XRD detectors, laser shadowgraphy, x-ray time-gated and time-integrated imaging, and time-integrated spatially resolved and time-gated spatially integrated spectrometers for harder (K-shell) and softer (L-shell) x-ray regions. Previous studies of SS X-pinches which relied mostly on x-ray diagnostics [1,5] were expanded here to include shadowgraphy. The typical shadowgraphy of explosion of the X-pinch recorded at 50 ns after current start is shown in Fig. 1b. The typical time-integrated spatially-resolved K-shell spectrum shown in Fig. 1c includes both "hot" Heα and "cold" Kα lines of Fe, Cr, and Ni. The presence of Heα lines localized in the middle indicates the hot thermal plasmas (Te>1.5 keV) in/near wire cross-point. The "cold" Kα lines spread towards the anode and represent the signatures of much cooler thermal plasmas (< 100eV) with non-thermal electron beams. The use of alloyed SS wires instead of pure Fe wires helps to study simultaneously three different materials in similar plasma conditions and lower opacity effects on spectra due to small concentrations in an original alloy.

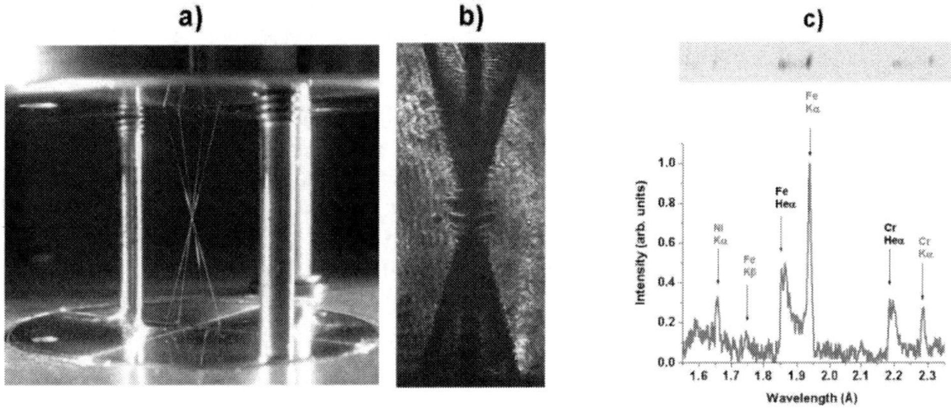

Figure 1. Stainless Steel X-pinch (shot 1591): a) Load before the shot. b) Laser shadow image. c) X-ray image of the spatially-resolved spectrum recorded by the LiF crystal (top) and lineout of the time-integrated spatially-resolved spectrum which includes He-like lines of Fe and Cr as well as characteristic "cold" Kα lines of Fe, Cr, and Ni.

The results of the experiment with combined DPWA are shown in Fig. 2. The wires ablated very little yet at 14 ns after the current start as it seen on ICCD image (Fig. 2c) and predicted by WADM [8] calculations (Fig. 2d). Then, as ablation progresses, faster ablation of Alumel wires results in higher density and lower velocity (up to 100 km/s) of the ablation flow and slower ablation of SS wires results in lower density and higher velocity (up to 120 km/s) of the ablation flow (Fig. 2e,f).

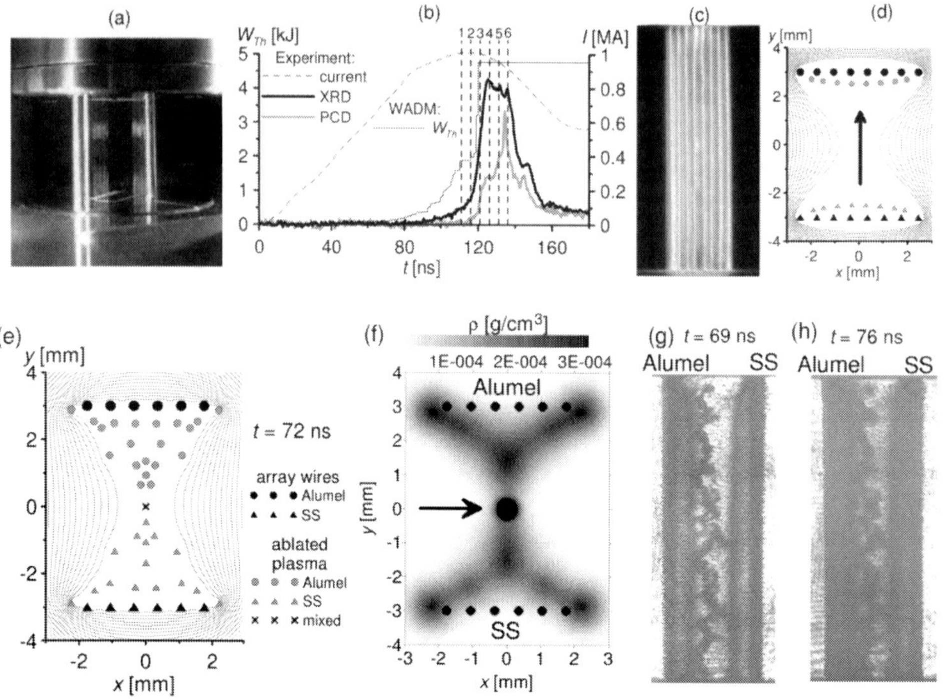

Figure 2. Alumel/Stainless Steel DPWA (shot 1774): (a) Load before the shot (side view). (b) Experimental current pulse, x-ray detector signals and the positions of MCP for time-gated spectra, and thermalized kinetic energy W_{Th} calculated by the WADM. (c) ICCD camera image of the load taken at 14 ns (side view perpendicular to wire planes). Positions of array wires and current filaments associated with the ablated plasmas (top view from the anode side) according to the WADM at 14 ns (d) and at 72 ns (e). View direction of ICCD camera is shown by black arrow in plot (d). Contours represent the intensity lines of the global magnetic field. (f) 2D mass density distribution of the ablated plasma reconstructed from the WADM simulation at 72 ns. An arrow in plot (f) shows the view direction of the laser shadow images (side view along wire planes) recorded at 69 ns (g) and 76 ns (h). Time is referred to the current pulse start.

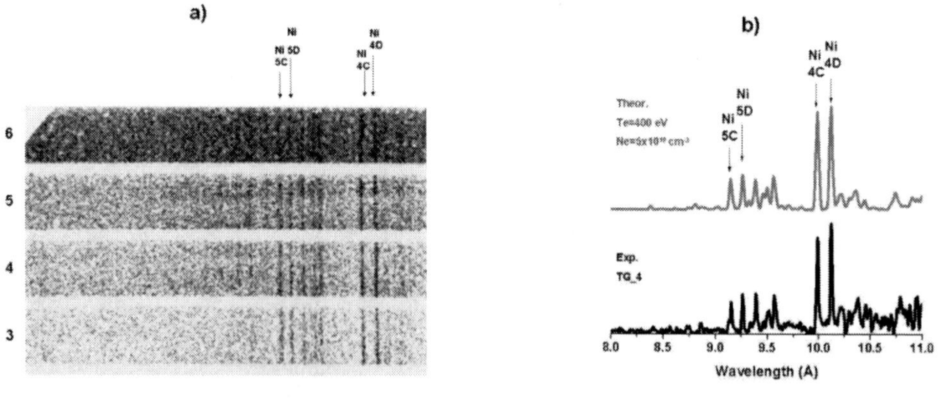

Figure 3. Alumel/Stainless Steel DPWA (shot 1774). Time-gated spectra images (a) and the spectrum recorded by the MCP frame 4 fit with non-LTE modelling (b).

Both plasma jets are supersonic with Mach number ~ 6 (sound speed is ~ 20 km/s). The disbalance of velocities of two plasma jets moving from the Alumel and SS sides and resultant 3D hydrodynamic instabilities in high-beta plasma may explain leaning of the precursor plasma column toward the Alumel wire side, which is clearly seen on shadowgraphy images taken at 69 and 76 ns after current start (see Fig. 2g and h, respectively). Although as compared to real astrophysical objects (see, for example, the discussion in Ref. [9]) the dimensional characteristics such as the length, time and density are very much different, the jet velocities and Mach numbers are well in the range.

Four time-gated spectra recorded by MCP frames 3-6 just before and during x-ray burst (see MCP locations in Fig. 2b and spectra in Fig. 3) are shown in Fig. 3a. The first three spectra (3-5) before x-ray burst express very small changes in line intensities and hence can be described by similar plasma parameters. These spectra include the diagnostically important Ne-like Ni lines (nC and nD) due to transitions from high Rydberg states n=4 and 5 and F-like spectral features between them. The synthetic spectrum calculated at Te=400 eV and Ne=5×10^{19} cm^{-3} reproduces well the experimental spectra recorded by frames 3-5 (see Fig. 3b). The spectrum at the peak of x-ray burst shows more continuum radiation which might indicate the presence of electron beams.

3. Conclusion

The new results of X-pinch and double planar wire array experiments are presented. The wire material was chosen to produce plasmas from astrophysically relevant Fe and Ni ions. Plasma jets from X-pinch and in addition from DPWA experiments were observed through laser shadowgraphy. The resonance He-like lines of Fe from hot plasmas as well as "cold" Kα lines from much cooler plasma with Fe, Cr, and Ni neutrals and low-ionized ions were recorded and analyzed.

Acknowledgment

This work was supported by NNSA under DOE Cooperative Agreements DE-FC52-06NA27588, DE-FC52-06NA27586, and in part by DE-FC52-06NA27616.

References

[1] Kantsyrev V L, Rudakov L I, Safronova A S, Fedin D A, Ivanov V V, Velikovich A L, Esaulov A A, Chuvatin A S, Williamson K, Ouart N D, Nalajala V, Osborne G, Shrestha I, Yilmaz M F, Shivaji P, Laca P J and Cowan T E 2006 *IEEE Trans. Plasma Science* **34** 194 and 2295

[2] Kantsyrev V L, Rudakov L I, Safronova A S, Esaulov A A, Chuvatin A S, Coverdale C A, Deeney C, Williamson K M, Yilmaz M F *et al* 2008 *Physics of Plasmas* **15** 030704

[3] Safronova A, Kantsyrev V, Ouart N, Yilmaz F, Fedin D, Astanovitsky A, LeGalloudec B, Batie S, Brown D, Nalajala V, Shrestha I, Pokala S, Cowan T, Jones B, Coverdale C, Deeney C *et al* 2006 *Journal of Quantitative Spectroscopy and Radiative Transfer* **99** 560

[4] Safronova A S, Kantsyrev V L, Neill P, Safronova U I, Fedin D A, Ouart N D, Yilmaz M F Osborne G C, Shrestha I, Williamson K M, Hoppe T, Harris C, Beiersdorfer P and Hansen S 2008 *Canadian Journal of Physics* **86** 267

[5] Safronova A S, Kantsyrev V L, Esaulov A A, Ouart N D, Safronova U I, Shrestha I and Williamson K M 2009 *European Physical Journal Special Topics* **169** 155

[6] Phillips KJH 2004 *The Astrophysical Journal* **605** 921

[7] Page MJ, Soria R, Zane S, Wu K and Starling RLC 2004 *Astronomy and Astrophysics* **422** 77

[8] Esaulov A A, Kantsyrev V L, Safronova A S, Williamson K M, Shrestha I, Osborne G C, Yilmaz M F, Ouart N D and Weller M E 2009 *High Energy Density Physics* **5** 166

[9] Nicolaï Ph, Stenz C, Kasperczuk A, Pisarczyk T, Klir D, Juha L., Krousky E, Masek K, Pfeifer M, Rohlena K, Skala J, Tikhonchuk V, Ribeyre X, Galera S, Schurtz G, Ullschmied J, Kalal M, Kravarik J, Kubes P *et al* 2008 *Physics of Plasmas* **15** 082701

Simulating thermo-mechanical interaction between a xenon gas-filled chamber and tungsten first-wall armor for the LIFE reactor design using the BUCKY 1-D radiation hydrodynamics code

T A Heltemes, A E Prochaska and G A Moses

Fusion Technology Institute, University of Wisconsin — Madison, 1500 Engineering Dr., Madison WI 53706, USA

E-mail: `taheltemes@wisc.edu`

Abstract. The BUCKY 1-D radiation hydrodynamics code has been used to simulate the dynamic thermo-mechanical interaction between a xenon gas-filled chamber and tungsten first-wall armor with an indirect-drive laser fusion target for the LIFE reactor design.

Two classes of simulations were performed: (1) short-time (0–2 ms) simulations to fully capture the hydrodynamic effects of the introduction of the LIFE indirect-drive target x-ray and ion threat spectra and (2) long-time (2–70 ms) simulations starting with quiescent chamber conditions characteristic of those at 2 ms to estimate xenon plasma cooling between target implosions at 13 Hz.

The short-time simulation results reported are: (1) the plasma hydrodynamics of the xenon in the chamber, (2) dynamic overpressure on the tungsten armor, and (3) time-dependent temperatures in the tungsten armor. The ramifications of local thermodynamic equilibrium (LTE) vs. non-LTE opacity models are also addressed.

1. Introduction

The BUCKY 1-D radiation hydrodynamics code has been used to model the chamber conditions subsequent to the implosion of an indirect-drive DT target for the LIFE reactor study [1]. This analysis is crucial for estimating the survivability of a tungsten-armored oxide dispersion strengthened (ODS) ferritic steel first wall material. The results presented here address the long-term survivability of the tungsten wall, which is evaluated by determining the maximum temperature excursion of the tungsten surface facing the xenon chamber fill gas. The design criteria for survivability is set at 2400°C, the temperature limit for arresting cyclic stress crack propagation through the tungsten armor [2]. Other chamber design issues and feasibility of rep-rated ICF reactors are discussed in detail in Ref. 2.

2. Simulation Setup

2.1. Ion and X-ray Source

The LIFE indirect-drive target integrated yield is given to be 37.5 MJ — 4.5 MJ in x-rays, 3.8 MJ in ions and 29.2 MJ in neutrons and gamma rays [3].

For an indirect-drive target, the majority of the mass is located in the hohlraum surrounding the DT capsule. For the LIFE reactor, the hohlraum has a total mass of 230 mg. The chamber simulation used this mass at a Maxwellian temperature of 20 keV to create a kinetic energy distribution of the gold ion population. This ion source was launched at $t = 0$ seconds in the simulation.

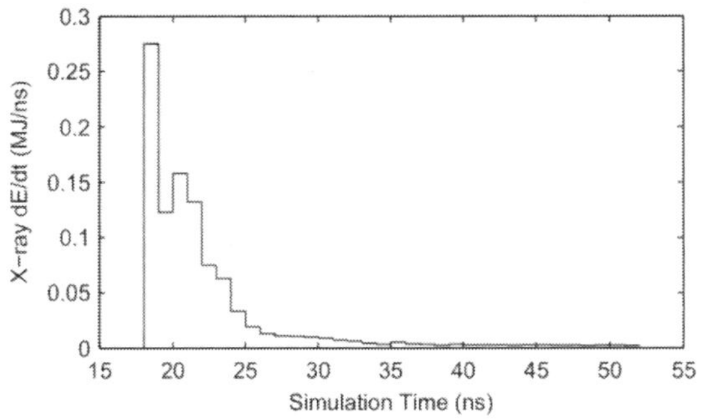

Figure 1. The time-differential of the LIFE x-ray threat spectra normalized to a 1 MJ x-ray yield.

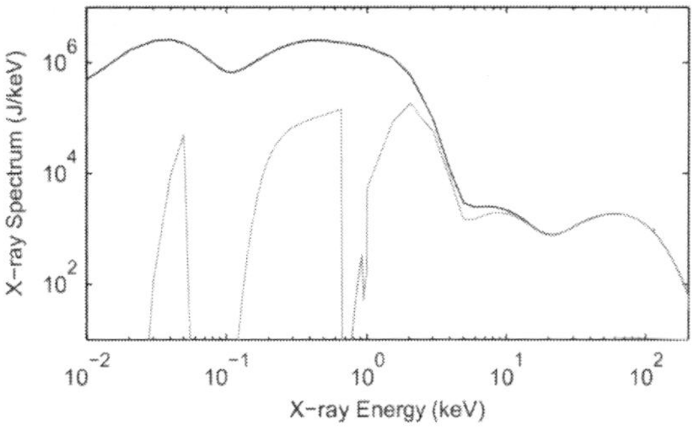

Figure 2. LIFE x-ray threat spectra as emitted (black) and at the tungsten armor surface (red).

The x-ray spectra was divided into 300 photon energy groups, ranging from 0.01 keV to 200 keV. Figure 1 shows the x-ray power vs. time, normalized to a 1 MJ total x-ray yield. Figure 2 shows the x-ray photon energy spectrum as emitted by the target and at the xenon-tungsten interface.

2.2. Chamber and Armor Characterization

The chamber fill gas is xenon at 226.6 Pa (1.7 torr) with an ambient temperature of 600°C [3]. The radius of the gas-armor interface is 2.5 m from the center of the target chamber. The simulation results reported here include a local thermodynamic equilibrium (LTE) EOS and opacity model and a non-LTE coronal model for the xenon chamber gas generated by the IONMIX code [4]. The chamber gas was partitioned into 450 hydrodynamic zones and a mass-matching boundary condition was used on the ten zones closest to the gas-armor interface.

The first-wall armor consists of a 308 μm layer of tungsten over a first-wall consisting of ODS low-activation ferritic steel at an ambient temperature of 600°C [3]. For the simulation results reported here, the tungsten-steel interface is treated as a fixed-temperature boundary condition of 600°C. YAC non-LTE data are used for tungsten armor opacities [5]. The armor material was partitioned into 50 finite-difference zones with logarithmic spacing and first-zone thickness of 100 nm.

For the solid and liquid phases of tungsten, the temperature-dependent values of thermal conductivity were obtained from the ITER Material Properties Handbook. The temperature dependent values for the specific heat capacity of tungsten were obtained from the NIST Standard Reference Database in the form of the Shomate equation. Phase changes were incorporated into the specific heat data by dividing the latent heats of fusion and vaporization by 232 K and inserting the resulting top hat functions into the specific heat data set with the midpoints on the melting and vaporization temperatures, respectively.

3. Simulation Results

3.1. Short-time Simulation

BUCKY was used to simulate the chamber for times 0 to 2 ms to capture the threat spectra and hydrodynamic effects on the chamber armor (figure 3). The dynamic surface temperature response and overpressure of the tungsten armor are reported in figure 4. The surface temperature response is driven primarily by the arrival of the unattenuated prompt x-rays of the target threat spectra (figure 2). The secondary rise in surface temperature observed at later simulation times is due to re-radiation from the xenon plasma. The overpressure data show chronologically distinct features, such as (i) the rise in overpressure due to x-ray heating of the xenon gas near the surface of the tungsten at approximately 20 ns, (ii) the steep rise in overpressure due to the arrival of the Marshak wave at approximately 4 μs, (iii) the arrival of the hydrodynamic shock wave driven by the ion energy deposition in the gas near the center

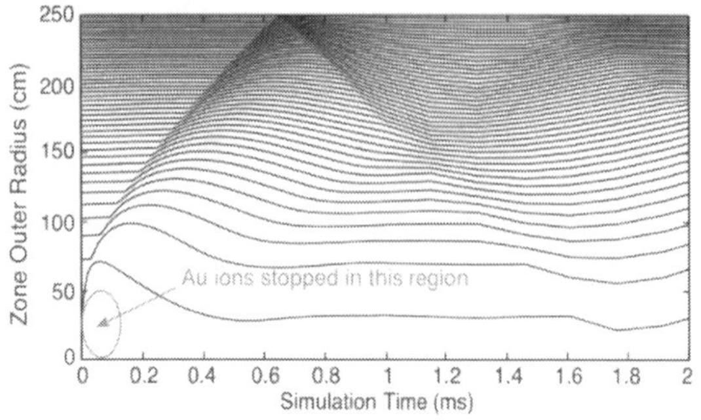

Figure 3. The hydrodynamic zone radius vs. time (RT) plot of the xenon fill gas for the short-time simulation. For clarity only zones $1+10\times(n-1)$ $(n=1,2\ldots45)$ are plotted.

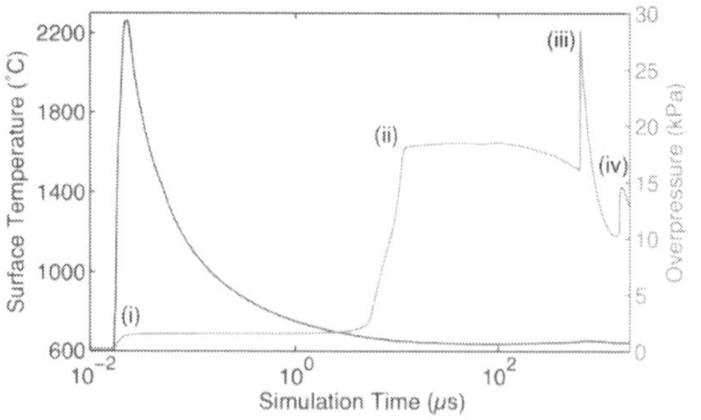

Figure 4. The overpressure and temperature at the xenon-tungsten interface.

of the chamber at approximately 0.7 ms, and (iv) the arrival of a second hydrodynamic shock, which is a rebound shock as the gas "sloshes" around at approximately 1.9 ms. This is an artificial shock inherent to 1-D simulations, where multi-dimensional non-uniformites cannot defocus shocks rebounding from the solid interface.

The choice of opacity model (LTE vs. non-LTE) for xenon has an impact on the plasma properties and its ability to attenuate ion and x-ray energy. The plasma opacity model has the greatest impact in the hydrodynamic zones near the center of the chamber, as shown in the average charge state and ion temperature (figure 5), which are the values of the first hydrodynamic zone. Although the plasma properties are vastly different, the effect at the armor interface is minimal. The differing opacity models yielded no change in the x-ray heating temperature spike and the re-radiation temperature rise was increased by 50°C when using the non-LTE opacity model.

3.2. Long-time Simulation

The BUCKY results at the end of the 2 ms LTE simulation were used as initial conditions for an idealized 1-D thermodynamic analysis of the evolution of the xenon gas and tungsten surface temperatures to 70 ms. This analysis was predicated on the assumption that at 2 ms the xenon gas is at a uniform temperature with no hydrodynamic motion. A second was that the xenon gas and tungsten were in quasi-steady-state, allowing the use of classical thermal transport equations. The quasi-steady-state assumption is that the radiative heat transfer rate of the xenon gas is equal to the heat transfer rate via conduction through the tungsten layer. The radiative heat transfer rate for the sphere of xenon gas is given by

$$\dot{Q}_R = 4\pi r_s^2 \varepsilon\sigma\left(T_g^4 - T_s^4\right) \tag{1}$$

where r_s^2 is the radius of the inner surface of the tungsten region, ε is the emissivity of the body, σ is the Stefan-Boltzmann constant, T_g^4 is the temperature of the xenon gas and T_s^4 is the temperature of the tungsten at the surface exposed to the xenon gas. The approximate heat transfer rate via conduction through a thin spherical shell is given by

$$\dot{Q}_C = 4\pi\left(\frac{r_s r_b}{r_b - r_s}\right) k_W \left(T_s - T_b\right) \tag{2}$$

where r_b is the radius of the outer surface of the tungsten region, k_W is the thermal conductivity of the tungsten layer and T_b is the temperature of the outer surface of the tungsten layer. Using (1) and (2) the

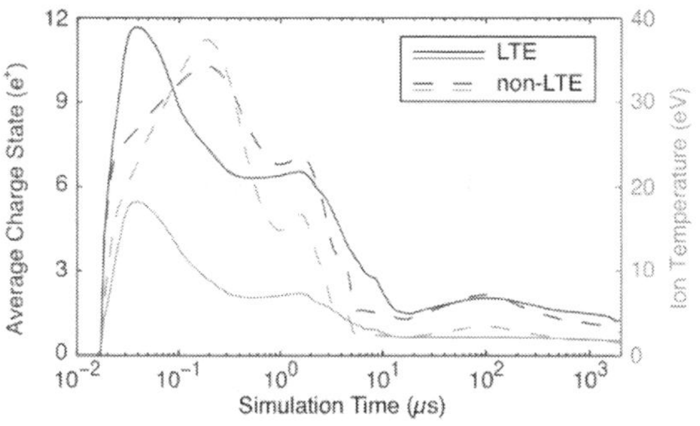

Figure 5. The central-zone values for the average ionic charge state (black) and plasma ion temperature (red).

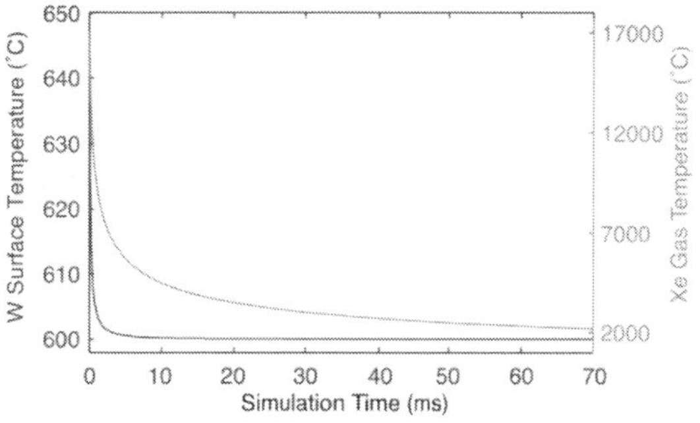

Figure 6. The long-time simulation of the tungsten surface temperature (black) and xenon gas (red) for the LIFE reactor chamber design.

value for ε can be calculated by

$$\varepsilon = \frac{k_W}{\sigma}\frac{r_b/r_s}{r_b - r_s}\frac{T_s - T_b}{T_g^4 - T_s^4}. \tag{3}$$

Using the temperatures reported by BUCKY at 2 ms, the value for ε is 3.16062×10^{-3}. A simple 1-D finite-difference algorithm using (1) and (2) was used to calculate the chamber cooling. The calculation was performed using the NIST/ITER thermophysical data for tungsten, an initial gas temperature of 1.5 eV and initial surface temperature of 642.4°C. For xenon temperatures in excess of 6000 K, the specific heat evaluated at 6000 K was used. Figure 6 shows the temperature evolution of the xenon gas and tungsten surface calculated with a fixed Δt of 10 μs.

4. Conclusion

The temperature rise in the tungsten surface due to the x-ray flash brings the temperature at the interface to 2260°C, below the maximum design limit of 2400°C. This result confirms that the specified 1.7 torr of xenon chamber fill gas is sufficient to protect a 2.5 m tungsten-armored ODS ferritic steel first wall. However, an analysis of the 70 ms temperature shows that while the tungsten armor returns to the 600°C initial condition, the xenon gas temperature only relaxes to 2200°C. This result requires that analysis be conducted over many target shots to ascertain the self-consistent cyclic steady-state values for the xenon gas and tungsten armor.

Acknowledgments

This work has been supported by Lawrence Livermore National Laboratory, contract number B5769261590-03.

References

[1] Peterson R R, MacFarlane J J and Moses G A 1995 BUCKY — a 1-D radiation hydrodynamics code for simulating inertial confinement fusion high energy density plasmas UWFDM 984 Fusion Technology Institiute, University of Wisconsin — Madison

[2] Raffray A R and The HAPL Team 2005 *J. Nucl. Mat.* **347** 178–191

[3] Latkowski J, Loosmore G and Amendt P 2008 LIFE X-ray source and chamber parameters private communications with LLNL

[4] MacFarlane J J 1989 *Comp. Phys. Commun.* 259

[5] Yuan J and Moses G A 2006 *J. Quant. Spect. Rad. Trans.* **99** 697–711

Journal of Physics: Conference Series **244** (2010) 032033 doi:10.1088/1742-6596/244/3/032033

Laboratory experiments on cluster/aerosol formation by colliding ablation plumes

Y Hirooka[1,4], K A Tanaka[2], H Sato[2], K Ishihara[2] and A Sunahara[3]

[1]National Institute for Fusion Science, 322-6 Oroshi, Toki, Gifu 509-5292, Japan
[2]Osaka University, 2-1 Yamadaoka, Suita, Osaka 565-0871, Japan
[3]Institute for Laser Technology, 2-6 Yamadaoka, Suita, Osaka 565-0871, Japan

E-mail: hirooka.yoshihiko@nifs.ac.jp

Abstract. First-of-a-kind experiments on cluster/aerosol formation by colliding ablation plumes have been conducted, radiating Al, Cu and C with 3ω-YAG laser at power densities between 2~30 J/cm^2/pulse. Visible spectroscopy indicates that the excitation light intensities of Cu and Al plumes are not necessarily be doubled in collision, but can rather be weakened due to atomic and molecular reactions. For colliding C plumes, Swan band radiation has been observed, indicative of C_2 and/or C_2^+ formation, and ion mass spectrometry has identified C_n^+-clusters, including C^+, C_2^+, C_3^+, C_4^+ and C_5^+. From ICCD camera observations, C plumes generated at power densities above ~15 J/cm^2/pulse tend to split into two components with respective velocities, only the slow component of which appears to be interactive to form clusters. Nano structures like CNT have been identified in deposits from colliding C plumes.

1. Introduction

It is predicted that, along with DT-pellet implosions in inertial fusion power reactors, target chamber walls will repeatedly be exposed to intense pulses o14MeV neutrons, X-rays, high-energy unburned fuel particles and pellet debris such as hydrocarbon fragments. As a result, wall materials will be subjected to thermo-physical erosion due to evaporation, sputtering and ablation, etc.

Eroded materials will either be re-condensed elsewhere on the wall after travelling across the chamber or collide with each other perhaps in the centre or the symmetric axis region to form clusters which can grow into aerosol to float until it is pumped out. These processes directly affect the wall lifetime as well as reactor performance in terms of pulse repetition rate because laser can be reflected, scattered or refracted by floating aerosol, which hinders subsequent implosions. Despite its critical importance, this chamber clearing issue has not clearly been addressed in the ICF research community. The present work is intended to investigate fundamental aspects of materials ablation behavior with the particular emphasis on cluster/aerosol formation by colliding plumes.

2. Experimental

A new setup shown in figure 1 has been put together for the Laboratory Experiments on cluster/Aerosol Formation by Colliding Ablation Plumes (to be referred to as LEAF-CAP in the remainder of the paper). In this setup, a 3ω-converted YAG laser beam (355nm, 320mJ/pulse, 6ns, 10Hz) is split into two equal-power beams, each line-focused into ~0.1cm x ~1cm to

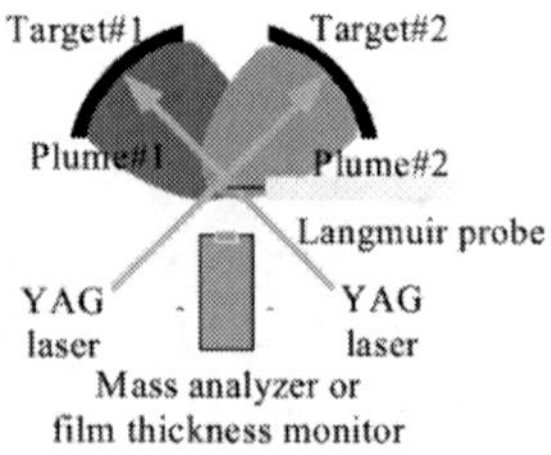

Figure 1 The LEAF-CAP setup.

radiate two targets at room temperature in vacuum (~10^{-6}Torr), i.e. a double target setup. These targets are in the form of rectangular disk (~5cm x ~1cm x ~0.5cm) but arc-shaped on the laser-facing side, and are positioned in such a way that two ablation plumes collide with each other in the centre-of-arc region. Used as target materials in the LEAF-CAP experiments are Al, Cu and C (isotropic graphite).

3. Results and discussion

3.1. CCD camera observations

Shown in figure 2 are some of the typical CCD camera images of laser ablation plumes of Al: (a) and (b), Cu: (c) and (d), and C: (e) and (f), all observed at 2.2J/cm^2/pulse in the single and double target setups, respectively, in the LEAF-CAP experiments. Notice that Al and Cu plume emission tails seen in the single target setup appear to be cut in the middle, as shown in figures (b) and (d), when two self-plumes collide with each other in the double target setup. Interestingly, although the plasma density tends to be doubled in the centre-of-arc region [1], when plumes collide with each other, the excitation light intensity is not necessarily doubled but can be weakened in some cases, contrary to expectation, presumably due to complicated atomic and molecular reactions. In contrast, C plumes rather strongly radiate in the centre-of-arc region when they collide with each other, as can be seen in figure (f), indicative of an exothermic reaction although they are barely visible individually.

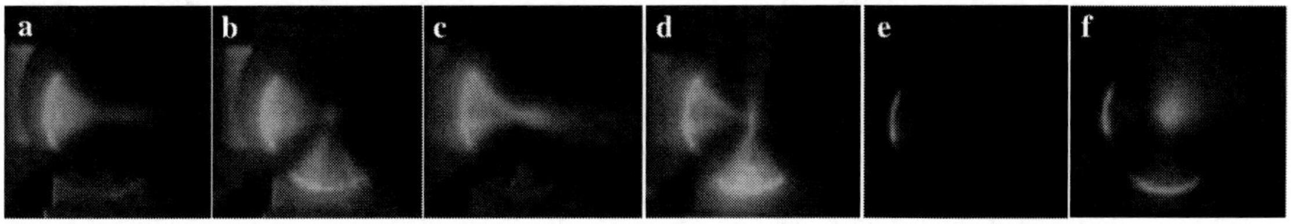

Figure 2 Ablation plumes of Al: (a, b), Cu: (c, d) and C: (e, f) in the single and double target setups, where the distance from the target surface and plume colliding center is about 1.5cm.

3.2. Visible spectroscopy

Visible spectroscopy measurements have provided data, essentially consistent with these CCD camera observations. In the case of Al, the intensity of Al-I approximately is doubled at 2.2J/cm^2/pulse (not shown here), whereas in the case of Cu, as shown in figure 3(a), the intensity of Cu-I at the same power density becomes smaller in the double target setup. In the case of C, the Swan band structure is clearly identified in spectra, indicating the formation of C_2 and/or C_2^+ when plumes collide with each other. Presumably related to C_n and/or C_n^+cluster formation to be described in 3.3, however, the line radiation marked with **?** at around 580nm is yet to be identified in these spectra.

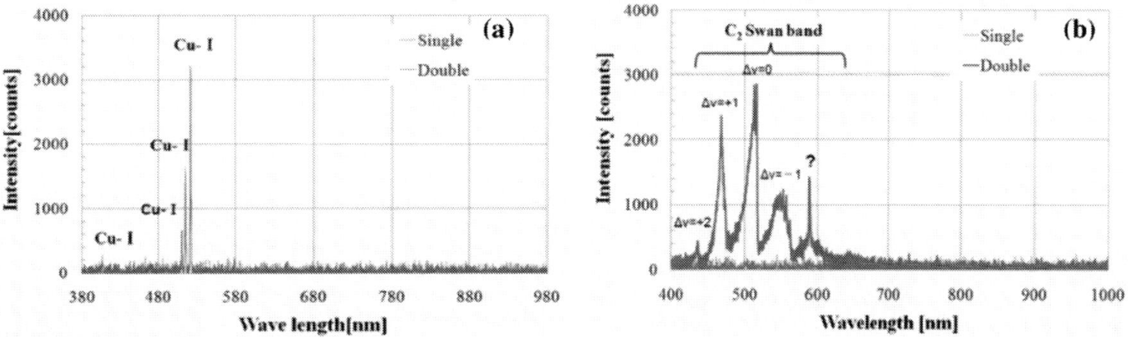

Figure 3 Visible spectroscopy of (a) Cu plumes at 2.2J/cm^2/pulse and (b) C plumes at 6.5 J/cm^2/pulse in the single and double target setups.

3.3. Ion mass spectrometry

As shown in figure 1, a qudrapole mass analyzer is employed, but used without hot filament in order that ions in colliding ablation plumes can directly be identified. A comparison between the mass spectra taken for Al and C plumes, all generated at 2.2J/cm^2/pulse, in the single and double target setups are shown in figure 4. The formation of Al^+ and Al^{++} is exhibited when Al plumes collide with each other although no sign of cluster formation is seen. In the case of C plume, C_1^+, C_2^+, C_3^+, C_4^+ and C_5^+ are clearly identified in mass spectra, which is believed to be due to their strong interaction potential, i.e. C-C bonding strength. Interestingly, C_n-cluster ion formation has been observed mainly at low laser power densities where the ion energy is also relatively low. Details will be described next.

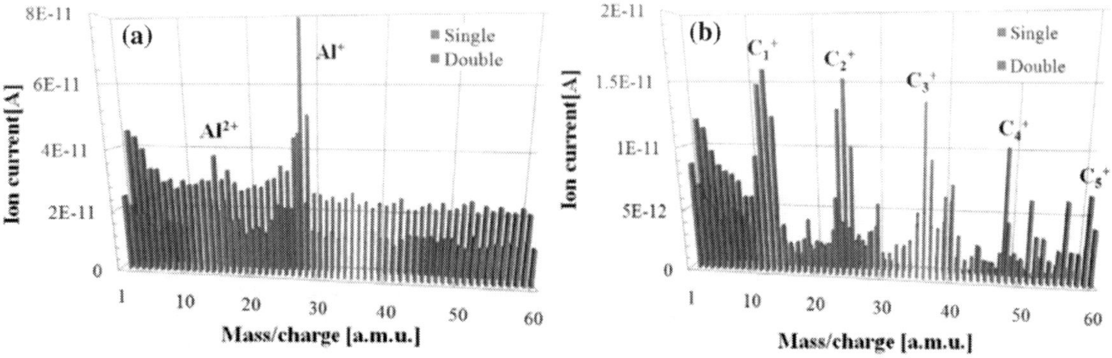

Figure 4 Ion mass spectra of (a): Al and (b): C plumes in the single and double target setups.

3.4. Plasma parameter and deposition rate measurements

For plasma parameter measurements, a linearly moveable Langmuir probe is inserted into the centre-of-arc region. Also, a quartz crystal film thickness monitor is positioned at about 2.5cm from the centre-of-arc, replacing the quadrapole mass analyzer. Data have been taken from C plumes generated at laser power densities between 2 and 13J/cm^2/pulse. As shown in figure 5 (a), the plasma density appears to increase linearly on the order of 10^{10} 1/cc, finally reaching ~3 x 10^{11}1/cc, as the laser power density increases. Along with the increase in plasma density, the deposition rate also increases due to the relation: $\Gamma = (n_o + n_i)v_c$, where Γ is the deposition rate, n_o is the neutral density, n_i is the ion density, and v_c is assumed to be the velocity of C_1^0 or C_1^+instead of C_n^0 or C_n^+ for simplicity. Meanwhile, the electron temperature increases from 0.5 to 3eV, as shown in figure 5 (b). Under adiabatic expansion conditions, the electron temperature at the point-of-origin for plume generation can be as high as ~15eV [1], where $n_o << n_i$ readily holds [2], so that the deposition rate may be given by: $\Gamma \cong n_i v_c$. As plotted in figure 5 (b), the ion temperature has been found to vary from 0.1 to 0.3eV, based on the relation: $kT_i = m_c v_c^2$, where m_c is assumed to be the mass of C_1^0 or $C_1^+$for simplicity.

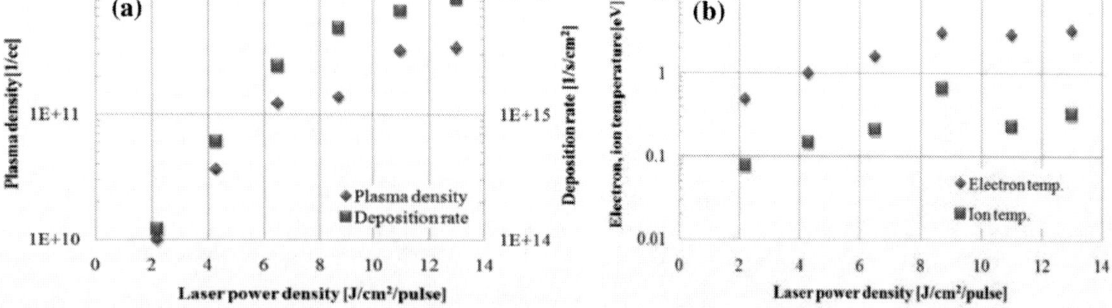

Figure 5 Plasma parameters and deposition rates measured for C plumes at 3-13J/cm^2/pulse.

3.5. ICCD camera observations

Shown in figure 6 are a series of the ICCD camera images taken for C plumes generated at 29J/cm^2/pulse in the double target setup. Notice that there are two distinctive components with their respective velocities, as seen in figure 6 (a). The kinetic energy estimated from these ICCD image data for the fast component is $E_f \approx 10$eV and that for the slow component is $E_s \approx 1$eV. One possible mechanism to explain this phenomenon is as follows [3]. Up on materials ablation, electrons tend to be emitted first, which will then drag ions out of the energy deposited volume. Multiply charged ions, if they are formed, will be attracted to these electrons more strongly than singly charged ions, leading to the separation of plumes. Here, one might conjecture that the formation of multiply charged ions would require power densities above a certain threshold. The threshold energy density in the present work seems to be around 15J/cm^2/pulse, which is relatively in agreement with earlier studies [4].

Interestingly and perhaps most importantly from the ICF reactor operation point of view, the fast components don't seem to collide but to miss each other, as can be seen in figure 6 (b), whereas the slow components apparently collide to interact, slowing down even further with strong radiation, as shown in figures 6 (c, d, e, and f). After the slow down, the directional kinetic energy of the combined plumes falls down on the order of 0.1eV, related to the deposition rate data analysis described in 3.4. Corroborating these findings, the ion-ion collision mean free paths for the fast and slow components are of the order of 10cm and 0.1cm, respectively, both estimated from λ_{ii}[cm]=$10^{12}T_i^2$[eV]/n_e[1/cc] [5].

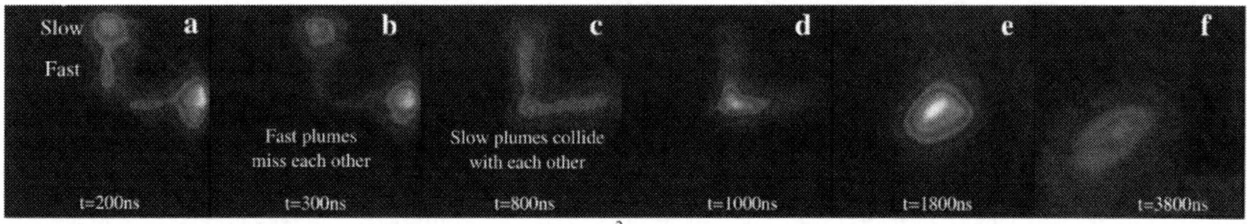

Figure 6 ICCD images of C plumes at 29J/cm^2/pulse in the double target setup.

3.6. Ablation deposits analysis

Carbon deposits generated after repeated colliding plume experiments have been analyzed with TEM. As shown in figure 7, data indicate the presence of nano structures like CNT (for carbon nano tube). From elemental analysis, other than carbon, no impurities have been detected in these deposits. The formation of these nano structures apparently favour low power density ablation conditions although details are still unclear at this point.

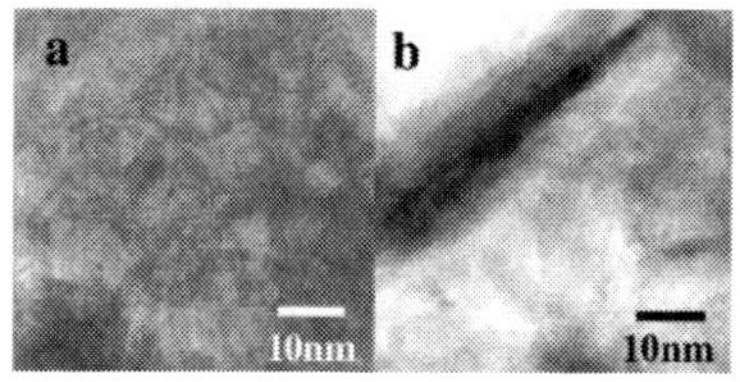

Figure 7 CNT-like nano structures.

4. Conclusion

To address the ICF chamber clearing issue, the first-of-a-kind demonstration has been performed, using a laboratory-scale experimental setup in which two laser ablation plumes collide with each other. Chosen as the target materials are Al, Cu and C. Data indicate that under identical conditions only colliding C plumes exhibit the formation of clusters. Also, cluster formation has been found to favour relatively low kinetic energy plumes. From these findings, one predicts that materials re-condensation as well as cluster/aerosol formation will most likely occur in ICF target chambers with carbon armors.

References

[1] H Sato, M S Thesis "Dynamics of Colliding Ablation Plume", Osaka Univ. Japan (2010).
[2] NIFS A&M database: http://dpc.nifs.ac.jp/ionfracdata/ (after Amaud and Rothenflug 1985).
[3] J F Ready, "Effects of high power laser radiation", Academic Press (1971).
[4] S S Harilal et al., J.Appl. Phys. **80**, 3561.
[5] P C Stangeby, "Boundary plasma physics", IoP Publishing (2000).

Current status of the SBS PCM approach to self-navigation of lasers on injected IFE pellets

Milan Kalal, Michaela Martinkova and Ondrej Slezak
Faculty of Nuclear Sciences and Physical Engineering, Czech Technical University in Prague, Brehova 7, 115 19 Prague 1, Czech Republic
E-mail: `kalal@fjfi.cvut.cz`

Hong Jin Kong, Jin Woo Yoon and Jae Sung Shin
Department of Physics, Korea Advanced Institute of Science and Technology, Daejeon, Korea

Elena Rostislavovna Koresheva and Sergei Aleksandrovich Startsev
P.N. Lebedev Physical Institute of Russian Academy of Sciences, Moscow, Russia

Abstract. Current status of SBS PCM based IFE approach proposed recently as an alternative to the IFE classical approach is presented. This technology is of particular importance to the direct drive scheme taking care of automatic self-navigation of every individual laser beam on the injected pellets with no need for any final optics adjustment. Conceptual design of one typical laser driver is shown and its features discussed. In comparison with the earlier design an upgraded scheme was developed with the low energy illumination laser beam (glint) entering the reactor chamber through the same entrance window as used by the corresponding high energy irradiation laser beam. Results of experimental verification of this improved design are reported. In these experiments for the fist time a complete setup including the pellet (realized by the static steel ball) was employed. The pellet survival conditions in the period between its low energy illumination and subsequent high energy irradiation were studied and the upper limits on the allowed energies absorbed were found for both DD and DT fuels.

1. Introduction

One of the very difficult challenges to deal with in the *direct* drive inertial fusion energy (IFE) integrated approach is connected with the need of simultaneous irradiation of thermonuclear pellets by many dozens of powerful laser beams inside the reactor chamber. Sophisticated *tracking* of injected pellets' trajectories is necessary for prediction of the place most suitable for interaction with the driver beams in order to achieve necessary *irradiation symmetry* and subsequent *fuel compression*. For the direct drive scheme the following set of parameters is being currently considered: pellets $\sim 4\,\mathrm{mm}$ in diameter should be delivered into the *virtual* sphere $\sim 5\,\mathrm{mm}$ in diameter located around the center of the reactor chamber $\sim 10\,\mathrm{m}$ in diameter. Combined precision of tracking and aiming should be $\sim 20\,\mu\mathrm{m}$. Navigation technologies developed so far (despite their gradual progress) are still well outside of the required margin even in the case of *evacuated* reactor chambers - as some time consuming adjustment of heavy *final optics* for every shot and every laser beam is always necessary. This last fact is also partially responsible for the very tight margin ($\pm 500\,\mu$m) on the pellets successful delivery into the above mentioned virtual sphere. This value was prescribed among other reasons in order to *minimize* the necessary final optics adjustment.

In reality there are some additional *serious obstacles* further complicating this direct drive IFE scheme - even putting in doubts its *practical feasibility*. Among the most serious ones is the *insufficient predictability* of the injected pellets' trajectories resulting from their expected interaction with remnants of previous fusion explosions due to the considered $5\div10$ Hz repetition rate. This might be one of the reasons why the *indirect* drive scheme seems to be currently considered a more serious IFE candidate as the corresponding *hohlraum* targets are by three orders of magnitude heavier compared to their direct drive counterparts thus allowing for much more reliable prediction of their trajectories.

In order to deal with these direct drive IFE scheme laser navigation difficulties a new approach was recently proposed employing the *stimulated Brillouin scattering* (SBS) *phase conjugating mirror* (PCM) technique [1, 2]. In the first public presentation of this novel approach [3] it was predicted that a *fully automatic* self-aiming of every individual laser beam on the injected pellets with no need for any final optics adjustment could be achieved. This idea was undergoing gradual improvement in its theoretical design [4]–[6] and subsequently it started to be tested also experimentally proving the principle [7, 8].

2. SBS PCM based IFE approach - Current Design

Current design of this approach is illustrated in the Fig. 1 where one particular laser channel is displayed during the *three distinct stages* of its functioning:

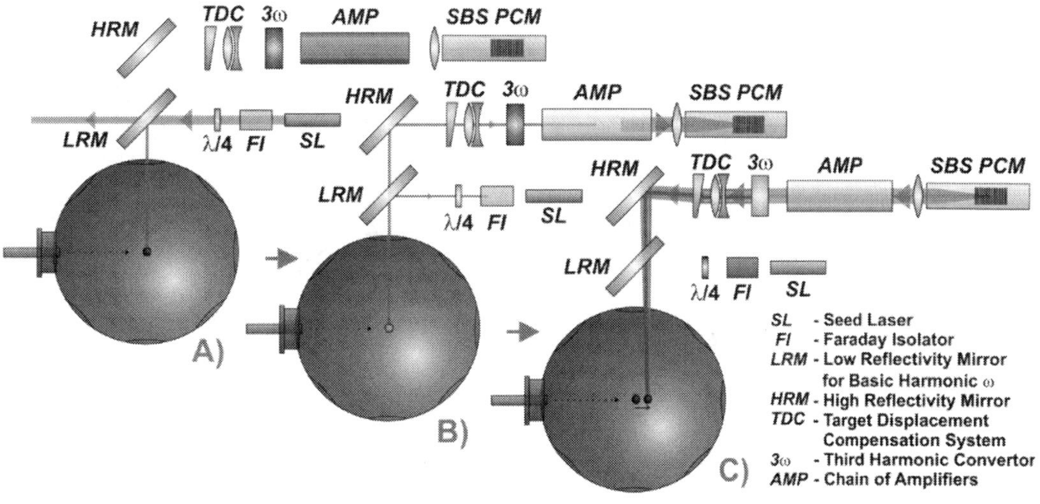

Figure 1. Three distinct stages of every individual laser channel functioning.

A) at the right moment (determined by careful tracking) when the injected pellet is approaching its best interaction position, a low energy *seeding* laser pulse (glint - *red* line) is sent to *illuminate* the pellet; B) reflected seeding laser pulse is collected by the focusing optics and amplified on its way to the SBS PCM cell; C) amplified pulse is reflected by the SBS PCM cell, amplified once again, converted to higher harmonic (*blue* line) and automatically aimed at the moving pellet by the *target displacement compensation system* (TDC) for its final high power *irradiation*. TDC is a completely *passive* system having its optical components *appropriately* designed for every individual channel taking advantage of their index of refraction dependence on the wavelength.

Typical displacements for the pellet injection speeds 100 m/s and 1 μs delay times corresponding to 300 m distance traveled by the laser beam outside the reactor chamber would be 100 μm. It should be noted that in comparison with the previous design presented in Ref. [4] the *new scheme* of *illumination* was developed with the seeding laser beam entering the reactor chamber through the *same* entrance window as used by the corresponding *irradiation* beam.

This SBS PCM based IFE approach has several important *advantages* over the *classical* one. Even if the injected pellets will inevitably reach for every shot slightly *different* (and difficult to predict with the accuracy required) positions within the prescribed area, their subsequent *displacement* from the position in which they will be *illuminated* into the position in which they will be *irradiated* should always be the *same* (provided, of course, that the injection speed will not vary substantially from shot to shot). Therefore, optical elements especially designed for taking care of every individual beam shift can be introduced once for all. Featuring *no moving parts* this technique can significantly *simplify* design of lasers and beam transport optics allowing for substantial increase in the number of laser beams employed. Every laser beam can operate as an independent sub-beam with much lower energy per pulse thus making the required *repetition rate* easier to achieve. With many laser beams available any *shape* of the final irradiating pulse can be realized by considering neighboring sub-beams as creating a required pulse shape when combined together on the pellet surface using different *delays* and *amplifications* of individual sub-beams. More details concerning these issues as well as the *illuminating* and *irradiating* schemes can be found in Ref. [5].

3. SBS PCM based IFE approach - Experimental Verification

Schematics of the experimental setup used for verification of the proposed design is shown in the Fig. 2.

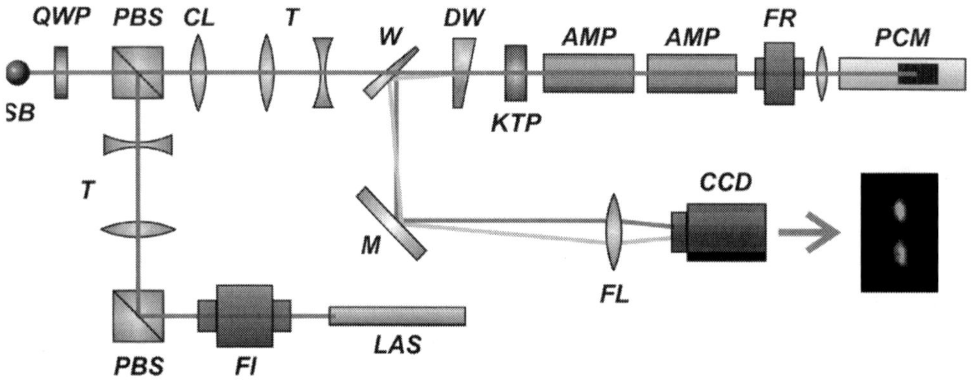

Figure 2. *AMP* - amplifier, *CCD* - CCD camera, *CL* - collecting lens, *DW* - dispersion wedge, *FI* - Faraday isolator, *FR* - Faraday rotator, *KTP* - KTP crystal, *LAS* - laser, *PBS* - polarizing beamsplitter, *SB* - steel ball, *T* - telescope 4x, *W* - wedge

Compared to the successful experiments performed earlier [7, 8] which confirmed the self-navigation principle (the change of the trajectory achieved by the incorporated conversion to the *second* harmonic - *green* line) in these new experiments the *complete* individual laser channel setup was assembled - including the *pellet* realized by the static *steel ball*. It should be noted that during these tests the laser channel was operating with much lower energies compared to the real ones (so far just below 1 J before *irradiating* the target). Combining description of the Fig. 1 with the Fig. 2 and its captions the results presented above should be self-explanatory.

4. Concluding remarks

In this paper an important upgrade to the IFE scheme employing SBS PCM approach was presented and results from experiments proving the principle of self-navigation of lasers on injected pellets were reported. In these experiments the complete laser channel setup already containing all basic components was used - including pellets realized at this stage of development by stationary steel balls. These were deliberately chosen as capable to withstand much higher illumination energies compared to the real pellets. The higher illumination energies became necessary due to the insufficient amplification available in the laser channel (with only two amplifiers employed).

In this context the injected pellet survival conditions in the period between its low energy *illumination* and subsequent high energy *irradiation* were studied. The upper limits on the illumination energies to be used in real experiments for currently considered direct drive pellets (4 mm in diameter, 45 μm thick polystyrene shell, 200 μm thick fuel layer) were calculated. It was found that the absorbed energy which leads to the cryogenic layer temperature increase by 1 K (from 17 K to 18 K) in the area of cryolayer/shell wall contact during 1 μs is $\sim$ 6 mJ in the case of DD and $\sim$ 14 mJ in the case of DT. Knowledge of these energies is crucial for proper design of individual laser channels - in particular their total amplification needed for high quality SBS PCM reflection to take place. The energies found are about ten times higher than those tentatively estimated in our very first laser channel design [7]. This is by all means the move in the right direction.

Encouraged by all these findings a new round of experiments is under preparation when the pellets (still substituted by the steel balls) would be already moving with the speed comparable to that used during the IFE injection.

5. Acknowledgments

This work has been supported by IAEA RC No.: 13781 & 13758; by the Ministry of Education, Youth, and Sports of the Czech Republic Research Project LC528 & by the grant KONTAKT No. ME933; by the Czech Technical University in Prague grant No.: CTU0916314; and by the Grant Agency of the Czech Republic grant No.: 202/08/H057. It was also supported by the Nuclear Research & Development Program of the Korea Science and Engineering Foundation (KOSEF) grant funded by the Korean Government (MEST) - grant code: M20090078160.

References

[1] Zeldovich B Y, Popovichev V I, Ragulsky V V and Faizullov F S 1972 Connection between the wave fronts of the reflected and exciting light in stimulated Mandelshtam-Brillouin scattering *Sov. Phys. JETP* **15** 109

[2] *Phase Conjugate Laser Optics* 2003 Eds: A Brignon and J-P Huignard (New York: Wiley-Interscience)

[3] Kalal M, Kong H J and Alexander N B 2007 Consideration of SBS PCM technique for self-aiming of laser fusion drivers on IFE targets - proposal and feasibility study *3rd International Conference on the Frontiers of Plasma Physics and Technology* March 5-9, 2007, Bangkok, Thailand

[4] Kalal M, Martinkova M, Slezak O, Kong H J and Alexander N B 2007 SBS PCM technique and its possible role in achieving IFE objectives *J. Phys.: Conf. Ser.* **112** 032049

[5] Ostermeyer M, Kong H J et al. 2008 Trends in stimulated Brillouin scattering and optical phase conjugation *Las. Part. Beams* **26** 297

[6] Kalal M, Kong H J, Slezak O and Yoon J W 2008 Some issues in development of SBS PCM based IFE driver, *3rd Workshop on SBS and Phase Conjugation* August 25-26, 2008, Harbin, China

[7] Kalal M, Kong H J, Martinkova M, Slezak O and Yoon J W 2008 Current status of designing SBS PCM based IFE driver *30th European Conference on Laser Interaction with Matter*, August 31-September 5, 2008, Darmstadt, Germany

[8] Kalal M, Kong H J, Martinkova M, Slezak O and Yoon J W 2009 SBS PCM Technique Applied for Aiming at IFE Pellets: First Tests with Amplifiers and Harmonic Conversion *J. Kor. Phys. Soc.* (accepted for publication)

Integrated process modeling for the laser inertial fusion Energy (LIFE) generation system

W R Meier[1], T M Anklam[1], A C Erlandson[1], R R Miles[1], A J Simon[1], R Sawicki[1] and E Storm[1]

[1]Lawrence Livermore National Laboratory, P.O. Box 808, Livermore, CA 94551

E-mail: meier5@llnl.gov

Abstract. A concept for a new fusion-fission hybrid technology is being developed at Lawrence Livermore National Laboratory. The primary application of this technology is base-load electrical power generation. However, variants of the baseline technology can be used to "burn" spent nuclear fuel from light water reactors or to perform selective transmutation of problematic fission products. The use of a fusion driver allows very high burn-up of the fission fuel, limited only by the radiation resistance of the fuel form and system structures. As a part of this process, integrated process models have been developed to aid in concept definition. Several models have been developed. A cost scaling model allows quick assessment of design changes or technology improvements on cost of electricity. System design models are being used to better understand system interactions and to do design trade-off and optimization studies. Here we describe the different systems models and present systems analysis results. Different market entry strategies are discussed along with potential benefits to US energy security and nuclear waste disposal. Advanced technology options are evaluated and potential benefits from additional R&D targeted at the different options is quantified.

1.0 Introduction

Fusion-fission hybrid technology based on Laser Inertial Fusion Energy (LIFE) is being developed at Lawrence Livermore National Laboratory. The primary application of this technology is base-load electrical power generation using energy multiplication in a fission blanket with depleted uranium (DU) as the beginning of life heavy metal (HM). An integrated process model (IPM) for a LIFE power plant has been developed that includes cost and performance algorithms for the major subsystems of the plant, including the laser, fusion target fabrication, fusion-fission chamber (including the tritium and fission fuel blankets), heat transfer and power conversion systems, and other balance of plant systems. The model has been developed in Visual Basic with an Excel spreadsheet user interface in order to allow experts in various aspects of the design to easily integrate their individual modules and provide a convenient, widely accessible platform for conducting the system studies. In this paper we describe key aspects of the IPM and report on the results of our systems analyses for a particular design based on indirect-drive, hot-spot ignition (HSI) targets and a fission blanket using depleted uranium fuel. The cost of electricity (COE) is used as the figure of merit for design optimization and to reveal key cost and performance sensitivities. Additional details on the LIFE plant are given in Refs. 1-5.

2.0 Integrated process model (IPM)

The integrated process model (IPM) includes detailed cost and performance models for the major subsystems of the power plant. Key features are summarized here with a more detail description of the model presented in Ref. 6.

Laser system capital costs are scaled from NIF project costs, with adjustment for N'th of a kind costing, diode pumping, active cooling and a more compact architecture. NIF is a one-of-a-kind R&D facility whereas this paper is addressing an N'th-of-a-kind power plant. NIF unit costs have been adjusted to give credit for future manufacturing learning, account for non-recurring costs, such as supply chain development, and adjusted for beam line architecture differences between NIF and a LIFE power plant. Additional costs for systems needed for high pulse rate operation, such as the diodes and power conditioning, and active cooling are added to arrive at the total LIFE laser cost.

Target costs are based on a conceptual design of an automated target production facility (target factory) including estimates for buildings, process equipment, fixed and variable operating and maintenance costs. The sum of the annual capital-related charges, operation and maintenance (O&M) and feed materials gives a unit cost of \$0.25/target. For a plant operating at 10 Hz and a capacity factor of 85%, 2.7×10^{8} targets are needed per year resulting in an annual target cost of ~\$67M.

The costs for the nuclear island (NI) and balance of plant (BOP) are derived various fission reactor studies [7,8]. The LIFE plant is similar to the liquid metal reactor in that neither technology requires the high pressure containment systems and structures typical of light water reactors, but both require intermediate coolant loops. Costs for the LIFE fusion target chamber and vacuum vessel, which are very different from a nuclear reactor pressure vessel, are calculated directly based on estimated raw material costs and an assumed fabrication cost multiplier. Additional allowances are included for LIFE unique systems such as tritium recovery, storage and special materials such as the molten salt coolants.

We have adopted the approach as reported in the MIT study for calculating the COE [9,10]. The power plant project is assumed to be financed with a combination of equity and debt financing yielding a nominal, tax adjusted cost of money of 7.8%. Annual costs are summed (accounting for the benefits of accelerated depreciation) and then deflated to 2007 dollars. This results in an effective fixed charge rate on capital investment of ~7.7%. The overnight construction cost, accounting for home and field office engineering, owners cost, etc. increases the direct capital cost by 59%. At this point we are assuming a uniform 10%, across the board contingency. Interest during construction is based on a 5 year construction period and increases the capital investment by ~10% (in 2007\$). Therefore the total capital cost (in 2007\$) is ~1.9 times the direct capital cost. In calculating the COE we assume a constant plant availability of 85%.

3.0 Base case results

As a point of comparison, we have selected a nominal base case design. The laser energy on target is 1.86 MJ and produces a target yields of 51 MJ (target gain = 27). The overall energy multiplication in the tritium breeding and fission blankets (including fission product decay heat) is 5.35, so at 10 Hz, the thermal power is 2690 MW_t. The molten salt coolant with outlet temperature of 650 C drives a Brayton power cycle with a thermal conversion efficiency of 46%, generating 1234 MW_e gross electric power. The 9.5% efficiency laser requires 196 MW_e and we assume that 3% (37 MW_e) of the gross electric is required for in-plant auxiliary power needs. The plant net power is 1000 MW_e.

The COE for this design is ~\$65/MWh (i.e., 65 mils/kWh). The top-level contributions to the COE are: capital 69%; operations and maintenance (O&M), 17%; and fuel, 14%. The contributions to the capital cost (\$4.2B total) are: power block (chamber, heat transfer system and power conversion system), 40%; balance of plant, 34%; laser, 18%; and special materials (e.g., molten salt), 8%.

4.0 Results for other LIFE options and other base-load technologies

Figure 1 gives COE results for three other LIFE designs: a design using a fissile blanket with a higher fission blanket gain, a pure fusion design, and a pure fusion design with a fast ignition target. All are designed to produce 1000 MWe net power. COEs are all within 10% of the base case LIFE with a DU blanket. The COEs for pure fusion designs are competitive because we have assumed a 20% cost savings in the balance of plant and a higher thermal conversion efficiency (50% vs. 46%) expected without the fission blanket.

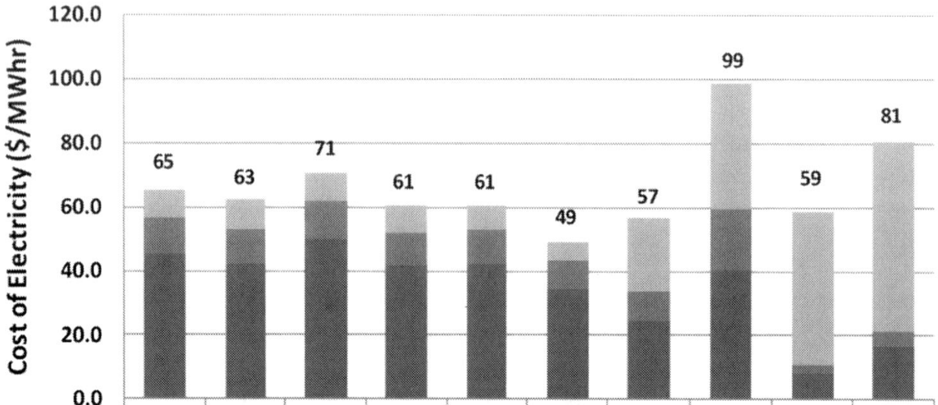

	LIFE-DU	LIFE-Fissile Fuel	LIFE-Pure Fusion	LIFE-Fusion /F Ign	LWR	LWR (46% eff)	Coal	Coal w/CCS	Gas	Gas w/CCS
Fuel	8.9	9.6	8.8	8.8	7.9	5.7	23.1	39.0	47.6	59.1
O&M	11.1	10.4	12.0	10.1	10.2	8.4	8.9	18.9	2.7	4.7
Capital	45.4	42.5	49.8	41.7	42.6	34.8	24.8	40.7	8.5	16.8

Figure 1. Comparison of LIFE to alternative base-load technologies.

We also show other technologies including a LWR, fission with higher conversion efficiency, and coal and gas, each with and without carbon sequestration. We conclude that LIFE costs compare favourably with other carbon-free base load technologies.

5.0 Design Space Studies

Next we examine the COE as a function of target yield, rep-rate and net electric power.

Figure 2 shows the COE as a function of rep-rate for net powers of 1000, 1300 and 1600 MW_e. At each point, the laser energy and target yield are adjusted to keep the net power fixed. As indicated, the COE is relatively insensitive to rep-rate with 5-10 Hz near optimal for all three power levels. Larger plants have significantly lower COEs reflecting more economical use of the capital investment in the laser and the economy of scale benefits of the larger nuclear island and BOP.

Figure 3 shows the COE as a function of target yield, again for three net electric powers. At very low yields, the target gain is low and the laser recirculating power is high, resulting in a high COE. COE decreases with increasing yield, with yields in the range of 50-75 MJ near optimal for this range of plant sizes. The markers on each line correspond to rep-rates of 20, 15, 10 and 5 Hz moving from left (low yield) to right.

6.0 Summary

We are developing LIFE as a carbon-free, base-load power technology that with adequate R&D funding and successful development of key technologies could enter the commercial market ~2030. The LIFE Integrated Process Model is being used to guide conceptual designs to attractive operating space and to identify high leverage R&D. Scoping-level cost estimates indicate that LIFE can be competitive with fission power and fossil plants with carbon sequestration. Large plants (>1000 MW_e)

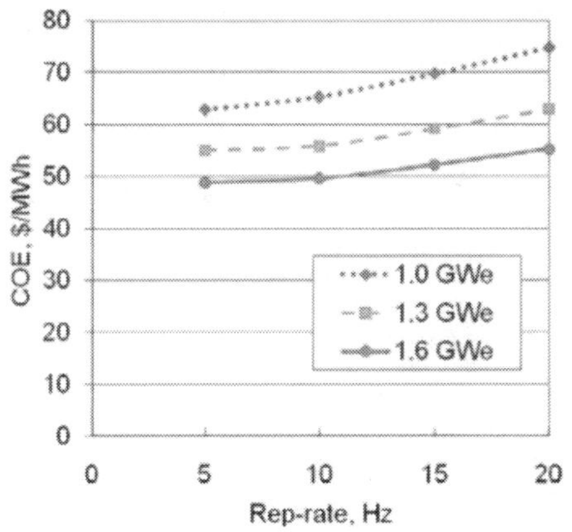

Figure 2. COE vs. rep-rate for net powers of 1.0, 1.3 and 1.6 GWe.

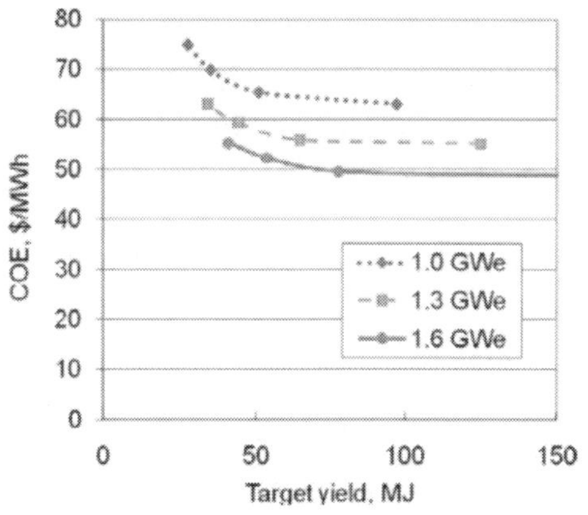

Figure 3. COE vs. target yield for net powers of 1.0, 1.3 and 1.6 GWe.

are most cost effective. Fission blanket energy multiplication allows operation at modest target yields (~50-75 MJ), reducing required laser energy and cost. The COE is relatively insensitive to rep-rate for fixed net power and rep-rates of ~10 Hz or less are optimal. With estimated cost reductions relative to NIF, laser capital cost is only ~17% of the LIFE plant capital cost. Assuming LIFE nuclear island and BOP costs are similar to fission plants, they account for 74% of total capital cost. We are now pursuing innovative lower-cost, high-efficiency laser architectures and higher-temperature, high efficiency chamber concepts to optimize the economics of LIFE.

Acknowledgements
This work performed under the auspices of the U.S. Department of Energy by Lawrence Livermore National Laboratory under Contract DE-AC52-07NA27344.

References
[1] Moses E et al., A sustainable nuclear fuel cycle based on laser inertial fusion-fission energy (LIFE), *Fusion Science and Technology*, **56**, 2, p. 547 (2009).
[2] Meier W R et al., Systems modeling for the laser fusion-fission energy (LIFE) power plant, *Fusion Science and Technology*, **56**, 2, p. 647 (2009).
[3] Kramer K J et al., Neutron transport and nuclear burnup analysis for the laser inertial confinement fusion-fission energy (LIFE) engine, *Fusion Science and Technology,* **56**, 2, p. 625 (2009).
[4] Abbott R P et al., Thermal and mechanical design aspects of LIFE engine, *Fusion Science and Technology,* **56**, 2, p. 618 (2009).
[5] Caird J et al., Nd:glass laser design for laser ICF fission energy (LIFE), *Fusion Science and Technology,* **56**, 2, p. 607 (2009).
[6] Anklam T et al., Economic Evaluation of Electrical Power Generation Using Laser Inertial LLNL report LLNL-MI-416420 (2009).
[7] U.S. advanced liquid metal reactor program – Actinide recycle system: 1994 capital and busbar cost estimates, General Electric Report GEFR-00940/UC-87Ta (1995).
[8] Delene J G et al., Nuclear energy cost data base - a reference data base for nuclear and coal-fired power plants power generation cost analysis, DOE/NE-0095 (Sept. 1988).
[9] Deutch J M et al., The future of nuclear power – An interdisciplinary MIT study, ISBN 0-615-12420-8 (2003). http://web.mit.edu/nuclearpower/
[10] Deutch J M et al., Update of the 2003 MIT future of nuclear power study, MIT report (2009). http://web.mit.edu/nuclearpower/

ADVANCES IN PROPOSED D-CLUSTER INERTIAL CONFIMENT FUSION TARGET

George H. Miley,[1] Xiaoling Yang,[1] Hora Heinrich,[2] Kirk Flippo,[3] Sandrine Gaillard,[3] Dustin Offermann,[3] and D. Cort Gautier[3]

[1]Department of Nuclear, Plasma and Radiological Engineering, University of Illinois, Urbana, IL 61801 USA; [2]Department of Theoretical Physics, University of New South Wales, Sydney, Australia; [3]Plasma Physics Group, Los Alamos National Laboratory Los Alamos, NM 87545 USA

ghmiley@uiuc.edu

Abstract. Our recent research has developed a technique for imbedding ultra high density deuterium "clusters" (D cluster) in Palladium (Pd) thin film. Experiments have shown that in Pd these condensed matter state clusters approach metallic conditions, exhibiting super conducting properties. This deuterium cluster is achieved through electrochemically loading-unloading deuterium into a thin metal film, such as Palladium (Pd). During the loading process, Palladium lattice expands significantly due to invasion of deuterium into the interstitial sites. With the large enough stress, some linear lattice imperfections, called dislocations, form at / transformation interface. These dislocation defects form a strong potential trap causing deuterium to condense. In the present study, a new method employing nano-structuring of the Pd is proposed to significantly improve the site density over the target volume, suggesting that a sizable region of the compressed target deuterium can reach densities an order of magnitude higher than possible with prior target designs. This improved cluster packing fraction will enable a significant increase of the fusion reaction burn density, hence the target burn-up efficiency.

1. Introduction

Current state-of the art ICF target technology employs cryogenic D-T filled targets. The targets are expensive and complex, requiring several hours of laser heating and cooling cycles and tritium decay smoothing during preparation. While such operation appears feasible for single shot experiments, the extension to rapid pulsing and later to low cost commercial target manufacture and delivery may be a show stopper for useful ICF power. Clearly a non-cryogenic design is a key goal to seek.

Recent superconducting quantum interference devices (SQUID) measurement has shown ultra-dense states of deuterons with many more than 100 deuterons within a crystal defect in a palladium crystal are possible, and a superconductive state of these clusters was demonstrated in these experiments [1, 2]. Similar ultra-dense state of deuterons was seen at surface defects of iron oxide resulted in ion energies of 630 eV through Mass spectrometry measurements [3]. It may well be assumed that both cluster states are of the same nature though the states are concentrated at the surface in the iron oxide case due to the catalytic generation in contrast to the Pd samples with localization in the bulk volume [2]. In both cases their existence was confirmed by the LENR process [4] which likewise should be valid including when an inverted Rydberg state is present. [3] A very important application would be using these clusters to achieve non-cryogenic targets for inertial confinement fusion. In principle this could also provide a compressed fuel density up to about 1000 times solid state density. [2][5][6].

2. Formation of Ultra-High-Density Deuterium Cluster in Palladium Foil

To date, we have demonstrated more than 100 deuterions located in the dislocation loops of Palladium thin film at room temperature. [1, 2] This high density deuterium cluster is achieved through electrochemically loading-unloading H/D into a thin metal film, such as Palladium (Pd). During the loading process, metal lattice expands significantly due to invasion of H/D into the interstitial sites. The Palladium lattice expands from 3.889Å to 3.894Å (-phase) when hydrogen is physically adsorbed (PdH0.015), and then further to 4.025Å (-phase) when chemical bonding is formed (PdH0.6). With the large enough stress, some linear lattice imperfections, called dislocations (Figure 1), form at / transformation interface. The diameter of the dislocation defects is around 2 burgers factor, and the length of them depends on the dimensions of the Palladium film. After loading, the loosely bonded hydrogen is unloaded by applying positive current to metal film, while the formed cluster materials remain in place due to their high binding energy. Additional dislocations and clusters are formed upon reloading. Thus this loading and unloading process is repeatedly done in order to achieve abundant and uniform distribution of dislocation loops. After about 6-10 cycles, the packing fraction of clusters saturates.

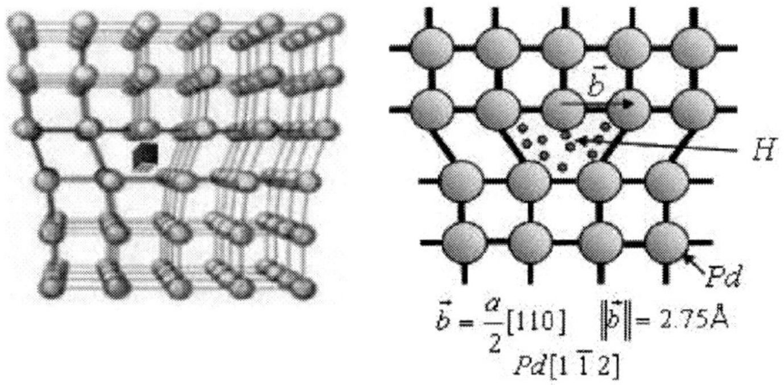

Figure 1 Scheme of edge dislocation loops in Pd containing condensed H/D.

The following experiment results briefly described the evidence of the UHD H clusters:

I. Temperature Programmed Desorption (TPD) Analysis

TPD measurement involves heating the hydrogen loaded sample while contained in a vacuum and simultaneously detecting released hydrogen in the vacuum by means of a mass analyzer. This heating and simultaneous detection result in a hydrogen partial pressure peak in the pressure-versus-time plot. The binding energy of the bound species and the total hydrogen desorbed could be calculated from the recorded curve. In our TPD experiment, the sample was heated from 20 °C to 700 °C in baseline vacuum of 10^{-8} Torr (see figure 2). The calculated binding energy of H/D based on the Garlick-Gibson model is estimated as 0.65±0.10 eV, [2] which is consistent with the result of Kirchheim for hydrogen trapping at dislocation core sites in cycled Pd. Thus, with known dislocation density 2×10^{11} cm^{-2}, [7] the local concentration of hydrogen within dislocation loops is calculated to be [H]/[Pd] ~ 1.8. The TPD measurement for pure Pd/PdO sample showed only negligible peak.

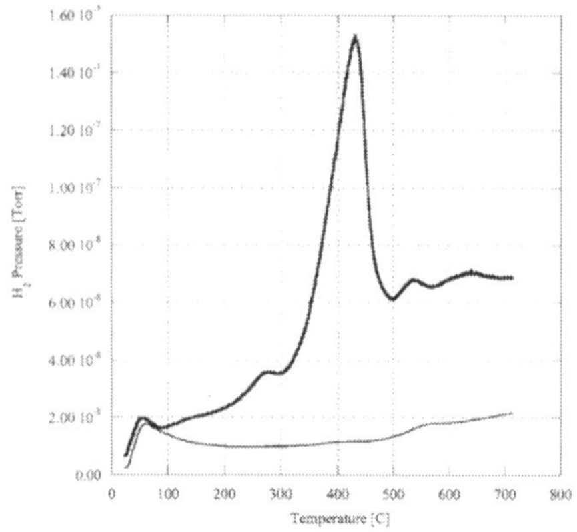

Figure 2 Thermal desorption measurements of Pd/PdO:Hx (thick solid line) and Pd/PdO (thin solid line). [1]

II. Superconductivity measurement

Another proof of the existence of UHD D cluster is its demonstrated superconductive properties in direct resistance measurement and SQUID measurement. After eliminating the temperature-independent part of resistance caused by dislocation formation, cluster material showed much lower resistance than pure Pd. This enhanced conductivity is best seen in the ratio of resistances – the larger temperature coefficient for the Pd:H cluster sample is an indication of a more metal-like behavior. Also, SQUID measurement of Pd/PdO:Hx showed enhanced diamagnetic feature when compared with the measurement of pure Pd/PdO sample. Both the resistance measurement and SQUID measurement indicate that hydrogen is condensed into a metal-like phase at dislocation cores. [1]

3. New Nano-Structure Film (NSF) Electrodes/Targets

Recently, a new type of electrode, the nano-structured film (NSF) electrode, was developed to create cluster formation in nanostructures manufactured on a Pd or Ni thin film electrode. The concept is to mimic the dislocation loops, but achieve a larger volumetric concentration or "packing fraction" than in the previous electrodes. This method used follows one originally developed by R.N. Rhonda of the International Nickel Plating Company. SEM image of plated Pd on Ni Microfelt or Ni foam are shown in Figure 3. The structures provided are thought to provide the conditions needed for cluster formation. Some preliminary electrolytic studies have been encouraging, but much more needs to be studied to understand the true promise of this new electrode design. If high volumetric densities of cluster sites can be created via methods outlined earlier, a high reaction rate per cc should result given a competitive power cell. The approach starts with creation of a micro-fiber felt sandwich structure. Palladium is then sputtered onto the Ni substrate prior to encapsulating it into the sandwich electrode configuration. Many local defects occur at interfaces in this configuration, giving a high volumetric density of potential cluster sites. This work is currently in progress, and preliminary results to data are encouraging.

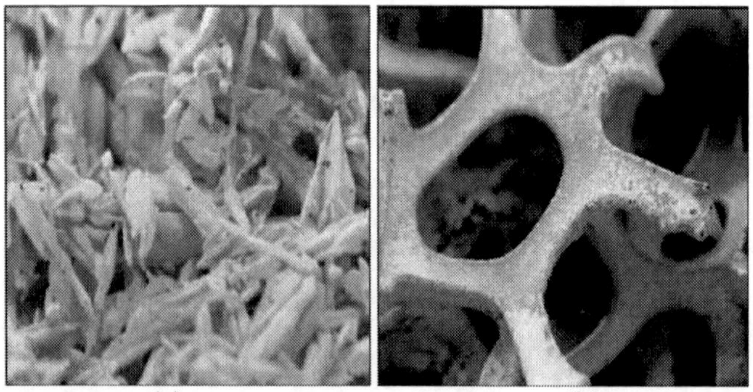

(a) (b)

Journal of Physics: Conference Series **244** (2010) 032036 doi:10.1088/1742-6596/244/3/032036

Figure 3 (a) Auger Scan of Ni Microfelt and (b) Ni foam electrodes [White color indicates Pd.].

4. Observation of Defects and Ultra-High-Density Deuterium Cluster in Other Materials

Encouraged about the result [1, 2] that SQUID measurements confirmed the localization of more than 100 nuclei of hydrogen in a crystal defect within a palladium crystal, an estimation was given [5] how many such defects may be generated in a crystal. A density of such defects in crystals may well be more than 10^{19} cm^{-3} before the crystal is breaking. This is the experience from the property in silicon crystals where such densities of doped atoms are well known. Similar density of defects with voids has been produced in silicon too by very intense electron bombardment. [8, 9, 10, 11] If these voids are then in an average distance of 10 atoms in the crystals for laser fusion targets based on LiH or similar crystals at room temperature, the targets with a density of fusion fuel of 1000 times solid state density may be possible.

A necessary condition, however, is that the clusters in the crystal volume should have the density of 10^{24} cm^{-3} These densities have been confirmed [3] from clusters which were produced in crystal defects at the surface of iron oxide by catalytic processes of inverted Rydberg states. These densities were proved from the Coulomb repulsion energy of deuterons in such states as time-of-flight measurements showed 630 eV w. This corresponds to a distance of the deuterons of 2.3 pm within the inverted Rydberg clusters. The repulsion process is given, when laser irradiation of 546 nm wave length removes the neutralizing electrons between the deuterons [3].

The mechanism of the electron emission by the laser radiation has been analyzed by inclusion of the correspondence principle of electromagnetic interaction [12]. A virtual quiver motion of the electrons in the quantum state in the inverted Rydberg cluster arrives at a quiver elongation of 2.3 pm at a laser intensity of 1.01×10^{10} W/cm^2. This intensity was just measured as the lowest threshold [3] for the measurement of the 630 eV deuterons.

A further conclusion of this analysis is that the state of the deuterons in these clusters represents an effective Bose-Einstein condensation at room temperature. [13] A further result is for a hydrogen atom, the virtual quiver motion of the electron within the laser field with an elongation of the Bohr radius corresponds to the well known ionization energy of 13.6 eV. This energy has the unique value of $^2mc^2/2$ expressed by the fine structure constant and the rest mass energy of the electrons mc^2 with the electron mass m and the vacuum speed of light c.

5. Summery

It may be concluded that the state of the deuterium clusters with more than 100 deuterons measured with SQUID in single crystal defects [1, 2] is similar to the ultra-density clusters [3]. These were measured by time-of-flight spectrometry during laser irradiation of surface defects which are considered as inverted Rydberg states D(-1). Following these exciting findings, new nano-manufacturing methods are now being explored to achieve larger packing fractions of these ultra-high-density deuterium cluster materials. Further, the extension of the technology to other materials, such as low Z materials, like Li and Be, is needed in order to reduce the cost and Bremsstrahlung losses. If these goals can be achieved, a remarkable new type of high performance non-cryogenic ICF target would become a reality.

References

[1] A. Lipson, B.J. Heuser, C. Castanov, G. Miley, B. Lyakov, A. Mittin. Transport and magnetic anomalies below 70 K in a hydrogen cycles Pd foil with a thermally grown oxide, Phys. Rev. B 72 (21): 212507-10, 2005

[2] G. H. Miley and X. Yang, (2009), Deuterium Cluster Target for Ultra-High Density Fusion, Fusion Science and Technology, 56 (1), 395-400

[3] Shahriar Bardei, Patrick U. Andersson, Leif Holmlid. (2009) High-energy Coulomb explosions in ultra-dense deuterium: Time-of-flight-mass spectrometry with variable energy and flight length, International Journal of Mass Spectrometry, 282 (1-2), 70-76.

[4] H. Hora, G.H. Miley. Maruhn-Greiner maximum from uranium fission for confirmation of low

energy nuclear reactions LENR via a compound nucleus with double layer magic numbers. Journal of Fusion Energy. 26 (2007) 349-354 & 347.

[5] L. Holmlid, H. Hora, G.H. Miley, X. Yang, (2009) Ultrahigh-density deuterium of Rydberg matter clusters for inertial confinement fusion targets. Laser and Particle Beams 27 529-532.

[6] Patrick U.. Andersson, Leif Holmlid. Ultra-dense deuterium: a possible nuclear fuel for inertial confinement fjusion (ICF). Physics Letters A doi:101.1016/j.physleta.2009.06.046.

[7] B.J. Heuser, J.S. King, SANS measurements of deuterium-dislocation trapping in deformed single crystal Pd, J. Alloys Compd. 261 (1997) 225

[8] H. Hora, Shift of Absorption of Evaporated Silicon Layers by Bombardment with 75 keV, Naturwissenschaften 48, 641 (1961)

[9] H. Hora, Change of Silicon by Intensive Bombardment with 50 keV Electrons, Zeitschrift f. angew. Phys. 14, 9-12 (1962).

[10] HORA, H. (1983). Stresses in silicon crystals from ion-implanted amorphous regions Appl. Phys. A 32, 1–5.

[11] GOLDSMID, H.J., HORA, H. & PAUL, G.L. (1984). Anomalous heat conduction of ion-implanted amorphous layers in silicon crystals using a laser-probe technique. Phys. Stat. Sol. (a) 81, K127–K130.

[12] H. Hora and P.H. Handel, New Experiments and theoretical development of the Quantum Modulation of Electrons (Schwarz-Hora effect), Advances in Electronics and Electron Physics, P.W. Hawkes ed. (Acad. Press, New York, 1987) Vol. 69, p.55-113., Chapter III.

[13] Yeong E. Kim, Theory of Bose–Einstein condensation mechanism for deuteron-induced nuclear reactions in micro/nano-scale metal grains and particles, Naturwissenschaften. 96(7):803-11, 2009A reference

Target developments program to prepare LMJ campaigns

R. Collier, F. Bachelet, R. Botrel, O. Breton, C. Chicanne, Ch. Dauteuil, F. Durut, E. Fleury, L. Guillot, C. Hermerel, L. Jeannot, O. Legaie, G. Legay, M. Martin, B. Reneaume, M. Theobald, O. Vincent-Viry

Commissariat à l'Energie Atomique, Direction des Applications Militaires, Valduc, F-21120 Is-sur-Tille, France.

remy.collier@cea.fr

Abstract. To carry out laser plasma experiments on CEA laser facilities, a R&D program was set up and is still under way to deliver complex targets. For a decade, specific developments are also dedicated to "Ligne d'Intégration Laser" (LIL) in France and Omega facilities (USA). To prepare the targets intended for the first experiments on the Laser "Mégajoule" (LMJ) facility, new developments are required, such as cocktail hohlraum fabrication, gas barrier coating and foam shells developments. For fusion experiments on LMJ, an important program is also under way to elaborate the Cryogenic Target Assembly (CTA), to fill and transport the CTA and to study the conformation process of the deuterium-tritium (DT) layer.

1. Introduction

Target developments started in France with first laser facilities four decades ago. The targets complexity has increased to meet ever more demanding experimental requirements. To achieve the production of these specific targets, materials and micro technologies research programs have started. Since 1999, this program involves CEA scientists, engineers and technicians to realize specific targets, used to carry out laser plasma experiments mainly on LIL and Omega facilities. This article presents an illustration of these micro technology fabrications through some complex laser targets.

The current know-how is not sufficient for some steps of the experimental program planned on the LMJ [1], [2]. New target developments are required such as cocktail hohlraums, DT gas shells and foam shells.

Among the LMJ targets, one of the most complex targets is the CTA [3]. An important program is dedicated to its manufacture, its filling (cryogenic targets filling station or IRCC) and its transportation (moving cryostat or VTCC) at cryogenic temperature. The CTA has to meet severe specifications and involves a lot of challenging tasks for its manufacture. To fill CTAs by permeation with DT, the IRCC needs to meet strict thermal, mechanical and dimensional specifications. Today, many parts of the installation have been successfully prototyped, such as the permeation cell, the cryogenic gripper, the DT purification process, and the cryogenic DT pressure process. After DT filling, to obtain a good combustion yield, a very homogenous DT ice layer with a low roughness at 1.5 K below the DT triple point, are also required. The conformation of the ice is studied in two specific cryostats: the integrating sphere and the study filling station (SFS).

2. Target developments for LIL and Omega plasma experiments

Various techniques are fully operational to fabricate target elements: Chemical Vapor Deposition (CVD), Physical Vapor Deposition (PVD), Glow Discharge Polymer (GDP), precision machining, laser machining [4], and electrochemical deposition. Characterization means are also needed for each step of target element manufacturing and after the assembling. The following paragraphs describe two complex targets delivered to LIL and Omega facilities.

2.1. Equation of state targets
Recent targets were realized to study the Equation of State (EOS) of boron on LIL facility [5]. These targets were composed of a gold spherical hohlraum, with two steps at 90°: one was the studied material i.e. boron, and the other was aluminum as a reference. Machining of these materials needed very specific fabrication processes to obtain thin films of 50 μm with 20 nm roughness. Gold cones were also added to shield diagnostic measurement from laser light.

2.2. Multilayers GDP shells
New designs have been developed to reduce hydrodynamic instabilities propagation during target implosion [6]. In this field, multilayer germanium doped and undoped CH_x coating were developed with GDP techniques. These films are realized by controlling the germanium flow in the GDP coater in order to realize many 1 μm thick layers as shown on picture 1. A specific Labview™ software was developed to pilot the mass flow controllers that allow us to obtain these multilayers.

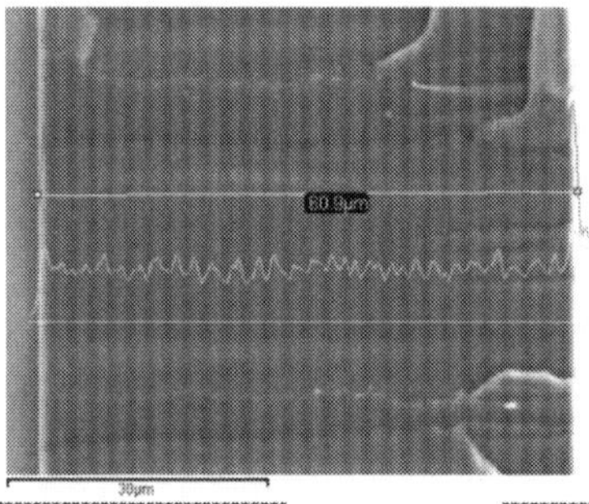

Figure 1. Picture of the cross section of a 60 μm thick coating with 1 μm thick germanium doped and undoped layers. The red curve shows the germanium concentration and the blue one, the oxygen concentration (Energy Dispersion Spectroscopy).

3. New target developments for LMJ
Before reaching the ignition on LMJ, different types of plasma experiments will have to be carried out. Several of these will need new kinds of targets which have not been required up till now. This part of the paper describes three key developments realized.

3.1. Cocktail hohlraums
In order to decrease the wall absorption of hohlraums during laser matter interaction, a thick layer of depleted uranium with gold (DU alloy), can be deposited on the inner surface of the hohlraums. In order to achieve such a coating, sputtering of massive targets of DU and gold directly in the hohlraums is used, the vacuum chamber being placed in a box gloves under nitrogen atmosphere.

Two major technological problems have to be solved to produce such pieces. First, the composition has to be controlled through sputtering parameters as pressure, power of sputtering and target – substrate distance. Gold – molybdenum (Mo) alloys have been synthesised on LMJ hohlraum with a thickness homogeneity better than 5%. The second problem is the rapid delamination of DU alloy in contact with water, air or hydrogen. To protect the DU alloy, a thin coating of dense gold is sputtered on DU alloy. The studies on the effective protection of gold on praseodymium (Pr) deposits have demonstrated that a layer of 0.2 μm of gold is sufficient to prevent Pr from any degradation. The first deposition of DU is to be fabricated to begin the studies of DU alloy aging in aggressive atmosphere (DT).

3.2. Gas barriers for DT gas shells

Gas hohlraum at room temperature with a DT filled CH_x shell would be needed on LMJ before cryogenic shots. The D_2 gas retention in CH_x shells is around 7 min (half life time - $T_{1/2}$). After including thin layers (100 – 300 nm) of metal oxides (such as TiO_x, SiO_2), the current D_2 gas retention reached for Omega targets is 1 month ($T_{1/2}$), which is currently experienced on Omega facility, but not sufficient with DT. Work is going on to control the deposition parameters to increase gas retention up to 3 months to be compatible with LMJ experiments.

3.3. Foam shells to study implosion

The specification required for 2 mm diameter "CH" foam shells are tight: sphericity > 99%, concentricity > 99.9%, porosity of the foam < 1 µm. Two densities are also required: 0.1 and 0.25 with metallic atoms as dopants such as ytterbium and titanium. The process uses two main steps: triple injection (followed by final polymerization and solvent exchanges) and drying by CO_2 supercritical dryer. The results obtained meet the main requirements (doping of Yb and Ti including). For sphericity and concentricity, good results (99.6% and 96% respectively) are obtained (15 experiments, hundreds of microballoons) and work is under way to reach the final goal on concentricity.

4. CTAs, filling station and conformation for ignition on LMJ

The CTA has to meet many requirements [7], among them the Contact Thermal Resistance (CTR) at cryogenic temperature when CTA is connected to a cryogripper. Satisfactory CTR results have been obtained by removing the "Beilby layer" [8] using electrolytic polishing (CTR < 1 K/W).

The developments of graded germanium doped CH_x shells is another priority [9]. The last developments focus on bumps characterization at the outer surface of CH_x shells by Digital Holography Microscope (DHM). The DHM allows the 3D mapping of the shell outer surface (pictures scale is 250 x 250 μm^2). The surface is mapped by recording images along equators using two rotation axis (72 pictures by equator and 12 equators to map a shell of 2100 µm diameter). Pictures are then stitched to rebuild pictures of entire equators and datas are reconstructed in 3D to identify bumps redundancy. The dimensions of each bump are determined by mathematical functions (wavelet or spherical function). At the end of the process, many results are available: number of bumps, bumps dimensional characteristics, rate covering surface, and rate redundancy surface [10].

The research program for operational filling facilities which will deliver CTAs filled by permeation in DT to LMJ has been already described [11]. Glove boxes have been received for most of them.

4.1. Layering studies

Layering studies are focused on crystal growing and thermal quenching in spherical integrating sphere. Experiments in a complete CTA have started in the Study Filling Station (SFS).

4.1.1. *Spherical integrating sphere.* Experiments on layering of solid deuterium in integrating sphere help to learn about the solid crystallization behaviour. Easier to handle since it is not radioactive, deuterium needs the use of a volumetric heating by an infrared laser. To growth a perfect crystal, a thermal stable environment is required at the triple point. It starts from a small crystal seed. The crystal growth speed has to be dominated by the latent heat of fusion. For these experiments the laser power fluctuations have to be at a sufficiently low level since they act directly on the shell temperature, and so the crystal growth.

Following the layering, nominal temperature is obtained by a rapid cooling. The basis of this concept, originally developed at CEA [12], is to cool down the solid faster that the roughness increases. A range of temperature speed drop has been explored. We determined that the minimal slope is 2 K/min. Roughness is keeping its initial value as far as the temperature is going down. It has been possible to keep a smooth layer at 2.5 K below the triple point. When temperature is stabilized (1.5 K under triple point), roughness stays in the requirements during about 5 seconds (figure 2).

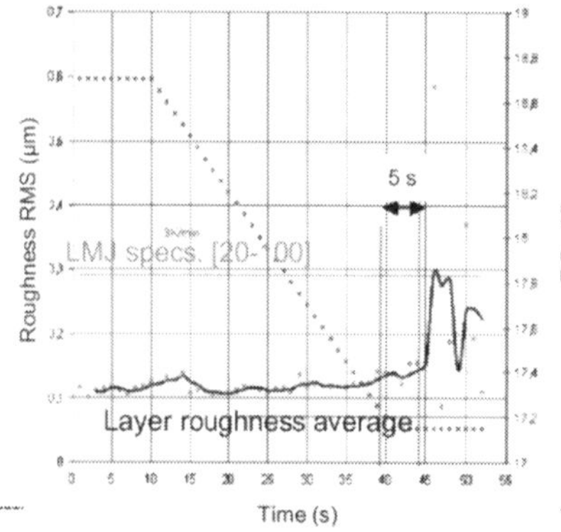

Figure 2. D_2 roughness and temperature during a rapid cooling at 3 K/min. When temperature is stabilized, 5 s are allowed for the LMJ laser shot.

4.1.2. *SFS.* The whole permeation filling system for the LMJ cryotarget is now fully operational [13]. First experiments have been successfully done with deuterium in a 70 µm thick CH capsule. Preliminary layerings have been done. Work is now focussed on optimizing the infra-red laser injection in order to reach a uniform volumetric heating in this cylindrical geometry.

5. Conclusion

The CEA covers since a long time all the fields of research and development in targets for laser plasma experiments. This know-how allows delivering targets for current laser facilities such as LIL and Omega. New fields of developments are required for some parts of the LMJ experimental program, and are well under way, one of the most important among them being a significant program on the cryogenic target assembly. The most part of the tritium workshop is now under completion.

References

[1] Ch. Lion, et al. this proceeding.
[2] J. Ebrardt, et al. this proceeding.
[3] C. Cherfils, et al. this proceeding.
[4] S. Bednarczyk, et al., "Excimer Laser Ablation of 1D, 2D and Double Faces Sinusoidal Profiles on Amorphous Hydrogenated Carbon, Brominated Polystyrene and Polyimide Polymers Foils," *Fusion Sci. and Technol.* 49 813 (2006).
[5] A. Casner, et al. this proceeding.
[6] L. Masse, et al. this proceeding.
[7] P. Baclet, et al., "Research Program for the Fabrication of the Cryogenic Target to Reach Ignition on the LMJ, Results and Propects," *Fusion Sci. Technol.* 49 565 (2006).
[8] Metallography 14:277-294 (1981).
[9] M. Theobald, et al., "Thickness, doping accuracy, and roughness control in graded germanium doped CHx microshells for LMJ," *Fusion Sci. Technol.* 55 438 (2009).
[10] F. Sandras, et al., "Characterization of the microshell surface using holographic measurements", *Fusion Sci. Technol.* 55 389 (2009).
[11] R. Collier, et al., "Overview of the filling station for LMJ cryogenic targets," *Fusion Sci. Technol.* 51 758 (2007).
[12] M. Martin, et al., "A way to reach the cryogenic's temperature and roughness requirements of LMJ facility," *Fusion Sci. Technol.* 51 747 (2007).
[13] F. Bachelet, et al., "The cryogenic studying ans filling facilities for the LMJ targets," *Fusion Sci. Technol.* 55 237 (2009).

Laser Machining for Fabrication of Targets used in the FIREX-I project

T. Fujimura[1], M. Nakai[1], A. Iwamoto[2], N. Hayashi[1], M. Tanabe[1], Yu. Izawa[1], K. Nagai[2], H. Homma[1], H. Nishimura[1], T. Norimatsu[1] and H. Azechi[1]
[1] Institute of Laser Engineering (ILE), Osaka University, Osaka 565-0871, Japan
[2] National Institute for Fusion Science (NIFS), Toki, Gifu 509-5292, Japan
[3] Chemical Resources laboratory, Tokyo Institute of Technology, Yokohama, Kanagawa 226-8503, Japan

E-mail: t-fujimura@ile.osaka-u.ac.jp

Abstract. This paper reports on way to fabricate a gas-tight targets dedicated for the first stage of Fast Ignition Realization Experiment (FIREX-I) at the Institute of Laser Engineering (ILE), Osaka University. It was found that a Ti;sapphire laser machining can be used to fabricate the target. The performance of the laser machining using a fs Ti;sapphire laser was examined on shell materials. The conditions for accurate machining were determined. Michelson interferometer with two different wavelengths which imitates a white light interferometer is an excellent tool for confirming the gas-tightness of the target after assembly.

1. Introduction

The first stage of Fast Ignition[1, 2] Realization Experiments (FIREX-I) [3] is proceeding in the Institute of Laser Engineering (ILE). In the FIREX-I project, unique targets were proposed as shown in Figure 1.[4] Resorcinol-Formaldehyde[5] low density foam and poly-palaxylylene gas barrier are shell material candidates [6]. The CD or CDBr are poly-styrene (PS) based materials are also shell material candidates. Gas-tightness is necessary for these targets since hydrogen fuel or tracer gas are planned to be fed into the shell. Precise machining to make a hole on the shell for inserting the guide cone and the gas feeder are needed. A technique of minute assembling these parts is also needed. A system to confirm the gas-tightness is necessary before starting cryogenic layering.

In a previous work, hole boring on PS shell was carried out by dissolving PS with solvent. Control of the hole diameter was, however, quite difficult and sometimes cracks was a few microns wide would develop. Nd;YAG laser machining has been tested to make a hole on the foam shell.[7] It was useful in the case of a RF foam shell without a gas barrier because the RF foam absorbs the second harmonics of the Nd;YAG laser and it easily decomposes. In the case of a foam shell with a gas barrier and PS shells, the Nd;YAG laser machining with ns is pulses inappropriate because the thermal propagation can not be ignored when the intensity of the ns-pulse laser increases to generate a plasma. Laser machining using a Ti;sapphire fs laser was tried to provide precise laser machining. Multi photon absorption happens and thermal damage does not propagated onto not irradiated areas[8]. Laser power dependence of ablation depth and residual debris around the cutting trace was studied.

It is necessary to control the amount of epoxy resin around the laser guide cone. It is better that theamount of epoxy resin be as little as possible from the point of implosion uniformity, but it is difficult to make a gas-tight target as the amount of epoxy resin decreases. A two wavelength

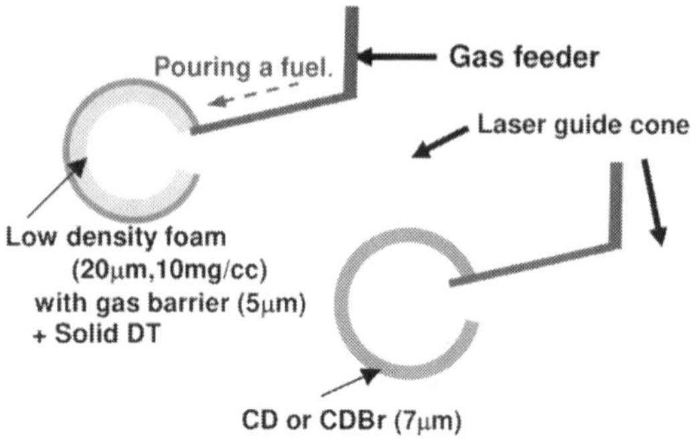

Figure 1. A schematic view of cryogenic targets planned to be used in the FIREX-I project.

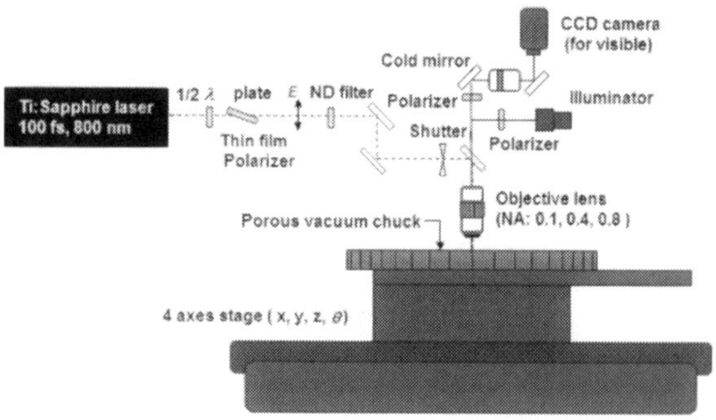

Figure 2. A schematic view of a set-up of a fs laser machining.

interference system for checking a gas-tightness after assembly was developed.

2. A fs laser machining

Figure 2 shows the set-up for fs laser machining. A Ti;sapphire laser was used. The wavelength, the pulse width and the repetition rate were 800 nm, 100 fs and 1 kHz, respectively. The laser power was controlled by rotating a 1/2 wave retardation plate. The resolution of the x axis, y axis and z axis of the stage was 100 nm, 100 nm and 1.0 μm respectively. The tests of fs laser machining were done on the RF foam plate to confirm that thermal propagation does not play an important role (Fig.3 (A)). The stage ran at 3.0 μm/sec, the spot diameter was 9.0 μm and the average intensity was 1.6×10^{15} W/cm^2. Figure 3 (A) shows the cutting width and was mostly same as the diameter of the laser spot. It was found that thermal propagation did not damage any non irradiated area. Laser machining of the foam

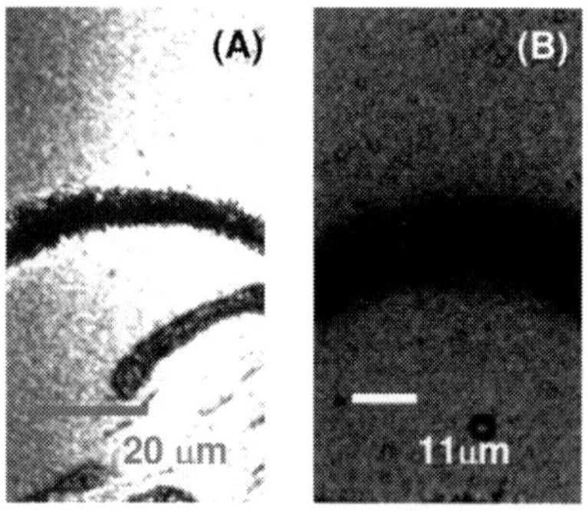

Figure 3. The laser microscopic images of laser machining. (A) Laser machining to the RF foam plate and (B) to the gas barrier on the foam plate. Figure 3 (B) shows the machining of the RF foam plate with 5.0 μm gas barrier

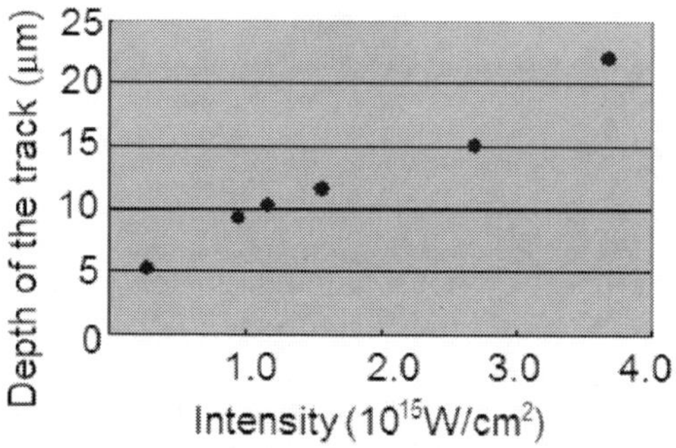

Figure 4. Laser power dependence of ablation depth of a PS sheet. The stage ran at 3.0 μm/sec. The depth was measured by a laser microscope.

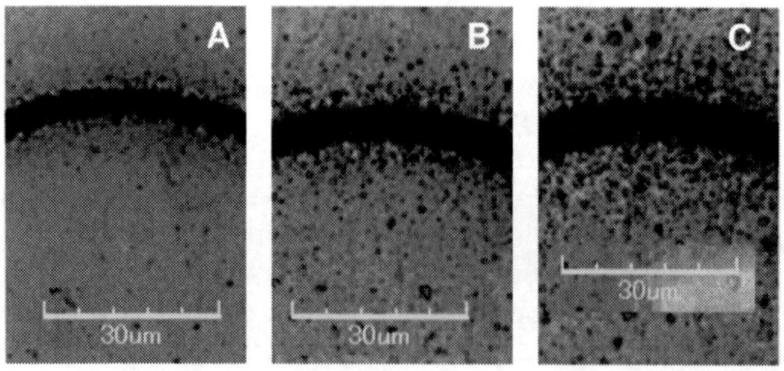

Figure 5. Laser power dependence of residual debris around the cutting trace. The laser intensities were (A) 1.3×10^{15}, (B) 1.7×10^{15} and (C) 3.0×10^{15} (W/cm^2) each.

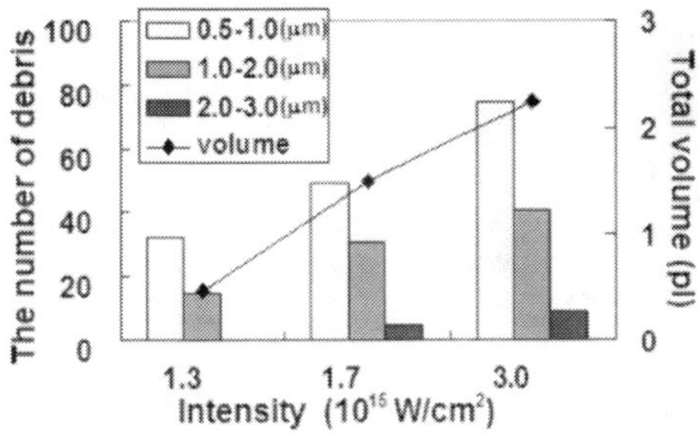

Figure 6. Results of image analysis. The number and the diameter of debris in a 30 μmx30 μm square are shown by bars. The point and line shows the total volume of debris in 30μm around the hole and then the debris is assumed as an ideal sphere.

plate with a gas barrier was tested for determining the minimum laser intensity to cut the gas barrier. It was possible to machine the foam with the gas barrier. The minimum intensity to make a hole on the 5.3 μm gas barrier was 3.1×10^{14} W/cm^2. It was also confirmed that there is no damage on the gas barrier when the laser (the intensity was 1.5×10^{14} W/cm^2) was focused 50μm above the surface. This indicated that the inner surface of the opposite side of the shell should not be damaged. There was little debris around the cutting edge. The dependence of laser intensity and ablation depth was measured on a PS plate. Figure 4 shows the average depth of the track as a function of the intensity. The table shows that the laser intensity needed to cut a 7μm-thick PS shell was 0.8×10^{15} W/cm^2. Figures 5 show the relationship between laser power and the debris from the fs laser machining on the PS plates. The analysis was done by using Image-J (National Institute of Health, USA) (Fig. 6). In the case of a laser intensity 1.3 $\times10^{15}$ W/cm^2 of, the average particle diameter was 0.73 μm, whereas in the case of a laser intensity 1.7 $\times10^{15}$ W/cm^2 of, the average particle diameter was 1.4 μm. As the laser intensity decreased, the particle size and the number decreased. Compared with the amount of epoxy resin (about 2.0 pl) for assembling, the total volume of the debris is so small that it can be ignored in the case of 1.3 $\times10^{15}$ W/cm^2 in laser intensity. The relationship between the total debris volume and intensity was not linear. As the laser intensity increased by 30%, the volume of debris increased by a factor of three. Judging from these results, it should be better to laser machine the targets with an intensity lower than 1.3 $\times10^{15}$ W/cm^2.

3. Confirmation of gas-tightness

After laser machining, a shell, a laser guide cone and a gas feeder were assembled by using a XYZ manipulator. A carbon fibre (the diameter is 1μm) was used as a brush the epoxy resin to minimizing the amount of epoxy. After assembling, it is necessary to confirm the gas-tightness of the target and that the gas feeder is not plugged. Figure 7 shows the set-up to check the gas-tightness of the assembled target. A Michelson interferometer was used. Two beams with different wavelengths from HeNe lasers were interfered for imitating a white light interferometer. One can identify the interference line where there is no phase difference between reference beams and object beams. SF_6 gas was used to fill the shell. SF6 has several advantages; molecular weight of SF_6 gas is high, so permeation of SF_6 gas through the gas barrier is small. SF_6 gas has a high refractive index and is not

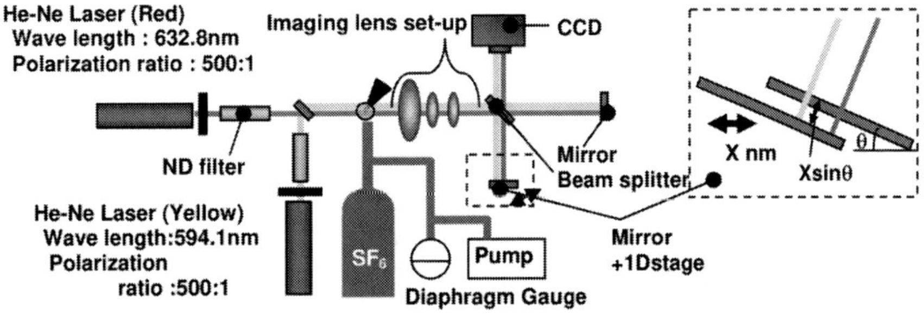

Figure 7. A schematic view of the Michelson Interferometer the gas-tightness.

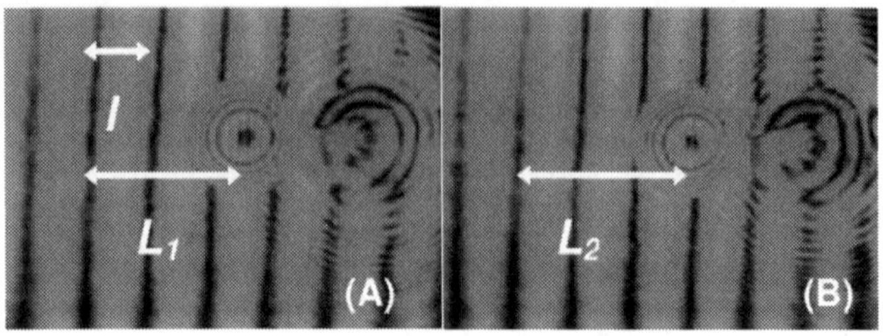

Figure 8. Interferometer image of the PS shell target. The target was evacuated (A) and 1.0×10^5 Pa of SF_6 gas was in the target (B).

corrosive. The target was evacuated and the interference image was taken (Fig. 8 (A)). The red laser was used as a basic interferometer. The phase of the reference beam was adjusted to make the image darkest at the center of the shell. The same procedure was done after filling SF_6 gas in the shell (Fig. 8 (B)). The object beam and the reference beam were inclined to each other to make interference fringes. The interval l between the fringes was measured. The distances (L_1 and L_2) between the center of the shell and certain dark line were measured in the both cases. The yellow laser was used to confirm that the same dark line was selected. The refractive indexes of the gas in the shell was calculated by using the following equation (1)

$$(n-1)=\frac{(L_2-L_1)}{l\cdot d}\lambda\,, \tag{1}$$

where d is a inner diameter of the shell and λ is a wavelength of the red laser. Since there is correlation between the gas pressure and the refractive index, the gas pressure was easily calculated by using Lorents-Lorenz formula. When the L_1 and L_2 are different, the gas feeder is not blocked. The correct value of the pressure should be measured when there are no leaks on the target. It was demonstrated that this instrument can be utilized to check the gas-tightness of the target room temperature.

Summary

This paper reports the way to make a gas-tight targets dedicated for the first stage of FIREX-I. The conditions for machining holes accurately on PS shells and foam shells with gas barriers were investigated using a fs laser. The fs laser machining contributes to fabricate the gas-tight targets. The utility of a two laser interference system for checking a gas-tightness was confirmed.

Acknowledgements

This work was supported by Research Fellowships of the Japan Society for the Promotion of Science for Young Scientists. The stuffs of TFT and EMP group in ILE were gratefully acknowledged.

References

[1] Yamanaka T 1983 *Institute of Laser Engineering Internal Report* Osaka University 5–6
[2] Tabak M, Hammer J, Glinsky E M, Kruer L W, Wilks C S, Woodworth J, Campbell E M, Perry D M and Mason R J 1994 *Phys. Plasmas* **1** 1626-34
[3] Azechi H and the FIREX project 2006 *PlasmaPhys. Control. Fusion* **48** B267-75
[4] Norimatsu T *et al.* 2003 *Fusion Science and Technol* **43** 339-45
[5] Nagai K *et al*. 2005 *Nucl.Fusion* **45** 1277-83
[6] Yang H, Nagai K, Nakai M and Norimatsu T 2008 *Laser Part. Beam* **26** 449-53
[7] Fujimura T *et al.* 2007 *Fusion Science and Technol.* **51** 677-81
[8] Obara M, Arai T and Midorikawa K 1998 *Laser Application Technology* 162-64 (in Japanese)

Study on possible fuel layering sequence for FIREX target

A Iwamoto[1], T Fujimura[2], M Nakai[2], K Nagai[3], T Norimatsu[2], H Azechi[2], R Maekawa[1] and H Sakagami[1]

[1]National Institute for Fusion Science, 322-6 Oroshi, Toki, Gifu 509-5292, Japan
[2]Institute for Laser Engineering, Osaka University, 2-6 Yamadaoka, Suita, Osaka 565-0871, Japan
[3]Chemical Resources Laboratory, Tokyo Institute of Technology, 4256 Nagatsuda, Midori-ku, Yokohama, Kanagawa 226-8503, Japan

iwamoto.akifumi@LHD.nifs.ac.jp

Abstract. A new procedure of fuel layering for the Fast Ignition Realization Experiment (FIREX) target is proposed. A conical laser guide heating technique was experimentally demonstrated in principle as the followings. It employed the target consisting of a polystyrene (PS) shell, a fill tube and a conical laser guide. At first, liquid fuel was fed into the shell and existed around the conical laser guide because the surface tension of the fuel must cause it. Then, it was solidified. The laser light provided a heat source to the conical laser guide so that the solid fuel was moved to the other interior of the shell. This process resulted in missing solid fuel around the conical laser guide. To fill the vacant space, liquid fuel was added as temperature was raised to the melting point. After the liquid fuel addition, temperature was lowered to the solidification point again. During this process, most of the solid fuel could survive.

1. Introduction

Development of a cryogenic target is one of the important issues for the FIREX project. A foam shell method [1] is a possibility. Its layering process must be modified to the FIREX target based on its configuration. Collaboration research by the Institute of Laser Engineering (ILE), Osaka University and the National Institute for Fusion Science (NIFS) has been conducted to develop it. However, some issues: a foam shell fabrication, residual voids in a foam-solid fuel layer, characterization of the layer and so on are still remaining. The foam shell with a 500 μm diameter has been developed at ILE [2]. However, more several years should be required to complete it. The characterization has also been studied using a dummy foam target [3]. We preliminary demonstrated the system to characterize the layer of solid fuel soaking the foam material. Thus, the development of the foam shell method strongly depends on the foam shell fabrication. To ensure supplying solid deuterium (D_2) fuel targets within several years, alternative methods should be considered.

To date, applying volumetric heating method has been studied for the FIREX target [4]. Assuming a D_2 and normal hydrogen (n-H_2) mixture fuel, the ortho-para conversion of n-H_2 was considered as the heat source. Its heating effect was tested using n-H_2 only. Solid H_2 could be moved from the warmer to the cooler part in a shell. The ortho-para conversion, however, can generate much lower heat power than other heat sources such as the beta-layering for a DT fuel [5] and the infrared heating for a D_2 fuel [6]. Therefore, longer duration is required to complete fuel layering, which might be impracticable. Furthermore, according to calculations, the target temperature profile was not spherical isotherms because the conical laser guide played a role of a heat exchanger. Target temperature control is essential. On the other hand, the infrared heating has been studied for the OMEGA target [7]. Forming D_2 fuel targets for central ignition was demonstrated. This technique is specialized to the target with spherical symmetry. Thus applying the volumetric heating process to the FIREX target is worth considering, whereas it must be more complicated than that to the OMEGA target.

Conical laser guide heating is proposed as a new heat source. This external heat source has a possibility to supply sufficient energy for fuel redistribution and should provide a simpler layering procedure than the conventional heat sources. The new idea is experimentally demonstrated in principle using a dummy target.

2. Proposed layering procedure by conical laser guide heating technique

We propose the new layering procedure applying the conical laser guide heating technique. Figure 1 shows the schematic of the sequence. (Step 1) Liquid fuel is filled around the conical laser guide owing to its surface tension and then solidified. (Step 2) Solid fuel moving requires temperature gradient in the shell. We employed the laser heating technique as an external heat source. The conical laser guide is irradiated with laser light to provide a heat source. The solid fuel would sublimate around the conical laser guide and condense somewhere. In natural convection cooling, the shell surface has uniform heat transfer. Therefore, the solid fuel could uniformly redistribute to the other interior of the shell. This process results in missing a solid fuel around the conical laser guide. (Step 3) After the redistribution, as temperature increases to the melting point, additional liquid fuel is supplied to the vacant space. This process time is as short as possible so that most of the solid fuel can survive. The Steps 2 and 3 are repeated until the completion of fuel layering.

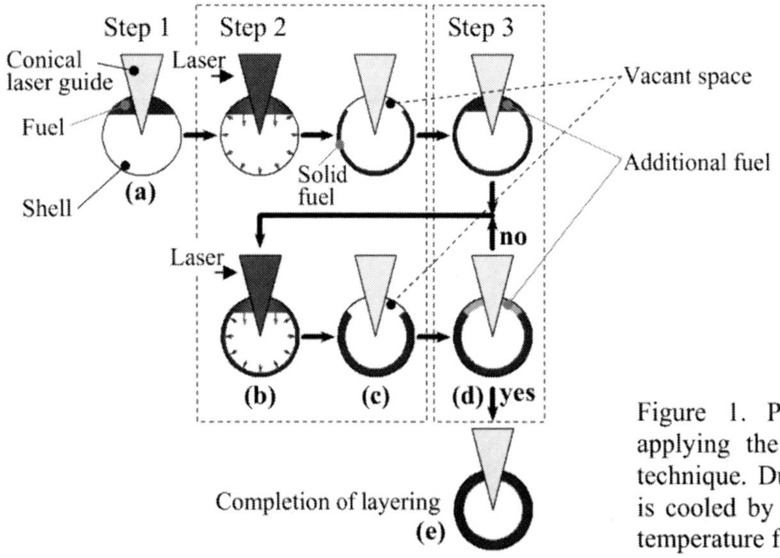

Figure 1. Poroposed layering sequence applying the conical laser guide heating technique. During the sequence, the target is cooled by gaseous helium with suitable temperature for each step.

3. Demonstration of the new technique

3.1. Experimental details

Figure 2 shows the photograph of the dummy target consisting of a PS shell, a gold conical laser guide and a glass fill tube. A 2 mm diameter shell was employed for easy observation of the layering processes. The conical laser guide was attached vertically to the shell. The target was put in the dedicated apparatus [8] and cooled by ambient gaseous helium (GHe) in a target chamber with viewing windows. A temperature controllable thermal shield was equipped in it. To heat the conical laser guide through a view window, a 5 mW He-Ne laser was applied. The laser light had too much energy to transmit moderate heating power on the conical laser guide. Therefore, it passed though two neutral density filters of ND20 and ND30 in series to reduce its energy. A shutter was utilized to intermittently irradiate with the laser light. Open and close time were 7 and 5sec, respectively. As a surrogate fuel, n-H_2 was used. Its ortho-para conversion might become a volumetric heat source. It was experimentally confirmed that the effect was negligible small.

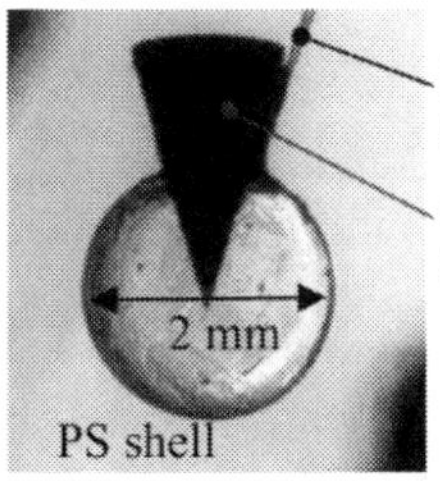

Figure 2. Dummy target with PS shell.

3.2. Layering demonstration
The new technique was experimentally demonstrated in principle. Figures 3 show photographs on each step in the sequence corresponding to that in Figure 1. To supply liquid H_2 (LH_2), the shield temperature was controlled at 13.40 K, which was not equal to the GHe and target temperatures. Gaseous H_2 (GH_2) pressure was increased across the saturated vapour pressure and LH_2 started to appear around the conical laser guide. Its supply was stopped just before the meniscus was inverted by gravity. The LH_2 layer should be contiguous around the shell. Then, the temperature was lowered to 13.20 K; the transition to a solid state was occurred. After the solidification, crystalline solid H_2 was observed as shown in figure 3(a). To ensure the solid state during the laser heating process, the temperature was set at 13.10 K. Then the conical laser guide was being heated. The photograph of the first Step 2 was missed; therefore, figure 3(b) represents the second laser heating process after the first fuel addition. Solid H_2 was forming a uniform layer except for around the conical laser guide in figure 3(c). To add LH_2 to the vacant space, the temperature was raised to the melting temperature of 13.40 K; after LH_2 addition, lowered to 13.20 K for solidification. During this process, most of the solid H_2 could survive. Figure 3(d) shows successful LH_2 addition. This process has another function that the solid surface is made smooth. The solid layer in figure 3(d) has a smoother surface than that in figure 3(c). If the fuel quantity does not meet the specification for a laser experiment, the sequence of Steps 2 and 3 is repeated. Eventually, a thick layer was formed as shown in figure 3(e). In this case, LH_2 was added four times and the whole process required within 8 hours. Thickness of the solid layer is 132~223 μm. Conspicuous variation is beside the conical laser guide, which might be caused by the surface energy difference in LH_2 addition. The inserted fill tube and applied glue must make the surface energy non-uniform. If the thinnest and thickest thicknesses, both locating around the conical laser guide, are ignored, the variation is reduced to 155~192 μm. Defects on the interior surface are also estimated at the maximum of ~30 μm. This demonstration does not satisfy the required specification on the FIREX target at all. However, the conical laser guide heating technique has the possibility to realize a uniformly thick layer for a non-spherical symmetric target.

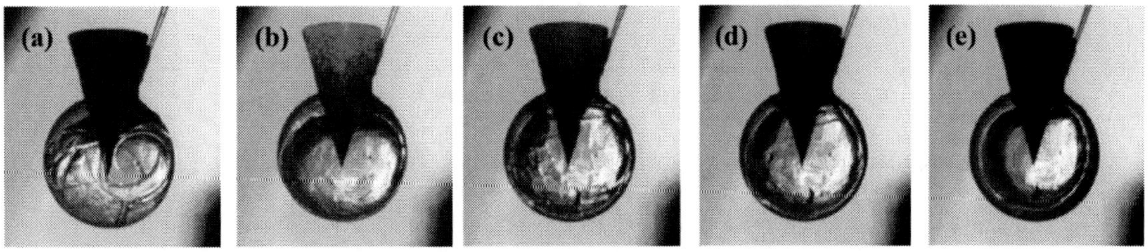

Figures 3. Photographs on each step corresponding to that in figure 1.

4. Calculation

The temperature distribution during the laser heating process (Step 2) is roughly evaluated using the ANSYS code. Steady state condition is considered to simulate maximum temperature raise. The 2D axisymmetric model is shown in figure 4. Ambient GHe temperature is assumed at 13.10 K where comes from the experiment. Uniform surface heat transfers of 23.6×10^{-6} W/mm^{2}K for the shell and 0.44×10^{-6} W/mm^{2}K for the laser guide are applied. Radiation from room temperature and the latent heat of sublimation are not considered. The conical laser guide is irradiated by the laser light with the heat flux of 31.7×10^{-9} W/mm^2, which is evaluated according to the laser energy and its frequency, the irradiating area and the absorption coefficient of gold.

Figure 5 shows the calculated temperature profile. The temperature gradually decreases from the conical laser guide to the shell bottom. The difference of ~0.7 mK is created in the solid layer. It must be enough for the solid fuel to move from the warmer to the cooler part. Compared to the melting temperature of 13.20 K from the experiment, the calculated maximum temperature of 13.102 K is much lower. The sublimation-condensation process was obviously below the melting temperature.

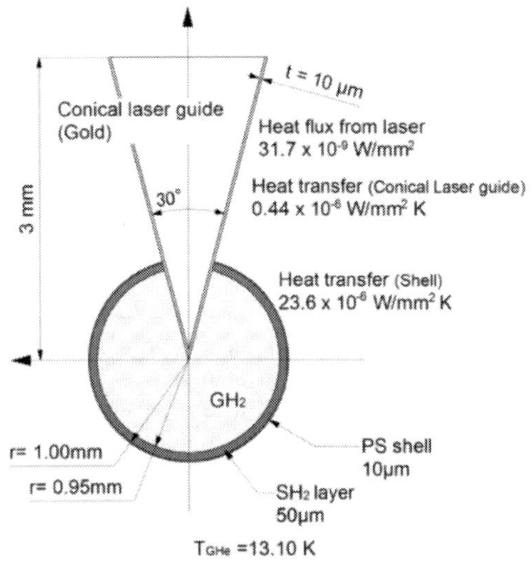

Figure 4. 2D axisymmetric model of target.

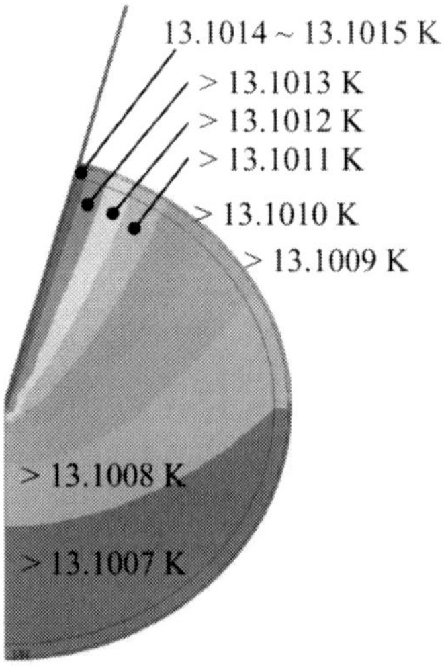

Figure 5. Temperature profile in laser heating

5. Summary

The new technique of the conical laser guide heating was proposed, which employs no volumetric heat source such as the IR heating. Fuel layering was experimentally demonstrated in principle. It has the possibility to realize a uniformly thick layer with a D_2 fuel for the FIREX target.

References

[1] Foreman L R and Hoffer J F 1988 *Nucl.Fusion* **28** 1609-12.
[2] Nagai K, et al. 2009 *Nucl. Fusion* **49** 095028.
[3] Fujimura F, Iwamoto A, Nakai M, Norimatsu T, Nagai K, Yang H, Kimura Y, Sakagami H, Mito T and Mima K 2008 *J. Phys.: Conf. Ser.* **112** 032066.
[4] Iwamoto A, et al. "Temperature Control in a Cryogenic Target with a Conical Laser Guide for Fuel Layering" to be published in *Fusion Sci. Technol.*
[5] Hoffer J K and Foreman L R 1988 *Phys. Rev. Lett* **60** 1310-13.
[6] Collins G W, Bittner D N, Monsler E, Letts S, Mapoles E R and Bernat T P 1996 *J. Vac. Sci. Technol.* **A14** 2897-2900.
[7] Harding D R, et al. 2006 *Phys. Plasmas* **13** 056316.
[8] Iwamoto A, Maekawa R, Mito T, Okamoto M, Motojima O, Sugito S, Okada K, Nakai M, Norimatsu T and Nagai K 2006 *Fusion Eng. Des.* **81** 1647-52.

Improvements in ICF Target Fabrication through High Precision Assembly and Nondestructive Characterization

K A D Obrey[1], D Schmidt, B M Patterson, R D Day, A C Valdez, D Capelli, R Perea, R B Randolph, D Hatch, F Garcia, and D Honnell
MST-7, Los Alamos National Laboratory, MS E549, Los Alamos, NM, USA

Email: defriend@lanl.gov

Abstract. Current ICF and HED targets are fielded on Omega, Z, and Trident; future campaigns will also be fielded on NIF. NIF will field less than 2 shots per day. With such few experiments, target fabrication and alignment accuracy, enhanced metrology and advanced component machining will be even more important. Future target designs are also becoming more complex and more stringent in terms of both manufacturing accuracy and precision. Several steps have been taken to improve the fabrication and characterization of targets, such as instituting an automated assembly station with 3 μm tolerances, utilizing non-destructive characterization tools for rapid component metrology and target assembly, and advancing machining capabilities. Recapitalization of target fabrication infrastructure is continuous.

1. High Precision Assembly

We constructed an automated assembly station to improve target assembly accuracy. This station has a 5-axis automated arm combined with triple theta station that can manipulate any object precisely for accurate assembly, figure 1. The finest stages produces 3 μm complete tolerance stack in XYZ directions and angular accuracies and repeatability of 0.0001° for all angular stations [1].

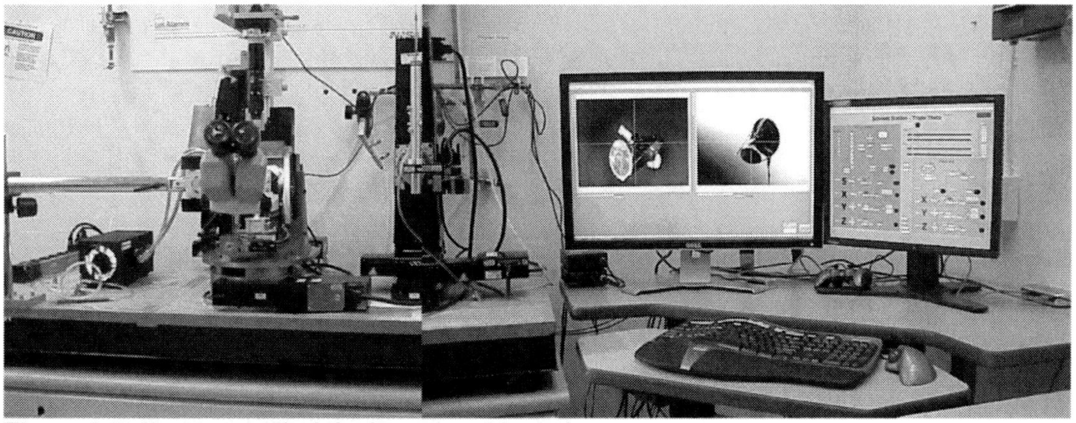

Figure 1. Fully automated triple-theta assembly station.

This station was first implemented in October 2007 for the Symergy campaign, however, the station was operated in manual mode. The Symergy campaign uses either a plastic or Be sputtered capsule mounted inside of a gold hohlraum, figure 2a. The centering specification for the capsule was set at 30 μm. When operating the station in manual mode, 60% of the 12 targets met the 30 μm centering specification. Instituting the automated feature for the January 2009 Symergy campaign, all assembled targets met the 30 μm centering specification. In fact, most of the targets fell well below 20 μm centering.

a) b)

Figure 2. a) January 2009 symergy target fielded at Omega and b) first-ever Be capsule in hohlraum fielded March 2008.

The March 2008 Symergy campaign fielded the first-ever sputtered Be capsule, figure 2b. This targets had a humagen fill-tube connected to the sputtered capsule. The location of this fill-tube made assembly quite difficult, and because of this only half of the targets with the fill-tube met the 30 μm specification.

2. Unique Nondestructive Characterization

A General Atomics (GA) manufactured sputtered Be Ar/Cu capsule was analyzed by a variety of nondestructive methods at LANL. Combination of confocal micro-x-ray fluorescence (confocal MXRF) and 3D tomography developed at LANL gives heretofore-unattainable information [2]. Confocal MXRF provides analytical information on the elemental content and elemental location within a target. No sample preparation is required when performing confocal MXRF; the target is not damaged or destroyed. 3D tomography is used for higher resolution non-destructive imaging based upon x-ray absorption, but no elemental information is measured.

In figure 3, the dotted line is extracted from the reconstructed slice (red arrow) on the tomography image. A low-x-ray absorption region (low intensity) seen on the outer surface is due to the presence of Be with only minor contaminants. High x-ray absorption regions (high intensity) located on inner portion of the shell cross section due to Cu doping of the Be. The colored solid lines of the profile are the x-ray scatter and elemental fluorescence intensity collected on the confocal MXRF. The iron and scatter profiles are seen throughout the capsule. The copper and nickel regions are located in the inner shell of the capsule.

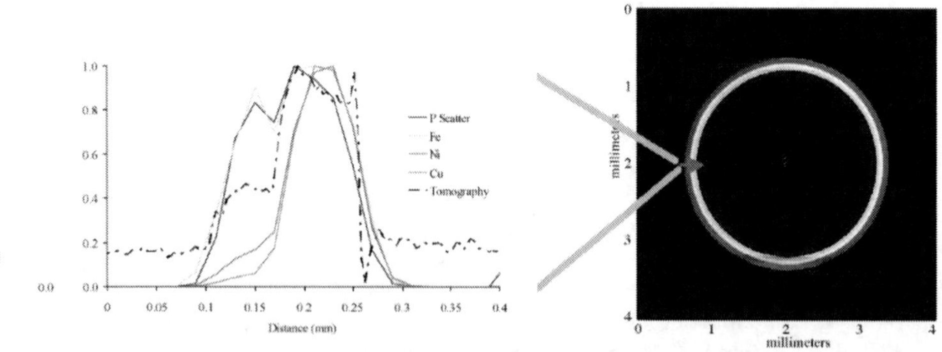

Figure 3. Combination of confocal micro-x-ray fluorescence (confocal MXRF) and 3D tomography used to analyze a GA manufactured sputtered Be/Cu-Ar capsule.

Journal of Physics: Conference Series **244** (2010) 032040 doi:10.1088/1742-6596/244/3/032040

2.1.1. *NIF Double Shell.* Double shells carefully assembled capsules of multiple nested shells of various materials. Confocal MXRF provides elemental information on double shells without destructive characterization. Figure 4a is an XRF spectrum of the double shell. The Ge doped GDP plastic shell and the gold shell are identified. When using confocal XRF, the different elementally doped plastic shells can be identified without any spectral contamination by the other, Figure 4a and imaged, figure 4b. Because double shell manufacturing is labor intensive, non-destructive techniques that provide useful information are very important.

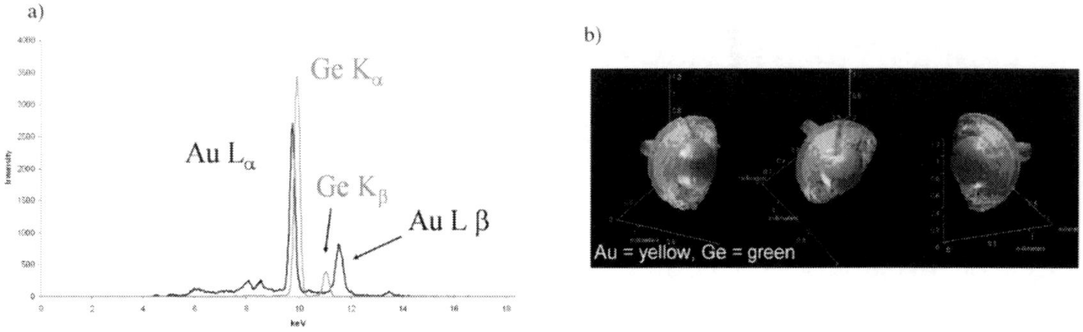

Figure 4. a) XRF spectra of a NIF double shell and b) confocal MXRF images of the NIF double shell both show the Ge and Au layers present.

3. Precision Micro-Machining

The targets used in Hydrodynamic experiments are used to assess the behavior of materials under heavily shocked conditions and require precisely machined and characterized surfaces with unusual features [3]. The particular target highlighted here is a disk that is 27 mm in diameter. The prominent feature of the target is a complex set of sine waves that are merged together to form the configuration shown in figure 5a. A photograph of a practice part, machined out of wax is shown in figure 5b. The fielded parts are made out of tin.

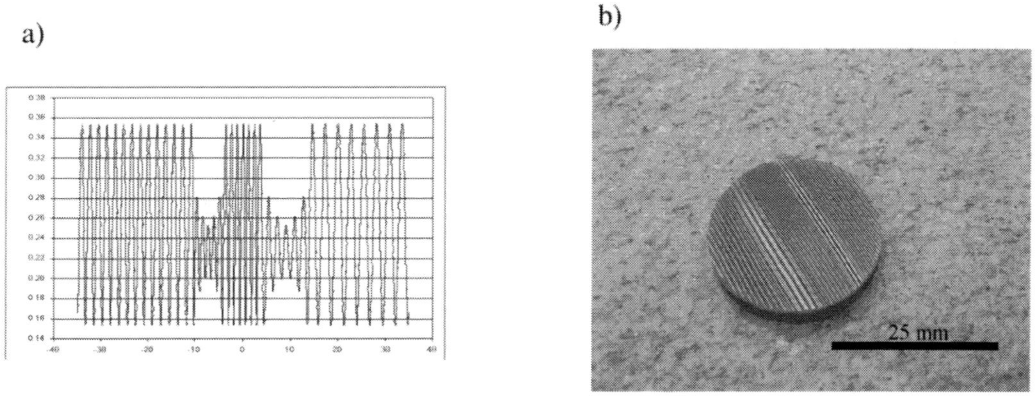

Figure 5. a) Plot of sine-wave set that must be produced on tin surface and b) final practice part with sine wave set on surface for Gas Gun target.

3.1.1. *Neutron Pinhole.* A 3x3 neutron pinhole was fabricated in 2006 [4] and has been used for experiments at the Omega laser facility. A second-generation neutron pinhole with enhanced capability is currently being fabricated. A detail drawing of this new design is shown in figure 6a and a representation of the pinhole is depicted in figure 6b.

The enhanced pinhole array increases the number of triangular pinholes to 20 from nine for the previously fabricated array; it also adds three penumbrae. The penumbrae will be made in a similar fashion, but the holes must be contoured in a manner similar to the sine wave target described earlier.

a)

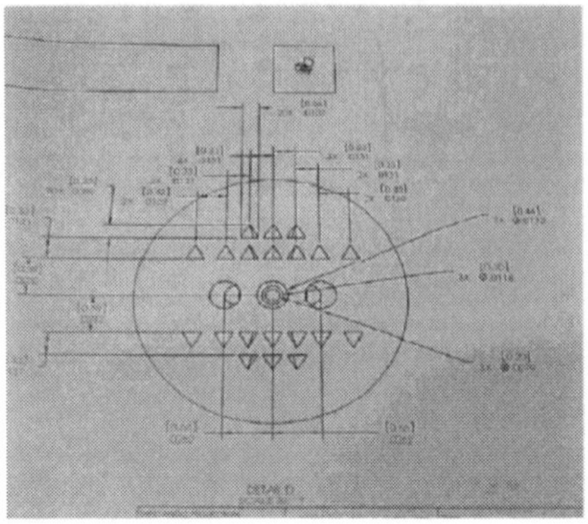

b)

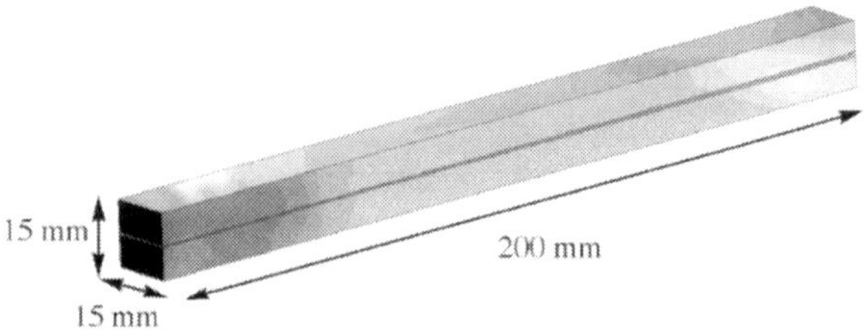

Figure 6. a) Drawing of the neutron pinhole and b) representation of the assembled neutron pinhole.

4. References

[1] Schmidt DW 2009 *Fusion Sci. Technol.* **55** 290

[2] Patterson BM, Obrey KAD, Havrilla GJ, Nikroo A, Huang HB 2009 *Fusion Sci. Technol.* **55** 417

[3] Day RD, Fierro F, Garcia FP, Hatch DJ, Randolph RB, Reardon P, Rivera G 2009 *Fusion Sci. Technol.* **55** 301

[4] Day RD, Brooks PM, Edwards RL, Garcia FP, Grim GP, Nobile A, Schmidt DW, Snow RC 2007 *Fusion Sci. Technol.* **51** 776

Journal of Physics: Conference Series **244** (2010) 032041 doi:10.1088/1742-6596/244/3/032041

A SCATTERED-NEUTRON DETECTOR FOR AREAL DENSITY MEASUREMENT

Y Arikawa[1], T Murata[2], S Fujino[3], H Yoshida[4], K Yamanoi[1], T Nakazato[1], T Shimizu[1], N Sarukura[1], M Nakai[1], T Norimatsu[1], H Azechi[1], K Kamada[5], Y Usuki[5], T Suyama[6], A Yoshikawa[7], N Satoh[8] and H Kan[8],

[1] Institute of Laser Engineering, Osaka Univ., 2-6 Yamadaoka, Suita, Osaka, 565-0871,Japan
[2] Tokai Univ., Kawayou, Minamiaso, Aso, Kumamoto, 869-1404, Japan,
[3] Kyushu Univ., 744 Nishikyo Motooka, Fukuoka, 819-0395, Japan,
[4] Ceramic Research Center of Nagasaki, Hiekoba, Hasami, Higashisonogi, Nagasaki, 859-3726, Japan,
[5] Furukawa Co., Ltd., Kannondai, Tsukuba, Ibaraki, 1-25-13, Japan
[6] Tokuyama Corporation Shibuya, 3-3-1, Shibuya, Tokyo, 150-8383, Japan
[7] Tohoku Univ, 2-1-1 Katahira, Aoyou, Sendai, Miyagi, 980-8577, Japan
[8] Hamamatsu Photonics K.K., 5000 Hiraguchi, Hamakitaku, Hamamatsu, Shizuoka, 434-8601
Corresponding author's e-mail address : arikawa-y@ile.osaka-u.ac.jp

Abstract. Experimental results are presented on the neutron scintillating properties of a custom-designed Praseodymium-doped Lithium-6 glass for a new deuterium-deuterium (DD) fusion scattered-neutron detector. Luminescence was observed at 278 nm-wavelength, and time-resolved measurement yielded ~5.4 ns decay time for neutron excitation. Actual time-of-flight data in laser fusion experiments at the GEKKO XII facility at the Institute of Laser Engineering (ILE), Osaka University reveal that it can clearly discriminate fusion primary-neutrons from the x-ray signals. This material promises the realization of more accurate DD fusion scattered-neutron diagnostics. Design work for a scattered-neutron detector is being conducted for the Fast Ignition Realization Experiment (FIRE-X) at ILE.

1. Introduction

In both for direct and indirect drive approaches, a high-density implosion is an essential issue. Implosion experiments with deuterium fuel areal densities (ρR, g/cm^2) over 0.2 g/cm^2 have been successfully achieved [1], and over 1 g/cm^2 implosion in ignition experiments are expected to be achieved soon at the National Ignition Facility. In this research area, the scattered-neutron method promises a way to accurately measure ρR. Primary neutrons having 2.45 MeV energy and down-scattered-neutrons in the plasma core having energies down to 0.27 MeV are generated in DD fusion reactions. The calculated neutron spectrum for single neutron elastic scattering normalized with respect to the total number of neutrons is shown in Figure 1. Additionally, the case for multiple neutron scattering has been previously reported in [2]. The ratio of the primary-neutrons (around 2.45 MeV energy) and scattered-neutrons around back scattering edge (0.27 – 0.6 MeV energy) shows a nearly linear response to the ρR [2]. This linearity will not be affected so much by the multiple scattering. Experimentally, scattered-neutron measurements have some technical difficulties and currently, a gated neutron imaging detector using a Lithium-6 glass scintillation fiber has been proposed as a promising option [2]. The advantages of Lithium-6 as a scintillator material are as follows: (1) The ^{6}Li(n,t)α reaction, with Q = 4.8 MeV, produces enough light output for scattered-neutrons having lower energy than the primary and (2) the resonant peak of the reaction cross section coincides with the back scattering edge of the neutron spectrum as shown in Figure 1; thereby allowing for increased detection efficiency. However the conventional Cerium (Ce^{3+})-doped Lithium-6 glass scintillator has an inherent problem. It has a relatively long decay profile; with its scintillation fall time from 90% to 10% of the peak value equal to 120 ns. Moreover, it has a significantly intense "after glow" [3]. This long decay time translates into a technical problem should it be used as a diagnostics tool in the GEKKO XII target chamber at ILE. With the chamber's radius at 85 cm, the scintillator must be positioned at about 30 cm away from the target in order to avoid the unwanted interference signals from the primary-neutrons scattered from the target chamber wall. This configuration entails the scintillator to have a response time of ~20 ns or better. In this work, experimental results are presented on the custom-developed, fast response Praseodymium (Pr^{3+})-doped lithium glass as a scintillator material. The detailed description about glass manufacturing process and basic characteristics are discussed elsewhere [4]. Furthermore, we also report on-going designing works of the scattered-neutron detector for the FIRE-X in ILE using this newly-developed material.

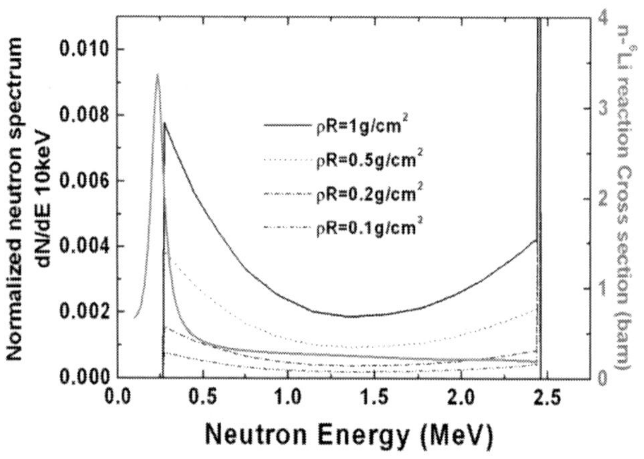

Figure 1 Calculated neutron spectrum originating from DD fusion implosion.

2. Material development

The Pr^{3+} ion with a higher emission cross section in the deep ultraviolet region (~270 nm) was chosen as the dopant material [3] over the slower, albeit more widely-used Ce^{3+} ion [4, 5]. Additionally, high Li-density fluoro-Lithium glass was preferred as the host material over UV-transparent, Li-doped fluoride crystals such as LiCAF[6], primarily due to its ease-of-preparation and design flexibility. The newly developed glass sample, which was named as APLF80+3Pr, has a composition of $20Al(PO_3)_3$-80LiF+PrF_3 (in mol) as starting materials; with 95.5 % ^{6}Li for the enriched material. The Lithium concentration (including residual ^{7}Li) of the finished glass sample was measured using an atomic absorption photometer to be 7.98 w%, 31.6 mmol/cc. This is the highest reported value for conventional ^{6}Li glass scintillators such as KG2 [5].

3. Scintillation characteristics

Figure 2 shows the photoluminescence spectra of the sample, compared with a KG2 conventional scintillator. Strong emission was observed at the design wavelength of ~ 278 nm. Furthermore, the host material was found to have good transmission in the vacuum ultraviolet region down to 180 nm and no absorption at the luminescence region was observed. Moreover, photon yield from a Cf-252 neutron source (broad energy spectrum around 1 MeV) was measured to be 310 photons for all solid angle at the peak of the charge distribution. Gamma rays from Cf-252 spontaneous fission was shielded with 3 cm thick lead, and we assumed the dominant pulse distribution were from thermal neutron since ^{6}Li(n,T)He has much higher cross section for thermal neutrons. The fluorescence lifetime using a Cf-252 neutron source was measured to be 5.4 ns as shown in Fig. 3 where the profile of a GS2, conventional lithium glass scintillator, is also shown for comparison. The APLF80+3Pr lifetime is shown to be sufficiently fast for our requirements.

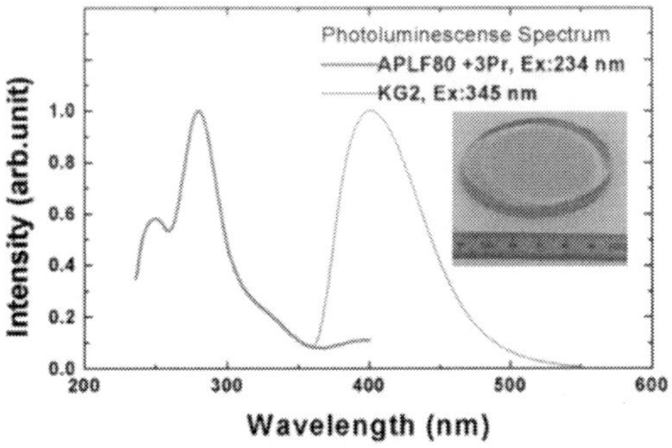

Figure 2 Photoluminescence spectra of the APLF80 + 3Pr and KG2

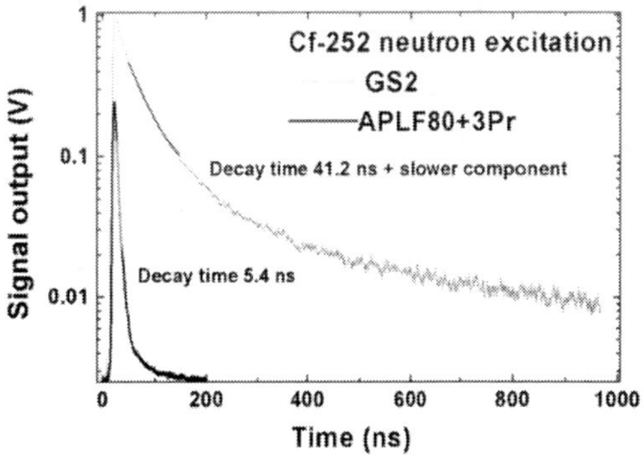

Figure 3 The scintillation decay profiles of the APLF80 + 3Pr and GS2 using Cf-252 neutron excitation.

In addition, fusion-originated neutrons were successfully observed at the GEKKO XII fusion experiment using the scintillator, as shown in Fig. 4. A deuterated polystyrene spherical shell target was irradiated by the GEKKO XII laser facility. Initially, a conventional plastic scintillator was used to establish a baseline for the number of primary neutrons and this was measured to be 5×10^{5}. In this experiment, a 24 mm-diameter, 10 mm-thick APLF80+3Pr sample was positioned at about 10 cm from the target and fluorescence was transmitted to the radiation shielded PMT using a UV fiberoptic bundle. Although scattered-neutrons were not observed since the neutron yield was too low (we estimated 6 primary-neutrons in this signal by ^{6}Li density and cross section), the scintillator's sufficiently fast response characteristics were evident. A dashed- line box in figure figure 4 also indicates the expected temporal location of the scattered neutron signal for a sufficiently intense neutron yield.

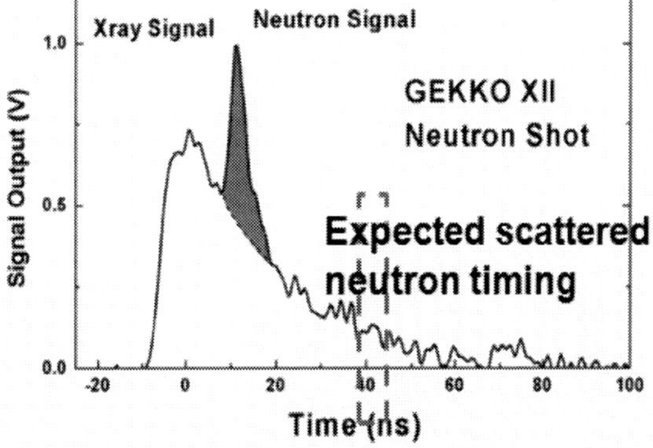

Figure 4 Detected primary neutron signal from the fusion- neutron observation experiment at the GEKKO XII facility using the APLF80+3Pr scinitllator..

4. Detector designing works

A scattered-neutron detector, based on the conventional design constructed with a multi channel scintillation array and a time-gated image intensified CCD camera [2, 6] for the FIRE-X system has been designed at ILE. The neutron yield (Ny) of the first phase of the fast ignition experiment is expected to be about 10^7 - 10^8, and ρR is expected to be the same as Refs. [7, 8] for a ~0.15 g/cm^2 CD target, which is equivalent to 0.021 g/cm^2 for a pure deuterium target. The scintillator array having a total size of 4 cm×4 cm×4cm is designed to be positioned at a distance of 30 cm from the target. The predicted scattered-neutron count is about 3 for a fusion shot with an Ny of 10^7 and ρR value of 0.021 g/cm^2, and about 30 counts for Ny of 10^8. The scintillator is composed of 400 channels; each having a pixel size of 2mm×2mm×40mm, to reduce strong unwanted signals from the x-ray or primary-neutrons. The signal to noise ratio (S/N) of this design is estimated by scaling the S/N values derived from the results in Fig. 4 at the 40 ns region. With the assumption that the signal from a scattered-neutron is 70%, that of a primary-neutron, the estimated S/N in intensity at the 40 ns region is 8.75 for Ny of 10^7 and 0.875 for Ny of 10^8. As such, it can be deduced that this design is feasible to be used for initial testing at the FIREX facility. In the future, the objective is to develop smaller structure scintillation fiber array designs using this new material to further improve the dynamic range of future scintillation systems.

5. Summary

We have succeeded in developing a fast-response ^{6}Li glass scintillator material suitable for scattered-neutron diagnostics of the inertial confinement fusion plasma experiments at the ILE, Osaka University. The scintillator exhibited a decay time of 5.4 ns for neutron excitation. Fusion-originated neutrons were successfully observed using the GEKKO XII laser at the Institute of Laser Engineering, Osaka University, and these results could pave the way for more accurate scattered-neutron detector systems for DD fusion experiments. In this regard, actual detector design and development for this new material, has commenced.

Acknowledgement

This work was supported by the Japan Society for the Promotion of Science under the contracts of Grant-in-Aid for Scientific Research (S) No. 18106016, Grant-in-Aid on Priority Area No. 16082204, Open Advanced Research Facilities Initiative, and Research Fellowship for Young Scientists No. 3273.

References

[1] Sangster T. C, *et. al.*, 2008, *Physic. Rev. Lett.* **100** 185006
[2] Izumi N, *et. al.*, (2003), *Rev. Sci. Instrum.* **74**, 1722–1725.
[3] Fairley E. J. *et. al.*, 1978, *Nucl. Instrum. Methods*, **150**, 159–163.
[4] Arikawa Y, *et. al.*, 2009, *Rev. Sci. Instrum.* **80**, 113504.
[5] Spowart A. R, 1976, *Nucl. Instrum.* Methods **135**, 441–453.
[6] L. Disdier, *et. al.*, 2003, Rev. Sci. Instrum. **74**, 3
[7] Kodama R, *et. al.*,2001, *Nature*, **412**, 798,
[8] Kodama R, *et. al.*,2002, *Nature,* **418**, 933

Update on recent results of LIL experiments

A. Casner[1], E. Alozy[1], S. Brygoo[1], M. Casanova[1], L. Chauvel[2], J. MChevalier[2], T. Chies[2], P. Combis[1], R. Courchignoux[2], S. Darbon[1], C. Debonnel[1], S. Depierreux[1], A. Duval[1], C. Esnault[1], J.E. Fuchs[1], H. Graillot[2], O. Henry[2], G. Huser[1], JP Jadaud[1], C. Labaune[3], A. Geille[2], P. Le Gourrierec[2], P. Loiseau[1], O. Lutz[2], C. Meyer[2], D T Michel[1,3], J L Miquel[1], M. Naudy[1], P. Nicolaï [4], D. Raffestin[2], P. Renaudin[1], C. Reverdin[1], Ph. Romary[2], C. Rousseaux[1], G. Thiell[2], V. Tikhonchuk[4], J.L. Ulmer[1], B. Villette[1], L. Videau[1] and R. Wröbel[1]

[1]CEA, DAM, DIF, F-91297 Arpajon, France.
[2]CEA, CESTA, F-33114 Le Barp, France.
[3]LULI, Ecole Polytechnique, F-91128 Palaiseau, France.
[4]CELIA, Université Bordeaux I, 351 cours de la Libération, F-33405 Talence, France.

E-mail: alexis.casner@cea.fr

Abstract. We present an overview of recent experiments fielded on the LIL facility. A key issue for mega-joule class laser facilities is shrapnel fragment generation. A specific collector was therefore developed to capture debris in aerogel and dedicated shots were done to quantitatively evaluate target fragmentation phenomena. The LIL panel of transmitted and backscattered light diagnostics is well suited for laser-plasma interaction (LPI) experiments. This include gas-filled hohlraum configurations relevant to Indirect Drive ignition targets in order to test the specific LMJ longitudinal SSD technique, as well as LPI experiments with foam targets for laser beam smoothing in underdense plasma in the context of Direct Drive. After Visar commissioning a campaign was dedicated to boron Equation of State (EOS).

1. The LIL plasma diagnostics suite.

LIL commissioning as a quadruplet prototype for LMJ ended in 2002 (see Ref. [1] for laser characteristics). Even if the LIL is still partially used as an optical testbed the facility is now mainly devoted to laser-plasma experiments [2,3]. A suite of 13 plasma diagnostics is installed around the target chamber. x-ray imaging is performed with multi-Kirkpatrick-Baez (KB) gated imagers and KB streaked microscopes, duplicated in hard (1 – 5 keV) and soft (< 1 keV) versions [4]. Radiative temperature laws for hohlraum shots is monitored with DMX, a time-resolved absolutely calibrated broadbrand spectrometer [5]. High resolution spectrometers [6] allow detailed spectroscopic studies [2]. In the x-ray and UV domains several diagnostics are used to get a complete energy balance. In

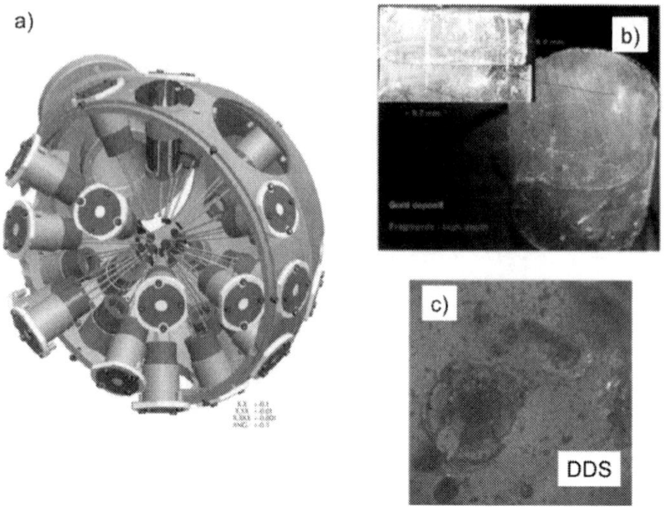

Figure 1: (a) Sketch of the capturing sphere used on fragmentation shots. (b) and (c) Examples of aerogel catcher and a glass sample (thickness 2 mm) severely damaged after a shot on a gold halfraum surrounded by aluminium rings.

particular the backscattered light inside the focusing aperture is collected with a elliptical Spectralon diffuser and stimulated Raman scattering (SRS) and stimulated Brillouin scattering (SRS) are analyzed with 2 spectrometers coupled to streak cameras. An exhaustive description of the LIL diagnostics can be found in [7,8].

2. Target fragmentation experiments.

Shrapnels and debris mitigation procedures are primordial to minimize the operations costs of MJ class laser facilities [9]. Specific shots were therefore done in order to benchmark integrated simulations of targets fragmentation. A capturing sphere (see Fig. 1(a)) allows varagel collectors [10] or various material to surround the irradiated target. As an example the fragmentation of a gold halfraum with aluminium rings induces severe damage on a 2 mm thick glass sample (similar to a LMJ Disposable Debris Shield) and many big solid fragments (with a size > 100 μm and high penetration depth) were collected (see Fig. 1(b) and (c)). The aluminium rings were in fact designed to work as shrapnels generators. An automatic detection algorithm for the number and size of the collected fragments will allow a quantitative comparison with target fragmentation simulations.

3. Laser-plasma interaction experiments.

3.1. LPI in gas-filled hohlraums

Due to LIL/LMJ focusing by gratings [1] the speckle pattern essentially moves along the beam direction and SSD smoothing is mainly longitudinal. To test its efficiency preliminary LPI experiments were performed with gas-filled targets (long open tube and closed halfraum) which emulate internal and external LMJ beams propagation (Fig. 2a)). The gas-filling pressure (1 atm neopentane) was also chosen to create plasma parameters relevant to LMJ target design: $n_e \sim 0.1\ n_C$ and $T_e / T_i \sim 2$. The LIL pulse shaping capability was harnessed to create a 6 ns pulse divided in a heating part (3 ns long) and an interaction part (2 ns long) at moderate laser intensity ($I \sim 6\ 10^{14}$ W. cm^{-2}).

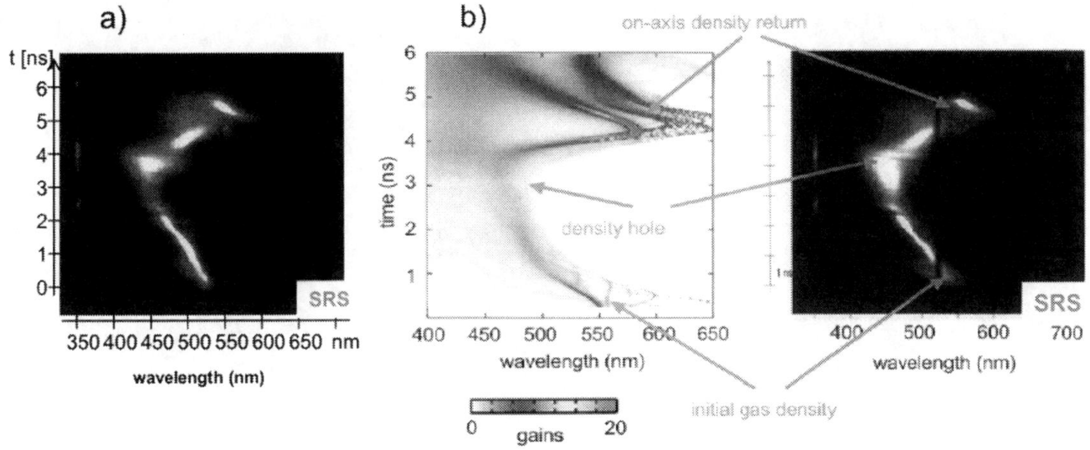

Figure 2: (a) Experimental Raman spectrum for a closed gas-filled halfraum used to emulate LMJ external cone propagation. (b) Comparison of experimental and postprocessed Raman spectrum.

Experimental results were highly reproducible for similar shots as illustrated by the experimental spectrum on Fig. 2 (a) and (b) and the measured reflectivities (SBS,SRS) were only moderate, less than 10 %. Even if the halfraum hydrodynamics is complex experimental SRS spectra compare well with FCI2 simulations postprocessed with the linear gain code PIRANAH [11] (see (Fig.2b)).

3.2. LPI in the context of Direct Drive: beam smoothing in underdense plasma

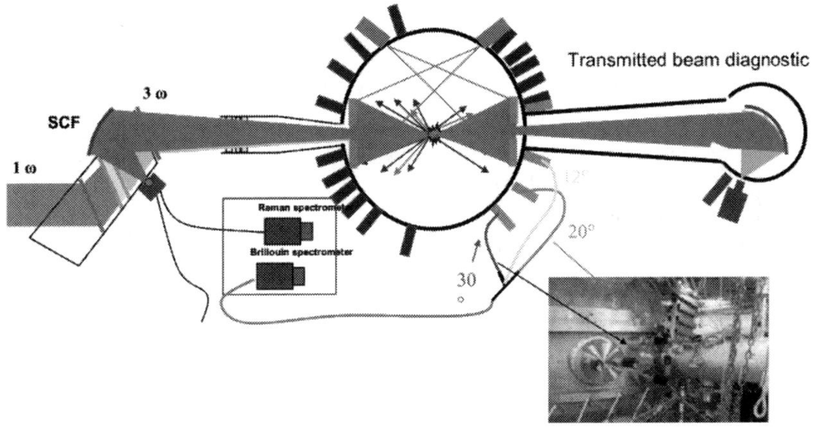

Figure 3: Scheme of the diagnostic designed for time-resolved spectral measurements of the near forward transmitted light.

LPI experiments performed on LIL by the academic community in the context of Direct Drive demonstrated laser beam smoothing in underdense plasma in the multikilojoule regime [3]. A following of this campaign concentrates on time-resolved spectral measurements of the near forward transmitted light [12]. Measurements at 2 different angles are time-multiplexed and transported to the SBS spectrometer through 20m of low-losses optical fiber (see Fig. 3). The measured spectral red-shift of the transmitted light through the hot (T_e ~2 keV) millimeter-long underdense plasma is in accordance with theoretical predictions. The effect of the SSD was also tested. At maximum SSD bandwidth SBS and SRS losses were reduced by a factor of ~2. This validates the efficiency of LMJ longitudinal SSD in this regime of moderate laser intensities ($I < 6\ 10^{14}$ W. cm^{-2}).

4. Equation of state experiments

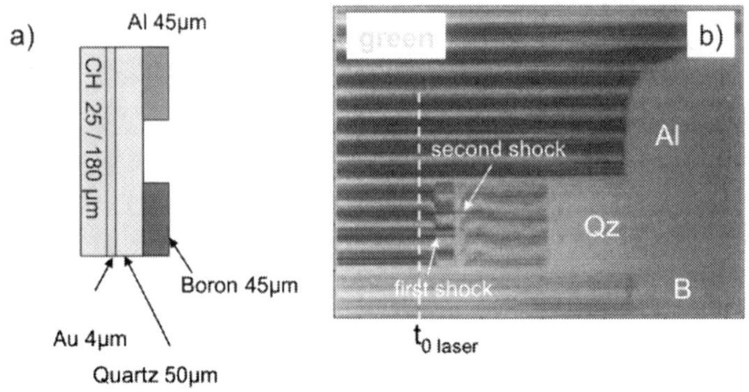

Figure 4: (a) Schematic diagram of the target used for Boron EOS experiments. (b) Visar interferogram (2 ω) obtained on a Indirect Drive shot.

A VISAR associated with a streak optical pyrometer was commissioned in 2008 and boron Equation of State (EOS) experiments were then fielded either in DD or ID. CH was used for pusher and quartz and aluminium were the reference materials (see Fig. 4(a) for target description). A typical interferogram for ID shots is shown on Fig. 4b). 2 shocks are visible in the quartz , the first shock being due to the heating of the 4 μm gold layer by M-band radiation [13]. EOS points up to 4 Mbar were obtained in ID. For DD shots preheat in quartz was observed and data analysis is underway. Dedicated phase plates will be furbished for future DD EOS campaigns.

5. Conclusion

Well-diagnosed plasma experiments are conducted on the LIL facility. Quantitative results are obtained either in the framework of LMJ program , academic DD approach or HED physics. In the future years a PW beam now under construction [14] will be focalized into the target chamber and will extend the experimental capabilities of the facility.

References

[1] Di Nicola J.M. et al., *J. Phys. IV,* **133**, 595 (2006).
[2] Schurtz G. et al., *Physical Review Letters* **98,** 095002 (2007).
[3] Depierreux S et al., *Physical Review Letters* **102,** 105005 (2009).
[4] Rosch et al., *Rev. Sci. Instrum.* **78**, 033704 (2004).
[5] Bourgade J. L. et al., *Rev. Sci. Instrum.* **72**, 1173 (2001).
[6] Reverdin C. et al., *Rev. Sci. Instrum.* **77**, 10, 033704 (2004).
[7] Videau L. et al., *Plasma Phys. Controlled Fusion* **50**, 124017 (2007).
[8] LeBreton J. P. et al., *Rev. Sci. Instrum.* **78**, 033704 (2004).
[9] Eder D. et al., *J. Phys. Conf. Series* **112,** 032023 (2008).
[10] Lescoute E. et al., *APS Shock compression of condensed matter* (2009).
[11] Loiseau P. et al, *37th Anomalous Absorption Conference* (2007).
[12] Labaune C et al., this conference.
[13] Brygoo S. et al., *APS Shock compression of condensed matter* (2009).
[14] Blanchot N. et al., this conference.

Journal of Physics: Conference Series **244** (2010) 032043 doi:10.1088/1742-6596/244/3/032043

Sensitivity of Capsule Implosion Symmetry due to Laser Beam Imbalance in a Scale 0.2 Hot Hohlraum at OMEGA

N.D. Delamater, D.C. Wilson, G.A. Kyrala, A. Seifter, N.M. Hoffman, Los Alamos National Laboratory, P.O. Box 1663, MS-T085, Los Alamos, NM 87545
ndd@lanl.gov

Abstract: Results are shown from recent experiments at the Omega laser facility, using 40 Omega beams driving the hohlraum with 3 cones from each side and up to 19.5 kJ of laser energy. Beam phasing is achieved by decreasing the energy separately in each of the three cones, by 3 kJ, for a total drive energy of 16.5 kJ. This results in a more asymmetric drive, which will vary the shape of the imploded symmetry capsule core from round to oblate or prolate in a systematic and controlled manner. These results show the sensitivity of capsule implosion symmetry for implosions in "high temperature" (275 eV) hohlraums at Omega. Dante measurements confirmed the predicted peak drive temperatures of 275 eV. Implosion core time dependent x-ray images were obtained from framing camera data which show the expected change in symmetry due to beam imbalance and which also agree well with post processed hydro code calculations.

1. Introduction

Symmetry capsules are planned as a diagnostic of implosion symmetry at varying times during the NIF drive[1]. A suitably designed symmetry capsule samples the drive symmetry during the implosion, whose duration varies for symmetry capsules of different shell thickness. Our capsules use Ge-doped plastic shells with shell thickness varying from 25 μm to 55 μm and filled with 36 atm d-^{3}He gas. We present data from Omega experiments using symmetry capsule implosions in vacuum gold hohlraums 1900x1200 μm, and 70% laser entrance hole, which is approximately a 0.2 NIF scale ignition hohlraum and reaches temperatures of 275 eV similar to those during the NIF drive. Implosion core time dependent x-ray images were obtained which show the expected change in symmetry due to beam balance; and also agree well with post-processed 2D Lagrangian rad-hydro code calculations. Limb brightened images were also seen for some of the capsules, also in agreement with calculation. The 0.2 scale hot hohlraum gives measured hohlraum drive temperatures from the Dante diagnostic in excellent agreement with simulation.

2. Hohlraum Energetics

Previously[2], we described the results of this experiment which were related to the proton spectroscopy aspect, where proton spectra is measured from the capsule implosion and related to the measurement of the capsule unablated shell mass for a measurement of the ρr of the shell. In this paper, we describe how symmetry is varied in the capsule implosion in a controlled manner, which agrees with calculation. The experiment controls the implosion symmetry by varying the relative energies in each of the laser cones. The nominal (maximum) energies are achieved when the 21 °, 42 ° and 59 ° laser cones each have energies of 490 J/beam, and for the 40 beams into the hohlraum give an energy of about 19.6 kJ. The pointing was optimized to give a round implosion with full energy.

The full energy shots gave peak hohlraum drive temperatures of about 275 eV and round implosions. In order to change the implosion drive symmetry, we reduced systematically the overall energy in each of the cones. For example, with the 59 ° cone, we applied 300 J/beam and 490 J/beam elsewhere for a total of 16.6 kJ. For the 42 ° cone, we applied 200 J/beam and 490 J/beam for the other two cones for a total energy of 16.6 kJ. For the 21° cone, we applied 200 J/beam and 490 J/beam for the other cones, giving a total energy of 16.6 kJ. For each configuration we measured the hohlraum drive with Dante and found a temperature drop of about 15 eV with the reduced laser energies.

3. Symmetry measurements

In the experiment, we obtained implosion symmetry data for implosions of the symmetry capsules with shell thicknesses from 25 μm to 55 μm for both full energy and reduced energy configurations. The implosion core x-ray image was observed using framing camera data filtered with 10 μm Be, so that the energy range is about 1-2 keV x-rays. We obtained images on a number of shots, and show here some results using symmetry capsules of shell thickness 25, 35, 45 and 55 μm. Figure 1 gives the x-ray images for full energy and reduced energy in the 21° cone for capsules of 35 μm shell thickness. The corresponding simulated x-ray image from post-processing the hydro calculation is also shown in Figure 1. The simulated images were taken at peak x-ray emission. The agreement in shape and size is good. Note that the implosion core goes from round with full energy to slightly oblate with reduced energy. Also note that a faint limb brightening is seen in both the data and simulations. This is due to the Ge-doping in the CH shell. The limb brightened image aids in the image analysis in determining the accurate shape of the implosion core. The images from both of the reduced energy shots show qualitatively a less round image as can be seen from the square contours in the images which indicate P4 asymmetry. The image shape can be described by the eccentricity, e=a/b, where a and b are the major and minor axis diameters. The a/b from the simulated image is 1.0, 0.86 and 0.96 for the full energy, lower 21 and lower 59 ° cones. Thus the effect of beam asymmetry is weak but measureable for the 35 μm shell thickness.

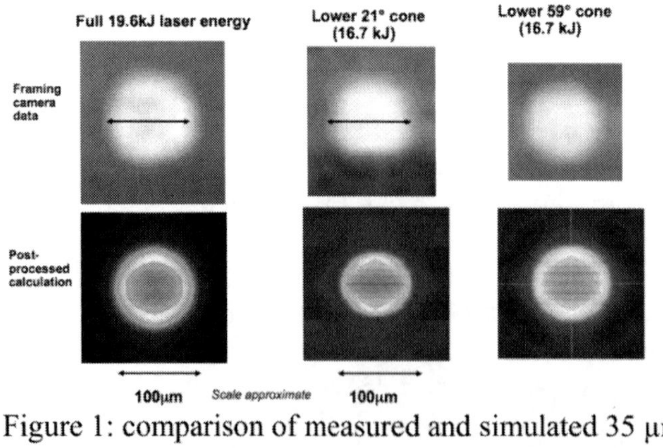

Figure 1: comparison of measured and simulated 35 μm capsule implosion x-ray images.

In Figure 2 we show similar comparison of measured and simulated 25 μm thick shell imploded core images. Limb brightening is also apparent for these images and we can see a slight shape change from full energy shot to reduced energy in the 59 ° cone. The simulated images have an a/b of 1.08 and 1.0 for the full energy and lower 59 ° cone shots respectively. The effect of beam imbalance is slight for the 25 μm shell thickness.

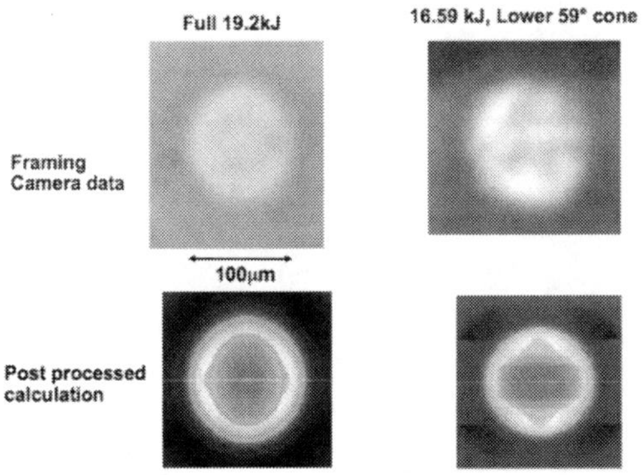

Figure 2: comparison of measured and simulated 25 μm capsule implosion x-ray images.

In Figure 3, results from the 45 μm shell implosions with reduced laser energy in the 42 and 59 ° cones are shown. Both framing camera data at peak emission and post-processed simulations at peak emission are shown. The simulated images have a/b of 0.84 and 0.94 for the lower 42 and lower 59 ° cone shots respectively. The framing camera data shows more of a symmetry change than calculated. Thus the effect of beam imbalance is significant for the capsules with shell thickness of 45 μm.

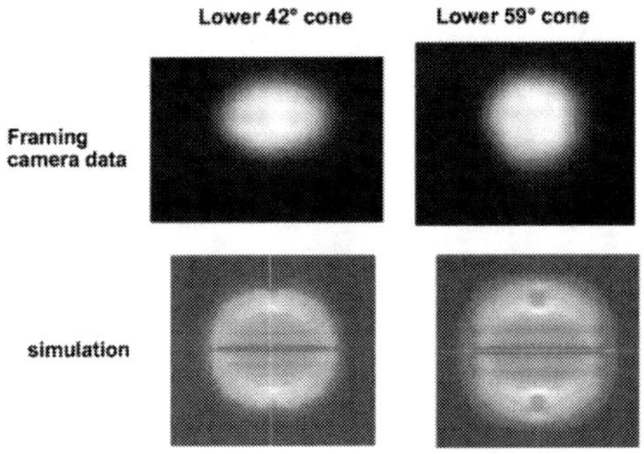

Figure 3: comparison of measured and simulated 45 μm capsule implosion x-ray images.

In Figure 4, results from the implosion of the thick 55 μm shell capsules are shown. These capsules generate the higher degree of convergence than the other thinner shell capsules, allowing instability growth to disrupt the symmetry of the implosion. The framing camera data showed a strong asymmetry developing during the implosion with the reduced energy in the 21 ° cone, and this was also well modeled in the post-processed implosion simulation. Thus the effect of a beam imbalance is significant for the thickest shell targets with 55 μm shell thickness. The thick shell targets in general have the longest implosion times, larger capsule convergence and are subject to turbulent mix and shell breakup during the implosions. Thus these targets would be expected to be most sensitive to the imposed laser beam imbalances.

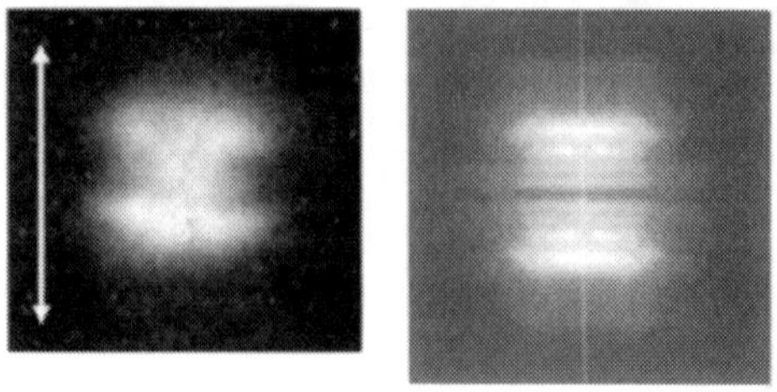

Figure 4: comparison of measured and simulated 55 μm capsule implosion x-ray images showing strong asymmetry developing in the implosion.

4. Conclusion

In conclusion, we have shown results from Omega experiments on symmetry in high temperature hohlraums, which are relevant for further NIF experiments. We show good agreement for hohlraum energetics between data and simulation[2]. We show qualitative agreement for implosion symmetry between data and simulation; as indicated by size and shape of the implosion core, so that we can "tune" the shape of the implosion in agreement with modeling simply by varying the energy in each of the laser cones entering the hohlraum. The capsules with 45 and 55 μm shell thickness are especially sensitive to beam imbalance. The effect of beam imbalance on the 25 and 35 μm shell capsules is less dramatic, though measureable.

References:
[1] J.D. Lindl, et.al., Physics of Plasmas, **11**, 339-491 (2004).
[2] N. Delamater, D.C. Wilson, G.A. Kyrala, et.al., Rev. Sci. Instr., **79**, 10E526 (2008).

X-ray Thomson Scattering for measuring Dense Beryllium Plasma Collisionality

T. Döppner[1], C. Fortmann[1,2], P.F. Davis[3], A.L. Kritcher[1,3], O.L. Landen[1], H.J. Lee[4], R. Redmer[5], S.P. Regan[6], S.H. Glenzer[1]

[1]Lawrence Livermore National Laboratory, P.O. Box 808, Livermore, CA 94551, USA
[2]University of California Los Angeles, Department of Physics, Los Angeles, USA
[3]University of California Berkeley, Berkeley, USA
[4]Stanford Linear Accelerator Center, Menlo Park, USA
[5]University of Rostock, Department of Physics, Rostock, Germany
[6]Laboratory for Laser Energetics, University of Rochester, Rochester, USA

E-mail: doeppner1@llnl.gov

Abstract. We are developing a target platform that utilizes short-pulse (10 ps) generated hot electrons (~200 keV) to isochorically heat solid density beryllium up to temperatures of several 10 eV. We use x-ray Thomson scattering to characterize the plasma conditions. X-rays from a Cl Ly-α line source at 2.96 keV are scattered off the plasma in forward direction where the inelastically scattered signal is sensitive to plasma oscillations. Besides Landau-damping the strong energy down-shifted plasmon signal is also broadened by electron-ion collisions which, in turn, allows to infer the collision rate and thus the conductivity in these plasmas. A precise knowledge of the collisionality in the parameter regime we are aiming at with these experiments is important to correctly model the conditions encountered during capsule implosions at the National Ignition Facility.

1. Introduction

Accurate characterization of warm dense matter is important for high energy density physics experiments [1], for inertial confinement fusion (ICF) [2], and for astrophysical studies. Plasmas in this regime are typically at solid density and above with temperatures on the order of a few eV to a few tens of eV. One plasma property of particular interest is the thermal conductivity. A precise knowledge of this quantity is important, for example, to correctly model the hydrodynamic instability sensitivity of capsule implosions on the National Ignition Facility [3, 4]. So far, benchmarking of models that calculate the conductivity for beryllium and deuterium is lacking. Predictions of various models for conditions relevant for ICF vary by up to a factor of six [5, 6, 7, 8].

X-ray Thomson scattering (cf. Ref. [9] for a recent review) has been developed to access plasmas at solid density and greater, approaching electron densities of $n_e \approx 10^{24}$ cm^{-3}. The characterization of isochorically heated [10, 11] and laser shock-compressed matter [1, 12] has shown the capability to reliably measure electron temperature and density. This also enabled the validation of models that calculate ion-ion structure factors [13] and radiation hydrodynamic codes, by testing shock-timing predictions. Here, we will outline how x-ray Thomson scattering can be utilized to measure the collisionality in warm dense matter.

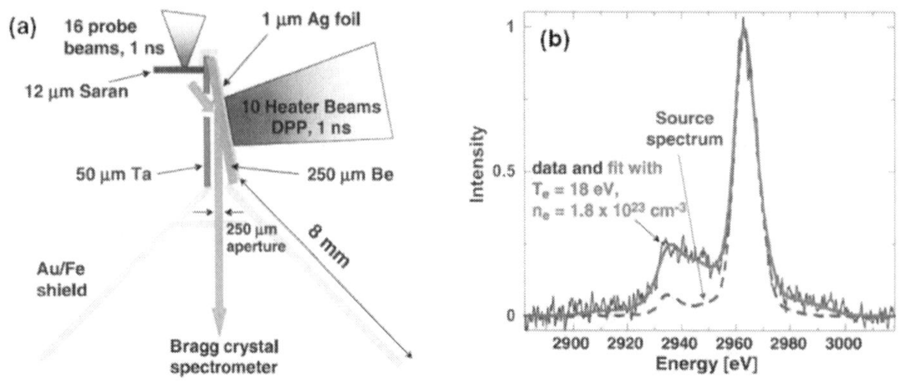

Figure 1. (a) Target design; (b) Measured x-ray Thomson scattering spectrum of isochorically heated beryllium. The best fit is obtained with $n_e = 1.8 \times 10^{23}$ cm^{-3} and $T_e = 18$ eV (solid line). The source spectrum is show for comparison (dashed line). Adapated from Ref. [14].

To achieve this, the experiment has to be in the collective scattering regime where x-rays are scattered from plasmons, i.e. electron plasma oscillations. The collectivity parameter, defined as $\alpha = 1/(k\lambda_{scr})$ with λ_{scr} the screening length, and $k = 4\pi(E_0/hc)\sin(\theta/2)$ the momentum transfer in the scattering process, enables to discriminate between the collective ($\alpha > 1$) and the non-collective scattering regimes ($\alpha < 1$). α mainly depends on the x-ray probe energy E_0, the scattering angle θ, and the plasma conditions. The inelastic collective scattering signal consists of two plasmon features, up- and down-shifted by $\hbar\omega_{res}$ from the quasi-elastic Rayleigh signal. The up-shifted plasmon is reduced by a factor of $\exp\{-(\hbar\omega_{res})/(k_B T_e)\}$ compared to the down-shifted feature as given by detailed balance [14] which is a powerful tool to measure T_e since it is based on first principles. In addition, the plasmon shift $\hbar\omega_{res}$ depends on n_e and T_e which, for solid density Be, is described in Ref. [15].

2. Collective scattering experiment

First we describe a successful collective scattering experiment that achieved sufficiently high temperature plasmas allowing to observe an up-shifted plasmon signal. The experiment was performed at the Omega laser facility at the Laboratory for Laser Energetics (LLE) at the University of Rochester. Fig. 1a shows a schematic of the target and the laser beam configuration. To isochorically heat a 250 μm thick Be foil with Ag L-shell radiation (3.2 - 3.6 keV), 10 drive beams with a total energy of 4.7 kJ at 351 nm in a 1 ns pulse width are incident on a 1 μm silver foil glued to the Be. We use distributed phase plates (DPP, type SG4) to achieve a smooth beam spot with a diameter of 800 μm yielding a drive intensity of 7×10^{14} W cm^{-2}. To generate the Cl Ly-α x-ray probe at 2.96 keV, 16 beams with a total energy of 7.4 kJ at 351 nm are focused to a 150 μm spot on a 12 μm Saran foil. The x–ray photons are collimated by a 50 μm thick, rectangular Ta pinhole in front of the Be which defines the scattering angle to 40°.

The plasma conditions are probed ~0.5 ns after the end of the heater beams when the highest plasma temperatures are predicted. Because of the 250 μm mean free path of 3 keV x-rays in Be, an 11° tilt is introduced so that scattered x-rays can exit at the rear surface with minimal reabsorption. The scattered signal is collected using a high–efficiency Bragg spectrometer with a highly oriented pyrolytic graphite (HOPG) crystal. For detection we use an x–ray framing camera with a 180 ps gate. The measured signal is read out by a CCD camera fiber coupled to the framing camera. Au shields are used to block the direct line of sight of the spectrometer to the plasmas generated by the heater and the probe beams.

The measured scattering spectrum is shown in Fig. 1b. Note that the source spectrum (dashed line) shows a red–shifted satellite in addition to the main Cl Ly-α line. However, the down-shifted inelastically scattered plasmon signal is much stronger than this naturally occurring He–like satellite. In addition to the down–shifted plasmon signal a clear inelastically scattered, up–shifted signal is observed with a strength that is governed by the detailed balance relation. The plasmon signals are quite broad because the collectivity parameter $\alpha = 1.22$ is close to the transition to non-collective scattering.

To infer density and temperature, synthetic scattering spectra were generated and fit to the experimental data. To model the spectra we applied the random phase approximation (RPA) for the electron structure factor [16], and included the measured instrument function. The best agreement between the measured scattering signal was obtained for $n_e = 1.8\times10^{23}$ cm^{-3} and an electron temperature $T_e = 18$ eV (cf. Fig. 1b). A sensitivity analysis yields error bars of 15% and 20% for T_e and n_e, respectively [14].

3. New target platform to measure Be collisionality

There are two drawbacks to the experiment described in the previous section. First, isochoric heating based on x-ray converter foils limits the attainable plasma temperatures. Secondly, the heater beams launch a shock wave into the foil which can lead to density inhomogeneities. In the experiment described above, probing the latter was avoided by using a sufficiently thick Be foil, which in turn limits the heating efficiency at the rear surface. In addition, due to optics damage thresholds the heater pulse width has to be on order of 1 ns to maximize pulse energy and peak power and thus plasma heating. However, this leads to significant undesirable rarefaction at the rear surface where the plasma conditions are probed [14].

In order to mitigate these issues we are developing a new target platform that utilizes fast electrons produced by powerful short-pulse laser beams to isochorically heat matter. If laser light interacts with a solid at intensities of 10^{18} W/cm^2 and above, at the critical surface up to 30% of the laser pulse energy is converted into high energy electrons. These are very efficiently (close to 100%) absorbed within the target, which is heated on a time scale of order 10 ps [17].

The LLE in Rochester recently commissioned the OmegaEP laser, capable of delivering up to 1 kJ in 10 ps at 1054 nm on target that can be combined with additional laser beams to create the x-ray probe for Thomson scattering. We are proposing an experiment that employs the short pulse beam to heat a 200 x 200 x 250 μm^3 Be cube, and up to 14 long pulse beams at 351 nm to create the Cl Ly-α x-ray probe at 2.96 keV. 250 J pulse energy within 10 ps and at best focus (50 μm) yields $\sim10^{18}$ W/cm^2 on target, creating fast electrons with $T_{hot} \approx 200$ keV that are capable of penetrating 250 μm Be. Assuming a conversion efficiency of 20% into hot electrons we expect to heat the Be cube to about 35 eV. The front surface will be much hotter at several 100 eV, and has to be shielded from the scattering experiment.

To record the scattering data, we have developed a highly-efficient spectrometer utilizing a cylindrical curved HOPG crystal in von-Hamos geometry with an image plate detector. The time resolution is determined by the 200 ps probe-beam pulse length. The scattering signal will compete with bremsstrahlung created by the fast electrons circulating in the Be, and a general high energy background due to the short-pulse environment. While we estimate the bremsstrahlung at most on same order of magnitude as the scattering signal, the latter is hard to quantify. Since we are expecting a scattering signal of ~10,000 photons per resolution element, even a tenfold stronger background would still give an acceptable signal-to-noise ratio of 10.

In order for the width of the plasmon signal to be sensitive to collisions, the scattering has to be well in the collective regime, i.e. $\alpha > 1.7$. To achieve this for given plasma parameters ($T_e \approx 20$ eV, $n_e \approx 2.8 \times 10^{23}$ cm^{-3}) and given x-ray probe energy (2.96 keV), we will observe the signal at scattering angles of $\theta \leq 30°$. To account for effects from electron-ion collisions, we use the Born-Mermin approximation (BMA), that is the dielectric function is modified via the

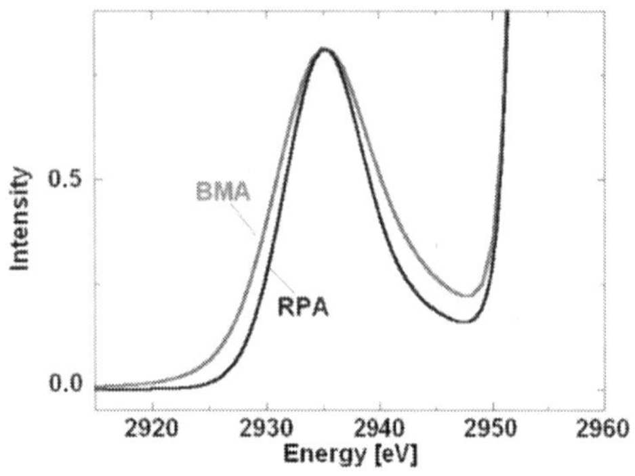

Figure 2. Simulated scattering spectra with (BMA) and without (RPA) collisions for solid Be ($n_e = 2.8\times10^{23}$ cm^{-3}) isochorically heated to 20 eV, assuming a 2.96 keV x-ray probe with 7 eV instrument function, and a scattering angle of 25°, resulting in a collectivity parameter α = 2.23.

Mermin approach, and the collision frequency is calculated in Born-approximation, for details see Refs. [18, 19]. Fig. 2 shows synthetic scattering spectra obtained with BMA (with collisions) and RPA (no collisions). For clarity, only the strong down-shifted plasmon signal is shown. Its width is broadened by 30% when collisions are included.

In conclusion, the proposed experiment will enable to benchmark models that calculate collision rates. It will further allow to infer the conductivity in solid denisty Be and link it to electron density and plasma temperature which can be extracted from the same scattering data with high accuracy. In addition to validating plasma physics models, precise conductivity measurements are important for improving our understanding on how to correctly simulate the conditions encountered during capsule implosions at the National Ignition Facility.

Acknowledgments
This work was performed under the auspices of the U.S. Department of Energy by the Lawrence Livermore National Laboratory, through the Institute for Laser Science and Applications, under contract DE-AC52-07NA27344. The authors also acknowledge support from Laboratory Directed Research and Development Grants No. 08-LW-004 and 08-ERI-002. C.F. and R.R. were supported by the Deutsche Forschungsgemeinschaft via the Sonderforschungsbereich 652.

References
[1] Kritcher A L, Neumayer P, Castor J et al. 2008 *Science* **322** 69
[2] Lindl J D, Amendt P, Berger R L et al. 2004 *Phys. Plasmas* **11** 339
[3] Hammel B A, Edwards M J, Haan S W et al. 2008 *J. Phys.: Conf. Series* **112** 022007
[4] Hammel B A, Haan S W, Clark D et al. 2009 *High Energy Dens. Phys.*, in print, doi: 10.1016/j.hedp.2009.12.005
[5] Lee Y T and More R M 1984 *Phys. Fluids* **27** 1273
[6] Chabrier G, Saumon D and Potekhin A Y 2006 *J. Phys. A: Math. Gen.* **39** 4411
[7] Sterne P A, Hansen S B, Wilson B G and Isaacs W A 2007 *High Energ. Dens. Phys.* **3** 278
[8] Recoules V, Lambert F, Decoster A, Canaud B and Clerouin J 2009 *Phys. Rev. Lett.* **102** 075002
[9] Glenzer S H and Redmer R 2009 *Rev. Mod. Phys.* **81** 1625
[10] Glenzer S H, Gregori G, Lee R W et al. 2003 *Phys. Rev. Lett.* **90** 175002
[11] Glenzer S H, Landen O L, Neumayer P et al. 2007 *Phys. Rev. Lett.* **98** 065002
[12] Lee H J, Neumayer P, Castor J et al. 2009 *Phys. Rev. Lett.* **102** 115001
[13] Kritcher A L, Neumayer P, Brown C et al. 2009 *Phys. Rev. Lett.* **103** 245004
[14] Döppner T, Landen O L, Lee H J et al. 2009 *High Energy Dens. Phys.* **5** 182
[15] Döppner T, Davis P F, Kritcher A L et al. 2009 *Proc. SPIE* **7451** 74510H
[16] Gregori G, Glenzer S H, Rozmus W, Lee R W and Landen O L 2003 *Phys. Rev. E* **67**
[17] Nilson P M, Theobald W, Myatt J F et al. 2009 *Phys. Rev. E* **79** 016406
[18] Höll A, Bornath T, Cao L et al. 2007 *High Energy Dens. Phys.* **3** 120
[19] Fortmann C, Bornath T, Redmer R et al. 2009 *Laser Part. Beams* **27** 311

COMPARISON OF TRIATURE DOPPLER VELOCIMETRY AND VISAR

Cenobio H. Gallegos,[1] Bruce Marshall,[2] Matthew Teel,[1] Vincent T. Romero,[1] Abel Diaz,[1] and Michael Berninger[1]

[1]National Security Technologies, LLC, Los Alamos Operations
182 East Gate Drive, Los Alamos, NM 87544, USA
[2]National Security Technologies, LLC, Special Technologies Laboratory
5520-B Ekwill Street, Santa Barbara, CA 93111, USA

E-mail: gallegch@nv.doe.gov

Abstract. Triature Photon Doppler Velocimetry (TDV) is an adaptation of Photonic Doppler Velocimetry (PDV) that rejects common-mode data noise after splitting PDV three ways, with each signal 120° out of phase from each other. Testing has demonstrated that the TDV also improves temporal resolution from the typical five-nanoseconds of PDV to a subnanosecond range. This paper compares the temporal response of TDV with that of PDV and VISAR [velocity interferometer system for any reflector] in an experiment with a subnanosecond (~120-picosecond rise time) shock source.

Laboratory tests were performed using a high-power laser on targets of copper and aluminum. A fast VISAR with a single-point PDV and a prototype TDV were used. A special probe that combined PDV, TDV, and fast VISAR made simultaneous velocity measurements. Breakout velocities of 1.3 km/second on copper and 2.5 km/second on aluminum were observed, where TDV resolved rise times of ~200 ps. This resolution was better than that of a fast VISAR, which can achieve ~500 ps temporal resolution. Test methods and results are presented.

1. Introduction

Photonic Doppler velocimetry (PDV) was introduced by Ted Strand of Lawrence Livermore National Laboratory. Data are taken by mixing the Doppler-shifted light from a dynamic target (shifted light) with the base laser light (unshifted light), producing a heterodyned frequency in GHz range. This frequency is directly proportional to the velocity of the target that can be extracted with a fast Fourier transform analysis (FFT). Because of FFT's inherent properties, the temporal resolution is 5 nanoseconds at best (requires a full cycle to extract a frequency).

Triature Doppler velocimetry (TDV) uses a quadrature analysis concept, using a portion of a cycle of three 120° out-of-phase, identical signals to extract a frequency. This improves the temporal resolution to better then 1 nanosecond.

The tests discussed here were performed with the Sandia National Laboratories fast VISAR system. Many experiments we conduct are light-starved; this condition limits the type of detectors available to experimenters that will achieve the maximum temporal resolution of VISAR systems. These comparison tests utilized avalanche photodiode (APD) detectors with a bandwidth from DC to 450 MHz and a sensitivity of 15kV/W; their temporal resolution is ≈800 picoseconds.

A PDV/VISAR probe was constructed, by NSTec, to focus a 1550 nm (PDV) single-mode fiber and a 532 nm (VISAR) 100-micron core fiber on the same plane with a magnification of 0.5. The focal points were ≈0.25mm apart, as shown in Figure 1.

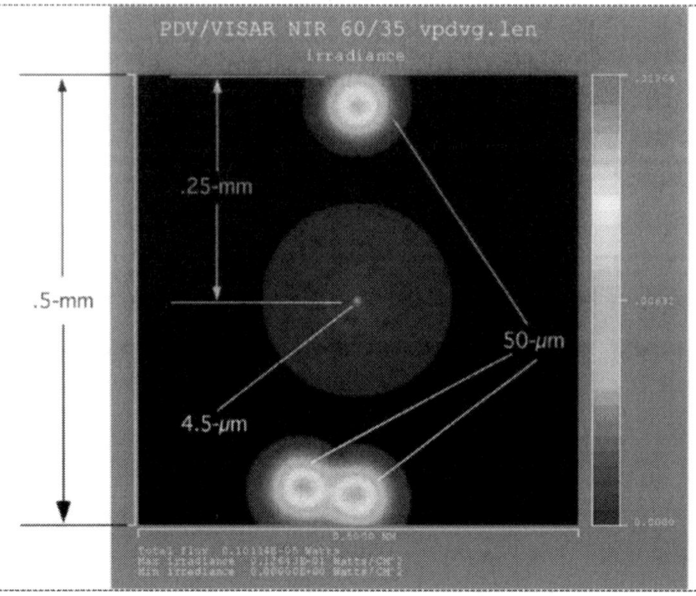

Figure 1. Irradiance of PDV/VISAR probe at the target. At the center is the PDV send/receive fiber. Surrounding it is the VISAR send/receive fiber. The probe magnification is 0.5.

Laser-induced shock was accomplished with a Continuum laser with a 120 mJ pulse at a 145 ps rise and 300 ps FWHM. The targets were 10-micron copper on glass or 10-micron aluminum on glass; the targets produced velocities of ≈1 km/sec and ≈2.5 km/sec, respectively. A two-lens imaging system was placed at the output of the laser to convert the Gaussian shape of the laser pulse to a flat-top pulse at the target, so that velocity was evenly distributed over a 1-mm-diameter area. This assured that velocities were identical for the PDV and VISAR, which enabled comparison. Figure 2 is a map of a series of experiments with the velocities measured using a PDV system shown in Table 1.

Initial calibrations of the PDV and VISAR were performed using a high-current calibration source that consisted of a 5 μF capacitor charged to 4.6 kV, generating approximately 14.5 kA. The current was focused onto a 0.25-inch copper strip, producing a shock of ≈3.5 km/sec with a slow rise time of about 0.5 μsec as seen in Figure 3.

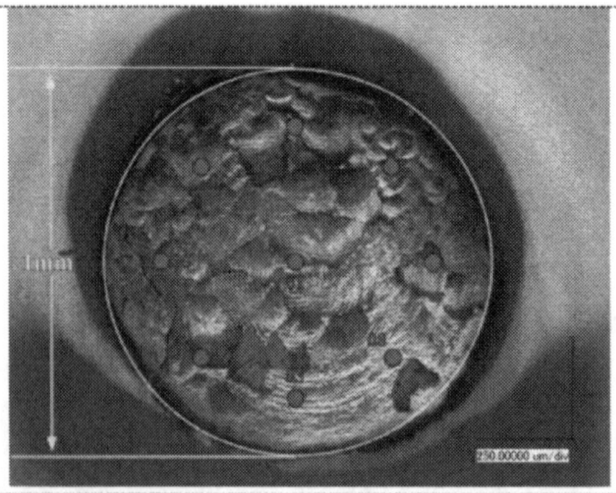

Figure 2. Velocity map of Table 1.

Table 1. Velocity at each position of Figure 2 (each position is a separate shot).

Copper Target				*Aluminum Target*			
Shot /Position	***Velocity***	***Shot /Position***	***Velocity***	***Shot /Position***	***Velocity***	***Shot /Position***	***Velocity***
1	379 m/sec	6	303 m/sec	1	1.1 km/sec	6	1.2 km/sec
2	378 m/sec	7	341 m/sec	2	1.3 km/sec	7	1.3 km/sec
3	303 m/sec	8	341 m/sec	3	1.5 km/sec	8	1.3 km/sec
4	379 m/sec	9	870 m/sec	4	1.4 km/sec	9	0.95 km/sec
5	455 m/sec	**Average**	**416 m/sec**	5	1.7 km/sec	**Average**	**1.31 km/sec**

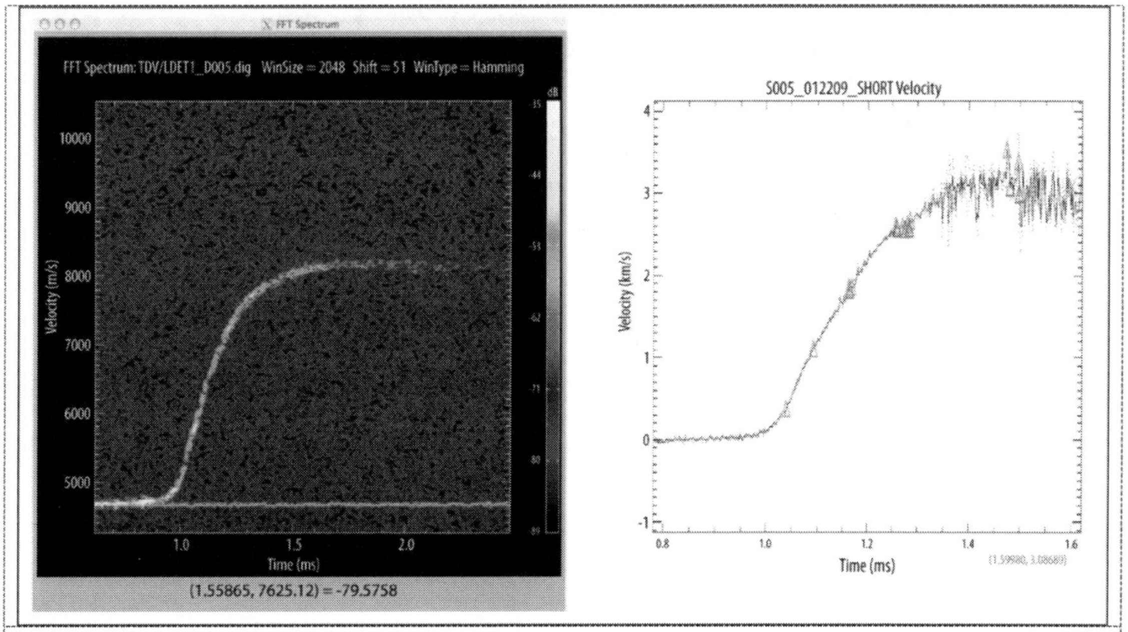

Figure 3. High-current calibration source PDV and VISAR data.

2. Results from TDV/VISAR comparison

The TDV system use MITEQ detectors that operate from 100 kHz to 20 GHz. We discovered data loss at the initial breakout at lower frequencies. This made it extremely difficult to determine the precise time of breakout and obtain a precise rise time for the data.

Optical up-conversion moves the data base line into the operating range of the detectors. This is accomplished by using a tunable laser and mixing it with the base laser, so producing a heterodyned signal that can be adjusted to the desired frequency. The mixed signal can be additive or subtractive; data are distributed above or below the base line, as shown in Figure 4.

Figure 5 is a comparison of the TDV (red) and the VISAR data (blue) at long and short time scales for an aluminum target. In the long time scale graph the data compare quite well. For the short time scale, the TDV data rise time was near 200 ps and the VISAR data at ~20 ps (one resolution element of the digitizer). The VISAR appears to have a fast rise time because the lower bandwidths miss fringes. In order to obtain the correct velocity, fringes are added by the analysis software. However, since it is impossible to know exactly where the fringes should be added in time, fringes are added in the same time period, and a false rise time results, but the correct end velocity is achieved.

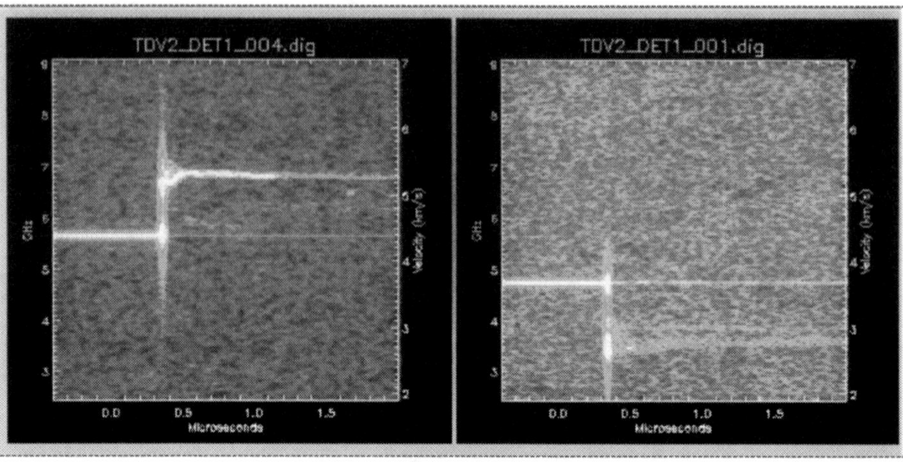

Figure 4. PDV up-converted data, where the tuneable laser is additive (left) and subtractive (right) of the base laser.

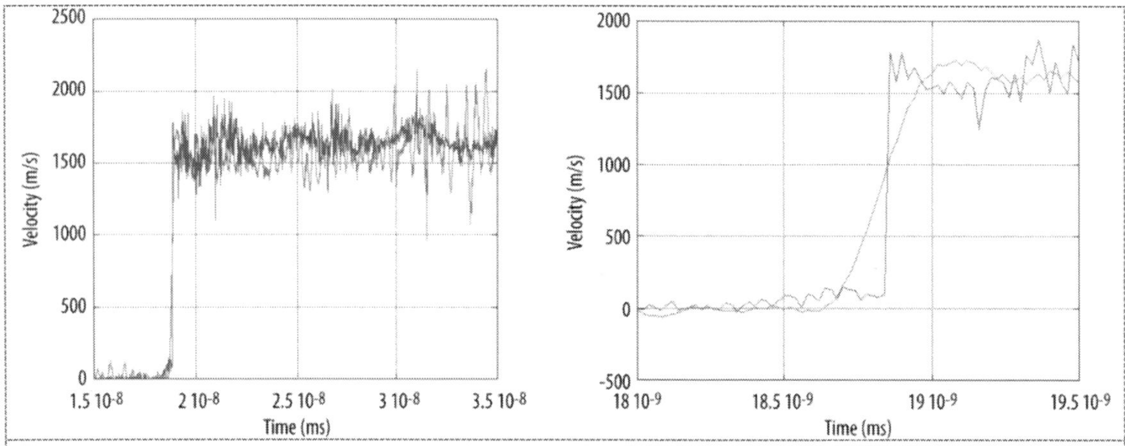

Figure 5. TDV (red) and VISAR data (blue) at a long and short time scale.

3. Conclusion

The Positive Light laser is a repeatable flat-top, high-power light source that produces fast velocities for testing velocimeter systems. Optical up-conversion works well with PDV and TDV to eliminate base line noise to determine accurate data breakout. VISAR resolution is limited by detector and recording bandwidth. TDV provides better than 0.5 nanosecond resolution.

This manuscript has been authored by National Security Technologies, LLC, under Contract No. DE-AC52-06NA25946 with the U.S. Department of Energy. The United States Government retains and the publisher, by accepting the article for publication, acknowledges that the United States Government retains a nonexclusive, paid-up, irrevocable, worldwide license to publish or reproduce the published form of this manuscript, or allow others to do so, for United States Government purposes.

Journal of Physics: Conference Series **244** (2010) 032046 doi:10.1088/1742-6596/244/3/032046

Target debris collection studies for inertial confinement fusion (ICF) experiments

G.P. Grim, T. N. Archuleta, P.A. Bradley, M.M. Fowler, A.C. Hayes, G. Jungman, A.W. Obst, R.S. Rundberg, D.J. Vieira, Y. Q. Wang, and J.B. Wilhelmy,

Los Alamos National Laboratory, PO Box 1663, Los Alamos, NM 87545, USA

E-mail: gpgrim@lanl.gov

Abstract. At the recently completed National Ignition Facility (NIF) at Lawrence Livermore National Laboratory, the initial set of diagnostics to be deployed are focused on measuring neutrons and γ's generated by $d(t,n)\alpha$ reactions in the imploded capsule. Although valuable for understanding pre-ignition experiments, this abbreviated diagnostic suite provides an incomplete picture of the plasma conditions obtained. Prompt radiochemical techniques, based on induced neutron and charged particle reactions within the imploded target, provide a novel and interesting new perspective. To enable these techniques requires the collection and assay of activated target material. In Nov. 2008, experiments were performed using the Omega Laser at the University of Rochester to study the efficiency of collecting debris from directly driven targets. Results from these experiments indicate that target debris was successfully collected, and the debris thermalization and transport scheme enhanced the debris collection up to 347% over direct collection.

1. Introduction

Prompt radiochemistry may be used in ICF experiments to study hydrodynamical mixing of the ablator and fuel[1, 2], as well as implosion dynamics, such as ablator velocity, through measurement of ^{9}Li production in Be targets, or ablator radius and thickness, by measuring suitably normalized production of (n,p) or (n,2n) reactions on ablator contaminants such as Si, or Cu. To enable these measurements, techniques must be developed to ensure efficient collection of activated target material from the implosion, as well as, in-situ assay of lifetimes as short as 100 ms.

Collection of activated debris products from ICF implosions has had some degree of success in the past[3, 4, 5, 6]. To lay a foundation for understanding these issues in the NIF environment, we have performed a set of debris collection experiments on the Omega laser at the University of Rochester, Rochester, NY. The goals were to gain understanding of the dynamics of debris generation and transport in an ICF experiment, and to develop strategies to optimize the collection of this debris. Below we present results from these experiments.

2. Target debris collection

In experiments performed using the Nova laser, it was observed that debris generated by the ablated and highly compressed target posses kinetic energies too great to collect on materials directly viewing the target[3]. At the Omega laser, this situation may be ameliorated by moving materials away from the target. At the NIF, it is anticipated that X-rays generated by laser target interactions for 1.1 MJ shots, will ablate the target chamber wall, at least along the direction of the laser entrance holes[7]. Thus, developing a strategy for addressing this phenomena is important to enable reliable debris collection in ICF experiments.

2.1. Debris thermalization and transport experiment

Fig. 1(a) illustrates the architecture of a debris collection experiment performed at the Omega laser in Nov. 2008. Debris striking a "reflector" foil, in close proximity to the target, will thermalize while ablating a sufficiently thick layer of the ablator, which will entrain the target material and transport it as a thermal vapor to a "collector" foil, elsewhere in the chamber. In the experiments fielded, a 5.03 cm diameter reflector foil was positioned 14 cm from target chamber center (TCC) and was contoured to have a 40.6 cm radius of curvature. A 10 cm collector opposed the reflector at a distance of 65 cm from TCC. To facilitate a number of different assay techniques, different collector materials were used and varied from shot to shot. These materials consisted of Si, Sn, Ta, and Au. The fractional solid angles subtended from TCC by the reflector and collectors was 0.81% and 0.15% of 4π.

Plastic CH targets were used in the experiments. All capsules were specified to have an outer diameter of 875 μm and a 15 μm wall thickness. The capsules were filled with 15 atmospheres of equimolar DT gas. To measure the collected debris we manufactured 90% enriched ^{13}CH targets and used the $^{13}C/^{12}C$ ($\delta^{13}C$) ratio as our primary signature. Direct assay of ^{13}C was possible for concentrations above 1E14 cm^{-2}. The target masses were nominally 39.4 μg for ^{13}C enriched targets, and 36.6 μg for the natural abundance targets. If the target debris were distributed uniformly, the CH concentrations impinging the reflector and collector surfaces should be 6.9E14 cm^{-2} and 3.2E13 cm^{-2} respectively.

To determine the materials present on the surface of the collectors, we used Rutherford backscattering spectrometry (RBS), nuclear reaction analysis (NRA) and time-of-flight secondary ion mass spectrometry (ToF-SIMS). To measure $\delta^{13}C$, the (d,p) reaction on C was used. Details of this analysis are being published in a separate manuscript, but the primary signature for C isotopes results from the characteristic protons energies produced in $^{12}C(d,p)^{13}C$ and $^{13}C(d,p)^{14}C$ reactions. These energies are 3.25 MeV for the $^{12}C(d,p)^{13}C$ reaction and 6.19 MeV for the $^{13}C(d,p)^{14}C$ reaction. Collector foils were exposed to a 4 mm^2, 1.4 MeV deuteron beam produced in the Ion Beam Materials Laboratory (IBML) at Los Alamos National Laboratory. The range of 1.4 MeV deuterons in Sn is approximately 10 microns.

Fig. 1(b) shows an example of the data collected from the NRA analysis. The data are proton spectra from the (d,p) reaction produced by 240 μC of deuteron beam impinging a 0.25 mm thick Sn foils. A 25 μm Al foil was placed in front of the surface barrier detector which further reduced the characteristic proton energies for ^{12}C and ^{13}C signatures to 2.7 MeV and 5.9 MeV, respectively. The relationship between the number of counts in each carbon peak, ^{12,13}A, and the respective concentrations, $^{12,13}n$, is given by the formula:

$$\frac{^{13}A}{^{12}A} = \frac{\sigma_{13}(d,p)}{\sigma_{12}(d,p)}\frac{^{13}n}{^{12}n}, \tag{1}$$

where, $\sigma_{12,13}(d,p)$, are the (d,p) reaction cross sections on the respective carbon isotopes. Due to the simultaneous measurement the geometric and intrinsic detection efficiencies are identical for both reactions. A "clean" Sn foil was used to determine the $\sigma_{13}(d,p)/\sigma_{12}(d,p)$ ratio by assuming the concentration ratio was equal to 0.011. Absolute concentration calibration was performed by implanting 40 keV ^{13}C ions from 1E14 cm^{-2} to 5E16 cm^{-2} on the surface of a

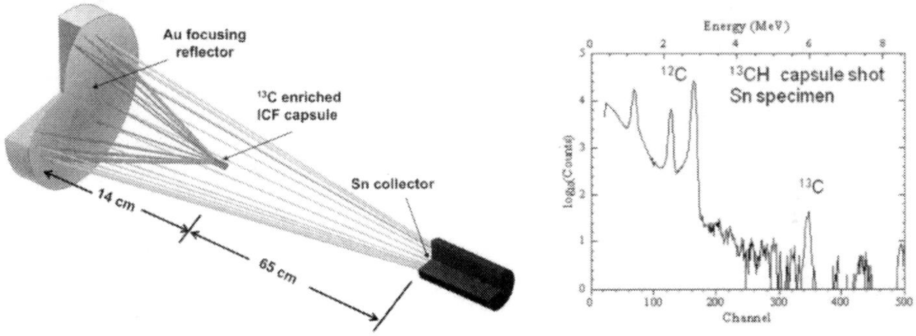

(a) Debris collector architecture for direct drive experiments.

(b) Sn foil exposed to three enriched ^{13}C targets.

Figure 1. Sketch of the target debris collection scheme deployed on the Omega laser facility and the NRA analysis of a Sn foil. A 5.03 cm diameter Au "reflector" foil, with 40.6 cm radius of curvature is located 14 cm from the target. The center radial of the foil points through target chamber center to a "collector" foil located 65 cm from the target. Proton spectra from the (d, p) reaction produced by 240 μC of beam on a Sn foil exposed to three successive enriched ^{13}CH target implosions.

clean Sn foil. The range of 40 keV C ions in Sn is nominally ~20 nm, thus to determine potential contributions from bulk carbon by the deutron beam, (d, p), measurements were performed on the back surface of the foils and used to correct for this effect. The (d, p) analysis technique simultaneously measures both ^{12}C and ^{13}C, thereby allowing for a direct measurement of δ^{13}C after correction with the cross section ratio above. The calibration studies indicate that the practical ^{13}C detection limit for this technique is ~1E14 cm^{-2}, provided that particle energy interference from other light isotopes can be avoided.

All implosions were generated by direct drive, using 23 kJ of laser energy delivered in a 1 ns square pulse. The data in Fig. 1(b) show a clear ^{13}C signal is present in foil. Visual inspection of the pre and post shot foils shows a visibly significant presence of Au. Due to the proximity of the mass of Sn to that of Au, RBS analysis could not be used on this foil, but a Si wafer positioned 65 cm from TCC, and opposing a similar reflector foil, was observed to have a Au concentration of 3.8E15 cm^{-2} present on the surface exposed to the Au reflector.

Table 1 summarizes the NRA data collected and analyzed to date. The data presented in the first row summarizes the calibration data. The δ^{13}C ratio was assumed to be 0.0110, and the shown carbon concentrations were based on the ^{13}C calibration data. Foil Sn-4 was measured on its front and back surfaces. The front surface was exposed to three successive enriched ^{13}CH target implosions. The back surface was protected from direct exposure through a seal and enclosure. The δ^{13}C ratio observed on the back surface was consistent with natural abundance carbon, while the front surface observation was 29.4% larger. The concentrations of ^{12}C and ^{13}C have decreased, from back to front, by 36.7% and 17.9% This decrease in concentrations levels is due to heating of the foils by X-rays and target debris ions arriving ahead of the reflected thermal debris. Further, this decrease indicates that the carbon present on the foils is consistent with a few monolayer surface contamination, probably due to processing and preparation of the foils. This same behavior is observed in TOF-SIMS analysis of SiOH concentrations on the surface of clean Si wafers used in the experiments. The concentration of SiOH decreased by a factor of 50 from wafers used in the experiment to those not used. Finally, to confirm that the increased ^{13}C observed on Sn-4 was not due to some other source, a natural abundance ^{nat}CH

Sample	$^{13}C \times 10^{14}$ cm^{-2}	$^{12}C \times 10^{16}$ cm^{-2}	$\delta^{13}C$
Natural Sn Foil	4.60 ± 0.29	4.14 ± 0.009	0.0110 ± 0.0007
Sn-4 ^{13}C exposure (front)	4.50 ± 0.29	2.91 ± 0.007	0.0154 ± 0.0010
Sn-4 no ^{13}C exposure (back)	5.48 ± 0.32	4.60 ± 0.009	0.0119 ± 0.0007
Sn-2 $^{nat.}CH$ exposure	3.48 ± 0.25	2.99 ± 0.008	0.0116 ± 0.0009

Table 1. Summary of the (d,p) NRA data for the various Sn foils analyzed.

target was imploded and debris collected on foil Sn-2. The measured $\delta^{13}C$ on the surface of the exposed foil is consistent with natural abundance. Thus, the excess ^{13}C observed on foil Sn-4 must be due to the ^{13}C in the enriched ^{13}CH target used in the experiment.

Using the measured ^{12}C concentration on Sn-4 to determine the amount of background ^{13}C present, leaves an excess ^{13}C concentration of $1.27 \pm 0.08 \times 10^{14}$ cm^{-2} from the target implosions. Under the assumption that the target debris is uniformly distributed, this implies an increase in collection efficiency due to reflection of between 47% and 340%. This range of values is due to the lack of data confirming whether the debris is from one, two, or three shots that the foil was exposed to. Since the concentration level is consistent with monolayer deposition on the foils, and the observed back-to-front concentration decreases for both C and SiOH are independent of the number of shots, it is likely that the increase due to reflection is at the higher end of the quoted range.

3. Summary

We have performed a debris collection experiment using directly driven targets at the Omega laser, based on thermalization, entrainment and focused transport to a collector substrate. We have observed statistically significant concentrations of target debris in the data analyzed and in enhancement of our collection efficiency over direct collection from 47% to 347%.

4. Acknowledgments

We'd like to acknowledge the staff of the Laboratory for Laser Energetics, and the support of the Los Alamos National Laboratory LDRD program.

References

[1] A. C. Hayes, G. Jungman, J. C. Solem, P. A. Bradley, and R. S. Rundberg, *"Prompt beta spectroscopy as a diagnostics for mix in ignited NIF capsules,"* Mod. Phys. Lett. A, **21**, 1029-1040 (2006)

[2] G.P. Grim et al., *"Prompt radiochemistry at the National Ignition Facility (invited),"* Rev. Sci. Instrum. **79**, (2008) 10E503-1-5

[3] S.G. Prussin, S.M. Lane, M.C. Richardson, and S.G. Noyes, *"Debris collection from implosion of microballoons,"* Rev. Sci. Instrum **57** (1986) 1734-1736

[4] H. Nakaishi, N. Miyanaga, M. Yamanaka, T. Yamanaka, M. Takagi, S. Nakai, K. Kobayashi, and I. Kimura, *"Experimental studies on debris collection for radiochemistry in inertial confinement fusion."* Rev. Sci. Instrum. **61** 2623-2627 (1990)

[5] Stephen M. Lane and Michael B. Nelson, *"A technique for shell compression measurements of laser fusion targets by neutron activation rubidium tracer,"* Rev. Sci. Instrum. **61** 3298-3300 (1990)

[6] J-P Garçonnet, C. Bayer, D. Juraszek, D. Schirmann, and A. Bertin, *"Radiochemistry measurements on Phebus laser,"* Rev. Sci. Instrum **63** 4866-4870 (1992)

[7] D. Eder et. al. *Assessment and mitigation of radiation, EMP, debris and shrapnel impacts at mega-joule class laser facilities,"* To be published in same conference journal proceedings.

The Sixth International Conference on Inertial Fusion Sciences and Applications IOP Publishing
Journal of Physics: Conference Series **244** (2010) 032047 doi:10.1088/1742-6596/244/3/032047

ICF Gamma-Ray Reaction History Diagnostics

H W Herrmann[1], C S Young[1], J M Mack[1], Y H Kim[1], A McEvoy[1], S Evans[1], T Sedillo[1], S Batha[1], M Schmitt[1], D C Wilson[1], J R Langenbrunner[1], R Malone[2], M I Kaufman[2], B C Cox[2], B Frogget[2], E K Miller[3], Z A Ali[4], T W Tunnell[2], W Stoeffl[5], C J Horsfield[6] and M Rubery[6]

[1]Los Alamos National Laboratory, P.O Box 1663, M/S E526, Los Alamos, NM 87545, USA
[2] National Security Technologies, Los Alamos, NM 87544, USA
[3] National Security Technologies, Special Technologies Laboratory, Santa Barbara, California, 93111 USA
[4] National Security Technologies, Livermore, CA, 94550, USA
[5] Lawrence Livermore National Laboratory, Livermore, CA 94550, USA
[6] Atomic Weapons Establishment, Aldermaston, Reading, RG7 4PR, UK

E-mail: herrmann@lanl.gov

Abstract. Reaction history measurements, such as nuclear bang time and burn width, are fundamental components of diagnosing ICF implosions and will be employed to help steer the National Ignition Facility (NIF) towards ignition. Fusion gammas provide a direct measure of nuclear interaction rate (unlike x-rays) without being compromised by Doppler spreading (unlike neutrons). Gas Cherenkov Detectors that convert fusion gamma rays to UV/visible Cherenkov photons for collection by fast optical recording systems have established their usefulness in illuminating ICF physics in several experimental campaigns at OMEGA. In particular, bang time precision better than 25 ps has been demonstrated, well below the 50 ps accuracy requirement defined by the NIF. NIF Gamma Reaction History (GRH) diagnostics are being developed based on optimization of sensitivity, bandwidth, dynamic range, cost, and NIF-specific logistics, requirements and extreme radiation environment. Implementation will occur in two phases. The first phase consists of four channels mounted to the outside of the target chamber at ~6 m from target chamber center (GRH-6m) coupled to ultra-fast photo-multiplier tubes (PMT). This system is intended to operate in the 10^{13}-10^{17} neutron yield range expected during the early THD campaign. It will have high enough bandwidth to provide accurate bang times and burn widths for the expected THD reaction histories (> 80 ps fwhm). Successful operation of the first GRH-6m channel has been demonstrated at OMEGA, allowing a verification of instrument sensitivity, timing and EMI/background suppression. The second phase will consist of several channels located just inside the target bay shield wall at 15 m from target chamber center (GRH-15m) with optical paths leading through the cement shield wall to well-shielded streak cameras and PMTs. This system is intended to operate in the 10^{16}-10^{20} yield range expected during the DT ignition campaign, providing higher temporal resolution for the expected burn widths of 10-20 ps associated with ignition. Multiple channels at each phase will allow for increased redundancy, reliability, accuracy and flexibility. In addition, inherent energy thresholding capability combined with this multiplicity will allow exploration of interesting gamma-ray physics well beyond the ignition campaign.

Journal of Physics: Conference Series **244** (2010) 032047 doi:10.1088/1742-6596/244/3/032047

1. Introduction

A Gamma Reaction History (GRH) diagnostic has been developed for the National Ignition Campaign and thoroughly tested on OMEGA. The GRH is a follow-on design to the Gas Cherenkov Detector (GCD-1) which has been productive on OMEGA for the past 10 years [1]. Figure 1 depicts the GRH currently installed on the outside of OMEGA target chamber [2,3]. It is designed to detect the 16.7 MeV gamma-rays which are emitted in approximately one of every 10,000 DT fusion reactions. These gammas are converted, primarily through Compton scattering, to relativistic electrons inside an aluminum γ-to-e$^-$ convertor located within the front flange. The electrons traveling faster than the speed of light in a pressurized gas generate UV/Visable Cherenkov radiation which is collected with reflective optics into an ultra-fast photomultiplier tube (PMT). Varying the pressure of the gas, typically CO2 or SF6, varies the index of refraction and hence the Cherenkov threshold defined by $v_e > c/n$ (where v_e is the electron velocity, c is the speed of light, and n is the index of refraction of the gas). The electrical PMT signal is sent to a Mach Zehnder modulator in the electronics enclosure for optical transmission to digitizers with minimal loss of bandwidth [4]. A complete system impulse response with a single-stage, microchannel plate (MCP) PMT has been measured to be 93 ps fwhm with 62 ps 10%-to-90% rise time; fast enough to measure the 80-150 ps burn widths expected from the THD campaign planned for the Spring of 2010 at NIF. Fiber optic connectors allow timing fiducials and calibration signals to be injected directly into the optical path and provide a dry-run capability for verifiable reliability.

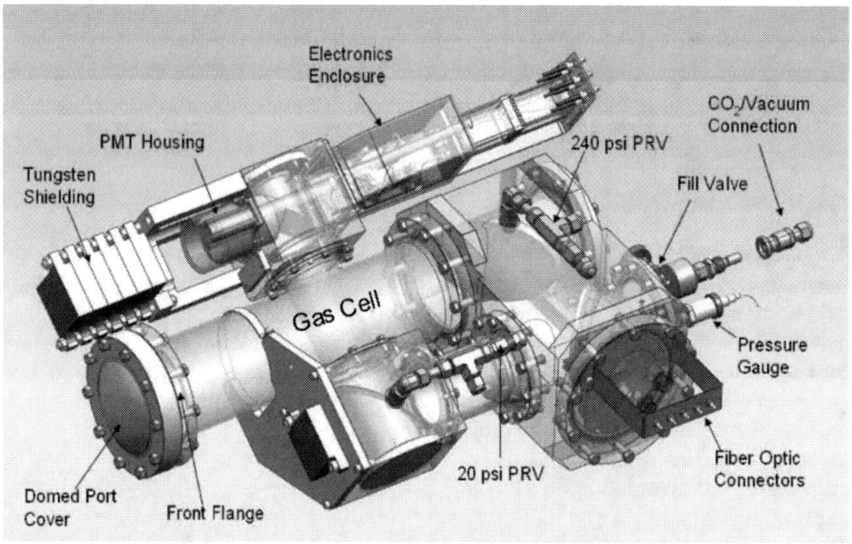

Figure 1: First GRH-6m channel, currently installed on OMEGA.

For the NIF, the GRH-6m will consist of 4 gas cell channels run in parallel for increased redundancy and reliability, and the ability to run at several gamma-ray threshold energies to diagnose various gamma-rays of interest (e.g., t(d,γ) at 16.7 MeV, t(h, γ) at 19.8 MeV, ^{12}C(n,n') γ at 4.4 MeV). These detectors will be capable of measuring fusion bang times relative to the t_0 of the composite laser pulse to within 50 ps accuracy and burn widths to within 5-10 ps accuracy at yields of 1×10^{13} to 1×10^{17} at NIF. These parameters will be key in diagnosing shell velocity and the various possible failure modes in attempts at ICF ignition on the NIF. The GRH may also prove useful as a means of providing a time-resolved measure of ablator ρR indicative of ablator mass remaining and ablator/fuel mix based on the ratio of ^{12}C(n,n') γ to t(d,γ). Efforts are underway to reduce the threshold to below 4 MeV for this purpose. A later implementation intended for diagnosing ignition shots will consist of Cherenkov detectors located at ~15m (GRH-15m) with the light coupled outside the shield wall to well shielded streak cameras for improved temporal response (~10 ps) and higher yields (>10^{16} DT-n).

2. Performance Tests at OMEGA

2.1. Background and Sensitivity

A gamma-ray measurement from an imploded DT-filled plastic capsule with the GRH pressure at 100 psia CO_2 (~6 MeV Cherenkov threshold) is shown in Figure 2. Also shown is a background measurement from a previous shot with the gas cell evacuated resulting in no Cherenkov production, but still exhibiting direct gamma interaction with the PMT (this background is also present when the PMT input window is optically blocked with black tape). DT fusion gammas and n-γ from DT fusion neutrons passing through the imploding capsule (primarily $^{12}C(n,n')\gamma$ from CH capsules) interact directly with the PMT causing the "precursor" peak approximately 4.1 ns before bang time. The speed-of-light transit time from target chamber center (TCC) through the optical delay designed into the GRH is calculated to be 4.26 ns longer than the direct line-of-sight (LOS) transit time from TCC to the microchannel plate (MCP) in the PMT (i.e., directly through the tungsten shielding). Other experiments using the GRH at OMEGA have shown that 6 inches of tungsten shielding is adequate to eliminate direct LOS gammas (and x-rays), and so the precursor from these shots with 9 inches of shielding are most likely from gammas originating at TCC and reaching the MCP on scattering paths delayed approximately 160 ps relative to a direct LOS path, or an extra 4.8 cm. The precursor from NIF indirect-drive shots is expected to be considerably larger due to n-γ from the hohlraum and thermo-mechanical package, but the optical delay provides ample time for this signal to subside before the main Cherenkov peak. Additional small peaks after the precursor are likely from n-γ generated by interaction of DT neutrons with mass near TCC. Similar peaks have been correlated with various TIM-based diagnostics as they are moved about TCC. Mass located closer than 22 cm from TCC may generate n-γ signal before or at bang time due to the slower transit time of the neutrons. The slight increase in background starting ~1 ns before bang time is most likely n-γ off the four fixed x-ray camera snouts located at 16.5 cm and the GCD at 20 cm from TCC. Current background levels are adequate for bang time and burn width measurements, and are expected to be lower at NIF due to an improved shielding design and less close-in mass near TCC. The DT neutron yield on shot 55073 was 1.27e13. A comparable signal can be expected from a 1.2e14 yield shot on NIF after taking into account the factor of 10 solid angle reduction in repositioning the GRH γ-to-e^- convertor from 1.87 m at OMEGA to 6.1 m at NIF. Approximately 650 DT fusion gammas were detected in this shot, based on detector sensitivity calculations using the ITS ACCEPT code.

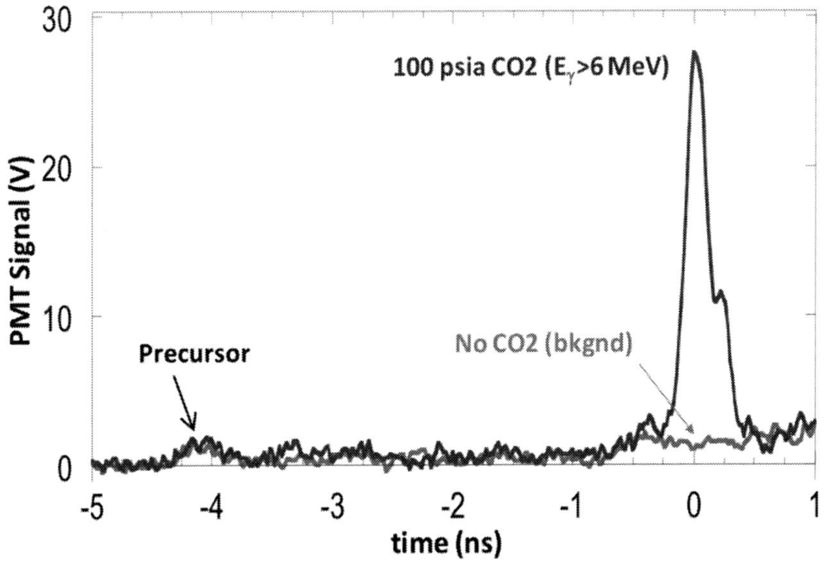

Figure 2: GRH data from high-yield shots with pressurized CO_2 (blue, shot 55073) and without (red, shot 55071) using a high gain ($6x10^5$), 2-stage MCP PMT (Photek PMT210). Time zero placed at peak of fusion reactivity.

2.2. Absolute Bang Time and Burn Width

The GRH was recently absolutely timed using the x-rays generated from 100 ps laser pulses directed onto gold ball targets. The γ-to-e⁻ convertor in the GRH was replaced with a BC-422 x-ray sensitive scintillator and thin aluminium x-ray windows replaced thick vacuum/pressure flanges for transmission above ~20 keV. The x-rays/scintillator photons have essentially the same transit time through the system as the γ-rays/relativistic electrons/Cherenkov photons. Thus the x-ray pulse can be used to generate a t_0 time stamp on the oscilloscope from which the optical fiducial is calibrated.

Figure 3 shows an x-ray shot and a high yield shot overlaid by lining up the timing fiducials. The difference between the rising edge of the x-ray pulse and the peak of the fusion signal is representative of the absolute bang time. Small correction factors are applied to the raw bang time to take instrument responses into account. A more detailed description of bang time determination will be published elsewhere. The fwhm of the Cherenkov peak in Figure 3 is 130 ps fwhm. Subtracting the 92 ps system response in quadrature results in a fusion burn width of ~91 ps, consistent with a full deconvolution and forward folding of Gaussian pulse shapes with the system impulse response to match the raw data.

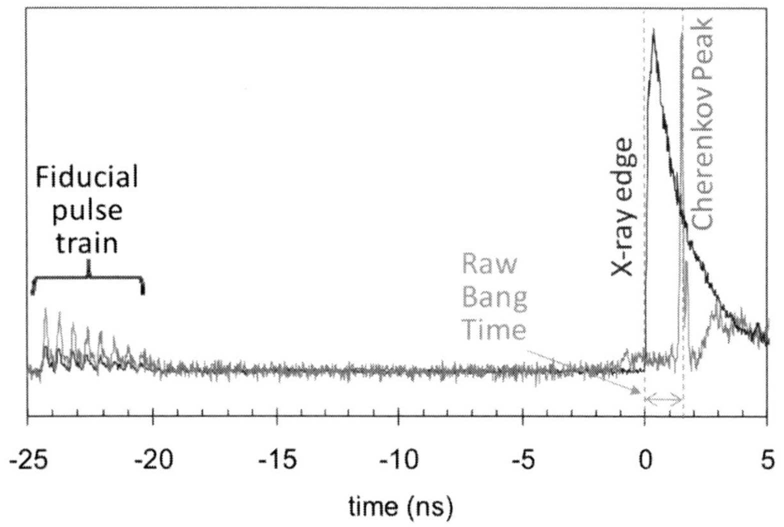

Figure 3: Response to x-ray pulse from a gold ball target (blue, shot 55061) used to calibrate the optical timing fiducial, overlaid with a Chernkov signal from a high yield DT shot (red, shot 55078). A single-stage MCP PMT (Photek PMT110) was used for fast time response. Time zero placed at x-ray signal initiation representing t_0 of the laser pulse.

3. Conclusion

The first channel of the 4-channel Gamma Reaction History diagnostic to be located 6 m from TCC on NIF has undergone testing at OMEGA. It has demonstrated adequate sensitivity, background suppression and absolute timing to meet bang time and burn width requirements. The use of four channels operating in parallel will improve reliability and accuracy, and allow exploration of gamma ray spectroscopy during the THD campaign.

References

[1] Mack J M , et al., *2006 Rad. Phys. Chem.* **75** 551 and references therein
[2] Herrmann H W, et al., 2008 *Rev. Sci. Instrum.* **79** 10E531
[3] Malone R M, et al., 2008 *Rev. Sci. Instrum.* **79** 10E532; also these proceedings
[4] Miller K M, et al., these proceedings

Journal of Physics: Conference Series **244** (2010) 032048 doi:10.1088/1742-6596/244/3/032048

Experimental evaluation of neutron induced noise on gated x-ray framing cameras

N Izumi, G Stone, C Hagmann, C Sorce, D K Bradley, M Moran, O L Landen, P Springer, W Stoeffl, R Tommasini, H W Hermann[1], G A Kyrala[1], V Y Glebov[2], J Knauer[2], T C Sangster[2], and J A Koch

Lawrence Livermore National Laboratory, Livermore, CA, USA

[1]Los Alamos National Laboratory, Los Alamos, NM, USA

[2]Laboratory for Laser Energetics, Rochester, NY, USA

izumi2@llnl.gov

Abstract. A micro-channel plate based temporally-gated x-ray camera (framing camera) is one of the most versatile diagnostic tools of inertial confinement fusion experiments; particularly for observation of the shape of x-ray self emission from compressed core of imploded capsules. However, components used in an x-ray framing camera have sensitivity to various kinds of secondary radiation induced by neutron. On early low-yield capsule implosions at the National Ignition Facility (NIF), the expected neutron production is about 5×10^{14}. Therefore, the expected neutron fluence at a framing camera located ~ 150 cm from the object is 2×10^{9} neutrons/cm^2. To obtain gated x-ray images in such harsh neutron environments, quantitative understanding of neutron-induced backgrounds is crucial.

1. Introduction

Because of their long mean free path, 14-MeV neutrons produced by the t(d,n)α reaction penetrate x-ray shielding along the detector line of sight, causing various interactions in materials surrounding the sensitive region of the x-ray framing camera such as elastic recoil, (n,n')γ, (n,p), (n,α). These various kinds of secondary ionizing radiation or visible light produced by them then stimulate the sensitive layer of the detector and induce background noise. To understand mechanisms of this neutron induced background, we have fabricated a test camera head and exposed it to strong DT neutron fluxes at the Omega laser facility.

2. Gated and time integrated components of x-ray framing camera

Fig. 1. shows a typical setup of pinhole imaging with an x-ray framing camera at the National Ignition facility. A pinhole array is located typically 8~20 cm from the capsule. The x-ray image is projected on to a micro-channel plate (MCP) surface (~150 cm from the object). Then secondary electrons produced on the MCP entrance are amplified in its pores. Once amplified electrons exit the MCP, they are accelerated by phosphor potential (to 3~6 kV) and converted to optical light on a phosphor. The optical signal from the phosphor is transferred to a recording device (charge-coupled-device (CCD) or a photographic film) by a fiber-optic plate (FOP) [1].

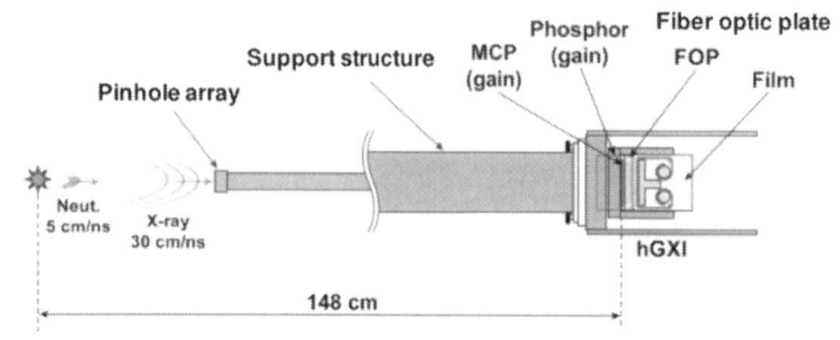

Fig. 1. Typical setup of an x-ray framing camera for gated pinhole x-ray imaging. The MCP gain is turned off on arrival of 14 MeV neutrons. However time integrating components following the MCP (phosphor, FOP, and photographic film) remain sensitive at the neutron arrival time.

To avoid image blur due to fast motion of an object, temporal gating of the x-ray camera is crucial. The x-ray sensitivity of the camera is switched by supplying a high voltage pulse on arrival of x-ray signal (~5ns after the x-ray emission on the object). The 14 MeV neutrons produced in the compressed plasma travels ~5.1 cm/ns and hit the detector ~24 ns after the x-ray arrival. Therefore there is delay between the x-ray and the neutron arrival. This delay helps to reduce the background because the MCP gain is turned off when neutrons hit the detector. However time-integrating components after the MCP (phosphor, photographic film) are still sensitive to excitation by ionizing radiation induced by neutron [2-4]. Therefore, the neutron sensitivity of those time integrating components is important.

3. Possible mechanisms of fast neutron induced background

Fig. 2 shows possible mechanisms of neutron induced background on time integrating components of the camera. First, neutron produces various charged particles in the sensitive layer of the photographic film. Protons and other light nuclei recoiled by 14 MeV neutrons deposit energy in the emulsion layer. Because of their longer mean free path, protons produced in the FOP by (n,p) reactions can also reach the sensitive layer and contribute to the background exposure. X-rays induced by those charged particles excite the film also. The 14-MeV neutrons have enough energy to cause nuclear excitation of materials surrounding the film. Gamma-rays produced by (n,n')γ reactions cause Compton scattering around the detector and then high energy knock-on electrons excite the sensitive layer.

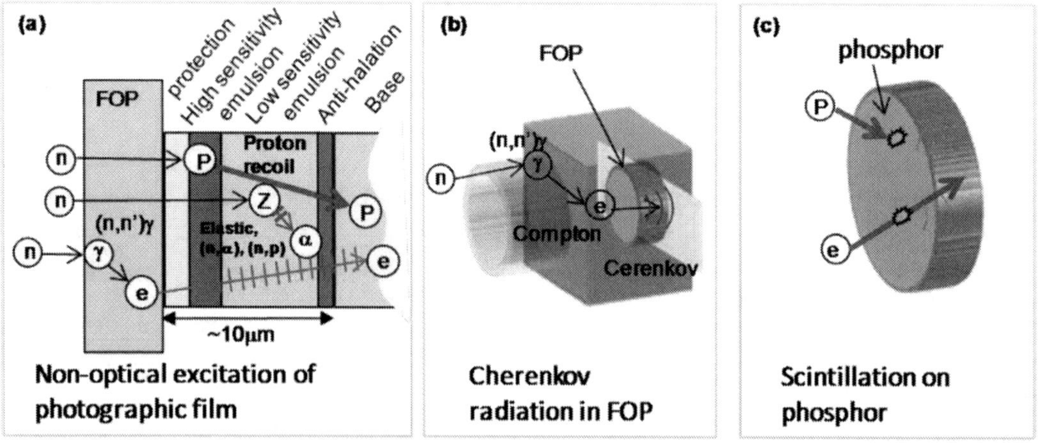

Fig. 2. Possible mechanisms of neutron induced backgrounds on an x-ray framing camera. (a) Non-optical excitation induced by neutron (charged particles and x-ray) directly exposes the sensitive layer of the photographic films. (b) High energy electrons induced by neutron interaction can produce Cherenkov radiation and scintillation in the FOP. (c) Neutron induced secondary charged particles excite phosphor layer.

Second, those high energy electrons cause Cherenkov radiation in the FOP and optically excite the photographic films. Third, the charged particles produced by neutron interaction excite the phosphor layer and scintillation on the phosphor makes background exposure of the film.

4. Experiments

To understand the contribution of the different excitation mechanisms, we assembled a test camera head with the time integrating components (phosphor, FOP, and photographic film). Figure 3(a) shows exploded view of the test camera head. The film cassette was located 43.2_cm from the OMEGA chamber center. A photographic film (Kodak TMAX 3200) was set in an aluminium cassette with various samples. Figure 3(b) is showing a film exposure data when neutron flux was 3×10^{9} neutrons/cm^2. The optical density of the film was converted to equivalent optical exposure (erg/cm^2) using a sensitometry test sample (a piece of the film which is pre-exposed to known light fluence, and developed simultaneously with the data film).

A part of film which is exposed to all the background mechanisms (region 1 of figure 3 (b)) showed optical density corresponding to 0.035erg/cm^2. In order to separate optical and non-optical background, visible light signal was partially blocked with thin black paint on the FOP surface. As shown in Fig. 3. (c), region under the light block (region 2) has exposure of 0.019erg/cm^2. Therefore about 50% of the recorded background was non-optical exposure. To differentiate the optical exposure from the phosphor and the FOP, a part of the phosphor coating was removed from the FOP (region 3). Difference of exposure with and without phosphor was ~ 0.004 erg/cm^2. Therefore about 40% of the background comes from the FOP and 10% is from the phosphor as optical excitation. A plastic (1mm thick) and a tantalum (0.8mm thick) sample were placed in front of the film to enhance the neutron induced ionizing radiation. As expected, exposure of the region covered with plastic (region 4: 0.07 erg/cm^2) was higher than the region covered with aluminium (region 5: 0.018 erg/cm^2). This enhancement is due to energy deposition from recoiled protons. The region covered with the tantalum disc (region 6: 0.03erg/cm^2) was also higher than the region covered by aluminium because of x-ray fluorescence of tantalum excited by neutron induced charged particles. In addition, four different materials were attached to the back side of the film; Pb, Cu, Al, and plastic. No significant difference was observed between the plastic (region 2: 0.019 erg/cm^2) and the aluminium backing (region 7: 0.018 erg/cm^2).

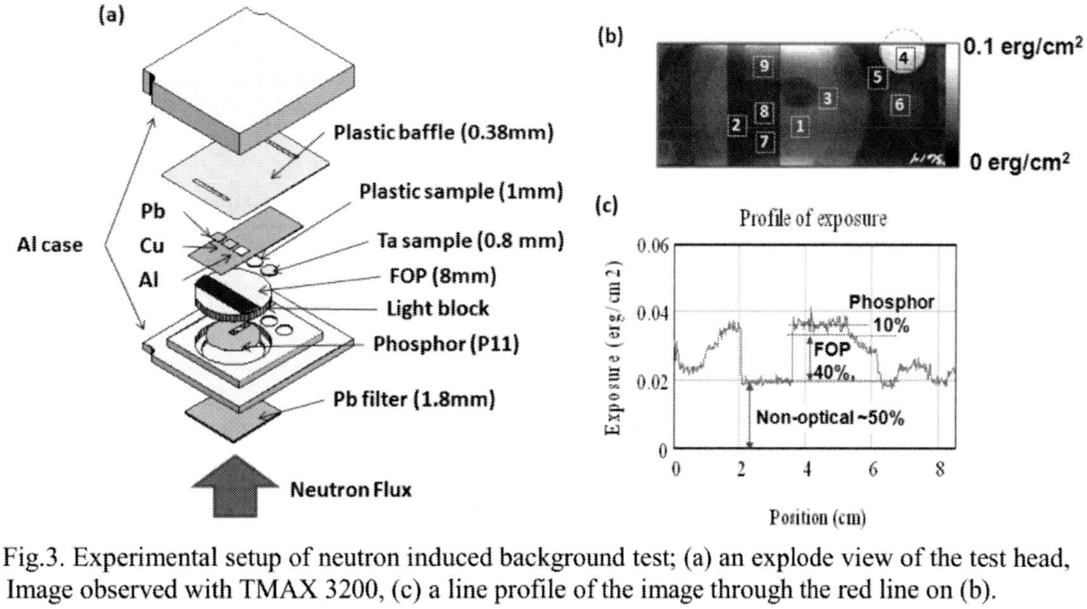

Fig.3. Experimental setup of neutron induced background test; (a) an explode view of the test head, (b) Image observed with TMAX 3200, (c) a line profile of the image through the red line on (b).

This is explained by the direction of the recoiling protons produced in the plastic sample. Since neutrons come predominantly from the center of the target chamber, most of the protons recoil away from the film and cause little exposure. The regions backed by the Cu foil (region 8: 0.023 erg/cm^2) and the Pb foil (region 9: 0.028 erg/cm^2) were higher than the region backed by the Al foil because x-ray fluorescence form those material were higher than that of aluminium.

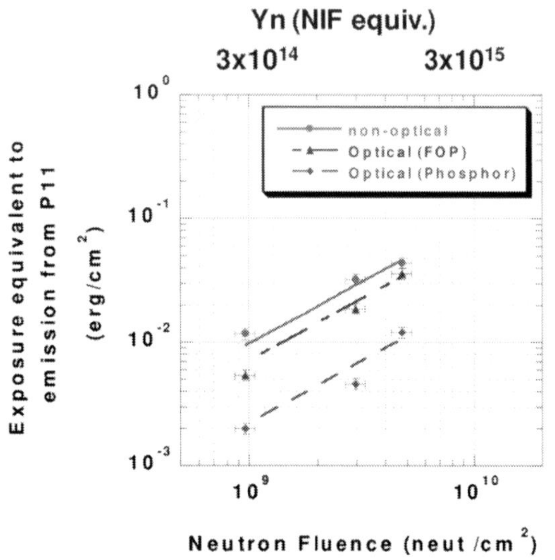

Fig. 4. Observed background exposure versus neutron fluence. The background exposure from each mechanism scales with neutron fluence.

Fig. 4 shows the observed background versus 14 MeV neutron fluence inferred from independent neutron production measurements. Background exposure of each component showed linear dependence on neutron fluence. This confirms that the background exposure was neutron induced.

5. Discussion

To achieve good signal to noise in measurements, the optical output for the x-ray signal has to be stronger than the neutron induced background. For the case of the low-yield NIF implosion experiments (Yn: 5×10^{14}, neutron fluence at the detector: 2×10^{9} neutrons/cm^2), the expected background exposure on a Kodak TMAX3200 film is 0.023erg/cm^2. Our framing camera operating under standard conditions (phosphor potential: +3kV) is driven into non-linear response regime at its optical output of 0.2erg/cm^2 [5] and then the expected maximal signal to background ratio is 8.6:1. With applying higher phosphor potential, it is possible to increase the linear response regime by a factor 2 and thereby the expected signal to background ratio can be raised to 17:1.

Acknowledgements

The authors acknowledge the support of the Omega crew, J. Emig, J. Tellinghuisen, S. Evans, V. Rekow, N. Hill, J. Armstrong, G. Pien, S. Alvarez, M. Moran, K. Loughman, S. Dent, S. Lucas, O. Gotchev, and W. Shmayda. Prepared by LLNL under Contract DE-AC52-07NA27344

References

[1] J. D. Kilkenny, Laser and Particle Beams 9, 49 (1991).
[2] R. A. Lerche, Rev. Sci. Instrum. 68, 628 (1997).
[3] P. A. Jaanimagi, R. Boni, and R. L. Keck, Rev. Sci. Instrum. 72, 801 (2001).
[4] J. L. Bourgade, et al., Rev. Sci. Instrum. 75, 4204 (2004).
[5] O. L. Landnen, et al., Proc. SPIE 2002, 2 (1993).

Simulation of Radiation Backgrounds associated with the HEXRI Diagnostics at the National Ignition Facility

H Khater, L Dauffy, R Tommasini, M Eckart and D Eder

Lawrence Livermore National Laboratory, 7000 East Avenue, Livermore, CA 94550, USA

E-mail: khater1@llnl.gov

Abstract. Experiments resulting in a significant neutron yield are scheduled to start in 2010 at the National Ignition Facility (NIF). A wide range of diagnostics will be used to measure several parameters of implosion such as the core and fuel shape, temperatures and densities, and neutron yield. Accurate evaluations of the neutron and gamma backgrounds are important for several diagnostics. Several Monte-Carlo simulations were performed to identify the expected signal to background ratios at several potential locations for the High Energy X-ray Imager (HEXRI) diagnostics. Gamma backgrounds were significantly reduced by using tungsten collimators. The collimators resulted in the reduction of the gamma background at the HEXRI scintillators by more than an order of magnitude during the first 40 ns following a Tritium-Hydrogen-Deuterium (THD) shot.

1. Introduction

The National Ignition Facility (NIF) at Lawrence Livermore National Laboratory is the world's largest and most powerful laser system for inertial confinement fusion. The NIF is a 192 laser beam facility that is capable of producing 1.8 MJ, 500 TW of ultraviolet light. Experiments resulting in a significant neutron yield are scheduled to start in 2010 at the NIF. Several experiments utilizing Tritium-Hydrogen-Deuterium (THD) and Deuterium-Tritium (D-T) targets are scheduled as part of the National Ignition Campaign (NIC). The THD and D-T targets have different spectra and the THD spectrum will depend on the fraction of deuterium in the target. A wide range of diagnostics will be used to measure several parameters of implosion such as the core and fuel shape, temperatures and densities, and neutron yield. Accurate evaluations of the neutron and gamma backgrounds are important for several diagnostics, such as the High Energy X-ray Imager (HEXRI). Several sources of neutron and gamma backgrounds will impact the accuracy of the diagnostics measurements. Fusion neutrons generated by fuel burn and secondary neutrons resulting from the fusion neutrons interaction with structures present inside and outside the Target Chamber (TC) contribute to the neutron background. Inside the TC, neutrons interact with the Cryogenic Target Positioner (CRYO TARPOS), the Target Positioner (TARPOS), and three other Diagnostics Instruments Manipulators (DIM). Neutrons also interact with the aluminum TC, the TC concrete shield, and the TC port covers. In the meantime, x-rays emitted from implosion, x-rays resulting from laser plasma interaction (LPI) of NIF beams with the hohlraum, and gamma-rays induced by neutron interactions with different structures inside and outside the TC contribute to the gamma background. A detailed model of the NIF facility and all structures inside the TC has been developed. Several Monte-Carlo simulations were performed

Journal of Physics: Conference Series **244** (2010) 032049 doi:10.1088/1742-6596/244/3/032049

to identify the expected signal to background (S/B) ratios at several potential locations for the HEXRI diagnostics. This paper updates and expands on our previous published work [1].

2. Computational Model

A detailed 3-D model of the NIF facility has been developed using the MCNP [2] radiation transport code. The TC is made of a 10-cm-thick aluminum wall surrounded by 40-cm of borated concrete. The TC includes 48 indirect-drive and 24 direct-drive laser beam ports as well as 120 diagnostic ports. All diagnostic and direct-drive ports are assumed to be unshielded and only covered with ~ 2"-thick port covers made of aluminum alloy. Indirect-drive ports are connected to fully modeled Final Optics Assemblies (FOA). The Target Bay (TB) shielding walls are made of 6'-thick concrete, and the typical thickness of a concrete Switchyard wall is 3'–3". The detailed TB model used in this analysis is shown in Figure 1.

Figure 1. Target Bay model.

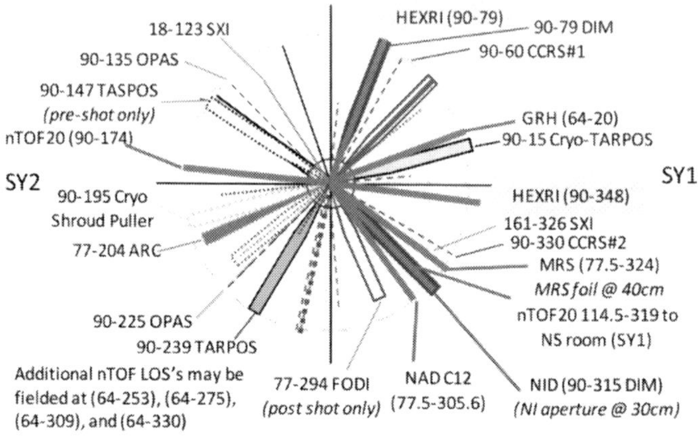

Figure 2. Potential locations of ignition/implosion diagnostics.

Figure 2 shows the NIF TC with potential locations for ignition diagnostics. The two target positioners use the (90°, 15°) and the (90°, 239°) TC ports. Two equatorial DIMs use ports (90°, 315°) and (90°, 45°), and a polar DIM uses the (0°, 0°) port. In this paper we present our evaluation of the radiation background associated with the HEXRI diagnostics. We examined the possibility of installing the HEXRI detectors outside the TC ports, namely at (90°, 79°) and (90°, 348°). As shown in Figure 2, additional diagnostics like neutron activation diagnostics (NAD), neutron imaging (NI), magnetic recoil spectrometer (MRS), gamma reaction history (GRH), and neutron-time-of-flight (nToF) will be installed at different locations inside and outside the TC. Evaluation of the radiation backgrounds associated with these diagnostics will be presented in future publications.

3. Radiation Backgrounds associated with the High Energy X-ray Imager (HEXRI) System

The HEXRI system is used to provide a sequence of two-dimensional radiographs of imploding targets. The radiographs are created by firing four beamlets of the NIF-ARC petawatt laser system on gold backlighters, and bremsstrahlung emitted by the backlighters is used to image the target. The backlighters are installed on the (90°, 239°) target positioner (TARPOS) and located at a distance of 5 cm from the Target Chamber Centre (TCC). The radiographs are recorded by scintillators located outside the (90°, 79°) and (90°, 348°) TC ports at a distance of 6 m from TCC and optically relayed to a gated CCD camera.

Several sources of radiation background have an impact on the signal recorded by HEXRI. Direct neutron impact on the collected signal has been eliminated by gating the HEXRI to ≤ 40 ns. Neutrons can only travel to within 2 m from TCC during this time and do not impact directly the collected signal. On the other hand, two gamma background sources do interfere with the collected signal. The first gamma source is the hard x-rays generated by LPI of the NIF laser with the hohlraum. These x-rays are partially shielded by restricting the portion of the hohlraum seen by the detector by carefully placing multiple collimators near the hohlraum. The collimators are made of platinum and are placed at distances of 4 mm and 5 cm from the TCC. The second gamma source is generated by the neutrons interactions with masses in the vicinity of the target. Gammas generated (mostly through n,γ reactions) within a 2 m distance from TCC will contribute to the background signal. Contribution to the collected signal was minimized through the use of 1-cm thick conical tungsten collimators. The collimators extended from a distance of 2.5 m from the TCC to the scintillators outside the TC ports. Figure 3 shows the HEXRI detector setup.

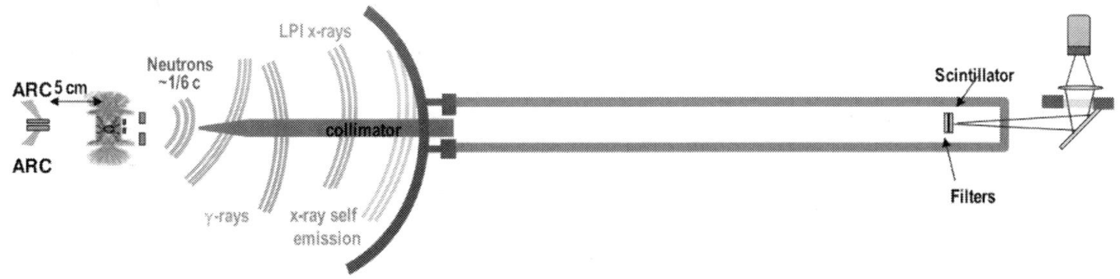

Figure 3. HEXRI detector setup.

Several models for near TCC masses were added to the NIF facility model. Photon production and neutron scattering within a two-meter radius from the TCC significantly impact the radiation backgrounds seen by the HEXRI detectors. Models for the hohlraum, the target positioners, the two equatorial DIMs, the polar DIM, MRS foil and NI collimator were included. The TC model showing the near TCC masses is shown in Figure 4. Design of the tungsten collimators was based on an evaluation of the contribution to the gamma background signal due to different near TCC masses. As shown in Figure 5, the two target positioners are the main contributors to the background signal. The tungsten collimators were designed to minimize contributions from these masses.

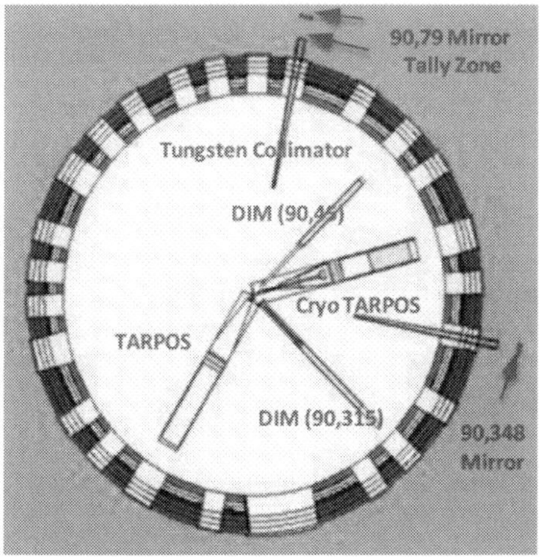

Figure 4. Masses near TCC.

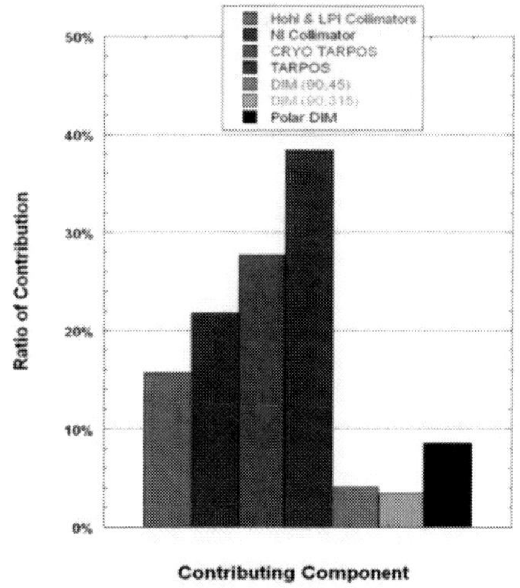

Figure 5. Contributions to background without collimators.

The radiation backgrounds were estimated for a THD target with 2% deuterium yielding a neutron yield of $2x10^{14}$. Figures 6 and 7 show the neutron and gamma fluences expected during the first 40 ns following the shot, respectively. As shown in the figures, the tungsten collimators were effective in minimizing the gamma streaming through the TC ports.

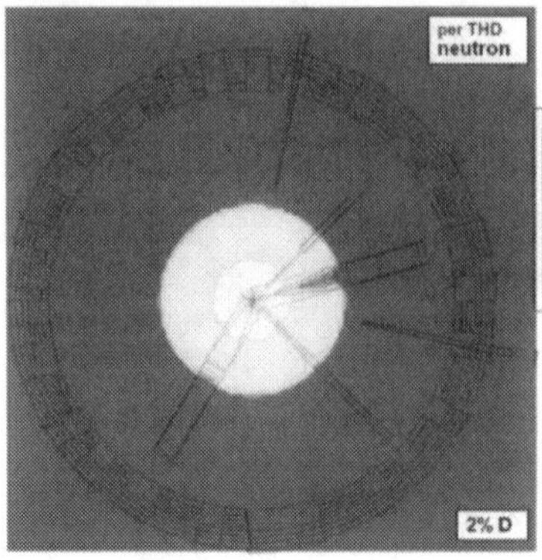

Figure 6. Neutron fluence during first 40 ns.

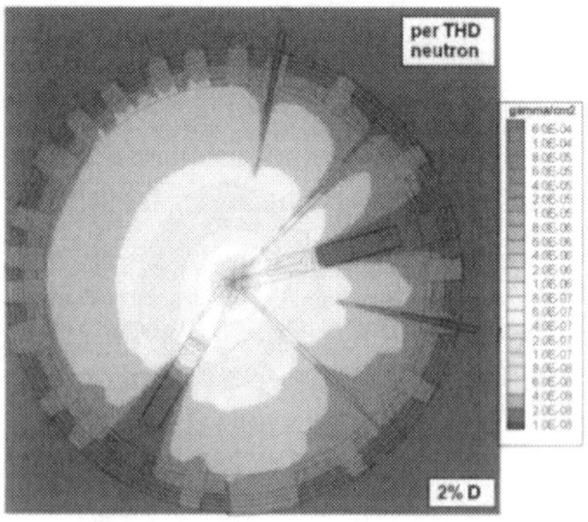

Figure 7. Gamma fluence during first 40 ns.

The impact of using the tungsten collimators on the recorded signal is further illustrated in Figure 8. The collimators were able to reduce the gamma background due to n,γ reactions by more than an order of magnitude. Finally, as shown in Figure 9, the set of collimators limited the background to signal (B/S) ratios from all sources to less than 10% of the estimated recorded signal.

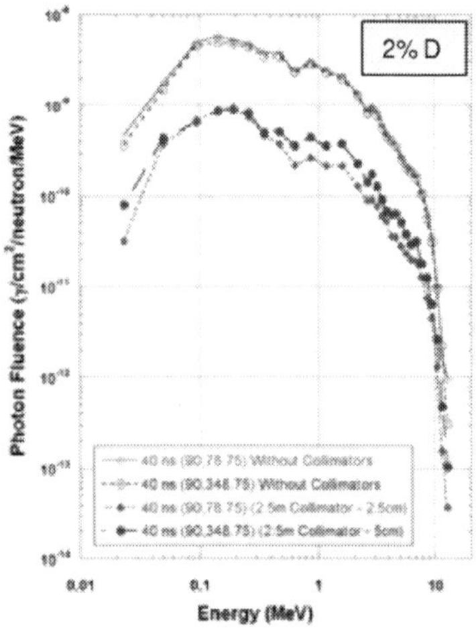

Figure 8. Gamma spectrum due to n,γ reactions.

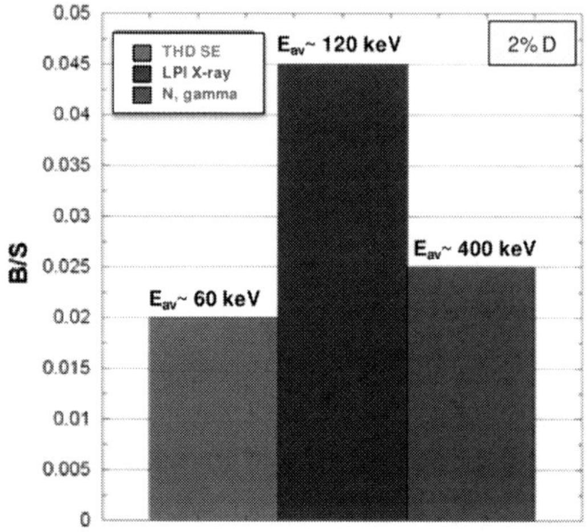

Figure 9. Gamma B/S ratios.

4. Conclusions

Detailed simulation of the radiation environment in NIF allowed for good understanding of the important issues related to performance of the HEXRI system. TARPOS and CRYO TARPOS are the largest contributors to the HEXRI gamma background. The use of tungsten collimators reduced the gamma background experienced by HEXRI detectors by more than an order of magnitude during THD shots. The current setup of the HEXRI system provides adequate reduction of radiation backgrounds.

Acknowledgement

This work was performed under the auspices of the U.S. Department of Energy by Lawrence Livermore National Laboratory under Contract DE-AC52-07NA27344.

References

[1] L. Dauffy et al., "*Gamma Background Calculation for the HEXRI Diagnostics at the National Ignition Facility,*" Journal of Physics: Conference Series 112 (2008) 032085

[2] X-5 Monte Carlo Team, "*MCNP - A General Monte Carlo N-Particle Transport Code, Version 5,*" Los Alamos National Laboratory, LA-UR-03-1987 (2005)

Measurement of DT Fusion and Neutron-induced Gamma-rays using Gas Cherenkov Detector

Y Kim[1], H W Herrmann[1], S Evans[1], T Sedillo[1], J R Langenbrunner[1], C S Young[1], J M Mack[1], A McEvoy[1], C J Horsfield[2], M Rubery[2], Z Ali[3], and W Stoeffl[4]

[1]Los Alamos National Laboratory, Los Alamos, NM 87545, USA
[2]Atomic Weapons Establishment, Aldermaston, Reading, RG7 4PR, UK
[3]National Security Technologies, Livermore, CA, 94550, USA
[4]Lawrence Livermore National Laboratory, Livermore, CA 94550, USA

E-mail: yhkim@lanl.gov

Abstract. A secondary gamma experiment was carried out using a Gas Cherenkov Detector (GCD) at the OMEGA laser facility. The primary experimental objective was to simulate neutron-induced secondary gamma production (n-γ) from a NIF implosion capsule, hohlraum, and thermo-mechanical package. The high-band width of the GCD enabled us to detect time delayed and Doppler broadened n-γ signals from five different puck materials (Si, SiO_2, Al, Al_2O_3, Cu) placed near target chamber center. These measurements were used for MCNP & ITS ACCEPT code validation purposes. By a simple change of the GCD CO_2 gas pressure the system can effectively eliminate signals induced by n-γ reactions and thereby allow quality measurements of DT fusion γ-rays that are produced at NIF (National Ignition Facility).

1. Introduction

A Gamma Reaction History (GRH) diagnostic which measures 16.75 MeV DT fusion γ-rays (DT-γ) has been developed by Los Alamos National Laboratory in collaboration with Lawrence Livermore National Laboratory, Atomic Weapons Establishment, and National Security Technologies [1]. The fast time response of the GRH will make it an important instrument for fusion studies at NIF. A complete system impulse response with a single-stage, micro-channel plate (MCP) photomultiplier tube (Photek PMT110), Mach Zehnder transmission line, and 12 GHz digitizer (Tektronix TDS5124) has been measured to be 93 ps fwhm with 62 ps 10%-to-90% rise time; fast enough to measure the 80-150 ps burn widths expected from the THD campaign planned for the Spring of 2010 at NIF. The GRH signal performance could possibly be compromised by an interference of various neutron-induced secondary γ-signals (n-γ) with the primary DT-γ output. For instance, indirect-drive NIF implosions produce 14 MeV DT neutrons that induce the production of n-γ from 1) implosion capsule (e.g., ^{12}C, ^{16}O, Be, etc.), 2) hohlraum (e.g., Au, U, etc.), and 3) thermo-mechanical package (TMP) (e.g., Si, Al, etc.). Experimental characterization of the GRH response to these various gamma rays is necessary to ensure accurate measurements of NIF fusion reaction history.

The primary experimental objective was to simulate n-γ production from a NIF implosion capsule, hohlraum and TMP materials. A secondary gamma experiment (a.k.a., "hockey puck" experiment) was carried out using a Gas Cherenkov Detector (GCD) at the OMEGA laser facility. Three group of pucks were placed between TCC and GCD; 1) SiO_2, Al_2O_3, and BeO pucks for glass and Be implosion capsule n-γ, 2) Au and Cu pucks for hohlraum n-γ, 3) Al and Si pucks for TMP n-γ. This paper describes the n-γ characteristics measured by the GCD and how the GCD can minimize DT-γ interference from n-γ.

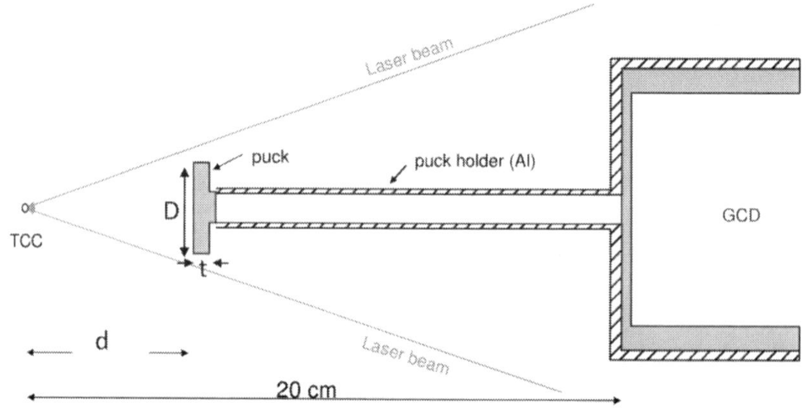

Figure 1: A schematic drawing of secondary gamma experiment using a GCD at OMEGA (puck used: Si, SiO_2, Al, Al_2O_3, and Cu).

2. Experimental Methods

The GCD used for these studies is described in greater detail elsewhere [2]. Briefly, fusion gammas are converted to relativistic electrons primarily through Compton scattering and pair production in a beryllium converter on the front of the detector. Electrons traveling faster than the speed of light in the gas then produce Cherenkov radiation as they travel through a pressurized CO_2 gas cell. The pressure of CO_2 in the gas cell determines the index of refraction and thus speed of light in the medium, hence establishing a variable energy thresholding capability. This UV/visible light is then collected onto an ultra-fast PMT and the signal is recorded on high bandwidth recorders. Figure 1 shows a schematic of the GCD converter and associated secondary γ-ray puck. Pucks were placed between TCC and GCD at a distance d = 5.8 cm (or 5.665 cm, depending on the type of puck). The pucks had a diameter D=30 mm (or 25.4 mm), and a thickness t= 5 mm (or 6.35 mm). A puck holder made of Aluminum was used to hold a puck and attached to the nose corn of the GCD. The GCD nose cone was located 20 cm away from TCC.

3. Experimental Results

Figure 2(a) shows time-dependent gamma-ray signals from DT-filled plastic capsule implosions measured without puck and puck holder to serve as a baseline (black curve, shot 54449, DT-n yield = 2.67e13), and with a Si puck in the holder (red curve, shot 54453, DT-n yield = 2.50e13). 60 laser beams were used with a total energy of 23 kJ of UV light on target. Silicon puck (D = 25.4 mm, t = 6.35 mm) was placed 5.665 cm away from TCC for shot 54453. The CO_2 gas pressure in the GCD was fixed at 100 psia, which corresponds to a threshold energy (E_{th}) of ~ 6.3 MeV. Both γ-signals are normalized by neutron-yield and a MCP gain value of 521. Time zero is placed at the highest peak of signals (peak B) for convenience.

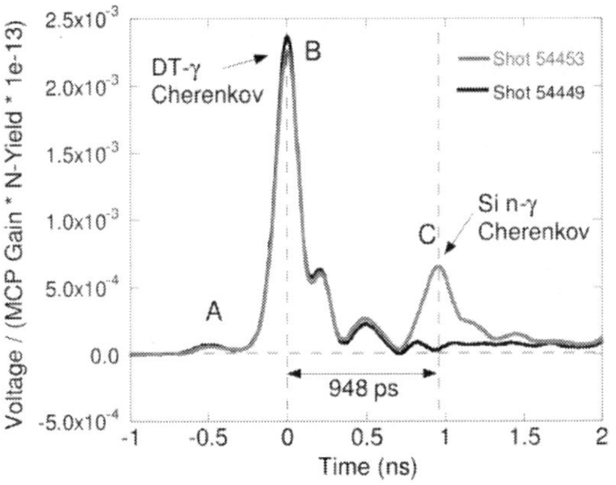

Figure 2a: Simultaneous measurement of fusion gamma and secondary Si gamma signals

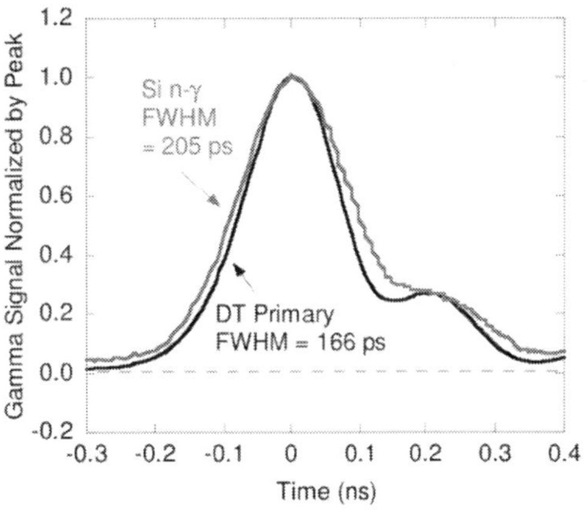

Figure 2b: Secondary Si n-γ signal overlaid with primary DT-γ

Without puck (black curve), reproducible DT-γ signals are measured and served as baseline for n-γ study. As explained in ref 2, a precursor (peak A) and a DT-γ (peak B) are detected. With puck (red curve), new signal peak (peak C) was observed at the 948 ps later. The time interval measured between peaks B and C (= 948 ps) is quite consistent with the expected time-of-flight difference between DT-γ and 14 MeV neutrons (~ 163 ps/cm) incident on the Si puck (= 163 ps/cm × 5.665 cm = 923 ps). Si n-γ signal shape was compared with that of DT-γ by normalizing each peaks (B and C) with their own peak height and overlaying peak C and B. Shown in Fig. 2b, full-width half maximum (FWHM) of the Si n-γ signal was 205 ps, which is wider than that of DT-γ signal (166 ps). The measured Si n-γ FWHM was quite close to the estimated value (~ 207 ps = $(166^2 + 16^2 + 123^2)^{1/2}$ ps) by taking account of 1) measured DT-γ FWHM (= 166 ps), 2) neutron Doppler broadening time (~ 16 ps at 5.665 cm) and 3) geometrical temporal spreading as the neutrons penetrate the puck (~ 123 ps). High-band width of the GCD enabled us to detect time delayed and Doppler broadened n-γ signals from the puck.

Four more puck materials (SiO_2, Al_2O_3, Al, and Cu) were tested with E_{th} kept at 6.3 MeV. Two shots were carried out for each puck. Each n-γ signal is normalized by corresponding neutron-yield and MCP gain. The baseline-subtracted n-γ signals are overlaid and shown in the vicinity of peak C in Figure 3. The silicon puck produced the strongest secondary signal ratioed to the DT-γ signal (n-γ/DT-γ ~ 27 %). The n-γ/DT-γ of SiO_2 and Al_2O_3 were similar at ~12%. The n-γ/DT-γ of Al was ~ 9 %. Copper produced the weakest n-γ above 6.3 MeV at n-γ/DT-γ ~ 7 %. Measured n-γ signals for the five puck materials are used as simulation codes validation purpose such as a MCNP (Monte Carlo N-Particle) code for n-γ production and an ITS ACCEPT code for GCD spectral sensitivity.

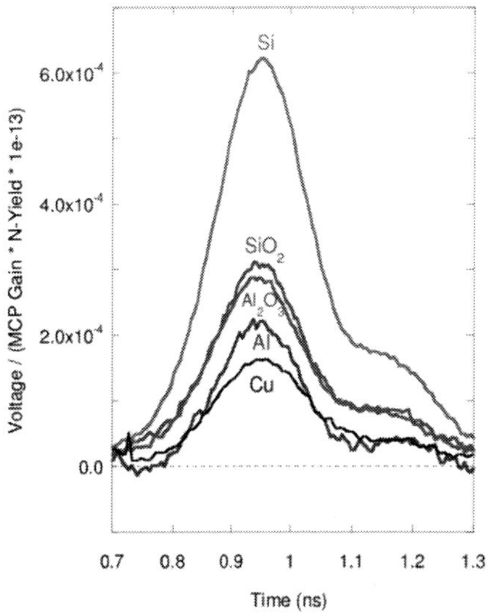

Figure 3: n-γ material sensitivity measured by GCD.

Using the Si puck, we changed the GCD threshold energy from 6.3 MeV (= 100 psi) to 8 MeV (= 64.5 psi) and 10 MeV (= 42.3 psi) to investigate relative sensitivity of the GCD to DT-γ at 16.7 MeV and n-γ's primarily above E_{th}. Figure 4 shows three measurements of DT-γ and Si n-γ as a function of GCD E_{th}. Precursors (peak A) are not affected by changing E_{th}

of GCD due to the fact that precursor is caused by prompt DT-γ and capsule n-γ, or their relativistic electrons, shining directly onto the PMT without incurring the 0.5 ns delay associated with a folded path in the GCD Cassegrainian optics. DT Cherenkov gammas (peak B) are reduced as the overall sensitivity of the GCD drops with increasing E_{th}. However, the Si n-γ signals are reduced much more significantly than the DT-γ signal. At E_{th} = 6.3 MeV, n-γ/DT-γ was about 27 %. At E_{th} = 8 MeV, the ratio dropped to ~ 9 %. At E_{th} = 10 MeV, the secondary signal was very small and the ratio was < 3 %. This threshold energy scan result demonstrates that secondary gammas can be thresholded out by changing gas pressure of the GCD. The GCD sensitivity to n-γ was simulated using MCNP for the Si n-γ spectrum (red curve in Fig. 4b) and using ITS ACCEPT for the GCD responses to various γ-ray energies (black curve in Fig. 4b) for varying E_{th}. Secondary n-γ spectrum decreases as energy increases, with little n-γ occurring above 10 MeV. Therefore, secondary gammas can be thresholded out by increasing E_{th}, while the sensitivity to DT-γ at 16.75 MeV is only marginally reduced.

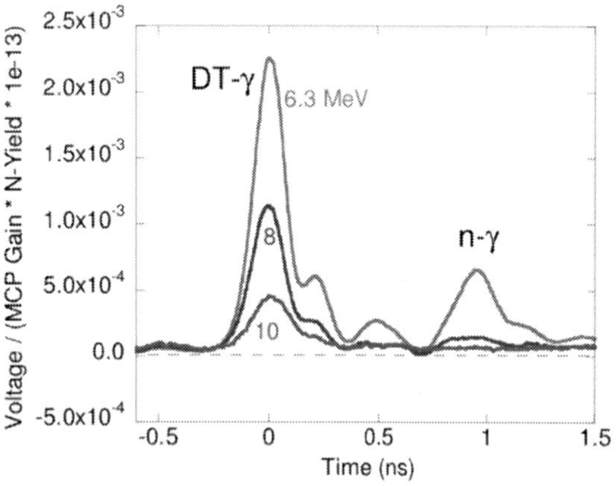

Figure 4a: Three measurements of DT-γ and n-γ as a function of GCD threshold energy.

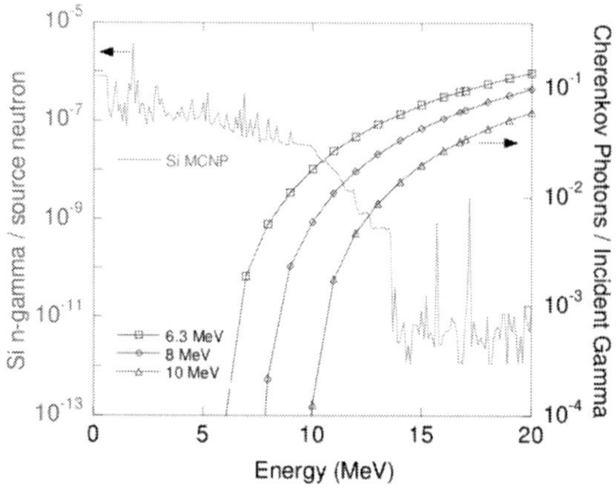

Figure 4b: Si n-γ spectrum simulated by MCNP (red curve) and GCD sensitivity as a function of threshold energy (black curves).

4. Conclusion

Characteristics of n-γ signals were experimentally studied using a GCD. High-band width of the GCD enabled us to detect time delayed and Doppler broadened n-γ signals from five different puck materials. Excellent GCD sensitivity was shown by measuring five n-γ signals, which were used for MCNP & ITS ACCEPT code validation purposes. By a simple change of GCD gas pressure, the system can effectively eliminate signals induced by n-γ reactions and thereby allow quality measurement of DT fusion γ-rays.

References

[1] Herrmann H W, et al., these proceedings
[2] Mack J M , et al., *2006 Rad. Phys. Chem.* **75** 551 and references therein

Investigations into Reconstruction Techniques for the National Ignition Facility Neutron Imaging System

E. Loomis[1], **G. Grim**[1], **C. Wilde**[1], **D.C. Wilson**[1], **M. Wilke**[1], **J. Finch**[2], **G. Morgan**[1], **I. Tregillis**[1], **D. Clark**[1]

[1]Physics Division, Los Alamos National Laboratory, Los Alamos, NM 87544, USA
[2]Global Nuclear Fuel, Wilmington, NC 28402, USA

E-mail: loomis@lanl.gov

Abstract. Neutron imaging is currently being developed as a primary diagnostic for inertial fusion studies at the National Ignition Facility (NIF). It is an attractive diagnostic for measuring asymmetries in the burn region and will be able to operate at neutron fluences found during ignition scale implosions. The most straightforward technique for imaging of the spatial distribution of deuterium-tritium (DT) fusion neutrons utilizes a simple pinhole aperture, which blocks all neutrons outside of the solid angle defined by the pinhole and results in a blurred image at the detector. We are currently investigating source image reconstruction techniques from detector images. Source reconstructions from Monte Carlo neutron transport (MCNP) calculations are shown to emulate hydrodynamic simulations with imposed Legendre asymmetries to high accuracy.

1. Introduction

With the completion of the National Ignition Facility (NIF), full-scale ignition attempts will soon be made to prove the viability of inertial confinement fusion (ICF) as a demonstration of controlled thermonuclear fusion opening the door for future fusion-based energy sources. Prior to these ignition attempts, ICF designs must rely upon the results from smaller scale experiments and complex numerical simulations to predict performance at the NIF. Some open questions standing in the way of ignition relate to implosion symmetry and our ability to optimize fusion burn.

Unfortunately, spatially-resolved x-ray diagnostics will not be able to operate at the neutron yields expected for ignition shots. The NIF neutron imaging system (NIS) is currently being developed to overcome this problem by directly imaging the spatial variation of deuterium-tritium (DT) fusion neutrons originating from the implosion [1, 2, 3]. (A description of the NIF-NIS design can be found elsewhere in these proceedings [Wilke].)

The basic idea of neutron imaging is that source neutrons originating outside of the field of view (FOV) of the aperture or originating in the FOV, but directed away from the aperture, are attenuated so that they make a negligible contribution to the recorded signal. For neutron imaging, the source fluence incident outside of the aperture opening must be attenuated using several centimeters of high-Z material given the large mean free path of neutrons in solid density matter. Image reconstruction techniques are then needed to reconstruct the source geometry from recorded images given knowledge of the neutron transport through the aperture.

The choice of aperture geometry has a great effect on image quality and often the type of reconstruction method needed. Penumbral apertures [1] have the advantage of higher signal-to-noise (SNR) ratio over that of pinhole imaging; however, pinholes can achieve a better spatial resolution. Reconstruction techniques designed for the former aperture type have used autocorrelation methods [4, 5] among others, which are limited to spatially invariant aperture point spread functions (PSF) or require the source to be centered in the FOV. This kind of precision alignment will be a difficult task at the NIF, which warrants the development of flexible and robust reconstruction techniques that are not limited to these conditions. Furthermore, it is expected that the NIF will produce several orders of magnitude higher neutron yields than any current laser facility along with improved background shielding so arrays of pinhole apertures should provide adequate SNR. Ring apertures [5] may be an alternative, but they must satisfy the ring delta function requirement [6] to high accuracy and may be limited by manufacturability.

For general imaging techniques not limited to the aformentioned conditions, formation of an image at the detector (in this case, scintillator) can be described mathematically by the product of a (4-dimensional) matrix containing the PSF from all points in the FOV to all points in the detector plane times a matrix containing the actual object (neutron source) (see Fig. 1).

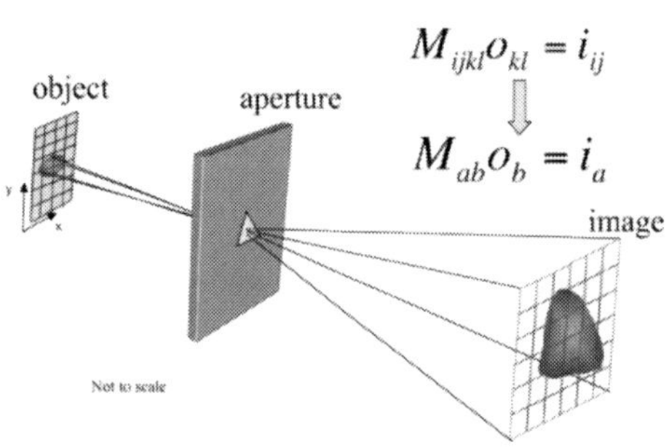

Figure 1. Schematic of image formation from neutron source

The image plane at the NIF is displaced by 28 m from the source plane resulting in a ~100 times magnification. Christensen et al. [7] applied a method based on singular value decomposition (SVD) to the linear system described in Fig. 1. The technique demonstrated the ability to reconstruct asymmetric neutron sources from experiments performed at the Omega Laser, however, the reconstructions were limited to ~6 μm pixels in the source plane for a ~185 μm × 185 μm FOV. Our current goal is to have 2 μm pixels with an equivalent or larger FOV to improve the attainable resolution with this technique. This paper describes the progress made to reach that goal. In the next section we discuss the reconstruction techniques currently being pursued along with the procedure used to form the reconstructed object. The techniques are demonstrated using numerical simulations of ICF implosions.

2. Procedure for NIF-NIS source reconstruction and relevant mathematics

The goal of reconstruction is to form the matrix $\mathbf{o}$ from knowledge of the transfer matrix $\mathbf{M}$ and detected image $\mathbf{i}$. The problem with this is that $\mathbf{M}$ cannot be inverted directly because it is extremely ill-conditioned and possibly rank deficient. Alternative methods can be used to approximate the matrix inversion, two of which have been implemented for the NIF-NIS. Both are based on finding solutions to the normal equations, which are the result of minimizing the residual in a least squares sense [8] giving

$$\mathbf{M}^T\mathbf{M}\mathbf{o} = \mathbf{M}^T\mathbf{i} \tag{1}$$

or the damped least squares (regularization) equivalent

$$\left(\mathbf{M}^T\mathbf{M} + \lambda^2\mathbf{L}^T\mathbf{L}\right)\mathbf{o} = \mathbf{M}^T\mathbf{i} \tag{2}$$

where the second term inside the parentheses damps high frequency noise in the solution.

To perform a reconstruction, knowledge of the spatial variation of the PSF is required. This information is obtained by propagating point sources of neutrons through the pinhole and onto a detector using a Monte Carlo particle transport code (MCNP). A single point source at any location in the source plane contributes a non-zero fluence to all points in the detector plane so if the calculation is repeated for all points in the source plane, a 4-dimensional transfer matrix M_{ijkl} is generated. The next step is to create the image i_{ij} of the object to be reconstructed. In the future, this will be the recorded image from implosion experiments, but to prove the technique, known objects are currently simulated using MCNP.

Once M_{ijkl} and i_{ij} have been formed, they are concatenated and sent to either reconstruction routine, which are described next. Once the reconstruction for o_b has been performed the array is converted back to o_{kl} resulting in the reconstructed source. The numerical methods are outlined below and focus is placed on their specific implementation for neutron image reconstruction.

2.1. SVD

The solution to Eqn. 1 in terms of the singular values and singular vectors is [9, 10, 11]

$$\mathbf{o} = \sum_{j=1}^{r} (1/\sigma_j)\, \mathbf{u}_j^T \mathbf{i} \mathbf{v}_j \tag{3}$$

where σ_j are the singular values of $\mathbf{M}$. A similar equation can be found for the regularized solution in [9]. The idea is to cut-off this expansion prior to reaching small singular values, which would otherwise result in noise amplification.

In our implementation, we use the CLAPACK (v. 3.1.1.1 [12]) function DGESVD to perform the decomposition and then reconstruct the solution using Eqn. 3. Calls to the DGESVD function require significant memory resources so all arrays are allocated dynamically. Upon output of the solution, the number of terms kept is chosen to optimize the reconstruction by balancing the norm of the solution and the norm of the residual [9].

2.2. Iterative

Besides the SVD, an iterative scheme has also been implemented that uses a biconjugate gradient (BCG) algorithm [13]. The BCG is the generalization of CG [8] for non-symmetric and non-positive definite systems. In our implementation, the BCG recursion is applied to the system in Eqn. 2 since it will reduce to Eqn. 1 by setting $\lambda = 0$.

The BCG is guaranteed to converge in N iterations where N is the dimension of the square matrix $\mathbf{M}$, although a stopping condition is placed on the recursion if the residual norm reaches a small enough value. This property of the BCG allows for much faster reconstructions and less memory requirements than for the SVD for large systems.

3. Simulated Reconstructions

We have recently adapted MCNP to accept source definitions of simulated implosions from radiation hydrodynamics calculations. With this tool, simulations of the NIF-NIS from the imploding capsule to the scintillator plane can be performed followed by reconstructions of known objects. Implosions have been simulated and used for image reconstruction consisting of low-mode Legendre asymmetries as shown in Fig. 2 a). After running the MCNP transport calculation the detector image shown in Fig. 2 b) is formed and is used in the reconstruction.

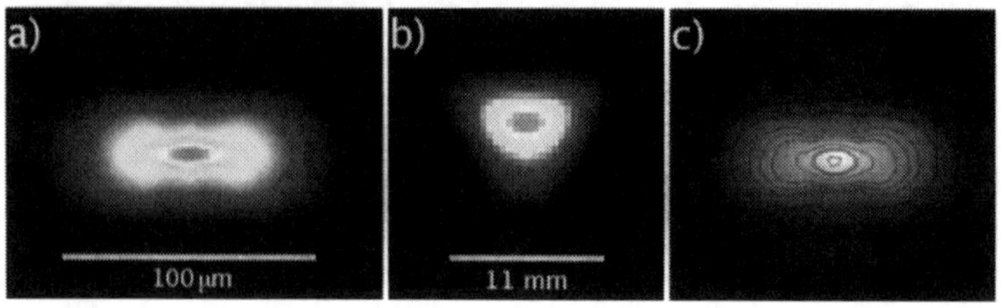

Figure 2. a) Simulated source with low-mode Legendre asymmetries b) Image of source at detector and c) contour plot of source reconstruction using 1000 term SVD expansion.

Note the triangular shape of this image, which is due to the triangular geometry of the current NIF pinhole. The transport calculations and reconstructions were performed using 74 $\times$ 74 pixels in both the source and image planes and the FOV of the source plane was 185 $\times$ 185 μm^2.

The original source shown in Fig. 2 a) was reconstructed using the SVD routine keeping 1000 terms in the expansion (BCG with 100 iterations gives very similar result) and the reconstruction is shown as a contour plot in c), which is seen to compare well with the input source. We are currently pursuing the development of a figure of merit for closeness of the reconstruction.

4. Conclusions

In this paper, we have demonstrated the ability to reconstruct objects from pinhole neutron images given knowledge of neutron transport through the aperture. We have found that SVD and BCG methods give very similar reconstructions although the BCG method is more efficient for large problems. The inclusion of noise in the reconstructions has begun with promising early results, but further tests and development are needed.

Acknowledgments

This work was performed for the U.S. Department of Energy for NNSA campaign 10 (Steve Batha, program manager). Eric Loomis would like to thank Valentina Staneva of Johns Hopkins University for helpful math discussions.

References

[1] Ress D, Lerche R, Ellis R, Lane S and Nugent K 1988 *Science* **241** 956–958
[2] Morgan G, Berggren R, Bradley P, Cverna F, Faulkner J, Gobby P, Oertel J, Swenson F, Tegtmeier J, Walton R, Wilke M, Wilson D and Disdier L 2001 *Review of Scientific Instruments* **72** 865–868
[3] Grim G, Morgan G, Wilke M, Gobby P, Christensen C and Wilson D 2004 *Review of Scientific Instruments* **75** 3572–3574
[4] Rouyer A 2003 *RSI* **74** 1234–1239
[5] Disdier L, Rouyer A, Lantuejoul I, Landoas O, Bourgade J, Sangster T, Glebov V and Lerche R 2006 *Phys Plasmas* **13** 056317
[6] Simpson R, Barrett H, Subach J and Fisher H 1975 *Optical Engineering* **14** 490–494
[7] Christensen C, Barnes C, Morgan G, Wilke M and Wilson D 2003 *Review of Scientific Instruments* **74** 2690–2694
[8] Strang G 1986 *Introduction to applied mathematics* (Wellesley: Wellesley-Cambridge Press)
[9] Hansen P 1992 *SIAM review* **34** 561–580
[10] Loan C V 1976 *SIAM J. Numer. Anal.* **13** 76–83
[11] Leon S 2002 *Linear Algebra: with applications* (New Jersey: Prentice Hall)
[12] Anderson E, Bai Z, Bischof C, Blackford S, Demmel J, Dongarra J, Du Croz J, Greenbaum A, Hammarling S, McKenney A and Sorensen D 1999 *LAPACK Users' Guide* 3rd ed (Philadelphia, PA: Society for Industrial and Applied Mathematics) ISBN 0-89871-447-8 (paperback)
[13] Saad Y 2000 *Iterative Methods for Sparse Linear Systems, 2nd edition* (online)

The Sixth International Conference on Inertial Fusion Sciences and Applications IOP Publishing
Journal of Physics: Conference Series **244** (2010) 032052 doi:10.1088/1742-6596/244/3/032052

Design and Construction of a Gamma Reaction History Diagnostic for the National Ignition Facility

R M Malone[1], B C Cox[1], S C Evans[2], B C Frogget[1], H W Herrmann[2], M I Kaufman[1], Y H Kim[2], J M Mack[2], K D McGillivray[3], M Palagi[3], W Stoeffl[4], A Tibbitts[1], T W Tunnell[1] and C S Young[2]

[1]National Security Technologies, LLC, P.O. Box 809, Los Alamos, NM 87544, USA
[2]Los Alamos National Laboratory, P.O. Box 1663, M/S E526, Los Alamos, NM 87545, USA
[3]National Security Technologies, LLC, P.O. Box 98521, Las Vegas, NV 89193, USA
[4]Lawrence Livermore National Laboratory, 7000 East Avenue, Livermore, CA 94550, USA

E-mail: malonerm@nv.doe.gov

Abstract. Gas Cherenkov detectors have been used to convert fusion gammas into photons to record gamma reaction history measurements. These gas detectors include a converter, pressurized gas volume, relay collection optics, and a photon detector. A novel design for the National Ignition Facility (NIF) using 90° off-axis parabolic mirrors efficiently collects signal from fusion gammas with 8-ps time dispersion. Fusion gammas are converted to Compton electrons, which generate broadband Cherenkov light (response is from 250 to 700 nm) in a pressurized gas cell. This light is relayed into a high-speed detector using three parabolic mirrors. The relay optics collect light from a 125-mm-diameter by 600-mm-long interchangeable gas (CO_2 or SF_6) volume. The parabolic mirrors were electroformed instead of diamond turned to reduce scattering of the UV light. All mirrors are bare aluminum coated for maximum reflectivity. This design incorporates a 4.2-ns time delay that allows the detector to recover from prompt radiation before it records the gamma signal. At NIF, a cluster of four channels will allow for increased dynamic range, as well as different gamma energy thresholds.

1. Introduction

The original Gas Cherenkov Detector (GCD) was fielded at the Laboratory for Laser Energetics (LLE) OMEGA laser facility to record high-energy gamma rays emitted during deuterium-tritium experiments [1]. Measuring fusion gammas provides precise bang time and burn width for understanding the fusion process. The original GCD used a compact optical layout making use of a Cassegrain telescope to image the back of the converter (Be or Al) onto a detector. However, light is generated through the volume of gas, not just at the converter, and this reduced collection efficiency. Because the GCD was light starved, it was moved close to the gamma source. Its position required penetration inside the OMEGA target vacuum chamber. Because of where the GCD detector is located, prompt gammas induce a precursor signal that is 0.5 ns ahead of the Cherenkov light. Tungsten shielding reduces the amplitude of this precursor signal.

The new gamma reaction history (GRH) diagnostic is external to the target chamber, requiring no chamber penetrations [2] [3]. There is no requirement to position the GRH at precise position from the gamma source. Just as in the original GCD, gammas from fusion reactions are converted to Compton electrons that are converted into Cherenkov photons. The Cassegrain optics are replaced with three 90° off-axis parabolic (OAP) mirrors to relay the light into a high-speed detector.

Because light is collected from source locations throughout a gas volume, the GRH detector is positioned at the stop position rather than at an image position. The stop diameter and its position are independent of the light-generation locations throughout the gas cell. This parabolic mirror relay is functioning as a non-imaging system. Because of the UV wavelength requirement, the parabolic mirror surface must be smooth to limit scattering.

2. Optical and Optomechanical designs

Figure 1 shows how light is collected into the detector. This example ray tracing shows five γs originating from a point at the target chamber center (TCC) that hit five field positions on the converter, producing Compton electrons at −2, 0, and +2 degrees. Here, Cherenkov light is emitted along each of these electron trajectories, and the detector collects light within a ±1-degree cone centered about each electron trajectory. The high-speed detector has a 1-cm collection area. The combination of OAP2 and OAP3 are used to demagnify the first stop diameter. Due to the 10 to 1 demagnification of the first stop onto the second stop (detector), the detector collection angle is limited to ±3 degrees of the source light, which can be summed up from any combination of electron and photon angles (shown here are ±2-degree electrons and ±1-degree photons).

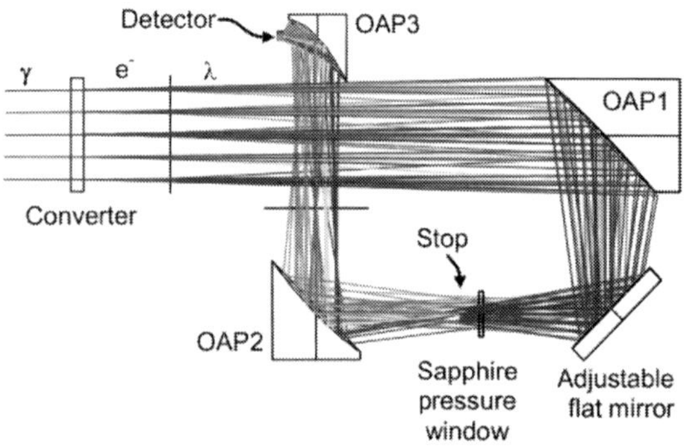

Figure 1. Cherenkov light is generated throughout the gas cell. Shown is optical ray tracing from one of many source planes.

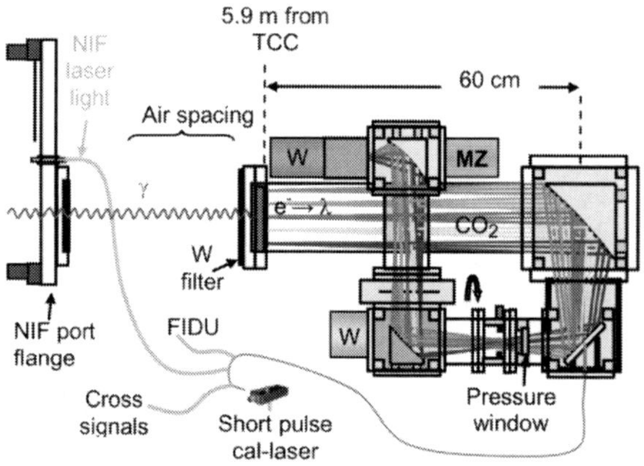

Figure 2. This new GRH diagnostic is external to the NIF target chamber. Optical fibers provide dry run and calibration signals.

Figure 2 shows the mounting hardware supporting the OAPs. The support structure is not shown. The Al converter measures 5 inches in diameter by 15-mm thick. Tungsten (W) shielding protects both the detector and the pressure window from line-of-sight and scattered radiation. A Mach-Zehnder (MZ) modulator transmits the data by optical fibers to a remote location and preserves signal bandwidth [4]. There is an air gap spacing between the chamber port and the converter housing. The OAP layout delays the Cherenkov light by 4.2 ns to enable the detector to recover from prompt gammas. Expected burn width of fusion is less than 1.5 ns. The temporal response of the optical system is limited by the dispersion of the sapphire pressure window to 8 ps over the wavelength range of 250 to 700 nm. Gammas will produce Cherenkov light in the sapphire pressure window, but the light goes the wrong direction due to the layout of the OAPs. Figure 3 shows a 3-D perspective of the mounting structures. The port cover is removable to exchange convertors or replace with an X-ray scintillator.

Optical ray tracings demonstrate how light can be collected from different angled trajectories of the Compton electrons as they traverse the gas volume. A Monte Carlo physics model of the conversion process from gammas to Cherenkov photons is used to generate photon trajectories. The collection efficiencies for different gamma energies that produce different electron angles are evaluated. Figure 4 shows an example of the ray counting from a set of photon trajectories produced inside a 100 psi CO_2 gas volume. Of the 10,000 photons that are generated by this Monte Carlo run, 524 usable photons make it into the detector. This demonstrates the challenge of collecting photons.

Inside the GRH enclosure, optical fibers insert light in two directions through a hole at the center of the turning mirror (see figure 2). This provides both calibration and dry run signals. A drive laser timing fiducial (FIDU) points directly at the detector (forward direction). Three fibers point towards the converter (backwards direction) and reflect their light off the converter to simulate a dry run light source starting from the converter plane. An additional target chamber light monitor signal collected by an optical fiber is multiplexed with the FIDU signal.

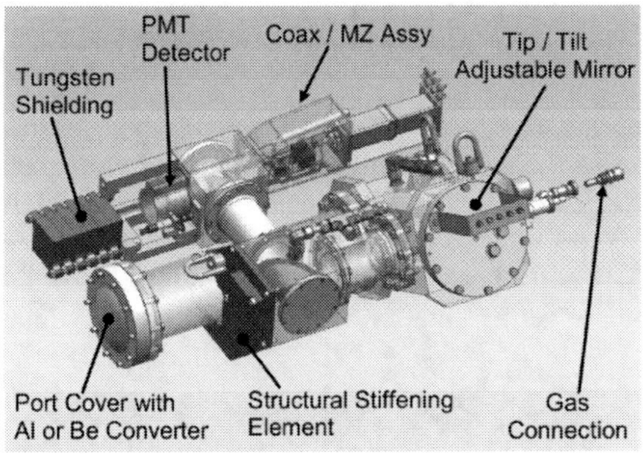

Figure 3. Component labeling of the GRH used at the LLE OMEGA. A Mach-Zehnder modulator transmits data to a remote location.

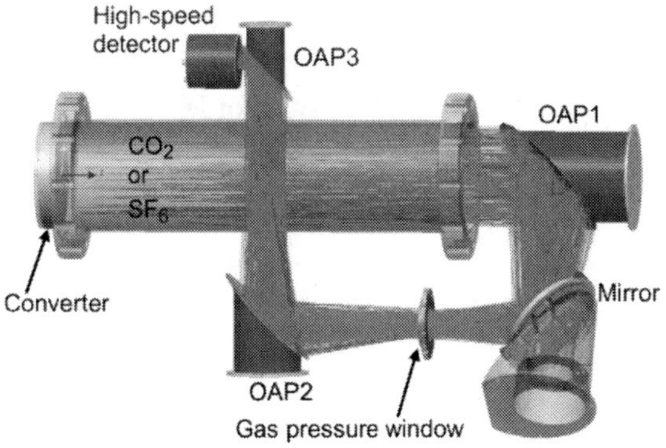

Figure 4. Monte Carlo rays derived from simulation code show that we collect 5.2% of Cherenkov light.

The problem with using relay optics with a volume source of light is that as the object position moves, the image position also moves. However, the stop position never moves and its diameter does not change, as its diameter is related to the full angle cone of light collected. The stop position is the best location to place a detector that is collecting light from many object source planes.

The parabolic mirrors were electroformed nickel mirrors instead of diamond-turned aluminum mirrors to reduce scattering of the UV light [5]. All mirrors are bare aluminum coated for maximum reflectivity at wavelengths below 400 nm. Because the amount of light scattered from a surface varies as the square of the ratio of surface roughness to wavelength, diamond turned OAPs will not work in the UV. We confirmed this formula by photographing the scattering at two different wavelengths on prototypes of both the electroformed nickel mirrors and the diamond-turned mirrors.

The optical alignment operation is fairly simple. Crosshairs are placed at the converter position. The detector is replaced with a grid pattern. The turning mirror has tilt adjustments. The hole in the turning mirror is tilted to line up with the cross hairs and grid pattern, sighted by eye. If alignment is not perfect, then a different set of rays are collected at the detector. The variation of ray count with misalignments is shown in figure 5, and demonstrates alignment sensitivity for the detector and the turning mirror. Figure 5 shows that the best z-axis position of the detector needs further optimization.

Double O-ring seals are used for both pressure and RF. The GRH enclosure prevents high-frequency electrical noise from affecting the photomultiplier tube (PMT) detector and associated electronics. The electrical noise can originate both from the environment and also from electron generation inside the gas cell. This enclosure has no welded parts. The external noise was blocked by making the enclosure electrically continuous with conductive O-rings. Chomerics Cho-seal 1285, a silicon O-ring impregnated with silver-plated aluminum, was used. The RF cavity resonances within this enclosure are damped with a magnetically loaded absorbing rubber. All enclosure surfaces were treated with a chemical conversion coating per MIL-D05541 class 3, which provided anti-corrosion protection and maximum electrical conductivity. Inside wall surfaces of the GRH were made to absorb light using a black Aeroglaze 300 series coating for vacuum systems, which does not outgas. Viton O-rings provide the pressure seal.

3. GRH installation on NIF

The initial NIF installation will have four GRH units, each set to a different gas pressure for detecting gammas ranging from 3.5 MeV to 20 MeV. For example, 200 psi of SF_6 gives a 3.5-MeV Cherenkov threshold, and 100 psi of CO_2 gives a 7-MeV Cherenkov threshold. [2] Each of the four units is attached to a support frame, as in figure 6. This frame is attached to a NIF chamber port. The four detectors are arranged into the middle of the cluster so they experience the same shielding environment. There is an 8-inch air separation between the chamber port and converter housing. The converter can be replaced with an X-ray scintillator when we record timing signals produced by the NIF drive lasers hitting a gold ball instead of a fusion capsule. The complete system attached to the port weighs 1500 lb. Design requirement for initial installation at NIF is to detect gammas emitted from fusion burns producing 10^{14} to 10^{16} neutrons and a system response time of 100 ps.

Numerous recording hardware components must be calibrated and maintained. Several high-bandwidth oscilloscopes with different time records archive the MZ signals [4]. Optical fibers feed light signals into each GRH unit. They supply both dry run and calibration signals. Another fiber looks through a small window at the center of this NIF port. This chamber light fiber provides an exact timing FIDU of the drive laser hitting target chamber center and, with a suitable fiber loop delay, this FIDU is added to the GRH data record.

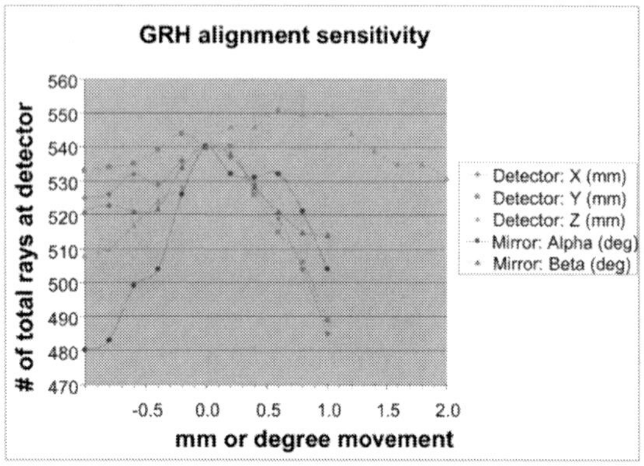

Figure 5. Optical illumination program optimizes the light collection efficiency of the detector.

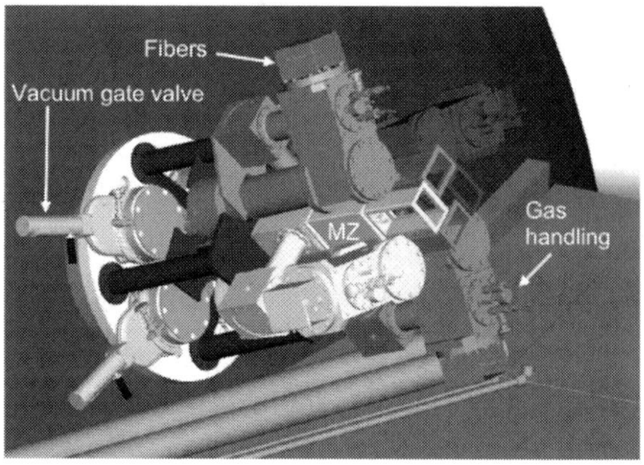

Figure 6. NIF will have four GRH units, each with a different gas pressure, for detecting gammas of several energy levels.

Acknowledgments
This manuscript has been authored by National Security Technologies, LLC, under Contract No. DE-AC52-06NA25946 with the U.S. DOE. The United States Government retains and the publisher, by accepting the article for publication, acknowledges that the United States Government retains a nonexclusive, paid-up, irrevocable, worldwide license to publish or reproduce the published form of this manuscript, or allow others to do so, for United States Government purposes.

References

[1] Mack J M, *et al., Nucl. Instrum. Methods Phys. Res.* A **513** 566
[2] Herrmann H W, *et al., Rev. Sci. Instrum.* **79** 10E531; also these proceedings
[3] Malone R M, *et al., Rev. Sci. Instrum.* **79** 10E532
[4] Miller K M, *et al.,* these proceedings
[5] Kaufman M I, *et al., Proc. SPIE* **7424** 74240M

Journal of Physics: Conference Series **244** (2010) 032053 doi:10.1088/1742-6596/244/3/032053

Coulomb explosions of deuterium clusters studied by compact design of Nomarski interferometer

Michaela Martinkova and Milan Kalal

Faculty of Nuclear Sciences and Physical Engineering, Czech Technical University in Prague, Brehova 7, 115 19 Prague 1, Czech Republic

E-mail: martimi2@fjfi.cvut.cz

Yong Joo Rhee

Korean Atomic Energy Research Institute, Daejon, 305-600, Korea

Abstract. Interactions of high-intensity femtosecond lasers with deuterium clusters leading to Coulombic explosions and subsequent production of *fusion neutrons* attracted in recent years considerable attention. In order to maximize the neutron yield finding a dependence of clusters *size* and their *spatial distribution* on experimental conditions became very important. In this paper a possibility to measure the deuterium clusters spatial distributions experimentally was analyzed. In combination with experiments recently performed in the Laboratory of Quantum Optics at the *Korea Atomic Energy Research Institute* (KAERI) *interferometry* was identified as the diagnostics suitable for such measurements.

1. Introduction

In recent years interactions of high-intensity *femtosecond* laser pulses with *deuterium clusters* leading to their *Coulombic explosions* and subsequent production of *fusion neutrons* attracted considerable attention [1]–[4]. From experiments performed it became clear that in order to *optimize* fusion neutrons production it would be very important to find a dependence of clusters *size* and their *spatial distribution* in the expanding deuterium gas on experimental conditions, in particular the *backing gas pressure* and *temperature*. This information would provide a deeper insight into the process of clusters *creation*, their *ionization*, and fusion neutrons production *efficiency*. For this purpose various diagnostics were employed, e.g. *Rayleigh scattering* or Taylor-Sedov *blast wave*. However, these diagnostics were able to determine the information required only a *qualitative* way. Therefore, some new approach would need to be looked for and applied. Such an ambitious project would need some rather *sophisticated diagnostics* to be implemented.

2. Preliminary experiments performed at KAERI

Several years ago experiments aimed to repeat/verify the results reported earlier were performed also in the Laboratory of Quantum Optics at the *Korea Atomic Energy Research Institute* (KAERI) using the Ti:Saphire laser system (30 fs, $I > 10^{16}\,\mathrm{W/cm^2}$). Successful production of fusion neutrons was achieved [5]–[7]. In these experiments *optical interferometry* was employed mainly to *visualize* the experimental conditions of the interaction process. This was followed by a series of experiments where interferometry was already used as a *potentially valuable diagnostics* in providing important information concerning the *clusters spatial distributions*. In order to test its limits the experimental setup illustrated in the Fig. 1 was used.

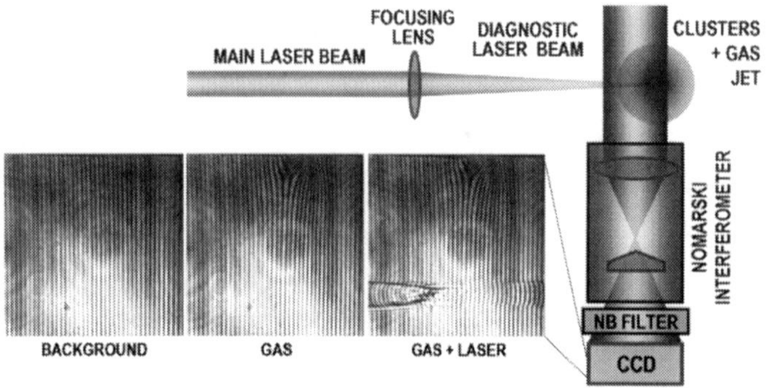

Figure 1. Experimental setup for testing sensitivity of interferometry diagnostics. The gas is injected into the vacuum chamber from the top. Very close resemblance of various structures in the signal free/low areas of individual interferograms is clearly visible.

It was confirmed that even if the diagnostic laser beam cross section was not exactly homogeneous and along its trajectory to CCD through various optical elements gradually accumulated some local disturbances, its structure from shot to shot remained practically the same. This feature allowed for a very effective usage of the *reference* shots (either the background ones or those without the action of the main laser) to be incorporated into FFT based interferogram analysis [8] –[10], thus providing results with a degree of accuracy potentially sufficient for clusters spatial profiles reconstruction. One example of the results obtained is demonstrated in the Fig. 2.

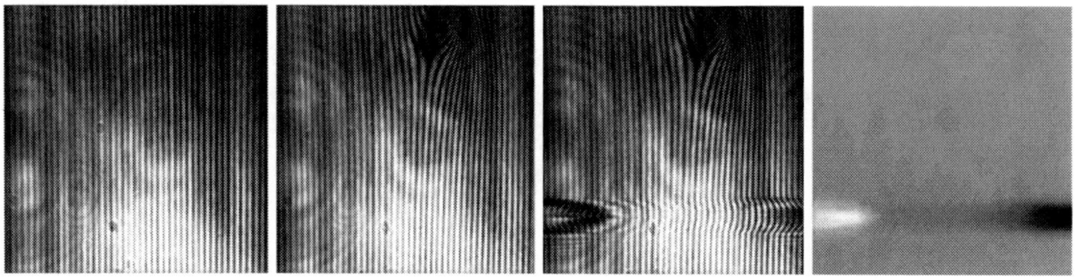

Figure 2. Interferograms from the left: background (vacuum); partially clusterized D_2 gas (injected from the top); laser interaction with partially clusterized D_2 gas. On the right: reconstructed phase shift of laser produced plasma (*dark* and *bright* regions - corresponding to positive and negative phase shifts - are the result of the *compact* interferometer design (for more details about this particular design see Ref. 10).

3. Theoretical analysis of clusters distribution measurement by interferometry

Encouraging results of the experiments performed at KAERI so far triggered further interest into preparation of a new series of experiments dedicated solely to the application of *optical interferometry* for determination of the actual *clusters spatial distributions*. In this Section some specific comments to this topic will be presented. It should be understood from the very beginning that this diagnostics does not provide any means how to determine the *actual sizes* of the clusters. Only some *average* values can be worked with contributing to the total *phase shift* of the probing part of the diagnostic beam along its path through the expanding gas/clusters jet according to the clusters respective spatial distribution.

Let's consider the situation when the injected deuterium gas is expanding into vacuum. In the *ideal* case of the *pure* gas (without any clusters created) the density profile in the cross-section perpendicular to the gas jet axis of symmetry at the distance z from the nozzle can be described by some function $f_1(r, z)$ where r represents the distance from the axis of symmetry. Now let's take into consideration that in the *real* case a certain part of the gas will participate in generation of clusters. Let's denote this density as $f_2(r, z)$ thus giving rise to the corresponding clusters density $f_C(r, z)$. When N would be the *effective* number of deuterium molecules in one cluster then $f_C(r, z) = f_2(r, z)/N$. Let's also presume that clusters and gas have *different* values of their respective *molar refractivity*. Under these assumptions the resulting *phase shifts* can be obtained from the following expressions:

$$\varphi_1(y, z) = k_1 A[f_1(r, z)], \tag{1}$$

$$\varphi_2(y, z) = k_1 A[f_1(r, z)] - (Nk_1 - k_C)A[f_C(r, z)], \tag{2}$$

where $\varphi_1(y, z)$ is the phase shift corresponding to the *pure* gas; $\varphi_2(y, z)$ is the phase shift of the gas/clusters *mixture*; k_1 and k_C are appropriate constants specific for the deuterium gas and the clusters, respectively; $A[f(r)]$ stands for the *Abel* integral transformation of the function $f(r)$ [11]. Provided that both the phase shifts $\varphi_1(y, z)$ and $\varphi_2(y, z)$ would be known, as well as the parameters k_1, k_C, and N, the resulting clusters spatial distribution can be found from the expression:

$$f_C(r, z) = \frac{A^{-1}[\varphi_1(y, z) - \varphi_2(y, z)]}{Nk_1 - k_C}, \tag{3}$$

where $A^{-1}[\varphi(y)]$ denotes the *Abel inversion* of the function $\varphi(y)$.

This theoretical analysis requires some important comments to be made from the point of its practical applicability. The first comment concerns the actual *knowledge* of various parameters needed for such analysis to be performed. The experimentally measured phase shift $\varphi_2(y, z)$ should pose no major problems. Also, the algorithm used for these data analysis should guarantee the required level of precision. One example of such analysis if provided in the Fig. 3. Somewhat different situation concerns the phase shift $\varphi_1(y, z)$. However, $\varphi_1(y, z)$ can actually be determined only *indirectly* as for the appropriate experimental conditions the partial clusterization already occurs thus the measured quantity would always be $\varphi_2(y, z)$. Hence, ways must be found to obtain this $\varphi_1(y, z)$ from measurements performed under slightly different conditions when no clusterization would occur combined with necessary *extrapolation* based on suitable theoretical analysis. As for the remaining parameters assembled in the denominator of the formula (3) they play only the role of the *multiplicative* factor. Therefore, if only the *relative* spatial distribution of clusters would be of some concern (and indeed in many cases such knowledge might be quite sufficient) they could be selected rather arbitrarily. For the knowledge of their *absolute* values, however, these parameters would need to be determined. Out of them k_1 is the easiest to obtain as it is directly related to D_2 gas molar refractivity. The average number of atoms in one cluster N can be approximately evaluated using the Hagena number [4]. The remaining parameter k_C related to molar refractivity of deuterium clusters is the most difficult to find. It might need to be obtained experimentally.

4. Conclusion

It this paper it was demonstrated that for the experimental conditions available in KAERI the optical interferometry diagnostics should be suitable for determination of deuterium clusters density spatial distribution assuming the effective size of clusters averaged over their size distribution (at least in the case when knowledge of relative density profiles would be sufficient).

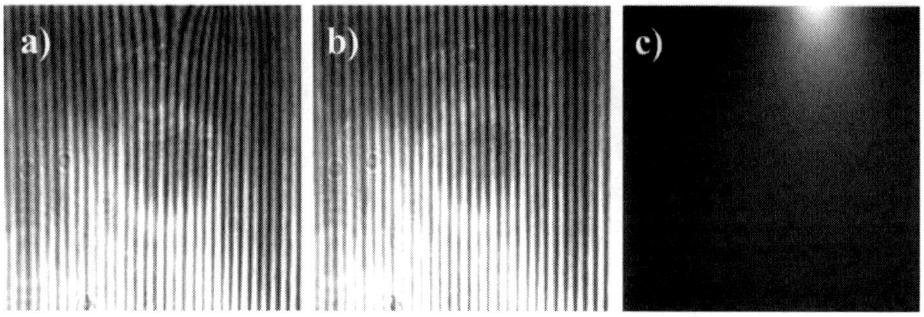

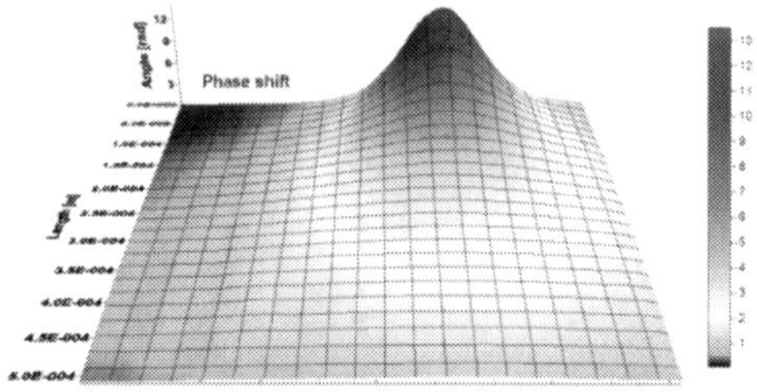

Figure 3. Interferograms of (a) partially clusterized D_2; (b) background (vacuum); (c) 2D reconstructed phase shift corresponding to (a) using (b) as a reference. 3D graph represents the same data as in (c).

5. Acknowledgments
This work has been supported by the IAEA RC No.: 13781; by the CTU in Prague grant No.: CTU0916314; by the Grant Agency of the Czech Republic grant No.: 202/08/H057; by the Ministry of Education, Youth, and Sports of the Czech Republic Research Project LC528, by the grant KONTAKT No. ME933; and the Narration of Prof. Stanislav Hanzl. This work has been also supported in part by the EU FP7 joint R&D project "Collaboration on HiPER project" funded by Korean MEST through NRF.

References
[1] Zweiback J, Cowan T E, Hartley J H, Howell R, Wharton K B, Crane J K, Smith R A and Ditmire T 2002 Detailed study of nuclear fusion from femtosecond laser-driven explosions of deuterium clusters *Phys. Plasmas* **9** 3108

[2] Ditmire T, Zweiback J, Yanovsky V P, Cowan T E, Hartley J H, Hays G and Wharton K B 1999 Nuclear fusion from explosions of femtosecond laser-heated deuterium clusters *Nature* **398** 489

[3] Ditmire T, Zweiback J, Yanovsky V P, Cowan T E, Hays G, and Wharton K B 2000 Nuclear fusion in gases of deuterium clusters heated with a femtosecond laser *Phys. Plasmas* **7** 1993

[4] Holkundar A R and Gupta N K 2008 Role of radial ninuniformities in the interaction of an intense laser with atomic clusters *Phys. Plasmas* **15** 123104

[5] Nam S, Han J, Rhee Y J, Cha Y H, Kwon D H, Lee Y W, Mun J H, Lee S, Kwon S and Cha H 2008 High-Energy Ion Emission from Deuterium Clusters by Using an Intense Femtosecond Laser *J. Kor. Phys. Soc.* **52** 1020

[6] Liu H J, Zheng Z J, Gu Y Q, Zhang B H, Rhee Y J, Nam S M, Han J M, Rhee Y W, Yea K H, Chen J B, Wang H B, Jiao Ch Y, He Y L, Wen T S, Wen X L and Chen M 2007 Deuterium Cluster Fusion Induced by the Intense Femtosecond Laser Pulse *Chinese Phys. Lett.* **24** 494

[7] Lee S, Kwon S, Lee K, Cha Y H, Kwon D H, Nam S, Yea K H, Lee Y W, Jeong Y U, Rhee Y J, Cha H and Lee J Y 2007 Effect of the Prepulse Width on the Neutron Generation in a Femtosecond Deuterated Polystyrene Plasma *J. Kor. Phys. Soc.* **51** 1695

[8] Nugent K A 1985 Interferogram Analysis Using an Accurate Fully Automatic Algorithm *Appl. Opt.* **24** 3101

[9] Kalal M, Luter-Davis B and Nugent K A 1988 Phase-amplitude imaging - The fully automated analysis of megagauss magnetic field measurements in laser-produced plasmas *J. Appl. Phys.* **64** 3845

[10] Kalal M, Slezak O, Martinkova M and Rhee Y J 2009 Compact Design of Nomarski Interferometer and its Application in Diagnostics of Coulomb Explosions of Deuterium Clusters *J. Kor. Phys. Soc.* **56** 287

[11] Kalal M and Nugent K A 1988 Abel Inversion Using Fast Fourier Transforms *Appl. Opt.* **27** 1956

The Sixth International Conference on Inertial Fusion Sciences and Applications IOP Publishing
Journal of Physics: Conference Series **244** (2010) 032054 doi:10.1088/1742-6596/244/3/032054

Design of High-Neutron-Yield, Polar-Drive Targets for Diagnostic Activation Experiments on the NIF

P W McKenty[1], R S Craxton[1], F J Marshall[1], T C Sangster[1], J A Marozas[1], A M Cok[1], M J Bonino[1], D R Harding[1], D D Meyerhofer[1,2], R L McCrory[1,2], J D Kilkenny[3], A Nikroo[3], J Fooks[3], M L Hoppe[3], J M Edwards[4], A J MacKinnon[4], D H Munro[4] and R J Wallace[4]

[1] Laboratory for Laser Energetics, University of Rochester, 250 E. River Road, Rochester, NY 14623, USA
[2] Departments of Mechanical Engineering and Physics & Astronomy, University of Rochester, NY 14623, USA
[3] General Atomics Corporation, La Jolla, CA, USA
[4] Lawrence Livermore National Laboratory, Livermore, CA, USA

E-mail: pmck@lle.rochester.edu

Abstract. Polar-drive (PD) target implosions have been designed for neutron diagnostic development on the NIF. These experiments use thin, room-temperature glass shells filled with low pressures of DT. Initial target implosions on the NIF will produce DT yields in the range of a few 10^{14} neutrons. The predicted yields are consistent with earlier data (10^{14} neutrons at 30 kJ) and recent PD scoping experiments performed on OMEGA. The experiments will use existing x-ray-drive phase plates with judicious repointing and defocusing to drive the implosions as uniformly as possible. These implosions have been modeled with three codes: *LILAC*, to optimize the 1-D design; *SAGE*, to optimize the pointing uniformity; and *DRACO*, to predict the yield from 2-D implosion simulations. Current simulation results indicate that the required yields will be obtained using up to 200-kJ UV light formed into a 1500-ps Gaussian pulse. Large-diameter glass shells (~1500-μm OD) are under development and fabrication at General Atomics. As tritium and environmental conditions evolve, similar target designs, with larger diameters and higher laser energies, are expected to produce thermonuclear yields approaching 10^{16} neutrons.

The recent dedication of the National Ignition Facility [1] (NIF) represents a landmark achievement for both the U. S. Department of Energy and its Inertial Confinement Fusion (ICF) program. As the NIF's young experimental campaigns begins to blossom, a strong emphasis has been placed on commissioning critical ignition diagnostics. Primary in this exercise is commissioning the neutron-detection and ion-temperature diagnostics. These diagnostics require a reliable and variable platform of neutrons to ensure their accurate calibration. Such a platform was demonstrated several years ago at the Omega Laser Facility [2] when the highest neutron yield (1.4×10^{14}) for an ICF experiment was obtained [3] using simple, room-temperature, gas-filled glass microballons.

These early OMEGA experiments laid the groundwork for a study prepared by Cok *et al.* [4], which examined the use of energy-scaled targets implementing polar drive (PD) to produce diagnostic neutrons for the NIF. An advantage of the PD configuration is its ability to uniformly drive a variety of room-temperature glass microballons [Fig. 1(a)], using simple Gaussian laser pulses [Fig. 1(b)] by means of minor adjustments to the NIF final optics [Fig. 1(c)]. The results of this study, shown in Fig. 1(d), clearly indicate a robust experimental platform capable of producing neutrons in the range from 1×10^{14} up to 1×10^{16} for energies from 100 kJ to 1 MJ.

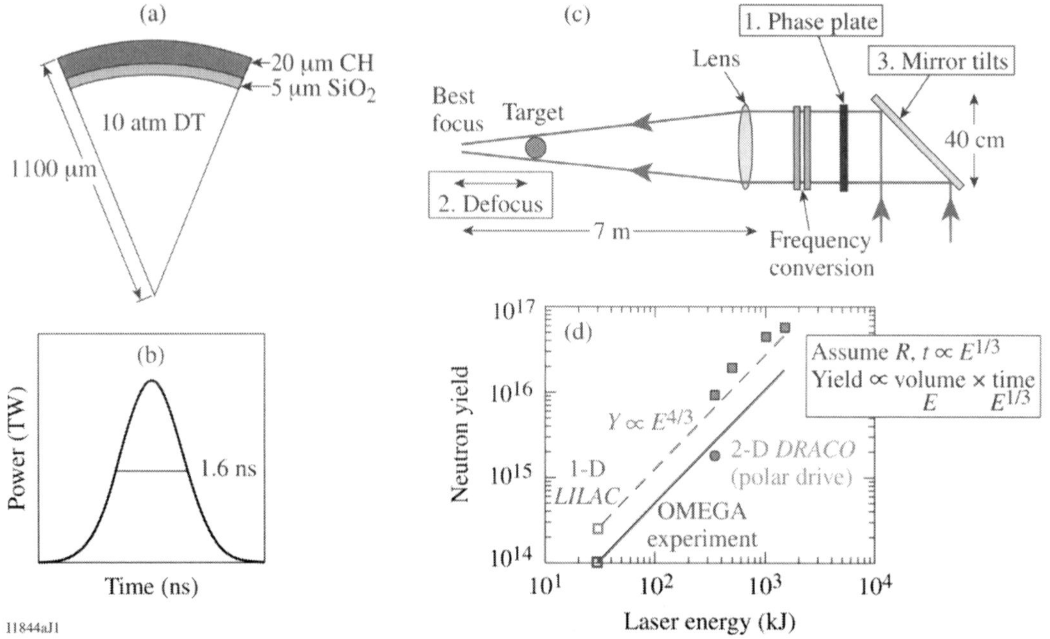

Figure 1. (a) Schematic of a typical glass microballoon target used in these simulations. (b) Laser pulse shape with incident energy of 350 kJ. (c) Final optics on the NIF used for polar drive: the existing x-ray-drive phase plates are used, the laser spot is defocused, and the individual beams are tilted to optimize laser uniformity. (d) Predicted target yields follow a simple energy scaling that is consistent with experimental results on OMEGA outlined in Ref. 3.

The current manufacturing capability for large-diameter, thin-walled glass shells is summarized in Fig. 2. From this map it can be seen that the original target design outlined in Ref. 4 had to been modified to fit within these limitations. The current design calls for a 4-μm shell with a 1500-μm diameter, which is represented by a star on the left edge of the shaded area in Fig. 2. Many of these shells were fabricated and sent to OMEGA for testing. The PD scheme, using 40 defocused and re-pointed beams, was implemented on OMEGA and a comparison of target performance using the PD scheme and energy-equivalent 60-beam symmetric illumination was performed.

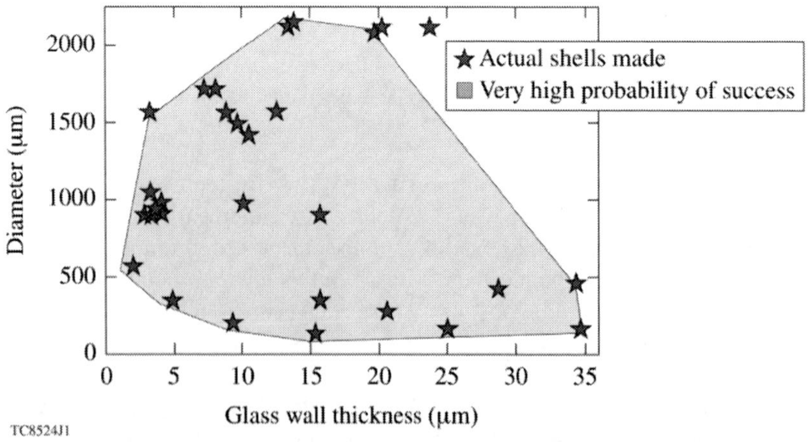

Figure 2. A plot showing the current parameter space of current glass microballoon target fabrication capabilities is used to determine the target designs described in this paper. Initial experiments on the NIF will use targets from the left-hand side of the space for diameters of 1500 μm, while future target designs will be chosen from the top of the region.

Figure 3(a) shows an image of one of the targets at target chamber center during these experiments. The experimental results, outlined in Fig. 3(b), indicate that the PD scheme can deliver approximately 60% of the clean symmetric one-dimensional (1-D) yield. One should also note the relative reproducibility of the platform, with each set of implosions lying within 10% of their individual mean.

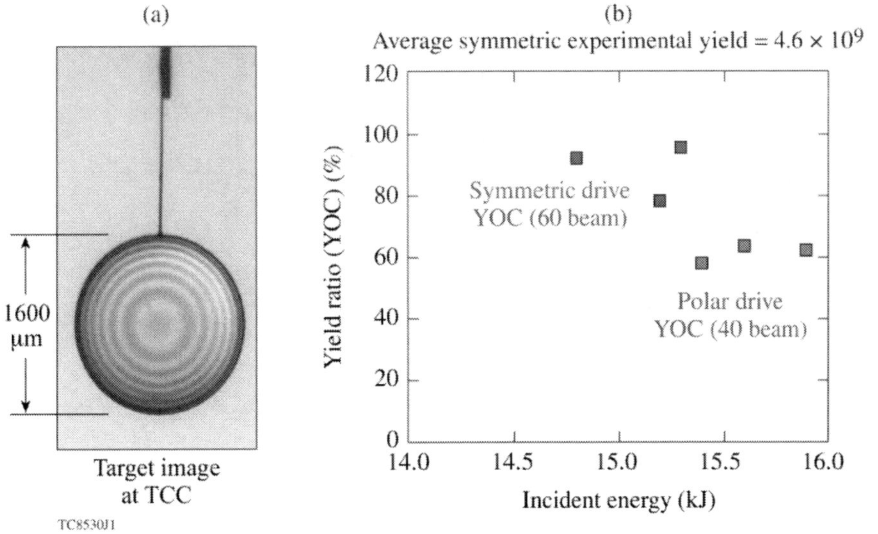

Figure 3. (a) Image of a target at target chamber center for the OMEGA experiments. (b) Experimental results for the yield ratio (yield relative to 1-D prediction) for targets driven using 60 beams in the symmetric configuration (blue) and 40 beams in the PD configuration (red) indicate that the PD concept can reliably produce 60% of symmetric calculations.

Initial experiments on the NIF are tentatively planned for late October 2009, while additional shots, implementing DT fills, are scheduled on OMEGA for the following month. With these results in hand, detailed designs will be developed using targets that sit at the upper limits of diameter given in Fig. 2 for higher-energy experiments with input energies approaching 1.0 MJ. These experiments are anticipated to produce neutron yields approaching 1×10^{16} and will serve to validate the robustness of the PD experimental platform in producing a reliable and reproducible source of neutrons over several orders of magnitude.

Acknowledgment

This work was supported by the U.S. Department of Energy Office of Inertial Confinement Fusion under Cooperative Agreement No. DE-FC52-08NA28302, the University of Rochester, and the New York State Energy Research and Development Authority. The support of DOE does not constitute an endorsement by DOE of the views expressed in this article.

References

[1] Campbell E M and Hogan W J 1999 *Plasma Phys. Control. Fusion* **41** B39–56
[2] Boehly T R *et al* 1995 *Rev. Sci. Instrum.* **66** 508–10
[3] Soures J M *et al* 1996 *Phys. Plasmas* **3** 2108–12
[4] Cok A M, Craxton R S and McKenty P W 2008 *Phys. Plasmas* **15** 082705

Journal of Physics: Conference Series 244 (2010) 032055 doi:10.1088/1742-6596/244/3/032055

Mach-Zehnder Fiber-optic Links for Reaction History Measurements at the National Ignition Facility

E Kirk Miller[1], H W Herrmann[2], W Stoeffl[3] and C J Horsfield[4]

[1] National Security Technologies, LLC, Special Technologies Laboratory, Santa Barbara, California, USA

[1] Email: millerek@nv.doe.gov

[2] Los Alamos National Laboratory, Los Alamos, New Mexico, USA

[3] Lawrence Livermore National Laboratory, Livermore, California, USA

[4] Atomic Weapons Establishment, Aldermaston, UK

Abstract. We present the details of the analog fiber-optic data link that will be used in the chamber-mounted Gamma Reaction History (GRH) diagnostic at the National Ignition Facility (NIF) located at the Lawrence Livermore Laboratory in Livermore, California. The system is based on Mach-Zehnder (MZ) modulators integrated into the diagnostic, with the source lasers and bias control electronics located remotely to protect the active electronics. A complete recording system for a single GRH channel comprises two MZ modulators, with the fiber signals split onto four channels on a single digitizer. By carefully selecting the attenuation, the photoreceiver, and the digitizer settings, the dynamic range achievable is greater than 1000:1 at the full system bandwidth of greater than 10 GHz. The system is designed to minimize electrical reflections and mitigate the effects of transient radiation darkening on the fibers.

1. Introduction

Traditionally, time-domain signals have been transmitted from high-bandwidth detectors to data recorders, such as oscilloscopes or digitizers, using coaxial cables. With coaxial cables, transmission bandwidth is maximized by (a) using high-quality cable, (b) keeping cable lengths as short as possible and (c) equalizing the cables, at the expense of signal amplitude. With short-pulse experiments like laser-driven inertial-confinement fusion (ICF) entering a new regime of energy and radiation emission, recording systems must be located 100 feet or more from the detectors [1]. At this length, preserving bandwidth above a few gigahertz for detectors with modest voltage output is not possible, so the National Ignition Facility (NIF) will use analog fiber-optic links to transmit data from the Gamma Reaction History (GRH) detectors to the digitizers [2].

The fiber links at NIF will initially be based on Mach-Zehnder (M-Z) modulators, commonly used in digital telecommunications because of their high bandwidth (> 10 GHz) and compatibility with low-loss single-mode optical fiber. M-Z transmission links are well suited to analog signal transmission and recording of pulsed power diagnostics for a few key reasons:

- Bandwidth is preserved over arbitrarily long fibers (kilometers or more).
- The M-Z modulator can tolerate large voltage pulses (> 500 V) without damage, thus optically isolating sensitive digitizer inputs from detector outputs.
- Modern digitizers, unlike their tube-based counterparts, have high-gain input amplifiers that are well matched to high-bandwidth photoreceivers which typically output less than one volt.

2. NIF Gamma Reaction History Mach-Zehnder System

Engineering an M-Z system for pulsed-power experimentation requires careful consideration of several details in order to optimize the use of commercially available components into a modestly priced system. In particular, we have optimized the NIF GRH system for low-noise, signal accuracy and tolerance to electromagnetic pulse (EMP) effects. The important steps to system optimization were:

- Eliminating switching power noise in the laser diodes, as off-the-shelf laser modules are typically designed for digital telecommunications systems which are insensitive to noise below 10 MHz
- Controlling the modulator bias before the shot with no RF signal on the input using components specifically designed for analog fiber-optic systems
- Shielding and isolating DC bias voltage from EMP by shunting high frequencies to ground through a 50 Ohm resistor using a bias tee

The system for the NIF GRH diagnostic has a modular design, with the active electronics (laser diode, bias controller, and recording system) on the diagnostic mezzanine, and only the modulator located at the target chamber wall. The default configuration for the NIF GRH M-Z system will use two M-Z modulators per detector, with a 10x difference in attenuation to provide a dynamic range of 1000:1 at the full system bandwidth. The system schematic is shown in Figure 1(a), with the corresponding digitizer channel coverage shown in part (b), as measured during the OMEGA high-yield campaign in July 2009.

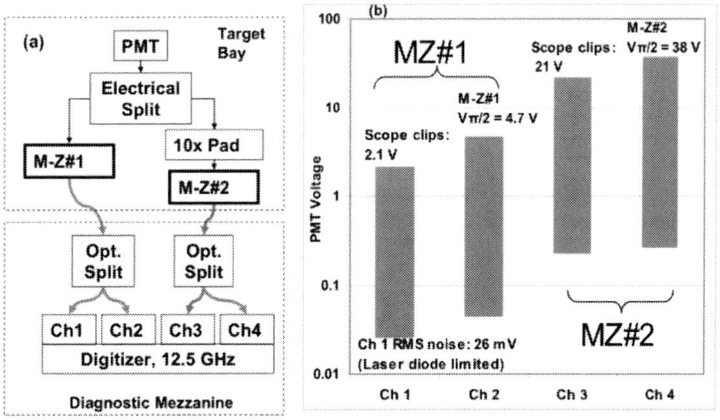

Figure 1: NIF GRH M-Z system. The schematic is shown in (a), with two M-Z modulators running to four digitizer channels. The system dynamic range is shown in (b), as measured during the high-yield shot campaign at OMEGA in July 2009.

Each M-Z optical signal is split using fiber couplers, and recorded on two photoreceivers attached to the digitizer. The noise in the laser diode ultimately determines the noise floor of the system; that noise level, however, is below the digitizer noise only if the signal is kept fully on scale. To achieve high dynamic range, one channel records the signal at the level of the laser noise and clips when the signal rises above a certain level, while the other channel records the full modulation amplitude of the

M-Z, though with higher noise floor due to the digitizer. This strategy is repeated for the second M-Z, giving overlapping coverage across three decades of signal level.

Four digitizer traces from one OMEGA shot are shown in Figure 2. The photomultiplier tube used on this shot has higher sensitivity and somewhat slower response than the tube more commonly used on the GRH for burn width measurements. However, this shot data shows the intrinsic dynamic range of the recording system, going from the 26 mV noise floor up to 27.5 V at the peak of the gamma signal. The data are composited using a noise-weighted spline fit, so that the traces with the lowest noise are most heavily weighted in the fit, until they go off-scale [3].

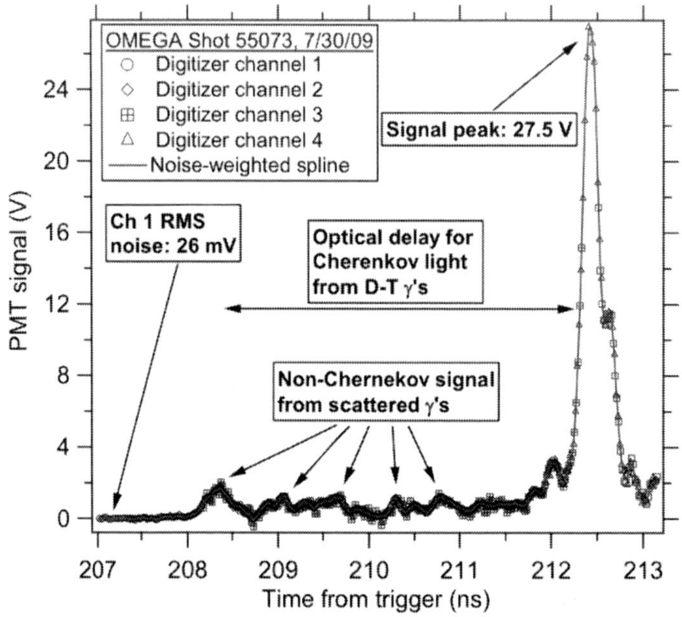

Figure 2: M-Z composite data from a high-yield shot at OMEGA, showing dynamic range of the M-Z recording system. The delay is engineered into the GRH to separate the Cherenkov signal from the prompt, scattered gamma signal.

3. Timing Measurement Considerations

One of the top-level requirements of the GRH diagnostic at NIF is precise measurement of so-called "bang time," which is the timing of the peak fusion reaction rate relative to the laser drive. For quantitative error estimation, we must consider the effect of peak width and noise level on the estimate of timing.

One strategy for determining peak time to better than one digitizer period is to fit a function to data, for example a Gaussian. With real data, such a strategy is always executed under circumstances of (a) finite sample intervals in the recorded data and (b) noise in the record. For additive white Gaussian noise on a Gaussian peak, the uncertainty in peak location due to least-squares fitting can be expressed as:

$$RMS\_Error \approx 0.007 * \underline{N} * \sqrt{\underline{W}}$$

Where $\underline{N}$ is the ratio of the RMS noise to the peak height, and $\underline{W}$ is the full-width-half-max of the peak, measured in digitizer samples.

Simulations were performed to verify this conclusion using 100 noisy Gaussian records at values of $\underline{N}$ from 0 to 20% (1% increments) and $\underline{W}$ from 4 to 20 (increments of 1). The results from the value $\underline{N}$=0.1 are shown in Figure 3(a), and good fits were obtained for all values of $\underline{N}$. We analyzed peaks in the M-Z records from the May 2009 high-yield campaign at OMEGA with this formalism, using peaks from the timing fiducial, precursor peaks, the gamma peak and the first peak after the gamma peak.

The results show very good agreement in the slope of the curve, though with some noise due to the much smaller number of samples, since each data point in (a) represents 100 simulated curves.

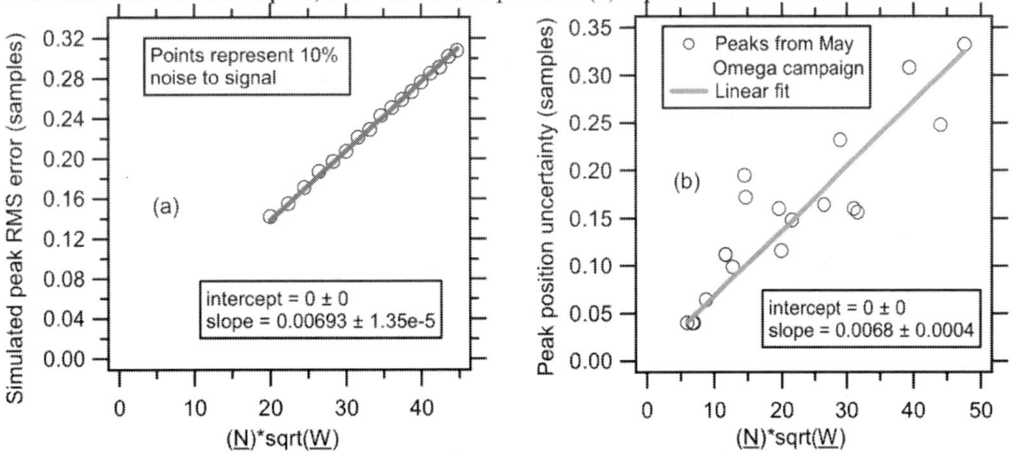

Figure 3: Peak uncertainty from Gaussian fitting. Simulation results for $\underline{N}$=0.1 are shown in (a), and measured data from the OMEGA high-yield campaign in May 2009 are shown in (b).

4. Conclusion

The Mach-Zehnder transmission and recording system for the NIF GRH diagnostic provides dynamic range of greater than 1000:1 at the full system bandwidth of 12 GHz. The system has a modular design based on low-noise electronics located on the diagnostic mezzanine and the optical modulator co-located with the detector on the target chamber wall. Results at OMEGA over the last three years show that the system is very reliable and that measurement of peak timing is robust, even at the subsample-interval timescale.

Acknowledgments

The authors wish to thank Mark May, John Galbraith, Monty Wood, Joe Mack, Carl Young, Tom Waltman and Jerry Blair for valuable discussions on reaction history measurements. Craig Halvorson, Bruce Marshall and Mark Lowry provided specific Mach-Zehnder expertise and encouragement. Fielding at OMEGA and the Idaho LINAC was made possible by Scott Evans, Tom Sedillo, Zaheer Ali, Greg Macrum, Yongho Kim, Dave Drews, Michael Rubery, Morris Kaufmann, Aric Tibbits, Bob Malone, Jim Tellinghuisen and Ian McKenna.

This manuscript has been authored by National Security Technologies, LLC, under Contract No. DE-AC52-06NA25946 with the U. S. Department of Energy.

References

[1] Miller, E. K., Macrum, G. S., McKenna, I. J., Herrmann, H. W., Mack, J. M., Young, C. S., Sedillo, T. J., Evans, S. C. and Horsfiled, C. J., 2007 *IEEE Trans. Nuc. Sci.***6-1** 2457

[2] Malone, R. M., Herrmann, H. W., Stoeffl, W., Mack, J. M., and Young C. S., 2008 *Rev. Sci. Instrum.* **79** 10E532

[3] Blair, J., 2007 *IEEE Instrumentation and Measurement Technology Conference Proceedings 2007 (Warsaw)*1

X-ray monochromatic high-speed imager for FIREX fast ignition research

Hiroaki Nishimura, Minoru Tanabe, Takashi Fujiwara, Shinsuke Fujioka, Hiroyuki Shiraga, and Hiroshi Azechi

Institute of Laser Engineering, Osaka University, 2-6 Yamadaoka, Suita, Osaka 565-0871 Japan

nishimu@ile.osaka-u.ac.jp

Abstract. A novel x-ray imager, consisting of two toroidally bent Bragg crystals and a 2D image sampler using a conventional x-ray streak camera, has been demonstrated for use in fast ignition resaerch. Sequential and 2D monochromatic x-ray images of laser-imploded core plasma were successfully obtained with a temporal resolution of 20 ps, a spatial resolution of 31 μm, and a spectral resolution of over 200, simultaneously.

1. Introduction

Fast ignition has been recognized as a pathway to efficient thermonuclear burn in laser-driven inertially confined fusion (ICF) plasmas [1]. This scheme differs from central ignition by using separate lasers for core plasma formation and additional heating processes with a large energy peta-watt laser. Only when the heating laser is injected within a plasma confinement time (typically less than 100 ps) [2], adequate energy transfer can be achieved. The scale of the core plasma is several tens of micrometers. Thus, high spatial and temporal resolutions, typically 10 μm and 10 ps, and seamless data acquisition are simultaneously required to investigate phenomena occurring in these short-lived dense plasmas. In addition, x-ray spectroscopy in the keV range can apply to diagnose plasma parameter such as electron temperature and density of the ICF plasmas.

Various x-ray spectroscopic and imaging methods have been used to measure electron temperature and density profiles of ICF plasmas [3-5]. Spatially, temporally, and spectrally resolved two-dimensional 2D images have been obtained by using a monochromatic x-ray framing camera [3]. However, its temporal resolution of 35 ps and frame interval of 50 ps do not allow for detailed observation of fast-ignitor plasmas.

We report a new type of monochromatic x-ray camera, called a monochromatic sampling image x-ray streak camera: M-SIXS. This method achieves high spatial, temporal, and spectral resolutions providing temporally continuous x-ray images of the core plasma.

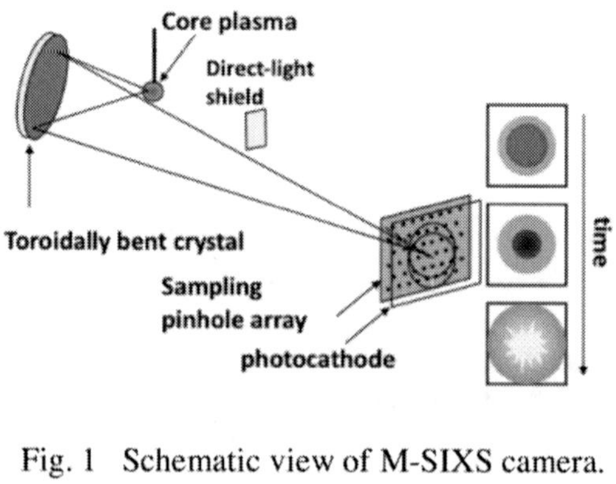

Fig. 1 Schematic view of M-SIXS camera.

2. Monochromatic sampling image x-ray streak camera

A basic M-SIXS consists of a toroidally bent Bragg crystal and a 2D sampling image x-ray streak camera SIXS [6]. Figure 1 shows a schematic view of a M-SIXS system. The use of a bent-crystal imaging device has several advantages in the keV x-ray region, such as high photon collection efficiency and high spectral and spatial resolutions, compared to an x-ray pinhole imager. Because of reflection optics, system geometry blocks energetic x rays and γ rays emitted from the fast-ignitor plasma with a thick direct shield. The toroidally bent crystal is bent to two different bend radii [7]. In order to avoid spherical aberration in the output image, toroidally bent crystals of two different radii are adopted. For a toroidal surface, the focal lengths f_m for the meridional plate and f_s for the sagittal plate are given by their respective direction radius R_m or R_s: $f_m=R_m \sin\theta_B/2$ and $f_s=R_s/2\sin\theta_B$. Here θ_B is the Bragg angle for the wavelength of the x ray. These focal lengths must be equal, providing the relation $R_s/R_m=\sin^2\theta_B$. When the image magnification is larger than 1, the spectral range covered by the crystal is primarily determined by the variation of the Bragg angle compared with the rocking-curve width [8]. In practice, a spatial resolution of less than 10 μm has been attained [3].

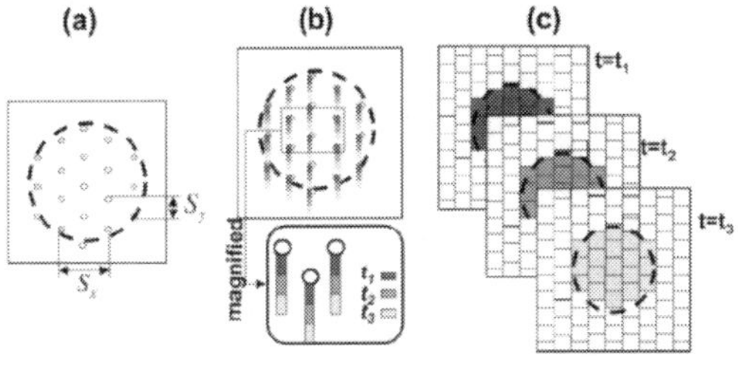

Fig. 2 Processes of image sampling and reconstruction. (a) Static x-ray image, (b) streaked x-ray image of the sampling points, and (c) reconstructed 2D images.

SIXS applies a 2D image sampling technique to an x-ray streak camera (XSC). This is suitable for observation of rapidly changing small x-ray sources. As shown in Fig. 1, an image of the object is focused on the photocathode, and then sampled by placing a periodic pinhole-array set in front of the photocathode. Figure 2 shows the image sampling and reconstruction processes. A static image, obtained without supplying a sweep voltage to the streak tube, is shown in Fig. 2(a), providing the absolute position of each sampling point. A streaked image of sampling points is shown in Fig. 2(b). Time-resolved 2D images can be reconstructed in such that the sampled images for a specific time are re-arranged as shown in Fig. 2(c). Spatial non-uniformity of photocathode sensitivity and dispersion of pinhole size are considered in the reconstruction process.

According to sampling theory, the sampling distances for image sampling Sx, Sy should be Sx, $Sy \leq (1/2)M\Delta r_{\text{imager}}$ in order to cover all of spatial frequencies in the original image information, where M and Δr_{imager} are the image magnification and the spatial resolution of the imager. If the sampling distance is larger than $(1/2)M\Delta r_{\text{imager}}$, some of the original image information between the sampling points is lost. Spatial resolution of the system, as determined by the sampling distances, is defined as $\Delta r_x=2S_x/M$ or $\Delta r_y=2S_y/M$. The size of the sampling pinhole does not affect the spatial resolution, because the sampling distance is much larger than the pinhole diameter. Temporal resolution is determined by both pinhole size and XSC sweep speed, and the window of available time is determined by the pinhole spacing in the sweep direction [6].

3. Experiment

Two x-ray photon energies, 3.27 keV corresponding to Cl^{15+} Heβ and 3.51 keV corresponding to Cl^{16+} Lyβ, were chosen to derive an electron temperature profile of the laser imploded-core plasma, supposing a case that chlorine is doped as a seed material in the fusion capsules. Specifications of the toroidally bent crystals dedicated to these two lines are listed in Table I. A100- μm-thick beryllium (Be) filter was set in front of each crystal to shield them from plasma debris. The crystals were mounted on miniature goniometers and *x*-*y*-*z*-translation stages to allow alignment to the correct tilts and distances. The initial alignment of the system was made optically, and then the final confirmations were made using the x-ray from laser-produced plasma. Small angular deviations between the optical and x-ray images have been investigated separately with an x-ray tube [9]. The measured angular

TABLE I. Specifications of two crystals.

X-ray energy (keV)	3.27	3.51
Bragg crystal	Silicon (220)	Quartz (11.2)
Bragg angle (deg)	80.7	76.4
Bent radii (mm)	200/195.8	200/189.5
Spectral window (eV)	11.7	26.2
Image magnification	25.8	26.2
Target-crystal (mm)	102.5	100.9
Crystal-detector (mm)	2648.2	2640.7

discrepancies were less than 1 mrad.

A sampling pinhole array was placed in front of the photocathode (13.2x17.3 mm^2) with a separation distance of 5 mm. The 55x34 pinholes were fabricated on a 25-μm-thick nickel substrate by a photolithographic method. The diameter of the pinholes was 30 ± 3 μm. The x and y sampling distances between the pinholes were 400 μm. Although the spatial resolutions of the both toroidally bent crystals were 10 μm, the sampling distance was three times larger than $(1/2)M\Delta r_{imager}$, so some of original image information between the sampling points was lost. For the setup adopted in this experiment, the resultant spatial resolution was 31 μm and the temporal resolution was 20 ps. In addition, the pinhole separation along the sweep direction was 800 μm, creating an observable time-window of 600 ps. The photocathode was made of a 0.03-μm-thick gold, coated on a 3-μm-thick plastic film. A 100-μm-thick Be filter was placed in front of the photocathode in order to absorb scattered visible light and soft x rays.

A proof-of-principle experiment was performed at the GEKKO XII laser facility. A deuterated plastic CD shell target was irradiated with 12 beams of 527 nm in wavelength, totaling 4 kJ of energy in a Gaussian pulse with a 1.3 ns full width at half maximum. To smoothen the laser intensity profile on the shell surface, random phase plates were installed for each laser beam. The laser beams were focused at d/R=-5, where R is the shell radius and d is the distance from the shell center to the laser focus point. d is defined as negative when the focus point is beyond the shell center. The CD plastic shell target was 481 μm in diameter and 5.23 μm in wall thickness.

Figure 3 shows a streaked monochromatic x-ray image of the laser-imploded-core plasma at 3.27 keV. Figure 4(a) shows the reconstructed 2D x-ray images of the core plasma at 3.27 keV. The outer frame sizes of the 2D images are 120x120 μm^2 and the frame interval is 20 ps. The time origin was defined as the time corresponding to half of the x-ray intensity peak in the streaked image shown in Fig. 3. Similarly, the reconstructed images for 3.51 keV were obtained, as shown in Fig. 4 (b). The two images, for 3.27 and 3.51 keV, were simultaneously recorded on the same photocathode for the same shot. The time origin for 3.51 keV corresponds to that of 3.27 keV. X-ray intensities for the two images gradually increase in time and reach their peak at 160 ps. The imploded core is about 50 μm in diameter.

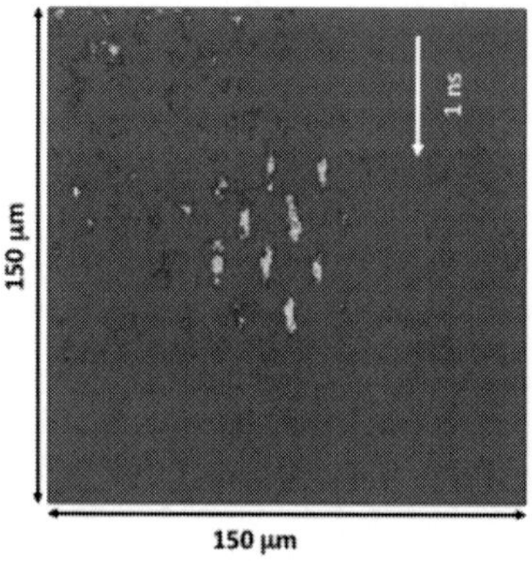

Fig. 3 Streaked monochromatic x-ray image of the laser imploded core plasma at 3.27 keV.

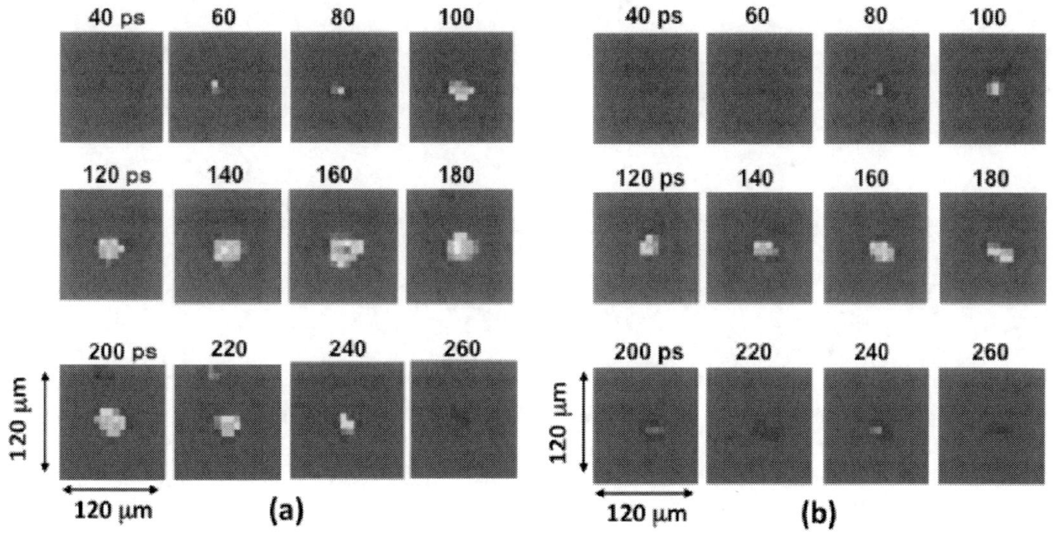

Fig. 4 Reconstructed monochromatic x-ray images of the core plasma for (a) 3.27 keV and for (b) 3.51 keV x ray. Frame interval is 20 ps.

4. Summary

A new monochromatic x-ray imager has been proposed, and its feasibility was successfully demonstrated with a temporal resolution of 20 ps, a spatial resolution of 31 μm, and a spectral resolution over 200, simultaneously. The system provided seamless images with high temporal resolution. The spatial resolution has not yet been sufficiently optimized for practical use in implosion experiments. A combination of higher sweep-speed, larger image magnification, and narrower sampling distances should lead to a temporal resolution better than 10 ps and a spatial resolution better than 10 μm. By using a chlorine-doped shell target, we plan to obtain temporal evolution of electron temperature profiles of fast-ignitor plasmas created with Gekko-XII and LFEX laser facilities.

References

[1] M. Tabak, J. Hammer, M. E. Glinsky, W. L. Kruer, S. C. Wilks, J. Woodworth, E. M. Campbell, M. D. Perry, and R. J. Mason, Phys. Plasmas **1**, 1626 (1994).

[2] R. Kodama, H. Shiraga, K. Shigemori, Y. Toyama, S. Fujioka, H. Azechi, H. Fujita, H. Habara, T. Hall, Y. Izawa *et al.*, Nature London **418**, 933 (2002).

[3] K. Fujita, H. Nishimura, I. Niki, J. Funakura, I. Uschmann, R. Butzbach, E. Förster, M. Nakai, M. Fukao, A. Sunahara *et al.*, Rev. Sci. Instrum. **72**, 744 (2001).

[4] J. A. Koch, J. T. W. Barbee, N. Izumi, R. Tommasini, R. C. Mancini, L. A. Welser, and F. J. Marshall, Rev. Sci. Instrum. **76**, 073708 (2005).

[5] N. Miyanaga, Y. Aoki, H. Shiraga, K. Shimada, K. Fujimoto, M. Heya, and M. Nakasuji, Rev. Sci. Instrum. **68**, 817 (1997).

[6] H. Shiraga, M. Nakasuji, M. Heya, and N. Miyanaga, Rev. Sci. Instrum. **70**, 620 (1999).

[7] E. Förster, K. Gäbel, and I. Ushmann, Laser Part. Beams **9**, 135 (1991).

[8] I. Uschmann, E. Förster, K. Gäbel, G. Hölzer, and M. Ensslen, J. Appl. Crystallogr. **26**, 405 (1993).

[9] M. Vollbrecht, I. Uschmann, E. Förster, K. Fujita, Y. Ochi, H. Nishimura, and K. Mima, J. Quant. Spectrosc. Radiat. Transf. **58**, 965 (1997).

Lasnex simulations of NIF vacuum hohlraum commissioning experiments

R. E. Olson[1], L. J. Suter[2], J. L. Kline[3], D. A. Callahan[2], M. D. Rosen[2], K. Widmann[2], E. A. Williams[2], D. E. Hinkel[2], N. B. Meezan[2], G. A. Rochau[1], A. L. Warrick[2], S. H. Langer[2], C. Thomas[2], S. N. Dixit[2], E. L. Dewald[2], M. B. Schneider[2], J. D. Moody[2], P. Michel[2], R. J. Wallace[2], O. L. Landen[2], J. Edwards[2], B. J. MacGowan[2], S. H. Glenzer[2]

[1]Sandia National Laboratories, Albuquerque, New Mexico 871815
[2]Lawrence Livermore National Laboratory, Livermore, California 94550
[3]Los Alamos National Laboratory, Los Alamos, New Mexico 87545

e-mail: reolson@sandia.gov

Abstract. Lasnex calculations and x-ray flux measurements are presented for a series of NIF vacuum hohlraum experiments that were among the first targets shot on NIF as part of the facility commissioning. An important result is that the hohlraum x-ray fluxes are significantly higher than predicted by pre-shot Lasnex calculations employing the baseline "configuration managed" physics packages used in the NIF ignition target calculations. A possible explanation for the high-flux vacuum hohlraum result has been explored via post-shot calculations in which non-baseline emissivity and heat conduction models are used.

1. Introduction

Among the first targets shot at NIF during the facility commissioning were empty, or "vacuum" gold hohlraums. Although these cylindrical targets were relatively simple, the on-target laser energies were up to 20x the maximum energy used in previous hohlraum experiments at Nova and Omega, and the targets had approximately 5x the surface area and more than 10x the volume of previous vacuum hohlraums heated to comparable radiation temperatures. An important goal of the experimental series was to validate the Lasnex code and associated physics packages at the increased size and energy regimes now available for hohlraum experiments at the NIF. An additional goal was to qualify the NIF Dante soft x-ray spectrometer [1].

Two types of vacuum hohlraums were used. "Scale 0.7" hohlraums were 6.40 mm long, 3.55 mm in diameter, with 2.65 mm diameter laser entrance holes (LEH), and "scale 0.9" hohlraums were 8.44 mm long and 4.55 mm in diameter with 2.95 mm diameter LEH. The hohlraum walls were 25 micron thick Au. Depending on the experiment, the hohlraums were heated with either half the NIF beams (actually 88 or 96 beams) or all the NIF beams (actually 184 or 192 beams). In all experiments, the laser pulse shape was nominally 2 ns square, and the laser power was approximately equally divided between the two LEH's. The peak on-target laser intensities ranged from 1.2×10^{14} to 2.8×10^{14} w/cm^2. The laser energies, peak intensities, and peak radiation temperatures for each experiment are indicated in Fig. 1.

The x-ray flux measurements were made with the NIF Dante [1] looking into the lower LEH at an angle of 37.4 degrees relative to the hohlraum axis. Other diagnostics used in these experiments included a Static X-ray Imager (SXI) [2], hard x-ray spectrometer (FFLEX) [3], Near Backscatter Imager (NBI) [4], and Full Aperture Backscatter (FABS) [4]. Further details concerning the experimental setup and measurements of the NIF vacuum hohlraum series are provided in Ref. 5.

The Lasnex calculations were done in 2D, with the laser beams approximated as cones entering the LEH at angles of 30 and 50 degrees for the 88 or 96 beam experiments and at 23.5, 30, 44.5, and 50 degrees (relative to the hohlraum axis) for the 184 and 192 beam experiments. In all calculations and experiments, the outer cone energy was a factor of 2 larger than the inner cone energy. The Lasnex output was post-processed to provide the simulated time- and spectral-resolved flux that would be seen by Dante at the angle of 37.4 degrees relative to the hohlraum axis.

2. X-ray Flux Measurements and Pre-Shot Lasnex Simulations

The NIF Dante [1] consists of an array of 18 K- and L-edge filtered photo-cathodes ("XRDs") with some channels also employing grazing-incidence mirrors. As described in Ref. 6, the NIF Dante filters, mirrors, and XRDs are calibrated with a random flux error of 5-18% (depending on the energy channel), with a total flux uncertainty of about 7%. A new data unfold code is being developed and tested as a part of the NIF Dante qualification effort. It is anticipated that, within the next few months the new code will be incorporated into the NIF Shot Analysis and Data Visualization (SADV) website. In the meantime, however, an existing "desktop" Dante unfold routine has been employed here as a data analysis tool. The "desktop" code has been used routinely by Sandia National Laboratory in the analysis of Omega hohlraum data (eg, Ref. 7). It is based upon the iterative "test-spectrum" method described in Ref 8, and its Omega Dante results have been cross-compared with similar unfold codes that exist at other laboratories. Inputs into the unfold code include raw digitizer "scope" files, collimator and solid angle measurements, cable compensation measurements, attenuator and signal divider calibrations, and response functions (which have been derived from the filter, mirror, and XRD calibrations [6]). The total flux vs. time unfolds shown in Figs 2-6 were done using a version of the Sandia desktop code that was modified for use with NIF Dante data files. As such, the work is somewhat preliminary. At the time of this conference, the NIF Dante is new, the NIF Dante SADV analysis routine is not yet qualified, and all aspects of the measurements and analysis are still being closely scrutinized.

The NIF ignition campaign includes a number of different types of experiments (eg., hohlraum energetics, shock timing, symmetry, convergent ablation). During the planning phases of the ignition campaign, a "configuration managed" Lasnex setup was established to ensure that the different program elements would all be using physics assumptions that are consistent with the baseline ignition target design calculations [9]. For the vacuum hohlraum experiments, the pre-shot Lasnex simulations were done using the "configuration managed" Lasnex setup. The pre-shot Lasnex Dante flux predictions are shown as the dashed curves labeled "Lasnex (CM)" in Figs. 2-6. As can be seen in the figures, the peak measured Dante fluxes are ~20-50% higher than the pre-shot predictions.

3. Post-Shot Lasnex Simulations

Although the Dante x-ray flux measurements are much higher than predicted by CM-Lasnex calculations, the results were not entirely unexpected. It has been known for some time that detailed configuration NLTE emissivity models that produce more radiation per unit volume than the default Lasnex average atom NLTE model will increase the efficiency of the hohlraum [10]. Furthermore, it is expected that the increase in hohlraum x-ray conversion efficiency will become more pronounced at larger scale sizes due to the larger volume-to-surface ratio [11]. It has also been known that the default Lasnex electron flux limiter of 0.05 can lead to an underestimate of the heat conduction losses from the Au coronal plasma, and that a flux limiter of 0.15 produces a better match to experimental data in Au sphere x-ray conversion efficiency measurements at Omega [12]. The plasma emissivity in CM-Lasnex is calculated using the average-atom XSN package [13]. Historically, XSN has been used in integrated computational simulations because it is fast enough for in-line use. However, its accuracy is severely limited. Upon obtaining the results shown in Fig. 2-6, a series of post-shot Lasnex calculations were run using recently-developed methods [14] for constructing atomic models

The first vacuum hohlraum experiments at NIF

NIF shot number	# beams & energy	laser intensity	Peak Tr
Scale 0.7:			
N090709	96 beams, 150 kJ,	1.2×10^{14} w/cm^2	Peak T_r ~ 235 eV
N090711	88 beams, 308 kJ,	2.8×10^{14} w/cm^2	Peak T_r ~ 290 eV
N090713	88 beams, 306 kJ,	Au:B liner;	Peak T_r ~ 290 eV
N090728	184 beams, 635 kJ,	2.7×10^{14} w/cm^2	Peak T_r ~ 345 eV
N090808	96 beams, 233 kJ,	1.9×10^{14} w/cm^2	Peak T_r ~ 265 eV
Scale 0.9:			
N090817	192 beams, 591 kJ,	2.5×10^{14} w/cm^2	Peak T_r ~ 310 eV
N090828	192 beams, 588 kJ,	cryogenic;	Peak T_r ~ 310 eV

Fig. 1 NIF vacuum hohlraum shot summary.

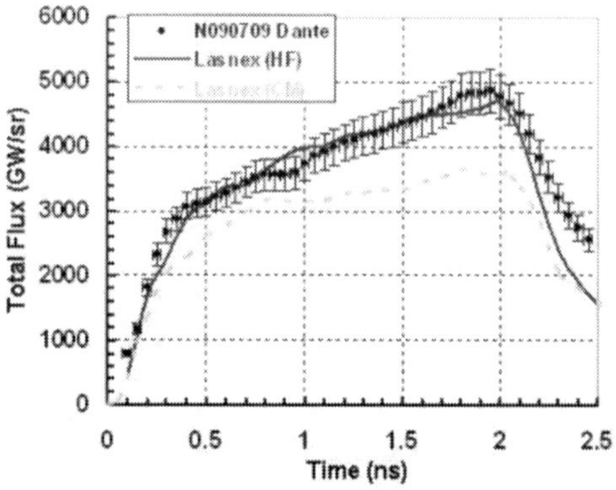

Fig. 2 Scale 0.7, 150 kJ, peak T_r~235 eV.

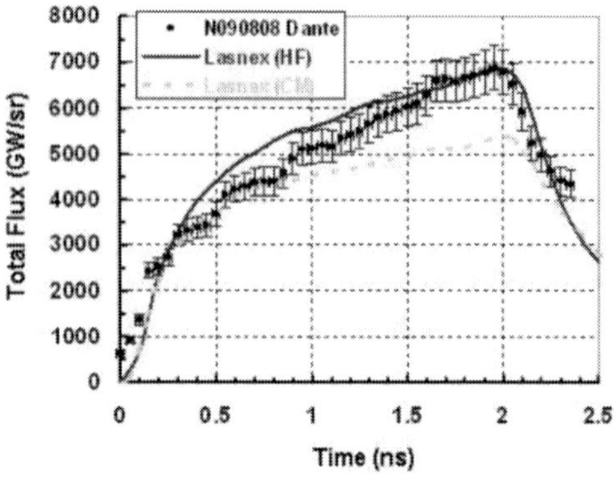

Fig. 3 Scale 0.7, 233 kJ, peak T_r~265 eV.

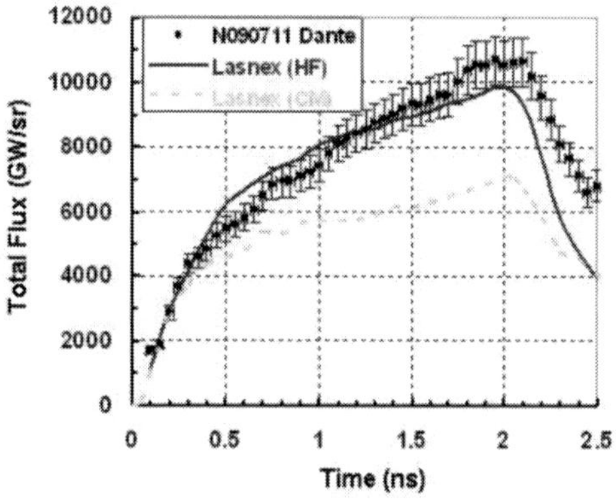

Fig. 4 Scale 0.7, 308 kJ, peak T_r~290 eV.

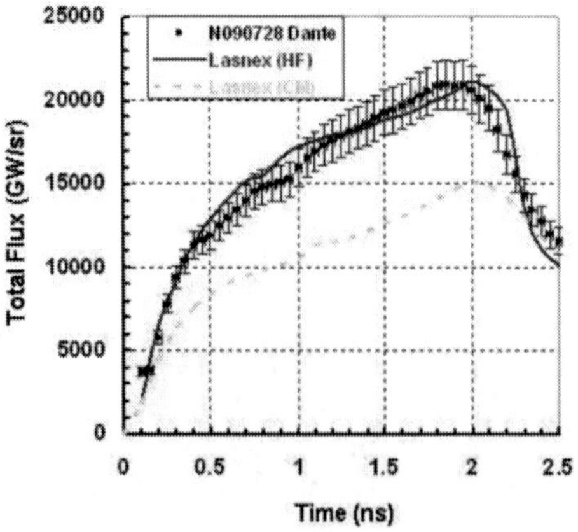

Fig. 5 Scale 0.7, 635 kJ, peak T_r~345 eV.

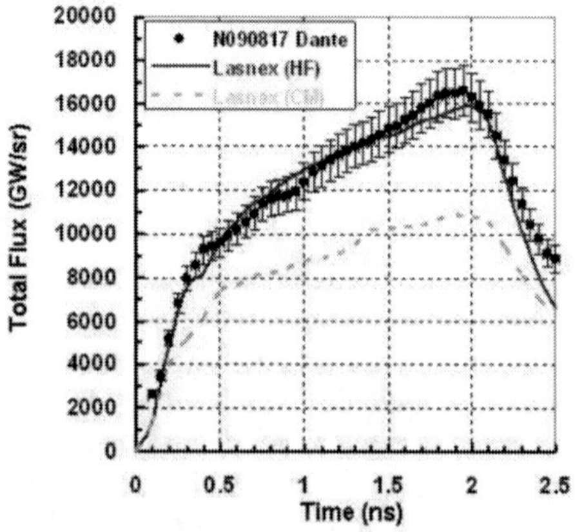

Fig. 6 Scale 0.9, 591 kJ, peak T_r~310 eV.

that are compact and fast enough for routine use in integrated simulations while providing accuracy comparable to far more detailed models. In addition, the electron heat conduction model was altered to use an electron flux limiter of 0.15. These Lasnex "high flux" calculations are shown as the solid blue curves labeled "Lasnex (HF)" in Figs 2-6. As can be seen in the figures, it is possible to explain the seemingly high Dante x-ray flux measurements across the entire range of the 240-340 eV peak Tr vacuum hohlraum experiments with the new "high flux" Lasnex model.

4. Conclusions

The x-ray fluxes measured in the first vacuum hohlraum experiments at NIF are significantly higher than predicted by pre-shot Lasnex calculations employing the default "configuration managed" physics models. It has been demonstrated via post-shot Lasnex calculations that there is a reasonable explanation for the high flux measurements involving the emissivity and heat conduction in the hot coronal plasma. Due to the significant increase in scale size, it seems possible that the hohlraum x-ray conversion efficiency at NIF is higher than in previous hohlraum experiments done at the Omega and Nova scales, and that the high x-ray fluxes indicated by Dante are feasible.

At this point in time, the NIF Dante diagnostic and the associated unfold code are new, and all aspects of the measurements and interpretations are being closely scrutinized. A new SADV Dante data unfold code is in the process of being qualified. NIF experiments are currently being planned to corroborate the high flux measurements with other diagnostic measurements. These include capsule bang time, shock timing, and ablation rate, as well as dedicated measurements of Au x-ray conversion efficiency.

References

[1] E. L. Dewald *et al.*, Rev. Sci. Instrum. **75**, 3759 (2004).
[2] M. D. Landon *et al.*, Rev. Sci. Instrum. **72**, 698 (2001).
[3] E. L. Dewald *et al.*, these proceedings.
[4] J. D. Moody *et al.*, these proceedings.
[5] J. L. Kline *et al.*, these procedings.
[6] K. M. Campbell *et al.*, Rev. Sci. Instrum. **75**, 3768 (2004).
[7] R. E. Olson *et al.*, Phys. Plasmas **11**, 2778 (2004).
[8] R. L. Kauffman *et al.*, Rev. Sci. Instrum. **66**, 678 (1995).
[9] D. A. Callahan *et al.*, these proceedings.
[10] L. J. Suter et al., Bull. Am. Phys. Soc. 53, 89 (2008).
[11] L. J. Suter et al., these proceedings.
[12] E. L. Dewald et al., Phys. Plasmas 15, 072706 (2008).
[13] W. A. Lokke and W. H. Grasberger, LLNL rpt. UCRL-52276, (1977).
[14] H. A. Scott and S. B. Hansen, High Energy Density Physics 6, 39 (2010).

Acknowledgements

LLNL is supported under Contract DE-AC52-07NA27344. LANL is operated by LANS LLC under Contract DE-AC5206NA25396. Sandia is a multiprogram laboratory operated by the Sandia Corporation, a Lockheed Martin Company, for the U.S. Department of Energy under Contract No. DE-AC04-94AL85000.

Journal of Physics: Conference Series **244** (2010) 032058 doi:10.1088/1742-6596/244/3/032058

A Reduced Model for the ICF Gamma-Ray Reaction History Diagnostic

M J Schmitt[1], D C Wilson[1], N M Hoffman[1], J R Langenbrunner[1], H W Herrmann[1], Y H Kim[1], C S Young[1], S C Evans[1], C J Cerjan[2], Wolfgang Stoeffl[2], D H Munro[2], L S Dauffy[2], K M Miller[3], C J Horsfield[4] and M S Rubery[4]

[1]Los Alamos National Laboratory, MS F699, Los Alamos, NM 87545 USA
[2]Lawrence Livermore National Laboratory, Livermore, CA 94550 USA
[3]National Securities Technologies, Livermore, CA 94550 USA
[4]Atomic Weapons Establishment, Aldermaston, Reading, RG7 4PR, UK

mjs@lanl.gov

Abstract. An analytic model for the gamma reaction history (GRH) diagnostic to be fielded on the National Ignition Facility is described. The application of the GRH diagnostic for the measurement of capsule rho-R during burn using 4.4 MeV carbon gamma rays is demonstrated by simulation.

1. The forward model

A forward model for the gamma reaction history (GRH) diagnostic [1] has been developed to simulate the output voltage waveform for typical low yield shots on NIF. The photo-electronic signal has the form

$$N_e = \eta_{\lambda e}\eta_{e\lambda}\left\{\delta t\eta_{\gamma e}R_{\gamma n}N_{ntot}\left(\frac{D_A}{4R_A}\right)^2\sqrt{\frac{4\ln 2}{\pi t_{FWHM}^2}}e^{-4\ln(2)(t-t_B-t_{TOFC}-t_{\gamma e}-t_{gas})^2/t_{FWHM}^2}+\bar{N}_\sigma\left(t-t_{gas}\right)\right\}$$

where N_e is the electron number as a function of time, N_{ntot} is the integrated neutron yield of the shot (typically $10^{14}-10^{16}$) and the rest of the system variables are defined in Table 1. When the sampling bin probability is greater than unity, we use a Gaussian noise model given by

$$\bar{N}_\sigma = INT\left\{\left[\delta t\eta_{\gamma e}R_{\gamma n}N_{ntot}\left(\frac{D_A}{4R_A}\right)^2\sqrt{\frac{4\ln 2}{\pi t_{FWHM}^2}}e^{-4\ln(2)(t-t_B-t_{TOFC}-t_{\gamma e}-t_{gas})^2/t_{FWHM}^2}\right]^{1/2} erfi(rand)\right\}$$

and if we are in a Poisson regime when the probability is less than one, we use

$$\bar{N}_\sigma = INT\left[\delta t\eta_{\gamma e}R_{\gamma n}N_{ntot}\left(\frac{D_A}{4R_A}\right)^2\sqrt{\frac{4\ln 2}{\pi t_{FWHM}^2}}e^{-4\ln(2)(t-t_B-t_{TOFC}-t_{\gamma e}-t_{gas})^2/t_{FWHM}^2} erfi(rand)\right]$$

where *erfi(rand)* is the inverse error function of a random number between 0 and 1, (yielding a Gaussian distributed noise amplitude with a mean of zero and standard deviation of unity) and *INT* generates an integer of the argument. For an integrated neutron shot yield, $N_{ntot} = 2x10^{14}$ (i.e. a low yield THD shot), a peak gamma flux rate of over 10^3 γ's/ps will occur at $t=t_B+t_{TOFC}$ yielding Gaussian statistics for a ~5ps sample time during the main part of the pulse. The value for the conversion efficiency of gamma rays (striking the converter plate) into productive electrons that enter the detector gas, $\eta_{\gamma e}$, has been determined using the ACCEPT code [2].

Similarly, the electrons in the gas generate productive Cherenkov photons with an efficiency, $\eta_{e\lambda}$, given by the ACCEPT code. The Cherenkov photons hit the microchannel plate (MCP) and generate photoelectrons in the MCP with efficiency, $\eta_{\lambda e} \approx 0.17$. This signal (with noise) is further amplified by the gain of the single stage MCP tube and circuit, G_{MCP}. However, the temporal response of the MCP to a single photon is neither instantaneous nor prompt. For the Photek/AWE system, the response function is dominated by a $t_r = 135$ps FWHM Gaussian transfer function followed by a ringing tail, as shown in Figure 1. This response width will be reduced to 90ps for the NIF system (shown in black in Figure 1).

Table 1. GRH model parameters.

Variable Name	Definition	Nominal value
$R_{\gamma n}$	Branching ratio of DT γ's versus neutrons	$2x10^{-5}$
t_B	Implosion bang time	0
t_{FWHM}	Implosion full width half max burn width	100 ps
D_A	GRH convertor plate diameter	12.7 cm
R_A	Target chamber radius	590 cm
$t_{TOFC} = R_A/c$	Time of flight delay to GRH converter plate	19.7 ns
$\eta_{\gamma e}$	Efficiency of γ's to Compton e^-	$8x10^{-3}$
$t_{\gamma e}$	Conversion time delay in convertor plate	50 ps
$\eta_{e\lambda}$	Conversion efficiency of e^- to Cherenkov photons	25
t_{gas}	Average electron and photon time delay in the gas cell	2 ns
$\eta_{\lambda e}$	Efficiency of Cherenkov photons to MCP circuit e^-	0.17
G_{MCP}	Electronic gain	10,000
δt	Digitization time	$4.89x10^{-12}$ s
R_Ω	Oscilloscope load resistance	50 Ω
V_{rms}	Electronic noise rms amplitude	30 mV

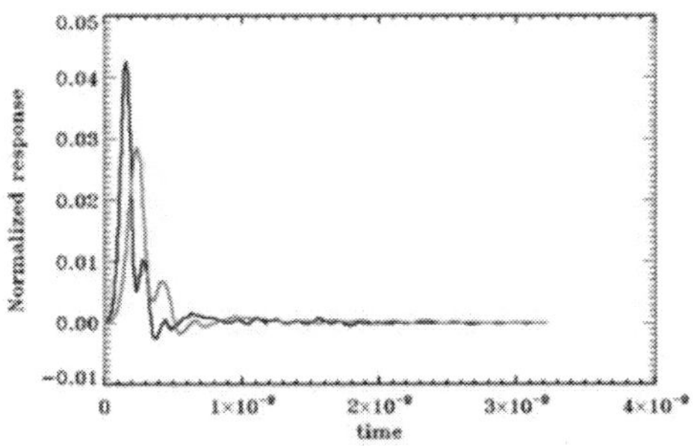

Figure 1. Measured GRH response function for 135ps (red) and a scaled 90 ps response.

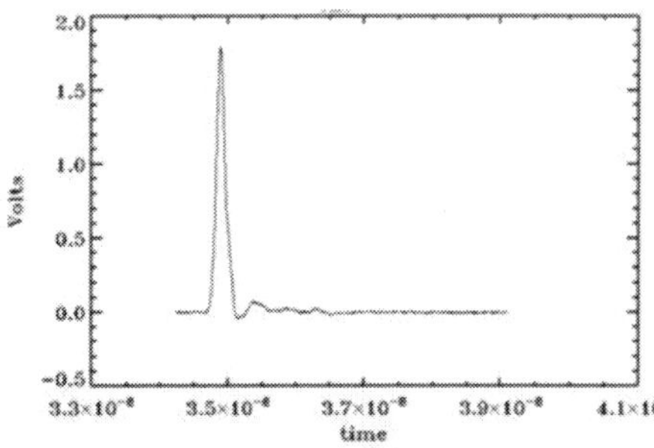

Figure 2. Simulated GRH voltage output using the 90 ps PMT response curve of Figure 1.

The binned temporal electron number density signal, N_s, out of the MCP is the convolution of the signal into the MCP and the MCP response function, *H(t)*, given by

$$N_s(t) = \int_{-\infty}^{\infty} d\tau N_e(\tau) H(t-\tau) = N_e(t) * H(t)$$

where the "*" represents the convolution. Since N_s represents the electron signal for each δt sampling bin, we can calculate the current and voltage as

$$V(t) = I(t)R_\Omega = qR_\Omega \frac{N_s}{\delta t} \simeq 1.6x10^{-6} N_s .$$

Simulated output for a 100ps Gaussian burn profile is given in Figure 2.

1.1 Arbitrary Dark Current Noise Model

Noise in the amplification chain is modelled by adding a scaled $1/f$ noise spectrum to the temporal signal. The model generates a temporal realization using a frequency transform given by

$$N_{dc}(t) = FFT^{-1}\left\{\sum_{f=f_1}^{f_c} \sqrt{S_N(f)}\left[erfi(rand1)e^{i2\pi(rand2)} + c.c.\right]\right\}$$

where FFT^{1} represents the inverse Fourier transform, f_c=0.5/δt is the cut off frequency (a.k.a. the Nyquist critical frequency) and S_N is the power spectrum of the electronic noise. The variables *rand*1 and *rand*2 are two independent *uniformly* distributed (from 0 to 1) random numbers and erfi represents the inverse error function used to generate a Gaussian distributed amplitude.

A $1/f$ model routine using this form has been added to the GRH model. The spectral terms are S_0=0,

$$S_n\Big|_{n=1}^{n=n_{\max}/2-1} = (n\Delta f)^{-1/2}\, erfi(rand(n-1))e^{i2\pi rand(n)},$$

$$S_{n_{\max}/2} = (n_{\max}\Delta f/2)^{-1/2}\, erfi(rand(n_{\max}/2-1))e^{i2\pi rand(n_{\max}/2)}$$

and $S_{n_{\max}-n}\Big|_{n=1}^{n=n_{\max}/2-1} = S_n^*$ where $rand(n)$ is now assumed to be an array of length n containing uniformly distributed random numbers. To calibrate this noise, the mean, μ, and standard deviation, σ, of each temporal realization of noise is calculated. For our discrete temporal noise realizations, these are defined $\mu = \frac{1}{n_{\max}}\sum_{n=0}^{n_{\max}-1} N_{dc}(n\delta t)$ and $\sigma = \sqrt{\frac{1}{n_{\max}}\sum_{n=0}^{n_{\max}-1}\left[N_{dc}(n\delta t)-\mu\right]^2}$.

First the mean is subtracted out. Then the noise amplitude is set by dividing out the actual standard deviation of each temporal realization and multiplying this by the desired noise level, V_{rms}.

1.2 Background gamma model

Background gamma rays from (n,n'γ) reactions occurring in materials surrounding the burning capsule will produce a secondary gamma signal observed by the GRH diagnostic. Most troublesome are low-Z materials close to the capsule. The capsule hohlraum is a laminated structure of radius R_{hohl} containing an Al cylinder. We have calculated this signal using the MCNP code, but we will not show the results in this paper. An analytical model also has been developed but has yet to be validated. We do not expect the hohlraum to be a significant GRH degradation source for NIF.

2. De-convolution Model

The inverse model first factors out all the conversion efficiencies of the detection chain and then performs a de-convolution to extract an approximation to the neutron production rate from the ICF shot. From the previous derivation of the forward model, the factor relating the voltage measure at the oscilloscope to the total neutron production rate of the shot is given by

$$F_{total} = \frac{V(t)}{\frac{dN_n}{dt}} = R_{\gamma n}\left(\frac{D_A}{4R_A}\right)^2 \eta_{\gamma e}\eta_{e\lambda}\eta_{\lambda e}G_{MCP}qR_\Omega = 1.575\text{x}10^{-24}$$

Note that we are ignoring the addition of noise and the transfer function, $H(t)$, of the electronic chain. However, we are assuming that the noise has zero mean and that the impulse response function of the electronic chain has been normalized to unity (times the amplifier gain, G_{MCP}, which we have included).

Besides just applying the scaling factor F^{-1}, a de-convolution can be performed to obtain a more accurate time profile of the gamma generation rate. Our initial model follows that of Horsfield. [3]. Assuming a forward model of the form

$$R(t) = F_{total} H(t) * E(t)$$

where R is the measure temporal response profile, H is the transfer function of the GRH diagnostic and E is the actual gamma rate of the ICF shot. We can obtain an estimate of the gamma rate using the following

$$E_{est1}(t) = \left[2R - R(t) * H(t)\right] F_{total}^{-1}$$

Followed by the iterative technique

$$E_n(t) = E_{n-1} - \alpha\left(E_{n-1} * H - E_{n-1}\right)$$

where this can be iterated many times and α is a factor less than one to maintain stability.

3. Extension to 4.4 MeV (n,n'γ) carbon gamma rays

Neutrons generated by D-T fusion in the ICF capsule will interact with the residual shell material on their way out of the capsule. These collisions generate 4.4 MeV gamma rays when the neutrons collide with carbon atoms in the plastic CH shell. Measurement of these gammas can provide a time-dependent measurement of the areal density, a.k.a. the rho-R, of the capsule while nuclear burn is occurring. Simulations [4] of the generated 4.4 MeV gamma rays have been calculated as a function of capsule rho-R and the results have been plotted in Figure 3. These results show a strong correlation between the number of 4.4 MeV gammas and capsule rho-R. LANL will be pursuing this method to improve the rho-R diagnostic capabilities of the GRH instrument for NIF.

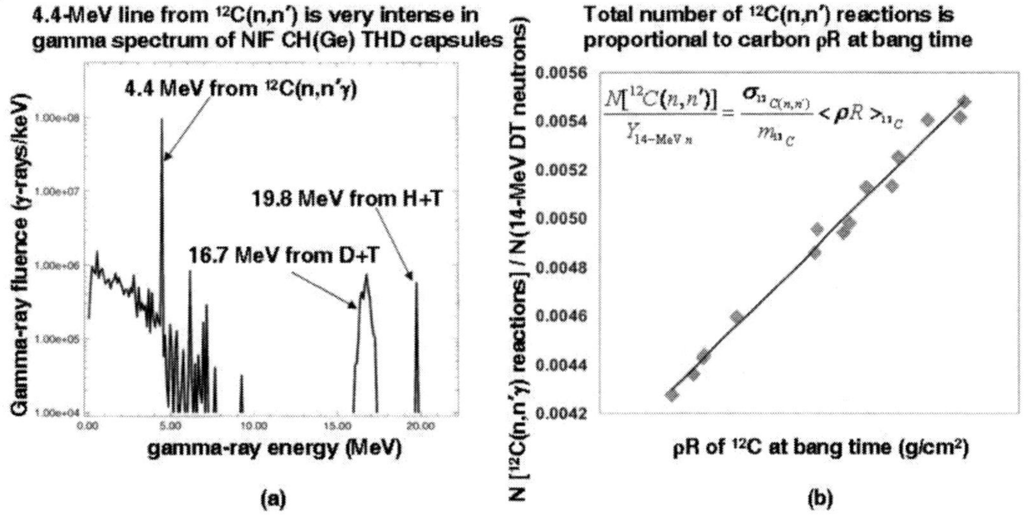

Figure 3. Gamma ray spectrum from the capsule and (b) correlation of 4.4MeV gamma rays for a ~25% change in rho-R from a set of rad-hydro calculations using nominal THD capsule parameters.

References

[1] Herrmann H W, et al., these proceedings.
[2] Mack J M , et al., *2006 Rad. Phys. Chem.* **75** 551 and references therein.
[3] Horsfield C J, private communication.
[4] Hoffman N M, private communication.

The Sixth International Conference on Inertial Fusion Sciences and Applications IOP Publishing
Journal of Physics: Conference Series **244** (2010) 032059 doi:10.1088/1742-6596/244/3/032059

MEASUREMENT OF FAST NEUTRON BY USING IMAGING PLATE

K. Shimada[1,2], N. Nakanii[1,2], I. Murata[1], H. Habara[1,2], K. A. Tanaka[1,2]

[1]*Graduate school of Engineering, Osaka University, 2-1 Yamadaoka, Suita, Osaka, Japan*

[2]*Institute of Laser Engineering, Osaka University, 2-6 Yamadaoka, Suita, Osaka, Japan*

E-mail; shimada-kazu@ile.osaka-u.ac.jp

Abstract. Measurements of yield and spatial distribution of fast neutrons are of extreme importance in the inertial confinement fusion research to estimate the spatial distribution of temperature in core plasma. In experiments using an ultra intense laser light such as fast ignition research, strong electromagnetic pulses generated by the laser irradiation make it very difficult to use a detector with an electric circuit. In order to avoid this problem, we have developed several detectors to measure fast electrons, X-ray, and so on under these environments using an imaging plate (IP). Here, we demonstrate the simple fast neutron detection method by using a stack of IPs. In the experiment using a monochromatic 14MeV neutron source, the IP has linear response to quantity of incident neutrons within the fluence from 10^4 to 10^9. The minimum neutron yield measureable by the detector is estimated to 10^3 or less depending on detection efficiency.

1. Introduction

In inertial confinement fusion (ICF) research, a small deuterium-tritium (DT) shell target is compressed to high density by high-power lasers. At the center of the highly compressed fuel, the ion temperature reaches several kiloelectron-Volts. Fusion reactions are then initiated in this hot spot, resulting in the production of energetic neutrons and alpha particles. Measurement of yield and spatial distribution of fast neutrons is one of key diagnostics to obtain fusion gain and heating efficiency. However in Fast ignition approach, in which an ultrashort-pulse laser is injected into highly compressed and high density core to obtain much higher fusion gain, the strong electromagnetic pulse (EMP) generated by the laser irradiation makes it very difficult to use detectors with an electric circuit such as a charge-couple device (CCD). Imaging plates (IPs) are widely used under such environments to measure fast electrons, X-ray, and so on [1,2] because IPs have less response to EMP. In addition, IPs are very cheap and are reusable, and are also easy to process as digitized data at high speeds. Here, we propose a neutron detection method using an IP stack with polyethylene as a converter from neutrons to protons. We performed a calibration experiment at OKTAVIAN, a 14MeV intense Neutron Source Facility at Osaka University. In section 2, we describe the design of the neutron measurement method. In section 3, the details of calibration experiment of this method in OKTAVIAN are explained. And section 4 is the summary.

2. Neutron detection method

A diagram of the fast neutron detection with IP stack is shown in Fig.1. The IP stack consists of a 0.3mm thick polyethylene sheet, three IP layers, and 1.4mm thick Teflon sheets. Each sheet is cut out with the size of $5 \times 5\text{cm}^2$. In the front of the stack, the polyethylene sheet is used as a neutron-proton converter because IP has little sensitive to neutrons whereas in polyethylene the incident neutron can create a recoil proton with almost same energy via elastic scattering. Then the first IP layer can record the incident recoil protons. The second and third IP layers only measure background noise of gamma ray because Teflon sheet in front of these layers contains less hydrogen, and the Teflon thickness is thicker than 14MeV recoil proton range. Background noise of gamma ray on the first IP layer is removed by substituting the back side IP layers if there are the signal in the same position with similar intensity. We use Fuji BAS-TR image plate because this image plate has no protective layer including hydrogen in order to eliminate the recoil proton signal on the later layers. This design was primarily adopted from Ref. 3, which used image plate in neutron radiography.

The sensitivity and the detection efficiency of this stack are calculated by considering the elastic scattering of fast neurons. The polyethylene sheet is thin enough in order that a single collision with a fast neutron can be expected. The efficiency of a single collision with a fast neutron in polyethylene is given

$$\varepsilon = \frac{N_H \sigma_H}{N_H \sigma_H + N_C \sigma_C} \{1 - \exp\left(-(N_H \sigma_H + N_C \sigma_C)\, d\right)\} \qquad (1)$$

where N_H and N_C are the density of hydrogen and carbon atoms, σ_H and σ_C are the elastic scattering cross section of hydrogen and carbon, and d is the thickness of the polyethylene sheet. When 0.3mm thick polyethylene is supposed, the ε calculated this equation is 0.0017.

The energy of recoiled proton is given as a function of a scattering angle of recoil in the following equation:

$$E_R = E_n \cos^2\theta \qquad (2)$$

where E_R, E_n and θ are recoiled proton kinetic energy, incident neutron kinetic energy and scattering angle of the recoiled proton in the laboratory coordinate system. Recoil proton energy distribution is almost constant from zero to incident neutron energy because hydrogen scattering is isotropic in laboratory system. The energy of recoil protons scattered by a single collision with the fast neutrons decreased in polyethylene via small angle scatterings. According to the calculation of SRIM 2008, only 70% of recoil protons reach the first later of IP because low energy proton is stopped in polyethylene. From this calculation and the efficiency of a single collision with a fast neutron, about 0.1% of incident neutron number is detected in IP as the signal of protons.

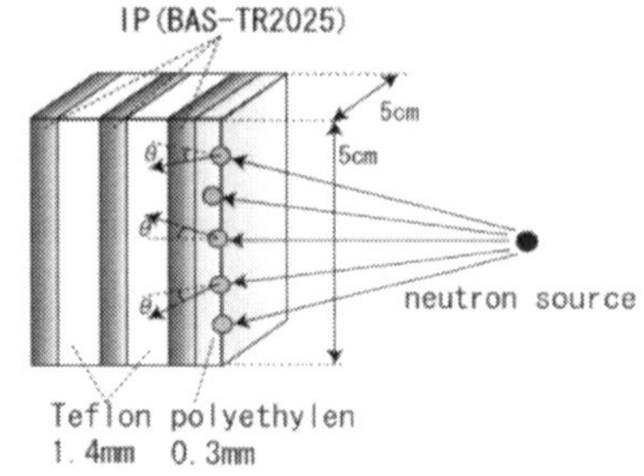

Figure.1 IP stack design to detect 14MeV neutrons

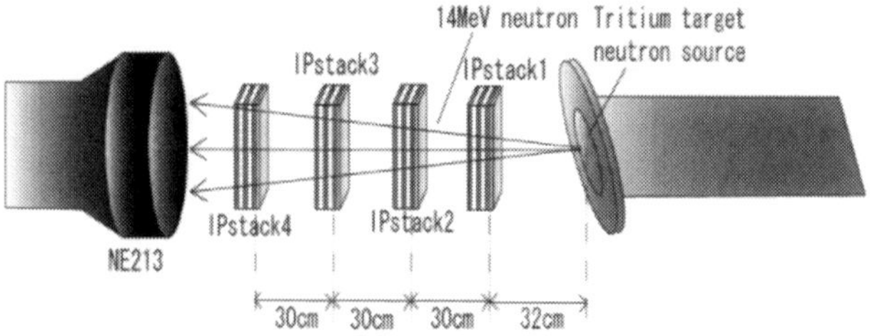

Figure.2 Experimental setup to calibrate neutron detection method using IP in OKTAVIAN

3. Calibration experiment

We calibrated this method using an intense neutron source, OKTAVIAN at Osaka University. At the OKTAVIAN facility, the 14MeV neutrons are produced via the T(d, n)^{4}He reaction by bombarding a Ti-T rotating target with a d$^+$ beam of about 5mA and 300kV. The yield of 14MeV neutrons is measured by a NE213 liquid scintillator [4], which is calibrated by the ^{24}Al(n, α)^{24}Na activation foil method.

The experimental setup is shown Fig. 2. Four sets of IP stack are used at different distance from the neutron source in order to measure the signal linearity to neutron number because the neutrons are a point source. After IP stacks were exposed for about 660 seconds, IPs were read out using an IP reader (Model BAS 1800: Fuji Photo Film Co. Ltd.) with the spatial resolution of 50μm in 16 bit data depth. The fast neutron yield of $2.0 \times 10^7 \mathrm{n/s}$ was measured by NE213. IPs have 10^6 pixels for the IP size of $5 \times 5\mathrm{cm}^2$ and a spatial resolution of 50μm. The number of IP pixel is much larger than the recoil proton number detected with IP. Therefore, counting the signal of protons can estimate the incident neutron numbers like single photon counting technique [5]. Fig. 3 shows histogram of IP signals for the first (solid line), second (dotted lines), and third layers (dashed lines). In the figure, the horizontal and vertical axes represent PSL value in each pixel and the number of pixels, respectively. The first IP layer has larger number of pixels with high PSL value compared with that in second and third IP layers. On the other hand, there is not so much difference between the signals of second and third IP layers. It clearly shows that the recoil protons are detected only in the first IP layer whereas the backside IP layers detect only background gamma noise. Taking into account of the neutron numbers from 4 sets of stacks with different distances from the source, the integrated signal count exceeding background PSL level shows a linear response to the number of incident neutrons as shown in Fig. 4. From the detection efficiency by equation (1) and SRIM results, one incident proton into IP may spread into around 20 pixels, which well agrees with the previous measurement using heavier ions [6]. And the average PSL value per one 14MeV neutron is calculated to $1.2 \pm 0.1 \times 10^{-3}$PSL/neutron, which shows good agreement to all of detected neutron number in the experiment.

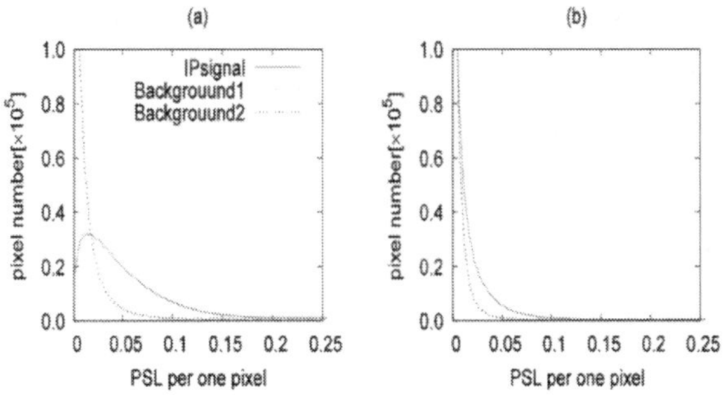

Figure.3 (a), (b) Signal of IP set in the place of the distance 32cm, 62cm, respectively

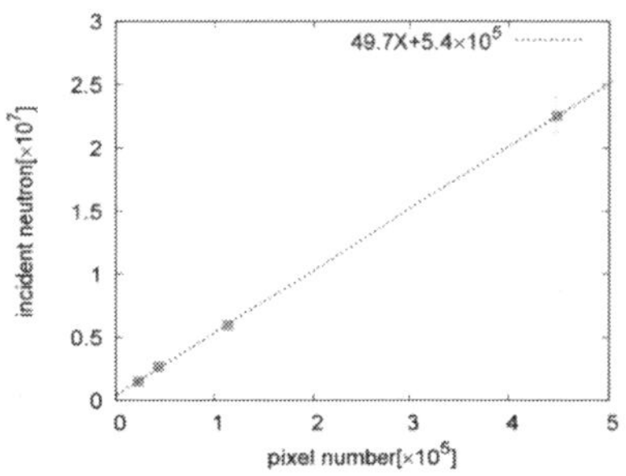

Figure.4 Dependence of number of incident neutron to signal count

4. Conclusion

A new neutron detection technique has been demonstrated using a stack of imaging plates (IP). Detection efficiency is estimated about 0.001 by SRIM simulation for our experimental conditions. Dynamic range to the incident neutron number is ideally from 10^4 to 10^9 when using IP of the $5 \times 5\text{cm}^2$ size. In addition IP signal has linear response to number of neutrons. Neutron detection system which is less influenced to EMP has been established using IP. It is useful in experiments using an ultra intense laser light such as fast ignition research.

References

[1] K.A. Tanaka, et al., Rev. Sci. Instrum. **76**, 01507 (2005)

[2] C.D. Chen, et al., Rev. Sci. Instrum. **79**, 10E305 (2008)

[3] M. Matsubayashi, et al., Nuclear Instruments and Methods in Physics Research **A** 463 (2001) 324–330

[4] M.A. Al-Ohali, et al., Nuclear Instruments and Methods in Physics Research **A** 396 (1997) 388–393

[5] C. Stoeckl, et al., Rev. Sci. Instrum. **75**, 10 (2004).

[6] G. Kanase et al, IEEE *Nucl. Sci. Sympo.* 1998, N12-6

Journal of Physics: Conference Series **244** (2010) 032060 doi:10.1088/1742-6596/244/3/032060

High-speed monochromatic x-ray imager for electron temperature mapping of fast igniter plasmas

Minoru Tanabe, Takashi Fujiwara, Shinsuke Fujioka, Nobuo Hayashi, Hiroaki Nishimura, Takeshi Fujimura, Mitsuo Nakai, Keiji Nagai, Hirofumi Homma, Hiroyuki Shiraga, Takayoshi Norimatsu, Hiroshi Azechi, and Kunioki Mima

Institute of Laser Engineering, Osaka University 2-6 Yamada-oka, Suita, Osaka 565-0871 Japan

E-mail: mtanabe@ile.osaka-u.ac.jp

Abstract. We report on experiments aimed at obtaining laser-imploded core plasma temperature maps using two color, monochromatic x-ray imager. This instrument uses two toroidally bent Bragg crystals and a high-speed sampling imager to provide spatial resolution of 25 μm, temporal resolution of 20 ps, and spectral resolution over 200, simultaneously. Using a deuterated plastic shell target filled with $CHClF_2$ gas, time-resolved monochromatic x-ray images in the Cl-He_β and Cl-Ly_β were successfully obtained.

1. Introduction

In fast ignition scheme [1], high-density fuel compression and additional heating with a high intensity ultrashort pulse laser are very important. Dynamics of laser-imploded core plasma and energy deposition by fast particles must be studied in order to achieve efficient fusion burn. The fuel is compressed to several tens of micrometers, and the heating process is found to occur in a temporal window less than 100 ps [2]. Highly spatial and temporal resolutions, typically 10 μm and 10 ps, and seamless data acquisition are simultaneously required to diagnose the core plasma. In addition, spectral information in the keV range can be used as a diagnostic of plasma parameters such as electron temperature and density.

Monochromatic x-ray imaging with a bent crystal [3] represents an indispensable tool for obtaining narrowband x-ray images covering x-ray line emissions from dopant in a target. Moreover, this monochromatic imaging geometry blocks energetic x rays emitted from fast-ignitor plasma with a thick high Z substrate direct shield. In previous researches, time-resolved monochromatic images have been obtained with monochromatic x-ray framing camera [4], and electron temperature and density profiles of the core plasma were derived [5]. In order to obtain higher temporal resolution in continuous x-ray images, a monochromatic x-ray sampling image x-ray streak camera (M-SIXS) has been developed [6]. In this paper, the feasibility of the M-SIXS in implosion experiments is reported.

2. Experimental arrangements

The M-SIXS consists of a bent Bragg crystal and a two-dimensional (2D) sampling image x-ray streak camera (SIXS) [7]. The instrument geometry of the M-SIXS for this research is schematically shown in Fig. 1. To minimize astigmatism and obtain high photon-collection

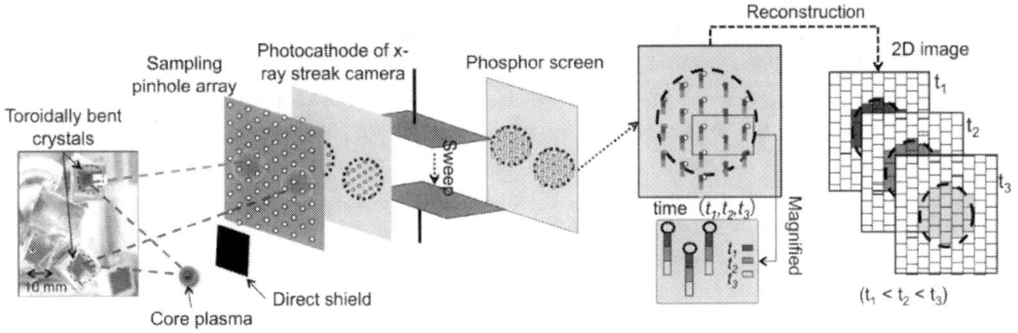

Figure 1. Schematic of the M-SIXS operational geometry, and process of image sampling and reconstruction.

efficiency, toroidally bent crystal [3] was used. In the separate experiments, a spatial resolution better than 10 μm has been confirmed with the toroidally bent crystal [4]. $CHClF_2$ gas was chosen as a dopant, and Cl^{15+} He_β and Cl^{16+} Ly_β lines were observed to derive electron temperatures. Specifications of two toroidally bent crystals in this research are listed in Table 1. A 100-μm-thick beryllium shield was set in front of each crystal.

Table 1. Specifications of two toroidally bent crystals.

X-ray energy (keV)	3.27	3.51
Bragg crystal	Silicon (220)	Quartz (11.2)
Bent radii (mm)	200/195.8	200/189.5
Spectral window (eV)	11.7	26.2
Image magnification	26.0	26.3
Target-Crystal (mm)	102.5	100.9
Crystal-Detector (mm)	2667	2659

SIXS is a 2D sampling image technique with an x-ray streak camera (XSC) [7]. Two monochromatic x-ray images of the core plasma are focused on a gold photocathode which is 13.2 mm in height and 17.3 mm in width, and then the images are sampled by placing a periodic pinhole-array in front of the photocathode. A separation distance between the pinhole-array and the photocathode was 5 mm. The 41 × 67 pinholes were fabricated on 25-μm-thick nickel substrate. The pinhole size was 30±3 μm. The sampled pinhole images are swept with high voltage in the streak tube, and the streaked image of sampling points can be obtained on a phosphor screen, as illustrated in Fig. 1. The spatial resolution of the system Δr_s, as determined by the sampling distances S_x, S_y, is defined as $\Delta r_s = 2S_x/M$ or $\Delta r_s = 2S_y/M$ [6; 7], where M is the image magnification. As the x and y sampling distances were set to 330 μm, the spatial resolution in this study was 25 μm. However, according to the sampling theory, the sampling distances for image sampling S_x, S_y should be $S_x, S_y \leq (1/2)M\Delta r_{imager}$ to cover all of the spatial frequencies in the original image information, where Δr_{imager} is the spatial resolution of the imager. As the spatial resolutions for two crystals were 10 μm, the sampling distances are 2.5 times larger than $(1/2)M\Delta r_{imager}$, thus some of the original image information between the sampling points was lost. Temporal resolution is determined by both the pinhole diameter and the sweep speed, and the observable time-window is determined by both the pinhole space in the sweep direction and the sweep speed. For the set-up adopted in this experiment, the temporal resolution and the observable time-window were estimated to be 20 ps and 500 ps, respectively.

A 6.2 μm in thickness, 490 μm in diameter deuterated plastic shell was used. As a tracer, $CHClF_2$ gas of 0.2 atm was filled in the shell. It was supported with as 20-μm-outer-diameter and 15-μm-inner-diameter glass capillary through which the tracer gas was filled. The capillary is struck on the shell, glued with a paste to avoid the leakage of the fill gas. The gas fill procedure is described as follows. First, the inside of the shell was evacuated. Second, a buffer gas tank of approximately 200 cm^3 in volume located outside a target chamber was filled with $CHClF_2$ gas. Pressure in the buffer tank was measured with a diaphragm pressure gauge directly installed in the buffer tank. After suitable gas pressure was attainted, a valve between the buffer tank and the gas cylinder was closed, then a valve between the buffer tank and the shell was opened slowly. The $CHClF_2$ gas was flowed to the shell inside through the capillary. Finally, the valve between the shell and the buffer tank was kept open at least 1 hours before the laser shot to equalize the pressures in the buffer tank and the shell inside.

Implosion experiment was performed at the GEKKO XII laser facility. The shell target was irradiated with 12 beams of 526 nm in wavelength and 3.7 kJ in total energy at a Gaussian pulse with 1.3 ns full width half maximum. To homogenize laser intensity distribution on the shell surface, random phase plates for each beam were used.

3. Experimental results

Figure 2 shows typical data from the two color M-SIXS. The two color x-ray images at 3.27 keV and 3.51 keV were simultaneously recorded on the same photocathode of the XSC. The absolute position, namely the center of the shell, was recorded in a separate laser shot. Moreover, the absolute position of each sampling point, spatial non-uniformity of the photocathode sensitivity, and dispersion of the pinhole size were provided with a static image, which obtained without supplying a sweep voltage to the streak tube. Using these experimental data, the 2D images for two color x rays were reconstructed from the streaked monochromatic images shown in Fig. 2.

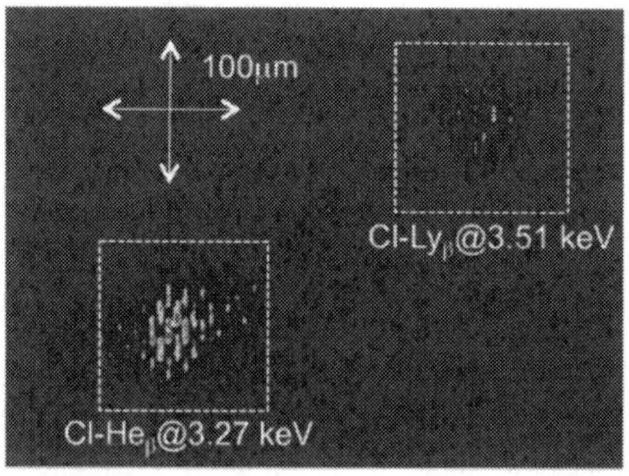

Figure 2. Streaked monochromatic x-ray images of the chorine doped core plasma at 3.27 keV (left) and 3.51 keV (right). The two color images were recorded on the photocathode of the XSC for the same laser shot.

Figure 3(a) shows reconstructed monochromatic x-ray images at 3.27 keV. The reconstructed images for 3.51 keV is also shown in Fig. 3(b). They are shown with same color scale corrected for the x-ray filter transmission and crystal reflectivity. The time interval is 20 ps and the overall time duration of x-ray emissions is about 400 ps. First frame images were defined as the onset of the line emission. Each frame image at 3.27 keV and 3.51 keV corresponds to the same time. The emission intensity gradually increases as time advances, and reached the peak at 360 ps after the onset of the emission. One can clearly see that the x-ray image for 3.27 keV has once a weak peak at 200 ps and a strong peak at 360 ps from the onset of the emission. The similar phenomenon is obtained with a time-resolved x-ray spectrograph. It is of great interest to study such plasma dynamics in detail for fast ignitor research.

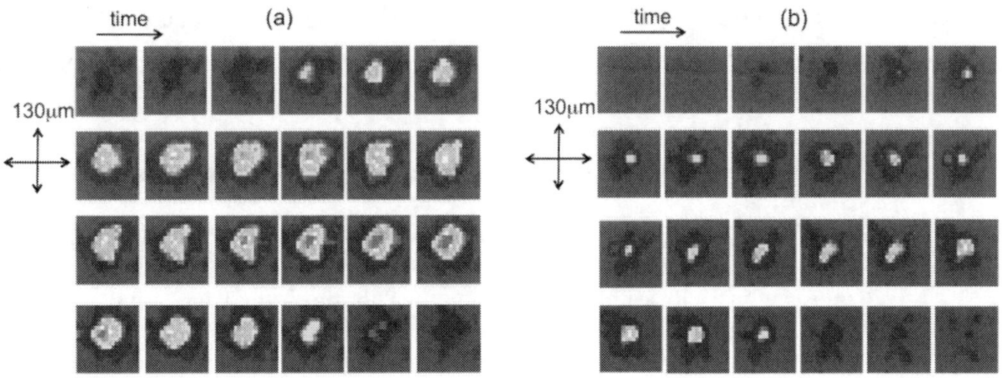

Figure 3. Reconstructed monochromatic x-ray images of the chorine doped core plasma for 3.27 keV x ray (a) and for 3.51 keV x ray (b). Frame interval is 20 ps. Two color x-ray images are shown with same color scale.

4. Summary

Feasibility of the M-SIXS for deriving electron temperature mapping of the core plasma has been demonstrated with a temporal resolution of 20 ps, a spatial resolution of 25 μm, and an energy resolution over 200 using a deuterated plastic shell target containing $CHClF_2$ gas. The two color monochromatic images at 3.27 and 3.51 keV have been successfully recorded for the same shot. Temporal evolution of electron temperature profiles is undertaken, and will be obtained in the near future.

Acknowledgments

This work was performed under the auspices of the Ministry of Education, Culture, Sports, Science and Technology of Japan (MEXT), under the project" Advanced Diagnostics for Burning Plasma (code 442)" /" Temperature and density mapping of imploded burning plasma with monochromatic c-ray imaging (code 20026005)" .

References

[1] Basov N G, Gus'kov S Y and Feokistov L P 1992 *J. Sov. Laser Res.* **13** 396
[2] Kodama R, Shiraga H, Shigemori K *et al.*, 2002 *Nature* **418** 934
[3] Förster E, Gäbel K and Ushmann I 1991 *Laser Part. Beams* **9** 135
[4] Fujita K, Nishimura H, Niki I *et al.*, 2001 *Rev. Sci. Instrum.* **72** 744
[5] Ochi Y, Golovkin I, Mancini *et al.*, L 2003 *Rev. Sci. Instrum.* **74** 1683
[6] Tanabe M, Fujiwara T, Fujioka S *et al.*, 2008 *Rev. Sci. Instrum.* **79** 10E908
[7] Shiraga H, Nakasuji M, Heya M *et al.*, 1999 *Rev. Sci. Instrum.* **70** 620

Journal of Physics: Conference Series **244** (2010) 032061 doi:10.1088/1742-6596/244/3/032061

Penumbral imaging with multi-penumbral-apertures and its heuristic reconstruction for nuclear reaction region diagnostics

Tatsuki Ueda[1], Shinsuke Fujioka[1], Shinya Nozaki[2], Yen-Wei Chen[3], Hiroaki Nishimura[1]

[1] Institute of Laser Engineering, Osaka University, 2-6, Yamada-oka, Suita, Osaka, 565-0871,Japan

[2] Transdisplinary Research Organization for Subtropics and Island Studies, 1 Senbaru Nishihara, Okinawa, 903-0213 Japan

[3] College of information Science and Engineering, Ritsumeikan University, 1-1-1 Nojihigasi Kusatsu, Shiga, 525-8577, Japan

E-mail: ueda-t@ile.osaka-u.ac.jp

Abstract. Imaging of nuclear reaction region is important to clarify heating mechanism in a fast-ignition plasma. The nuclear reaction region can be identified by hard x-ray and neutron images, which are emanated from the heated region. We proposed a novel penumbral imaging that is suitable for imaging quanta having strong penetrating power, such as hard x ray and neutron. Using multiple penumbral apertures arranged with *M*-sequence leads to two orders of magnitude higher detection efficiency than that with a single aperture. In addition, a heuristic method was introduced to a image reconstruction procedure for reducing artifacts caused by noise in a penumbral image. A proof-of-principle experiment indicates that the proposed imaging is superior to the conventional one.

1. Introduction

The fast ignition scheme is investigated to achieve fusion ignition and burn with a relatively small laser facility. Institute of Laser Engineering of Osaka University launches FIREX (Fast Ignition Realization Experiment) project[1] to demonstrate fusion ignition with this scheme. A high-intense short-pulse laser must heat locally a fusion fuel for achieving a high gain of fusion energy with the fast ignition scheme. Physical processes dominates the local heating can be understand by identifying the nuclear reaction region in a fast ignition plasma. Advanced penumbral imaging technique has been proposed to obtain two-dimensional images of hard x-rays and/or neutrons emitted from the nuclear reaction region. Penumbral imaging[2] is suitable for imaging quanta having strong penetrating power, such as hard x-rays and neutrons. Penumbral imaging is one of encoded imaging techniques. A reconstruction or decoding process is required for obtaining an objective image from the pneumbral image. In a conventional penumbral imaging, a reconstructed objective image is distorted mainly due to noises and aperture shape uncertainties. A heuristic method is introduced to the image reconstruction procedure for reducing the image distortion. It was found in the previous study that neutron yield of 10^{11} is required for the conventional penumbral imaging.[3] This yield is two orders of magnitude larger than that predicted for the FIREX project. We used multi-apertures for the penumbral imaging. Signal intensity increases with multi-penumbral-apertures arranged in *M*-matrix compared to that with single one, signal increment is proportional to number of the apertures. We carried out a proof-of-principle experiment for the proposed imaging technique, which is a combination of the heuristic reconstruction method and the multi-penumbral-apertures.

2. Heuristic image reconstruction of penumbral image

The penumbral imaging[4, 5, 6] uses aperture, whose diameter is larger than objective size. The penumbral image consists of a uniformly bright region surrounded by a penumbra. Spatial information of the objective is contained in the penumbral region. Deconvolution operation is used for the image reconstruction in the conventional penumbral imaging. Weiner filter is used to reduce amplitude of noise.[6] In the case that the amplitude of noise is much lower than that of signal, the use of Weiner filter is a simple and straightforward technique to reduce noises, however, in the contrary case, noise is difficult to be removed by the Weiner filter. Noise remained in a penumbral image is amplified by the deconvolution process.

Heuristic method[7, 8, 9], which has high tolerance to noise, has been adapted to the reconstruction process. In the heuristic method, the optimal objective image is estimated by minimizing the mean square difference between an obtained penumbral image and an estimated one. The heuristic method was compared with the conventional one. Figure 1(a) is the penumbral image obtained in an experiment. Signal to noise ratio (SNR) of the penumbral image is about 17.7. The objective image reconstructed by the conventional method is shown in Fig. 1(b). Although SNR of the penumbral image is sufficiently high, artifacts are overlapped on the image reconstructed by deconvolution process. The clear image without artifacts was obtained with the heuristic method as shown in Fig. 1(c). The heuristic method clearly excludes artifacts those appear in the conventional reconstructed image.

3. Proof-of-principle experiment of Uniformly redundant array of penumbral aperture

Uniformly redundant array of penumbral aperture (URPA) was proposed to increase image intensity.[10] Penumbral apertures are arrayed in a two-dimensional M-matrix. There is a *G*-matrix corresponding to a *M*-matrix. A convolution of the *M*- and the *G*-matrix generates the perfect delta function. Multiple penumbral images arrayed in a *M*-matrix are unified automatically by convoluting with a G-matrix, and signal intensity is easily enhanced. The concept of the URPA imaging technique is shown in Fig. 2. The multiple penumbral images arrayed in the M-matrix are recorded on a detector as shown in Fig. 2(c). The unified penumbral image (Fig. 2(e)) is obtained by convoluting the multiple penumbral images (Fig. 2(c)) with the G-matrix (Fig. 2(d)) and the object image (Fig. 2(f)) is reconstructed from the unified penumbral image (Fig. 2(e)).

The proposed scheme was applied to imaging 2-3 keV x-rays. A Nd: YAG laser was used in this experiment. Energy, pulse duration, and wavelength of the laser were, respectively, 1.5 J, 3.5 ns, and 1064 nm. Penumbral apertures (400 μm in diameter) were fabricated on a 25 μm-thick tantalum substrate. A separation distance between apertures was 500 μm. The distance between the aperture and the object was 211 mm and the distance between the aperture and the detector was 916 mm, thus camera magnification was 4.3. A 100 μm-thick beryllium was put in front of the aperture as an x-ray

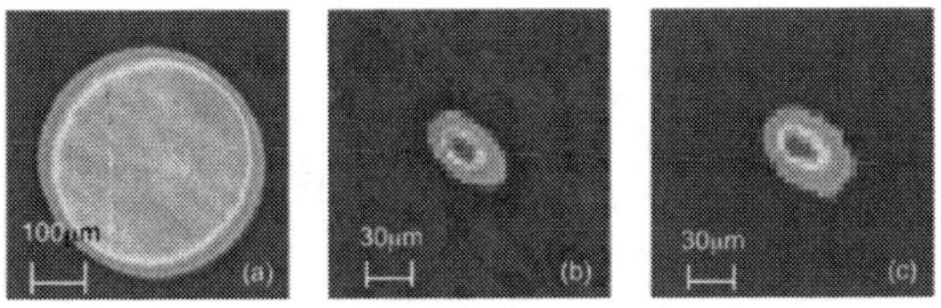

Figure 1. Comparison between the heuristic method and the conventional one. (a) penumbral image for comparison, (b) reconstructed image with the conventional reconstruction, (c) that of with the heuristic reconstruction.

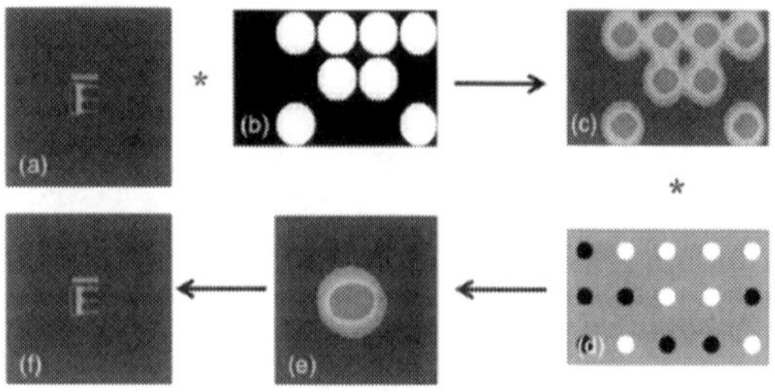

Figure 2. Basic concept of the URPA. (a) the objective image, (b) the multi-penumbral-apertures, (c) the multiple penumbral images arrayed with the M-matrix, (d) the G-matrix corresponding to the M-matrix, (e) the unified penumbral image, (f) the objective image reconstructed from the unified penumbral image.

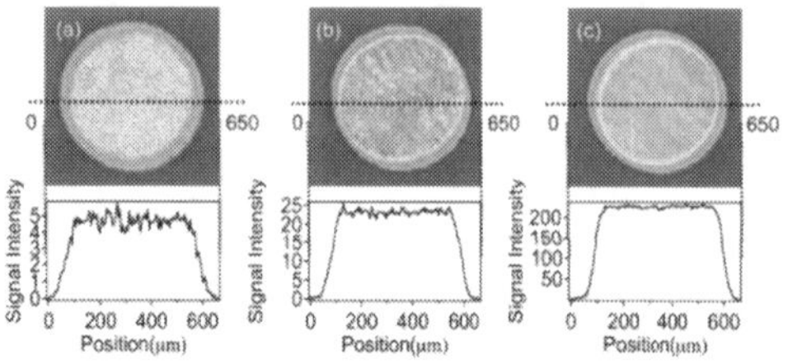

Figure 3. Comparison between the single aperture and URPA.
(a) the penumbral image of the single aperture and that of line profile, (b) the unified penumbral image with 3 x 5 URPA and that of line profile, (c) the unified penumbral image with 7 x 9 URPA and that of line profile.

filter to exclude low energy photons whose energy is less than 2 keV. A planar gold foil was irradiated by a laser pulse, and x-ray images from laser-produced plasma were imaged with the URPA. Three kinds of apertures were used in the experiment; (1) single penumbral aperture, (2) 3×5 URPA and (3) 7×9 URPA. The penumbral image with single aperture is shown in Fig. 3(a). Unified penumbral images obtained with 3×5 URPA or 7×9 URPA are shown in Figs. 3(b) and (c) and, corresponding line profiles are shown below them. SNRs of the unified penumbral images were evaluated to be 2.7, 6.3 and 17.7 for single, 3 x 5, 7 x 9 URPA, respectively. SNRs of unified penumbral images increased in proportion to square root of the number of apertures.

4. Combination of heuristic image reconstruction and URPA imaging

The proposed imaging technique was compared with the conventional penumbral one. The reconstructed image of the conventional penumbral imaging is shown in Fig. 4(a). A single aperture and the reconstruction procedure based on decombolution and a Wiener filter were used for Fig 4(a). The reconstructed image is strongly distroted by artifacts because of low SNR. The reconstructed image of the proposed imaging technique is shown in Fig. 4(b). The high quality reconstructed image was obtained by using the proposed imaging technique from an object whose signal intensity is not sufficient.

We evaluate spatial resolution of the proposed imaging technique. For this evaluation, an experiment was performed on Gekko-XII laser facility, Osaka University. Configurations of the imaging system were the same as that in the proof-of-principle experiment. A cross slit (slit width 100 μm) was backlit by x rays, and backlit slit was imaged with URPA. Chlorided plastic was irradiated by Gekko-XII laser beams to produce a backlight x-ray source, which emits 2.7 keV x rays. The reconstructed image is shown in Fig. 5(a). Modulation transfer function (MTF) of the imaging system was evaluated from the slit image as shown in Fig 5(b). Spatial resolution was 25 μm in the present system, here spatial resolution was defined as the wavelength corresponds to 0.1 of MTF.

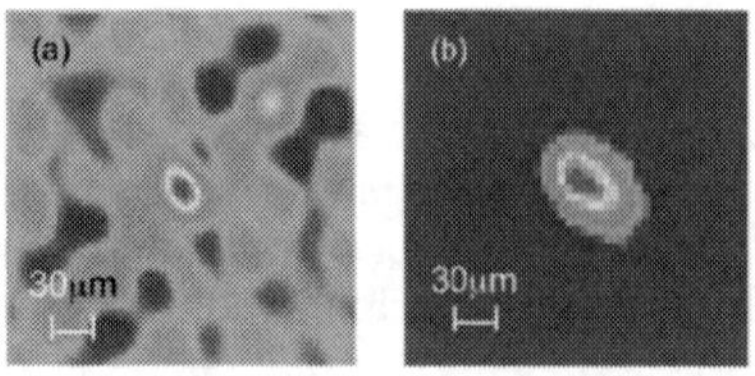

Figure 4. Comparison between the proposed penumbral imaging and the conventional penumbral imaging. (a) the reconstructed image with the single aperture and the deconvolution reconstruction, (b) that with the 7 x 9 URPA and the heuristic one.

Journal of Physics: Conference Series **244** (2010) 032061 doi:10.1088/1742-6596/244/3/032061

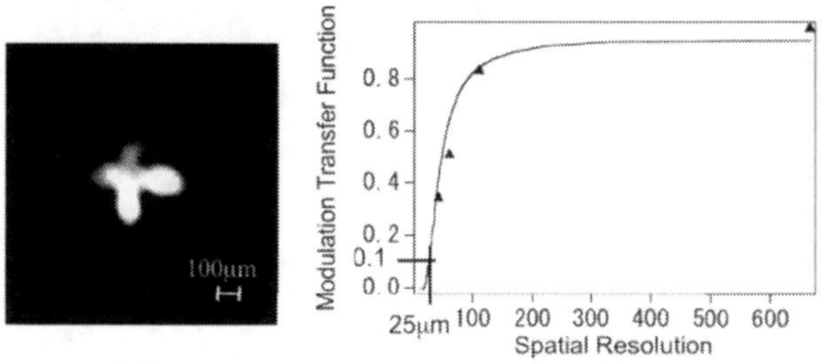

(a) (b)

Figure 5. Spatial resolution was evaluated by Modulation Transfer Function.
(a) the reconstructed image of backlight cross slit, (b) the relationship between Modulation Transfer Function and spatial resolution.

Spatial resolution is limited by several factors, i.e. x-ray diffraction, unroundness of the apertures, error of aperture positions, detector resolution and SNR of the penumbral image. Diffraction Spread (Δx) on the detector caused by x-ray diffraction is given by

$$\Delta x = 2.4 \frac{\lambda}{D} d, \quad (1)$$

here λ is the wavelength of the backlight x-ray source, D the aperture diameter, d the distance between the target and the aperture. Unroundness of the aperture distorts a reconstructed image, and error of the aperture position blurs a reconstructed image. Spatial resolution limited by x-ray diffraction was only 0.6 μm in the present system. Standard deviation of the aperture unroundness was 1.47 μm, by which spatial resolution was limited to 1.8 μm. Standard deviation of the aperture position errors was 2.6 μm, by which the spatial resolution is limited to 1.8 μm. Spatial resolution of the detector was 10.5 μm. As the result of the above evaluation, SNR of the penumbral image limits dominantly the spatial resolution in the present experiment.

Acknowledge

This work was performed under the auspices of the Ministry of Education, Culture, Sports, Science and Technology of Japan (MEXT), under the project "Advanced Diagnostics for Burning Plasma (code 442)"/"Advanced penumbral image technique based on heuristic image reconstruction procedure (code 20026004).

Reference

[1]Azechi H and the FIREX project 2006 *PlasmaPhys. Control. Fusion* **48** B267-75.
[2] K. A. Nugent, B. Luther-Davis, April 1984 , Opt. Commun, vol, 49, no.6, pp. 393-396.
[3] Y.-W. Chen et al., Dec. 1995, *IEICE Trans. Electron.*, vol.E78-c, no.12, pp. 1787-1792.
[4] Y.-W. Chen, R. Kodama, Z. Nakao, Jun. 1998, *IEEE Trans. Nucl. Sci.*, vol.45, no.3, pp. 992-996.
[5] Y.-W. Chen, Z. Nakano, I. Nakamura et al., Feb. 1997, *IEICE Trans. Electron.*, vol.E80-c, no.2, pp. 346-351.
[6] X.-H. Han, H. Lin, S. Y. Dai, J. Li, Sep. 2002, *Proc. of SPIE,* vol.6768, 67872O, pp. 67872O-1 - 67872O-6.
[7] S. Nozaki, Y.-W. Chen, Z. Nakano, S. Fujioka, and H. Shiraga, *Rev. Sci. Instrum.*, vol.73, no.9, pp. 3198-3204.
[8] S. Nozaki, Y.-W. Chen, Z. Nakano, R. Kodama, and H. Shiraga, March 2003, *Rev. Sci. Instrum.*, vol.74, no.3, pp. 2240-2244.
[9] S. Nozaki, Y.-W. Chen, Oct. 2004, *Rev. Sci. Instrum.*, vol.75, no.10, pp. 3980-3982.
[10] Y.-W. Chen, H. Yamamoto, S. Nozaki, Oct. 2004, *Rev. Sci. Instrum.*, vol.75, no.10, pp. 4017-4019.

Radiation power losses and opacity of mid-Z impurities

D Benredjem[1], A Calisti[2], J C Pain[3], F Gilleron[3]

[1] Laboratoire Aimé Cotton, Campus d'Orsay, 91400 Orsay, France
[2] PIIM, Université de Provence, Marseille, France
[2] CEA, DAM, DIF, F-91297 Arpajon Cedex, France

E-mail: djamel.benredjem@lac.u-psud.fr

Abstract. We present results on opacity and radiation losses of mid-Z impurities in ICF conditions. Two approaches were investigated. The first one is based on a detailed calculation where the atomic database is provided by the MCDF code. Then a lineshape code (PPP) based on a fast algorithm was adapted to the calculation of emissivity and opacity profiles. The second approach is a statistical one. It involves high-order moments of the radiative power losses. Atomic calculations were performed with the Cowan code. The Normal Inverse Gaussian and the Generalized Gaussian distributions were then used to calculate the radiation power loss profile.

The knowledge of opacity and radiative power losses (RPL) is very important when one deals with radiative properties of ICF and astrophysical plasmas. Many experiments have been performed in the recent years on iron (see for example Ref. [1]), aluminum (see for example Ref. [2]) and heavier elements such as gold [3]. In Ref. [1], the absorption of the $2p-3d$ transition array in Fe V to Fe X has been measured. Concerning aluminum, recent results have been obtained at the LULI facility [2] by irradiating a very thin foil (≈tens of nm) with a subpicosecond laser pulse.

In this work we use a detailed line calculation. Two approaches were investigated. In the first one the PPP line profile code [4, 5] developed at the University of Marseille is adapted to opacity and RPL calculations. This approach is interesting in the sense that it accounts for all major line broadening mechanisms. We have explored a second approach which is based on statistical distributions. This approach involves high-order moments of the RPL. The code based on the second approach is much faster than the PPP-like code.

The line shape of a radiative line results from a competition between homogeneous and inhomogeneous broadening and it can be written as a sum of L Lorentzians (see Ref. [6]):

$$I(\omega) = \sum_{q=1}^{L} \frac{c_q(\omega - f_q) + a_q\gamma_q}{(\omega - f_q)^2 + \gamma_q^2}, \tag{1}$$

where $(a_q + ic_q)$ are the generalized intensities and $(f_q + i\gamma_q)$ are the generalized frequencies. The profile can then be seen as resulting from the contribution of L two-level systems. It takes the following form:

$$I(\omega) = \sum_{q=1}^{L} a_q\Psi_q, \tag{2}$$

where Ψ_q is the profile of the Stark component q belonging to the E1 radiative line. We have $a_q = S_q/e^2$, where S_q designates the effective line strength, in atomic units.

Considering that each two-level system contributes to the emission and absorption processes, it is possible to define an opacity and RPL for each one of them.

1. Opacity

The opacity of an ion of net charge i is given by

$$k_i(\nu) = \sum_{u-l} \sum_{q=1}^{L} N_{l,q} \frac{(2\pi)^2}{3\hbar c} f_q S_q \Psi_q(\nu), \tag{3}$$

where the first sum involves all radiative transitions between upper levels labeled u and lower levels labeled l. The second sum accounts for all the Stark components associated with the radiative line $u-l$. $N_{l,q}$ is the population density of the lower level of the 2-level system. For each radiative transition there are L Stark transitions, and each one is broadened due to electron-ion collisions and thermal Doppler detuning. To account for the various ionic stages, the opacity is written as

$$k(\nu) = \sum_i p_i k_i(\nu), \tag{4}$$

where the p_i's are the ionic fractions. We have $\sum_i p_i = 1$.

We are interested in the transition array $1s-2p$. The required atomic data are given by the MCDF code [7]. The atomic basis used for the ion of charge i, *i.e.*, Al^{i+}, is formed by the following superconfigurations : $(1)^2(2)^{R-2}$, $(1)^1(2)^{R-1}$, $(1)^1(2)^{R-2}(3)^1$, where $R = Z - i$. The notation $(n)^r$ represents the ensemble of all the configurations $(nl_1)^{r_1}(nl_2)^{r_2}\cdots$ with the condition $\sum_j r_j = r$. For example, $(2)^2$ represents the three configurations $2s^2$, $2s2p$ and $2p^2$. We focus on the spectral range 7.7–8.5 Å which is covered by the Al^{4+} to Al^{11+} inner shell $n=1$ to 2 transitions and the Al^{4+} $n=1$ to 3 transitions. The electron density is estimated to be 5×10^{22} cm^{-3} and the electron temperature to 50 eV [2].

The transmission is given by the Beer's law:

$$T(\nu) = \exp\left[-k(\nu)\frac{\rho_l}{\rho}\right], \tag{5}$$

where ρ_l and ρ are the areal density and the density of the absorbing sample, respectively. The free-free and free-bound continuum absorptions are neglected.

There are three possibilities for the distribution of population among the various ion levels: i. the distribution is assumed in local thermodyamic equilibrium (LTE), ii. it is given by the statistical weights, iii. it is obtained by using a collisional-radiative code (NLTE distribution). Our code is versatile and can treat each possibility without significant difference between computing times.

Figure 1 shows a comparison of the measured and the calculated transmissions of the $1s-2p$ transition array. Most transmission peaks are attributed to one ionic species. The agreement is good for the peak positions, intensities and widths. The ion Stark broadening is negligible for the $1s-2p$ transition array. It is worth noting that for transition arrays involving higher configurations such as $1s-3p$, the ion Stark linewidth is important and can be used to predict the electron density [2].

2. Radiative power losses

In this section, we present the calculated RPL involving all the transitions between states of the ground superconfiguration (SC) of an ion which is generally of the type $(n)^r$. Owing to

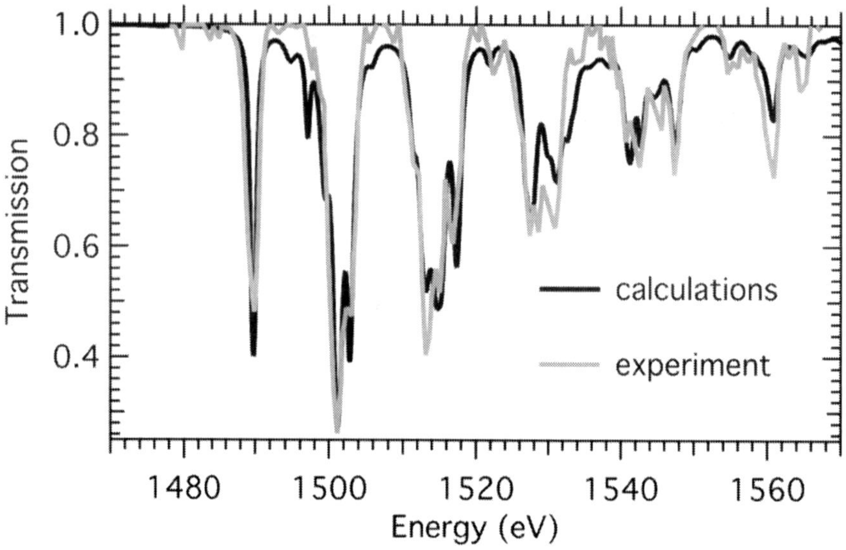

Figure 1. Transmission in aluminum as a function of the photon energy. The electron density and temperature are respectively 5×10^{22} cm^{-3} and 5 eV.

numerical constraints inherent to detailed line calculations, we restrict ourselves to $n = 3$ and $r = 4$. In fact, we are limited by the atomic data calculations required for the RPL calculations. For higher r values, atomic data calculations are made very complex by the large dimensions of the matrices to be diagonalized.

The radiative power losses of an ion i is given by

$$\epsilon_i(\nu) = \sum_{u-l} \sum_{q=1}^{L} N_{u,q} \frac{2}{3} \left(\frac{2\pi}{c}\right)^3 f_q^4 S_q \Psi_q(\nu), \tag{6}$$

where $N_{u,q}$ is the population density of the upper level of a 2-level system. As in the previous section, the RPL involving all ionic stages are given by

$$\epsilon(\nu) = \sum_i p_i \epsilon_i(\nu). \tag{7}$$

We have calculated the RPL profile of Ge^{18+} and Au^{65+} in the $(3)^4$ SC. The non relativistic configurations involved are $3s^2 3p^2$, $3s^2 3p3d$, $3s^2 3d^2$, $3s3p^3$, $3s3p^2 3d$, $3s3p3d^2$, $3s3d^3$, $3p^4$, $3p^3 3d$, $3p^2 3d^2$, $3p3d^3$, $3d^4$. There are 33,884 E1 lines contributing to the RPL.

For such SCs the precise RPL profile calculated by Eqs. (6) and (7) necessitates long calculation times. We then have explored a statistical approach. This approach is based on the Normal Inverse Gaussian (NIG) and Generalized Gaussian (GG) distributions [8, 9] which require high-order moments of the integrated RPL.

Figure 2.a shows the RPL profile of Ge^{18+} obtained by the two approaches. Due to modest density and Z values, the ion Stark broadening is negligible. The Doppler line broadening overcomes the electron (impact) broadening due to the high temperature value. We have introduced an apparatus width of 1 eV. The NIG distribution, which takes into account the first four moments, only provides an envelope of the precise profile. It is not able to reproduce

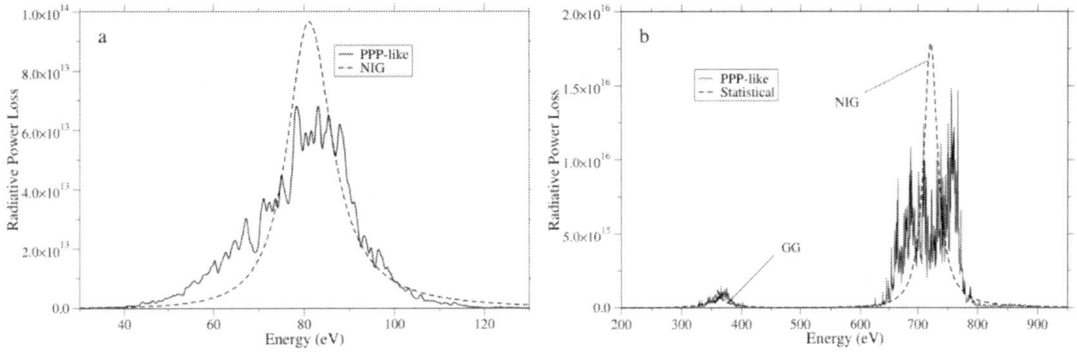

Figure 2. Radiative power losses of ions in the superconfiguration $(3)^4$ as a function of the photon energy. The RPL values have to be multiplied by the ground level population density. Full curve: precise profile obtained by using the PPP code, Dashed curve: RPL profile obtained with statistical distributions (NIG or NIG+GG). Figure 2.a: Ge^{18+}, $N_e = 10^{20}$ cm^{-3} and $T_e = 1000$ eV, Figure 2.b: Au^{65+}, $N_e = 10^{24}$ cm^{-3} and $T_e = 12$ keV.

the individual peaks. The Gram-Charlier expansion series [8] cannot be used since it exhibits negative features, even if very high-order moments are introduced in the distribution.

In Figure 2.b we show the RPL profile of Au^{65+} for the same SC. We have set $N_e = 10^{24}$ cm^{-3} and $T_e = 12$ keV. We have checked that despite a high density, the ion Stark broadening is negligible. For the same reason as above, the Doppler broadening overcomes the electron broadening. The main contribution to the RPL is around 700 eV. A smaller contribution emerges in the 350 eV energy range due to the spin-orbit interaction. In order to account for the entire spectrum, both statistical distributions (NIG and GG) are used.

3. Conclusions and perspectives

We have calculated the transmission of an aluminum plasma created by two femtosecond beams and a picosecond x-ray backlighter source. The agreement between measured and calculated transmission is good for peak positions, widths and magnitudes. All broadening mechanisms are taken into account, but it appears that the ion Stark brodening is very small for the $1s - 2p$ transition array. We have also calculated the RPL of germanium and gold plasma by using the same approach. We concentrated on the $(3)^4$ superconfiguration. Due to long calculation times a statistical approach was explored. The statistical approach provides an appropriate envelope for the germanium case but not for gold.

References

[1] Chenais-Popovics C, Merdji H, Missala T, Gilleron F, Gauthier J C, Blenski T, Perrot F, Klapisch M and Eidmann K 2000 *ApJ* **127** 275
[2] Audebert P, Renaudin P, Bastiani-Ceccotti S, Geindre J P, Chenais-Popovics C, Tzortzakis S, Nagels-Silvert S, Shepherd R, Matsushima I, Gary S, Girard F, Peyrusse O and Gauthier J C 2005 *Phys. Rev. Lett.* **94** 025004
[3] Eidman K, Bar-Shalom A, Saemann A and Winhart G 1998 *Europhys. Lett.* **44** 459
[4] Calisti A, Khelfaoui F, Stamm R, Talin B and Lee R W 1990 *Phys. Rev. A* **42** 5433
[5] Talin B, Calisti A, Godbert L, Stamm R, Lee R L and Klein L 1995 *Phys. Rev. A* **5** 1918
[6] Benredjem D, Calisti A and Mossé C 2009 *High Energy Density Physics* **5** 187
[7] Grant I P, McKenzie B J, Norrington P H, Mayers D F and Pyper N C 1980 *Comput. Phys. Comm.* **21** 207
[8] Pain J C, Gilleron F, Bauche J and Bauche-Arnoult C 2009 doi:10.1016/j.hedp.2009.05.001
[9] Gilleron F, Pain J C, Bauche J and Bauche-Arnoult C 2008 *Phys. Rev. E* **77** 026708

Journal of Physics: Conference Series **244** (2010) 042002 doi:10.1088/1742-6596/244/4/042002

Analytical Expressions for Radiative Opacities of Low Z Plasmas

J G Rubiano[1,2,3], P Martel[1,2], JM Gil[1,2], R. Rodriguez[1,2], R. Florido[1,2], M A Mendoza[1] and E Minguez[2]
[1]Departamento de Física, Universidad de Las Palmas de GC, Las Palmas de GC, Spain
[2]Instituto de Fusión Nuclear (DENIM), Universidad Politécnica de Madrid, Madrid, Spain

E-mail: jgarcia@dfis.ulpgc.es

Abstract. In this work we obtain analytical expressions for the radiative opacity of several low Z plasmas (He, Li, Be, and B) in a wide range of temperatures and densities. These formulas are obtained by fitting the proposed expression to mean opacities data calculated by using the code ABAKO/ RAPCAL. This code computes the radiative properties of plasmas, both in LTE and NLTE conditions, under the detailed-level-accounting approach. It has been successfully validated in the range of interest in previous works.

1. Introduction

The accurate determination of the radiative properties of hot dense matter is needed in several research fields such as ICF target physics analysis, magnetic fusion or astrophysics, and the radiative opacity is a key factor to be determined. Thus, ICF target simulation involves the use of hydrodynamic radiation codes that require the input of thousands of spectrally resolved opacity data for each point of a fine mesh of temperatures and densities and determining them takes a large computation time. A usual approach consists of weighting these opacities in a small number of multigroups opacities or even only one group using then the Rosseland and Planck mean opacities. Even in this case, the accurate computation of these quantities is too huge to be used in-line with a hydrodynamic code.

For these reasons, analytical formulas giving mean radiatives properties versus temperature and density of the plasma are a useful tool. In this work, our main goal is to obtain analytical expressions for the Rosseland and Planck mean radiative opacities of several low Z plasmas in a wide range of temperature and densities. These formulas are obtained by fitting the free parameters of the proposed expression to data calculated by the code ABAKO/ RAPCAL [1]. This code computes the radiative properties of plasmas such as the spectrally resolved opacity or the Rosseland and Planck mean opacities, both in LTE and NLTE conditions, under the detailed-level-accounting approach. The code has been successfully tested by comparing its results with reference codes in the fourth and fifth Kinetic Code Comparison Workshops [2, 3]. It also has demonstrated to be useful for diagnostic of experiments [4].

---

[3] To whom any correspondence should be addressed.

2. Opacity calculation

We have carried on a detailed analysis of the mean opacity for helium, lithium, beryllium and boron plasma, using high quality atomic data obtained using the FAC code [5]. This code provides atomic magnitudes calculated into the DLA approach, using appropriate coupling schemes and including configuration interaction. The calculation of ionic charge state distributions and level populations is performed by solving a collisional-radiative steady-state model (CRSS). The level populations are computed with a reasonable accuracy for plasmas of any element in a wide range of conditions embracing Coronal Equilibrium (CE), non-LTE and LTE situations. The spectrally resolved opacity is obtained as a sum of the bound-bound, bound free, free-free and scattering contributions

$$\kappa(\nu) = \frac{1}{\rho}\left(\mu_{bb}(\nu) + \mu_{bf}(\nu) + \mu_{ff}(\nu) + \mu_{scatt}(\nu)\right) \tag{1}$$

The bound–bound spectrum includes all the allowed transitions in the dipole approximation along with all the detailed atomic levels considered. Line profiles incorporate Doppler, natural and electron-impact widths. The bound-free and free-free processes are calculated using the Kramer's semiclassical expressions. The Rosseland and Planck mean opacities are then computed by:

$$\frac{1}{\kappa_{Rosseland}} = \int_0^\infty d\nu B'(\nu,T)/\kappa(\nu) \qquad \kappa_{Planck} = \int_0^\infty d\nu B(\nu,T)\left[\kappa(\nu) - \kappa_{scatt}(\nu)\right] \tag{2}$$

where κ_{scatt} is the absorption coefficient contribution by scattering (Thomson cross section), and $B(\nu,T)$ and $B'(\nu,T)$ are the Planck weighting function and its derivative.

3. Results and discussion

Analytical expressions can be found in the literature [6–8] for the opacity of low-Z plasmas in a wide range of temperatures and densities. The expression most commonly used to model Rosseland and Planck mean opacities is a power law depending on temperature and density,

$$k_{R,P} = e^a \rho^b T^c \tag{3}$$

where T is the temperature (eV) and ρ is the mass density (g·cm^{-3}).The definition of the parameter a can change depending on the author. These formulas are usually fit to match LTE data but there are no values for a wide range of density and temperature conditions where NLTE assumptions are important. In a logarithmic representation, equation (3) is a plane surface over the entire range of temperatures and densities. This geometrical form can give a good fit for highly ionized plasmas, but it fails at low temperatures where the atomic structure leads to local modulations of the mean opacity.

In order to extend its range of application, we have modified this formula including a multiplicative term depending on the temperature and the electron density, thus,

$$\mu_{R,P}\left(cm^{-1}\right) = \rho\kappa_{R,P} = e^{a_0} T^{a_1} \rho^{a_2} f(x,y) \tag{4}$$

with

$$f(x,y) = \left[\exp a_3 xy + a_4 x^2 + a_5 y^2\right] \tag{5}$$

where x = log(T) and y=log(ρ). We have fit the mass absorption coefficient μ instead of the radiative opacity because it has a smoother behavior. This new expression is a quadratic fit to surface which gives more flexibility in relation to a linear one.

In figure 1, it is shown the average ionization for a beryllium plasma in the range 1-10^3 eV for several densities from 10^{-9} to 10^{-1} g·cm^{-3}. At low temperatures the behaviour of the mean ionization depends strongly on the plasma density, thus under 10 eV we found huge variations in the average ionization. In the range 15 – 100 eV the low density curves trend to mach and for temperatures above 300 eV the fully striped ion dominates for all the cases studied. This complexity found in the ionization at low temperature has its counterpart in the mean opacity, and it is necessary to perform different fits for each range of temperatures.

Journal of Physics: Conference Series **244** (2010) 042002 doi:10.1088/1742-6596/244/4/042002

In figure 2 we plot the behavior of the Planck mean absorption coefficient (a) and its fit (b) for boron plasmas in the temperature range between 1 and 10^3 eV and density from 10^{-10} to 10^{-1} g cm^{-2}. Figures 3(a) and 3(b) show the Rosseland mean for beryllium plasmas under the same conditions. The fit has been performed using a quadratic response surface model and since it is not possible fit the whole surface in the entire range of temperature, a piecewise fit has been done. One can appreciate that performing a linear fit over the entire range of temperatures and densities will yield a poor representation of the data, whereas equation 4 provides a better representation of the behavior of the Planck and Rosseland means at low temperatures.

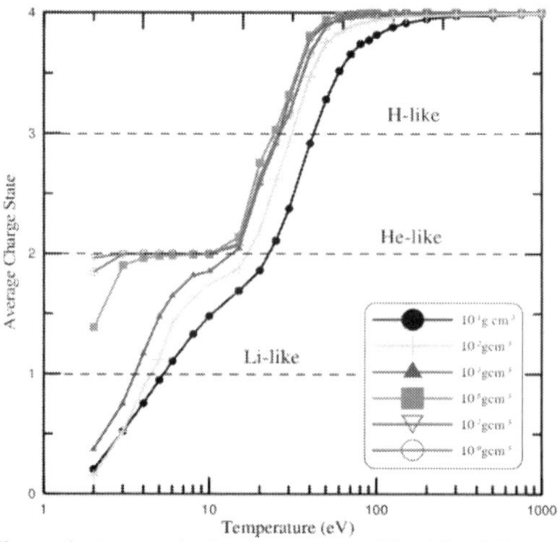

Figure 1. Average ionization computed by ABAKO vs. temperature at several densities for a Beryllium plasma

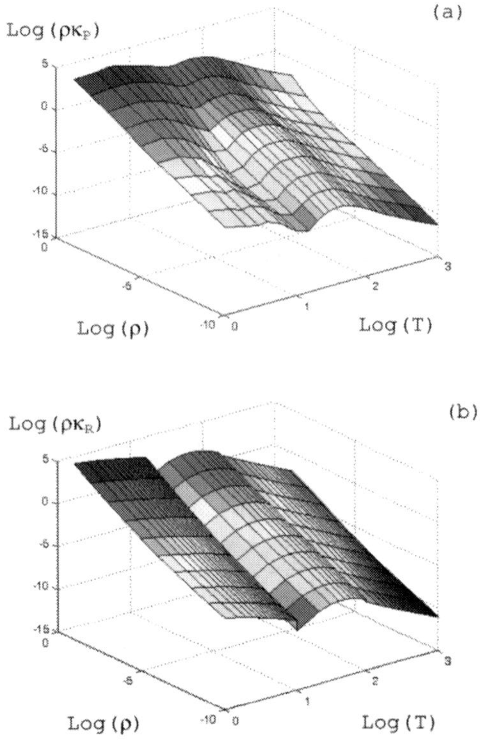

Figure 2. (a) Planck absorption coefficient (cm^{-1}) for a boron plasma (b) fit using equations (4)-(5).

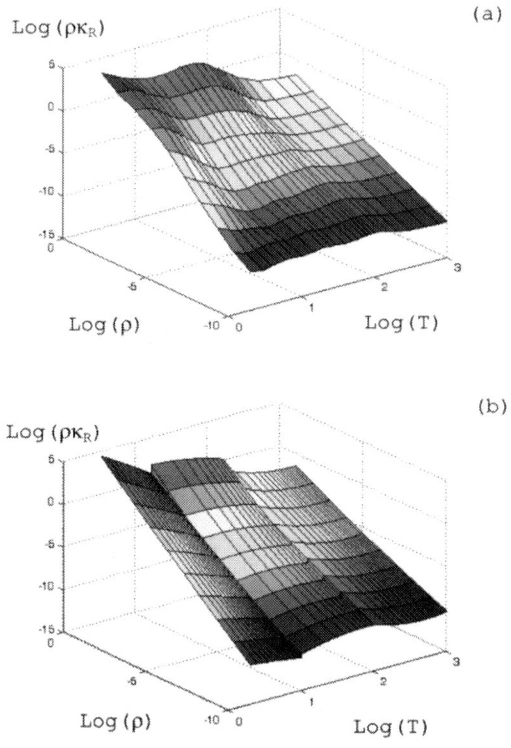

Figure 3. (a) Rosseland absorption coefficient (cm^{-1}) for a beryllium plasma (b) fit using equations (4)-(5).

Values of the parameters obtained for several low Z elements are given in Table 1. These results should be used with caution because the accuracy of the fit increases with the temperature, thus for the range 10^2 – 10^3 eV we find local deviations around 10-20 % in the opacities but in the range 1 -10 eV they can deviate by factor 5 from the calculated values. Nevertheless this fit could be a useful tool in the study of phenomena in hydrodynamics involving radiative transport, when these local deviations can be assumed.

Table 1. Fit coefficients for He, Li, B and Be Plasma.

	Planck			**Rosseland**		
			Z=2			
	2≤T≤5 eV	**5<T≤ 10 eV**	**10<T≤10³ eV**	**2≤T≤10 eV**	**10<T≤ 10² eV**	**10² <T≤ 10³ eV**
a_0	1.5483E+01	-6.5726E+00	2.7636E+01	1.5179E+01	2.9428E+01	1.3474E+00
a_1	-5.8973E+00	2.1957E+01	-4.5688E+00	2.8546E+00	-5.8835E+00	-4.5316E-01
a_2	1.3979E+00	1.2924E+00	2.2334E+00	2.2316E+00	2.9015E+00	1.1514E+00
a_3	2.2544E-02	1.8327E-01	1.6779E-02	-6.2619E-02	-2.6930E-01	-2.3174E-02
a_4	3.1109E+00	-5.5971E+00	5.1518E-02	-1.2338E+00	-7.2351E-03	0.0000E+00
a_5	1.1628E-02	2.0203E-02	3.9707E-02	1.4569E-02	2.3920E-02	0.0000E+00
			Z=3			
	1≤T≤10 eV	**10<T≤ 10² eV**	**10²<T≤10³ eV**	**1≤T≤10 eV**	**10<T≤ 10² eV**	**10² <T≤ 10³ eV**
a_0	2.1195E+01	9.3986E+00	2.7807E+01	1.9424E+01	1.5594E+01	2.4996E+01
a_1	-9.5624E+00	4.2568E+00	-5.1315E+00	-7.7632E+00	1.5201E+00	-7.6016E+00
a_2	2.1810E+00	1.5861E+00	1.7701E+00	2.2338E+00	2.8647E+00	1.4683E+00
a_3	-3.8400E-01	2.8961E-02	-3.7040E-03	-1.0297E-01	-2.4069E-01	-4.3706E-02
a_4	2.4094E+00	-1.0845E+00	1.0344E-01	2.0365E+00	-8.6555E-01	5.5563E-01
a_5	1.2152E-02	2.3232E-02	2.5038E-02	1.7773E-02	2.5594E-02	7.0353E-03
			Z=4			
	1 ≤T≤ 10 eV	**10 <T≤ 10² eV**	**10² <T≤ 10³ eV**	**1 ≤T≤ 10 eV**	**10 <T≤ 10² eV**	**10² <T≤ 10³ eV**
a_0	2.5021E+01	-1.3335E+01	3.1069E+01	2.1479E+01	1.0988E+01	4.2232E+01
a_1	-1.1426E+01	1.5940E+01	-6.1691E+00	-5.2281E+00	1.8834E+00	-1.2768E+01
a_2	1.9024E+00	1.4265E+00	1.4141E+00	2.0529E+00	2.6111E+00	1.5055E+00
a_3	-8.7274E-02	-2.4457E-02	-4.7942E-03	-1.9886E-01	-1.3254E-01	-4.8302E-02
a_4	2.6164E+00	-2.5665E+00	1.8157E-01	-1.5990E-01	-5.5587E-01	9.4243E-01
a_5	1.5436E-02	1.1775E-02	1.4057E-02	4.6793E-03	3.0221E-02	7.5590E-03
			Z=5			
	1 ≤T≤ 20 eV	**20<T≤2×10²eV**	**2×10²<T≤10³eV**	**1 ≤T≤ 10 eV**	**20<T≤2×10²eV**	**2×10²<T≤10³eV**
a_0	1.6109E+01	-2.0046E+01	3.0288E+01	1.7740E+01	8.6753E+00	5.2951E+01
a_1	-8.8648E-01	1.6572E+01	-5.6685E+00	-5.7576E-01	2.1382E+00	-1.5937E+01
a_2	1.4934E+00	1.2575E+00	1.1586E+00	1.9185E+00	2.5659E+00	1.1639E+00
a_3	4.5994E-02	-2.7165E-02	-2.0608E-03	-1.2073E-01	-1.3522E-01	-1.2894E-02
a_4	-2.8942E-01	-2.3022E+00	1.2952E-01	-9.2101E-01	-5.0343E-01	1.1695E+00
a_5	1.5892E-02	5.6522E-03	6.2114E-03	3.6289E-03	2.8486E-02	3.1661E-03

Acknowledgments

This work has been supported by a Research Project of the Spanish Ministry of Science and Innovation ENE2008-06668-C02/FTN and also by the ''Keep in Touch'' Project of the European Union.

References

[1] Rodriguez R, Gil JM, Florido R, Rubiano JG, Martel P and Mínguez E 2008 *Laser and Particle Beams*, **26**, 433–448.

[2] Rubiano JG, Florido R, Bowen C, Lee RW and Ralchenko Yu 2007 *High Energy Density Physics*, **3**, 225-232.

[3] Fontes CJ, Abdallah, J, Bowen C, Lee RW and Ralchenko Yu 2009 *High Energy Density Physics*, **5**, 15–22.

[4] Mínguez E, Florido R, Rodriguez R, Gil JM, Rubiano JG, Mendoza M A, Suárez D and Martel P, *High Energy Density Physics*, Article in Press. doi:10.1016/j.hedp.2009.05.016 (2009)

[5] Gu. M F 2003 *Astrophys. J.* **582**, 1241-1250.

[6] Lindl J, *Physics of Plasmas*, **2**, 11, 39-71

[7] Tsakiris G D and Eidman K 1987 *J. Quant. Spectrosc. Radiat. Transfer*, **38**,353-368.

[8] Mínguez E, Martel P, Gil JM, Rubiano JG and Rodriguez R. 2002 *Fusion Engineering and Design*, **60**, 17–25.

Opacity calculations for ICF target physics using the ABAKO/RAPCAL code

E. Mínguez[a], R. Florido[ab], R. Rodriguez[ab], J.M. Gil[ab], J.G. Rubiano[ab], M. A. Mendoz[ab], D. Suárez[b], P. Martel[ab]

[a] Instituto de Fusión Nuclear, Universidad Politécnica de Madrid, Madrid, 28006, Spain.
[b] Departamento de Física, Universidad de las Palmas de Gran Canaria, Las Palmas de Gran Canaria 35017, Spain

E-mail: emilio.minguez@upm.es

Abstrat: In this work we present a set of atomic models (called ABAKO/RAPCAL), and its validation with experiments and with other NLTE models. We consider that our code permits the diagnosis and the determination of opacity data. A review of calculations and simulations for the validation of this set is presented.As an interesting product of these calculations, we can obtain accurate analytical formulas for Rosseland and Planck mean opacities. These formulas are useful for the use as input data in hydrodinamic simulations of targets where the computation task is so hard that in line computation with sophisticated opacity codes is prohibitive. Analytical opacities for several Z-plasmas are presented in this work.

1.- Introduction.

In the plasmas generated in ICF targets the calculation of radiative properties is a difficult task due to the huge number of levels and transitions involved, even for low Z elements. In particular, the understanding of ICF plasmas requires emissivities and opacities both for hydro-simulations and diagnostics.To assist in these calculations, the code ABAKO/RAPCAL has been developed [1] which can determine populations, radiative properties, spectrally resolved opacities and emissivities, and perform K-shell spectroscopic diagnostics to infer both the average electron density and the temperature of the plasma in LTE and non – LTE (NLTE).

The atomic structure model used in this code follows a relativistic detailed configuration-accounting (DCA) scheme, where relativistic electronic configurations are determined by solving the monoelectronic Dirac equation, using parametric analytical potentials developed by our group in the last years [2-6]. The non –local thermodynamic equilibrium (NLTE) system of rate equations is solved assuming steady-state, and analytical formulas are used to compute rate coefficients. The code provides good results in a wide range of temperatures and densities, reproducing the Coronal and Saha results in the limits of low and high density, respectively. Optically thin and thick plasmas can also be treated.

2.- Spectroscopic diagnostic.

ABACO/RAPCAL has been used to analyze several laser produced plasmas. In this way, an experimental study to measure the opacity and emissivity of bound-bound transitions in laser-shocked dense hot aluminium plasma was carried out at the LULI laser facility [7]. A database of emergent intensities in the photon energy from 1700 to 2400 eV was computed, producing a 20 x 20 grid of electron temperatures and densities. An optically 80 μm thick aluminium plasma was assumed, which matches with the thickness of the aluminium layer of the target.

The electron temperature and density for a given spectra is performed by researching in the database the synthetic spectrum that yields the best fit to the data. Following in this way with the application of this procedure, each of the spectral line outs in the recorded film results in a spatial profile of electron temperature and density.

3.- Relativistic screening hydrogenic model.

In ABAKO/RAPCAL, analytical potentials are used to solve the Dirac equation avoiding self-consistency, Detailed multifrequency apacities are determined creating a database for any element. However a simpler approach consists in calculate energy levels using the analytical expressions of the hydrogenic atom.

This approach has been widely used in plasma physics and is the basic for a family of closely related models called screened hydrogenic model (SHN). This relativistic model [8] is able to obtain fast and accurate computation of detailed opacities with a reduced computational time, and it has the adventage that can be coupled to hydrodynamic codes.

Planck and Rosseland mean opacities obtained with our model are compared (Table 1) with those from TOPS Opacities [9] given by Los Alamos National Laboratory, for several elements. Inspite of its simplicity, the model gives the magnitude order of the values correctly.

Table 1: Comparison of k_R and k_P with TOPS (LANL)

	ρ (g/cm3)	T (eV)	k_R (cm²/g)	TOPS k_R (cm²/g)	K_P (cm²/g)	TOPS k_P (cm²/g)
Al	0.27	100.0	1.028E+03	1.414E+03	3.241E+03	5.265E+03
	0.27	125.0	3,585E+02	4.965E+02	1.528E+03	2.446E+03
	0.27	150.0	1.442E+02	2.024E+02	7.785E+02	1.166E+03
Ar	1.00	200.0	5.820E+02	7.075E+02	1.492E+03	2.429E+03
	1.00	500.0	1.278E+01	1.839E+01	1.085E+02	1.395E+02
	1.00	1000.0	4.054E+00	6.242E+00	2.250E+01	5.406E+01
Fe	1.00	500.0	5.467E+01	7.493E+01	2.171E+02	2.813E+02
	1.00	1000.0	2.114E+00	3.322E+00	2.492E+01	2.872E+01
	10.00	1000.0	1.207E+01	2.192E+01	8.525E+01	1.140E+02
	50.00	1000.0	2.979E+01	6.239E+01	2.430E+02	2.690E+02

4.- NLTE analytical mean opacities.

In the simulation of ICF targets, radiation hydrodynamics codes need the use of thousands of spectrally resolved opacity points for each temperature and density mesh point. A usual approach consists of weighting these opacities in only one group, using the Rosseland and Planck mean opacities. For this reason, analytical expressions for the Rosseland and Planck mean radiative opacities of several low Z plasmas in a wide range of temperature and densities are determined. These formulas are obtained fitting the analytical expression proposed to mean opacities data computed by using the code ABAKO / RAPCAL.

Analytical expressions used to model Rosseland and Planck mean opacities were proposed to match LTE data, but there are no values for a wide range of density and temperature conditions where NLTE assumptions are important . In order to extend those expressions for NLTE, a new one is proposed

$$\mu_{R,P}\left(cm^{-1}\right) = \rho\kappa_{R,P} = e^{a2} T^{a1} \rho^{a2} f(x,y)$$

with

$$f(x,y) = \left[\exp a_3 xy + a_4 x + a_5 y\right]$$

where x = log (T) and y = log (ρ).

The variation of Planck and Rosseland mean absorption coefficient in cm^{-1} for lithium in the range of 1 to 10^{-3} eV for temperature and 10^{-10} to 10^{-1} g cm^{-2} for density are shown in figure 1.

5. – Conclusions

ABAKO /RAPCAL can determine spectrally resolved opacities under LTE and NLTE conditions, that can be uses as a database for radiation hydrodynamic codes to simulate ICF targets. Also, we have used this code to measure, experimentally, opacity an emissivities of hot dense plasmas created with laser.

With the analytical potentials used in ABACO/RAPCAL, we have developed a new relativistic screening hydrogenic models that give successful results and can be compled to radiation hydrodynamic codes for getting on-line detailed opacities.

Finally, we have obtained NLTE analytical mean opacities, using data from ABACO/RAPCAL fitting with an analytical formula coefficients most important elements used in ICF targets.

Acknowledgements

This work has been supposted by a Reseach Projet of the Spanish Ministry of Science and Innovation ENE 2008-06668-CO2/FTN

References

[1] R. Florido, R. Rodriguez, J.M. Gil, J.G. rubiano, P. Marte, E. Mínguez, R. Mancini. *Physical Review E* (accepted for publiction).

[2] P. Martel, L. Doreste, E. Mínguez, J. M. Gil, J. Quant. *Spectrosc. Radiat. Transfer* 54 (1995) 621-636.

[3] P. Martel, J. G. Rubiano, J. M. Gil, L. Doreste, E. Mínguez. J. Quant. *Spectrosc. Radiat. Transfer 60* (1998) 623-633.

[4] J. M. Gil, P. Marte, E. Mínguez, J. G. Rubiano, R. Rodriguez, F. H. Ruano, J. Quant. *Spectrosc. Radiat. Transfer 75* (2002) 539-557.

[5] R. Rodriguez, J. G. rubiano, J. M. Gil, P. Marte, E. Mínguez, R. Florido, , J. Quant. *Spectrosc. Radiat. Transfer 75* (2002) 723-739.

[6] R. Rodriguez, R. Florido, J.M. Gil, J. G. Rubiano, P. Martel, E. Mínguez. Laser Part. *Beams 26* (2008) 433-448.

[7] R. Schott, F. Philippe, P. Ângelo, E. Dufour, A. Poquerusse, E. Dalmier, P. Sauvan, E. Sauvan, E. Mínguez , J. M. Gil, J. G. Rubiano, R. Rodriguez, P. Martel, R.Mancini, J. Quant. *Spectrosc. Radiat. Transfer 81* (2003) 441-450

[8] J. G. Rubiano, R. Rodriguez, J.M. Gil, F. H. Ruano, P. Martel, E. Mínguez, J. Quant. *Spectrosc, Radiat. Transfer 72* (2002) 575-588.

[9] LANL T-4 Opacities – *Los Alamos National Laboratory.*

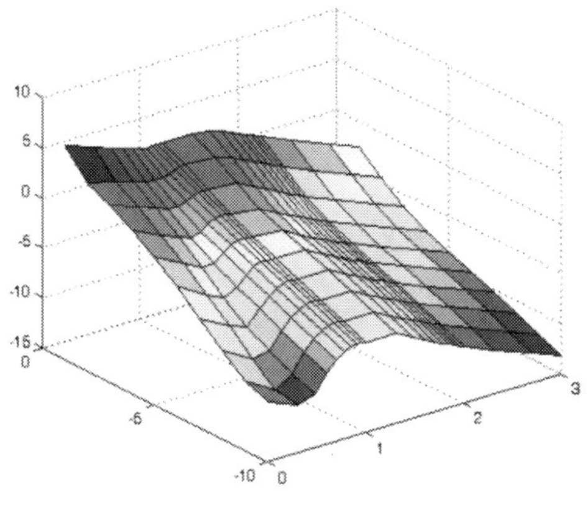

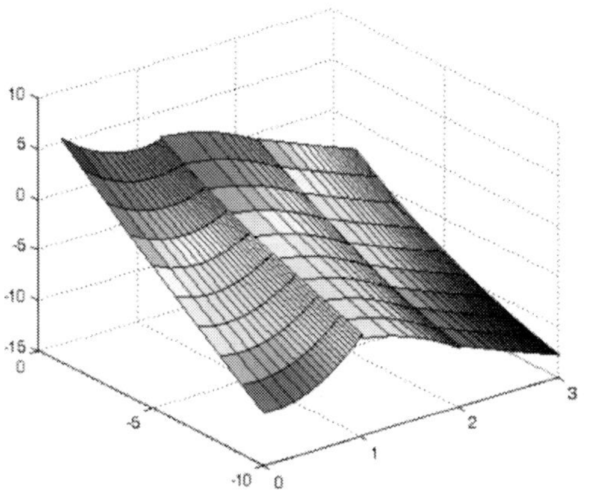

(a) (b)

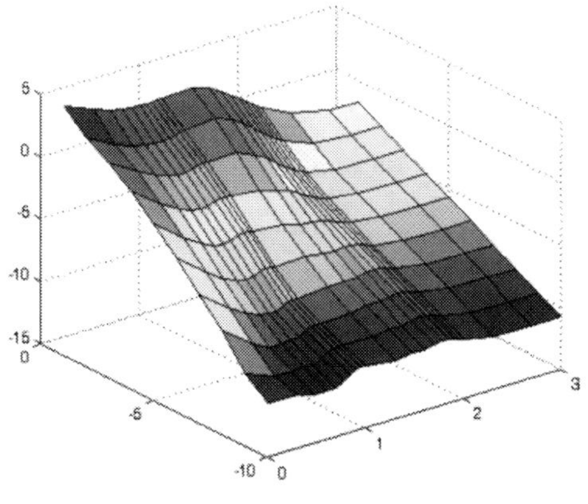

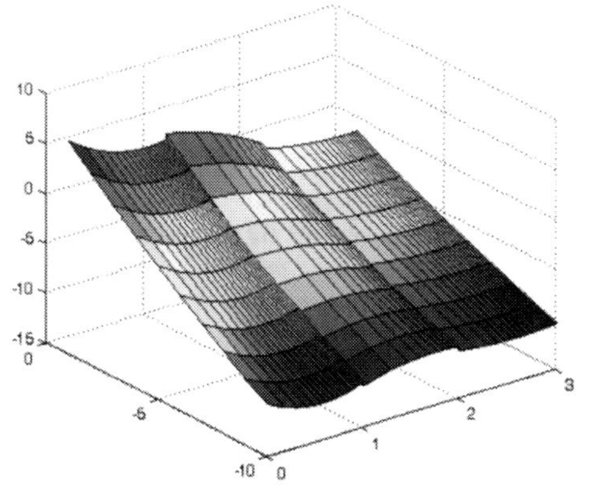

(c) (d)

Figure 1. For a Li plasma: (a) Planck absorption coefficients in cm^{-1} with ABAKO, (b) fit using analytical expressions,(c) Rosseland absorption coefficients in cm^{-1} with ABAKO, (d) fit using analytical exppressions.

Time-duration dependence from the simultaneous measurements of iron K- and L-shell radiation from laser produced plasmas

N D Ouart[1], A S Safronova[1], A Ya Faenov[2,3], T A Pikuz[2,3], S V Gasilov[3], S Stagira[4], F Calegari[4], M Nisoli[4], S De Silvestri[4], L Poletto[5] and P Villoresi[5]

[1] Physics Dept. MS220, University of Nevada, Reno, 1664 N. Virginia St., Reno, NV 89557, USA
[2] Kansai Photon Science Institute, Japan Atomic Energy Agency, 8-1 Umemidai, Kizugawa-city, Kyoto 619-0215, Japan
[3] Joint Institute of High Temperatures, Russian Academy of Sciences, Izhorskaya ul. 13/19, Moscow 127412, Russia
[4] National Laboratory for Ultrafast and Ultraintense Optical Science - CNR - INFM, Department of Physics, Politecnico, Milano, Italy
[5] Laboratory for Ultraviolet and X-Ray Optical Research - CNR - INFM, D.E.I. - Universita di Padova, Padova, Italy

E-mail: nouart@unr.nevada.edu

Abstract. High resolution X-ray spectra of plasmas produced by the interaction of Ti:Sa laser pulses (duration from 60 fs to 1 ps, and energy from 15 mJ to 128 mJ) with a solid Fe target are investigated experimentally and theoretically. Interestingly, both L-shell and K-shell Fe spectra were simultaneously measured while changing the time-duration of the laser experiments. In particular, the L-shell Fe radiation from F-, Ne-, and Na-like ions between ~14-16.3 Å were observed from the first order reflection with a mica crystal. Simultaneously measured alongside the L-shell radiation was the Fe cold Kα lines (~1.94 Å), from 8th order reflection. Two non-LTE Fe kinetic models have been developed to account for the L-shell radiation from the Fe ions and for the K-shell radiation from the low ionization stages emitting the Fe cold Kα lines. Preliminary analysis of the simultaneously measured Fe L- and K-shell radiations from changing the time-duration of the experiments shows varying relative line intensities from different ionization stages. Resulting plasma parameters as well as the opacity of the strongest resonance line, 3C, and their dependence on changing the time-duration will be discussed.

1. Introduction

Studies of different laboratory and astrophysical plasmas, such as tokamaks, astrophysical, and ICF applications, involve the comprehensive investigation of K- and L-shell iron spectra. Previous work was mainly focused on the separate study and measurements of L-shell [1] or K-shell iron spectra [2]. Also, it is very important to measure and analyze them simultaneously, especially for the case of fs-scale interactions with solid targets when fast electrons are produced and ionize solid targets up to the M-shells [3]. In particular, it is interesting to investigate the ionization balance and spectra of iron for the case of < 100 fs laser pulses interactions with a solid target. In the present paper, we are reporting such investigation for the first time.

2. Experimental Setup

The experiments were performed using a Ti:Sapphire (Ti:Sa) laser system from a national laboratory for ultrafast and ultraintense optical science (ULTRAS-CNR-INFM, Politecnico di Milano, Italy), which generates 800 nm pulses with 10 Hz repetition rate, maximum energy of 130 mJ, and minimum duration of 60 fs. The duration of laser pulses was varied between 60 fs and 2 ps by changing the distance between gratings in the compressor stage. The laser pulse contrast can be varied between ~ 10^{-3} - 10^{-5} in the nanosecond time scale and ~ 10^{-2} - 10^{-4} in the ps scale. Iron slabs were used as targets. The laser beam was focused on a target surface at an incidence angle of 15° by an off-axis parabola mirror in a spot between 10–20 μm in diameter (the Rayleigh length of the focusing system in vacuum is about 200 μm). The peak intensity varies in the range 4×10^{15} – 6×10^{17} W/cm^2 for the pulse duration between 1 ps and 60 fs. The target was continuously shifted in order to shoot on a fresh surface each time. Laser radiation focusing and plasma expansion were controlled by an X-ray pinhole camera with a 15-μm aperture diameter. The 1:1 plasma image was recorded on a RAR-2492 film in the X-ray spectral range with photon energy greater then 0.4 keV. In this experiment, the laser pulse contrast was intentionally reduced to about 10^{-2} in the 100-ps scale by inverting a dielectric steering mirror between the regenerative and the high power amplifiers; in this way, a weak prepulse reflected from the substrate surface was generated.

X-ray radiation from the plasma was measured by the focusing spectrometer with spatial resolution (FSSR) spectrometer [4, 5] based on the spherically bent mica crystal (2d = 19.91 Å). The crystal curvature radius is R = 100 mm and the aperture size is 8x26 mm^2. The X-ray radiation was registered by a CCD camera (Roper Scientific, 1300 x1300-cells matrix) with the cell size of 20 μm. The CCD-matrix was protected from the visible and vacuum ultraviolet VUV radiation by three layers of the 1μm polypropylene film coated with a 0.2-μm Al layer. The spectrometer was adjusted to the Bragg reflection at the angle θ = 49.8° to observe the first order reflections in the vicinity 14.1-16.2 Å capturing the 3-2 transitions of L-shell Fe with the spectral resolution $\lambda/\delta\lambda$ ~ 5000. Simultaneously measured in 8th order reflections was radiation from the Fe Kα line in the spectral range 1.75 – 2 Å. The spectral measurements were performed with the one-dimensional spatial resolution about 50 μm transverse to the plasma expansion direction.

3. Experimental Results and Modeling of Line Radiation

X-ray spectra from the plasmas were obtained from the interaction of a Ti:Sa laser with a solid Fe target. The obtained spectra, especially for the case of heating by 60 fs laser pulses, were noisy due to the interaction of fast electrons and gamma rays with the CCD. Therefore x-ray spectra were cleaned with a special procedure to remove the noise [6]. The solid line in figure 1 shows an experimental spectrum with the L-shell lines from hotter plasma (1st order reflection) and cold Kα lines from cooler plasma (8th order reflection). The L-shell lines were produced by the prepulse plasma. The laser radiation interacted with the plasma producing hot electrons which heated the solid dense iron target producing the cold Kα lines. A total of nine experiments were analyzed. Each experiment had either best contrast, ps-prepulse, or ns-prepulse with main pulse durations of 60 fs, 600 fs, or 1 ps. All experimental spectra showed increasing intensities of the cold Kα lines with decreasing laser main pulse duration. The L-shell lines also decreased with decreasing main pulse duration, but only slightly with experiments with the best contrast.

In order to model radiation from the experiments, two non-LTE kinetic models were used. The first model accounted for the L-shell radiation from the hotter multiply-charged ions and has been used to model radiation from the implosion of stainless steel x-pinches [7] and wire arrays [8]. An example of a synthetic spectrum accounting for the L-shell radiation is shown in figures 1 and 2. Some L-shell spectral features seen are the F-like lines labels F1, F2, and F3 which are $2p^4\,3d \rightarrow 2p^5$ transitions and the Ne-like lines labeled 3C and 3D which are $2p^5\,3d \rightarrow 2p^6$ transitions. The relative ratios of lines from adjacent ionization stages, such as F2 to 3D, were used to constrain the electron temperature. For instance, the ratio of F2 to 3D is lower in figure 1 then in figure 2 and therefore the electron temperature is cooler in figure 1. The unlabeled lines between 3C and 3D are mainly Na-like satellite lines and were helpful to constrain electron density as well as the electron temperature.

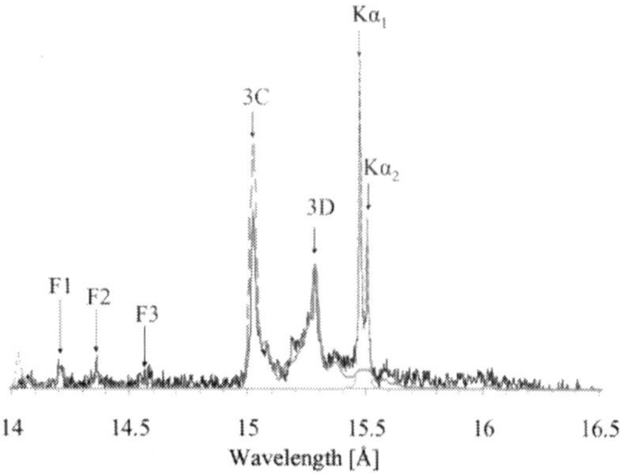

Figure 1. An experimental spectrum from the interaction of a Ti:Sa laser system with a ps-prepulse and 60 fs main pulse and an iron target (solid line). The result of modeling the L-shell lines (long dash line) with electron temperature of 170 eV and electron density of 10^{21} cm^{-3}. The result of modeling the cold Kα lines (short dash line) with electron temperature 20 eV, electron density of 10^{22} cm^{-3} and a 0.5% fraction of hot electrons.

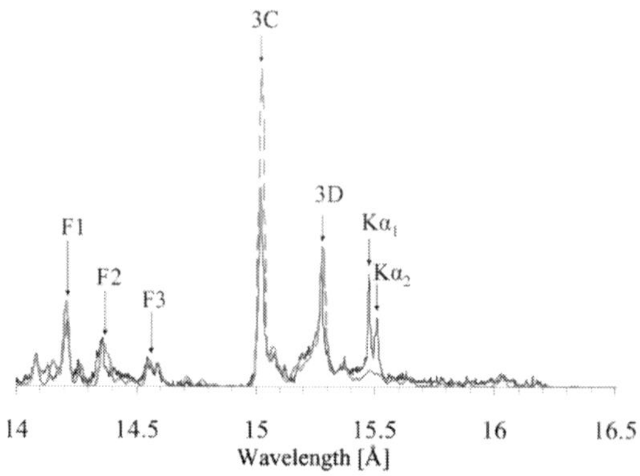

Figure 2. An experimental spectrum from the interaction of a Ti:Sa laser system with a ns-prepulse and 1 ps main pulse and an iron target (solid line). The results of modeling the L-shell lines (long dash line) with electron temperature 195 eV and electron density of 10^{21} cm^{-3}.

The second model was recently developed to account for the cold Kα radiation by the excitation of inner shell states from the interaction of hot electrons with the solid dense iron target. An example of a synthetic spectrum accounting for the cold Kα lines is shown is figure 1. These lines originate from cooler denser plasma from inner shell excitations with hot electrons in lower ionization stages. It was found in modelling that increasing the fraction of hot electrons not only shifted the ionization balance to higher ionization stages, but also tended to broaden the ionization balance. This tendency constrained the fraction of hot electrons to < 1% to reproduce the experimental spectra.

L-shell modelling of spectra with the best contrast and ps-prepulse showed that the electron temperature increased while the electron density decreased with increasing main pulse duration. However, spectra produced with ns-prepulse showed no appreciable difference in electron density and a small change in electron temperature with the different main pulse durations. In figures 1 and 2, there is a difference in intensity of 3C lines between the synthetic spectrum and experimental spectrum which is due to opacity. The 3C line has the largest radiative decay rate of the Ne-like resonance lines of Fe and therefore the effects of opacity should occur first in this line compared to the other Ne-like lines. In figure 2 there is also a difference in the F1 line intensities between the synthetic spectrum and experimental spectrum. Again, this difference is due to opacity. The F1 line has the largest radiative decay rate of the F-like resonance lines. The difference between the ratio of the theoretical to the experimental 3C line intensity correlates with increasing Kα intensity.

4. Summary

X-ray spectra were obtained from the interaction of a Ti:Sa laser with a solid iron target. Simultaneous measurements of L-shell emission (1st order reflection) from the hotter plasma and the cold Kα (8th order reflection) emission from the cooler plasma were obtained. The L-shell lines were produced by the prepulse plasma. The laser radiation interacted with the plasma producing hot electrons which heated the solid dense iron target producing the cold Kα lines. Radiation was modelled from the different regions with two non-LTE kinetic models. The first model accounted for the L-shell radiation from the hotter multiply charged ions, while the second accounted for the cold Kα radiation by the excitation of inner shell states from the interaction of hot electrons with the solid dense iron target. All experimental spectra showed increasing intensities of the cold Kα lines with decreasing laser main pulse duration. The L-shell lines also decreased with decreasing main pulse duration, but only slightly with experiments with the best contrast. L-shell modelling of the experimental spectra with the best contrast and ps-prepulse showed that the electron temperature increased while the electron density decreased with increasing main pulse duration. However, spectra produced with ns-prepulse showed no appreciable difference in electron density and a small change in electron temperature with the different main pulse durations. A difference in the intensity of the 3C line between synthetic spectra and experimental spectra was found and is due to opacity. This difference was found to correlate with strong cold Kα emission.

Acknowledgments
This work was partly supported by the DOE/NNSA Cooperative agreement DE-FC52-06NA27588, by fellowship support with the National Physical Science Consortium with Sandia National Labs., by the Ministry of Education, Science, Sports, and Culture of Japan(MEXT), Grant-in-Aid for Kiban A (20244065, 2008), by the Russian Foundation for Basic Research (RFBR) (09-02-92482-MNKC_a), and by the Russian Academy of Sciences (RAS) Presidium Program of basic research 12.

References
[1] May M J, Beiersdorfer P, Dunn J, Jordan N, Hansen S, Osterheld A L, Faenov A Ya, Pikuz T A, Skobelev I Yu, Flora F, Bollanti S, Di Lazzaro P, Murra D, Reale A, Reale L, Tomassetti G, Ritucci A, Francucci M, Martellucci S and Petrocelli G 2005 Accurate wavelength measurements and modeling of Fe xv to Fe xix spectra recorded in high-density plasmas between 13.5 and 17 Å *The Astrophysical Journal Supplement Series* **158** 230–41

[2] Doschek G A 2006 High-resolution X-ray spectra of solar flares *Advances in Space Research* **38** 1482-9

[3] Hansen S B, Faenov A Ya, Pikuz T A, Fournier K B, Shepherd R, Chen H,Widmann K, Wilks S C, Ping Y, Chung H K, Niles A, Hunter J R, Dyer G and Ditmire T 2005 Temperature determination using Kα spectra from M-shell Ti ions *Physical Review E* **72** 036408

[4] Faenov A Ya, Pikuz S A, Erko A I, Bryunetkin B A, Dyakin V M, Ivanenkov G M, Mingaleev A R, Pikuz T A, Romanova V M and Shelkovenko T A 1994 High-performance X-ray spectroscopic devices for plasma microsources investigations *Physica Scripta* **50** 333-8

[5] Skobelev I Yu, Faenov AYa, Bryunetkin B A, Dyakin V M, Pikuz T A, Pikuz S A, Shelkovenko T A, Romanova V M and Mingaleev A R 1995 Investigating the emission properties of plasma structures with X-ray imaging spectroscopy methods *Jour. Exp. Theor. Phys.* **81** 692-717

[6] Gasilov S V, Faenov AYa, Pikuz T A, Villoresi P, Poleto L, Stagira S, Calegari F, Vozzi C and Nisoli M 2007 Extracting ion emission lines from femtosecond-laser plasma x-ray spectra heavily contaminated by spikes *Journ. of Applied Physics.* **102** 063303

[7] Safronova A S, Kantsyrev V L, Esaulov A A, Ouart N D, Safronova U I, Shrestha I and Williamson K M 2009 X-ray spectroscopy and imaging of stainless steel X-pinches with application to astrophysics *Eur. Phys. J. Special Topics* **169** 155-8

[8] Safronova A S, Kantsyrev V L, Esaulov A A, Ouart N D, Yilmaz M F, Williamson K M, Shrestha I , Osborne G C, Greenly J B, Chandler K M, McBride R D, Chalenski D A, Hammer D A, Kusse B R and LePell P D 2008 Spectroscopy and implosion dynamics of low wire number nested arrays on the 1 MA COBRA generator *Phys. Plasmas* **15** 033302

Journal of Physics: Conference Series **244** (2010) 042005 doi:10.1088/1742-6596/244/4/042005

Spectroscopic Studies of Tin Plasma Using Laser Induced Breakdown Spectroscopy

Nek M Shaikh[1,2], Y Tao[1], R A Burdt[1], S Yuspeh[1], N Amin[1,3] and M S Tillack[1]
[1]Center for Energy Research, University of California, San Diego, 9500 Gilman Drive, La Jolla, CA, USA 92093-0417
[2]Institute of Physics, University of Sindh, Jamshoro, Pakistan 76080
[3]Department of Physics, University of Agriculture, Faisalabad, Pakistan

E-mail: neksheikh@yahoo.com

Abstract. Laser-induced Sn plasma generated at different laser intensities has been characterized using visible emission spectroscopy. A CO_2 laser pulse 85 ns in duration is used to generate plasma from a planar Sn sample in a vacuum of 10^{-5} torr. The plasma electron temperature is inferred by the Boltzmann plot method from singly ionized Sn emission lines, and plasma electron density is inferred using Stark broadened profiles. Electron temperature is measured in the range of (0.53 - 1.28) eV, and electron density is measured in the range of (9.19×10^{15} - 7.45×10^{16}) cm^{-3}, as the laser intensity is varied from (1×10^{10} to 2.5×10^{10}) W/cm^2. The plasma shielding effect has been observed within the laser intensities of (2×10^{10} – 2.5×10^{10}) W/cm^2.

1. Introduction

Laser-induced breakdown spectroscopy (LIBS) is a powerful and flexible tool for qualitative and quantitative elemental studies in a wide class of applications from material science to medicine. The ablation process using long pulse duration lasers (> 1 ns) is divided into three stages. In the first stage, the laser light interacts with the solid resulting in rapid ionization of the target surface into plasma on a time scale short compared with the pulse duration. In the second stage, the laser light is efficiently absorbed by the plasma which expands isothermally. In the third stage, after the end of the laser pulse, the resultant plasma plume expands quasi-adiabatically in a medium, which can include vacuum or a background gas, with or without applied fields [1,2].

Laser produced Sn plasma is the most promising candidate for the next generation extreme ultraviolet (EUV) light source used in the semiconductor lithography industry to produce microchips with feature size of 32 nm or smaller. It is considered as a potential EUV source due to its high conversion efficiency from laser to EUV light, and also due to the ability to control the ionic plasma debris [3,4]. Here we use the LIBS technique to study laser produced Sn plasma.

Most of the previous studies [5-7] using visible emission spectroscopy of laser-produced Sn plasma have used 1064 nm Nd:YAG laser radiation, and there have been minimal efforts to study the visible emissions from Sn plasma produced by a CO_2 laser with 10.6 μm wavelength. Here we measure the visible spectrum radiated by laser-produced Sn plasma using a CO_2 laser at intensities from 1×10^{10} to 2.5×10^{10} W/cm^2. The spectral line intensities allow the deduction of electron temperature and electron density in space and time and thus give information regarding the fundamental plasma physics involved.

2. Experimental setup

The experimental setup shown in figure 1 is one variation of the LIBS technique in practice. The laser used in the experiments consists of a master oscillator and power amplifier. The gas mixture ratio of the oscillator is (CO_2:N_2: He = 20:4:76 %), optimized to remove the long μs tail and enahnce the peak power of the oscillator to 1 MW. The pulse duration of the laser from the oscillator is shortened by an air-breakdown plasma shutter. The plasma shutter is triggered by free electrons from an air-breakdown plasma induced by a Q-switched Nd:YAG laser and pumped by the oscillator itself.

A 2 mm thick 98.4% pure Sn sample was mounted in the vacuum chamber on a 3-dimensional motorized stage which was translated in the focal plane before each shot to provide a fresh surface. The vacuum was maintained below 10^{-5} torr throughout the experiments. The CO_2 laser with pulse duration of 84 ns was focused on the target surface at normal incidence by an F/10 meniscus lens. The optical emission from the plasma plume was collected normal to the target surface with a 1:1 imaging system. The optical system consisted of a collimating and focusing lens used to image the visible plasma radiation to the entrance slit of a 0.5m Czerny–Turner spectrograph (Acton Pro, Spectra-Pro 500i). The spectrograph was equipped with three gratings: 150, 600 and 2400 grooves/mm. The output of the spectrograph was coupled to an intensified charge-coupled device (PI MAX, Model 512 RB) camera that was operated with vertical binning of the 512 × 512 pixel array to provide spectral intensity versus wavelength. A programmable timing generator enables the acquisition of time resolved plasma spectra by controlling the delay between the laser pulse arrival and the detector system, as well as the intensification (exposure) time. The highest resolution grating (2400 groove/mm) was used for better resolution when measuring the Stark broadened transition lines; these lines being used to estimate the electron number density. For the measurement of the excitation temperature, the 600 groove/mm grating was used.

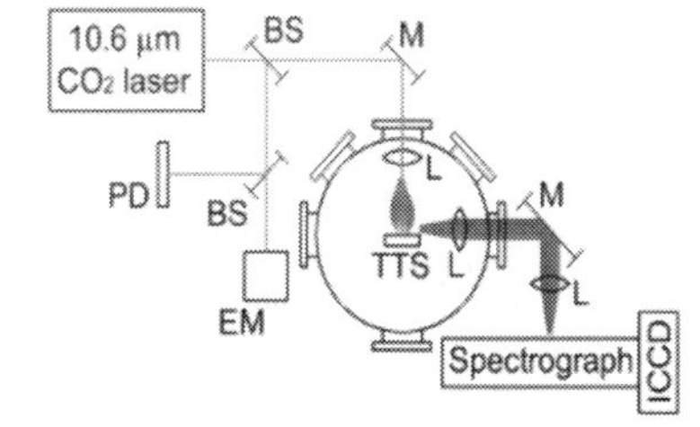

Figure 1. Experimental arrangement – BS-beam sampler, M-mirror, PD-photodiode, EM-laser energy monitor, TTS-target translation stage, L-lens, ICCD – intensified charge-coupled device camera, Spectrograph – Acton 500i SpectraPro.

3. Result and discussion

A sample of the emission spectrum covering the spectral region from 554 – 581 nm belonging to neutral (Sn I) and singly ionized (Sn II) is reproduced in Figure 2. In order to study the effects of the laser intensity on the emission characteristics of the laser produced Sn plasma in detail, the intensities of the Sn (II) lines at 556.19 nm (6d $^2D_{5/2} \rightarrow$ 6p $^2P_{3/2}$), 558.89 nm (4f $^2F_{5/2} \rightarrow$ 5d $^2D_{3/2}$), and 579.92 nm (4f $^2F_{7/2} \rightarrow$ 5d $^2D5_{3/2}$) were measured at various laser intensities. Figure 3 shows the dependence of the normalized intensities on the laser intensity. It is found that emission intensity of the Sn (II) transition lines at 556.19 nm, 558.89 nm and 579.92 nm increases by a factor of 13.0, 9.0, and 7.7 times with the increases of laser intensities from 1×10^{10} W/cm^2 to 2.5×10^{10} W/cm^2, whereas at the higher values of laser intensity from 2×10^{10} W/cm^2 to 2.5×10^{10} W/cm^2 a negligible variation is observed. As the emission intensity is associated with the density and temperature of the ablated material, so at the higher CO_2 laser intensity, which produces the dense plasma generated by the leading edge of the laser pulse, prevents light from reaching the surface. Therefore, most of the energy in the remaining part of the laser pulse will be absorbed by material in front of the surface prevent the tailing part of the laser pulse. Thus the laser energy delivered at high intensity is less effective in causing vaporization than the energy delivered at lower intensity.

In addition, we have recorded the spatial variation of the three transition lines from the Sn plasma along the normal to the target surface. In this experiment, the plasma is generated at a constant laser intensity of 1.5×10^{10} W/cm^2. Figure 4 shows the variation of the optical emission of Sn ions at different positions along the normal axis. It is found that emission intensity of the Sn (II) transition lines are all weaker close the surface up to a distance of 2 mm where a peak is reached, and then decay

Journal of Physics: Conference Series **244** (2010) 042005 doi:10.1088/1742-6596/244/4/042005

from 2 mm to the limit of our experimentation (5 mm). Up to 2 mm from the target surface, the intensities of the ionic lines are increasing because the collision process is dominant in this region which re-excites such ionic levels. Past 2 mm, the electron- ion recombination is fast as to result in a decrease of the emission intensities.

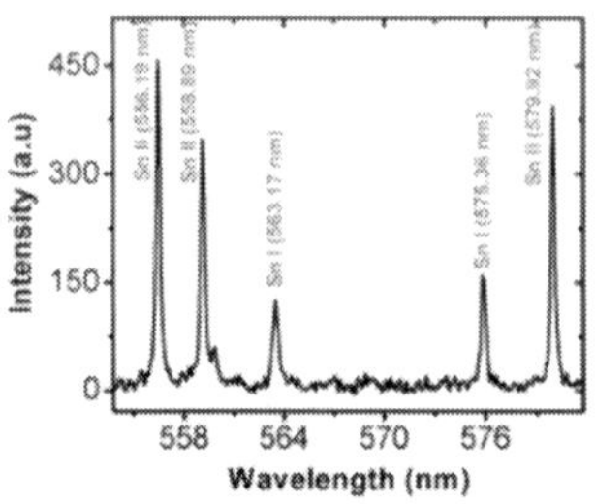

Figure 2. Emission Spectrum of the Sn plasma.

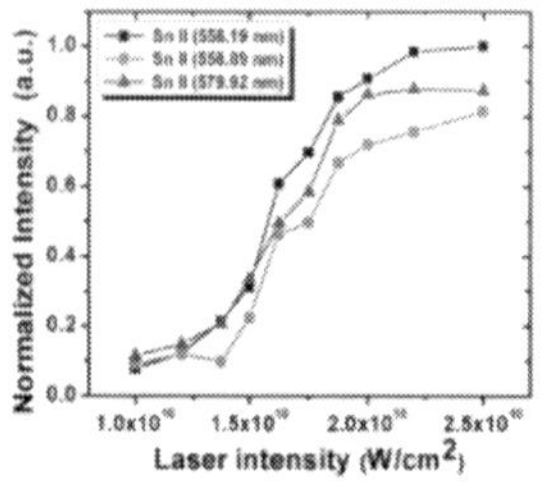

Figure 3. Variation in the integrated intensity of the Sn II transitions with laser intensity.

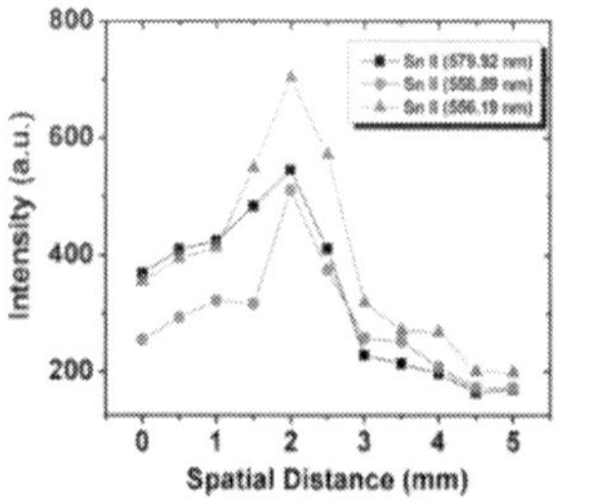

Figure 4. Spatial variation in the integrated intensity of the Sn II transitions.

3.1. Temperature and Electron Number Density

The key parameters of laser ablated plume are electron density and electron temperature. These parameters have been estimated under the assumption of local thermodynamic equilibrium as discussed in our previous work [8]. Five Sn II lines at 533.23 nm (6d $^2D_{3/2} \rightarrow$ 6p $^2P_{1/2}$), 556.19 nm (6d $^2D_{5/2} \rightarrow$ 6p $^2P_{3/2}$), 558.89 nm (4f $^2F_{5/2} \rightarrow$ 5d $^2D_{3/2}$), 645.35 nm (6p $^2P_{3/2} \rightarrow$ 6s $^2S_{1/2}$) and 684.4 nm (6p $^2P_{1/2} \rightarrow$ 6s $^2S_{1/2}$) have been employed in the Boltzmann plot method under the assumption of local thermodynamic equilibrium (LTE).

The shape and width of the spectral lines emitted by the plasma are governed by the collisional processes perturbing the emitting atoms and ions. Hence, the plasma density can be inferred from the profile of line spectra. The FWHM of the Stark broadened profile of Sn (II) lines at 556.19 nm has been used to estimate the electron number density [5,8]. The FWHM of the 556.19 nm is obtained by fitting the lorentizan profile shown in figure 5. The Stark broadening parameter ω was taken from Ref. [7]. Spectroscopic constants of Sn (II) lines used for the estimation of temperature have been taken from the NIST database [9]. The minimum criteria for LTE proposed by McWhirter [10] have also been verified and found to be the minimum electron number density $N_e \geq 2.2\times10^{15}$ cm^{-3} to justify the LTE assumption.

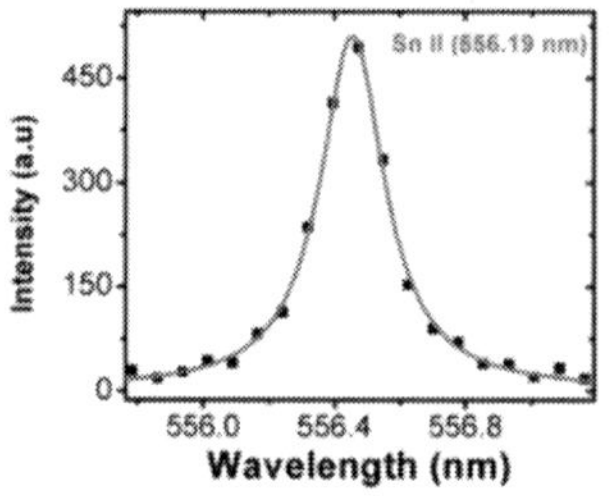

Figure 5. Lorentian fit on the Sn II transition line at 556.19 nm

Figure 6 shows the estimated values of the plasma temperature and number density of the Sn plasma dependence on the laser intensity. It is found that the T_e in the range of (0.53 - 1.28) eV, increases by the factor of 2.4 and the N_e is in the range of (9.19×10^{15} - 7.45×10^{16}) cm^{-3}, increases by the factor of 8 with the increase in the laser intensity from 1×10^{10} to 2.5×10^{10} W/cm^2. With increase in laser irradiance, the temperature and number density are found to increase up to the 2×10^{10} W/cm^2 laser irradiance and then to saturate. The saturation in temperature and number density above the 2×10^{10} W/cm^2 laser irradiance level is presumably due to plasma shielding, i.e., absorption and/or reflection of the laser photons by the plasma plume.

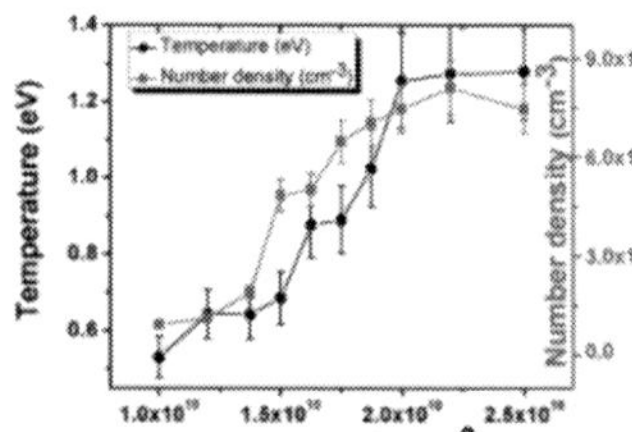

Figure 6. Dependence of T_e and N_e of the Sn plasma with laser intensity.

In the Previous reported work of O'Shay *et. al.*[6] measured higher temperatures and electron densities of Sn plasma generated by a 1.064 μm Nd:YAG laser than the work presented here. The laser parameters such as wavelength, spot size, pulse lengths and laser intensity used in Ref. 6 are 1.064 μm, 58 μm, 10 ns and 3.8 $\times 10^{11}$ W/cm^2 respectively, whereas in the present work we used as 10.6 μm, 200 μm, 84 ns, (1 - 2.5) $\times 10^{10}$ W/cm^2 respectively. The analytical properties of the laser ablated plasma are highly influenced by these parameters. Besides the longer pulse duration, higher spot size and lower intensity in the present work, the most important factor that influence the lower values of the temperature and number density is the laser wavelength, which directly affects the critical electron density (N_{cr}). The critical electron density at which the plasma becomes opaque to laser radiation is inversely proportional to the square of the incident laser wavelength. Thus critical electron density for 1.064 μm, Nd: YAG laser is two orders of magnitude greater than the CO_2 laser, whereas the optical depth of the 10.6 μm CO_2 laser in the target surface is less than the 1.064 μm Nd:YAG laser. This difference in laser absorption density explains the lower measured electron density in CO_2 laser-produce plasma plumes.

4. Summary

The effects of the CO_2 laser intensity on the emission, T_e and N_e of the Sn plasma have been studied. The variation in T_e and N_e with the laser irradiance shows that both the parameters increase with the increase in the laser intensity. It is found that emission intensity of the Sn (II) transition lines at 556.19 nm, 558.89 nm and 579.92 nm increases by the factor of 13.0, 9.0 and 7.7 and the T_e and N_e increases by the factor of 2.4 and 8 with the increases of laser intensities from 1×10^{10} W/cm^2 to 2.5×10^{10} W/cm^2, whereas at the higher values of laser intensity from 2×10^{10} W/cm^2 to 2.5×10^{10} W/cm^2 a negligible variation is observed. Plasma shielding effect has been observer within the laser intensity of (2×10^{10} – 2.5×10^{10}) W/cm^2.

Acknowledgement

Dr. N. M. Shaikh and Dr. N. Amin are grateful to Higher Education Commission (HEC) of Pakistan for providing the scholarships for the post doctoral research work. Dr. N. M. Shaikh is also thankful to University of Sindh, Jamshoro, Pakistan for the grant of sabbatical leave.

References

[1] Singh R K, Narayan J1990, *J. Phys. Rev.* B **41,** 8843
[2] Russo R E, Mao X L, Liu C and Gonzalez J 2004 *J. Anal. At. Spectrom.* **19** 1084
[3] White J, O'Sullivan G, Zakharov S, Choi P, Zakharov V, Nishimura H, Fujioka S and Nishihara K 2008 *Appl. Phys. Lett.* **92** 151501
[4] Tillack M S, Sequoia K L and Tao Y 2008 *J. of Phys.: Conference Series* **112** 042060
[5] Harilal S S, O'Shay B, and Tillack M S, Mathew M V 2005 *J. Appl. Phys.* **98** 013306
[6] O'Shay B, Najmabadi F, Harilal S S and Tillack M S 2007 *J of Phys.: Conference Series* **59** 773
[7] Alonso-Medina A and Colo´n C 2008 *Astrophysical Journal* **672** 1286
[8] Shaikh N M, Hafeez S, Kalyar M A, Ali R and Baig M A 2008 *J. Appl. Phys.* **104** 103108
[9] Handbook of Basic Atomic Spectroscopic Data, NIST, http://physics.nist.gov/PhysRefData/Handbook/Tables/tintable4.htm
[10] McWhirter R W P 1965 *Plasma Diagnostic Techniques* ed R H Huddlestone and S L Leonard (New York: Academic) chapter 5

Proton Beam Weibel Instability Simulations of Energy Transfer in Gamma-ray Bursts

S Davis[1,2], V Tikhonchuk[1], E d'Humières[1], S Weber[1]

[1] Centre Lasers Intenses et Applications, Université Bordeaux 1, Talence, France
[2] Traditional Fulbright – Aquitaine Research Scholar, USA Department of State

E-mail: davis@celia.u-bordeaux1.fr

Abstract. Proton beam Weibel instability may explain key ingredients for Gamma-Ray Bursts (GRBs): amplified magnetic field, relativistic electron production and a power law energy distribution. By using particle-in-cell simulations, we demonstrate that a part of the directed kinetic energy of a proton beam is converted into kinetic energy of electrons, ions, electric and magnetic fields. A finite lifetime for the magnetic field is demonstrated, indicating a maximum for the GRB duration. An experiment probing the GRB environment is proposed.

1. Introduction

Historically, Gamma-ray bursts (GRBs) - short, intense pulses of γ rays, were discovered serendipitously by orbiting Vela satellites in 1963. Today, GRB are known to consist of both γ rays and any lower-energy emission that occurs simultaneously with them. The x-ray emission usually accompanies the γ-ray emission as a low-energy tail [1]. GRBs are observed almost daily with satellites and ground-based instruments. Their isotropic distribution across the sky indicates a cosmological origin. GRB observations in the radio, infrared, optical and x-ray domain have been reported [1]. The measured energy fluxes are of the order of 10^{-11} J/cm^2 along with estimated distances up to 3 Gpc (distance of 10^{23} km corresponding to the optical red shift $Z \sim 5$), providing an estimated isotropic electromagnetic energy release of $\sim 10^{46}$ J. The total GRB duration is typically less than 100 s. Some of them are accompanied by a weaker afterglow that continues for as long as a few days. GRBs consist of a series of ms pulses with the majority of photons between 100 keV and 1 MeV; some up to GeV. GRB spectra have a broad power law tail, $dN_\gamma/d\varepsilon \sim \varepsilon^{-s}$. Breaks in GRB spectra indicate a structure of the source of radiation consisting of jets and dense bulk material moving with relativistic velocities [2].

The GRB origin is related to the explosion of a central engine with energy release in the form of two relativistic jets. As reviewed in [1], the central engine is jet-like consisting of a baryon–dominated or a lepton–dominated relativistic plasma flow. The fireball (internal-external shocks) model [1, 3] describes the GRB phenomenology and its principle parameters, although it does not define the inner engine that produces GRBs [1]. In this investigation, we assume baryon flux dominated jets from the central engine. The jets have relativistic velocities with an effective relativistic factor $\Gamma \sim 10 - 100$. The jet deceleration, heating of electrons and ions with magnetic field generation is due to: (1) the internal collisions of dense globules in the flying ejecta and (2) the interaction of jets with the interstellar medium (ISM). The electromagnetic radiation is explained by the synchrotron emission of relativistic electrons in a strong magnetic field and by Compton scattering. The fireball model relies on the hypothesis of energy equipartition: the total energy released in a form of directed kinetic energy of ions is evenly distributed between electrons, ions and magnetic field.

The fact that the jets are relativistic allows reconciling a very short duration GRB recorded on the Earth with a very long slowing down length that cannot be less than $10^{10} - 10^{12}$ km. The difference between the slowing down time of $\sim 10^5$ s and the GRB duration of ~ 100 s is explained by relativistic compression by a factor of $2\Gamma^2$ of a signal emitted from the source moving towards the observer. Moreover, such a large propagation length makes it evident that the slowing down process has to be collisionless and the directed kinetic energy of the jet is transformed into the energy of electrons, ion and electromagnetic fields via plasma instabilities. The Weibel-type instabilities are considered to be viable candidates [4]. However, theoretical arguments [5] predict a protracted distance for the proton energy exchange with the mechanisms for strong electron heating and strong magnetic field generation remaining unclear. Recent numerical simulations with particle-in-cell (PIC) codes address the issues of the development of ion Weibel instabilities in relativistic proton streams with formation of collisionless shocks [6 - 8]. However, the simulation results are contradictory and incomplete. Hence, the problem of energy dissipation by collisionless plasma flows within the context of the Fireball Model remains unresolved. Furthermore, it is advantageous to define the conditions whereby the main features of the model are verified to scale under controlled conditions in laboratory experiments.

The objective of the present paper is to propose a methodology and a preliminary design of an experiment whereby the collisionless slowing down processes of dense streams of high energy plasmas can be studied and test the major hypothesis of the fireball model. We focus exclusively on interacting shocks that expand into the surrounding ISM, and we investigate relativistic electrons accelerated within the shocks due to energy transfer from proton streams. The methodology incorporates the slowing down of relativistic proton streams, the repartition of the dissipated energy between the electrons, protons and magnetic fields; the electron heating mechanism, efficiency and spectrum of electromagnetic emission. In addition to its astrophysical goals, such an experiment is of interest for inertial fusion, whereby the method of fast ion ignition relies on the energy deposition of more than 10 kJ into the compressed fuel core at a time scale of a few ps. The role of collective processes in the transport and dissipation of such extremely high ion currents remains to be determined.

2. Scheme of laboratory experiment on the collisionless dissipation of fast ion beams

The experiments on ion acceleration with high intensity laser beams demonstrate a possibility of efficient proton acceleration to energies of several tens of MeV with pulse duration of several ps and a total charge of a few μC [9]. These proton beams have a very low emittance of less than 0.01 mm· mrad and can be transported over distances of a few mm. The proton density in such beams may exceed $10^{18} - 10^{19}$ cm^{-3}. We chose to investigate collisionless dissipation processes of such beams via interaction with plasma of comparable density. The estimates and numerical simulations shown below demonstrate that the collisionless shock might have a thickness of a few mm. Therefore, the experiment is feasible with present day plasmas and with lasers having pulse energy of a few kJ and petawatt power. The processes of energy conversion can be measured from the energy spectrum of the transmitted ion beam, from interferometric observations of the density distribution in the interaction zone, and from the spectrum and energy of soft x-ray emission.

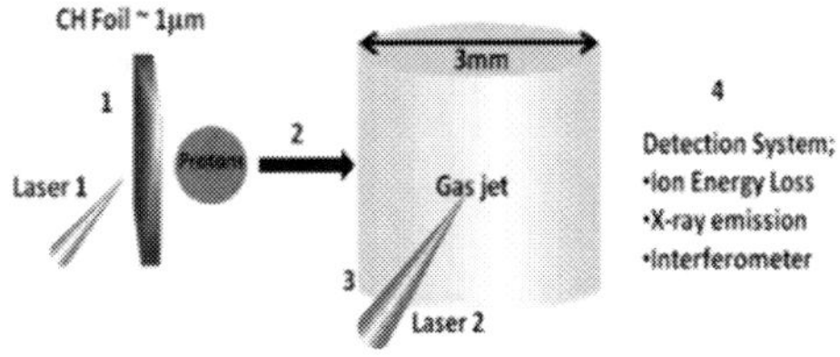

Figure 1. Schematic of the experiment to probe the collective effects in ion beam collisions: 1 – a proton bunch is created from a thin foil irradiated by a high intensity short laser pulse, 2 – transport zone of the proton bunch to the interaction site, 3 – target plasma created by ionization of a gas jet with an auxiliary laser beam, and 4 – diagnostic equipment: ion energy detector, interferometer, x-ray detector.

Although the experiment proposed in Figure 1 is non-relativistic, it verifies the conditions of the proton Weibel instability and allows us to investigate electron heating and magnetic field generation that are the key descriptors of the GRB Fireball Model. Measurements for the proton beam slowing down length, x-ray emission and energy spectrum can be performed. The ion beam is created via interaction of a high intensity laser pulse with a thin solid foil made of plastic or metal. The laser and target parameters are discussed in the next section. A special target design may be used for controlling the proton energy spectrum, number of protons and their divergence. The secondary plasma is placed at a distance of ~ 1 mm, which is sufficiently far to avoid any parasitic coupling between the primary and secondary targets while maintaining a high ion density. The secondary plasma is created by an auxiliary laser or an x-ray flash. The gas density and the size of the interaction region are chosen in order to optimize collective processes and to make visible the collisionless shock. The characteristics of the proton beam and its interaction with the secondary plasma are discussed below.

3. Ion acceleration in laser plasma interaction

Studies over the last ten years demonstrate efficient proton acceleration with laser pulses in thin solid foils. The model of target normal sheath acceleration (TNSA) [10] explains it as a three-step process: first, the laser pulse energy is converted into hot electrons with the energy of the order of the laser ponderomotive potential at the front side of the target. Then, these hot electrons cross the foil and create a strong electrostatic potential at the rear side. Because of an extremely steep density profile, the sheath electric field exceeds the values of 1 MV/μm and effectively ionizes and accelerates the ions from the surface. In our simulations with a one-dimensional version of the particle-in-cell code PICLS [11] we chose a 1 μm wavelength laser pulse with intensity of 10^{20} W/cm^2, and pulse duration of 0.3 ps. It interacts with a 5 μm thick CH foil (density 4×10^{22} cm^{-3}) and accelerates electrons to the average energy of 16 MeV. The energy spectrum of protons accelerated from the rear side of the foil, their spatial and angular distributions are shown in Figure 2.

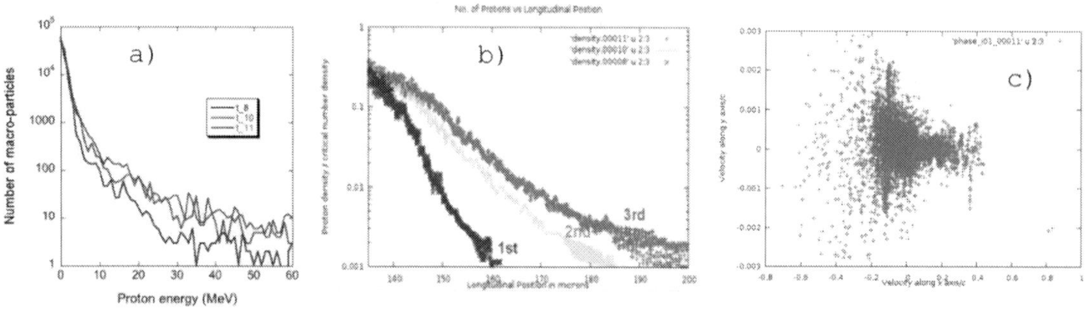

Figure 2. Characteristics of the proton beam accelerated from the rear side of a 5 μm CH foil: a) energy spectrum at three time moments of 1, 1.3, and 1.5 ps; b) spatial distribution of fast protons for the same three time moments; the density is normalized to the critical density $n_c = 10^{21}$ cm^{-3}; c) velocity distribution of protons in the plane of laser polarization at 1.5 ps; the proton velocities in the normal direction, v_x, and in the target plane, v_y, are normalized to the light velocity, c.

The protons in the interval from 10 to 30 MeV have a very small divergence of less than 1% (Fig. 2c). Their density just after the acceleration is ~ 10^{19} cm^{-3} and they are spread over the layer of thickness of ~ 10 μm (Fig. 2b). Assuming that they are accelerated from the spot of 100 μm in diameter, we estimate the total number of protons to be ~ 10^{12} in the source region. These protons are spreading ballistically along the propagation direction. We estimate that at a distance of ~ 1 mm from the source the proton bunch will have a thickness of 100 μm and a density of ~ 10^{18} cm^{-3}. These parameters are inputs for modeling of the proton bunch interaction with the secondary plasma.

4. Fast ion beam interaction with secondary plasma

The collisionless interaction of two plasmas is most efficient if both plasmas have similar densities. The simulations were conducted with the two-dimensional version of PICLS [11] in the center of mass reference system. Two quasineutral plasmas were colliding with equal bulk velocities $v_i = 0.2c$. The transverse temperature of the target plasma moving to the left is 100 eV. It plays role of the ISM in the GRB model. The transverse temperature of the plasma moving to the right is 10 keV. It plays the role of the external jet. The length of the simulation box in the propagation direction (x) is 1.2 μm, than is, about four ion inertia length, c/ω_{pi}, the length in the perpendicular direction is c/ω_{pi} with periodic boundary conditions. The proton-to-electron mass ratio is 1836, the initial densities of both plasmas is 10^{18} cm^{-3}. Initially, each plasma occupies one-half of the simulation box.

The simulation results are presented in Figure. 3. The electron Weibel instability is excited immediately as the interpenetration begins and produces electron kinetic energy dissipation only. This results in minor electron heating and generation of a weak magnetic field with the energy of the order of the initial electron kinetic energy. Large energy exchange begins after a few ion plasma periods followed by a rapid increase of the magnetic field, shock structure formation and coalescence of small-scale electron current filaments into larger scale structures with magnetic field of 400 – 500 T.

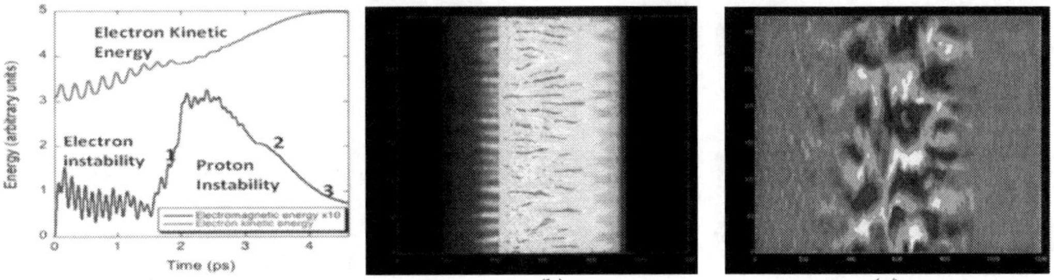

(a) (b) (c)

Figure 3. Simulation results of interaction of two colliding plasmas: a) temporal evolution of the electron energy and the magnetic field energy, b) proton density spatial distribution at the time of 3 ps when the ion Weibel instability is excited. Black is zero density and red is maximum density, which exceeds the average density by a factor of 3. c) The spatial structure of magnetic field (perpendicular to the simulation plane) at 3 ps. Orange – red color represents zero, bright yellow is for positive magnetic fields and blue is for negative magnetic fields; the maxima correspond to the fields of 500 T. The box size is 4 c/ω_{pi} in the horizontal direction and c/ω_{pi} in the vertical direction.

Similar results have been reported in previous publications [6, 8], which use smaller ion-to-electron mass ratio and different boundary conditions. More importantly, the collisionless shock structure found in our simulations has a finite extent of only a few ion inertial lengths. The continuing fusion of the current filaments leads to overall decrease of the magnetic energy and the interaction ceases after 4 – 5 ps, that is, after ~ 10 ion plasma periods. Consequently, the energy dissipated in the plasma collision is about 3% of the ion directed kinetic energy, which agrees with the number reported in Ref. [8]. However, this is more than 10 times less than reported in Ref. [7]. This might be related to a lower proton energy and smaller simulation box in our case. The energy in our simulations is deposited mainly in the electron component. The electrons display an almost isotropic distribution function with a power law energy distribution $dN_\varepsilon/d\varepsilon \sim \varepsilon^{-3}$, up to the energy of 150 MeV.

5. Conclusions

An experiment for studying the collisionless mechanisms of fast ion dissipation in laboratory plasma is proposed. The laser pulse interaction with thin solid target enables creation of a dense proton beam with the energy flux exceeding 10^{17} W/cm^2 and the density more than 10^{18} cm^{-3}. The proton Weibel instability can develop on the length of 1 mm and generate a collisionless shock necessary for GRB emissions. About 2% of the directed proton energy is transformed into electrons and generated a flash of electromagnetic energy. Simulations of similar physical processes at the GRB scale are ongoing.

References

[1] Piran T 2004 *Rev. Mod. Phys.* **76** 1143
[2] Davis S P 1995 *Measurements of Time Dilation in Gamma-Ray Bursts by Analysis of Temporal Structure* Ph D dissertation
[3] Piran T 1999 *Phys. Rept.* **314** 575
[4] Kazimura Y *et al.* 1997 *Ap. J.*, **498** L183
[5] Lyubarsky Y & Eichler D 2006 *Ap. J.* **647** 1250
[6] Dieckmann M E *et al.* 2008 *Plasma Phys. Contr. Fusion* **50** 065020
[7] Spitkovsky A 2008 *Ap. J.* **673** L39
[8] Kato T N & Takabe H 2008 *Ap. J.* **681** L93
[9] Snavely R A *et al.* 2000 *Phy Rev Lett* **85** 2945
[10] Wilks S C *et al.* 2001 *Phys. Plasmas* **8** 542
[11] Sentoku Y & Kemp A J 2008 *J. Comp. Phys.* **227** 6846

Journal of Physics: Conference Series **244** (2010) 042007 doi:10.1088/1742-6596/244/4/042007

Understanding the implications of the data from recent high-energy-density Kelvin-Helmholtz shear layer experiments

O.A. Hurricane[1], J.F. Hansen[1], E.C. Harding[2], R.P. Drake[2], H.F. Robey[1], B.A. Remington[1], C.C. Kuranz[2], M.J. Grosskopf[2], R.S. Gillespie[2], and H. Park[1]

[1]Lawrence Livermore National Laboratory, 7000 East Avenue, Livermore, CA 94550, USA
[2]Department of Atmospheric, Oceanic, and Space Sciences, University of Michigan, 2455 Hayward St., Ann Arbor, Michigan 48109-2143, USA

E-mail: `hurricane1@llnl.gov`

Abstract. The first successful high energy density Kelvin-Helmholtz (KH) shear layer experiments (O.A. Hurricane, et al., Phys. Plasmas, **16**, 056305, 2009; E.C. Harding, et al., Phys. Rev. Lett., **103**, 045005, 2009) demonstrated the ability to design and field a target that produces an array of large diagnosable KH vortices in a controlled fashion. Data from these experiments vividly showed the complete evolution of large distinct eddies, from formation to apparent turbulent break-up. Unexpectedly, low-density bubbles/cavities comparable to the vortex size ($\sim 300-400$ μm) appeared to grow up in the free-stream flow above the unstable material interface. In this paper, the basic principles of the experiment will be discussed, the data reviewed, and the progress on understanding the origin of the above bubble structures through theory and simulation will be reported on. (IFSA 1.10.096)

In May 2009, our team fielded the first successful high-energy-density-physics (HEDP) Kelvin-Helmholtz (KH) experiments [1, 2] on the Omega laser at the University of Rochester. These experiments proved out the unique conceptual design [3] that relied upon shock driven baroclinic vorticity production and also showed that vivid high quality data (see Fig. 1) could be obtained on KH in a HEDP environment.

The basic configuration consists of a stack of opaque high density plastic and low density foam all of which is contained in a shock tube of rectangular cross-section, made from Be so as to be able to radiograph through it with x-rays of a few keV energy (see Fig. 2) – details of the target design can be found in [1]. Laser energy (4 kJ in a 1 ns pulse for this case) is delivered to an 820 μm diameter spot on an ablator covering the low density foam part of the target (on the left of Fig. 2). In this way, a strong shock is launched into the low density foam such that the pressure gradient at the leading edge of the shock would essentially be at right angles to the density gradient at the interface of the two dissimilar materials thus maximizing $\nabla P \times \nabla \rho$. The interface between the two materials is perturbed by a sinusoidal contour with amplitude ($a = 30$ μm) and wavelength ($\lambda = 400$ μm) chosen such that a number of large vortices would develop nonlinear structure in the expected field of view during the experiment. By in large, the data from our May 2008 experiments were consistent with expectations based upon two-dimensional (2D) simulation using the CALE [4] code.

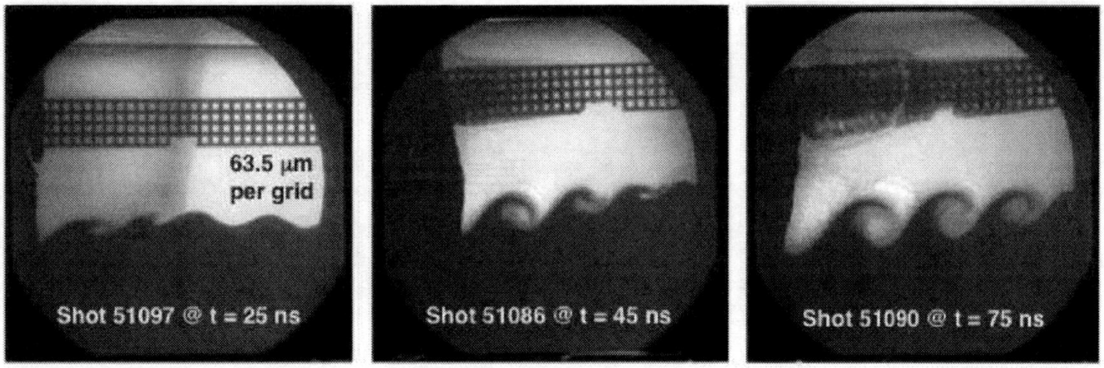

Figure 1. From left to right, radiographic data from Omega shots 51097, 51086, and 51090 are shown. These three images show the time development of the KH instability at 25 ns, 45 ns, and 75 ns respectively. In the left frame, the vorticity producing shock wave is visible in the low density (100 mg/cc) carbon foam. Wave crest begin to develope immediately after passage of the shock wave and grow into full blown vortices (middle frame). At late time (right frame), the spiral arms of the vortices appear to begin to diffuse away presumably the result of turbulence onset.

The images shown in Fig. 1 are simply converted into datum of vortex height versus time [2] that can be compared with simulation and theory. In Fig. 3 an updated comparison of the vortex height data is shown against a revised simulation result and theory. The data shown in Fig. 3 are identical to those shown in Ref. [1, 2], but the simulation result shown here superceeds that presented previously. Here the simulation used to produce the data shown in Fig. 3 has been corrected to include the actual as-shot Be shock tube thickness of 500 μm rather than the 200 μm thickness used for the simulations shown in [1, 2] and a more accurate method of determining the vortex height from the simulation has also been used.

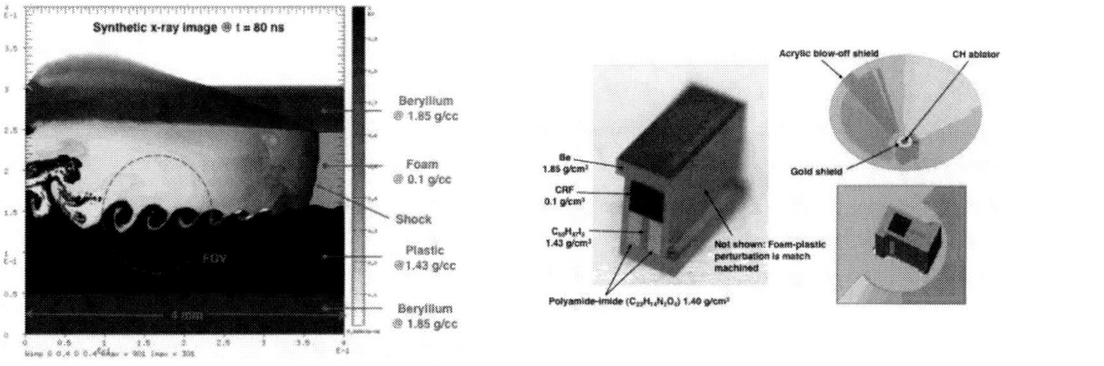

Figure 2. Left: A simulated radiograph result of a simulation of the target performance is shown at t = 80 ns. The materials that compose the target are annotated in the image. The field-of-view (FOV) accessible in the actual experiment is shown as the dashed red circle. Right: an annotated picture of the actual target. Note that the Be tube sides are 200 μ m thick, while the top and bottom sections of the Be tube are 500 μ m thick.

The vortex model theory shown in Fig. 3 comes from using the expression for the fluid circulation, Γ, derived in Ref. [3] (with values P = 1.62 Mbar, ρ_H = 1.43 g/cc, ρ_L = 0.1 g/cc, and $\gamma = 5/3$) in combination with the differential equations for the flow field

$$\frac{dx}{dt} = \frac{\Gamma}{4\lambda} \frac{\sinh\left(\frac{2\pi y}{\lambda}\right)}{\cos^2\left(\frac{\pi x}{\lambda}\right) + \sinh^2\left(\frac{\pi y}{\lambda}\right)}; \quad \frac{dy}{dt} = \frac{\Gamma}{4\lambda} \frac{\sin\left(\frac{2\pi x}{\lambda}\right)}{\cos^2\left(\frac{\pi x}{\lambda}\right) + \sinh^2\left(\frac{\pi y}{\lambda}\right)} \tag{1}$$

where a these differential equations imply vortex growth up to a saturation of the vortex amplitude to a value of $y_{max} = \cosh^{-1}(3)\lambda/2\pi \approx 0.281\lambda$ [3] the full vortex height then being $h_{max} = 2y_{max}$. Since Eqs. (1) trace out the trajectory that a massless particle would follow starting from some initial point (x_0, y_0) at $t = 0$, the full vortex height as plotted in Fig. 3 is then twice the value of the envelope of solutions to Eqs. (1) using the (x, y) locations that trace out the initial interface (see Fig. 4). An attempt to include the added complication of flow in the direction of vortex growth, due to the effect the transmitted shock, is shown in Fig. 3 as the stretched vortex model and is arrived at by adding a constant y-velocity of 2 μm/ns (from simulation) to the vortex model solution.

At late-time, the simulation and the stretched vortex model (which uses simulation derived values) both over predict the data. The simulation does exhibit the same change in growth rate at around $t = 38$ ns that the data shows. Inspection of the simulation at $t = 38$ ns indicates that this is the time at which a shock traveling in the $-\hat{y}$ direction, that was reflected from the top of the Be shock tube, impacts the chain of vortices slightly compressing them.

The late-time over-prediction of the simulation is likely explained by the fact that the simulation is 2D, while the target itself is 3D. That is in 2D, the simulation the post-shock expansion of the shock-tube would under-estimate the real decay in the post-shock flow that results from the shock-tube expanding in 3D. Circumstantial evidence that supports this 3D shock-tube expansion hypothesis, is that the earlier 2D simulation [1] that uses a thinner Be shock-tube wall thickness than the simulation presented here is closer to the data at late time. An actually 3D simulation would be necessary to fully prove this hypothesis and some action in that direction is underway.

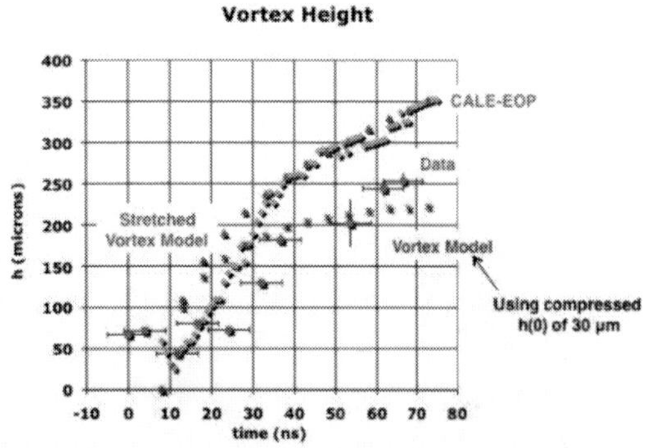

Figure 3. Vortex height versus time is shown for the experiment (red squares), the simulation (blue diamonds), and theory (purple asterisks and green triangles). The data and simulation both show a period of post-shock amplitude compression between 10 - 15 ns and subsequent amplitude growth. A distinct change is growth-rate is seen around 38 ns.

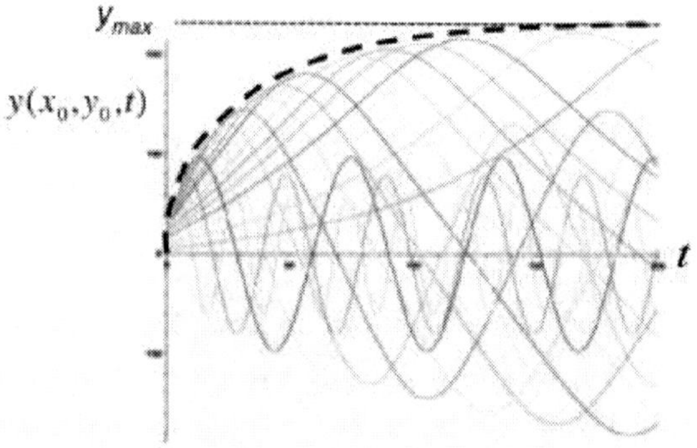

Figure 4. A family of solutions to the vortex model is shown and the envelope of these solutions is given by the dashed curve.

Bubbles located around the crests of the largest vortices at late-time (e.g. see the largest vortex of the rightmost frame of Fig. 1) in the data were unexpected. Several speculations about the origins of these bubble were put forth in the initial reports on these experiments [1, 2]. In spite of a strong superficial resemblance, further investigation of the trans-sonic shocklet [5] in the context of the present experiment now seems an unlikely explanation as the free-stream flow in the vicinity of the large vortices at late time is too low according simulations. Attempts to create a transonic effect in the simulations by assuming an ideal-gas equation of state and varying γ from $\gamma = 1.001$ to $\gamma = 6.0$ were also unsuccessful.

Cavitation-like behavior as an explanation for the observed bubbles is presently being investigated throught the use of simulations that allow for multi-phase equations of state for the carbon foam used in the experiments. A more mundane effect that is related to the cavitation that occurs when an object penetrates from one fluid across and interface to another fluid [6] is presently the leading explanation for the bubbles observed in the radiographs.

That is the simulation indicates that albator plasma may be entrained into the free-stream flow, due to the impulsive way in which this experiment is driven, and that the ablator material may be located in a bubble-like structure around the location of the largest vortex at 75 ns into the experiment. Moreover, as false-color simulated radiographs clearly show, the bubble of ablator material would have about 1/2 to 1/4 the optical depth of the surrounding post-shock carbon foam material. The morphology of the simulated bubble of albator material is not, however, a perfect match to the observation but it is the closest simulated feature seen so far.

While some features seem in our HEDP KH experiment are yet to be fully explained, we have demonstrated to viability of the target design concept and demonstrated that high quality data on a KH unstable shear layer can be obtained. Simulation and theory are in fair agreement with the data obtained from this experiment with regard to the overall height of the vortex layer, especially considering the fact that the simulation uses a simple laser source and lacks the 3D expansion of the shock tube.

In late 2009, we plan of fielding a follow-up set of HEDP KH experiments on the Omega laser that will examine the density scaling of the instability growth and simultaneously test the effect of the density on the development of the low density bubbles that appeared in the free-stream of the initial trials of this platform.

Further extensions of this platform to study multi-mode KH under HEPD conditions is easily accomplished by judicious choice of initial interface perturbation. Future experiments on NIF could create sustained steady flows with higher Reynolds numbers and use larger targets (for better diagnosis) both of which would enhance the usefulness of this platform.

Acknowledgments

This work has benefitted from the input of Dr.'s Andy Cook, Paul Dimotakis, George Langstaff, Aaron Miles, Paul Miller, Phil Sterne, Alan Wan, and Charlie Verdon. This work was performed under the auspices of the U.S. Department of Energy Lawrence Livermore National Laboratory under contract No. DE-AC52-07NA27344 and support by a National Laser User Facilities Grant.

[1] O.A. Hurricane, J.F. Hansen, H.F. Robey, B.A. Remington, M.J. Bono, E.C. Harding, R.P. Drake, C.C. Kuranz 2009 *Phys. Plasmas*, **16**, 056305

[2] E.C. Harding, J.F. Hansen, O.A. Hurricane, R.P. Drake, H.F. Robey, C.C. Kuranz, B.A. Remington, M.J. Bono, M.J. Grosskopf, R.S. Gillespie 2009 *Phys. Rev. Lett.*, **103**, 045005

[3] O.A. Hurricane 2008 High Energy Den. Phys., doi:10.1016/j.hedp.2008.02002

[4] R. T. Barton 1985 *Development of a multimaterial two-dimensional, arbitrary Lagrangian-Eulerian mesh computer program,* in *Numerical Astrophysics*, J.M. Centrella, J.M. LeBlanc, R.L. Bowers, eds., (Jones and Bartlett Publishers, Boston), p. 482.

[5] P.E. Dimotakis 1991 *AIAA 91-1724, Proc. 22nd Fluid Dyn., Plasma Dyn., & Lasers Conf.*

[6] F. Ronald Young 1999 *Cavitation,* (Imperial College Press, London), p. 203-4

Journal of Physics: Conference Series **244** (2010) 042008 doi:10.1088/1742-6596/244/4/042008

Laboratory Experiment to Study Collisionless Shock

Y. Kuramitsu[1], Y. Sakawa[1], T. Morita[2], H. Aoki[2], H. Tanji[3], S. Dono[3], J. N. Waugh[4], C. D. Gregory[5], M. Koenig[5], N. Woolsey[4], and H. Takabe[1]

[1] Institute of Laser Engineering, Osaka University, 2-6 Yamadaoka, Suita, Osaka, 565-0871 Japan

[2] Graduate School of Science, Osaka University, 1-1 Machikaneyama-cho,Toyonaka, Osaka 560-0043, Japan

[3] Graduate School of Engineering, Osaka University, 2-1, Yamadaoka, Suita, Osaka 565-0871, Japan

[4] Department of Physics, University of York, York, YO10 5DD, UK

[5] Laboratoire pour l'Utilisation des Lasers Intenses, UMR 7605, CNRS - CEA - Université Paris VI - Ecole Polytechnique, 91128 Palaiseau Cedex, France

E-mail: `kuramitsu-y@ile.osaka-u.ac.jp`

Abstract. We report the experimental results of collisionless shock formation in counterstreaming plasmas produced by a high-power laser system. The experiment was performed with Gekko XII HIPER laser system at the Institute of Laser Engineering. In order to model collisionless shocks in the universe, supersonic counterstreaming plasma flows were generated using a CH double-plane target. By using the self-emission measurements, we observed the emission increase toward the shock through the downstream. We also observed the density jump associated with the emission increase. The width of the transition region is shorter than the ion-ion mean-free-path calculated from the measured plasma velocity and density.

1. Introduction

Collisionless shocks are abundant in space and astrophysical plasmas, e.g., the Earth's bow shock, interplanetary traveling shocks, and supernova remnants. It is considered that the first order Fermi acceleration or the diffusive shock acceleration (DSA) is a standard theory for nonthermal acceleration of energetic particles or cosmic rays in the universe. In many circumstances, space and astrophysical plasmas are collisionless, where the particle transport and relaxation are caused by wave-particle interactions. In space plasmas the electric and magnetic field, and plasma distribution functions, those are essential to understand the acceleration processes, are directory obtained by *in situ* observations. The exponential growth of the proton and electron fluxes toward the interplanetary shock was observed *in situ* [1] . This is predicted by the DSA, and the above observation is recognized as an evidence of the DSA. On the other hand, in astrophysical plasmas there is no way to directly measure the key quantities to discuss the particle acceleration. Recently, the X-ray images of SN1006 showed the filamentary structures with the profiles of exponential growth [2]. This indicates that the DSA can produce the nonthermal cosmic acceleration. However, there are many unknown parameters to prove the scenario of the DSA. A laboratory experiment can be an alternative approach to study the collisionless shocks and the particle acceleration. In this paper we report experimental

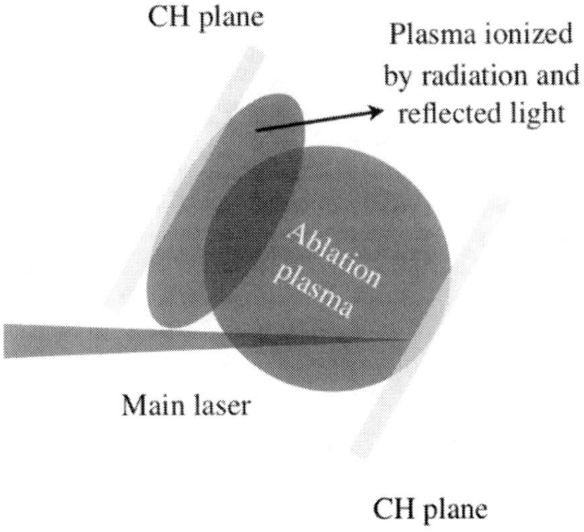

Figure 1. Schematic of double plane target. The thickness of both planes are 60 μm, and the separation between the planes were 4.5 mm. The target normal is 30 degrees from the laser axis.

results of collisionless shock formations in counterstreaming plasmas produced by a high-power laser system, Gekko XII (GXII). Irradiating double CH (plastic) plane targets, we produced counterstreaming plasmas in order to excite a collisionless shock. By using gated optical imager and streaked optical pyrometry, the self-emission increase was observed during the two plasmas interactions. Furthermore, by using shadowgraphy a filamentary structure, which corresponds to a sudden density change, was simultaneously observed at the same timing and position of the self-emission increase. The mean-free-pathes of the ions, estimated from the plasma density and the counterstreaming relative velocity, were larger than the shock transitions. These results strongly indicate the collisionless shock formation in the laboratory plasmas.

2. Experiment

The experiment was performed by GXII laser at the Institute of Laser Engineering at the Osaka university. The laser conditions were; the wavelength was 351 nm, the pulse length was 500 ps, the focal spot diameter was 300μm, the pulse energy was $\sim$ 100 J. To produce counterstreaming plasmas, a CH double-plane target was irradiated with the laser. Figure 1 shows schematic drawing of the configuration of the laser and the double plane target. The both plane were 60μm thick and the distance between the planes were 4.5 mm. The target normal is aligned 30 degrees from the laser propagation axis, so that the laser shot the right plane first then produced the ablation plasma at the front side of the right plane. The left plane is also ionized by the plasma, radiations and reflected light from the left plane.

The counterstreaming plasmas were observed by the transverse diagnostics with the probe laser and with the self-emition. The optical probe conditions were; the wavelength was 532 nm, and the pulse length was $\sim$ 14 ns. With the probe laser we obtained the density information of the two-dimensional snapshot by interferometry and shadowgraphy (250 ps gate width). We also measured the time evolution of the central plasma by streaked interferometry. Using 450 nm interference filters the self-emission was gated at the wavelength. We obtained the two-dimensional snapshot of the self-emission by gated optical imager (GOI) (1.6 ns gate width) and of the time evolution by self-emission streaked optical pyrometer (SOP). The slits of the streak cameras were also rotated in order to measure the time evolutions of the central plasmas.

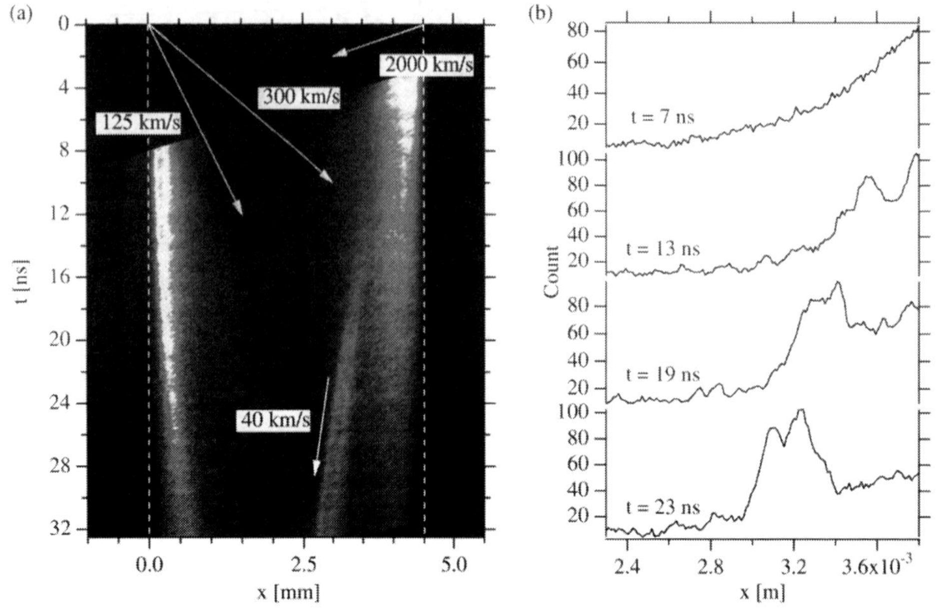

Figure 2. (a) SOP image. The laser timing corresponds to the top of the image. (b) The line profiles of the sections correspond to the time at $t = 7, 13, 19$, and 23 ns.

3. Results

The two counterstreaming plasmas were seen in streaked interferogram (not shown). The density of the plasmas, obtained from interferogram before the plasmas occupied the hole field of view, $t \lesssim 5$ ns, was from $\sim 1 \times 10^{18}$ at the tip of the plasma to $\gtrsim 1 \times 10^{19}$ at the opaque region to the probe light. The plasma density is considered to decrease as the plasmas expand. Figure 2 (a) shows an SOP image where $t = 0$ is the main laser timing and the vertical dotted lines corresponds to the nominal target positions. The dashed arrows represent the fast plasma velocities estimated from the streaked interferogram taken on the same shot as Fig. 2 (a). In Fig. 2 (a) the fast plasma seen in streaked interferogram was not seen, but the plasma from the left plane went through the plasma from the right plane since the mean-free-paths of the fast counterstreaming ions are estimated as the order of meters, and then started to interact with the plasma staying around the right plane at $t \sim 12$ ns. The two stream interaction enhanced the self-emission at the interaction front and the sharp structure or shock propagated from right to left at 40 km/s. Figure 2 (b) shows the line profiles of 2 (a) at $t = 7, 13, 19$, and 23 ns. At $t = 7$ ns the profile shows the emission increase gradually toward the right plane since there was no two stream interaction. At $t = 13$ ns, after the two plasmas interacted, a structure was formed at $x = 3.4$ mm. As time passed, the profiles became steeper and steeper, clearly showing the shocked structure. The shock transition width was typically $\sim 100\mu$m. The mean-free-path of ions is written as $\lambda_{ii} = 2\pi\epsilon_0^2 m_i^2 \langle v_i \rangle^4 / (n_i Z^4 e^4 \ln \Lambda)$, where ϵ_0 is the vacuum dielectric constant, m_i is the ion mass, $\langle v_i \rangle$ is the average ion velocity, n_i is the ion density, Ze is the ion charge, $\ln \Lambda = \ln(\lambda_D 4\pi\epsilon_0 m_i \langle v_i \rangle^2 / (Z^2 e^2))$, and λ_D is the Debye length [3]. Using the relative velocities of the counterstreaming ions 165 km/s $\leq \langle v_i \rangle \leq$ 340 km/s, the mean-free-path for the counterstreaming flow is estimated as 0.4 mm $\leq \lambda_{ii} \leq$ 4 mm with $Z = 3.5$, a mass number $A = 6.5$, averaged values for a proton-carbon ion plasma, $n_e = Zn_i = 10^{18}\text{cm}^{-3}$, and $T_e = 100$ eV. These are larger than the shock transition width.

Figures 3 (a) and 3 (b)–3 (d) show the snapshots of the shock structures by self-emission GOI

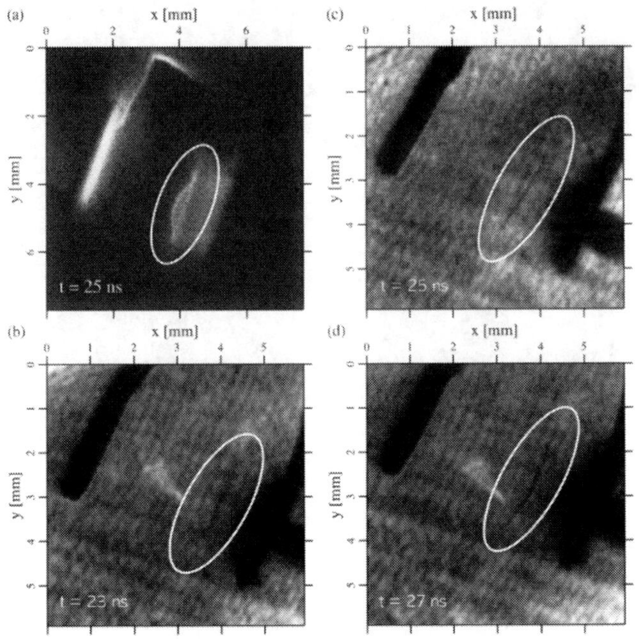

Figure 3. (a) GOI snap shot at $t = 25$ ns. (b)–(d) Shadowgraphs at $t = 23, 25, 27$ ns.

at $t = 25$ ns taken on the same shot as Fig. 2 (a), and by shadowgraphy at $t = 23, 25, 27$ ns taken on different shots from Fig. 2 (a), respectively. In Fig. 3 (a) one can clearly see a filamentary emission increase or a bow-shock like structure in front of the right plane (marked by the circle). Besides the shocked self-emission structure, we observed thin structures correspond to the sudden density change in shadowgraphs in three different timing $t = 23, 25, 27$ ns in Figs. 3 (b)–3 (d) (marked by the circles). Furthermore, the thin structures in shadowgraphs moved toward the left at about the same velocity as the shock velocity observed in SOP of Fig. 2 (a). The self-emission increase observed by GOI and SOP associated with the large density change.

4. Conclusions

Two counterstreaming plasmas were created by irradiating the double-CH targets by GXII laser. The self-emission measurements showed the steepened structures as the counterstreaming interactions evolved. The shadowgraphs showed the thin structures corresponding to the large second derivative of the electron density. The position and the propagation speed of the thin structures were represented by the structure observed by self-emission, indicating the shock transition. The shock transition width is less than the mean-free-path of the counterstreaming ions. Our results strongly indicate the collisionless shock formation in counterstreaming laboratory plasmas.

Acknowledgments

The authors would like to thank T. N. Kato, S. Matsukiyo, and T. Hada for their helpful comments.

References

[1] Shimada N, Terasawa T, Hoshino M, Naito T, Matsui H, Koi T and Maezawa K 1999 Ap&SS **264** 481–488

[2] Bamba A, Yamazaki R, Ueno M and Koyama K 2003 ApJ **589** 827–837

[3] Spitzer L 1962 *Physics of Fully Ionized Gases* (Interscience)

The Sixth International Conference on Inertial Fusion Sciences and Applications IOP Publishing
Journal of Physics: Conference Series **244** (2010) 042009 doi:10.1088/1742-6596/244/4/042009

Jet Formation in Counterstreaming Plasmas Produced by High-Power Laser Beams

Y. Kuramitsu[1], Y. Sakawa[1], J. N. Waugh[2], C. D. Gregory[3], T. Morita[4], S. Dono[5], H. Aoki[4], H. Tanji[5], M. Koenig[3], N. Woolsey[2], and H. Takabe[1]

[1] Institute of Laser Engineering, Osaka University, 2-6 Yamadaoka, Suita, Osaka, 565-0871 Japan
[2] Department of Physics, University of York, York, YO10 5DD, UK
[3] Laboratoire pour l'Utilisation des Lasers Intenses, UMR 7605, CNRS - CEA - Université Paris VI - Ecole Polytechnique, 91128 Palaiseau Cedex, France
[4] Graduate School of Science, Osaka University, 2-1, Yamadaoka, Suita, Osaka 565-0871, Japan
[5] Graduate School of Engineering, Osaka University, 2-1, Yamadaoka, Suita, Osaka 565-0871, Japan

E-mail: `kuramitsu-y@ile.osaka-u.ac.jp`

Abstract. We report the experimental results of a plasma jet formation in counterstreaming plasmas created by irradiating a double CH-plane target with a high-power laser system. The time evolution of the jet collimation was obtained over several shots, each shot taking a shadowgraphy image at a different timing. Since no jet formation was observed when a single CH-plane target was irradiated, the presence of the counterstreaming plasma, as an ambient plasma, is essential to the jet plasma collimation.

1. Introduction

Astrophysical jets have been observed in various astrophysical phenomena, such as active galactic nuclei (AGN), X-ray binary systems, and young stellar objects (YSO), and they are often highly collimated over large distances. It has been considered that there are three possible factors to affect the jet collimation: (1) the presence of ambient medium, (2) the radiative cooling of jets, and (3) the presence of external magnetic fields. Since it is impossible to locally observe such astrophysical phenomena, laboratory experiments can be an alternative approach to investigate and prove the physics in the jet formation and collimation. By irradiating geometrically designed targets with high-power lasers, plasma jets have been generated in laboratory experiments. The effects of the radiative cooling was investigated by using conical targets with different atomic numbers [1]. The scaling parameters relevant to astrophysical conditions have been obtained from direct measurements [2]. The effects of the presence of ambient medium was studied by comparing jets propagation in vacuum with in an ambient medium of neutral gas [3]. In astrophysical circumstances, however, a plasma jet is considered to interact with a surrounding plasma. While a plasma particle interacts with a neutral gas particle via collision, a plasma particle can interact with a plasma particle via the electric and magnetic field besides the collision. In this article we report the jet formation and collimation in counterstreaming laboratory plasmas. The two counterstreaming plasmas were created by irradiating a double

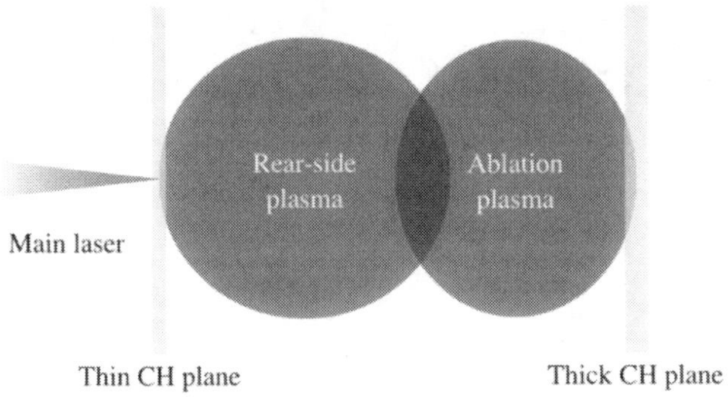

Figure 1. Schematic of the double-plane target. The thickness of the left and right planes are 10 μm and 60 μm, respectively. The separation between two planes is 4.5 mm.

CH-plane target with high-power laser beams. In the counterstreaming plasmas, plasma jet formation was observed by using shadowgraphy. The time evolution of the jet collimation were obtained over shots with different timing. When a single CH-plane target was irradiated, no jet formation was observed; the counterstreaming plasma from the other plane enable the jet plasma to collimate due to two-streaming interactions.

2. Experiment

The experiment was performed by using Gekko (GXII) HIPER laser system. We shot CH (plastic) double-plane targets, where the first and the second planes were 10 μm and 60 μm thick with the separation of 4.5 mm. Figure 1 shows the schematic of the double-plane target. The main laser pulses were focused on the spot of 300 μm diameter with the wavelength of 527 nm and the pulse length of 500 ps. The pulse energy was $\sim$ 120 J per beam. We used four beams per one shot. The first CH plane was irradiated with the laser beams. The first plane was designed to be thin enough so that the plasma is formed at the rear-side of the plane. Then the transmitted light and the radiation from the first plane ablate the second CH plane, resulting in formation of the ablation plasma in front of the second plane.

We used transverse diagnostics in order to observe the spatial and temporal evolution of plasma flows. For two-dimensional transverse images, we used interferometry and shadowgraphy with an optical probe (the wavelength was 532 nm, the pulse length was $\sim$ 14 ns) by using two ICCD cameras with the minimum gate widths of 250 ps. The interferometer was a Normarski type, and the image separation was vertical to the plasma flow axis in order to avoid the overlapping the two planes. Interferometry gives the plasma electron density, and the shadowgraphy gives a picture of the plasma produced, where the refractive index varies rapidly and where the density is opaque to the probe beam. For the on-axis time evolution of plasmas, we used self-emission streaked optical pyrometry (SOP) with wavelength gated at 450 nm, and streaked interferometry. We also used two X-ray pinhole cameras from different angles to obtain the focal spot images.

3. Results

Figrues 2 (a) and 2 (b) show a streaked interferogram and an SOP image taken on the same shot, respectively. The spatial and temporal scales of the two streak images are set to be identical. The horizontal and vertical dotted lines coorespond to the main laser timing $t = 0$ and the nominal target positions, respectively. The main laser came from the left then produced a plasma flow behind the left plane. The streaked interferometry observed the relatively low-density ($\sim 10^{18}$ cm^{-3}), rapidly expanding ($\sim$ 400 km/s) plasma at the rear-side of the left plane (dashed line 1), and simultaneously observed the fast ablation plasma from the right plane (dashed line 2) with a similar velocity ($\sim$ 400 km/s) by the transmitted laser pulse and the radiation from the left plane. These two plasmas streamed toward each other. The SOP

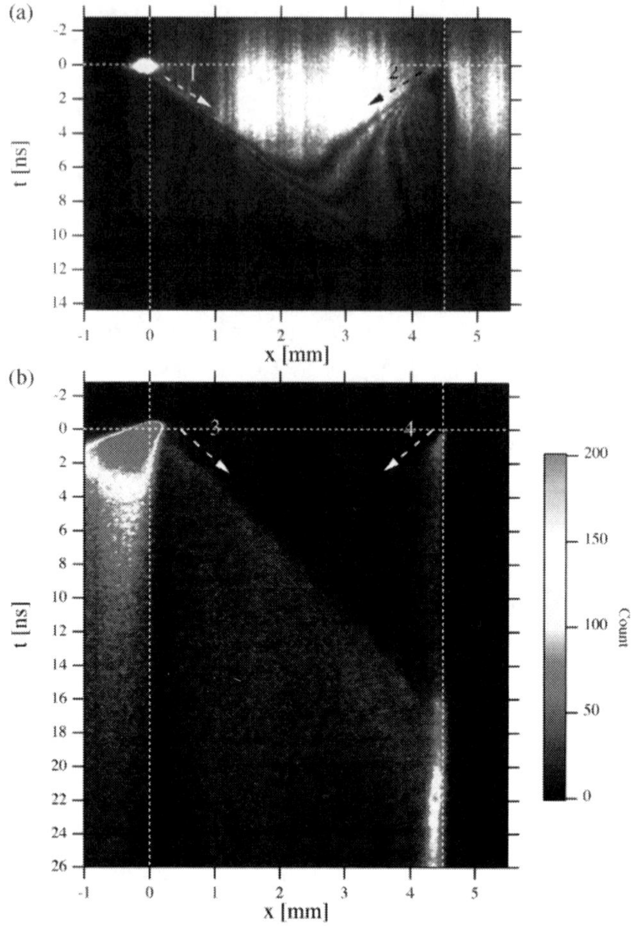

Figure 2. (*a*) Streaked interferogram. The dashed lines 1 and 2 indicate flow velocities of ± 400 km/s, respectively. (*b*) Self-emission streaked optical pyrometer image. The dashed lines 3 and 4 indicate flow velocities of ± 290 km/s, respectively.

observed part of the slower ($\sim$ 290 km/s) and denser ($\gtrsim 10^{19}$ cm^{-3}) plasma at the rear-side of the left plane (dashed line 3) that followed the fast plasma, and also simultaneously observed the corresponding slower, denser component of the ablation plasma from the right plane (dashed line 4) propagating at a similar velocity ($\sim$ 290 km/s).

Figure 3 shows the time series of the plasma propagation obtained over shots by shadowgraphy with different delay from the main laser timing. For the comparison purpose, the bottom right image shows the shadowgram at 15.5 ns for a single-plane target. The black shadows correspond to high-density plasma where the probe laser cannot go through. The plasma from the right plane is too low to be opaque in the shadowgrams. In early times, the plasmas from the left plane showed bumpy shape, implying that the main four laser pulses are not well aligned at a spot. The high-density plasmas from the left plane flow showed some directivity to the main laser axis; it is not isotropic expansion. The counterstreaming flows started to collide around $\sim$ 6 ns (see Fig 2 (a)), and as time passed the tip of the plasma from the left plane started to collimate from $t = 9.5$ ns. The collimated plasma flow or jet became more clear as time evolved and reached the right plane at 15.5 ns. These jet formation was not seen when the single-plane target was irradiated. The lateral size of the jet in Fig. 3 at t = 15.5 ns is only 0.12 mm, which is an order of magnitude smaller than that of the dense plasma at the same timing when the single plane target was irradiated.

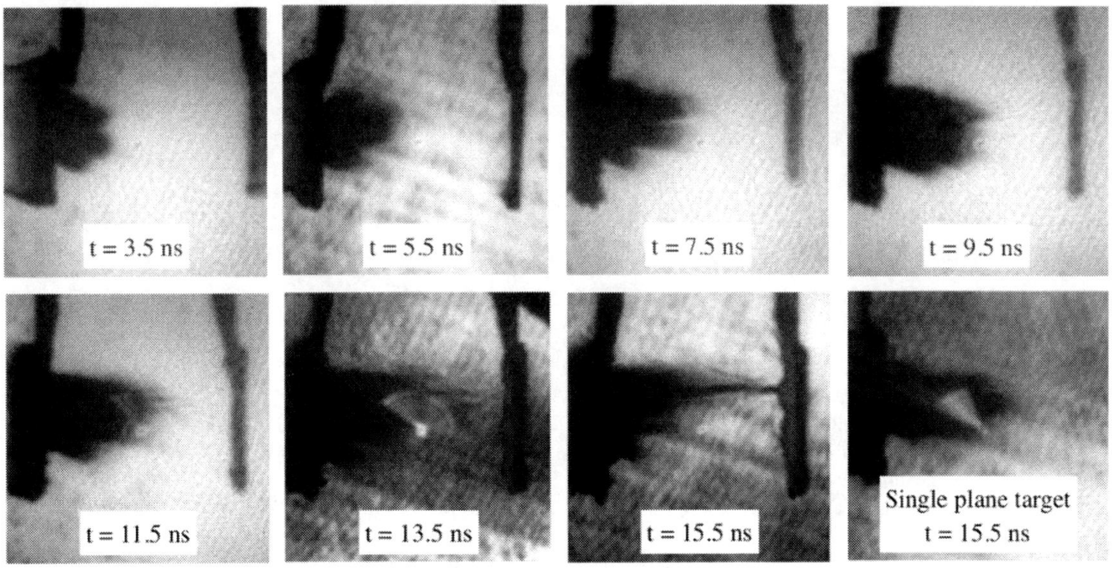

Figure 3. Time series of shadowgrams with different delay from the main laser. The bottom right image shows the single-plane target at 15.5 ns.

4. Conclusion

We have presented the jet formation and propagation in the counterstreaming plasmas created by irradiating the double plane targets with the high-power laser system, GXII. The streaked interferogram clearly shows the formation of the counterstreaming plasmas. By using shadowgrapy we observed the time evolution of the plasma propagation in the countersteaming plasma. In the presence of the counterstreaming plasma, the plasma can be collimated, thus the jet formation is possible. The effect of surrounding plasma is more eminent than neutral media reported before [3]. When we shot the single plane target or when there is no surrounding plasma, we did not observe the jet formation. Therefore, the presence of the ambient plasma is essential to the jet formation. There are two major candidates to account for the jet collimation in the counterstreaming plasmas. One is the plasma confinement by the shock wave and another is magnetic field enhanced in the counterstreaming plasmas. In order to confirm this, the magnetic field measurement is necessary in the future experiments. In the future we will discuss the mechanism of the jet formation and collimation in the counterstreaming plasmas.

Acknowledgments

The authors would like to thank S. Okajima, T. N. Kato, A. Mizuta, and N. Ohnishi for their helpful comments.

References

[1] Shigemori K, Kodama R, Farley D R, Koase T, Estabrook K G, Remington B A, Ryutov D D, Ochi Y, Azechi H, Stone J and Turner N 2000 Phys. Rev. E **62** 8838–8841

[2] Loupias B, Koenig M, Falize E, Bouquet S, Ozaki N, Benuzzi-Mounaix A, Vinci T, Michaut C, Rabec Le Goahec M, Nazarov W, Courtois C, Aglitskiy Y, Faenov A Y and Pikuz T 2007 *Physical Review Letters* **99** 265001–+

[3] Gregory C D, Howe J, Loupias B, Myers S, Notley M M, Sakawa Y, Oya A, Kodama R, Koenig M and Woolsey N C 2008 ApJ **676** 420–426

Collisionless shock generation by a high-power laser

T. Morita[1], Y. Sakawa[2], Y. Kuramitsu[2], S. Dono[3], H. Aoki[1], H. Tanji[3], T. Kato[2], Y. T. Li[4], Y. Zhang[4], X. Liu[4], J. Y. Zhong[5], N. Woolsey[6], H. Takabe[1,2], and J. Zhang[7]

[1]Graduate School of Science, Osaka University, 1-1 Machikane-yama, Toyonaka, Osaka 560-0043, Japan
[2]Institute of Laser Engineering, Osaka University, 2-6 Yamada-oka, Suita, Osaka 565-0871, Japan
[3]Graduate School of Engineering, Osaka University, 2-6 Yamada-oka, Suita, Osaka 565-0871, Japan
[4]Beijing National Laboratory for Condensed Matter Physics, Institute of Physics, Chinese Academy of Sciences Beijing 100190, China
[5]The National Astronomical Observatories, Chinese Academy of Sciences, Beijing 100012, China
[6]Department of Physics, University of York, Heslington, York, YO10 5DD, United Kingdom
[7]Shanghai Jiao Tong University, Shanghai 200240, China

E-mail: moritat@ile.osaka-u.ac.jp

Abstract. We report the experimental generation of an electrostatic collisionless shock in counter-streaming plasmas without an external magnetic field. The collisionless counter-streaming plasmas were created by a high-power laser. The expansion of plasmas was measured with interferometry and shadowgraphy. Large density jump was observed in both diagnostics. The density jump is the collisionless shock because the width of this jump is much shorter than the ion-ion mean-free-path calculated from flow velocities and densities. Particle-in-cell simulations suggest the formation of the electrostatic collisionless shock in the counter-streaming plasmas. In our experiment, considering the shock width, the observed shock is the electrostatic shock.

1. Introduction

Collisionless shocks have been studied for many years[1, 2]. The collisionless shock is an important subject in not only plasma physics but also astrophysics. Cosmic rays whose energy is less than 10^{15} eV are speculated to be accelerated around the collisionless shock surface in supernova remnants in our galaxy. There is, however, no experimental evidence with a model experiment even for the formation scenario of the collisionless shock and resultant particle acceleration. It is demonstrated that in the situation where counter-streaming plasmas are produced, collisionless shocks can be produced two different kinds of collisionless shocks can be produced. One is essentially one-dimensional and electrostatic shock[3, 4], and the other is collisionless shock based essentially on the multi-dimensional physics, namely mediated by nonlinearity of Weibel instability[5, 6].

In this paper, we present the experimental observation of the electrostatic collisionless shock generated in the counter-streaming plasmas without an external magnetic field using a high-power laser.

2. Experimental setup

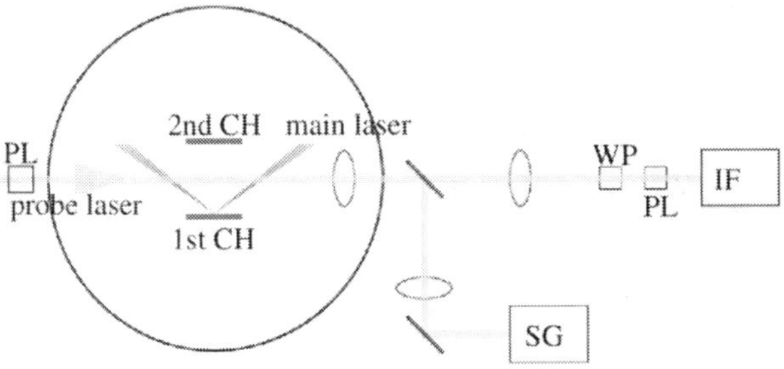

Figure 1. The top view of experimental setup. SG and IF represent the ICCD cameras for shadowgraphy and interferometry, respectively. WP is the Wollaston prism and PL is the polarizer.

The experiment was performed with Shenguang-II laser facility (Shanghai, China). There are eight main beams in which the energy is about 260 J/beam in 351 nm (3ω) and pulse width is 1 ns. Probe laser is a short pulse laser with the width of 70 ps in 527 nm (2ω). Figure 1 shows the top view of the experimental setup. Our targets consist of two thin CH foils. The size of each foil is 2 mm $\times$ 2 mm $\times$ 100 μm, and two foils are separated by 4.5 mm. Four beams were focused on the surface of one side of the CH foil with the incident angle of 60 degrees from the target normal direction with 150 μm diameter at intensity $\simeq 6 \times 10^{15}$ W/cm^2. In the optical path for Nomarski interferometry, a Wollaston prism was placed near the focal spot of the probe laser. This prism can split incident light into two orthogonal linearly polarized beams. Two polarizer were placed before the target chamber and after the Wollaston prism so that the two beams interfere at the photoelectric surface of an ICCD camera. In the shadowgraphy, the incident probe beam passed through the plasmas was directly detected by the other ICCD camera. In our experiment, four beams were used to ablate only one of the CH foils (first CH). The plasma from the other CH foil (second CH) was created by the radiation and/or the plasma from the first CH. In this way counter-streaming plasma flows are created between two foils and interact near the surface of the second CH. Interferometry and shadowgraphy were taken at same region and timing, and hence, the information of density profile and derivative of density gradient can be compared.

3. Result

Phase difference $\delta\theta$ between two rays (one passes through plasmas and the other passes through vacuum) is expressed by optical path length l as $\delta\theta = \omega/(2\mathrm{n_c}c)\int \mathrm{n_e}(l)\mathrm{d}l$[7], where ω is the frequency of the probe beam, c is the speed of light, $\mathrm{n_c}$ is the critical density and $\mathrm{n_e}$ is the electron density. Assuming that plasmas are created axially symmetrically, $\mathrm{n_e}$ can be calculated by Abel inversion method. Abel inversion is calculated with Bockasten's method[8]. The range of $\mathrm{n_e}$ which can be calculated from interferometry is about $1 \times 10^{18} - 5 \times 10^{19}$ cm^{-3}.

Figures 2(a) and 2(b) show the interferogram and shadowgraph respectively, measured at 9 ns after the peak of the main laser (t = 9 ns). The horizontal axis x represents the distance from the surface of the first CH. In Fig. 2(a), the interference fringes suddenly shift and a large density jump exists near the second CH. Sudden brightness changes are observed in the shadowgraph at the same position where the density jump exits. The arrows in Fig. 2(a) and 2(b) indicate the x position of sudden brightness changes. The density profile calculated from Fig. 2(a) is shown in Fig. 2(c). Figure 2(d) shows the density profile at y = 3.5 mm in Fig. 2(c). A large density jump is observed at $x \simeq 3.1$ mm. The width of this jump is about 100 μm.

Figure 3 shows the density profile (Fig. 3(a)) and the shadowgraph (Fig. 3(b)) measured at t = 5ns. Some fine structures vertical to the flow direction are observed near the second CH in Fig. 3(b) at $x \simeq 3.6$ mm. This means that there is a very large density gradient. The arrows in Fig .3(a) and 3(b) show the horizontal position of the large density gradients. Assuming that the plasma from the second CH is created by the radiation and both plasma flows start moving at t = 0 with a constant velocity, flow velocity v_{first} of the plasma with

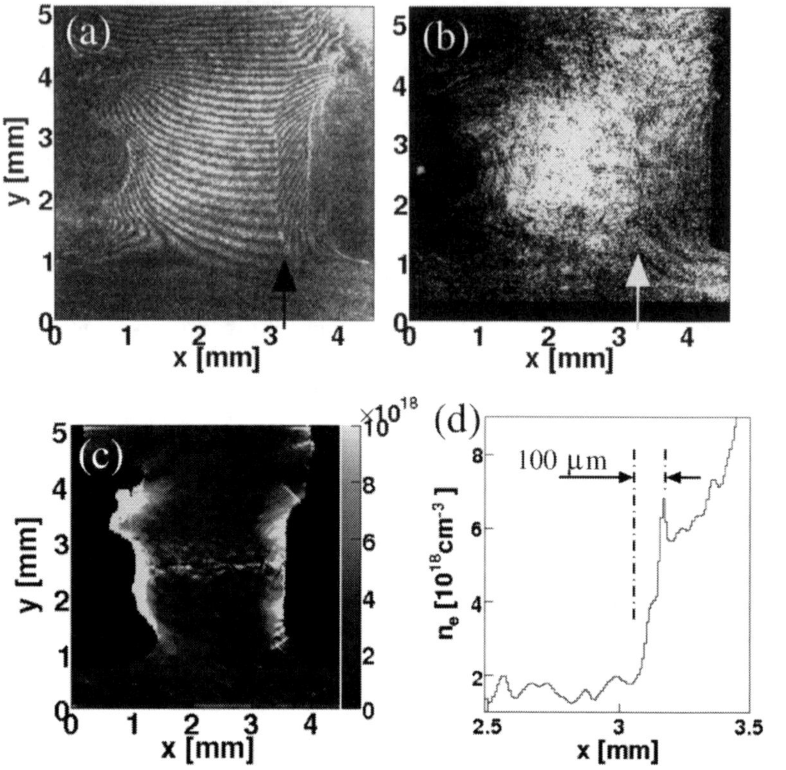

Figure 2. (a) The interferogram and (b) shadowgraph measured at t = 9 ns. x = 0 and 4.5 mm are the surface of the first and second CH foils, respectively. Four beams were focused on the first CH at $x = 0$ and $y \simeq 2.5$ mm. (c) The density profile calculated from (a). (d) The density profile measured at y = 3.5 mm.

$n_e \simeq 2 \times 10^{18}$ cm^{-3} from first CH is roughly estimated as $v_{\text{first}} = \Delta x/\Delta t \simeq 2.0$ mm/3.0 ns $\simeq$ 830 km/s, and the velocity from the second CH ($n_e \simeq 2 \times 10^{18}$ cm^{-3}) is $v_{\text{second}} \simeq 0.7$ mm/3.0 ns $\simeq$ 230 km/s from the data at $t = 3$ ns (not shown). The ion-ion mean-free-path λ_{ii} is expressed as $\lambda_{\text{ii}} = m_i^2 v^4/(8\pi n_i Z^4 e^4 \ln\Lambda)$[9], where coulomb logarithm is calculated with following formula with reduced mass $m_{\text{r}} = (1/m_{\text{first}} + 1/m_{\text{second}})^{-1} = m_i/2$, $\ln\Lambda = \ln(4\pi\epsilon_0\lambda_{\text{D}} m_{\text{r}} v^2/(Z^2e^2))$, where m_{first} and m_{second} are the ion masses coming from the first and second CH, respectively, and λ_{D} is the Debye length. From the values of relative velocity $\simeq$ 1060 km/s and the electron density $n_e = 8 \times 10^{18}$ cm^{-3}, λ_{ii} is calculated as $\lambda_{\text{ii}} = 35$ mm for the electron temperature $T_e = 1$ eV and 25 mm for $T_e = 1000$ eV. It is difficult to estimate T_e in our diagnostics, however λ_{ii} does not strongly depend on T_e. Since the width of the measured density jump at $t = 9$ ns ($\simeq 100$ μm) is much shorter than λ_{ii}, the counter-streaming plasmas created in our experiment is collisionless and this density jump is the collisionless shock.

4. Discussion and conclusion

Particle-in-cell simulations show two possibilities for generation of collisionless shock in counter-streaming plasma. One is electrostatic shock[3, 4] and the other is "Weibel-mediated" shock[5, 6]. In latter case, the dissipation mechanism is provided by the magnetic field generated by the Weibel instability. The width of density transition region is evaluated as $W \simeq 100c/\omega_{\text{pi}}$ where ω_{pi} is the ion plasma frequency. In the experiment, c/ω_{pi} is evaluated as $\simeq 110$ μm for $n_e \simeq 8 \times 10^{18}$ cm^{-3}, and $W \simeq 11$ mm which is much larger than the observed structure. In the case of electrostatic shock, the width of transition region is evaluated as $0.5c/\omega_{\text{pe}}$[3]. The electron inertial length evaluated from the experiment is $\simeq 2$ μm in $n_e \simeq 8 \times 10^{18}$ cm^{-3}. The width of shock structure is larger than this electron inertial length, but this is due to the resolution of experimental instruments. From the narrowest fringe in interferogram, the resolution of interferogram is about 50 μm. As a result, the width of shock structure $\simeq 100$ μm

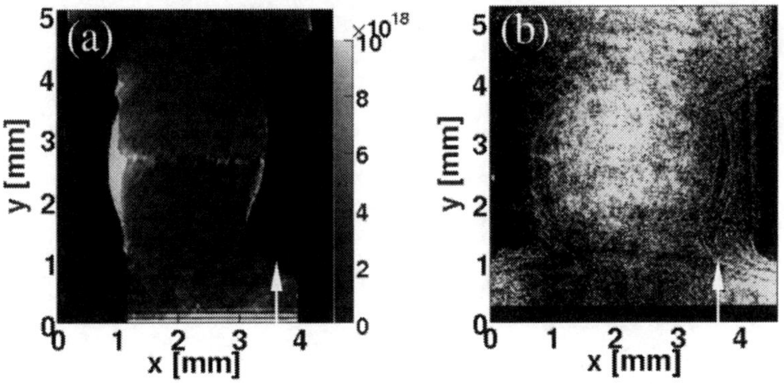

Figure 3. (a) The density profile and (b) shadowgraph measured at $t = 5$ ns. The surfaces of the two CH foils are at $x = 0$ and 4.5 mm. The first CH was irradiated at $x = 0$ and $y \simeq 2.5$ mm.

is reasonably regarded as the electrostatic collisionless shock.

In summary, counter-streaming collisionless plasma flows were created with Shenguang-II laser facility. The large density gradient was observed in both the interferogram and shadowgraph at the same position. The width of the density jump was less than 100 μm, which is much shorter than the ion-ion mean-free-path, and the jump is the collisionless shock.

5. Acknowledgments

The authors acknowledge the help of the technical staffs in "Shenguang-II" laser facility for the realization of this experiment and the helpful discussion with C. Gregory and J. Waugh. This work is supported by the National Basic Research Program of China (973 Program) (Grant No.2007CB815100) and NIFS/NINS under the project of Formation of International Network for Scientific Collaborations.

References

[1] B. Lembege and J. M. Dawson, Phys. Rev. Lett. **62**, 2683 (1989); J. Denavit, Phys. Rev. Lett. **69**, 3052 (1992); L. O. Silva *et al.*, Phys. Rev. Lett. **92**, 015002 (2004); M. Chen *et al.*, Phys. Plasmas **14**, 053102 (2007);

[2] R. J. Taylor *et al.*, Phys. Rev. Lett. **24**, 206 (1970); H. Ikezi *et al.*, Phys. Fluids, **16**, 2167 (1973); H. Ikezi *et al.*, Phys. Rev. Lett. **36**, 794 (1976); N. Izumi *et al.*, Phys. Rev. E **65**, 036413 (2002); M. S. Wei *et al.*, Phys. Rev. Lett. **93**, 155003 (2004); H. Habara *et al.*, Phys. Rev. E **70**, 046414 (2004); A. R. Bell *et al.*, Phys. Rev. A **38**, 1363 (1988); C. Courtois *et al.*, Phys. Plasmas **11**, 3386 (2004); N. C. Woolsey *et al.*, Plasma Phys. Cont. Fusion **46**, B397 (2004); L. Romagnani *et al.*, Phys. Rev. Lett. **101**, 025004 (2008).

[3] T. Kato *et al.*, submitted to Phys. Plasmas (2009).

[4] G. Sorasio, M. Marti, R. Fonseca, and L. O. Silva, Phys. Rev. Lett. **96**, 045005 (2006).

[5] T. Kato and H. Takabe, Astrophys. J. Lett. **681**, L93 (2008).

[6] H. Takabe *et al.*, Plasma Phys. Cont. Fusion **50**, 124057 (2008).

[7] I. H. Hutchinson, Principles of Plasma Diagnostics, Cambridge University Press (2005).

[8] K. Bockasten, J. Opt. Soc. Am. **51**, 943 (1961); R. Illingworth *et al.*, J. Phys. E: Sci. Instrum. **14**, 147 (1981).

[9] L. Spitzer, Jr., Physics of Fully Ionized Gases (2nd ed.), Interscience, New York (1962).

Journal of Physics: Conference Series **244** (2010) 042011 doi:10.1088/1742-6596/244/4/042011

Laboratory studies of multi-material radiative astrophysical jets propagation in plasmas

Ph Nicolaï[1], C Stenz[1], V Tikhonchuk[1], A Kasperczuk[2], T Pisarczyk[2], L Juha[3], E Krousky[3], K Masek[3], M Pfeifer[3], K Rohlena[3], J Skala[3], J Ullschmied[4], V Kmetik[4], M Kalal[5], D Klir[6], J Kravarik[6], P Kubes[6], P Pisarczyk[7], E Tabakhoff[8]

[1]CELIA, Université Bordeaux 1-CNRS-CEA, 33405 Talence cedex, France
[2]Institute of Plasma Physics and Laser Microfusion, 00-908 Warsaw, Poland
[3]Institute of Physics AS CR, Na Slovance 2, 182 21 Prague 8, Czech Republic
[4]Institute of Plasma Physics AS CR, Za Slovankou 3, 182 00 Prague 8, Czech Republic
[5]Czech Technical University in Prague, FNSPE, Brehova 7, 115 19 Prague 1, Czech Republic
[6]Czech Technical University in Prague, FEL, Technicka 2, 166 27 Prague 6, Czech Republic
[7]Warsaw University of Technology, ICS, 00-665, Warsaw, Poland
[8]LULI, Ecole Polytechnique, CNRS, 91128 Palaiseau, France

E-mail: nicolai@celia.u-bordeaux1.fr

Abstract. This work reports on results of a series of experiments dedicated to studying of plasma jets created by a simple method of matter ablation of multi-layer targets. The collimation and X-ray emission of multi-material jets strongly depend on the atomic number of materials and have been measured with two complementary diagnostics, in the optical and soft X-ray domains. A possibility of multi-material laboratory jet production scaled to astrophysical conditions is demonstrated.

1. Introduction

The physical origins of fine structuring of localized high speed flows and their dynamics are the astrophysical problems that can be resolved by using appropriately scaled laboratory experiments. The interaction experiments at the PALS laser facility [1] demonstrate a reliable formation of supersonic jets from metallic targets. Until now, we produced mono-material jets [2, 3], and studied their interaction with an ambient plasma [4]. The dimensionless numbers of our jets have values close to those seen in young stellar objects. These are the Mach number, the jet-plasma density ratio and the dissipative processes described by the cooling parameter. Nevertheless, the structure of astro-jets is much more complex than that seen in our experiments. The astrophysical flows may show the presence of several shocks with variable velocities in the jet body inducing collisions and generating fragmentation. The bow shock, which propagates through the gas, may break up into clumps and the jet itself may appear to be clumpy both along and laterally to the axis. These features may be related to time-variability in the flow ejection velocity. To verify this hypothesis, we experimentally produced multi-material jets by using multi-layer targets. As the ablated matter velocity depends on the material, a multi-layer target induces a strongly inhomogeneous velocity profile in the jet body that may break out and generate a shock. Internal jet structures have been observed by means of a three-frame imaging system in the optical domain (interferograms) and with a four-frame camera in the soft x-ray domain.

2. Experimental conditions

The experiment has been carried out at the PALS iodine laser facility [1]. Figure 1 presents the schematic experimental setup. A laser beam at the wavelength of 0.438 μm was focused by an asperical lens in a focal spot of 600 μm, on a multi-layer target. The laser energy on target was of $\sim$ 30 J and the pulse duration was 250 ps. In front of it, a helium gas puff was placed at the distance of 3.5 mm. A 15 bars backing pressure supersonic gas nozzle was used to create a cylindrical column with a density of neutral atoms 4×10^{18} cm^{-3}. The gas axis was parallel to the solid target surface and it was aligned in such a way that the plasma jet was incident perpendicularly to it. The gas density profile corresponds to a plateau of maximum density with the radius of 1.5 mm surrounded by a 1.5 mm smooth gradient. The gas presence allows to visualise the jet propagation and due to its low atomic number, it does not affect the X-ray emission.

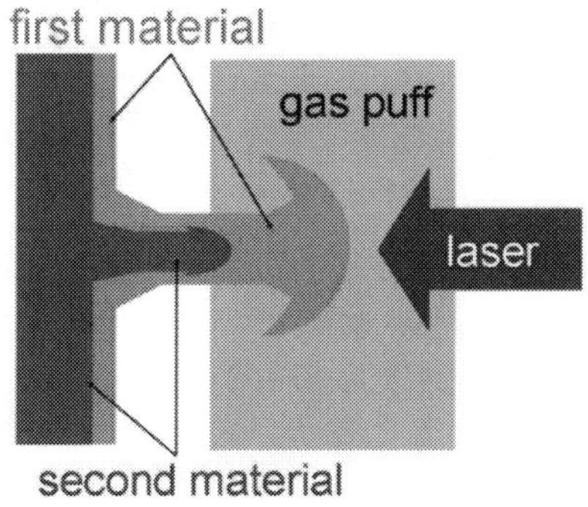

Figure 1. Cartoon that shows the interaction geometry.

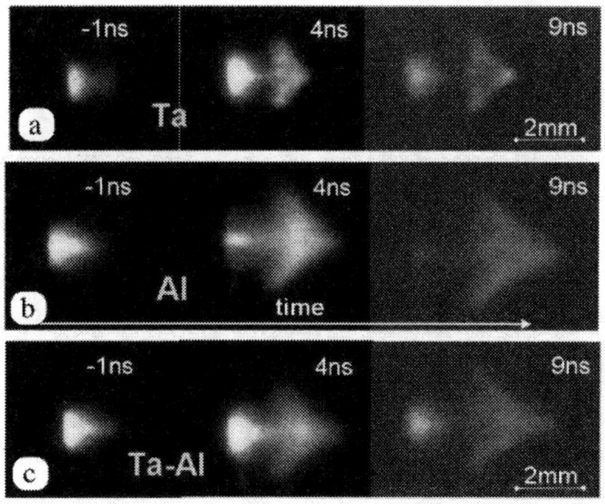

Figure 2. X-ray images of jets launched (a) from a tantalium target, (b) from an aluminium target and (c) from a 65 nm layer of aluminium set on a massive tantalium slab after -1 ns, 4 ns and 9 ns.

The X-ray diagnostic had an exposure time of 4 ns, the line of vision was parallel to the target plane (perpendicular to the jet propagation), and the pinhole diameter was 80 μm. The soft X-ray emission in the energy range 10 – 1000 eV was transferred to the microchannel plate detector coupled with the CCD type digital camera Nikon D1X (4028×1324 pixels, 12 bits). The second main diagnostic was a three-frame interferometric/shadowgraphic system, illuminated by an auxiliary laser beam at the wavelength of 0.438 μm. Each channel was equipped with its own interferometer of the folding wave type [5] with an objective focal length of 250 mm and a wedge angle of 3°. The images were recorded by high resolution 10 bits dynamics CCD cameras Pulnix TM-1300, with the matrix of 1300×1030 pixels. The exposure time of the interferometer was $\sim$ 250 ps.

3. Experimental results

Figure 2 presents 2D soft X-ray images obtained from tantalium and aluminium targets and from a thin layer of aluminium (65 nm) covering a massive tantalium target. As already mentioned,

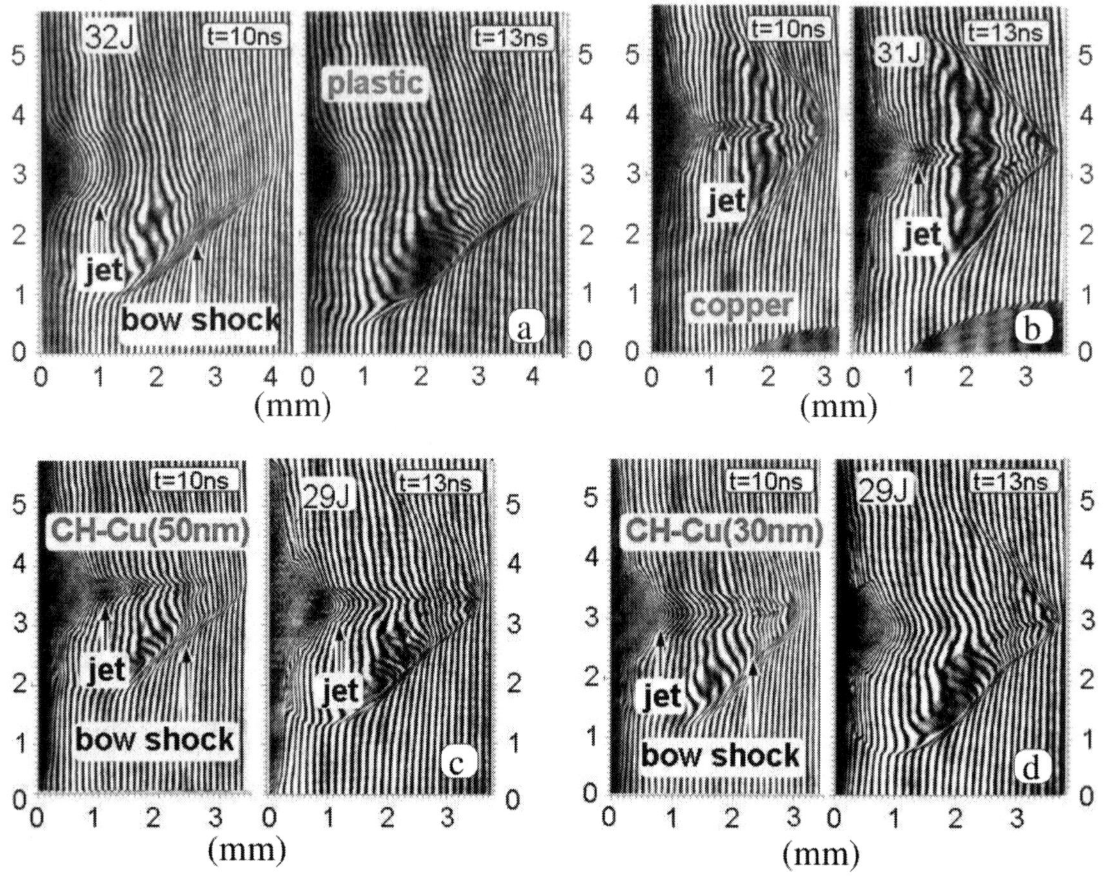

Figure 3. Interferograms of the interaction of jet launched from a plastic slab (a), from a copper slab (b) from a plastic slab covered with a 50 nm layer of copper (c)or with a 30 nm layer of copper (d) with helium gas at 15bars at 10 ns and 13 ns.

the helium gas emission does not appear in images. Only the jet is visible. The luminosity of different parts of the jet depends on its material composition. Due to a lower ablation velocity, the tantalium jet is shorter than the one of aluminium and the interaction zone with the gas (jet head) is smaller. After 9 ns, these differences are clearly visible. Another difference between a high and a low atomic number targets concerns the jet foot emission. In the aluminium case, this emission is very weak and only the interaction zone, heated by the reverse schock continues to emit X-rays. In the tantalium target, the foot emission remains bright and is still visible after 9 ns. Note that each image has its own luminosity scale. One can compare dimensions, shape and relative X-ray emissions between the foot and the head of the jet but not the absolute X-ray intensities between two images. The jets launched from multi-material target combine characteristics of both tantalium and aluminium jets. The jet head and body correspond to the first ablated material (aluminium) whereas the jet foot is characteristic of tantalium emission. These images provide a clear evidence of presence of two components in the plasma jet.

Figure 3 shows typical interferograms of jets obtained with different materials. Interferograms present a slight asymmetry in the shock front. The gas velocity $\sim$ 10 km/s = 1 μm/ns has no influence on the interaction process in the time scale of 10 ns. This effect can be explained by the laser energy distribution in the focal spot which is not perfectly symmetric. Figs. 3-a and b are obtained, respectively, from massive plastic and copper targets. Some features are similar to those seen in X-rays images: the jet is longer and the interaction zone is larger for the plastic target than for the copper one. In addition, this diagnostic allows to observe the jet collimation. The jet made of Cu has a diameter less than 300 μm, even after 13 ns, whereas for a plastic target, after 10 ns, the jet is much wider. Similar features of monomaterial jets have been already observed and explained in previous experiments [3]. It has been shown that jet formation depends on the ratio between the radiation cooling time and the expansion time. If the radiation time is shorter than the expansion time, a jet is formed, otherwise a more spherical blow-off structure is produced. A short radiation time is related to specific laser conditions, such as short pulse duration, large focal spot diameter and a specific concave focal spot shape [4]. However the most important condition for the jet formation is a high atomic number target. This has also been observed in other experiments [6, 7], where the radiation cooling was identified as a condition for jet collimation.

Figures 3-c and 3-d show interferograms obtained from plastic-copper multi-layer targets. For a copper layer thickness of 50 nm, the jet maintains its shape almost 10 ns. Indeed, at this time, the jet is mainly composed of the first ablated material, copper, and collimation is not affected by presence of plastic. Only after 13 ns, the jet base appears wider and its shape changes. The use of a thinner copper layer enhances this effect. With a 30 nm thick copper layer, as early as at 10 ns, the jet collimation is degraded. After 13 ns, the jet width comes close to the one of plastic jet. So, the interferometric diagnostic confirms the presence of both materials in the jet and the modification of jet collimation. Moreover, the plastic plasma propagates faster than that of copper. Therefore, the velocity profile should correspond to the shock wave. For the reverse case (aluminium coated on to tantalum), it should correspond to the rarefaction wave. However, the variations in density are not strong enough to be observed using the present diagnostics.

4. Conclusion

Multi-material plasma jets have been produced and observed by means of two complementary diagnostics. By comparing the features of jets launched from mono-material targets, it is possible to discriminate each component in multi-material jets. This experiment is a first step to the creation of flows with internal shocks. The use of layers with different atomic numbers induces ablation velocity variation and, in addition, allows a better observation of hydrodynamic discontinuities in soft X-rays. These shocks could be scalable to astrophysical objects, morphologically and in terms of dimensionless parameters such as Mach number and radiation cooling time [8, 4]. Studies of complex internal structures will improve our understanding of astrophysical jets and open new capabilities to study their evolution.

5. References

[1] Jungwirth K, Cejnarova A, Juha L, Kralikova B, Krasa J, Krousky E, Krupickova P, Laska L, Masek K, Mocek T, Pfeifer M, Prg A, Renner O, Rohlena K, Rus B, Skala J, Straka P, Ullschmied J et al. 2001, *Phys. Plasmas* **8**, 2495.

[2] Kasperczuk A, Pisarczyk T, Borodziuk S, Ullschmied J, Krousky E, Masek K, Rohlena K, Skala J, Hora H 2006, *Phys. Plasmas* **13**, 062704.

[3] Nicolai Ph, Tikhonchuk V T, Kasperczuk A, Pisarczyk T, Borodziuk S, Rohlena K, Ullschmied J 2006, *Phys. Plasmas* **13**, 062701.

[4] Nicolai Ph, Stenz C, Kasperczuk A, Pisarczyk T, Klir D, Juha L, Krousky E, Masek K, Pfeifer M, Rohlena K, Skala J, Tikhonchuk V, Ribeyre X, Galera S, Schurtz G, Ullschmied J, Kalal M, Kravarik J, Kubes P, Pisarczyk P, Schlegel T 2008, *Phys. Plasmas* **15**, 082701.

[5] Borodziuk S, Kasperczuk A, Pisarczyk T, Gus'kov S, Ullschmied J, Kralikova B, Rohlena K, Skala K, Kalal M and Pisarczyk P 2004, *Opt. Appl.* **34**, 31.

[6] Farley D R, Estabrook K G, Glendinning S G, Glenzer S H, Remington B A, Shigemori K, Stone J M, Wallace R J, Zimmerman G B, Harte J A 1999, *Phys. Rev. Lett.* **83**, 1982.

[7] Lebedev S V, Chittenden J P, Beg F N, Bland S N, Ciardi A, Ampleford D, Hughes S, Haines M G, Frank A, Blackman E G, Gardiner T 2002 , *Astrophys. J.* **564**, 113.
[8] Ryutov D, Drake R P, Kane J, Liang E, Remington B A, Wood-Vasey W M 2001, *Astrophys. J. Phys.* **518**, 821.

A MODEL EXPERIMENT OF A DOUBLE-CONE TARGET USING A GAP TARGET

Y. Sakawa[1), H. Nakamura[1), S. Oshima[1), M. Hatakeyama[2), N. Kageiwa[2), S. Hino[2), S. Tanimoto[2), M. Tanabe[2), H. Habara[2), H. Homma[1), T. Norimatsu[1), T. Jitsuno[1), H. Cai[1), W. Zhou[2), T. Johzaki[1), A. Sunahara[3), H. Nagatomo[1), H. Nishimura[1), K. A. Tanaka[1),2)], K. Mima[1), and H. Azechi[1)

1) Institute of Laser Engineering, Osaka University, 2-6 Yamadaoka, Suita 565-0871, Japan
2) Grad. School of Engineering, Osaka Univ., 2-1 Yamadaoka, Suita 565-0871, Japan
3) Institute for Laser Technology, 2-6 Yamadaoka, Suita 565-0871, Japan

E-mail: sakawa-y@ile.osaka-u.ac.jp

Abstract. We have conducted preliminary experiments to prove the vacuum-shielding effect using an Al-Cu double-foil (both Al and Cu foils are 1 mm x 1 mm x 10 μmt) target with a vacuum-gap. Particle-in-cell simulation results have shown that the double-cone confines the electrons for hundreds of femtoseconds by the sheath electric field inside the vacuum-gap [T. Nakamura, *et al.*, Phys. Plasmas **14**, 103105 (2007)]. Cu-Kα intensity decreased with the vacuum-gap, i.e., the number of fast electrons that reached the Cu foil decreased with the gap. In the stack measurement, the number of electrons detected in the target surface direction increased with the gap. These results indicate that when the double-cone target is used, fast electrons created by an ignition laser can be reflected from the vacuum-gap, move along the cone surface, and be transferred towards the cone tip.

1. Introduction

In the Fast Ignition scheme with a re-entrant cone, an efficient heating of a dense core with fast electrons is required. It is known that a cone target is surrounded by coronal plasma generated during the implosion, and electrons created by an ignition laser can escape freely from cone sides to the surrounding coronal plasma. As a result, an energy flux to the core plasma through the cone tip decreases. Nakamura et al. have suggested using a double-cone target to prevent the electrons from escaping aside [1]. Their two-dimensional (2D) Particle-in-cell (PIC) simulation results have shown that the double-cone confines the electrons for hundreds of femtoseconds (fs) by the sheath electric field inside the vacuum-gap with an immobile ion background [1]. However, since the ions are accelerated into the gap by the electrostatic fields, the plasma expansion could be significant for picoseconds (ps) time scale, and the sheath electric field inside the gap could be reduced to a very low level in several hundreds of femtoseconds.

Recently, Cai et al. have demonstrated 2.5D simulations using the PIC framework ASCENT (assembly system for computational experiment), in order to clarify whether the double cone is still effective in confining the high-energy electrons after 1 ps, and how much energy can be confined and focused to the compressed core by using the double-cone [2]. In the simulation, a P-polarized laser pulse (1.06 μm in wavelength, 1.2 x 10^{19} W/cm^2 in intensity, 67 fs in rise-time, and flat-top pulse shape for 1.5 ps in pulse duration) has been irradiated on a double-cone target, in which the inner-cone wall has been isolated from the background corona plasma due to a vacuum-gap of 3 μm. It has been found that the number of high-energy electrons escaping from the side-walls of the cone is greatly reduced by the vacuum-gap inside the wing of the double-cone. Two mechanisms to confine high-energy electrons have been found; the sheath electric field at the rear-side of the inner-cone wing and the quasistatic magnetic field inside the vacuum-gap. The quasistatic magnetic field has been due to a localized supply of high-energy electrons, originally produced at the inner-cone and the cone tip. In addition to this electron current coming from the cone tip, an opposite-directional surface-current moving along the inner-surface of the outer-cone has given rise to a large quasistatic magnetic field inside the gap. The quasistatic fields have continued to confine the high-energy electrons for longer than a few picoseconds. It has been shown that whereas about 30% of the total input laser energy escapes into the surrounding corona plasma from the side-walls as high-energy electrons in the single-cone case, the escaped energy is reduced to about 11% percent in the double-cone case [2].

We have conducted preliminary experiments with double-foil targets with a vacuum-gap to prove the effect of sheath electric filed using a high-power laser system. In the experiment, since the laser intensity is nearly an order of magnitude lower and pulse duration is by a factor of three lower than those used in PIC simulation [2], generation of quasistatic magnetic field is expected to be small.

2. Experimental

The experiments were performed with Gekko MII laser system (Nd: Glass laser, wavelength = 1.05 μm, pulse duration = 400 - 500 fs, energy = 4 - 13 J on target, laser intensity = (1 - 4) x 10^{18} W/cm^2, S- or P-polarized) at the Institute of Laser Engineering, Osaka University. In order to clarify the vacuum-shielding effect, an Al-Cu double-foil (both Al and Cu foils are 1 mm x 1 mm x 10 μmt) target with a vacuum-gap was used. The gap width ΔW was varied as 0, 10, 20, and 100 μm with four Al spaces (100 μm x 100 μm x ΔW) located at four corners of the target. The laser beam was irradiated on an Al foil side with 20 (S-20) or 70 (S-70) degrees from the target normal direction for S-polarized and 60 degrees (P-60) for P-polarized cases. Figure 1 shows schematic diagram of the experimental setup for S-20. An X-ray pinhole camera (XPHC) was used to measure the incident laser spot-size on the Al foil at 38.5 degrees from the axis of the laser direction. Cu K-shell emission was measured from the rear-side of the Cu foil using an x-ray spectrometer located at 48.5 degrees from the laser axis. A stuck of imaging plates (IPs) was placed behind the target, 35 mm and 15 mm from the target for S- and P-polarized cases, respectively, in order to measure the transport of fast electrons. An Al foil was placed in front of the IPs to block the laser light.

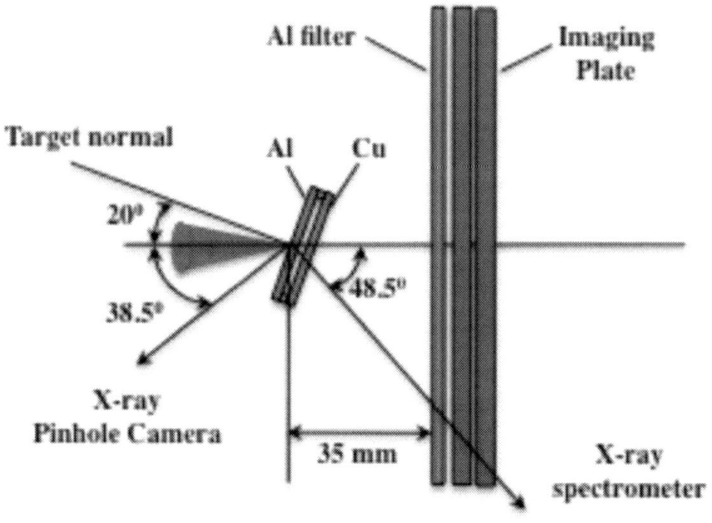

Fig. 1 Schematic diagram of the experimental setup for S-20. Polarization of incident laser beam is perpendicular to the sheet.

3. Results and discussion

In the rear-side x-ray spectroscopy, Cu-Kα1. Kα2, and Kβ lines were observed. We focus only on the peak intensity of Cu-Kα1 line. Figure 2(a) shows x-ray intensity from the front-side Al foil measured by XPHC as a function of laser energy E_L for S-20. We see that the x-ray intensity increases monotonically with the laser energy independent of the gap width ΔW. Figure 2(b) represents the rear-side Cu-Kα intensity for S-20. Whereas the Cu-Kα intensity increases by increasing the laser energy, the intensity at a given laser energy is lower with the vacuum-gap. The decrease in Cu-Kα intensity with the vacuum-gap is much clear for S-70 as shown in Fig. 2(c); Cu-Kα intensity for the gap-targets with $E_L > 10$ J is even lower than that for no-gap targets with $E_L < 8$ J. For S-70, because of the location of the XPHC, the front-side emission from Al foil was not measured. The measured reduction of the rear-side Cu-Kα intensity with the vacuum-gap indicates the number of fast electrons that reached the Cu foil decreased with the gap.

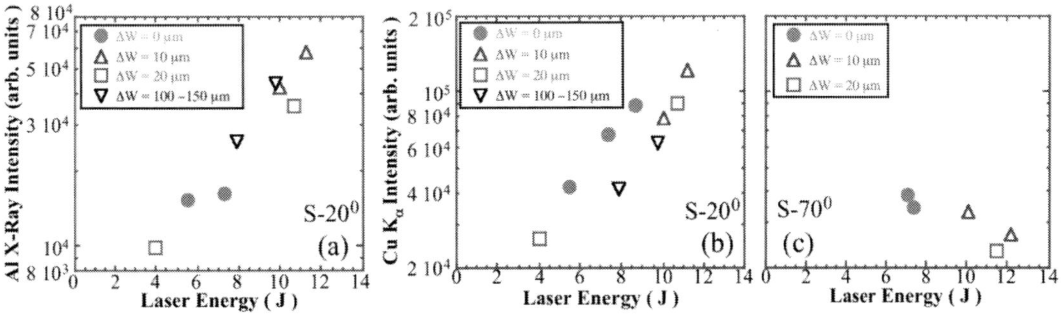

Fig. 2 (a) X-ray intensity from the front-side Al foil measured by XPHC for S-20, and rear-side Cu-Kα intensities for (b) S-20 and (c) S-70.

Kα radiation is emitted by inner-shell transition when an L-shell electron fills a vacancy in the K-shell via radiative decay. The lifetime of the K-shell vacancy induced by a fast electron is less than 10^{-12} s. Cu K-shell collisional-ionization cross-section as a function of the incoming electron energy has the maximum around 27 keV, and energy range of electrons for the half-maximum is roughly 10 – 270 keV [3]. Therefore, we can estimate the flux of the tens to few hundreds of keV electrons from the Cu-Kα intensity measurements.

Figures 3(a) and 3(b) show two-dimensional (in vertical and horizontal directions) intensity profiles of IP for ΔW = 0 and 20 μm, respectively, for S-70. The vertical angle $\theta_V = 0$ and the horizontal angle $\theta_H = 0$ correspond to the axis of the incident laser beam. The horizontal angle $\theta_H = -20$ and 0 degrees correspond to the directions of specular reflection and the target surface. Figure 3(c) shows signal intensity profiles as a function of horizontal angle θ_H for ΔW = 0 [Fig. 2(a)] and 20 μm [Fig. 3(b)] at $\theta_V = -10$ degrees. We see that signal intensity profile for ΔW = 0 shows two peaks; in the directions of specular reflection ($\theta_H = -20$ degrees) and axis of the incident laser beam ($\theta_H = 20$ degrees). On the other hand, for ΔW = 20 μm, the signal intensity peaks in the directions of target surface ($\theta_H = 0$). Similar results were obtained for P-60, i.e., the signal intensity shows two peaks in the directions of specular reflection and axis of the incident laser beam for ΔW = 0, and one peak in the direction of target surface for ΔW = 20 μm. Note that the signal intensities in the directions of specular reflection and axis of the incident laser beam for ΔW = 20 μm are nearly identical to those for ΔW = 0. On the other hand, the signal intensity in the directions of target surface increased nearly by a factor of three by adding ΔW from 0 to 20 μm. Therefore, in the stack measurement, the number of electrons detected in the target surface direction increased with the vacuum-gap.

When Al-Cu no-gap targets are irradiated by the laser beam, a sheath electric field is created in the rear-side of the Cu foil, and low-energy electrons that cannot escape the sheath field are reflected back to the target and may transported towards the target surface direction. When a vacuum-gap is added to the Al-Cu interface, sheath electric fields are created at the vacuum-gap at the rear-side of Al foil in addition to the rear-side of the Cu foil. Therefore, there are two sheath fields, and the reflection of the electrons at the sheaths are enhanced for the gap-target compared with that for the no-gap target.

These results indicate that when the double-cone target is used, fast electrons created by an ignition laser can be reflected from the vacuum-gap, move along the cone surface, and be transferred towards the cone tip.

4. Summary and Conclusions

We have conducted preliminary experiments to prove the vacuum-shielding effect using an Al-Cu double-foil target with a vacuum-gap. Cu-K α intensity decreased by increasing the vacuum-gap width ΔW, i.e., the number of fast electrons that reached the Cu foil decreased with the gap. In the stack measurement, the number of electrons detected in the target surface direction increased with a vacuum-gap. These results indicate that when the double-cone target is used, fast electrons created by an ignition laser can be reflected from the vacuum-gap, move along the cone surface, and be transferred towards the cone tip.

Acknowledgments

The authors would like to thank colleagues at the ILE in particular GMII laser group for laser operation and TMS group for target fabrication.

References

[1] T. Nakamura, *et al.*, Phys. Plasmas **14**, 103105 (2007).
[2] H. Cai, *et al.*, Phys. Rev. Lett. **102**, 245001 (2009).
[3] C. Hombourger, J. Phys. B: At. Mol. Opt. Phys. **31**, 3693 (1998).

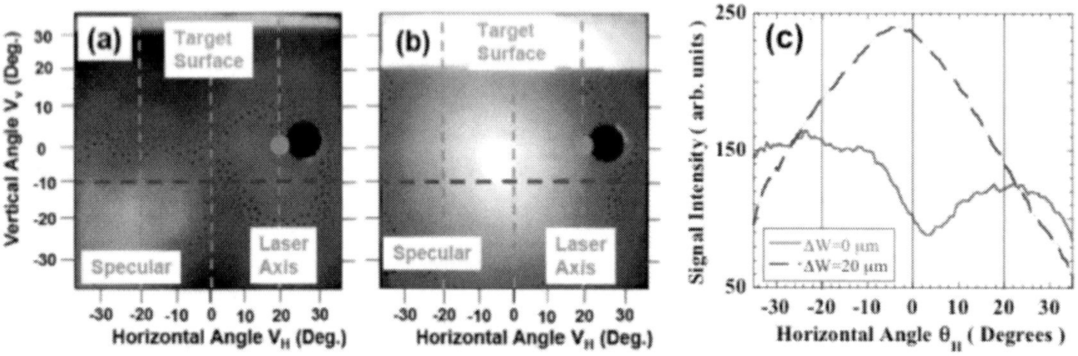

Fig. 3 Two-dimensional (in vertical and horizontal directions) intensity profiles of IP for (a) $\Delta W = 0$ and (b) 20 μm, for S-70. The vertical angle $\theta_V = 0$ and the horizontal angle $\theta_H = 0$ (indicated by a solid circle) correspond to the axis of the incident laser beam. The horizontal angle $\theta_H = -20$ and 0 degrees correspond to the directions of specular reflection and the target surface, respectively. Laser energies were (a) 14.8 J and (b) 22.9 J. (c) Signal intensity profiles as a function of horizontal angle θ_H for $\Delta W = 0$ [Fig. 3(a)] and 20 μm [Fig. 3(b)] at $\theta_V = -10$ degrees.

X-ray Spectroscopy of Non-thermal Equilibrium Laboratory Photo-ionized Plasma

Norimasa Yamamoto[1], Shinsuke Fujioka[1], Hiroaki Nishimura[1], David Salzmann[1], Feilu Wang[2], Yutong Li[3], Yi Zhang[3], Shoujun Wang[3], Quan-Li Dong[3], Jiayong Zhong[2], Jie Zhang[3], Yong-Joo Rhee[4], Yong-Woo Lee[4], Jae Min Han[4], Duckhee Kwon[4], Gang Zhao[2], Kunioki Mima[1] and Hideaki Takabe[1]

[1]Institute of Laser Engineering, Osaka University, Suita, Osaka 565-0871, Japan
[2]National Astronomical Observatories, Chinese Academy of Sciences, Beijing, China,
[3]Institute of Physics, Chinese Academy of Science, Beijing, China
[4]Korea Atomic Energy Research Institute, Yuseong, Daejeon, Korea

yamamoto-no@ile.osaka-u.ac.jp

Abstract. Photo-ionized silicon plasma was studied using 500-eV Planckian radiation from implosion core plasma generated with Gekko-XII laser to simulate a astrophysical photo-ionized plasma. Three major features were observed in an energy range from 1.80 to 1.88 keV and origins of the features were replicated with a time-dependent collisional-radiative model including photo-ionization processes: The line at 1.865 keV is the resonance line by inner-shell photo-ionization of Li-like Si ions and that at 1.840 keV is the sum of satellite lines attributed to the inner-shell photo-ionization of Be-like ions. Emission lines in the soft x-ray/extreme ultra-violet (EUV) region (i.e., 140 to 300 eV) from the low-temperature Si plasma in prior to the photo-ionization were also identified to extract the intial temperature and density of the pre-formed Si.

1. Introduction

Photo-ionized plasma in astronomy is generated around x-ray binary star, consisting of a compact star and a companion star as Cygnus X-3 [1] and Vela X-1 [2]. In general, the observed x-ray spectra with a x-ray telescope (i.e. *Chandra*, *XMM-Newton*, *Suzaku* etc.) are analyzed to fit the theoretical spectra. Therefore, simulation experiments of astronomy-like plasmas in laboratory are highly important for validation of theoretical models. Creation of astronomy-like photo-ionized plasma are more difficult than the generations of collisional-ionization plasmas (i.e. solar, star etc.), because the creation high radiation temperature x-ray sources in laboratories is very difficult. Even in recent experiments, temperature of radiation source are around one hundred eV [3], was not sufficient to mimic similar situations observed in universe since the radiation temperate of the photo-ionized plasma is on the order of 1-2 keV.

In our work, x-ray sources like x-ray binary star are generated with laser driven-implosion. The compressed core plasma is known to be one of the most high radiation sources. Then astronomy-like photo-ionized plasmas are generated and x-ray spectra of the plasma were measured.

2. Experiments

Target layout for the present experiments is shown in Figure 1(a). A plastic-made spherical shell for the core plasma formation was set at the center of target-chamber, and a Si plate for photo-ionization plasma was set beside 1.2 mm away from the shell. Diameter and thickness of the shell were 505 μm and 6.4 μm, respectively. The Si plate was 1-μm-thick silicon coated on a 25-μm-thick plastic sheet and was in a square shape of 500 × 500 μm^2. The Si plate was irradiated with a 1-J, Nd-YAG laser pulse in prior to the core plasma generation. The pulse width of the Nd:YAG laser was 10 ns. The implosion plasma was generated with GXII laser of ~350 J/beam (in total 4.0 kJ for 12 beams) with a pulse width of 1.2 ns in a Gaussian profile. A Ta-made slit was inserted between these two targets to choose a portion of the expanding Si plasma and to avoid debris restraints from the imploded core. The implosion plasma was observed to be like a black-body radiation source of 500 eV as shown in Figure 1(b). This means that the temperature of the x-ray source is high enough to ionize the pre-formed Si ions emitting x-ray spectra with a peak at1.4 keV. It was experimentally checked that only when both the Si plasma and the implosion core were simultaneously generated were intense x-ray from the photo-ionized Si plasma observed [4]. X-ray emission from these plasmas were measured with an x-ray streak camera, a transmission grating spectrometer, a flat-field EUV spectrometer, and a Bragg crystal (RbAP) spectrometer attached with an x-ray CCD camera.

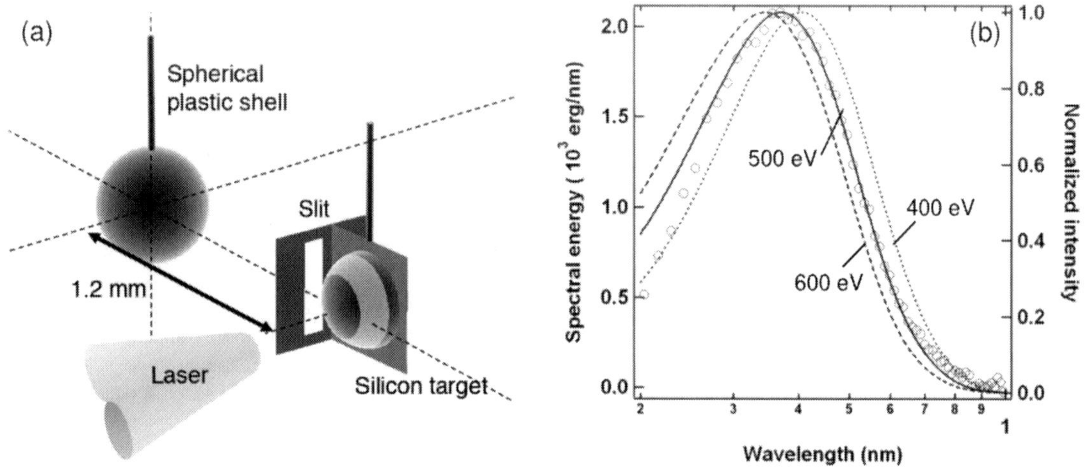

Figure 1(a). Experimental layout of the Si plate and the plastic shell. (b) Comparison of x-ray spectra from the implosion core plasma and Planckian spectra for various radiation temperatures. Best fit is seen for 500 eV.

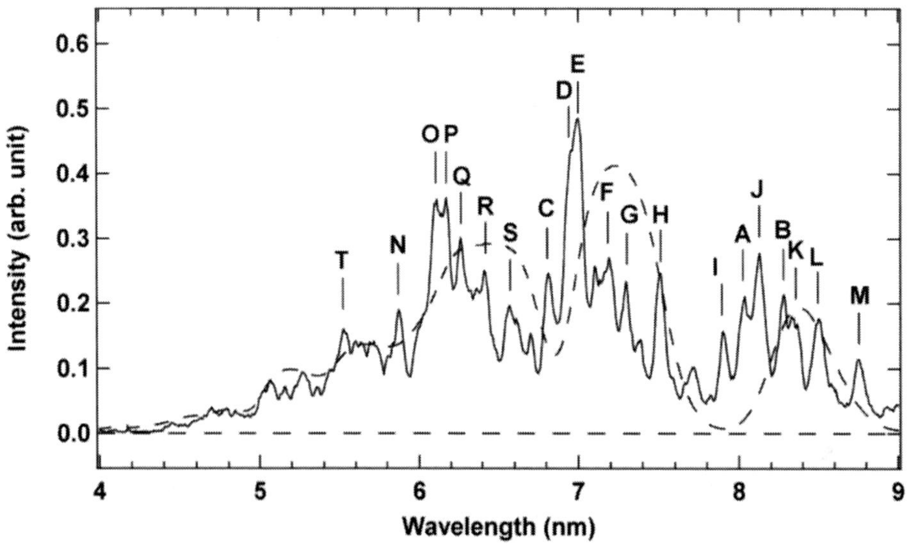

Figure 2. Comparison of the measured spectra (solid line) from the pre-formed Si plasma with the calculated spectra (dashed line) by FLYCHK with the best-fit parameter. Letters shown on each line correspond to those listed in Table 1.

3. Soft x-ray spectra from pre-formed Si plasma and photo-ionized Si plasma

3.1. Identifications of lines from pre-formed Si plasma

Soft x-ray spectra from the pre-formed Si plasma were measured with a flat field grating spectrometer. The spectra are shown in Figure 2. Lines in the range of 4 – 9 nm (300 – 140 eV) are identified by NIST database [5] and the identifications were checked by a collisional-radiative model without photo-ionization process [6]. These wavelengths can only be roughly determined because of accumulations of many lines in the spectra. Line identifications are summarized in Table 1. The measured soft x-ray spectral lines of pre-formed Si plasma are identified to be Si VI – Si IX. Lines of A – H and O – T are derived from *2p – 3d* transitions. Lines of I – M are by *2p – 3s* transition and line of N is by *2s – 3p* transition. For the estimation of electron temperature and density, the measured spectra were fit using the calculated spectra of FLYCHK code [7]. Since spectral resolution of FLYCHK is not sufficient in this energy range, the measured spectra were fit by structures of the calculated spectra. The estimated electron temperature and density were 28 eV and 8×10^{19} cm^{-3}, respectively.

Table 1. Line identification table. Wavelengths were estimated by energy levels data of NIST database.

Spectrum	*letter*	*λ (nm)*	*Transitions*
Si VI	*A*	8.04 – 8.07	$2s^2\,2p^5\,{}^2P - 2s^2\,2p^4\,({}^1D)\,3d\,{}^2S, {}^2P, {}^2D, {}^2F$
	B	8.30 – 8.35	$2s^2\,2p^5\,{}^2P - 2s^2\,2p^4\,({}^3P)\,3d\,{}^2S, {}^2P, {}^2D, {}^2F$
Si VII	*C*	6.81 – 6.87	$2s^2\,2p^4\,{}^3P - 2s^2\,2p^3\,({}^2P)\,3d\,{}^3P, {}^3D$
	D	6.94 – 6.98	$2s^2\,2p^4\,{}^3P - 2s^2\,2p^3\,({}^2D)\,3d\,{}^3S, {}^3P$
	E	7.01 – 7.03	$2s^2\,2p^4\,{}^3P - 2s^2\,2p^3\,({}^2D)\,3d\,{}^3D, {}^1P$
	F	7.14 – 7.23	$2s^2\,2p^4\,{}^1D - 2s^2\,2p^3\,({}^2D)\,3d\,{}^1P, {}^1D, {}^1F$
	G	7.31 – 7.34	$2s^2\,2p^4\,{}^3P - 2s^2\,2p^3\,({}^4S)\,3d\,{}^3D$
	H	7.52	$2s^2\,2p^4\,{}^1S - 2s^2\,2p^3\,({}^2D)\,3d\,{}^1P$
	I	7.92	$2s^2\,2p^4\,{}^3P - 2s^2\,2p^3\,({}^2P)\,3s\,{}^3P$
	J	8.15 – 8.20	$2s^2\,2p^4\,{}^3P - 2s^2\,2p^3\,({}^2D)\,3s\,{}^3D$
	K	8.41	$2s^2\,2p^4\,{}^1D - 2s^2\,2p^3\,({}^2D)\,3s\,{}^1D$
	L	8.52 – 8.57	$2s^2\,2p^4\,{}^3P - 2s^2\,2p^3\,({}^4S)\,3s\,{}^3S$
	M	8.80 – 8.88	$2s\,2p^5\,{}^3P - 2s\,2p^4\,({}^4P)\,3s\,{}^3P$
Si VIII	*N*	5.89	$2s^2\,2p^3\,{}^4S - 2s\,2p^3\,({}^5S)\,3p\,{}^4P$
	O	6.14 – 6.19	$2s^2\,2p^3\,{}^2D - 2s^2\,2p^2\,({}^1D)\,3d\,{}^2D, {}^2F$
	P	6.26	$2s^2\,2p^3\,{}^2P - 2s^2\,2p^2\,({}^1D)\,3d\,{}^2S$
	Q	6.28 – 6.33	$2s^2\,2p^3\,{}^2P, {}^2D - 2s^2\,2p^2\,({}^3P, {}^1D)\,3d\,{}^2P, {}^2D$
	R	6.37 – 6.39	$2s^2\,2p^3\,{}^2D - 2s^2\,2p^2\,({}^3P)\,3d\,{}^2F$
	S	6.54 – 6.55	$2s^2\,2p^3\,{}^2P - 2s^2\,2p^2\,({}^3P)\,3d\,{}^4D$
Si IX	*T*	5.53 – 5.54	$2s^2\,2p^2\,{}^3P - 2s^2\,2p3d\,{}^3D$

3.2. Identifications of laboratory photo-ionized plasma spectral lines

The measured spectra of Figure 3(a) were analyzed with a time-dependent collisional-radiative model including photo-ionization processes [8]. In the measured spectra, the line at 1.865 keV is the resonance lines ($1s^2\,{}^1S_0 - 1s2p\,{}^1P_1$) and that at 1.840 keV is a sum of the Li-like satellite lines for the resonance line. The model calculation shows that the resonance line is attributed to the photo-ionization of $1s^22p$ states of Li-like Si ions, and they are also produced by photo-ionization of the pre-formed Si plasma consisting mostly of N- and O-like ions. In the same manner, the satellite lines are to the inner-shell photo-ionization of $1s^22s2p$ and $1s^22p^2$ states of Be-like ions. Ion densities of these p-states increase by collisions between s- and p-states. However the line at 1.853–4 keV in the measured spectra could not appear in the calculated spectra for the present experimental conditions. A value of the x-ray energy corresponds to the intercombination line ($1s^2\,{}^1S_0 - 1s2p\,{}^3P_{1,2}$). Figure 3(b) and 3(c) show the observed spectra from Cygnus X-3 and Vela X-1 with Chandra. Spectra of Cygnus

X-3 seem to consist of two strong broad lines, and spectra of Vela X-1 consist of two strong lines and one weak line. In astronomy, these lines are believed to be those from He-like ions lines, resonance, inter-combination, and forbidden lines ($1s^2\ ^1S_0 - 1s2s\ ^3S_1$).

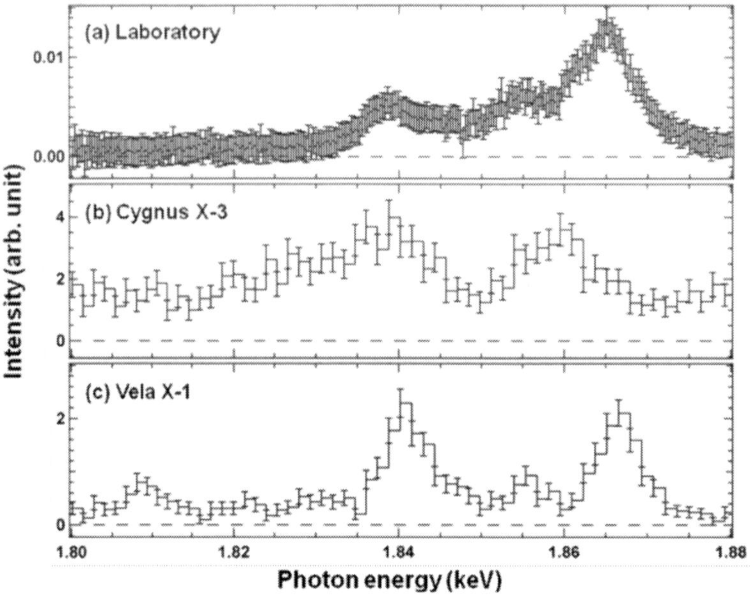

Figure 3. Comparison of (a) the measured spectra from the present work, (b) Cygnus X-3 and (c) Vela X-1 with *Chandra*.

4. Conclusions

We generated astronomy-like photo-ionized plasma with GXII laser. The radiation source derived from implosion fits well with a black-body radiation of 500 eV. Soft x-ray spectra from the pre-formed Si plasma generated with 1J Nd:YAG laser were identified and electron temperature and density of the pre-formed plasma were estimated to be 28 eV and 8×10^{19} cm^{-3}, respectively. When the radiation source and pre-formed Si plasma were simultaneously generated, photo-ionized Si plasma were generated and strong x-rays from the plasma were measured. Then origins of lines in the measured spectra were identified by the theoretical model.

Acknowledgements

This work was partially supported by the Grant-in-Aid for Young Scientists (B) 21740397 of Japan Society for the Promotion of Science.

References

[1] F. Paerels et al., Astrophysical J. **533** (2000) L135-138.
[2] S. Watanabe et al., Astrophysical J. **651** (2006) 421-437.
[3] F. Wang et al., Phys. Plasmas **15** (2008) 073108-073114.
[4] S. Fujioka et al., Nature Physics advanced online publication, 18 October 2009 (doi:10.1038/nphys1402).
[5] T. Shirai, J. Sugar, A. Musgrove and W. L. Wiese, J. Phys. Chem. Ref. Data, Monograph **8** (2000). (NIST online Atomic database ; http://www.physics .nist.gov/PhysRefData/ASD/).
[6] N. Yamamoto et al., Astrophyscal J. **689** (2008) 646-652.
[7] H. K. Chung et al., High Energy Density Physics **1** (2005) 3.
[8] F. Wang et al., to be published, Astrophyical J. (2009).

Journal of Physics: Conference Series **244** (2010) 042014 doi:10.1088/1742-6596/244/4/042014

Measurement of The Dynamic Response of Compressed Hydrogen by Inelastic X-Ray Scattering

K Falk[1], A P Jephcoat[2,3], B J B Crowley[4,1], R R Fäustlin[5], C Fortmann[6], F Y Khattak[7], A K Kleppe[2], D Riley[8], S Toleikis[5], J Wark[1], H Wilhelm[2], G Gregori[1]

[1]Dept. of Physics, Clarendon Laboratory, University of Oxford, Oxford, OX1 3PU, UK
[2]Diamond Light Source, Harwell Science and Innovation Campus, Chilton, OX11 0DE, UK
[3]Dept. of Earth Sciences, University of Oxford, Oxford, OX1 3PR, UK
[4]AWE PLC, Aldermaston, Reading, RG7 4PR, UK
[5]DESY, Notkestr. 85, D-22607 Hamburg, Germany
[6]Institut für Physik, Universität Rostock, 18051Rostock, Germany
[7]Dept. of Physics and Astronomy, University of California, Los Angeles CA 90095
[8]Dept. of Physics, Kohat University of Science and Technology, Kohat-26000, Pakistan
[9]School of Mathematics and Physics, Queens University of Belfast, Belfast BT7 1NN, UK

E-mail: k.falk1@physics.ox.ac.uk

Abstract. Measurement of the dynamic properties of hydrogen and helium under extreme pressures is a key to understanding the physics of planetary interiors. The inelastic scattering signal from statically compressed hydrogen inside diamond anvil cells at 2.8 GPa and 6.4 GPa was measured at the Diamond Light Source synchrotron facility in the UK. The first direct measurement of the local field correction to the Coulomb interactions in degenerate plasmas was obtained from spectral shifts in the scattering data and compared to predictions by the Utsumi-Ichimaru theory for degenerate electron liquids.

1. Introduction

A central feature of planetary research and laboratory astrophysics is the creation of extreme high pressure, high temperature states of hydrogen and helium. At solid densities and high pressures (above a few GPa) degeneracy and strong inter-particle coupling start to play an important role in determining the equation of state (EOS) of the hydrogen, which is still largely unknown in this regime. To date, most experimental efforts have concentrated on creating high density and temperature conditions in fast shocks created by laser ablation [1, 2]. State-of-the-art EOS measurements of laser-driven shocks on pre-compressed H and He samples have been demonstrated using Velocity Interferometer System for Any Reflector (VISAR) diagnostics [3, 4]. While the laser driven dynamic compression reaches higher pressures, diamond anvil cells (DAC) that compress the samples statically can be used to create steady state conditions which are more representative of the planetary interiors than laser-shock experiments [5]. X-ray Thomson scattering (XRTS) [6] is capable of providing more EOS information, such as electron temperature, density and ionization state, and VISAR is less applicable at higher pressures while XRTS can still be used in such conditions, making it the preferred method for measurements of the dynamic response of highly compressed H.

Journal of Physics: Conference Series **244** (2010) 042014 doi:10.1088/1742-6596/244/4/042014

2. Experimental Setup

This proof-of-principle experiment was conducted on the I15 (Extreme Conditions) beamline at Diamond Light Source, UK. The hydrogen gas loaded inside a 0.1 mm thick gasket with diameter of 250 μm was compressed using a diamond anvil cell and probed by monochromatic 19,952 eV X-rays from the synchrotron, figure 1. The inelastic X-ray scattering spectra from hydrogen DAC's at P_1 = 2.8 GPa (liquid phase) and P_2 = 6.4 GPa (solid phase) were observed in Von Hamos geometry using a curved highly annealed pyrolytic graphite (HAPG) crystal [7] coupled to a photon-counting detector (Pilatus 100k) with scattering angle of 32.2±1.1°, corresponding to a regime where the magnitude k of the scattering wavevector is comparable to the Thomas-Fermi screening wavenumber of the electrons (k_{TF} = 1.7 x 10^{10} m^{-1}). The DAC was positioned with 20 μm accuracy, corresponding to an uncertainty of 10 μm ($\sim$ 1 eV) at the detector plane. Uncertainties in the synchrotron energy (as determined using a Pd filter) were also found to be within 1 eV.

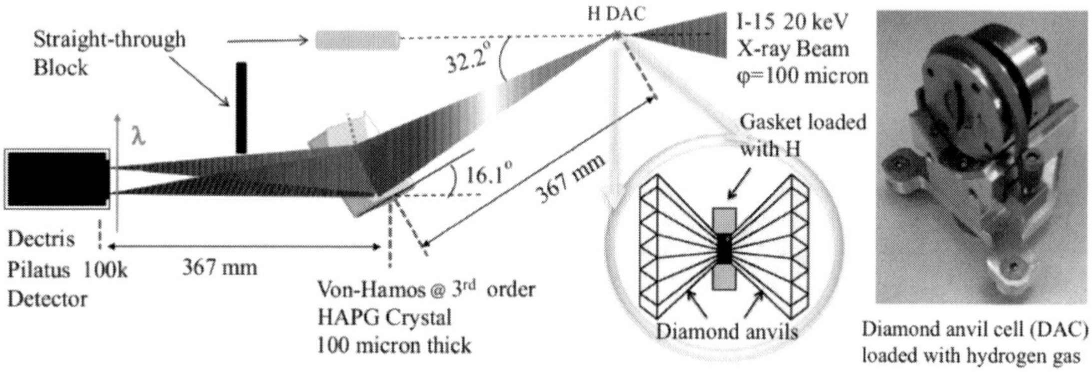

Figure 1. Apparatus arrangement.

The pressure within the anvil cell was measured through the well known relationship for fluorescence of ruby under pressure [8] (a small ruby probe was buried inside the anvil cell together with the hydrogen sample and the fluorescence was measured using an optical laser and a calibrated spectrometer). Since the pressure and temperature of the hydrogen in the anvil cell are known (room temperature, T = 300 K) the electron density could be determined from the equation of state (EOS) data for compressed hydrogen experimentally determined by ref. [5] and was found to be 6.0 x 10^{22} cm^{-3} for P_1 and 7.9 x 10^{22} cm^{-3} at P_2 at which hydrogen is fully degenerate (see table 1).

Table 1. Physical parameters in compressed hydrogen. The Wigner-Seitz radius given by $d = (3/(4\pi n_e))^{1/3}$ is the inter-particle spacing at electron density n_e and a_B is the Bohr radius.

	P_1 = 2.8 GPa (liquid)	P_2 = 6.4 GPa (solid)
Brueckner parameter, $r_s = \frac{d}{a_B}$:	2.99	2.74
Fermi energy, $E_F = \frac{\pi^2\hbar^2}{2m_e}\left(\frac{3n_e}{\pi}\right)^{2/3}$:	5.6 eV	6.7 eV
Degeneracy parameter, $\theta = \frac{k_B T}{E_F}$:	0.0046	0.0039
Plasma frequency, $\omega_p = \left(\frac{n_e e^2}{\varepsilon_0 m_e}\right)^{1/2}$:	9.12 eV	10.4 eV

3. Reduction Analysis

The weak H scattering compared to the diamond is a result of the small sample volume needed to achieve high compression of the hydrogen gas within the larger anvils and the fact that each C atom has more electrons to scatter from than the H atoms. In order to compensate for the dominant Compton scattering from the diamond anvils, we retrieved the hydrogen scattering data in the form of a ratio of spectra from hydrogen at P_1 and P_2 through a careful reduction analysis procedure using the measured signal referenced to the empty cell.

The ratio of the 2.8 GPa to 6.4 GPa spectra was computed using the following equations. The empty cell scattering spectrum corresponds to:

$$I_0 = B \cdot S_{anvil} \tag{1}$$

And the scattering spectrum of hydrogen loaded DAC at pressure P:

$$I_P = A_P \cdot (S_{anvil} + S_H^P) \tag{2}$$

Where A and B are calibration constants, which depend on the absolute response of the measuring apparatus and their ratio is a well defined quantity given by the ratio of average background noise levels of the empty and H loaded DAC spectra, and S_{anvil} and S_H^P are the dynamic structure factors from the diamond anvils only and hydrogen at pressure P, respectively:

$$\frac{S_H^{P_1}}{S_H^{P_2}} = \frac{(B/A_{P_1})(I_{P_1}/I_0) - 1}{(B/A_{P_2})(I_{P_2}/I_0) - 1} \tag{3}$$

Since the quantity on the RHS of equation (3) is directly obtained from the experimental data using the measured spectra from the empty cell and the DAC's loaded at the two pressures, the ratio $S_H^{P_1}/S_H^{P_2}$ is thus well defined.

4. Local Field Correction

Since we only obtained the ratio of the scattering signals of hydrogen at the two different pressures, the dynamic structure at a given pressure has to be retrieved from this ratio using a known result for the other pressure. Thus, in the next step, the experimentally determined ratio was multiplied by the spectrum calculated for P_2 obtaining the response at 2.8 GPa (blue line in figure 2). P_2 has been chosen, because at the higher pressure we have polycrystalline solid hydrogen (hexagonal close packed) [9]. The electrons are effectively free as they satisfy the Bloch functions for electronic motion inside the periodic potentials within the crystalline lattice of localized ions. Since, in our experiment, k lies outside the Brillouin zone boundary, the lattice does not influence scattering from the electrons. This phase is also more degenerate (lower degeneracy factor $\Theta = k_B T/E_F$) resulting from higher Fermi energy making the effects of finite temperature less important. Therefore the calculation of the scattering spectrum for P_2 can be justifiably carried out using the Random Phase Approximation (RPA) at T = 0 K [10]. The electron-electron dynamic structure factor for electrons interacting via Coulomb forces is given by the Fluctuation Dissipation Theorem, for $k_B T \ll \hbar\omega$:

$$S_H^P(k,\omega) = \frac{\varepsilon_0 \hbar k^2}{\pi e^2 n_e} \cdot \mathrm{Im}\left[\frac{-1}{\varepsilon(k,\omega)}\right] \tag{4}$$

where n_e is the electron density, e the electron charge, ε_0 the permittivity of free space, k is the magnitude of the change in wavevector due to the scattering and ω is the inelastic energy exchange. The dielectric function $\varepsilon(k,\omega)$ relates to the electron susceptibility $\chi(k,\omega)$ as $\varepsilon(k,\omega) = 1 - (e^2/\varepsilon_0 k^2) \cdot \chi(k,\omega)$ [13]. The electron susceptibility can also be written as:

Journal of Physics: Conference Series **244** (2010) 042014 doi:10.1088/1742-6596/244/4/042014

$$\chi(k,\omega) = \frac{\chi_0(k,\omega)}{1 + \frac{e^2}{\varepsilon_0 k^2} G(k,\omega)\chi_0(k,\omega)} \tag{5}$$

where χ_0 is the Lindhard susceptibility at T = 0 K and $G(k,\omega)$ is the local field correction [13], which is a measure of the effects of strong coupling due to higher order interaction terms and incorporates the short-range correlations, both neglected by the RPA. For $G(k,\omega) = 0$, $\chi(k,\omega) = \chi_0(k,\omega)$ and the RPA is reproduced. While the dynamic response for liquid hydrogen is obtained from Eq. (3) using the RPA approximation for the solid hydrogen, equation (3) can also be directly applied to retrieve the liquid response if the local field is known. In case of liquid hydrogen, the theoretical value of the local field correction can be estimated using the Utsumi-Ichimaru equations in the static approximation to be $G(k,\omega) = G(k) = 0.91$ [11], and the spectrum calculated using this value of G corresponds to the dashed green line in figure 2.

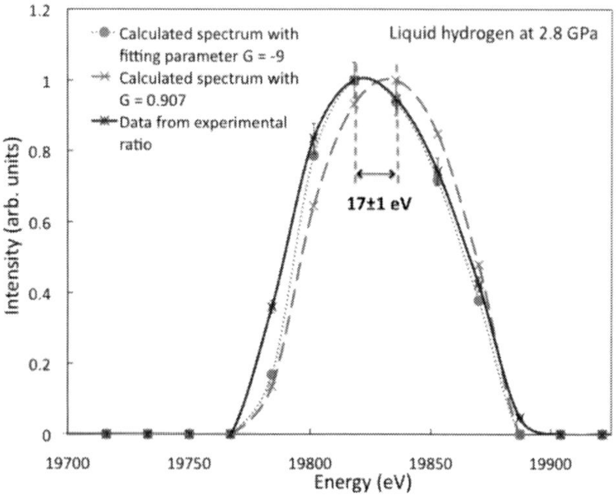

Figure 2. Normalized plots of the electron-electron feature of the scattering spectrum for H at 2.8 GPa.

There is a noticeable shift of $\sim 17 \pm 1$ eV between the peaks of the calculated spectrum using the Utsumi-Ichimaru theory and the data obtained from experiment (assuming a nearly free electron gas for the solid H), which is a result of our assumptions breaking down in the case of liquid hydrogen, where the electrons are no longer free and short range correlations can become important. The calculated response overlaps with the experimental data for $G(k = 5.59 \pm 0.19$ x 10^{10} m^{-1}, $\omega \sim 130$ eV$) = -9 \pm 1.5$ (red dots, figure 2). However, since $\omega \sim 130$ eV is significantly larger than plasma frequency ω_p, static approximations, such as the one of Utsumi-Ichimaru, may not be applicable here. Our results are the first experimental measurement of local field effects across the liquid to solid phase transition in hydrogen. They indicate a much stronger effect of correlations than suggested by fluid theories. This is important since the local field correction can be linked to dynamic properties of the compressed material, namely the isothermal compressibility and conductivity [11]. Building upon the results presented in this study, we would propose to expand and improve the XRTS method at free-electron X-ray laser facilities (FEL), where laser-driven DAC's [4] can be both heated and further compressed to the warm dense matter regime relevant to the planetary interior regime, and studied with a high-power coherent X-ray source.

References

[1] Glenzer S H and Redmer R 2009 *Rev. Mod. Phys.* **81** 1625-63
[2] Hicks D G et al. 2009 *Phys. Rev. B* **79** 014112

[3] Loubeyre P et el. 2009 *High Pressure Research* **24** 25-31
[4] Eggert J et el. 2008 *Phys. Rev. Lett.* **100** 124503
[5] Mao H et al. 1988 *Science* **239** 4844
[6] Glenzer et al. 2003 *Phys. Rev. Lett.* **90** 17
[7] Legall H et al. 2009 *J. Appl. Cryst.* **42** 572-9
[8] Syassen K 2008 *High Pressure Research* **28** 75-126
[9] Gregoryanz E et al. 2003 *Phys. Rev. Lett.* **90** 17
[10] Arista N R and Brandt W 1984 *Phys. Rev. A* **29** 3
[11] Ichimaru S 1982 *Rev. Mod. Phys* **54** 4
[12] Gregori G et al. 2007 *High Energy Density Physics* **3** 99-108
[13] Gregori G et al. 2003 *J. Phys. A: Math. Gen.* **36** 5971-80

X-ray Thomson Scattering Measurement in shock-compressed Beryllium

H J Lee, T Döppner[1], P Neumayer[1], J Castor[1], R W Falcone, C Fortmann[2], B A Hammel[1], A L Kritcher, O L Landen[1], R W Lee[1], D D Meyerhofer[3], D H Munro[1], R Redmer[2], S P Regan[3], S Weber[1], R Wallace[1], S H Glenzer[1]

University of California Berkeley, Berkeley, CA 94709, USA
[1] Lawrence Livermore National Laboratory, Livermore, CA, USA
[2] Institut für Physik, Universität Rostock, D-18051 Rostock, Germany
[3] Laboratory for Laser Energetics, Rochester, NY, USA

Email: haelee@slac.stanford.edu

Abstract. We have measured x-ray scattering spectra of shock-compressed beryllium. In these studies, 6 keV x-rays have been produced to perform spectrally resolved measurements of the plasmas employing both non-collective and collective scattering at the Omega laser facility. Laser beams with intensities of $10^{14}<I<10^{15}$ Wcm^{-2} and 3-4 ns pulse width have irradiated beryllium to launch a strong shock and compress the target homogeneously. The x-ray scattering measurements of the shock-compressed matter have shown inelastic Compton and Plasmon scattered spectra indicating a dense Fermi-degenerate plasma state with a Fermi energy above 30 eV with temperatures in the range of 10–15 eV. These findings reflect compression by a factor of 3 in agreement with radiation-hydrodynamic modeling.

Introduction

X-ray Thomson scattering technique has shown great success to determine plasma properties in dense matters [1-4]. As plasma becomes denser, one reaches regimes that are found in high energy density physics conditions, planetary interiors, and inertial fusion schemes. Since the study of dense plasmas has been severely limited due to the lack of an appropriate x-ray source and a lack of robust diagnostic technique, direct and accurate measurements of thermodynamic and physical properties by x-ray Thomson scattering diagnostics are promising to test dense plasma modeling and to address fundamental physics questions in dense matter [5].

In x-ray Thomson scattering study, the dense plasma created by intense nanosecond laser is irradiated by unpolarized x-rays with energy E_0 (or wavelength λ_0) and scattered photons are detected by a Bragg crystal spectrometer coupled to CCD camera. During the scattering process, the incident photons transfer the momentum $\hbar k$ and the energy $\hbar\omega = (\hbar^2 k^2/2m_e)$ to the electrons. The magnitude of the k-vector is given by $k=4\pi\,(E_0/hc)\sin(\theta/2)$ with θ, scattering angle. Here, since k determines the scaling length by $\lambda^*\sim 2\pi/k$, the ratio to the screening length λ_s distinguishes the scattering regime; collective and non-collective regime. The dimensionless scattering parameter α is defined as

$$\alpha = \frac{1}{k\lambda_s} \sim \frac{\lambda^*}{\lambda_s}$$

When the scattering length is larger than the screening length ($\alpha>1$), spatial and temporal correlation of electrons become important and scattered radiation reflects the collective fluctuation. In case of $\alpha<1$, the probe radiation is scattered from individual particle fluctuation of its velocity. Experimentally, the screening length by k can be determined through changing E_0 (6.18 keV) and θ (25° and 90°).

Forward and backward scattering

X-ray Thomson scattering experiment is understood in both forward and backward scattering geometry; that is collective and non-collective scattering regime. In the forward scattering geometry, the density fluctuation in the dense plasma behaves collectively and oscillates at a frequency

$$\omega_{pl}^2 = \omega_p^2 + 3k^2 v_{th}^2\left(1 + 0.088 n_e \Lambda_e^3\right) + \left(\frac{\hbar k^2}{2m_e}\right)^2$$

where $\omega_p=(n_e e^2/\varepsilon_0 m_e)^{1/2}$ is the plasma frequency, $v_{th}=(k_B T_e/m_e)^{1/2}$ is the thermal velocity and $\Lambda_e=\hbar/(2\pi m_e k_B T_e)^{1/2}$ is the thermal wave length [2]. Here small value of k works in this regime. In such a regime strong intensity of collective motion governs the spectrum of fluctuations, that is a Plasmon feature.

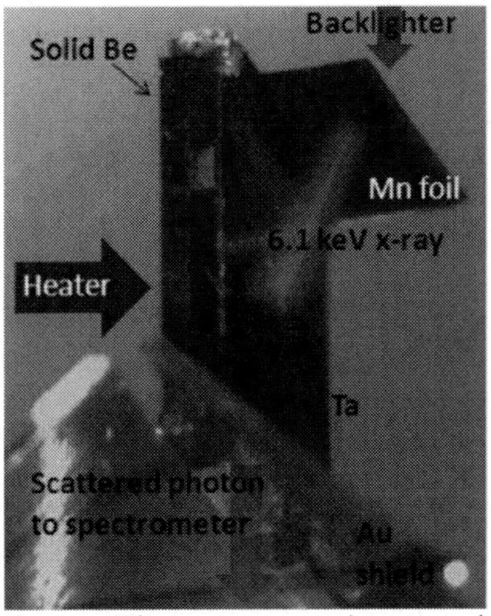

Figure 1. A real target picture of x-ray Thomson scattering experiment. X-ray scattered spectra are observed in the downward direction with a gated highly oriented pyrolytic graphite crystal spectrometer.

In the backward scattering geometry, individual particle-like fluctuation is the major contribution and Compton feature is strong in this regime. The frequency shift of a scattered photon by a free electron is determined by the Compton and the Doppler effects

$$\omega = \frac{-\hbar k^2}{2m_e} \pm kv$$

In this Compton scattering regime, the scattering process is non-collective and the spectrum shows the Compton down-shifted response that is broadened by the thermal motion of the electrons and that reflects the electron velocity in the direction of the scattering vector k. In shock-compressed Fermi-degenerate plasmas, the width of Compton scattering spectrum is proportional to the Fermi-energy, $E_F^{1/2}$, and hence to $n_e^{1/3}$[5]. The intensity ratio of Compton to elastic peak in Fermi-degenerated plasma is sensitive to the ion temperature.

The total electron density fluctuation function can be described as total dynamic structure factor and further explained as scattering cross section. The total electron dynamic structure factor $S(k,\omega)$ includes bound electron and ion contributions in addition to the Compton scattering component by free electrons.

$$S(k,\omega) = |f(k) + q(k)|^2 S_{ii}(k,\omega) + Z_f S_{ee}^0(k,\omega) + Z_b \int d\omega' \bar{S}_{ce}(k,\omega-\omega') S_s(k,\omega')$$

Here, the $S_{ii}(k,\omega)$ denotes dynamic ion-ion structure factor, ion form factor $f(k)$ is for scattering from the tightly bound electrons, $q(k)$ is for screening cloud of valence electrons surrounding the ions, $S_{ee}^0(k,\omega)$ is the dynamic structure factor of free electrons, Z_f is ionization state, and the last term is for the bound-free contribution.

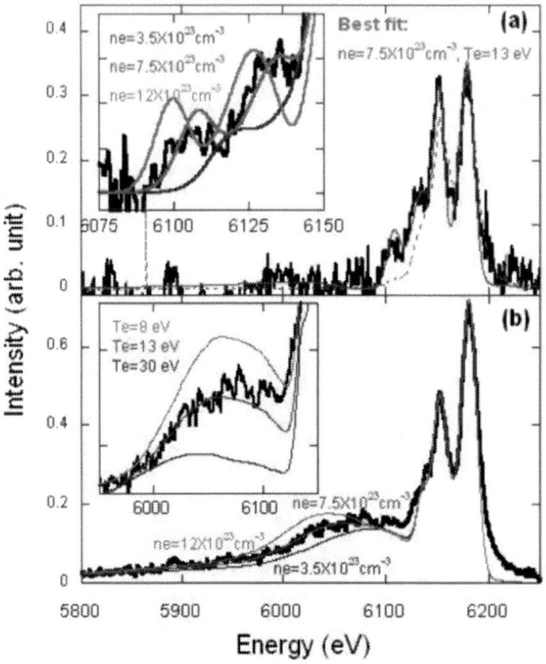

Figure 2. (a) Plasmon and (b) Compton scattering spectrum from compressed Fermi-degenerate beryllium. Sensitivity analysis shows the electron density 7.5×10^{23} cm^{-3} with an error bar of $\pm$ 6 % and electron temperature 13$\pm$3 eV.

X-ray Thomson scattering on shock-compressed plasmas

With powerful lasers in Omega laser facility, shock-compression resulting in a Fermi-degenerate plasma was achieved and measured in both collective and non-collective scattering regime. Figure 1 shows a real target picture of forward scattering. Twelve heater beams smoothed with phase plates irradiate 250 μm thick solid Be in ~1 mm diameter focal spot. Laser beams with intensities of $10^{14} < I < 10^{15}$ Wcm^{-2} and 3-4 ns pulse width launch a strong shock and compress the target homogeneously. Delayed backlighter beams are focused on Mn foil to produce 6.18 keV x-ray that can penetrate through compressed matter reaching $n_e \sim 10^{24}$ /cm^3. Ta sheet with about 400 μm aperture between Be and Mn foil limits the scattering angle 25° ± 7° and confine the intersection area to the compressed Be. (Change of the incident angle of x-rays by putting Mn foil parallel to Ta sheet makes scattering angle of 90° ± 7° available.) The scattered photons are collected using graphite Bragg crystal spectrometer coupled to x-ray framing camera and charge coupled device. Two gold shields block huge bremsstrahlung from Be and Mn foils so that only scattered photons from dense plasma are detected.

Radiation hydrodynamic calculations indicate that these irradiation conditions compress a target homogeneously at pressure in the range of 20-35 Mbar and shock breaks out about 4.5 ns. At 4.5ns, the calculation shows compression by a factor of 3 with $7\times10^{23} < n_e < 8\times10^{23}$ cm^{-3} for 3×10^{14} Wcm^{-2}. The simulated temperature is between $12 < T_e < 14$ eV in the dense beryllium.

Figure 2 (a) and (b) show x-ray scattering spectra from compressed beryllium from forward and backward scattering geometry, respectively. In Fig. 2 (a) inelastic Plasmon feature due to a collective free electron motion is clearly stronger than satellite elastic response of Mn x-ray source spectrum in dash line. Sensitivity analysis using the theoretical form factor $S(k,\omega)$ (the inset of Fig. 2 (a)) provides that the position of Plasmon response is sensitive to the electron density of dense matter and an accurate measurement of the electron density of $n_e = 7.5\times10^{23}$ cm^{-3} with an error bar of ± 6 %.

The Compton spectrum of Fig. 2 (b) shows parabolic shape in compressed dense state indicating Fermi-degenerate plasma. The Compton scattering spectrum directly reflects the electron distribution function. For the analysis, we assume T_e=T_i and Z=2 consistent with calculations and with the measurements from isochorically heated Be [2]. The comparison with theoretical calculations shows that the width of Compton peak is sensitive to the electron density change and the intensity of the peak is sensitive to the electron temperature. Density and temperature obtained in this way are n_e = 7.5 x 10^{23} cm^{-3} and T = 13 eV for a Fermi temperature of E_F = 30 eV; for a Fermi-degenerate system the width of the Compton spectrum yields the Fermi energy, $E_F=\hbar^2(3\pi^2 n_e)^{2/3}/(2m_e)$. These results are well agreed with the radiation hydrodynamic calculations. Structure factors according to the experimental conditions are summarized in table 1.

time (ns)	P (Mbar)	ne (×10^{23}/cc)	Te (eV)	Z	Sii (f+q)^2	f	q	Sii	α	k (Å^{-1})	θ
4.6	29	7.5	13	2	1.783	1.67	0.2	0.5	0.48	4.43	90
4.4	29	7.5	13	2	0.778	1.97	1.4	0.07	1.5	1.35	25

Table 1. Experimental conditions (scattering angle and vector) and analyzed results.

Conclusion

In this paper, we have demonstrated x-ray Thomson scattering diagnostics by measuring shock-compressed beryllium in both collective and non-collective regime. The agreement with radiation hydrodynamic calculation indicates the accurate and direct measurement of the degeneracy and adiabat in these single-shocked foils. This method allows application to a number of dense matter questions including future applications to measure degeneracy and adiabat, e.g., in ICF and high energy density physics studies.

Acknowledgments

This work was performed under the auspices of the U.S. Department of Energy by Lawrence Livermore National Laboratory under contract No. DE-AC52-07NA27344 and supported by the National Laboratory User Facility program.

References

[1] S. H. Glenzer *et al.*, Phys. Rev. Letters **90**, 175002 (2003).
[2] S. H. Glenzer *et al.*, Phys. Rev. Letters **98**, 065002 (2007).
[3] A. L. Kritcher *et al.*, Science **322**, 69 (2008).
[4] T. Doppner *et al.*, High Energy Density Physics **5**,182 (2009).
[5] O. L. Landen *et al.*, J. Quant. Spectrosc. Radiat. Transfer **71**, 465 (2001).
[6] H. J. Lee *et al.*, Phys. Rev. Letters **102**, 115001 (2009).

Designfor solid-state Rayleigh-Taylor experiments in tantalum at Omega

S.M. Pollaine, B.A. Remington, H.S. Park, S.T. Prisbrey, R.M. Cavallo

Lawrence Livermore National Laboratory, L-477, Livermore, CA

pollaine@llnl.gov

Abstract. We have designed an experiment for the Omega - EP laser facility to measure the Rayleigh-Taylor (RT) growth rate of solid-state Ta samples at ~1 Mbar pressures and very high strain rates, 10^7-10^8 s^{-1}. A thin walled, hohlraum based, ramp-wave, quasi-isentropic drive has been developed for this experiment. Thick samples (~50 um) of Ta, with a pre-imposed sinusoidal rippled on the driven side, will be accelerated. The ripple growth due to the RT instability is greatly reduced due to the dynamic material strength. We will show detailed designs, and a thorough error analysis used to optimize the experiment and minimize uncertainty.

This work was performed under the auspices of the Lawrence Livermore National Security, LLC, (LLNS) under Contract No. DE-AC52-07NA27344.

1. Introduction

There are large uncertainties in our understanding of material strength at high strain rates and pressures. Models agree where data exists, below strain rates of 10^5 s^{-1}, but diverge at strain rates larger than 10^5 s^{-1}. Somewhere between 10^5 s^{-1} and 10^8 s^{-1}, the movement of dislocations switches from the thermal activation regime to the phonon drag regime, with a corresponding change in the slope of d ln(strength) / d ln(strain rate). We have chosen to measure material strength by its effect on the growth rate of Rayleigh-Taylor instabilities [1]. The yield strength acts like an effective viscosity that slows down the Rayleigh-Taylor growth rate [2]. Classical growth, with out strength, has a growth rate given by $\gamma = \sqrt{(gkA)}$. In the presence of strength, the growth rate is $\gamma = \sqrt{(gkA+\nu^2k^4)}-\nu k^2$ with an effective viscosity of $\nu=Y/(\rho\varepsilon\sqrt{6})$ For comparison, liquid Al has a viscosity of 0.01 poise, while a strength of 3 kb at a strain rate of 10^7 s^{-1} has an effective viscosity of 120 poise.

To keep the material solid, we need to keep it close to the isentrope while bringing it up to the desired pressure. Thus we use a technique developed by Barnes [3,4], in which we shock up a reservoir to high pressure, let it rarify across a vacuum gap, and have it gently stagnate against the sample, driving it along a quasi-isentrope.

2. Experiment

We have designed a 1 Mb Ta Rayleigh-Taylor Omega experiment driven by a thin-walled hohlraum. Fig. 1 shows the hohlraum with the laser rays from 3 distinct cone angles, along with the backlighter. The hohlraum wall consists of a 1 μm layer of Au backed by 160 μm of CH. 19.2 kJ of 0.35 μm light enter the hohlraum in 1 ns, bringing the internal temperature up to 135 eV and dropping to about 50 eV after the laser turns off (fig. 2). This temperature drives our target, which consists of a reservoir made of 25 μm Be and 200 μm C_4H_3Br, a 400 μm gap, and a target consisting of 12 μm CH_2 preheat shield, 50 μm Ta, and 500 μm LiF tamper. The interface between the CH_2 heat shield and the Ta has a pre-imposed ripple with amplitude 1.5 μm and wavelength 50 μm to seed the Rayleigh-Taylor growth. The Ta reaches a pressure of 1.3 Mb about 35 ns after the start of the laser pulse.

The hohlraum had only 1 μm of Au, instead of a thick Au wall, because previous hohlraum shots in March 2008 and May 2008 showed that the temperature of the hohlraums stayed unexpectedly high after the laser turned off. This elevated temperature launched successive shocks into the target. The thinner wall becomes more transparent after about 10 ns, reducing the temperature and the late time shocks.

On June 1, 2009, we got our first backlit image from a Rayleigh-Taylor shot, taken 70 ns after the start of the laser pulse. The ripples grew by a factor of 2± 0.3, less than expected.

3. Error analysis

To analyze the data, we need to measure the drive, the ripple contrast, the MTF of the radiography image, and the spectrum of x rays that produced the image. We found that the growth factor is roughly inversely proportional to the Ta strength, as shown in Fig. 3 so that the percent error in growth factor is the same as the percent error in strength.

The drive temperature is measured with Dante[7], and these measurements are good to about 1.2 eV, or 3.5% in flux. Late time temperature measurements are good to about 2.6 eV out of 40 eV, or about 30% in flux. Collectively, these errors add up to about 9% uncertainty in the drive. This can be reduced by measuring the velocity of the back surface with a VISAR[8] interferometer. Velocity can be measured to about 1%, resulting in a 2% error in the drive. Although we plan to measure velocity simultaneously with the x-ray radiography of the ripple growth, to date we have used separate drive shots to measure the drive, correcting for changes in the overall laser energy. This introduces an error due to fluctuations from shot to shot of about 4% in drive, which is still less than the uncertainty from the temperature measurements.

The MTF is about 75% at a wavelength of 50 μm. An uncertainty of 10% in MTF translates into a 10% uncertainty in growth factor, and 10% uncertainty in material strength.

Fig. 4 shows 2 possible spectra for the backlighter source. Each spectrum is multiplied by the transparency of the target as a function of photon energy, then by the response function of the detector. The derivative of the final spectrum with respect to Ta thickness is plotted as a function of Ta thickness in fig. 5. Since we're using the contrast in x-ray image to deduce the growth of the RT ripples, any error in how much a Ta ripple causes a difference in x-ray brightness directly impacts our estimate of ripple growth. Fig. 5 shows that there's about a 6% difference between the two spectra. Furthermore, any error in Ta thickness affects the spectrum as well, with each extra micron of Ta reducing the overall x-ray ripple contrast by 2.6%. Thus a 3 μm error in Ta thickness results in an 8% error in growth factor.

Table I below lists many parameters of the experiment that affect the estimate of the growth factor, the extent of their affect, the uncertainty of each parameter, and the contribution of that uncertainty to the total uncertainty in the growth factor, and thus the material strength. If the errors were all uncorrelated, then the total uncertainty in growth factor would just be the quadrature sum of the individual errors, about 16%. However, several of the errors are correlated, and these correlations must be taken into account. The three hohlraum temperatures, peak, mid time and late time, are correlated since any problems with a Dante channel will affect all three temperatures. Although the correlations are not 100%, we will assume they are to be conservative in our overall estimate. Thus we will add these three uncertainties linearly to get 9% overall uncertainty due to hohlraum temperature

The other main correlations are with the Ta thickness. Ta thickness has two effects on the overall error budget. A thicker Ta sample has more mass, thus moves more slowly, so the RT ripple grows more slowly. A thicker Ta sample also affects the spectrum of the backlighter used to measure the RT ripple. In addition, the glue thickness between the Ta and LiF tamper is estimated by measuring the thickness of the entire target – the CH_2 heat shield, the Ta, the glue and the LiF thickness is

Journal of Physics: Conference Series **244** (2010) 042016 doi:10.1088/1742-6596/244/4/042016

measured, and the CH_2, Ta and LiF thicknesses are subtracted to get the glue thickness. Thus the error in glue thickness is negatively correlated with the Ta thickness. Therefore, these three error contributions from the Ta thickness are linearly added to get (1.14+1.08+2.6)*3 = 14.5%. The remaining errors are assumed independent, and are added in quadrature. This yields a total error estimate in the growth factor of $\sigma = 22.6\%$. The probability of the error in growth factor, and thus in yield strength, being greater than x% is then $\mathrm{erfc}(x/\sigma/\sqrt{2})$. Thus there is a 27% chance that the error in yield strength is greater than 25%, as shown in fig. 6.

There are several things we can do to reduce this overall error of 22.6%. With more careful measurements, we could reduce the MTF error from 10% to 5%. We plan to use simultaneous VISAR measurements with our Rayleigh-Taylor growth measurements, which will eliminate the uncertainty due to temperature and shot-to-shot fluctuations in drive. We also will be imaging Ta steps of 35, 45, 55 and 65 μm thickness so we can interpolate the Ta thickness variations from the x-ray image of the ripple. This replaces the uncertainty in backlighter spectrum with the uncertainty in Ta step thickness and density of about 5-10%. With these improvements, the overall uncertainty reduces to 16.6%. Then the chance that the error in yield strength is greater than 25% reduces from 27% to 13%. Fig. 6 shows how the error probability improves with these changes.

Table I. **Affect on growth factor, expected error, and total error of growth factor due to each uncertainty**

Description	$\frac{\partial\sigma}{\partial x_i}(\frac{\%}{unit})$	δx_i	$\delta\sigma_i$ (%)
Be thickness (μm)	-0.2	0.5	0.1
Glue thickness (μm)	-0.15	2	0.3
C_4H_3Br thickness (μm)	-0.14	1	0.14
Gap thickness (μm)	-0.08	5	0.4
CH_2 thickness (μm)	-0.27	2	0.54
Ta thickness (μm)	-1.14	3	3.4
Glue thickness (μm)	+1.08	2	2.2
LiF thickness (μm)	-0.011	2	0.02
Ripple amplitude (μm)	0.75	0.15	0.1
Peak Tr (eV)	0.83	1.15	0.95
Mid-time Tr (eV)	1.2	1.15	1.4
Late-time Tr (eV)	2.6	2.55	6.6
Spatial nonuniformity (%)	6.9	1	6.9
Timing errors (ns)	1.1	1	1.1
EOS errors (EOP/LEOS)	2.0	na	2.0
1.3*strength surface layer (μm)	0.24	1	0.24
Thermal conductivity (%)	-0.008	100	0.8
Ripple contrast (PSL)	12.	~.02	0.24
MTF error (%)	1.0	10	10
Photon statistics			2.3
Backlighter spectrum			5.8
Spectrum, Ta thickness (μm)	-2.6	3	8
VISAR error (%)	0.98	2	2.0
VISAR drive ≠ RT drive (%)	0.71	6	4.2
Heat shield strength			0
Grain structure (anisotropy)			0.4
Model dependence			3.2

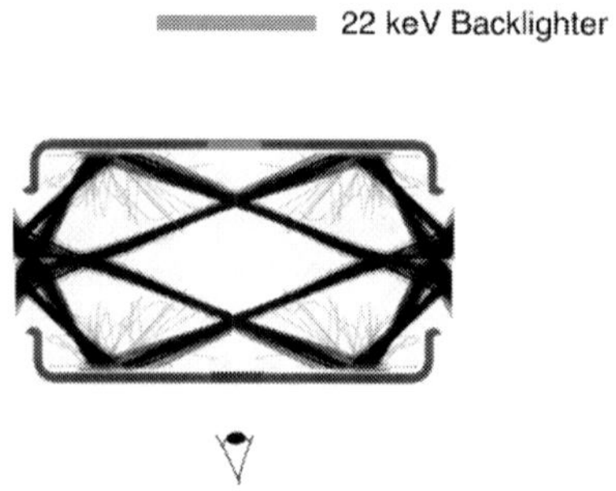

Figure 1 Hohlraum with backlighter and laser rays

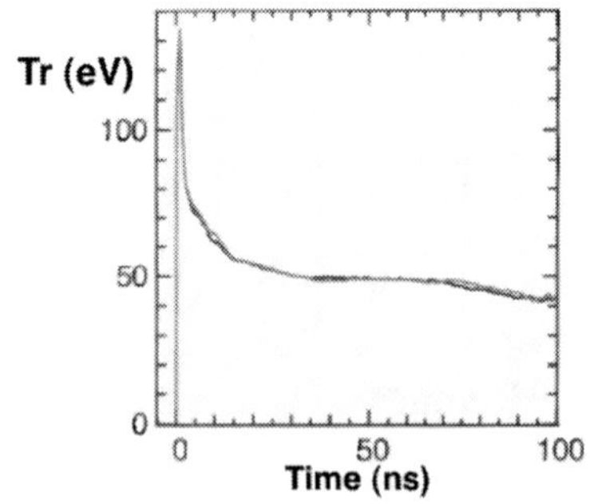

Figure 2 Hohlraum radiation temperature vs time

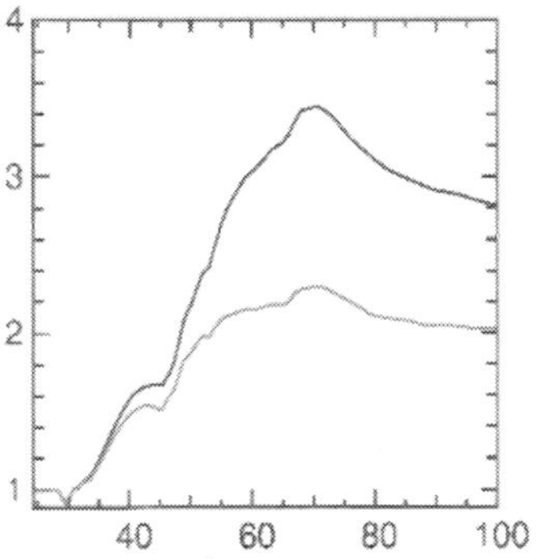

Figure 3 Growth factor vs. time (ns) for Y_0 = 15.4 kb (black) and 23.1 kb (red) in the Steinberg-Guinan

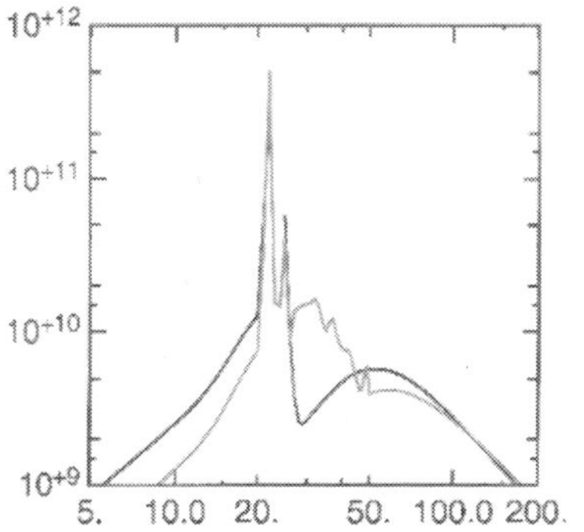

Figure 4. Two possible backlighter spectra, intensity vs. energy (keV)

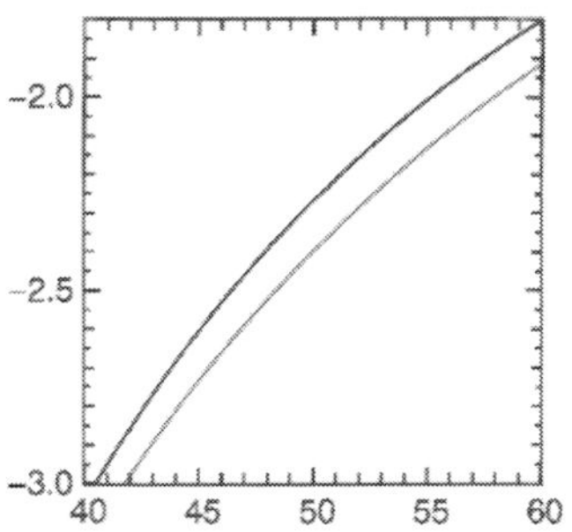

Figure 5. Change in detector response per micron of Ta for the two spectra of fig. 4.

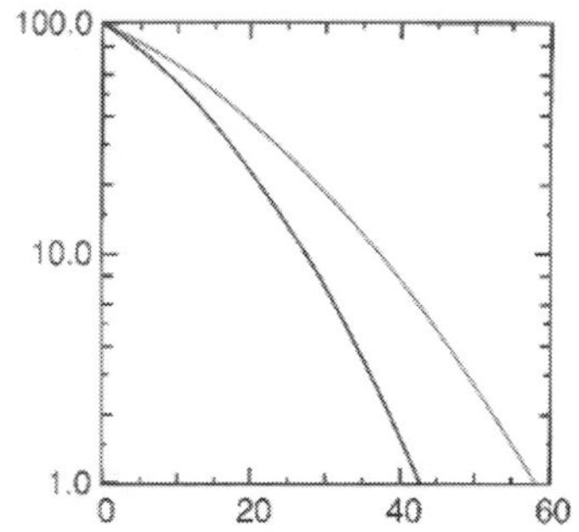

Figure 6. Probability that total error in strength is greater than the value on the x-axis (%) for a total error σ = 22.6% (red) and 16.6% (black).

References

[1] K.T. Lorenz, J.J. Edwards, S.G. Glendinning, A.F. Jankowsky, J. McNaney, S.M. Pollaine, and B.A. Remington, "Accessing ultrahigh-pressure, quasi-isentropic states of matter", *Physics Plasmas* **12**, 056309 (2005).

[2] J.D. Colvin, M. Legrand, B.A. Remington, G. Schurtz, S.V. Weber, "A Model for Instability Growth in Accelerated Solid Metals", *J. Appl. Phys.* **93**, 5287 (2003)

[3] J.F. Barnes, P.J. Blewett, R.G. McQueen, K.A. Meyer, and D. Venable, *J. Appl. Phys.* **45**, 727 (1974).

[4] M.J. Edwards et al, *Phys. Rev. Lett.* **92**, 75002 (2004)

[5] D.J. Steinberg, S.G. Cochran and M.W. Guinan, "A Constitutive Model for Metals Applicable at High Strain Rate", *J. Appl. Phys.* **51**,1498 (1980).

[6] D.L. Preston, D.L. Tonks, and D.C. Wallace, "Model of plastic deformation for extreme loading conditions", *J. Appl. Phys.* **93,** 211 (2003).

[7] H.N. Kornblum, R.L. Kauffman and J.A. Smith, Measurement of 0.1–3-keV X-rays from laser plasmas, *Rev Sci Instrum* **57** (1986), pp. 2179–2181, and Dewald E L *et al, Rev. Sci. Instrum.* **75,** 3759 (2004).

[8] P.M. Celliers, D.K. Bradley, G.W. Collins, D.G. Hicks, T.R. Boehly, and W.J. Armstrong, *Rev. Sci. Instrum.* **75**, 4916 (2004).

Journal of Physics: Conference Series **244** (2010) 042017 doi:10.1088/1742-6596/244/4/042017

Inferring the Electron Temperature and Density of Shocked Liquid Deuterium Using Inelastic X-Ray Scattering

S P Regan[1], P B Radha[1], T R Boehly[1], T Doeppner[2], K Falk[3], S H Glenzer[2], V N Goncharov[1], G Gregori[3], O L Landen[2], R L McCrory[1,4], D D Meyerhofer[1,4], P Neumayer[2], T C Sangster[1], and V A Smalyuk[1]

[1] Laboratory for Laser Energetics, University of Rochester, 250 East River Road, Rochester, NY 14623-1299
[2] Lawrence Livermore National Laboratory, Livermore, CA 94550, USA
[3] Oxford University, Oxford OX1 3PU, UK
[4] also Departments of Mechanical Engineering and Physics, University of Rochester, Rochester, NY 14623-1299

E-mail: sreg@lle.rochester.edu

Abstract. An experiment designed to launch laser-ablation-driven shock waves (10 to 70 Mbar) in a planar liquid-deuterium target on the OMEGA Laser System and to diagnose the shocked conditions using inelastic x-ray scattering is described. The electron temperature (T_e) is inferred from the Doppler-broadened Compton-downshifted peak of the noncollective $(\alpha_s = 1/k\lambda_D > 1)$ x-ray scattering for $T_e > T_{Fermi}$. The electron density (n_e) is inferred from the downshifted plasmon peak of the collective ($\alpha_{scatter} > 1$) x-ray scattering. A cylindrical layer of liquid deuterium is formed in a cryogenic cell with 8-μm-thick polyimide windows. The polyimide ablator is irradiated with peak intensities in the range of 10^{13} to 10^{15} W/cm^2 and shock waves are launched. Predictions from a 1-D hydrodynamics code show the shocked deuterium has a thickness of ~0.1 mm with spatially uniform conditions. For the drive intensities under consideration, electron density up to ~5×10^{23} cm^{-3} and electron temperature in the range of 10 to 25 eV are predicted. A laser-irradiated saran foil produces Cl Ly$_\alpha$ emission. The spectrally resolved x-ray scattering is recorded at 90° for the noncollective scattering and at 40° for the collective scattering with a highly oriented pyrolytic graphite (HOPG) crystal spectrometer and an x-ray framing camera.

1. Introduction

An experimental platform to study inelastic x-ray scattering from shocked deuterium is being developed on the 60-beam, 30-kJ, 351-nm OMEGA Laser System [1]. Inelastic x-ray scattering [2,3] is a powerful diagnostic for equation-of-state (EOS) measurements in inertial confinement fusion (ICF). Laser-ablation-driven shock waves are launched into planar liquid-deuterium targets creating warm dense matter with conditions similar to an imploding shell of thermonuclear fuel in ICF during the shock propagation and acceleration phases. Diagnosing the plasma conditions of the shocked deuterium is important because the minimum energy needed to achieve ignition (E_{min}) depends on the plasma conditions in the shell [4]: The ratio of the pressure in the shell to the Fermi-degenerate pressure is defined as the shell's adiabat (α) and $E_{min} \sim \alpha^{1.8}$ [4]. A saran foil is irradiated with a group of tightly focused beams having an overlapped intensity of ~10^{16} W/cm^2 and producing Cl Ly$_\alpha$ emission at 2.96 keV. The electron density (n_e) and temperature (T_e) are inferred from the spectral line shapes of the inelastic x-ray scattering of the Cl Ly$_\alpha$ emission from the shocked deuterium. The T_e and average ionization (Z) of the shocked deuterium are inferred from the noncollective scattering (i.e., scattering from the electrons) [2] recorded at a 90° scattering angle. The n_e of the shocked deuterium is inferred from the collective scattering (i.e., scattering from electron plasma waves or plasmons) [3] recorded at a 40° scattering angle. The design of the experiment, including predictions of the shocked plasma conditions from the 1-D hydrodynamics code *LILAC* [5], is presented in Sec. 2. A brief discussion of the inelastic x-ray scattering is given in Sec. 3. Future experiments combining inelastic x-ray-scattering observations with shock-velocity measurements [6–9] are described in Sec. 4.

2. Experimental configuration

The experimental configuration is shown in Fig. 1 for the $\theta = 90°$ scattering geometry. The 8-μm-thick plastic ablator containing the planar layer of liquid deuterium is irradiated with a constant intensity UV laser drive with 10^{14} W/cm^2. The laser drive, formed with six pairs of beams staggered in time as shown in the lower right of Fig. 1, is uniform over a 0.5-mm diameter. A laser-ablation driven shock wave is launched through the liquid deuterium creating warm dense matter. Twelve tightly focused beams irradiate a saran backlighter with 10^{16} W/cm^2 generating Cl Ly$_\alpha$ emission (λ_0 = 4.188 Å), which is scattered at θ = 90° and detected with an x-ray framing camera (XRFC) outfitted with a HOPG crystal spectrometer [2,3]. The timing of the backlighter beams is also shown. The integration time of the x-ray scattering measurements is ~200 ps.

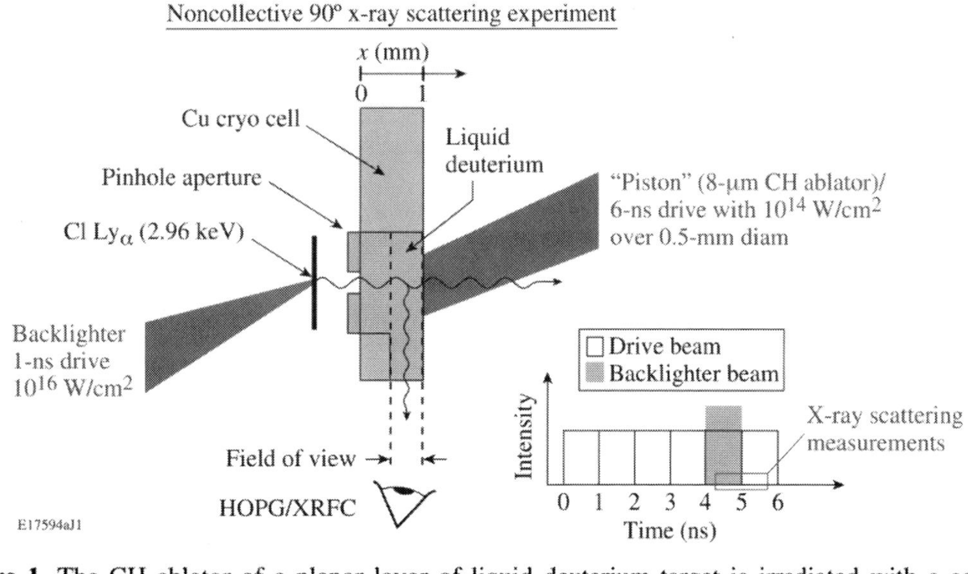

Figure 1. The CH ablator of a planar layer of liquid deuterium target is irradiated with a constant intensity (10^{14} W/cm^2) 6-ns UV laser drive. A laser-ablation-driven shock wave is launched through the liquid deuterium creating warm dense matter. Twelve tightly focused beams irradiate a saran backlighter with 10^{16} W/cm^2. Cl Ly$_\alpha$ emission scattered at 90° is detected with an x-ray framing camera (XRFC) outfitted with a HOPG crystal spectrometer. The timing of the drive and backlighter beams as well as the x-ray scattering observations are shown in the lower right.

The 1-D hydrodynamics code *LILAC* [5] is used to model the plasma conditions of the warm dense matter presented in Fig. 2. Laser absorption via inverse bremsstrahlung is calculated using a ray-trace algorithm. Electron thermal conduction is calculated using a flux-limited diffusion model. Radiative transport is calculated using the Los Alamos Astrophysical tables. The *SESAME* tables are used to determine the EOS. Uniform conditions are predicted in the ~100-μm-thick shocked region with $n_e = 2.0 \times 10^{23}$ cm^{-3}, $\rho \sim 0.8$ g/cm^3, $T_e = 22$ eV, $Z = 1$, and $P = 12$ Mbar.

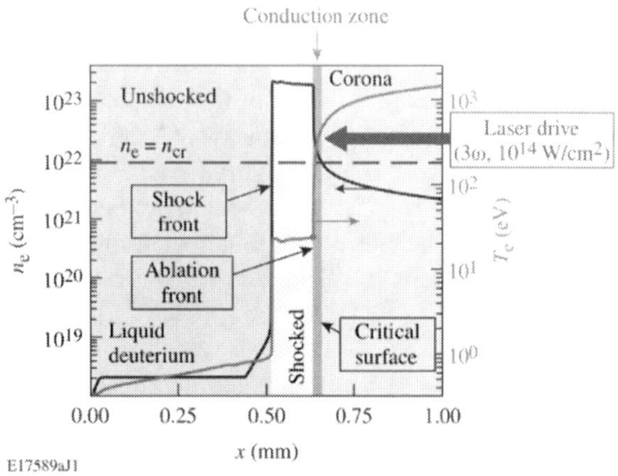

Figure 2. A laser-ablation-driven shock wave is launched in a liquid-deuterium target with a constant intensity laser drive of ~10^{14} W/cm^2. The shocked conditions at t = 5 ns are predicted with the 1-D hydrodynamics code *LILAC* [5]. Uniform conditions in the shocked region (100-μm-thick, 500-μm-diam cylinder) with $n_e = 2.0 \times 10^{23}$ cm^{-3}, $\rho \sim 0.8$ g/cm^3, T_e = 22 eV, and $P = 12$ Mbar are predicted.

3. Inelastic x-ray scattering

Inelastic x-ray scattering is defined as collective or noncollective depending on the scattering parameter $\alpha_s = 1/k\lambda_D$, where the probe wave number $k = 4\pi/\lambda_0 \sin(\theta/2)$ and λ_D is the Debye length. If $\alpha_s < 1$ the scattering is from the free electrons and it is referred to as noncollective [2,3]. Figure 3 shows the calculated x-ray scattering of Cl Ly$_\alpha$ emission ($\lambda_0 = 4.188$ Å) from electrons in the shocked deuterium for fixed electron density $n_e = 2.0 \times 10^{23}$ cm^{-3} and average ionization $Z = 1$, with the electron temperature varied from $T_e = 18$ eV to $T_e = 26$ eV. The scattering angle is $\theta = 90°$ and the scattering parameter is $\alpha_s = 0.61$. The Compton peak (inelastic) is on the low energy side of the Rayleigh peak (elastic) at 2.96 keV. The width of the Compton-downshifted $\left(\Delta E_C = \hbar^2 k^2/2m_e\right)$, Doppler-broadened component indicates the electron temperature and the ratio of the inelastic-to-elastic peaks indicates the average ionization (Z) of the plasma. All of the T_e-sensitive features shown in Fig. 3 can be resolved with the gated x-ray spectrometer. If $\alpha_s > 1$ the scattering is from the electron plasma waves or plasmons and is referred to as collective. Figure 4 shows the calculated x-ray scattering of Cl Ly$_\alpha$ emission from plasmons for fixed electron temperature $T_e = 22$ eV and average ionization $Z = 1$, with the electron density varied from $n_e = 1.7 \times 10^{23}$ cm^{-3} to $n_e = 2.3 \times 10^{23}$ cm^{-3}. To lowest order, the energy-downshifted plasmon feature is related to the electron plasma frequency $\omega_{pe} = \left(e^2 n_e/4\pi\varepsilon_0 m_e\right)^{1/2}$. Again these n_e-sensitive differences can be resolved in the experiment.

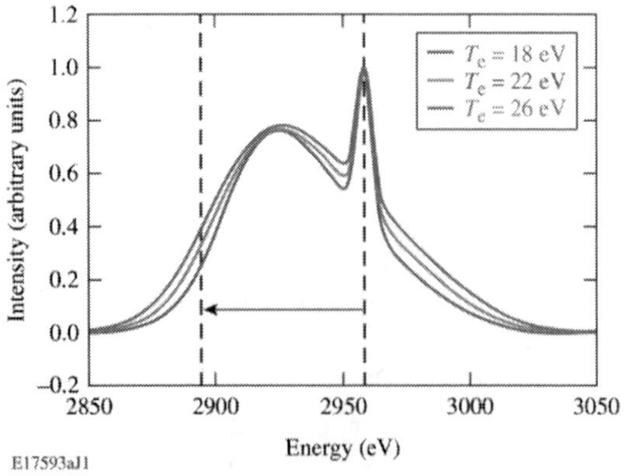

Figure 3. For shocked deuterium the x-ray scattering of Cl Ly$_\alpha$ emission at 2.96 keV (λ_0 = 4.188 Å) from electrons has been calculated for the following conditions: The electron density and ionization are kept fixed at $n_e = 2.0 \times 10^{23}$ cm^{-3} and Z =1, while the electron temperature is varied from T_e = 18 eV to T_e = 26 eV. The scattering angle is $\theta = 90°$ and the scattering parameter is $\alpha_s = 0.61$. The Compton peak is on the low energy side of the Rayleigh peak at 2.96 keV.

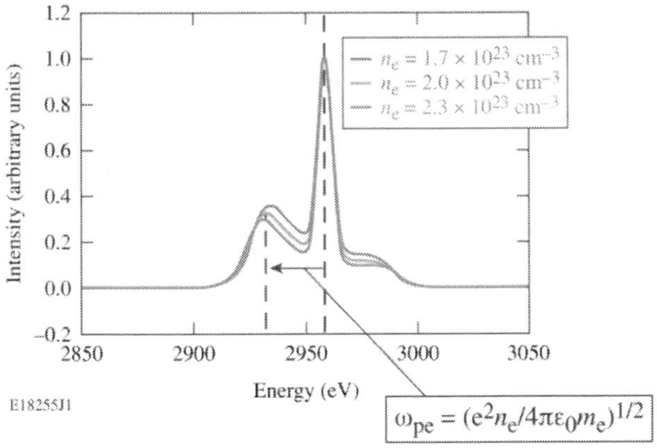

Figure 4. For shocked deuterium the x-ray scattering of Cl Ly_α emission at 2.96 keV (λ_0 = 4.188 Å) from plasmons has been calculated for the following conditions: The electron temperature and ionization are kept fixed at T_e = 22 eV and Z = 1, while the electron density is varied from $n_e = 1.7 \times 10^{23}$ cm^{-3} to $n_e = 2.3 \times 10^{23}$ cm^{-3}. The scattering angle is $\theta = 40°$ and the scattering parameter is $\alpha_s = 1.26$.

4. Future directions

Typical EOS measurements diagnose the shock velocity and infer the velocity of the particles [6–9]. Experiments on OMEGA will combine inelastic x-ray-scattering observations with shock-velocity measurements for $P < 10$ Mbar; consequently, P, ρ, n_e, T_e, and Z of the shocked deuterium will be measured. They will also involve multiple shock waves launched by shaped-laser-pulse drives. The comparison with visible diagnostic measurements may only be possible for laser drive intensities as low as ~10^{13} W/cm^2, which have negligible levels of target preheat. Systematic differences between the plasma conditions inferred by the two diagnostic techniques for the shocked deuterium will be studied. Ideally, accuracies of the plasma condition measurements ~10% are needed for this research.

5. Conclusions

An experiment designed to launch laser-ablation-driven shock waves (10 to 70 Mbar) in a planar liquid-deuterium target on the OMEGA Laser System [1] and diagnose the shocked conditions using inelastic x-ray scattering [2,3] has been designed. Inelastic x-ray scattering is a powerful diagnostic for high-pressure ($P > 10$ Mbar) EOS research, which is inaccessible to optical shock-velocity measurements [6–9].

Acknowledgment

This work was supported by the U.S. Department of Energy Office of Inertial Confinement Fusion under Cooperative Agreement No. DE-FC52-08NA28302, the University of Rochester, and the New York State Energy Research and Development Authority. The support of DOE does not constitute an endorsement by DOE of the views expressed in this article. Partial support from the Engineering and Physical Sciences Research Council (Grant No. EP/G007187/1) and the Science and Technology Facilities Council of the United Kingdom is also acknowledged.

References

[1] Boehly T R *et al.* 1997 *Opt. Commun.* **133** 495–506
[2] Glenzer S H *et al.* 2003 *Phys. Rev. Lett.* **90** 175002; Gregori G *et al.* 2003 *Phys. Rev. E* **67** 026412; Sawada H *et al.* 2007 *Phys. Plasmas* **14** 122703
[3] Glenzer S H *et al.* 2007 *Phys. Rev. Lett.* **98** 065002
[4] Herrmann M C *et al.* 2001 *Nucl. Fusion* **41** 99–111; Betti R *et al.* 2002 *Phys. Plasmas* **9** 2277–2286
[5] Delettrez J *et al.* 1987 *Phys. Rev. A* **36** 3926–3934
[6] Collins G W *et al.* 1998 *Science* **281** 1178–1181
[7] Eggert J *et al.* 2008 *Phys. Rev. Lett.* **100** 124503
[8] Celliers P M *et al.* 2004 *Rev. Sci. Instrum.* **75** 4916–4929
[9] Celliers P M *et al.* 2005 *J. Appl. Phys.* **98** 113529

Journal of Physics: Conference Series **244** (2010) 042018 doi:10.1088/1742-6596/244/4/042018

Hugoniot and temperature measurements of liquid hydrogen by laser-shock compression

T Sano[1], N Ozaki[2], T Sakaiya[3], K Shigemori[1], M Ikoma[4], T Kimura[2], K Miyanishi[2], T Endo[2], A Shiroshita[1], H Takahashi[3], T Jitsui[2], Y Hori[4], Y Hironaka[1], A Iwamoto[5], T Kadono[1], M Nakai[1], T Okuchi[6], K Otani[7], K Shimizu[8], T Kondo[3], R Kodama[2], and K Mima[1]

[1] Institute of Laser Engineering, Osaka University, Suita, Osaka 565-0871, Japan
[2] Graduate School of Engineering, Osaka University, Suita, Osaka 565-0871, Japan
[3] Department of Earth and Space Science, Osaka University, Toyonaka, Osaka 560-0043, Japan
[4] Department of Earth and Planetary Sciences, Tokyo Institute of Technology, Ookayama, Meguro, Tokyo 152-8551, Japan
[5] National Institute of Fusion Science, Toki, Gifu 509-5292, Japan
[6] Institute for Study of the Earth's Interior, Okayama University, Misasa, Tottori 682-0193, Japan
[7] Advanced Research Center for Beam Science, Institute for Chemical Research, Kyoto University, Uji, Kyoto 611-0011, Japan
[8] KYOKUGEN, Center for Quantum Science and Technology under Extreme Conditions, Osaka University, Toyonaka, Osaka 560-8531, Japan

E-mail: `sano@ile.osaka-u.ac.jp`

Abstract. Hydrogen at high pressure in the fluid state is of great interest for target design of inertial confinement fusion and understanding the interior structure of gas giant planets. In this work, we successfully obtained the Hugoniot data for liquid hydrogen up to 55 GPa under laser-driven shock loading using impedance matching to a quartz standard. The shocked temperature was determined simultaneously by the brightness temperature. The compression and temperature along the principal Hugoniot are in good agreement with theoretical models. High reflectivity of hydrogen was observed at 40 GPa, which suggests the fluid becomes conducting.

The properties of hydrogen at high pressure and high density are of great scientific interest. The equation of state (EOS) of hydrogen at these conditions is essential for modeling of the interior structure of gas giant planets [1]. The EOS of hydrogen isotopes has important practical applications for inertial confinement fusion [2]. Chemical free-energy models [3] and *ab initio* simulations [4] have been used to predict the properties of warm dense hydrogen, but the results vary widely and have not converged yet. Therefore accurate experimental EOS data for hydrogen are required for evaluation of the theoretical models and for further understanding of the fundamental nature of hydrogen. The principal Hugoniot for liquid deuterium has been measured up to 220 GPa using laser-driven shock waves [5]. For the case of liquid hydrogen, on the other hand, the Hugoniot was studied experimentally only to 10 GPa by a gas gun and explosive method [6, 7]. The metallization of hydrogen on the Hugoniot is expected to occur at much higher pressure. In this work, we carried out laser-shock experiments of liquid hydrogen

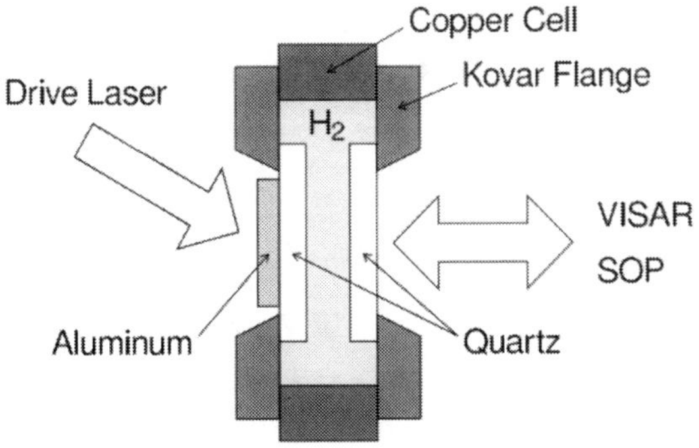

Figure 1. Sketch of the cryogenic hydrogen target. The quartz plates were glued to kovar flanges attached to a copper cell filled with liquid molecular hydrogen at 15 K. The thickness of the quartz was 50 μm for both of them. The laser-side quartz was deposited by 40 μm aluminum to reduce the x-ray fluence from ablation plasma.

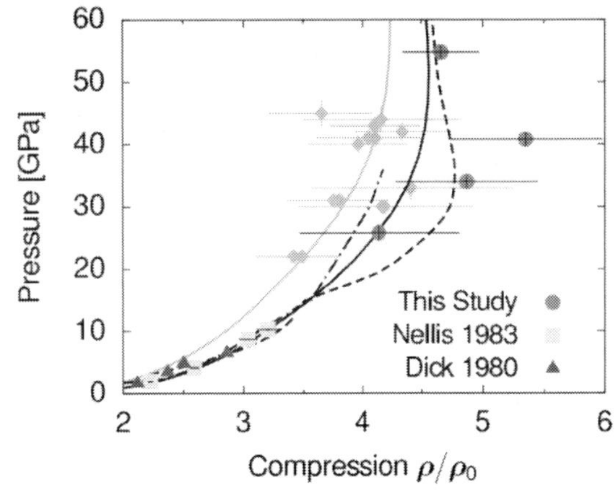

Figure 2. Pressure P versus compression ρ/ρ_0 for the principal Hugoniot of liquid hydrogen. Data are from Dick & Kerley [6] (blue triangles), Nellis et al. [7] (green squares), and this work (red circles). Also shown are theoretical predictions from the model EOS of Kerley [3] (solid line), quantum molecular dynamics simulations [4] (dashed line), and the linear mixing model [10] (dot-dashed line). For reference, the Hugoniot data for liquid deuterium are shown by gray diamonds, and gray curve is the EOS model for deuterium.

to pressures exceeding 10 GPa in order to make a quantitative comparison of the hydrogen Hugoniot around the metal transition with the deuterium data.

The experiment was performed on the GEKKO/HIPER laser facility at the Institute of Laser Engineering, Osaka University. Laser energies between 0.8 and 1.4 kJ were delivered using a nominally square pulse of 2.5 ns in duration. The laser focal spot of 600 μm in diameter was smoothed using Kinoform phase plates. This resulted in average laser intensities between 5 and 8×10^{13} W/cm^2. Figure 1 shows the experimental setup and target arrangement. The cryogenic hydrogen targets consisted of two α-quartz plates, by which the hydrogen layer was sandwiched. In this target, the quartz was designed to be used not only as a window material but also as a reference standard for the impedance-matching measurement.

Shock velocities in the laser-side quartz U_{sQ} and hydrogen U_{sH} were measured using a line-imaging velocity interferometer system for any reflector (VISAR) [8]. At the high pressure involved in these experiments, shock waves in the initially transparent quartz and hydrogen are reflecting. Then the VISAR can provide a direct time-resolved measurement of shock velocities in these media. Two VISARs were run concurrently on each shot to resolve the 2π phase-shift ambiguities at shock breakout. The velocity sensitivities were 4.14 and 14.53 km/s/fringe for quartz, and 5.71 and 20.05 km/s/fringe for liquid hydrogen. The time-resolved VISAR measurements allowed velocities to be tracked continuously during transit through both the standard and sample. Shock velocities immediately before and after the quartz-hydrogen interface were adopted for the impedance-matching analysis. The quartz Hugoniot was previously established to have a linear U_s-U_p relation [9]. Impedance-matching occurs at the intersection of the hydrogen Rayleigh line and the release isentrope of quartz. Because of

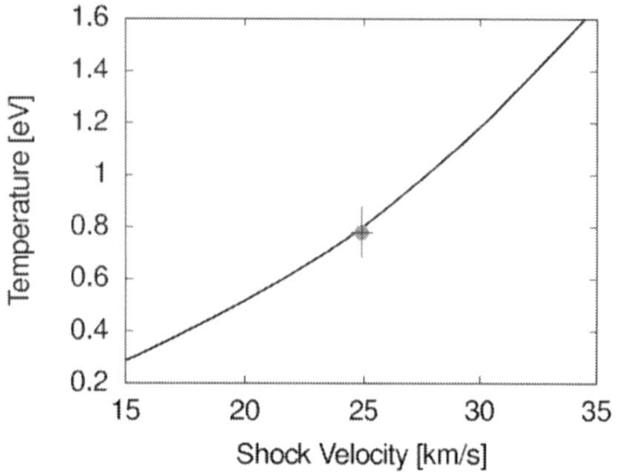

Figure 3. Shocked temperature T versus shock velocity in hydrogen U_{sH}. Solid curve shows the prediction by the SESAME model[3].

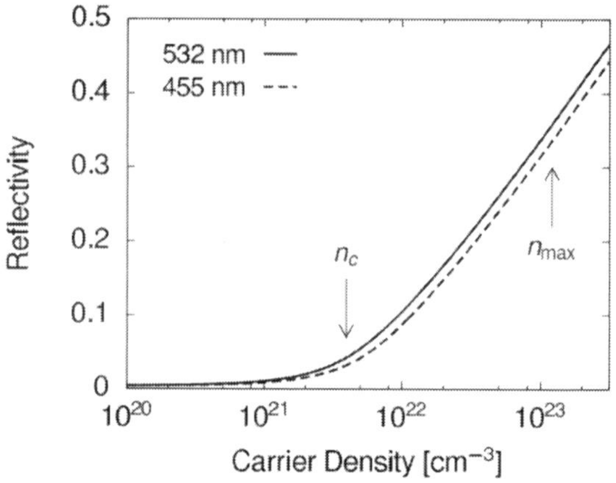

Figure 4. Drude reflectivity plotted as a function of carrier density at the VISAR wavelength 532 nm (solid line) and SOP wavelength 455 nm (dashed line). The difference in the reflectivity between these two wavelengths is at most 2%. The critical density n_c and maximum density n_{max} of electrons are indicated by the arrows.

the low shock impedance of hydrogen, quartz release curves from several hundreds of GPa down to very low pressure are needed. In this work, the quartz release is approximated by a reflection of the principal Hugoniot in the P-U_p plane.

Figure 2 depicts the principal Hugoniot in the P-ρ plane. These results are shown along with the previous experimental data [6, 7] and theoretical models calculated for hydrogen [3, 4, 10]. The pressure range of hydrogen obtained in this work was 25-55 GPa. The highest pressure is more than 5 times higher than the previous data. The compression ρ/ρ_0 of hydrogen ranges from 4.1 to 5.3, which is mostly comparable to the models but slightly softer at 30-40 GPa. The compression of deuterium has been measured as 3.4-4.4 at the same pressure range [11, 5], so that hydrogen exhibits systematically higher compression compared to deuterium. This trend is qualitatively consistent with theoretical predictions [3] and can be seen in the earlier data below 10 GPa [7].

Using a streaked optical pyrometer (SOP), the temperature T of shocked hydrogen was measured simultaneously [12, 13]. We extracted the temperature by fitting the absolute spectral radiance to a gray body Plank spectrum with T as a fit parameter. The radiance was obtained in a single spectral band centered at 455 nm with a bandwidth 38 nm. The emissivity ε is related to the reflectivity R of the shock front through $\varepsilon(\lambda) = 1 - R(\lambda)$. For the reflectivity at the SOP wavelength, we adopted the reflectivity measured at the VISAR wavelength assuming a weak dependence of R on the wavelength. Absolute reflectivity is determined by comparing the shock reflectivity to that from the aluminum surface which has a known value. The obtained R of shock front in hydrogen was 16$\pm$2% at the shock velocity $U_{sH} \approx 25$ km/s corresponding to the pressure $P_{\rm H} \approx 40$ GPa. The result of shocked temperature was 0.78 ± 0.10 eV at $U_{sH} = 24.9 \pm 0.5$ km/s (see Fig. 3). Theoretical models predicted that Hugoniot temperatures of liquid hydrogen are higher by a factor of ~ 1.3 than those of liquid deuterium at the same pressure [3]. Deuterium temperature was measured as 0.5-0.7 eV around 40 GPa by previous shock experiments [14]. Although the uncertainty in temperature measurements is still large, the hydrogen T obtained in this work is higher than the deuterium data. At this temperature, molecular hydrogen begins to dissociate and the fraction reaches to $\sim 30\%$ according to the Kerley's model [3].

Drude-type models are often applied to parameterize the optical properties of liquid metal. Within the Drude description, the complex index of refraction is given by $\hat{n}_s^2 = 1 - (\omega_p^2/\omega^2)(1 + i/\omega\tau_e)^{-1}$ where $\omega_p = (4\pi n_e e^2/m_e)^{1/2}$ is the plasma frequency, n_e is the carrier density, e is the electron charge, m_e is the electron mass, and $\omega = 2\pi c/\lambda$ is the optical frequency. Here the electron relaxation time is assumed to be $\tau_e = R_0/v_F$ where R_0 is the interparticle spacing and v_F is the electron Fermi velocity. We adopt $R_0 = 0.126$ nm corresponding to a density 0.4 g/cm^3. Figure 4 shows the Drude reflectivity of liquid metallic hydrogen derived as a function of n_e. High reflectivity is produced when the carrier density exceeds the critical density n_c defined by $\omega_p^2(n_c) = \omega^2$. The critical density is $n_c = 3.9 \times 10^{21}$ cm^{-3} at $\lambda = 532$ nm. Taking $R = 16\%$, the carrier density is given by 1.8×10^{22} cm^{-3}, which is about 15% of total hydrogen number density.The Fermi energy estimated from this carrier density is 2.5 eV, which is higher than the shock temperature. Based on the Drude model, the reflectivity increases with the carrier density and saturates at $R = 36\%$ for the fully ionized case. It will be interesting to confirm the saturated value of the reflectivity by future experiments.

The hydrogen Hugoniot obtained in this work is strongly dependent on the quartz EOS. Although the shock Hugoniot is relatively well determined, the EOS data along the release curves are still model dependent. For comparison, we calculated a release isentrope from the Kerley 7360 model for α-quartz [15] and obtained the corresponding Hugoniot data, $U'_{p\mathrm{H}}$ and P'_{H}. At higher pressure, the mirror curve is higher than Kerley's model. But just above 50 GPa, they cross and the Kerley model gives a larger particle velocity. The Hugoniot pressure P'_{H} is higher by 2% and the density ρ'_{H} larger by 6% compared to the mirror approximation case. Apparently, improvement of the quartz EOS is essential for the further development of Hugoniot measurements using quartz standards.

In summary, we have obtained the Hugoniot data P-ρ-T for liquid hydrogen in an unexplored range of pressure up to 55 GPa. The results demonstrate that the hydrogen Hugoniot cannot be scaled by density from the deuterium data. As for the study of planetary interiors, the hydrogen EOS data at much higher pressure are required since the transition to metallic hydrogen is anticipated to be at $P \sim$ 200-400 GPa in Jupiter [1]. However, the hydrogen temperature must be kept lower because the Hugoniot temperature at this pressure range is too high to reproduce Jupiter's conditions. Therefore off-Hugoniot measurements of hydrogen by means of reflection shocks [16] and/or precompressed samples [17] will be an important next step.

References

[1] Saumon D and Guillot T 2004 *Astrophys. J.* **609** 1170–1180
[2] Lindl J 1995 *Phys. Plasmas* **2** 3933–4024
[3] Kerley G I 2003 *Sandia National Laboratories, Technical Rep. SAND2003-3613*
[4] Holst B, Redmer R and Desjarlais M P 2008 *Phys. Rev. B* **77** 184201
[5] Hicks D G *et al.* 2009 *Phys. Rev. B* **79** 014112
[6] Dick R D and Kerley G I 1980 *J. Chem. Phys.* **73** 5264–5271
[7] Nellis W J *et al.* 1983 *J. Chem. Phys.* **79** 1480–1486
[8] Celliers P M *et al.* 2004 *Rev. Sci. Instrum.* **75** 4916–4929
[9] Hicks D G *et al.* 2005 *Phys. Plasma* **12** 082702
[10] Ross M unpublished data
[11] Knudson M D *et al.* 2004 *Phys. Rev. B* **69** 144209
[12] Miller J E *et al.* 2007 *Rev. Sci. Instrum.* **78** 034903
[13] Ozaki N *et al.* 2009 *Phys. Plasma* **16** 062702
[14] Collins G W *et al.* 2001 *Phys. Rev. Lett.* **87** 165504
[15] Kerley G I 1999 *Kerley Publishing Services Rep. KPS99-4*
[16] Boehly T R *et al.* 2004 *Phys. Plasmas* **11** L49–L52
[17] Loubeyre P *et al.* 2004 *High Pressure Res.* **24** 25–31

The Sixth International Conference on Inertial Fusion Sciences and Applications IOP Publishing
Journal of Physics: Conference Series **244** (2010) 042019 doi:10.1088/1742-6596/244/4/042019

Target Design for High Energy Density Physics Experiment using Intense Ion Beams

Toru Sasaki[1], Takashi Kikuchi[1], Nob. Harada[1] and Kazuhiko Horioka[2]

[1]Department of Electrical Engineering, Nagaoka University of Technology, Kamitomioka 1603-1, Nagaoka, Niigata, Japan

[2]Department of Energy Sciences, Tokyo Institute of Technology, Nagatsuta 4259, Midori-ku, Yokohama, Japan

E-mail: sasakit@vos.nagaokaut.ac.jp

Abstract. A new approach for high energy density physics experiments using intense ion beams is presented. To make dense matter plasma as the interior of Jupiter, we use a diamond anvil cell for isentropic compression and the intense ion beams for isochoric heating. Results indicated that the temperature of $10\rho_s$ (ρ_s: solid density) pre-compressed target can be achieved 4000 K irradiated by argon ion beams. We also investigated a radiation hydrodynamics (RH) target for understanding optical properties. Making well-defined RH plasma, we calculated the spherical target using intense ion beams. The results indicated that the spherical target plasma could be point-spot like temperature distribution and target size was about 0.4 mm full width of half maximum.

1. Introduction

High energy density physics (HEDP) is of interest in the efficient implosion of the inertial confinement fusion, the originating Solar Systems evaluated by the interior of giant planets as the Jupiter and the Saturn, and the generating process of highly energetic particles in astrophysical phenomena as the Gamma-ray burst, and so on [1]. Understanding these behavior and/or structure, we should have exact data on the equation of state (EOS) and the transport properties of electron and radiation in high energy density (HED) state. HED state of matter, which is categorized as a warm dense matter (WDM), a hot dense matter (HDM), and a radiation dominated matter (RDM), has a potential of abundant scientific discovery including the properties of plasma.[2]

Intense ion beams, which are provided by the induction systems, are opening up an attractive branch for HEDP [3, 4, 5, 6, 7, 8, 9, 10]. Accelerators producing opportunely tailored energies of intense ion beams are promising energy drivers for inertial fusion energy and are also expected to be able to provide a useful tool for creating HED matters. Compared to the high power devices, the intense ion beams using induction technology [11] have many advantages; in uniformity of energy deposition, in large volume of the samples compared to diagnostic resolution, in virtue of the long stopping range, in an advantageous environment for diagnostics, in an ability to heat a variety of target materials in a condition with high repetition rate and multiple beam lines.

For the HEDP study based on the ion beams irradiation, the target should be in equilibrium and as well-defined as possible to accurately diagnose the physical parameters. In this study,

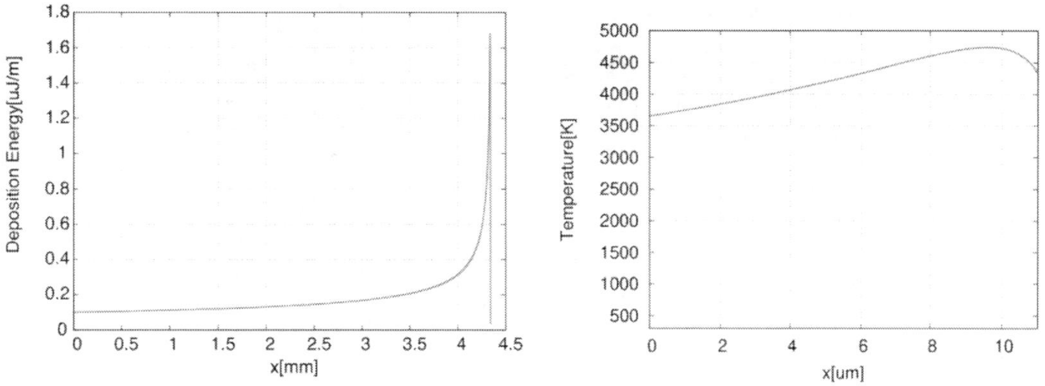

(a) Deposition energy profile in diamond irradiated by argon ion beams.

(b) Compressed hydrogen irradiated by the argon ion beams with 2×10^{10} particles.

Figure 1. Estimation of the ion beam heated DAC. (a) deposition energy profile in DAC irradiated by argon ion beam which consists of the providable parameter of KEK Digital Accelerator. (b) estimated temperature using SESAME[12] and deposition energy profile around Bragg peak.

we propose a method to make a HED state of matter based on providable beam parameters in KEK digital accelerator. In this paper, low entropy target and spherical point-spot like radiator are discussed using intense ion beams.

2. Low entropy target for condense plasma experiments based on intense ion beams

Low entropy target plasma is expected to be applicable for planetary science. Making a critical parameter for exploring the structure of Jupiter, we have to generate a warm dense hydrogen with 6000 K and 200 GPa. The critical parameter region of hydrogen in Jovian interior is estimated to be about 10 ρ_s (ρ_s: solid density) from SESAME[12] and QEOS[13]. Generating highly compressed hydrogen, a shock compression driven by ultra short pulse laser or pulsed-power device has been practical method. The shock compression methods are too dissipative to make the state of Jovian interior. A scheme with multiple shock compression may be useful for formation of the WDM state. However, the shock relation intrinsically includes the EOS, that is, the adiabatic constant depends on the state of matters. Then the estimated condition formed by the multiple shock compression should become quite unclear.

Diamond-anvil cell (DAC), which consists of two opposing diamonds with a sample compressed between the culets, can be generated extreme pressure condition, statically. Additional heating method of DAC as laser-heated or electron cyclotron resonance (ECR) heated DAC is also to create extreme pressure (P > 100 GPa) and temperature (T > 3000 K). However, caused by the broken anvils, approached parameters of both laser-heated and ECR heated DAC depended on the diamonds properties. Pulsed intense ion beams heated DAC, which are programed to heat the sample by the Bragg peak, can provide the uniform samples and the lower damaged anvils compared to the both heating methods in DAC sample.

Figure 1 (a) shows a deposition energy profile of diamond target irradiated by 4.8 GeV argon ion beams, which is providable beam parameter of KEK digital accelerator. As shown, the estimated stopping range is about 4.4 mm and the deposition energy of target is concentrated from 4.2 mm to 4.4 mm. The result indicated that the concentrated deposition energy can heat the compressed sample without the diamond heating. We estimated the temperature of 10 ρ_s

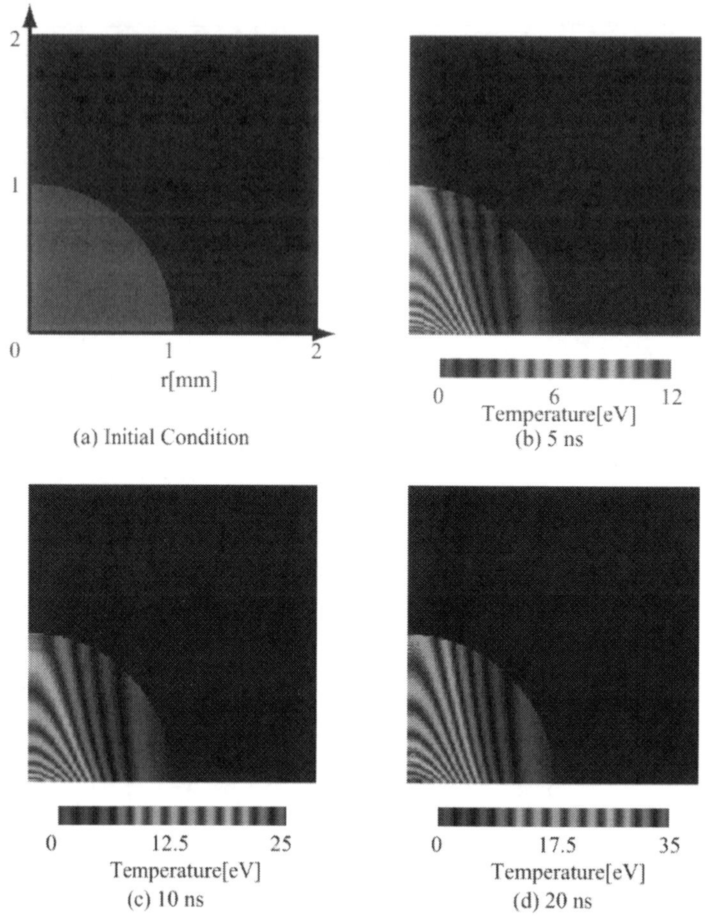

Figure 2. Time evolution of spherical target irradiated by Argon ion beams with 2×10^{13} per bunch.

hydrogen target irradiated by the argon ion beams with 2×10^{10} particles as shown in Fig. 1 (b). It reveals that the temperature of compressed hydrogen target can be uniformly about 4300 K with 10 μm in the target length.

3. Quasi-spherical plasma for radiation hydrodynamics experiments based on intense ion beams

The hydrodynamics dominated by radiative energy flux and pressure plays a crucial role in not only for exploring astrophysical phenomena but also for evaluating the efficiency of hohlraum target for ICF driven by HIBs or lasers [1, 14] Especially, the explosion process in supernova is strongly affected by the radiation transfer in a wide range of parameters from optical thick to thin condition. Then, we cannot determine directly the origin of radiation from the astronomical observation due to the unclarified radiative effects. In order to study such processes, we should evaluate the energy transfer in dense high temperature matter. In a region from HDM to RDM, we should consider both the hydrodynamic energy flux and the radiation flux. The focusing beams as KEK Digital accelerator with PS-ring are able to produced both HDM and RDM conditions[2].

We calculated the hydrodynamics of target with two dimensional spherical geometry, which is assumed aluminum solid target with 1 mm in radius. The detail of calculation is shown in Ref. [10]. In this calculation, the beam radius is set to be 1 mm with a Gaussian distribution. The beam duration is 100 ns, and the beam particle number is 2×10^{13} argon ions with monotonic particle energy (1.67 GeV) which is considered to the stopping range in the target.

Figure 2 shows preliminary results of the evolutions of temperature at each time. As shown in Fig. 2, the point-spot like temperature distribution is able to be generated by the light ion beams. The radius of generated point-spot plasma is 0.4 mm full width of half maximum at center of target. It might indicated that the more smaller spherical plasma is created by irradiation of smaller beam radius.

4. Summary

We proposed HED experiment based on intense ion beams as possible parameters of KEK digital accelerator. New target plasmas, which are low entropy target and point-spot like spherical radiation source, were generated by the programed intense ion beams. To make dense matter plasma as the interior of Jupiter, we use a diamond anvil cell for isentropic compression and the intense ion beams for isochoric heating. It indicated that the temperature of $10\rho_s$ pre-compressed target can be achieved 4000 K irradiated by the argon ion beams. We also investigated the radiation hydrodynamics based on the intense ion beams. To generate the well-defined plasma, we calculated the spherical target using intense ion beams with 1 mm in radius. The results indicated that the spherical target plasma could be point-spot like temperature distribution and the target size was about 0.4 mm full width of half maximum.

Acknowledgements

A part of this work was performed under the Joint Research of Institute of Laser Engineering, Osaka University.

References

[1] Drake R P 2005 *High-Energy-Density Physics Fundamentals, Inertial Fusion, and Experimental Astrophysics* (Springer)
[2] Sasaki T, Asai T, Takahashi T, Kikuchi T, Nakajima M, Horioka K and Takayama K 2009 *Journal of Plasma and Fusion Research SERIES* **8** 174
[3] Dewald E, Constantin C, Udrea S, Jacoby J, Hoffmann D, Niemann C, Wieser J, Tahir N A, Kozyreva A, Shutov A and Tauschwitz A 2002 *Laser and Particle Beams* **20**
[4] Kozyreva A, Basko M, schlegel T, Tauschwiz A and Hoffmann D H H 2003 *Physical Review E* **68** 056406
[5] Grisham L R 2004 *Physics of Plasmas* **11** 5727–5729
[6] Tahir N A, Hoffmann D H H, Kozyreva A, Tauschwitz A, Shutov A, Maruhn J A, Spiller P, Neuner U, Jacoby J, Roth M, Bock R, Juranek H and Redmer R 2001 *Physical Review E* **63** 016402
[7] Tahir N A, Deutsch C, Fortov V E, Gryaznov V, Hoffmann D H H, Kulish M, Lomonosov I V, Mintsev V, Ni P, Nikolaev D, Piriz A R, Shilkin N, Spiller P, Shutov A, Temporal M, Ternovoi V, Udrea S and Varentsov D 2005 *Physical Review Letters* **95** 035001
[8] Barnard J J, Armijo J, More R M, Friedman A, Kaganovich I, Logan B G, Marinak M, Penn G, Sefkow A, Santhanam P, Stoltz P, Veitzer S and Wurtele J 2007 *Nuclear Instruments and Methods in Physics Research A* **577** 275
[9] Barnard J J, Armijo J, Bailey D S, Friedman A, Bieniosek F M, Henestroza E, Kaganovich I, Leung P T, Logan B G, Marinak M M, More R M, Ng S F, Penn G E, Perkins L J, Veitzer S, Wurtele J S, Yu S S and Zylstra A B 2009 *Nuclear Instruments and Methods in Physics Research Section A* **606** 134
[10] Sasaki T, Kikuchi T, Nakajima M, Kawamura T and Horioka K 2009 *Nuclear Instruments and Methods in Physics Research Section A* **606** 161
[11] Takayama K, Arakida Y, Dixit T, Iwashita T, Kono T, Nakamura E, Otsuka K, Shimosaki Y, Torikai K and Wake M 2007 *Physical Review Letters* **98** 054801
[12] Lyon S P and Johnson J D 1998 *T-1 Handbook of the SESAME Equation of State Library* (LA-CP-98-100)
[13] More R M, Warren K H, Young D A and Zimmerman G B 1988 *Physics of Fluids* **31** 3059
[14] Castor J I 2004 *Radiation Hydrodynamics* (Cambridge University Press)

Journal of Physics: Conference Series **244** (2010) 042020 doi:10.1088/1742-6596/244/4/042020

Theoretical study of density condition of solid metallic hydrogen for fuel pellet

Kazunori Shibata[1] **and Ryosuke Kodama**[1,2]

[1] Photon Pioneers Center, Osaka University, 2-1 Yamadaoka Suita, Osaka 565-0871 JAPAN

[2] Graduate school of engineering, Osaka University, 2-1 Yamadaoka Suita, Osaka 565-0871 JAPAN

E-mail: ka-shibata@eie.eng.osaka-u.ac.jp

Abstract. Hydrogen fuel pellet is often used in the laser fusion. In many cases, the hydrogen state as a fuel pellet is kept in low temperature in order to achieve high density solid or liquid states. Such states are very delicate and not easy to handle. On the other hand, a solid metallic state is said to possess similar high density. Additionally, it could be metastable in ambient temperature and utilized as a maintenance free fuel pellet. We have focused on the solid atomic hydrogen and studied theoretically the density condition of the metallization. In this study, we have picked up seven crystalline structures and compared the transition density for each structure. It is found that the density becomes higher as its structural filling factor becomes larger.

1. Introduction

Hydrogen target is often used as a fuel pellet for the laser driven inertial fusion research. As a target, deuterium and tritium are mixed and kept in solid or liquid state to obtain high initial density. Typically, the density is about 0.2 g/cc. Such a high density state is maintained in low temperature because it is not kept in ambient temperature and pressure. On the other hand, as a state of high density hydrogen, a solid metallic state is expected to exist. The solid metallic hydrogen is supposed to have similar or higher density compared with the hydrogen state as a fuel pellet. If there is a metastable solid metallic state in ambient temperature and pressure, it can be used as a new fuel pellet [1] which is very easy to handle. In this paper, we focus on the solid atomic hydrogen and investigate the metal-insulator transition (MIT) density for several structures. The solid metallic hydrogen was first predicted over 70 years ago [2] and challenges to make it are still continued. The solid metallic state is thought to exist in the low temperature (up to 1000K) and high pressure (more than 350GPa) region. In the limit of high pressure (and about room temperature), monatomic solid metallic hydrogen is said to be exist. At the present time, the highest pressure achieved in room temperature is about 320GPa [3]. However, there was no evidence for the metallization and the dissociation of a hydrogen molecule. The extrapolation of this experiment and other recent studies [4] indicate that about 400GPa is necessary to make the solid metallic hydrogen. It is uncertain which of the metallization and the dissociation of hydrogen molecule is takes place at lower pressure. This is because to execute experimental research on ultra-high pressure region (more than about 300GPa) in room temperature is quite difficult[5]. Unfortunately, numerical simulations in such density and temperature region are also difficult. At the moment, in many cases, solid molecular hydrogen is research subject of the numerical studies on the metal-insulator transition (MIT) [6]. Few calculations of the MIT of solid atomic hydrogen were done in body centered cubic structure [7]. In this study, we consider the MIT of solid atomic hydrogen. We have assumed several crystal structures and figured out the transition densities for each structure. We have discussed the relationship between the densities and the structures.

2. Solid atomic hydrogen

In this study, we consider a crystal of solid atomic hydrogen and discuss metal-insulator transition of the material. All the electrons are bounded in 1s state. We consider inter-electronic interaction only in the case that two electrons are in the same site. Such two electrons are both in 1s state and possesses opposite spin. We ignore the interaction if two electrons are in different sites and also, we ignore the overlap integral for different sites. The model Hamiltonian is Hubbard Hamiltonian and it is given by

$$\hat{H} = -E_{1s}\sum_{i}\hat{n}_{i\sigma} - t\sum_{\langle i,j\rangle}\hat{C}^{\dagger}_{j\sigma}\hat{C}_{i\sigma} + U\sum_{i}\hat{n}_{i\uparrow}\hat{n}_{i\downarrow}. \quad (1)$$

Here, $\hat{C}^{\dagger}_{i\sigma}$ is creation and $\hat{C}_{i\sigma}$ is annihilation operator for an electron in i site and possesses spin σ. $\hat{n}_{i\sigma} = \hat{C}^{\dagger}_{i\sigma}\hat{C}_{i\sigma}$ is number operator. E_{1s} is bound energy of 1s electron. The first term is negligible because this term always brings the eigenvalue $-E_{1s}N$ (N is particle number). t is hopping energy which represents an energy that an electron moves toward a neighbor atom by the Coulomb attractive potential of the atom in the neighbor site.

$$t = \frac{e^2}{4\pi\varepsilon_0}\int \phi^*_{1s}(\vec{r}-\vec{R}_j)\frac{1}{|\vec{r}-\vec{R}_j|}\phi_{1s}(\vec{r}-\vec{R}_i)d\vec{r} = \frac{e^2}{4\pi\varepsilon_0 a_B}(\alpha+1)e^{-\alpha}. \quad (2)$$

Here, $\phi_{1s}(\vec{r}-\vec{R}_i)$ is 1s wave function bounded in site i. a is distance of first nearest neighbors and $\alpha = a/a_B$. a_B is Bohr radius. U is so called Hubbard U which represents the energy of the electronic interaction. This value is given by a mean value,

$$U = \langle\psi_U|\hat{V}_{int}|\psi_U\rangle = \frac{5}{8}\frac{e^2}{4\pi\varepsilon_0 a_B} = 17.0(eV). \quad (3)$$

Here, $\psi_U = \phi_{1s}(\vec{r}_1)\phi_{1s}(\vec{r}_2)$ and $\hat{V}_{int} = \frac{e^2}{4\pi\varepsilon_0|\vec{r}_1-\vec{r}_2|}$. Comparing t and U, we obtain $t/U = 1.6(\alpha+1)e^{-\alpha}$. The density region we are concerned is about $\alpha > 3.3$(see, fig. 2(a)) and then, $t/U < 0.254$. It indicates the third term of the Hamiltonian is dominant for the MIT. It means, the MIT is Mott transition. By using this Hubbard model, we consider below seven structures to compare each other. The seven structures are simple cubic (sc), body centered cubic (bcc), face centered cubic (fcc), hexagonal close-packed (hcp), diamond structure, wurtzite structure and graphite structure.

3. Way of calculations

In order to treat the Mott MIT in the solid atomic hydrogen, we use Coherent Potential Approximation (CPA)[8]. In this approximation, single electron Green function $G(\omega)$ and its self energy $\Sigma(\omega)$ for an electron of energy ω is decided through two equations.

$$\Sigma(\omega) = \frac{nU}{1-(U-\Sigma(\omega))G(\omega)}, G(\omega) = \int\frac{\rho_0'(E)}{\omega - E - \Sigma(\omega) + i\delta}. \quad (4)$$

Here, $\rho_0'(E)$ is the density of state (DOS) without interaction between electrons. The DOS is obtained by the eigenvalues of first and second term of the Hamiltonian (1). $n = 1/2$

represents paramagnetic state and δ is infinitesimal. After finishing the calculation, we consider $\delta \to +0$ and we can obtain the density of state with interaction between electrons as $\rho(\omega) = -(1/\pi)\lim_{\delta\to+0} \mathrm{Im}G(\omega)$. By definition, $\rho_0'(E)$ varies with structures. This is because possible direction and number of the hopping is different from the structures and the eigenvalues are changed. To simplify our discussion, we normalize the DOS so that the normalized DOS $\rho_0(x)$ satisfies $\int \rho_0(x)dx = 1$ and the energy is also normalized and restricted $-1 \leq x \leq 1$. In fig. 1, we show a rough sketch of $\rho_0(x)$ for the case of $N \to \infty$. In this limit, $\rho_0(x)$ between fcc-hcp and diamond-wurtzite are almost the same (of course, these $\rho_0(x)$ are different in the case of finite N).

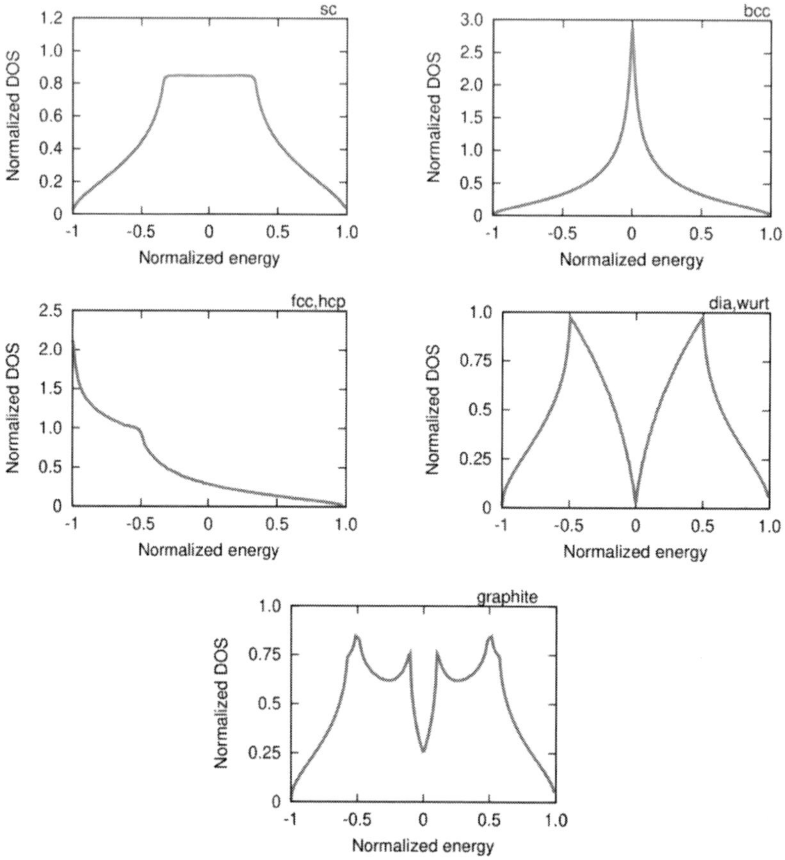

Figure 1. The density of state without interaction for each structure. Particle number $N \to \infty$ is shown. In this limit, the DOSs of fcc-hcp and diamond-wurtzite are corresponding each other.

The relationship between the band width W and the hopping energy t also varies with the structures and it is given by $W = 12t$ for sc, $W = 16t$ for bcc, fcc and hcp, $W = 8t$ for diamond and wurtzite, and $W = 2(\sqrt{10}+1)t$ for graphite structure. The order of the proportional coefficient is like this; (a) diamond=wurtzite<graphite<sc<bcc=fcc=hcp.

4. Calculation result

We calculate and obtain the electronic states and the energy gap. Figure 2(a) represents the relationship between $\alpha = a/a_B$ and the energy gap for every structure we treat. Recall the sequence (a), we can say that large band width contributes to the metallization at larger distance of the neighboring atoms. Because the density n is proportional to α^{-3}, it can be expect that the transition density becomes larger as opposite to the sequence (a). However, this is not obvious (and as we show soon, this is incorrect) because the proportional coefficient between n and α^{-3} is also different from each structure. The coefficient (we denote it as c) is related to the structural filling factor and is given by $n = c \times m_p/(a_B\alpha)^3$ (m_p is mass of a proton). It is calculated as $3\sqrt{3}/8$ for diamond and wurtzite, $4\sqrt{3}/9$ for graphite, 1 for sc, $3\sqrt{3}/4$ for bcc, and $\sqrt{2}$ for fcc and hcp. The order is like this; (b) diamond=wurtzite<graphite<sc<bcc<fcc=hcp. The sequence (b) is almost the same as the sequence (a). However, the sequence (b) conflicts with the sequence (a) because the density for same α becomes larger as the sequence (b). In fig. 2(b), we show the density dependence of the energy gap. The density for the energy gap disappears is same as the sequence (b). It indicates the contribution of the filling factor (sequence (b)) is more effective than that of the broadening of the band width (sequence (a)). The value of the MIT of each structure is n=0.20 for diamond and wurtzite, n=0.24 for graphite, n=0.25 for sc, n=0.28 for bcc, and n=0.30 for fcc and hcp (unit is g/cc).

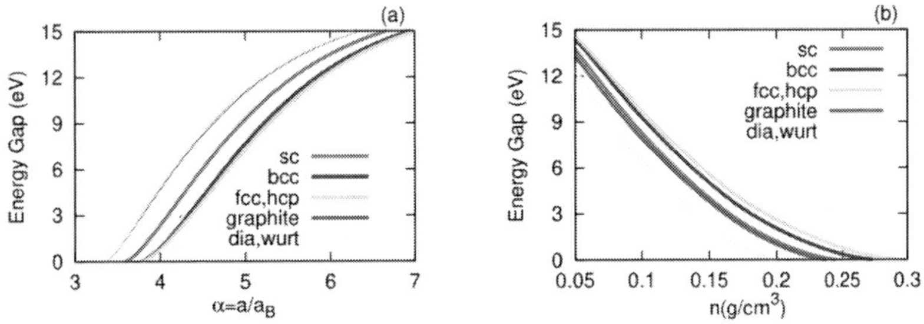

Figure 2. (a) The relationship between the energy gap and $\alpha = a/a_B$. As is obvious from the curves, energy gap decreases as the density becomes higher (smaller α). (b) The density dependence of the energy gap for each structure.

5. Conclusion

In this study, we have investigated the structure dependence of the density when the MIT takes place in solid atomic hydrogen. We also calculated the transition density for each structure. As a result, the transition density varies with the structure and the density becomes higher for the structure of larger filling factor. This is because the filling factor is the dominant factor to decide the transition density. The transition density we calculated in this study is similar to the transition density region obtained for bcc structure[9]. This means the assumptions and parameters we used here are reliable at least qualitatively. If we consider deuterium or tritium, the expected MIT density becomes twice or three times larger.

References

[1] Nellis W J 2000 *Sci. Am.* **May** 84
[2] Wigner E and Huntington H B 1935 *J. Chem. Phys.* **3** 764-70
[3] Loubeyre P, Occelli F and LeToullec R 2002 *Nature* **416** 613-17
[4] Takezawa T, Nagao K and Nagara H 2001 *J. Low. Temp. Phys.* **123** 315-29
[5] Narayana C, Luo H, Oroloff J and Ruoff A L 1998 *Nature* **393** 46-9
[6] Stadele M and Martin R M 2000 *Phys. Rev. Lett.* **84** 6070-3
[7] Li J L, Rignanese G M, Chang E K, Blase X and Louie S G 2002*Phys. Rev. B* **66** 035102
[8] Soven P 1967 *Phys. Rev.* **156** 809-813
[9] Pfrommer B G and Louie S G 1998 *Phys. Rev. B* **58** 12680-3

The Sixth International Conference on Inertial Fusion Sciences and Applications IOP Publishing
Journal of Physics: Conference Series 244 (2010) 042021 doi:10.1088/1742-6596/244/4/042021

Outline of experimental schemes for measurements of thermophysical and transport properties in warm dense matter at GSI and FAIR

Anna Tauschwitz[1,2], Mikhail Basko[3], Vladimir Efremov[4], Igor Iosilevskiy[4], Joachim Jacoby[1], Joachim Maruhn[1], Paul Neumayer[2,5], Vladimir Novikov[6], Andreas Tauschwitz[5], Frank Rosmej[7,8]

[1] University of Frankfurt am Main, Max-von-Laue-Str. 1, 60438 Frankfurt, Germany
[2] ExtreMe Matter Institute EMMI, GSI Helmholtzzentrum für Schwerionenforschung, Planckstr. 1, 64291 Darmstadt, Germany
3 Institute for Theoretical and Experimental Physics, Bolshaya Cheremushkinskaya 25, 117218 Moscow, Russia
[4] Joint Institute for High Temperatures, Izhorskaya 13/19, 125412 Moscow, Russia
[5] GSI Helmholtzzentrum für Schwerionenforschung, Planckstr. 1, 64291 Darmstadt, Germany
[6] Keldysh Institute of Applied Mathematics, Miusskaya sq. 4, 125047 Moscow, Russia
[7] Université Pierre et Marie Curie, UMR7605, case 128, 4 Place Jussieu, 75252 Paris Cedex 05, France
[8] Centre de Recherche LULI, École Polytechnique, route de Saclay, 91128 Palaiseau, France

E-mail: an.tauschwitz@gsi.de

Abstract. Different experimental schemes for investigation of warm dense matter produced with intense energetic ion beams are presented. The described target configurations allow direct measurements of thermophysical and transport properties of warm dense matter without hydrodynamic recalculations. The presented experiments will be realized at the current GSI synchrotron SIS-18 and the future FAIR facility in the framework of the WDM-collaboration.

1. Introduction

Intense energetic ion beams are a promising tool for production and investigation of matter at high energy density. Using a beam from the SIS-18 heavy ion synchrotron at GSI a specific energy of several kJ/g can be deposited in an extended volume of condensed matter. Extreme states of matter at nearly solid densities and thermal energies comparable to the Fermi energy are referred to as warm dense matter (WDM). Understanding of WDM is difficult because of the complexity of first principle theories and the lack of well-defined experimental measurements [1]. For precise thermophysical experiments using heavy ion beams appropriate target schemes are indispensable.

In this paper we address three experimental schemes for investigation of WDM with intense ion beams. These schemes constitute currently the central experimental activities of the WDM-collaboration which is part of the core program of the international FAIR facility [2].

2. Experimental schemes

2.1. Isochoric heating

In the case of isochoric heating the density of a substance, ρ_0 remains constant and the energy deposited by the ion beam is equal to the internal energy of the heated matter, ϵ. Then any measured quantity is a function of a well defined thermodynamic state (ρ_0, ϵ). The available ion pulses are not short enough, however, and the conditions for isochoric heating are not satisfied. Therefore, a dynamic confinement scheme was developed to realize isochoric heating of matter in cylindrical [3] as well as in spherical geometry [4]. Dynamically confined targets provide the conditions of quasi-isochoric heating by employing a thin low-Z tamper which is as well heated by the ion beam to produce a confining pressure on the core target material. The use of a low-Z tamper allows to apply X-ray scattering diagnostics to investigate the core target material [4].

An ion-beam target configuration in spherical geometry in drawn Fig. 1(a). Fig. 1(b) shows the normalized mean density variation in a target consisting of a hydrogen core surrounded by a carbon shell. It is seen that the target can be optimized to minimize the density variation during the heating phase or to obtain a mean density after irradiation equal to the initial one. In the calculation the hydrogen core has a radius of 400 μm. The ion beam has 100 ns length and consists of 8×10^{10} uranium ions.

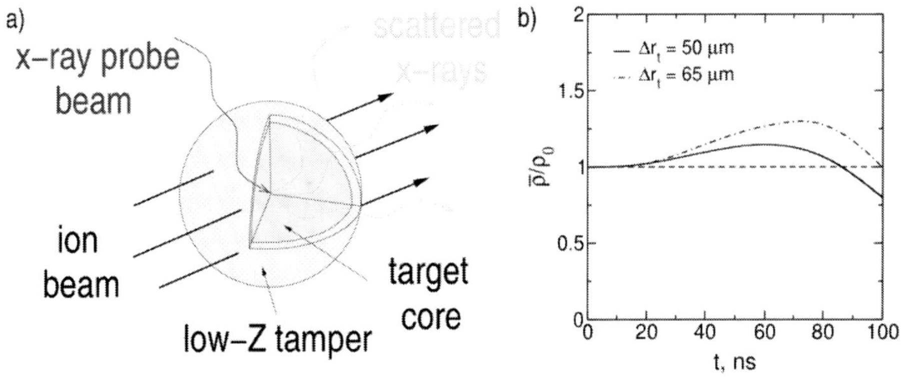

Figure 1. a) Dynamically confined target in spherical geometry. b) Mean density variation during the ion beam heating for two different tamper thicknesses.

2.2. Isothermal expansion

Heating of thin foils with intense heavy-ion beams allows the creation of isolated samples of warm dense matter suitable for experimental determination of frequency-dependent opacities [5]. The proposed measurements will help to benchmark atomic physics models in the WDM regime. For the material temperatures of about 1 eV the measurements have to be performed in the XUV-range. In the experiments at the GSI facility, a backlighter source is driven by a table-top laser as shown in Fig. 2(a). The higher ion beam intensities available at FAIR will allow to use a second ion-beam heated foil as a backlighter.

For precise opacity measurements a constant target temperature is essential. Figure 2(b) shows the density and temperature distribution in the initially 200 nm thick Bi foil irradiated with an ion beam. For the calculation an ion beam consisting of 10^{10} uranium ions with a pulse length of 100 ns is assumed. After 45 ns the target foil is expanded resulting in a gaussian density profile and a constant temperature of about 1 eV, which is appropriate for opacity measurements. Simulation results agree with an analytical treatment of the isothermal expansion of the heated foils given in [6].

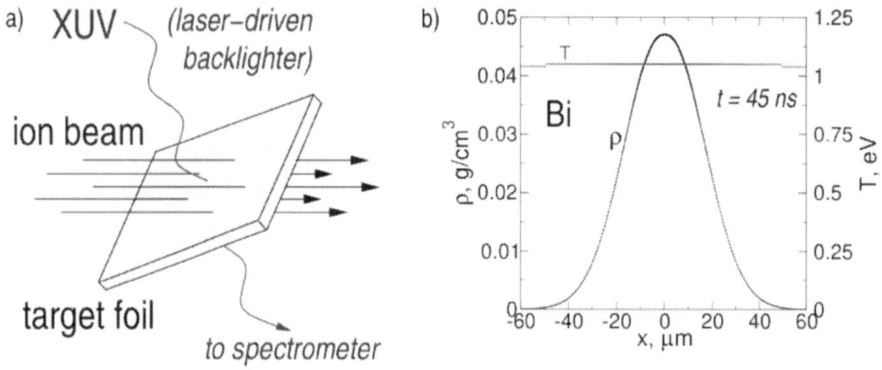

Figure 2. a) Ion-beam target and backlighter configuration for opacity measurements. b) Target density and temperature in the ion beam heated Bi foil.

2.3. *Quasi-static heating*

It was suggested to measure equation of state data using ion beam heating of foams which are highly dispersed porous samples. The main goal is to detect the moment t_x when the pores of the heated sample close [7,8]. At the moment t_x the density is determined by the mean density of the cold sample and the enthalpy, H by the ion beam energy deposition. Hence, the caloric expansion coefficient $\alpha_p = (\partial\rho/\partial H)_p$ is obtained. Since at $t = t_x$ the heated sample is fairly homogeneous, a measurement of the surface temperature defines the thermal expansion coefficient $\alpha_p^{'} = (\partial\rho/\partial T)_p$ and the heat capacity $c_p = (\partial H/\partial T)_p$. The proposed experimental scenario can be applied, e.g. to a number of high boiling temperature metals in the temperature range of 1000–7000 K and will help to determine their critical points, which are highly uncertain for those metals [7]. Measurements with certain compounds can also confirm or disprove predicted non-congruent evaporation [7-9].

It is proposed to replace the complicated three-dimensional structure of a foam sample by a stack target consisting of n separated thin foils [10] as shown in Fig. 3(a). The gap between the foils determines the mean density of the target, $\rho_{00} = \rho_0 l_0/(l_0 + \Delta l_0)$. Similar to foam samples, the porosity of the stack target can be defined as $\rho_{00}/\rho_0 < 1$. At $t < t_x$ the foils expand quasi-statically, if the sound propagation time is short on the time scale of the energy deposition. The later can be achieved if the foils are sufficiently thin. If the expansion is quasi-static, the density, the pressure, and the internal energy (and temperature) are constant in space. The surface velocity of the expanding foils is proportional to the foil thickness, $u_l \propto l$. When the foils merge, the launched acoustic waves increase the surface velocity of the entire target by $2u_l$, when escaping from the target surface.

The evolution of the surface velocity for a stack target consisting of $n = 10$ foils in presented in Fig. 3(b). For the hydrodynamic calculations a stack made of aluminum and an energy deposition of 10 kJ/g in a 100 ns pulse are assumed. This corresponds to about 10^9 ions of uranium. The initial foil thickness is $l_0 = 5\,\mu$m, the gap between the foils is $\Delta l_0 = 1.1\,\mu$m. The expected sharp increase of the surface velocity will help to detect the homogenization time t_x.

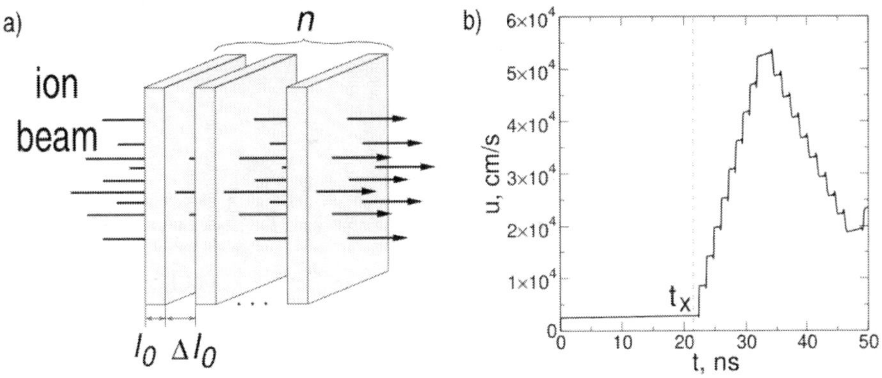

Figure 3. a) Stack target configuration. b) Surface velocity of a stack target consisting of 10 foils irradiated with an ion beam.

3. Summary

The described experimental schemes allow the determination of thermophysical and transport properties of WDM directly from experimental data without the necessity of intermediate hydrodynamic recalculations. The proposed schemes make use of isochoric heating, isothermal expansion and quasi-static heating of the irradiated samples. The presented experiments can be started using the existing heavy ion synchrotron SIS-18 at GSI. Use of the future FAIR facility will extend the accessible temperature range and improve the accuracy of the experiments due to the higher ion beam intensity.

References

[1] R. W. Lee, S. J. Moon, H.-K. Chung, W. Rozmus, H. A. Baldis, G. Gregori, R. C. Cauble, O. L. Landen, J. S. Wark, A. Ng, S. J. Rose, C. L. Lewis, D. Riley, J.-C. Gauthier, and P. Audebert *J. Opt. Soc. Am. B* 2003 **20** 770

[2] *WDM - radiative properties of warm dense matter produced by intense heavy ion beams*, FAIR Baseline Technical Report 2006, vol 5. Available at `http://www.gsi.de/fair/reports/btr.html`

[3] A. Kozyreva, M. Basko, F. Rosmej, T. Schlegel, A. Tauschwitz, D.H.H. Hoffmann *Phys. Rev. E* 2003 **68** 056406

[4] An. Tauschwitz, J.A. Maruhn, D. Riley, G. Shabbir Naz, F.B. Rosmej, S. Borneis, A. Tauschwitz *High En. Dens. Phys.* 2007 **3** 371

[5] An. Tauschwitz, V.G. Novikov, A. Tauschwitz, F.B. Rosmej, J. Abdallah, E. Onkels, J. Jacoby, J.A. Maruhn 2009 *Appl Phys B* **95** 13

[6] R.A. London and M.D. Rosen *Phys. Fluids* 1986 **29** 3813

[7] I. Iosilevskiy, V. Gryaznov *J. Nucl. Mater.* 2005 **344** 30

[8] I. Iosilevskiy *High Energy Density Physics with Intense Ion and Laser Beams* May 2007, GSI-2007-2

[9] I. Iosilevskiy, V. Gryaznov, E. Yakub, C. Ronchi, V. Fortov *Contr. Plas. Phys.* 2003 **43** 316

[10] An. Tauschwitz, V. Efremov, J.A. Maruhn, F.B. Rosmej, A. Tauschwitz *Nucl. Instr. Meth. B* 2009 **267** 2449

The Sixth International Conference on Inertial Fusion Sciences and Applications IOP Publishing
Journal of Physics: Conference Series **244** (2010) 042022 doi:10.1088/1742-6596/244/4/042022

Ultraintense laser interaction with nanoscale targets: a simple model for layer expansion and ion acceleration

B. J. Albright[1], L. Yin[1], B. M. Hegelich[1,2], K. J. Bowers[1,7], C. Huang[1], A. Henig[2,4], J. C. Fernández[1], K. A. Flippo[1], S. A. Gaillard[1,3], T. J. T. Kwan[1], X. Q. Yan[4,5], T. Tajima[2,6], D. Habs[2,4]

[1]Los Alamos National Laboratory, Los Alamos, NM 87545, USA
[2]Department für Physik, Ludwig-Maximilians-Universität München, 85748 Garching, Germany
[3]Forschungszentrum Dresden-Rossendorf, P.O. Box 510119, Dresden, Germany
[4]Max-Planck-Institut f. Quantenoptik, 85748 Garching, Germany
[5]State Key Lab of Nuclear Physics and Technology, Peking University Beijing 100871, China
[6]Photo-Medical Research Center, JAEA, Kyoto 619-0215, Japan

E-mail: balbright@lanl.gov

Abstract. A simple model has been derived for expansion of a thin (up to 100s of nm thickness) target initially of solid density irradiated by an ultraintense laser. In this regime, ion acceleration mechanisms, such as the Break-Out Afterburner (BOA) [1], emerge with the potential for dramatically improved energy, efficiency, and energy spread. Ion beams have been proposed [2] as drivers for fast ignition inertial confinement fusion [3]. Analysis of kinetic simulations of the BOA shows the period of enhanced acceleration occurs between times t_1, when the target becomes relativistically transparent to the laser, and t_2, when the target becomes classically underdense and the enhanced acceleration terminates. A simple model for target expansion has been derived that contains early, one-dimensional (1D) expansion of the target and three-dimensional (3D) expansion at late times. The model assumes expansion is slab-like at the instantaneous ion sound speed and requires as input target composition, laser intensity, laser spot area, and the efficiency of laser absorption into electron thermal energy.

1. Introduction

Laser-ion accelerators have been proposed as drivers for fast ignition inertial confinement fusion [2]. If a laser pulse with very low pre-pulse is employed, an ultrathin, target of solid density with thickness as small as a few nm may be used. Such targets enable new acceleration mechanisms, such as the Break Out Afterburner [1, 4], and Radiation Pressure Acceleration/Phase-Stable Acceleration [5] to be accessed. These mechanisms may lead to order-of-magnitude improvement in beam energy and efficiency over target-normal sheath acceleration [6], while at the same time maintaining high beam quality (low $\Delta\mathcal{E}/\mathcal{E}$) as well as accelerate mid-Z ion species like carbon, as in recent studies of ion-driven fast ignition [7].

[7] Guest scientist. Present address: D. E. Shaw Research LLC, 120 West 45th St. 39th Floor, New York, NY 10036, USA.

In this paper, we focus on the BOA, which relies on a transition of the target from opaque to relativistically underdense as the target expands and its electrons are heated. As the laser penetrates the target, at time denoted t_1, rapid acceleration of target ions ensues until time t_2, when the target becomes classically underdense ($n_e = n_{cr}$). Optimal energy of such beams is obtained when peak intensity of the laser pulse falls between t_1 and t_2.

To guide experimental design, it is fruitful to have a simple model for the dynamics of the expanding target subject to a laser drive. Such a model is presented here, along with its application to a setting relevant to those obtained at the LANL Trident facility. It should be noted that a self-similar analysis of a similar problem has been reported by Yan et al. [8].

2. Model of Layer Expansion

We consider a target of thickness ℓ_0 irradiated by an ultraintense ($I > 1.4\times 10^{18}$) laser with spot area A. We further assume laser energy absorption at efficiency ε by target electrons (of initial density n_0) and that the electrons have a single temperature T that evolves in time. The target will exhibit 1D longitudinal expansion initially, until instantaneous target thickness $\ell(t) \sim \sqrt{A}$ and lateral expansion of the plasma must be taken into account. We allow the plasma to expand at the ion sound speed $c_s = (5Z_iT/3m_i)^{1/2}$; here Z_i denotes ion core charge.

According to the First Law of Thermodynamics, laser heating should balance the increase in electron internal energy, PdV work, and losses (radiative and conductive). In time dt,

$$\varepsilon AIdt = \frac{3}{2}n_0A\ell_0dT + nTdV - (P_{\mathrm{Br}} - P_{\nabla T})Vdt, \tag{1}$$

where $n(t)$ is instantaneous electron density of the slab, V is the expanding plasma volume, and $P_{\nabla T}$ and P_{Br} are power densities from thermal conduction and radiation losses respectively. This expression is valid for optically thin (to thermal xrays) targets, as appropriate for solid density or lower, fully ionized plasma at sub-micron thickness and $T >\sim 1$ keV. At lower temperature (though still fully ionized), the plasma is opaque and bremsstrahlung is radiated only from the outer optical depth of plasma; the above expression is an upper bound to such losses.

Lateral thermal conduction can be ignored generally for sub-micron target thickness. Radiative power loss requires more care: in optically thin plasma [9], $d\log T/dt = P_{\mathrm{Br}}/(nk_BT) = 3.4\times 10^{-15} n_{[cm^{-3}]} Z_i T_{[keV]}^{-1/2}$ s^{-1}. For $n \sim 10^{23}$ cm^{-3} and $\Theta \sim$keV, the radiative cooling time is ~ 100 ps, well exceeding the pulse length; later, n is lower, T is higher, and relative losses to radiation are lower still. Consequently, we ignore radiation power losses in our model to leading order.

To accommodate the transition from 1D to 3D expansion, we use the ansatz

$$n(t) = n_0\frac{\ell_0}{\ell}\left[\frac{A + C\ell_0^2}{A + C\ell^2}\right], \tag{2}$$

where

$$\ell(t) = \ell_0 + \int_0^t dt'\, c_s(t') \tag{3}$$

and

$$V \sim \ell(A + C\ell^2), \tag{4}$$

where C is a numerical coefficient of order unity. At early time, $\ell^2 \ll A$, the bracketed term in (2) is approximately unity, and $dV \approx Ad\ell$, so the expansion is 1D; at late time, $n \sim \ell^{-3}$ and $dV \sim \ell^2 d\ell$, as appropriate for 3D expansion.

It is convenient to express this equation in scaled units: we normalize time $s \equiv ct/\ell_0$, distance $z = \ell/\ell_0$, and temperature $\Theta \equiv T/m_ec^2$ and write scaled sound speed as $c_s = c\Theta^{1/2}X_I$, where

$X_i \equiv [(5/3)(m_e/m_p)(Z_i/A_i)]^{1/2} = 0.030\,(Z_i/A_i)^{1/2}$ for ion mass $m_i = A_i m_p$. We define a normalized intensity $\tilde{I} = \varepsilon I/n_0 m_e c^3$ and let $\delta = \ell_0^2/A \ll 1$. If we assume the target foil matter beneath the laser spot expands at late time into a hemisphere ($C = 2\pi/3$), the evolution obeys

$$z' = X_i \Theta^{1/2} \tag{5}$$

$$\Theta' = \frac{2}{3}\left(\tilde{I} - X_i \frac{\Theta^{3/2}}{z}\left[\frac{(1+C\delta)(1+3C\delta z^2)}{1+C\delta z^2}\right]\right), \tag{6}$$

with primes indicating d/ds. (A different expansion profile would entail a different value of C; though the values of t_1 and t_2 would change somewhat with different values of C, the essential behavior of the model would remain the same). The initial conditions are that $z(0) = 1$ and $\Theta(0) = \Theta_0$. In the $\delta \to 0$ limit, this reduces to the second-order, nonlinear ODE for Θ reported in Henig et al. [10].

Times t_1 and t_2 are obtained readily from the model: $n(t_1) = \langle\gamma\rangle\, n_{\rm cr}$ ($n_{\rm cr} = m_e \omega_0^2/4\pi e^2$ is the critical density in CGS units); $n(t_2) = n_{\rm cr}$. The average relativistic factor of the electrons follows from the expression for internal energy of the electrons $\frac{3}{2}T = m_e c^2(\langle\gamma\rangle - 1)$, from which

$$\langle\gamma\rangle = 1 + \frac{3}{2}\Theta. \tag{7}$$

As described in Ref. [10], maximum ion energy occurs when the peak of the laser intensity falls between t_1 and t_2, a point graphically illustrated in the left panel of the Figure.

3. Application to Recent Trident Experiment and VPIC Kinetic Simulations

We consider the application of this model to the expansion of a diamond-like-carbon (DLC) layer heated via a laser pulse of 1 micron light similar to that attained at the LANL Trident User Facility. We assume that the laser pulse has sine-squared intensity profile (in time) with FWHM of 540 fs and peak intensity 2×10^{20} W/cm^2 focused into a spot size 7 microns in diameter with negligible pre-pulse or pedestal. The DLC layer has $n_0/n_{\rm cr} = 821$ and is assumed to absorb the laser pulse with 50% efficiency. These experimental parameters are similar to those reported by Henig et al. [10].

The Figure shows the evolution of the target layer for a particular choice of target thickness ($\ell_0 = 80$ nm) near the optimal thickness. The thick, solid curve is $n_e/n_{\rm cr}$ as a function of time for a 3D expanding layer. The dashed curve is $\langle\gamma\rangle$ as a function of time and the time $t_1 \approx 400$ fs is where the two cross. The time $t_2 \approx 700$ fs is when $n_e/n_{\rm cr} = 1$. The dotted curve is the laser intensity profile (arb. units.) and the peak of the profile, at 540 fs, is between t_1 and t_2. The dot-dashed curve is the same as the solid curve, but for a 1D expansion ($A \to \infty$ limit); for this set of target parameters, the pulse terminates before t_2.

The right panel of the Figure shows average electric field envelope amplitude $\langle E\rangle$ averaged between t_1 and t_2 for the same laser and target parameters as in the left panel. Analysis of a suite of 1D, 2D, and 3D kinetic plasma simulations of the BOA using the VPIC particle-in-cell code [11] as well as analytic theory [8, 12] show that in the BOA, peak ion beam energy scales linearly with $\langle E\rangle$. In this case, the peak ion energy is determined to be near $\ell_0 = 80$ nm.

4. Conclusions

This simple model cannot capture the full dynamics of the expanding layer. Kinetic simulations of BOA [4] show that electron phase space possesses complicated structure and may not be represented well by an electron gas that instantaneously relaxes to a single, equlibrium temperature. Moreover, relativistic nature of the electron gas is not treated rigorously and the model employs an ad hoc assumption about laser absorption. Finally, though the model

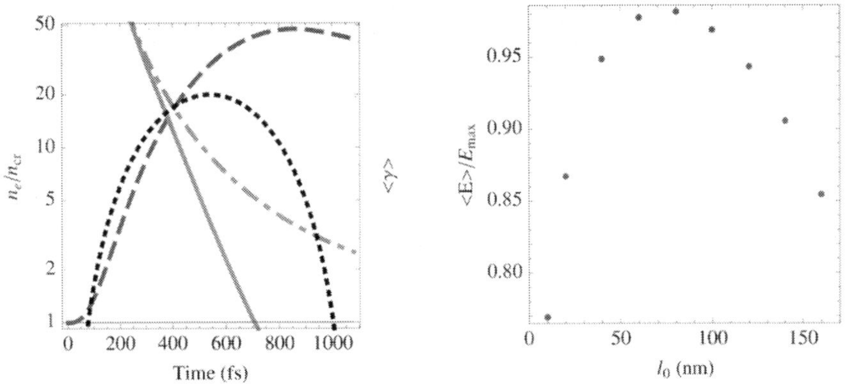

Figure 1. Left panel: Density n_e/n_{cr} of electrons in a DLC layer as a function of time for a 20 μm diameter laser spot (solid curve) and one of infinite size (dot-dashed) for a laser of peak intensity 2×10^{20} W/cm^2. The sine-squared pulse profile (arb. units) is shown by the dotted curve. The dashed curve is $\langle\gamma\rangle$. The time t_1, when the BOA starts, is around 400 fs, where the solid and dashed curves cross; $t_2 = 700$ fs for 3D expansion (where the solid curve crosses $n_e/n_{cr} = 1$. Right panel: Laser electric field amplitude $\langle E\rangle / E_{max}$ averaged between t_1 and t_2.

recovers the appropriate limiting dynamics of an expanding 1D and 3D gas, the transition is treated heuristically. Recent work by Yan et al. remedies many of these issues [8]. Despite its shortcomings, the simple model presented here yields insight into the dynamics of layer expansion, which lie at the heart of the BOA.

Acknowledgments

Work performed under the auspices of the U.S. D.O.E. by the LANS LLC, Los Alamos Nat. Lab. and supported by LDRD and by DFG through Transregio SFB TR18 and the DFG Cluster of Excellenence Munich-Centre for Advanced Photonics (MAP). A. Henig acknowledges support from IMPRS-APS; X. Q. Yan, from the Humboldt Foundation and NSFC(10855001).

References

[1] L. Yin et al., Lasers Part. Beams **24**, 291 (2006)
[2] M. Roth et al., Phys. Rev. Lett. **86**, 436 (2001); M. Temporal, J. J. Honrubia, and S. Atzeni, Phys. Plasmas **9**, 3098 (2002);
[3] M. Tabak et al., Phys. Plasmas **1**, 1626 (1994)
[4] L. Yin et al., Phys. Plasmas **14**, 056706 (2007)
[5] A. Macchi et al., Phys. Rev. Lett. **94**, 165003 (2005); A.P.L. Robinson, et al., New J.Physics **10**, 013021 (2008); X. Q. Yan et al., Phys. Rev. Lett. **100**, 135003 (2008); X.Q.Yan et al., Phys. Rev. Lett. **103**, 135001 (2009); A Henig et al. Phys. Rev. Lett., **103**, 245003 (2009)
[6] Maksimchuk, A. et al. Phys. Rev. Lett. **84**, 4108 (2000); Snavely, R. et al., Phys. Rev. Lett. **85**, 2945 (2000); Hatchett, S. et al., Phys. Plasmas **5**, 2076 (2000); Hegelich, M. et al., Phys. Rev. Lett. **89**, 085002 (2002); Roth, M. et al., Phys. Rev. Spec. Topic Accelerators Beams **5**, 061002 (2002)
[7] J. C. Fernández et al., J. Physics: Conf. Series **112**, 022051 (2008); B. J. Albright et al., J. Physics: Conf. Series **112**, 022029 (2008); J. C. Fernández et al., Nuclear Fusion **49** 065004 (2009)
[8] X. Q. Yan et al., Appl. Phys. B, DOI 10.1007/s00340-009-3707-5 (2009).
[9] J. D. Huba 2000 *NRL Plasma Formulary* NRL/PU/6790-00-426 (Washington, DC: Naval Research Laboratory) p 57.
[10] A. Henig et al., Phys. Rev. Lett. **103**, 045002 (2009).
[11] K. J. Bowers et al., Phys Plasmas **15**, 055703 (2008).
[12] B. J. Albright et al., Phys. Plasmas **14**, 094502 (2007).

Journal of Physics: Conference Series **244** (2010) 042023 doi:10.1088/1742-6596/244/4/042023

Investigation of high intensity laser proton acceleration with underdense targets

E. d'Humières[1,*], J. L. Feugeas[1], P. Nicolaï[1], S. Gaillard[2], T. Cowan[2,3], Y. Sentoku[2] and V. Tikhonchuk[1].

[1]Centre Lasers Intenses et Applications, Université de Bordeaux-CNRS-CEA, 33405 Talence, France.

[2]Department of Physics, University of Nevada, Reno, Nevada 89557, USA.

[3]Forschungszentrum Dresden-Rossendorf, P.O. Box 510119, 01300 Dresden, Germany.

dhumieres@celia.u-bordeaux1.fr

Abstract. In the last few years, intense research has been conducted on laser-accelerated ion sources and their applications. Recently, experiments have shown that a gaseous target can produce proton beams with characteristics comparable to those obtained with solid targets. In underdense laser proton acceleration, volume effects dominate the acceleration, while in target normal sheath acceleration, the electric field value is directly related to the electron surface density. Using Particle-In-Cell simulations, we have studied in detail the effect of an underdense density gradient on proton acceleration with high intensity lasers. Underdense laser ion acceleration strongly depends on the length, the shape and the amplitude of the density gradient and on the laser pulse shape. The accelerated proton beam characteristics in the shock-like regime are very promising.

1. Introduction

The production of energetic ions demonstrated recently [1,2] by focusing intense laser pulses on solid foils has triggered a number of theoretical and experimental studies, relevant for the fast ignitor scheme in the inertial confinement fusion concept [3], medical applications [4,5,6], and dense plasma diagnostics and probing [7]. The dominant acceleration mechanism in these experiments was the Target Normal_Sheath Acceleration (TNSA) mechanism [8]. Another promising way to accelerate ions to high energies is to use underdense or near-critical density targets. Compared to solid targets where laser absorption is limited to the target surface, the laser pulse inside low density plasmas heats electrons on a large volume leading to higher laser absorption. This acceleration regime is also advantageous for several applications as less debris are produced. It is also more adapted to high repetition rate lasers. The laser contrast, which can be problematic with thin solid targets, is less detrimental as gas jets are transparent for laser intensities below 10^{13} W/cm^2. Ion acceleration due to the interaction of a very high intensity laser with underdense targets has already been studied with PIC simulations [9] but the simulated plasmas never reproduced the gaseous targets available for experiments. The first experiments of ion acceleration via laser interaction with an underdense target showed radial acceleration [10]. Recent experiments showed that strong longitudinal proton acceleration can be achieved using high intensity lasers [11]. The longitudinally accelerated protons are in this case far more energetic than the transverse accelerated protons. The exact accelerating mechanism leading to the acceleration of energetic ions observed in [11] has not yet been described in detail and therefore is the center of an ongoing debate [12]. For an underdense target, is it not easy to produce a sharp gradient plasma experimentally. It is therefore important to understand the dependence of ion acceleration on the density gradient to optimize it. In this article, we show that even using underdense plasmas, protons can be accelerated efficiently in term of energy, charge, and collimation. Using Particle-In-Cell simulations, we studied in detail the interaction of a high intensity laser pulse with an underdense plasma. In section 2, we describe the shock-like regime of underdense laser ion acceleration. We discuss the dependence on laser and target parameters of the shock-like regime in Section 3. Section 4 addresses the possibility to study the shock-like regime experimentally by coupling a Particle-In-Cell code with a hydrodynamic code.

2. Detailed description of the shock-like accelerating mechanism

In this section, we demonstrate the efficiency of the shock-like regime of laser underdense ion acceleration using 2D Particle-In-Cell simulations performed with the code PICLS [13]. The wavelength of the incident pulse is 1 μm, its pulse duration 23 fs and its irradiance 3.2×10^{19} W/cm^2. The full-width-at-half-maximum (FWHM) of the focal spot is 6 μm. The pulse interacts with the target in normal incidence. Its electric field is in the simulation plane (p-polarization). The spatial and temporal profiles are truncated Gaussians. The plasma is composed of protons and electrons with a 0.08 n_c maximum density, a cosine-square density profile with a 100 microns FWHM in the x-direction and uniform in the y-direction. Protons are accelerated with a shock-like structure [14] as shown in Fig. 1.c. After 3.3 ps, the maximum proton energy saturates at 8.4 MeV (Fig. 1.a) showing the efficiency of this acceleration mechanism compared to the TNSA mechanism with similar laser parameters. Figs. 1.a and 1.b also demonstrate that the number of accelerated protons and their divergence are comparable with TNSA-accelerated beams. The number of accelerated particles is calculated using the conservative hypothesis that the characteristic dimension in the third direction corresponds to the laser pulse FWHM.

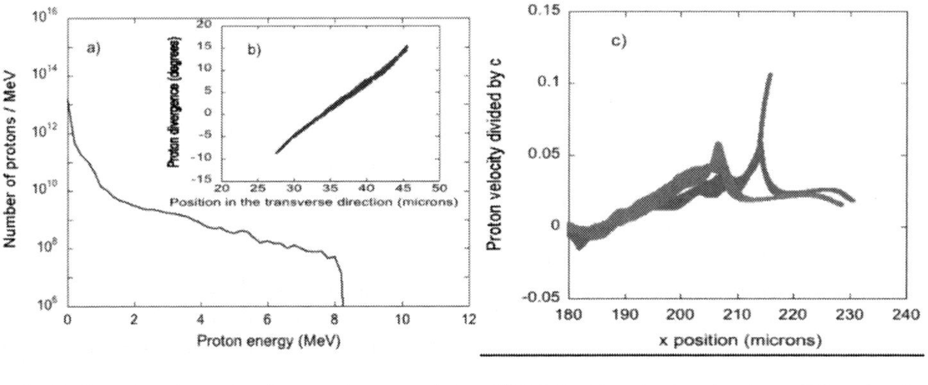

Figure 1: (a,b) Simulated proton beam characteristics 3285 fs after the beginning of the simulation. (c) Proton phase space 1.47 ps after the beginning of the simulation (red) and 1.83 ps after the beginning of the simulation (green).

The shock-like mechanism starts as a strong asymmetrical Coulomb explosion in the decreasing density gradient and evolves into wave breaking driven by the strong electric field present in this region. In 1D, this field is created by energetic electrons accelerated during the laser pulse propagation. In 2D and 3D, this field is also enhanced by the electric field generated due to the fast evolving strong magnetic field gradients in this region. If the ion wave launched during this process is fast enough and crosses other ions, these ions can be reflected on this ion wave and accelerated at twice the velocity of the incoming ion wave. This is clearly visible in Fig. 2.a in the strong acceleration phase between t=1.5 ps and t=1.8 ps (same times as in Fig. 1.c). The velocity increase during this phase is not correlated to the variation in maximum electric field. This acceleration process is therefore referred to as the shock-like regime of underdense laser ion acceleration.

3. Dependence of the shock-like mechanism on laser and target parameters

The shock-like regime depends strongly on the characteristics of the density gradient. For sharp density gradients, most energetic protons are accelerated at the back surface, similarly as in TNSA. For small density gradients, the most energetic protons are accelerated close to the back surface and wave breaking is observed. For intermediate density gradients, the most energetic protons are accelerated by a shock-like mechanism in the decreasing density ramp. If the density profile becomes too long, wave breaking is again the dominant mechanism. The plasma scale length plays an important role in the optimization process as it controls laser absorption in energetic electrons responsible for the strong accelerating longitudinal electric field. Other target parameters like the maximum density and the overall length of the plasma also influence the efficiency of the process as they act on the initial asymmetrical Coulomb explosion and the topology of the mentioned magnetic field. Laser intensity, pulse duration and laser focal spot width also play a role on the efficiency of the acceleration process. For each set of laser parameters and target density and shape, an optimum configuration exists. Fig 2.b shows the evolution of the maximum proton energy in the case of 1D simulations of the interaction of a 6×10^{19} W/cm^2, 23 fs laser pulse with a 33 microns FWHM cosine-square target.

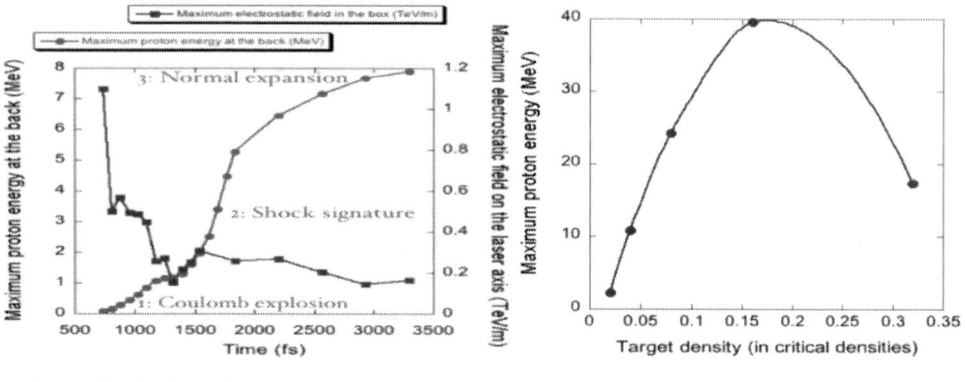

Figure 2: (a) Evolution of the maximum longitudinal electrostatic field (blue) and of the maximum proton energy (red) as a function of time. (b) Evolution of the maximum proton energy as a function of target density.

4. Coupling of a Particle-In-Cell code and a hydrodynamic code to assess the possibility of inestigating the shock-like regime in experiments

As mentioned above, part of the optimization of this acceleration process comes from achieving plasmas with particular densities and gradients. For this, it is possible to explode a thin foil using a long pulse laser. We have performed 2D hydrodynamic simulations of such plasma production using the code CHIC [15]. Figure 3.a shows the result of the interaction of a 5×10^{13} W/cm^2, 1 ns laser pulse focused to 10 microns FWHM focal spot with CH foils with different initial thicknesses after 800 ps of interaction. Using variable initial foil thickness or interaction duration allows to investigate the interaction of a short pulse with targets presenting different maximum density, length and density gradients. We used these results in PICLS to simulate in 2D the interaction of a 5×10^{19} W/cm^2, 350 fs laser pulse with a 10 microns FWHM focal spot on the exploded foil to accelerate protons to high energies. After 500 ps of interaction in the hydrodynamic simulation, the target has approximately a cosine-square density profile with a 120 microns FWHM. The maximum proton energy measured 3.3

ps after the beginning of the PIC simulation is 32 MeV (Fig. 3.b). This energy is comparable to the maximum proton energies obtained using similar laser parameters with solid targets.

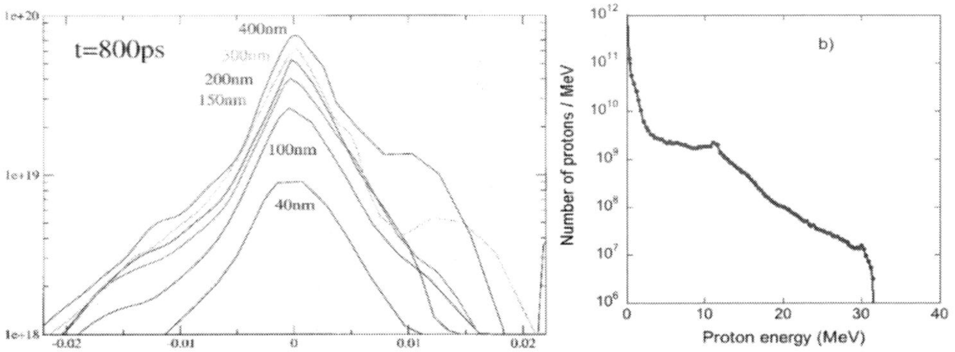

Figure 3: (a) Electron density profile at the center of the simulation box for various initial CH target thicknesses after 800 ps (the x-axis is in cm and the y-axis is in cm^{-3}). (b) Proton energy spectrum in the forward direction after 3.3 ps using an initial thickness of 400 nm.

5. Conclusions

PIC simulations were used to study proton acceleration with underdense targets in various density gradient conditions. The propagation of a high intensity laser pulse in an underdense target can lead to the production of energetic proton beams. The most efficient acceleration observed in such targets is the result of the formation of a shock-like structure in the decreasing density gradient. This regime is strongly dependent on the density gradient length and shape and on the shape of the laser pulse. It features a very high ratio of maximum proton energy over laser energy, a low divergence and a high number of accelerated protons but is nowadays difficult to achieve experimentally with short pulse lasers because it requires short, dense and sharp underdense plasmas. A setup including a long pulse laser to explode a thin solid foil is proposed to study this regime experimentally. This mechanism could open important perspectives for the optimization of high-energy proton beams.

References

[1] E. Clark *et al.*, Phys. Rev. Lett. **84**, 670 (2000).
[2] R. A. Snavely *et al.*, Phys. Rev. Lett. **85**, 2945 (2000).
[3] M. Roth *et al.*, Phys Rev ST-AB **5**, 061002 (2002).
[4] S. Fritzler *et al.*, Appl. Phys. Lett. **83**, 3039 (2003).
[5] V. Malka *et al.*, Med. Phys. **31**, 6 (2004).
[6] K. W. D. Ledingham *et al.*, J. Phys. D: Appl. Phys. **37**, 2341 (2004).
[7] M. Borghesi *et al.*, Phys. Plasmas **9**, 2214 (2002).
[8] S. C. Wilks *et al.*, Phys. Plasmas **8**, 542 (2001).
[9] T. ZH. Esirkepov *et al.*, JETP Lett. **70**, 82 (1999), M. Yamagiwa *et al.*, Phys. Rev. E **60**, 5987 (1999), Y. Sentoku *et al.*, Phys. Rev. E **62**, 7271 (2000).
[10] K. Krushelnick *et al.*, Phys. Rev. Lett. **83**,737 (1999).
[11] L. Willingale *et al.*, Phys. Rev. Lett. **96**, 245002 (2006), A. Yogo *et al.*, Phys. Rev. E **77**, 016401 (2008), P. Antici *et al.*, New J. Phys. **11**, 023038 (2009).
[12] S. V. Bulanov *et al.*, Phys. Rev. Lett. **98**, 049503 (2007), L. Willingale *et al.*, Phys. Rev. Lett. **98**, 049504 (2007).
[13] Y. Sentoku and A. Kemp, J. Comput. Phys. **227**, 6846 (2008).
[14] V.T. Tikhonchuk et al., Plasma Phys. and Control. Fusion **47**, 869 (2005), F. Peano et al., Phys. Plasmas **14**, 056704 (2007).
[15] J. Breil et al, J. Comp. Phys. **224**, 785 (2007).

Journal of Physics: Conference Series **244** (2010) 042024 doi:10.1088/1742-6596/244/4/042024

Broadening of the energy spectrum of ions accelerated by laser-driven shocks in over-dense slab plasmas

Quan-Li Dong[1,*], Min-Qing He[1], Zheng-Ming Sheng[1,2], Min Chen[1], Shou-Jun Wang[1], and Jie Zhang[1,2]

[1] Institute of Physics, Chinese Academy of Sciences, Beijing 100190, China

[2] Department of Physics, Shanghai Jiaotong University, Shanghai, 200240, China

E-mail: * qldong@aphy.iphy.ac.cn

Abstract. We investigated numerically the broadening mechanisms of the energy spectrum of ions accelerated by electrostatic shocks in over dense plasmas irradiated by intense ultra-short laser pulses. It was found that the width of the shock-accelerated ion spectrum could be ineluctably broadened by two mechanisms, which are the continuously decreasing speed of the shock front due to the energy dissipation into ions, and the further acceleration by the sheath field at target rears when energetic ions arrive there. Other effects, such like that of driving laser pulse duration, were also studied.

1. Introduction

It has been shown that ions located deep in plasmas can be accelerated by electrostatic shock waves driven by intense laser pulses. The formation of collisionless electrostatic shocks (CLESs) and their capabilities of accelerating ions were studied both theoretically and experimentally in plasma stream collisions or intense laser-plasma interactions [1-5]. With particle-in-cell PIC simulations it was found that the shock acceleration of ions can even dominate the sheath acceleration at the target rear when the target thickness is under a certain threshold, which is determined by the intensity of the incident laser pulses [2]. The accelerated ions are expected to present a plateau-shaped energy spectrum, which is proven recently by spherical mass-limited target experiments that manifest converging shocks producing collimated proton beams. Detailed studies of the shock formation and propagation, and its acceleration of ions have indicated that such a plateau shape is caused by two mechanisms: the energy dissipation from the shock front, and the indistinguishable acceleration by the sheath field at the target rear when the shock-accelerated ions arrive [3, 4]. Any proposal trying to produce mono-energetic ion beams with laser-driven shocks should take into consideration of the above two processes.

2. Accelerated ion spectrum broadening due to the energy dissipation of the electrostatic shocks

The speed of the shock front decreases as the wave propagates since the shock front loses energy to ions in the acceleration process. As the result, the ion energy spectrum widens as the shock propagates over longer distances. It was expected that by using a sandwich target, which consists in the central part an ultra-thin layer of heavier ions, mono-energetic heavy ion beam could be produced [4]. This target configuration was once brought out trying to avoid the ion energy spectrum broadening due to the decreasing shock speed by selecting a *single* shock speed with the thinner heavy ion layer positioned in the target middle area. The position of the heavy ion layer could help to select different shock speed. The energy spread of the heavy ion beam is still, however, determined by the thickness of the central layer. Fig.1 (a) and (b) show the temporal evolutions of the heavy ion distribution in their phase space when sandwich targets applied with central layers of different thickness. For both cases, the electron density is set as n_e=20n_c, the ion density of the Left and the Right light ion (helium) layers are set as n_{i1L}=n_{i1R}=20n_c, and the heavy ion (carbon) density is set as 5n_c with the charge set as 4^+. In the case with a thicker heavy ion layers as in Fig.1(a), the thickness of the Left and the Right layer of light ions were set as l_{i1L}=0.625λ_0, l_{i1R}=1.625λ_0 and that of the central heavy ion layer l_{i2} =0.25λ_0, respectively. In the second case with a thinner heavy ion layer, as in Fig.1 (b), l_{i1L}=0.625λ_0 l_{i1R}=1.75λ_0 and l_{i2} =0.125λ_0. The bold lines in Fig.1 (b) show the positions of the light and the heavy ion layers. With the heavy ion layer as thin as l_{i2} =0.125λ_0, the shock break through at t=130τ_0, while the breaking out time for the central layer of 0.25λ_0 is t=175τ_0, as shown by the red crosses in Fig.1 (a) and (b). Here λ_0 and τ_0 are respectively the wavelength and the period of the applied laser pulse shaped as $\sin^2(t)$ with the duration for both cases as 80τ_0. For the thicker heavy ion layer, the shock speed decreasing behavior is clearly shown when the shock propagates in the heavy ion layer, while for thinner heavy ion layer, such phenomenon is unconspicuous. From this point of view, the sandwich target works partially as expected to select a single shock speed. This can not, however, ensure the mono-energy ion beam generation, because after the shock's breakthrough the heavy ion layer, the shock-accelerated ion bunch might begin to explode due to the coulomb repulsion between ions, as shown in the heavy ion phase space, Fig.1 (a) and (b). This process causes the expansion of the ion bunch. A large extension in space will deteriorate the mono-energy quality of the ion bunch as it arrives at the target rear, where the sheath field is spatially and temporally variable. This phenomenon is discussed below in Section 3. Another important phenomenon in the simulations is that not all heavy ions experiencing the shock front could be accelerated. It has been pointed out by Chen and He in their simulation studies [3, 4] that the shock can not trap all the ions in its propagating path, resulting into a lower energy tail in the spectrum of shock-accelerated ions. To push all ions in the central heavy ion layer, the laser pulse needs to propagate through to the end of the layer. Of course, such ion acceleration should be attributed to that by the laser front, not by a shock. The effects on the laser pulse duration are discussed below in Section 4.

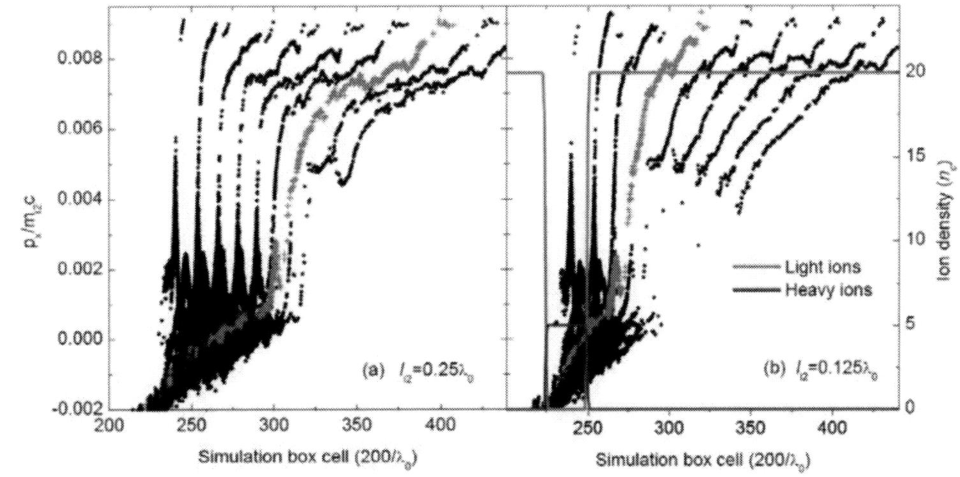

Figure 1. The temporal evolutions of the heavy ion distributions in phase space when sandwich targets applied with central heavy ion layers of different thickness. The time is between t=80 τ_0 and 205 τ_0 from the left to the right in the graph with the step of 15 τ_0. (a) n_{i1L}=n_{i1R}=4n_{i2}=20n_c, l_{i1L}=0.625λ_0 and l_{i1R}=1.625λ_0, l_{i2} =0.25λ_0, a=5, T=80τ_0. (b) l_{i1L}=0.625λ_0, l_{i1R}=1.75λ_0 and l_{i2} =0.125λ_0, other parameters are the same to that in (a). The red crosses in both cases are phase-space plots right after shocks breaking through the central heavy ion layers.

3. Further acceleration of ions in the sheath field at target rears

It has been reported that the shock-accelerated ions can be further accelerated in the sheath field formed in the expanding plasma part at target rears [2]. The solid line in Fig.2 (a) indicates the temporal evolution of the longitudinal electric field experienced by one accelerated test ion. The increasing part of the electric field after $60\tau_0$ is the sheath field at the target rear. The dashed line in Fig.2 (a) is the longitudinal momentum of the accelerated ion [3]. It is obvious that the ion was accelerated between $25\tau_0$ and $45\tau_0$, after which period, it propagates with quasi-constant velocity through the unperturbed region of the plasma, and finally arrived at the target rear, where it is further accelerated. Fig.2 (b) shows the temporal evolution of the shock accelerated heavy ion energy spectrum in simulations with sandwich targets. The ion energy spectrum presents a narrow width around the peak of 0.26 MeV just after the shock breaks through the heavy ion layer at 130 τ_0. After a long time propagation in the bulk plasma, the earlier accelerated part of the heavy ion bunch arrives at the sheath field at the target rear at 310 τ_0, and continues to be accelerated to higher energies, as indicated by the shift of the peak to ~0.3 MeV. The width of the spectrum is also broadened as expected. In the contrast, at this time, those ions at the relatively lower energy tail of the heavy ion bunch still keep their relatively lower velocity. The situation is changed at 400 τ_0, when the ion bunch tail arrives at the target rear, where those ions are also accelerated to form the second energy spectrum peak at 0.16 MeV. Both peaks are broadened at later time of 450 τ_0 and 500 τ_0, which phenomenon appears 400 τ_0 after the laser pulse finishes. The further acceleration by the sheath field at the target rear after the shock appears to be unavoidable. As Silva *et al.* pointed out that the energy spectrum of ions accelerated by shocks evolves into a plateau, which was demonstrated by Henig *et al.* as the significant feature of this accelerating mechanism in experiments.

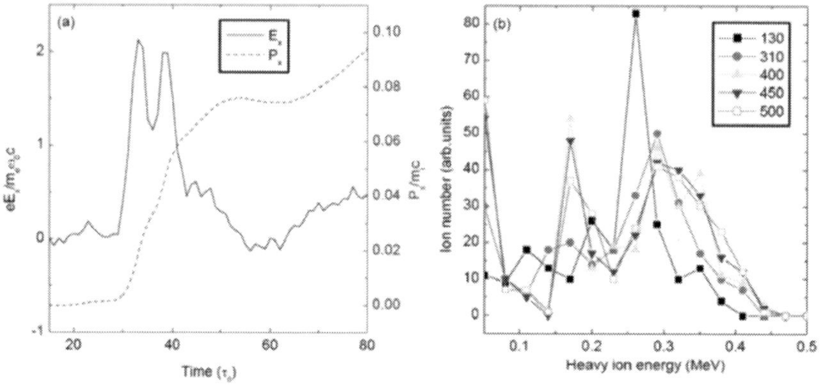

Figure 2. The temporal evolution of the longitudinal electric field experienced by one accelerated test ion (solid line in (a)). The dashed line is the longitudinal momentum of the accelerated ion [3]. (b) The temporal evolution of the heavy ion energy spectrum. The further acceleration of heavy ions by the sheath field at target rear broadens the spectrum long after the finish of the shock acceleration process.

4. The duration effects of driving laser pulses on the energetic ion spectrum

During the pulse duration, the laser continues to heat the plasma to a higher temperature, which in turn determines the shock speed and the penetration depth of the laser pulse into the plasma. As mentioned above, to push forward the whole central heavy ion layer in sandwich targets, the laser pulse need to be long enough or the plasma density is low enough so that the intense part of the laser pulse can bore into the plasma until the end of the layer. With the same target configuration, we studied the pulse duration effects on the heavy ion acceleration in the central layer of the sandwich targets. Fig.3 (a) and (b) give the heavy ion distributions at different times in the phase space when the pulse duration are

130 τ_0 and 180 τ_0, respectively. Fig.3 can also be compared with Fig.1 (b), which is the phase space of heavy ions when the laser pulse is 80 τ_0, showing a real shock-acceleration process. For the longest laser pulse situation, the whole heavy ion layer is pushed forward at 115 τ_0. However, the quality of the heavy ion energy spread worsens as the whole layer propagate into the depth of the slab plasma due to effects of the existing light ion layer that are partially accelerated by the laser pulse.

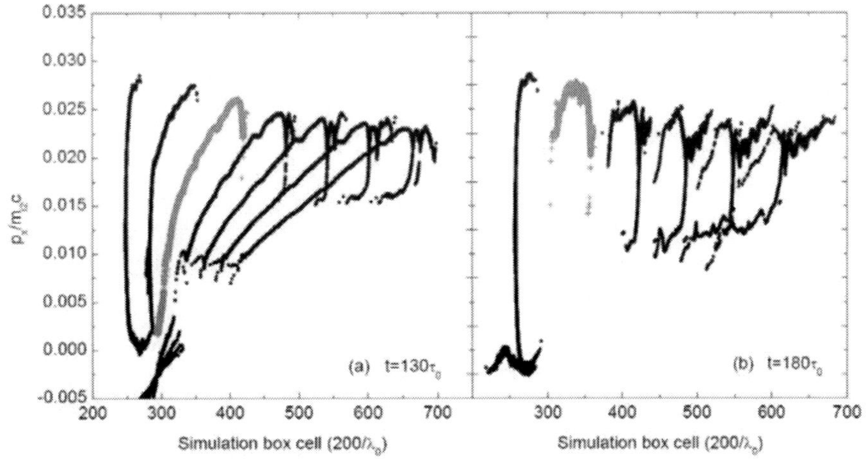

Figure 3. The temporal evolutions of the heavy ion distributions in phase space with sandwich targets. Laser pulses with different durations were used, 130 τ_0 for (a), and 180 τ_0 for (b). Other simulation parameters are the same to those of Fig.1 (b). The red crosses are for the time when the heavy ion layers completely leave their original positions.

5. Summary

We have shown two kinds of mechanisms that broaden the ion spectrum of the ions accelerated by laser-driven shocks. By introducing the expected heavy ions as the central ultra-thin layer of sandwich targets, it is possible to select a *single* shock speed so that the energetic ion spectrum can *partially* avoid broadening due to the energy dissipation from the shock front to the accelerated ions. But the following exploding effects within the ion bunch, and the further acceleration of the shock-accelerated ion bunch that takes place long after the shock acceleration process, will ineluctably worsen the mono-energy quality of the ion beam in experiments. The collision between energetic ions and the background ions is also found to have effects (not described here). Analysis only of the characteristics of the early shock acceleration process might come into being a misleading conclusion. Even with a longer intense laser pulse that propagates through to the end of the central ultra-thin heavy ion layer of the sandwich targets, it is not easy to obtain a mono-energy ion beam.

6. Acknowledgments

This work is jointly supported by the Natural Science Foundation of China (Grant No. 60621063, 60678007,10774184), the CAS project KJCX2-YW-T01, and the National Basic Research Program of China No.2007CB815101.

References:

[1] Denavit J 1992 Phys. Rev. Lett. **69** 3052
[2] Silva L O, Marti M, Davies J R *et al.* 2004 Phys. Rev. Lett. **92** 015002
[3] He Q-M, Dong Q-L, Sheng Z-M *et al.* 2007 Phys. Rev. E **76** 035402(R)
[4] Chen M, Sheng Z-M, Dong Q-L *et al.* 2007 Phys. Plasmas **14** 113106
[5] Henig A, Kiefer D, Geissler M *et al.* 2009 Phys. Rev. Lett. **102** 095002

Journal of Physics: Conference Series **244** (2010) 042025 doi:10.1088/1742-6596/244/4/042025

Control of proton energy in ultra-high intensity laser-matter interaction

A. Maksimchuk[1], S. S. Bulanov[1], A. Brantov[2], V. Yu. Bychenkov[2], V. Chvykov[1], F. Dollar[1], D. Litzenberg[3], G. Kalintchenko[1], T. Matsuoka[1], S. Reed[1], V. Yanovsky[1], and K. Krushelnick[1]

[1]*FOCUS Center and Center for Ultrafast Optical Science, University of Michigan, Ann Arbor, 48109, USA*
[2]*P. N. Lebedev Physics Institute, Russian Academy of Sciences, Moscow 119991, Russia*
[3]*Department of Radiation Oncology, University of Michigan, Ann Arbor, 48109*

tolya@umich.edu

Abstract. Recent breakthroughs in short pulse laser technology resulted in (i) generation of ultra-high intensity ($2x10^{22}$ W/cm^2) and (ii) ultra-high contrast (10^{-11}) short pulses at the Hercules facility of the University of Michigan, which has created the possibility of exploring a new regime of ion acceleration – the regime of Directed Coulomb Explosion (DCE). In this regime of sufficiently high laser intensities and target thicknesses approaching the relativistic plasma skin depth it is possible to expel electrons from the target focal volume by the laser's ponderomotive force allowing for direct laser ion acceleration combined with a Coulomb explosion. That results in greater than 100 MeV protons with a quasi-monoenergetic energy spectrum. The utilization of beam shaping, namely, the use of flat-top beams, leads to more efficient proton acceleration due to the increase of the longitudinal field. According to the results of 2D PIC simulations a 500 TW laser pulse with a super-Gaussian beam profile interacting with 0.1 micron aluminium-hydrogen foil is able to produce monoenergetic protons with the energy up to 240 MeV.

1. Introduction

With the development of short-pulse high-intensity lasers it has become possible to generate beams of energetic particles from laser plasma interactions such as relativistic electrons and multi-MeV ions [1]. These beams of ions have a wide array of applications such as a compact source of radioactive isotopes [2], a table-top fission source [3] and as a ion injector source [4]. Beams of energetic protons can be used to generate quasi-homogeneous warm dense matter by the isochoric heating of solid density foils [5] or can be used as a radiography source to detect and study the evolution of electric fields in laser-matter interactions [6]. The possibilities to use beams of laser driven ions for hadron radiation therapy [7], for the fast ignitor research [8] and even for generation of exotic particles [9] have also been discussed.

Most proton beam applications require repetitive and controllable quasi-monoenergetic beams with energies that are an order of magnitude above current laser accelerated protons. In the majority of the experiments to date, protons from laser-solid interactions were produced in the target normal sheath acceleration (TNSA) regime [10] from foils with thicknesses orders of magnitude greater than the plasma skin depth. This generates proton beams with energies up to several to tens of

MeV but with an exponential energy spectrum in the number of accelerated protons followed by a cutoff.

Here we propose to explore a new regime of ion acceleration using the interaction of ultra-relativistic, ultra-short and ultra-clean laser pulses with ultra-thin membranes at focused intensities $\sim 10^{22}$ W/cm^2 to achieve controllable quasi-monoenergetic beams of protons with energies in the 100-200 MeV range in the Directed Coulomb Explosion regime [11,12].

2. Directed Coulomb Explosion Regime

When an intense, prepulse-free, ultra-short laser pulse, I~10^{22} W/cm^2, directly interacts with an ultra-thin foil, it ionizes the target in a few femtoseconds maintaining the integrity of overdense plasma. It is reasonable to assume that under the action of the intense laser pulse, the electrons are evacuated from the foil region with transverse dimensions of the order of the focal spot diameter. If the laser field is much stronger than the Coulomb attraction field, the ions cannot retain electrons near the backside of the target, which leads to ion acceleration in the so-called Coulomb explosion regime [13]. In order to expel all the electrons and achieve Coulomb Explosion the following condition on laser electromagnetic vector-potential, a = 0.85 (I [W/cm^2] λ^2[μm] 10^{-18})$^{1/2}$ and foil thickness l must be satisfied

$$a > \pi \frac{N_e}{n_{cr}} \frac{l}{\lambda} \qquad (1)$$

Here, N_e is the electron density, n_{cr} is the plasma critical density, and λ is the laser wavelength. In order to estimate the typical energy of the accelerated ions in the Coulomb explosion regime, we assume that all the electrons produced by the ionization in the focal spot region are forced to leave the foil. In this case, the electric field near the positively charged layer is, $E_0 = 2\pi N_i Z_i e l$, where N_i is the heavy ion density in the foil, and Z_i is the heavy ion electric charge. The size of the region where this estimation for E_0 is valid is of the order of the focal spot size, d, in both transverse and longitudinal directions providing a one-dimensional regime of ion acceleration. When the ions leave this region, the Coulomb explosion regime becomes three-dimensional, leading to an immediate drop in ion acceleration efficiency [14]. Thus the maximum energy of the protons can be estimated as

$$E_{\max} = \pi N_i Z_i e^2 l d \ . \qquad (2)$$

It follows from Eq. (2) that $E_{\max}$ increases linearly with foil thickness. However, the foil thickness, l, in equation (2) must satisfy the Coulomb explosion regime condition, $l/\lambda < a n_{cr} / \pi N_e$. Correspondingly, one may conclude that for a given laser intensity and plasma density there is an optimum foil thickness for which the accelerated ions reach maximum possible energy. This was confirmed in 2D PIC simulations [11] (see Fig. 1a). Further increasing of foil thickness does not provide an ion energy increase.

For very thin targets laser light just transmits through a foil without pushing it. The transparency condition has the form $l/\lambda < \pi a N_e / n_{cr}$, that coincides with the Coulomb explosion condition (1). When a foil is thicker, laser light reflects and accelerates foil by radiation pressure [16]. In this case the laser field acts as a piston driving a flow of heavy ions tearing across the foil. Thus the regime described here is an efficient combination of the radiation pressure and Coulomb explosion effects and we call it a Directed Coulomb Explosion regime (DCE).

We used the 2D PIC code REMP [17] to numerically study the acceleration of ions in high-intensity laser interactions with an ultra-thin two-layer aluminum-hydrogen foil. In the simulations presented here, the uniform grid mesh size is λ/200. The pulse is linearly polarized along the z-axis. The temporal and spatial profiles of the pulse are Gaussian if not stated otherwise. The following parameters were used in the simulations: laser power of 150-500 TW, pulse duration of 30 fs, and a spot size of 1.0λ (FWHM). The aluminum layer thickness was varied from l=0.0125λ to l=0.2λ. The electron density of the foil is $400n_{cr}$, where n_{cr} is the critical density of the plasma. The hydrogen layer thickness was 0.05λ with an electron density of $10n_{cr}$.

The calculated spectrum of the accelerated protons is shown in Fig. 1b. The spectrum has quasi-monoenergetic features and is peaked at 140 MeV, with FWHM of 10 MeV or with the energy spread $\Delta E/E_{peak} = 6\%$. The width of the peak is mainly due to the Coulomb repulsion of protons and can be varied by changing the thickness of hydrogen layer, giving an additional parameter that controls the acceleration of protons in laser-foil interactions. Proton spectra at lower power exhibit similar features but the peaks are positioned at lower energies, i.e. at 45 MeV for 250 TW pulse and at 20 MeV for 150 TW. We estimated than ~4.10^8 protons are contained in the monoenergetic peak at 500 TW.

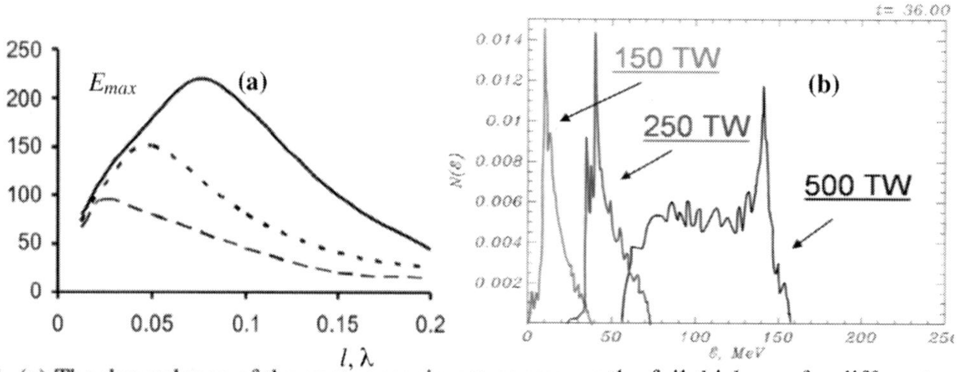

Fig. 1. (a) The dependence of the proton maximum energy on the foil thickness for different values of the laser pulse power: 150 TW (dashed curve), 300 TW (dotted curve), 500 TW (solid curve). (b) The spectra of protons accelerated from the double layer foil for 150 TW, 250 TW and 500 TW Gaussian pulses.

3. Proton Acceleration in DCE regime with Super-Gaussian Beams

Since proton acceleration is mainly due to the charge separation electric field, an increase in the acceleration effectiveness can be achieved by the enhancement of this field effect. Further optimization of low Z ion beam quality (maximum energy increase and energy spread decrease) can be achieved by employing the shaping of laser beams.

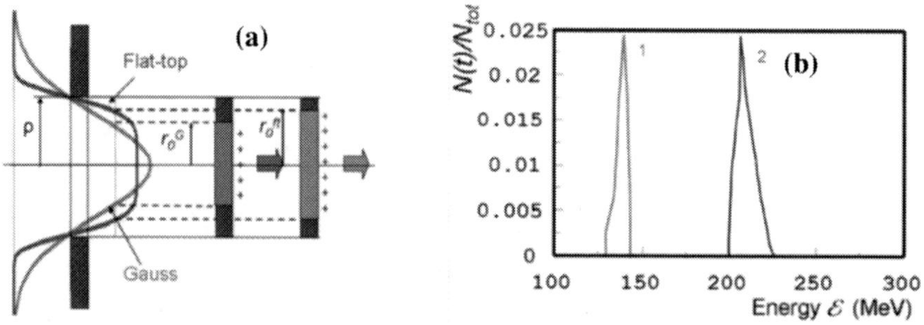

Fig.2. a) Principal scheme of foil fraction acceleration by Gaussian (radius of the area with all electrons expelled is r_0^G) and flat-top beams (radius of the area with all electrons expelled is r_0^{ft}); b) Spectra of protons that are accelerated inside an angle of 10 degrees to the target normal by Gaussian (1) and flat-top (2) beams (a 500 TW laser pulse interacts with a 0.1λ thick aluminium foil with a 50 nm second layer of hydrogen).

The utilization of flat-top (super-Gaussian) beams enhances the longitudinal field for the same laser power, leading to maximum energy increases up to 45 % above the value generated by the conventional Gaussian beams for low Z ions according to the theoretical estimates [17]. This effect is due to the fact that flat-top beams evacuate electrons from a larger area on the foil, thus generating a stronger longitudinal electric field, as well as more effectively preventing the electrons from returning to the evacuated region due to a higher ponderomotive force (see Fig. 2a).

Our 2D PIC simulations show that the resulting proton spectrum demonstrates the same peaked structure as in the Gaussian beam case (Fig. 2b), but the maximum energy is increased by 45 % up to 250 MeV for 500 TW laser pulses, as it was predicted by the theoretical estimate of the effect [17]. Another advantage of using super-Gaussian beams is that the relative width of the high-energy peak in proton spectrum is smaller. This is mainly due to larger maximum energy, since the Coulomb explosion of the moving proton layer determines the absolute value of the energy spread in proton spectrum. This fact gives us one more parameter of laser-foil interaction that controls the energy spread: the initial thickness of the hydrogen layer.

5. Summary

We suggested accelerating protons to energies greater than 100 MeV with quasi-monoenergetic spectral features in the regime of Directed Coulomb Explosion using prepulse free interactions of a ultra-relativistic laser with ultra-thin membranes. The repetitive Ti:Sapphire Hercules laser at the University of Michigan recently set the world record for the highest focused intensity of $2x10^{22}$ W/cm^2 [17] and ASE intensity contract of 10^{-11} [18] and is capable to demonstrate this new regime of ion acceleration.

This work was supported by the NSF grant FOCUS (PHY-0114336).

References

[1] E. L. Clark *et al., Phys. Rev. Lett.* **84**, 670 (2000); A. Maksimchuk *et al., ibid.* **84**, 4108 (2000); R. A. Snavely *et al., ibid.* **85**, 2945 (2000).
[2] M. I. K. Santala *et al., Appl. Phys. Lett.* **78**, 19 (2001); K. Nemoto *et al., ibid.* **78**, 595 (2001).
[3] K. W. D. Ledingham, P. McKenna, R. P. Singhal, *Science* **300**, 1107 (2003).
[4] K. Krushelnick *et al., IEEE Trans. on Plasma Science* **28**, 1184 (2000).
[5] P. K. Patel *et al., Phys. Rev. Lett.* **91**, 125004 (2003).
[6] M. Borghesi *et al., Phys. Plasmas* **9**, 2214 (2002).
[7] S. V. Bulanov *et al., Plasma Phys. Rep,* **28**, 453 (2002).
[8] M. Roth *e. al., Phys. Rev. Lett.* **86**, 436 (2001); V. Yu. Bychenkov *et al. Plas. Phys. Rep.* **27**, 1017 (2001).
[9] V. Yu. Bychenkov *et al., JETP Lett.* **74**, 586 (2001).
[10] S. C. Wilks *et al., Phys. Plasmas* **8**, 542 (2001).
[11] S. S. Bulanov *et al., Med. Phys.* **35**, 1770 (2008).
[12] S. S. Bulanov *et al., Phys. Rev. E* **78**, 026412 (2008).
[13] S. V. Bulanov *et al., Phys. Lett. A* **299**, 240 (2002); E. Fourkal *et al., Phys. Rev. E* **71**, 036412 (2005).
[14] V. Yu. Bychenkov and V. F. Kovalev, *Quant. Electronics* **35**, 1143 (2005).
[15] T. Zh. Esirkepov *et al., Phys. Rev. Lett.* **92**, 175003 (2004).
[16] T. Zh. Esirkepov *et al. Comp. Phys. Commun.* **135**, 144 (2001).
[17] V. Yanovsky *et al., Optics Express* **16**, 2109 (2008).
[18] V. Chvykov *et. al., Opt. Lett.* **31**, 1456 (2006).

Synchrotron x-ray radiation from laser wakefield accelerated electron beams in a plasma channel

T. Matsuoka[1], S. Kneip[2], C. McGuffey[1], C. Palmer[2], J. Schreiber[2], C. Huntington[3], Y. Horovitz[1,5], F. Dollar[1], V. Chvykov[1], G. Kalintchenko[1], A.G.R. Thomas[1], V. Yanovsky[1], K. Ta Phuoc[4], S. P. D. Mangles[2], Z. Najmudin[2], A. Maksimchuk[1], and K. Krushelnick[1]

[1]FOCUS Center and Center for Ultrafast Optical Science, University of Michigan, Ann Arbor, 48109, USA

[2]The Blackett Laboratory, Imperial College London, London, SW7 2BZ, UK

[3]Atmospheric Oceanic and Space Sciences, Space Physics Research Lab., University of Michigan, Ann Arbor, MI 48109 USA

[4]Laboratoire d'Optique Appliquee, ENSTA, Ecole Polytechnique, Palaiseau, 91761, France

[5] Present address: Dynamical Experiments Group, Propulsion Division, Soreq NRC, Yavnee 81800, Israel.

matsuoka@eecs.umich.edu

Abstract. Synchrotron x-ray radiation from laser wakefield accelerated electron beams was characterized at the HERCULES facility of the University of Michigan. A mono-energetic electron beam with energy up to 400 MeV was observed in the interaction of an ultra-short laser pulse with a super-sonic gas jet target. The experiments were performed at a peak intensity of 5×10^{19} W/cm^2 by using an adaptive optic. The accelerated electron beam undergoes a so called "betatron" oscillation in an ion channel, where plasma electrons have been expelled by the laser ponderomotive force, and, therefore, emits synchrotron radiation. We observe broad synchrotron x-ray radiation extending up to 30 keV. We find that this radiation is emitted in a beam with a divergence angle as small as 12×4 mrad^2 and can have a source size smaller than 3 microns and a peak brightness of 10^{22} photons/mm^2/mrad^2/second/0.1% bandwidth, which is comparable to currently existing 3rd generation conventional light sources. This opens up the possibility of using laser-produced "betatron" sources for many applications that currently require conventional synchrotron sources.

1. Introduction

Laser wakefield acceleration (LWFA) is an active research field since it can potentially enable compact, energy tunable and cost-effective sources of high-energy electron beams [1-3]. One attractive application of those electron beams is the use as an x-ray source for imaging applications. "Betatron" motion of electrons in a plasma with respect to x-ray generation has been studied experimentally [4-5] and theoretically [6]. In this article, the experimental characterization of a synchrotron x-ray source from LWFA is reported.

2. Experimental configurations

The experimental geometry is shown in Figure 1. The HERCULES facility of the University of Michigan [7] provided an interaction pulse ($\lambda = 800$ nm, 30 fs) which irradiates supersonic He gas-jets. A 10 cm laser beam was focused by a 1 m focal length off-axis parabolic mirror after wavefront correction with a deformable mirror to a spot size of 10 μm (FWHM) containing 40 % of the energy. The laser beam wavefront was corrected at a low energy (~ 0.35 mJ) by the method described in [8]. A focused peak intensity of 8×10^{19} W/cm^2 can be reached for 100 TW laser power. For these experiments the temporal contrast at the nanosecond time scale was 10^{-8}. A probe pulse was split-off from the interaction pulse by a 2 μm thick pellicle and sent through the plasma orthogonally to the interaction pulse in order to measure the plasma density via a shearing interferometer.

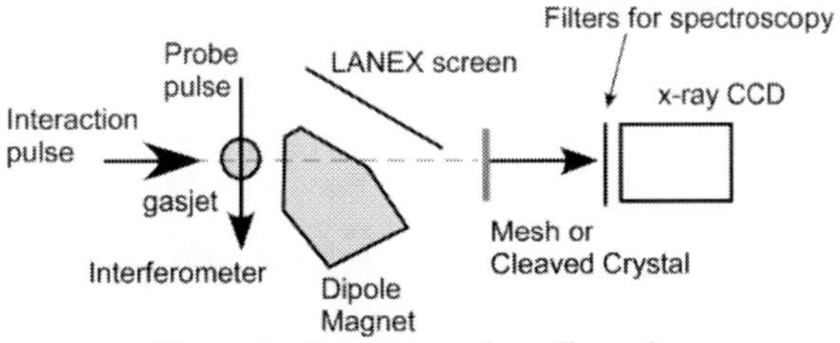

Figure 1. Experimental configuration

Accelerated electrons were deflected by a dipole magnet and sent to a LANEX (Kodak) scintillating screen and the signal was read by a CCD camera. A tracking code using the measured magnetic field map was used to relate the deflection of the electron beam on the LANEX screen to an electron energy. X-rays from the plasma were detected by an x-ray CCD camera. In some experiments, a mesh, a cleaved crystal or filters were inserted between the interaction region and the x-ray CCD camera in order to measure the x-ray beam divergence, source size or x-ray spectrum respectively.

3. "Betatron" motion in electron spectrum

Electron spectra as a function of plasma density are shown in Figure 2. The length of the plasma was approximately 2.5 mm. At a fixed laser power (30 TW), the electron spectra could be categorized into three types depending on the plasma density. First, a broad spectrum with multiple peaks is observed at high densities (e and f). Second, the spectra showed a sinusoidal shape in space, potentially due to "betatron" motion of the accelerated electron beam, and are observed at medium densities (b, c and d). Third, quasi-monoenergetic spectra are observed at the lowest densities (a). Observed "betatron" motion in the electron spectrum is similar to the one reported in the reference of 9. In the reference, the experimentally observed "betatron" motion of accelerated electrons was interpreted by 2D-PIC simulations as a consequence of off-axis injection in the radial direction, i.e., orthogonal to the laser propagation direction. The experiments and simulations showed that the asymmetrical laser intensity profile in radial direction might lead to an asymmetrically shaped wake and off-axis electron injection. More studies are required in order to fully understand the source of the "betatron" motion.

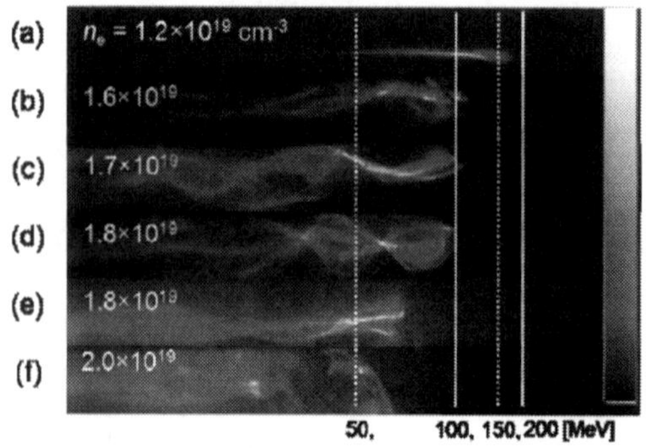

Figure 2. (color online) Image of electron spectrum as a function of plasma density. The color intensity scale is normalized by the peak intensity for each individual image.

4. X-ray characterization

A mesh was set behind the magnet and x-rays originating from the interaction region cast a shadow of the mesh on the x-ray CCD camera as shown in Figure 3. The size of the plasma was 5 mm in diameter and the plasma density was 5×10^{18} cm^{-3}. In the particular shot shown below, the laser power was 70 TW on target which yields an intensity of 5.3×10^{19} W·cm^{-2}. The x-ray image shows a 12x4 mrad^2 beam divergence which corresponds to a K parameter ($K \sim \theta\gamma$) of $K_x = 5$ and $K_y = 1.5$, where θ, γ, x and y are the beam divergence of x-rays, Lorentz factor of the electron beam, horizontal direction and vertical direction, respectively. Here, the electron beam energy was simultaneously measured to be 220 MeV was used.

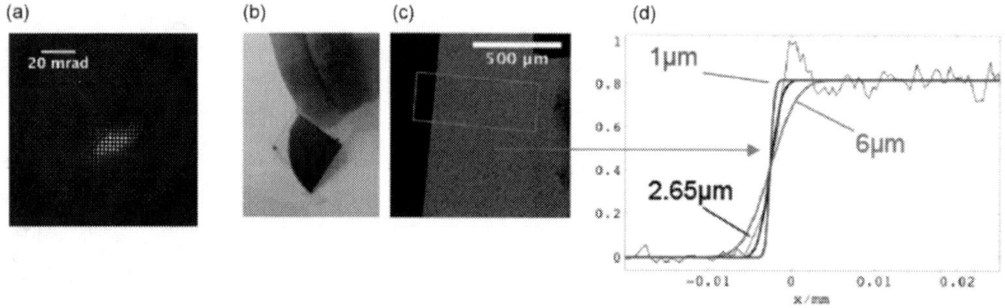

Figure 3. (color online) (a) Image of a mesh illuminated by the synchrotron x-ray radiation from a LWFA electron beam. The beam divergence is as small as 12x4 mrad^2. (b) InSb cleaved crystal, (c) Image of the crystal illuminated by the synchrotron x-ray radiation from a LWFA electron beam. (d) Line outs of the x-ray image with the experimental profile. Fitting curves by use of various source sizes are overlaid.

To quantify the source dimensions more accurately, penumbral images were taken by replacing the mesh by a cleaved edge crystal (Figure 3b). The image of the half-shadow due to the edge was obtained at a magnification of 30 (Figure 3c). The half-shadow is the result of the convolution of the source function and the aperture function. If they are assumed to be a Gauss function and a step function, respectively, the convolution takes the shape of an error function. Fitting error functions to the lineout across the half-shadow yields the best fit for the source size. Preliminary results reveal an unprecedented small FWHM x-ray source size smaller than 2.7 μm (Figure 3d) which changes little with plasma density [10].

In order to obtain the x-ray spectrum, the x-ray yield was measured through a set of metal foil filters of different material and thickness. Various spectral distributions were compared with experimental results until the agreement between predicted and measured x-ray yield is satisfactory. The actual spectrum can be assumed to have a synchrotron-like shape [6], and we consequently vary the critical energy E_{crit} to maximize the agreement between measured and predicted x-ray yield. We used up to six filters simultaneously per shot with $1/e$ cut off energies from keV to tens of keV allowing for a reliable reconstruction of the critical energy E_{crit}. Figure 4 shows various typical synchrotron spectra from a range of conventional light sources together with our experimental result. Based on spectral measurement and x-ray beam divergence measurements, the highest total photon number observed in our experiment is of the order 10^8 photons per shot. The peak brightness is normalized to the source size, the divergence, 0.1 % bandwidth and the x-ray pulse duration, which is assumed to be the laser pulse duration. It is due to the unprecedented small source size and good collimation of the x-ray beam that the source yields a brightness comparable to currently existing 3rd generation conventional light sources. Small source size and divergence in turn are the merit of quasi-monoenergetic electron beams and their unique injection and acceleration dynamics.

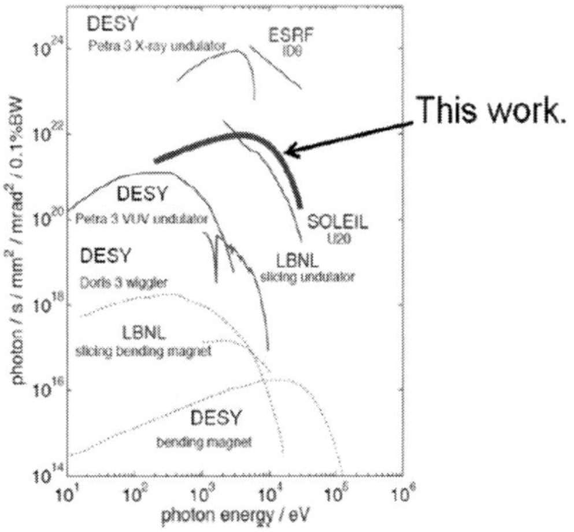

Figure 4. (color online) Normalized x-ray peak brightness for a typical shot for a selection of conventional 2nd (green dashed lines) and 3rd (blue lines) generation light sources [11].

5. Summary

Synchrotron x-rays from mono-energetic electrons were fully characterized. The x-ray divergence was found to be as small as 12x4 mrad^2 and the x-ray source size can be smaller than 3 μm. A peak brightness of more than 10^{22} [ph/s/mm^2/mrad^2/0.1%BW] comparable to 3rd generation light sources was inferred from the above mentioned observations.

References

[1] S. P. D. Mangles, et. al, Nature **431**, 535 (2004)
[2] C. G. R. Geddes, et. al, Nature **431**, 538 (2004)
[3] J. Faure, et. al, Nature **431**, 541 (2004)
[4] A. Rousse, et. al, Phys Rev Lett **93**, 135005 (2004)
[5] S. Kneip, et. al, Phys Rev Lett **100**, 105006 (2008)
[6] E. Esarey et. al, Phys. Rev. E **65**, 056505 (2002)
[7] V. Yanovsky, et. al, Opt Express **16**, 2109 (2008)
[8] S. W. Bahk, et. al, Opt Lett **29**, 2837 (2004)
[9] Y. Glinec, et. al, A Letters Journal Exploring The Frontiers of Physics **81**, 64001-p1 (2008)
[10] S. Kneip, et. al, Proc. of SPIE Vol. **7359**, 73590T-1 (2009)
[11] A. Rousse, et al, European Physical Journal D **45**, 391(2007)

This work was supported by the US National Science Foundation through the Physics Frontier Center FOCUS (Grant No PHY-0114336) and NSF/DNDO Grant No.-0833499. S. Kneip would like to acknowledge the Euroleap network for financial assistance.

Journal of Physics: Conference Series **244** (2010) 042027 doi:10.1088/1742-6596/244/4/042027

Industrial Applications of Laser Neutron Source

S. Nakai[1], K. Mima[1)5)] ,Y. Kato[1], K. Tanaka[2], Y. Ikeda[3] H.Azechi[4], K. Miyanaga[4], M. Nakai[4], M. Perlado[5], R. Gonzalez Arrabal[5]

1) Graduate School for the Creation of New Photonics Industies,
1955-1 Kurematsu-cho, Nishiku, Hamamatsu, Shizuoka, 431-1202, Japan
2) Graduate School of Engineering, Osaka University,
2-1 Yamada-oka, Suita, Osaka 565-0871, Japan
3) Life Science Division, Japan Atomic Energy Agency
2-4 Shirakata Shirane Tokai, Naka, Ibaraki, 319-1106, Japan
4) Institute of Laser Engineering, Osaka University,
2-6 Yamada-oka, Suita, Osaka 565-0871, Japan
5) Universidad Politecnica de Madrid,
C/José Gutiérrez Abascal, 2 , Madrid, 28006 Spain

s.nakai@gpi.ac.jp

Abstract. The industrial applications of the intense neutron source have been widely explored because of the unique features of the neutron-matter interaction. Usually, intense neutron sources are assembled with fission reactors or high energy ion accelerators. The big size and high cost of these systems are the bottle neck to promote the industrial applications of intense neutrons. In this paper, we propose the compact laser driven neutron source for the industrial application. As the first step of our project for the versatile applications of laser driven neutron source, Li-neutron and/or Li-proton interactions have been investigated for the application to the development of Li battery.

1. Introduction

The high density and intensity laser-matter interaction physics have been well investigated being directly related to the laser fusion research. Various physical processes have been clarified to generate neutrons, high energy electrons and ions, and intense x-ray and -ray by using relatively small laser energy with high intensity short pulses.

The high power pulse lasers with high efficiency and rep-rate operation are progressing rapidly aiming to develop the laser driver for the fusion power plant. The technology of the system consists of wide range of innovative engineering fields such as high power laser diode for pumping, new laser host materials of glass, ceramic, and/or crystal, precise pulse shaping and compression, frequency conversion, phase control and beam pointing. The developments of fusion driver are leading the front edge of industrial laser technologies. To promote the industrial application of the laser fusion technology, especially laser produced neutrons and protons, a specified committee: "Committee of Industrial Application of Laser Neutron Source" has started with the support of IFE forum getting the collaborations of academia, and related industries through out Japan. The collaborative research with Universidad Politecnica de Madrid has also started. This work is being performed partly under the IAEA program "Coordinated Research Project on Pathway to Energy from Inertial Fusion – an Integrated Approach".

2. Generation of Neutrons, Particle Beams by High Power Laser

Neutron generation by short pulse laser of order of ns to fs have been well investigated for many years. The physical model of typical examples is shown in Fig. 1. They are ①gas cluster target, ②thin film target, ③spherical pellet target for implosion. There are many optimized structures of each target such as density control, composite material and Z number optimization, layered structure etc. The pulse width and shape are also optimized according to the type of the targets and structures, of which features are also shown in Fig. 1.

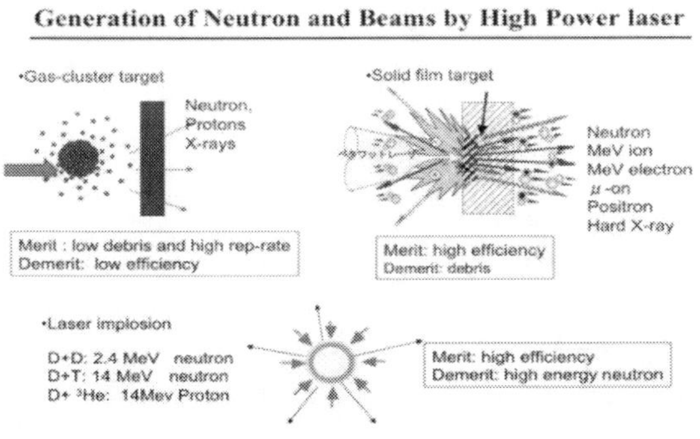

Fig. 1

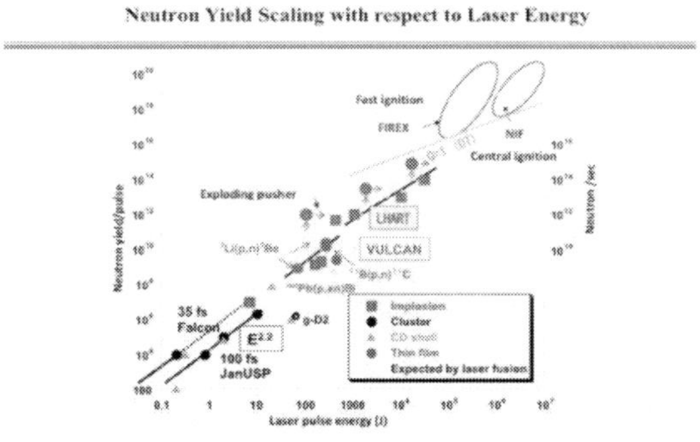

Fig. 2

Fig. 2 shows the compiled data of neutron generation as a function of injected laser energy with different pulse width depending on the related physical processes. The data points are basically single shot results. The rep-rate operation is needed for the industrial application as a neutron source. For this purpose, repeatable target feeding and high average power pulse laser must be developed. The right vertical axis shows the neutron fluence per second when the neutron production shots are repeated at 10 Hz to 1 kHz. It should be noted that fusion ignition can enhance the neutron production beyond the line Q=1 where the fusion energy is equal to the incident laser energy, and also it is expected to be demonstrated in near future of 2010~2015.

3. Industrial Applications of Laser Neutron Source

The applications of neutron sources in science and industry are discussed in the previous report in IFSA2007[1]. It spans over wide fields of material science and technology, nuclear energetics, medical, and new methodology in diagnostics, especially in measurements of light elements such as Hydrogen and Li.

The required source neutron flux is evaluated from the necessary neutron energy and flux density at the sample of the specific application, and the design of energy moderation and guiding of neutrons.

For the diagnostics of Li-ion battery and fuel cell, thermal neutron flux of 10^{11} ~10^{12} /sec is required. For the medical applications like BNCT, the flux is 10^{12} ~10^{13} /sec. For semiconductor doping (NTD), the flux is 10^{13} ~10^{14} /sec. For material processing like annihilation of radio-activity in which long life radio-active elements in a used nuclear fuel are transmuted by fusion neutron , fusion material test, and fusion- fission hybrid in which a under critical fission reactor is driven by fusion neutron irradiation, the flux higher than 10^{15} /sec is required.

In the following section, the laser driven nuclear processes are discussed for the applications to the new diagnostics which are necessary for the developments of Li ion battery and Hydrogen fuel system.

4. Application of Laser Produced Neutron and Ion to Li-ion Battery Diagnostics

4.1 Diagnostics

In the R&D of the lithium ion battery, it is necessary to diagnose the damage of electrode and the dynamics of lithium and negative ions in the electrodes and electrolyte. Neutrons and ions are appropriate and unique tools for the Li-ion battery diagnostics. However, because the conventional irradiation facilities are huge and nowadays only specialized nuclear research site can be used, the speed of material developments for the advanced battery is limited. Neutron and ion sources driven by lasers could provide a compact combined system for this purpose. The applications of laser driven neutron and ion for observing electrodes, lithium, negative ions are discussed here. The images of diagnostic systems are the facility at University of Texas in the reference [2)] for neutron diagnostics and the Fig. 3 for proton diagnostics. Here, an intense short pulse laser irradiates a solid target or cluster to generate mono-energy protons. Laser produced MeV protons are guided and directly injected onto the Li-ion battery to probe Li-ion depth profile, negative ion depth profile and so on. Otherwise, the protons are injected into liquid Li metal target to generate low energy neutrons. The neutrons are decelerated to thermal neutrons that are used for the Li battery diagnostics as discussed in the reference [4] or in some other new methods.

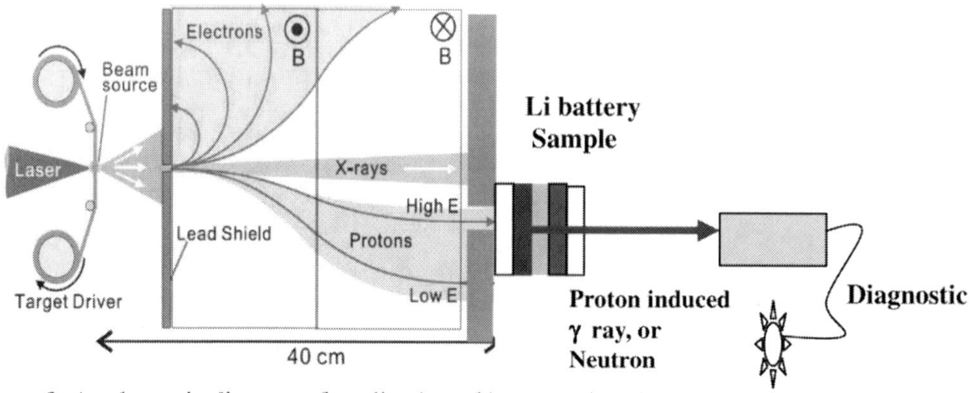

Figure 3 A schematic diagram of application of laser produced proton beam to

4.2 Laser neutron diagnostics

On the other hand, neutrons generated by ultra intense short pulse laser can be also applied for the Li battery diagnostics. The thermal neutron scattering, excitation, and reactions of light elements are used for the diagnostics. As an interesting example, we describe the application of neutron absorption reaction. The reaction: $^{6}Li + n \rightarrow t(2.7MeV) + \alpha\,(2.1MeV)$ could be used for measuring Li depth profile by observing the emitted t and/or α energy spectrum. Namely, the t and α energy spectra correspond to the Li number density. The lowering of energy of t or α corresponds to depth of the place where the reaction takes place. In this way, the Li depth profile can be measured[3)]. Multi-MeV protons generated by laser are used for generating sub-MeV neutrons through p-Li reactions.

4.3 Laser produced proton diagnostics and mono energy proton generation

[Proton diagnostics]

When MeV protons are injected into Li- battery, protons react with light nuclei like Li, F, and so on to produce γ rays and/or neutrons. Since these nuclear processes are threshold reaction, we can deduct ion density depth profile from the proton beam energy dependence of γ ray yield and/or neutron spectrum. The proton induced γ•ray diagnostics has been known as the PIGE (Proton Induced Gamma Emission) [4)]. Energy spectrum of neutrons induced by the reaction ^{7}Li (p. n) ^{7}Be can be used for diagnostics of Li depth profile, since neutron energy depends on proton energy. Although the energy spread is the critical issue, the laser produced protons are unique as a probe beam because of very low transverse emmitance and short pulse. The energy spread could be overcome by introducing energy selector as shown in Figure 3. Let's take a quasi mono-energy proton beam with $(\Delta E/E)_{FWHM} \sim 0.1$[5)]. For obtaining a beam of 1% energy spread, 10% of the total flux can be used. In the case of Li ion battery, the width of a typical sample is 200 μm. Therefore, the spatial resolution obtained by this proton beam diagnostics like PIGE could be a few μm. This resolution is not very good, but interested in the field of Li ion battery community.

[Mono energy proton generation by structured clusters]

Journal of Physics: Conference Series **244** (2010) 042027 doi:10.1088/1742-6596/244/4/042027

Mono-energy protons can be generated by irradiating an ultra intense short pulse laser on clusters of hydrogen-high Z atom mixture. When the fraction of proton charge is much smaller than the total cluster charge and the Z/M of the high Z ion is small enough compared with that of proton, then we can assume that protons are accelerated in a Coulomb field of the charged high Z sphere. Namely, the final velocity of proton (the proton velocity at r = ∞) is given by

$$E = Mv_f^2/2 = Zn_0 e^2 (3R^2 - r_0^2)/(6\varepsilon_0),$$

where R is the initial size of the cluster and r_0 is the initial radial position of the proton layer. So, the energy difference between protons at r_0 and $r_0 + dr_0$ is given by $dE = -Zn_0 e^2 r_0/(3\varepsilon_0)\, dr_0$. The proton number in a spherical shell of inner radius r_0 and outer radius $r_0 + dr_0$ is

$$n_p 4\pi r_0^2 d r_0 = - 4\pi r_0^2 n_p (d r_0/dE)\, dE, \quad where\ r_0 = (3R^2 - 6E\varepsilon_0/Zn_0e^2)^{1/2}.$$

When $N(E)\, dE$ is number of proton in the energy range of $(E, E+dE)$, then, $N(E)$ is given by

$$N(E) = 2\pi n_p (6\varepsilon_0/Zn_0e^2)^{3/2}(R^2 Zn_0e^2/2\varepsilon_0 - E)^{1/2},$$

where $R^2 Zn_0e^2/3\varepsilon_0 = E_{min} < E < E_{max} = R^2 Zn_0e^2/2\varepsilon_0$, otherwise $N(E) = 0$.
In this case, the FWHM of the proton energy spectrum is 30%.
For the nuclear reaction diagnostics by proton, this proton energy spectrum is too broad and energy selection is required. However, the proton beam brightness for the probing is 3 times higher than conventional laser produced proton beam. On the other hand, this proton can be used for producing low energy neutrons. In the ^{7}Li(p,n) ^{7}Be reaction, the maximum neutron energy is 1MeV when maximum and minimum proton energy are 3MeV and 2MeV. Proton energy spectrum from a high Z cluster in which hydrogen is weakly doped. When the total charge of cluster is 1.25×10^{-10} C, then, $E_{max} = R^2 Zn_0e^2/2\varepsilon_0 = 2.75$ MeV and $E_{mini} = 1.83$ *MeV*. The cluster radius R is 0.1 μm. This neutron source could be useful for a compact thermal neutron diagnostic system because sub-MeV neutrons have large scattering.
The application of intense neutron source is becoming increasingly important in developing new methodology in research and engineering. The laser driven neutron and ion beam sources will provide a compact and versatile facility for the industrial applications.

5 Summary

We are now getting into the new phase of development in the Laser Inertial Fusion Energy (LIFE) with the accumulated results of scientific and engineering researches and developments through out the world for many years. The most challenging and significant issue toward LIFE is the high power laser development of rep-rate operation with high efficiency, compact, and low cost for construction and maintenance. The progress of laser technology is opening wider fields of industrial applications in manufacturing, medical, biological and energy engineering.

The application of intense neutron source is becoming increasingly important in developing new methodology in science and engineering. The laser driven neutron and ion beam sources provide the compact and versatile facility for the industrial application. The feasibility has been examined theoretically and a new project for the industrial application has been proposed.

References:

[1] S. Nakai et al IFSA2007 Proceeding Journal of Physics:Conference Series 112(2008)042070
[2]K.Unlu, B.W.Wehring, J. Radioanalytical and Nucl. Chem. 217(1997)273
[3]S.M. Whitney, et al. J.Radioanalytical Nucl. Chem.282(2009)173
[4]R.Mateus, A.PJesus, B.Braizinh, J. Cruz, J.V. Pinto, J.P.Ribeiro, Nuclear instruments and methods B 190(2002)117-121
[5] For an example, B.M.Hegelich, *Nature* **439**, (2006) 441-444

Journal of Physics: Conference Series **244** (2010) 042028 doi:10.1088/1742-6596/244/4/042028

XUV Emission from Autoionizing Hole States Induced by Intense XUV-FEL at Intensities up to 10^{17} W/cm^2

F B Rosmej[1,2*], E Galtier[1], D Riley[3], T Dzelzainis[3], P Heinmann[4], F Y Khattak[5], R W Lee[4], B Nagler[6], A Nelson[4], T Tschentscher[7], S M Vinko[4], T Whitcher[4], S Toleikis[7], R Fäustlin[7], R Soberierski[8], L Juha[9], M Fajardo[10], J S Wark[6], J Chalupsky[9], V Hajkova[9], J Krzywinski[11], M Jurek[8], M Kozlova[9]

[1]Université Pierre et Marie Curie UPMC, UMR 7605, LULI, case 128, 4 Place Jussieu, 75252 Paris Cedex 05, France
[2]Ecole Polytechnique, Laboratoire pour l'Utilisation des Lasers Intenses LULI, Physique Atomique dans les Plasmas Denses PAPD, 91128 Palaiseau, France
[3]School of Mathematics and Physics, Queen's University, Belfast BT7 1NN, UK
[4]Lawrence Livermore National Laboratory, USA
[5]Department of Physics, Kohat University, Kohat, NWFP, Pakistan
[6]Clarendon Laboratory, University of Oxford, South Parks Road OX1 3PU, UK
[7]HASYLAB DESY, Hamburg, Germany
[8]Institute of Physics, Polish Academy of Sciences, Poland
[9]Institute of Physics, AS CR, Prague 8, Czech Republic
[10]Instituto Superior Técnico, U. Técnica de Lisboa, Portugal
[11]SLAC, USA

*E-mail: frank.rosmej@upmc.fr

Abstract. Aluminium targets were irradiated with 92 eV radiation from FLASH Free Electron Laser at DESY at intensities up to 10^{17} W/cm^2 by focussing the beam on target down to a spot size of ~1 µm by means of a parabolic mirror. High resolution XUV spectroscopy was used to identify aluminium emission from complex hole-states. Simulations carried out with the MARIA code show that the emission characterizes the electron heating in the transition phase solid-atomic. The analysis allows constructing a simple model of electron heating via Auger electrons.

1. Introduction

The study of dense strongly coupled plasmas DSCP and warm dense matter WDM is of intense interest because such matter is difficult to handle theoretically, with electrostatic and thermal energies being of equivalent importance, rather than one being a perturbation of the other as in solid state physics or classical plasma physics. Moreover in this regime experimental data is not easy to obtain and one of the key issues is the ability to create uniform well-defined samples of matter near solid density but with temperatures of ~ 1-100 eV. A possible way forward is the use of the new generation Free Electron Lasers (XUV-FEL and X-FEL) to irradiate solid samples [1, 2].

Free electron XUV- and X-ray lasers allow volumetric heating because the plasma frequency of the solid is smaller than the laser frequency. The radiation penetrates therefore deep into the solid and the analysis of the data does not suffer from critical surface effects. Moreover, in regimes where thermal energies are too low to create short wavelength radiation that can exit a high density sample without reabsorption, the emission from photoionized core states, allows to get information on high energy density matter at low temperatures [3].

We report on one of the first experiments with a high-intensity 4th generation light source FLASH (XUV-Free-Electron-Laser at DESY in Germany) where the solid is effectively heated and destroyed after intense photoionization of the target. Finally we present an analysis of the characteristic XUV transitions from the target.

2. Experimental setup

The present experiment was carried out using the FLASH XUV-Free Electron Laser Facility at DESY in Germany. Solid Al-foils were irradiated with intense pulses of $\lambda = 13.5$ nm radiation operating at 5 Hz. The pulse length was $\tau \sim 15$ fs and the beam was focussed down to about 1 μm diameter with a multilayer parabolic optic (multilayer Mo-Si) with a focal length of 269 mm to achieve intensities up to 10^{17} Wcm^{-2}. A 3 mm aperture in the beam limited its size and the pulse energies varied between 5 and 10 μJ on target. The best focus was established by analysis of PMMA (Polymethylmethacrylate) ablation using a range of target positions relative to the focus optic and a range of pulse energies. This analysis established that the best focus was ~ 1 μm FWHM. Further, the reflectivity of the optic was measured after the experiment and found to be 48%. The spectrometer employed a 1200 lines/mm Hitachi grating with variable line spacing to create a flat spectral focal plane on a CCD camera positioned on a vacuum flange, providing a spectral coverage from ~ 10-30 nm. An Al edge filter was used to establish that the spectral resolution achieved was ~ 0.1 nm thus providing an effective spectral resolution of about $\lambda/\delta\lambda$ ~150 in the relevant spectral range. The samples were composed of 10 μm thick Al foils that were continuously moved transverse to the FEL beam across the focus to expose a fresh surface each shot. Figure 1 shows an experimental spectrum obtained by integrating 2000 shots–at irradiation intensities of about 10^{16} W/cm^2. The dominant spectral feature (Figure 1) is identified as atomic Al IV lines: $1s^22s^22p^6$-$1s^22s^22p^5$ $(^2P^o)3s$ ^1P and ^3P at 16.1 nm and 16.2 nm, respectively by comparison with transition energies obtained from Hartree-Fock calculations [4]. The transitions at shorter wavelengths are related to transitions $1s^22s^22p^6$-$1s^22s^22p^5 3d$ ^1P, ^3P and ^3D at 13.0 nm, 13.1 nm and 13.2 nm respectively.

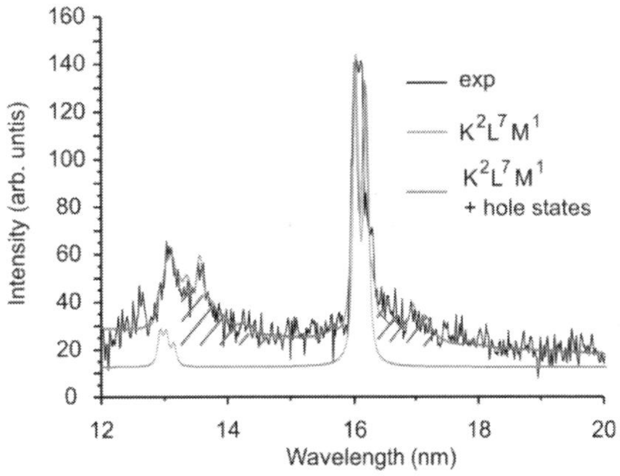

Figure 1. Aluminum emission (solid black line) at best focus. The dominant spectral features are Al IV resonance line emission (blue solid line). The hatched areas indicates potential emission from hole states. The spectral fit (solid green line) has been obtained from MARIA-code simulations [6] and the application of a genetic algorithm [7].

3. Analysis of the XUV Al target emission

A comparison between the experimental data in Figure 1 (black solid line) and the simulation of the Al IV lines (blue solid line) shows large discrepancies indicated by the hatched area. It is known that transitions located at the red wing of resonance line transitions might be due to screened transitions, so called satellite transitions, firstly investigated in detail for the He-like resonance line of highly charged ions [5]. The observed transitions on the red wavelength wing of the Al IV resonance line transitions might therefore originate from XUV-transitions into the L-shell of heated Al while spectator electrons are present in the M-shell:

I. $K^2L^7M^3 \rightarrow K^2L^8M^2 + h\nu_{Al\,II}$
II. $K^2L^7M^2 \rightarrow K^2L^8M^1 + h\nu_{Al\,III}$

In order to understand the effective wavelengths shift due to several spectator electrons, we have performed detailed atomic structure calculations by means of the Multi-Configuration Hartree-Fock method [4]. As the number of line transitions I. and II. is too large to be analysed in table form, we represent the transitions by spectral simulations using

$$I(\omega)=\sum_{i,j} g_j A_{ji}\, e^{-(E_j-E_0)/kT_e}\, \Phi_{ji}(\omega) \quad (1)$$

g_j is the statistical weight, A_{ji} the radiative decay, E_j and E_0 are the energies of level j and the ground state respectively, Φ_{ij} is the line profile. Figure 2 shows the spectral distribution of M-L transitions for different number of electrons in the M-shell: x=1, 2, 3. The five vertical dotted lines 1-5 indicate the central positions of dominating emission groups.

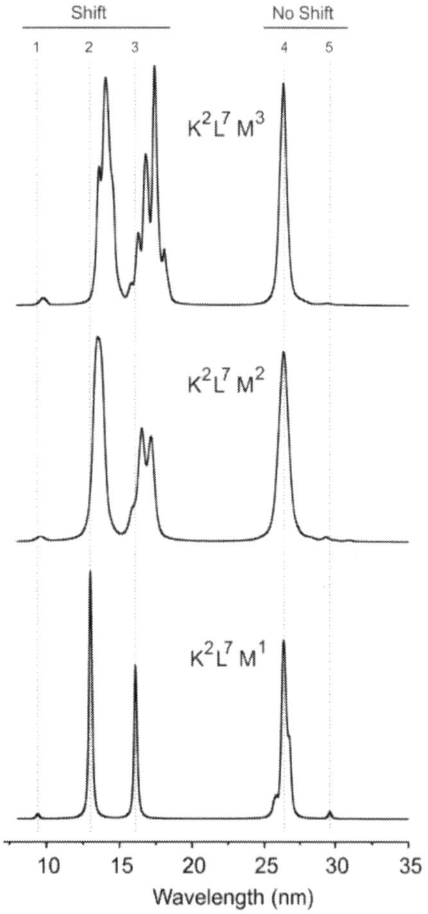

Figure 2: Spectral distribution of transitions originating from 3-2 transitions with different number of screening electrons.

The spectral distribution of transitions originating from $K^2L^7M^2$ and $K^2L^7M^3$ configuration indicate rather strong shifts to the red for the first 3 groups (dotted lines 1-3 at about 9.4 nm, 13.0 nm and 16.1 nm, respectively). As screening electron and optical electrons are in the same M-shell, the shifts are rather strong (about 0.5-0.6 nm for each additional M-electron). In contrast to the dotted lines 1-3 in Fig. 2 are the dotted lines 4 and 5 (at 26.3 nm and 29.6 nm). Almost all transitions, independent of the number of electrons in the M-shell, are located near these dotted lines. The atomic structure calculations indentifies inner-shell transitions with $\Delta n=0$:

III. $1s^22s^12p^6M^X \rightarrow 1s^22s^22p^5M^X + h\nu$

For these transitions, the screening effect from M-electrons is not large and barely visible in the spectral distribution. The simulation presented in Figure 2 (solid curve indicated as $K^2L^7M^1$ + hole states) show that the discrepancies indicated by the hatched area can be attributed to the transitions I. and II. because the shifted transitions (originating from configurations $K^2L^7M^2$ and $K^2L^7M^3$) just fall into the requested wavelength interval.

Figure 3 indentifies a large temperature sensitivity of the spectral distribution of hole states from about 1 eV – 50 eV. Based on these sensitivities the electron temperature is estimated from the fit presented in Figure 1 (solid curve indicated as $K^2L^7M^1$ + hole states) to be about 25±5 eV.

We note that there is a substantial difference in the temporal evolution of the Al IV emission compared to those of Al II and Al III of channels I and II. Transition probabilities of Al IV lines are 10^9 to 10^{11} s^{-1} [4], so that the self-emission occurs about 0.1 ns after excitation (taking into account radiation transport effects). This time is much longer than the 15 fs FEL pulse and corresponds to a recombination phase. As the Al IV resonance lines are sensitive to radiative recombination, a low temperature recombining phase may considerably contribute to the time integrated spectra. The situation for the Al II and Al III emission is dramatically different. The configurations $K^2L^7M^2$ and $K^2L^7M^3$ are autoionizing, their characteristic time scales being ≤ 100 fs [4]. We therefore based our analysis on hole states from Al II, III rather than resonance line emission from Al IV.

A temperature T_e, which is much higher than melt, enables one to construct a qualitative model of electron heating. After photoionization of the L_{II} and L_{III} shells an excess energy of ~19 eV will be distributed amongst the 4 conduction band electrons. Assuming rapid thermalization on sub fs-scale [8] and accounting for the 12 eV Fermi energy for Al, a T_e of ~8 eV is obtained. This T_e represents an excited transient state that will start relax after ~40 fs by either radiative decay or by autoionization. As the branching ratio favours autoionization by 99.8 % the energy of the decaying electron is shared with the three remaining conduction band electrons. Simulations show that the energy difference, e.g. between the configuration $K^2L^8M^3$ and $K^2L^7M^4$ is about 70 eV which is equivalent to the kinetic energy of the Auger electrons [4]. Assuming that all the Al atoms in the intense FEL beam are photoionized this excess 70 eV will rapidly thermalize with the 8 eV electrons-resulting in T_e of ~22 eV. This model is in good agreement with the spectroscopic analysis of channels I and II presented in Figure 1. As saturated absorption has been observed [9] (which means Al becomes transparent to the laser intensity due to the photoionization of all $2p^6$-electrons) the spectroscopically determined T_e of ~25 eV indicates that Auger electron heating plays an important role in the laser-matter interaction.

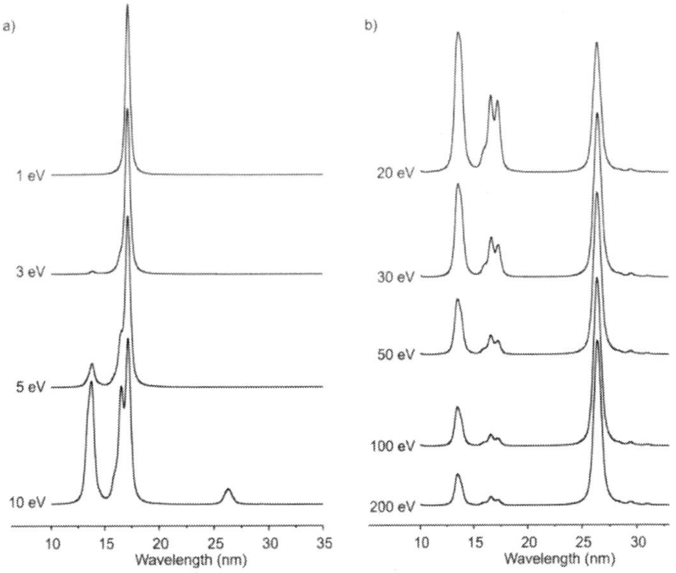

Figure 3: Electron temperature sensitivity of $K^2L^7M^2$-$K^2L^8M^1$ +hν transitions.

Acknowledgement
The support of the LULI-Research Center at Palaiseau, France, is greatly acknowledged.

References

[1] Lee R W, Moon S J, Chung H K et al. 2003, JOSA B **20** 770.
[2] Fajardo M, Zeitoun P, Gauthier J-C 2004, EJD **29** 69.
[3] Rosmej F B, Lee R W, Europhysics Letters **77** 24001.
[4] Cowan R D 1980, The Theory of Atomic Structure, Berkeley.
[5] Gabriel A H 1972, Mon. Not. R. Astron. Soc. **160**, 99.
[6] Rosmej F B, Europhysics Letters **55**, 472 (2001).
[7] Galtier E, Rosmej FB, Riley D et al. 2009, Proceedings of the 36th EPS Conference on Plasma Physics, Sofia, Bulgaria, ECA Vol. **33E**, P-2.034.
[8] Fann S W, Storz R, Tom H W K, Bokor J 1992, Phys. Rev. B **46**, 13592.
[9] Nagler B, Zastrau U, Fäustlin R et al 2009, Nature Physics **5**, 693.

Journal of Physics: Conference Series **244** (2010) 042029 doi:10.1088/1742-6596/244/4/042029

Laser-produced molybdenum plasma analysis with a Collisional - Radiative model included in the hydrodynamic code med103

Shou-Jun Wang[1, †], **Quan-Li Dong**[1, ‡], **Lei Zhang**[1] **and Jie Zhang**[1, 2]

[1] Beijing National Laboratory for Condensed Matter Physics, Institute of Physics, Chinese Academy of Sciences, Beijing 100190, China
[2] Department of Physics, Shanghai Jiao Tong University, Shanghai 200240, China

E-mail: † sjwang@aphy.iphy.ac.cn, ‡ qldong@aphy.iphy.ac.cn

Abstract. The Collisinal-Radiative (CR) model with fine atomic data included in the hydrodynamic code MED103 is used for the soft x-ray spectra analysis of laser-produced molybdenum plasmas. The charge distribution drawn from the soft x-ray spectra can be used to determine the pre-plasma conditions for the plasma-based molybdenum soft x-ray lasers (SXRLs).

1. Introduction

X-ray spectroscopy is a powerful tool in the study of plasmas both in the universe and the laboratory. The laboratory plasmas generated by high intense lasers, z-pinch machines and tokamak devices, etc., usually have quite different conditions. Such complicated plasmas impose high requirements on the theoretical analysis of the spectra recorded with various spectrometers equipped with gratings or crystals. In order to extract the physics underlying the phenomenon from the recorded spectrographs, one usually turns to certain radiation-hydrodynamics codes. The Collisional-Radiative model (CR model) has been proven as the most successful model to analyze the laboratory plasma which is usually in non-local thermodynamic equilibrium (non-LTE). Much work has been done in the development of various codes with CR model, for the wide range of applications in plasma experimental design and spectral analysis [1]. International workshops on non-LTE kinetics have led to a significant progress on the CR models and hence to a better agreement with well characterized measurements [2].

In this paper, we present the application of the Collisional-Radiative (CR) model code on the analysis of laser-produced Molybdenum plasmas. By comparing the calculated soft x-ray spectra with that measured in experiments, it is able to deduce the charge distribution of the laser-produced plasmas, and the contributions of different physical process can also be obtained with further detailed analysis.

2. Experiments

The experiment was carried out on the XL-II Ti:sapphire laser facility at the Institute of Physics, Chinese Academy of Sciences. The laser typically delivers pulses of energy about 500 mJ in 200 picoseconds with central wavelength at 800nm. The uncompressed beam was directed into the target chamber and focused onto a molybdenum slab target using a single $f/8$ spherical lens. By adjusting the lens position from the target we could vary the focal spot size from tens to hundreds of micrometers monitored by a cross-slit camera filtered with 1.2 μm thickness aluminum. The x-ray image was recorded on a back-illuminated CCD. In some shots, a serial of neutral density filters were inserted into the beam path to change the laser energy focused on the target, so that the difference of the two intensity-changing methods could also be studied. The plasma x-ray emission was observed 45° off the target normal by using an on-axis flat field spectrometer equipped with a Hitachi 1200 lines/mm variably spaced spherical grating, which was located 321 mm from the target and set to a grazing angle of 3°. Another 2048 × 512 pixel, back-illuminated CCD camera was coupled to the focal plane of the grating to record the spectrograph. Figure 1 shows the typical corrected on-axis spectra of the plasma created by a 400-mJ laser pulse with the average intensity of 3.5×10^{12} W/cm^2. The well-resolved line is imposed on a strong recombination continuum emission.

3. Data Analysis

3.1.1. Calibration and spectra intensity correction of CCD camera

A long-year-used CCD camera usually has the layer of silicon oxide grown on the silicon surface and the dead layer, which affect the quantum efficiency QE dramatically and trouble the spectra analysis in the soft x-ray region. We calibrated our CCD camera by using the method of Li's [3].

The interaction efficiency of a CCD can be written as

$$\eta_{\text{int}}(\lambda) = \exp[-\alpha_{ox}(\lambda)t_{ox}] \times \{1 - \exp[-\alpha(\lambda)t]\},$$

where t and t_{ox} are the thickness of the active region and that of the oxide layer. α and α_{ox} are the related absorption coefficients [4].

The reflectivity of the surface is neglected. The collection efficiency is

$$\eta_{coll}(\lambda) = \exp[-\alpha(\lambda)t_d],$$

with t_d the thickness of the dead layer.

The quantum efficiency QE of the device is the production of the two above, i.e.,

$$QE(\lambda) = \eta_{\text{int}}(\lambda)\eta_{coll}(\lambda).$$

With this semi-empirical method we can get the calculated quantum efficiency curves by simply assuming the layer thickness then comparing the silicon L-edge jump with the experiment spectra to find the fitted thickness for the QE. In the data correction we used $t_{ox} = 10\,nm$, $t = 10\,\mu m$ and $t_d = 30\,nm$.

The calculated QE can be taken as the semi-empirical calibration of the CCD camera with which we corrected the spectrograph of the molybdenum plasma emissions in soft x-ray region 8nm ~ 22nm, shown as the red line in Figure 1.

3.1.2. Theoretical Model

To investigate the physical progress, the Collisional-Radiative (CR) model was employed to simulate the results by using the same laser conditions with the experiment. In the Collisional-Radiative model, atomic level populations are calculated by solving multi-level atomic rate equations self-consistently, which is a complicated and time-consuming progress. For the molybdenum simulation 1579 levels from Mo9 to Mo15 was used, the others were treated as H-like. The model considers important atomic processes and their inverse processes: collisional excitation and deexcitation for radiationless bound-bound transitions, collisional ionization and recombination, autoionization and electron capture. For radiative process, the code includes the spontaneous emission for bound-bound transitions and the radiative recombination, while an escape factor is introduced for the re-absorption of the emission instead of considering the photo-excitation and the photo-ionization processes, since there is no external field applied in the experiments except the laser pulse. In the model the time dependent rate equations are resolved within the hydro code.

The rates of the collisional excitation and the de-excitation are from the method of van Wyngaarden et al. [5] with the collision strength calculated with the Cowan code [6].

The collisional ionization rate is calculated with Lotz's [7] model. The rate coefficient from an ion i to ion i+1 is given by

$$R_{i,i+1}/N_e = 2.97\times10^{-6}\left(\xi/I_n\sqrt{T_e}\right)E_I(U),$$

here the ξ is the number of electrons in the outer shell of the ion being ionized. I_n is the ionization potential of the shell in eV. The E_I is the exponential integral of the first kind with $U = I_n/T_e$. For a Maxwellian electron distribution, the collisional recombination is calculated according to the principle of detail balance:

$$R_{i+1,i}/N_e = 1.66\times10^{-22}\left(g_i/g_{i+1}\right)(e^U/T_e^{3/2})R_{i,i+1}$$

where the g_i and g_{i+1} are the statistical weights of the respective states and $R_{i,i+1}$ is the ionization rate coefficient.

In laser-generated molybdenum plasmas, strong continuum emission due to the recombinations was observed to dominate in measured soft x-ray rang. The atomic process contributes to the continuum emission is the radiative recombination. The rate is given by

$$R_{i+1,i} = \left(n_i/n_{1+1}\right)^* 4\pi\int\alpha_{i,i+1}\left(2hv^3/c^2\right)e^{-hv/kT_e}\frac{dv}{hv}$$

The $\left(n_i/n_{1+1}\right)^*$ given by the Saha-Boltzmann equation

$$\left(n_i/n_{1+1}\right)^* = 1.66\times10^{-22}\left(g_i/g_{i+1}\right)(e^U/T_e^{3/2}).$$

The $\alpha_{i,i+1}$ is the radiative recombination cross-section for the species of interest, which assumed given by the hydrogenic formula allowing for the non-hydrogenic ionization potential, I_n, of the ion of charge Z, i.e.

$$\alpha_{i,i+1} = 4.12\times10^{26}(I_n^{5/2}/Z)(1/v^3)$$

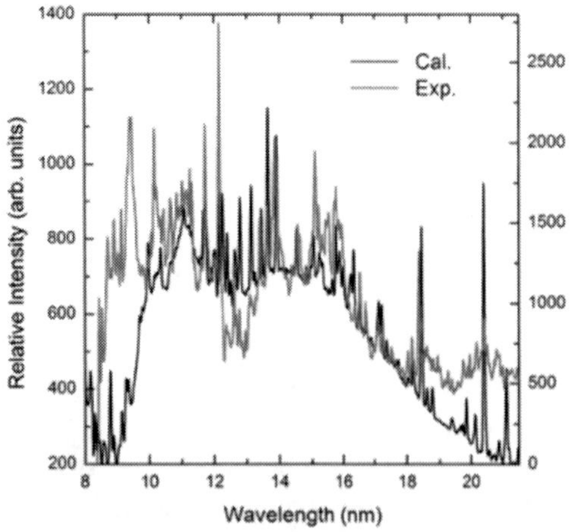

Figure 1. Comparison of the Mo plasma spectra recorded with CCD corrected by detector response and simulated one with CR model code.

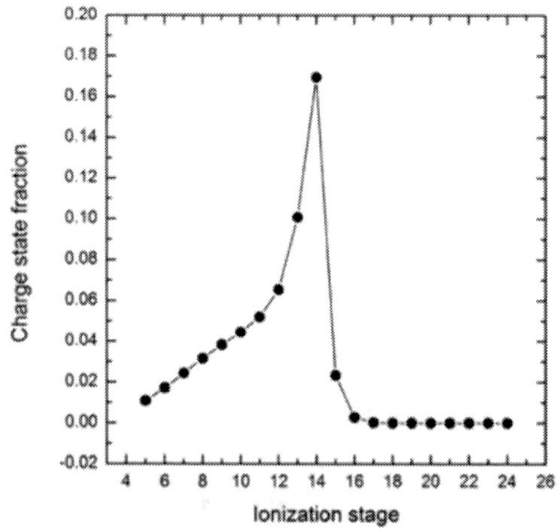

Figure 2. Charge state fraction of the average ionization stage between the electron density of 5×10^{19} cm^{-3} and 5×10^{20} cm^{-3} from the hydrodynamic code MED103.

To investigate the physical progress from the spectra of molybdenum plasmas, we did the CR model simulation using the same experiment laser conditions of 3.5×10^{12} W/cm^2, also the results are time-integrated and space-integrated as in the experiments. The comparison of the measured spectra after semi-empirical calibration, and the synthetic one calculated with CR model, is shown in Figure 1. A good recombination continuum shape agreement is seen over the whole wavelength scale. However, the intensity at both wings of the spectra does not match with the measurement quite well. The intensity mismatch at the wings of Figure 1 might be caused by the fact that the experiment has a Gaussian laser intensity distribution at the focus, while the numerical simulations assume only a single intensity. Furthermore, for a long distance transfer the wave front cannot be the perfect Gaussian shape anymore, which means the spectrum recorded in our experiment might come from sparks in the laser focus. At both the shorter and the longer wavelength end of the spectra, the emissions measured are obviously intense than calculated one. For the shorter wavelength, the continuum is emitted from highly ionized stages generated in the center of plasma by the laser intensity more than 3.5×10^{12} W/cm^2. On the other side, the continuum at the longer wavelength is contributed by the lowly ionized particles in the plasma of low density. The synthetic spectrum from the CR model code is good enough to reproduce what had taken in the laboratory. Figure 2 shows the charge state fraction of the ionization stage which is averaged during the laser pulse interaction, where the electron density located at the range between 5×10^{19} cm^{-3} and 5×10^{20} cm^{-3}, the electron temperature is above 100eV in the simulation. Since the measured soft x-ray emission is mostly from the high density and high temperature plasma part in the experiments, the calculation is comparable to that in the experiments. The molybdenum atomic level population distributions extracted from the measured spectra can be used as the pre-plasma analysis for the Nickel-like molybdenum soft x-ray lasers.

4. Conclusion

A Collisinal-Radiative model code with fine atomic data included in the hydrodynamic code was developed. A good recombination continuum shape agreement between the simulation and the semi-empirical calibrated Mo plasma experiment spectrum was obtained over the whole wavelength scale. The intensity mismatch at both wings of the two spectra was attributed to the intensity distribution of the laser focus, which is a Gaussian one in experiments, and a flat-top one assumed in the numerical calculations. With such a time dependent CR model code, plasmas information can be extracted from the measured spectra directly.

Acknowledgments

This work is jointly supported by the National Natural Science Foundation of China (under grants 60621063, 60678007, 10774184), the CAS project KJCX2-YW-T01, and National Basic Research Program of China (973 Program) (grant No. 2007CB815101).

References

[1] Chung H K, Chen M H, Morgan W L, Ralchenko Y and Lee R W 2005 *High Energy Density Physics* **1** 3
[2] Bowen C, Lee R W and Ralchenko Y, 2006 *Journal of Quantitative Spectroscopy and Radiative Transfer* **99** 102
[3] Li Y, Tsakiris G D and Sigel R, *Rev. Sci. Instrum.* 1995 **66** 80
[4] Palik E A, *Handbook of Optical Constants of Solids* (Academic, London, 1985)
[5] Van Wyngaarden W L, Bahdra K and Henry R J W, 1979 *Phys. Rev. A* **20** 1409
[6] Cowan R D, *The Theory of Atomic Structure and Spectra*, (University of California Press, Berkeley, 1981)
[7] W. Lotz, *Z. Phys.* 1967 **206**, 205; 1968 **216**, 241; 1969 **220**, 486

X-ray source for microstructure imaging under air conditions based on fs laser plasma

S A Pikuz Jr[1,2], O V Chefonov[1], S V Gasilov[1], P S Komarov[1], A V Ovchinnikov[1], I Yu Skobelev[1], S Yu Ashitkov[1], M V Agranat[1] and A Ya Faenov[1,3]

[1]Joint Institute for High Temperatures RAS, Moscow 125412, Russia
[2]Moscow Institute of Physics and Technology, Dolgoprudny 141700, Russia
[3]Kansai Photon Science Institute JAEA, Kizugawa-city, Kyoto 619-0215, Japan

spikuz@gmail.com

Abstract. X-ray emission and possibility of obtaining radiography images with laser plasma X-ray source operating at atmospheric pressure and room temperature is investigated. Table top 40 fs 100 uJ laser and Cu and Fe bulk metal targets were used to produce X-ray emission in spectral range of 3-10 keV. The spectra, yield and size of X-ray source were measured under different experimental conditions. X-ray photons of up to 10 keV energy are produced effectively even by relatively low laser intensity < 10^{15} W/cm^2. The source is applied for absorption imaging provided the sensitivity on 10 μm sample thickness and 30 μm spatial resolution of microstructures. The ability to perform phase contrast measurements is demonstrated. The advantages of the proposed method are in smart setup of investigated object within air environment; spectral range tunability; and sensitivity to low-contrast or micron thin details.

1. Introduction

The development of methods and tools for sensitive and inexpensive diagnostics of low-contrast objects and microstructures becomes important aim for the purposes of material science, medicine and high energy density physics. For radiography applications it is important to use a source with very high brightness that usually provided by synchrotrons. However, recent achievements in femtosecond laser plasma interaction allow to substitute synchrotron sources by compact inexpensive table-top laser facilities. It was shown that femtosecond laser plasma provides intense short-pulsed flow of several keV X-ray photons (see for ex. [1-5]) and hundreds keV fast ions [6, 7], which energies are enough to penetrate through a number of microscale objects. To provide the diagnostics of weakly-absorbed samples the methods of phase-enhanced imaging are commonly used where the size of backlighting source is desirable for [8, 9].

In the case of laser focusing in vacuum it is easily possible to provide the laser intensity of 10^{17}-10^{19} W/cm^2 generating hot dense plasma x-ray source of several microns size. But it is impossible or ineffective to put the sample for radiography inside vacuum environment for a number of microscale objects and life species. Then the process of laser propagation and focalization in ambient gas and following X-ray generation becomes quite more complicated. Recent experiments [10,11] shows the interaction of powerful fs laser pulses with solid target in gas environment causes X-ray generation. The use of micro objective allows to create the plasma blow in air very close to solid target surface. In this case the some part of laser radiation is able to penetrate through plasma blow, which is enough to create dense hot plasma at solid target surface as a source of X-rays. In the same time, air plasma blow can be considered as a source of fast electrons providing another mechanism of x-ray photon generation at solid target surface.

The aims of present work are: 1) to determine the parameters of X-ray source created on solid target surface standing in air environment; 2) to apply the source for convenient radiography of microstructures in absence of vacuum chamber using inexpensive laser source of 100 uJ energy.

2. Experimental setup

The experiment was performed using a Ti:Sapph laser system consisting of main oscillator (Coherent, Mira) and regenerative amplifier (Coherent, Evolution). The scheme of the experiment is shown in Figure 1. The laser delivers of 100 uJ energy during 40 fs pulse with 1 kHz repetition rate Pulse contrast of 10^3 is provided near the pulse pedestal (due to ASE), and of 10^6 in the picosecond scale. Also there is typical prepulse in the nanosecond scale due to the pockels cells leakage with 10^3 times lower intensity then the main pulse. Laser radiation was focused by the microfocus objective with N_A = 4 to the bulk cylinder-shape solid targets with incidence angle of 45°. To measure the focal spot profile and laser energy inclined glass plate were used to deliver a part of laser beam to calibrated photodiode and imaging system. Intensity distribution in the focal spot was measured for the laser energy below and over ionization threshold in air. In the case of lower energy the focal spot has Gaussian profile with 3 μm diameter, but in the case of gas breakdown the plasma blow is about 20 μm wide (see Figure 2), that corresponds to laser intensity of around $6x10^{14}$ W/cm^2. In order to provide the best focusing conditions during long time of target irradiation Fe and Cu cylinder-shape targets were rotated continuously.

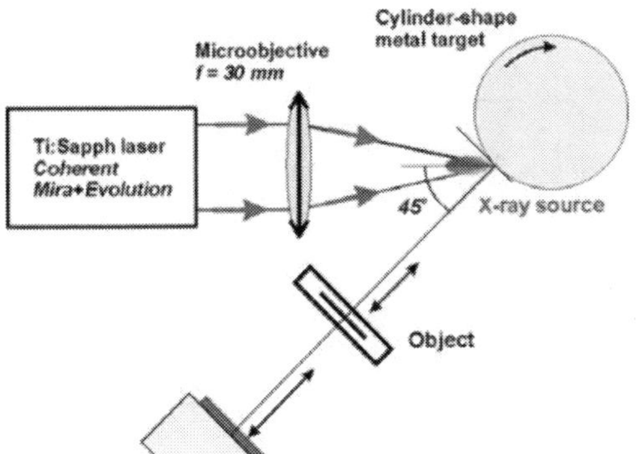

Figure 1. Experimental setup. Laser beam was focused by the micro objective on the continuously refreshed target surface standing in air. Imaged object and CCD positions changed in order to provide different magnification and phase-contrast imaging

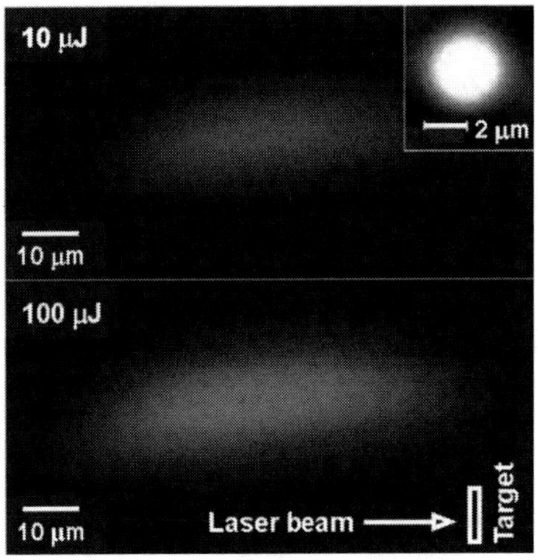

Figure 2. Imaging in visible light of plasma blow in air near solid target surface obtained along laser beam axis for 10 and 100 uJ laser energy. The image of laser focal spot at target surface for the energy below air breakdown is shown in the insertion.

For characterization of emitted X-rays emitted we used back illuminated Roper Scientific X-ray CCD camera with 20 μm pixel size. CCD was set to observe the plasma in the direction close to the normal incidence on the target. Single photon counting regime [12] was employed to estimate X-ray plasma emission spectrum. Entering aperture of the camera was protected by 15 μm thick Al filter, while the minimum distance to the camera during all measurements was 20 cm at least. In this case, due to the X-ray absorption in air, only photons with energy more than 2 keV can be detected by the camera. X-ray source size was measured using the penumbra effect in the images of test objects consisting either of two parallel thin wires or knife-like edge or pinhole with the 30 μm diameter.

3. Results and discussion

The experiments were done to investigate the mechanism and spectral distribution of X-ray generation on solid target in air conditions. We did not expect large number of electrons potent to produce characteristic emission with energies higher than 10 keV when the laser intensity of 10^{14}-10^{15} W/cm^2 is applied. However X-ray generation in Cu and Fe bulk target was examined to produce the emission in 5-10 keV spectral region which is still useful for imaging on many objects. The demand energy of X-ray radiation is also limited by air absorption in the gap between target and detector and corresponds to be more than 2-3 keV.

The spectrum of Cu target radiation registered by CCD matrix on 30 cm from the target is shown in Figure 3. The data demonstrates quite large photon flux obtained during 1 sec of exposition. Here Cu K_α and K_β spectral lines have a good contrast comparing with adjacent bremsstrahlung background. Total number of photons emitted in K_α line is estimated to be about 2×10^6 per second per steradian, and $\sim 2 \times 10^5$ in the K_β line, which give us conversion efficiency of around 10^{-8}.

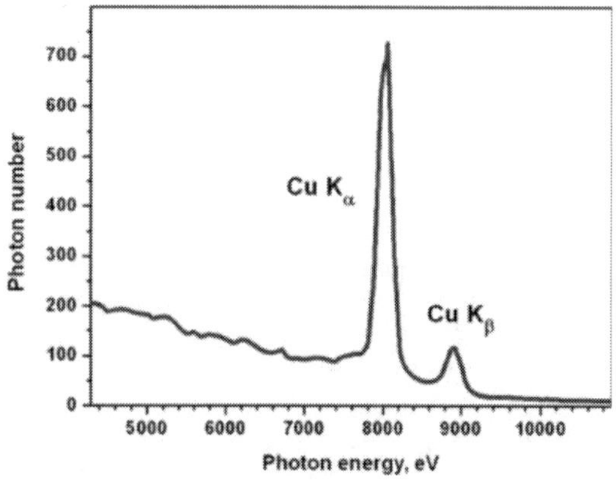

Figure 3. X-ray spectrum of the source measured by CCD in single photon counting regime placed at 300 mm in air during 1 sec exposure time.

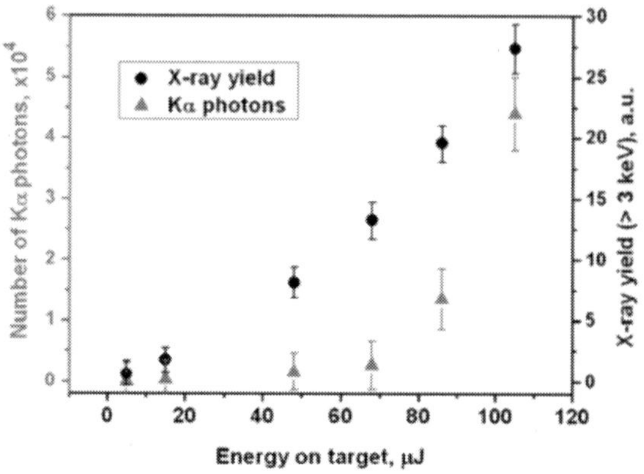

Figure 4. The dependence of overall (3-10 keV range) and K_a radiation of Cu target in air depending on laser energy. Data acquired by CCD matrix during 60 sec.

The measurements were done for the dependence of X-ray generation on implied laser energies in the range 10 - 100 μJ (see Figure 4). We were confident that all of this energy is focused in the measured spot size because the intensity does not greatly exceed air ionization threshold. We observed almost linear growth of total X-ray flux with energies more than 3 keV, while noticeable amount of K_α X-ray began to be emitted only for laser energies more than 60 μJ. Source size was also measured for the same energy range by the knife-like method. Intensity profile near the edge-air interface was use to calculate shape and size of the X-ray emission zone. It was found to have the same size of 50 μm under overall laser energy range.

The source was applied for imaging of thin films and low-contrast biological creatures. In Figure 5 the image of an insect are shown obtained using Cu plasma source. The data with probe radiation filtering by 120 μm aluminum foil was obtained additionally to provide the observation in harder x-rays to get the sensitivity on thick part of the object. The trace of insect leg image demonstrates its hollow internal structure and gives us an evidence of phase contrast effect representing the fringes at the leg edge. Similarly the details and features on insect wings having the thickness less than 10 μm are well distinguished that verifies the ability of the suggested imaging technique and also may be explained by phase contrast effect.

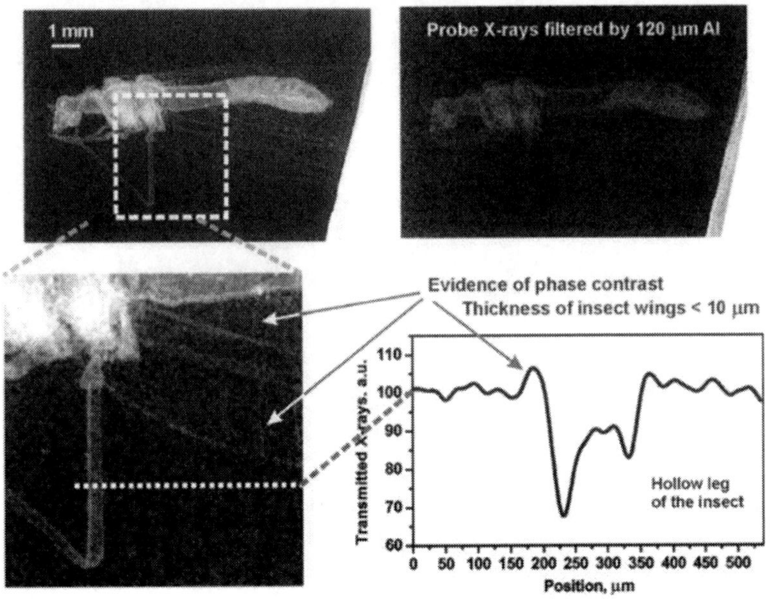

Figure 5. X-ray shadowgraphy images of the insect obtained by use of Cu fs-laser plasma source. Both source and object are located in air environment at distance of 100 mm between.

The increase of laser pulse energy up to several mJ on the target should cause the photon generation of tens of keV energies. Thus, the source and application method will be convenient for analysis and quality control of thin membranes and microstructures as well as of mm-thick structures and biomedical samples alive in air environment.

4. Conclusion

The approach to provide a common X-ray source for convenient radiography imaging by use of table-top femtosecond laser is successfully examined. Short laser pulse propagation and focalization in air at room pressure causes the gas breakdown before reaching the target surface. X-ray photons of up to 10 keV energy are produced effectively even by relatively low laser intensity < 10^{15} W/cm^2. The spectral distribution and geometry parameters of the source are determined. The source is experimentally proven to be applied for absorption and phase contrast imaging of different microstructures situated at atmosphere environment. The advantages of the proposed method are in smart setup of investigated object within air environment; spectral range tunability and sensitivity to weakly-absorbed or micron-thin details.

The work is carried out under the agreement with Russian Federal Government #02.740.11.0236 and partially supported by RFBR grant 09-02-92482-MNKS_a, CRDF BRHE Grant #Y5-11-02 and RAS Presidium Program for fundamental researches No. 27.

References

[1] Toth R, Fourmaux S, Ozaki T et al. 2007 *Physics of Plasmas* **14** 053506
[2] Zhavoronkov N, Gritsai Y, Bargheer M et al. 2005 *Appl. Phys. Lett.* **86** 244107
[3] Ewald T, Schwoerer H, Sauerbrey R 2002 *Europhys. Lett.* **60** 710
[4] Calegari F, Valentini G, Vozzi C et al 2007 *Optics Letters* **32** 2593
[5] Fukuda Y, Faenov A Ya, Pikuz T et al. 2008 *Appl. Phys. Lett.* **92** 121110
[6] Repsilber T, Borghesi M, Gauthier J-C et al. 2005 *Appl. Phys. B* **80** 905
[7] Andreev A A, Komarov V A, Platonov K Yu, Charukhchev A V 2007 *Tech. Phys. Lett.* **33** 239
[8] Wilkins S W, Gureyev T E, Gao D, Pogany A, Stevenson A W 1996 *Nature (London)* **384** 335
[9] Chakera J A, Ali A, Tsui Y Y, Fedosejevs R 2008 *Appl. Phys. Lett.* **93** 261501
[10] Gordienko V M, Dzhidzhoev M S, Zhvania I A, Makarov I A 2007 *Quantum Electron.* **37** 599
[11] Hou B, Mordovanakis A, Easter J, Krushelnik K, Nees J A 2008 *Appl. Phys. Lett.* **93** 201503
[12] Fourment C, Arazam N, Bonte C et al. 2009 *Rev. Sci. Instrum.* **80** 083505

AUTHOR INDEX

AUTHOR INDEX

AUTHOR INDEX

AUTHOR INDEX

AUTHOR INDEX

AUTHOR INDEX

AUTHOR INDEX

AUTHOR INDEX

AUTHOR INDEX

AUTHOR INDEX

AUTHOR INDEX

Institute of Physics
Dirac House, Temple Back
Bristol BS1 6BE UK

CURRAN ASSOCIATES INC
proceedings.com